PRODUCT AND PROCESS DESIGN PRINCIPLES
Synthesis, Analysis, and Evaluation

Second Edition

Warren D. Seider

Department of Chemical and Biomolecular Engineering
University of Pennsylvania

J. D. Seader

Department of Chemical and Fuels Engineering
University of Utah

Daniel R. Lewin

Department of Chemical Engineering
Technion—Israel Institute of Technology

WILEY

John Wiley and Sons, Inc.

Executive Editor   Bill Zobrist
Acquisitions Editor   Wayne Anderson
Marketing Manager   Katherine Hepburn
Media Editor   Martin Batey
Senior Production Editor   Valerie A. Vargas
Senior Designer   Kevin Murphy
Cover Designer   Carol Grobe
Outside Production Services   Ingrao Associates

This book was set in Times Roman by UG / GGS Information Services, Inc. and printed and bound by Hamilton Printing. The cover was printed by Phoenix Color.

This book is printed on acid-free paper.

Library of Congress Cataloging-in-Publication Data:

Seider, Warren D.
    Product and process design principles: synthesis, analysis, and evaluation / by Warren
D. Seider, J. D. Seader, Daniel R. Lewin. — 2nd ed.
        p.   cm.
    Rev. ed. of: Process design principles. ©1999.
    ISBN 0-471-21663-1 (acid-free paper)
    ISBN 0-471-45247-5 (WIE)
    1. Chemical Processes   I. Seader, J. D.   II. Lewin, Daniel R.   III. Seider, Warren D.
Process Design Principles.   IV. Title.
TP155.7.S423   2003
660′.2812—dc21

                                                                2002192252

Printed in the United States of America

10 9 8 7 6 5 4 3

To the memory of my parents, to Diane, and to Benjamin, Deborah, Gabriel, and Joe

To the memory of my parents, to Sylvia, and to my children

To my parents, Harry and Rebeca Lewin, to Ruti, and to Noa and Yonatan

To the memory of Richard R. Hughes, a pioneer in computer-aided simulation and optimization, with whom two of the authors developed many concepts for carrying out and teaching process design.

About the Authors

Warren D. Seider is Professor of Chemical Engineering at the University of Pennsylvania. He received a B.S. degree from the Polytechnic Institute of Brooklyn and M.S. and Ph.D. degrees from the University of Michigan. Seider has contributed to the fields of process analysis, simulation, design, and control. He coauthored *FLOWTRAN Simulation—An Introduction* in 1974 and has coordinated the design course at Penn for over 20 years involving projects provided by many practicing engineers in the Philadelphia area. He has authored or coauthored over 80 journal articles and authored or edited six books. Seider was the recipient of the AIChE (American Institute of Chemical Engineers) Computing in Chemical Engineering Award in 1992. He served as a Director of AIChE from 1984 to 1986 and has served as chairman of the CAST Division and the Publication Committee. He helped to organize the CACHE (Computer Aids for Chemical Engineering Education) Committee in 1969 and served as its chairman. Seider is a member of the Editorial Advisory Board of *Computers and Chemical Engineering.*

J. D. Seader is Professor of Chemical Engineering at the University of Utah. He received B.S. and M.S. degrees from the University of California at Berkeley and a Ph.D. from the University of Wisconsin. From 1952 to 1959, he designed processes for Chevron Research and directed the development of one of the first computer-aided process design programs. From 1959 to 1965, he conducted rocket engine research for Rocketdyne on all of the engines that took astronauts to the moon. Before joining the faculty at the University of Utah in 1966, Seader was a professor at the University of Idaho. He is the author or coauthor of 109 technical articles, seven books, and four patents. Seader is coauthor of the section on distillation in the 6th and 7th editions of *Perry's Chemical Engineers' Handbook*. He is coauthor of *Separation Process Principles*, published in 1998. Seader was Associate Editor of *Industrial and Engineering Chemistry Research* for 12 years, starting in 1987. He was a founding member and trustee of CACHE for 33 years, serving as Executive Officer from 1980 to 1984. For 20 years, he directed the use by and distribution to 190 chemical engineering departments worldwide of Monsanto's FLOWTRAN process simulation computer program. Seader served as Chairman of the Chemical Engineering Department at the University of Utah from 1975–1978, and as a Director of AIChE from 1983 to 1985. In 1983, he presented the 35th Annual Institute Lecture of AIChE. In 1988, he received the Computing in Chemical Engineering Award of the CAST Division of AIChE.

Daniel R. Lewin is Professor of Chemical Engineering and the director of the Process Systems Engineering (PSE) research group at the Technion, the Israel Institute of Technology. He received his B.Sc. from the University of Edinburgh and his D.Sc. from the Technion. His research focuses on the interaction of process design and process control and operations, with emphasis on model-based methods. He has authored or coauthored over 90 technical publications in the area of process systems engineering, as well as the first edition of this textbook, and the multimedia CD that accompanies it. Professor Lewin has been awarded a number of prizes for research excellence, and twice received the Jacknow Award in recognition of teaching excellence at the Technion. He served as Associate Editor of the *Journal of Process Control* and is a member of the International Federation of Automatic Control (IFAC) Committee on Process Control and Optimal Control.

Preface

OBJECTIVES

A principal objective of this textbook and accompanying CD-ROM, referred to here as courseware, is to describe modern strategies for the design of chemical products and processes. Since the early 1960s, the emphasis in undergraduate education has been on the engineering sciences. In recent years, however, more scientific approaches to product and process design have been developed, and the need to teach students these approaches has become widely recognized. Consequently, this courseware has been developed to help students and practitioners better utilize the modern approaches to process design and the growing importance of the design by chemical engineers of new products that require the selection of an appropriate chemical or chemical mixture or involve chemical reactions and/or the separation of chemical mixtures. Like workers in thermodynamics; momentum, heat, and mass transfer; and chemical reaction kinetics, product and process designers apply the principles of mathematics, chemistry, physics, and biology. Designers, however, utilize these principles, and those established by engineering scientists, to create chemical products and processes that satisfy societal needs while returning a profit. In so doing, designers emphasize the methods of synthesis and optimization in the face of uncertainties, often utilizing the results of analysis and experimentation prepared in cooperation with engineering scientists.

In this courseware, the latest design strategies are described, most of which have been improved significantly with the advent of computers, numerical mathematical programming methods, and artificial intelligence. Since most curricula place little emphasis on design strategies prior to design courses, this courseware is intended to provide a smooth transition for students and engineers who are called upon to design creative new products and processes.

The first edition of this textbook focused on the design of commodity chemical processes. While this material has been updated and augmented to include recent developments, the second edition, with a new title, broadens this focus to include the design of chemical products. Strategies for product design have been added for both configured industrial and consumer products. Also, numerous examples of the process design of specialty chemicals are included based on batch, rather than continuous, processing. In addition, the second edition has expanded coverage on the design of process units, the estimation of capital costs, profitability analysis, and optimization. Of particular note is the addition of more than 100 equations for the estimation of purchase costs for most kinds of process equipment.

This courseware is intended for seniors and graduate students, most of whom have solved a few open-ended problems but have not received instruction in a systematic approach to product and process design. To guide this instruction, the subject matter is presented in five parts. As discussed in Chapter 1, Figure 1.2 shows how these parts relate to the entire design process and to each other. All of the parts are presented at the senior level.

In this second edition, the scope of Part One is broadened to include the idea-generation stage in designing new chemical products, and the search for chemicals or chemical mixtures that have the desired properties and performance. When a process is required to produce the chemicals, the steps in process design are covered, expanded beyond those presented in the first edition. When configured industrial or consumer products are needed, with or without a new chemical process, the steps in detailed product design are covered with many examples in Part Three. Because many specialty chemicals are manufactured in small quantities, the second edition has been expanded to include the design of batch processes.

For process design, the coverage is similar to that in the first edition, but expanded to include new methods and more coverage of solids processing. The emphasis throughout the text, and especially in Part One, on process invention, and Part Two, on detailed process synthesis, is on the steps in process creation and the development of a base-case design(s). For the former, methods of preparing the synthesis tree of alternative flowsheets are covered. Then, for the most promising flowsheets, a base-case design(s) is developed, including a detailed process flow diagram, with material and energy balances. The base-case design(s) then enters the detailed design stage, in which the equipment is sized, cost estimates are obtained, a profitability analysis is completed, and optimization is carried out, as discussed in Part Three.

Throughout this courseware, various methods are utilized to perform the extensive calculations and provide graphical results that are visualized easily, including the use of computer programs for simulation and design optimization. The use of these programs is an important attribute of this courseware. We believe that our approach is an improvement over an alternative approach that introduces the strategies of process synthesis *without computer methods*, emphasizing heuristics and *back-of-the-envelope* calculations. We favor a blend of heuristics and analysis using the computer. Since the 1970s, many faculty have begun to augment the heuristic approach with an introduction to the analysis of prospective flowsheets using simulators, such as ASPEN PLUS, HYSYS.Plant, PRO/II, CHEMCAD, FLOW-TRAN, BATCH PLUS, and SUPERPRO DESIGNER. Today, most schools use one of these simulators, but often without adequate teaching materials. Consequently, the challenge for us, in the preparation of this courseware, has been to find the proper blend of modern computational approaches with simple heuristics.

In the chapters on the design of commodity chemical processes, emphasis is placed on the synthesis of processes that operate at steady state and present no unusual control problems. For these processes, dynamic simulators, such as ASPEN DYNAMICS and HYSYS.Plant, are useful for studying startup, shutdown, upsets, and the performance of alternative control systems. Dynamic analysis often suggests designs that are easier to implement and control. As processes become more integrated, to achieve more economical operation, their responses to disturbances and setpoint changes become more closely related to the design integration, and consequently, the need to assess their controllability gains importance. To introduce several methods, Part Four is intended for readers who have studied linear control theory for single-input, single-output (SISO) controllers (usually in a first course in process control). Emphasis is placed on the methods for assessing the controllability of processes designed to operate at a steady state, with the consideration of frequency-dependent measures only when necessary. Control systems are designed for the most promising processes, and the ability of the processes to reject typical disturbances is evaluated using dynamic simulation. In summary, Part Four is intended to show that, to achieve more profitable designs, it is important to consider *plantwide control* during process design. This is accomplished, qualitatively and then quantitatively, using the simpler strategies for multiple-input, multiple-output (MIMO) control.

FORMAT OF COURSEWARE AND THE MULTIMEDIA CD-ROM

This courseware takes the form of a conventional textbook accompanied by computer programs to be utilized by the reader in various aspects of his or her design studies. As the design strategies have been elucidated during the development of this courseware, fewer specifics have been provided in the chapters concerning the software packages involved. Instead, a multimedia CD-ROM had been developed to give many examples of the simulator input and output, with frame-by-frame instructions, to discuss the nature of the models provided for the processing units, with several example calculations presented as well. The CD-ROM uses voice, video, and animation to introduce new users of the steady-state simulators to the specifics of ASPEN

PLUS and HYSYS.Plant. These include several tutorials that provide instruction on the solution of problems for courses in material and energy balances, thermodynamics, heat transfer, separations, and reactor design. In many cases, students will have been introduced to the process simulators in these courses. Also, video segments show portions of a petrochemical complex in operation, including distillation towers, heat exchangers, pumps and compressors, and chemical reactors. The CD-ROM includes over 90 files that contain the solutions for many examples using either programs in the Aspen Engineerintg Suite (e.g., ASPEN PLUS, Aspen IPE, and BATCH PLUS) or HYSYS.Plant, as well as problems solved using GAMS, an optimization package, and the MATLAB scripts in Chapter 21. The files are referred to in each example and can easily be used to vary parameters and to explore alternative solutions. Furthermore, a Web site is maintained at www.wiley.com/college/seider, in which revisions will be entered on a regular basis, especially as they relate to new releases of the commercial software.

As observed in the Table of Contents for the textbook, Chapters 8 and 9, and Sections 16.7 and 17.8 are provided in PDFs (portable document format) on the CD-ROM, with only the objectives of the chapters and sections presented in the textbook. Furthermore, Appendix II provides a list of design projects, whose detailed statements are provided in the file, Design Problem Statements.pdf, on the CD-ROM. These involve the design of chemical processes in several industries. Many are derived from the petrochemical industry, with much emphasis on environmental and safety considerations, including the reduction of sources of pollutants and hazardous wastes, and purification before streams are released into the environment. Several originate in the biochemicals industry, including fermentations to produce pharmaceuticals, foods, and chemicals. Others are involved in the manufacture of polymers and electronic materials. Each design problem has been solved by groups of two or three students at the University of Pennsylvania, with copies of their design reports available by Interlibrary Loan from the Towne Library at the University.

ADVICE TO STUDENTS AND INSTRUCTORS

In the use of this textbook and CD-ROM, students and instructors are advised to take advantage of the following six features:

Feature 1

The textbook is organized around the key steps in product and process design shown in Figure 1.2. These steps reflect current practice and provide a sound sequence of instruction, yet with much flexibility in permitting the student and instructor to place emphasis on preferred subjects.

Students can study the chapters in Part One in sequence. Although they provide many examples and exercises, the CD-ROM can be referred to for details of the process simulators. Chapters in Parts Two, Three, and Four can be studied as needed. There are many cross-references throughout the text—especially to reference materials needed when carrying out designs. For example, students can begin to learn heuristics for heat integration in Chapter 3, learn algorithmic methods in Chapter 10, learn the strategies for designing heat exchangers and estimating their costs in Part Three (Chapters 13 and 16), and learn the importance of examining the controllability of heat-exchanger networks in Part Four.

Instructors can begin with Part One and design their courses to cover the other chapters as desired. Because each group of students has a somewhat different background depending on the subjects covered in prior courses, the textbook is organized to give instructors much flexibility in their choice of subject matter and the sequence in which it is covered. Furthermore, design instructors often have difficulty deciding on a subset of the many subjects to be covered. This book provides sufficiently broad coverage to permit the instructor to emphasize certain subjects in lectures and homework assignments, leaving others as reference materials to be used by the students when carrying out their design projects. In a typical situation, when teaching the students to generate design alternatives, select a base-case design, and carry out its analysis, the textbook enables the instructor to place emphasis on one or more of the following subjects: synthesis of chemical reactor networks (Chapter 6), synthesis of separation trains (Chapter 7), energy efficiency (lost work analysis and heat and power integration—Chapters 9 and 10), process unit design (e.g., heat exchangers—Chapter 13), and controllability assessment (Part Four).

Feature 2

This textbook introduces the key steps in product design with numerous examples.

Students can begin in Sections 1.1 and 1.2 to learn about industrial and consumer products, the idea generation stage, methods of stimulating innovation in product design, and the special attributes of pharmaceutical product design. In Chapter 2, they can learn to find chemicals and chemical mixtures having desired properties and performance; that is, to carry out molecular-structure design. Chapter 3 shows how to synthesize a pharmaceutical process for the manufacture of tissue plasminogen activator (tPA), and Chapter 4 introduces the methods of batch process simulation for the process. Then, they can turn to Chapter 12 to learn how to optimize the design and scheduling of batch processes. Finally, Chapter 19 expands on the steps for product design and shows how to utilize them in designing several configured industrial and consumer products, including a hemodialysis device, a solar desalination unit, a hand warmer, silicon coated chips, automotive fuel cells, and environmentally-safe refrigerants.

Instructors can create a course in product design using the materials and exercises referred to in the preceding paragraph. The product designs in Chapter 19 can be expanded upon and/or used as the basis of design projects for student design teams.

Feature 3

Process synthesis is introduced mostly using heuristics in Part One (Chapters 3 and 5), whereas Part Two provides more detailed algorithmic methods for chemical reactor network synthesis, separation train synthesis, the synthesis of reactor-separator-recycle networks, heat and power integration, mass integration, and the optimal design and sequencing of batch processes.

This feature enables the student to begin carrying out process designs using easy to understand rules of thumb when studying Part One. As these ideas are mastered, the student can learn algorithmic approaches that enable him or her to produce better designs. For example, Chapter 3 introduces two alternative sequences for the separation of a three-component mixture (in the vinyl chloride process), whereas Chapter 7 shows how to generate and evaluate many alternatives for the separation of multicomponent mixtures, both ideal and nonideal.

This organization provides the instructor the flexibility to emphasize those subjects most useful for his or her students. Part One can be covered fairly quickly, giving the students enough background to begin work on a design project. This can be important at schools where only one semester is allotted for the design course. Then, as the students are working on their design projects, the instructor can take up more systematic, algorithmic methods, which can be applied to improve their designs. In a typical situation, when covering Part One, the instructor would not cover nonideal separations, such as azeotropic, extractive, or reactive distillations. Consequently, most students would begin to create simple designs involving reactors followed by separation trains. After the instructor covers the subject matter in Chapter 7, the students would begin to take advantage of more advanced designs.

Feature 4

Process simulators, steady state, dynamic, and batch, are used throughout the textbook (ASPEN PLUS, HYSYS.Plant, CHEMCAD, PRO/II, BATCH PLUS, and SUPERPRO DESIGNER). This permits access to large physical property, equipment, and cost databases and the examination of aspects of numerous chemical processes. Emphasis is placed on the usage of simulators to obtain data and perform routine calculations throughout.

Through the use of the process simulators, students learn how easy it is to obtain data and perform routine calculations. They learn effective approaches to building up knowledge about a process through simulation. The CD-ROM provides the students with the details of the methods used for property estimation and equipment modeling. They learn to use simulators intelligently and to check their results. For example, in Chapter 3, examples show how to use simulators to assemble a preliminary database and to perform routine calculations when computing heat loads, heats of reaction, and vapor/liquid equilibria. Then, in Chapter 4, two examples show how to use the simulators to assist in the synthesis of the toluene hydrodealkylation and monochlorobenzene separation processes. Virtually all of the remaining chapters show examples of the use of simulators, as well as Aspen IPE and a profitability analysis spreadsheet, to obtain additional information, including equipment sizes, costs, profitability analyses, and the performance of control systems.

Because the book and CD-ROM contain so many routine examples of how the simulators are useful in building up a process design, the instructor has time to emphasize other aspects of process design. Through the examples and multimedia instruction on the CD-ROM, with emphasis on ASPEN PLUS and HYSYS.Plant, the students obtain the details they need to use the simulators effectively, saving the instructor class time, as well as time in answering detailed questions as the students prepare their designs. Consequently, the students obtain a better understanding of the design process and are exposed to a broader array of concepts in process design. In a typical situation, when creating a base-case design, students use the examples in the text and the encyclopedic modules and tutorials on the CD-ROM to learn how to obtain physical property estimates, heats of reaction, flame temperatures, and phase distributions. Then, students learn to create a reactor section, using the simulators to perform routine material and energy balances. Next, they create a separation section and eventually add recycle. Thanks to the coverage of the process simulators in Part One and the CD-ROM, the instructor need only review the highlights in class. Note that at many universities the students receive prior instruction, often using the CD-ROM, in prior core chemical engineering courses.

Feature 5

Part Three includes chapters that provide instruction and examples of the design of heat exchangers, multistage and packed towers, and pumps, compressors, and expanders. In addition, Chapter 16 provides guidelines for selecting processing equipment and equations for estimating the purchase costs of a broad array of equipment items. Furthermore, Section 16.7 shows how to use the Aspen Icarus Process Evaluator (IPE), with the process simulators, to estimate purchase costs and the total permanent investment for a chemical plant.

Students can use the chapters in Part Three when carrying out their design projects. In this book, most of the information they need is provided for estimating equipment sizes, purchase costs, and operating costs, and for carrying out profitability analyses.

Instructors can use the chapters on equipment design to supplement the subjects covered in earlier courses, selecting topics most appropriate for their students.

Feature 6

To our knowledge, this book is the first design text to emphasize the importance of assessing plantwide controllability. Modern computing tools are enabling practitioners and students to be more aware that processes selected on the basis of steady-state economics alone often perform poorly and are less profitable.

When studying Chapter 20, students learn that they can begin to screen processes during process synthesis by selecting the variables to be measured and adjusted and beginning to formulate control loops, but *without* detailed controller design. Then, in Chapter 21, well-established methods enable students to screen alternative processes using standard linear approximations, again *without* detailed controller design. Also, in Chapter 21, for the most promising processes, the controllers are designed and simulations are carried out to show that the linear analyses are effective. Throughout Part Four (Chapters 20 and 21), alternatives for heat-exchanger networks; heat-integrated distillation towers; stirred-tank reactor designs; and processes with reactors, separators, and recycle loops are compared to show that there are significant differences in controllability and resiliency.

From the instructor's perspective, Chapter 20 provides basic introductory material that can be taught with little effort and little control background. Chapter 21 is more advanced. Although some instructors will cover it thoroughly, others may prefer to select from among the case studies to introduce their students to the effectiveness of controllability and resiliency analysis for linearized systems. Then, even if time is not available for the students to tune controllers and run dynamic simulations, the case studies in Chapter 21 can be used to show the effectiveness of this approach during process design. In a typical situation, the instructor would use the alternative heat-exchanger networks, heat-integrated distillation towers, reactor sequences, and/or processes with recycle to show the importance of considering controllability in the design process.

ONE- OR TWO-SEMESTER COURSES

In one semester, it is possible to emphasize process design by covering Part One, many topics in Parts Two and Three, and Chapter 20. Students solve homework exercises and take midsemester and final exams but do not work on a comprehensive design project. The latter is reserved for a design project course in a second semester. Alternatively, many departments teach process design concepts in a single course, which includes a comprehensive process design project. For such a course the same materials can be covered, somewhat less thoroughly, or a subset of the subjects can be covered. The latter often applies in departments that cover design-related topics in other courses. For example, many departments teach economic analysis before the students take a process design course. Other departments teach the details of equipment design in courses on transport phenomena and unit operations. This textbook and CD-ROM are well suited for these courses because they provide much reference material that can be covered as needed.

For departments that teach a course in process design followed by a course in product design, this textbook provides instruction in most of the topics covered in both courses.

ACKNOWLEDGMENTS

In the preparation of this courseware, several graduate and post-doctoral students made significant contributions, including Charles W. White III, George J. Prokopakis, Joseph W. Kovach III, Tulio R. Colmenares, Miriam L. Cygnarowicz, Alden N. Provost, David D. Brengel, Soemantri Widagdo, Amy C. Sun, Roberto Irrizary-Rivera, Leighton B. Wilson, James R. Phimister and Pramit Sarma, at the University of Pennsylvania; and Oren Weitz, Boris Solovyev, and Eyal Dassau, at the Technion. Tulio developed the lecture notes, with many examples, that are the basis for Chapter 10 on "Heat and Power Integration." Soemantri coauthored the review article on azeotropic distillation in which many of the concepts in Section 7.6, "Sequencing Azeotropic Distillation Columns," are presented. He also provided much advice on product design based upon experiences at 3M Corporation. James was the teaching assistant during the semester in which many new concepts were introduced. Subsequently, he wrote introductory material on the use of GAMS, which appears in the file, GAMS.pdf, on the CD-ROM. In addition, Holger Nickisch, who received bachelors degrees in chemical engineering and business at Penn, created the spreadsheet, Profitability Analysis-1.0.xls, which is discussed in Section 17.8 and provided on the CD-ROM. The successes in our process design courses are closely related to the many contributions of these students. Their help is very much appreciated.

Students at the Technion, Eyal Dassau, Joshua Golbert, Garry Zaiats, and Daniel Schweitzer, and students at the University of Pennsylvania, Murtaza Ali, Scott Winters, Diane M. Miller, Michael DiTillio, Christopher S. Tanzi, Robert C. Chang, Daniel N. Goldberg, Matthew J. Fucci, and Robyn B. Nathanson, implemented the multimedia CD-ROM and assisted in many other ways. Their efforts are also appreciated. In this regard, seed money for the initial development of the CD-ROM was provided by Dean Gregory Farrington, University of Pennsylvania, and is acknowledged gratefully.

Several colleagues at the University of Pennsylvania and industrial consultants from local industry in the Philadelphia area were very helpful as these materials evolved, especially Arnold Kivnick, Leonard A. Fabiano, Scott L. Diamond, Robert M. Busche (Bio-en-gene-er Associates, Wilmington, DE), F. Miles Julian, Robert F. Hoffman (Rowan University), Robert Nedwick (Penn State University—formerly ARCO Chemical Company), Robert A. Knudsen (Lyondell), and David Kolesar (Rohm & Haas).

Four faculty, Michael E. Hanyak, Jr. (Bucknell), Daniel W. Tedder (Georgia Tech), Dale E. Briggs (Michigan), and Colin S. Howat (Kansas), reviewed a preliminary version of the first edition. Three additional faculty, John T. Baldwin (Texas A&M), William L. Luyben (Lehigh), and Daniel A. Crowl (Michigan Tech), reviewed a preliminary version of the second edition. In addition, Professor Ka Ng (Hong Kong University of Science and Technology), Dr. Soemantri Widagdo (3M Corporation), Professor Costas Maranas (Penn State), and Professor Luke Achenie (Connecticut) reviewed selected chapters. Their suggestions and critiques were extremely helpful. In addition, Lorenz T. Biegler (CMU) provided helpful suggestions concerning the organization of the initial version of Chapter 18, "Optimization of Process Flowsheets." Some of the material for this chapter was adopted from Chapter 13 by L. T. Biegler in *FLOWTRAN Simulation—An Introduction* (Seader et al., 1987).

The cooperation of Jila Mahalec, Vladimir Mahalec, Herbert I. Britt, Atilla Forouchi, Sanjay Patnaik, Lawrence Fry, Lorie Roth, Robert L. Steinberger, Siva Natarajan, Bahram Meyssami, and Lawrence B. Evans at Aspen Technology; and Bill Svrcek, Rich Thomas, and James Holoboff at Hyprotech (now part of Aspen Technology), has been especially valuable and is acknowledged with appreciation.

It is of special note that during the preparation of the first edition, Professors Christodoulos A. Floudas (Princeton) and William L. Luyben (Lehigh) provided W.D. Seider an opportunity to lecture in their classes and utilize some of these materials as they were being developed. Their interactions and insights have been very helpful. This cooperation, and some of the work on this project, was facilitated in part by NSF Grant No. EEC-9527441 from the Combined Research and Curriculum Development Program.

Throughout the development of the first and second editions, and the CD-ROM, A. Wayne Anderson, the Editor for College Publishing at John Wiley & Sons, was extremely helpful. Wayne's excellent advice and guidance is very much appreciated.

It is important to acknowledge the secretarial support, provided in the most efficient and effective manner, by John Linscheid, who made it possible to prepare the first edition. Finally, W.D. Seider received two Lady Davis Visiting Professorships at the Technion during the spring of 1996 and 2002, and D.R. Lewin was a Visiting Professor at the University of Pennsylvania during the summer of 1997. Financial support in connection with these sabbatical leaves enabled them to work on the manuscript and is very much appreciated.

W. D. Seider
J. D. Seader
D. R. Lewin
May 2003

Contents

Part One

PRODUCT AND PROCESS INVENTION—HEURISTICS AND ANALYSIS

In five chapters, Part One describes the key steps in *product and/or process invention*, one of the most creative activities carried out by chemical engineers. It begins with the *primitive problem statement* and describes the formation of the design team, its idea generation phase and gathering of information, and the creation of more specific problems, with emphasis on the consideration of environmental, safety, and ethical issues. Then, for *product design*, it shows how to find chemicals and chemical mixtures having desired properties and performance. For *process design*, step-by-step, it shows how the preliminary database is created and how a *synthesis tree* of potential flowsheets is generated with very little analysis. For the most promising flowsheets, the design team creates a *base-case design*, including a detailed process flow diagram, with material and energy balances, and a more complete *process integration*. For the base-case design(s), a more detailed database is assembled, pilot-plant testing is undertaken, and a process simulation model is normally developed before detailed design begins, the steps of which are described in Part Three of this text.

Throughout Chapters 1 and 3, the role of the *process simulators* is deemphasized. Rather, emphasis is placed on the steps in process synthesis and the development of the base-case design. Although from time to time these chapters refer to simulation results, a formal discussion of the role of process simulators during process creation is not attempted until Chapter 4 (and the accompanying multimedia CD-ROM). Chapter 3 introduces the steps in process synthesis without discussing many of the key heuristics used in making decisions. A more formal treatment is reserved for Chapter 5.

Chapter 1

The Design Process

1.0 OBJECTIVES

This chapter introduces the broad array of design decisions that confront chemical engineers and the steps involved in developing a product or process design. Special emphasis is placed on the growing importance of protecting the environment and ensuring safe and reliable chemical products, as well as manufacturing facilities, considerations that are prominent in the minds of product and process design teams. Because computers are utilized in so many aspects of the design process, the role of the computer is also highlighted, to introduce the kinds of computer tools available and where they are used.

After studying this chapter, the reader should

1. Be knowledgeable about the kinds of design decisions that challenge product and process design teams.
2. Have an appreciation of the key steps in carrying out a product and/or process design. The part-by-part and chapter-by-chapter organization of this book teaches how to implement these steps.
3. Be aware of the many kinds of environmental issues and safety considerations that are prevalent in the design of new chemical products and processes.
4. Appreciate the importance of maintaining high ethical principles in product and process design.
5. Understand that chemical engineers use a blend of hand calculations, spreadsheets, mathematical computer packages, and process simulators to design products and processes.

Although you will not solve any design problems in this chapter, you will obtain the background information that will be expanded upon and referred to throughout the remaining chapters of this text.

1.1 DESIGN OPPORTUNITIES

The design of chemical products begins with the identification and creation of potential opportunities to satisfy societal needs and to generate profit. Thousands of chemical products are manufactured, with companies like Minnesota Mining and Manufacturing (3M) having developed over 50,000 chemical products since being founded in 1904. Who has not used their Scotch Magic Tape™? The scope of chemical products is extremely broad. They can be roughly classified as: (1) basic chemical products, (2) industrial products, and (3) consumer products. As shown in Figure 1.1a, basic chemical products are manufactured from natural resources. They include commodity and specialty chemicals (e.g., commodity chemicals—ethylene, acetone, vinyl chloride; specialty chemicals—difluoroethylene, ethylene glycol

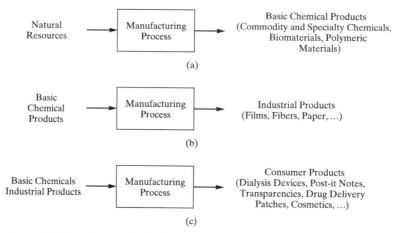

Figure 1.1 Manufacture of chemical products.

monomethyl ether, diethyl ketone), biomaterials (e.g., pharmaceuticals, tissue implants), and polymeric materials [e.g., ethylene copolymers, polyvinyl chloride (PVC), polystyrene]. The manufacture of industrial products begins with the basic chemical products, as shown in Figure 1.1b. Industrial products include films, fibers (woven and nonwoven), and paper. Finally, as shown in Figure 1.1c, consumer products are manufactured from basic chemical and industrial products. These include dialysis devices, hand warmers, Post-it notes, ink-jet cartridges, detachable wall hangers, solar desalination devices, transparencies for overhead projectors, drug delivery patches, fuel cells, cosmetics, detergents, pharmaceuticals, ...

Many chemical products are manufactured in small quantities and the design of a product focuses on identifying the chemicals or mixture of chemicals that have the desired properties, such as strength, stickiness, porosity, and permeability, to satisfy specific consumer needs. For these, the challenge is to create a product that can be protected by patents and has sufficiently high market demand to command an attractive selling price. After the chemical mixture is identified, it is often necessary to design a manufacturing process.

Other chemical products, often referred to as commodity chemicals, are required in large quantities. These are often intermediates in the manufacture of specialty chemicals and industrial and consumer products. These include ethylene, propylene, butadiene, methanol, ethanol, ethylene oxide, ethylene glycol, ammonia, nylon, and caprolactam (for carpets), together with solvents like benzene, toluene, phenol, methyl chloride, and tetrahydrofuran, and fuels like gasoline, kerosene, and diesel fuel. These are manufactured in large-scale processes that produce billions of pounds annually in continuous operation. Since they usually involve small well-defined molecules, the focus of the design is on the process to produce these chemicals from various raw materials.

Design projects have many points of origin. Often they originate in the research labs of chemists, biochemists, and engineers who seek to satisfy the desires of customers for chemicals with improved properties for many applications (e.g., textiles, carpets, plastic tubing). In this respect, several well-known products, such as Teflon (polytetrafluoroethylene), were discovered by accident. At DuPont, a polymer residue that had accumulated in a lab cylinder of tetrafluoroethylene was found to provide a slippery surface for cookware capable of withstanding temperatures up to 250°C, among many similar applications. In other cases, an inexpensive source of a raw material(s) becomes available and process engineers are called on to design processes that use this chemical, often with new reaction paths and methods of separation. Other design problems originate when new markets are discovered, especially in the developing countries of Southeast Asia and Africa. Yet another source of design projects are engineers themselves who often have a strong feeling that a new chemical or route to produce an existing chemical may be very profitable, or that a market exists for a new chemical product.

When a new chemical product is envisioned, the design project can be especially challenging, as much uncertainty often exists in the chemical(s) or mixture of chemicals best suited for the product, as well as the product configuration, the modes of manufacture, and the market demand. For the manufacture of a commodity chemical, the design project is usually less comprehensive, as the focus is usually on the design of the manufacturing facility or chemical process.

Design Team

To address a given problem statement, or to create one, a small design team may be assembled, often involving a chemical engineer, a chemist, perhaps a biochemist, and a marketing person, as appropriate. Other expertise is added to the team to address specific areas. The team undertakes the design problem following a sequence of well-recognized steps in product and process design, as described in the next section.

1.2   STEPS IN PRODUCT AND PROCESS DESIGN

In examining the steps commonly practiced in designing a chemical product or process, it is important to keep in mind that design problems are open ended and may have many solutions that are attractive and near optimal. Furthermore, no two designers design a complex product or process following exactly the same steps. In fact, to capture the know-how of experienced designers and better understand the design process, cognitive scientists recommend that the designer's steps be tracked so that others can learn to apply them when working on the design of similar products and processes.

Design is the most creative of engineering activities, with many opportunities to invent imaginative new products and processes. It is also the essence of engineering, differentiating an engineer from a scientist. Whereas chemical engineers engaged in product or process development exercise creativity in formulating experiments and theories to uncover and explain the mechanisms of processing operations (often involving complex reaction kinetics with heat and mass transfer in various flow fields), chemical engineers engaged in product and process design face different challenges including (1) determining the composition of chemical mixtures to provide desired properties, (2) creating complex flowsheets and selecting operating conditions to produce desired products with a high degree of yield and selectivity, little recycle, and low utility costs, and (3) creating configured industrial and consumer products. The creation of product and process designs are rarely straightforward and routine; rather they involve innovative approaches that lead to more profitable products and processes that are environmentally sound, and operationally safe. Frequently, they result in patents that give the company protection in the United States for 17 years.

Turn next to Figure 1.2, in which the steps in designing chemical products and processes are shown. Figure 1.2a shows just the steps, each one of which is expanded in Figure 1.2b and described next. Beginning with a potential opportunity, the design team creates and assesses a so-called primitive problem. When necessary, the team seeks to find chemicals or chemical mixtures that have desired properties and performance. Then, when a process is required to produce the chemicals, process *creation* (or invention) is undertaken. When the gross profit is favorable, a base-case design is developed. This is the subject of Part One. In parallel, algorithmic methods are employed to find better process flowsheets, as discussed in Part Two. Also, in parallel, plantwide controllability assessment is undertaken to eliminate processes that are difficult to control, as discussed in Part Four. When the process looks promising and/or when a configured industrial or consumer product is to be designed, the design team carries out detailed design, equipment sizing, and optimization, as

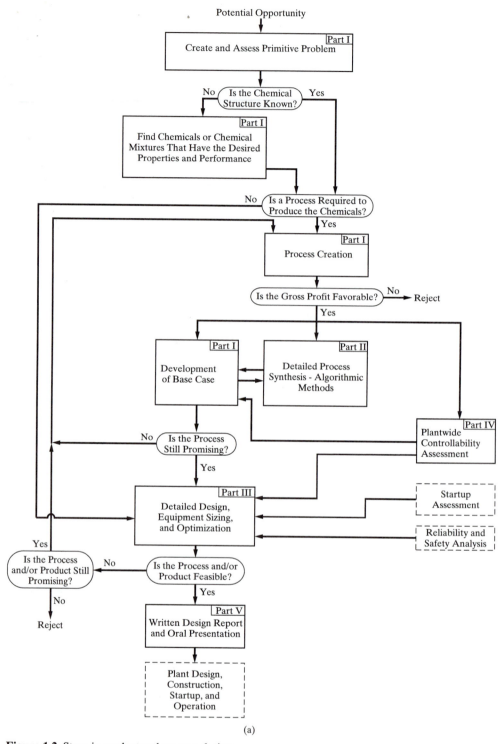

(a)

Figure 1.2 Steps in product and process design.

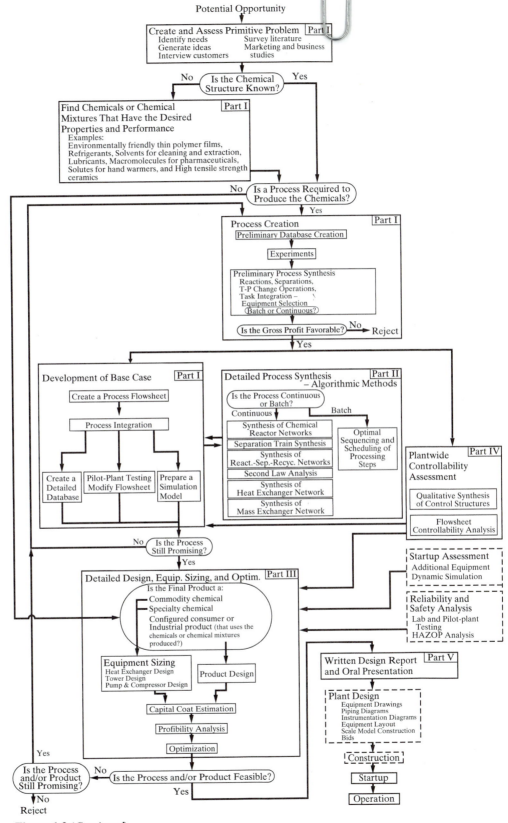

Figure 1.2 (*Continued*).

discussed in Part Three. Finally, Part Five focuses on the written design report and the oral presentation. As the design process is unfolded in this text, reference to Figure 1.2 will be made often.

In the paragraphs that follow, emphasis is placed on the first step of Part One, followed by a brief introduction to the other steps in Part One.

Create and Assess Primitive Problem

Product and process designs begin with a potential opportunity, often a gleam in the eye of an engineer. Usually, the opportunity arises from a customer need, often identified by interviewing customers. Given an array of needs, an effort is made to arrive at specifications for the product; for example, desired density, viscosity, and latent heat of crystallization for a solvent to be used in a hand warmer. In many cases, a design team engages in a formal session to identify needs and generate ideas for the product. This involves brainstorming to arrive at concepts that are potentially promising in satisfying the needs. Having generated several ideas, an attempt is made to select the most promising from among them using the principles of thermodynamics, chemical kinetics, heat and mass transfer, etc. In so doing, designers and design teams create and assess *primitive problems* that are most worthy of further research and development. These steps are discussed thoroughly in *Chemical Product Design* (Cussler and Moggridge, 2001). For many consumer products, the assistance of mechanical engineers is required to provide sufficient strength in the final product. For example, an inexpensive competitor to Scotch Magic Tape™ is almost useless because, although it has good adhesive properties, a desired length of it cannot be withdrawn from the roll because the tape tears prematurely.

Typical Opportunities and Primitive Problems

As examples, consider the following three potential opportunities and the kinds of primitive design problems generated, with all three considered in more depth later in Chapters 3 and 19:

EXAMPLE 1.1   *Dialysis Device*

Consider the possibility of designing an inexpensive (say, less than $10) throw-away product for patients with temporary or permanent kidney failure; a device that provides the only treatment for patients with end-stage renal disease (ESRD), whose kidneys are no longer capable of their function. This treatment, which is required three times per week, for an average of 3–4 hours per dialysis, was performed on more than 200,000 patients in the United States in 1996. ∎

EXAMPLE 1.2   *Drug to Counter Degradation of Blood*

In the manufacture of pharmaceuticals, consider the possible production of plasminogen activators, which are powerful enzymes that trigger the proteolytic (breaking down of proteins to form simpler substances) degradation of blood clots that cause strokes and heart attacks. Since the mid-1980s, Genentech, a U.S. company, has manufactured tissue plasminogen activator (tPA), which they currently sell for $2,000 per 100-mg dose, with annual sales of $300 MM/yr (MM in American engineering units is thousand-thousand or 1 million). Given that their patent will expire soon, Genentech has developed a next generation, Food and Drug Administration (FDA)-approved, plasminogen activator called TNK-tPA, which is easier and safer for clinicians to use. With a rapidly growing market, the question arises as to whether an opportunity exists for another company to manufacture a generic (i.e., without a brand name) form of tPA that can compete favorably with TNK-tPA. ∎

EXAMPLE 1.3   *Vinyl Chloride Manufacture*

Consider the need to manufacture vinyl chloride,

$$
\begin{array}{ccc}
H & & Cl \\
\diagdown & & \diagup \\
& C = C & \\
\diagup & & \diagdown \\
H & & H
\end{array}
$$

a monomer intermediate for the production of polyvinyl chloride,

$$
\begin{array}{ccc}
CHCl & CHCl & CHCl \\
\diagup \diagdown \; \diagup & \diagdown \; \diagup & \diagdown \; \diagup \diagdown \\
CH_2 & CH_2 & CH_2
\end{array}
$$

an important polymer (usually referred to as just vinyl) that is widely used for rigid plastic piping, fittings, and similar products. An opportunity has arisen to satisfy a new demand for vinyl-chloride monomer, on the order of 800 million pounds per year, in a petrochemical complex on the Gulf Coast, given that an existing plant owned by the company produces 1 billion pounds per year of this commodity chemical. Because vinyl-chloride monomer is an extremely toxic substance, it is required that all new facilities be designed carefully to satisfy governmental health and safety regulations.   ∎

Clearly, potential opportunities and primitive problems are generated regularly in the fast-paced corporate, government, and university research environments. In product design, especially, it is important to create an environment that encourages the identification of consumer needs, the generation of product ideas, and the selection from among these ideas of the most promising alternatives, as discussed in the next subsection.

Selecting Alternatives—Assessing the Primitive Problem

Normally, the designer or a small design team generates many potential ideas for products and processes as potential solutions for the primitive problem, particularly if these individuals are well familiar with the existing products or situation. Ideas may also come from potential customers, who may be frustrated with the existing products or situation. The ideas are best generated in a noncritical environment, when "brainstorming." Often, the best ideas may initially be those that might otherwise receive the most criticism. All of the ideas are collected, organized, discussed, and carefully assessed. Cussler and Moggridge (2001) present extensive lists of ideas for several products. From the list of ideas, a selection of the most promising alternatives is made, based upon technical or marketing considerations; for example, thermodynamics or advertising potential. Consider the following design alternatives that are typical of those selected from ideas for the three primitive problems above, noting that it is best at first to consider in depth at least two, and possibly three, alternatives.

EXAMPLE 1.4   *(Example 1.1 Revisited)*

To satisfy the need for a dialysis product, preliminary designs for the following two alternatives might be generated

Alternative 1 Design a hemodialysis device capable of causing urea (CH_4ON_2), uric acid ($C_5H_4O_3N_4$), creatinine ($C_4H_7ON_3$), phosphates, and other low-molecular-weight metabolites to transfer by diffusion from the blood to a dialysate, to be selected. The device should also permit glucose and salts to be transferred by diffusion in the opposite direction. It should be designed to be discarded after a single treatment at a selling price less than \$10.00. The cost and/or performance of the new product should compare favorably with the C-DAK 4000 artificial kidney of Althin CD Medical, Inc. (acquired by Baxter International, Inc. in March 2000). In 1992, 60 million of these disposable, sterilized membrane modules, which weigh less than 100 grams each, were sold at \$5.00 to \$6.00 each.

Alternative 2. Design a peritoneal dialysis device (which uses the serous membrane that lines the cavity of the abdomen of a mammal) having similar attributes as the hemodialysis device. ∎

EXAMPLE 1.5   *(Example 1.2 Revisited)*

In assessing the primitive problem to produce plasminogen activators, two promising alternatives might be generated by a design team:

Alternative 1. While a generic form of tPA may not compete well against TNK-tPA in the United States, it may be possible to market a low-cost generic tPA in foreign markets, where urokinase and streptokinase are low-cost alternatives, which sell for only $200/dose, but are associated with increased bleeding risks. Market analysis suggests that a maximum production rate of 80 kg/yr would be appropriate over the next five years.

Alternative 2. Given the possibility that lower health care reimbursements are received by hospitals in the United States, it may be reasonable to develop a similar process that competes favorably with TNK-tPA in the United States.

Other promising alternatives are likely to arise, often initiated by the successes of a research laboratory. ∎

EXAMPLE 1.6   *(Example 1.3 Revisited)*

To satisfy the need for an additional 800 million lb/yr of vinyl chloride, the following four plausible alternatives might be selected for further consideration.

Alternative 1. A competitor's vinyl-chloride plant, which produces 2 MMM (billion) lb/yr of vinyl chloride and is located about 100 miles away, might be expanded to produce the required amount, which would be shipped by truck or rail in tank car quantities. In this case, the design team projects the purchase price and designs storage facilities. This might be the simplest solution to provide the monomer required to expand the local PVC plant.

Alternative 2. Purchase and ship, by pipeline from a nearby plant, chlorine from the electrolysis of NaCl solution. React the chlorine with in-house ethylene to produce the monomer and HCl as a by-product.

Alternative 3. Because the existing company petrochemical complex produces HCl as a byproduct in many processes (e.g., in chloroform and carbon tetrachloride manufacture) at a depressed price because large quantities are produced, HCl is normally available at low prices. Reactions of HCl with acetylene, or ethylene and oxygen, could produce 1,2-dichloroethane, an intermediate that can be cracked to produce vinyl chloride.

Alternative 4. Design an electrolysis plant to produce chlorine. One possibility is to electrolyze the HCl, available from within the petrochemical complex, to obtain H_2 and Cl_2. React chlorine, according to alternative 2. Elsewhere in the petrochemical complex, react hydrogen with nitrogen to form ammonia or with CO to produce methanol. ∎

These are typical of the alternatives that might be selected from a large number of ideas and which serve as a base on which to begin the engineering of a product or a process. At this stage, the alternatives require further, more detailed study, and hence it is important to recognize that, as the engineering work proceeds, some alternatives will be rejected and new alternatives may be generated. This is a crucial aspect of design engineering. On the one hand, it is important to generate large numbers of ideas leading to a few promising alternatives. On the other hand, to meet the competition with a product designed and manufactured in a timely fashion, it is important to winnow the alternatives that might require too extensive an engineering effort to be evaluated; for example, a process that requires exotic materials of construction or extreme conditions of temperature and/or pressure.

Literature Survey

When generating alternative specific problems, design teams in industry have access to company employees, company files, and the open literature. These resources provide helpful leads to specific problems, as well as information about related products, thermophysical property and transport data, possible flowsheets, equipment descriptions, and process models. If the company has been manufacturing the principal products, or related chemicals, information available to the design team provides an excellent starting point, enabling the team to consider variations to current practice very early in the design cycle. In spite of this, even when designing a next-generation product or plant to expand the production of a chemical product, or retrofitting a plant to eliminate bottlenecks and expand its production, the team may find that many opportunities exist to improve the processing technologies. Several years normally separate products, plant start-ups, and retrofits, during which technological changes are often substantial. For example, the recent shift in distillation, particularly under vacuum conditions, from trays to high-performance packings. For this reason, it is important to make a thorough search of the literature to uncover the latest data, flowsheets, equipment, and models that can lead to an improved product and more profitable design. Several literature resources are widely used by design teams. These include the Stanford Research Institute (SRI) Design Reports, encyclopedias, handbooks, indices, and patents, many of which are available electronically, with an increasing number available on the World Wide Web (WWW, Internet).

SRI Design Reports   SRI, a consortium of several hundred chemical companies, publishes documentation of many chemical processes in considerable detail. While their reports provide a wealth of information, most are written under contract for clients, and consequently, are not available to the public. Yet, some materials are available online, by subscription, to the public. Furthermore, most industrial consultants have access to these reports and may be able to provide helpful information to student design teams, especially those who carry out some of the design work in company libraries.

Encyclopedias   Three very comprehensive, multivolume encyclopedias contain a wealth of information concerning the manufacture of most chemicals. Collectively, these encyclopedias describe the uses for the chemicals, the history of manufacture, typical process flowsheets and operating conditions, and related information. The three encyclopedias are: the *Kirk-Othmer Encyclopedia of Chemical Technology* (1991), the *Encyclopedia of Chemical Processing and Design* (McKetta and Cunningham, 1976), and *Ullmann's Encyclopedia of Industrial Chemistry* (1988). For a specific chemical or substance, it is not uncommon for one or more of these encyclopedias to have 5 to 10 pages of pertinent information, together with literature references for more detail and background. Although the encyclopedias are updated too infrequently to always provide the latest technology, the information they contain is normally very helpful to a design team when beginning to assess a primitive design problem. Many other encyclopedias may also be helpful, including the *McGraw-Hill Encyclopedia of Science and Technology* (1987), *Van Nostrand's Scientific Encyclopedia* (Considine, 1995), the *Encyclopedia of Fluid Mechanics* (Cheremisinoff, 1986), and the *Encyclopedia of Material Science and Engineering* (Bever, 1986).

Handbooks and Reference Books   Several key handbooks and reference books are well known to chemical engineers. These include *Perry's Chemical Engineer's Handbook* (Perry and Green, 1997), the CRC *Handbook of Chemistry and Physics* (the so-called *Rubber Handbook*, published annually by CRC Press, Boca Raton, FL) (Lide, 1997), *JANAF Thermochemical Tables* (Chase, 1985), *Riegel's Handbook of Industrial Chemistry* (Kent, 1992), the *Chemical*

Processing Handbook (McKetta, 1993a), the *Unit Operations Handbook* (McKetta, 1993b), *Process Design and Engineering Practice* (Woods, 1995a), *Data for Process Design and Engineering Practice* (Woods, 1995b), the *Handbook of Reactive Chemical Hazards* (Bretherick, 1990), and the *Standard Handbook of Hazardous Waste Treatment and Disposal* (Freeman, 1989), among many other sources.

Indexes   To search the current literature, especially the research and technology journals, several indexes are extremely helpful. These provide access to a broad spectrum of journals, including electronic access to issues since the late 1970s, with rapidly improving search engines. These indexes provide links to most of the articles published during this period, including kinetics data, thermophysical property data, and much related information for many chemicals. Of primary interest to a design team, are the *Applied Science and Technology Index* (with electronic access to 350 journals since 1983), the *Engineering Index* (with access to 4,500 journals, technical reports, and books, electronically since 1985), *Chemical Abstracts* (one of the most comprehensive scientific indexing and abstracting services in biochemistry, organic chemistry, macromolecular chemistry, physical and analytical chemistry, and applied chemistry and chemical engineering—available electronically with entries since 1907), and the *Science Citation Index* (with access to 3,300 journals since 1955, available electronically since 1980, with searches that indicate where the author's work has been cited).

Patents   These are important sources with which the design team must be aware to avoid the duplication of designs protected by patents. After the 17 years that protect patented products and processes in the United States are over, patents are often helpful in the design of next-generation processes to produce the principal chemicals, or chemicals that have similar properties, chemical reactions, and so on. However, many patents withhold important know-how that may be vital to success. Since a patent is legal property, like a house or a car, it, and perhaps the know-how, can be owned, bought, and sold. Often patents are licensed for fees on the order of 3–6 percent of gross sales. This can be important when a design team decides to incorporate a patented chemical product in its design. Patents from the United States, Great Britain, Germany, Japan, and other countries, are available in major libraries, from which copies can be ordered. Also, most patents can be accessed using the World Wide Web. U.S. patents, those issued and some applications, are available for download, with no charge, at the Internet site, www.uspto.gov/patft.

Auxiliary Studies

While creating and assessing a primitive design problem, design teams often initiate studies of (1) technical feasibility, (2) marketing, and (3) business considerations. For a promising product, the technical feasibility study identifies existing and potentially new manufacturing methods, the advantages and disadvantages of each method, and eventually (as the design proceeds), the reasons for selecting a specific method. In the marketing analysis, other manufacturers are identified, as well as plant capacities, price histories of the raw materials and products, regulatory restrictions, and principal uses of the product. Marketing considerations often far outweigh technical considerations. Many products or process designs are rejected by management for marketing reasons. For each promising design, business objectives and constraints are normally considered, usually in a business study. These include plant capacity, product quality, likely size of first plant expansion, mechanical completion and startup dates, maximum capacity available, maximum operating costs as a function of capacity, seasonal demand changes, inventory requirements, and minimum acceptable return on investment.

Stimulating Innovation in Product Design

The invention and commercialization of new products, and chemical products in particular, can benefit by corporate organization to encourage interactions between researchers, marketers, salespeople, and others. In this regard, companies like 3M and General Electric (G.E.) are noted for their corporate policies that seek to maintain a climate in which innovation flourishes, several examples of which are discussed next.

Fifteen Percent Rule   At 3M, managers are expected to allow employees 15 percent of their time to work on projects of their own choosing. This rule, which has become a fundamental part of the 3M culture, is assessed nicely by Bill Coyne, a research and development manager: "The 15 percent part of the Fifteen Percent Rule is essentially meaningless. Some of our technical people use much more than 15 percent of their time on projects of their own choosing. Some use less than that; some use none at all. The number is not so important as the message, which is this: the system has some slack in it. If you have a good idea, and the commitment to squirrel away time to work on it, and the raw nerve to skirt your lab manager's expressed desires, then go for it" (Gundling, 2000).

Tech Forum   This terminology, which is used at 3M, is typical of organizational structures designed to encourage technical exchange and a cross-fertilization of ideas between persons working in many corporate divisions at widely disparate locations. At 3M, the Tech Forum is organized into chapters and committees, with the chapters focused on technology, including the Physics Chapter, the Life Sciences Chapter, and the Product Design Chapter. Seminars are held related to their own areas of technology, presented by outside speakers or 3M employees. Some chapters do not have a technical focus. For example, the Intellectual Property Chapter is primarily targeted at patent attorneys. Also, at 3M, the Tech Forum hosts a two-day "Annual Event" at the St. Paul headquarters, with each of the 3M labs invited to assemble a booth. Since the company rewards labs when other divisions use their technology, employees have an incentive to participate in this internal trade show.

Stretch Goals   Another 3M policy, intended to *stretch* the pace of innovation, is the rule that at least 30 percent of annual sales should come from products introduced in the last four years. The policy has recently been refined to establish an even greater sense of urgency such that "10 percent of sales should come from products that have been in the market for just one year" (Gundling, 2000).

Process Innovation Technology Centers   Since 75 percent of manufacturing at 3M is done internally, two technology centers are provided. One is staffed with chemical engineers and material scientists to help researchers scale-up a new idea from the bench to production, with a focus on core technologies. The other handles the development and scale-up for key manufacturing process technologies such as coating, drying, and inspection and measurement. The latter is staffed primarily with chemical and mechanical engineers and software development personnel. These centers work closely with researchers and engineers involved in product development and equipment design.

Six Sigma   At companies like Motorola, Allied Signal, General Electric, DuPont, and many others, emphasis has been placed on quality control in manufacturing. Recently, General Electric has created teams to implement a 6 Sigma strategy for quality control (Coe, 2000). For each product and process, a team seeks to reduce the deviations from outside of critical-to-quality (CTQ) specifications to less than 3.4 events per million opportunities; that is, to reduce variance in the CTQ specifications such that the upper and lower acceptance limits lie six standard deviations on either side of the target value. As enthusiasm for this strategy has been growing, several publications have appeared, including Rath and Strong's *Six Sigma Pocket Guide* (2000), and papers introducing the approach (e.g., Wheeler, 2002) and discussing its application to the operation of an ethylene plant (Trivedi, 2002). More recently, emphasis has been placed on applying the analysis during product and process design, with the objective of achieving more reliable designs; see, for example, Rath and Strong's *Design for Six Sigma Pocket Guide* (2002). Further discussion of this analysis is included in Section 19.1 on "Steps in Designing Industrial and Consumer Products."

Pharmaceutical Products

Special considerations are needed for the design of pharmaceutical products. As the design team creates and assesses the primitive problem, it is important to be aware of the typical *development cycle* or time line for the discovery and development of a new chemical entity, as discussed thoroughly by Pisano in *The Development Factory* (1997). The four key steps are examined next.

Discovery   This exploratory research is intended to identify molecules that will prove safe and effective in the treatment of disease. This step involves working backwards to isolate classes of compounds or specific molecular structures that are likely to have the desired therapeutic effect, like blocking a particular enzyme that may cause elevated blood pressure. Most of the work involves literature and patent searches, isolation or synthesis of test tube quantities, and testing on laboratory animals. This is an iterative process that usually involves the exploration of thousands of compounds to locate a handful that are sufficiently promising for further development. Increasingly, it involves methods of genomic analysis, with laboratory testing using microfluidic devices, and the application of data mining techniques to locate the most promising proteins, and cells within which they can be grown, from numerous databases of laboratory data.

Preclinical Development   During this phase, a company seeks to obtain sufficient data on a drug to justify the more expensive and risky step of testing in humans. First, the drug is injected into animal species to determine its toxicity. In addition, pharmacological and pharmoacokinetic studies are undertaken to quantify the main and side effects and the speeds of adsorption and metabolism. In parallel, formulations are devised for administering the drug (e.g., in tablets, capsules, injections, or cream). This phase ends with the preparation and filing of an Investigational New Drug (IND) application that is filed with the FDA in the United States. This application seeks approval to begin testing the drug on humans. Note that about 50 percent of the potential drugs are eliminated for some reason during this phase. While the preclinical development is underway, the process research phase is initiated, in which alternative synthetic routes are considered and evaluated on a laboratory scale.

Clinical Trials   These trials are administered over three phases, each of which has a duration of one–two years. In Phase 1 trials, the drug is tested in multiple doses on healthy volunteers to determine whether there are significant side effects and to identify maximum tolerable doses. When Phase 1 is successful, Phase 2 trials involve afflicted patients. Both the drug and *placebo* treatments are administered to a control group, where the patients are unaware of whether they have received the drug, a placebo, or a substitute drug. During Phase 1 and especially during Phase 2, the development of a pilot plant facility is accelerated, as the demand for test quantities increases, leading into Phase 3 trials. During the latter phase, the drug is administered to thousands of patients at many locations over several years. The intent is to confirm the safety and efficacy of the drug over long-term use, as compared with existing drugs. When successful, the data are submitted to the FDA. Note that when approval is granted, only the expanded pilot plant, constructed for Phase 3, is permitted to produce the drug for commercial distribution.

Approval   Together with the data from the clinical trials, an application is prepared for the FDA requesting permission to sell the drug. The FDA evaluates the application during a period which can last up to two years.

Summary   In summary, extensive work to find the molecules, usually proteins, that have the appropriate therapeutic properties begins in the step to *Create and Assess the Primitive Problem*. This work continues into the step to *Find Chemicals or Chemical Mixtures Having Desired Properties and Performance*, which is introduced in the next section. Then, as Phases 1 and 2 of the Clinical Trials proceed, process design is undertaken to produce large quantities of the drug, first for Phase 3 testing and then for commercial operation, as discussed in Chapter 3, "Process Creation," in Section 3.4 for a plant to produce tissue plasminogen activator.

Find Chemicals or Chemical Mixtures Having Desired Properties and Performance

For those primitive problems in which desired properties and performance have been specified, it is often necessary to identify chemicals or chemical mixtures that meet these specifications. Examples include (1) thin polymer films to protect electronic devices, having a high glass-transition temperature and low water solubility, (2) refrigerants that boil and condense at desired temperatures and low pressures, while not reacting with ozone in the earth's stratosphere, (3) environmentally friendly solvents for cleaning, for example, to remove ink pigments, and for separations, as in liquid–liquid extraction, (4) low-viscosity lubricants, (5) proteins for pharmaceuticals that have the desired therapeutic effects, (6) solutes for hand warmers that remain supersaturated at normal temperatures, solidifying at low temperatures when activated, and (7) ceramics having high tensile strength and low viscosity for processing. Often design problems are formulated in which the molecular structure is manipulated, using optimization methods, to achieve the desired properties. For this purpose, methods of property estimation are needed, which often include group contribution methods, and increasingly, molecular simulations (using molecular dynamics and Monte-Carlo methods). The search for molecular structure is often iterative, involving heuristics, experimentation, and the need to evaluate numerous alternatives in parallel, especially in the discovery of pharmaceutical proteins, as discussed in Chapter 2.

For some chemical products, like creams and pastes, the specification of desired properties is a key to successful product design. Creams and pastes are colloidal systems that contain immiscible liquid phases, as well as solid particles. As described by Wibowo and Ng (2001), the first step in their design involves the identification of *product quality factors*, including *functional* quality factors (e.g., protects, cleans, and decorates the body, delivers an active pharmaceutical ingredient, . . .), *rheological* quality factors (e.g., pours easily, spreads easily on the skin, does not flow under gravity, stirs easily, coats uniformly, . . .), *physical* quality factors (e.g., remains stable for an extended period, melts at a specified temperature, releases an ingredient at a controlled rate, . . .), and *sensorial* quality factors (e.g., feels smooth, does not feel oily, appears transparent, opaque, or pearlescent, does not cause irritation, . . .). Given these specifications, the second step involves *product formulation*, which involves selection of the ingredients, the emulsion type (if applicable), the emulsifier (if necessary), and determination of the product microstructure. Then, the process creation step and product evaluation steps follow, as discussed below. In a second paper, Wibowo and Ng (2002) expand upon these steps, and apply them for several products, including dry toner, laundry detergent, shampoo, and cosmetic lotion.

When specifying desired properties and selecting potential chemicals and chemical mixtures, design teams must be acutely aware of environmental and safety issues and regulations. These issues are so important that they are discussed separately in Sections 1.3 and 1.4, and throughout the book. See, for example, the selection of environmentally friendly refrigerants and solvents in Chapter 2 and the need to avoid producing and storing hazardous intermediates in Section 5.2.

Process Creation

For those primitive problems for which a process must be designed to produce the chemical products, the design team begins the process creation step. As shown in Figure 1.2, this involves the assembly of a preliminary database that is comprised of thermophysical property data, including vapor–liquid equilibrium data, flammability data, toxicity data, chemical prices, and related information needed for preliminary process synthesis. In some cases, experiments are initiated to obtain important missing data that cannot be accurately estimated, especially when the primitive problem does not originate from a laboratory study. In this regard, experimental reaction data are always required, as is experimental separation data when mixtures to be separated are moderately to highly nonideal. Then, preliminary process synthesis begins with the design team creating flowsheets involving just the reaction, separation, and temperature- and pressure-change operations. Process equipment is selected in a so-called *task-integration* step. This latter step involves the selection of the operating mode; that is, continuous, batch, or semicontinuous. Only those flowsheets that show a favorable gross profit are explored further; the others are rejected. In this way, detailed work on the process is avoided when the projected cost of the raw materials exceeds that of the products. These steps are described in detail in Chapter 3, in which flowsheets of operations are synthesized to address alternative 2 for the problem of increasing the production of vinyl chloride, and alternative 1 for the problem of producing tissue plasminogen activator (tPA).

Product Type

Returning to Figure 1.1, the manufacturing processes shown differ significantly depending on the chemical product type. For the manufacture of basic chemical products, the process flowsheet involves chemical reaction, separation, pumping, compression, and similar operations, as listed in Table 1.1. On the other hand, for the manufacture of industrial products, extrusion, blending, compounding, and stamping are typical operations. Note also that the focus on quality control of the product shifts from the control of the physical, thermal, chemical, and rheological properties to the control of optical properties, weatherability, mechanical strength, printability, and similar properties. This book focuses on the manufacture of basic chemical products. Processes are not discussed for the manufacture of industrial products. Later, configured industrial and consumer products are discussed in Chapter 19, where emphasis is placed on their *design*, but not on the manufacturing process, which includes parts making, parts assembly and integration, and finishing.

Development of Base Case Process

To address the most promising flowsheet alternatives for the manufacture of basic chemicals, the design team is usually expanded or assisted by specialized engineers, to develop *base-case designs*. This usually involves the development of just one flow diagram for each favorable process. As described in Section 3.5, the design team begins by creating a detailed process flow diagram, accompanied by material and energy balances, and a list of the major equipment items. A material balance table shows the state of each stream; that is, the temperature, pressure, phase, flow rate, and composition, plus other properties as appropriate. In

Table 1.1 Process Operations for Three Types of Chemical Products

Product Type	Process Operation					
	1 Raw Materials Handling	2 Conversion	3	4	5 Quality Control	6 Packaging
Basic chemical products	Feeding, pumping, compressing	Chemical reaction	**Purification:** Separation	—	**Intrinsic: quality:** Physical, thermal, chemical, rheological, properties	**Shipped** tank cars, pipelines
Industrial products	Feeding, pumping, web handling, drying dehumidifying	Extrusion, blending, compounding	**Primary forming:** Die/profile extrusion, pultrusion, molding	**Secondary forming:** Sealing, stamping, thermal/light treatment, orientation	**Functional quality:** Optical weatherability, mechanical strength, printability	**Packaged** cartons, sacks, rolls
Consumer products	Web/fiber handling, belt conveying, discrete handling	Parts making includes die cutting, molding	**Parts assembly and integration**	**Finishing**	**Functional (see above) and use quality:** durability, lifetime, ease of use, effectiveness	**Packaged** individual containers, plastic wrapped, vials, bottles, cans

many cases, the material and energy balances are performed, at least in part, by a computer-aided process simulator, such as ASPEN PLUS, HYSYS, CHEMCAD, and PRO/II for commodity chemicals, and BATCH PLUS and SUPERPRO DESIGNER for specialty chemicals, especially pharmaceuticals. Then, the design team seeks opportunities to improve the designs of the process units and to achieve more efficient process integrations for the production of commodity chemicals, applying the methods of *heat and power integration*, for example, by exchanging heat between hot and cold streams, and *mass integration* to minimize raw materials and wastes.

For each base-case design, three additional activities usually take place in parallel. Given the detailed process flow diagram, the design team refines the preliminary database to include additional data such as transport properties and reaction kinetics, feasibilities of the separations, matches to be avoided in heat exchange (i.e., forbidden matches), heuristic parameters, equipment sizes and costs as a function of throughput, and so on. This is usually accompanied by pilot-plant testing to confirm that the various equipment items will operate properly and to refine the database. If unanticipated data are obtained, the design team may need to revise the flow diagram. In some cases, equipment vendors run tests, as well as generate detailed equipment specifications. To complement these activities, a simulation model is prepared for the base-case design. Process simulators are often useful in generating databases because of their extensive data banks of pure-component properties and physical property correlations for ideal and nonideal mixtures. When not available, simulation programs can regress experimental data taken in the laboratory or pilot plant for empirical or theoretical curve fitting.

As shown in Figure 1.2, in developing a base-case design, the design team checks regularly to confirm that the process remains promising. When this is not the case, the team often returns to one of the steps in process creation or redevelops the base-case design.

Finally, before leaving this topic, the reader should note that *process creation* and *development of a base-case process* are the subjects of Part One of this book, entitled "Product and Process Invention—Heuristics and Analysis" (Chapters 1–5.)

Detailed Process Synthesis Using Algorithmic Methods

While the design team develops one or more base-case designs, detailed process synthesis may be undertaken using algorithmic methods as described in Part Two. For continuous processes, these methods: (1) create and evaluate chemical reactor networks for conversion of feed to product chemicals (Chapter 6), separation trains for recovering species in multicomponent mixtures (Chapter 7), and reactor–separator–recycle networks (Chapter 8), and (2) locate and reduce energy usage (Chapter 9), create and evaluate efficient networks of heat exchangers with turbines for power recovery (Chapter 10), and networks of mass exchangers (Chapter 11) to reduce waste. For batch processes, these methods create and evaluate optimal sequences and schedules for batch operation (Chapter 12). With the results of these methods, the design team compares the base case with other promising alternatives, and in many cases identifies flowsheets that deserve to be developed along with, or in place of, the base-case design. More specifically, Chapter 9 discusses second law analysis, which provides an excellent vehicle for screening the base-case design or alternatives for energy efficiency. In this analysis, the lost work is computed for each process unit in the flowsheet. When large losses are encountered, the design team seeks methods to reduce these losses. Then, in Chapter 10, algorithmic methods are utilized to synthesize networks of heat exchangers, turbines, and compressors to satisfy the heating, cooling, and power requirements of the process. These methods, which place emphasis on the minimization of utilities such as steam and cooling water, are used by the design team to provide a high degree of heat and power integration in the most promising processes.

Industrial Chemicals

In the manufacture of industrial chemicals, such as films, fibers, and paper, processes are synthesized involving extrusion, blending, compounding, sealing, stamping, and related operations, as summarized in Table 1.1. These processes normally involve large throughputs, and consequently, are usually continuous, with some discrete operations that occur at high frequency, such as sealing and stamping. Methods of process synthesis rely heavily on heuristics and are not as well developed as for the manufacture of basic chemical products. For these processes, the emphasis is on the unit operations that include single- and twin-screw extrusion, coating, and fiber spinning, subjects beyond the scope of this book. Excellent coverage of these processes is provided in the *Principles of Polymer Processing* (Tadmor and Gogos, 1979), and in Chapter 12 of *Process Modeling* (Denn, 1986).

Plantwide Controllability Assessment

An assessment of the controllability of the process is initiated after the detailed process flow diagram has been completed, beginning with the qualitative synthesis of control structures for the entire flow diagram, as discussed in Chapter 20. Then measures are utilized that can be applied before the equipment is sized in the detailed design stage, to assess the ease of controlling the process and the degree to which it is inherently resilient to disturbances. These measures permit alternative processes to be screened for controllability and resiliency with little effort and, for the most promising processes, they identify promising control structures. Subsequently, control systems are added and rigorous dynamic simulations are carried out to confirm the projections using the approximate measures discussed previously. This is the subject matter of Chapter 21, which together with Chapter 20, comprise Part Four of this book on "Plantwide Controllability Assessment."

Detailed Design, Equipment Sizing, and Optimization—Configured Product Design

Depending on the primitive design problem, the detailed design involves equipment sizing of units in a new process for commodity and specialty chemicals, that is, *detailed process design*, and/or the determination of the product configuration for a configured industrial or consumer product (that uses the chemicals or chemical mixtures produced), as shown in Figure 1.2.

Commodity and Specialty Chemicals

For a new process to produce commodity or specialty chemicals, after completing the base-case design, the design team usually receives additional assistance in carrying out the detailed process design, equipment sizing and capital-cost estimation, profitability analysis, and optimization of the process. These topics are covered in separate chapters in Part Three. Although these chapters describe several methods, having a range of accuracy, for computing equipment sizes, cost estimates, and profitability analyses, it is important to recognize that the more approximate methods are often sufficient to distinguish between alternatives during product conception and process creation (Part One), and detailed process synthesis (Part Two). Throughout these parts of the book, references are made to the approximate methods included in Chapters 16 and 17.

While carrying out these steps, the design team formulates startup strategies to help identify the additional equipment that is usually required. In some cases, using dynamic simulators, the team extends the model for the dynamic simulation of the control system and tests startup strategies, modifying them when they are not implemented easily. In addition, the team often prepares its recommendations for the initial operating strategies after startup has been completed.

Another crucial activity involves a *formal* analysis of the reliability and safety of the proposed process, as discussed in Section 1.4. Note that, as discussed earlier, in Section 1.4, and throughout the book, these considerations must be foremost throughout the design process. If not accomplished earlier during process creation, detailed design, and controllability analysis, formal safety analysis usually involves laboratory and pilot-plant testing to confirm that typical faults (valve and pump failures, leaks, etc.) cannot propagate through the plant to create accidents, such as explosions, toxic clouds of vapor, or fires. Often, HAZOP (*Haz*ard and *Op*erability) analyses are carried out to check systematically all of the anticipated eventualities.

When the detailed process design stage is completed, the economic feasibility of the process is checked to confirm that the company's profitability requirements have been met. If this proves unsatisfactory, the design team determines whether the process is still promising. If so, the team returns to an earlier step to make changes that it hopes will improve the profitability. Otherwise, this process design is rejected. Methods for and examples of HAZOP analysis, together with risk assessment, are not presented in this textbook. Instead, the reader is referred to the text by Crowl and Louvar (1990) and the following books developed by the Center for Chemical Process Safety of the American Institute of Chemical Engineers:

1. *Safety, Health, and Loss Prevention in Chemical Processes: Problems for Undergraduate Engineering Curricula—Student Problems* (1990).
2. *Guidelines for Hazard Evaluation Procedures, Second Edition with Worked Examples* (1992).
3. *Self-Study Course: Risk Assessment* (2002).

The latter reference is particularly noteworthy for instructors because it provides a Power-Point file that can be integrated into a safety lecture.

Configured Industrial or Consumer Products

When the primitive design problem leads to a configured industrial or consumer product, much of the design activity is centered on the three-dimensional structure of the product. Typical chemically related, industrial and consumer products, many of which are discussed in this book, include, for example, hemodialysis devices, solar desalination units, automotive fuel cells, hand warmers, multilayer polymer mirrors, integrated circuits, germ-killing surfaces, insect repelling wristbands, disposable diapers, ink-jet cartridges, transparencies for overhead projectors, sticky wall hangers, and many others. In many cases, the product must be configured for ease of use and to meet existing standards, as well as to be easily manufactured. Increasingly, when determining the product configuration, distributed-parameter models, involving ordinary and partial differential equations, are being created. Simple discretization algorithms are often used to obtain solutions, as well as finite-element packages like the *Finite-element Toolbox* in MATLAB and the FEMLAB and FLUENT packages.

As discussed in Chapter 19, the design team carries out a detailed literature search, with care to explore the patent literature, to determine that a similar product does not exist. When not available, the product invention phase begins in which ideas are generated and screened, with detailed three-dimensional designs developed. For the most promising inventions, prototypes are built and tested on a small-scale. Only after the product passes these initial tests, do the design engineers focus on the process to manufacture the product in commercial quantities. This scale-up often involves pilot-plant testing, as well as consumer testing, before a decision is made to enter commercial production. Clearly, the methods of capital-cost estimation, profitability analysis, and optimization of Chapters 16–18 are applied before and during the design of the manufacturing facility, although when the product can be sold at a high price, due to large demand and limited production, detailed profitability analysis and optimization are less important. For these products, it is most important to reduce the design and manufacturing time to capture the market before competitive products are developed.

Written Design Report and Oral Presentation

Reporting and presentation are key steps in selling company management on the need to proceed with the design. As discussed in Chapter 22, the written design report is a work in progress as the design evolves. The design report is the basis for the final, detailed design of the product and/or plant, suitable for manufacturing and/or construction.

Plant Design, Construction, Startup, and Operation

In creating the plant design for a chemical process and/or the manufacturing plant for a product, much detailed work is done, often by contractors, using many mechanical, civil, and electrical engineers. For processes that produce commodity and specialty chemicals, they complete equipment drawings, piping diagrams, instrumentation diagrams, the equipment layout, the construction of a scale model, and the preparation of bids. For the manufacture of products, the focus shifts to devices that cast and shape component parts, robots that assemble the parts, finishing, and packaging and labeling, as shown in Table 1.1. Then, the construction phase is entered, in which engineers and project managers play a leading role. The design team may return to assist in plant startup and operation. Note that the final design and construction activities are placed in dashed boxes in Figure 1.2 because they are usually not the responsibilities of chemical engineers.

Summary

With the preceding description of Figures 1.1 and 1.2, and Table 1.1, the reader should have gained a good appreciation of the subjects to be learned in product and process design and how this text is organized to describe the design methodologies in the context of many kinds of design problems. Having completed the coverage on creating and assessing the primitive problem, for the next steps, beginning with molecular structure design, the reader should turn to Chapter 2. Before leaving this chapter, however, discussions are provided on environmental protection in Section 1.3, important safety considerations in Section 1.4, engineering ethics in Section 1.5, and the role of the computer in product and process design in Section 1.6. These are key concerns of a design team throughout the design process, and consequently, these sections provide background for many of the discussions in the chapters that follow.

1.3   ENVIRONMENTAL PROTECTION

One of the most significant changes that has occurred since the late 1970s throughout the manufacturing and transportation sectors within the United States, and those of many other industrialized nations, is the transformation of environmental protection from a secondary to a primary issue. Because of the Environmental Protection Agency (EPA), with the cooperation of the U.S. Congress, tighter environmental regulations have been legislated and enforced over this period. This has resulted in a noticeable improvement in air quality (especially in urban areas), a reduction in water pollution, and considerable progress in the remediation of many waste dumps containing toxic chemicals. In short, the United States and many other industrialized nations are rapidly increasing their emphases on maintaining a clean environment. To bring this about, in recent years, large investments have been made by the chemical process industries to eliminate sources of pollution. These have increased the costs of manufacturing which, in turn, have been transmitted to consumers through increased costs of end products. Because most producers are required to satisfy the same regulations, the effect has been to translate the costs to most competitors in an evenhanded manner. Problems have arisen, however, when chemicals are produced in countries that do not have strict environmental standards and are subsequently imported into the United States at considerably lower prices. Issues such as this are discussed regularly at international conferences on the environment, which convene every two or three years with the objective of increasing the environmental standards of all countries. In this section, several of the more pressing environmental issues are reviewed, followed by a discussion of many environmental factors in process design. Then, a few primitive problem statements are reviewed, with reference to the more complete statements provided on the CD-ROM that accompanies this book. For more comprehensive coverage in these areas, the reader is referred to the discussions of "Environmental Protection, Process Safety, and Hazardous Waste Management" in *Frontiers in Chemical Engineering* (Amundson, 1988); *Environmental Considerations in Process Design and Simulation* (Eisenhauer and McQueen, 1993); *Pollution Prevention for Chemical Processes* (Allen and Rosselot, 1997); and *Green Engineering: Environmentally Conscious Design of Chemical Processes* (Allen and Shonnard, 2002). Early efforts to protect the environment focused on the removal of pollutants from waste gas, liquid, and solid streams. Effort has now shifted to *waste minimization* (e.g., waste reduction, pollution prevention).

Environmental Issues

At the risk of excluding many key environmental issues, the following are singled out as being closely related to the design and operation of chemical processes.

Burning of Fossil Fuels for Power Generation and Transportation

Because fossil fuels are the predominant sources of power worldwide, their combustion products are a primary source of several pollutants, especially in the urban centers of industrialized nations. More specifically, effluent gases from burners and fires contain sizable concentrations of SO_2, the nitrogen oxides (NO_x), CO, CO_2, soot, ash, and unburned hydrocarbons. These, in turn, result in many environmental problems, including acid rain (principally concentrated in H_2SO_4), smog and hazes (concentrated in NO_x), the accumulation of the so-called *greenhouse gas* (CO_2), volatile toxic compounds (e.g., formaldehyde, phenol), and organic gases (e.g., CO), which react with NO_x, especially on hot summer days, altering the O_3 level. As the adverse impacts of pollutants on animals, plant life, and humans are being discovered by scientists and engineers, methods are sought to reduce their levels significantly. In some cases, this is accomplished by one of several methods, such as separating the sources (e.g., sulfur compounds) from fuels; adjusting the combustion process (e.g., by reducing the temperature and residence time of the flame to produce less NO_x); separating soot, ash, and noxious compounds from effluent gases; reacting the effluent gases in catalytic converters; or through the use of algae to consume (through photosynthesis) large quantities of CO_2 in flue gases (a recently proposed technique now under study). As a rule of thumb, it should be noted that the cost of cleaning combustion products is approximately an order of magnitude less than the cost of removing contaminants from fuel. This is an important heuristic, especially when designing processes that are energy intensive, requiring large quantities of fuel.

Handling of Toxic Wastes

In the chemical and nuclear power industries, large quantities of toxic wastes are produced annually, largely in wastewater streams, which in 1988 amounted to 97% of the wastes produced, as shown in the pie chart of Figure 1.3. While a small portion is incinerated (on the order of 3% in the late 1980s), the bulk is disposed of in or on the land, with a variety of methods having been introduced over the past century to bury these wastes. Since the late 1960s, many of the burial sites (e.g., Love Canal, Times Beach) have threatened the health of nearby residents and, more broadly, have threatened to contaminate the underground water supply throughout entire states and countries. In this regard, studies by the state of California have shown that aqueous waste streams from the processing of electronic materials are posing widespread threats to the groundwater in California's Silicon Valley. In fact, this area has a leading number of sites on the U.S. National Priority List of toxic waste dumps (which is comprised of approximately 10,000 sites throughout the United States). In process design, it is essential that facilities be included to remove pollutants from wastewater streams. The design of mass-exchange networks (MENs) for this and other purposes is the subject of Chapter 11.

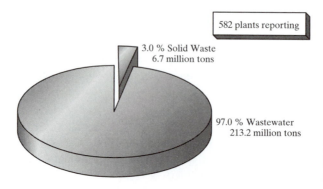

582 plants reporting

3.0 % Solid Waste
6.7 million tons

97.0 % Wastewater
213.2 million tons

Figure 1.3 Hazardous waste generation in the United States in 1988 (Eisenhauer and McQueen, 1993).

Bioaccumulated Chemicals

Probably the most well-known cases of chemicals that have been discovered to *bioaccumulate* in the soil and plant life are the insecticide DDT (1,1-bis(4-chlorophenyl)-2,2,2-trichloroethane; $C_{14}H_9Cl_5$) and the solvent PCBs (polychlorinated biphenyls). DDT was sprayed in large quantities by low-flying airplanes to kill insects and pests throughout the 1950s. Unfortunately, although effective for protecting crops, forests, and plant life, toxic effects in birds, animals, and humans were strongly suspected, as discussed in Section 1.5. Consequently, DDT was banned by the U.S. EPA in 1972. Its effect, however, will remain for some time due to it having bioaccumulated in the soil and plant life.

Toxic Metals and Minerals

In this category, major changes have taken place since the late 1960s in response to the discoveries of the toxic effects of lead, mercury, cadmium, and asbestos on animals and humans. After lead poisoning (accompanied by brain damage, disfigurement, and paralysis) was related to the ingestion of lead-based paints by children (especially in older buildings that are not well maintained), the EPA banned lead from paints as well as from fuels. In fuel, tetraethyl lead had been used as an octane enhancer throughout the world. It was subsequently replaced by methyl *tertiary*-butyl ether (MTBE), which is also being replaced due to reports that it can contaminate groundwater. Mercury, which has been the mainstay of manometers in chemistry laboratories, has similarly been found to be extremely toxic, with disastrous effects of accidental exposure and ingestion reported periodically. In the case of asbestos, its toxic effects have been known since the late 1940s, yet it remains a concern in all buildings built before then. Gradually, as these buildings are being renovated, sheets of asbestos insulation and asbestos ceiling tiles are being removed and replaced by nontoxic materials. Here, also, the incidents of asbestos poisoning are associated most often with older buildings that have not been well maintained.

Summary

As the adverse effects of these and other chemicals becomes better understood, chemical engineers are being called on to satisfy far stricter environmental regulations. In many cases, these regulations are imposed to be safe even before sufficient data are available to confirm toxic effects. For these reasons, chemical companies are carefully reexamining their existing processes and evaluating all proposed plants to confirm that they are environmentally sound, at least insofar as meeting the regulations imposed, or anticipated to be imposed, by the environmental regulation agencies.

Environmental Factors in Process Design

The need to retrofit existing plants and to design environmentally-sound new plants has required chemical engineers to become far more proficient in accounting for environmentally related factors. In this section, a few of the better recognized factors are discussed. Additional coverage is included related to purges in Section 5.3, energy conservation in Chapters 9 and 10, wastewater treatment in Chapter 11, and fuel cells for automobiles in Section 19.9. More complete coverage may be found in the comprehensive textbook, *Green Engineering: Environmentally Conscious Design of Chemical Processes* (Allen and Shonnard, 2002).

Reaction Pathways to Reduce Byproduct Toxicity

The selection of reaction pathways to reduce byproduct toxicity is a key consideration when a design team creates specific problems from a primitive problem, and during preliminary process synthesis, when the reaction operations are positioned. As the reaction operations are determined by chemists and biochemists in the laboratory, the toxicity of all of the chemicals, especially chemicals recovered as byproducts, needs to be evaluated. For this purpose, companies have toxicity laboratories and, in many cases, large repositories of toxicity data. One useful source, especially for students at universities, is the *Pocket Guide to Chemical Hazards* of the National Institute for Occupational Safety and Health (NIOSH, 1987). Clearly, when large quantities of toxic chemicals are anticipated, other reaction pathways must be sought; when these cannot be found, design concepts are rejected, except under unusual circumstances.

Reducing and Reusing Wastes

Environmental concerns have caused chemical engineers to place even greater emphasis on recycling, not only unreacted chemicals but also product and byproduct chemicals. In so doing, design teams commonly anticipate the *life cycles* of their products and byproducts, paying special attention to the waste markets, so as to select the appropriate waste quality. Stated differently, the team views the proposed plant as a producer of engineering scrap and attempts to assure that there will be a market for the chemicals produced after their useful life is over. Clearly, this is a principal consideration in the production of composite materials and polymers. In this connection, it is important to plan on producing segregated wastes when they are desired by the waste market, and in so doing, to avoid overmixing the waste streams.

Avoiding Nonroutine Events

To reduce the possibilities for accidents and spills, with their adverse environmental consequences, processes are often designed to reduce the number of transient operations, clean-up periods, and catalyst regeneration cycles. In other words, emphasis is on the design of a process that is easily controlled at or near a nominal steady state, with reliable controllers and effective fault-detection sensors.

Materials Characterization

Often, waste chemicals are present in small amounts in gaseous or liquid effluents. To maintain low concentrations of such chemicals below the limits of environmental regulations, it is important to use effective and rapid methods for measuring or deducing their concentrations from other measurements. In this regard, the design team needs to understand the effect of concentration on toxicity, which can vary significantly in the dilute concentration range. Yet another consideration is to design the plant to use recycled chemicals—that is, someone else's waste. When this is accomplished, it is necessary to know the range of compositions within which the waste chemicals are available.

Design Objectives, Constraints, and Optimization

Environmental objectives are normally not well defined because economic objective functions normally involve profitability measures, whereas the value of reduced pollution is not easily quantified by economic measures. As a consequence, design teams often formulate mixed objective functions that attempt to express environmental improvements in financial terms. In other cases, the team may settle for the optimization of an economic objective function, subject to bounds on the concentrations of the solutes in the waste streams. It is important to assess whether the constraints are *hard* (not allowed to be violated) or *soft* (capable of being violated under unusual circumstances). Emphasis must be placed on the formulation of each constraint and the extent to which it must be honored.

Regulations

As mentioned previously, some environmental regulations can be treated as constraints to be satisfied during operation of the process being designed. When a mathematical model of the proposed process is created, the design team can check that these constraints are satisfied for the operating conditions being considered. When an objective function is formulated, the design variables can be adjusted to obtain the maximum or minimum while satisfying the constraints. Other regulations, however, are more difficult to quantify. These involve the expectations of the public and the possible backlash should the plant be perceived as a source of pollution. In a similar vein, constraints may be placed on the plant location, principally because the local government may impose zoning regulations that require chemical plants to be located in commercial areas, beyond a certain distance from residential neighborhoods. To keep these regulations from becoming too prohibitive, chemical companies have a great incentive to gain public confidence by satisfying environmental regulations and maintaining excellent safety records, as discussed in Section 1.4.

Intangible Costs

Like the regulations imposed by local governments, some of the economic effects of design decisions related to the environment are very difficult to quantify. These include the cost of liability when a plant is found to be delinquent in satisfying regulations, and in this connection, the cost of legal fees, public relations losses, and delays incurred when environmental groups stage protests. Normally, because these costs cannot be estimated reliably by a design team, mixed objectives are not formulated and no attempts are made to account for them in an optimization study. Rather, the design team concentrates on ensuring that the regulations will be satisfied, thereby avoiding legal fees, public relations losses, and the complications associated with public demonstrations.

Properties of Dilute Streams

Most pollutants in the effluent and purge streams from chemical plants are present in dilute concentrations. Furthermore, since the regulations often require that their concentrations be kept below parts per million or parts per billion, reliable and fast analysis methods are needed to ensure that the regulations are satisfied. Beyond that, it is often important to understand the impact of the concentration on the kinetics of these species in the environment—for example, the rates of chemical reaction of organic species, such as CO, with NO_x in the atmosphere to produce O_3, and the rate at which other reaction byproducts are formed. With this knowledge, a company can help regulatory agencies arrive at concentration limits more scientifically and, in some cases, at limits that are less restrictive, and cost companies, and the consumers of their products, less in the long run. Note that, in urban smog, high concentrations of ozone often create problems for people with respiratory ailments.

Properties of Electrolytes

Many aqueous streams contain inorganic compounds that dissociate into ionic species, including acids, bases, and salts, often in dilute concentrations. These electrolytic solutions commonly occur in the manufacture of inorganic chemicals (e.g., soda ash, Na_2CO_3), in the strong solvents used in the pulp and paper industry, in the aqueous wastes associated with the manufacture of electronic materials (e.g., silicon wafers, integrated circuits, photovoltaic films), and in many other industries. Strong electrolytes dissociate into ionic species whose interactions with water and organic molecules are crucial to understanding the state of a mixture—that is, the phases present (vapor, water, organic liquid, solid precipitates, etc.) at a given temperature and pressure. Hence, when designing processes that involve electrolytes, a design team needs to include the properties of ionic species in its thermophysical properties database. Fortunately, to provide assistance for designers, databases and facilities for estimating the thermophysical properties of a broad base of ionic species, over an increasing range of temperatures and pressures, are available in process simulators.

Environmental Design Problems

Since the late 1970s, the number of design projects focusing on the solution of environmental problems has increased significantly. These, in turn, are closely related to environmental regulations, which have become increasingly strict. Although it is beyond the scope of this book to provide a comprehensive treatment of the many kinds of designs that have been completed, it is important that the reader gain a brief introduction to typical design problems. This is accomplished through the design projects listed in Table 1.2. As can be seen, a large fraction of the design projects are concerned with air quality; others involve water treatment; two involve soil treatment; one involves the conversion of waste fuel to chemicals; and one proposes the use of a biochemical conversion to consume solid waste and produce ethanol fuel. On the multimedia CD-ROM that accompanies this book, the primitive problem statements for these design projects, as they were presented to student groups, are reproduced in unabridged form. See the file Design Problem Statements.pdf. Keep in mind that, as the de-

Table 1.2 Environmental Design Projects

Project	Location in Book
Air quality	
R134a Refrigerant	App. II—Design Problem A-II.7.1
Biocatalytic desulfurization of diesel fuel	App. II—Design Problem A-II.7.2
Sulfur recovery using oxygen-enriched air	App. II—Design Problem A-II.7.3
California smog control	App. II—Design Problem A-II.7.4
Zero emissions from a THF plant	App. II—Design Problem A-II.7.5
Volatile organic compound (VOC) abatement—thermal incineration, catalytic incineration, or adsorption, for ozone control	App. II—Design Problem A-II.7.6
Recovery and purification of HFC-125 (pentafluoroethane) by distillation	App. II—Design Problem A-II.7.7
CO_2 fixation by microalgae to mitigate the greenhouse effect	App. II—Design Problem A-II.7.8
H_2 generation for reformulated gasoline	App. II—Design Problem A-II.7.9
Water treatment	
Effluent remediation from wafer fabrication	App. II—Design Problem A-II.8.1
Recovery of germanium from optical fiber manufacturing effluents	App. II—Design Problem A-II.8.2
Solvent waste recovery	App. II—Design Problem A-II.8.3
Soil treatment	
Phytoremediation of lead-contaminated sites	App. II—Design Problem A-II.9.1
Soil remediation and reclamation	App. II—Design Problem A-II.9.2
Miscellaneous	
Fuel processor for 5 kW PEM fuel cell unit	App. II—Design Problem A-II.10.1
Combined-cycle power generation	App. II—Design Problem A-II.10.2
Production of low-sulfur diesel fuel	App. II—Design Problem A-II.10.3
Waste-fuel upgrading to acetone and isopropanol	App. II—Design Problem A-II.10.4
Conversion of cheese whey (solid waste) to lactic acid	App. II—Design Problem A-II.10.5
Ethanol for gasoline from corn syrup	App. II—Design Problem A-II.10.6

signs proceeded, the design teams often upgraded the information provided, and in some cases created *specific problems* that were not anticipated by the originator of the primitive problem statement.

A closer look at Table 1.2 shows that the projects address many aspects of air quality control. Two alternative approaches to sulfur removal from fuels are proposed, one involving desulfurization of the fuel, the other the recovery of sulfur from its combustion products. One is concerned with NO_x removal from combustion products, and three involve the recovery of hydrocarbons from effluent gases. One explores the interesting possibility of growing algae by the photosynthesis of CO_2 from combustion gases as a vehicle for reducing the rate at which CO_2 is introduced into the atmosphere. Under water treatment, the projects involve the recovery of organic and inorganic chemicals from aqueous waste streams. Two alternative approaches to soil treatment are proposed, including the use of phytoremediation; that is, using plants to absorb lead and other heavy metals. All of the projects involve chemical reactions, and consequently, the design teams are comprised of chemical engineers, chemists, and biochemists. In this respect, it seems clear that chemistry and biology are the key ingredients that qualify chemical engineers to tackle these more challenging environmental problems.

1.4   SAFETY CONSIDERATIONS

A principal objective in the design and operation of chemical processes is to maintain safe conditions for operating personnel and inhabitants who live in the vicinity of the plants. Unfortunately, the importance of meeting this objective is driven home periodically by accidents, especially accidents in which lives are lost and extensive damage occurs. To avoid this, all companies have extensive safety policies and procedures to administer them. In recent years, these have been augmented through cooperative efforts coordinated by technical societies, for example, the Center for Chemical Plant Safety of the American Institute of Chemical Engineers, which was formed in 1985, shortly after the accident in Bhopal, India, on December 3, 1984. In this accident, which took place in a plant partially owned by Union Carbide and partially owned locally, water (or some other substance—the cause is still uncertain) accidentally flowed into a tank in which the highly reactive intermediate, methyl isocyanate (MIC) was stored, leading to a rapid increase in temperature accompanied by boiling, which caused toxic MIC vapors to escape from the tank. The vapors passed through a pressure-relief system and into a scrubber and flare system that had been installed to consume the MIC in the event of an accidental release. Unfortunately, these systems were not operating, and approximately 25 tons of toxic MIC vapor were released, causing a dense vapor cloud that escaped and drifted over the surrounding community, killing more than 3,800 civilians and seriously injuring an estimated 30,000 more.

Like Section 1.3 on environmental issues, this section begins with a review of two safety issues that are considered by many design teams, followed by an introduction to many of the design approaches for dealing with these issues. For more comprehensive coverage of these areas, the reader is referred to *Chemical Process Safety: Fundamentals with Applications* (Crowl and Louvar, 1990); *Plant Design for Safety—A User-Friendly Approach* (Kletz, 1991); and a collection of student problems, *Safety, Health, and Loss Prevention in Chemical Processes: Problems for Undergraduate Engineering Curricula—Student Problems* (American Institute of Chemical Engineers, 1990).

The U.S. Chemical Safety and Hazard Investigation Board (CSB), established by the Clean Air Act Amendments of 1990, is an independent federal agency with the mission of ensuring the safety of workers and the public by preventing or minimizing the effects of chemical incidents. They attempt to determine the root and contributing causes of chemical accidents. Their web site at http://www.csb.gov is a very useful source of brief and detailed accident reports.

Safety Issues

Of the many potential safety issues, two are singled out for coverage here because they must be confronted often in the design of chemical, petroleum, and petrochemical plants and in other plants in which exothermic reactions and operations occur at elevated pressures.

Fires and Explosions

In organic chemical processes, it is not uncommon for sizable concentrations of flammable chemicals to accumulate in air or pure oxygen with the possibilities of ignition or even explosion. For this reason, laboratory studies have been carried out to determine the flammability limits for many of the common organic chemical vapors. These limits at 25°C and 1 atm are listed for many chemicals in Table 1.3, where the LFL is the lower flammability limit (that is, the volume percent of the species in air below which flammability does not occur) and the UFL is the upper flammability limit (above which flammability does not occur). Within these limits, flames and explosions can occur and, consequently, design teams must be careful to keep the concentrations outside the flammability range. In addition, Table 1.3 provides autoignition temperatures, above which a flammable mixture is capable of extracting enough energy from the environment to self-ignite. At lower temperatures, an ignition source must be present. The flash point, given in the second column of Table 1.3, is the lowest temperature at which sufficient vapor exists in air to form an ignitable mixture. At the flash point, the vapor burns, but only briefly as insufficient vapor is formed to sustain combustion.

Table 1.3 pertains to pure chemicals. For mixtures, the flammability limits are often estimated using the Le Chatelier equation, an empirical equation that must be used with caution:

$$\text{LFL}_{\text{mix}} = \frac{1}{\sum_{i=1}^{C} (y_i/\text{LFL}_i)} \quad \text{UFL}_{\text{mix}} = \frac{1}{\sum_{i=1}^{C} (y_i/\text{UFL}_i)} \quad (1.1)$$

where LFL_i and UFL_i are the flammability limits of species i, y_i is the mole fraction of species i in the vapor, and C is the number of chemical species in the mixture, excluding air.

To extend the flammability limits to elevated temperatures and pressures, the following equations have been developed:

$$\text{LFL}_T = \text{LFL}_{25} \left(1 - \frac{0.75(T - 25)}{\Delta H_c} \right) \quad (1.2a)$$

$$\text{UFL}_T = \text{UFL}_{25} \left(1 + \frac{0.75(T - 25)}{\Delta H_c} \right) \quad (1.2b)$$

and

$$\text{UFL}_p = \text{UFL} + 20.6(\log P + 1) \quad (1.3)$$

where T is the temperature (in °C), ΔH_c is the net heat of combustion (in kcal/mol at 25°C), P is the pressure (in MPa absolute), and UFL is the upper flammability limit at 101.3 kPa (1 atm). The lower flammability limit is not observed to vary significantly with the pressure. These equations, plus others to estimate the flammability limits for species not listed in Table 1.3, are presented by Crowl and Louvar (1990), with a more complete discussion and references to their sources. With this kind of information, the process designer makes sure that flammable mixtures do not exist in the process during startup, steady-state operation, or shutdown.

Toxic Releases and Dispersion Models

In chemical processing, it is desirable to avoid working with chemicals that are toxic to animals, humans, and plant life. This is an important consideration as design teams select from

Table 1.3 Flammability Limits of Liquids and Gases

Compound	Flash Point (°F)	LFL (%) in air	UFL (%) in air	Autoignition temperature (°F)
Acetone	0.0[a]	2.5	13	1,000
Acetylene	Gas	2.5	100	
Acrolein	−14.8	2.8	31	
Acrylonitrile	32	3.0	17	
Aniline	158	1.3	11	
Benzene	12.0[b]	1.3	7.9	1,044
Butane	−76	1.6	8.4	761
Carbon monoxide	Gas	12.5	74	
Chlorobenzene	85[b]	1.3	9.6	1,180
Cyclohexane	−1[b]	1.3	8	473
Diborane	Gas	0.8	88	
Dioxane	53.6	2.0	22	
Ethane	−211	3.0	12.5	959
Ethyl alcohol	55	3.3	19	793
Ethylene	Gas	2.7	36.0	914
Ethylene oxide	−20[a]	3.0	100	800
Ethyl ether	−49.0[b]	1.9	36.0	180
Formaldehyde		7.0	73	
Gasoline	−45.4	1.4	7.6	
Heptane	24.8	1.1	6.7	
Hexane	−15	1.1	7.5	500
Hydrogen	Gas	4.0	75	1,075
Isopropyl alcohol	53[a]	2.0	12	850
Isopropyl ether	0	1.4	7.9	830
Methane	−306	5	15	1,000
Methyl acetate	15	3.1	16	935
Methyl alcohol	54[a]	6	36	867
Methyl chloride	32	8.1	17.4	1,170
Methyl ethyl ketone	24[a]	1.4	11.4	960
Methyl isobutyl ketone	73	1.2	8.0	860
Methyl methacrylate	50[a]	1.7	8.2	790
Methyl propyl ketone	45	1.5	8.2	941
Naphtha	−57	1.2	6.0	550
Octane	55.4	1.0	6.5	
Pentane	−40	1.51	7.8	588
Phenol	174	1.8	8.6	
Propane	Gas	2.1	9.5	
Propylene	−162	2.0	11.1	927
Propylene dichloride	61	3.4	14.5	1,035
Propylene oxide	−35	2.3	36	869
Styrene	87[b]	1.1	7.0	914
Toluene	40	1.2	7.1	997

[a]Open-cup flash point

[b]Closed-cup flash point

Sources: Martha W. Windholtz, Ed., *The Merck Index: An Encyclopedia of Chemicals, Drugs, and Biologicals*, 10th ed. (Merck, Rahway, NJ, 1983), p. 1124; Gressner G. Hawley, Ed., *The Condensed Chemical Dictionary*, 10th ed. (Van Nostrand Reinhold, New York, 1981), pp. 860–861; Richard A. Wadden and Peter A. Scheff, *Engineering Design for the Control of Workplace Hazards* (McGraw-Hill, New York, 1987), pp. 146–156.

among the possible raw materials and consider alternate reaction paths, involving inter-mediate chemicals and byproducts. In some cases, decisions can be made to work with non-toxic chemicals. However, toxicity problems are difficult to avoid, especially at the high concentrations of chemicals in many process streams and vessels. Consequently, the potential for a release in toxic concentrations during an accident must be considered care-fully by design teams. In so doing, a team must identify the ways in which releases can occur; for example, due to the buildup of pressure in an explosion, the rupture of a pipeline due to surges at high pressure, or the collision of a tank car on a truck or train. It is also important for the team to select protective devices and processing units, to assess their potential for failure, and, in the worse case, to model the spread of a dense, toxic vapor. Given the potential for the rapid spreading of a toxic cloud, it is often necessary to find an alternative design, not involving this chemical, rather than take the chance of ex-posing the surrounding community to a serious health hazard. Although it is beyond the scope of this discussion, it should be noted that dispersion models are developed by chemical engineers to predict the movement of vapor clouds under various conditions—for example, a continuous point release, at steady state, with no wind; a puff with no wind; a transient, continuous point release with no wind; as well as all of the previously mentioned factors with wind. These and other models are described by Crowl and Louvar (1990) and de Nevers (1995), accompanied by example calculations.

Design Approaches toward Safe Chemical Plants

In the previous discussion of two important safety issues, design approaches to avoid acci-dents have been introduced. This section provides a more complete enumeration without dis-cussing implementational details, which are covered by Crowl and Louvar (1990).

Techniques to Prevent Fires and Explosions

One method of preventing fires and explosions is *inerting*—that is, the addition of an inert gas to reduce the oxygen concentration below the minimum oxygen concentration (MOC), which can be estimated using the LFL and the stoichiometry of the combustion reaction. An-other method involves avoiding the buildup of static electricity and its release in a spark that can serve as an ignition source. Clearly, the installation of grounding devices, and the use of antistatic additives that increase conductivity, reducing the buildup of static charges, can help to reduce the incidence of sparking. In addition, explosion-proof equipment and instru-ments are often installed; for example, explosion-proof housings that do not prevent an ex-plosion, but that do absorb the shock and prevent the combustion from spreading beyond the enclosure. Yet another approach is to ensure that the plant is well ventilated, in most cases constructed in the open air, to reduce the possibilities of creating flammable mixtures that could ignite. Finally, sprinkler systems are often installed to provide a rapid response to fires and a means to contain them effectively.

Relief Devices

In processes where pressures can build rapidly, especially during an accident, it is crucial that the design team provide a method for relieving the pressure. This is accomplished using a variety of relief devices, depending on whether the mixtures are vapors, liquids, solids, or combinations of these phases. In some cases the vessels can be vented to the atmosphere; in other cases they are vented to containment systems, such as scrubbers, flares, and con-densers. The devices include relief and safety valves, knock-out drums, rupture disks, and the like. Relief system design methodologies are presented in detail in the AIChE publica-tion, *Emergency Relief System Design Using DIERS Technology* (1992).

Hazards Identification and Risk Assessment

Hazards identification and risk assessment are key steps in process design. As shown in Figure 1.2, they are normally carried out in connection with the preparation of the final design. In these steps, the plant is carefully scrutinized to identify all sources of accidents or hazards. This implies that the design team must consider the propagation of small faults into catastrophic accidents, an activity that is complicated by the possibility of two or more faults occurring either simultaneously or in some coordinated fashion. At some point, especially after the economics satisfy the feasibility test, the design team normally prepares a HAZOP study mentioned on page 19, in which all of the possible paths to an accident are identified. Then, when sufficient probability data are available, a fault tree is created and the probability of the occurrence for each potential accident is computed. Clearly, this requires substantial experience in operating comparable facilities, which is generally available in the large chemical companies.

Material Safety Data Sheets

A process design should be accompanied by a Material Safety Data Sheet (MSDS) for every chemical appearing in the process. These sheets, which are developed by chemical manufacturers and kept up to date under OSHA (Occupational Safety and Health Agency of the federal government) regulations, contain safety and hazard information, physical and chemical characteristics, and precautions on safe handling and use of the chemical. The MSDSs, which usually involve several pages of information are available on the Internet at:

> http://hazard.com/msds/
> http://www.ilpi.com/msds/

1.5 ENGINEERING ETHICS

In 1954, the National Society of Professional Engineers (NSPE) adopted the following statement, known as the Engineers' Creed:

> *As a Professional Engineer, I dedicate my professional knowledge and skill to the advancement and betterment of human welfare.*
>
> *I pledge:*
>
> *To give the utmost of performance;*
> *To participate in none but honest enterprise;*
> *To live and work according to the laws of man and the highest standards of*
>   *professional conduct;*
> *To place service before profit, the honor and standing of the profession before*
>   *personal advantage, and the public welfare above all other considerations.*
> *In humility and with need for Divine Guidance, I make this pledge.*

In 1977, a similar statement was approved by the Accreditation Board for Engineering and Technology (ABET), as follows:

> *Engineers uphold and advance the integrity, honor, and dignity of the engineering profession by:*
>
>   *I. Using their knowledge and skill for the enhancement of human welfare;*
> *II. Being honest and impartial, and serving with fidelity the publics, their*
>   *employees;*
> *III. Striving to increase the competence and prestige of the engineering*
>   *profession; and*
> *IV. Supporting the professional and technical societies of their disciplines.*

These two statements have to do with ethics, also called *moral philosophy*, which is derived from the Greek *ethika*, meaning character. Thus, ethics deals with standards of conduct or morals. Unfortunately, there are no universal standards; only the ethics of Western civilization are considered in detail here. There is a movement toward the development of global ethics, which is described briefly at the end of this section.

Engineering ethics is concerned with the personal conduct of engineers as they uphold and advance the integrity, honor, and dignity of engineering while practicing their profession. This conduct of behavior has obligations to (1) self, (2) employer and/or client, (3) colleagues and co-workers, (4) public, and (5) environment.

Specific examples of these obligations are given in individual codes of ethics adopted by the various engineering societies (e.g., AIChE, ASCE, ASME, and IEEE) and by the NSPE. The following is the Code of Ethics adopted by the American Institute of Chemical Engineers (AIChE):

Members of the American Institute of Chemical Engineers shall uphold and advance the integrity, honor, and dignity of the engineering profession by: being honest and impartial and serving with fidelity their employers, their clients, and the public; striving to increase the competence and prestige of the engineering profession; and using their knowledge and skill for the enhancement of human welfare. To achieve these goals, members shall:

1. *Hold paramount the safety, health, and welfare of the public in performance of their professional duties.*
2. *Formally advise their employers or clients (and consider further disclosure, if warranted) if they perceive that a consequence of their duties will adversely affect the present or future health or safety of their colleagues or the public.*
3. *Accept responsibility for their actions and recognize the contributions of others; seek critical review of their work and offer objective criticism of the work of others.*
4. *Issue statements and present information only in an objective and truthful manner.*
5. *Act in professional matters for each employer or client as faithful agents or trustees, and avoid conflicts of interest.*
6. *Treat fairly all colleagues and co-workers, recognizing their unique contributions and capabilities.*
7. *Perform professional services only in areas of their competence.*
8. *Build their professional reputations on the merits of their services.*
9. *Continue their professional development throughout their careers, and provide opportunities for the professional development of those under their supervision.*

A more detailed code of ethics for engineers was adopted initially by the NSPE in July 1964. Since then, it has been updated 22 times and will probably continue to receive updates. The 1996 version is shown in Figure 1.4. Some idea of the direction in which engineering ethics is moving may be gleaned from the following changes made since 1996, as taken from the NSPE Web site at:

http://www.nspe.org.

February 2001—The NSPE Board approved the following change to the Code of Ethics: Deletion of Section III.1.e. "Engineers shall not actively participate in strikes, picket lines, or other collective coercive action.

*July 2002—The NSPE Board approved the following changes to the Code of
Ethics: New Section II.1.e. Engineers shall not aid or abet the unlawful practice of
engineering by a person or firm. Old Section II.1.e. was renumbered as new Section
II.1.f.*

It is important for an engineer, or one preparing for entry into the profession, to develop
the ability to address, in an ethical fashion, significant workplace problems that may involve
difficult choices. For this purpose, the Online Ethics Center (OEC) for Engineering and Science (formerly the World Wide Web Ethics Center for Engineering and Science), was established in 1995 under a grant from the National Science Foundation (NSF). The Center,
located at Case-Western Reserve University, provides very extensive educational resources,
including more than 100 case studies, at the Web site:

http://onlineethics.org/

Figure 1.5 provides just a sample of Center case studies dealing with public safety and welfare. The Center also sponsors conferences, addresses the ABET Readiness Committee call
for a "*Guide to Ethics for Dummies,*" and provides sample student assignments, from fresh-
man to senior level, on practical ethics for use by instructors.

A breach of ethics or a courageous show of ethics is frequently newsworthy. An example
of a breach of ethics is that of an MIT student who used university hardware to distribute
commercial software over the Internet. Recent examples of a more courageous show of
ethics, which are presented as case studies by the WWW Ethics Center for Engineering and
Science, include

1. The attempts by Roger Boisjoly to avert the *Challenger* space disaster.
2. The emergency repair by William LeMessurier of structural supports for the Citicorp
   Tower in New York City.
3. The campaign of Rachel Carson for control of the use of pesticides.

The work of Rachel Carson has had a significant impact on stirring the world to action on
environmental protection, beginning with concerted efforts on college campuses. Carson was
a U.S. Fish and Wildlife Service biologist who, in 1951, published *The Sea Around Us*,
which won the National Book Award. In 1962, Carson's book *Silent Spring* was published.
In that book, she criticized the widespread use of chemical pesticides, fertilizers, and weed
killers, citing case histories of damage to the environment. In particular, she cited the disappearance of songbirds (thus, the title of the book) due to the use of synthetic, chlorine-
containing pesticide DDT (previously discussed in Section 1.3), which kills insects by
acting as a nerve poison. Synthetic pesticides were developed during a period of great economic development after World War II in an attempt to reduce insect-caused diseases in
humans and to increase food production. More specifically, the use of DDT practically
eliminated the anopheles mosquito that had caused malaria in many countries in Asia,
Africa, and South and Central America. Carson claimed that the problems created by DDT
were worse than the problems it solved. DDT disrupted reproductive processes and caused
bird eggs to be infertile or deformed. Because DDT breaks down very slowly in the soil,
its concentration builds up in the food chain as larger organisms eat smaller ones. Even
though no adverse effects of DDT on humans have been found, its use in the United States
was banned in 1972. However, it is still manufactured in the United States, and it is still
used in parts of the world for malaria control.

Concern over the environment has led to much interest in the development of global
ethics. Considerable information on this subject is available on the Web site of The Institute
for Global Ethics, which is located at Camden, Maine:

http://www.globalethics.org

Preamble

Engineering is an important and learned profession. As members of this profession, engineers are expected to exhibit the highest standards of honesty and integrity. Engineering has a direct and vital impact on the quality of life for all people. Accordingly, the services provided by engineers require honesty, impartiality, fairness and equity, and must be dedicated to the protection of the public health, safety, and welfare. Engineers must perform under a standard of professional behavior that requires adherence to the highest principles of ethical conduct.

I. Fundamental Canons

Engineers, in the fulfillment of their professional duties, shall:

1. Hold paramount the safety, health, and welfare of the public.
2. Perform services only in areas of their competence.
3. Issue public statements only in an objective and truthful manner.
4. Act for each employer or client as faithful agents or trustees.
5. Avoid deceptive acts.
6. Conduct themselves honorably, responsibly, ethically, and lawfully so as to enhance the honor, reputation, and usefulness of the profession.

II. Rules of Practice

1. Engineers shall hold paramount the safety, health, and welfare of the public.

a. If engineers' judgment is overruled under circumstances that endanger life or property, they shall notify their employer or client and such other authority as may be appropriate.
b. Engineers shall approve only those engineering documents that are in conformity with applicable standards.
c. Engineers shall not reveal facts, data, or information without the prior consent of the client or employer except as authorized or required by law or this Code.
d. Engineers shall not permit the use of their name or associate in business ventures with any person or firm that they believe are engaged in fraudulent or dishonest enterprise.
e. Engineers having knowledge of any alleged violation of this Code shall report thereon to appropriate professional bodies and, when relevant, also to public authorities, and cooperate with the proper authorities in furnishing such information or assistance as may be required.

2. Engineers shall perform services only in the areas of their competence.

a. Engineers shall undertake assignments only when qualified by education or experience in the specific technical fields involved.
b. Engineers shall not affix their signatures to any plans or documents dealing with subject matter in which they lack competence, nor to any plan or document not prepared under their direction and control.
c. Engineers may accept assignments and assume responsibility for coordination of an entire project and sign and seal the engineering documents for the entire project, provided that each technical segment is signed and sealed only by the qualified engineers who prepared the segment.

3. Engineers shall issue public statements only in an objective and truthful manner.

a. Engineers shall be objective and truthful in professional reports, statements, or testimony. They shall include all relevant and pertinent information in such reports, statements, or testimony, which should bear the date indicating when it was current.
b. Engineers may express publicly technical opinions that are founded upon knowledge of the facts and competence in the subject matter.
c. Engineers shall issue no statements, criticisms, or arguments on technical matters that are inspired or paid for by interested parties, unless they have prefaced their comments by explicitly identifying the interested parties on whose behalf they are speaking, and by revealing the existence of any interest the engineers may have in the matters.

4. Engineers shall act for each employer or client as faithful agents or trustees.

a. Engineers shall disclose all known or potential conflicts of interest that could influence or appear to influence their judgment or the quality of their services.
b. Engineers shall not accept compensation, financial or otherwise, from more than one party for services on the same project, or for services pertaining to the same project, unless the circumstances are fully disclosed and agreed to by all interested parties.
c. Engineers shall not solicit or accept financial or other valuable consideration, directly or indirectly, from outside agents in connection with the work for which they are responsible.
d. Engineers in public service as members, advisors, or employees of a governmental or quasi-governmental body or department shall not participate in decisions with respect to services solicited or provided by them or their organizations in private or public engineering practice.
e. Engineers shall not solicit or accept a contract from a governmental body on which a principal or officer of their organization serves as a member.

5. Engineers shall avoid deceptive acts.

a. Engineers shall not falsify their qualifications or permit misrepresentation of their or their associates' qualifications. They shall not misrepresent or exaggerate their responsibility in or for the subject matter of prior assignments Brochures or other presentations incident to the solicitation of employment shall not misrepresent pertinent facts concerning employers, employees, associates, joint venturers, or past accomplishments.
b. Engineers shall not offer, give, solicit, or receive, either directly or indirectly. any contribution to influence the award of a contract by public authority, or which may be reasonably construed by the public as having the effect of intent to influencing the awarding of a contract. They shall not offer any gift or other valuable consideration in order to secure work. They shall not pay a commission, percentage, or brokerage fee in order to secure work, except to a bona fide employee or bona fide established commercial or marketing agencies retained by them.

Figure 1.4 NSPE code of ethics for engineers—1996 version.

III. Professional Obligations

1. Engineers shall be guided in all their relations by the highest standards of honesty and integrity.

a. Engineers shall acknowledge their errors and shall not distort or alter the facts.
b. Engineers shall advise their clients or employers when they believe a project will not be successful.
c. Engineers shall not accept outside employment to the detriment of their regular work or interest. Before accepting any outside engineering employment they will notify their employers.
d. Engineers shall not attempt to attract an engineer from another employer by false or misleading pretenses.
e. Engineers shall not actively participate in strikes, picket lines, or other collective coercive action.
f. Engineers shall not promote their own interest at the expense of the dignity and integrity of the profession.

2. Engineers shall at all times strive to serve the public interest.

a. Engineers shall seek opportunities to participate in civic affairs, career guidance for youths, and work for the advancement of the safety, health, and well-being of their community.
b. Engineers shall not complete, sign, or seal plans and/or specifications that are not in conformity with applicable engineering standards. If the client or employer insists on such unprofessional conduct, they shall notify the proper authorities and withdraw from further service on the project.
c. Engineers shall endeavor to extend public knowledge and appreciation of engineering and its achievements.

3. Engineers shall avoid all conduct or practice that deceives the public.

a. Engineers shall avoid the use of statements containing a material misrepresentation of fact or omitting a material fact.
b. Consistent with the foregoing, engineers may advertise for recruitment of personnel.
c. Consistent with the foregoing, engineers may prepare articles for the lay or technical press, but such articles shall not imply credit to the author for work performed by others.

4. Engineers shall not disclose, without consent, confidential information concerning the business affairs or technical processes of any present or former client or employer, or public body on which they serve.

a. Engineers shall not, without the consent of all interested parties, promote or arrange for new employment or practice in connection with a specific project for which the engineer has gained particular and specialized knowledge.
b. Engineers shall not, without the consent of all interested parties, participate in or represent an adversary interest in connection with a specific project or proceeding in which the engineer has gained particular specialized knowledge on behalf of a former client or employer.

5. Engineers shall not be influenced in their professional duties by conflicting interests.

a. Engineers shall not accept financial or other considerations, including free engineering designs, from material or equipment suppliers for specifying their product.
b. Engineers shall not accept commissions or allowances, directly or indirectly, from contractors or other parties dealing with clients or employers of the engineer in connection with work for which the engineer is responsible.

6. Engineers shall not attempt to obtain employment or advancement or professional engagements by untruthfully criticizing other engineers, or by other improper or questionable methods.

a. Engineers shall not request, propose, or accept a commission on a contingent basis under circumstances in which their judgment may be compromised.
b. Engineers in salaried positions shall accept part-time engineering work only to the extent consistent with policies of the employer and in accordance with ethical considerations.
c. Engineers shall not, without consent, use equipment, supplies, laboratory, or office facilities of an employer to carry on outside private practice.

7. Engineers shall not attempt to injure, maliciously or falsely, directly or indirectly, the professional reputation, prospects, practice, or employment of other engineers. Engineers who believe others are guilty of unethical or illegal practice shall present such information to the proper authority for action.

a. Engineers in private practice shall not review the work of another engineer for the same client, except with the knowledge of such engineer, or unless the connection of such engineer with the work has been terminated.
b. Engineers in governmental, industrial, or educational employ are entitled to review and evaluate the work of other engineers when so required by their employment duties.
c. Engineers in sales or industrial employ are entitled to make engineering comparisons of represented products with products of other suppliers.

8. Engineers shall accept personal responsibility for their professional activities, provided, however, that engineers may seek indemnification for services arising out of their practice for other than gross negligence, where the engineer's interests cannot otherwise be protected.

a. Engineers shall conform with state registration laws in the practice of engineering.
b. Engineers shall not use association with a nonengineer, a corporation, or partnership as a "cloak" for unethical acts.

9. Engineers shall give credit for engineering work to those to whom credit is due, and will recognize the proprietary interests of others.

a. Engineers shall, whenever possible, name the person or persons who may be individually responsible for designs, inventions, writings, or other accomplishments.
b. Engineers using designs supplied by a client recognize that the designs remain the property of the client and may not be duplicated by the engineer for others without express permission.
c. Engineers, before undertaking work for others in connection with which the engineer may make improvements, plans, designs, inventions, or other records that may justify copyrights or patents, should enter into a positive agreement regarding ownership.
d. Engineers' designs, data, records, and notes referring exclusively to an employer's work are the employer's property. Employer should indemnify the engineer for use of the information for any purpose other than the original purpose.

Figure 1.4 (*Continued*)

- **Suspected Hazardous Waste**
  A supervisor instructs a student engineer to withhold information from a client about the suspected nature of waste on the client's property, to protect what the supervisor takes to be the client's interest.
- **Clean Air Standards and a Government Engineer**
  An engineer defies immediate supervisor because he believes supervisor's instruction would pose an environmental health hazard.
- **The Responsibility for Safety and the Obligation to Preserve Client Confidentiality**
  Tenants sue their building's owner, and the owner employs an engineer who finds structural detects not mentioned in the tenant's lawsuit. Issues of public safety versus client confidentiality.
- **Code Violations with Safety Implications**
  Engineer discovers deficiencies in a building's structural integrity, and it would breach client confidentiality to report them to a third party.
- **Whistleblowing City Engineer**
  An engineer privately informs other city officials of an environmental threat, a problem her supervisor has ordered her not to disclose.
- **Safety Considerations and Request for Additional Engineering Personnel**
  An engineer is concerned for worker safety during construction but yields to his client's objections at the cost of an on-site representative.
- **Engineer's Dispute with Client Over Design**
  A client believes an engineer's designs are too costly; but the engineer fears that anything less may endanger the public.
- **Do Engineers Have a Right to Protest Shoddy Work and Cost Overruns?**
  An engineer who is employed by a government contractor objects to a subcontractor's poor performance and is ignored and silenced by management.
- **Change of Statement of Qualifications for a Public Project**
  An engineering firm takes measures to remedy a deficit in a particular area of expertise needed to successfully compete for and carry out a public project.
- **Knowledge of Damaging Information**
  An engineer has a conflict between honoring an agreement to an employer and reporting a hazard to protect the public interest.

Figure 1.5 Engineering ethics cases in the Ethics Center for Engineering & Science (http://www.cwru.edu/affil/wwwethics).

The Institute exists because many believe in their statements:

1. *Because we will not survive the twenty-first century with the twentieth century's ethics.*
2. *The immense power of modern technology extends globally. Many hands guide the controls and many decisions move those hands. A good decision can benefit millions, while an unethical one can cripple our future.*

The Institute strongly believes that education in ethics must begin at the middle- and high-school level. Accordingly, they provide instructional materials suitable for that level. They also stress the concept of moral courage, which they are in the process of defining. In a recent white paper, the Institute makes the following statements:

Moral courage is different from physical courage. Physical courage is the willingness to face serious risk to life or limb instead of fleeing from it.

Moral courage is not about facing physical challenges that could harm the body. It's about facing mental challenges that could harm one's reputation, emotional well-being, self-esteem, or other characteristics. These challenges, as the term implies, are deeply connected with our moral sense—our core moral values.

Moral courage, . . , has four salient characteristics:

- *It is the courage to be moral—to act with fairness, respect, responsibility, honesty, and compassion even when the risks of doing so are substantial.*
- *It requires a conscious awareness of those risks. The sleepwalker on the ridgepole is not courageous unless, waking up, he or she perceives the danger and goes forward anyway.*

- *It is never formulaic or automatic, but requires constant vigilance against its opposite (moral timidity) and its counterfeit (moral foolhardiness).*
- *It can be promoted, encouraged, and taught through precept, example, and practice.*

The teaching of engineering ethics to senior engineering students can be difficult, especially when students raise questions from their personal experiences. Years ago, a student in a senior design class, at an appointment in the instructor's office, asked the following question: "Two weeks ago, I accepted an offer of employment from a company that had set a deadline for accepting the offer. Yesterday, I received a better offer from another company for a better job opportunity at a higher starting salary. What should I do?" At that time, the instructor was inclined to tell the student to stand by the commitment to the first company. Several years later the tables were turned. A senior student told the instructor that he had accepted an offer with an excellent company and then rejected two other offers that he had received. One month later, the student informed the instructor that the company, to which he had committed, reneged on the offer because of a downturn in the economy. Furthermore, job offers from the two other companies that had made offers were no longer available. From then on, when asked, the instructor recited these two episodes and told students to look out for their own best interests. If they got a better offer after accepting an earlier offer, renege on the first and take the second. Was the instructor giving ethical advice?

1.6   ROLE OF COMPUTERS

Many calculations in process design do not require detailed algorithms because they involve simple equations and graphic procedures that can be carried out quickly without the complications of computers. In some circles, designers take pride in making quick and effective decisions using heuristics and back-of-the-envelope calculations. Indeed, in the earliest steps in Figure 1.2, calculations are often quite approximate and the sources of data not very extensive. However, it does not take long for design teams to seek some computational assistance, particularly with process simulators, if for no other reason than to access their extensive physical property data banks. As more data are obtained and flowsheets become more complicated, designers use a combination of computer resources involving spreadsheets, mathematical packages, and process simulators, both steady-state and dynamic. In this section, the objective is to introduce briefly these three computational aids, with emphasis on their role in the design process. Table 1.4 is a listing of the more widely used computer programs that have been found useful in process design. A much more extensive list of chemical engineering software is *The 1997 CEP Software Directory* of the AIChE, which describes more than 1,700 commercial computer programs. An up-to-date searchable version of the CEP Software Directory is provided on the Web site:

http://www.cepmagazine.org/features/software/.

As of March 2002, the list contained more than 1,800 programs. The search can be made by keyword, title, description, or vendor. The list can also be browsed by vendor, title, or by one of more than 35 categories.

Table 1.4 does not include traditional procedural programming languages such as FORTRAN, C, C++, Java and Visual Basic. Although many chemical engineering departments continue to provide instruction in one or more of these languages, recent surveys have concluded that few chemical engineers in industry write their own computer programs. Instead, they use the higher-level languages listed in Table 1.4 to solve their problems. Consequently, many departments now require students to learn the use of such mathematical packages as spreadsheets, MATLAB, POLYMATH, MATHCAD, and MAPLE; together with process simulators such as ASPEN PLUS, CHEMCAD, and HYSYS.Plant.

Table 1.4 Computer Programs Useful in Process Design

1. Spreadsheets (commercial programs):
   Corel Quattro Pro 10 in Word Perfect Office 2002 (http://www.corel.com)
   Lotus 1-2-3 in Smart Suite 9.8 (http://www.lotus.com)
   Microsoft Excel XP (http://www.microsoft.com)
2. Symbolic mathematics (commercial programs):
   DERIVE 5 (http://www.derive.com)
   MACSYMA (http://www.mathtools.net/MATLAB/Symbolic/Other/)
   MAPLE 8 (http://www.maplesoft.com)
   MATHCAD 11 (http://www.mathcad.com)
   MATHEMATICA 4.2 (http://www.wolfram.com)
3. Numerical mathematics (commercial programs):
   MACSYMA 2.4 (http://www.scientek.com/macsyma/mxmain.htm)
   MAPLE 8 (http://www.maplesoft.com)
   MATHCAD 11 (http://www.mathcad.com)
   MATHEMATICA 4.2 (http://www.wolfram.com)
   MATLAB 6.5 (http://www.mathworks.com)
   POLYMATH 5.1 (http://www.polymath-software.com)
   TK SOLVER 4 (http://www.uts.com)
4. Ordinary and Partial Differential Equations by numerical methods (public domain software that can be downloaded from the Internet site at http://netlib2.cs.utk.edu):
   COLMOD An automatic continuation code for solving stiff boundary value problems.
   COLNEW A general purpose code for solving mixed-order systems of boundary value problems using spline collocation and Newton's method.
   ODEPACK A collection of codes for solving stiff and nonstiff systems of initial value problems.
   PDECOL A code for solving coupled systems of partial differential equations of at most second order, in one space and one time dimension, or two space dimensions, using collocation on finite elements.
5. Ordinary and Partial Differential Equations by numerical methods (commercial programs):
   FEMLAB 2.3 (http://www.femlab.com)
   MATLAB 6.5 (http://www.mathworks.com)
6. Optimization (commercial programs):
   LINDO APT 2.0 (http://www.lindo.com)
   Microsoft Excel 2002 (http://www.microsoft.com)
   GAMS (http://www.gams.com)
7. Bifurcation analysis (public domain program):
   AUTO97 (http://indy.cs.concordia.ca/auto) A code for tracking by continuation the solution of systems of nonlinear algebraic and/or first-order ordinary differential equations as a function of a bifurcation parameter (available only for UNIX-based computers)
8. Process simulation, synthesis of distillation trains and heat-exchanger networks
   Aspen Engineering Suite 11.1 (includes ASPEN PLUS, ASPEN DYNAMICS, ASPEN PINCH, ASPEN SPLIT, BATCH PLUS, etc.) and HYSYS.Plant
   Aspen Technology, Inc., Ten Canal Park, Cambridge, MA 06141
    Phone: 617-577-0100
    FAX: 617-577-0303
    http://www.aspentec.com
   CHEMCAD 5.2.1
    Chemstations, 2901 Wilcrest Drive, Suite 305, Houston, TX 77042
    Phone: 713-978-7700
    FAX: 713-978-7727
    http://www.chemstations.net
   Process Engineering Suite (includes PRO/II, HEXTRAN, etc.)
    Simulation Sciences, Inc., 601 South Valencia Ave., Suite 100 Brea, CA 92823
    Phone: 714-579-0412
    FAX: 714-579-7927
    http://www.simsci.com

Table 1.4 (*Continued*)

SUPERPRO DESIGNER 4.9 (includes BatchPro, EnviroPro, and BioPro Designer)
  Intelligen, Inc., 2326 Morse Avenue, Scotch Plains, NJ 07076
  Phone: 908-654-0088
  FAX: 908-654-3866
  http://www.intelligen.com/SuperPro.htm
9. Stand-alone distillation packages
    CHEMSEP 4.3 (http://www.chemsep.org)
      Calculates continuous, multicomponent, multistage distillation, absorption, and stripping columns by equilibrium-stage and
      rate-based methods
    MULTIBATCHDS 2.0 (http://www.che.utexas.edu/cache/multibatch.html)
      Calculates multicomponent batch distillation

Spreadsheets

In the world of the personal computer, spreadsheets such as Excel, Lotus 1-2-3, and Quattro Pro have become as easy to use as word processors and communication packages. Most engineers can enter tables of data and program the spreadsheets to evaluate arithmetic expressions with very little effort. Whole columns and rows are manipulated and graphed, and results are stored and annotated without the complex formatting instructions that accompany procedural languages like FORTRAN. For this reason, many engineers have switched from procedural languages to spreadsheets, even for the implementation of iterative algorithms. More recently, spreadsheets have added *solvers* for the solution of problems in nonlinear equations and optimization. In the design area, spreadsheets are widely used for profitability analysis, as discussed in Section 17.8. Given the total installed cost of the equipment and the unit costs of the products, byproducts, and the raw materials, as well as the utilities, spreadsheets compute the total capital investment, the cost sheet, and various profitability measures, such as the return on investment and the net present value. Results from spreadsheets are readily plotted in any of a variety of graphs, and for the more difficult computational problems, a spreadsheet can be linked to a procedural language. For example, Visual Basic for Applications (VBA) is available as a macro language for Microsoft Excel.

Mathematical Packages

Many kinds of engineering calculations can be carried out quickly and efficiently with symbolic and numeric mathematical packages. Examples are the analysis of linear systems (MATLAB), symbolic manipulations in algebra and calculus (MATHEMATICA, MAPLE), numerical integration of ordinary and partial differential equations (FEMLAB and MATLAB), and solution of mathematical programs in optimization (GAMS and LINDO). Optimization capabilities include the solution of linear programs (LPs), nonlinear programs (NLPs), mixed-integer linear programs (MILPs), and mixed-integer nonlinear programs (MINLPs). In this book, several optimization problems are solved with GAMS, which is introduced on the multimedia CD-ROM that accompanies this book. A less important, but growing, analysis in design involves the controllability of a process, which is often assessed by linearizing about a steady state. In this book, MATLAB is used for this purpose. For those willing to write brief FORTRAN main programs, public domain software is available for the numerical solution of ordinary and partial differential equations (COLMOD, COLNEW, ODEPACK, and PDECOL) and for bifurcation analysis of nonlinear systems (AUTO).

Process Simulators

Computer-aided process design programs, often referred to as *process simulators, flowsheet simulators*, or *flowsheeting packages*, are widely used in process design. The facilities of the major simulators are more comprehensive than required for just material and energy balances of a flowsheet with recycle and design specifications. As seen beginning in Chapter 4, these packages are comprised of data banks, physical property models, and equipment operation and sizing models. The extensive data banks contain data on the thermophysical and transport property constants for thousands of chemicals. The simulators contain many models of reactors and unit operations, so-called simulation models that can be used to calculate material and energy balances. Other models compute equipment sizes and costs, and profitability measures. Consequently, the process simulators are used to carry out many kinds of calculations throughout the design process, beginning with the early steps in process creation.

Six major process simulators are widely used in the chemical process industries today. These are ASPEN PLUS, BATCH PLUS and HYSYS.Plant (Aspen Technology, Inc.), PRO/II (Simulation Sciences, Inc.), CHEMCAD (ChemStations, Inc.), and SUPERPRO DESIGNER (Intelligen, Inc.). In this book and the associated multimedia CD-ROM, coverage is provided of ASPEN PLUS and HYSYS.Plant, which are the two most widely used process simulators. It should be mentioned that once the principles of process simulation are understood, it is a relatively easy matter to switch from one simulator to another.

In connection with the process simulators, it is a common criticism that their data banks and models are oriented toward petrochemical processes; that is, processes involving vapors, non-electrolyte liquids, and organic molecules having low molecular weights. Although this has been true, it is important to recognize that in recent years facilities have been added to handle solids, aqueous streams with electrolytes, as well as polymerization and fermentation reactions. When process simulators are applicable, they are, by far, the most effective way to carry out process design.

Computational Guidelines

With the broad array of computational packages available, especially to universities where they can be licensed at relatively low cost, a design team can lose valuable time in learning to use a package when the calculations can be completed rapidly using simple equations and graphics. Computer packages should be used only when they can be applied easily and when their rigorous models for equipment and thermophysical properties are justified. Normally, ease of use is based on experience, which grows rapidly with exposure. As engineers and students gain familiarity with computer packages, they are often used very effectively throughout the design process, providing a real advantage to a design team. There is one basic premise, however, that must be observed: In process design, to permit a company to be competitive, it is crucial that deadlines be met. To the extent that computer packages can accelerate the completion of the project, possibly with more rigorous results, they should be used. However, the deadlines must be met, with or without the computer. When computational difficulties are encountered, it is important that the design team find other methods for obtaining results, even if they are more approximate.

1.7   SUMMARY

Having studied this chapter, the reader should

1. Be well acquainted with the steps in designing chemical products and processes as presented in Figure 1.2, which also serves as a road map to the five parts and the chapters in this book.
2. Have examined the primitive design problem statements presented herein and gained familiarity with some of the problem statements on the multimedia CD-ROM that accompanies this book.
3. Be knowledgeable about the principal issues in environmental protection and the many safety considerations that are foremost in the minds of product and process designers. The reader should also have some familiarity with the many design methods used to protect our environment and provide safe chemical processes.

4. Understand that it is crucial for engineers to maintain high ethical principles, especially as they relate to protecting the public against environmental and safety problems. At the minimum, the reader should be familiar with the codes of ethics presented herein and recognize that engineers are often confronted with difficult choices that must be resolved using high ethical standards.

5. Have gained an appreciation of the kinds of computer tools regularly used by process design teams and at least an initial concept of the roles of the various computer packages in process design.

REFERENCES

Allen, D.T., and K.S. Rosselot, *Pollution Prevention for Chemical Processes*, John Wiley & Sons, New York (1997).

Allen, D.T., and D.R. Shonnard, *Green Engineering: Environmentally Conscious Design of Chemical Processes*, Prentice-Hall, Englewood Cliffs, New Jersey (2002).

American Institute of Chemical Engineers, *Safety, Health, and Loss Prevention in Chemical Processes: Problems for Undergraduate Engineering Curricula—Student Problems*, AIChE, New York (1990).

American Institute of Chemical Engineers, *Emergency Relief System Design Using DIERS Technology,* AIChE, New York (1992).

American Institute of Chemical Engineers, *Guidelines for Hazard Evaluation Procedures, Second Edition with Worked Examples*, AIChE, New York (1992).

American Institute of Chemical Engineers, *Self-Study Course: Risk Assessment*, AIChE, New York (2002).

Amundson, N.R., Ed., *Frontiers in Chemical Engineering: Research Needs and Opportunities*, National Research Council, National Academy Press, Washington, DC (1988).

Bever, M.B., Ed., *Encyclopedia of Materials Science and Engineering*, Pergamon Press, Oxford (1986).

Bretherick, L., *Handbook of Reactive Chemical Hazards*, Butterworth, London (1990).

Chase, M.W., Ed., *JANAF Thermochemical Tables*, 3rd ed., Parts 1 and 2, in *J. Phys. Chem. Ref. Data*, **14** (Suppl. 1) (1985).

Cheremisinoff, N.P., Ed., *Encyclopedia of Fluid Mechanics*, Gulf Publishing Co., Houston (1986).

Coe, J.T., *Unlikely Victory: How General Electric Succeeded in the Chemical Industry*, AIChE, New York (2000).

Considine, D.M., Ed., *Van Nostrand's Scientific Encyclopedia*, 8th ed., Van Nostrand, New York (1995).

Crowl, D.A., and J.F. Louvar, *Chemical Process Safety: Fundamentals with Applications*, Prentice-Hall, Englewood Cliffs, New Jersey (1990).

Cussler, E.L., and G.D. Moggridge, *Chemical Product Design*, Cambridge University Press, Cambridge (2001).

de Nevers, N., *Air Pollution Control Engineering*, McGraw-Hill, New York (1995).

Denn, M.M., *Process Modeling*, Longmans, New York (1986).

Eisenhauer, J., and S. McQueen, *Environmental Considerations in Process Design and Simulation*, Energetics, Inc., Columbia, Maryland (1993).

Freeman, H.M., Ed., *Standard Handbook of Hazardous Waste Treatment and Disposal*, McGraw-Hill, New York (1989).

Gundling, E., *The 3M Way to Innovation: Balancing People and Profit*, Kodansha International, Tokyo (2000).

Kent, J.A., Ed., *Riegel's Handbook of Industrial Chemistry*, 9th ed., Van Nostrand Reinhold, New York (1992).

Kirk-Othmer Encyclopedia of Chemical Technology, 4th ed. Wiley-Interscience, New York (1991).

Kletz, T., *Plant Design for Safety—A User-Friendly Approach*, Hemisphere, Washington, DC (1991).

Lide, D.R., Ed., *Handbook of Chemistry and Physics*, 78th ed. CRC Press, Boca Raton, Florida (1997).

McGraw-Hill Encyclopedia of Science and Technology, 6th ed. McGraw-Hill, New York (1987).

McKetta, J.J., Ed., *Chemical Processing Handbook*, Marcel Dekker, New York (1993a).

McKetta, J.J., Ed., *Unit Operations Handbook*, Marcel Dekker, New York (1993b).

McKetta, J.J., and W.A. Cunningham, Eds., *Encyclopedia of Chemical Processing and Design*, Marcel Dekker, New York (1976).

National Institute for Occupational Safety and Health, *Pocket Guide to Chemical Hazards*, NIOSH, Cincinnati, Ohio (1987).

Perry, R.H., and D.W. Green, Ed., *Perry's Chemical Engineer's Handbook*, 7th ed. McGraw-Hill, New York (1997).

Pisano, G.P., *The Development Factory: Unlocking the Potential of Process Innovation*, Harvard Business School Press, Cambridge (1997).

Rath and Strong, *Six Sigma Pocket Guide*, Rath and Strong/AON Management Consulting, Lexington, Massachusetts (2000).

Rath and Strong, *Design for Six Sigma Pocket Guide*, Rath and Strong/AON Management Consulting, Lexington, Massachusetts (2002).

Tadmor, Z., and C.G. Gogos, *Principles of Polymer Processing*, Wiley, New York (1979).

Trivedi, Y.B., "Applying 6 Sigma," *CEP*, 76–81, July, 2002.

Ullmann's Encyclopedia of Industrial Chemistry, 5th ed., VCH, Deerfield Beach, Florida (1988).

Ulrich, K.T., and S.D. Eppinger, *Product Design and Development*, 2nd ed., McGraw-Hill, New York (2000).

Wheeler, J.M., "Getting Started: Six-Sigma Control of Chemical Operations," *CEP*, 76–81, June, 2002.

Wibowo, C., and K.M. Ng, "Product-Oriented Process Synthesis and Development: Creams and Pastes," *AIChE J.*, **47** (12), 2746–2767 (2001).

Wibowo, C., and K.M. Ng, "Product Centered Processing: Manufacture of Chemical-Based Consumer Products," *AIChE J.*, **48** (6), 1212–1230 (2002).

Woods, D.R., *Process Design and Engineering Practice*, Prentice-Hall, Englewood Cliffs, New Jersey (1995a).

Woods, D.R., *Data for Process Design and Engineering Practice*, Prentice-Hall, Englewood Cliffs, New Jersey (1995b).

Chapter 2

Molecular Structure Design

2.0 OBJECTIVES

As discussed in Section 1.2, the design of many chemical products, like creams and pastes, begins with specifications of their desired properties and performance. Given these specifications, which are based upon technical, marketing, and business considerations, a design team seeks to invent new chemicals or chemical mixtures that have these desired properties and performance, including density, boiling point, viscosity, average molecular weight and solubility in solvents, among many others. In this chapter, several methods for *molecular structure design* are introduced. These methods select from among numerous permutations of atoms and molecular groups to identify molecules, and mixtures of molecules, that satisfy specifications for properties and performance. For this purpose, methods of property estimation, including semiempirical and group-contribution methods, are introduced. Methods for Monte-Carlo and molecular dynamics estimation of properties are also introduced, with emphasis on their promising role in product design, but are not used in the examples presented. To satisfy property specifications, where appropriate, optimization algorithms are covered. These adjust the number and position of preselected atoms and molecular groups to minimize the difference between the property estimates and specifications. Where property estimation techniques are not effective, the role of experimental methods, with emphasis on the discovery of pharmaceuticals, is covered. Examples and discussions show how to use the methods of molecular structure design to locate:

    a. polymers that have desired properties, such as density, glass-transition temperature, and water absorption,

    b. refrigerants that boil and condense at desired temperatures and low pressures, while not reacting with ozone,

    c. solvents that convey solids, such as ink pigments and paints, evaporating rapidly, while being non-toxic and environmentally friendly,

    d. solvents for liquid–liquid extraction, extractive distillation, or azeotropic distillation,

    e. macromolecules as pharmaceuticals, such as proteins that function as antibodies, which are Y-shaped multidomain proteins that bind specific antigens or receptors with exquisite selectivity,

    f. solutes for hand warmers that remain supersaturated at low temperatures, crystallizing exothermically, only when activated.

The design of other chemical products, not covered in this chapter, include lubricants having low viscosity that withstand high engine temperatures, and ceramics having high tensile strength and low density.

As shown in Figure 1.2, molecular structure design is often a key step in *product design*. In this chapter, its iterative nature is emphasized, often involving heuristics, experimentation, and the need to evaluate numerous alternatives in parallel, as in the discovery of pharmaceuticals. Typical work processes in industry are covered.

After studying this chapter, the reader should

1. Be aware of typical considerations in specifying the physical properties and performance of potential chemical products.
2. Know how to set up a search for chemicals and chemical mixtures that satisfy specifications for physical properties.
3. Understand the role of group-contribution methods, and other molecular modeling techniques, in estimating properties during molecular structure design.
4. Know how to apply optimization methods to locate molecular structures having the desired properties.
5. Appreciate the role of parallel experimentation in searching for pharamaceuticals.
6. Be aware of the many kinds of chemical products discovered using molecular structure design.

2.1   INTRODUCTION

Often, the search for new chemical products is motivated by a desire to improve the capabilities and performance of existing products. Increasingly, lighter products are sought, having more strength, that are biodegradable, safer to manufacture, less toxic, and more environmentally friendly, to mention several of the common objectives. In the idea-generation phase of product design (see Section 1.2), these kinds of objectives are identified, and as the designer(s) focuses on the most promising ideas, more quantitative specifications are established for the properties of a chemical or chemical mixture.

Often, product development is closely related to the discoveries of a research and development group. As an example, consider the search for improved liquid solvents, which are ubiquitous in the processing and conveyance of chemicals. A key concern, noted by Brennecke and Maginn (2001), is that, due to a narrow liquidus range (difference between boiling and freezing points), most solvents are quite volatile at typical processing conditions. Since about 20 million tons of volatile organic compounds (VOCs) are estimated to be discharged into the atmosphere annually in connection with U.S. industrial operations (Allen and Shonnard, 2002), this is of considerable environmental concern. Furthermore, with solvents estimated to comprise two-thirds of industrial emissions and one-third of VOC emissions nationwide, chemical engineers have been challenged to develop processes that sharply reduce emissions. In response, Brennecke and Maginn suggest that *ionic liquids*, which involve organic salts that sharply reduce the vapor pressure of liquids, may be worthy of development as environmentally-friendly products. With this in mind, an aim of a product design team might be to explore the effect of various cations and anions, at various concentrations, on the vapor pressure, and to estimate emissions at typical operating temperatures. This could lead to new ionic solvent products, designed for specific applications. To satisfy emissions regulations, salt–solvent combinations would be sought that reduce emissions at low cost, involving small salt concentrations. Promising possibilities include quaternary ammonium salts, commonly used for phase-transfer catalysis and for gas separations, for example, to recover water vapor or carbon dioxide. In these salts, the organic "R" group can be adjusted to affect such properties as the water solubility of the ionic solvent.

When designing these products, either measurements of the vapor pressure as a function of salt concentration, or property estimation methods, are needed. As discussed in the next section, the latter are often available for the mixtures involved. For example, the calculation of vapor–liquid equilibria for aqueous electrolytes is carried out commonly using property estimation systems, such as those provided by the OLI electrolyte engine (Aspen, 1999; Zemaitis et al., 1986) and the ASPEN PLUS simulator (Chen et al., 2001). Recently, methods have been developed to calculate equilibria for organic electrolyte solutions, which should be applicable for many ionic solvents, and are implemented in ASPEN PLUS (Getting Started, 1999).

Like ionic liquids, many chemical products are designed using property estimation methods, as discussed in the next section. These include polymer membranes and refrigerants. However, for pharmaceuticals, the properties of proteins are normally not estimated. Rather, they are determined experimentally in the laboratory as discussed next.

Pharmaceutical Product Design

Returning to Section 1.2, it is the *discovery* step of the *development cycle* in which the molecular structure design is carried out. As discussed by Pisano (1997), two distinct strategies have evolved for drugs that are synthesized chemically as compared with those derived through recombinant biotechnology methods.

Synthetic Chemical Drugs

These drugs are synthesized through a sequence of chemical reactions that either add or subtract atoms. Consequently, synthesis of potential molecules begins with identification of the starting materials and selection of the reactions, which comprise the *synthetic route*. Initially, much work is done conceptually, outside of the laboratory, where chemists explore routes using journal articles and computer simulations, to the extent possible. Gradually, a tree of alternative routes leading to potentially attractive drugs evolves. As the tree is pruned, several alternatives are modeled more carefully to locate the most promising routes and reject the least attractive. Then, heuristics are applied to select those routes that have desirable characteristics; for example, the fewest reaction steps, with high selectivity and yields, involving nonhazardous byproducts, easy separations to recover the desired products, and safe implementation.

Normally, the limitations of theory, such as the inability to predict kinetic rates and conversions, require the chemist to carry out small-scale experiments for the most promising routes. Mostly, experiments are performed one reaction at a time, rather than several in parallel, to better understand the kinetics, energy requirements, solvents needed, etc.

Gradually, an attractive route is developed, with a pilot plant designed to manufacture sufficient quantities of product for preclinical trials, while the search continues for more attractive routes. The scale-up from laboratory to pilot-plant quantities involves many of the considerations introduced in the next chapter on process creation; that is, *process synthesis*. As illustrated in Section 3.4, for the manufacture of tissue plasminogen activator (tPA), the flowsheet of process operations is synthesized and the equipment is selected, together with operating strategies that involve batch processing almost entirely. Then, as described in Section 4.5 and subsequent chapters, chemical engineers on the design team implement the methods of process simulation, equipment sizing, and cost estimation.

The search for attractive routes is best characterized as iterative, with new leads continually appearing due to successes and failures in the laboratory and pilot plant. While some theory leads to the initial chemical routes, automated methods of route synthesis are not yet practical, especially for the synthesis of proteins having on the order of 500–600 amino acids. In fact, chemists have been severely limited in the choice of smaller target molecules, which can be synthesized with far fewer chemical reaction steps. To generate proteins, which have become key therapeutic drugs, chemists and biochemists have turned to cell cultures.

Genetically Engineered Drugs

As discussed by Pisano (1997), the key breakthrough came when Herbert Cohen and Stanley Boyer, of the University of California at San Francisco, invented a means of inserting genes into bacterial cells. By *expressing* the gene for a protein into the DNA of a bacterial or mammalian cell, the latter becomes capable of producing that protein. For example, as discussed in Section 3.4, the tPA gene can be isolated from human melanoma cells, inserted into Chinese hamster ovary (CHO) cells, which then generate the tPA protein.

To genetically engineer a drug, chemists and biochemists begin by identifying target proteins; that is, proteins that have the desired therapeutic properties, such as a monoclonal antibody or insulin. Then, the gene sequence that *codes* for the protein must be identified, together with a host cell to be used for growing the protein. Identification of the desired therapeutic properties usually begins with knowledge of a disease, leading the chemist or biochemist to work backward to find a protein that inhibits a chemical reaction involved in that disease. This was the approach used at Eli Lilly to find Prozac, a serotonin inhibitor, for treating depression.

At the heart of process development is the need to precisely measure the quantity and purity of protein expression. These measurements are crucial for determining the rate and conversion of protein growth, and the purification yield, which are the basis for the design of cultivation, fermentation, and separation equipment. Initially, process researchers focus on identifying the most promising cell lines, cells that can produce the desired protein, that have a high rate of production and require the least expensive nutrients for growth. Many iterations are usually required in this search, involving many prospective cell lines and operating conditions. For this purpose, automated lab benches are often employed, permitting hundreds and thousands of cell clones to be evaluated experimentally in parallel.

2.2   PROPERTY ESTIMATION METHODS

Theoretical approaches to molecular structure design require accurate estimates of physical and transport properties. These are derived commonly from the principles of thermodynamics and transport phenomena, and using molecular simulations. Since the literature abounds with estimation methods, reference books and handbooks are particularly useful sources. One of the most widely used, *Properties of Gases and Liquids* (Poling et al., 2001), provides an excellent collection of estimation methods and data for chemical mixtures in the vapor and liquid phases. For polymers, *Properties of Polymers* (van Krevelen, 1990) provides a collection of group-contribution methods and data for a host of polymer properties.

In recent years, property information systems have become widely available in computer packages. Some are available on a stand-alone basis, such as PPDS2 (1997), while others are available within the chemical process simulators, such as ASPEN PLUS, HYSYS.Plant, PRO/II, CHEMCAD, BATCH PLUS, and SUPERPRO DESIGNER. Commonly, constants and parameters are stored for a few thousand chemical species, with programs provided to estimate the property values of mixtures, and determine the constants and parameters for species that are not in the data bank using estimation methods or the regression of experimental data. Virtually all of the property systems estimate the properties of mixtures of organic chemicals in the vapor and liquid phases. Methods are also provided for electrolytes and some solids, but these are less predictive and less accurate.

Computer Data Banks

Data banks for the pure species may be viewed as a collection of *data records*, each containing the constants and parameters for a single chemical [e.g., the critical properties (T_c, P_c, v_c), the normal boiling point (T_{nbp}), vapor pressure coefficients, heat capacity coefficients, acentric factor, etc.]. One such data bank, which is utilized in ASPEN PLUS, is that compiled by Poling et al. (2001) (also Reid et al., 1977, 1987) in Appendix A of the *Properties of Gases and Liquids*. This data bank, which is referred to as the ASPENPCD (ASPEN PLUS Pure Component Data Bank), contains data for 472 chemicals, using data solely from Reid et al. (1977). Another data bank, known as PURECOMP, originates from the DIPPR® (Design Institute for Physical Property Data, sponsored by the AIChE) data bank, with information supplemented by Aspen Technology, Inc. (e.g., the UNIFAC group contributions) and parameters from the ASPENPCD data bank. It is an updated DIP-PRPCD data bank and has superseded the ASPENPCD data base as the main source of parameters for pure components. For Version 11 of ASPEN PLUS, the PURECOMP data base was renamed PURE11. The PURECOMP data bank contains data for over 1,727 chemicals (mostly organic) not including the ionic species in electrolytes. In addition, ASPEN PLUS provides access to the AQUEOUS data bank for over 900 ionic species, to be used for electrolytes.

When the constants and parameters are not stored for a chemical species, most of the property information systems have programs for the regression of experimental data (e.g., tables of vapor pressures, liquid densities, and heat capacities as a function of temperature). Finally, when insufficient experimental data are available, programs are often provided to estimate the properties based upon the molecular structure, using group- and bond-contribution methods, often utilizing limited data (e.g., the normal boiling point). These are particularly useful in the early stages of product and process design before a laboratory or pilot-plant study is initiated.

Property Estimation

Each of the property information systems has an extensive set of subroutines to determine the parameters for vapor pressure equations (e.g., the extended Antoine equation), heat capacity equations, etc., by regression and to estimate the thermophysical and transport properties. The latter subroutines are called to determine the state of a chemical mixture (phases at equilibrium) and its properties (density, enthalpy, entropy, etc.) When calculating phase equilibria, the fugacities of the species are needed for each of the phases. A review of the phase equilibrium equations, as well as the facilities provided by the process simulators for the calculation of phase equilibria, is provided on the CD-ROM that accompanies this book (see *ASPEN→Physical Property Estimation* and *HYSYS→Physical Property Estimation*).

As mentioned above, when a data record for a pure species cannot be located in one of its data banks, each of the property information systems permits the designer to enter the missing constants and parameters. Furthermore, methods are provided to estimate the constants and parameters when the designer cannot provide these. This is especially important when laboratory and pilot-plant data are not available.

Usually, bond- or group-contribution methods are used to estimate the constants and parameters for pure species, with the designer providing the molecular structure of the chemical species, as shown, for example, for trifluoropropylene:

$$H-\underset{①}{C}=\underset{②}{C}-\underset{③}{C}-F$$

Here, all atoms, with the exception of hydrogen, are numbered and the bonds associated with each carbon atom and its adjacent numbered atoms can be specified as follows:

	Atom 1		Atom 2	
Number	Type	Number	Type	Bond Type
1	C	2	C	Double Bond
2	C	3	C	Single Bond
3	C	4	F	Single Bond
3	C	5	F	Single Bond
3	C	6	F	Single Bond

Using bond- and group-contribution techniques, as discussed by Poling et al. (2001) and Joback and Reid (1987), many properties can be estimated, including the critical volume, normal boiling point, liquid density and heat of vaporization at the normal boiling point, and ideal-gas heat capacity coefficients. Similarly, using the UNIFAC group contribution method, the activity coefficients of trifluoropropylene in solution with other chemical species can be estimated for use in computing phase equilibria.

Polymer Property Estimation

As mentioned above, van Krevelen (1990) presents semi-empirical, group-contribution methods and data for each group in a polymer "repeating unit." Data are provided to estimate a host of polymer properties, including the density, specific heat, glass-transition temperature, water absorption, and refractive index. For a specific property, these are in one of two forms:

$$p\{n\} = \frac{\sum_{i=1}^{N} A_i n_i}{\sum_{i=1}^{N} B_i n_i} \tag{2.1}$$

and

$$p\{n\} = \left(\frac{\sum_{i=1}^{N} A_i n_i}{\sum_{i=1}^{N} B_i n_i} \right)^d \tag{2.2}$$

were n_i is the number of groups of type i in the polymer "repeating unit," N is the number of types of groups in the repeating unit, A_i is the contribution associated with group i, B_i is the molecular weight of group i, and d is an exponent for each property to be estimated. Note that the denominator summation is either the molecular weight or the specific volume of the repeat unit. In most cases, these property estimates lie within 5 to 10 percent of experimental values, which are often sufficiently close to permit the selection of repeating units to meet property specifications.

EXAMPLE 2.1

Estimate the glass-transition temperature of polyvinyl chloride, T_g, with repeating unit, —(CH$_2$CHCl)—, using the following group contributions (van Krevelen, 1990):

Group	A_i	B_i
—CH$_2$—	2,700	14
—CHCl—	20,000	48.5

SOLUTION

$$T_g = \frac{2,700 \times 1 + 20,000 \times 1}{14 \times 1 + 48.5 \times 1} = 363 \text{ K}$$

This compares fairly well with the experimental value of 356 K (van Krevelen, 1972, page 114). ∎

Caution. When using property estimation methods, especially group- and bond-contribution methods, care must be taken to avoid large differences from experimental values, especially when the molecules, temperatures, and pressures, are substantially different from those used to estimate the parameters of the methods.

Microsimulation

Two methods, of a more fundamental nature than group- and bond-contribution methods, are being used increasingly to improve estimates of thermophysical and transport properties. These involve molecular dynamics and Monte-Carlo simulations, with small collections (typically 100–10,000) of interacting molecules, and are commonly referred to as *microsimulations*. In general, Monte-Carlo simulations are numerical statistical methods that utilize sequences of random numbers. The name Monte Carlo was coined during the Manhattan Project of World War II, which resulted in the development of the atomic bomb. Monte Carlo is the capital of Monaco, a world center for games of chance. A simple example of a Monte-Carlo simulation is the evaluation of the integral of a complex function, $y = f\{x\}$ over the interval $\{x_1, x_2\}$. A plot of $f\{x\}$ against x over the specified interval is prepared. A range of y is selected from 0 to y_1, where y_1 is greater than the largest value of y in the interval. A random-number generator is then used to select pairs of y–x values within the range of the rectangle bounding the x-interval and the y-range. Suppose that 900 pairs are selected. If 387 pairs fall under the $f\{x\}$ curve of the plot, then the value of the integral is:

$$\frac{387}{900}(y_1 - 0)(x_2 - x_1)$$

Molecular Dynamics

This method involves the numerical integration of the equations of motion ($F = ma$) for each of the molecules, subject to intermolecular forces, in time. The molecules are positioned arbitrarily in a simulation cell, that is, a three-dimensional cube, with initial velocities also specified arbitrarily. Subsequently, the velocities are scaled so that the summation of the kinetic energies of the molecules, $3NkT/2$, gives the specified temperature, T, where N is the number of molecules and k is the Boltzmann constant. Note that after many collisions with the walls and the other molecules, the relative positions and velocities of the molecules are independent of the initial conditions.

   During the simulation, the force on each molecule is calculated as the sum of the forces of interaction with all of the surrounding molecules. These are the dispersion forces, also referred to as the London and van der Waals forces, which depend on the separation distance, r, between two molecules, as shown in Figure 2.1. These are represented by the dimensionless form of the commonly used Lennard-Jones pair potential, $U\{r\}/\epsilon$:

$$\frac{U\{r\}}{\epsilon} = 4\left[\left(\frac{\sigma}{r}\right)^{12} - \left(\frac{\sigma}{r}\right)^6\right] \tag{2.3}$$

which expresses the intermolecular potential between two molecules as a function of the distance, r, between them. In this equation, $U\{r\}$ is the intermolecular potential energy, ϵ is the maximum energy of attraction between a pair of molecules, and σ is the collision diameter of the molecules. Differentiating Eq. (2.3) with respect to r, the negative gradient of the dimensionless intermolecular potential, is the dimensionless force, $F\{r\}/(\epsilon/\sigma)$, between them:

$$\frac{F\{r\}}{\epsilon/\sigma} = -24\left[2\left(\frac{\sigma}{r}\right)^{13} - \left(\frac{\sigma}{r}\right)^7\right] \tag{2.4}$$

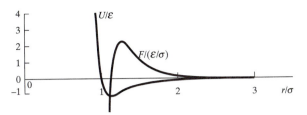

Figure 2.1 Lennard-Jones pair potential and force.

Then, for each molecule i, the summation of all forces acting on it is computed and its equation of motion:

$$\frac{d^2 r_i}{dt^2} = \frac{F_i}{m_i} \tag{2.5}$$

is integrated across the time step. Values of ε and σ for many molecules and methods of estimating them for other molecules are given by Bird et al. (2002). To obtain property estimates, the set of differential equations, one for each molecule, is integrated using picosecond time steps over several hundred thousand time steps. Then, time averages are computed to give properties such as the configurational energy, the pressure, and the self-diffusion coefficient.

With the increased availability of software (e.g., DISCOVER in CERIUS² by Materials Studio, ETOMICA: Kofke and Mihalick, 2002) and faster computers, these simulations are being carried out more routinely for the estimation of thermophysical and transport properties, as well as for the calculation of phase equilibria. The estimates are tuned to match experimental data by adjusting the energy and size parameters.

Monte-Carlo Methods

In Monte-Carlo simulations, the energy of the molecular system is minimized by randomly moving the molecules in accordance with a desired probability distribution. After each move, the energy of each molecule is computed. When the total energy is reduced, the move is accepted and the molecules are redistributed. Moves are continued until equilibrium is achieved. As for molecular dynamics simulations, potential functions are provided. After convergence, the thermophysical properties, at equilibrium, are computed by averaging. Monte-Carlo methods, which are particularly effective for the calculation of thermophysical properties, including phase equilibria, are considered in detail by Rowley (1994).

2.3   OPTIMIZATION TO LOCATE MOLECULAR STRUCTURE

Molecular structure design relies on accurate property estimation methods. When sufficiently accurate, the atoms and groups in the molecular structure are adjusted to minimize the sum of the squares of the differences between the property estimates and the specified values:

$$\min_n \sum_{j=1}^{P} (p_j\{n\} - p_j^{spec})^2 \tag{2.6}$$

where p_j^{spec} is specified by the designer for property j in an array of P target properties. Often this minimization is carried out subject to specified bounds; for example:

$$p_j^L \le p_j\{n\} \le p_j^U \tag{2.7}$$

and

$$n_i \in \{n_i^L, n_i^U\}, \quad i = 1, \dots, N \tag{2.8}$$

where n_i is the number of groups of type i in molecule j, and N is the number of types of molecular groups in molecule j. It is also necessary to assure that when a new molecular group is added to a molecule that the number of free attachments available for bonding is zero, and when added to the repeating unit of a polymer, the number of free attachments is two. This can be checked by computing the number of free attachments:

$$f = \sum_{i=1}^{N} (v_i - 2)n_i + 2 \qquad (2.9)$$

where v_i is the valence, or number of free bonds, associated with molecular group i. The molecular group can be added to a molecule when $f = 0$ and to a repeating unit when $f = 2$. To locate the molecular structure at the optimum, the objective function (2.6) is combined with the constraints (2.7) $-$ (2.9) into a mixed-integer nonlinear program (MINLP), which is solved using a mathematical programming solver, such as GAMS. In the next three subsections, these steps are illustrated for polymer, refrigerant, and solvent designs.

Polymer Design

Having discussed the estimation of polymer properties in the previous section, the methods of polymer design are described in connection with the design of a polymer film in Example 2.2.

EXAMPLE 2.2

A polymer film is needed to protect an electronic device. Since the device will operate at temperatures below 60°C and must be protected by a fairly dense layer, which absorbs small concentrations of water, a design team has prepared the following product quality specifications: (1) density = 1.5 g/cm³, (2) glass-transition temperature = 383 K (50 degrees above the operating temperature), and (3) water absorption = 0.005 g/g polymer. As stated initially by Derringer and Markham (1985), candidate molecular groups, together with their group contributions, are:

i	Group	Y_i	V_i	H_i	M_i
1	—CH₂—	2,700	15.85	0.000033	14
2	—CO—	27,000	13.40	0.11	28
3	—COO—	8,000	23.00	0.075	44
4	—O—	4,000	10.00	0.02	16
5	—CONH—	12,000	24.90	0.75	43
6	—CHOH—	13,000	19.15	0.75	30
7	—CHCl—	20,000	29.35	0.015	48.5

where M_i, V_i, Y_i, and H_i are the contributions for group i in estimating the molecular weight, M, the molar volume, V, the glass-transition temperature, T_g, and the water absorption, W, according to:

$$M = \sum_{i=1}^{7} M_i n_i \quad \text{g/mol}$$

$$V = \sum_{i=1}^{7} V_i n_i \quad \text{cm}^3/\text{mol}$$

$$Y = \sum_{i=1}^{7} Y_i n_i \quad \text{K(g/mol)}$$

$$H = \sum_{i=1}^{7} H_i n_i \quad \text{mol H}_2\text{O/mol polymer}$$

$$\rho = M/V \quad \text{g/cm}^3$$

$$T_g = Y/M \quad \text{K}$$

$$W = 18H/M \quad \text{g H}_2\text{O/g polymer}$$

Formulate the mixed-integer nonlinear program and use GAMS to obtain the optimal solution. For an introduction to GAMS, see the file GAMS.pdf on the CD-ROM that accompanies this book.

SOLUTION

Using the objective function in Eq. (2.6), with relative differences, the nonlinear program is:

$$\min_{\substack{\text{w.r.t.}\\ n}} \left(\frac{\rho - \rho^{spec}}{\rho^{spec}}\right)^2 + \left(\frac{T_g - T_g^{spec}}{T_g^{spec}}\right)^2 + \left(\frac{W - W^{spec}}{W^{spec}}\right)^2$$

s.t.:
$$M = \sum_{i=1}^{7} M_i n_i$$
$$V = \sum_{i=1}^{7} V_i n_i$$
$$Y = \sum_{i=1}^{7} Y_i n_i$$
$$H = \sum_{i=1}^{7} H_i n_i$$
$$\rho = M/V$$
$$T_g = Y/M$$
$$W = 18H/M$$
$$0 \le n_i \le 7 \qquad i = 1, \ldots, 7$$
$$1 \le \rho \le 1.5$$
$$298 \le T_g \le 673$$
$$0 \le W \le 0.18$$

Note that Eq. (2.9) is not included because each group has just two attachments.

GAMS Program

This mixed-integer nonlinear program is coded in GAMS as follows:

```
VARIABLES
     n1, n2, n3, n4, n5, n6, n7, TG, RHO, W, H, V, Y, M, STG,
     SRHO, SW, Z;
POSITIVE VARIABLES
     STG, SRHO, SW, W, H, V, Y, TG, M, RHO;
INTEGER VARIABLES
     n1, n2, n3, n4, n5, n6, n7;

     n1.LO = 0;   n1.UP = 7;
     n2.LO = 0;   n2.UP = 7;
     n3.LO = 0;   n3.UP = 7;
     n4.LO = 0;   n4.UP = 7;
     n5.LO = 0;   n5.UP = 7;
     n6.LO = 0;   n6.UP = 7;
     n7.LO = 0;   n7.UP = 7;

     M.LO = 14;
     V.LO = 10;
     H.LO = 3.3E-5;
     Y.LO = 2700;

     W.LO = 0;        W.UP  = 0.18;
     TG.LO = 298;     TG.UP  = 673;
     RHO.LO = 1;      RHO.UP  = 1.5;

     STG.L = 383; SRHO.L = 1.50; SW.L = .005;
EQUATIONS
     SPEC1, SPEC2, SPEC3, MOLWEIGHT, GLASSTMP, YTOT, HTOT,
     VTOT, DENSITY, ABSORBANCE, OBJ;
```

```
OBJ..  Z  =E= ((SQR((STG - TG)/STG)) + (SQR((SRHO - RHO)/SRHO))
                  + (SQR((SW - W)/SW)));

SPEC1..  SRHO =E= 1.5;

SPEC2..  STG  =E= 383;

SPEC3..  SW   =E= .005;

MOLWEIGHT..  M  =E= n1*(14) + n2*(28) + n3*(44) + n4*(16)
                    + n5*(43) + n6*(30) + n7*(48.5);

YTOT..       Y  =E= n1*(2700) + n2*(27000) + n3*(8000)
                    + n4*(4000) + n5*(12000) + n6*(13000)
                    + n7*(20000);

HTOT..       H  =E= n1*(3.3E-5) + n2*(0.11) + n3*(0.075)
                    +n4*(0.02) + n5*(0.75) + n6*(0.75)
                    + n7*(0.015);

VTOT..       V  =E= n1*(15.85) + n2*(13.40) + n3*(23)
                    + n4*(10) + n5*(24.9) + n6*(19.15)
                    + n7*(29.35);

GLASSTMP..   TG  =E= (Y/M);

DENSITY..    RHO =E= (M/V);

ABSORBANCE.. W  =E= ((18*H)/M)

MODEL GROUPS /ALL/;
SOLVE GROUPS USING MINLP MINIMIZING Z;
OPTION DECIMALS = 4;
DISPLAY TG.L, RHO.L, W.L, n1.L, n2.L, n3.L, n4.L, n5.L, n6.L,
n7.L, Z.L;
```

All variables in GAMS must be declared. Then, n_i, $i = 1, \ldots, 7$, are declared as integer variables, with lower and upper bounds specified. The remaining variables are real and are declared positive, with bounds specified for ρ, T_g, and W. Note that each equation is assigned a name, including the objective function, which is named Z. The SOLVE statement indicates that the MINLP is to minimize Z using the MINLP solver. Finally, the variables to be displayed in the solution are identified, with the .L suffix indicating the level (final) value computed.

GAMS Solution

```
*     VARIABLE TG.L =     384.6847
*     VARIABLE RHO.L    = 1.4889
*     VARIABLE W.L      = 0.0049
*     VARIABLE n1.L     = 3.0000
*     VARIABLE n2.L     = 0.0000
*     VARIABLE n3.L     = 0.0000
*     VARIABLE n4.L     = 0.0000
*     VARIABLE n5.L     = 0.0000
*     VARIABLE n6.L     = 0.0000
*     VARIABLE n7.L     = 6.0000
*     VARIABLE Z.L      = 0.0007
```

At the minimum, the repeat unit has three —CH_2— groups and six —CHCl— groups; that is, —$[(CH_2)_3(CHCl)_6]$—. The objective function, $Z = 0.0007$ and the three properties lie within 2% of specifications. ∎

Refrigerant Design

In 1937, Thomas Midgely, working for General Motors, published the first comprehensive study of the design of small molecules for refrigerants. Through examination of the periodic table, he concluded that inert gases are too light (having very low boiling points) for most applications, and that the metals are impractical due to their potential for freezing when the refrigeration system is shut down. Consequently, he concentrated on compounds involving the C, N, O, S, H atoms, and the halogens, F, Cl, Br, and I. Although compounds containing F tend to be more flammable, he considered their relatively low toxicity to be an overriding advantage. While compounds containing Cl are less flammable, they are more toxic, but he considered them not sufficiently toxic to be excluded. Compounds containing Br and I were considered to be far too toxic. Midgely also recognized the desirability of refrigerants having: (1) a large latent heat of vaporization, to reduce their throughput when removing a specified heat duty; (2) a low viscosity, to reduce the recirculation power; and (3) a low freezing point, to reduce the possibility of freezing. His work led to the development of a number of refrigerants containing C, Cl, and F atoms, called Freons. Further work by DuPont led to additional Freon refrigerants containing H atoms as well.

In the years that followed, with the increasing usage of refrigerants for home refrigerators and air-conditioning systems, especially automobile air conditioners, the production of Freon 21, $CHCl_2F$, as well as other CHClF forms, grew rapidly. However, because the concentration of these compounds increased to parts-per-billion in the stratosphere, and chlorine atoms were found to react with ozone, decreasing significantly the earth's ozone layer, CFCs (compounds containing Cl) were banned in the Montreal protocol of 1987. This led to a search for new refrigerant products having comparable properties, but excluding chlorine. One possibility, Freon 11, CF_3H, was rejected because it boils at too low a temperature, $-82°F$. Another, HFC-134a, CFH_2CF_3, has become a popular alternative. Another legacy left by Midgely was tetraethyl lead. In 1921, he calculated that adding lead to gasoline would make it burn better and prevent engine knock. Not until a century later, by which time more than five trillion gallons of leaded gasoline had been burned by automobiles, did public health concerns overcome industrial resistance and usher in the current era of unleaded gasoline.

Consider, next, the problem of designing a new refrigerant product, given the temperatures at which heat is to be absorbed by the evaporator and rejected from the condenser of a refrigerator. Note that the design of a conventional refrigerator is discussed in Sections 9.6 and 9.9 and in most books on engineering thermodynamics. Beginning with a selection of k molecular groups, each of which can appear in a candidate refrigerant n times, up to n_{max} times, the number of distinct molecular designs is:

$$\sum_{n=2}^{n_{max}} C\{k, n\} = \sum_{n=2}^{n_{max}} \frac{(k + n - 1)!}{n!(k - 1)!} \tag{2.10}$$

where $C\{k,n\}$ is the number of combinations of k groups taken n at a time (Joback and Stephanopoulos, 1989). Clearly, this number can become very large, on the order of millions, for as few as 10 molecular groups. To illustrate the problem of selecting from among such a large number of combinations, consider the next example. Note that a less restrictive formulation was solved initially by Joback and Stephanopoulos (1989). Subsequently, Gani and co-workers (1991) excluded oxygen atoms and added restrictions that limit the scope of the search for new molecules. Yet another formulation, which includes oxygen atoms, was provided by Duvedi and Achenie (1996). Note also that chlorine is included in the search to show that molecules containing chlorine are the most desirable when the ozone layer is disregarded. In the example that follows, a measure used by Duvedi and Achenie (1996) to estimate the ozone depletion potential (ODP) is shown to lead to different refrigerants.

EXAMPLE 2.3

It is desired to design a refrigerant that can absorb heat at temperatures as low as 30°F (-1.1°C) and reject heat at temperatures as high as 110°F (43.3°C). For the design, consider 13 molecular groups: CH_3, CH_2, CH, C, OH, O, NH_2, NH, N, SH, S, F, and Cl. When combining these groups, compounds with double or triple bonds, which tend to polymerize, should be avoided. Also, compounds involving both nitrogen and halides should be avoided, as these tend to be explosive. Desirable refrigerants should have: (1) a vapor pressure, $P^s\{-1.1°C\} > 1.4$ bar, to assure that leaks are from the refrigeration system (rather than vacuum operation, into which air and water vapor can leak), (2) $P^s\{43.3°C\} < 14$ bar, to keep the compression ratio from exceeding 10, (3) an enthalpy of vaporization, $\Delta H^v\{-1.1°C\} > 18.4$ kJ/mol, to reduce the amount of refrigerant needed (where 18.4 kJ/mol is the latent heat of vaporization of Freon 12, the refrigerant banned in 1987), and (4) a liquid heat capacity, $c_{pl}\{21.1°C\} < 32.2$ cal/(mol K), to reduce the amount of refrigerant that flashes across the valve (where 32.2 cal/(mol K) is the heat capacity of liquid Freon 12). Note that 21.1°C is the average of the extreme temperatures.

SOLUTION

The estimation methods by Duvedi and Achenie (1996) are used:

1. Normal boiling point and critical properties (Joback and Reid, 1987).

$$T_b = 198.2 + \sum_{i=1}^{N} T_{b_i} n_i$$

$$T_c = T_b \left[0.584 + 0.965 \sum_{i=1}^{N} T_{c_i} n_i - \left(\sum_{i=1}^{N} T_{c_i} n_i \right)^2 \right]^{-1}$$

$$P_c = \left(0.113 + 0.0032 n_A - \sum_{i=1}^{N} P_{c_i} n_i \right)^{-2}$$

where the temperatures and pressures are in K and bar, and n_A is the total number of atoms in the molecule.

2. Vapor pressure—Riedel–Plank–Miller method (Reid et al., 1977).

$$\ln P_r^s = \frac{-G[1 - T_r^2 + k(3 + T_r)(1 - T_r)^3]}{T_r}$$

$$G = 0.4835 + 0.4605h$$

$$h = T_{b_r} \frac{\ln P_c}{1 - T_{b_r}}$$

$$k = \frac{\left[\dfrac{h}{G} - (1 + T_{b_r}) \right]}{(3 + T_{b_r})(1 - T_{b_r})^2}$$

where T_r and P_r are the reduced temperature and pressure.

3. Liquid heat capacity—Chueh and Swanson method (Reid et al., 1987).

$$c_{pl} = 0.239 \sum_{i=1}^{N} c_{pl_i} n_i$$

where c_{pl} is in cal/mol K.

4. Latent heat of vaporization.

At normal boiling point—Vetere modification of Kistiakowsky eqn. (Duvedi and Achenie, 1996):

$$\Delta H_b^v = S_{vb} T_b$$

$$S_{vb} = 44.367 + 15.33 \log T_b + 0.39137 T_b/M + 0.00433 T_b^2/M$$
$$-5.627 \times 10^{-6} T_b^3/M$$

At other temperatures (Reid et al., 1987):

$$\Delta H^v\{T\} = \Delta H_b^v \left[\frac{1 - T/T_c}{1 - T_b/T_c} \right]^n$$

$$n = \left[\frac{0.00264(\Delta H_b^v)}{RT_b} + 0.8794 \right]^{10}$$

where the latent heat of vaporization is in J/mol and M is the molecular weight.

The group contributions for use in the above equations from Joback and Reid (1987) are:

Group	Valence	T_c	P_c	V_c	T_b	n_i	c_{pl}	M
—CH$_3$	1	0.0141	−0.0012	65	23.58	4	36.8	15.04
—CH$_2$—	2	0.0189	0	56	22.88	3	30.4	14.03
—CH=	3	0.0164	0.002	41	21.74	2	21	13.02
=C=	4	0.0067	0.0043	27	18.25	1	7.36	12.01
—OH	1	0.0741	0.0112	28	92.88	2	44.8	17.01
—O—	2	0.0168	0.0015	18	22.42	1	35	16
—NH$_2$	1	0.0243	0.0109	38	73.23	3	58.6	16.03
—NH—	2	0.0295	0.0077	35	50.17	2	43.9	15.02
—N=	3	0.0169	0.0074	9	11.74	1	31	14.01
—S—	2	0.0119	0.0049	54	68.78	1	33	32.07
—SH	1	0.0031	0.0084	63	63.56	2	44.8	33.08
—F	1	0.0111	−0.0057	27	−0.03	1	17	19
—Cl	1	0.0105	−0.0049	58	38.13	1	36	35.45

Using a mixed-integer nonlinear program with various objective functions, Duvedi and Achenie (1996) found three compounds that satisfy the specified constraints. These are:

Compound	ΔH^v, kJ/mol at −1.1°C	c_{pl}, cal/mol-K at 21.1°C	P^s, bar at −1.1°C	P^s, bar at 43.3°C
CCl$_2$F$_2$	18.76	27.1	2.94	10.67
CF$_3$OH	19.77	24.7	2.69	13.57
CH$_3$Cl	20.37	17.4	2.39	8.72

ΔH^v is reported at the lowest temperature; that is, the temperature at which evaporation occurs in a refrigerator, while c_{pl} is reported at the average temperature. Note that the values of ΔH^v and P^s are computed using experimental T_b (CCl$_2$F$_2$ = 244.2K, CF$_3$OH = 251.48K, CH$_3$Cl = 249.1K) because the group contribution method is not sufficiently accurate. Note also that differences in ΔH^v from those reported by Duvedi and Achenie (1996) are due to the differences in T_b and simplifications in the methods for estimating S_{vb}.

Since CH$_3$Cl contains chlorine, which depletes ozone in the earth's stratosphere, Example 2.4 repeats the search using the ozone depletion potential.   ∎

EXAMPLE 2.4

Redesign the refrigerant molecules using the ozone depletion potential (ODP) defined for molecules having one carbon atom:

$$\text{ODP} = 0.585602 n_{Cl}^{-0.0035} \, e^{M/238.563},$$

and having two carbon atoms

$$\text{ODP} = 0.0949956 n_{Cl}^{-0.0404477} \, e^{M/83.7953}$$

Repeat the search in Example 2.3 by minimizing the ODP.

SOLUTION

Using a mixed-integer nonlinear program with various objective functions, Duvedi and Achenie (1996) found two compounds that satisfy the specified constraints. These are:

Compound	ΔH^v, kJ/mol at −1.1°C	c_{pl}, cal/mol at 21.1°C	P^s, bar at −1.1°C	P^s, bar at 43.3°C	ODP
SF$_2$	18.3	16.0	3.84	13.9	0
CH$_3$CHF$_2$	20.6	21.9	2.08	7.91	0

Note that the latent heat of vaporization of SF_2 is sufficiently close to 18.4 kJ/mol to be acceptable, given the approximate estimation methods. Also, neither molecule contains chlorine, and hence, the ODP is zero. ∎

Solvent Design

Organic solvents play a key role in many aspects of chemical processing and in the delivery of chemicals to consumers. In chemical processing, solvents are often used: (1) to mobilize solids, frequently dissolving them; (2) to clean equipment, as in removing grease and grime; and (3) in cleaning clothing, as in dry cleaning. In contrast, in the delivery of chemicals to consumers, solvents often convey particles onto surfaces in coatings, such as in paint and printing ink.

Until the past decade, the solvent market was dominated by a few principal products, solvents known for their ability to "dissolve most anything" (Kirschner, 1994). These included acetone, mixed xylenes, and 1,1,1-trichloroethane, which are manufactured in large-scale processes by the major chemical companies. For environmental and health reasons, over the past decade, there has been a gradual shift away from these solvents. The U.S. Environmental Protection Agency maintains a Toxic Release Inventory (TRI), which includes acetone, 1,1,1-trichloroethane and other common solvents, whose emissions into the air have been gradually reduced. Other solvents are included on the hazardous air pollutants (HAP) list of the 1990 Clean Air Act; for example, 1,1,1-trichloroethane which, like Freon refrigerants, accumulates in the stratosphere and destroys ozone. Furthermore, other solvents, like monomethylether, monoethylether, and their acetates, have been associated with high miscarriage rates.

Chemical companies are increasingly challenged to reduce the usage of these targeted solvents, and consequently, a host of solutions are being sought, including shifts toward: (1) aqueous solvents, where possible, (2) more concentrated paints and coatings, containing less solvent, and (3) hot-melt, ultraviolet-cured, and waterborne adhesives. For example, in the manufacture of cosmetics and personal care products, solvents are being dropped from some formulations due to their high content of volatile organic compounds (VOCs) and are being replaced, for example, by water-based hair sprays and solid deodorant sticks.

In meeting the challenge, chemical companies are designing a growing number of environmentally-friendly solvents, that is, *engineered* or *designer* solvents, that satisfy the specifications for each application; and end users are altering their usage patterns. As a result, new solvents are appearing gradually, designed as specialty chemicals, to replace the use of commodity solvents; for example, diacetone alcohol cleaner is a replacement for acetone in the shipbuilding industry. In some cases, the cleaning methods themselves are changing; for example, a one-step, vapor-degreasing process is replaced by a two-step process involving a dip-tank rinse followed by drying. Another example includes the recycling and reuse of cleaning solvents. In the dry-cleaning business, because the principal solvent, perchloroethylene, is suspected to be a carcinogen and appears on the list of HAPs, used solvent is being filtered and recycled, refrigerated condensers are being installed to recover vapor emissions, and new water and steam-cleaning (wet) processes are being developed.

The search for each new *specialty* solvent can be viewed as a product design problem. For this purpose, design strategies have been evolving, some of which have been computerized, at least partially. In this section, two examples are presented in which: (1) a new solvent is designed as a replacement for 1,1,1-trichloroethane, for cleaning surfaces, in the lithographic printing industry (Sinha et al., 1999), and (2) a solvent is selected to remove a solute from a mixture in a liquid–liquid extraction process (Pretel et al., 1994; Gani et al., 1991). Like the previous examples on the design of polymers and refrigerants, initially desired properties are selected, together with a set of candidate molecular groups and target property values. Then, chemical structures are determined whose property estimates, using group-contribution methods, are closest to the target properties.

Property Estimation

For solvent design, in addition to the normal boiling point, liquid density, and the latent heat of vaporization; solubility and related properties, as well as health and safety properties, must be estimated. Estimation methods for these properties are discussed next.

Solubility and Related Measures.   For the design of solvents to clean surfaces, to apply coating resins, and to swell cured elastomers, the Hansen solubility parameter:

$$\delta_T = \sqrt{\delta_D^2 + \delta_P^2 + \delta_H^2} \tag{2.11}$$

provides a useful measure of solvent performance. As defined in Eq. (2.11), this parameter is comprised of three solubility parameters: (1) δ_D, to account for nonpolar (dispersive) interactions, (2) δ_P, to account for polar interactions, and (3) δ_H, to account for hydrogen-bonding interactions. These three contributions may be estimated using group-contribution methods:

$$\delta_D = \frac{\sum_{i=1}^{N} n_i F_{D_i}}{V_0 + \sum_{i=1}^{N} n_i V_i} \tag{2.12}$$

$$\delta_P = \frac{\sqrt{\sum_{i=1}^{N} n_i (1{,}000 \, F_{P_i})}}{V_0 + \sum_{i=1}^{N} n_i V_i} \tag{2.13}$$

$$\delta_H = \sqrt{\frac{\sum_{i=1}^{N} n_i (-U_{H_i})}{V_0 + \sum_{i=1}^{N} n_i V_i}} \tag{2.14}$$

where n_i is the number of groups of type i in the solvent molecule, N is the number of group types in the solvent molecule, and F_{D_i}, F_{P_i}, U_{H_i}, and V_i are the contributions associated with group i. The latter are tabulated for common groups by van Krevelen and Hoftyzer (1976—F_{D_i}, F_{P_i}), Hansen and Beerbower (1971—U_{H_i}), and Constantinou and Gani (1994—V_i). Note that the constant associated with the molar volume prediction for liquids is $V_0 = 12.11 \text{ cm}^3/\text{mol}$.

Given estimates of these three solubility parameters, a solvent is likely to dissolve a solute when:

$$4(\delta_D - \delta_D^*)^2 + (\delta_P - \delta_P^*)^2 + (\delta_H - \delta_H^*)^2 \le (R^*)^2 \tag{2.15}$$

where R^*, referred to as the radius of interaction, as defined by Hansen (1969) and δ_D^*, δ_P^*, and δ_H^*, are parameters related to the solute. Note that the left-hand side of inequality (2.15) is the distance between the solute and solvent molecules, a measure of the solute-solvent interaction.

For other applications, like the selection of solvents for the liquid–liquid extraction of solutes from mixtures, solubility and related measures are determined on the basis of the liquid-phase activity coefficients, γ_{ij}, for solute-solvent pairs. Usually, for screening purposes, it is sufficient to estimate the liquid-phase activity coefficient at infinite dilution, γ_{ij}^∞, using group-contribution methods.

When considering solvent S for extraction of solute A from species B, Pretel and co-workers (1994) use the UNIFAC group-contribution method to obtain estimates of four solvent properties:

$$\text{Solvent Selectivity} = S_s = \beta = \frac{x_{A,S}}{x_{B,S}} = \frac{\gamma_{B,S}^\infty MW_A}{\gamma_{A,S}^\infty MW_B} \tag{2.16}$$

$$\text{Solvent Power} = S_P = x_{A,S} = \frac{1}{\gamma_{A,S}^{\infty}} \frac{MW_A}{MW_S} \tag{2.17}$$

$$\text{Solute Distribution Coefficient} = m = K_D = \frac{x_{A,S}}{x_{A,B}} = \frac{\gamma_{A,B}^{\infty}}{\gamma_{A,S}^{\infty}} \frac{MW_B}{MW_S} \tag{2.18}$$

$$\text{Solvent Loss} = S_l = x_{S,B} = \frac{1}{\gamma_{S,B}^{\infty}} \frac{MW_S}{MW_B} \tag{2.19}$$

where x are mass fractions. Clearly, a desirable solvent will have large selectivity, solvent power, distribution coefficient, and low solvent loss, as discussed in Example 2.6 below.

Health and Safety Measures. Several empirically-defined properties are useful in restricting the selection of solvents to those having low impacts on health and safety. These are presented next.

    Bioconcentration Factor. This factor, which is related to the likelihood of a solvent accumulating in and harming living tissue, was correlated by Veith and Konasewich (1975) as:

$$\log_{10} \text{BCF} = 0.76 \log_{10} K_{ow} - 0.23 \tag{2.20}$$

where K_{ow} is the octanol-water partition coefficient, which is expressed as:

$$\log_{10} K_{ow} = \sum_{i=1}^{N} n_i \chi_i^0 + 0.12 \sum_{i=1}^{N} n_i \chi_i^1 \tag{2.21}$$

where χ_i^0 and χ_i^1 are the *fragment* and *factor* of group i, respectively, as tabulated by Hansch and Leo (1979).

    Toxicity Measure. The lethal concentration of a solvent, LC_{50}, a useful measure of toxicity, has been correlated by Konemann (1981) as:

$$\log_{10} LC_{50} = -0.87 \log_{10} K_{ow} - 0.11 \tag{2.22}$$

    Flash Point. The flash-point temperature is a measure of the explosive potential of vapor mixtures in air. For paraffins, aromatics, and cycloparaffins, it has been correlated as a function of the normal boiling point (Butler et al., 1956; Lyman et al., 1981):
$$T_f = 0.683 T_b - 119 \tag{2.23}$$
where the temperatures are in kelvin.

EXAMPLE 2.5

In a lithographic printing process, ink is conveyed to an impression plate by means of a train of rubber rollers known as "blankets." These blankets must be cleaned regularly since their cleanliness is crucial for the production of high quality images. It is desired to replace the current solvent, 1,1,1-trichloroethane, with an environmentally-friendly solvent, having the ability to rapidly dissolve dried ink and having a short drying time; that is, with a small latent heat of vaporization, and consequently, a short drying time and low utility costs for vaporization. In addition, the solvent should cause negligible swelling of the blanket and be nonflammable. These are the desired product quality specifications.

SOLUTION

This solution is based upon that presented by Sinha and co-workers (1999). A set of 12 molecular groups is selected upon which the search for solvent molecules is based. These are: CH_3—, —CH_2—, Ar— (C_6H_5—), Ar= (C_6H_4=), —OH, CH_3CO—, —CH_2CO—, —COOH, CH_3COO—, —CH_2COO—, —CH_3O, and —CH_2O—. Note that chlorine is omitted to avoid ozone-depletion problems.

    Next, specifications are provided to define the desired properties of the solvent molecules to be designed. The ink residue is assumed to consist of phenolic resin, Super Bakacite 1001, for which the following parameters were estimated: $\delta_D^* = 23.3$ MPa$^{1/2}$, $\delta_P^* = 6.6$ MPa$^{1/2}$, $\delta_H^* = 8.3$ MPa$^{1/2}$, and $R^* = 19.8$ MPa$^{1/2}$. For the lithographic blanket, which is typically polyisoprene rubber, swelling is avoided when

$\delta_P > 6.3$ MPa$^{1/2}$. Furthermore, the bioconcentration factor is sufficiently low and the lethal concentration is sufficiently high when $\log_{10} K_{ow} < 4.0$. Finally, to assure that the solvent is liquid at ambient pressure, it is required that $T_b > 323$ K and $T_m < 223$ K, where T_m is the melting point temperature. Note that while no bounds are placed upon the standard latent heat of vaporization at 298 K, ΔH^v, it is minimized to reduce the drying time and the cost of heating utilities.

Group Contributions

The following group contributions have been taken from van Krevelen and Hoftyzer (1976), Hansen and Beerbower (1971), and Constantinou and Gani (1994).

Group	Valence	T_{bi}	T_{mi}	F_{Di}	F_{Pi}	U_{Hi}	H_{Vi}	V_i	χ_i^0	χ_i^1
CH$_3$—	1	0.8894	0.464	420	0	0	4.116	26.14	0.89	1
—CH$_2$—	2	0.9225	0.9246	270	0	0	4.65	16.41	0.66	1
Ar—	1	6.2737	7.5434	1,430	110	0	33.042	70.25	1.9	1
Ar=	2	6.2737	7.5434	1,430	110	0	33.042	70.25	1.67	1
—OH	1	3.2152	3.5979	210	500	−19,500	24.529	5.51	−1.64	1
CH$_3$CO—	1	3.566	4.8776	210	800	−2,000	18.99	36.55	−0.44	2
—CH$_2$CO—	2	3.8967	5.6622	560	800	−2,000	20.41	28.16	−0.67	2
—COOH	1	5.8337	11.563	409	450	−11,500	43.046	22.32	−1.11	1
CH$_3$COO—	1	3.636	4.0823	806	510	−3,300	22.709	45	−0.6	2
—CH$_2$COO—	2	3.3953	3.5572	609	510	−3,300	17.759	35.67	−0.83	2
—CH$_3$O	1	2.2536	2.9248	520	410	−4,800	10.919	32.74	−0.93	2
—CH$_2$O—	2	1.6249	2.0695	370	410	−4,800	7.478	23.11	−1.16	2

To estimate the normal boiling point, the melting point, and the standard latent heat of vaporization at 298 K:

$$T_b = T_{b0} \ln \left(\sum_{i=1}^{N} n_i T_{bi} \right)$$

$$T_m = T_{m0} \ln \left(\sum_{i=1}^{N} n_i T_{mi} \right)$$

$$\Delta H^v = \Delta H_0^v + \sum_{i=1}^{N} n_i H_{Vi}$$

where $T_{b0} = 204.2$ K, $T_{m0} = 102.4$ K, and $\Delta H_0^v = 6.829$ KJ/mol.

Sinha and co-workers (1999) formulate a mixed-integer nonlinear program, which minimizes ΔH^v to locate three compounds that satisfy the specified constraints. These are:

Compound	ΔH^v, kJ/mol	T_b, K	T_m, K	δ_P, MPa$^{1/2}$	$\log_{10} K_{ow}$
Methyl ethyl ketone	35.5	354.9	193.2	9.66	1.59
Diethyl ketone	40.1	385.7	206.6	8.21	2.37
Ethylene glycol monomethyl ether	47.6	387.4	200.2	11.5	−0.65

∎

EXAMPLE 2.6

It is desired to locate a solvent for the liquid–liquid extraction of ethanol from its azeotrope with water. This dehydration has been carried out principally by heterogeneous azeotropic distillation using benzene, now known to be a carcinogen, as an entrainer. If such a solvent can be located, liquid-liquid extraction could become the preferred processing technique.

SOLUTION

Potential molecular groups for the solvents are selected from among those in the UNIFAC VLE (vapor–liquid equilibrium) tables (Hansen et al., 1991). Solvents are sought that have the following properties: MW < 300, $T_b - T_{b,\text{furfural}} > 50$ K, $S_s > 7$ wt./wt., m > 1.0 wt%/wt%, and $S_l > 0.1$ wt%.

Pretel and co-workers (1994) estimate these properties, as well as the solvent power, S_P, and the solvent density, ρ_s, for many candidate solvents. They observe that the constraints are not satisfied for any of the candidates, and consequently, conclude that liquid–liquid extraction is not a favorable process for the dehydration of the ethanol–water azeotrope.   ∎

Solutes For Hand Warmers

In Section 19.4, hand warmers are discussed as examples of the design of configured consumer products. Early in the product design process, a design team identifies product quality specifications. As discussed, typically these include identifying a salt that is: (1) capable of remaining dissolved in a supersaturated solution at temperatures far below its equilibrium crystallization temperature, (2) capable of crystallization at a moderate temperature, such as 125°F, (3) having a large latent heat of crystallization, and (4) capable of being regenerated, that is, easily dissolved for reuse. In this case, an approach is described in Example 19.3 for screening several candidate salts using experimental data.

2.4   SUMMARY

This section has concentrated on the search for molecules to meet the thermophysical and transport property specifications being considered by a design team for a new chemical product(s). Emphasis has been placed on the use of group contribution methods for the property estimates. Molecular simulation methods, which are gaining favor, are introduced briefly, but are not used in the examples presented. Through examples for the design of polymer repeating units, refrigerants, solvents for removal of printing ink, and solvents for liquid–liquid extraction, optimization methods are employed to locate the best molecules.

REFERENCES

Allen, D.T., and D.R. Shonnard, *Green Engineering: Environmentally Conscious Design of Chemical Processes*, Prentice-Hall, Englewood Cliffs, New Jersey (2002).

Aspen OLI User Guide, Version 10.2, Aspen Technology Inc., Cambridge, MA (1999).

Bird, R.B., W.E. Stewart, and E.N. Lightfoot, *Transport Phenomena*, 2nd ed., John Wiley & Sons, New York (2002).

Brennecke, J.F., and E.J. Maginn, "Ionic Liquids: Innovative Fluids for Chemical Processing," *AIChE J.*, **47**(11), 2384–2389 (2001).

Butler, R.M., G.M. Cooke, G.G. Lukk, and B.G. Jameson, "Prediction of Flash Points for Middle Distillates, *Ind. Eng. Chem.*, **48**, 808–812 (1956).

Chen, C.-C., C.P. Bokis, and P. Mathias, "Segment-based Excess Gibbs Energy Model for Aqueous Organic Electrolytes," *AIChE J.*, **47**(11), 2593–2602 (2001).

Constantinou, L., and R. Gani, "New Group Contribution Method for Estimating Properties of Pure Compounds," *AIChE J.*, **40**, 1697–1710 (1994).

Dare-Edwards, M.P., "Novel Family of Traction Fluids Derived from Molecular Design," *J. Synth. Lubr.*, **8**(3), 197 (1991).

Derringer, G.C., and R.L. Markham, "A Computer-Based Methodology for Matching Polymer Structures with Required Properties," *J. Appl. Polymer Sci.*, **30**, 4609–4617 (1985).

Duvedi, A.P., and L.E.K. Achenie, "Designing Environmentally Safe Refrigerants Using Mathematical Programming," *Chem. Eng. Sci.*, **51**(15,) 3727–3729 (1996).

Gani, R., B. Nielsen, and A. Fredenslund, "A Group Contribution Approach to Computer-aided Molecular Design," *AIChE J.*, **37**(9), 1318–1332 (1991).

Getting Started Modeling Processes with Electrolytes, Version 10.2, Aspen Technology, Inc., Cambridge, MA (1999).

Giannelis, E.P., "Molecular Engineering of Ceramics. Chemical Approaches to the Design of Materials," *Eng.: Cornell Q.*, **23**(2), 15 (1989).

Hansch, C., and A.J. Leo, *Substituent Constants for Correlation Analysis in Chemistry and Biology*, Wiley, New York (1979).

Hansen, C.M., "The Universality of the Solubility Parameter," *Ind. Eng. Chem. Prod. Res. Dev.*, 2–11 (1969).

Hansen, C.M., and A. Beerbower, "Solubility Parameters," in A. Standen, Ed., *Kirk-Othmer Encyclopedia of Chemical Technology*, Wiley-Interscience, New York (1971).

Hansen, H.K., P. Rasmussen, Aa. Fredenslund, M. Schiller, and J. Gmehling, "Vapor–liquid Equilibria by UNIFAC Group Contribution: 5. Revision and Extension," *Ind. Eng. Chem. Res.*, **30**, 2352 (1991).

Joback, K.G., and R.C. Reid, "Estimation of Pure-Component Properties from Group Contributions," *Chem. Eng. Commun.*, **57**, 233–243 (1987).

Joback, K.G., and G. Stephanopoulos, "Designing Molecules Possessing Desired Physical Property Values," in J.J. Siirola, I.E. Grossmann, and G. Stephanopoulos, Eds., Proceedings of *Foundations of Computer-aided Process Design (FOCAPD'89)*, pp. 363–387, AIChE New York (1989).

Kirschner, E.M., "Environment, Health Concerns Force Shift in Use of Organic Solvents," *Chem. Eng. News*, 13–20, June 20 (1994).

Kofke, D.A., and B.C. Mihalick, "Web-based Technologies for Teaching and Using Molecular Simulation," *Fluid Phase Equil.*, **194–197**, 327–335 (2002).

Konemann, H., "Quantitative Structure–Activity Relationships in Fish Toxicity Studies. 1. Relationship for 50 Industrial Chemicals," *Toxicology*, **19**, 209–221 (1981).

Lewin, D.R., W.D. Seider, J.D. Seader, E. Dassau, J. Golbert, D. Goldberg, M.J. Fucci, and R.B. Nathanson, *Using Process Simulators in Chemical Engineering: A Multimedia Guide for the Core Curriculum*, Version 2.0, John Wiley & Sons, New York (2003).

Lyman, W.J., W.F. Reehl, and D. H. Rosenblatt, *Handbook of Chemical Property Estimation Methods*, McGraw-Hill, New York (1981).

Maranas, C.D., "Optimal Computer-Aided Molecular Design: A Polymer Design Case Study," *Ind. Eng. Chem. Res.*, **35**, 3403–3414 (1996).

Midgely, T., "From Periodic Table to Production," *Ind. Eng. Chem.*, **29**, 241–244 (1937).

Pisano, G.P., *The Development Factory: Unlocking the Potential of Process Innovation*, Harvard Business School Press, Cambridge (1997).

Poling, B.E., J.M. Prausnitz, and J.P. O'Connell, *Properties of Gases and Liquids*, 5th ed., McGraw-Hill, New York (2001).

PPDS2 for Windows: User Manual and Reference Guide, Nat'l. Eng. Lab., E. Kilbridge, Glasgow, UK (1997).

Pretel, E.J., P. Araya Lopez, S.B. Bottini, and E.A. Brignole, "Computer-aided Molecular Design of Solvents for Separation Processes," *AIChE J.*, **40**(8), 1349–1360 (1994).

Reid, R.C., J.M. Prausnitz, and T.K. Sherwood, *The Properties of Gases & Liquids*, 3rd ed., McGraw-Hill, New York (1977).

Reid, R.C., J.M. Prausnitz, and B.E. Poling, *The Properties of Gases & Liquids*, 4th ed., McGraw-Hill, New York(1987).

Rowley, R.L., *Statistical Mechanics for Thermophysical Property Calculations*, Prentice-Hall, Englewood Cliffs, New Jersey (1994).

Sinha, M., L.E.K. Achenie, and G.M. Ostrovsky, "Environmentally Benign Solvent Design by Global Optimization," *Comput. Chem. Eng.*, **23**, 1381–1394 (1999).

van Krevelen, D.W., *Properties of Polymers: Correlation with Chemical Structure*, Elsevier, Amsterdam (1972).

van Krevelen, D.W., *Properties of Polymers*, 3rd ed., Elsevier, Amsterdam (1990).

van Krevelen, D.W., and P.J. Hoftyzer, *Properties of Polymers: Their Estimation and Correlation with Chemical Structure,* Elsevier, Amsterdam (1976).

Veith, G.D., and D.E. Konasewich, "Structure–activity Correlations in Studies of Toxicity and Bioconcentration with Aquatic Organisms," Proceedings of Symposium in Burlington, Ontario—Canada Center for Inland Water (1975).

Zemaitis, J.F., D.M. Clarke, M. Rafal, and N.C. Scrivner, *Handbook of Aqueous Electrolyte Thermodynamics*, DIPPR, AIChE, New York, NY (1986)

EXERCISES

2.1 For the polymer film in Example 2.2, use GAMS to locate the repeat units having the second and third lowest values of the objective function.

2.2 Many companies and municipalities are reluctant to handle chlorine, either in processing or in incinerating wastes. Resolve Example 2.2 without the —CHCl— group.

2.3 For an electronic device designed to operate at higher temperatures near a furnace, a high glass-transition temperature, 423 K, is required. Resolve Example 2.2 with this constraint.

2.4 It is desired to find a refrigerant that removes heat at $-20°C$ and rejects heat at $32°C$. Desirable refrigerants should have $P^s\{-20°C\} > 1.4$ bar, $P^s\{32°C\} < 14$ bar, $\Delta H^v\{-20°C\} > 18.4$ kJ/mol, and $c_{pl}\{6°C\} < 32.2$ cal/mol-K. For the candidate groups, CH_3, CH, F, and S, formulate a mixed-integer nonlinear program and use GAMS to solve it. Use the group contribution method in Section 1 of the solution to Example 2.3 to estimate T_b. Hint: maximize the objective function, $\Delta H^v\{-20°C\}$.

2.5 Using the group contributions in Example 2.5, determine whether methyl propyl ketone, methyl butyl ketone, and methyl isobutyl ketone are suitable solvents.

Chapter 3

Process Creation

3.0 OBJECTIVES

This chapter covers many of the steps under the blocks of "Process Creation" and "Development of Base Case" in Figure 1.2, which provides an overview of the steps in designing new chemical products and processes. These are two of the major blocks on which design teams concentrate much of their effort.

After studying this chapter, the reader should

1. Understand how to go about assembling design data and creating a preliminary database.
2. Be able to implement the steps in creating flowsheets involving reactions, separations, and $T-P$ change operations. In so doing, many alternatives are identified that can be assembled into a synthesis tree containing the most promising alternatives.
3. Know how to select the principal pieces of equipment and to create a detailed process flow diagram, with a material and energy balance table and a list of major equipment items.
4. Understand the importance of building a pilot plant to test major pieces of equipment where some uncertainty exists.
5. Have an initial concept of the role of a process simulator in obtaining data and in carrying out material and energy balances. This subject is expanded upon in Chapter 4.

3.1 INTRODUCTION

This chapter begins with the steps often referred to as *process creation*, which are implemented by a design team in seeking a solution to each of the design alternatives prepared when assessing the primitive design problem, as discussed in Section 1.2. It describes the components of the preliminary database and suggests several sources, including the possibility of carrying out laboratory experiments. Then, using the database, it shows how to create a synthesis tree, with its many promising flowsheets, for consideration by the design team. This is accomplished first for the design of a continuous process to produce a commodity chemical, vinyl chloride, and subsequently, for the design of a batch process to produce a pharmaceutical.

For each of the most promising alternatives in the synthesis tree, a base-case design is created. Because this is central to the work of all design teams, the strategy for creating a detailed process flow diagram is covered and the need for pilot-plant testing is discussed.

3.2 PRELIMINARY DATABASE CREATION

Having completed an initial assessment of the primitive design problem and having conducted a literature search, the design team normally seeks to organize the data into a compact database, one that can be accessed with ease as the team proceeds to create process flowsheets and develop a base-case design. At this stage, several specific problems are being considered, involving several raw materials, the desired products, and several byproducts and reaction intermediates. For these chemicals, basic thermophysical properties are needed, including molecular weight, normal boiling point, freezing point, critical properties, standard enthalpy and Gibbs free energy of formation; and vapor pressures, densities, heat capacities, and latent heats as a function of temperature. If chemical reactions are involved, some rudimentry information concerning the rates of the principal chemical reactions, such as conversion and product distribution as a function of space velocity, temperature, and pressure, is often needed before initiating the process synthesis steps. When necessary, additional data are located, or measured in the laboratory, especially when the design team gains enthusiasm for a specific processing concept. In addition, the team needs environmental and safety data, including information on the toxicity of the chemicals, how they affect animals and humans, and flammability in air. Material Safety Data Sheets (MSDSs) will be available for chemicals already being produced but will have to be developed for new chemicals. Also, for preliminary economic evaluation, chemical prices are needed. Additional information, such as transport properties, detailed chemical kinetics, the corrosivity of the chemicals, heuristic parameters, and data for sizing equipment, is normally not needed during process creation. It is added by the design team, after a detailed process flow diagram has been created, and before work on the detailed design of the equipment commences.

When the data are assembled, graphs are often prepared with curves positioned to provide a good representation, especially for experimental data with scatter. Alternatively, the coefficients of equations, theoretical or empirical, are computed using regression analysis programs. This is especially common for thermophysical property data, such as the vapor pressure, P^s, as a function of the temperature, T, and vapor–liquid equilibrium data, as discussed later in this section.

If molecular structure design is previously carried out, as discussed in Chapter 2, most of the pertinent data will have been collected. This is especially the case for protein pharmaceuticals where automated lab benches are often employed, permitting hundreds and thousands of cell clones to be evaluated experimentally in parallel, as discussed in Section 2.1.

Thermophysical Property Data

For basic properties, such as molecular weight, normal boiling point, melting point, and liquid density (often at 20°C), the CRC *Handbook of Chemistry and Physics* (CRC Press, Boca Raton, FL, annual) provides a compilation for a large number of organic and inorganic compounds. In addition, it provides vapor pressure data and enthalpies and free energies of formation for many of these compounds, as well as selected properties, such as the critical temperature, for just a few of these compounds. Similar compilations are provided by *Perry's Chemical Engineers' Handbook* (Perry and Green, 1997), the *Properties of Gases and Liquids* (Poling et al., 2001), and *Data for Process Design and Engineering Practice* (Woods, 1995). In addition, extensive databases for more than 1,000 compounds are provided by process simulators (e.g., ASPEN PLUS, HYSYS.Plant, CHEMCAD, PRO/II, BATCH PLUS, and SUPERPRO DESIGNER), as discussed in Section 2.2. These are extremely useful as they are accessed by large libraries of programs that carry out material and energy balances, and estimate equipment sizes and costs.

Because phase equilibria are important in most chemical processes, design teams usually spend considerable time assembling data, especially vapor–liquid and liquid–liquid equilibrium data. Over the years, thousands of articles have been published in which phase equilibria data are provided. These can be accessed by a literature search, although the need to search the literature has largely been negated by the extensive compilation provided in

Vapor–Liquid Equilibrium Data Collection (Gmehling et al., 1980). In this DECHEMA data bank, which is available both in more than 20 volumes and electronically, the data from a large fraction of the articles can be found easily. In addition, each set of data has been regressed to determine interaction coefficients for the binary pairs to be used to estimate liquid-phase activity coefficients for the NRTL, UNIQUAC, Wilson, etc., equations. This database is also accessible by process simulators. For example, with an appropriate license agreement, data for use in ASPEN PLUS can be retrieved from the DECHEMA database over the Internet. For nonideal mixtures, the extensive compilation of Gmehling (1994) of azeotropic data is very useful.

In this section, space is not available to discuss the basics of phase equilibrium; for this material, the reader is referred to many excellent thermodynamics books (e.g., Smith et al., 1997; Kyle, 1984; Balzhiser et al., 1972; Sandler, 1997; Walas, 1985; de Nevers, 2002). Yet process designers usually need to work with phase equilibria data to obtain reasonable predictions for phase conditions and separations of specific mixtures in the temperature and pressure ranges anticipated. This usually requires data regression using models that are best suited for the compositions, temperatures, and pressures under study. Consequently, in this section, two examples are presented in which methods of data regression are needed. To assist the reader, a review of the basics of phase equilibrium is presented on the CD-ROM that accompanies this textbook, *Using Process Simulators in Chemical Engineering: A Multimedia Guide for the Core Curriculum (ASPEN → Physical Property Estimation;* Lewin et al., 2003), in which the equations are derived, the data banks are summarized, and many of the phase equilibrium models are tabulated and discussed briefly.

EXAMPLE 3.1

This example involves vapor–liquid equilibrium (VLE) data for the design of a distillation tower to dehydrate ethanol. A portion of the *T–x–y* data for an ethanol–water mixture, measured at 1.013 bar (1 atm) using a Gillespie still (Rieder and Thompson, 1949), is shown in Figure 3.1a. Here, it is desired to use regression analysis to enable the UNIQUAC equation to represent the data accurately over the entire composition range.

SOLUTION

Using ASPEN PLUS and data from the DECHEMA data bank, with the details described on the multimedia CD-ROM that accompanies this textbook (*ASPEN → Physical Property Estimation → Equilibrium Diagrams → Property Data Regression*), the *x–y* diagram in Figure 3.1b is obtained, which compares the data points with a curve based on the following built-in interaction coefficients retrieved from the VLE-IG data bank: $a_{E,W} = 2.0046$, $a_{W,E} = -2.4936$, $b_{E,W} = -728.97$, and $b_{W,E} = 756.95$. Then the data regression system is used with the Rieder and Thompson data and much better agreement between the data and the VLE estimates is obtained, as shown in Figure 3.1c. Note that the data regression system adjusts the interaction coefficients to $a_{E,W} = 3.8694$, $a_{W,E} = -3.9468$, $b_{E,W} = -1,457.2$, and $b_{W,E} = 1,346.8$.

Clearly, data regression is needed to obtain a rigorous design for the distillation. Furthermore, in this case, the UNIQUAC equation represents the nonidealities of this polar mixture quite well. When the Peng–Robinson (Reid et al., 1987) equation is used instead, as shown on the multimedia CD-ROM, the data are not represented as well after the data regression is completed. ∎

EXAMPLE 3.2

A second example is provided in which vapor–liquid equilibrium data for a CH_4–H_2S mixture are utilized in connection with the design of a natural gas expander plant. In this case, a portion of the *P–x–y* data, measured by Reamer et al. (1951), is shown in Figure 3.2a, and regression analysis is used to enable the Soave–Redlich–Kwong (SRK) equation to represent the data better.

SOLUTION

ASPEN PLUS is used with the SRK equation:

$$P = \frac{RT}{V - b} - \frac{a}{V(V + b)}$$

where

$$a = \sum_{i=1}^{2} \sum_{j=1}^{2} x_i x_j (a_i a_j)^{0.5} (1 - k_{ij})$$

$$b = \sum_{i=1}^{2} x_i b_i$$

$$a_i = f\left[T, T_{ci}, P_{ci}, \omega_i\right]$$

$$b_i = f\left[T_{ci}, P_{ci}\right]$$

$$k_{ij} = k_{ji}$$

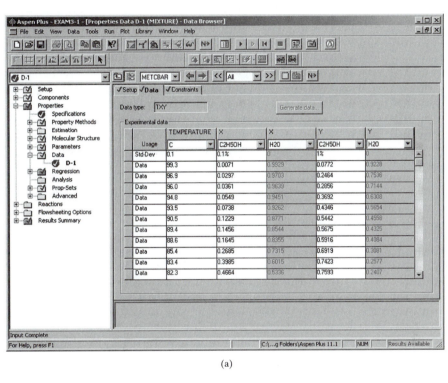

(a)

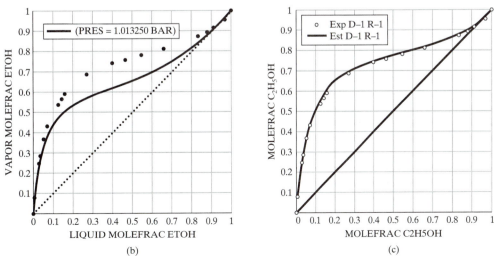

(b) (c)

Figure 3.1 Regression of ethanol–water data using UNIQUAC: (a) VLE data (Rieder and Thompson, 1949); (b) x–y diagram before regression; (c) x–y diagram after regression.

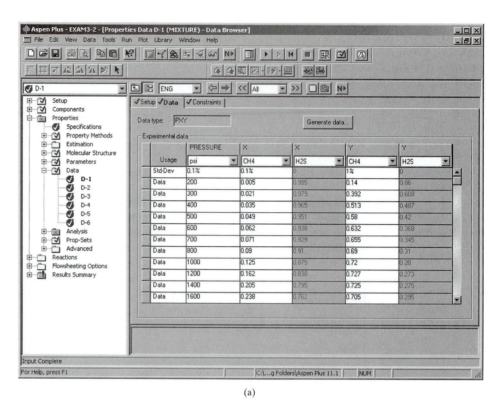

(a)

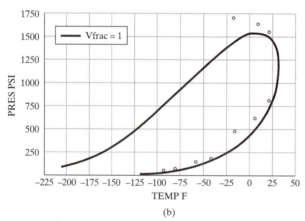

(b)

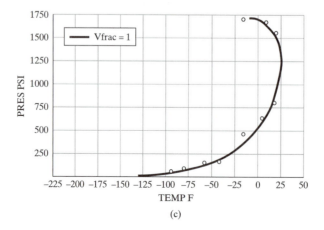

(c)

Figure 3.2 Regression of CH_4–H_2S data using the Soave–Redlich–Kwong equation: (a) VLE data (Reamer et al., 1951); (b) phase envelope before regression—75.2 mol% CH_4; (c) phase envelope after regression—75.2 mol% CH_4.

and P is the pressure, T is the absolute temperature, V is the molar volume of the mixture, a_i and b_i are pure-component constants, T_{ci}, P_{ci}, and ω_i are the critical temperature and pressure and the acentric factor for species i, and k_{ij} and k_{ji} are the binary interaction coefficients. Using the built-in parameters retrieved from the EOS-LIT data bank, the P–T phase envelope in Figure 3.2b is obtained, with the details described on the multimedia CD-ROM (*ASPEN → Physical Property Estimation → Equilibrium Diagram → Property Data Regression*). Then the data regression system is used to adjust three of the parameters to $k_{CH_4,H_2S} = -0.11399$, $\omega_{CH_4} = -0.3344$, and $\omega_{H_2S} = 0.04377$. The result is a significant improvement of the phase envelope when compared with the experimental data, as shown in Figure 3.2c. Note, especially, the improvement in the critical region at the elevated pressures. ∎

Environmental and Safety Data

As mentioned in Section 1.3, design teams need toxicity data for raw materials, products, byproducts, and intermediates incorporated in a process design. In toxicology laboratories operated by chemical companies and governmental agencies, such as the U.S. Environmental Protection Agency (EPA) and the U.S. Food and Drug Administration (FDA), tests are run to check the effects of various chemicals on laboratory animals. The chemicals are administered in varying dosages, over differing periods, and in different concentrations, stimulating effects that are measured in many ways, including effects on the respiratory system, the skin, and the onset of cancer. In most cases, the results are provided in extensive reports or journal articles. In some cases, chemicals are difficult to classify as toxic or nontoxic.

Already it is well known that a number of common chemicals are toxic to humans and need to be avoided. One source of information on these chemicals is the Toxic Chemical Release Inventory (TRI), which is maintained by the U.S. EPA, and includes over 600 chemicals. A list of these chemicals is available at the Internet site:

http://www.epa.gov/tri/chemical/index.htm

Another source is provided by the ratings of the National Fire Protection Association (NFPA), which are tabulated for many chemicals in *Data for Process Design and Engineering Practice* (Woods, 1995). The first of three categories is titled "Hazard to Health" and is rated from 0 to 4, with 0 meaning harmless and 4 meaning extremely hazardous.

As seen in Table 1.3 and discussed in Section 1.4, data on the flammability of organic compounds are tabulated and, for those compounds not included in the table, methods are available to estimate the data. In addition, tables of flammability data are also available for aerosols and polymers in *Perry's Chemical Engineers' Handbook* (Perry and Green, 1997). The NFPA ratings provide a less quantitative source for many chemicals under "Flammability Hazard," which is the second of the three categories (also rated from 0 to 4).

Chemical Prices

Economics data are often related to supply and demand, and consequently they fluctuate and are much more difficult to estimate. Most companies, however, carry out market studies and have a basis for projecting market size and chemical prices. In view of the uncertainties, to be safe, economic analyses are often conducted using a range of chemical prices to determine the sensitivity of the results to specific prices.

One widely used source of prices of commodity chemicals is the *Chemical Market Reporter*, a biweekly newspaper that provides up-to-date prices for chemicals of commerce and is available in most university libraries. It should be noted, however, that these prices may not reflect the market situation at a particular location; nevertheless, they provide a good starting point. Obviously, to obtain better estimates, at least for the immediate future, the manufacturers of the chemicals should be contacted directly. Lower prices than those listed can often be negotiated. Articles on chemicals of commerce in trade magazines can be searched for on the web site, http://www.findarticles.com. To obtain chemical prices, search for the Chemical Market Reporter. Also, www.chemicalmarketreporter.com can be used with a library user name and password.

In some cases, it may be desirable to estimate the prices of utilities, such as steam, cooling water, and electricity, during process creation. Here also, appropriate prices can be obtained from local utility companies. As a start, however, values are often tabulated, as provided in Table 17.1.

Summary

To the extent possible, using the literature, company files, computer data banks, and similar sources, the design team assembles a preliminary database for use in preliminary process synthesis, the subject of Section 3.4. Typically, the database contains thermophysical property data, rudimentary reaction-rate data, data concerning toxicity and flammability of the chemicals, and chemical prices. In cases where data cannot be located, estimation methods are often available. However, when the results are sensitive to the estimates, conclusions must be drawn with caution. In most cases, when a process looks promising, an experimental program is initiated, as discussed in the next section. Note that other kinds of data are normally not necessary until the detailed process flow diagram has been developed for the base-case design, and the design team is preparing to complete the detailed design of the equipment items. Note also that when molecular structure design has been used to select the chemical product, experimental data and/or theoretical estimates are usually available in data banks, especially in drug development.

3.3   EXPERIMENTS

Many design concepts are the result of extensive experiments in the laboratory, which provide valuable data for the design team. Often, however, laboratory experiments are carried out in small vessels, using small quantities of expensive solvents, and under conditions where the conversion and selectivity to the desired product are far from optimal. For this reason, as a design concept becomes more attractive, it is common for the design team to request additional experiments at other conditions of compositions, temperatures, and pressures, and using solvents that are more representative of those suitable for large-scale production. In cases where no previous in-house experimental work has been done, laboratory programs are often initiated at the request of the design team, especially when estimates of the rates of reaction are not very reliable. When chemical reactions involve the use of catalysts, it is essential that experiments be conducted on catalyst life using feedstocks that are representative of those to be used for large-scale production, and that may contain potential catalyst poisons.

Laboratory experiments may also be necessary to aid in the selection and preliminary design of separation operations. The separation of gas mixtures requires consideration of absorption, adsorption, and gas permeation, all of which may require the search for an adequate absorbent, adsorbent, and membrane material, respectively. When nonideal liquid mixtures are to be separated, laboratory distillation experiments should be conducted early because the possibility of azeotrope formation can greatly complicate the selection of adequate separation equipment, which may involve the testing of one or more solvents or entrainers. When solids are involved, early laboratory tests of such operations as crystallization, filtration, and drying are essential.

Clearly, as data are obtained in the laboratory, they are tabulated and usually regressed, to allow addition to the preliminary database for use by the design team in preliminary process synthesis, the subject of the next section.

3.4   PRELIMINARY PROCESS SYNTHESIS[1]

Design teams use many kinds of processing operations to carry out chemical reactions and to separate products and byproducts from each other and from unreacted raw materials. In

---

[1]Adapted from Myers and Seider (1976), Chap. 3.

many respects, one of the greatest challenges in process design involves the synthesis of configurations that produce chemicals in a reliable, safe, and economical manner, and at high yield with little or no waste. Until recently, this part of the design process, often referred to as *process synthesis*, in which many kinds of process operations are configured into flowsheets, was performed from experience gained in similar processing situations, with little formal methodology.

Thanks to research over the past 30 years, coupled with methods of decision-tree analysis and mathematical programming, synthesis strategies have become more quantitative and scientific. In Part Two of this text, a primary objective is to cover many of the modern strategies for synthesizing process flowsheets. The objective of this introductory section, however, is simply to show some of the steps and decision processes, mostly by example. After examining two case studies, involving the synthesis of a vinyl chloride process and of a process to manufacture tissue plasminogen activator (tPA), the reader should have a good appreciation of the principal issues in process synthesis.

As discussed earlier, preliminary process synthesis occurs after an alternative processing concept has been created. Having defined the concept and assembled the preliminary database, usually with some experimentation, the design team sets out to synthesize a flowsheet of process operations to convert the raw materials to the desired products. First, it decides on the *state* of the raw materials, products, and byproducts, before assembling different configurations of the process operations.

To introduce this approach, this section begins by reviewing the concept of the chemical state, followed by a review of the principal operations, before covering several of the key steps in process synthesis, and utilizing them to create the vinyl chloride and tPA processes. Throughout this development, it should be clear that the synthesis or invention of a chemical process involves the generation and solution of a large combinatorial problem. Here, intuition and experience are as important to the design team as to the composer or artist. The emphasis in this section is on the use of *heuristics*, or rules of thumb for the synthesis step. However, throughout this text, it will be evident, especially in Part Two, that many quantitative methods of synthesis, combined with optimization, are available to the design team to generate the most promising process flowsheets.

Chemical State

As the first step in process synthesis, the design team must decide on raw material and product specifications. These are referred to as *states*. Note that the state selections can be changed later with modifications to the flowsheets. To define the state, values of the following conditions are needed:

1. Mass (flow rate)
2. Composition (mole or mass fraction of each chemical species of a unique molecular type)
3. Phase (solid, liquid, or gas)
4. Form, if solid phase (e.g., particle size distribution and particle shape)
5. Temperature
6. Pressure

In addition, some well-defined properties, such as the intrinsic viscosity, average molecular weight, color, and odor of a polymer, may be required. These are often defined in connection with the research and marketing departments, which work to satisfy the requests and requirements of their customers. It is not uncommon for a range of conditions and properties to be desired, some of which are needed intermittently by various customers as their downstream requirements vary. When this is the case, care must be taken to design a process that is sufficiently flexible to meet changing demands.

Figure   3.3 Process   synthesis problem.

For commodity chemicals, of the above conditions, the scale (i.e., production level or flow rate) of the process is a primary consideration early in the design process. Working together with the marketing people, the scale of the process is determined on the basis of the projected demand for the product. Often the demographics of the most promising customers have an important impact on the location of the plant and the choice of its raw materials. As the scale and the location are established, the composition, phase, form, temperature, and pressure of each product and raw material stream are considered as well. When the desired states of these streams have been identified, the problem of process synthesis becomes better defined. As shown in Figure 3.3, for the production of vinyl chloride, it remains to insert the process operations into the flowsheet.

It is noteworthy that once the state of a substance is fixed by conditions 1–6, all physical properties (except for the form of a solid), including viscosity, thermal conductivity, color, refractive index, and density, take on definite values. Furthermore, the state of a substance is independent of its position in a gravitational field and its velocity. Although there are other conditions (magnetic field strength, surface area) whose values are needed under certain conditions, the six conditions listed above are usually sufficient to fix the state of a substance.

Process Operations

Throughout the chemical engineering literature, many kinds of equipment, so-called *unit operations*, are described, including distillation columns, absorbers, strippers, evaporators, decanters, heat exchangers, filters, and centrifuges, just to mention a few. The members of this large collection, many of which are listed in Tables 4.1 and 4.2, in connection with process simulators, all involve one or more of these basic operations:

1. Chemical reaction
2. Separation of chemical mixtures
3. Phase separation
4. Change of temperature
5. Change of pressure
6. Change of phase
7. Mixing and splitting of streams or batches
8. Operations on solids, such as size reduction and enlargement

Since these are the building blocks of nearly all chemical processes, it is common to create flowsheets involving these basic operations as a first step in process synthesis. Then, in a *task integration* step, operations are combined where feasible. In the remainder of this section, before considering the steps in process synthesis, each of the basic operations is considered in some detail.

Chemical reaction operations are at the heart of many chemical processes. They are inserted into a flowsheet to effect differences in the molecular types between raw material and product streams. To this end, they involve the chemistry of electron transfers, free-radical exchanges, and other reaction mechanisms, to convert the molecular types of the raw materials into products of other molecular types that have the properties sought by a company's customers. Clearly, the positioning of the reaction operations in the flowsheet involves many considerations, including the degree of conversion, reaction rates, competing side reactions, and the existence of reactions in the reverse direction (which can result in constraints on the conversion at equilibrium). These, in turn, are related closely to the temperature and pressure at which the reactions are carried out, the methods for removing or supplying energy, and the catalysts that provide competitive reaction rates and selectivity to the desired products. In the next subsections, many of these issues are considered in the context of process synthesis. These are revisited throughout the text, especially in Sections 5.2, 5.3, and 5.5 and Chapter 6.

Separation operations appear in almost every process flowsheet. They are needed whenever there is a difference between the desired composition of a product or an intermediate stream and the composition of its source, which is either a feed or an intermediate stream. Separation operations are inserted when the raw materials contain impurities that need to be removed before further processing such as in reactors, and when products, byproducts, and unreacted raw materials coexist in a reactor effluent stream. The choice of separation operations depends first on the phase of the mixture and second on the differences in the physical properties of the chemical species involved. For liquid mixtures, when differences in volatilities (i.e., vapor pressure) are large, it is common to use vapor–liquid separation operations (e.g., distillation), which are by far the most common. For some liquid mixtures, the melting points differ significantly and solid–liquid separations, involving crystallization, gain favor. When differences in volatilities and melting points are small, it may be possible to find a solvent that is selective for some components and not others, and to use a liquid–liquid separation operation. For other mixtures, particularly gases, differences in absorbability (in an absorbent), adsorbability (on an adsorbent; e.g., activated carbon, molecular sieves, or zeolites) or differences in permeability through a membrane may be exploited with adsorption and membrane separation operations. These and many other separation operations are considered throughout this text, especially in Chapters 5 and 7. The first example of process synthesis that follows introduces some of the considerations in the positioning of distillation operations, and Section 5.4 and Chapter 14 contribute to create the underpinnings for a comprehensive treatment of the synthesis of separation trains in Chapter 7. Many separation operations require phase-separation operations, which may be accomplished by vessels called flash drums for vapor–liquid separation, by decanters for liquid–liquid separation, and by filters and centrifuges for liquid–solid separation.

The need to change temperatures usually occurs throughout a chemical process. In other words, there are often differences in the temperatures of the streams that enter or leave the process or that enter or leave adjacent process operations, such as reaction and separation operations. Often a process stream needs to be heated or cooled from its *source* temperature to its *target* temperature. This is best accomplished through heat exchange with other process streams that have complementary cooling and heating demands. Since the energy crisis in 1973, numerous strategies, some of which are covered in Chapter 10, have been invented to synthesize networks of heat exchangers that minimize the need for heating and cooling utilities, such as steam and cooling water. In the example of process synthesis, heating and cooling operations are inserted into the flowsheet to satisfy the heating and cooling demands, and a few of the concepts associated with heat integration are introduced. Then, in Section 5.5 and Chapter 13, additional concepts are presented to accompany Chapter 10 on heat and power integration.

The positioning of pressure-change operations such as gas compressors, gas turbines or expanders, liquid pumps, and pressure-reduction valves in a process flowsheet is often ignored in the early stages of process design. As will be seen, it is common to select the pressure levels for reaction and separation operations. When this is done, pressure-change operations will be needed to decrease or increase the pressure of the feed to the particular operation. In fact, for processes that have high power demands, usually for gas compression, there is often an opportunity to obtain much of the power through integration with a source of power, such as turbines or expanders, which are pressure-reduction devices. In process synthesis, however, where alternative process operations are being assembled into flowsheets, it has been common to disregard the pressure drops in pipelines when they are small relative to the pressure level of the process equipment. Liquid pumps to overcome pressure drops in lines and across control valves and to elevate liquid streams to reactor and column entries often have negligible costs. Increasingly, as designers recognize the advantages of considering the controllability of a potential process while developing the base-case design, the estimation of pressure drops gains importance because flow rates are controlled by adjusting the pressure drop across a valve. In the first example of process synthesis, some of the important considerations in positioning the pressure-change operations in a flowsheet are introduced. These are developed further in Sections 5.6, 9.9, 10.8, and Chapter 15.

Often there are significant differences in the phases that exit from one process operation and enter another. For example, hot effluent gases from a reactor are condensed, or partially condensed, often before entering a separation operation, such as a vapor–liquid separator (e.g., a flash vessel or a distillation tower). In process synthesis, it is common to position a phase-change operation, using temperature- and/or pressure-reduction operations, such as heat exchangers and valves.

The mixing operation is often necessary to combine two or more streams and is inserted when chemicals are recycled and when it is necessary to blend two or more streams to achieve a product specification. In process synthesis, mixing operations are inserted usually during the distribution of chemicals, a key step that is introduced in the first example of process synthesis and expanded upon in Section 5.3. Because the impact of mixing on the thermodynamic efficiency and the utilization of energy is often very negative, as discussed in Chapter 9, it is usually recommended that mixer operations not be introduced unless they are necessary—for example, to avoid discarding unreacted chemicals. In this regard, it is noteworthy that mixing is the reverse of separation. Although there is an energy requirement in separating a stream into its pure constituents, mixing can be accomplished with no expenditure of energy other than the small amount of energy required when an agitator is used to speed up the mixing process. In cases where the streams are miscible and of low viscosity, mixing is accomplished easily by joining two pipes, avoiding the need for a mixing vessel. Splitting a stream into two or more streams of the same temperature, pressure, and composition is also readily accomplished in the piping.

Synthesis Steps

Given the states of the raw material and product streams, process synthesis involves the selection of processing operations to convert the raw materials to products. In other words, each operation can be viewed as having a role in eliminating one or more of the property differences between the raw materials and the desired products. As each operation is inserted into a flowsheet, the effluent streams from the new operation are closer to those of the required products. For example, when a reaction operation is inserted, the stream leaving often has the desired molecular types, but not the required composition, temperature, pressure, and phase. To eliminate the remaining differences, additional operations are needed. As separation operations are inserted, followed by operations to change the temperature, pressure, and phase, fewer differences remain. In one parlance, the operations are inserted with the goal of reducing the differences until the streams leaving the last operations are identical in state to the required products. Formal, logic-based strategies, involving the proof of theorems that assert that all of the differences have been eliminated, have been referred to as *means–end analysis*. In process synthesis, these formal strategies have not been developed beyond the synthesis of simple processes. Rather, an informal approach, introduced by Rudd, Powers, and

Siirola (1973) in a book entitled *Process Synthesis*, has been adopted widely. It involves positioning the process operations in the following steps to eliminate the differences:

Synthesis Step	Process Operations
1. Eliminate differences in molecular types	Chemical reactions
2. Distribute the chemicals by matching *sources* and *sinks*	Mixing
3. Eliminate differences in composition	Separation
4. Eliminate differences in temperature, pressure, and phase	Temperature, pressure, and phase change
5. Integrate tasks; that is, combine operations into *unit processes* and decide between continuous and batch processing	

Rather than discuss these steps in general, it is probably more helpful to describe them as they are applied to the synthesis of two processes, in this case, the vinyl chloride and tPA processes, which are synthesized in the next subsections.

Several general observations, however, are noteworthy before proceeding with the examples. First, like the vinyl chloride and tPA processes to be discussed next, most chemical processes are built about chemical reaction and/or separation operations. Consequently, the steps involved in synthesizing these processes are remarkably similar to those for the manufacture of other chemicals. As the syntheses proceed, note that many alternatives should be considered in the application of each step, many of which cannot be eliminated before proceeding to the next steps. The result is that, at each step, a new set of candidate flowsheets is born. These are organized into *synthesis trees* as the steps are applied to create the vinyl chloride and tPA processes. The synthesis trees are compact representations of the huge combinatorial problem that almost always develops during process synthesis. As will be seen, approaches are needed to eliminate the least promising branches as soon as possible, to simplify the selection of a near-optimal process flowsheet. These approaches are further refined in subsequent chapters. The decision between continuous and batch processing is introduced briefly next, before proceeding with the two examples.

Continuous or Batch Processing

When selecting processing equipment in the task-integration step, the production scale strongly impacts the operating mode. For the production of commodity chemicals, large-scale continuous processing units are selected, whereas for the production of many specialty chemicals as well as industrial and consumer chemical products, small-scale batch processing units are preferable. The choice between continuous or batch, or possibly semicontinuous, operation is a key decision. See Section 12.1 for a more complete discussion of this subject.

Example of Process Synthesis: Manufacture of Vinyl Chloride

As introduced in Section 1.2, during description of a typical primitive design problem, vinyl chloride (chloroethylene) is a monomer intermediate for the production of polyvinyl chloride, an important plastic that is widely used for rigid plastic piping, fittings, and similar products. Over the years, large commercial plants have been built, some of which produce over 1 billion lb/yr. Hence, polyvinyl chloride, and the monomer from which it is derived, is referred to commonly as a commodity chemical that is produced continuously, rather than in batch, virtually everywhere. Historically, vinyl chloride was discovered in 1835 in the laboratory of the French chemist Regnault, and the first practical method for polymerizing vinyl chloride was developed in 1917 by the German chemists Klatte and Rollett (Leonard, 1971). Vinyl-chloride monomer is an extremely toxic substance, and therefore, industrial plants that manufacture it or process it must be designed carefully to satisfy governmental health and safety regulations.

In this example, only the production of the monomer is considered, with a focus on the specific problems identified by the design team as alternatives 2 and 3 in Section 1.2. The objective is to create several promising flowsheets, as candidate solutions to these specific problems, to be inserted later into Figure 3.3. In addition to data from the chemistry laboratory, two patents (Benedict, 1960; B.F. Goodrich Co., 1963) play a key role in process synthesis. These were located by the design team during their literature search and entered into the preliminary database. When appropriate, they will be referred to in connection with the synthesis steps that follow.

Step 1. Eliminate Differences in Molecular Type

For the manufacture of vinyl chloride, data from the chemistry laboratory focus on several promising chemical reactions, involving the chemicals shown in Table 3.1. Note that since vinyl chloride has been a commodity chemical for many years, these chemicals and the reactions involving them are well known. For newer substances, the design team often begins to carry out process synthesis as the data are emerging from the laboratory. The challenge, in these cases, is to guide the chemists away from those reaction paths that lead to processes that are costly to build and operate, and to arrive at designs as quickly as possible, in time to capture the market before a competitive process or chemical is developed by another company.

Returning to the manufacture of vinyl chloride, the principal reaction pathways are as follows.

1. Direct Chlorination of Ethylene

$$C_2H_4 + Cl_2 \rightarrow C_2H_3Cl + HCl \tag{3.1}$$

This reaction appears to be an attractive solution to design alternative 2 in Section 1.2. It occurs spontaneously at a few hundred degrees Celsius, but unfortunately does not give a high yield of vinyl chloride without simultaneously producing large amounts of byproducts such

Table 3.1 Chemicals That Participate in Reactions to Produce Vinyl Chloride

Chemical	Molecular weight	Chemical formula	Chemical structure
Acetylene	26.04	C_2H_2	H—C≡C—H
Chlorine	70.91	Cl_2	Cl—Cl
1,2-Dichloroethane	98.96	$C_2H_4Cl_2$	
Ethylene	28.05	C_2H_4	
Hydrogen chloride	36.46	HCl	H—Cl
Vinyl chloride	62.50	C_2H_3Cl	

as dichloroethylene. Another disadvantage is that one of the two atoms of expensive chlorine is consumed to produce the byproduct hydrogen chloride, which may not be sold easily.

2. Hydrochlorination of Acetylene

$$C_2H_2 + HCl \rightarrow C_2H_3Cl \tag{3.2}$$

This exothermic reaction is a potential solution for the specific problem denoted as alternative 3. It provides a good conversion (98%) of acetylene to vinyl chloride at 150°C in the presence of mercuric chloride ($HgCl_2$) catalyst impregnated in activated carbon at atmospheric pressure. These are fairly moderate reaction conditions, and hence, this reaction deserves further study.

3. Thermal Cracking of Dichloroethane from Chlorination of Ethylene

$$C_2H_4 + Cl_2 \rightarrow C_2H_4Cl_2 \tag{3.3}$$
$$\frac{C_2H_4Cl_2 \rightarrow C_2H_3Cl + HCl}{C_2H_4 + Cl_2 \rightarrow C_2H_3Cl + HCl} \quad \text{(overall)} \tag{3.4}$$
$$\tag{3.1}$$

The sum of reactions (3.3) and (3.4) is equal to reaction (3.1). This two-step reaction path has the advantage that the conversion of ethylene to 1,2-dichloroethane in exothermic reaction (3.3) is about 98% at 90°C and 1 atm with a Friedel–Crafts catalyst such as ferric chloride ($FeCl_3$). Then, the dichloroethane intermediate is converted to vinyl chloride by thermal cracking according to the endothermic reaction (3.4), which occurs spontaneously at 500°C with conversions as high as 65%. The overall reaction presumes that the unreacted dichloroethane is recovered entirely from the vinyl chloride and hydrogen chloride and recycled. This reaction path has the advantage that it does not produce dichloroethylene in significant quantities, but it shares the disadvantage with reaction path 1 of producing HCl. It deserves further examination as a solution to design alternative 2.

4. Thermal Cracking of Dichloroethane from Oxychlorination of Ethylene

$$C_2H_4 + 2HCl + \tfrac{1}{2}O_2 \rightarrow C_2H_4Cl_2 + H_2O \tag{3.5}$$
$$\frac{C_2H_4Cl_2 \rightarrow C_2H_3Cl + HCl}{C_2H_4 + HCl + \tfrac{1}{2}O_2 \rightarrow C_2H_3Cl + H_2O} \quad \text{(overall)} \tag{3.4}$$
$$\tag{3.6}$$

In reaction (3.5), which *oxychlorinates* ethylene to produce 1,2-dichloroethane, HCl is the source of chlorine. This highly exothermic reaction achieves a 95% conversion of ethylene to dichloroethane at 250°C in the presence of cupric chloride ($CuCl_2$) catalyst, and is an excellent candidate when the cost of HCl is low. As in reaction path 3, the dichloroethane is cracked to vinyl chloride in a pyrolysis step. This reaction path should be considered also as a solution for design alternative 3.

5. Balanced Process for Chlorination of Ethylene

$$C_2H_4 + Cl_2 \rightarrow C_2H_4Cl_2 \tag{3.3}$$
$$C_2H_4 + 2HCl + \tfrac{1}{2}O_2 \rightarrow C_2H_4Cl_2 + H_2O \tag{3.5}$$
$$\frac{2C_2H_4Cl_2 \rightarrow 2C_2H_3Cl + 2HCl}{2C_2H_4 + Cl_2 + \tfrac{1}{2}O_2 \rightarrow 2C_2H_3Cl + H_2O} \quad \text{(overall)} \tag{3.4}$$
$$\tag{3.7}$$

This reaction path combines paths 3 and 4. It has the advantage of converting both atoms of the chlorine molecule to vinyl chloride. All of the HCl produced in the pyrolysis reaction is consumed in the oxychlorination reaction. Indeed, it is a fine candidate for the solution of design alternative 2.

Table 3.2 Assumed Cost of Chemicals Purchased or Sold in Bulk Quantities

Chemical	Cost (cents/lb)
Ethylene	18
Acetylene	50
Chlorine	11
Vinyl chloride	22
Hydrogen chloride	18
Water	0
Oxygen (air)	0

Given this information, it seems clear that the design team would reject reaction path 1 on the basis of its low *selectivity* with respect to the competing reactions (not shown) that produce undesirable byproducts. This leaves the other reaction paths as potentially attractive to be screened on the basis of the chemical prices. Although it is too early to estimate the cost of the equipment and its operation, before the remaining process operations are in place, the design team normally computes the *gross profit* (i.e., the profit excluding the costs of equipment and the operating costs) for each reaction path and uses it as a vehicle for screening out those that cannot be profitable. To illustrate this process for the production of vinyl chloride, Table 3.2 provides a representative set of prices for the principal chemicals, obtained from a source such as the *Chemical Marketing Reporter*, as discussed earlier. The gross profit is computed as the income derived from the sales of the products and byproducts less the cost of the raw materials. It is computed by first converting to a mass basis, as illustrated for reaction path 3:

	C_2H_4	+	Cl_2	=	C_2H_3Cl	+	HCl
lbmol	1		1		1		1
Molecular weight	28.05		70.91		62.50		36.46
lb	28.05		70.91		62.50		36.46
lb/lb of vinyl chloride	0.449		1.134		1		0.583
cents/lb	18		11		22		18

Then, the gross profit is $22(1) + 18(0.583) - 18(0.449) - 11(1.134) = 11.94$ cents/lb of vinyl chloride. Similar estimates are made for the overall reaction in each of the reaction paths, it being assumed that complete conversion can be achieved without any side reactions (not shown), with the results shown in Table 3.3.

Table 3.3 Gross Profit for Production of Vinyl Chloride (Based on Chemical Prices in Table 3.2)

Reaction path	Overall reaction	Gross profit (cents/lb of vinyl chloride)
2	$C_2H_2 + HCl = C_2H_3Cl$	−9.33
3	$C_2H_4 + Cl_2 = C_2H_3Cl + HCl$	11.94
4	$C_2H_4 + HCl + \frac{1}{2}O_2 = C_2H_3Cl + H_2O$	3.42
5	$2C_2H_4 + Cl_2 + \frac{1}{2}O_2 = 2C_2H_3Cl + H_2O$	7.68

Even without the capital costs (for construction of the plant, purchase of land, etc.) and the operating costs (for labor, steam, electricity, etc.), the gross profit for reaction path 2 is negative, whereas the gross profits for the other reaction paths are positive. This is principally because acetylene is very expensive relative to ethylene. The fairly high price of HCl also contributes to the inevitable conclusion that vinyl chloride cannot be produced profitably using this reaction path. It should be noted that the price of HCl is often very sensitive to its availability in a petrochemical complex. In some situations, it may be available in large quantities as a byproduct from another process at very low cost. At a much lower price, reaction path 2 would have a positive gross profit, but would not be worthy of further consideration when compared with the three reaction paths involving ethylene. Turning to these paths, all have sufficiently positive gross profits, and hence are worthy of further consideration. It is noted that the price of HCl strongly influences the gross profits of reaction paths 3 and 4, with the gross profit of reaction path 5 midway between the two. Before proceeding with the synthesis, the design team would be advised to examine how the gross profits vary with the price of HCl.

Figure 3.4 shows the first step toward creating a process flowsheet for reaction path 3. Each reaction operation is positioned with arrows representing its feed and product chemicals. The *sources* and *sinks* are not shown because they depend on the *distribution of chemicals*, the next step in process synthesis. The flow rates of the external sources and sinks are computed assuming that the ethylene and chlorine sources are converted completely to the vinyl chloride and hydrogen chloride sinks. Here, a key decision is necessary to set the scale of the process, that is, the production rate at capacity. In this case, a capacity of 100,000 lb/hr (~800 million lb/yr, assuming operation 330 days annually—an operating factor of 0.904) is dictated by the primitive problem statement in Section 1.2. Given this flow rate for the product (principal sink for the process), the flow rates of the HCl sink and the raw material sources can be computed by assuming that the raw materials are converted to the products according to the overall reaction. Any unreacted raw materials are separated from the reaction products and recycled. By material balance, the results in Figure 3.4 are obtained, where each flow rate in lbmol/hr is 1,600.

Similar flowsheets, containing the reaction operations for reaction paths 4 and 5, would be prepared to complete step 1 of the synthesis. These are represented in the synthesis tree in Figure 3.9, which will be discussed after all of the synthesis steps have been completed. Note that their flowsheets are not included here due to space limitations, but are requested in Exercise 3.5 at the end of the chapter. As the next steps in the synthesis are completed for reaction path 3, keep in mind that they would be carried out for the other reaction paths as well. Note, also, that only the most promising flowsheets are developed in detail, usually by an expanded design team or, in some cases, by a competitive design team.

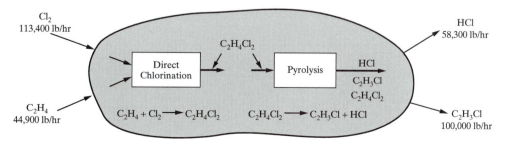

Figure 3.4 Reaction operations for the thermal cracking of dichloroethane from the chlorination of ethylene (reaction path 3).

Step 2. Distribute the Chemicals

In step 2, where possible, the sources and sinks for each of the chemical species in Figure 3.4 are matched so that the total mass flow into a reactor equals the total mass flow out. This often entails the introduction of mixing operations to eliminate differences in flow rates when a single sink is supplied by two or more sources. In other cases, a single source is divided among several sinks. To achieve the distribution of chemicals in Figure 3.5, the ethylene and chlorine sources are matched with their sinks into the chlorination reactor. It is assumed that ethylene and chlorine enter the reactor in the stoichiometric ratio of 1:1 as in reaction (3.3). Because the raw materials are in this ratio, no differences exist between the flow rates of the sources and sinks, and hence, no mixers are needed. Flow rates of 113,400 lb/hr of chlorine and 44,900 lb/hr of ethylene produce 158,300 lb/hr of dichloroethane. When it is desired to have an excess of one chemical in relation to the other so as to completely consume the other chemical, which may be toxic or very expensive (e.g., Cl_2), the other raw material (e.g., C_2H_4) is mixed with recycle and fed to the reactor in excess. For example, if the reactor effluent contains unreacted C_2H_4, it is separated from the dichloroethane product and recycled to the reaction operation. Note that the recycle is the source of the excess chemical, and the flow rate of the external source of C_2H_4 for a given production rate of dichloroethane is unaffected. This alternative distribution of chemicals is discussed further in Section 5.3 and illustrated in Figure 5.1. Returning to the distribution of chemicals in Figure 3.5, note that, at reactor conditions of 90°C and 1.5 atm, experimental data indicate that 98% of the ethylene is converted to dichloroethane, with the remainder converted to unwanted byproducts such as trichloroethane. This loss of yield of main product and small fraction of byproduct is neglected at this stage in the synthesis.

Next, the dichloroethane source from the chlorination operation is sent to its sink in the pyrolysis operation, which operates at 500°C. Here only 60% of the dichloroethane is converted to vinyl chloride with a byproduct of HCl, according to reaction (3.4). This conversion is within the 65% conversion claimed in the patent. To satisfy the overall material balance, 158,300 lb/hr of dichloroethane must produce 100,000 lb/hr of vinyl chloride and 58,300 lb/hr of HCl. But a 60% conversion only produces 60,000 lb/hr of vinyl chloride. The additional dichloroethane needed is computed by mass balance to equal $[(1 - 0.6)/0.6] \times 158{,}300$ or 105,500 lb/hr. Its source is a recycle stream from the separation of vinyl chlo-

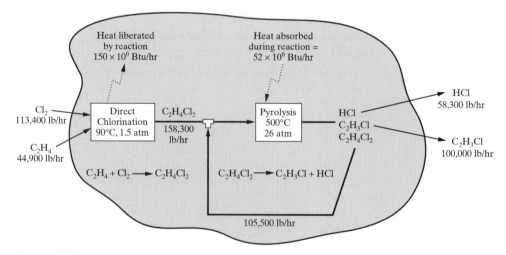

Figure 3.5 Flowsheet showing a distribution of chemicals for thermal cracking of dichloroethane from chlorination of ethylene (reaction path 3).

ride from unreacted dichloroethane, from a mixing operation, inserted to combine the two sources, to give a total of 263,800 lb/hr. The effluent stream from the pyrolysis operation is the source for the vinyl chloride product, the HCl byproduct, and the dichloroethane recycle. To enable these chemicals to be supplied to their sinks, one or more separation operations are needed and are addressed in the next synthesis step.

Figure 3.5 also shows the heats of reaction for the two reaction steps. These are computed at the temperatures and pressures of the reaction operations from heats of formation and heat capacities as a function of temperature. There are many sources of this data, especially the process simulators that are discussed in Chapter 4. When a simulator, such as ASPEN PLUS, is used, it is convenient to define each of the reaction operations and to perform an energy balance at reactor conditions. The simulators report the rate at which heat must be transferred to or from the reactor to achieve exit conditions from given inlet conditions or, if operated adiabatically, the exit conditions for no heat transfer, as discussed on the multimedia CD-ROM (*ASPEN → Chemical Reactors* and *HYSYS → Chemical Reactors*) that accompanies this text. For reaction path 3, the chlorination operation provides a large source of energy, 150 million Btu/hr, but at a low temperature, 90°C, whereas the pyrolysis operation requires much less energy, 52 million Btu/hr, at an elevated temperature, 500°C. Since this heat source cannot be used to provide the energy for pyrolysis, other uses for this energy should be sought as the synthesis proceeds. These and other sources and sinks for energy are considered during task integration in step 5.

As for the pressure levels in the reaction operations, 1.5 atm is selected for the chlorination reaction to prevent the leakage of air into the reactor to be installed in the task integration step. At atmospheric pressure, air might leak into the reactor and build up in sufficiently large concentrations to exceed the flammability limit. For the pyrolysis operation, 26 atm is recommended by the B.F. Goodrich patent (1963) without any justification. Since the reaction is irreversible, the elevated pressure does not adversely affect the conversion. Most likely, the patent recommends this pressure to increase the rate of reaction and, thus, reduce the size of the pyrolysis furnace, although the tube walls must be thick and many precautions are necessary for operation at elevated pressures. The pressure level is also an important consideration in selecting the separation operations, as will be discussed in the next synthesis step.

Referring to Figure 3.9, at the "Distribution of Chemicals" level, two branches have been added to the synthesis tree to represent the two distributions in connection with reaction path 3. Each of these branches represents a different partially completed flowsheet, that is, Figures 3.5 and 5.1. Other distributions arise in connection with reaction paths 4 and 5. These are represented using dashed lines in the synthesis tree.

Step 3. *Eliminate Differences in Composition*

As mentioned earlier, for each distribution of chemicals, the needs for separation become obvious. In Figure 3.5, for example, it is clear that the pure effluent from the chlorination reaction operation needs no separation, but the effluent from the pyrolysis operation is a mixture that needs to be separated into nearly pure species. Here, the source of the three species in the effluent is at a composition far different from the compositions of the three sinks: vinyl chloride product, HCl byproduct, and the dichloroethane for recycle. To eliminate these composition differences, one or more separation operations are needed.

One possibility is shown in Figure 3.6, in which two distillation towers in series are inserted into the flowsheet. Distillation is possible because of the large volatility differences among the three species. This can be seen by examining the boiling points in Table 3.4, which can be obtained from vapor pressure data in the preliminary database, or from a process simulator. In the first column, HCl is separated from the two organic chemicals. In

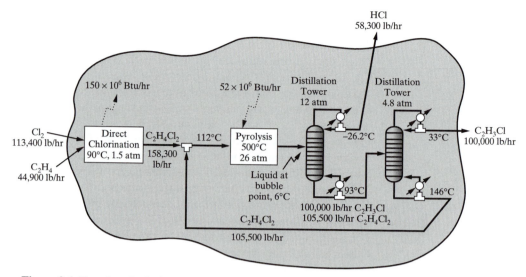

Figure 3.6 Flowsheet including the separation operations for the vinyl-chloride process.

the second column, vinyl chloride is separated from dichloroethane. At 1 atm, the boiling point of HCl is very low, −84.8°C, and hence if HCl were recovered at 1 atm as the distillate of the first tower, very costly refrigeration would be necessary to condense the reflux stream. At 26 atm (the pyrolysis reaction pressure), HCl boils at 0°C, and much less costly refrigeration could be used. The B.F. Goodrich patent recommends operation at 12 atm without any justification. At this pressure, HCl boils at −26.2°C and the bottoms product, comprised of vinyl chloride and dichloroethane, with trace quantities of HCl, has a bubble point of 93°C, which can be calculated by a process simulator. The bottoms product at this reduced temperature and pressure is farther away from the critical points of vinyl chloride–dichloroethane mixtures at the bottom of the distillation column. It is likely, therefore, that B.F. Goodrich selected this lower pressure to avoid operation in the critical region where the vapor and liquid phases approach each other and are much more difficult to disengage (i.e., have small flooding velocities and require very large diameters and tray spacings). Furthermore, low-pressure steam is adequate for the reboiler. When this distillation tower is inserted into the flowsheet, the conditions of its feed stream, or sink, need to be identified. If the feed is a saturated liquid, the temperature is 6°C at 12 atm, with a mild refrigerant required for cooling. A preferable feed temperature would be 35°C or higher, which could be achieved by completing the cooling and partial condensation of the pyrolysis reactor effluent with cooling water, but the introduction of vapor into the column would increase the refrigeration load of the condenser at −26.2°C. Upon making this specification, key differences (temperature, pressure, and phase) appear between the effluent from the pyrolysis operation and the feed to the distillation column. These are eliminated in the next synthesis step by inserting tempera-

Table 3.4 Boiling Points and Critical Constants

Chemical	Normal boiling point (1 atm, °C)	Boiling point (°C)			Critical constants	
		4.8 atm	12 atm	26 atm	T_c(°C)	P_c(atm)
HCl	−84.8	−51.7	−26.2	0	51.4	82.1
C_2H_3Cl	−13.8	33.1	70.5	110	159	56
$C_2H_4Cl_2$	83.7	146	193	242	250	50

ture and pressure change operations, with each temperature specification leading to a some-what different flowsheet.

After the first distillation operation is inserted into the flowsheet, the second follows naturally. The bottoms from the HCl-removal tower is separated into nearly pure species in the second tower, which is specified at 4.8 atm, as recommended by the B.F. Goodrich patent. Under these conditions, the distillate (nearly pure vinyl chloride) boils at 33°C and can be condensed with inexpensive cooling water, which is available at 25°C. The bottoms product boils at 146°C, and hence, the vapor boilup can be generated with medium-pressure steam, which is widely available in petrochemical complexes.

Alternative separation operations can be inserted into Figure 3.5. When distillation is used, it is also possible to recover the least volatile species, dichloroethane, from the first column, and separate HCl from vinyl chloride in the second column. Yet another possibility is to use a single column with a side stream that is concentrated in the vinyl chloride product. Absorption with water, at atmospheric pressure, can be used to remove HCl. The resulting vapor stream, containing vinyl chloride and dichloroethane, could be dried by adsorption and separated using distillation. With so many alternatives possible, the process designer needs time or help to select the most promising separation operations. As mentioned previously, this topic is considered in detail in Chapter 7.

Furthermore, as before, the synthesis tree in Figure 3.9 is augmented. In this case, the new branches represent the different flowsheets for the alternative separation operations. Clearly, as each step of the synthesis is completed, the tree represents many more possible flowsheets.

Step 4. Eliminate Differences in Temperature, Pressure, and Phase

When the reaction and separation operations are positioned, the states of their feed and product streams are selected. This is accomplished usually by adjusting the temperature and pressure levels to achieve the desired reaction conversions and separation factors. Subsequently, after the flowsheets have been created, these are often adjusted toward the economic optimum, often using the optimizers in the process simulators discussed in Chapter 18. In this synthesis step, however, the states are assumed to be fixed and operations are inserted to eliminate the temperature, pressure, and phase differences between the feed sources, the product sinks, and the reaction and separation operations.

Figure 3.7 shows one possible flowsheet. It can be seen that liquid dichloroethane from the recycle mixer at 112°C and 1.5 atm undergoes the following operations:

1. Its pressure is increased to 26 atm.
2. Its temperature is raised to the boiling point, which is 242°C at 26 atm.
3. Dichloroethane liquid is vaporized at 242°C.
4. Its temperature is raised to the pyrolysis temperature, 500°C.

Note that an alternative flowsheet would place operations 1 and 2 after operation 3. However, this is very uneconomical, as the cost of compressing a vapor is far greater than the cost of pumping a liquid because the molar volume of a vapor is so much greater than that of a liquid (typically, a factor of 100 times greater). For a more complete discussion of this observation, which is just one of many design heuristics or rules of thumb, see Section 5.7.

In addition, the hot vapor effluent from the pyrolysis operation (at 500°C and 26 atm) is operated upon as follows:

1. Its temperature is lowered to its dew point, 170°C at 26 atm.
2. The vapor mixture is condensed to a liquid at its bubble point, 6°C at 12 atm, by lowering the pressure, cooling, and removing the latent heat of condensation.

Finally, the dichloroethane recycle stream is cooled to 90°C to avoid vaporization when mixed with the reactor effluent at 1.5 atm.

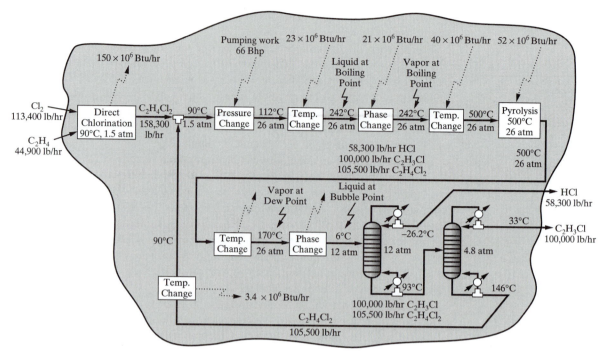

Figure 3.7 Flowsheet with temperature-, pressure-, and phase-change operations in the vinyl-chloride process.

Branches to represent the two new flowsheets are added to the synthesis tree in Figure 3.9 after this synthesis step has been completed.

Step 5. Task Integration

At the completion of step 4, each of the candidate flowsheets has a complete set of operations that eliminates the differences between the raw materials and products. Still, with the exception of the distillation operations, specific equipment items are not shown. The selection of the processing units, often referred to as unit operations, in which one or more of the basic operations are carried out, is known as *task integration*. To assist in this selection, the reader is referred to *Chemical Process Equipment* (Walas, 1988).

Figure 3.8 shows one example of task integration for the vinyl chloride process. At this stage in process synthesis, it is common to make the most obvious combinations of operations, leaving many possibilities to be considered when the flowsheet is sufficiently promising to undertake the preparation of a base-case design. As you examine this flowsheet, with the descriptions of the process units that follow, see if you can suggest improvements. This is one of the objectives in Exercise 3.3. Throughout the chapters that follow, techniques are introduced to obtain better integration for this and other processes that manufacture many other chemicals.

1. *Chlorination reactor and condenser.* The direct chlorination operation in Figure 3.7 is replaced by a cylindrical reaction vessel, containing a rectifying section, and a condenser. A pool of liquid dichloroethane, with ferric chloride catalyst dissolved, fills the bottom of the vessel at 90°C and 1.5 atm. Ethylene is obtained commonly from large cylindrical vessels, where it is stored as a gas at an elevated pressure and room temperature, typically 1,000 psia and 70°F. Chlorine, which is stored commonly in the liquid phase, typically at 150 psia and 70°F, is evaporated carefully to remove the viscous liquid (taffy) that contaminates most

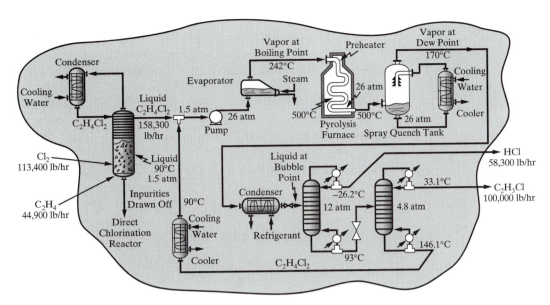

Figure 3.8 Flowsheet showing task integration for the vinyl-chloride process.

chlorine produced by electrolysis. Chlorine and ethylene in the vapor phase bubble through the liquid and release the heat of reaction as dichloroethane is produced. This heat causes the dichloroethane to vaporize and rise up the rectifying section into the condenser, where it is condensed with cooling water. Note that heat is needed to drive the reboiler in the first distillation column at 93°C, but the heat of reaction cannot be used for this purpose unless the temperature levels are adjusted. How can this be accomplished?

Most of the condensate is mixed with the effluent from the recycle cooler to be processed in the pyrolysis loop. However, a portion is refluxed to the rectifying section of the column, which has several trays, to recover any of the less volatile species (e.g., trichloroethane) that may have vaporized. These *heavies* accumulate at the bottom of the liquid pool and are removed periodically as impurities.

2. *Pump.* Since the pressure-change operation involves a liquid, it is accomplished by a pump, which requires only 66 Bhp, assuming an 80% efficiency. The enthalpy change in the pump is very small and the temperature does not change by more than 1°C.

3. *Evaporator.* This unit, in the form of a large kettle, with a tube bundle inserted across the bottom, performs the temperature- and phase-change operations. Saturated steam that passes through the tubes condenses as the dichloroethane liquid is heated to its boiling point and vaporized. The large vapor space is provided to enable liquid droplets, entrained in the vapor, to coalesce and drop back into the liquid pool, that is, to disengage from the vapor which proceeds to the pyrolysis furnace.

4. *Pyrolysis furnace.* This unit also performs two operations: It preheats the vapor to its reaction temperature, 500°C, and it carries out the pyrolysis reaction. The unit is constructed of refractory brick, with natural gas-fired heaters, and a large bundle of Nickel, Monel, or Inconel tubes, within which the reaction occurs. The tube bundle enters the coolest part of the furnace, the so-called *economizer* at the top, where the preheating occurs.

5. *Spray quench tank and cooler.* The quench tank is designed to rapidly quench the pyrolysis effluent to avoid carbon deposition in a heat exchanger. Cold liquid (principally dichloroethane) is showered over the hot gases, cooling them to their dew point, 170°C. As the gases cool, heat is transferred to the liquid and removed in the adjacent cooler. The warm liquid, from the pool at the base of the quench vessel, is circulated to the cooler, where it is contacted with cooling water. Any carbon that deposits in the quench vessel settles to the bottom and is bled off periodically. Unfortunately, this carbon deposition, as well as the corrosive HCl, is anticipated to prevent the use of the hot effluent gases in the tubes of the evaporator, which would have to be serviced often to remove carbon and replace corroded tubes. Note that coke formation in the pyrolysis products is discussed by Borsa et al. (2001). Consequently, large amounts of heat are transferred to cooling water and the fuel requirements for the process are high. As noted later in the section on pilot-plant testing, the design team is likely to measure the rate of carbon deposition and, if it is not very high, may decide to implement a design with a feed/product heat exchanger.

6. *Condenser.* To produce a saturated liquid at 6°C, the phase-change operation is carried out by a condenser that transfers heat to a mild refrigerant. Then the pressure is lowered to 12 atm across a valve.

7. *Recycle cooler.* To prevent vapor from entering the pump, when the recycle stream is mixed with effluent from the dichlorination reactor, the recycle stream is cooled to 90°C (below the boiling point of dichloroethane at 1.5 atm) using cooling water.

This completes the task integration in Figure 3.8. Can you suggest ways to reduce the need for fuel and hot utilities such as steam?

Synthesis Tree

Throughout the synthesis of the vinyl chloride process, branches have been added to the synthesis tree in Figure 3.9 to represent the alternative flowsheets being considered. The bold branches trace the development of just one flowsheet as it evolves in Figures 3.3–3.8. Clearly, there are many alternative flowsheets, and the challenge in process synthesis is to find ways to eliminate whole sections of the tree without doing much analysis. By eliminating reaction paths 1 and 2, as much as 40% of the tree is eliminated in the first synthesis step.

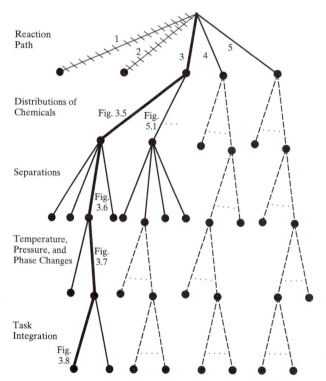

Figure 3.9 Inverted synthesis tree for the production of vinyl chloride.

Similar screening techniques are applied by the design team in every step, as discussed throughout this book.

To satisfy the objective of generating the most promising flowsheets, care must be taken to include sufficient analysis in each synthesis step to check that each step does not lead to a less profitable flowsheet or exclude the most profitable flowsheet prematurely. For this reason, it is common practice in industry to mix these synthesis steps with analysis using the simulators introduced in the next chapter.

Heuristics

It is important to keep in mind that, when carrying out the steps in preliminary process synthesis, the resulting synthesis tree is closely related to any heuristics or rules of thumb used by the design team. In the vinyl chloride example, emphasis was placed on the synthesis steps, and not on the use of heuristics by the design team. An exception is the heuristic that it is cheaper to pump a liquid than compress a gas. Heuristics are covered more thoroughly in Chapter 5, where it will become clear that the synthesis tree can be improved significantly. See also *Conceptual Design of Chemical Processes* (Douglas, 1988) and Walas (1988), where many heuristics are presented.

Example of Process Synthesis: Manufacture of Tissue Plasminogen Activator (tPA)

As introduced in Section 1.2, tissue plasminogen activator (tPA) is a recombinant therapeutic protein comprised of 562 amino acids, as shown schematically in Figure 3.10. Note that tPA is produced using a recombinant cell, which results from a recombination of genes. To eliminate blood clots, tPA activates plasminogen to plasmin, an enzyme, which dissolves fibrin formations that hold blood clots in place. In this way, blood flow is reestablished once the clot blockage dissolves; an important effect for patients suffering from a heart attack (microcardial infarction) or stroke. This example shows the steps in synthesizing a process

Figure 3.10 Schematic of tissue plasminogen activator (tPA).

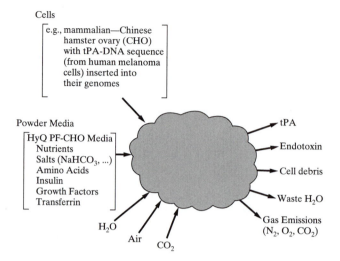

Cells

$$\left[\begin{array}{l} \text{e.g., mammalian—Chinese} \\ \text{hamster ovary (CHO)} \\ \text{with tPA-DNA sequence} \\ \text{(from human melanoma} \\ \text{cells) inserted into} \\ \text{their genomes} \end{array} \right.$$

Powder Media

$$\left[\begin{array}{l} \text{HyQ PF-CHO Media} \\ \text{Nutrients} \\ \text{Salts (NaHCO}_3, ...) \\ \text{Amino Acids} \\ \text{Insulin} \\ \text{Growth Factors} \\ \text{Transferrin} \end{array} \right.$$

H_2O

Air CO_2

→ tPA

→ Endotoxin

→ Cell debris

→ Waste H_2O

Gas Emissions
(N_2, O_2, CO_2)

Figure 3.11 Process synthesis problem.

to address the challenges posed by the specific problem posed in alternative 1 of Section 1.2; that is, to manufacture less expensive forms of tPA that can be sold for $200 per 100-mg dose. Note that it leads to a batch process, involving many small process units that must be scheduled for the manufacture of tPA, rather than a large-scale continuous process, as for the manufacture of vinyl chloride.

Stated differently, based upon extensive research in the biochemistry laboratory, the tPA gene was isolated from human melanoma cells, and the process synthesis problem in Figure 3.11 created. As shown, tPA is produced using mammalian [e.g., Chinese hamster ovary (CHO)] cells that have tPA-DNA as part of their genetic contents (genome). In an aerobic bioreaction operation, the tPA-CHO cells grow in a nutrient media, HyQ PF-CHO—Hyclone media, a blend of nutrients, salts (including $NaHCO_3$), amino acids, insulin, growth factors, and transferrin, specifically for growth of CHO cells. Other ingredients include sterilized water, air, and CO_2. In addition to tPA, endotoxins may be a contaminant of the product, which must be removed because they elicit a variety of inflammatory responses in animals. Other byproducts include cell debris, wastewater, and gas emissions, especially N_2 from air, unconsumed O_2 from air, and CO_2, which regulates the pH. An important source of data, in addition to that taken in the biochemistry laboratory, is a U.S. patent, filed by Genentech (Goeddel et al., 1988), which provides considerable qualitative and quantitative information.

Step 1. Eliminate Differences in Molecular Type

In the manufacture of a macromolecule like tPA through cell growth, a complex array of chemical reactions is often approximated by global reactions that are understood far less than the well-defined reactions for the manufacture of a simple monomer, like vinyl chloride. In terms of global reactions to manufacture tPA, two principal reaction paths are provided by the biochemist, as follows.

1. Mammalian Cells   Into CHO cells, the tPA-DNA sequence must be inserted and expressed. The resulting tPA-CHO cells are specially selected CHO cells, with many copies of tPA-DNA inserted into their genomes, and which secrete high levels of tPA. This tPA-DNA insertion step is summarized in the reaction:

tPA−DNA sequence + CHO cells → selected high expressing tPA−CHO cells   **(3.8)**

The product of this "catalyst preparation" is a master stock of tPA–CHO cells, which are prepared in the laboratory and stored in 1-mL aliquots at $-70°C$ to be used as inoculum for the bioreaction:

$$tPA-CHO \text{ cells} + HyQ \text{ PF}-CHO \text{ media} + O_2 \rightarrow \text{Increased cell numbers} \qquad (3.9)$$

As the cells grow in this aerobic cultivation at a rate of 0.39×10^6 cell/(mL-day), oxygen from air is consumed at the rate of 0.2×10^{-12} mol O_2/(cell-hr), and tPA is produced at the rate of 50 pico gram tPA/(cell-day). The latter is secreted gradually into the liquid media solution. Note that reaction (3.8) is carried out once during the research and development phase. Initially, 1–10 mg of tPA-DNA are added to 10^6 cells to produce a few tPA-CHO cells in many unmodified CHO cells. After careful selection, one tPA-CHO cell (the "founder" cell) is selected and amplified to yield about 10^6 cells/mL in 10–100 L. These cells are frozen in aliquots.

2. Bacterial Cells A promising alternative is to insert the tPA-DNA sequence into the genome of *Escherichia coli* (*E. coli*) cells, as summarized by the reaction:

$$tPA-DNA \text{ sequence} + E. coli \text{ cells} \rightarrow \text{selected high expressing } tPA-E. coli \text{ cells} \qquad (3.10)$$

Then, the tPA–*E. coli* bacteria cells, which are grown in the laboratory, are frozen in aliquots at $-70°C$ to be used as inoculum for the fermentation reaction:

$$tPA-E. coli \text{ cells} + \text{powder media} + O_2 \rightarrow \text{increased cell numbers} \qquad (3.11)$$

A batch fermentation of tPA–*E. coli* can produce 5–50 mg tPA/L-broth at harvest. *Escherichia. coli* may require disruption to release tPA, which is then more difficult to separate. Should a process be synthesized based upon this reaction path, reaction rate data from the laboratory will be needed. Unlike CHO cells, *E. coli* cells do not add sugar groups (glycosylation) to tPA. Like CHO cells, tPA–*E. coli* cells are produced and frozen during the research and development phase.

Returning to the reaction path with CHO cells, using laboratory data, the reaction operation is inserted onto the flowsheet, as shown in Figure 3.12. At a production rate of 80 kg/yr of tPA, the lab reports that the following ingredients are consumed and waste products are produced:

Ingredients	kg/yr	Wastes	kg/yr
tPA-CHO cells	small	Endotoxin	0.155
HyQ PF-CHO media	22,975	Cell debris	22,860
Water	178,250	Wastewater	178,250
Air	3,740	Gas emissions (N_2, O_2, CO_2)	4,036
CO_2	296		

The reaction operation provides sinks for tPA-CHO cells, from cold storage at $-70°C$, and HyQ PF-CHO media in water, air, and carbon dioxide. Its effluent is a source of tPA, at 112 kg/yr, endotoxin, cell debris, water, nitrogen, and carbon dioxide. When separated, these species are the sources for the product sinks from the flowsheet. Note that the combined cell growth and tPA production operation takes place at 37°C, 1 atm, and pH = 7.3. The latter is achieved by the $NaHCO_3$ in the powder media, with fine tuning by manipulation of the flow rate of CO_2.

Before accepting a potential reaction path, it is important to examine the gross profit; that is, the difference between the sales revenues and the cost of ingredients. To accomplish this, the sales price of tPA is projected (e.g., $200 per 100-mg dose), and the costs of ingredients are projected, with estimates often obtained from the suppliers. A typical list of

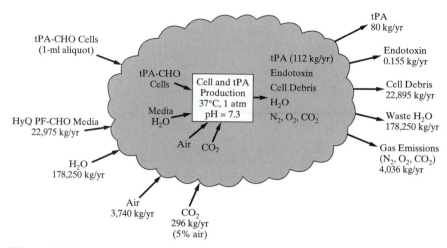

Figure 3.12 Reaction operations using mammalian CHO cells.

cost estimates is shown in Table 3.5. The cost of water for injection (WFI) is based upon estimates of the cost of sterilizing municipal water (12 cents/kg = 45 cents/gal = \$450/1,000 gal, which is far higher than the typical cost of process water = \$0.50/1,000 gal). The costs of sterilized air and carbon dioxide are for industrial cylinders of compressed gases. The cost of the tPA-CHO cells is not included, as it is associated with the cost of research, which is subsequently estimated as an operating cost.

Using these costs, the gross profit is estimated:

$$\text{Gross Profit} = 2,000,000 - 287.2 \times 233 - 2,228 \times 0.12 - 3.7 \times 1,447 - 46.8 \times 1,742$$
$$= \$1,846,000/\text{kg tPA}$$

Clearly, this is very high for tPA, a typical pharmaceutical. However, the gross profit does not account for the operating costs, which include the cost of research, the cost of utilities, and the investment cost, and are high for separations that involve expensive mass separating agents. With such a promising gross profit, the process synthesis proceeds at an accelerated pace.

Step 2. Distribute the Chemicals

In this step, the sources and sinks for each species in Figure 3.12 are matched so that the total mass flow rate into the reaction operation equals the mass flow rate out. This often entails the introduction of mixing operations, as illustrated in the previous example for vinyl chloride.

Table 3.5 Assumed Cost of Chemicals Produced or Sold

Chemical	kg/kg tPA	Cost (\$/kg)
tPA	1	2,000,000
HyQ PF CHO powder media	287.2	233
Water for injection (WFI)	2,228	0.12
Air	46.8	1,742
CO_2	3.7	1,447
tPA-CHO cells	—	[a]

[a]Not included in gross profit estimate—related to cost of research, an operating cost.

In this case, only one mixing operation is introduced, in which the HyQ PF-CHO powder media is mixed with water, as shown in Figure 3.13. Otherwise the sources and sinks are matched directly. However, the effluent from the cell growth, tPA production reactor must be separated before its species are matched with the product sinks.

Step 3. Eliminate Differences in Composition

For most distributions of chemicals, composition differences exist between streams to be separated and the sinks to which these species are sent. Clearly, in Figure 3.13, the effluent from the cell growth, tPA production reactor must be separated.

Many separation system possibilities exist, with one provided in Figure 3.14. Here, the reactor effluent is sent to a separator for recovery of the gas emissions from the liquid mixture, with the latter sent to a centrifuge to remove wet cell debris from the harvest media or clarified broth. Note that because proteins lose their activity at temperatures above ~0°C, the centrifuge, and all other separation operations, are operated at 4°C, slightly above the freezing point of water. The harvest media is mixed with arginine hydrochloride, an amino acid:

$$CH_2CH_2CH_2NH \overset{\overset{\displaystyle +NH_2}{\displaystyle \|}}{\underset{}{C}} NH_2$$

which prevents tPA from self-aggregating. Note that 45,870 kg/yr provides a concentration of 2.0 molar, which is sufficient to prevent aggregation.

The resulting mixture is sent to microfilters to remove large quantities of wastewater, which passes through the filters. For this step, alternate separators, like gel filtration and an Acticlean Etox resin (by Sterogene) should be considered. The retentate from the filter, which contains tPA, other proteins, endotoxin, arginine hydrochloride, and some water, is sent to an affinity chromatography operation. Here, tPA is selectively adsorbed on a resin

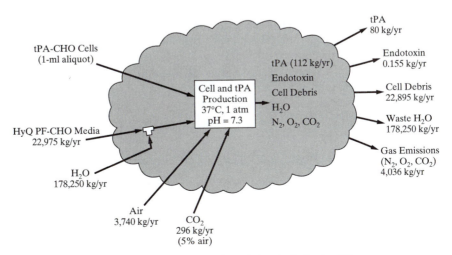

Figure 3.13 Flowsheet showing a distribution of chemicals for the tPA process.

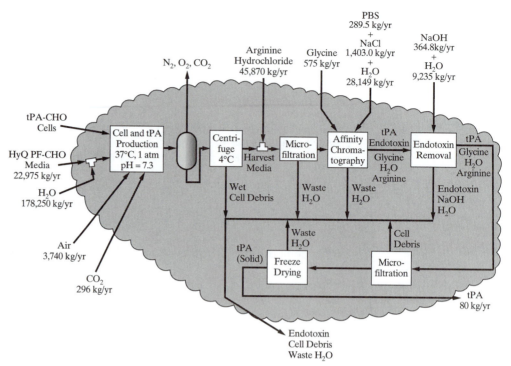

Figure 3.14 Flowsheet including the separation operations for the tPA process.

(e.g., CNBr-activated Sepharose, by Amersham Biotech). The resin is then eluted with glycine, an amino acetic acid:

$$H_2N-\underset{\underset{COOH}{|}}{\overset{\overset{H}{|}}{C}}-H$$

From lab measurements, 575 kg/yr of glycine are sufficient for the elution process. After the column is eluted, it is equilibrated with a mixture of 289.5 kg/yr of phosphate buffer solution (PBS) and 1,403.0 kg/yr of NaCl, with the quantities determined in the lab.

The resulting tPA solution is sent to an endotoxin removal column where the endotoxin is adsorbed selectively onto a resin (e.g., Acticlean Etox by Sterogene). This column is washed with a mixture of 364.8 kg/yr of NaOH and 9,235 kg/yr of water to remove the endotoxin. The effluent stream is microfiltered, to remove cell debris, which does not pass through the filter. Then, wastewater is removed in a freeze drying operation to provide tPA in powder form.

Step 4. Eliminate Differences in Temperature, Pressure, and Phase

In the manufacture of tPA, the ingredients are assumed to be available at 20°C, water is mixed with the HyQ PF-CHO powder media at 4°C, the cultivations (cell production operations) occur at 37°C, and the separations occur at 4°C. The exothermic heat of the cultivation is removed at 37°C. Only small pressure changes occur and can be neglected at this stage of process synthesis. Similarly, no phase-change operations are added to the flowsheet. Hence, only a few temperature change operations are added to Figure 3.14, with the resulting flowsheet shown in Figure 3.15.

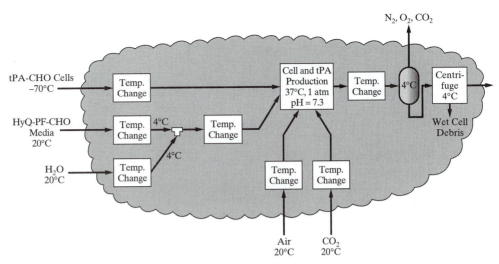

Figure 3.15 Flowsheet with the temperature-change operations in the tPA process.

Step 5. Task Integration

At this stage in the synthesis, various items of equipment are selected, often combining two or more adjacent operations into a single equipment item; that is, in *task integration*. The first key decision involves whether to operate in continuous or batch mode. For small throughputs, such as 80 kg/yr of tPA, the decision is nearly always to operate in batch mode. Choices of batch size and time are usually based upon the slowest operation, usually the cultivation process. For tPA, it is determined by the growth rate of tPA-CHO cells [0.39×10^6 cell/(ml-day)], the inlet and outlet concentrations (0.225×10^6 and 3×10^6 cell/mL), and the rate of tPA growth [50 pg tPA/(cell-day)]. To produce 1.6 kg of tPA per batch, 2.24 kg of tPA are produced by cultivation, allowing for losses in the separation process. At this production rate, 2.24 kg of tPA can be produced in 8 days in a 4,000-L batch (within a 5,000-L vessel). Allowing time for charging and cleaning, 14 days are reserved, and hence, 25 batches are produced annually, assuming 50 operating weeks. With two batch trains in parallel, 50 batches are produced annually; that is, 1.6 kg of tPA are produced per batch.

The flowsheet in Figure 3.16a begins in a 1-L laboratory cultivator, into which a 1-mL aliquot of tPA-CHO cells is charged from cold storage at $-70°C$. To this, HyQ PF-CHO media, water, air, and CO_2 are added. Cultivation takes place over five days to produce 1.2 kg/batch of inoculum, which is emptied from the cultivator and transferred to the plant in one day. This effluent inoculates three cultivators in series which carry out the cell and tPA production operation. The first is 40 L, with a 30-L batch that grows cells from 1.05×10^6 to 3×10^6 cell/mL in five days, with two additional days for loading and cleaning. The second is 400 L, with a 300-L batch that grows cells from 0.25×10^6 to 3×10^6 cell/mL in seven days, with 2.5 additional days for loading and cleaning. Finally, the third is 5,000 L with a 4,000-L batch that grows cells from 0.225×10^6 to 3×10^6 cell/mL in eight days, with six additional days for loading and cleaning. Note that gas emissions, containing N_2, O_2, and CO_2, are vented continuously from the cultivators. A 5,000-L mixing tank is installed to load and mix the powder media and water in two days. Note the tank jacket through which refrigerant is circulated. This vessel is followed by a microfilter, which sterilizes the mixture by removing bacteria, and a hot water heat exchanger. One last vessel, a 5,000-L holding tank, is provided to hold the contents of one cultivator batch (2.24, 457.17, 0.0031, 3,565 kg/batch of tPA, tPA-CHO cells, endotoxin, and water, respectively), in the event that the centrifuge is taken off-line for servicing. The effluent from the third cultivator is cooled to 4°C in the shell-and-tube heat exchanger, which is cooled by a refrigerant on the shell side.

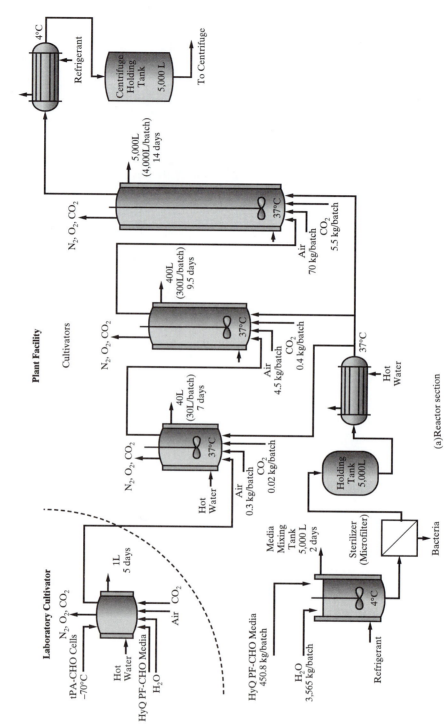

Figure 3.16 Flowsheet showing a task integration for the tPA process.

(a) Reactor section

Plant Facility

Cultivators

Laboratory Cultivator

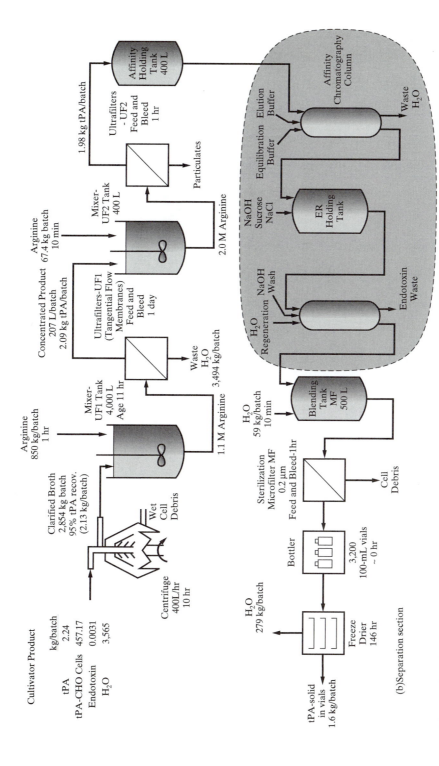

Cultivator Product

	kg/batch
tPA	2.24
tPA-CHO Cells	457.17
Endotoxin	0.0031
H₂O	3,565

Figure 3.16 (*Continued*)

(b) Separation section

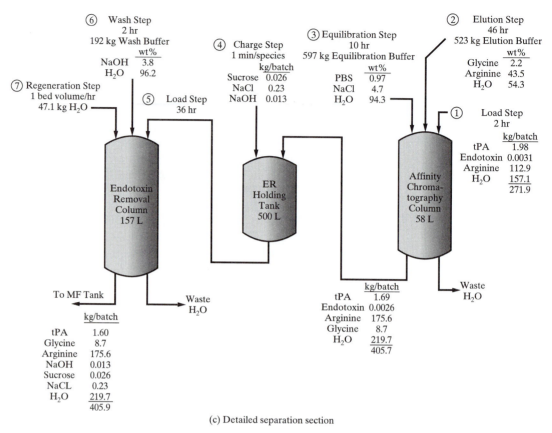

⑥ Wash Step
2 hr
192 kg Wash Buffer

	wt%
NaOH	3.8
H₂O	96.2

③ Equilibration Step
10 hr
597 kg Equilibration Buffer

	wt%
PBS	0.97
NaCl	4.7
H₂O	94.3

② Elution Step
46 hr
523 kg Elution Buffer

	wt%
Glycine	2.2
Arginine	43.5
H₂O	54.3

④ Charge Step
1 min/species

	kg/batch
Sucrose	0.026
NaCl	0.23
NaOH	0.013

⑦ Regeneration Step
1 bed volume/hr
47.1 kg H₂O

⑤ Load Step
36 hr

① Load Step
2 hr

	kg/batch
tPA	1.98
Endotoxin	0.0031
Arginine	112.9
H₂O	157.1
	271.9

Endotoxin
Removal
Column
157 L

ER
Holding
Tank
500 L

Affinity
Chroma-
tography
Column
58 L

To MF Tank

Waste
H₂O

Waste
H₂O

	kg/batch
tPA	1.60
Glycine	8.7
Arginine	175.6
NaOH	0.013
Sucrose	0.026
NaCL	0.23
H₂O	219.7
	405.9

	kg/batch
tPA	1.69
Endotoxin	0.0026
Arginine	175.6
Glycine	8.7
H₂O	219.7
	405.7

(c) Detailed separation section

Figure 3.16 (*Continued*)

Turning next to the separation section in Figure 3.16b. The centrifuge is designed to handle small batches, at a rate of 400 L/hr over 10 hr. It rotates at high speed with the wet cell mass (which contains all of the tPA-CHO cells, five wt% of the tPA, 20 wt% of the water, and none of the endotoxin fed to the centrifuge) thrown to the outside collection volume and removed. Note that at this stage in process synthesis recovery fractions are estimated using heuristics and experimental data when available. Also, since the endotoxin contaminant must be removed entirely, it is assumed to be entirely recovered (100%) in the effluent from the microfilters. The clarified broth (2,854 kg/batch) exits through the central tube overhead. It enters a mixing tank in which arginine hydrochloride is added to form a 1.1 molar solution, which is microfiltered to remove 3,494 kg/batch of wastewater. The concentrated product, at 207 L/batch and containing 98, 5.62, and 5.62 wt% of the tPA, arginine hydrochloride, and water fed to the microfilter, is mixed with 67.4 kg/batch of arginine in a second mixing vessel to give 2.0 molar arginine. This solution is microfiltered to remove particulate matter before being sent to the affinity holding tank. The effluent, which contains 95, 98, 100, and 98 wt% of the tPA, arginine, endotoxin, and water fed to the microfilter, is loaded into a 58-L affinity chromatography column, which adsorbs 100, 100, 2, and 2 wt% of tPA, endotoxin, arginine, and water, as shown in Figure 3.16c. Most of the adsorbed tPA, 1.69 kg/batch, is eluted with a stream containing glycine (523 kg/batch at 2.2, 43.5, and 54.3 wt% of glycine, arginine, and water, respectively) and sent to a 500-L holding tank (405.7 kg/batch containing 1.69, 8.7, 175.6, 0.0026, and 219.7 kg/batch of tPA, glycine, arginine, endotoxin, and water, respectively). Note that the elution buffer recovers 85 wt% of the tPA and endotoxin from the resin. The affinity chromatography column is equilibrated with an equilibration buffer (597 kg/batch containing 0.97, 4.7, and 94.3 wt% PBS, NaCl, and water, respectively). After a caustic and sucrose mix is added to the holding tank (0.013, 0.026, and 0.33 kg/batch of NaOH, sucrose, and NaCl, respectively) the mixture is loaded into the endotoxin removal column (406.0 kg/batch). In this 15.7-L column, the endotoxins are adsorbed, and removed, by washing with caustic (192 kg/batch containing 3.8 and 96.2 wt% NaOH and water, respectively), which is discarded. The endotoxin removal column is regenerated with 47.1 kg/batch of water, while the endotoxin-free solution (405.9 kg/batch containing 1.6, 8.7, 175.6, 0.013, 0.026, 0.23, and 219.7 kg/batch of tPA, glycine, arginine, NaOH, sucrose, NaCl, and water, respectively) is sent to a holding tank, where 59 kg/batch of water are added. After sterilization with a microfilter to remove cell debris, from which 99.7% of the tPA is recovered, the solution is sent to a bottler and 100-mL vials, each containing 100 mg of tPA, are conveyed to a freeze drier, where the water is evaporated.

It is important to recognize that the batch sizes in Figure 3.16 are representative. However, as discussed subsequently in Section 4.5 and Chapter 12, the batch times and vessel sizes are key design variables in scheduling and optimizing batch processes.

Synthesis Tree

Clearly, at each step in the synthesis of the process flowsheet, alternatives are generated and the synthesis tree fills in. For the tPA process, a schematic of a synthesis tree is shown in Figure 3.17. Note that the bold branch corresponds to the flowsheets in Figures 3.12–3.16. In design synthesis, the engineer strives to identify the most promising alternatives, eliminating the least promising alternatives by inspection, wherever possible. Initially, heuristic rules help to make selections. Eventually, algorithmic methods involving optimization can be introduced to check the heuristics and identify more promising alternatives, as discussed in Chapter 12. It should be emphasized, however, that the design window, beginning during Phases 1 and 2 of the clinical trials, is small, typically on the order of 12–16 months, before

Reaction Path   Fig. 3.12

Distribution of Chemicals   Fig. 3.13

Separations   Fig. 3.14

Temperature Changes   Fig. 3.15

Task Integration   Fig. 3.16

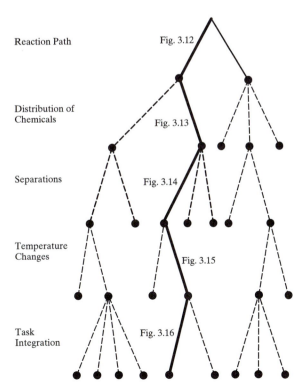

Figure 3.17 Inverted synthesis tree for the production of tPA.

Phase 3 begins (see Section 1.2). Consequently, emphasis is normally placed on the rapid development of a promising design, and less on design optimization. Stated differently, for high-priced pharmaceuticals, it is far more important to be first-to-market rather than achieve relatively small savings in the capital investment or operating expenses for the plant through design optimization.

Algorithmic Methods

Finally, before leaving this section on preliminary process synthesis, the limitations of the heuristic approaches should not be overlooked. Many algorithmic methods are very effective for the synthesis of alternative flowsheets, their analysis, and optimization. These methods are usually used by design teams in parallel with their work on the development of the base-case design, which is the subject of the next section. The algorithmic methods are easily implemented and are illustrated with many examples in Part Two of this text (Chapters 6–12).

3.5 DEVELOPMENT OF THE BASE-CASE DESIGN

At some point in the synthesis of alternative flowsheets, it becomes important to select one or two of the most promising alternatives for further development into the so-called *base-case design*(s). To accomplish this, the design team is usually expanded, mostly with chemical engineers, or assisted by more specialized engineers, as the engineering workload is increased significantly. With expanded engineering involvement, the design team sets out to create a detailed process flow diagram and to improve the task integration begun in preliminary process synthesis. Then, in preparation for the detailed design work to follow, a detailed database is created, a pilot plant is often constructed to test the reaction steps and the more important, less-understood separation operations, and a simulation model is commonly prepared. As the design team learns more about the process, improvements are made, especially changes in the flow diagram to eliminate processing problems that had not been envisioned. In so doing, several of the alternative flowsheets generated in preliminary process synthesis gain more careful consideration, as well as the alternatives generated by the algorithmic methods, in detailed process synthesis [which often continues as the base-case design(s) is being developed].

Flow Diagrams

As the engineering work on the base-case design proceeds, a sequence of *flow diagrams* is used to provide a crucial vehicle for sharing information. The three main types are introduced in this subsection, beginning with the simplest *block flow diagram* (*BFD*), proceeding to the *process flow diagram* (*PFD*), and concluding with the *piping and instrumentation diagram* (*P&ID*). These are illustrated for the vinyl-chloride process synthesized in the previous section (see Figure 3.8)—the so-called base-case design.

Block Flow Diagram (BFD)

The block flow diagram represents the main processing sections in terms of functional blocks. As an example, Figure 3.18 shows a block diagram for the vinyl-chloride process, in which the three main sections in the process, namely, ethylene chlorination, pyrolysis, and separation are shown, together with the principal flow topology. Note that the diagram also indicates the overall material balances and the conditions at each stage, where appropriate. This level of detail is helpful to summarize the principal processing sections and is appropriate in the early design stages, where alternative processes are usually under consideration.

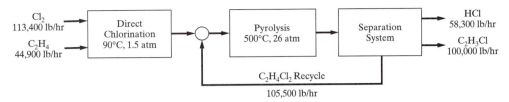

Figure 3.18 Block flow diagram for the vinyl-chloride process.

Process Flow Diagram (PFD)

Process flow diagrams provide a more detailed view of the process. These diagrams display all of the major processing units in the process (including heat exchangers, pumps, and compressors), provide stream information, and include the main control loops that enable the process to be regulated under normal operating conditions. Often, preliminary PFDs are constructed using the process simulators. Subsequently, more detailed PFDs are prepared using software such as AUTOCAD and VISIO, the latter having been used to prepare Figure 3.19 for the vinyl-chloride process. The conventions typically used when preparing PDFs are illustrated using this figure and are described next.

Processing Units.   Icons that represent the units are linked by arcs (lines) that represent the process streams. The drawing conventions for the unit icons are taken from accepted standards, for example, the ASME (American Society for Mechanical Engineers) standards (ASME, 1961). A partial list of typical icons is presented in Figure 3.20. Note that each unit is labeled according to the convention: U-XYY, where U is a single letter identifying the unit type (V for vessel, E for exchanger, R for reactor, T for tower, P for pump, C for compressor, etc.), X is a single digit, identifying the process area where the unit is installed, and YY is a two-digit number identifying the unit itself. Thus, for example, E-100 is the identification code for the heat exchanger that condenses the overhead vapors from the chlorination reactor. Its identification code indicates that it is the 00 item installed in plant area 1.

Stream Information.   Directed arcs that represent the streams, with flow direction from left to right wherever possible, are numbered for reference. By convention, when streamlines cross, the horizontal line is shown as a continuous arc, with the vertical line broken. Each stream is labeled on the PFD by a numbered diamond. Furthermore, the feed and product streams are identified by name. Thus, streams 1 and 2 in Figure 3.19 are labeled as the ethylene and chlorine feed streams, while streams 11 and 14 are labeled as the hydrogen chloride and vinyl-chloride product streams. Mass flow rates, pressures, and temperatures may appear on the PFD directly, but more often are placed in the stream table instead, for clarity. The latter has a column for each stream and can appear at the bottom of the PFD or as a separate table. Here, because of formatting limitations in this text, the stream table for the vinyl-chloride process is presented separately in Table 3.6. At least the following entries are presented for each stream: label, temperature, pressure, vapor fraction, total and component molar flow rates, and total mass flow rate. In addition, stream properties such as the enthalpy, density, heat capacity, viscosity, and entropy, may be displayed. Stream tables are often completed using a process simulator. In Table 3.6, the conversion in the direct chlorination reactor is assumed to be 100%, while that in the pyrolysis reactor is only 60%. Furthermore, both towers are assumed to carry out perfect separations, with the overhead and bottoms temperatures computed based on dew- and bubble-point temperatures, respectively.

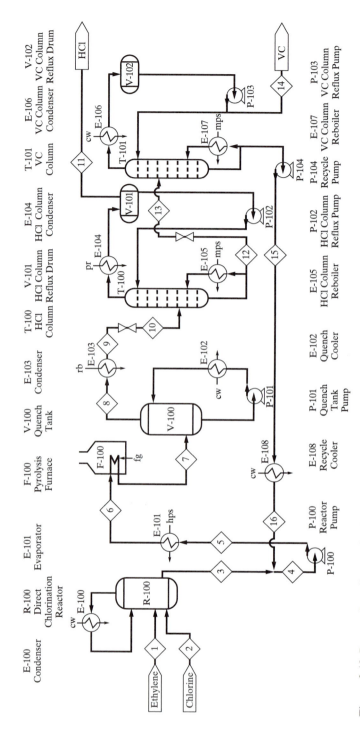

Figure 3.19 Process flow diagram for the vinyl-chloride process.

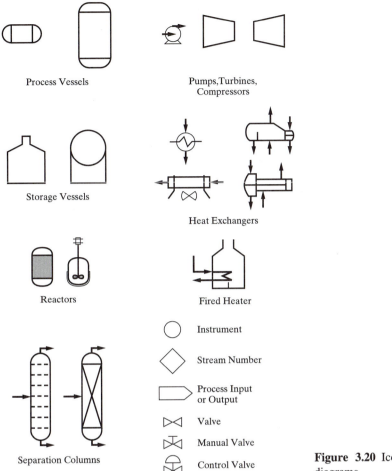

Process Vessels

Pumps,Turbines, Compressors

Storage Vessels

Heat Exchangers

Reactors

Fired Heater

Instrument

Stream Number

Process Input or Output

Valve

Manual Valve

Separation Columns

Control Valve

Figure 3.20 Icons in process flow diagrams.

Utilities. As shown in Figure 3.19, various utility streams are utilized for heating or cooling the process streams. For example, E-100, the overhead condenser for the direct chlorination reactor, which operates at 90°C, is cooled using cooling water (*cw*). The other cooling utilities are refrigerated brine (*rb*) and propane refrigerant (*pr*), each selected according to the temperature level of the required utility. Heating utilities are fuel gas (*fg*), high-pressure steam (*hps*), and medium-pressure steam (*mps*). A list of heating and cooling utilities, with temperature ranges, and the abbreviations commonly used on PFDs is presented in Table 3.7 (see also Table 17.1 and the subsection on *utilities* in Section 17.2).

Equipment Summary Table. This provides information for each equipment item in the PFD, with the kind of information typically provided for each type of unit shown in Table 3.8. Note that the materials of construction (MOC), and operating temperature and pressure, are required for all units. Also note that suggestions for the materials of construction are provided in Appendix III.

In summary, the PFD is the most definitive process design document, encapsulating much of the commonly referred to design information. As such, it is used and updated throughout much of process design. However, it lacks many details required to begin the construction engineering work for the plant. Many of these details are transmitted in the *Piping and Instrumentation Diagram.*

Table 3.6 Stream Summary Table for the Vinyl-Chloride Process in Figure 3.19

Stream Number	1	2	3	4	5	6	7	8
Temperature (°C)	25	25	90	90	91.3	242	500	170
Pressure (Atm)	1.5	1.5	1.5	1.5	26	26	26	26
Vapor fraction	1.0	1.0	0.0	0.0	0.0	1.0	1.0	1.0
Mass flow (lb/hr)	44,900	113,400	158,300	263,800	263,800	263,800	263,800	263,800
Molar flow (lbmol/hr)	1,600	1,600	1,600	2,667	2,667	2,667	4,267	4,267
Component molar flow (lbmol/hr):								
Ethylene	1,600	0	0	0	0	0	0	0
Chlorine	0	1,600	0	0	0	0	0	0
1,2-dichloroethane	0	0	1,600	2,667	2,667	2,667	1,067	1,067
Vinyl chloride	0	0	0	0	0	0	1,600	1,600
Hydrogen chloride	0	0	0	0	0	0	1,600	1,600

Stream Number	9	10	11	12	13	14	15	16
Temperature (°C)	6	6.5	−26.4	94.6	57.7	32.2	145.6	90
Pressure (Atm)	26	12	12	12	4.8	4.8	4.8	4.8
Vapor fraction	0.0	0.0	1.0	0.0	0.23	1.0	0.0	0.0
Mass flow (lb/hr)	263,800	263,800	58,300	205,500	205,500	100,000	105,500	105,500
Molar flow (lbmol/hr)	4,267	4,267	1,600	2,667	2,667	1,600	1,067	1,067
Component molar flow (lbmol/hr):								
Ethylene	0	0	0	0	0	0	0	0
Chlorine	0	0	0	0	0	0	0	0
1,2-dichloroethane	1,067	1,067	0	1,067	1,067	0	1,067	1,067
Vinyl chloride	1,600	1,600	0	1,600	1,600	1,600	0	0
Hydrogen chloride	1,600	1,600	1,600	0	0	0	0	0

Table 3.7 Heating and Cooling Utilities—Identifiers and Temperature Ranges

Identifier	Utility	Typical Operating Range (°F)
Hot Utilities—In increasing cost per BTU:		
lps	Low-pressure steam, 15 to 30 psig	250 to 275°F
mps	Medium-pressure steam, 100 to 150 psig	325 to 366°F
hps	High-pressure steam, 400 to 600 psig	448 to 488°F
fg	Fuel oils	
fo	Fuel gas	Process waste stream
po	Petroleum oils	Below 600°F
dt	Dowtherms	Below 750°F
Cold Utilities—In increasing cost per BTU:		
bfw	Boiler feed water	Used to raise process steam
ac	Air cooling	Supply at 85 to 95°F—temperature approach to process 40°F
rw	River water	Supply at 80 to 90°F (from cooling tower), return at 110°F
cw	Cooling water	Supply at 80 to 90°F (from cooling tower), return at 115 to 125°F
cw	Chilled water	45 to 90°F
rb	Refrigerated brine	0 to 50°F
pr	Propane refrigerant	−40 to 20°F

Table 3.8 Equipment Summary Specifications

Equipment type	Required specification
Vessel	Height, diameter, orientation, pressure, temperature, materials of construction (MOC)
Towers	Height, diameter, orientation, pressure, temperature, number of and type of trays, height and type of packing, MOC
Pumps	Driver type, flow, suction and discharge pressures, temperature, shaft power, MOC
Compressors	Driver type, inlet flow, suction and discharge pressures, temperature, shaft power, MOC
Heat exchangers	Type, area, duty, number of shell and tube passes, for both shell and tubes: operating temperature, pressure, pressure drop, and MOC
Fired heaters	Type, tube pressure and temperature, duty, radiant and convective heat transfer area, MOC

Piping and Instrumentation Diagram (P&ID)

This is the design document transmitted by the process design engineers to the engineers responsible for plant construction. It is also used to support the startup, operation of the process, and operator training. Consequently, it contains items that do not appear in the PFD, such as the location and type of all measurement and control instruments, the positioning of all valves, including those used for isolation and control, and the size, schedule, and materials of construction of piping. As a result, a number of interconnected P&IDs are prepared for a process that is represented on a single PFD. For more details on the preparation of P&IDs, the reader is referred to the books by Sandler and Luckiewicz (1993) and Ulrich (1984).

Calculations Supporting Flow Diagrams

As indicated when discussing the stream table (Table 3.6), and emphasized when synthesizing the vinyl-chloride process in the previous section, the material balances for the process streams are nearly complete after preliminary process synthesis. These are conducted by means of spreadsheets and by process simulators, as discussed in Chapter 4. At this stage, the design team checks the assumptions. It also completes the material and energy balances associated with heat addition and removal, without attempting to carry out heat and power integration. As indicated in the section on process integration that follows, the design team carries out heat and power integration just prior to the detailed design stage.

It should also be noted that, during the synthesis of the vinyl-chloride process, no attempt was made to complete calculations to determine the number of stages and reflux ratios for the distillation towers, and furthermore, perfect splits may be assumed. Hence, the condenser and reboiler heat duties are not yet known. The vapor stream, S1, is assumed to be saturated, pure dichloroethane, which releases its heat of vaporization, 143.1 Btu/lb, to cooling water, which is heated from 30° to 50°C. Both the direct chlorination reactor and the pyrolysis furnace are assumed to operate adiabatically, and natural gas is assumed to have a lower heating value of 23,860 Btu/lb (heat of combustion at 25°C). The liquid effluent from the quench is assumed to have a composition in vapor–liquid equilibrium at 150°C and 26 atm. The stream is cooled to 50°C with cooling water to release the heat necessary to cool the pyrolysis products from 500 to 170°C (4.66×10^7 Btu/hr). No attempt is made to calculate the amount of propane refrigerant necessary to remove the heat to cool the pyrolysis effluent to its bubble point at 6°C (5.20×10^7 Btu/hr); this calculation is completed during process integration, when the heat and power integration is completed.

These calculations could have been completed using the process simulators, which are commonly used to calculate the heats of reaction, enthalpy changes upon heating and cooling, and vapor–liquid equilibria, as well as to perform material and energy balances using approximate models involving specifications of split fractions in separators and conversions in chemical reactors. Note that a complete simulation is usually not justified until the design team is ready to begin the detailed design. Gradually, additional detail is added to the simulation model; for example, the number of stages and the reflux ratio are selected for the distillation columns, and the material and energy balances are completed with recycle streams that are not assumed to be pure. As mentioned previously, this is the subject matter of Chapter 4, in which the methods of building a simulation model are introduced. After studying Chapter 4, the reader should be able to prepare a simulation for the vinyl-chloride process (see Exercise 4.5) and prepare a more accurate representation of the flow diagram in Figure 3.19 and Table 3.6.

Process Integration

With the detailed process flow diagram completed, the task integration step, which was initiated in the preliminary process synthesis, is revisited by the design team. The assumptions are checked and opportunities are sought to improve the designs of the processing units, and to achieve a more efficient process integration. In the latter, attempts are made to match cold streams that need to be heated with hot streams that have cooling requirements, so as to reduce the need for external utilities such as steam and cooling water. In addition, where possible, power is extracted from hot streams at elevated pressures, so as to drive compressors and pumps. Also, when solvents, such as water, are used as mass separating agents, opportunities are sought to reduce the amount of solvent used through mass integration. Often, significant improvements can be made in the process design beyond those achievable in the preliminary process synthesis. The algorithmic methods in Chapter 10 for heat and power integration and in Chapter 11 for mass integration are commonly applied by the design team; they provide a systematic approach to minimizing the utilities, matching the hot and cold streams, inserting turbines (as a part of heat engines), minimizing the amount of solvent used, and so on.

Detailed Database

Having completed the process flow diagram (PFD), the design team seeks to check its key assumptions further and to obtain the additional information needed to begin work on the detailed design. As discussed earlier, this usually involves three activities in parallel, the first of which is to create a detailed database by refining and adding to the preliminary database. In the other two activities, a pilot plant is constructed to confirm that the equipment items operate properly and to provide data for the detailed data bank, and a simulation model is prepared to enable the team to project the impact of changes in the design and operation parameters, such as temperatures, pressures, reflux ratios, and the number of stages.

In creation of the detailed database, it is common to add transport and kinetics data, as well as data concerning the feasibility of the separations, the identity of any forbidden matches in heat exchange, heuristic parameters, and data for sizing the equipment. Each process requires somewhat different data, and hence it is inappropriate to generalize. However, it is instructive to examine the mix of data needed by a design team in connection with the vinyl chloride process in Figure 3.19.

Beginning with the chlorination reactor, data are needed to determine the impact of the concentrations of C_2H_4, Cl_2, and $FeCl_3$ catalyst in the $C_2H_4Cl_2$ pool on the intrinsic rate of the chlorination reaction (in $kmol/m^3$ hr). With these data, the team can determine the order of the reaction and its rate constant as a function of temperature, and eventually compute the residence time to achieve nearly complete conversion.

Similar data are required for the pyrolysis reactor. In this case, the intrinsic rate of reaction is needed as a function of concentration, temperature, and pressure. Furthermore, since the rate of reaction may be limited by the rate at which heat is transferred to the reacting gases, it is probably desirable to estimate the tube-side heat transfer coefficient, h_i, as a function of the Reynolds and Prandtl numbers in the tubes. The appropriate equations and coefficients, which are described in Chapter 13, would be added to the database.

In the vinyl-chloride process, because of the significant differences in the volatilities of the three principal chemical species, distillation, absorption, and stripping are prime candidates for the separators, especially at the high production rates specified. For other processes, liquid–liquid extraction, enhanced distillation, adsorption, and membrane separators might become more attractive, in which case the design team would need to assemble data that describe the effect of solvents on species phase equilibrium, species adsorption isotherms, and the permeabilities of the species through various membranes.

A key limitation in the flowsheets in Figures 3.8 and 3.19 is that the cold $C_2H_4Cl_2$ stream is not heated by the pyrolysis products because the rate of carbon deposition in such a feed/product heat exchanger is anticipated to be high, and would cause the heat exchanger to foul with carbon. As discussed above, the design team would normally apply the methods of heat and power integration to design a network of heat exchangers that would effect significant economies. Hence, it is important to learn more about the rate of carbon deposition. Before the team proceeds to the detailed design stage, it needs data to confirm the validity of this perception above—that is, to enable it to characterize the intrinsic rate of carbon deposition. If the rate is found to be sufficiently low, the team may decide to cool the hot pyrolysis products through heat exchange with the cold streams. For maintenance, to remove carbon deposits periodically, two heat exchangers could be installed in parallel, one of which would be operated while the other is being cleaned. This would provide substantial savings in fuel and cooling water utilities. On the other hand, if the rate of carbon deposition is high, the design team would avoid the exchange of heat between these two streams; that is, it would continue to consider the exchange of that heat to be a so-called *forbidden match*.

The additional data for sizing the equipment are typically maximum pressure drops, tube lengths, and baffle spacings in heat exchangers, surface tensions and drag coefficients for estimating the flooding velocities (to be used in determining the tower diameters), specifications for tray spacings in multistaged towers, and residence times in flash vessels and surge tanks. Examples of the use of this type of data for the detailed design of heat exchangers are provided in Chapter 13, and for the detailed design of a distillation tower in Chapter 14.

Pilot-Plant Testing

Clearly, as the detailed database is assembled, the needs for pilot-plant testing become quite evident. For the manufacture of new chemicals, a pilot plant can produce quantities of product suitable for testing and evaluation by potential customers. Very few processes that include reaction steps are constructed without some form of pilot-plant testing prior to doing detailed design calculations. This is an expensive, time-consuming step that needs to be anticipated and planned for by the design team as early as possible, so as to avoid extensive delays. Again, although it is inappropriate to generalize, the vinyl-chloride process provides good examples of the need for pilot-plant testing and the generation of data for detailed design calculations.

As mentioned in the previous subsection, kinetic data are needed for both the chlorination and pyrolysis reactors, as well as to determine the rate of carbon deposition. In all three cases, it is unlikely that adequate data can be located in the open literature. Consequently, unless sufficient data exist in company files, or were taken in the laboratory and are judged to be adequate, pilot-plant testing is needed. Generally, the pilot-plant tests are conducted by a development team working closely with the design team. As the data are recorded, regression analyses are commonly used to compute the coefficients of compact equations to be stored in the database.

As mentioned in connection with the need for laboratory experiments, pilot-plant tests also help to identify potential problems that arise from small quantities of impurities in the feed streams, and when unanticipated byproducts are produced, usually in small quantities, that have adverse effects, such as to impart an undesired color or smell to the product. When a catalyst is used, the impact of these species needs to be studied, and, in general, the useful life of the catalyst needs to be characterized. Pilot plants can also verify separation schemes developed during process design.

Process Simulation

As mentioned throughout the discussion of preliminary process synthesis and the creation of the process flow diagram, the process simulator usually plays an important role, even if a simulation model is not prepared for the entire flowsheet. When parts of a simulation model exist, it is common for the design team to assemble a more comprehensive model, one that enables the team to examine the effect of parametric changes on the entire process. In other cases, when the process simulators have not been used for design, a simulation model is often created for comparison with the pilot-plant data and for parametric studies.

High-speed PCs, which have excellent graphical user interfaces (GUIs), have replaced workstations as the preferred vehicle for commercial simulators, and are now finding widespread use throughout the chemical process industries. The use of simulators, which is the subject of the next chapter, has become commonplace in assisting the design team during process creation.

3.6   SUMMARY

Having studied this chapter, the reader should

1. Be able to create a preliminary database for use in preliminary process synthesis.
2. Understand the steps in preliminary process synthesis and be able to use them to develop other flowsheets for the manufacture of vinyl chloride and tPA (corresponding to the other branches of the synthesis trees in Figures 3.9 and 3.17), as well as for the manufacture of other chemicals.
3. Understand the steps taken by the design team in preparing one or more base-case designs. For the manufacture of vinyl chloride, or another chemical, you should be able to create a detailed process flow diagram and understand the need to complete the task integration step begun during preliminary process synthesis and carry out the process integration step. In addition, determine whether continuous or batch operation is more suitable.
4. Know how to prepare for the detailed design step, that is, to expand the database to include important kinetics data and the like, and to seek data from a pilot plant when necessary. You should also recognize the need for a model of the process, usually implemented by process simulators, to be covered in Chapter 4.

REFERENCES

ASME, *Graphical Symbols for Process Flow Diagrams*, ASA Y32.11, Amer. Soc. Mech. Eng., New York (1961).

Audette, M., C. Metallo, and K. Nootong, *Human Tissue Plasminogen Activator*, Towne Library, University of Pennsylvania, Philadelphia, Pennsylvania (2000).

Balzhiser, R.E., M.R. Samuels, and J.D. Eliassen, *Chemical Engineering Thermodynamics*, Prentice-Hall, Englewood Cliffs, New Jersey (1972).

Benedict, D.B., Process for the Preparation of Olefin Dichlorides, U.S. Patent 2,929,852, March 22 (1960).

Borsa, A.G., A.M. Herring, J.T. McKinnon, R.L. McCormick, and G.H. Ko, "Coke and Byproduct Formation during 1,2-Dichloroethane Pyrolysis in a Laboratory Tubular Reactor," *Ind. Eng. Chem. Res.*, **40**, 2428–2436 (2001).

B.F. Goodrich Co., Preparation of Vinyl Chloride, British Patent 938,824, October 9 (1963). *Instrument Symbols and Identification*, Instrument Society of America Standard ISA-S5–1, Research Triangle Park, North Carolina (1975).

de Nevers, N., *Physical and Chemical Equilibrium for Chemical Engineers*, Wiley-Interscience, New York (2002).

Douglas, J.M., *Conceptual Design of Chemical Processes*, McGraw-Hill, New York (1988).

Gmehling, J., *Azeotropic Data*, VCH Publishers, Deerfield Beach, Florida (1994).

Gmehling, J., U. Onken, W. Arlt, P. Grenzheuser, U. Weidlich, and B. Kolbe, *Vapor–Liquid Equilibrium Data Collection*, 13 Parts, DECHEMA, Frankfurt, Germany (1980).

Goeddel, D.V., W.J. Kohr, D. Pennica, and G.A. Vehar, *Human Tissue Plasminogen Activator*, U.S. Patent 4,766,075, August 23 (1988).

Kohn, J.P., and F. Kurata, Heterogeneous Phase Equilibria of the Methane–Hydrogen Sulfide System, *AIChE J.*, **4**(2), 211(1958).

Kyle, B.J., *Chemical and Process Thermodynamics*, Prentice-Hall, Englewood Cliffs, New Jersey (1984).

Leonard, E.C., Ed., *Vinyl and Diene Monomers, Part 3*, Wiley-Interscience, New York (1971).

Myers, A.L., and W.D. Seider, *Introduction to Chemical Engineering and Computer Calculations*, Prentice-Hall, Englewood Cliffs, New Jersey (1976).

Perry, R.H., and D.W. Green, Ed., *Perry's Chemical Engineers' Handbook*, 7th ed., McGraw-Hill, New York (1997).

Pisano, G. P., *The Development Factory: Unlocking the Potential of Process Innovation*, Harvard Business School Press, Boston (1997).

Poling, B.E., J.M. Prausnitz, and J. P. O'Connell, *Properties of Gases and Liquids*, 5th ed., McGraw-Hill, New York (2001).

Reamer, H.H., B.H. Sage, and W.N. Lacey, Phase Equilibria in Hydrocarbon Systems, *Ind. Eng. Chem.*, **43**, 976 (1951).

Rieder, R.M., and A.R. Thompson, Vapor–Liquid Equilibrium Measured by a Gillespie Still, *Ind. Eng. Chem.*, **41**(12), 2905 (1949).

Rudd, D.F., G.J. Powers, and J.J. Siirola, *Process Synthesis*, Prentice-Hall, Englewood Cliffs, New Jersey (1973).

Sandler, H.J., and E.T. Luckiewicz, *Practical Process Engineering*, XIMIX, Philadelphia, Pennsylvania (1993).

Sandler, S.I., *Chemical and Engineering Thermodynamics*, 2nd ed., Wiley, New York (1997).

Smith, J.M., H.C. Van Ness, and M.M. Abbott, *Chemical Engineering Thermodynamics*, 5th ed., McGraw-Hill, New York (1997).

Ulrich, G.D., *A Guide to Chemical Engineering Process Design and Economics*, Wiley, New York (1984).

Walas, S.M., *Phase Equilibria in Chemical Engineering*, Butterworth, London (1985).

Walas, S.M., *Chemical Process Equipment*, Butterworth, London (1988).

Woods, D.R., *Data for Process Design and Engineering Practice*, Prentice-Hall, Englewood Cliffs, New Jersey (1995).

EXERCISES

3.1 For an equimolar solution of *n*-pentane and *n*-hexane, compute:

a. The dew-point pressure at 120°F

b. The bubble-point temperature at 1 atm

c. The vapor fraction, at 120°F and 0.9 atm, and the mole fractions of the vapor and liquid phases

3.2 For the manufacture of vinyl chloride, assemble a preliminary database. This should include thermophysical property data, MSDSs for each chemical giving toxicity and flammability data, and the current prices of the chemicals.

3.3 Consider the flowsheet for the manufacture of vinyl chloride in Figure 3.8.

a. If the pyrolysis furnace and distillation towers are operated at low pressure (1.5 atm), what are the principal disadvantages? What alternative means of separation could be used?

b. For the process shown, is it possible to use some of the heat of condensation from the $C_2H_4Cl_2$ to drive the reboiler of the first distillation tower? Explain your response. If not, what process change would make this possible?

c. Consider the first reaction step to make dichloroethane. Show the distribution of chemicals when ethylene is 20% in excess of the stoichiometric amount and the chlorine is entirely converted. Assume that 100,000 lb/hr of vinyl chloride are produced.

d. Consider the first distillation tower. What is the advantage of cooling the feed to its bubble point at 12 atm as compared with introducing the feed at its dew point?

e. Why isn't the feed to the pyrolysis furnace heated with the hot pyrolysis products?

f. What is the function of the trays in the direct chlorination reactor?

g. Suggest ways to reduce the need for fuel and hot utilities such as steam.

3.4 a. To generate steam at 60 atm, two processes are proposed:

1. Vaporize water at 1 atm and compress the steam at 60 atm.

2. Pump water to 60 atm followed by vaporization.

Which process is preferred? Why?

b. In a distillation tower, under what circumstances is it desirable to use a partial condenser?

3.5 Synthesize a flowsheet for the manufacture of vinyl chloride that corresponds to one of the other branches in the synthesis tree in Figure 3.9. It should begin with reaction path 4 or 5.

3.6 Using the chemical engineering literature, complete the detailed database for the detailed design of the base-case process in Figure 3.19. When appropriate, indicate the kind of data needed from a pilot plant and how this data should be regressed.

Chapter 4

Simulation to Assist in Process Creation

4.0 OBJECTIVES

In Chapters 2 and 3, the emphasis was on finding molecules and chemical mixtures that have desired properties and on the creation of alternative process flowsheets that arise from the primitive design problem. The steps in generating the preliminary database, carrying out experiments, performing preliminary process synthesis, preparing a process flow diagram for the base-case design, and developing a detailed database and carrying out pilot-plant testing, prior to preparing the detailed design, were described in Chapter 3. For the production of vinyl-chloride monomer, a synthesis tree was generated and a base-case design was initiated. Throughout, emphasis was on calculations to obtain bubble- and dew-point temperatures, heats of reaction, and so on, and to satisfy material and energy balances, calculations carried out routinely by process simulators. Similarly, for the production of tissue plasminogen activator (tPA), the laboratory tests to locate the target protein, the gene sequence that codes for the protein, and the host cell to be used for growing the protein, were introduced in Chapter 2. Then, a synthesis tree was generated and a base-case design was initiated in Chapter 3. However, no instruction was provided on the use of the process simulators. This is the objective of the current chapter, which focuses on the basics of steady-state process simulation and describes the key role that process simulators play in assisting the design team in process creation. Four of the major process simulators are introduced for steady-state simulation: ASPEN PLUS by Aspen Technology, Inc.; HYSYS.Plant by Aspen Technology, Inc. (originally by Hyprotech, Ltd.); CHEMCAD by ChemStations, Inc.; and PRO/II by Simulation Sciences, Inc. A multimedia CD-ROM, *Using Process Simulators in Chemical Engineering: A Multimedia Guide for the Core Curriculum* (Lewin et al., 2003), provides detailed instructions on the use of the process simulators, with current emphasis on ASPEN PLUS and HYSYS.Plant. Chapter 4 concludes with an introduction to batch process simulation, placing emphasis on BATCH PLUS by Aspen Technology, Inc., and SUPERPRO DESIGNER by Intelligen, Inc.

After studying this chapter, and the associated CD-ROM, the reader should

1. Understand the role of process simulators in process creation and be prepared to learn about their roles in equipment sizing and costing, profitability analysis, optimization, and dynamic simulation in the chapters that follow.
2. For steady-state simulation, be able to create a simulation flowsheet, involving the selection of models for the process units and the sequence in which process units associated with recycle loops are solved to obtain converged material and energy balances.

3. Understand degrees of freedom in modeling process units and flowsheets, and be able to make design specifications and follow the iterations implemented to satisfy them. When using HYSYS.Plant, the reader will learn that its implementation of *bidirectional information flows* is very efficient in satisfying many specifications.

4. Learn the step-by-step procedures for using ASPEN PLUS and HYSYS.Plant. The CD-ROM covers many of these steps. Additional assistance is available by consulting the extensive user manuals distributed with the software.

5. Be able to use the process simulators systematically during process creation, following sequences similar to those illustrated later in this chapter for a toluene hydrodealkylation process. The reader will learn to simulate portions of the process (the reactor section, the distillation section, etc.) before attempting to simulate the entire process with its recycle loops. Many examples and exercises enable the reader to master these techniques.

6. Be able to use the batch process simulators to carry out material and energy balances, and to prepare an operating schedule in the form of a Gantt chart for the process.

4.1 INTRODUCTION

Having concentrated on the generation of process flowsheets in Chapter 3, this chapter focuses on the role of analysis, that is, the solution of the material and energy balances coupled with phase equilibria, and the equations of transport and chemical kinetics. The emphasis in this chapter is on finding suitable operating conditions for processes (temperatures, pressures, etc.). Computing packages that model the process units are introduced and utilized to model the highly integrated flowsheets commonly designed to achieve more profitable operation. As has been mentioned, these packages are referred to as *process simulators*, most of which are used to simulate potential processes in the steady state—that is, to determine the unknown temperatures, pressures, and component and total flow rates at steady state. More recently, these packages are being extended to permit the dynamic simulation of processes and their control systems as they respond to disturbances and changes in operating points. Chapter 4 concentrates on steady-state simulation and the scheduling of batch processes during process creation; Chapter 21 shows how to use the HYSYS.Plant dynamic simulator to confirm that a potential process is easily controlled as typical disturbances arise. Both Chapters 4 and 21 are accompanied by extensive coverage on the multimedia CD-ROM that accompanies this book, which explains how to use ASPEN PLUS and HYSYS.Plant. The Aspen Icarus Process Evaluator (Aspen IPE) package, provided by Aspen Technology, Inc., is used for cost estimation and an economics spreadsheet is used for profitability analysis. These topics are covered separately in Sections 16.7 and 17.8, respectively. Finally, the packages have extensive facilities for process optimization, with Chapter 18 and the multimedia CD-ROM concentrating on optimization in ASPEN PLUS and HYSYS.Plant.

Often during the synthesis steps and the creation of the base-case design, as mentioned when synthesizing the vinyl-chloride process in Sections 3.4 and 3.5, process simulators are utilized by the design team to calculate heat duties, power requirements, phase and chemical equilibria, and the performance of multistaged towers, among many other calculations. For the production of commodity chemicals, as the alternative flowsheets evolve, it is common to perform these calculations assuming operation in the steady state; hence, many *steady-state simulators* have become available to process engineers. For the production of specialty chemicals in batch processes, it is common to perform similar calculations using batch process simulators.

In this chapter, the principles behind the use of several widely used flowsheet simulators are introduced. For processes in the steady state, these include ASPEN PLUS, HYSYS.Plant, CHEMCAD, and PRO/II. For batch processes, these include BATCH PLUS and SUPER-PRO DESIGNER.

The multimedia CD-ROM that accompanies this book also explains how to use the dynamic simulators. Emphasis is placed on HYSYS.Plant. Using HYSYS.Plant, the design team can complete a steady-state simulation, add controllers, and activate the integrator to carry out a dynamic simulation. Similar facilities are provided in ASPEN DYNAMICS by Aspen Technology, Inc.

A primary objective of this chapter is to show how to use the process simulators during process synthesis to better define the most promising processes. After the basic principles are covered, a case study is presented in which the simulators are used to help synthesize the reactor and separation sections of a toluene hydrodealkylation process. Finally, a case study is presented in which the BATCH PLUS simulator is used to help synthesize the operating schedule for a tissue plasminogen activator (tPA) process. Many of the details concerning the process simulators are presented on the multimedia CD-ROM that accompanies this book. The latter covers the ASPEN PLUS and HYSYS.Plant simulators, with step-by-step audio instructions for completing the input forms. Some coverage of BATCH PLUS is also provided. In addition, the CD-ROM provides video segments from a large-scale petrochemical complex to illustrate some of the equipment being modeled, tutorials on the estimation and regression of physical property data, and .bkp and .hsc files for the ASPEN PLUS and HYSYS.Plant examples throughout the book.

4.2 PRINCIPLES OF STEADY-STATE FLOWSHEET SIMULATION

Given the detailed process flow diagram for the base-case design (e.g., Figure 3.19), or a process flow diagram after the task integration step in process synthesis, or even an incomplete flow diagram after one of the earlier steps, it is often possible to use a process simulator to solve for many of the unknown temperatures, pressures, and flow rates in the steady state. For an existing process, analysis using a process simulator is performed routinely to study potential changes in the operating conditions or the possibility of a retrofit to improve its profitability.

In this section, the objective is to cover the basics of steady-state simulation, with an introduction to ASPEN PLUS, HYSYS.Plant, CHEMCAD, and PRO/II. However, no attempt is made to show how to use these simulators when carrying out the step-by-step strategy in Chapter 3. This is accomplished in Section 4.3, in which a case study is presented involving the synthesis of a process to hydrodealkylate toluene by reaction with hydrogen to produce benzene and methane. Readers who have experience in using steady-state simulators may find it preferable to skim through these materials, to identify those sections that can add to their understanding, and to proceed to Section 4.3. Others, with little or no experience, are advised to study this section at least through the subsection on "Flash Vessel Control" before proceeding to the next section.

Process and Simulation Flowsheets

Process flowsheets are the *language* of chemical processes. Like a work of art, they describe an existing process or a hypothetical process in sufficient detail to convey the essential features.

Analysis, or simulation, is the tool chemical engineers use to interpret process flowsheets, to locate malfunctions, and to predict the performance of processes. The heart of analysis is the mathematical model, a collection of equations that relate the process variables, such as stream temperature, pressure, flow rate, and composition, to surface area, valve settings, geometrical configuration, and so on. The steady-state simulators solve for the unknown variables, given the values of certain known quantities.

There are several levels of analysis. In order of increasing complexity, they involve: material balances, material and energy balances, equipment sizing, and profitability analysis. Additional equations are added at each level. New variables are introduced, and the equation-solving algorithms become more complicated.

Fortunately, most chemical processes involve conventional process equipment: heat exchangers, pumps, distillation columns, absorbers, and so on. For these process units, the equations do not differ among chemical processes. The physical and thermodynamic properties and chemical kinetics constants differ, but not the equations. Hence, it is possible to prepare one or more equation-solving algorithms for each process unit to solve the material and energy balances and to compute equipment sizes and costs. A library of subroutines or models, usually written in FORTRAN or C, that automate such equation-solving algorithms is at the heart of process simulators. These subroutines or models are hereafter referred to as *procedures*, *modules*, or *blocks*. As discussed at the end of this section, in a small but growing class of simulators (e.g., gPROMS, ABACUS, and as options in ASPEN PLUS and HYSYS.Plant), equations that model a process unit are stored, rather than embedded in FORTRAN or C subroutines that solve the equations associated with the model for each process unit. Given the interconnecting streams, equations for the units in a process are assembled to be solved simultaneously by an equation solver such as the Newton–Raphson method.

To use a flowsheet simulator effectively, it is very helpful to distinguish between process flowsheets and the so-called *simulation flowsheets* associated with process simulators. These distinctions are drawn in the next subsections.

Process Flowsheets

A process flowsheet is a collection of icons to represent process units and arcs to represent the flow of materials to and from the units. The process flowsheet emphasizes the flow of material and energy in a chemical process, as illustrated in Figure 4.1.

Simulation Flowsheets

A simulation flowsheet, on the other hand, is a collection of simulation units to represent computer programs (subroutines or models) that *simulate* the process units and arcs to represent the flow of information among the simulation units. A simulation flowsheet emphasizes information flows. The analogy between the process flowsheet and the simulation flowsheet is illustrated by comparing Figures 4.1 and 4.2a. The latter has been prepared specifically for ASPEN PLUS. The simulation flowsheet may use blocks or icons to represent the process units. For ASPEN PLUS, Figure 4.2a uses blocks whereas Figure 4.2b uses icons. Figures 4.2c, 4.2d, and 4.2e for HYSYS.Plant, CHEMCAD, and PRO/II, respectively, use icons.

Several constructs appear in Figure 4.2:

1. The *arcs* represent the transfer of flow rates, temperature, pressure, enthalpy, entropy, and vapor and liquid fractions for each stream. The stream names can be thought of as the

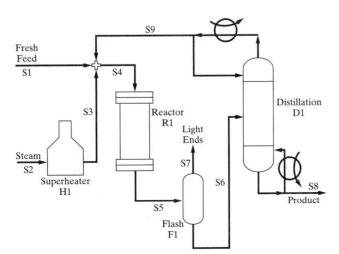

Figure 4.1 Process flowsheet.

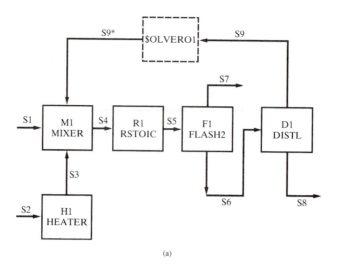

(a)

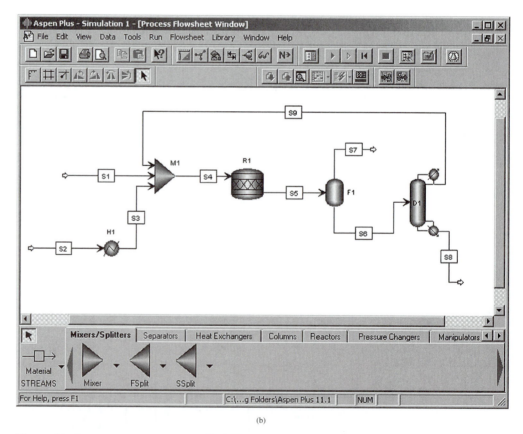

(b)

Figure 4.2 Simulation flowsheet: (a) ASPEN PLUS blocks; (b) ASPEN PLUS icons; (c) HYSYS.Plant icons; (d) CHEMCAD icons; (e) PRO/II icons.

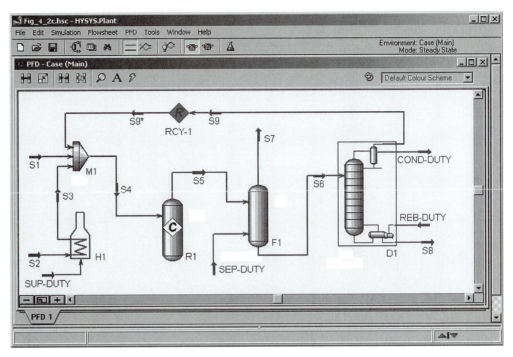

Unit Name	Model Name
H1	**Heater**
M1	**Mixer**
R1	**Reactor**
F1	**Separator**
D1	**Column**

(c)

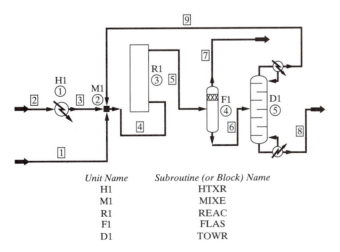

Unit Name	Subroutine (or Block) Name
H1	HTXR
M1	MIXE
R1	REAC
F1	FLAS
D1	TOWR

The CHEMCAD simulation flowsheet also assigns unique numbers, included in this figure, to the units.

(d)

Figure 4.2 (*Continued*)

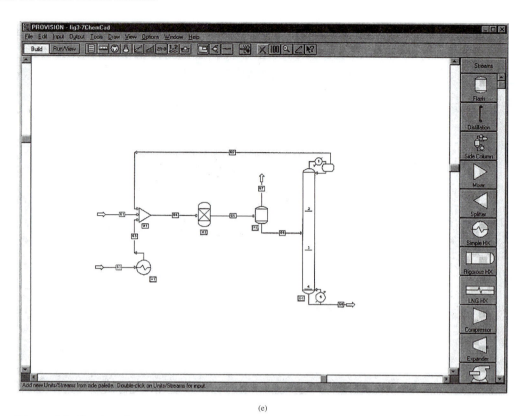

(e)

Figure 4.2 (*Continued*)

names of vectors that store stream variables in a specific order, as illustrated for ASPEN PLUS:

Vector Element		Vector Element (Continued)	
1 to C	chemical flow rates, kmol/s	C + 5	vapor fraction (molar)
C + 1	total flow rate, kmol/s	C + 6	liquid fraction (molar)
C + 2	temperature, K	C + 7	mass entropy, J/kg K
C + 3	pressure, MPa	C + 8	density, kg/m^3
C + 4	mass enthalpy, J/kg	C + 9	molecular weight, kg/kmol

where C is the number of chemical species.

2. The *solid-line rectangles* in Figure 4.2a, and the icons in Figures 4.2b–4.2e, represent *simulation units*. In Figure 4.2a, the upper character string provided by the user is a unique name of the simulation unit or *unit name*. The lower character string is the name of the subroutine or model, or so-called *block name* in many of the simulators. Although *model* or *block* are commonly used, the term *subroutine* is used throughout this book to emphasize that the models are computer codes. The equations to model a process unit involve equipment parameters, such as area, number of equilibrium stages, or valve settings. Although different values of the parameters characterize each occurrence of a process unit in the process flowsheet, the same subroutine or model is often used several times in a simulation flowsheet. In Figure 4.2c, for HYSYS.Plant, the unit names are in upper case, having been selected to be identical to those in Figure 4.2a, and the model names are tabulated separately in boldface (for emphasis, since HYSYS.Plant does not use upper case to represent its model names). In Figure 4.2d, for CHEMCAD, the unit names are the upper character string and the subroutine (or block) names are tabulated separately. In Figure 4.2e, for PRO/II, the unit names are shown, and the subroutine names are on the menu of icons.

3. The *dashed-line rectangle* in Figure 4.2a represents a mathematical convergence unit that uses a subroutine to adjust the stream variables in the information recycle loop because iterative calculations are necessary. These are discussed under "Recycle." Note that when entering a simulation flowsheet into most of the simulators, the mathematical convergence unit is not specified or shown in the flowsheet. Rather, it is positioned by the simulator unless the user intervenes. HYSYS.Plant is an exception, as shown in Figure 4.2c, in which the user positions the recycle convergence unit, RCY-1. In Figures 4.2d and 4.2e, CHEMCAD and PRO/II do not show the convergence unit, but it exists and is transparent to the user.

To convert from a process flowsheet to a simulation flowsheet, one replaces the process units with appropriate simulation units. For each simulation unit, a subroutine (or block, or model) is assigned to solve its equations. Each of the simulators has an extensive list of subroutines (or blocks, or models) to model and solve the process unit equations. In most cases, the models range from approximate to detailed and rigorous, with the most approximate models used during the initial steps of process synthesis and the more rigorous models gradually substituted as fewer flowsheets remain competitive. To make effective usage of the simulators, process engineers need to become familiar with the underlying assumptions in the models provided by each simulator. These are described in user manuals that accompany simulator software. It is an objective of this section, and especially the multimedia CD-ROM that accompanies this book, to discuss the principal models available. Partial lists for the four major simulators are provided in Table 4.1. In most simulators, new subroutines (or blocks, or models) may be programmed by a user and inserted into the library. These can call, in turn, upon the extensive libraries of subroutines and data banks provided by the process simulators for estimation of the thermophysical and transport properties (see the multimedia CD-ROM that accompanies this text), equipment sizes and costs (see Section 16.7), and so on.

Observe that a mixing unit, modeled using the MIXER subroutine in ASPEN PLUS, the **Mixer** model in HYSYS.Plant, and the MIXE and MIXER subroutines in CHEMCAD and PRO/II, is introduced to simulate the mixing of streams, even though mixing is often performed in an actual process in the pipeline. Similarly, a stream dividing unit, modeled using the FSPLIT subroutine in ASPEN PLUS, the **Tee** model in HYSYS.Plant, and the DIVI and SPLITTER subroutines in CHEMCAD and PRO/II, is needed to branch a flow to two or more destinations. Note that simulators often have units to duplicate a process stream or multiply the flow rate of a process stream. In ASPEN PLUS, the DUPL subroutine is used to prepare two identical copies (S1A and S1B) of stream S1, as shown in Figure 4.3a. In Figure 4.3b, the MULT subroutine multiplies the flow rate of stream S1 to give stream S1A. Dashed lines are used because these simulation units do not correspond to process units in a chemical plant. Duplication and multiplication are accomplished quite differently and with the same subroutine, SREF, in CHEMCAD. Referring to Figure 4.3c, streams 98 and 99 entering and leaving, respectively, unit R1, which is modeled by the SREF subroutine, are fictitious and are not streams in the actual process. The SREF subroutine requires that the user state the name of the stream to be duplicated or multiplied, for example, stream 5, and the name of the resulting duplicated stream, for example, stream 6.

The steady-state models in simulators do *not* solve time-dependent equations. They simulate the steady-state operation of process units (operation subroutines) and estimate the sizes and costs of process units (cost subroutines). Two other types of subroutines are used to converge recycle computations (convergence subroutines) and to alter the equipment parameters (control subroutines). These subroutines are discussed in this section.

Table 4.1 Unit Subroutines

(a) ASPEN PLUS [excluding solids-handling equipment—see Seider et al. (1999) Table A-IV.2]

Mixers and splitters	MIXER	Stream mixer
	FSPLIT	Stream splitter
Separators	SEP	Component separator—multiple outlets
	SEP2	Component separator—two outlets
Flash drums	FLASH2	Two-outlet flash drums
	FLASH3	Three-outlet flash drums
Approximate distillation	DSTWU	Winn–Underwood–Gilliland design
	DISTL	Edmister simulation
	SCFRAC	Edmister simulation—complex columns
Multistage separation	RADFRAC	Two and three phases, with or without reaction
(Equilibrium-based simulation)	MULTIFRAC	Ditto—with interlinked column sections
	PETROFRAC	Ditto—for petroleum refining
	ABSBR	Absorbers and Strippers
	EXTRACT	Liquid–liquid extractors
(Mass transfer simulation)	RATEFRAC	Two phases—mass transfer model for staged or packed columns
Heat exchange	HEATER	Heater or cooler
	HEATX	Two-stream heat exchanger
	MHEATX	Multistream heat exchanger
Reactors	RSTOIC	Extent of reaction specified
	RYIELD	Reaction yields specified
	RGIBBS	Multiphase, chemical equilibrium
	REQUIL	Two-phase, chemical equilibrium
	RCSTR	Continuous-stirred tank reactor
	RPLUG	Plug-flow tubular reactor
Pumps, compressors, and turbines	PUMP	Pump or hydraulic turbine
	COMPR	Compressor or turbine
	MCOMPR	Multistage compressor or turbine
	VALVE	Control valves and pressure reducers
Pipeline	PIPE	Pressure drop in a pipe
	PIPELINE	Pressure drop in a pipe
Stream manipulators	MULT	Stream multiplier
	DUPL	Stream duplicator

(b) HYSYS

Mixers and splitters	**Mixer**	Stream mixer
	Tee	Stream splitter
Separators	**Component Splitter**	Component separator—two outlets
Flash drums	**Separator**	Multiple feeds, one vapor and one liquid product
	3-Phase Separator	Multiple feeds, one vapor and two liquid products
	Tank	Multiple feeds, one liquid product
Approximate distillation	**Shortcut Column**	Fenske–Underwood design
Multistage separation (Equilibrium-based simulation)	**Column**	Generic multiphase separation, including absorber, stripper, rectifier, distillation, liquid–liquid extraction. Additional strippers and pump-arounds can be added. All models support two or three phases and reactions. Physical property models are available for petroleum refining applications.
Heat exchange	**Cooler/Heater**	Cooler or heater
	Heat Exchanger	Two-stream heat exchanger
	Lng	Multistream heat exchanger
Reactors	**Conversion Reactor**	Extent of reaction specified
	Equilibrium Reactor	Equilibrium reaction
	Gibbs Reactor	Multiphase chemical equilibrium (stoichiometry not required)
	CSTR	Continuous-stirred tank reactor
	PFR	Plug-flow tubular reactor
Pumps, compressors, and turbines	**Pump**	Pump or hydraulic turbine
	Compressor	Compressor

Table 4.1 (*Continued*)

(b) HYSYS.Plant (*Continued*)

	Expander	Turbine
	Valve	Adiabatic valve
Pipeline	**Pipe Segment**	Single/multiphase piping with heat transfer

(c) CHEMCAD

Mixers and splitters	MIXE	Stream mixer
	DIVI	Stream splitter
Separators	CSEP	Component separator—multiple outlets
	CSEP	Component separator—two outlets
Flash drums	FLAS	Two-outlet flash drums
	LLVF	Three-outlet flash drums
	VALV	Valve
Approximate distillation	SHOR	Winn–Underwood–Gilliland design
Multistage separation	SCDS, TOWR	Two and three phases, with or without reaction
(Equilibrium-based and mass	TPLS	Ditto—with interlinked column sections
transfer simulation)	EXTR	Liquid–liquid extractors
Heat exchange	HTXR	Heater or cooler
	HTXR	Two-stream heat exchanger
	LNGH	Multistream heat exchanger
	FIRE	Fired heater
Reactors	REAC	Extent of reaction specified
	EREA	Two-phase, chemical equilibrium
	GIBS	Multiphase, chemical equilibrium
	KREA	Continuous-stirred tank reactor
	KREA	Plug-flow tubular reactor
Pumps, compressors, and turbines	PUMP	Pump or hydraulic turbine
	COMP, EXPN	Compressor or turbine
Pipeline	PIPE	Pressure drop in a pipe
Stream manipulators	SREF	Stream multiplier
	SREF	Stream duplicator

(d) PRO/II

Mixers and splitters	MIXER	Combines two or more streams
	SPLITTER	Splits a single feed or mixture of feeds into two or more streams
Flash drums	FLASH	Calculates the thermodynamic state of any stream when two variables are given by performing phase equilibrium calculations
Distillation column	COLUMN	Splits feed stream(s) into its components based on temperature and pressure
		By default, a distillation column includes a condenser and reboiler
Heat exchanger	HX	Heats or cools a single process stream, exchanges heat between two process streams or between a process and utility stream
	HXRIG	Rates a TEMA shell-and-tube heat exchanger, rigorously calculating heat transfer and pressure drop
	LNGHX	Exchanges heat between any number of hot and cold streams; identifies zone temperature crossovers and pinch points
Reactors	REACTOR	Models simultaneous reactions defined by fraction converted
	EQUILIBRIUM	Models one reaction defined as an approach to equilibrium temperature or as a fractional approach to chemical equilibrium
	GIBBS	Simulates a single-phase reactor at minimum Gibbs free energy
	CSTR	Simulates a continuously fed, perfectly mixed reactor; adiabatic, isothermal, or constant volume
	PLUG	Simulates a tubular reactor exhibiting plug-flow behavior (no axial mixing or heat transfer)
Pumps, compressors, and turbines	PUMP	Increases P of a stream
	COMPRESSOR	Compresses the feed stream according to specifications
	EXPANDER	Expands stream to the specified conditions and determines the work produced
	VALVE	Simulates the pressure drop
	PIPE	Simulates the pressure drop in a pipe

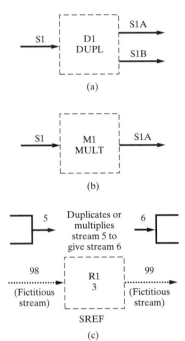

Figure 4.3 Stream manipulators: (a) duplication in ASPEN PLUS; (b) multiplication in ASPEN PLUS; (c) duplication and multiplication in CHEMCAD.

Each of the simulators has a similar syntax for specifying the topology of the simulation flowsheet. In ASPEN PLUS, for the simulation flowsheet in Figure 4.4a, the engineer draws the flowsheet in Figure 4.4b. Similarly, in HYSYS.Plant, CHEMCAD, and PRO/II, the simulation flowsheets are shown in Figures 4.4c–4.4e. Because the instructions for a new user of ASPEN PLUS are involved, new users are referred to the module *ASPEN → Principles of Flowsheet Simulation → Creating a Simulation Flowsheet* in the multimedia CD-ROM that accompanies this book. The ASPEN PLUS *Getting Started* manual is another good source of these instructions. For a new user of HYSYS.Plant, instructions are found in the multimedia CD-ROM by referring to the module *HYSYS → Principles of Flowsheet Simulation → Getting Started in HYSYS*.

When using the process simulators, it is important to recognize that, with some exceptions, most streams are comprised of chemical species that distribute within one or more solution phases that are assumed to be in phase equilibrium. The exceptions are streams involving so-called nonconventional components, which are usually solids such as coal, ash, and wood.

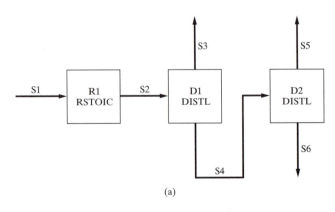

Figure 4.4 Acyclic flowsheet: (a) simulation flowsheet, (b) ASPEN PLUS flowsheet form; (c) HYSYS.Plant PFD; (d) CHEMCAD simulation flowsheet; (e) PRO/II simulation flowsheet.

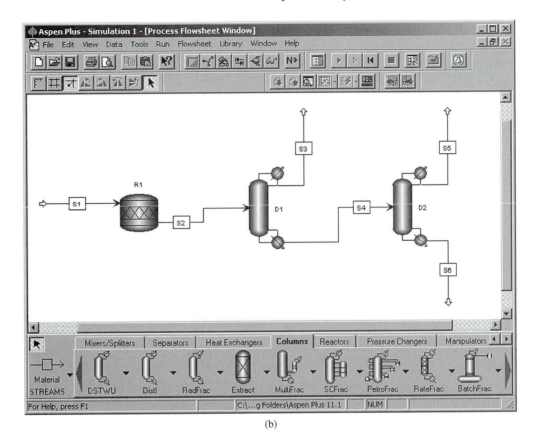

(b)

(c)

Figure 4.4 (*Continued*)

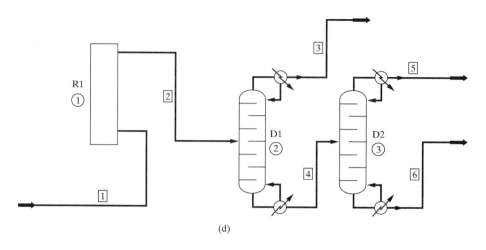

(d)

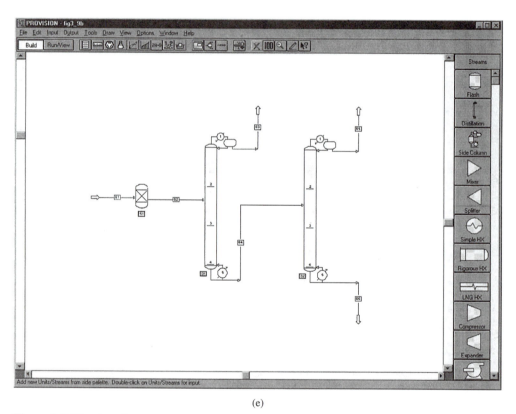

(e)

Figure 4.4 (*Continued*)

ASPEN PLUS has facilities for modeling these streams, but new users should not be concerned with these models until they have gained experience with streams in phase equilibrium.

For each stream in vapor–liquid equilibrium, there are $C + 2$ degrees of freedom, where C is the number of chemical species. These degrees of freedom can be satisfied by specifying C species flow rates (or $C - 1$ species mole fractions and the total flow rate) and two intensive variables such as the temperature, pressure, vapor fraction, or enthalpy. For example, when specifying the species flow rates for a stream and its pressure and temperature, all of the intensive properties are computed by solving the vapor–liquid equilibrium equations. These properties include the vapor fraction, enthalpy, and entropy. Alternatively, when the pressure and vapor fraction are specified, the remaining intensive properties are computed. Bubble-point and dew-point temperatures are computed by specifying the vapor fraction to be zero and unity, respectively.

Since phase equilibrium is a major segment of courses in thermodynamics, it does not seem appropriate to devote space in a chapter on process simulation to this subject. Yet it is important, when learning to use the process simulators, to understand how they apply the theory of phase equilibrium in modeling streams as well as so-called *flash vessels*, that is, vapor–liquid separators. On the multimedia CD-ROM, in the module *ASPEN → Separators → Phase Equilibria and Flash* and in the module *HYSYS → Separations → Flash*, concepts on phase equilibria and flash separations are reviewed. In addition, and perhaps of more use to many readers, these modules present the solution to a simulation of a flash vessel using ASPEN PLUS and HYSYS.Plant. In so doing, they show how to use *heat streams* in a process simulation. Note that the multimedia CD-ROM gives audio tutorials for completing the ASPEN PLUS input forms and examining the results, and for completing the HYSYS.Plant simulation flowsheet, its *Workbook* specsheets, and associated inputs. Readers using HYSYS.Plant should refer to the module *HYSYS → Principles of Flowsheet Simulation → Getting Started in HYSYS → Units Catalog* for a list of the unit models, with links to modules that provide detailed information.

Unit Subroutines

Table 4.1 lists the unit subroutines (or blocks, or models) in each of the four simulators. Several of the subroutines are referred to in the sections that follow, with descriptions of many on the multimedia CD-ROM and detailed descriptions in the user manuals and *Help* screens.

Degrees of Freedom

A degrees-of-freedom analysis (Smith, 1963; Rudd and Watson, 1968; Myers and Seider, 1976) is incorporated in the development of each subroutine (or block, or model) that simulates a process unit. These subroutines solve sets of $N_{\text{Equations}}$ involving $N_{\text{Variables}}$, where $N_{\text{Equations}} < N_{\text{Variables}}$. Thus, there are $N_{\text{D}} = N_{\text{Variables}} - N_{\text{Equations}}$ degrees of freedom, or input (decision) variables. Most subroutines are written for known values of the input stream variables, although HYSYS.Plant permits specification of a blend of input and output stream variables, or output stream variables entirely.

EXAMPLE 4.1

Consider the cooler in Figure 4.5, in which the binary stream S1, containing benzene and toluene at a vapor fraction of $\phi_1 = 0.5$, is condensed by removing heat, Q. Carry out a degrees-of-freedom analysis.

SOLUTION

At steady state, the material and energy balances are

$$F_1 x_{\text{B1}} = F_2 x_{\text{B2}} \tag{4.1}$$

$$F_1 x_{\text{T1}} = F_2 x_{\text{T2}} \tag{4.2}$$

$$F_1 h_1 + Q = F_2 h_2 \tag{4.3}$$

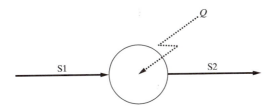

Figure 4.5 Schematic of a cooler.

S1 →

Mixer → S3

S2 →

Figure 4.6 Benzene–toluene mixer.

where F_i is the molar flow rate of stream i, x_{ji} is the mole fraction of species j in stream i, and h_i is the enthalpy of stream i, which can be expressed as

$$h_i = h_i[P_i, \phi_i, \underline{x}_i] \qquad i = 1, 2 \tag{4.4}$$

and $x_{Ti} = 1 - x_{Bi}$, $i = 1, 2$. Note that, in this case, the pressure, P, and vapor fraction, ϕ, accompany the mole fractions as the $C + 2$ intensive variables that provide the enthalpy and other intensive variables of each stream. For this model, $N_{\text{Equations}} = 7$, and $N_{\text{Variables}} = 13$ (F_i, h_i, P_i, ϕ_i, x_{Bi}, and x_{Ti}, $i = 1, 2$, and Q). Hence, $N_D = 13 - 7 = 6$, and one set of specifications is comprised of the variables of the feed stream (F_1, P_1, ϕ_1, x_{B1}) and P_2 and Q. In the process simulators, so-called heater and cooler subroutines are provided to solve the equations for specifications like these. ∎

EXAMPLE 4.2

Consider the mixer in Figure 4.6, in which binary streams S1 and S2, also containing benzene and toluene, are mixed isobarically to form stream S3. Carry out a degrees-of-freedom analysis.

SOLUTION

At steady state, its material and energy balances are

$$F_1 x_{B1} + F_2 x_{B2} = F_3 x_{B3} \tag{4.5}$$
$$F_1 x_{T1} + F_2 x_{T2} = F_3 x_{T3} \tag{4.6}$$
$$F_1 h_1 + F_2 h_2 = F_3 h_3 \tag{4.7}$$

Using temperature and pressure as the intensive variables, Eq. (4.4) becomes

$$h_i = h_i\{T_i, P, \underline{x}_i\} \qquad i = 1, 2, 3 \tag{4.8}$$

and $x_{Ti} = 1 - x_{Bi}$, $i = 1, 2, 3$. For this model, $N_{\text{Equations}} = 9$, and $N_{\text{Variables}} = 16$ (F_i, h_i, T_i, x_{Bi}, and x_{Ti}, $i = 1, 2, 3$, and P). Hence, $N_D = 16 - 9 = 7$, and a common set of specifications is comprised of the variables of the feed streams (F_1, x_{B1}, T_1, and F_2, x_{B2}, T_2) and P. ∎

Consider the information flows between a unit subroutine and the stream and equipment vectors in Figure 4.7 for the FLASH2 subroutine of ASPEN PLUS. These are typical of the subroutines (or models, or blocks) in all of the simulators. In ASPEN PLUS, FOR-TRAN subroutines that model the process units have access to vectors containing the *inlet* (*feed*) and *outlet* (*product*) *stream variables*, and *equipment parameters*, respectively. The equipment parameter vectors are created as the ASPEN PLUS forms described on the multimedia CD-ROM in the module *ASPEN → Separators → Phase Equilibria and Flash → Flash vessels → FLASH2* are completed by the user; in this case, specifications for the temperature and pressure are entered. Assume that the process consists only of a flash vessel, modeled by the FLASH2 subroutine. Then, the variables for the stream FEED are entered into its vector. Estimates of the enthalpy, vapor and liquid fractions, entropy, and density are computed by the property estimation system. After all of the forms have been completed, an ASPEN PLUS program is generated by ASPEN PLUS. This program is a compact representation of the specifications provided on the forms. It has many paragraphs, two of which are shown in Figure 4.7 (see the multimedia CD-ROM for the entire program). Next, ASPEN PLUS interprets the program, generates a calculation sequence (providing the order in which the simulation units are computed), and calls the appropriate subroutine (model) for each simulation unit.

During execution of a unit subroutine, the stream vectors and equipment parameters are accessed, from a so-called B vector in ASPEN PLUS, and changes are recorded when new

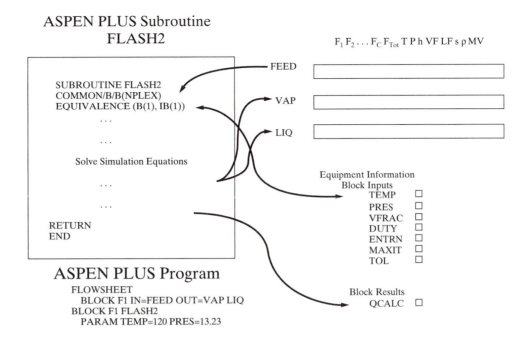

ASPEN PLUS Subroutine
FLASH2

$F_1 F_2 \ldots F_C F_{Tot}$ T P h VF LF s ρ MV

FEED

SUBROUTINE FLASH2
COMMON/B/B(NPLEX)
EQUIVALENCE (B(1), IB(1))
. . .

VAP

. . .

LIQ

Solve Simulation Equations

. . .

. . .

RETURN
END

Equipment Information
Block Inputs
TEMP ☐
PRES ☐
VFRAC ☐
DUTY ☐
ENTRN ☐
MAXIT ☐
TOL ☐

ASPEN PLUS Program

FLOWSHEET
    BLOCK F1 IN=FEED OUT=VAP LIQ
    BLOCK F1 FLASH2
        PARAM TEMP=120 PRES=13.23

Block Results
QCALC ☐

Subroutine Calls

CALL FLASH2

Figure 4.7 ASPEN PLUS unit subroutine—information transfer.

values are computed as the equations are solved. In most of the simulators, the unit subroutines take the variables of the feed streams as input and compute the variables of the product streams; most equipment parameters are specified, but some are computed and stored.

In the schematic of the FLASH2 subroutine in Figure 4.7, on the second line, the B vector, which contains the stream vectors and equipment parameters for all of the streams and simulation units, is referred to in the B common storage. When simulating the flash vessel, F1, the stream variables are taken from the FEED vector and two of the equipment parameters, TEMP, PRES, VFRAC, and DUTY, are taken from the subroutine inputs. As the flash equations are solved, the stream variables are stored in the VAP and LIQ vectors and the heat duty is stored as a parameter, QCALC.

The parameters to be entered by the user for each simulation unit are summarized in connection with its input form, and the associated *Help* information, as well as in Volume 1 of the *ASPEN PLUS Reference Manual*. Note that many default values are provided by ASPEN PLUS that are replaced easily by the user as desired. Upon completion of all of the unit computations, the latest values of all of the stream variables and equipment parameters are displayed on the forms or placed in a report file for printing.

For each unit, the vector of parameters computed by a unit subroutine is saved for display and printing, and to initiate iterative computations for subsequent executions of the subroutine.

In HYSYS.Plant, the models are programmed to reverse the information flow when appropriate, that is, to accept values for the variables of the product streams and to compute the variables of the feed streams. HYSYS.Plant implements the so-called *bidirectional information flow*, as described next.

Bidirectional Information Flow

In nearly all of the flowsheet simulators, the material and energy balances for the process units are solved given specifications for the inlet streams and the equipment parameters, along with selected variables of the outlet streams (e.g., temperatures and pressures). The unknown variables to be computed are usually those of the outlet streams (typically, the flow rates and compositions). The HYSYS.Plant simulator is a notable exception in that most combinations of specifications are permitted for each simulation model. With this flexibility, HYSYS.Plant can implement a *reverse information flow*, in which specifications are provided for the product streams and the unknown variables of the inlet streams are computed. More commonly, HYSYS.Plant implements a *bidirectional information flow*, involving the calculation of the unknown variables associated with the inlet and outlet streams. Whenever a stream variable is altered, the adjacent process units are recomputed. This causes the information to flow in parallel to the material streams, when a unit downstream is recomputed, or opposite to the material streams when a unit upstream is recomputed.

EXAMPLE 4.3   *(Example 4.1 Revisited)*

For the cooler in Figure 4.5, it is desired to specify the vapor fraction of the effluent stream, ϕ_2, the heat duty, Q, and the pressure drop. Can this be accomplished with bidirectional information flow?

SOLUTION

In the HYSYS.Plant simulator, bidirectional information flow is utilized to compute the vapor fraction of the feed stream, ϕ_1. This cannot be accomplished directly in the other process simulators, where, instead, an iteration loop is created in which a guess is provided for ϕ_1 and iterations are carried out until the specified value of ϕ_2 is obtained, as discussed in the next subsection. Note that for the heater or cooler model in most simulators, the vapor fraction can be specified for both streams and the heat duty computed. ∎

Control Blocks—Design Specifications

Occasionally, the need arises to provide specifications for variables or parameters that are not permitted by a unit subroutine (or block, or model). To accomplish this, all of the simulators provide a facility for iterative adjustment of the variables and parameters that are permitted to be specified so as to achieve the desired specifications. Guesses are made for the so-called manipulated variables. Then, the simulation calculations are performed and a *control* subroutine compares the calculated values with the desired specifications, which may be called *set points*. When significant differences, or errors, are detected, the control subroutine prepares new guesses, using numerical methods, and transfers control to repeat the simulation calculations. Since the procedure is analogous to that performed by feedback controllers in a chemical plant (which are designed to reject disturbances during dynamic operation), it is common to refer to these convergence subroutines as *feedback control* subroutines (Henley and Rosen, 1969).

In the HYSYS.Plant simulator, this is accomplished by the **Adjust** operation, in CHEMCAD by the CONT subroutine, and in PRO/II by the CONTROLLER subroutine. In ASPEN PLUS, the equivalent is accomplished with so-called *design specifications*. The latter terminology is intended to draw a distinction between simulation calculations, where the equipment parameters and feed stream variables are specified, and design calculations, where the desired properties of the product stream (e.g., temperature, composition, flow rate) are specified and the equipment parameters (area, reflux ratio, etc.) and feed stream variables are calculated. In HYSYS.Plant, the **Adjust** operation is used to adjust the equipment parameters and some feed stream variables to meet the specifications of the stream variables. Furthermore, the **Set** object is used to adjust the value of an attribute of a stream in proportion to that of another stream.

For assistance in the use of the **Adjust** and **Set** objects, the reader is referred to the module *HYSYS → Principles of Flowsheet Simulation → Getting Started in HYSYS → Convergence of Simulation* on the multimedia CD-ROM that accompanies this text. As was discussed in the subsection on bidirectional information flow, for *all* of its subroutines, HYSYS.Plant provides a *bidirectional information flow*, that is, when product stream variables are specified, the subroutines calculate most of the unknown inlet-stream variables. In CHEMCAD, a control unit, with one inlet stream and one outlet stream (which may be identical to the inlet stream), is placed into the simulation flowsheet using the CONT subroutine. The parameters of the control unit are specified so as to achieve the desired value of a stream variable (or an expression involving stream variables) or an equipment parameter (or an expression involving equipment parameters) by manipulating an equipment parameter or a stream variable. This is the *feed-backward mode*, which requires that the control unit be placed downstream of the units being simulated. The CONT subroutine also has a *feed-forward mode*.

As an example of using a feedback control subroutine in ASPEN PLUS, return to the benzene-toluene mixer in Example 4.2.

EXAMPLE 4.4   *(Example 4.2 Revisited)*

For an equimolar feed stream, S1, at 1,000 lbmol/hr and 100°F, the flow rate of a toluene stream, S2, at 50°F is adjusted to achieve a desired temperature of the mixer effluent (e.g., 85°F), as shown in Figure 4.8a. Convergence units for feedback control (design specifications) are shown on simulation flowsheets as *dotted circles* connected to streams and simulation units by *dotted arcs*. The arcs represent the information flow of stream variables to the control unit and information flow of adjusted equipment parameters to simulation units. Note that the control units of most simulators can adjust the flow rates of the streams. After the calculations by the MIXER subroutine are completed, the control subroutine samples the effluent temperature. It adjusts the flow rate of stream S2 when the specified temperature is not achieved and transfers to the MIXER subroutine to repeat the mixing calculations. This cycle is repeated until the convergence criteria are satisfied or the maximum number of iterations is exceeded.

Instructions to create a design specification using ASPEN PLUS for the mixing unit M1 are provided in the module *ASPEN → Principles of Flowsheet Simulation → Control Blocks—Design Specifications* on the multimedia CD-ROM that accompanies this text.

Based on the input specifications in this module, ASPEN PLUS generates the program in the module, and the simulator reports that

```
SEQUENCE USED WAS: $OLVER01 M1 (RETURN $OLVER01)
```

The iterative procedure used by $OLVER01 is initiated in the manner shown in Figure 4.9a. As indicated above, an initial guess for the manipulated variable (800 lbmol/hr), and the minimum and maximum values of the manipulated variable (0 and 2,000 lbmol/hr), are provided. Then, $OLVER01 adjusts the manipulated variable, using one of several convergence algorithms, until the convergence tolerances are satisfied with $F_2 = 402.3$ lbmol/hr. When the upper or lower bound is reached, a message is provided that convergence has not been achieved. For this example, the secant method was used to achieve convergence, with the iteration history displayed in Figure 4.9b.

For the benzene–toluene mixer, Figure 4.8b shows the HYSYS.Plant simulation flowsheet in which the **Adjust** operation manipulates the flow rate of stream S2 to achieve the desired temperature. ∎

Calculation Order

In most process simulators, the units are computed (simulated) one at a time. The calculation order is automatically computed to be consistent with the flow of information in the simulation flowsheet, where the information flow depends on the specifications for the chemical process. Usually, the variables of the process feed streams are specified and information flows parallel to the material flows. In other words, the calculations proceed from unit to unit, beginning with units for which all of the feed streams have been specified. For the

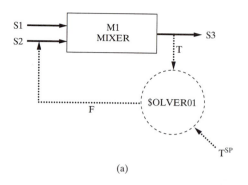

(a)

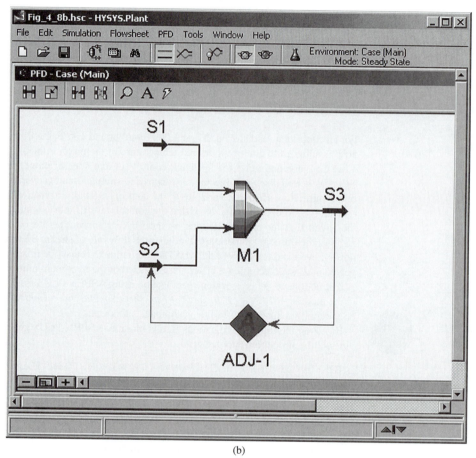

(b)

Figure 4.8 Feedback control—design specifications for the benzene–toluene mixer: (a) ASPEN PLUS blocks; (b) HYSYS.Plant icons.

flowsheet in Figure 4.4, the units are calculated in the order R1, D1, and D2, that is, starting from the feed end of the process. Before initiating the computations, ASPEN PLUS is provided with data for the variables of the feed stream, S1, and equipment parameters for the three units. The calculation orders for HYSYS.Plant, CHEMCAD, and PRO/II are the same. For HYSYS.Plant, the simulation flowsheet is shown in Figure 4.4c, using the **Conversion Reactor, Column**, and **Column** models, respectively. Similarly, the CHEMCAD simulation flowsheet is shown in Figure 4.4d, using the REAC, TOWR, and TOWR subroutines. Finally, the PRO/II simulation flowsheet is shown in Figure 4.4e, using the REACTOR, COLUMN, and COLUMN subroutines.

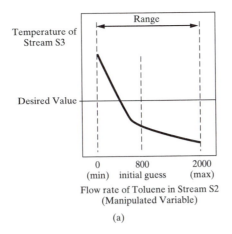

(a)

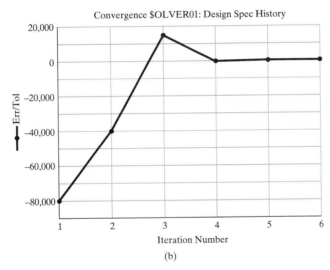

Figure 4.9 Graphical solution of the mixer control problem: (a) specifications for the manipulated variable; (b) ASPEN PLUS iteration history using the secant method.

(b)

After the subroutine (or model, or block) computations are completed, all of the stream variables and equipment parameters may be displayed or printed, as illustrated in the report files for ASPEN PLUS in the multimedia CD-ROM under the module *ASPEN → Principles of Flowsheet Simulation → Interpretation of Input and Output: Program Output.*

Recycle

Flowsheets are rarely acyclic, as in Figure 4.4. In process synthesis, most distributions of chemicals involve recycle streams as in Figure 4.1. For the simpler distributions, where the fractional conversions or the extents of reaction are known, the split fractions are specified, and no purge streams exist, as in the vinyl-chloride process (Figures 3.8 and 3.19), the flow rates of the species in the recycle streams can be calculated directly (without iteration).

When the reaction operations involve reversible reactions or competing reactions, the split fractions of the species leaving the separators are complex functions of the operating conditions such as the temperatures, pressures, and reflux ratios, and purge streams exist, then iterative calculations are necessary. In these cases, the simulation flowsheets usually contain information recycle loops, that is, cycles for which too few stream variables are known to permit the equations for each unit to be solved independently. For these processes, a solution technique is needed to solve the equations for all of the units in an information recycle loop.

One solution technique is to *tear* one stream in the recycle loop, that is, to guess the variables of that stream (Henley and Rosen, 1969; Myers and Seider, 1976; Westerberg et al., 1979). Based on *tear stream* guesses, information is passed from unit to unit until new values of the variables of the tear stream are computed. These new values are used to repeat the calculations until the convergence tolerances are satisfied. The variables of the tear streams are often referred to as *tear variables*.

In process simulators, recycle convergence units are inserted into the tear stream. These units can be represented by dashed rectangles, as illustrated in Figures 4.2a and 4.10a. In so doing, an additional stream vector is created. Convergence units use convergence subroutines to compare the newly computed variables (in the feed stream to the convergence unit) with guessed values (in the product stream from the convergence unit) and to compute new guess values when the two streams are not identical to within convergence tolerances.

In most process simulators, the convergence units are positioned automatically. Consider the flowsheet in Figure 4.10a. The process feed is stream S10, which the user would specify. Unit H1 could then be calculated. The set of units, M1, R1, D1, and D2 constitutes a recycle loop. A convergence unit must be placed somewhere in this loop. In a recycle loop, calculations begin with the streams leaving the convergence unit. Each of the units in the loop is then computed, returning to the convergence unit, where convergence is checked. When convergence is not achieved, the simulator repeats the loop calculations. Upon satisfying the convergence criteria, control is transferred to the unit following the recycle loop in the calculation order. In Figure 4.10a, that unit is D3. ASPEN PLUS names the recycle convergence units $OLVER01, $OLVER02, . . . , in sequence. The names of the convergence units are reported in the calculation sequence output, which is illustrated below for the flowsheet in Figure 4.10a:

```
SEQUENCE USED WAS: H1 $OLVER01 M1 R1 D1 D2 (RETURN $OLVER01) D3
```

Note that Figure 4.10a shows the simulation flowsheet with the recycle convergence unit, $OLVER01, inserted in stream S6. Here, S6* denotes the vector of guesses for the stream variables of the tear stream, and S6 denotes the vector of stream variables after the units in the recycle loop have been simulated. Although the ASPEN PLUS simulation flowsheet in Figure 4.10b does not show $OLVER01 and S6*, the user should recognize that they are implemented. The user can supply guesses for S6*, or they are supplied by the simulator.

All of the recycle convergence subroutines in simulators implement the successive substitution (direct iteration) and the bounded Wegstein methods of convergence, as well as more sophisticated methods for highly nonlinear systems where the successive substitution or Wegstein methods may fail or may be very inefficient. These other methods include the Newton–Raphson method, Broyden's quasi-Newton method, and the dominant-eigenvalue method (Wegstein, 1958; Henley and Rosen, 1969; Myers and Seider, 1976; Westerberg et al., 1979). Each of these five methods determines whether the relative difference between the guessed variables (e.g., for S6* in Figure 4.10a) and calculated variables (e.g., stream S6 in Figure 4.10a) are all less than a prespecified tolerance. If not, the convergence subroutine computes new guesses for its output stream variables and iterates until the loop is converged.

Consider the flowsheet in Figure 4.10. The variables for streams S1 and S10 are specified and the recycle stream (S6) has been selected as the tear stream. Let x^* be the value of a particular variable (element) of stream vector S6*, the stream output of convergence unit $OLVER01, and let $f\{x^*\}$ be the corresponding value for the corresponding calculated variable in stream S6, which enters $OLVER01, as determined by taking x^* and calculating the units M1, R1, D1, and D2 in that order. The value of x to initiate the next iteration is determined by $OLVER01 using one of the five mentioned convergence methods. When the method of successive substitutions is specified, the new guess for x is simply made equal to $f\{x^*\}$. A sequence of iterations may exhibit the behavior shown in Figure 4.11a. After a

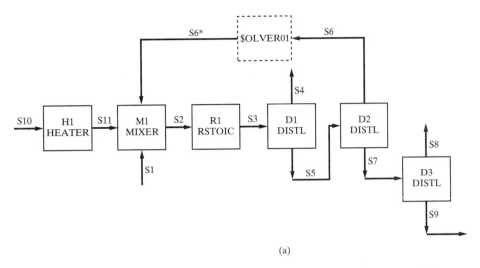

(a)

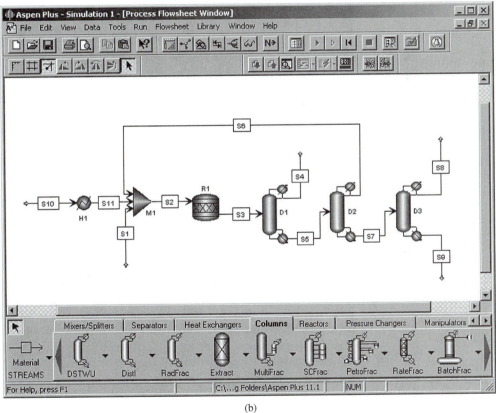

(b)

Figure 4.10 Process with recycle: (a) simulation flowsheet; (b) ASPEN PLUS simulation flowsheet.

number of iterations, the locus of iterates intersects the 45° line, giving the converged value of x in stream S6. When the slope of the locus of iterates $(f\{x\}, x)$ is close to unity (for processes with high recycle ratios), a large number of iterations may be required before convergence occurs.

Wegstein's method can be employed to accelerate convergence when the method of successive substitutions requires a large number of iterations. As shown in Figure 4.11b, the previous two iterates of $f\{x^*\}$ and x^* are extrapolated linearly to obtain the next value of x as

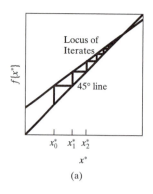

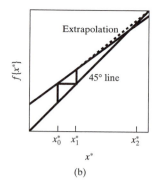

Figure 4.11 Convergence of a recycle loop: (a) successive substitution method; (b) Wegstein's method.

(a)

(b)

the point of intersection with the 45° line. The equation for this straight-line extrapolation is derived easily as

$$x = \left(\frac{s}{s-1}\right)x^* - \left(\frac{1}{s-1}\right)f\{x^*\} \tag{4.9}$$

where s is the slope of the extrapolated line. A more convenient form of Eq. (4.9) uses a weighting function defined by $q = s/(s-1)$, giving

$$x = qx^* + (1-q)f\{x^*\} \tag{4.10}$$

Thus, weights q and $1 - q$ are applied, respectively, to x^* and $f\{x^*\}$. Equation (4.10), with q defined by the slope, is usually employed when the slope is less than 1, such that $q < 0$. Typically, q is bounded between -20 and 0 to ensure stability and a reasonable rate of convergence. Wegstein's method reduces to the method of successive substitutions, $x = f\{x^*\}$, when $q = 0$.

When the tear stream is determined automatically by the process simulator, it is possible to override it. For example, ASPEN PLUS selects stream S2, but it can be replaced with stream S6. To do so, select *Convergence* from the *Data* pulldown menu. Then select *Tear*, which produces the *Tear Streams Specifications* form. Enter S6 as the tear stream. Other simulators permit the override in a similar manner.

Figure 4.12a shows a simulation flowsheet with two recycle loops for ASPEN PLUS. Flowsheets for CHEMCAD and PRO/II are identical except for the subroutine (or model) names for the units. Note that no recycle convergence units are shown. This is typical of the simulation flowsheets displayed by most process simulators. The flowsheet for HYSYS.Plant is an exception because the recycle convergence unit(s) are positioned by the user and appear explicitly in the flowsheet. For ASPEN PLUS, CHEMCAD, and PRO/II, to complete the simulation flowsheet, either one or two convergence units are inserted, as described below. Note that a single convergence unit suffices because stream S6 is common to both loops, as illustrated in Figure 4.12b. Stream S6 is torn into two streams, S6 and S6*, with guesses provided for the variables in S6*. Since no units are outside of the loops, all units are involved in the iterative loop calculations. The calculation sequence is

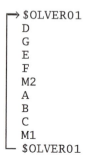

In ASPEN PLUS, the calculation sequence output is

```
SEQUENCE USED WAS:
  $OLVER01 D G E F M2 A B C M1
  (RETURN $OLVER01)
```

Note that this is the calculation sequence prepared by ASPEN PLUS. Alternatively, when the user prefers to provide guesses for the two recycle streams, S5 and S10, the simulation flowsheet in Figure 4.12c is utilized. To accomplish this in ASPEN PLUS, select *Convergence* from the *Data* pulldown menu. Then, select *Tear* which produces the *Tear Streams*

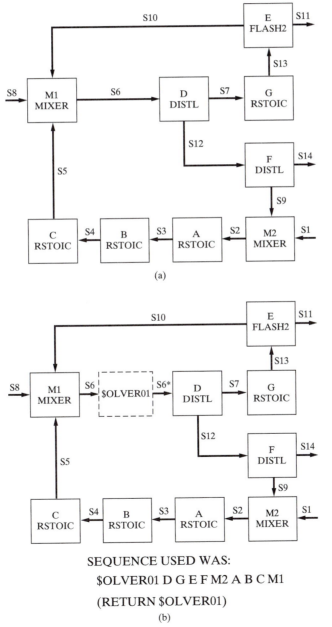

(a)

(b)

SEQUENCE USED WAS:

$OLVER01 D G E F M2 A B C M1

(RETURN $OLVER01)

Figure 4.12 Nested recycle loops: (a) Incomplete simulation flowsheet; (b) simulation flowsheet with a single tear stream and a single recycle convergence unit; (c) simulation flowsheet with two tear streams and a single recycle convergence unit; (d) simulation flowsheet with two tear streams and two recycle convergence units.

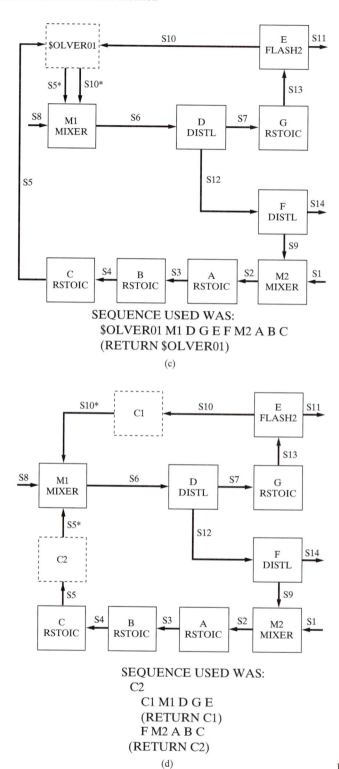

SEQUENCE USED WAS:
$OLVER01 M1 D G E F M2 A B C
(RETURN $OLVER01)

(c)

SEQUENCE USED WAS:
C2
C1 M1 D G E
(RETURN C1)
F M2 A B C
(RETURN C2)

(d)

Figure 4.12 (*Continued*)

Specifications form. Enter S5 and S10 as the tear streams. Then the calculation sequence output becomes

```
SEQUENCE USED WAS:
  $OLVER01 M1  D  G  E  F  M2  A  B  C
  (RETURN $OLVER01)
```

In this case, a single convergence unit, $OLVER01, checks for convergence and adjusts the guess values for streams S5 and S10 simultaneously.

Yet another sequence, shown in Figure 4.12d, can be programmed for ASPEN PLUS, with instructions for completing the ASPEN PLUS forms provided in in the module *ASPEN → Principles of Flowsheet Simulation → Recycle → Multiple Recycle Loops* on the multimedia CD-ROM that accompanies this text. This results in the calculation sequence output.

```
SEQUENCE USED WAS:
  C2
    C1 M1  D  G  E
    (RETURN C1)
    F  M2  A  B  C
  (RETURN C2)
```

In this sequence, the internal loop, C1, is converged during every iteration of the external loop, C2 (which includes C1). This may be efficient when the units outside C1 require extensive computations.

A more complex flowsheet, which contains three recycle loops, is shown in Figure 4.13a. Two calculation sequences are illustrated in Figure 4.13b and 4.13c. These involve the minimum number of tear streams, S5 and S8, and result in the following output from ASPEN PLUS:

Option 1

```
SEQUENCE USED WAS:
  CONV2  F  G
    CONV1  D  A  B  C
    (RETURN CONV1)
    E
  (RETURN CONV2)
```

Option 2

```
SEQUENCE USED WAS:
  CONV3  F  G  D  A  B  C  E
  (RETURN CONV3)
```

In both options, guesses are provided for the variables in streams S5 and S8. In option 1, the internal loop, CONV1, is converged during every iteration of the external loop, CONV2. In option 2, both loops are converged simultaneously. Note that the minimum number of tear streams may not provide for the most rapid convergence. An alternative solution procedure for this flowsheet involves three tear streams, for example, S7, S9, and S11 with one convergence unit.

When using ASPEN PLUS, the details of the convergence forms and the CONVERGENCE paragraph generated can be found in Chapter 17, Volume 2, of the *ASPEN PLUS User Guide*. See also the modules in *ASPEN → Principles of Flowsheet Simulation → Recycle* on the multimedia CD-ROM that accompanies this book. For HYSYS.Plant, the user can consult the modules under *HYSYS → Principles of Flowsheet Simulation → Getting Started in HYSYS → Convergence of Simulation → Recycle* on the CD-ROM, and for CHEMCAD and PRO/II, their user manuals.

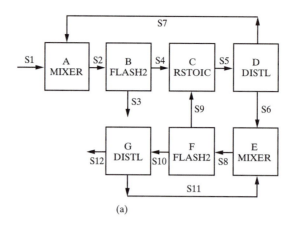

(a)

Option 1

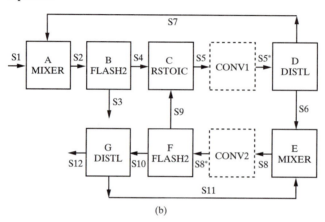

(b)

Option 2

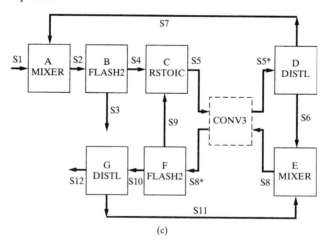

(c)

Figure 4.13 Three recycle loops: (a) incomplete simulation flowsheet; (b) simulation flowsheet with two tear streams and two recycle convergence units; (c) simulation flowsheet with two tear streams and one recycle convergence unit.

Recycle Convergence Methods

In the previous subsection, the successive substitution and Wegstein methods were introduced as the two methods most commonly implemented in recycle convergence units. Other methods, such as the Newton–Raphson method, Broyden's quasi-Newton method, and the dominant-eigenvalue method, are candidates as well, especially when the equations being solved are highly nonlinear and interdependent. In this subsection, the principal features of all five methods are compared.

For the recycle convergence unit in Figure 4.14, let

$$\underline{y} = \underline{f}\{\underline{x}^*\} - \underline{x}^* \tag{4.11}$$

where $\underline{x}^*$ is the vector of guesses for n recycle (tear) variables and $\underline{f}\{\underline{x}^*\}$ is the vector of the recycle variables computed from the guesses after one pass through the simulation units in the recycle loop. Clearly, the objective of the convergence unit is to adjust $\underline{x}^*$ so as to drive $\underline{y}$ toward zero.

Newton–Raphson Method

The Newton–Raphson second-order method can be written as

$$\underline{\underline{J}}\{\underline{x}^*\} \, \underline{\Delta x} = -\underline{y}\{\underline{x}^*\} \tag{4.12}$$

where $\underline{\Delta x} = \underline{x} - \underline{x}^*$. Substituting and rearranging, the new values of the recycle variables, $\underline{x}$ are

$$\underline{x} = \underline{x}^* - \underline{\underline{J}}^{-1}\{\underline{x}^*\}\underline{y}\{\underline{x}^*\} \tag{4.13}$$

In these equations, the Jacobian matrix:

$$\underline{\underline{J}}\{\underline{x}^*\} = \begin{bmatrix} \dfrac{\partial y_1}{\partial x_1} & \dfrac{\partial y_1}{\partial x_2} & \cdots & \dfrac{\partial y_1}{\partial x_n} \\[2ex] \dfrac{\partial y_2}{\partial x_1} & \dfrac{\partial y_2}{\partial x_2} & \cdots & \dfrac{\partial y_2}{\partial x_n} \\[2ex] \vdots & \vdots & & \vdots \\[2ex] \dfrac{\partial y_n}{\partial x_1} & \dfrac{\partial y_n}{\partial x_2} & \cdots & \dfrac{\partial y_n}{\partial x_n} \end{bmatrix}_{\underline{x}^*} \tag{4.14}$$

is evaluated at $\underline{x}^*$.

In each iteration of the Newton–Raphson method, when the guesses are close to the true values, the length of the error vector, $\| \underline{y} \|$, is the square of its length after the previous iteration; that is, when the length of the initial error vector is 0.1, the subsequent error vectors are reduced to 0.01, 10^{-4}, 10^{-8}, ... However, this rapid rate of convergence requires that n^2 partial derivatives be evaluated at $\underline{x}^*$. Since most recycle loops involve many process units, each involving many equations, the chain rule for partial differentiation cannot be implemented easily. Consequently, the partial derivatives are evaluated by numerical perturbation; that is, each guess, x_i, $i = 1, \ldots, n$, is perturbed, one at a time. For each

Figure 4.14 Recycle convergence unit.

perturbation, δx_i, $i = 1, \ldots, n$, a pass through the recycle loop is required to give y_j^p, $j = 1, \ldots, n$. Then the partial derivatives in the ith row are computed by difference:

$$\frac{\partial y_j}{\partial x_i} \cong \frac{y_j^p - y_j}{\delta x_i} \qquad j = 1, \ldots, n \tag{4.15}$$

This requires $n + 1$ passes through the recycle loop to complete the Jacobian matrix for just one iteration of the Newton–Raphson method; that is, for $n = 10$, eleven passes are necessary, usually involving far too many computations to be competitive.

Alternatively, so-called *secant methods* can be used to approximate the Jacobian matrix with far less effort (Westerberg et al., 1979). These provide a *superlinear* rate of convergence; that is, they reduce the errors less rapidly than the Newton–Raphson method, but more rapidly than the method of successive substitutions, which has a *linear* rate of convergence (i.e., the length of the error vector is reduced from 0.1, 0.01, 10^{-3}, 10^{-4}, 10^{-5}, . . .). These methods are also referred to as *quasi-Newton methods*, with Broyden's method being the most popular.

Method of Successive Substitutions

To compare the method of successive substitutions with the Newton–Raphson method, or the quasi-Newton methods, the former can be written:

$$\underline{x} = \underline{f}\{\underline{x}^*\} \tag{4.16}$$

Subtracting $\underline{x}^*$ from both sides:

$$\underline{x} - \underline{x}^* = \underline{f}\{\underline{x}^*\} - \underline{x}^* \tag{4.17}$$

or

$$-\underline{\underline{I}}\,\Delta x = -\underline{y}\{\underline{x}^*\} \tag{4.18}$$

Note that the Jacobian matrix is replaced by the identity matrix, and hence, each element of the $\underline{\Delta}x$ vector is influenced only by its corresponding element in the $\underline{y}$ vector. No interactions from the other elements of the $\underline{y}$ vector influence Δx_i.

Wegstein's Method

Rewriting Eq. (4.9) for n-dimensional vectors,

$$\underline{x} = \begin{bmatrix} \dfrac{s_1}{s_1 - 1} & & \\ & \ddots & \\ & & \dfrac{s_n}{s_n - 1} \end{bmatrix} \underline{x}^* - \begin{bmatrix} \dfrac{1}{s_1 - 1} & & \\ & \ddots & \\ & & \dfrac{1}{s_n - 1} \end{bmatrix} \underline{f}\{\underline{x}^*\} \tag{4.19}$$

and subtracting $\underline{x}^*$ from both sides,

$$\underline{x} - \underline{x}^* = \begin{bmatrix} \dfrac{1}{1 - s_1} & & \\ & \ddots & \\ & & \dfrac{1}{1 - s_n} \end{bmatrix} \underline{y}\{\underline{x}^*\} \tag{4.20}$$

or

$$-\underline{\underline{A}}\,\Delta x = -\underline{y}\{\underline{x}^*\} \tag{4.21}$$

where A is a diagonal matrix with the elements $1 - s_i$, $i = 1, \ldots, n$. Although Wegstein's method provides a superlinear rate of convergence, note that like the method of successive substitutions, no interactions occur.

Dominant-Eigenvalue Method

In the dominant-eigenvalue method, the largest eigenvalue of the Jacobian matrix is estimated every third or fourth iteration and used in place of s_i in Eq. (4.20) to accelerate the method of successive substitutions, which is applied at the other iterations (Orbach and Crowe, 1971; Crowe and Nishio, 1975).

Flash with Recycle Problem

To master the concepts of recycle analysis, it is recommended that the reader solve several of the exercises at the end of the chapter. Of these, the so-called flash with recycle problem (Exercise 4.1a) should be tackled first. Although it involves just one recycle loop, it demonstrates a very important principle. See if you can identify it!

Consider the simple process in Figure 4.15. For the three cases, compare and discuss the flow rates and compositions of the product streams. The model for the flash vessel is presented in the modules under *HYSYS → Separations → Flash* and *ASPEN → Separators → Phase Equilibria and Flash → Flash Vessels* on the multimedia CD-ROM, which include a narrated video of an industrial flash vessel, and the models for the mixer and splitter are rather straightforward. The pump model deserves special attention and is discussed in the modules under *HYSYS → Pumps, Compressors, & Expanders → Pumps* and *ASPEN → Pumps, Compressors, & Expanders → Pumps* on the multimedia CD-ROM, which include a narrated video of an industrial pump.

Note that the flash with recycle process is a good representation of a quench vessel, in which hot gases, typically from an exothermic reactor, are quenched by a cold liquid recycle. Quenches are often needed to provide rapid cooling of a reactor effluent by direct-contact heat transfer. Cold liquid is showered over hot, rising gases. As some of the liquid vaporizes, the latent heat of vaporization is absorbed, and cooling occurs. Quenches are particularly

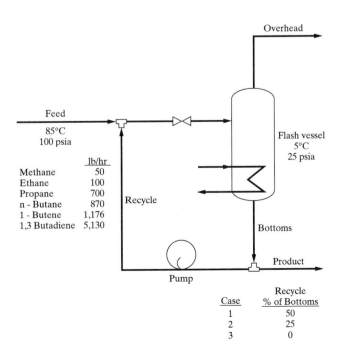

Figure 4.15 Flash with recycle process.

effective for the rapid cooling of organic vapors so as to avoid, or at least reduce, the deposition of solid carbon by chemical reaction. Any solid that is deposited can be bled with the condensate from the vessel bottoms rather easily. The alternative, shell-and-tube heat exchangers, often become fouled with solids and must be shut down periodically for cleaning.

Flash Vessel Control

Next, it is recommended that the reader solve a variation on the flash with recycle problem. In this variation (Exercise 4.1b), case 3 is modified so as to determine the flash temperature to obtain 850 lb/hr of overhead vapor.

Equation-Oriented Architectures

In the discussion thus far, unit subroutines (or blocks, or models) have been utilized to solve the equations that model the process units, given values for the degrees of freedom (i.e., a consistent set of specifications) associated with each process unit. The simulators determine a calculation sequence, which can be altered by the user, for proceeding from equipment subroutine to equipment subroutine in solving the equations associated with the entire process flowsheet. In most simulators, information flows in parallel with the flow of material and energy in the process flowsheet. In HYSYS.Plant, bidirectional information flow enables the simulator to return to execute a subroutine when one of its degrees of freedom has been altered, either upstream or downstream. When using subroutines, it is necessary to tear streams in recycle loops and perform iterative calculations. Similarly, when specifying degrees of freedom that require the calculation of equipment parameters, such as the area of a heat exchanger, iterative calculations are necessary to satisfy these so-called *design specifications*.

In contrast, several so-called *equation-oriented* simulators have been developed. These include gPROMS (Imperial College), ABACUS (M.I.T.), and as options in Version 11 of ASPEN PLUS and Version 3.0.1 of HYSYS.Plant (Aspen Technology, Inc.). In these simulators, libraries of equations are stored to represent the model associated with each process unit. Using the connectivity of the process flowsheet, that is, the streams that connect the process units, a set of equations is assembled for the entire flowsheet. Then, the degrees of freedom are determined by the simulator. The user is required to make enough specifications to satisfy the degrees of freedom. The simulator then solves the independent set of equations. Typically, a variation on the Newton–Raphson method is utilized and convergence is achieved when the residuals of the equations are sufficiently small.

To construct equation-oriented models for an entire process, it becomes important to identify specifications that are consistent, avoiding overspecifying or underspecifying subsets of the equations. When convergence is not achieved, facilities are provided to examine the values of selected variables and the residuals of selected equations. This requires well-designed programs that can display subsets of variables and equation residuals.

Clearly, equation-oriented simulators avoid iterations through subroutines in converging recycle loops and design specifications. Given good initial guesses, this is a major advantage. In ASPEN PLUS, good guesses can be provided by solving the unit subroutines in an initial pass through the flowsheet. Furthermore, ASPEN PLUS permits the creation of a hybrid simulation, in which subroutines are used to solve the equations associated with some process units while the equations associated with the remaining process units are solved simultaneously.

4.3   SYNTHESIS OF THE TOLUENE HYDRODEALKYLATION PROCESS

In this section, process simulators are utilized to assist in carrying out the steps introduced in Sections 3.4 and 3.5 for the synthesis of a process to hydrodealkylate toluene. This process was used actively following World War II, when it became favorable to convert large quantities of toluene, which was no longer needed to make the explosive TNT, to benzene for use in the manufacture of cyclohexane, a precursor of nylon. In this case, the design alternative generated from the *primitive problem*, as discussed in Section 1.2, involves the conversion of toluene to benzene and, for this purpose, the principal reaction path is well defined. It involves

$$C_7H_8 + H_2 \rightarrow C_6H_6 + CH_4 \tag{R1}$$

which is accompanied by the side reaction

$$2C_6H_6 \rightarrow C_{12}H_{10} + H_2 \tag{R2}$$

Laboratory data indicate that the reactions proceed irreversibly without a catalyst at temperatures in the range of 1,200–1,270°F with approximately 75 mol% of the toluene converted to benzene and approximately 2 mol% of the benzene produced in the hydrodealkylation reaction converted to biphenyl. Since the reactions occur in series in a single processing unit, just a single reaction operation is positioned in the flowsheet, as shown in Figure 4.16. The plant capacity is based on the conversion of 274.2 lbmol/hr of toluene, or approximately 200 MMlb/yr, assuming operation 330 days per year.

One distribution of chemicals involves a large excess of hydrogen gas to prevent carbon deposition and absorb much of the heat of the exothermic hydrodealkylation reaction. Furthermore, to avoid an expensive separation of the product methane from the hydrogen gas, a purge stream is utilized in which methane leaves the process, unavoidably with a comparable amount of hydrogen. Because the performance of the separation system, to be added in the next synthesis step, is unknown, the amount of hydrogen that accompanies methane in the purge stream is uncertain at this point in synthesis. Hence, the distribution of chemicals in Figure 4.17 is known incompletely. Note, however, that the sources and sinks of the chemicals can be connected and an estimate for the toluene recycle prepared based on the assumption of 75 mol% conversion and complete recovery of toluene from the effluent stream. Also, at 1,268°F and 494 psia, a typical operating pressure, the heat of reaction is 5.84 × 10⁶ Btu/hr, as computed by ASPEN PLUS using the RSTOIC subroutine and the Soave–Redlich–Kwong equation of state.

One selection of separation operations, shown in Figure 4.18, involves a flash separator at 100°F and a slightly reduced pressure, to account for anticipated pressure drops, at 484 psia. The liquid product is sent to a distillation train in which H_2 and CH_4 are recovered first, followed by C_6H_6 and then C_7H_8. Note that the pressures of the distillation columns have not yet been entered. These are computed to permit the usage of cooling water in the condensers; that is, the pressures are adjusted to set the bubble- or dew-point temperatures of the vapor streams to be condensed at 130°F or greater. This is accomplished using ASPEN PLUS for simulation of the distillation section, to be discussed shortly.

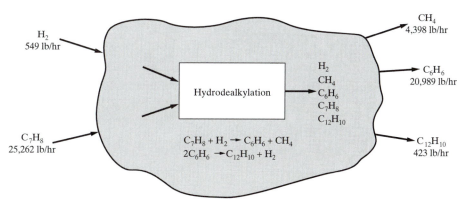

Figure 4.16 Reaction operation for the hydrodealkylation of toluene.

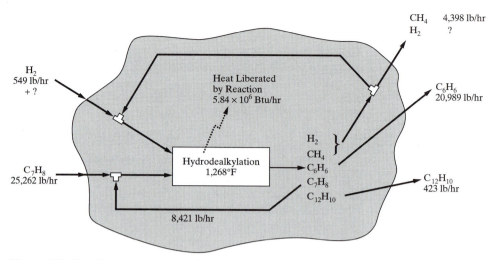

Figure 4.17 Distribution of chemicals for the hydrodealkylation of toluene.

The next synthesis step involves positioning operations to change the temperatures, pressures, and phases where differences exist between the reaction and separation operations, as well as the sources of the raw materials and sinks for the product chemicals. For this process, the toluene and hydrogen feed streams are assumed to be available at elevated pressure, above that required in the hydrodealkylation reactions. When this is not the case, the appropriate operations to increase the pressure must be inserted. One arrangement of the temperature-, pressure-, and phase-change operations is shown in Figure 4.19 for the reaction section only. Clearly, large quantities of heat are needed to raise the temperature of the feed chemicals to 1,200°F, and similarly large quantities of heat must be removed to partially condense the reactor effluent. These heat loads are calculated by ASPEN PLUS as discussed shortly.

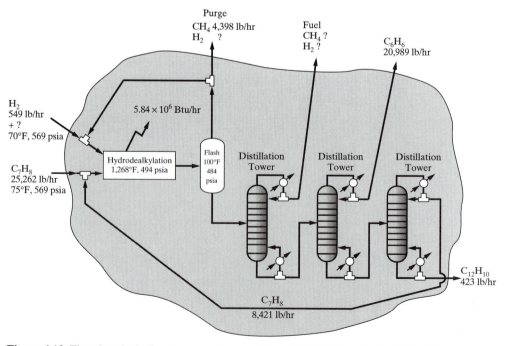

Figure 4.18 Flowsheet including the separation operations for the toluene hydrodealkylation process.

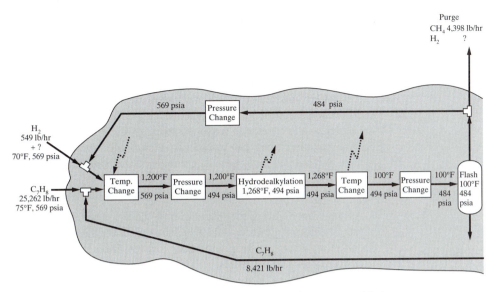

Figure 4.19 Reaction section for the toluene hydrodealkylation process with the temperature-, pressure-, and phase-change operations.

The next synthesis step involves task integration, that is, the combination of operations into process units. In one task integration, shown in Figure 4.20, reactor effluent is quenched rapidly to 1,150°F, primarily to avoid the need for a costly high-temperature heat exchanger, and is sent to a feed/product heat exchanger. There, it is cooled as it heats the mixture of feed and recycle chemicals to 1,000°F. The stream is cooled further to 100°F, the temperature of the flash separator. The liquid from the quench is the product of the reactor section, yet a portion of it is

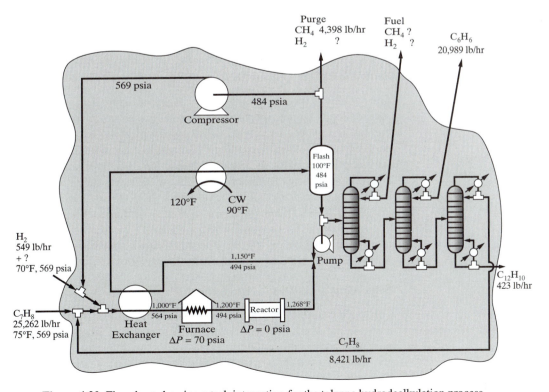

Figure 4.20 Flowsheet showing a task integration for the toluene hydrodealkylation process.

recycled to quench the reactor effluent. The vapor product is recycled after a portion is purged to keep methane from building up in the process. This recycle is compressed to the pressure of the feed chemicals, 569 psia. Returning to the feed/product heat exchanger, the hot feed mixture leaves at 1,000°F and is sent to a gas-fired furnace for further heating to 1,200°F, the temperature of the feed to the reactor. Note that the gases are heated in a tube bank that resides in the furnace, and hence a high pressure drop is estimated (70 psia). On the other hand, the hydrodealkylation reactions take place in a large-diameter vessel that has negligible pressure drop. Clearly, at a later stage in the process design, these pressure drops, along with pressure drops in the connecting pipes, can be estimated. Normally, however, small errors in the pressure drops have only a small impact on the equipment sizes and costs as well as the operating costs.

Process Simulation

As mentioned during the discussion of the synthesis steps, process simulators are very useful. They are used to calculate heats of reaction, heat added to or removed from a stream, power requirements for pumps and compressors, performance of a flash separator at various temperatures and pressures, and bubble- and dew-point temperatures associated with distillates and bottoms products, among many other quantities.

In this subsection, three simulations are suggested that, when carried out as exercises, show the more comprehensive role process simulators normally play during process synthesis. The first simulation involves the reactor section of the proposed process. It is intended to provide a better understanding of its performance. Note that several assumptions are made concerning the recycle streams, so as not to complicate the analysis. Then the separation section, involving three distillation towers, is examined, with specifications made for the flow rates and compositions of the product streams. Finally, after obtaining a better understanding of the performance of these two sections, the entire process is simulated. In this simulation, the flow rates and compositions of the recycle and purge streams are computed to satisfy material and energy balances. Of course, during any of these simulations, the specifications can be varied to gain a better understanding of the performance of the process. In Exercise 17.21, you will have an opportunity to use the Aspen Icarus Process Evaluator (Aspen IPE) to size all of the equipment, estimate its installation costs, and perform a profitability analysis.

Simulation of the Reactor Section

The conditions for this simulation are shown in Figure 4.21 and summarized in Exercise 4.2. As mentioned before, representative values are assumed for the flow rates of the species in the gas and toluene recycle streams. Also, typical values are provided for the heat transfer coefficients in both heat exchangers, taking into consideration the phases of the streams involved in heat transfer, as discussed in Section 13.3. Subroutines and models for the heat exchangers and reactor are described in the ASPEN and HYSYS modules on *Heat Exchangers* and *Chemical Reactors* on the multimedia CD-ROM that accompanies this text. In ASPEN PLUS and HYSYS.Plant, there are no models for furnaces, and hence it is recommended that you calculate the heat required using the HEATER subroutine and the **Heater** model, respectively. For estimation of the thermophysical properties, it is recommended that the Soave–Redlich–Kwong equation of state be used.

Simulation of the Distillation Section

The specifications for the distillation section are provided in Figure 4.22, and summarized in Exercise 4.3, in which three product streams are specified. The objective is to determine the tower pressures, number of equilibrium stages, and reflux ratios. In this problem, toluene and biphenyl are lumped together as a single product. Two configurations are examined for separating hydrogen and methane, as a single product, from benzene, and from toluene and biphenyl. Subsequently, the distillation column to separate toluene from biphenyl can be designed.

Species Flow Rates (lbmol/hr)	Feed	Recycle	Gas Recycle
H_2	0	0	2,045.9
CH_4	0	0	3,020.8
C_6H_6 (benzene)	0	3.4	42.8
C_7H_8 (toluene)	274.2	82.5	5.3
$C_{12}H_{10}$ (biphenyl)	0	1.0	0

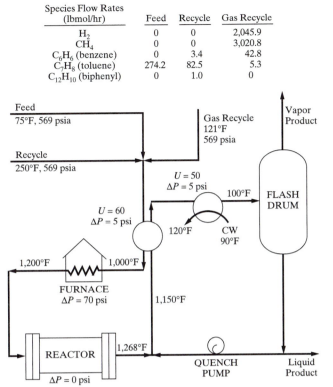

Figure 4.21 Reactor section of the toluene hydrodealkylation process.

An objective is to examine the two separation sequences shown. In the *direct sequence*, valves A and D are open, B and C are closed, and product 1 (H_2 and CH_4) is recovered in the distillate of the first tower. Alternatively, in the *indirect sequence*, valves B and C are open, and product 3 (C_7H_8 and $C_{12}H_{10}$) is recovered in the bottoms product of the first tower.

Using the flowsheet simulators, design calculations are needed to estimate the reflux ratio and the theoretical tray requirements for the two towers in each of the sequences. In ASPEN PLUS, this is accomplished with the DSTWU subroutine, which is described in the module *ASPEN → Separators → Distillation → FUG Shortcut Design* on the multimedia CD-ROM. In HYSYS.Plant, the **Shortcut Column** model is used, which is described in the modules under *HYSYS → Separations → Distillation → Shortcut Distillation Column* on the CD-ROM. The reflux ratio is set, arbitrarily, to 1.3 times the minimum and the column pressures are adjusted to obtain distillate temperatures greater than or equal to 130°F (to permit the use of cooling water for condensation). However, no column pressure is permitted to be less than 20 psia (to avoid vacuum operation). Total condensers are used, except a partial condenser is needed when methane is taken overhead.

The operating pressures, reflux rates, and numbers of trays are a good basis for comparison of the two sequences. It is preferable, however, to determine the capital and operating costs and to compare the sequences on the basis of profitability. For this purpose, Aspen IPE can be used to estimate the costs, as discussed in Section 16.7, and an economics spreadsheet can be used to carry out the profitability analysis, as discussed in Section 17.8.

Simulation of the Complete Process (Exercise 4.4)

Having completed the simulations of the reactor and distillation sections, the entire process in Figure 4.20 is simulated rather easily. For this simulation, it is recommended that the purge/recycle ratio be set initially to 0.25. Note that 0.25 is somewhat arbitrary for the

Flow Rate (lbmol/hr)

Species	Reactor Product Stream	Product 1	Product 2	Product 3
H_2	1.5	1.5		
CH_4	19.3	19.2	0.1	
C_6H_6	262.8	1.3	258.1	3.4
C_7H_8	84.7		0.1	84.6
$C_{12}H_{10}$	5.1			5.1

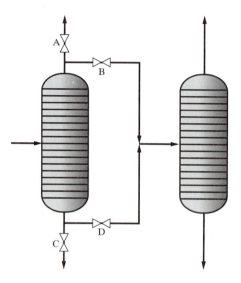

Figure 4.22 Toluene hydrodealkylation process— distillation section.

purge/recycle ratio, which should be adjusted to determine its impact on the recirculation rate, equipment sizes, power requirements, and so on; see Exercise 4.4 for this purpose. It is also recommended that the amount of hydrogen added to the process feed stream be adjusted to the amount of hydrogen leaving in the purge stream. This can be accomplished in ASPEN PLUS using a design specification. Also, the initial guesses for the recycle streams can be set equal to the values assumed when simulating the reactor section of the process. For the distillation columns, the RADFRAC subroutine can be used for simulation with the number of stages and the reflux ratio previously calculated by the DSTWU subroutine. See the module *ASPEN →* *Separators → Distillation → MESH Equations → RADFRAC* on the multimedia CD-ROM for an example using the RADFRAC subroutine. In HYSYS.Plant, the **Column** model is used, as described in the module *HYSYS → Separations → Distillation → Column Setup*.

As mentioned in Section 3.5, "Development of the Base-Case Design," the simulation model prepared for the complete process is often the source of the stream conditions in the PFD (e.g., Figure 3.19). Furthermore, as the design team completes the process integration step, the model can be improved to represent the more complete PFD.

In this section, several subroutines have been recommended for usage with ASPEN PLUS and HYSYS.Plant. These recommendations can be extended readily to permit the simulations to be carried out with CHEMCAD or PRO/II.

4.4 STEADY-STATE SIMULATION OF THE MONOCHLOROBENZENE SEPARATION PROCESS

Another process, which is considered throughout this text, involves the separation of a mixture consisting of HCl, benzene, and monochlorobenzene (MCB), the effluent from a reactor to produce MCB by the chlorination of benzene. As discussed in Chapter 7, when separating a light gaseous species, such as HCl, from two heavier species, it is common to vaporize the feed partially, followed by separation of the vapor and liquid phases in a *flash separator*. To obtain nearly pure HCl, the benzene and MCB can be absorbed in an absorber. Then, since benzene and MCB have significantly different boiling points, they can be separated by distillation. The process that results from this synthesis strategy is shown in Figure 4.23. Included on the diagram is the design basis (or specifications). Note that a portion of the MCB product is used as the absorbent.

As shown in the flowsheet, the feed is partially vaporized in the preheater, H1, and separated into two phases in the flash vessel, F1. The vapor from F1 is sent to the absorber, A1, where most of the HCl vapor passes through, but the benzene is largely absorbed using recycled MCB as the absorbent. The liquid effluents from F1 and A1 are combined, treated to remove the remaining HCl with insignificant losses of benzene and MCB, and distilled in D1 to separate benzene from MCB. The distillate rate is set equal to the benzene flow rate in the feed to D1, and the reflux ratio is adjusted to obtain the indicated MCB impurity in the distillate. The bottoms are cooled to 120°F in the heat exchanger, H2, after which one-third of the bottoms is removed as MCB product, with the remaining two-thirds recycled to the absorber. Note that this fraction recycled is specified during the *distribution of chemicals* in process synthesis, along with the temperature of the recycle, in an attempt to absorb benzene without sizable amounts of HCl. Furthermore, the temperature of stream S02 is specified to generate an adequate amount of vapor, three equilibrium stages are judged to be sufficient for the absorber (using the approximate Kremser–Brown equations), and the number of stages and the reflux ratio are estimated for the distillation column. Using the process

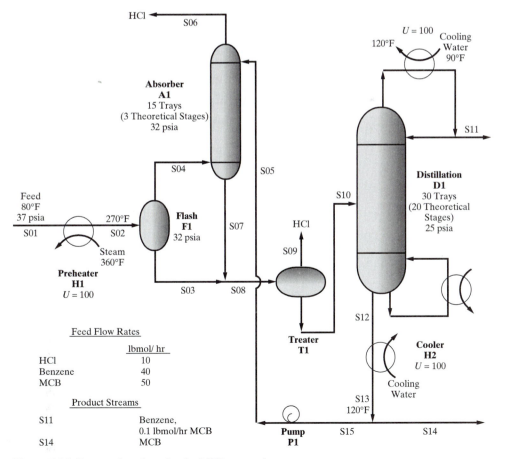

Figure 4.23 Process flowsheet for the MCB separation process.

simulators, these specifications are adjusted routinely to see how they affect the performance and economics of the process. Also, note that due to space limitations, a more complete, unit-by-unit description of the process and its specifications is reserved for the multimedia CD-ROM that accompanies this text. See the module *ASPEN → Principles of Flowsheet Simulation → Interpretation of Input and Output: Sample Problem.*

Use of Process Simulators

To determine the unknown temperatures and flow rates of the species, that is, to satisfy the material and energy balances, the MCB separation process is simulated in the steady state using ASPEN PLUS. This is accomplished by first creating an ASPEN PLUS simulation flowsheet, as illustrated in Figure 4.24. Then, the ASPEN PLUS forms are completed and the *Run* button is depressed, which produces the results shown in the modules under *ASPEN → Principles of Flowsheet Simulation → Interpretation of Input and Output* on the multimedia CD-ROM, which provides a unit-by-unit description of the input and the computer output. Then, parametric studies can be carried out, as recommended in Exercise 4.6.

Aspen IPE is also used to calculate equipment sizes and estimate capital costs for the MCB separation process in Section 16.7. Then, a profitability analysis is performed in Section 17.8. In Section 21.5, process controllers are added and their responses to various disturbances are computed using HYSYS.Plant in dynamic mode. Hence, for the MCB separation process, the process simulators have been used throughout the design process, although most design teams use a variety of computational tools to carry out these calculations.

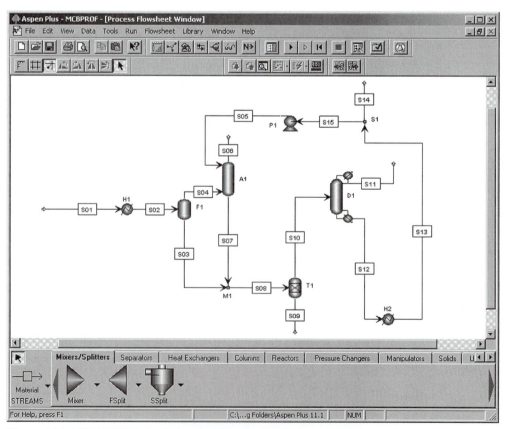

Figure 4.24 ASPEN PLUS simulation flowsheet for the MCB separation process.

4.5   PRINCIPLES OF BATCH FLOWSHEET SIMULATION

During the task integration step of process synthesis, as equipment items are selected, key decisions are made regarding whether they operate in continuous, batch, or semicontinuous modes, as discussed in Section 3.4. These decisions are based upon throughput and flexibility considerations. When the throughput is small, for example, on the laboratory scale, continuous operation is often difficult and impractical to maintain, it usually being simpler and more profitable to complete a batch in hours, days, or weeks. Even for larger throughputs, where multiple products are produced, with variably sized orders received regularly, batch processes offer the ease of switching from the production of one product to another; that is, flexibility, which is more difficult to achieve in continuous operation. These and other issues are discussed in more detail in Chapter 12.

As shown for the manufacture of tissue plasminogen activator (tPA) in Section 3.4, when batch operation is selected for an equipment item, either the batch time or batch size must be selected, with the other determined as a function of the throughput specification (e.g., 80 kg/yr of tPA). Furthermore, for a single-product plant involving a serial sequence of processing steps, when the product throughput is specified, the throughput for each process unit is determined, as shown in the synthesis of the tPA process in Section 3.4. In many cases, available vessel sizes are used to determine the size of a batch.

Given the process flowsheet and the specifics of operation for each equipment item, it is the role of batch process simulators, like BATCH PLUS, by Aspen Technology, Inc., and SUPERPRO DESIGNER, by Intelligen, Inc., to carry out material and energy balances, and to prepare an operating schedule in the form of a Gantt chart for the process. Then, after the equipment and operating costs are estimated, and profitability measures are computed, the batch operating parameters and procedures can be varied to increase the profitability of the design.

Process and Simulation Flowsheets

As in the steady-state simulation of continuous processes, it is convenient to convert from a *process flowsheet* to a *simulation flowsheet*. To accomplish this, it is helpful to be familiar with the library of models (or *procedures*) and *operations* provided by the simulator. For example, when using SUPERPRO DESIGNER to simulate two fermentation reactors in series, the process flowsheet in Figure 4.25a is replaced by the simulation flowsheet in Figure 4.25b. In BATCH PLUS, however, this conversion is accomplished without drawing the simulation flowsheet, since the latter is generated automatically on the basis of the recipe specifications for each equipment item.

In the simulation flowsheets, the arcs represent the streams that convey the batches from equipment item to equipment item. Each arc bears the stream name and represents the transfer of information associated with each stream; that is, the mass of each species per batch, temperature, pressure, density, and other physical properties.

The icons represent the models for each of the equipment items. Unlike for the simulation of continuous processes, these models involve a sequence of process operations, which are specified by the designer. Typically, these operations are defined as a *recipe* or *campaign* for each equipment item, and usually involve charging the chemicals into the vessel, processing the chemicals, removing the chemicals from the vessel, and cleaning the vessel. Note that in the SUPERPRO DESIGNER simulation flowsheet in Figure 4.25b, the microfiltration model represents both the microfilter and its holding tank in the process flowsheet, Figure 4.25a.

Equipment Models

Table 4.2 lists the equipment models (or procedures) and operations in each of the two simulators. Some of the models carry out simple material balances given specifications for the feed stream(s) and the batch (or vessel) size or batch time. Others, like the batch distillation

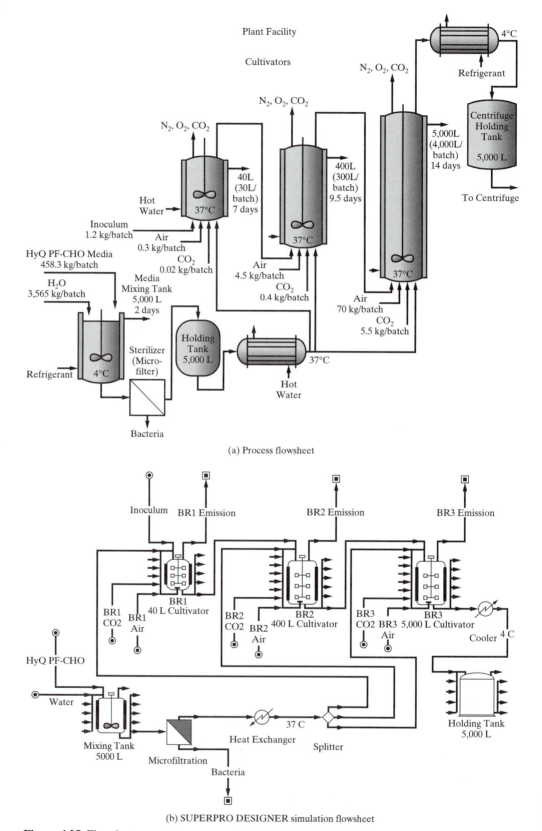

(a) Process flowsheet

(b) SUPERPRO DESIGNER simulation flowsheet

Figure 4.25 Flowsheets.

Table 4.2 Equipment Models

(a) BATCH PLUS equipment models		
Class	Mode	Type
Adsorption	Batch	Adsorption system
Agitator	Continuous	Agitator—3-blade retreat impeller, helical ribbon, paddle, propeller, turbine
Biotech	Batch	Autoclave, cell factory, diafilter, filter-depth, incubator, incubator-shaker, laminar flow hood, lyophilizer, microfilter, triblender, ultrafilter
	Continuous	Bead mill, homogenizer, sterilizer, transfer panel, valve
Centrifuge	Batch	Centrifuge, centrifuge—decanter, disk-stack, filter, horizontal basket, multichamber-bowl, tubular-bowl, vertical basket
Column	Batch	Column, column—chromatography
	Continuous	Column—continuous packed, continuous tray
Compressor	Continuous	Compressor, blower, fan
Conveyor	Continuous	Conveyor—pneumatic
Crystallizer	Batch	Crystallizer
	Continuous	Crystallizer—continuous
Dryer	Batch	Dryer, dryer—agitated pan, blender, conical, freeze, fluid bed, horizontal paddle, rotary, spray, tray
	Continuous	Dryer—continuous, fluid bed—continuous
Emission control	Either	Vapor emission vent
Evaporator	Continuous	Evaporator—long tube, thin film, wiped film
Extractor	Batch	Extractor
	Continuous	Extractor—continuous
Fermenter	Batch	Fermentor
	Continuous	Fermentor—continuous
Filling	Continuous	Filling system
Filter	Batch	Filter—agitated Nutsche, air, bag, belt, cross flow, dryer, in-line, pot, press, sparkler, tank sheet
	Continuous	Filter—continuous
Formulation and packaging	Batch	Blender, coater, high gear granulator, kneader, mill-hammer, screen, sifter
	Continuous	Classifier, extruder, filling system, granulator-fluid bed, mill—continuous, jet; tableting unit
Changing component for formulation	Continuous	Air distributor plate, agitator—impeller, blade; chopper, distributor plate, filter socks, nozzle, screen—mill
Generic	Batch	Generic batch
Heat exchanger	Batch	Condenser
	Continuous	Cooling tower, electric heater, fired heater, heat exchanger, heat exchanger plate, heat exchanger shell and tube, refrigeration unit
Heat transfer	Batch	Internal helical-coil, jacket—agitated conventional, baffled conventional, conventional, dimple, half-pipe coils
Hopper	Batch	Hopper, plate feeder
Instrument		Flow meter, moisture analyzer, scale, tester—hardness, friability, thickness; disintegration bath
Mixer	Batch	Mixer
	Continuous	Mixer—in-line
Piping	Batch	Piping
Pump	Continuous	Pump, pump—liquid ring vacuum, vacuum
Reactor	Batch	Reactor
	Continuous	Reactor—continuous
Scrubber	Batch	
Solid transport	Continuous	Screw conveyor, vacuum-pressure lock
Storage location	Batch	Inventory location, inventory location-vapor
Tank	Batch	Tank
Miscellaneous	Continuous	After burner, cyclone, demister, dust collector, ejector, hydrocyclone, steam jet

(Continued)

Table 4.2 (*Continued*)

(b) BATCH PLUS Operations

Batch operations	Age, centrifuge, charge, clean, cool, concentrate, crystallize, decant, distill, dry, evacuate, extract, filter, filter-in-place, heat, heat-to-reflux-and-age, line-blow, line-flush, open/close-vent, pH-adjust, pressurize, purge, QC-test, quench, quench-in-place, react, react-distill, start-sweep, stop-sweep, transfer, transfer-through-heat-exchanger, utilize, vent, wash-cake, yield-react
Chromatography operations	Elute-column, equilibrate-column, load-column, regenerate-column, wash-column
Continuous operations	Crystallize-continuously, Distill-continuously, Dry-continuously, Extract-continuously, filter-continuously, react-continuously
Biotech operations	Cell-disrupt, centrifuge-by-settling, depth-filter, diafilter, ferment, ferment-continuously, microfilter, sterilize, transfer-through-sterilizer, ultrafilter

(c) SUPERPRO DESIGNER Procedures (Equipment Models)

Group	Mode	Type
Vessel	Batch	Reactor, fermentor, seed fermentor, airlift fermentor
Continuous reaction	Continuous	Stoichiometric (CSTR, PFR, fermentor, seed fermentor, airlift fermentor)
	Continuous	Kinetics (CSTR, PFR, fermentor, seed fermentor)
	Continuous	Equilibrium (CSTR)
	Continuous	Environmental (Well-mixed aerobic, biooxidation, . . .)
Filtration	Batch	Microfiltration, ultrafiltration, reverse osmosis, diafiltration, dead end filtration, Nutsche filtration, plate and frame filtration, baghouse filtration, electrostatic precipitation
	Feed and Bleed (continuous)	Microfiltration, ultrafiltration, reverse osmosis
	Either	Rotary, vacuum filtration, air filtration, belt filtration, granular media filtration, baghouse filtration, electrostatic precipitation
Centrifugation	Batch	Basket centrifuge
	Either	Decanter centrifuge, disk-stack centrifuge, bowl centrifuge, Centritech centrifuge, cyclone, hydroclone
Homogenization	Either	High pressure, bead milling
Chromatography/ adsorption	Batch	Gel filtration, packed bed adsorption (PBA) chromatography, granular activated carbon (GAC)—liquid and gaseous stream
Drying	Batch	Tray drying, freeze drying
	Either	Spray drying, fluid bed drying, drum drying, rotary drying, sludge drying
Sedimentation	Either	Decanting (2-liquid phases), clarification, inclined plane (IP) clarification, thickener basin, dissolved air flotation tank, oil separator
Distillation	Batch	Shortcut batch distillation
	Either	Flash drum, shortcut distillation
Extraction	Either	Mixer-settler, differential column extractor, centrifugal extractor
Phase change	Either	Condensation for gas streams, multiple-effect evaporation, crystallization
Adsorption/stripping	Either	Absorber, stripper, degasifier
Storage	Batch	Hopper, equalization tank, junction box mixing
	Either	Blending tank, flat bottom tank, receiver, horizontal tank, vertical on legs tank, silo
Heat exchange	Either	Heating, electrical heating, cooling, heat exchanging (2-streams), heat sterilization
Mixing	Batch	Bulk flow (tumble mixer)
	Either	Bulk flow (2–9 streams), discrete flow (2–9 streams)
Splitting	Either	Bulk flow (2–9 streams), discrete flow (2–9 streams), component flow (2–9 streams)
Size reduction	Either	Grinding (bulk or discrete flow), shredding (bulk or discrete flow)
Formulation and packaging	Either	Extrusion, blow molding, injection molding, trimming, filling, assembly, printing, labeling, boxing, tableting
Transport (near)	Either	Liquid (pump)
		Gas (compressor, fan)
		Solids (belt conveyor—bulk or discrete flow, pneumatic conveyor—bulk or discrete flow, screw conveyor—bulk or discrete flow, bucket elevator—bulk or discrete flow)

Table 4.2 (*Continued*)

(c) SUPERPRO DESIGNER procedures (equipment models) (*Continued*)

Group	Mode	Type
Transport (far)	Either	By land (truck—bulk or discrete flow, train—bulk or discrete flow)
		By sea (ship—bulk or discrete flow)
		By air (airplane—bulk or discrete flow)

(d) SUPERPRO DESIGNER Operations

Absorb	Adsorb	Agitate	Assemble
Bio-oxidize	Bioreact	Centrifuge	Charge
Clarify	Clean-in-place (CIP)	Compress	Concentrate (batch)
Concentrate (feed & bleed)	Condense	Convert to bulk	Convert to discrete
Convey	Cool	Crystallize	Cyclone
Cycloning	Decant	Degasify	Diafilter
Distill	Dry	Dry cake	Elevate
Elute	Equalize	Equilibrate	Evacuate
Exchange heat	Extract/phase split	Extrude	Ferment (kinetic)
Ferment (stoichiometric)	Fill	Filter	Flash
Flotate	Gas sweep	Grind	Handle solids flow
Heat	Hold	Homogenize	Incinerate
Label	Load	Mix	Mix solids
Mold	Neutralize	Oxidize	Pack
Pass through	Precipitate	Pressurize	Print
Pump	Pump gas	Purge/inlet	Radiate
React (equilibrium)	React (kinetic)	React (stoichiometric)	Regenerate
Separate oil	Shred	Split	Steam-in-place (SIP)
Sterilize	Store	Store solids	Strip
Tablet	Thicken	Transfer in	Transfer out
Transport	Trim	Vaporize/concentrate	Vent
Wash	Wash cake		

models, integrate the dynamic MESH (Material balance, Equilibrium, Summation of mole fractions, Heat balance) equations, given specifications like the number of trays, the reflux ratio, and the batch time. Detailed documentation of the equipment models is provided in user manuals and help screens.

More specifically, a list of the BATCH PLUS equipment models is provided in Table 4.2a. These are organized under the *class* of model, with a list of *type* of equipment, and an indication of whether a model can be used in *batch*, *continuous*, or *either* mode. Similarly, for SUPERPRO DESIGNER, a list of *procedures* (equipment models) is provided in Table 4.2c. These are organized here as *groups* of equipment types.

For each equipment item, the engineer must specify the details of its operations. These include specifications for charging, processing, emptying, and cleaning. When using BATCH PLUS, these are specified in the steps in a recipe, with the equipment items defined as the steps are specified. A full list of the operations available is provided in Table 4.2b. Following this discussion, the results for a BATCH PLUS simulation of the reactor section of the tPA process are provided in Example 4.5, with detailed instructions provided for specifying the operations and equipment items provided in the tutorial *ASPEN → Tutorials → Batch Process Simulation → tPA Manufacture* on the multimedia CD-ROM that accompanies this text. In SUPERPRO DESIGNER, since the engineer provides a simulation flowsheet, the operations are specified unit-by-unit. Its list of operations is provided in Table 4.2d.

Combined Batch and Continuous Processes

Since it is possible to have adjacent equipment items operating in batch and continuous modes, it is important to understand the conventions used when preparing a mixed simulation with batch and continuous operations. In most cases, it is desirable to install a holding tank to moderate the surges that would otherwise occur.

In SUPERPRO DESIGNER, each flowsheet is defined by the engineer as either *batch* or *continuous*. In batch mode, stream results are reported on a per batch basis, even for streams associated with continuous processes in a batch flowsheet. Each equipment item is designated as operating in batch/semicontinuous or continuous mode. Scheduling information must be included for all items designated as operating as batch/semicontinuous. Semicontinuous units operate continuously while utilized, but are shut down between uses. Equipment items designated as continuous are assumed to operate at all times, and are excluded from operation schedules (and Gantt charts).

When a SUPERPRO DESIGNER flowsheet is defined to be in continuous mode, streams are reported on a per hour basis. Scheduling information is not required, and no overall batch time is calculated. Individual batch processes can be inserted into the flowsheet, with their batch and turnaround times specified.

In BATCH PLUS, every simulation is for an overall batch process, with stream values always reported on a per batch basis. Continuous operations, however, can be inserted. For these units, a feed is loaded, the vessel is filled to its surge volume, and an effluent stream immediately begins to transfer the product downstream. This differs from normal batch operation, which involves loading all of the feed and completing the processing steps before unloading. Specific units in BATCH PLUS, such as the *Fermenter*, can also operate as *fed-batch*. In such operations, a feed is added continuously to the batch while an operation is taking place.

With SUPERPRO DESIGNER and BATCH PLUS, caution must be exercised when introducing continuous operations into batch processes, as no warnings are provided when a continuous process unit is running dry. When a feed to a continuous unit runs dry, the simulator assumes that this unit is shut down and restarted when the feed returns. Clearly, such operation is infeasible for many units, such as distillation columns and chemical reactors. Consequently, when continuous processes are included, it is important to check the results computed by the batch simulators to be sure that unreasonable assumptions have not been made.

An advantage of adding continuous operations arises when the process bottleneck is transferred to the continuous unit. When a schedule is devised such that the continuous unit is always in operation, batch cycling is avoided.

EXAMPLE 4.5   *tPA Cultivators*

As discussed in Section 3.4, tPA-CHO cells are used to produce tPA. These cells are duplicated to a density of 3.0×10^6 cell/mL, after which the culture becomes too dense and the tPA-CHO cells die at a high rate. For this reason, engineers cultivate tPA-CHO cells in a sequence of bioreactors, each building up mass to a density of 3.0×10^6 cell/mL, with the accumulated cell mass used to inoculate the next largest reactor, until the desired cell mass is reached.

In this example, the objective is to determine the effective time between batches; that is, the *cycle time*, which is less than the total occupied time of a sequence of batch operations. The cycle time is smaller because while one batch is moving through the sequence, other batches are being processed simultaneously in other pieces of equipment both upstream and downstream. Therefore, the effective time between batches, or the cycle time, is determined by the equipment unit that requires the most processing time. This equipment unit is known as the *bottleneck*, and consequently, to reduce the cycle time, engineers seek to reduce the processing time of the bottleneck as much as possible. Usually, the bottleneck is associated with the largest process unit, often the main bioreactor, because these reactors involve the largest cultivation times. See Chapter 12 for a more complete discussion of the cycle time and bottleneck.

For this example, the BATCH PLUS simulator is used to determine the cycle time for a portion of the tPA process that involves just two cultivators, as shown in Figure 4.26. Initially, a mixing tank is charged with 3,565 kg of water and 458.3 kg of HyQ PF-CHO media, with a charge time of one hour. The material in the tank is cooled to 4°C for one day and aged for two days to allow for quality assurance testing. Then, this material is transferred to a 0.2-μm microfilter for sterilization, to remove bacteria over a two-hour period, and sent to a holding tank. Next, the first cultivator is charged with 1.2 kg of tPA-CHO cells in one hour. Then, 21.2 kg of material from the holding tank are heated in a heat exchanger to 37°C and added to the first cultivator in 0.5 day, after which cultivation takes place over the next five days. The yield from the cultivation is 15.3 wt% tPA-CHO cells, 0.01 wt% endotoxin, 84.7 wt% water, and 0.01 wt% tPA. The products of Cultivator 1 are fed to Cultivator 2 in 0.5 day. Then, 293.5 kg of media from the holding tank are heated to 37°C and fed to Cultivator 2 in 0.5 day, after which the cultivation takes place over seven days. Immediately after Cultivator 1 is emptied, it is cleaned-in-place using 60 kg of water over 20 hours. Note that to override the BATCH PLUS estimate, a charge time of 1 min should be entered. Then, it is sterilized at 130°C for two hours and cooled to 25°C (with one-hour heat-up and cool-down times). The yield of the cultivation in Cultivator 2 is 11.7 wt% tPA-CHO cells, 7.67×10^{-4} wt% endotoxin, 88.3 wt% water, and 0.039 wt% tPA. After this cultivation, the contents of Cultivator 2 are cooled in a heat exchanger to 4°C and transferred to the centrifuge holding tank over 0.5 day. After Cultivator 2 is emptied, it is cleaned-in-place using 600 kg of water over 20 hours, and sterilized using the procedure for Cultivator 1.

To determine the cycle time and the bottleneck unit, create a multiple batch Gantt chart using BATCH PLUS. Generate equipment content and capacity reports to determine the sizes of the equipment items. Examine the stream table report to monitor the production of tPA-CHO cells and tPA in the process.

SOLUTION

When using BATCH PLUS, as discussed step-by-step on the multimedia CD-ROM that accompanies this text (see *ASPEN → Tutorials → Batch Process Simulation → tPA Manufacture*), the materials are specified; that is, tPA-CHO cells, tPA, media, water, nitrogen, oxygen, and carbon dioxide. Then, each equipment item is entered with its recipe of operations. Note that there is no cultivator model in BATCH PLUS, and consequently, the *Fermenter* model is used in its place. Given this information, BATCH PLUS generates a recipe of operations for the process, shown in Figure 4.27a, and prepares a simulation flowsheet (using Microsoft VISIO). BATCH PLUS also generates a table including the per batch flow rates of each stream in the process in a time-dependent manner, a portion of which is shown

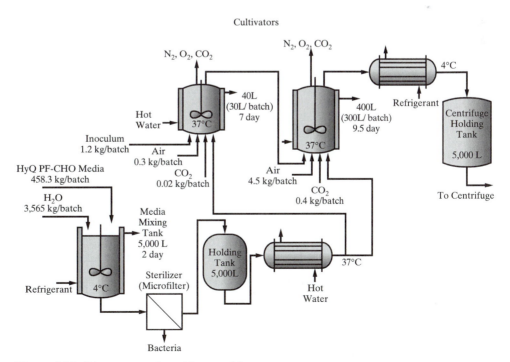

Figure 4.26 tPA reactor section with two cultivators.

in Figure 4.27b. Study of this report allows the monitoring of the growth of tPA-CHO cells, and their production of tPA, as they travel from vessel to vessel. The third column from the last in the report indicates that the final stream in the process contains 36.9 kg of tPA-CHO cells, 0.12 kg of tPA, 0.0024 kg of endotoxin, and 278.8 kg of water. For the simulation flowsheet and the entire stream table, see *ASPEN → Tutorials → Batch Process Simulation → tPA Manufacture* on the multimedia CD-ROM.

In addition, BATCH PLUS uses Microsoft EXCEL to prepare an *Equipment Contents Report*, which displays, for each vessel in the process, an inventory of the contents of the vessel during each step the vessel is utilized. This includes the mass of components, as well as overall liquid and solid volume and mass. Inspection of these reports allows estimation of required vessel sizes. The report for the Mixing Tank, shown in Figure 4.27c, indicates a maximum liquid and solid volume of 4,050 L after operation 1.1. It can, therefore, be concluded that the Mixing Tank unit must be larger than 4,050 L; for example, 5,000 L.

1.1. Charge Mixing Tank with 458.3 kg of Media. The charge time is 1 h. Charge Mixing Tank with 3565 kg of WATER. The charge time is 1 h.

1.2. Cool unit Mixing Tank to 4 C. The cooling time is 1 day.

1.3. Age the contents of unit Mixing Tank for 2 day.

1.4. Microfilter the contents of Mixing Tank in Microfilter. The mode of operation is Batch Concentration. Unspecified components go to the Permeate. The operation time is 2 h. The permeate stream is sent to Holding Tank.

1.5. Charge Fermenter 1 with 1.2 kg of tPA-CHO Cells. The charge time is 1 h.

1.6. Transfer contents of unit Holding Tank to Fermenter 1 through heat exchanger Heat Exchanger. The final stream temperature is 37 C. Transfer 21.2 kg of vessel contents. The transfer time is 0.5 day.

1.7. Ferment in unit Fermenter 1. The yield of tPA-CHO Cells in the Solid phase is 0.153, of Endotoxin in the Liquid phase is 0.0001, of tPA in the Liquid phase is 0.0001, of WATER in the Liquid phase is 0.847, of Media in the Liquid phase is 0, of Media in the Solid phase is 0 and of tPA-CHO Cells in the Liquid phase is 0. The fermentation time is 5 day. Continuously add 0.02 kg of CARBON-DIOXIDE. Continuously add 0.3 kg of AIR.

Start Parallel
    Series

1.8. Transfer contents of unit Fermenter 1 to Fermenter 2. Transfer 100% of vessel contents. The transfer time is 0.5 day.

1.9. Transfer contents of unit Holding Tank to Fermenter 2 through heat exchanger Heat Exchanger. The final stream temperature is 37 C. Transfer 293.5 kg of vessel contents. The transfer time is 0.5 day.

1.10. Ferment in unit Fermenter 2. The yield of tPA-CHO Cells in the Solid phase is 0.117, of Endotoxin in the Liquid phase is 7.67e-6, of tPA in the Liquid phase is 0.00039, of WATER in the Liquid phase is 0.883, of Media in the Solid phase is 0, of Media in the Liquid phase is 0 and of tPA-CHO Cells in the Liquid phase is 0. The fermentation time is 7 day. Continuously add 0.4 kg of CARBON-DIOXIDE. Continuously add 4.5 kg of AIR.

    Series

1.11. Clean unit Fermenter 1. Clean with 60 kg of WATER. The feed time is 1 min. Cleaning time is 20 h.

1.12. Sterilize the contents of Fermenter 1. The sterilization temperature is 130 C. The heat-up time is 1 h. Maintain the temperature for 2 h. The cool-down time is 1 h.

End Parallel

1.13. Transfer contents of unit Fermenter 2 to Centrifuge Holding Tank through heat exchanger Cooler. The final stream temperature is 4 C. Transfer 100% of vessel contents. The transfer time is 0.5 day.

1.14. Clean unit Fermenter 2. Clean with 600 kg of WATER. The feed time is 1 min. Cleaning time is 20 h.

1.15. Sterilize the contents of Fermenter 2. The sterilization temperature is 130 C. The heat-up time is 1 h. Maintain the temperature for 2 h. The cool-down time is 1 h.

(a)

Figure 4.27 BATCH PLUS simulation for Example 45. (a) Operations recipe for the process-BATCH PLUS output. (b) Last three columns of stream table (for the complete table, see *ASPEN → Tutorials → Batch Process Simulation → tPA Manufacture* on the CD-ROM. (c) Mixing Tank report (for Fermenter 1 and 2 reports, see the CD-ROM). (d) 3-batch Gantt chart.

Step Stream Table

Process (Version):	Reactor Growth Chain (1.0)		Key Input Intermediate:	Media	
Step (Version):	Step1 (1.0)		Key Output Intermediate:	tPA	
Simulation Date:	12/16/2001 17:10		Number of Batches:	1	
			Plan Quantity:	0.12	kg

BATCH PLUS Stream Label			1.13. Transfer-Through-Heat-Exchanger-107	1.14. Clean-111	1.14. Clean-112
Operation			1.13. Transfer-Through-Heat-Exchanger	1.14. Clean	1.14. Clean
Start Time		(min)	24,000.00	24,720.00	25,921.00
End Time		(min)	24,720.00	24,721.00	25,922.00
Total Time		(min)	720.00	1.00	1.00
From Unit			Cooler		Fermenter 2
To Unit			Centrifuge Holding Tank	Fermenter 2	
Stream Type			Intermediate	Input	Output
Mass - (kg)	**Per Batch**	**Mol Wt**			
Total			3159000	600.0000	800.0000
Endotoxin	18.02		0.0024		
tPA	18.02		0.1232		
AIR	28.95				
CARBON-DIOXIDE	44.01				
WATER	18.02		278.8288	600.0000	600.0000
Media	18.02				
tPA-CHO Cells	18.02		36.9456		
Total Mass		(kg)	315.90	600.00	600.00
Total Volume		(liter)	312.09	603.73	603.73
Mass Flowrate		(kg/min)	0.44	37.27	37.27
Volume Flowrate		(liter/h)	26.01	2,250.00	2,250.00
Composite Product Factor			2,565.12	4,872.03	4,872.03
Phase			Liquid1+Solid	Liquid1	Liquid1
Temperature		(C)	4.00	25.00	25.00
Average Density		(kg/Cubic m)	1,012.20	993.81	993.81
Average Viscosity		(cp)	1.35	0.92	0.92
Average Heat Capacity		(kJ/kg-K)	4.21	4.18	4.18
Average Molecular Weight			18.02	18.02	18.02

(b)

Step Equipment Contents

Process (Version):	Reactor Growth Chain (1.0)		Key Input Intermediate:	Media	
Step (Version):	Step1 (1.0)		Key Output Intermediate:	tPA	
Simulation Date:	12/16/01 17:10		Number of Batches:	1	
			Plan Quantity:	0.12	kg

Mixing Tank

Operation		START	1.1. Charge	1.2. Cool	1.3. Age	1.4. Microfilter
Time	(min)	0.00	60.00	1,500.00	4,380.00	4,500.00
Mass - (kg)	**Mol Wt**					
Total		5.8957	4,024,4221	4,024,5086	4,024,6086	6,3424
WATER	18.02		3,565.0000	3,565.0000	3,565.0000	
NITROGEN	28.01	4.5223	0.8607	1.0038	1.0038	4.8650
OXYGEN	32.00	1.3734	0.2614	0.3048	0.3048	1.4774
Media	18.02		458.3000	458.3000	458.3000	
Liquid+Solid Mass	(kg)	0.00	4,023.30	4,023.30	4,023.30	0.00
Liquid+Solid Volume	(liter)	0.00	4,048.34	3,968.36	3,968.36	0.00
Phase		Gas	Gas+Liquid1	Gas+Liquid1	Gas+Liquid1	Gas
Temperature	(C)	25.00	25.00	4.00	4.00	4.00
Pressure	(kPa)	101.33	101.33	101.33	101.33	101.33
Average Liq+Sol Density	(kg/Cubic m)	0.00	993.81	1,013.84	1,013.84	0.00
Average Liq+Sol Viscosity	(cp)	0.00	0.92	1.53	1.53	0.00
Average Liq+Sol Heat Capacity	(kJ/kg-K)	0.00	4.18	4.21	4.21	0.00
Average Liq+Sol Molecular Weight		0.00	18.02	18.02	18.02	0.00

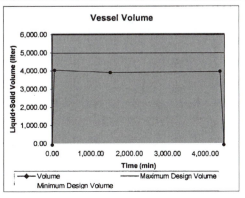

(c)

Figure 4.27 (*Continued*)

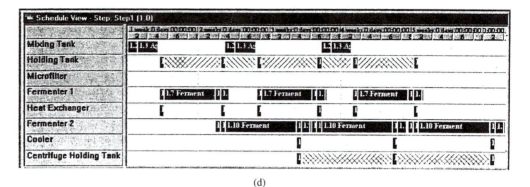

(d)

Figure 4.27 (*Continued*)

Similarly, the reports for Fermenters 1 and 2 are shown in *ASPEN → Tutorials → Batch Process Simulation → tPA Manufacture* on the CD-ROM, indicate maximum volumes of 22.8 L and 322 L after operations 1.7 and 1.8. On this basis, 40-L and 400-L vessels are selected for Fermenters 1 and 2.

Finally, when the Gantt chart prepared by BATCH PLUS is extended to show three batches, as shown in Figure 4.27d, the bottleneck of the process is determined quite easily. For each vessel, solid blocks show the time period during which it is in operation. Solid blocks are for the first, second, and third batches, respectively. The bottleneck is associated with the equipment unit that is utilized at all times; that is, for which the red, blue, and green blocks touch each other. Clearly, this unit determines the cycle time. Note that these results can be reproduced using the folder BATCH PLUS-EXAM 4-5 on the CD-ROM that accompanies this book. ∎

4.6   SUMMARY

Having studied this chapter and the accompanying material on the multimedia CD-ROM, the reader should

1. Be able to prepare a simulation flowsheet, beginning with a process flowsheet.
2. Be able to prepare a steady-state simulation using ASPEN PLUS and HYSYS.Plant, and be familiar with the capabilities of CHEMCAD and PRO/II.
3. Be able to set up simulations involving recycle loops and design specifications, which are usually implemented with iterative control algorithms.
4. Understand how to follow the calculation sequences implemented automatically by ASPEN PLUS, CHEMCAD, and PRO/II for recycle calculations and when satisfying design specifications. For recycle calculations, the reader should be able to alter the calculation sequence by specifying tear streams.
5. Understand the many places that process simulators are useful during process synthesis and when preparing the base-case design, as discussed in Sections 3.4 and 3.5.
6. Be able to prepare a simulation of a batch process using BATCH PLUS and SUPER-PRO DESIGNER.
7. Have completed several exercises involving steady-state simulation using one of the four simulators, ASPEN PLUS, HYSYS.Plant, CHEMCAD, and PRO/II, and involving batch process simulation using one of the two simulators, BATCH PLUS and SUPERPRO DESIGNER.

REFERENCES

Crowe, C.M., and M. Nishio, "Convergence Promotion in the Simulation of Chemical Processes—The General Dominant Eigenvalue Method," *AIChE J.*, **21**, 3 (1975).

Edmister, W.C., "Absorption and Stripping-Factor Functions for Distillation Calculation by Manual and Digital Computer Methods," *AIChE J.*, **3**(2), 165 (1957).

Henley, E.J., and E.M. Rosen, *Material and Energy Balance Computations*, Wiley, New York (1969).

Lewin, D.R., W.D. Seider, J.D. Seader, E. Dassau, J. Golbert, D.N. Goldberg, M.J. Fucci, and R.B. Nathanson, *Using Process Simulators in Chemical Engineering: A Multimedia Guide for the Core Curriculum*, Version 2.0, John Wiley & Sons, New York (2003).

Myers, A.L., and W.D. Seider, *Introduction to Chemical Engineering and Computer Calculations*, Prentice-Hall, Englewood Cliffs, New Jersey (1976).

Orbach, O., and C.M. Crowe, "Convergence Promotion in the Simulation of Chemical Processes with Recycle—The Dominant Eigenvalue Method," *Can. J. Chem. Eng.*, **49**, 509 (1971).

Rudd, D.F., and C.C. Watson, *Strategy of Process Engineering*, Wiley, New York (1968).

Seader, J.D., W.D. Seider, and A.C. Pauls, *FLOWTRAN Simulation—An Introduction*, 3rd ed., CACHE, Austin, Texas (1987).

Seider, W.D., J.D. Seader, and D.R. Lewin, *Process Design Principles: Synthesis, Analysis, and Evaluation*, John Wiley & Sons (1999)

Smith, B.D., *Design of Equilibrium Stage Processes*, McGraw-Hill, New York (1963).

Wegstein, J.H., "Accelerating Convergence of Iterative Processes," *Commun. Assoc. Comp. Machinery*, **1**(6), 9 (1958).

Westerberg, A.W., H.P. Hutchison, R.L. Motard, and P. Winter, *Process Flowsheeting*, Cambridge University Press, Cambridge (1979).

EXERCISES

4.1 *Flash with recycle* a. Consider the flash separation process shown in Figure 4.15. If using ASPEN PLUS, solve all three cases using the MIXER, FLASH2, FSPLIT, and PUMP subroutines and the RK-SOAVE option set for thermophysical properties. Compare and discuss the flow rates and compositions for the overhead stream produced by each of the three cases.

b. Modify case 3 of Exercise 4.1a so as to determine the flash temperature necessary to obtain 850 lb/hr of overhead vapor. If using ASPEN PLUS, a design specification can be used to adjust the temperature of the flash drum to obtain the desired overhead flow rate.

4.2 *Toluene hydrodealkylation process—reactor section.* As discussed in Section 4.3, toluene (C_7H_8) is to be converted thermally to benzene (C_6H_6) in a hydrodealkylation reactor. The main reaction is $C_7H_8 + H_2 \rightarrow C_6H_6 + CH_4$. An unavoidable side reaction occurs that produces biphenyl: $2C_6H_6 \rightarrow C_{12}H_{10} + H_2$.

The reactor section of the process is shown in Figure 4.21, as are the conditions for the feed and two recycle streams. The flow rate of the quench stream should be such that the reactor effluent is quenched to 1,150°F. Conversion of toluene in the reactor is 75 mol%. Two mole percent of the benzene present after the first reaction occurs is converted to biphenyl. Use a process simulator to perform material and energy balances with the Soave–Redlich–Kwong equation of state (RK-SOAVE option in ASPEN PLUS).

4.3 *Toluene hydrodealkylation process—separation section.* As discussed in Section 4.3, the following stream at 100°F and 484 psia is to be separated by two distillation columns into the indicated products.

Species	lbmol/hr			
	Feed	Product 1	Product 2	Product 3
H_2	1.5	1.5		
CH_4	19.3	19.2	0.1	
C_6H_6 (benzene)	262.8	1.3	258.1	3.4
C_7H_8 (toluene)	84.7		0.1	84.6
$C_{12}H_{10}$ (biphenyl)	5.1			5.1

Two different distillation sequences are to be examined, as shown in Figure 4.22. In the first sequence, H_2 and CH_4 are removed in the first column.

If using ASPEN PLUS, the DSTWU subroutine is used to estimate the reflux ratio and theoretical tray requirements for both sequences. In addition, the RK-SOAVE option is used. Specify a

reflux ratio equal to 1.3 times the minimum. Use design specifications to adjust the isobaric column pressures so as to obtain distillate temperatures of 130°F; however, no column pressure should be less than 20 psia. Also, specify total condensers except note that a partial condenser is used when H_2 and CH_4 are taken overhead.

4.4 Complete a simulation of the entire process for the hydrodealkylation of toluene in Figure 4.20. Initially, let the purge/recycle ratio be 0.25; then, vary this ratio and determine its effect on the performance of the process. Use a design specification to determine the unknown amount of hydrogen to be added to the feed stream (equal to that lost in the purge.) If using ASPEN PLUS, the distillation columns can be simulated using the RADFRAC subroutine with the number of stages and reflux ratio previously computed by the DSTWU subroutine. (To carry out an economic analysis of the process, see Exercise 17.21.)

4.5 a. Complete a steady-state simulation of the vinyl chloride process in Figure 3.19. First, create a simulation flowsheet. Assume that:

Cooling water is heated from 30 to 50°C.
Saturated steam is available at 260°C (48.4 atm).

If using ASPEN PLUS, use the UNIQUAC option set for thermophysical properties.

b. Carry out process integration and repeat the steady-state simulation.

4.6 For the monochlorobenzene separation process in Figure 4.23, the results of an ASPEN PLUS simulation are provided in the modules under *ASPEN → Principles of Flowsheet Simulation → Interpretation of Input and Output* on the multimedia CD-ROM. Repeat the simulation with:

a. 25% of MCB recycled at 130°F

Stream S02 at 250°F
15 theoretical stages in the distillation column

b. Other specifications of your choice

4.7 *Cavett Problem.* A process having multiple recycle loops formulated by R.H. Cavett [*Proc. Am. Petrol. Inst.*, **43**, 57 (1963)] has been used extensively to test tearing, sequencing, and convergence procedures. Although the process flowsheet requires compressors, valves, and heat exchangers, a simplified ASPEN PLUS flowsheet is shown in Figure 4.28 (excluding the recycle convergence units). In this form, the process is the equivalent of a four-theoretical-stage, near-isothermal distillation (rather than the

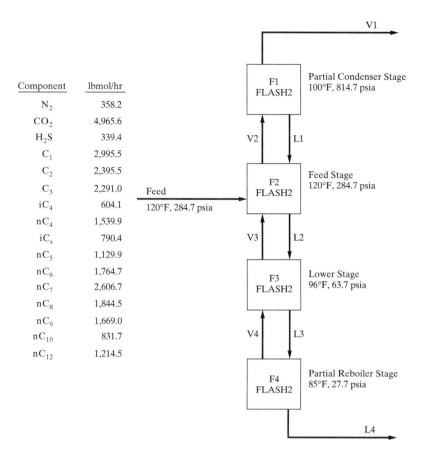

Component	lbmol/hr
N_2	358.2
CO_2	4,965.6
H_2S	339.4
C_1	2,995.5
C_2	2,395.5
C_3	2,291.0
iC_4	604.1
nC_4	1,539.9
iC_s	790.4
nC_5	1,129.9
nC_6	1,764.7
nC_7	2,606.7
nC_8	1,844.5
nC_9	1,669.0
nC_{10}	831.7
nC_{12}	1,214.5

Figure 4.28 Near-isothermal distillation process.

conventional near-isobaric type), for which a patent by A. Gunther [U.S. Patent 3,575,007 (April 13, 1971)] exists. For the specifications shown on the flowsheet, use a process simulator to determine the component flow rates for all streams in the process.

4.8 Use a process simulator to model a two-stage compression system with an intercooler. The feed stream consists of 95 mol% hydrogen and 5 mol% methane at 100°F and 30 psia; 440 lbmol/hr is compressed to 569 psia. The outlet temperature of the intercooler is 100°F and its pressure drop is 2 psia. The centrifugal compressors have an isentropic efficiency of 0.9 and a mechanical efficiency of 0.98.

Determine the power requirements and heat removed for three intermediate pressures (outlet from the first three stages): 100, 130, 160 psia. If using ASPEN PLUS, use the MCOMPR subroutine and the RK-SOAVE option.

4.9 Consider the ammonia process in which N_2 and H_2 (with impurities Ar and CH_4) are converted to NH_3 at high pressure (Figure 4.29). If using ASPEN PLUS, use the following subroutines:

Compressor	COMPR
Reactor	RSTOIC
Heat Exchanger	HEATER
High-Pressure Separator	FLASH2
Low-Pressure Separator	FLASH2
Recirculating Compressor	COMPR

You are given the feed stream and fraction purged in the splitter. Prepare a simulation flowsheet and, when applicable, show the calculation sequence prepared by the process simulator (if using ASPEN PLUS, complete SEQUENCE USED WAS:)

4.10 The feed (equimolar A and B) to a reactor is heated from 100°F to 500°F in a 1–2 parallel-counterflow heat exchanger with a mean overall heat transfer coefficient of 75 Btu/hr ft^2°F. It is converted to C by the exothermic reaction, A + B → C, in an adiabatic plug-flow tubular reactor (Figure 4.30). For a process simulator, prepare a simulation flowsheet and show the calculation sequence to determine:

a. Flow rates and unknown temperatures for each stream

b. Heat duty and area of the countercurrent shell-and-tube heat exchanger

4.11 Consider the simulation flowsheets in Figure 4.31, which were prepared for ASPEN PLUS. The feed stream, S1, is specified, as are the parameters for each process unit.

Complete the simulation flowsheets using sequences acceptable to ASPEN PLUS. If any of the streams are torn, your flowsheets should include the recycle convergence units. In addition, you should indicate the calculation sequences.

This problem is easily modified if you are working with HYSYS.Plant, CHEMCAD, or PRO/II.

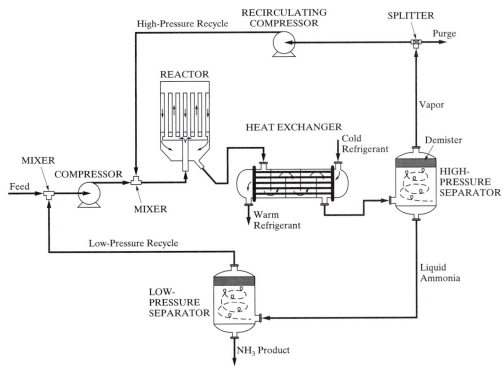

Figure 4.29 Ammonia reaction loop.

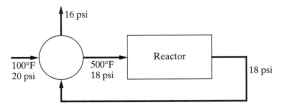

Figure 4.30 Reactor with feed/product heat exchanger.

4.12 Use a process simulator to determine the flow rate of saturated vapor benzene at 176.2°F and 1 atm to be mixed with 100 lbmol/hr of liquid benzene to raise its temperature from 25 to 50°F. Prepare a good initial estimate. *Note:* $\lambda_{NBP} = 13{,}200$ Btu/lbmol, $c_p = 0.42$ Btu/lb°F.

4.13 A distillation tower is needed to separate an equimolar mixture, at 77°F and 1 atm, of benzene from styrene. The distillate should contain 99 mol% benzene and 95 mol% of the benzene fed to the tower.

Use a process simulator to determine the minimum number of trays at total reflux (N_{min}), the minimum reflux ratio (R_{min}), and the theoretical number of trays at equilibrium when $R = 1.3R_{min}$.

4.14 Use a process simulator to determine the heat required to vaporize 45 mol% of a liquid stream entering an evaporator at 150°F and 202 psia and containing

	lbmol/hr
Propane	250
n-Butane	400
n-Pentane	350

Assume that the evaporator product is at 200 psia. Use the Soave–Redlich–Kwong equation of state.

4.15 For an equimolar mixture of *n*-pentane and *n*-hexane at 10 atm, use a process simulator to compute:

a. The bubble-point temperature

b. The temperature when the vapor fraction is 0.5

4.16 Hot gases from the toluene hydrodealkylation reactor are cooled and separated as shown in the flowsheet of Figure 4.32. In a steady-state simulation, can the composition of the recycle stream be determined without iterative recycle calculations? Explain your answer.

4.17 Given the feed streams and the parameters of the process units as shown in Figure 4.33, complete the simulation flowsheet for ASPEN PLUS and show the calculation sequence (i.e., complete the statement SEQUENCE USED WAS:). If any of the streams are torn, your flowsheet should include the stream convergence units.

The simulation flowsheet can be modified for HYSYS.Plant, CHEMCAD, or PRO/II and the exercise repeated.

4.18 Suppose 100 lbmol/hr of steam (stream STI) at 500°F and 1 atm heats 40 lbmol/hr of cold water (stream CWI) from 70°F

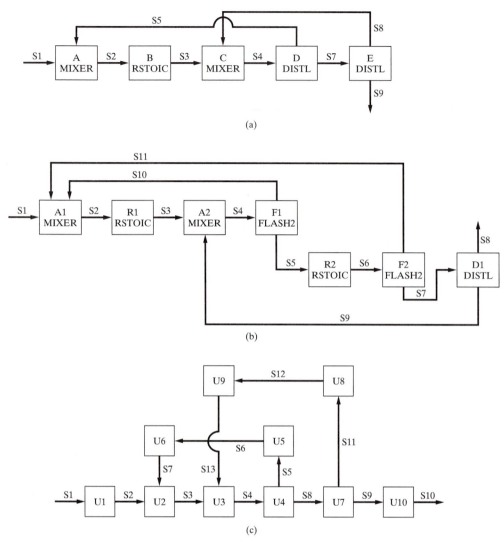

Figure 4.31 Interlinked recycle loops.

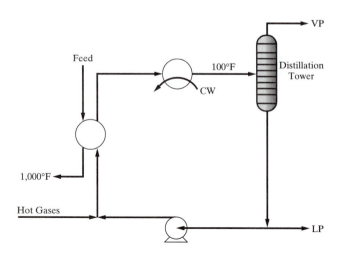

Figure 4.32 Combined quench/distillation process.

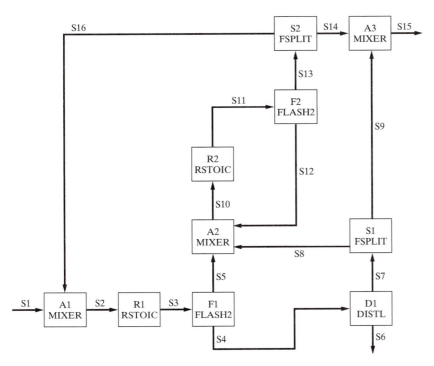

Figure 4.33 Incomplete simulation flowsheet.

to 120°F in a shell-and-tube heat exchanger. For the simulation flowsheet in ASPEN PLUS (Figure 4.34), determine the outlet temperature of the steam.

The problem can be modified for the usage of HYSYS.Plant, CHEMCAD, or PRO/II.

4.19 *Debottlenecking reactor train.* When the third tPA cultivator in Section 3.4 is added to the two cultivators in Example 4.5, as shown in Figure 4.25a, a significant time strain is placed on the process because the combined feed, cultivation, harvest, and cleaning time in this largest vessel is long and rigid. Consequently, the remainder of the process is designed to keep this cultivator in constant use, so as to maximize the yearly output of product. Note that, in many cases, when an equipment item causes a bottleneck, a duplicate is installed so as to reduce the cycle time.

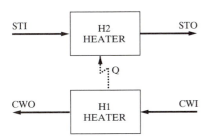

Figure 4.34 Simulation flowsheet with a heat stream.

For this exercise, the third cultivator is added to the simulation in Example 4.5, with the specifications for the mixer, filter, holding tank, heat exchanger 1, and first two cultivators identical to those in Example 4.5. After the cultivation is completed in Cultivator 2, its cell mass is transferred as inoculum to Cultivator 3 over 0.5 day. Then, the remaining media from the mixing tank is heated to 37°C and added over 1.5 days, after which cultivation takes place over eight days. Immediately after the transfer from Cultivator 2 to Cultivator 3, Cultivator 2 is cleaned-in-place using 600 kg of water over 20 hr. The yield of the cultivation in Cultivator 3 is 11.4 wt% tPA-CHO cells, 7.7×10^{-5} wt% endotoxin, 88.9 wt% water, and 0.0559 wt% tPA. When the cultivation is completed in Cultivator 3, its contents are cooled in a heat exchanger to 4°C and transferred to the centrifuge holding tank over one day, and Cultivator 3 is cleaned using 600 kg of water over 67 hr and sterilized using the procedure for Cultivators 1 and 2.

To eliminate an undesirable bottleneck(s), and reduce the cycle time to 14 days (total operation time of Cultivator 3), it may be necessary to add an equipment unit(s).

Print and submit the text recipes and 3-batch schedules for both the original process and the modified process, if debottlenecking is necessary, as prepared by BATCH PLUS.

4.20 *tPA Process simulation.* For the entire process flowsheet in Figure 3.16a,b,c of Section 3.4, complete a BATCH PLUS simulation. Print and submit the text recipes and 3-batch schedules for both the original process and the modified process, as prepared by BATCH PLUS.

Note that the operating times, batch sizes, and recovery percentages are shown in Figure 3.16a,b,c. Unfortunately, BATCH PLUS determines that 15,000 kg/batch of elution buffer are required to elute the affinity chromatography column. To circumvent this, after 523 kg have been fed, specify a "cut" of 404 kg. The latter is collected as the elution effluent while the difference is rejected in the wastewater. Also, BATCH PLUS does not model the selective adsorption of endotoxin without tPA, arginine, sucrose, glycine, NaCl, and NaOH. Consequently, to obtain the desired effluent streams, one approach is to adsorb endotoxin, while recovering the waste effluent (which contains tPA, arginine, sucrose, glycine, NaCl, and NaOH). Then, the elution buffer (500 kg/batch of water) is used to elute the endotoxin, which is rejected as wastewater.

Chapter **5**

Heuristics for Process Synthesis

5.0 OBJECTIVES

This chapter returns to the steps of preliminary process synthesis in Section 3.4, in which a strategy is recommended that involves assembling the process operations in a specific order, as follows:

1. Chemical reactions (to eliminate differences in molecular type)
2. Mixing and recycle (to distribute the chemicals)
3. Separation (to eliminate differences in composition)
4. Temperature, pressure, and phase change
5. Task integration (to combine operations into unit processes)

In Section 3.4, as the operations are inserted into alternative flowsheets to manufacture vinyl chloride, *rules of thumb* or *heuristics* are utilized. For example, when positioning the direct chlorination operation, it is assumed that because the reaction is nearly complete at 90°C, ethylene and chlorine can be fed in stoichiometric proportions. Furthermore, when positioning the pyrolysis operation, the temperature and pressure are set at 500°C and 26 atm to give a 60% conversion. These assumptions and specifications are based on many factors, not the least of which is experience in the manufacture of vinyl chloride and similar chemicals. In this case, a patent by the B.F. Goodrich Co. [British Patent 938,824 (October 9, 1963)] indicates the high conversion of ethylene and chlorine over a ferric chloride catalyst at 90°C and recommends the temperature and pressure levels of the pyrolysis reaction. The decision not to use ethylene in excess, to be sure of consuming all of the toxic chlorine, is based on the favorable conversions reported experimentally by chemists. In the distillation operations, the choice of the key components, the quality of the feed streams and the distillate products, and the pressure levels of the towers are also based on rules of thumb. In fact, heuristics like these and many others can be organized into an *expert system*, which can be utilized to synthesize sections of this and similar chemical processes.

Normally, design teams use heuristics when generating the alternatives that make up a synthesis tree, such as that shown in Figure 3.9. For the most part, heuristics are easy to apply; that is, they involve the setting of temperatures, pressures, excess amounts of chemicals, and so on. Often, they require little analysis in that simple material balances can be completed without iterations before proceeding to the next synthesis step. Consequently, several promising flowsheets are generated rapidly, with relatively little effort. Then, as described in Section 3.5, the emphasis of the design team shifts to the creation of a *base-case design*. The assumptions are checked, a process flow diagram is assembled (e.g., Figure

3.19), and a complete material and energy balance is carried out, often using the process simulators discussed in Chapter 4.

Clearly, the heuristics used by a design team to generate the synthesis tree are crucial in the design process. Section 3.4 provides just a brief introduction to these heuristics, and hence it is the objective of this chapter to describe the principal heuristics used in process design more thoroughly. A total of 53 heuristics are presented in Sections 5.2 through 5.9. In many cases, the heuristics are accompanied by examples. For quick reference, the heuristics are collected together in Table 5.2 at the end of this chapter. Additional guidance in the selection of equipment is given in Chapters 16 and 17 when determining equipment purchase and operating costs, Chapter 6 when designing reactors, Chapter 13 when sizing heat exchangers, Chapter 14 when sizing distillation towers, and Chapter 15 when sizing pumps, compressors, and gas expanders.

After studying this chapter and the heuristics in Table 5.2, the reader should

1. Understand the importance of selecting reaction paths that do not involve toxic or hazardous chemicals, and when unavoidable, to reduce their presence by shortening residence times in the process units and avoiding their storage in large quantities.
2. Be able to distribute the chemicals, when generating a process flowsheet, to account for the presence of inert species, to purge species that would otherwise build up to unacceptable concentrations, to achieve a high selectivity to the desired products, and to accomplish, when feasible, reactions and separations in the same vessels (e.g., reactive distillations).
3. Be able to apply heuristics in selecting separation processes to separate liquids, vapors, vapor–liquid mixtures, and other operations involving the processing of solid particles, including the presence of liquid and/or vapor phases.
4. Be able to distribute the chemicals, by using excess reactants, inert diluents, and cold (or hot) shots, to remove the exothermic (supply the endothermic) heats of reaction. These distributions can have a major impact on the resulting process integration.
5. Understand the advantages, when applicable, of pumping a liquid rather than compressing a vapor.

Through several examples and the exercises at the end of the chapter, the reader should be able to apply the heuristics in Table 5.2 when generating a synthesis tree. Also, he or she should obtain an appreciation of the role of heuristics and recognize that the heuristics covered are only a subset of many that are widely applied in chemicals processing.

5.1 INTRODUCTION

In a chapter on heuristics for process synthesis, it is important to emphasize that heuristics are used commonly by design teams to expedite the generation of alternative flowsheets in preliminary process synthesis. Then, as the alternatives are generated, or afterwards, it is common to perform material and energy balances, and related forms of analysis, often using a process simulator. Although this chapter is devoted to heuristics, in the remainder of this section, emphasis is placed on the kinds of analyses usually used to improve upon designs suggested by heuristics. It is important to understand the consequences of heuristics and to recognize, at least through one typical design, the way in which process simulators are used to explore and alter heuristics.

Return to the design of the toluene hydrodealkylation process, as it is presented in Section 4.3. In the reactor section, after heuristics are utilized to set (1) the large excess of H_2 in the hydrodealkylation reactor, (2) the temperature level of the quenched gases that enter the feed-product heat exchanger, and (3) the temperature in the flash vessel, the simulator is used to complete the material and energy balances and to examine the effects of these heuristics on the performance of the reactor section. In the distillation section, after heuristics are used to set (1) the quality of the feed, (2) the use of partial or total condensers, (3) the use of cooling water in the condensers, and (4) the ratio of the reflux ratio to the minimum reflux ratio (R/R_{min}), the simulator is used to determine the appropriate pressure levels in the columns, to estimate the number of stages and the position of the feed stage, to estimate the reflux ratio and, most important, to compute the distillate and bottoms products. The simulator provides an excellent vehicle for studying the effect of departures from the heuristics on the performance of the separation train.

Eventually, the two sections of the plant are combined and heuristics are used to set the purge-to-recycle ratio. Here, the simulator determines the recycle streams, which in the analysis heretofore were specified, once again using heuristic rules. Then the simulator provides an easy-to-use vehicle for studying the effect of the purge-to-recycle ratio on the performance of the process.

Having completed a simulation of the flowsheet, or possibly after working with the reactor and separation sections alone, the designer can estimate the capital and operating costs and can compute and optimize measures of the profitability, as discussed in Chapters 16–18. Furthermore, the simulator is used often by the engineer to study the effect of making small changes in the structure of the flowsheet (e.g., to recover toluene and biphenyl from the bottom of the first distillation tower). By examining their impact on the performance and profitability, the designer implements an *evolutionary* synthesis strategy, often using the process simulator. Some prefer to refer to this approach as process synthesis by *interactive analysis*. The basic approach is to examine sections of a plant one by one, generating alternative structures with heuristics and experimenting with them, retaining the most promising as operations are added to complete the flowsheets.

As mentioned above, in this chapter the focus is on the heuristics, because they are crucial in generating quickly the most promising structures. Subsequently, more systematic methods for generating many alternative flowsheets are considered in Part Two. Throughout this chapter, examples are provided to show how to use simulators to assist in evaluating the effect of the heuristics on the performance of the processes being designed. Even when the so-called *algorithmic* approaches to process synthesis are introduced in Part Two, the heuristics discussed in this chapter play an important role.

Eventually, the methods of mathematical programming are introduced in Chapters 10 and 18. Through the use of mixed-integer nonlinear programs (MINLPs), these methods are designed to optimize *superstructures* involving all of the potential streams and process units to be considered during the optimization, with many streams and process units turned off because they are associated with suboptimal solutions. When MINLPs can be formulated and solved rigorously, the use of heuristic rules can be sharply reduced or even eliminated. However, MINLPs are difficult to formulate in the early stages of process design because there are substantial uncertainties, and consequently, it is important to place emphasis on the rules of thumb needed to get started in the design process. As will be shown in this and subsequent chapters, these rules often lead to near-optimal designs.

5.2 RAW MATERIALS AND CHEMICAL REACTIONS

> ***Heuristic 1: Select raw materials and chemical reactions to avoid, or reduce, the handling and storage of hazardous and toxic chemicals.***

As discussed in Chapter 3, the selection of raw materials and chemical reactions is often suggested by chemists, biologists, biochemists, or other persons knowledgeable about the chemical conversions involved. In recent years, with the tremendous increase in awareness of the need to avoid handling hazardous and toxic chemicals, in connection with environmental and safety regulations (as discussed in Sections 1.3 and 1.4), raw materials and chemical reactions are often selected to protect the environment and avoid the safety problems that are evident in Material Safety Data Sheets (MSDSs). For example, recall that when the vinyl chloride process was synthesized in Section 3.4, the reaction of acetylene with HCl was rejected because of the high cost of acetylene. Today, in addition, this reaction path would be rejected on the basis of the high reactivity of acetylene and the difficulty of assuring safe operation in the face of unanticipated disturbances.

In connection with the handling of hazardous chemicals, the 1984 accident in Bhopal, India, in which water was accidentally mixed with the active intermediate, methyl isocyanate, focused worldwide attention on the need to reduce the handling of highly reactive intermediates. As discussed in Section 1.4, within an hour of the accident, a huge vapor cloud swept across Bhopal, leading to the death of over 3,800 victims in the vicinity of the Union Carbide plant. This accident, together with the discovery of polluted groundwaters adjacent to chemical plants, especially those that process nuclear fuels, have led safety and environment experts to call for a sharp reduction in the handling of hazardous chemicals.

For these reasons, societal needs are increasingly being formulated that call for new processes to avoid or sharply reduce the handling of hazardous chemicals. As an example, consider the manufacture of ethylene glycol, the principal ingredient of antifreeze. Ethylene glycol is produced commonly by two reactions in series:

$$C_2H_4 + \tfrac{1}{2}O_2 \rightarrow \overset{\displaystyle O}{\overset{\displaystyle \diagup\ \diagdown}{CH_2\!-\!CH_2}} \tag{R1}$$

$$\overset{\displaystyle O}{\overset{\displaystyle \diagup\ \diagdown}{CH_2\!-\!CH_2}} + H_2O \rightarrow \overset{\displaystyle OH\quad OH}{\underset{}{\overset{\mid\qquad\mid}{CH_2\!-\!CH_2}}} \tag{R2}$$

The first reaction involves the partial oxidation of ethylene over a Ag-gauze catalyst. Since both reactions are highly exothermic, they need to be controlled carefully. More important from a safety point of view, a water spill into an ethylene oxide storage tank could lead to an accident similar to the Bhopal incident. Yet it is common in processes with two reaction steps to store the intermediates so as to permit the products to be generated continuously, even when maintenance problems shut down the first reaction operation.

Given the societal need (or primitive design problem) to eliminate the storage of large quantities of reactive intermediates, such as ethylene oxide, four alternative processing concepts (or specific problems) are possible:

1. Eliminate the storage tank(s), causing intermittent interruptions in the production of ethylene glycol when the oxidation reaction shuts down.
2. Use costly chlorine and caustic (compared to oxygen from air) in a single reaction step:

$$CH_2\!=\!CH_2 + Cl_2 + 2NaOH(aq) \rightarrow \overset{\displaystyle OH\quad OH}{\underset{}{\overset{\mid\qquad\mid}{CH_2\!-\!CH_2}}} + 2\ NaCl \tag{R3}$$

This alternative requires more expensive raw materials, but completely avoids the intermediate.
3. As ethylene oxide is formed, react it with carbon dioxide to form ethylene carbonate, a much less active intermediate. This reaction

$$\overset{\displaystyle O}{\overset{\displaystyle \diagup\ \diagdown}{CH_2\!-\!CH_2}} + CO_2 \rightarrow \overset{\displaystyle O}{\overset{\displaystyle \|}{\overset{\displaystyle C}{\underset{\textstyle CH_2\!-\!CH_2}{\overset{\diagup\ \diagdown}{O\qquad O}}}}} \tag{R4}$$

occurs smoothly over a tetraethylammonium bromide catalyst. Ethylene carbonate can be stored safely and hydrolyzed to form the ethylene glycol product as needed.

4. Carry out reactions (R1) and (R4) consecutively over a Ag-gauze catalyst by reacting ethylene in a stream containing oxygen and carbon monoxide. To consider this as an alternative processing concept, laboratory or pilot-plant data on the rates of reaction are necessary.

In summary, there is an increasing emphasis on retrofitting processes to eliminate active intermediates, and in the design of new processes to avoid these chemicals entirely. Furthermore, the designers of new processes are being asked, with increasing frequency, to select raw materials and reactions accordingly. These have become important considerations in the early stages of process design.

5.3   DISTRIBUTION OF CHEMICALS

> ***Heuristic 2: Use an excess of one chemical reactant in a reaction operation to consume completely a valuable, toxic, or hazardous chemical reactant. The MSDSs will indicate which chemicals are toxic and hazardous.***

After the reaction operations are positioned in a process flowsheet, the sources of chemicals (i.e., the feed streams and reactor effluents) are distributed among the sinks for chemicals (i.e., the feed streams to the reaction operations and the products from the process). In this distribution, decisions are made concerning (1) the use of one chemical reactant in excess in a reaction operation, (2) the handling of inert species that enter in the feed streams, and (3) the handling of undesired byproducts generated in side reactions. For example, as we have seen in Figure 3.5, one distribution of chemicals for the vinyl chloride process involves stoichiometric amounts of ethylene and chlorine fed to the direct-chlorination reactor. Alternatively, an excess of ethylene can be utilized as shown in Figure 5.1. In this distribution, the reactor is designed to consume completely the hazardous and toxic chlorine, but the recovery of unreacted ethylene from the dichloroethane product is required. Clearly, an important consideration is the degree of the excess, that is, the ethylene/chlorine ratio. It governs the costs of separation and recirculation, and often plays a key role in the process economics. In many design strategies, this ratio is set using heuristics, with larger ratios used to assure consumption of the most hazardous chemicals. Eventually, as a *base-case design* evolves, the ratio is varied systematically, often using a process simulator. In mathematical programming strategies, it is treated as

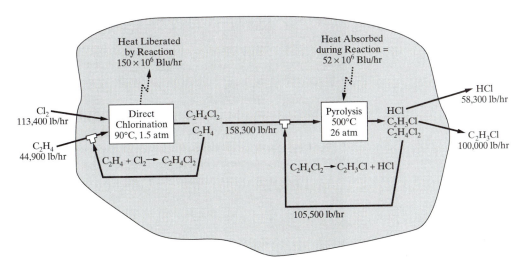

Figure 5.1 Distribution of chemicals for the production of vinyl chloride involving an excess of ethylene.

a *design variable*, to be varied during optimization, with a lower bound. Note that for exothermic reactions, the excess chemical often serves the useful function of absorbing the heat of reaction and thereby maintaining more moderate temperatures. This is an important approach to handling large heats of reaction and is considered with several common alternatives in Section 5.5, in the subsection on heat removal from exothermic reactors. An excess of one chemical reactant is also used to increase conversion of the other (limiting) reactant when the extent of reaction is limited by equilibrium. Also, side reactions can be minimized by using an excess of one reactant.

Inert Species

> *Heuristic 3:* ***When nearly pure products are required, eliminate inert species before the reaction operations when the separations are easily accomplished and when the catalyst is adversely affected by the inert, but not when a large exothermic heat of reaction must be removed.***

Often impure feed streams contain significant concentrations of species that are inert in chemical reaction operations. When nearly pure products are required, an important decision concerns whether impurities should be removed before or after reaction operations. As an example, consider the flowsheet in Figure 5.2a, in which two reaction operations have been positioned. An impure feed stream of reactant C contains the inert species D, and hence a decision is required concerning whether to remove D before or after reaction step 2, as shown in Figures 5.2b and 5.2c, respectively. Clearly, the ease and cost of the separations, that is, D from C, and D from E (plus unreacted A and C) must be assessed. This can

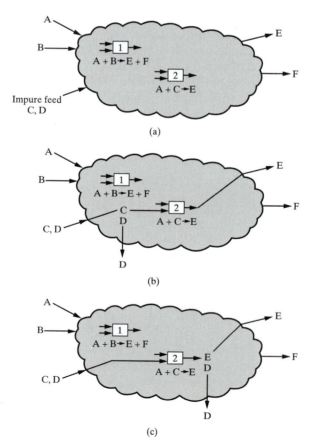

Figure 5.2 Partial distribution of chemicals showing the alternatives for removing inert species D: (a) reaction operations; (b) recovery before reaction; (c) recovery after reaction.

be accomplished by examining the physical properties on which the separations are based. For example, when considering distillation, estimates of the relative volatilities are used. When the mixtures are ideal, the relative volatility, α_{ij}, is simply a ratio of the vapor pressures ($\alpha_{ij} = P_i^s/P_j^s$). Otherwise, activity coefficients are needed ($\alpha_{ij} = \gamma_i P_i^s/\gamma_j P_j^s$). When the relative volatilities differ significantly from unity, an easy and inexpensive separation is anticipated. Similarly, when considering crystallization, the differences in the freezing points are examined, and for dense membrane separations, the permeabilities of the pure species are estimated as the product of the solubility in the membrane and the molecular diffusivity. Other considerations are the sizes of the reactors and separators, with larger reactors required when the separations are postponed. Also, for exothermic reactions the inert species absorb some of the heat generated, thereby lowering the outlet temperatures of the reactors.

EXAMPLE 5.1   *Distribution of Chemicals*

To satisfy the Clean Air Act of 1990, gasoline must have a minimum oxygen atom content of 2.0 mol%. In the 1990s, the most common source of this oxygen was methyl *tertiary*-butyl ether (MTBE), which is manufactured by the reaction

$$CH_3OH + iso-\text{butene} \rightleftharpoons MTBE$$

It is desired to construct an MTBE plant at your refinery, located on the Gulf Coast of Texas. Methanol will be purchased and *iso*-butene is available in a mixed-C_4 stream that contains

	Wt%
1-Butene	27
iso-Butene	26
1,3-Butadiene	47

During process synthesis, in the distribution of chemicals, a key question involves whether it is preferable to remove 1-butene and 1,3-butadiene before or after the reaction operation. In this example, distillation is considered, although other separation methods normally are evaluated as well. It should be noted that recently MTBE was found to contaminate ground water and, thus, in most locations is no longer the preferred source of oxygen.

SOLUTION

These hydrocarbon mixtures are only mildly nonideal, and hence it is satisfactory to examine the boiling points of the pure species or, better yet, their vapor pressures. These can be tabulated and graphed as a function of temperature using a simulator; for example, the following curves are obtained from ASPEN PLUS (and can be reproduced using the EXAM5-1.bkp file on the CD-ROM):

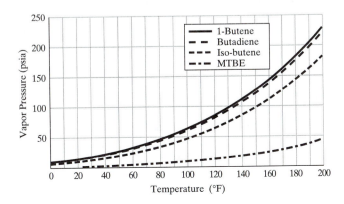

With respect to MTBE, the relative volatilities, $\alpha = P^s/P^s_{MTBE}$, at 200°F, are 5.13 (1-butene), 4.96 (1,3-butadiene), and 4.04 (iso-butene). Clearly, the relative volatilites of 1-butene and 1,3-butadiene are very close, but each differs significantly from the value for iso-butene. On this basis, the former two compounds can be separated by distillation before or after the reaction operation. Other considerations, such as their impact on the catalyst, the volumes of the reactors and distillation towers, and the temperature levels in the exothermic reactors, should be evaluated in making this decision.   ∎

EXAMPLE 5.2   *Positioning an Equilibrium Reaction Operation*

Consider the reaction and distillation operations for the isomerization of *n*-butane to isobutane according to the reaction

$$n\text{-}C_4H_{10} \rightleftharpoons i\text{-}C_4H_{10}$$

The feed to the process is a refinery stream which contains 20 mol% *iso*-butane. Show the alternatives for positioning the reaction and distillation operations.

SOLUTION

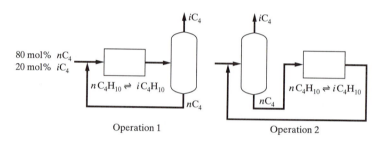

As shown in the diagram, either operation can be placed first. By positioning the distillation column first, a nearly pure feed is sent to the reaction operation, providing a higher conversion to *iso*-butane. The effectiveness of this configuration depends on the relative difficulty of achieving the distillation separation. To determine this, the two configurations should be simulated.   ∎

Purge Streams

> **Heuristic 4:** *Introduce purge streams to provide exits for species that enter the process as impurities in the feed or are formed in irreversible side reactions, when these species are in trace quantities and/or are difficult to separate from the other chemicals. Lighter species leave in vapor purge streams, and heavier species exit in liquid purge streams.*

Trace species, often introduced as impurities in feed streams or formed in side reactions, present special problems when the chemicals are distributed in a flowsheet. In a continuous process, these accumulate continuously unless a means is provided for their removal, either by reaction, separation, or through purge streams. Since the reaction or separation of species in low concentration is usually costly, purge streams are used when the species are nontoxic and have little impact on the environment. Purge streams are also used for removing species present in larger amounts when their separation from the other chemicals in the mixture is difficult. As an example, consider the distribution of chemicals in the ammonia process ($N_2 + 3H_2 \rightleftharpoons 2NH_3$) in Figure 5.3. Trace amounts of argon accompany nitrogen, which is recovered from air, and trace amounts of methane accompany hydrogen, which is produced by steam reforming ($CH_4 + H_2O \rightleftharpoons 3H_2 + CO$). After reforming, the carbon monoxide and unreacted methane and steam are recovered, leaving trace quantities of methane. Although nitrogen and hydrogen react at high pressures, in the range of 200–400

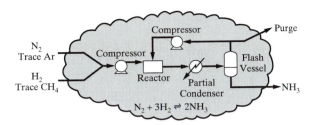

Figure 5.3 Ammonia reactor loop.

atm, depending on the process throughput, the conversion is low (usually in the range of 15–20 mol%), and large quantities of unreacted nitrogen and hydrogen are recirculated. The purge stream provides a sink for the argon and methane, which otherwise would build to unacceptable concentrations in the reactor bed, which is packed with reduced iron oxide catalyst. The light gases from the flash vessel are split into purge and recycle streams, with the purge/recycle ratio being a key decision variable. As the purge/recycle ratio increases, the losses of nitrogen and hydrogen increase, with an accompanying reduction in the production of ammonia. This is counterbalanced by a decrease in the recirculation rate. In the early stages of process synthesis, the purge/recycle ratio is often set using heuristics. Eventually, it can be varied with a process simulator to determine its impact on the recirculation rates and equipment sizes. Then it can be adjusted, also using a process simulator, to optimize the return on investment for the process, as discussed in Chapter 18. Note that the alternative of separating trace species from the vapor stream, thereby avoiding the purge of valuable nitrogen and hydrogen, may also be considered. These separations—for example, adsorption, absorption, cryogenic distillation, and gas permeation with a membrane—may be more expensive. Finally, it should be recognized that argon and methane are gaseous species that are purged from the vapor recycle stream. Other processes involve heavy impurities that are purged from liquid streams.

EXAMPLE 5.3   *Ammonia Process Purge*

In this example, the ammonia reactor loop in Figure 5.3 is simulated using ASPEN PLUS to examine the effect of the purge-to-recycle ratio on the compositions and flow rates of the purge and recycle streams. For the ASPEN PLUS flowsheet below, the following specifications are made:

Simulation Unit	Subroutine	T (°F)	P (atm)
R1	REQUIL	932	200
F1	FLASH2	−28	136.3

and the Chao–Seader option set is selected to estimate the thermophysical properties. Note that REQUIL calculates chemical equilibria at the temperature and pressure specified, as shown on the multimedia CD-ROM that accompanies this textbook (*ASPEN → Chemical Reactors → Equilibrium Reactors → REQUIL*).

The combined feed stream, at 77°F and 200 atm, is comprised of

	lbmol/hr	Mole Fraction
N_2	24	0.240
H_2	74.3	0.743
Ar	0.6	0.006
CH_4	1.1	0.011
	100.0	1.000

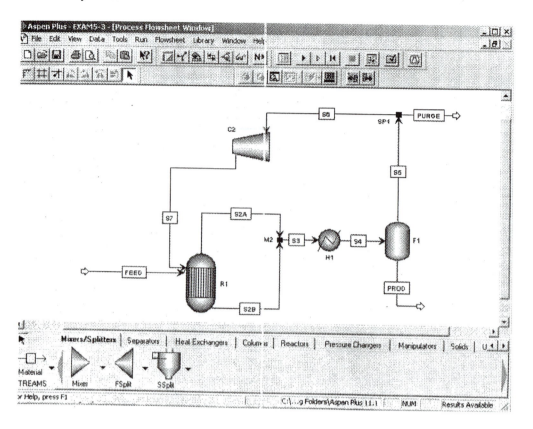

SOLUTION

Several variables are tabulated as a function of the purge/recycle ratio:

Purge/Recycle Ratio	PROD Flow Rate (lbmol/hr)	Recycle Flow Rate (lbmol/hr)	Purge Flow Rate (lbmol/hr)	Purge Mole Fraction Ar	Purge Mole Fraction CH_4
0.1	39.2	191.0	19.1	0.028	0.052
0.08	40.75	209.3	16.7	0.033	0.060
0.06	42.4	233.9	14.0	0.040	0.074
0.04	44.3	273.5	10.9	0.053	0.093
0.02	45.8	405.6	8.1	0.072	0.133

In all cases, the mole fraction of Ar and CH_4 in the purge are significantly greater than in the feed. As the purge/recycle ratio is decreased, the vapor effluent from the flash vessel becomes richer in the inert species and less H_2 and N_2 are lost in the purge stream. However, this is accompanied by a significant increase in the recycle rate and the cost of recirculation, as well as the reactor volume. Note that the EXAM5-3.bkp file on the CD-ROM can be used to reproduce these results. Although not implemented in this file, the purge/recycle ratio can be adjusted parametrically by varying the fraction of stream S5 purged in a *sensitivity analysis*, which is one of the *model analysis tools* found in most simulators. The capital and operating costs can be estimated and a profitability measure optimized as a function of the purge/recycle ratio. ∎

> **Heuristic 5:** *Do not purge valuable species or species that are toxic and hazardous, even in small concentrations (see the MSDSs). Add separators to recover valuable species. Add reactors to eliminate, if possible, toxic and hazardous species.*

In some situations, the recovery of trace species from waste streams is an important alternative to purging. This, of course, is the case when an aqueous stream contains trace quantities

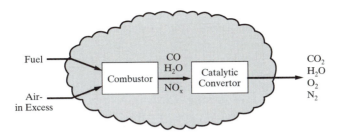

Figure 5.4 Catalytic conversion of combustion effluents.

of rare metals, as can occur when catalysts are impregnated on ceramic supports. In other situations—for example, in the handling of aqueous wastes—environmental regulations are such that trace quantities of organic and inorganic chemicals must be recovered or converted into an environmentally acceptable form. One process to treat aqueous streams in the vicinity of leaking tanks is *supercritical oxidation*, using acoustic waves or lasers to produce plasmas. In this process, the waste species [including chlorinated hydrocarbons, pesticides, phenols (e.g., *p*-nitrophenol), and esters] are oxidized at temperatures and pressures associated with supercritical water (Hua et al., 1995a,b). Yet another example involves the catalytic conversion of hydrocarbons and carbon monoxide in the exhaust gases from internal combustion engines. As illustrated in Figure 5.4, rather than purge the exhaust gases from a combustion engine, catalytic converters commonly convert carbon monoxide and nitrogen oxides to carbon dioxide and nitrogen, respectively. Again, the decision to insert a reaction step, rather than to separate or purge, in the early stages of process design is made often based on the availability of a catalyst and experience; that is, heuristics.

Recycle to Extinction

> **Heuristic 6:** *Byproducts that are produced in reversible reactions, in small quantities, are usually not recovered in separators or purged. Instead, they are usually recycled to extinction.*

Often small quantities of chemicals are produced in side reactions, such as the reaction of benzene to form biphenyl in the toluene hydrodealkylation process. When the reaction proceeds *irreversibly*, small quantities of byproducts must be separated away, as in Figure 4.20, or purged; otherwise they will build up in the process until the process must be shut down.

When the reaction proceeds *reversibly*, however, it becomes possible to achieve an equilibrium conversion at steady state by recycling product species without removing them from the process. In so doing, it is often said that undesired byproducts are *recycled to extinction*. It is important to recognize this when distributing chemicals in a potential flowsheet so as to avoid the loss of chemicals through purge streams or the insertion of expensive separation operations. Recycle to extinction, which is considered in more detail in Section 8.5, is most effective when the equilibrium conversion of the side reaction is limited by a small chemical equilibrium constant at the temperature and pressure conditions in the reactor.

EXAMPLE 5.4   *Reversible Production of Biphenyl*

This example follows the simulation of the complete toluene hydrodealkylation process at the end of Section 4.3 and is presented without a solution because it is the basis for Exercise 5.4. To recycle the biphenyl to extinction, the flowsheet in Figure 4.20 is modified to eliminate the last distillation column, and unreacted toluene from the second column is recycled with biphenyl. This is accomplished by the reversible reaction

$$2C_6H_6 \rightleftharpoons C_{12}H_{10} + H_2$$

Note that this eliminates one of the two waste streams from the process. The other, which loses large quantities of H_2, is the gas purge stream. To avoid this loss, the use of membrane or adsorption separators should be considered. ■

Selectivity

> *Heuristic 7: For competing reactions, both in series and parallel, adjust the temperature, pressure, and catalyst to obtain high yields of the desired products. In the initial distribution of chemicals, assume that these conditions can be satisfied. Before developing a base-case design, obtain kinetics data and check this assumption.*

When chemical reactions compete in the formation of a desired chemical, the reaction conditions must be set carefully to obtain a desirable distribution of chemicals. Consider, for example, the series, parallel, and series-parallel reactions in Figure 5.5, where species B is the desired product. For these and similar reaction systems, it is important to consider the temperature, pressure, ratio of the feed chemicals, and the residence time when distributing the chemicals. One example of series-parallel reactions occurs in the manufacture of allyl chloride. This reaction system, which involves three competing second-order exothermic reactions, is shown in Figure 5.5d, with the heats of reaction, ΔH_R, activation energies, E, and preexponential factors, k_0, in Table 5.1. Note that because $E_1/E_2 > 1$ and $E_1/E_3 < 1$, the conversion to allyl chloride is highest at intermediate temperatures. In the early stages of process synthesis, when distributing the chemicals, these considerations are helpful in setting the temperature, pressure, and the ratio of propylene/chlorine in the feed.

Table 5.1 Heats of Reaction and Kinetics Constants for the Allyl Chloride Process[a]

Reaction	ΔH_R (Btu/lbmol)	k_0 [lbmol/(hr ft^3 atm^2)]	E/R (°R)
1	−4,800	206,000	13,600
2	−79,200	11.7	3,430
3	−91,800	4.6×10^8	21,300

[a]Biegler and Hughes, 1983.

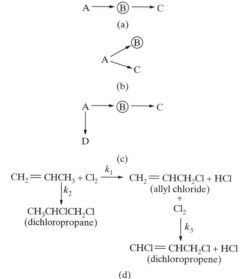

A ⟶ Ⓑ ⟶ C

(a)

A ⟨ Ⓑ
       C

(b)

A ⟶ Ⓑ ⟶ C
│
↓
D

(c)

$CH_2\!=\!CHCH_3 + Cl_2 \xrightarrow{\;k_1\;} CH_2\!=\!CHCH_2Cl + HCl$
                                              (allyl chloride)
$\Big\downarrow k_2$ $+$
$CH_3CHClCH_2Cl$ Cl_2
(dichloropropane) $\Big\downarrow k_3$

                              $CHCl\!=\!CHCH_2Cl + HCl$
                                   (dichloropropene)

(d)

Figure 5.5 Competing reactions: (a) series reactions; (b) parallel reactions; (c) series-parallel reactions; (d) exothermic allyl chloride reactions.

When selectivity is the key to the success of a process design, it is not uncommon to carry out an extensive analysis of the reactor alone, before distributing the chemicals, and proceeding with the synthesis of the flowsheet. In other cases, using simulation models, the distribution of chemicals is carried out as the process is optimized to achieve an economic objective.

EXAMPLE 5.5   *Selectivity of the Allyl Chloride Reactions*

To demonstrate the advantages of running the reactions at intermediate temperatures, show the rate constants for the three competing reactions as a function of temperature.

SOLUTION

As shown in the following graph, the rate constant of the desirable Reaction 1 is the largest relative to the rate constants of the other two reactions at the intermediate temperatures.

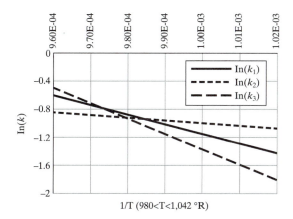

Often, adequate selectivities cannot be achieved by simply adjusting the temperature and pressure of the reactions. In these cases, the field of catalysis plays a crucial role. Chemists, biochemists, and chemical engineers work closely to design catalysts that permit the desired reactions to proceed more rapidly at lower temperatures while reducing the rates of side reactions. For this purpose, many successful reaction operations utilize crystalline zeolites, which are very effective *shape-selective catalysts*. In fact, it is possible to synthesize zeolite structures in which the cavities are just small enough to prevent the reactant molecules for the undesired side reactions from migrating to the reaction sites. Other commonly used catalysts involve the rare metals; for example, platinum, palladium, and rhodium. Clearly, when distributing chemicals in process synthesis, it is crucial to understand the relative rates of the competing reactions. Company laboratories are a key source of this information, as are patents and articles in the scientific journals. Process designers spend considerable time searching the extensive literature to locate appropriate catalysts.

Of course, many fine books have been written on the subject of catalysis. To study how to design reactors to achieve the desired selectivity, several outstanding textbooks are available; consider, for example, *Elements of Chemical Reaction Engineering* (Fogler, 1986) and *The Engineering of Chemical Reactions* (Schmidt, 1998).

Reactive Separations

> **Heuristic 8: *For reversible reactions especially, consider conducting them in a separation device capable of removing the products, and hence driving the reactions to the right. Such reaction-separation operations lead to very different distributions of chemicals.***

The last step in process synthesis recommended in Section 3.4 is *task integration*, that is, the combination of operations into process units. In the synthesis steps recommended there, reaction operations are positioned first, chemicals are distributed (as discussed earlier in this section), and separation operations are positioned, followed by temperature-, pressure-, and phase-change operations, before task integration occurs. In some cases, however, this strategy does not lead to effective combinations of reaction and separation operations, for example, reactive distillation towers, reactive absorption towers, and reactive membranes. Alternatively, when the advantages of merging these two operations are examined by a design team, a combined reaction-separation operation is placed in the flowsheet before chemicals are distributed, with a significant improvement in the economics of the design. Although the subject of reactive separations is covered in Section 7.5, a brief introduction to reactive distillation is provided next.

Reactive distillation is used commonly when the chemical reaction is reversible, for example,

$$a\text{A} + b\text{B} \rightleftharpoons c\text{C} + d\text{D}$$

and there is a significant difference in the relative volatilities of the chemicals at the conditions of temperature and pressure suitable for the reaction. In conventional processing, when a reversible reaction operation is followed by a distillation column, it is common to use an excess of a feed chemical to drive the reaction forward. Alternatively, when the reaction takes place in the gas phase, the pressure is raised or lowered, depending on whether the summation of the stoichiometric coefficients is negative or positive. An advantage of reactive distillation, as shown for the production of methyl acetate,

$$\text{MeOH} + \text{HOAc} \rightleftharpoons \text{MeOAc} + \text{H}_2\text{O}$$

in Figure 5.6, is that the product chemicals are withdrawn from the reaction section in vapor and liquid streams, thereby driving the reaction forward without excess reactant or changes in pressure. Since methanol is more volatile than acetic acid, it is fed to the bottom of the reaction zone, where it concentrates in the vapor phase and contacts acetic acid, which is fed at the top of the reaction zone and concentrates in the liquid phase. As methyl acetate is formed, it concentrates in the vapor phase and leaves the tower in the nearly pure distillate. The water product concentrates in the liquid phase and is removed from the tower in a nearly pure bottoms stream.

In summary, when the advantages of a combined operation (involving a reversible reaction and distillation, in this case) are clear to the design team, the operation can be in-

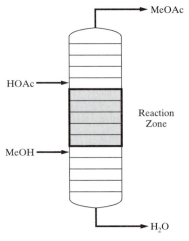

MeOH + HOAc ⇌ MeOAc + H₂O

Figure 5.6 Reactive distillation to produce methyl acetate.

serted into the flowsheet *before* the chemicals are distributed in process synthesis. This is a heuristic design procedure that can simplify the synthesis steps and lead to a more profitable design.

Optimal Conversion

Consider the case of a single reaction with a large chemical equilibrium constant such that it is possible to obtain a complete conversion. However, the optimal conversion may not be complete conversion. Instead, an economic balance between a high reactor section cost at high conversion and a high separation section cost at low conversion determines the optimum. Unfortunately, a heuristic for the optimal conversion is not available because it depends on many factors. This subject is considered in more detail in Chapter 8 on reactor–separator–recycle networks.

5.4 SEPARATIONS

Separations Involving Liquid and Vapor Mixtures

> *Heuristic 9:* **Separate liquid mixtures using distillation, stripping, enhanced (extractive, azeotropic, reactive) distillation, liquid–liquid extraction, crystallization and/or adsorption. The selection between these alternatives is considered in Chapter 7.**

> *Heuristic 10:* **Attempt to condense or partially condense vapor mixtures with cooling water or a refrigerant. Then, use Heuristic 9.**

> *Heuristic 11:* **Separate vapor mixtures using partial condensation, cryogenic distillation, absorption, adsorption, membrane separation and/or desublimation. The selection among these alternatives is considered in Chapter 7.**

The selection of separation processes is dependent on the phase of the stream to be separated and the relative physical properties of its chemical species. Liquid and vapor streams are separated often using the strategy recommended by Douglas (1988) in *Conceptual Design of Chemical Processes*. This strategy is reproduced here using the original figures, slightly modified, with the publisher's permission. It is expanded upon in Chapter 7.

When the reaction products are in the liquid phase, Douglas recommends that a liquid-separation system be inserted in the flowsheet, as shown in Figure 5.7. The liquid-separation system involves one or more of the following separators: distillation and enhanced distillation, stripping, liquid–liquid extraction, and so on, with the unreacted chemicals recovered in a liquid phase and recycled to the reaction operation.

For reaction products in the vapor phase, Douglas recommends that an attempt be made to partially condense them by cooling with cooling water or a refrigerant. Cooling water can cool the reaction products typically to 35°C, as shown in Figure 5.8. The usual objective is to obtain a liquid phase, which is easier to separate, without using refrigeration, which involves an expensive compression step. When partial condensation occurs, a liquid separation system is inserted, with a liquid purge added when necessary to remove trace inerts that concentrate

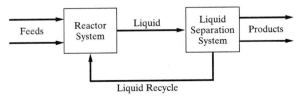

Figure 5.7 Flowsheet to separate liquid reactor effluents. (Reprinted with permission from Douglas, 1988.)

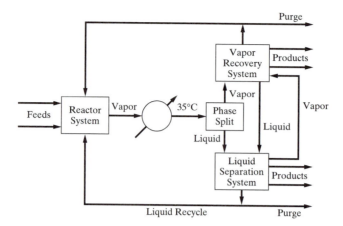

Figure 5.8 Flowsheet to separate vapor reactor effluents. (Modified and reprinted with permission from Douglas, 1988.)

in the liquid and are not readily separated. The vapor phase is sent to a vapor recovery system, which involves one or more of the following separations: partial condensation (at elevated pressures and cryogenic temperatures), cryogenic distillation, absorption, adsorption, membrane separation, and desublimation. Unreacted chemicals are recycled to the reactor section and vapor products are removed. A vapor purge is added when necessary to remove inerts that concentrate in the vapor and are not readily separated. Any liquid produced in the vapor recovery system is sent to the liquid recovery system for product recovery and the recycle of unreacted chemicals.

When the reactor effluent is already distributed between vapor and liquid phases, Douglas combines the two flowsheets, as shown in Figure 5.9. It should be recognized that the development of the separation systems for all three flowsheets involves several heuristics. First, certain separation devices, such as membrane separators, are not considered for the separation of liquids. Second, to achieve a partial condensation, cooling water is utilized initially, rather than compression and refrigeration. In this regard, it is presumed that liquid separations are preferred. An attempt is made to partially condense the vapor products, but no attempt is made to partially vaporize the liquid products. While these and other heuristics are based on considerable experience and usually lead to profitable designs, the designer needs to recognize their limitations and be watchful for situations in which they lead to suboptimal designs. Furthermore, for the separation of multicomponent streams, formal methods have been developed to synthesize separation trains involving vapors or liquids. These are covered in the next chapter.

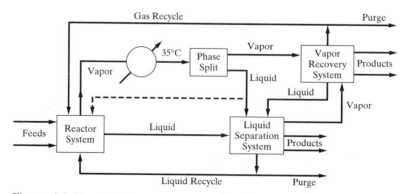

Figure 5.9 Flowsheet to separate vapor/liquid reactor effluents. (Modified and reprinted with permission from Douglas, 1988.)

Separations Involving Solid Particles

For streams that involve solid phases or species that crystallize or precipitate, additional considerations are necessary when selecting a separation system because several steps may be necessary due to the impossibility of removing dry solids directly from a liquid. When separating inorganic chemicals, in an aqueous solution especially, the stream is often cooled or partially evaporated to recover solids by crystallization, followed by filtration or centrifugation, and then drying. Often slurries are concentrated by settling, centrifugation, or filtration, before drying, as discussed in Section 7.7. Other devices for the removal of solid particles from gases and liquids are cyclones and hydroclones, respectively, as discussed in Section 5.9.

Crystallization occurs in three different modes. *Solution crystallization* applies mainly to inorganic chemicals, which are crystallized from a solvent, often water, with an operating temperature far below the melting point of the crystals. *Precipitation* is fast solution crystallization that produces large numbers of very small crystals. It usually refers to the case where one product of two reacting solutions is a solid of low solubility, for example, the precipitation of insoluble silver chloride when aqueous solutions of silver nitrate and sodium chloride are mixed together. In *melt crystallization*, two or more chemicals of comparable melting points are separated at an operating temperature in the range of the melting points. Crystallization is capable of producing very pure chemicals when conducted according to the following heuristics, noting that recovery by any mode of crystallization may be limited by a eutectic composition.

> **Heuristic 12:** *Crystallize inorganic chemicals from a concentrated aqueous solution by chilling when solubility decreases significantly with decreasing temperature. Keep the solution at most 1 to 2°F below the saturation temperature at the prevailing concentration. Use crystallization by evaporation, rather than chilling, when solubility does not change significantly with temperature.*

> **Heuristic 13:** *Crystal growth rates are approximately the same in all directions, but crystals are never spheres. Crystal growth rates and sizes are controlled by limiting the extent of supersaturation, $S = C/C_{saturation}$, where C is concentration, usually in the range $1.02 < S < 1.05$. Growth rates are influenced greatly by the presence of impurities and of certain specific additives that vary from case to case.*

> **Heuristic 14:** *Separate organic chemicals by melt crystallization with cooling, using suspension crystallization, followed by removal of crystals by settling, filtration, or centrifugation. Alternatively, use layer crystallization on a cooled surface, with scraping or melting to remove the crystals. If the melt forms a solid solution, instead of a eutectic, use repeated melting and freezing steps, called fractional melt crystallization, or zone melting to obtain nearly pure crystalline products.*

Prior to crystallization, it is common to employ evaporation to concentrate a solution, particularly an aqueous solution of inorganic chemicals. Because of the relatively high cost of evaporating water with its very large heat of vaporization, the following heuristics are useful for minimizing the cost.

> **Heuristic 15:** *Using multiple evaporators (called effects) in series, the latent heat of evaporation of water is recovered and reused. With a single evaporator, the ratio of the amount of water evaporated to the amount of external steam supplied to cause the evaporation is typically 0.8. For two effects,*

the ratio becomes 1.6; for three effects 2.4, and so forth. The magnitude of the boiling-point elevation caused by the dissolved inorganic compounds is a controlling factor in selecting the optimal number of effects. The elevation is often in the range of 3 to 10°F between solution and pure water boiling points. When the boiling-point rise is small, minimum evaporation cost is obtained with 8 to 10 effects. When the boiling-point rise is appreciable, the optimal number of effects is small, 6 or less. If necessary, boost interstage steam pressures with steam-jet or mechanical compressors.

Heuristic 16:   *When employing multiple effects, the liquid and vapor flows may be in the same or different directions. Use forward feed, where both liquid and vapor flow in the same direction, for a small number of effects, particularly when the liquid feed is hot. Use backward feed, where liquid flows in a direction opposite to vapor flows, for cold feeds and/or a large number of effects. With forward feed, intermediate liquid pumps are not necessary, whereas they are for backward feed.*

Solution crystallization produces a slurry of crystals and mother liquor, which is partially separated by filtration or centrifugation into a wet cake and a mother liquor. Filtration through a filter medium of porous cloth or metal may be carried out under gravity, vacuum, or pressure. Centrifugation may utilize a solid bowl or a porous bowl with a filter medium. Important factors in the selection of equipment include: (1) moisture content of the cake, (2) solids content of the mother liquor, (3) fragility of the crystals, (4) crystal particle size, (5) need for washing the crystals to replace mother liquor with pure water, and (6) filtration rate. Filtration rate is best determined by measuring the rate of cake thickness buildup using a small-scale laboratory vacuum leaf filter test with the following criteria: Rapid, 0.1 to 10 cm/s; Medium, 0.1 to 10 cm/min; Slow, 0.1 to 10 cm/hr.

Heuristic 17:   *When crystals are fragile, effective washing is required, and clear mother liquor is desired, use: gravity, top-feed horizontal pan filtration for slurries that filter at a rapid rate; vacuum rotary-drum filtration for slurries that filter at a moderate rate; and pressure filtration for slurries that filter at a slow rate.*

Heuristic 18:   *When cakes of low moisture content are required, use: solid-bowl centrifugation if solids are permitted in the mother liquor; centrifugal filtration if effective washing is required.*

Wet cakes from filtration or centrifugation operations are sent to dryers for removal of remaining moisture. A large number of different types of commercial dryers have been developed to handle the many different types of feeds, which include not only wet cakes, but also pastes, slabs, films, slurries, and liquids. The heat for drying may be supplied from a hot gas in direct contact with the wet feed or it may be supplied indirectly through a wall. Depending on the thickness of the feed and the degree of agitation, drying times can range from seconds to hours. The following heuristics are useful in making a preliminary selection of drying equipment:

Heuristic 19:   *For granular material, free flowing or not, of particle sizes from 3 to 15 mm, use continuous tray and belt dryers with direct heat. For free-flowing granular solids that are not heat sensitive, use an inclined rotary cylindrical dryer, where the heat may be supplied directly from a hot gas or indirectly from tubes, carrying steam, that run the length of the dryer and are located in one or two rings concentric to and located just inside the dryer rotating shell. For small, free-flowing particles of*

1 to 3 mm in diameter, when rapid drying is possible, use a pneumatic conveying dryer with direct heat. For very small free-flowing particles of less than 1 mm in diameter, use a fluidized-bed dryer with direct heat.

Heuristic 20: *For pastes and slurries of fine solids, use a drum dryer with indirect heat. For a liquid or pumpable slurry, use a spray dryer with direct heat.*

5.5   HEAT REMOVAL FROM AND ADDITION TO REACTORS

After positioning the separation operations, the next step in process synthesis, as recommended in Section 3.4, is to insert the operations for temperature, pressure, and phase changes. To accomplish this, many excellent algorithms for heat and power integration have been developed. These are presented in Chapter 10. The objective of this section, which considers several approaches to remove the heat generated in exothermic reaction operations and to add the heat required by endothermic reaction operations, is more limited. The subject is discussed at this point because several of the approaches for heat transfer affect the distribution of chemicals in the flowsheet and are best considered after the reaction operations are positioned. These approaches are discussed next, together with other common approaches to remove or add the heat of reaction. First, the methods for removing the heat generated by exothermic reaction operations are presented. Then, some distinctions are drawn for the addition of heat to endothermic reaction operations. For the details of heat removal from or addition to complex reactor configurations, the reader is referred to Section 6.2.

Heat Removal from Exothermic Reactors

Given an exothermic reaction operation, an important first step is to compute the *adiabatic reaction temperature*, that is, the maximum temperature attainable, in the absence of heat transfer. Note that this can be accomplished readily with any of the process simulators. Furthermore, algorithms have been presented for these iterative calculations by Henley and Rosen (1969) and Myers and Seider (1976), among many sources.

EXAMPLE 5.6   *Adiabatic Reaction Temperature*

Consider the reaction of carbon monoxide and hydrogen to form methanol:

$$CO + 2H_2 \rightarrow CH_3OH$$

With the reactants fed in stoichiometric amounts at 25°C and 1 atm, calculate the standard heat of reaction and the adiabatic reaction temperature.

SOLUTION

In ASPEN PLUS, the RSTOIC subroutine is used with a feed stream containing 1 lbmol/hr CO and 2 lbmol/hr H_2 and the PSRK method (Soave–Redlich–Kwong equation of state with Holderbaum–Gmehling mixing rules). To obtain the heat of reaction, the fractional conversion of CO is set at unity, with the product stream temperature at 25°C and the vapor fraction at 1.0. The latter keeps the methanol product in the vapor phase at 2.44 psia, and hence both the reactants and product species are vapor. The heat duty computed by RSTOIC is −38,881 Btu/hr, and hence the heat of reaction is $\Delta H_r = -38{,}881$ Btu/lbmol CO.

To obtain the adiabatic reaction temperature for complete conversion, the heat duty is set at zero and the pressure of the methanol product stream is returned to 1 atm. This produces an effluent temperature of 1,158°C (2,116°F), which is far too high for the Cu-based catalyst and the materials of construction in most reactor vessels. Hence, a key question in the synthesis of the methanol process, and similar processes involving highly exothermic reactions, is how to lower the product temperature. In most cases, the designer is given or sets the maximum temperature in the reactor and evaluates one of the heat-removal strategies described in this section.

Note that these results can be reproduced using the EXAM5-6.bkp file on the CD-ROM. Also, RSTOIC-type subroutines are described in Section 6.1. ■

> ***Heuristic 21:***   ***To remove a highly exothermic heat of reaction, consider the use of excess reactant, an inert diluent, or cold shots. These affect the distribution of chemicals and should be inserted early in process synthesis.***
>
> ***Heuristic 22:***   ***For less exothermic heats of reaction, circulate reactor fluid to an external cooler, or use a jacketed vessel or cooling coils. Also, consider the use of intercoolers between adiabatic reaction stages.***

To achieve lower temperatures, several alternatives are possible, beginning with those that affect the distribution of chemicals.

1. Use an excess of one reactant to absorb the heat. This alternative was discussed earlier in Section 5.3 and is illustrated in Figure 5.10a, where excess B is recovered from the separator and recirculated to the reaction operation. Heat is removed in the separator or by exchange with a cold process stream or a cold utility (e.g., cooling water).

EXAMPLE 5.7   *Excess Reactant*

Returning to Example 4.6, an excess of H_2 is specified such that the mole ratio of H_2/CO is arbitrarily 10 and the temperature of the reactor effluent stream is computed.

SOLUTION

Again, using the RSTOIC subroutine in ASPEN PLUS with a complete conversion of CO, the effluent temperature is reduced to 337°C (639°F), a result that can be reproduced using the EXAM5-7.bkp file on the CD-ROM. The Sensitivity command can be used to compute the effluent temperature as a function of the H_2/CO ratio, as in Exercise 5.8. ■

2. Use of an inert diluent, S. Figure 5.10b illustrates this alternative. One example occurs in the manufacture of methanol, where carbon monoxide and hydrogen are reacted in a fluidized bed containing catalyst. In the Marathon Oil process, a large stream of inert oil is circulated through the reactor, cooled, and recirculated to the reactor. Note that diluents like oil, with large heat capacities, are favored to maintain lower temperatures with smaller recirculation rates. The disadvantage of this approach, of course, is that a new species, S, is introduced, which, when separated from the reaction mix, cannot be removed entirely from the desired product. As in Alternative 1, heat is removed in the separator or by exchange with a cold process stream or a cold utility.

EXAMPLE 5.8   *Dodecane Diluent*

Returning to Example 5.6, 5 lbmol/hr of dodecane is added to the reactor feed (1 lbmol/hr CO and 2 lbmol/hr H_2) and the temperature of the reactor effluent stream is computed.

SOLUTION

In this case, the effluent temperature is reduced sharply to 77.6°C (171.7°F), a result that can be reproduced using the EXAM5-8.bkp file on the CD-ROM. The dodecane flow rate is adjusted to give an adequate temperature reduction. This is accomplished in Exercise 5.9. ■

3. Use of cold shots. In this alternative, as illustrated in Figure 5.10c, one of the reactants is cooled and distributed among several adiabatic reaction operations in series. The effluent from each stage, at an elevated temperature, is cooled with a cold shot of reactant B. In each stage, additional A is reacted until it is completely consumed in the last stage. The reader is

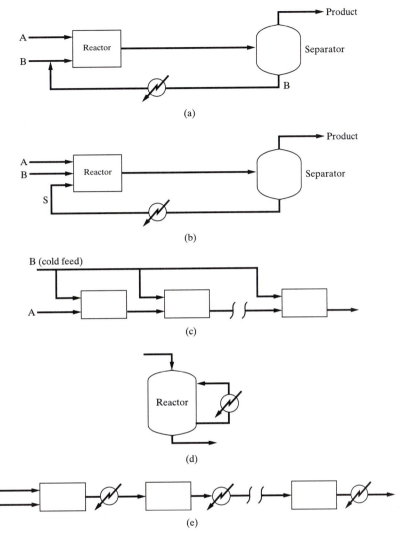

Figure 5.10 Flowsheets to remove the heat of reaction: (a) use of excess reactant; (b) use of inert diluent, S; (c) use of cold shots; (d) diabatic operation; (e) use of intercoolers.

referred to Example 6.3, which shows how to adjust the cold-shot distribution in an ammonia synthesis reactor to maximize the conversion of synthesis gas to ammonia.

The next two alternatives do not affect the distribution of chemicals and are usually considered for moderately exothermic reactions, later in process synthesis—that is, during heat and power integration, when opportunities are considered for heat exchange between high- and low-temperature streams.

4. Diabatic operation. In this case, heat is removed from the reaction operation in one of several ways. Either a cooling jacket is utilized or coils are installed, through which a cold process stream or cold utility is circulated. In some cases, the reaction occurs in catalyst-filled tubes, surrounded by coolant, or in catalyst-packed beds interspersed with tubes that convey a coolant stream, which is often the reactor feed stream, as illustrated for the ammonia reactor (TVA design) in Figure 5.11. Here, heat transfer from the reacting species in the catalyst bed preheats the reactants, N_2 and H_2, flowing in the cooling tubes, with the tube bundle designed to give adequate heat transfer as well as reaction. The design procedure is similar to that for the design of heat exchangers in Chapter 13. It is noted that for

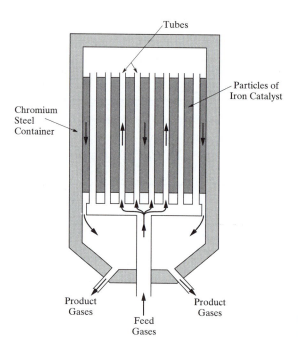

Figure 5.11 Tubular ammonia reactor. (TVA Design, from Baddour et al., 1965.)

large-capacity, ammonia-synthesis reactors, the optimal design usually calls for a series of adiabatic beds packed with catalyst, with cooling achieved using cold shots, as shown in Figure 5.10c. Most commercial ammonia reactors use a combination of these two cooling arrangements. Yet another alternative is to circulate a portion of the reacting mixture through an external heat exchanger in which the heat of reaction is removed, as shown in Figure 5.10d.

5. Use of intercoolers. As shown in Figure 5.10e, the reaction operation is divided into several adiabatic stages with heat removed by heat exchangers placed between each stage. Here, also, heat is transferred either to cold process streams that require heating or to cold utilities.

In all of these alternatives, the design team selects acceptable temperature levels and flow rates of the recirculating fluids. These are usually limited by the rates of reaction, and especially the need to avoid thermal runaway or catalyst deterioration, as well as the materials of construction and the temperature levels of the available cold process streams and utilities, such as cooling water. It is common to assign temperatures on the basis of these factors early in process synthesis. However, as optimization strategies are perfected, temperature levels are varied within bounds. See Chapters 10 and 18 for discussions of the use of optimization in process synthesis and optimization of process flowsheets, as well as Example 6.3 to see how constrained optimization is applied to design an ammonia cold-shot converter.

Heat Addition to Endothermic Reactors

> ***Heuristic 23:*** *To control temperature for a highly endothermic heat of reaction, consider the use of excess reactant, an inert diluent, or hot shots. These affect the distribution of chemicals and should be inserted early in process synthesis.*

> ***Heuristic 24:*** *For less endothermic heats of reaction, circulate reactor fluid to an external heater, or use a jacketed vessel or heating coils. Also, consider the use of interheaters between adiabatic reaction stages.*

For endothermic reaction operations, sources of the heat of reaction are needed. As in the case of exothermic reaction operations, the heat of reaction and the adiabatic reaction temperature can be computed using a simulator. The latter provides a lower bound on the temperature of the reactor effluent.

When adding heat to endothermic reaction operations, three approaches for heat addition affect the distribution of chemicals in the flowsheet and are best considered immediately after the reaction operations are positioned. Given that the feed stream is preheated, these approaches closely parallel the first three approaches for heat removal from exothermic reaction operations in Figures 5.10a–5.10c. When an excess of a reactant is utilized, the decrease in effluent temperature varies inversely with the degree of excess. Similarly, when an inert diluent is added, the effluent temperature is decreased inversely with the amount of diluent. For example, when ethylbenzene is pyrolyzed to produce styrene,

$$\text{Ethylbenzene} \rightarrow \text{Styrene} + \text{H}_2$$

superheated steam is added to provide the heat of reaction and keep the reaction temperature elevated. Of course, the addition of steam significantly increases the reactor volume and both the operating and installation costs. Finally, it is possible to add hot shots of reactants to a series of reactors, in a way similar to the cold shots added in Figure 5.10c.

The same two alternatives that do not affect the distribution of chemicals apply for the addition of heat to endothermic reaction operations. In this case, the diabatic operation in Figure 5.10d involves the addition of heat in ways similar to the removal of heat. Jackets, coils, and heat exchanger designs are very common. Also, interheaters between stages, as in Figure 5.10e, are used in many situations in place of intercoolers.

5.6 HEAT EXCHANGERS AND FURNACES

As discussed in the previous two sections, heat exchange is commonly used in conjunction with separation and reaction steps to change the temperature and/or phase condition of a process stream. When using a process simulator to perform heat exchange calculations, it is necessary to select a method of heat exchange from the following six possibilities, where all but the last two involve the separation, by a solid wall, of the two streams exchanging heat.

1. Heat exchange between two process fluids using a double-pipe, shell-and-tube, or compact heat exchanger.
2. Heat exchange between a process fluid and a utility, such as cooling water or steam, using a double-pipe, shell-and-tube, air-cooled, or compact heat exchanger.
3. High-temperature heating of a process fluid using heat from the products of combustion in a furnace (also called a fired heater).
4. Heat exchange within a reactor or separator, rather than in an external heat-exchange device such as a shell-and-tube heat exchanger or furnace, as described in Section 5.5.
5. Direct heat exchange by mixing the two streams that are exchanging heat.
6. Heat exchange involving solid particles.

The following heuristics are useful in selecting an initial basis for the heat exchange method and the operating conditions. Details of heat exchanger selection and design are presented in Chapter 13.

> **Heuristic 25:** *Unless required as part of the design of the separator or reactor, provide necessary heat exchange for heating or cooling process fluid streams, with or without utilities, in an external shell-and-tube heat exchanger using countercurrent flow. However, if a process stream requires heating above 750°F, use a furnace unless the process fluid is subject to chemical decomposition.*

Preliminary estimates of exit temperatures of streams flowing through a heat exchanger can be made with the following heuristics.

> *Heuristic 26: **Near-optimal minimum temperature approaches in heat exchangers
> depend on the temperature level as follows:
> 10°F or less for temperatures below ambient.
> 20°F for temperatures at or above ambient up to 300°F.
> 50°F for high temperatures.
> 250 to 350°F in a furnace for flue gas temperature above inlet process
> fluid temperature.***

As an example, suppose it is desired to heat 25,000 lb/hr of toluene at 100°F and 90 psia with 25,000 lb/hr of styrene at 300°F and 50 psia. Under these conditions, assume that both streams will be liquid, but this must be verified by flash calculations after the exit temperatures and pressures have been determined. From the previous two heuristics, use a shell-and-tube heat exchanger with countercurrent flow and a minimum approach temperature of 20°F. Let the average specific heats of the two streams be 0.43 Btu/lb-°F for toluene and 0.44 Btu/lb-°F for styrene. Initially it is not known to which end of the heat exchanger the 20°F minimum approach applies. Assume it applies at the toluene inlet end. If so, the styrene exit temperature is $100 + 20 = 120$°F. This gives a heat-exchanger duty, based on styrene of:

$$Q = 25,000(0.44)(300 - 120) = 1,980,000 \text{ Btu/hr}$$

Using this duty, the exit temperature of toluene, $T_{\text{toluene out}}$, can be computed from:

$$Q = 1,980,000 = 25,000(0.43)(T_{\text{toluene out}} - 100)$$

Solving, $T_{\text{toluene out}} = 284.2$°F. But this gives a temperature approach of $300 - 284.2 = 15.8$°F at the styrene inlet end, which is less than the minimum approach of 20°F. Therefore, the minimum approach must be applied to the styrene inlet end. Similar calculations give $T_{\text{toluene out}} = 280$°F and $T_{\text{styrene out}} = 124.1$°F. This corresponds to an approach temperature at the toluene inlet end of 24.1°F, which is greater than the minimum approach temperature and, therefore, is acceptable.

> *Heuristic 27: **When using cooling water to cool or condense a process stream,
> assume a water inlet temperature of 90°F (from a cooling tower) and a
> maximum water outlet temperature of 120°F.***

When cooling and condensing a gas, both sensible and latent heat can be removed in a single heat exchanger. However, because of the many two-phase flow regimes that can occur when boiling a fluid, it is best to provide three separate heat exchangers when changing a subcooled liquid to a superheated gas, especially when the difference between the bubble point and dew point is small. The first exchanger preheats the liquid to the bubble point; the second boils the liquid; the third superheats the vapor.

> *Heuristic 28: **Boil a pure liquid or close-boiling liquid mixture in a separate heat
> exchanger, using a maximum overall temperature driving force of 45°F
> to ensure nucleate boiling and avoid undesirable film boiling as
> discussed in Section 13.1.***

As discussed in detail in Section 13.1, the minimum approach temperature in a countercurrent-flow heat exchanger may occur at an intermediate location rather than at one of the two ends when one of the two streams is both cooled and condensed. If the minimum temperature approach is assumed to occur at one of the two ends of the heat exchanger, a smaller approach or a temperature crossover that violates the second law of thermodynamics may occur at an intermediate location. To avoid this situation, the following heuristic should be applied:

> *Heuristic 29: **When cooling and condensing a stream in a heat exchanger, a zone
> analysis, described in Section 13.1, should be made to make sure that
> the temperature difference between the hot stream and the cold stream***

is equal to or greater than the minimum approach temperature at all locations in the heat exchanger. The zone analysis is performed by dividing the heat exchanger into a number of segments and applying an energy balance to each segment to determine corresponding stream inlet and outlet temperatures for the segment, taking into account any phase change. A process simulation program conveniently accomplishes the zone analysis.

When using a furnace to heat and/or vaporize a process fluid, the following heuristic is useful for establishing inlet and outlet heating medium temperature conditions so that fuel and air requirements can be estimated.

Heuristic 30: *Typically, a hydrocarbon gives an adiabatic flame temperature of approximately 3,500°F when using the stoichiometric amount of air. However, use excess air to achieve complete combustion and give a maximum flue-gas temperature of 2,000°F. Set the stack gas temperature at 650 to 950°F to prevent condensation of corrosive components of the flue gas.*

Pressure drops of fluids flowing through heat exchangers and furnaces may be estimated with the following heuristics.

Heuristic 31: *Estimate heat-exchanger pressure drops as follows:*
1.5 psi for boiling and condensing.
3 psi for a gas.
5 psi for a low-viscosity liquid.
7–9 psi for a high-viscosity liquid
20 psi for a process fluid passing through a furnace.

Unless exotic materials are used, heat exchangers should not be used for cooling and/or condensing process streams with temperatures above 1,150°F. Instead, use the following heuristic for direct heat exchange.

Heuristic 32: *Quench a very hot process stream to at least 1,150°F before sending it to a heat exchanger for additional cooling and/or condensation. The quench fluid is best obtained from a downstream separator as in Figure 4.21 for the toluene hydrodealkylation process. Alternatively, if the process stream contains water vapor, liquid water may be an effective quench fluid.*

Streams of solid particles are commonly heated or cooled directly or indirectly. Heat transfer is much more rapid and controllable when using direct heat exchange.

Heuristic 33: *If possible, heat or cool a stream of solid particles by direct contact with a hot gas or cold gas, respectively, using a rotary kiln, a fluidized bed, a multiple hearth, or a flash/pneumatic conveyor. Otherwise, use a jacketed spiral conveyor.*

5.7 PUMPING, COMPRESSION, PRESSURE REDUCTION, VACUUM, AND CONVEYING OF SOLIDS

As mentioned in the previous section, it is common to consider the integration of all temperature- and pressure-change operations. This is referred to as *heat and power integration* and is covered in Chapter 10 after important thermodynamic considerations are presented first in Chapter 9. At this point, however, there are several important heuristics that are useful in determining what type of operations to insert into the flowsheet to increase or decrease pressure. Details of the equipment used to perform pressure-change operations are presented in Chapter 15.

Increasing the Pressure

Gases: To increase the pressure, the most important consideration is the phase state (vapor, liquid, or solid) of the stream. If the stream is a gas, the following heuristic applies for determining whether a *fan*, *blower*, or *compressor* should be used.

> **Heuristic 34:** *Use a fan to raise the gas pressure from atmospheric pressure to as high as 40 inches water gauge (10.1 kPa gauge or 1.47 psig). Use a blower or compressor to raise the gas pressure to as high as 206 kPa gauge or 30 psig. Use a compressor or a staged compressor system to attain pressures greater than 206 kPa gauge or 30 psig.*

In Figure 4.20 for the toluene hydrodealkylation process, the pressure of the recycle gas leaving the flash drum at 100°F and 484 psia is increased with a compressor to 569 psia, so that, after pressure drops of 5 psia through the heat exchanger and 70 psia through the furnace, it enters the reactor at a required pressure of 494 psia.

The following heuristic is useful for estimating the exit temperature, which can be significantly higher than the entering temperature, and the power requirement when increasing the gas pressure by a single stage of reversible, adiabatic compression.

> **Heuristic 35:** *Estimate the theoretical adiabatic horsepower (THp) for compressing a gas from:*
>
> $$\mathrm{THp} = \mathrm{SCFM}\left(\frac{T_1}{8{,}130a}\right)\left[\left(\frac{P_2}{P_1}\right)^a - 1\right] \qquad (5.1)$$
>
> *where SCFM = standard cubic feet of gas per minute at 60°F and 1 atm (379 SCF/lbmol), T_1 = gas inlet temperature in °R, inlet and outlet pressures, P_1 and P_2, are absolute pressures, and $a = (k - 1)/k$, with k = the gas specific heat ratio, C_p/C_v.*
>
> *Estimate the theoretical exit temperature, T_2, for a gas compressor from:*
>
> $$T_2 = T_1(P_2/P_1)^a \qquad (5.2)$$

For example, if air at 100°F is compressed from 1 to 3 atm (compression ratio = 3), using $k = 1.4$, the THp is computed to be 128 Hp/standard million ft³/day, with an outlet temperature = 306°F.

When using a compressor, the gas theoretical exit temperature should not exceed approximately 375°F, the limit imposed by most compressor manufacturers. This corresponds to a compression ratio of 4 for $k = 1.4$ and $T_2 = 375$°F. When the exit gas temperature exceeds the limit, a single gas compression step cannot be used. Instead, a multistage compression system, with intercoolers between each stage, must be employed. Each intercooler cools the gas back down to approximately 100°F. The following heuristic is useful for estimating the number of stages, N, required and the interstage pressures.

> **Heuristic 36:** *Estimate the number of gas compression stages, N, from the following table, which assumes a specific heat ratio of 1.4 and a maximum compression ratio of 4 for each stage.*

Final Pressure/Inlet Pressure	Number of Stages
<4	1
4 to 16	2
16 to 64	3
64 to 256	4

> *Optimal interstage pressures correspond to equal Hp for each*
> *compressor. Therefore, based on the above equation for theoretical*
> *compressor Hp, estimate interstage pressures by using approximately*
> *the same compression ratio for each stage with an intercooler pressure*
> *drop of 2 psi or 15 kPa.*

For example, in Exercise 4.8, a feed gas at 100°F and 30 psia is to be compressed to 569 psia. From the above table, with an overall compression ratio of 569/30 = 19, a 3-stage system is indicated. For equal compression ratios, the compression ratio for each stage of a 3-stage system = $19^{1/3}$ = 2.7. The estimated stage pressures are as follows, taking into account a 2 psi drop for each intercooler and its associated piping:

Stage	Compressor Inlet Pressure, psia	Compressor Outlet Pressure, psia
1	30	81
2	79	213
3	211	569

When compressing a gas, the entering stream must not contain liquid and the exiting stream must be above its dew point so as not to damage the compressor. To remove any entrained liquid droplets from the entering gas, a vertical knock-out drum equipped with a demister pad is placed just upstream of the compressor. To prevent condensation in the compressor, especially when the entering gas is near its dew point, a heat exchanger should also be added at the compressor inlet to provide sufficient preheat to ensure that the exiting gas is well above its dew point.

Liquids: If the pressure of a liquid is to be increased, a pump is used. The following heuristic is useful for determining the types of pumps best suited for a given task, where the head in feet is the pressure increase across the pump in psf (pounds force/ft^2) divided by the liquid density in lb/ft^3.

> **Heuristic 37:**   *For heads up to 3,200 ft and flow rates in the range of 10 to 5,000 gpm,*
> *use a centrifugal pump. For high heads up to 20,000 ft and flow rates*
> *up to 500 gpm, use a reciprocating pump. Less common are axial*
> *pumps for heads up to 40 ft for flow rates in the range of 20 to 100,000*
> *gpm and rotary pumps for heads up to 3,000 ft for flow rates in the*
> *range of 1 to 1,500 gpm.*

For liquid water, with a density of 62.4 lb/ft^3, heads of 3,000 and 20,000 ft correspond to pressure increases across the pump of 1,300 and 8,680 psi, respectively.

When pumping a liquid from an operation at one pressure, P_1, to a subsequent operation at a higher pressure, P_2, the pressure increase across the pump must be higher than $P_2 - P_1$ in order to overcome pipeline pressure drop, control valve pressure drop, and possible increases in elevation (potential energy). This additional pressure increase may be estimated by the following heuristic.

> **Heuristic 38:**   *For liquid flow, assume a pipeline pressure drop of 2 psi/100 ft of pipe*
> *and a control valve pressure drop of at least 10 psi. For each 10-ft rise*
> *in elevation, assume a pressure drop of 4 psi.*

For example, in Figure 3.8 the combined chlorination reactor effluent and dichloroethane recycle at 1.5 atm is sent to a pyrolysis reactor operating at 26 atm. Although no pressure drops are shown for the two temperature-change and one phase-change operations, they may be estimated at 10 psi total. The line pressure drop and control valve pressure drop may be estimated at 15 psi. Take the elevation change as 20 ft, giving 8 psi. Therefore, the total additional pressure increase is 10 + 15 + 8 = 33 psi or 2.3 atm. The required corresponding pressure increase across the pump (pressure change operation) is, therefore, (26 − 1.5) + 2.3 = 26.8 atm. For a liquid density of 78 lb/ft^3 or 10.4 lb/gal, the required pumping head is 26.8(14.7)(144)/78 = 730 ft. The flow rate through the pump is (158,300 + 105,500)/10.4/60 = 422 gpm. Using Heuristic 37, select a centrifugal pump.

The following heuristic provides an estimate of the theoretical pump Hp. Unlike the case of gas compression, the temperature change across the pump is small and can be neglected.

Heuristic 39: *Estimate the theoretical horsepower (THp) for pumping a liquid from:*

$$THp = (gpm)(Pressure\ increase,\ psi)/1,714 \qquad (5.3)$$

For example, the theoretical Hp for pumping the liquid in Figure 3.8, using the above data, is (422)(26.8)(14.7)/1,714 = 97 Hp.

Decreasing the Pressure

The pressure of a gas or liquid stream can be reduced to ambient pressure or higher with a single throttle valve or two or more such valves in series. The adiabatic expansion of a gas across a valve is accompanied by a decrease in the temperature of the gas. The exiting temperature is estimated from Eq. (5.2) above for gas compression. For a liquid, the exit temperature is almost the same as the temperature entering the valve. In neither case is shaft work recovered from the fluid. Alternatively, it is possible to recover energy in the form of shaft work that can be used elsewhere by employing a turbine-like device. For a gas, the device is referred to as an expander, expansion turbine, or turboexpander. For a liquid, the corresponding device is a power-recovery turbine. The following heuristics are useful in determining whether a turbine should be used in place of a valve.

Heuristic 40: *Consider the use of an expander for reducing the pressure of a gas or a pressure recovery turbine for reducing the pressure of a liquid when more than 20 Hp and 150 Hp, respectively, can be recovered.*

Heuristic 41: *Estimate the theoretical adiabatic horsepower (THp) for expanding a gas from:*

$$THp = SCFM\left(\frac{T_1}{8,130a}\right)\left[1 - \left(\frac{P_2}{P_1}\right)^a\right] \qquad (5.4)$$

Heuristic 42: *Estimate the theoretical horsepower (THp) for reducing the pressure of a liquid from:*

$$THp = (gpm)(Pressure\ decrease,\ psi)/1,714 \qquad (5.5)$$

In Figure 3.7, the pyrolysis effluent gas, following cooling to 170°C and condensation to 6°C at 26 atm, is reduced in pressure to 12 atm before entering the first distillation column. The flowsheet in Figure 3.8 shows the use of a valve, following the condenser, to accomplish the pressure reduction. Should a pressure recovery turbine be used? Assume a flow rate of 422 gpm. The pressure decrease is (26 − 12)(14.7) = 206 psi. Using Eq. (5.5), THp = (422)(206)/1,714 = 51, which is much less than 150 Hp. Therefore, according to the above heuristic, a valve is preferred. Alternatively, the pressure reduction step could be inserted just prior to the condenser, using an expander on the gas at its dew point of 170°C. The total flow rate is (58,300 + 100,000 + 105,500)/60 = 4,397 lb/min. The average molecular weight is computed to be 61.9, giving a molar flow rate of 71 lbmol/min. The corresponding SCFM (standard cubic feet per minute at standard conditions of 60°F and 1 atm) is (71)(379) = 26,900. Assume $k = 1.2$, giving $a = (1.2 − 1)/1.2 = 0.167$. With a decompression ratio of 12/26 = 0.462 and $T_1 = 797°R$, Eq. (5.4) above gives 1,910 THp, which is much more than 150 Hp. Therefore, according to the above heuristic, an expander should be used. The theoretical temperature of the gas exiting the expander, using the above Eq. (5.2) is 797(0.462)$^{0.167}$ = 701°R = 241°F = 116°C.

Pumping a Liquid or Compressing a Gas

When it is necessary to increase the pressure between process operations, it is almost always far less expensive to pump a liquid rather than compress a vapor. This is because the power required to increase the pressure of a flowing stream is

$$W = \int_{P_1}^{P_2} V \, dP \tag{5.6}$$

where V is the volumetric flow rate, which is normally far less for liquid streams—typically two orders of magnitude less. Hence, it is common to install pumps having approximately 10 Hp, whereas comparable compressors require approximately 1,000 Hp and are far more expensive to purchase and install. For these reasons, if the low-pressure stream is a vapor and the phase state is also vapor at the higher pressure, it is almost always preferable to condense the vapor, pump it, and revaporize it, rather than compress it, as illustrated in Figure 5.12. The exception to this heuristic is when refrigeration is required for condensation, which, as discussed in Chapter 9, involves extensive compression of the working fluid. If the low-pressure stream is a liquid and the high-pressure stream is a vapor, it is preferable to increase the pressure first with a pump and then vaporize the liquid, rather than vaporize the liquid and then compress it. This is the subject of Exercise 5.12 and the following example.

> **Heuristic 43: To increase the pressure of a stream, pump a liquid rather than compress a gas, unless refrigeration is needed.**

EXAMPLE 5.9   *Feed Preparation of Ethylbenzene*

Ethylbenzene is to be taken from storage as a liquid at 25°C and 1 atm and fed to a styrene reactor as a vapor at 400°C and 5 atm at 100,000 lb/hr. In this example, two alternatives are considered for positioning the temperature- and pressure-increase operations.

SOLUTION

1. Pump first. Using the PUMP and HEATER subroutines in ASPEN PLUS discussed on the multimedia CD-ROM that accompanies this book, 12.5 brake Hp are required to pump the liquid followed by 4.67×10^7 Btu/hr to vaporize the liquid and heat the vapor to 400°C.
2. Vaporize the liquid first. Using the HEATER and COMPR subroutines, discussed on the multimedia CD-ROM, 4.21×10^7 Btu/hr are required to vaporize the liquid and heat it to 349.6°C, followed by 1,897 brake Hp to compress the vapor to 5 atm at 400°C.

Clearly, the power requirement is substantially less when pumping a liquid. Note that these results can be reproduced using the EXAM5-9.bkp file on the CD-ROM. ∎

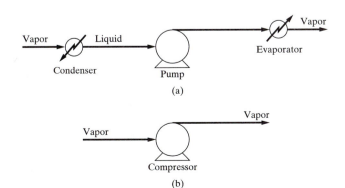

Figure 5.12 Alternatives for raising the pressure of a vapor stream: (a) pump a liquid; (b) compress the vapor.

Vacuum

When process pressures less than the ambient pressure are required, special devices and considerations are necessary. Vacuum operation is most common in crystallization, distillation, drying, evaporation, and pervaporation operations. A vacuum inside the equipment causes inleakage of ambient-pressure air. A vacuum system is used to remove this air together with any associated vapor in the process stream that passes through the equipment. For continuous processes, vacuums are predominantly in the range of 1 to 760 torr (0.13 to 101.3 kPa). In this range, it is common to use a vacuum pump, which compresses the gas from vacuum (suction pressure) to ambient pressure, or a jet-ejector system, which uses a flow of pressurized water or steam to mix with and remove the gas to create the vacuum. To design the vacuum system, it is necessary to estimate the inleakage of air, determine the total amount of gas (inleakage plus associated vapor) to be removed, and select an appropriate system for the vacuum level required. The following heuristics are useful. Details of vacuum equipment are presented in Section 16.6.

> *Heuristic 44:* *Estimate inleakage of air by:*
>
> $$w = kV^{0.667} \qquad\qquad (5.7)$$
>
> *where, w = lb/hr of air inleakage, V = ft³ of volume of the equipment under vacuum, and k = 0.2 for pressures greater than 90 torr, 0.15 for pressures between 21 and 89 torr, 0.10 for pressures between 3.1 and 20 torr, and 0.051 for pressures between 1 and 3 torr.*
>
> *Heuristic 45:* *To reduce the amount of gas sent to the vacuum system if its temperature is greater than 100°F, add a condenser using cooling water before the vacuum system. The gas leaving the condenser will be at a dew-point temperature of 100°F at the vacuum pressure.*
>
> *Heuristic 46:* *For pressures down to 10 torr and gas flow rates up to 10,000 ft³/min at the inlet to the vacuum system, use a liquid-ring vacuum pump. For pressures down to 2 torr and gas flow rates up to 1,000,000 ft³/min at the inlet to the vacuum system, use a steam-jet ejector system (one-stage for 100 to 760 torr, two-stage for 15 to 100 torr, and three-stage for 2 to 15 torr). Include a direct-contact condenser between stages.*
>
> *Heuristic 47:* *For a three-stage steam-jet ejector system used to achieve a vacuum of 2 torr, 100 pounds of 100 psig steam per pound of gas are required.*

Conveying Granular Solids

The movement of streams of granular solids horizontally and/or vertically is achieved with conveyors and elevators. When selecting the type of equipment, important considerations are stickiness, abrasiveness, particle size, and density of the solid particles.

> *Heuristic 48:* *If the solid particles are small in size, low in particle density, and are not sticky or abrasive, use pneumatic conveying with air at 1 to 7 ft³/ft³ of solids and 35 to 120 ft/s air velocity for distances up to 400 ft. Otherwise, for sticky and/or abrasive solids of any size and density, use a screw conveyer and/or bucket elevator for distances up to 150 ft. For solid particles of any size and shape, and not sticky, use a belt*

conveyor, with inclination up to 30° if necessary, for long distances up to a mile or more.

Changing the Pressure of Granular Solids

Continuous processes frequently involve granular solids, either alone, slurried in a liquid, or fluidized in a gas. In many cases, the streams containing the solids are at ambient pressure and are moved with ease by conveyors and elevators. However, in some cases, elevated pressure or vacuum may be required. If solid particles alone are being processed, they are placed in a closed hopper where the pressure is then adjusted to the required pressure. The solids are then conveyed under pressure or vacuum. For a slurry, the solids are placed in a hopper, from which they are discharged into a liquid at ambient pressure. The resulting slurry is then pumped to the desired pressure. A gas–solid particles mixture is formed by discharging solid particles from a hopper, through a rotary valve (also referred to as a rotary air-lock valve), into a gas stream at elevated pressure. For discharge pressures up to 15 psig, a standard rotary valve is used. For discharge pressures in the range of 15 psig to almost 100 psig, high-performance rotary valves are available. Rotary valves must be carefully designed to minimize or avoid gas leakage, prevent bridging of solids in the valve, and avoid wear of the vanes in the valve.

5.8   CHANGING THE PARTICLE SIZE OF SOLIDS AND SIZE SEPARATION OF PARTICLES

It is frequently necessary to change the particle size of solids to meet product specifications or change reaction or drying rate. Methods to accomplish changes in particle size, discussed in detail in *Chemical Process Equipment—Selection and Design* by S.M. Walas (1988) and in *Perry's Chemical Engineers' Handbook*, 7th Edition (1997), include crushing, grinding, and disintegration to reduce particle size; compression, extrusion, and agglomeration to increase particle size; and size separation devices to obtain a narrow range of particle size. Crushers and grinders are used with screens in closed circuits, wherein oversize material is recycled. Grindability is determined mainly by hardness as measured by the following Mohs scale, which ranges from 1 for the softest material to 10 for the hardest:

Material	Mohs Scale
Talc, $Mg_3SiO_{10}(OH)_2$	1
Gypsum, $CaSO_4 \cdot 2H_2O$	2
Calcite, $CaCO_3$	3
Fluorite, CaF_2	4
Apatite, $Ca_5(PO_4)_3(OH,F,Cl)$	5
Feldspar, Na, Ca, K, Al silicates	6
Quartz, SiO_2	7
Topaz, Al_2SiO_4 (F, OH)$_3$	8
Corundum, Al_2O_3	9
Diamond, C	10

Materials with a Mohs scale from 1 to 3 are considered soft and include graphite, many plastics, asphalt, sulfur, many inorganic salts, marble, and anthracite coal. Intermediate hardness extends from a Mohs scale of 4 to 6 and includes limestone, asbestos, and glass. Hard materials are characterized by a Mohs scale of 7 to 10 and include sand, granite, and emery. The following heuristics apply to particle-size reduction. Size of small particles is commonly stated in terms of screen mesh size according to the following U.S. standard (ASTM EII), where not all mesh sizes are given:

Mesh (Openings/inch)	Sieve Opening, mm
4	4.75
6	3.35
8	2.36
12	1.70
16	1.18
20	0.841
30	0.600
40	0.420
50	0.300
70	0.212
100	0.149
140	0.106
200	0.074
270	0.053
400	0.037

Heuristic 49: *Crushing of coarse solids. Use a jaw crusher to reduce lumps of hard, abrasive, and/or sticky materials of 4 inches to 3 feet in diameter to slabby particles of 1 to 4 inches in size. Use a gyratory crusher to reduce slabby materials of 8 inches to 6 feet in size to rounded particles of 1 to 10 inches in diameter. Use a cone crusher to reduce less hard and less sticky materials of 2 inches to 1 foot in diameter to particles of 0.2 inch (4 mesh) to 2 inches in diameter.*

Heuristic 50: *Grinding to fine solids. Use a rod mill to take particles of intermediate hardness as large as 20 mm and reduce them to particles in the range of 10 to 35 mesh. Use a ball mill to reduce particles of low to intermediate hardness of 1 to 10 mm in size to very small particles of less than 140 mesh.*

Heuristic 51: *Particle-size enlargement. Use compression with rotary compression machines to convert powders and granules into tablets of up to 1.5 inches in diameter. Use extruders with cutters to make pellets and wafers from pastes and melts. Use roll compactors to produce sheets from finely divided materials; the sheets are then cut into any desired shape. Use rotating drum granulators and rotary disk granulators with binders to produce particles in the size range of 2 to 25 mm.*

Heuristic 52: *Size separation of particles. Use a grizzly of spaced, inclined, vibrated parallel bars to remove large particles greater than 2 inches in diameter. Use a revolving cylindrical perforated screen to remove intermediate-size particles in the size range of 0.25 inch to 1.5 inches in diameter. Use flat, inclined woven screens (U.S. standard) that are vibrated, shaken or impacted with bouncing balls to separate small particles in the size range of 3 to 80 mesh. Use an air classifier to separate fine particles smaller than 80 mesh.*

5.9   REMOVAL OF PARTICLES FROM GASES AND LIQUIDS

Fine particles are most efficiently removed from dilute suspensions in gases and liquids by using centrifugal force in cyclones and hydroclones, respectively.

> *Heuristic 53:* *Use a cyclone separator to remove, from a gas, droplets or solid particles of diameter down to 10 microns (0.01 mm). Use a hydroclone separator to remove, from a liquid, insoluble liquid droplets or solid particles of diameter down to 5 microns (0.005 mm). However, small amounts of entrained liquid droplets are commonly removed from gases by vertical knock-out drums equipped with mesh pads to help coalesce the smallest droplets.*

5.10   SUMMARY

Having studied this chapter, the reader should

1. Be able to implement the steps in Section 3.4 for process synthesis more effectively, using the many heuristics presented herein, and summarized in Table 5.2. The examples and exercises should enable him or her to gain experience in their application.
2. Recognize the limitations of the heuristics in Table 5.2 and the role of the process simulator in permitting the systematic variation of parameters and the examination of alternative designs. The reader should also recognize that the heuristics listed are a subset of the many rules of thumb that have been applied by design teams in carrying out process synthesis.

Table 5.2 Heuristics in Chapter 5

	Heuristic
Reaction operations	
1	Select raw materials and chemical reactions to avoid, or reduce, the handling and storage of hazardous and toxic chemicals.
Distribution of chemicals	
2	Use an excess of one chemical reactant in a reaction operation to consume completely a valuable, toxic, or hazardous chemical reactant. The MSDSs will indicate which chemicals are toxic and hazardous.
3	When nearly pure products are required, eliminate inert species before the reaction operations when the separations are easily accomplished and when the catalyst is adversely affected by the inert, but not when a large exothermic heat of reaction must be removed.
4	Introduce purge streams to provide exits for species that enter the process as impurities in the feed or are formed in irreversible side reactions, when these species are in trace quantities and/or are difficult to separate from the other chemicals. Lighter species leave in vapor purge streams, and heavier species exit in liquid purge streams.
5	Do not purge valuable species or species that are toxic and hazardous, even in small concentrations (see the MSDSs). Add separators to recover valuable species. Add reactors to eliminate, if possible, toxic and hazardous species.
6	Byproducts that are produced in *reversible* reactions, in small quantities, are usually not recovered in separators or purged. Instead, they are usually **recycled to extinction**.
7	For competing reactions, both in series and parallel, adjust the temperature, pressure, and catalyst to obtain high yields of the desired products. In the initial distribution of chemicals, assume that these conditions can be satisfied. Before developing a base-case design, obtain kinetics data and check this assumption.
8	For reversible reactions especially, consider conducting them in a separation device capable of removing the products, and hence driving the reactions to the right. Such reaction-separation operations lead to very different distributions of chemicals.

(Continued)

Table 5.2 (*Continued*)

Separation operations— liquid and vapor mixtures	
9	Separate liquid mixtures using distillation, stripping, enhanced (extractive, azeotropic, reactive) distillation, liquid–liquid extraction, crystallization, and/or adsorption. The selection between these alternatives is considered in Chapter 7.
10	Attempt to condense or partially condense vapor mixtures with cooling water or a refrigerant. Then, use Heuristic 9.
11	Separate vapor mixtures using partial condensation, cryogenic distillation, absorption, adsorption, membrane separation and/or desublimation. The selection among these alternatives is considered in Chapter 7.
Separation operations— involving solid particles	
12	Crystallize inorganic chemicals from a concentrated aqueous solution by chilling when solubility decreases significantly with decreasing temperature. Keep the solution at most 1 to 2°F below the saturation temperature at the prevailing concentration. Use crystallization by evaporation, rather than chilling, when solubility does not change significantly with temperature.
13	Crystal growth rates are approximately the same in all directions, but crystals are never spheres. Crystal growth rates and sizes are controlled by limiting the extent of supersaturation, $S = C/C_{saturation}$, where C is concentration, usually in the range $1.02 < S < 1.05$. Growth rates are influenced greatly by the presence of impurities and of certain specific additives that vary from case to case.
14	Separate organic chemicals by melt crystallization with cooling, using suspension crystallization, followed by removal of crystals by settling, filtration, or centrifugation. Alternatively, use layer crystallization on a cooled surface, with scraping or melting to remove the crystals. If the melt forms a solid solution instead of a eutectic, use repeated melting and freezing steps, called fractional melt crystallization, or zone melting to obtain nearly pure crystalline products.
15	Using multiple evaporators (called effects) in series, the latent heat of evaporation of water is recovered and reused. With a single evaporator, the ratio of the amount of water evaporated to the amount of external steam supplied to cause the evaporation is typically 0.8. For two effects, the ratio becomes 1.6; for three effects 2.4, and so forth. The magnitude of the boiling-point elevation caused by the dissolved inorganic compounds is a controlling factor in selecting the optimal number of effects. The elevation is often in the range of 3 to 10°F between solution and pure water boiling points. When the boiling-point rise is small, minimum evaporation cost is obtained with 8 to 10 effects. When the boiling-point rise is appreciable, the optimal number of effects is small, 6 or less. If necessary, boost interstage steam pressures with steam-jet or mechanical compressors.
16	When employing multiple effects, the liquid and vapor flows may be in the same or different directions. Use forward feed, where both liquid and vapor flow in the same direction, for a small number of effects, particularly when the liquid feed is hot. Use backward feed, where liquid flows in a direction opposite to vapor flows, for cold feeds and/or a large number of effects. With forward feed, intermediate liquid pumps are not necessary, whereas they are for backward feed.
17	When crystals are fragile, effective washing is required, and clear mother liquor is desired, use: gravity, top-feed horizontal pan filtration for slurries that filter at a rapid rate; vacuum rotary-drum filtration for slurries that filter at a moderate rate; and pressure filtration for slurries that filter at a slow rate.
18	When cakes of low moisture content are required, use: solid-bowl centrifugation if solids are permitted in the mother liquor; centrifugal filtration if effective washing is required.
19	For granular material, free flowing or not, of particle sizes from 3 to 15 mm, use continuous tray and belt dryers with direct heat. For free-flowing granular solids that are not heat sensitive, use an inclined rotary cylindrical dryer, where the heat may be supplied directly from a hot gas or indirectly from tubes, carrying steam, that run the length of the dryer and are located in one or two rings concentric to and located just inside the dryer rotating shell. For small, free-flowing particles of 1 to 3 mm in diameter, when rapid drying is possible, use a pneumatic conveying dryer with direct heat. For very small free-flowing particles of less than 1 mm in diameter, use a fluidized-bed dryer with direct heat.
20	For pastes and slurries of fine solids, use a drum dryer with indirect heat. For a liquid or pumpable slurry, use a spray dryer with direct heat.

Table 5.2 (*Continued*)

Heat removal and addition	
21	To remove a highly exothermic heat of reaction, consider the use of excess reactant, an inert diluent, or cold shots. These affect the distribution of chemicals and should be inserted early in process synthesis.
22	For less exothermic heats of reaction, circulate reactor fluid to an external cooler, or use a jacketed vessel or cooling coils. Also, consider the use of intercoolers between adiabatic reaction stages.
23	To control temperature for a highly endothermic heat of reaction, consider the use of excess reactant, an inert diluent, or hot shots. These affect the distribution of chemicals and should be inserted early in process synthesis.
24	For less endothermic heats of reaction, circulate reactor fluid to an external heater, or use a jacketed vessel or heating coils. Also, consider the use of interheaters between adiabatic reaction stages.
Heat exchangers and furnaces	
25	Unless required as part of the design of the separator or reactor, provide necessary heat exchange for heating or cooling process fluid streams, with or without utilities, in an external shell-and-tube heat exchanger using countercurrent flow. However, if a process stream requires heating above 750°F, use a furnace unless the process fluid is subject to chemical decomposition.
26	Near-optimal minimum temperature approaches in heat exchangers depend on the temperature level as follows: 10°F or less for temperatures below ambient. 20°F for temperatures at or above ambient up to 300°F. 50°F for high temperatures. 250 to 350°F in a furnace for flue gas temperature above inlet process fluid temperature.
27	When using cooling water to cool or condense a process stream, assume a water inlet temperature of 90°F (from a cooling tower) and a maximum water outlet temperature of 120°F.
28	Boil a pure liquid or close-boiling liquid mixture in a separate heat exchanger, using a maximum overall temperature driving force of 45°F to ensure nucleate boiling and avoid undesirable film boiling as discussed in Section 13.1.
29	When cooling and condensing a stream in a heat exchanger, a zone analysis, described in Section 13.1, should be made to make sure that the temperature difference between the hot stream and the cold stream is equal to or greater than the minimum approach temperature at all locations in the heat exchanger. The zone analysis is performed by dividing the heat exchanger into a number of segments and applying an energy balance to each segment to determine corresponding stream inlet and outlet temperatures for the segment, taking into account any phase change. A process simulation program conveniently accomplishes the zone analysis.
30	Typically, a hydrocarbon gives an adiabatic flame temperature of approximately 3,500°F when using the stoichiometric amount of air. However, use excess air to achieve complete combustion and give a maximum flue-gas temperature of 2,000°F. Set the stack gas temperature at 650 to 950°F to prevent condensation of the corrosive components of the flue gas.
31	Estimate heat-exchanger pressure drops as follows: 1.5 psi for boiling and condensing. 3 psi for a gas. 5 psi for a low-viscosity liquid. 7–9 psi for a high-viscosity liquid. 20 psi for a process fluid passing through a furnace.
32	Quench a very hot process stream to at least 1,150°F before sending it to a heat exchanger for additional cooling and/or condensation. The quench fluid is best obtained from a downstream separator as in Figure 4.21 for the toluene hydrodealkylation process. Alternatively, if the process stream contains water vapor, liquid water may be an effective quench fluid.
33	If possible, heat or cool a stream of solid particles by direct contact with a hot gas or cold gas, respectively, using a rotary kiln, a fluidized bed, a multiple hearth, or a flash/pneumatic conveyor. Otherwise, use a jacketed spiral conveyor.

(Continued)

Table 5.2 (*Continued*)

Pressure increase operations	

34   Use a fan to raise the gas pressure from atmospheric pressure to as high as 40 inches water gauge (10.1 kPa gauge or 1.47 psig). Use a blower or compressor to raise the gas pressure to as high as 206 kPa gauge or 30 psig. Use a compressor or a staged compressor system to attain pressures greater than 206 kPa gauge or 30 psig.

35   Estimate the theoretical adiabatic horsepower (THp) for compressing a gas from:

$$\text{THp} = \text{SCFM}\left(\frac{T_1}{8{,}130a}\right)\left[\left(\frac{P_2}{P_1}\right)^a - 1\right] \tag{5.1}$$

where SCFM = standard cubic feet of gas per minute at 60°F and 1 atm (379 SCF/lbmol), T_1 = gas inlet temperature in °R, inlet and outlet pressures, P_1 and P_2, are absolute pressures, and $a = (k - 1)/k$, with k = the gas specific heat ratio, C_p/C_v.

Estimate the theoretical exit temperature, T_2, for a gas compressor from:

$$T_2 = T_1(P_2/P_1)^a \tag{5.2}$$

36   Estimate the number of gas compression stages, N, from the following table, which assumes a specific heat ratio of 1.4 and a maximum compression ratio of 4 for each stage.

Final Pressure/Inlet Pressure	Number of Stages
<4	1
4 to 16	2
16 to 64	3
64 to 256	4

Optimal interstage pressures correspond to equal Hp for each compressor. Therefore, based on the above equation for theoretical compressor Hp, estimate interstage pressures by using approximately the same compression ratio for each stage with an intercooler pressure drop of 2 psi or 15 kPa.

37   For heads up to 3,200 ft and flow rates in the range of 10 to 5,000 gpm, use a centrifugal pump. For high heads up to 20,000 ft and flow rates up to 500 gpm, use a reciprocating pump. Less common are axial pumps for heads up to 40 ft for flow rates in the range of 20 to 100,000 gpm and rotary pumps for heads up to 3,000 ft for flow rates in the range of 1 to 1,500 gpm.

38   For liquid flow, assume a pipeline pressure drop of 2 psi/100 ft of pipe and a control valve pressure drop of at least 10 psi. For each 10-ft rise in elevation, assume a pressure drop of 4 psi.

39   Estimate the theoretical horsepower (THp) for pumping a liquid from:

$$\text{THp} = (\text{gpm})(\text{Pressure increase, psi})/1{,}714 \tag{5.3}$$

Pressure decrease operations	

40   Consider the use of an expander for reducing the pressure of a gas or a pressure recovery turbine for reducing the pressure of a liquid when more than 20 Hp and 150 Hp, respectively, can be recovered.

41   Estimate the theoretical adiabatic horsepower (THp) for expanding a gas from:

$$\text{THp} = \text{SCFM}\left(\frac{T_1}{8{,}130a}\right)\left[1 - \left(\frac{P_2}{P_1}\right)^a\right] \tag{5.4}$$

42   Estimate the theoretical horsepower (THp) for reducing the pressure of a liquid from:

$$\text{THp} = (\text{gpm})(\text{Pressure decrease, psi})/1{,}714 \tag{5.5}$$

Pumping liquid or compressing gas	

43   To increase the pressure of a stream, pump a liquid rather than compress a gas, unless refrigeration is needed.

Table 5.2 (*Continued*)

Vacuum	
44	Estimate inleakage of air by:

$$w = kV^{0.667} \tag{5.7}$$

	where, w = lb/hr of air inleakage, V = ft^3 of volume of the equipment under vacuum, and k = 0.2 for pressures greater than 90 torr, 0.15 for pressures between 21 and 89 torr, 0.10 for pressures between 3.1 and 20 torr, and 0.051 for pressures between 1 and 3 torr.
45	To reduce the amount of gas sent to the vacuum system if its temperature is greater than 100°F, add a condenser using cooling water before the vacuum system. The gas leaving the condenser will be at a dew-point temperature of 100°F at the vacuum pressure.
46	For pressures down to 10 torr and gas flow rates up to 10,000 ft^3/min at the inlet to the vacuum system, use a liquid-ring vacuum pump. For pressures down to 2 torr and gas flow rates up to 1,000,000 ft^3/min at the inlet to the vacuum system, use a steam-jet ejector system (one-stage for 100 to 760 torr, two-stage for 15 to 100 torr, and three-stage for 2 to 15 torr). Include a direct-contact condenser between stages.
47	For a three-stage steam-jet ejector system used to achieve a vacuum of 2 torr, 100 pounds of 100 psig steam per pound of gas are required.
Conveying granular solids	
48	If the solid particles are small in size, low in particle density, and are not sticky or abrasive, use pneumatic conveying with air at 1 to 7 ft^3/ft^3 of solids and 35 to 120 ft/s air velocity for distances up to 400 ft. Otherwise, for sticky and/or abrasive solids of any size and density, use a screw conveyer and/or bucket elevator for distances up to 150 ft. For solid particles of any size and shape, and not sticky, use a belt conveyor, with inclination up to 30° if necessary, for long distances up to a mile or more.
Solid particle size change and separation	
49	Crushing of coarse solids. Use a jaw crusher to reduce lumps of hard, abrasive, and/or sticky materials of 4 inches to 3 feet in diameter to slabby particles of 1 to 4 inches in size. Use a gyratory crusher to reduce slabby materials of 8 inches to 6 feet in size to rounded particles of 1 to 10 inches in diameter. Use a cone crusher to reduce less hard and less sticky materials of 2 inches to 1 foot in diameter to particles of 0.2 inch (4 mesh) to 2 inches in diameter.
50	Grinding to fine solids. Use a rod mill to take particles of intermediate hardness as large as 20 mm and reduce them to particles in the range of 10 to 35 mesh. Use a ball mill to reduce particles of low to intermediate hardness of 1 to 10 mm in size to very small particles of less than 140 mesh.
51	Particle-size enlargement. Use compression with rotary compression machines to convert powders and granules into tablets of up to 1.5 inches in diameter. Use extruders with cutters to make pellets and wafers from pastes and melts. Use roll compactors to produce sheets from finely divided materials; the sheets are then cut into any desired shape. Use rotating drum granulators and rotary disk granulators with binders to produce particles in the size range of 2 to 25 mm.
52	Size separation of particles. Use a grizzly of spaced, inclined, vibrated parallel bars to remove large particles greater than 2 inches in diameter. Use a revolving cylindrical perforated screen to remove intermediate-size particles in the size range of 0.25 inch to 1.5 inches in diameter. Use flat, inclined woven screens (U.S. standard) that are vibrated, shaken, or impacted with bouncing balls to separate small particles in the size range of 3 to 80 mesh. Use an air classifier to separate fine particles smaller than 80 mesh.
53	Use a cyclone separator to remove, from a gas, droplets or solid particles of diameter down to 10 microns (0.01 mm). Use a hydroclone separator to remove, from a liquid, insoluble liquid droplets or solid particles of diameter down to 5 microns (0.005 mm). However, small amounts of entrained liquid droplets are commonly removed from gases by vertical knock-out drums equipped with mesh pads to help coalesce the smallest droplets.

REFERENCES

Baddour, R.F., P.L.T. Brian, B.A. Logeais, and J.P. Eymery, "Steady-State Simulation of an Ammonia Synthesis Converter," *Chem. Eng. Sci.*, **20**, 281 (1965).

Biegler, L.T., and R.R. Hughes, "Process Optimization: A Comparative Case Study," *Comput. Chem. Eng.*, **7**(5), 645 (1983).

Douglas, J.M., *Conceptual Design of Chemical Processes*, McGraw-Hill, New York (1988).

Fogler, H.S., *Elements of Chemical Reaction Engineering*, Prentice-Hall, Englewood Cliffs, NJ (1986).

B.F. Goodrich Co., "Preparation of Vinyl Chloride," British Patent 938,824 (October 9, 1963).

Henley, E.J., and E.M. Rosen, *Material and Energy Balance Computations*, Wiley, New York (1969).

Hua, I., R.H. Hochemer, and M.R. Hoffmann, "Sonochemical Degra-dation of *p*-Nitrophenol in a Parallel-Plate Near-Field Acoustical Proces-sor," *Environ. Sci. Technol.*, **29**(11), 2790 (1995a).

Hua, I., R.H. Hochemer, and M.R. Hoffmann, "Sonolytic Hydrolysis of *p*-Nitrophenyl Acetate: The Role of Supercritical Water," *J. Phys. Chem.*, **99**, 2335 (1995b).

Myers, A.L., and W.D. Seider, *Introduction to Chemical Engineering and Computer Calculations*, Prentice-Hall, Englewood Cliffs, New Jersey (1976).

Perry, R.H., and D.W. Green, Ed., *Perry's Chemical Engineers' Handbook*, 7th ed., McGraw-Hill, New York (1997).

Schmidt, L.D., *The Engineering of Chemical Reactions*, Oxford University Press, Oxford (1998).

Walas, S.M., *Chemical Process Equipment—Selection and Design*, Butterworth, Stoneham, Massachusetts (1988).

EXERCISES

5.1 For the production of ethylene glycol, how much is the gross profit per pound of ethylene glycol reduced when chlorine and caustic are used to avoid the production of ethylene oxide?

5.2 Consider ethyl-tertiary-butyl-ether (ETBE) as an alternative gasoline oxygenate to MTBE. While the latter appears to have the best combination of properties such as oxygen content, octane number, energy content, and cost, the former can be manufactured using ethanol according to:

$$C_2H_5OH + Iso-butene \rightleftharpoons ETBE$$

Since ethanol can be manufactured from biomass, it is potentially more acceptable to the environment.

a. Rework Example 5.1 for this process.

b. Is reactive distillation promising for combining the reaction and separation operations? If so, suggest a distribution of chemicals using a reactive distillation operation.

5.3 For the ammonia process in Example 5.3, consider operation of the reactor at 932°F and 400 atm. Use a simulator to show how the product, recycle, and purge flow rates, and the mole fractions of argon and methane, vary with the purge-to-recycle ratio. How do the power requirements for compression vary, assuming 3 atm pressure drop in the reactor and 1 atm pressure drop in the heat exchanger.

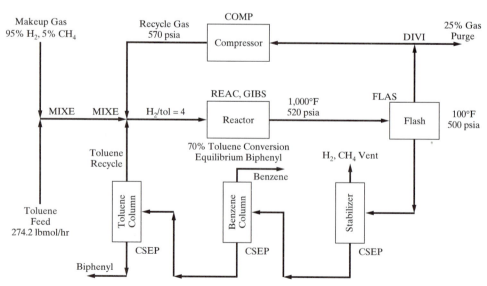

Figure 5.13 Flowsheet for the toluene hydrodealkylation process.

5.4 *Revamp of a toluene hydrodealkylation process.* This problem considers some waste-minimization concepts. Our operating toluene hydrodealkylation unit, shown in Figure 5.13, involves the hydrogenation of toluene to benzene and methane. An equilibrium side reaction produces a small quantity of biphenyl. To be more competitive, and eliminate waste, the process needs to be studied for a possible revamp. The customer for our small production of biphenyl has informed us that it will not renew its contract with us, and we have no other prospective buyer for biphenyl. Also, a membrane separator company believes that if we install their equipment, we can reduce our makeup hydrogen requirement. Make preliminary process design calculations with a simulator to compare the two alternatives below, and advise me of the technical feasibility of the second alternative and whether we should consider such a revamp further. For your studies, you will have to perform mainly material balance calculations. You will not make detailed distillation calculations, and liquid pumps need not be modeled. For the second alternative, calculate the required area in square feet of the membrane unit and determine if it is reasonable.

Alternative 1. Do no revamp and use the biphenyl for its fuel value.

Alternative 2. Eliminate operation of the toluene column and recycle the biphenyl (with the toluene) to extinction. This should increase the yield of benzene. Also, install a membrane separation unit to reduce hydrogen consumption.

Current plant operation: The current plant operation can be adequately simulated with CHEMCAD, using the equipment models indicated in the flow diagram. Alternatively, any other simulator can be used with appropriate models. Note that the flow diagram of the process includes only the reactor, separators, and recycle-gas compressor. The plant operating factor is 96% (8,410 hr/yr). The feedstock is pure toluene at a flow rate of 274.2 lbmol/hr, which is fixed for both alternatives, because any additional benzene that we can make can be sold. The makeup hydrogen is 95 mol% hydrogen and 5 mol% methane. Our reactor outlet conditions are 1,000°F and 520 psia. The hydrogen-to-toluene molar ratio in the feed to the reactor must be 4 to prevent coke formation. The toluene conversion is 70%. The biphenyl in the reactor effluent is the chemical equilibrium amount. The flash drum conditions are 100°F at 500 psia. The flash vapor is not separated into hydrogen and methane, but is purged to limit methane buildup in the recycle gas. The purge gas, which has fuel value, is 25% of the vapor leaving the flash vessel. Perfect separations can be assumed for the three columns. Based on this information, you can obtain the current plant material balance.

Alternative 1. Simulate the current plant operation. Note that the process has two recycle loops that must be converged. The SRK equation of state is adequate for K-values and enthalpies. From your converged material balance, summarize the overall component material balance in pounds per year (i.e., process feeds and products).

Alternative 2. Eliminate the toluene column and rerun the simulation. Since the biphenyl will be recycled to extinction, the benzene production should increase. Replace the stream divider, which divides the flash vapor into a purge and a gas recycle, with a membrane separation unit that can be modeled with a CSEP (black-box separator) unit.

For Alternative 2, the vendor of the membrane unit has supplied the following information:

Hydrogen will pass through the membrane faster than methane.

Vapor benzene, toluene, and biphenyl will not pass through the membrane.

The hydrogen-rich permeate will be the new recycle gas. The retentate gas will be used for fuel.

Tests indicate that the purity of the hydrogen-rich permeate gas will be 95 mol% with a hydrogen recovery of 90%. However, the pressure of the permeate gas will be 50 psia, compared to 500 psia for the recycle gas in the current plant operation. A pressure of 570 psia is required at the discharge of the recycle-gas compression system. Thus, a new compressor will be needed.

Run the revamped process with the simulator. From your converged material balance, summarize the overall component material balance in pounds per year (i.e., process feeds and products).

The membrane unit is to be sized by hand calculations on the basis of the hydrogen flux through the membrane. Tests by the vendor using a nonporous cellulose acetate membrane in a spiral-wound module indicate that this flux is 20 scfh (60°F and 1 atm) per square foot of membrane surface area per 100 psi hydrogen partial-pressure driving force. To determine the driving force, take the hydrogen partial pressure on the feed side of the membrane as the arithmetic average between the inlet and the outlet (retentate) partial pressures. Take the hydrogen partial pressure on the permeate side as that of the final permeate.

Summarize and discuss your results in a report and make recommendations concerning cost studies.

5.5 For the reaction system:

$$A + B \xrightarrow{k_1} C + A \xrightarrow{k_3} E$$
$$\searrow^{k_2} D + A \nearrow^{k_4}$$

select an operating temperature that favors the production of C. The preexponential factor and activation energy for the reactions are tabulated as follows:

Rxtn	k_0 (m⁶/kmol² · s)	E (kJ/kmol)
1	3.7×10^6	65,800
2	3.6×10^6	74,100
3	5.7×10^6	74,500
4	1.1×10^7	74,400

5.6 Propylene glycol mono-methyl-ether acetate (PMA) is produced by the esterification of propylene glycol mono-methyl ether (PM) in acetic acid (HOAc):

$$\underset{\text{PM}}{CH_3CH(OH)CH_2OCH_3} + \underset{\text{HOAc}}{CH_3COOH} \rightleftharpoons$$

$$\underset{\text{PMA}}{CH_3CH(OC(CH_3)O)CH_2OCH_3} + H_2O$$

Conventionally, the reaction takes place in a fixed-bed reactor followed by the recovery of PMA from water, and unreacted PM and HOAc. Prepare a potential distribution of chemicals for a reactive distillation process with the feed at 203°F and 1 atm.

5.7 For the following reactions, determine the maximum or minimum temperatures of the reactor effluents assuming:

a. Complete conversion

b. Equilibrium conversion

The reactants are available in stoichiometric proportions, at the temperature and pressure indicated.

	T_0 (°F)	P (atm)
a. $C_7H_8 + H_2 \rightarrow C_6H_6 + CH_4$	1,200	38.7
b. $SO_2 + \frac{1}{2}O_2 \rightarrow SO_3$	77	1.0
c. $CO + \frac{1}{2}O_2 \rightarrow CO_2$	77	1.0
d. $C_2H_4Cl_2 \rightarrow C_2H_3Cl + HCl$	932	26.0

Also, find the heats of reaction at the conditions of the reactants.

5.8 For Example 5.7, use a simulator to graph the effluent temperature of the methanol reactor as a function of the H_2/CO ratio.

5.9 For Example 5.8, use a simulator to graph the effluent temperature of the methanol reactor as a function of the dodecane flow rate.

5.10 Divide the methanol reaction operation in Example 5.6 into five consecutive stages in series. Feed the CO reactant entirely into the first operation at 25°C and 1 atm. Divide the H_2 reactant into five cold shots and vary the temperature of H_2 before dividing it into cold shots. Assuming that the reaction operations are adiabatic, determine the maximum temperature in the flowsheet as a function of the temperature of the cold shots. How does this compare with the adiabatic reaction temperature?

5.11 Repeat Exercise 5.10 using intercoolers instead of cold shots and an unknown number of reaction stages. The feed to the first reactor is at 25°C and 1 atm. Throughout the reactors, the temperature must be held below 300°C. What is the conversion of CO in the first reactor? How many reaction stages and intercoolers are necessary to operate between 25°C and 300°C?

5.12 *Alternatives for preparing a feed.* A process under design requires that 100 lbmol/hr of toluene at 70°F and 20 psia be brought to 450°F and 75 psia. Develop at least three flowsheets to accomplish this using combinations of heat exchangers, liquid pumps, and/or gas compressors. Discuss the advantages and disadvantages of each flowsheet, and make a recommendation as to which flowsheet is best.

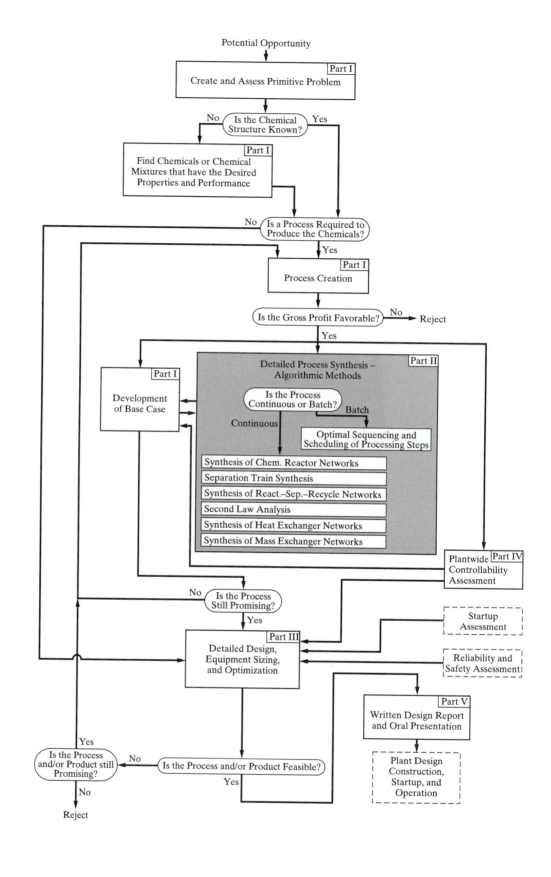

Part Two

DETAILED PROCESS SYNTHESIS—ALGORITHMIC METHODS

This part presents, in six chapters, the more formal approaches that most design teams use after creating a *synthesis tree* and while developing a *base-case design(s)*, as described in Part One. The basic idea is to use these approaches to generate more promising alternatives while developing a base-case design(s).

First, the focus is on chemical reactors in Chapter 6, in which reactor models are reviewed, followed by a discussion of configurations for heat transfer in exothermic and endothermic reactors. Then, *attainable region* analysis is introduced for the optimal design of reactor networks.

In Chapter 7, the focus is on separation processes, in which the criteria for the selection of separation processes and the choices of equipment are reviewed before systematic methods of process synthesis are covered. The latter begin with sequences of ordinary distillation columns, then general vapor–liquid separation processes, and subsequently sequences that include azeotropic distillation columns. Also covered are considerations in selecting separation systems for gas mixtures and for solid–fluid systems.

Chapter 8 extends the coverage in Chapters 6 and 7 to provide a treatment of reactor–separator–recycle networks, with emphasis on the best location of the separation section and the optimal reactor conversion.

Design teams may carry out some form of *second-law analysis*, especially when the process is energy-intensive—that is, it has high heating, cooling, and power demands. Their approach is normally to check the promising designs for energy efficiency. In so doing, they locate large sources of lost work, with the need for the utilization of additional fuel. They strive to reduce these losses, thereby increasing the thermodynamic efficiency and the profitability of the process. This is the subject of Chapter 9.

Second-law analyses identify the need to achieve improved heat and power integration of the process, that is, *process integration*. They also aid the design team in carrying out process integration, which is covered in Chapter 10. Using the methods introduced in that chapter, the design team determines the minimum utilities in a network of heat exchangers and selects optimal, or near-optimal, heat exchanger networks (HENs), *without* considering the details of heat exchanger design. The latter is the subject of Chapter 13, in Part Three on detailed design.

The coverage of heat and power integration is followed by a similar treatment of mass integration in Chapter 11. The focus is on determining the minimum usage of mass separating agents in a network of mass exchangers, and on the selection of optimal, or near-optimal, mass exchange networks (MENs), *without* considering the details of mass exchanger design. The latter is the subject of Chapter 14, in Part Three on the detailed design of multistage and packed towers.

For batch processes, which are predominant in the manufacture of specialty chemicals, Chapter 12 provides an introduction to the methods of optimizing the design and scheduling of these processes. Individual process units, combinations of reactors and separators, and multiproduct plants are considered.

To see how Part Two relates to the other four parts of this text and to the entire design process, see the figure on page 202.

Chapter 6

Reactor Design and Reactor Network Synthesis

6.0 OBJECTIVES

The design of the reactor section of the process flow diagram addresses the need to eliminate differences in molecular type. More specifically, it is desired to ensure sufficient yield and selectivity of the required product species, by appropriate selection of a single reactor or network of reactors. As shown in this chapter, particular care is required in heat exchange in endothermic and exothermic reactors, to sustain stable operation at the desired temperature level.

This chapter begins by describing the types of reactor models available in the flowsheet simulators, discussing also the applicability of the reactor types to model particular process circumstances. A section is then dedicated to discussing arrangements for heat transfer in exothermic and endothermic reactors, with examples from the processing industry. Finally, *attainable region* analysis is introduced, as a tool for the optimal design of reactor networks.

After studying this chapter, the reader should

1. Be familiar with the types of reactor models available in the simulators and their use in process simulation. Further assistance is provided on the CD-ROM that accompanies this book (*ASPEN → Chemical Reactors* and *HYSYS → Chemical Reactors*).
2. Be able to design a system for heat transfer in association with the reactor, to sustain an exothermic or endothermic reaction at its desired temperature level, and study the design using simulation.
3. Be able to determine if a reactor network should be considered and, if so, design it using the concept of the *attainable region*.

6.1 REACTOR MODELS

Chemical reactors, particularly for continuous processes, are often custom designed to involve multiple phases (e.g., vapor, liquid, reacting solid, and solid catalyst), different geometries (e.g., stirred tanks, tubular flows, converging and diverging nozzles, spiral flows, and membrane transport), and various regimes of momentum, heat, and mass transfer (e.g., viscous flow, turbulent flow, conduction, radiation, diffusion, and dispersion). There are so many configurations, involving different combinations of these attributes, that attempts to develop generalized reactor models have met with limited success.

Most of the process simulators provide four kinds of reactor models, including: (1) a stoichiometric model that permits the specification of reactant conversions and extents of reaction for one or more specified reactions; (2) a model for multiple phases (vapor, liquid, and solid) in chemical equilibrium, where the approach to equilibrium for individual reactions can be specified; (3) a kinetic model for a continuous-stirred-tank reactor (CSTR) that assumes perfect mixing of homogeneous phases (liquid or vapor); and (4) a kinetic model for a plug-flow tubular reactor (PFTR or PFR), for homogeneous phases (liquid or vapor) and assuming no backmixing (dispersion). These ideal models are used in the early stages of process synthesis, when the details of the reactor designs are less important, but reactor effluents and heat duties are needed.

The ideal reactor models are replaced by custom-made models as the details gain significance. For this purpose, all of the flowsheet simulators provide facilities for the insertion of user-generated models. These are refined often as the design proceeds and as reactor data from the laboratory or pilot plant are regressed, with some of the simulators providing facilities for estimating the parameters of kinetic models by nonlinear regression.

When working with the ideal reactor models, the reader should refer to available textbooks on reactor analysis and design, for example, *Elements of Chemical Reaction Engineering* by H.S. Fogler (1999), *The Engineering of Chemical Reactions* by L.D. Schmidt (1998), *Chemical Reaction Engineering* by O. Levenspiel (1999), and to the user manuals and tutorial segments of the flowsheet simulators. The following discussion of the ideal reactor models used in simulators is preceded by a brief review of reaction stoichiometry and reaction extent, which together provide the basis for calculation of the conversion reaction model in the simulators.

Reaction Stoichiometry

For most of the reactor models in the flowsheet simulators, it is necessary to provide R chemical reactions involving C chemical species:

$$\sum_{j=1}^{C} v_{ij} A_j = 0, \; i = 1, \ldots, R \tag{6.1}$$

where A_j is the chemical formula for species j and v_{ij} is the stoichiometric coefficient for species j in reaction i (negative for reactants, positive for products). As an example, for the manufacture of methanol, let the chemicals be ordered according to decreasing volatility, that is, (1) H_2, (2) CO, and (3) CH_3OH. The reaction can be written as $-2H_2 - CO + CH_3OH = 0$, and the stoichiometric coefficient matrix is

$$v_{ij} = \begin{matrix} H_2 \\ CO \\ CH_3OH \end{matrix} \begin{bmatrix} -2 \\ -1 \\ 1 \end{bmatrix}$$

Extent of Reaction

Consider a single reaction. In the stoichiometric reactor models, one specifies the fractional conversion, X_k, of key reactant k,

$$X_k = \frac{n_{k,in} - n_{k,out}}{n_{k,in}}, \tag{6.2}$$

where $n_{k,in}$ and $n_{k,out}$ are the moles of species k entering and leaving the reactor and $0 \le X_k \le 1$, or the extent (number of moles extent) of reaction i,

$$\xi_i = \frac{\Delta n_{ij}}{\nu_{ij}}, j = 1, \dots, C \qquad (6.3)$$

is specified. The molar flow rates of the components in the reactor effluent, $n_{j,out}$, are computed from the component molar flow rates in the reactor feed, $n_{j,in}$, by a component material balance equation that is consistent with the reaction stoichiometry. If a specification of the fractional conversion of the key component, k, is made with Eq. (6.2):

$$n_{j,out} = n_{j,in} - n_{k,in}X_k\left(\frac{\nu_j}{\nu_k}\right), j = 1, \dots, C \qquad (6.4)$$

If the extent of the reaction is specified:

$$n_{j,out} = n_{j,in} + \xi\nu_j, \quad j = 1, \dots, C \qquad (6.5)$$

For example, for the conversion of CO and H_2 to CH_3OH, assuming an initial feed of 100 kmol/hr CO and 600 kmol/hr H_2 and 70% conversion of CO (the key component), using Eq. (6.4), the molar flow rates of the three components in the reactor effluent are:

$$n_{H_2,out} = 600 - 100(0.7)(-2/-1) = 460 \text{ kmol/hr}$$
$$n_{CO,out} = 100 - 100(0.7)(-1/-1) = 30 \text{ kmol/hr}$$
$$n_{CH_3OH,out} = 0 - 100(0.7)(1/-1) = 70 \text{ kmol/hr}$$

The mole fraction of methanol in the reactor effluent is $y_{CH_3OH} = 70/(460 + 30 + 70) = 0.125$. If, instead, the extent of reaction is specified as 70 kmol/hr, Eq. (6.5) gives the same results.

For multiple reactions, the reactions must be specified as series or parallel. The former is equivalent to having reactors in series with the feed to each reactor, except the first, being the product from the previous reactor. Each reaction can have a different key reactant. For parallel reactions, it is preferable to specify the extent of reaction for each reaction, which results in:

$$n_{j,out} = n_{j,in} + \sum_{i=1}^{R} \xi_i\nu_{i,j}, \quad j = 1, \dots, C \qquad (6.6)$$

Equilibrium

A chemical reaction can be written as a general stoichiometric equation, in terms of reactants A, B, etc. and products R, S, etc.,

$$aA + bB + \cdots = rR + sS + \cdots \qquad (6.7)$$

In writing this equation, it is very important that, unless otherwise stated, each reactant and product is understood to be the pure component in a separate and designated phase: gas, liquid, or solid. A reaction is characterized by two important thermodynamic quantities, namely the heat of reaction and the Gibbs (free) energy of reaction. Furthermore, these two quantities are functions of temperature and pressure. Thermodynamic data are widely available in simulators and elsewhere, for more than a thousand components, for the calculation of these two quantities under standard state conditions, for example, at a reference temperature of 25°C and 1 bar with all components in a designated phase, usually as an ideal gas. The effect of temperature on the heat of reaction depends on the heat capacities of the reactants and products and the effect of temperature on those heat capacities. For many reactions, the effect of temperature on the heat of reaction is relatively small. For example, the reaction of CO and H_2 to form gaseous methanol,

$$CO + 2H_2 \rightleftharpoons CH_3OH, \tag{6.8}$$

has a standard heat of reaction, ΔH°_{rxn}, at 25°C of $-90,400$ kJ/kmol of methanol, while at 800°C, the heat of reaction, ΔH°_{rxn}, is $-103,800$ kJ/kmol, a relatively small change for such a large change in temperature. By contrast, the effect of temperature on the Gibbs energy of reaction can be very large. For example, for the same methanol formation reaction, the standard Gibbs energy of reaction, ΔG°_{rxn}, is $-25,200$ kJ/kmol at 25°C. Already, at 500°C, the Gibbs energy of reaction, ΔG°_{rxn}, has undergone a drastic change to $+88,000$ kJ/kmol.

Many reactions of industrial importance are limited by chemical equilibrium, with partial conversion of the limiting reactant and, with the rate of the reverse reaction equal to the rate of the forward reaction. For a specified feed composition and final temperature and pressure, the product composition at chemical equilibrium can be computed by either of two methods: (1) chemical equilibrium constants (K-values) computed from the Gibbs energy of reaction combined with material balance equations for a set of independent reactions, or (2) the minimization of the Gibbs energy of the reacting system. The first method is applicable when the stoichiometry can be specified for all reactions being considered. The second method requires only a list of the possible products.

For the first method, a chemical equilibrium constant, K, is computed for each independent stoichiometric reaction in the set, using the equation,

$$K = \frac{a_R^r a_S^s \cdots}{a_A^a a_B^b \cdots} = \exp\left(-\frac{\Delta G^\circ_{rxn}}{RT}\right) \tag{6.9}$$

where a_i is the component activity.

For a gas solution, the activity is given by

$$a_i = \overline{\phi}_i y_i P = \overline{\phi}_i P_i, \tag{6.10}$$

where $\overline{\phi}_i$ is the fugacity coefficient of component i in the gas mixture, equal to 1.0 for an ideal gas solution, and P_i is the partial pressure. In general, $\overline{\phi}_i$ is a function of T, P, and composition. At low to moderate pressures, $\overline{\phi}_i = 1.0$, so that the activity is equal to the partial pressure in bar. It is common to replace the activity in the equation for K with the above equation to give:

$$K = \frac{y_R^r y_S^s \cdots}{y_A^a y_B^b \cdots}\left(\frac{\overline{\phi}_R^r \overline{\phi}_S^s \cdots}{\overline{\phi}_A^a \overline{\phi}_B^b \cdots}\right) P^{r+s+\cdots-a-b-\cdots} = \frac{y_R^r y_S^s \cdots}{y_A^a y_B^b \cdots} K_\phi P^{r+s+\cdots-a-b-\cdots} = \frac{P_R^r P_S^s \cdots}{P_A^a P_B^b \cdots} K_\phi \tag{6.11}$$

where $K_\phi = 1.0$ for low to moderate pressures.

For a liquid solution, the activity is given by:

$$a_i = x_i \gamma_i \exp\left[\frac{\overline{V}_i}{RT}(P - P_i^s)\right] \tag{6.12}$$

where γ_i is the activity coefficient of component i in the liquid mixture and is equal to 1.0 for an ideal liquid solution, $\overline{V}_i$ is the partial molar volume of component i, and P_i^s is the vapor pressure of component i. The pressures are in bar. In general, γ_i is a function of T, P, and composition. For ideal liquid solutions at low to moderate pressures, $\gamma_i = 1.0$, so that the activity is equal to the mole fraction. It is common to replace the activities in Eq. (6.9) with Eq. (6.12) to give

$$K = \frac{x_R^r x_S^s \cdots}{x_A^a x_B^b \cdots}\left(\frac{\gamma_R^r \gamma_S^s \cdots}{\gamma_A^a \gamma_B^b \cdots}\right) = \frac{x_R^r x_S^s \cdots}{x_A^a x_B^b \cdots} K_\gamma \tag{6.13}$$

where $K_\gamma = 1.0$ for ideal liquid solutions at low to moderate pressures. Most textbooks on chemical thermodynamics present charts of $\log_{10} K$ as a function of temperature for many chemical reactions. The van't Hoff equation relates K to temperature by

$$\left(\frac{d \ln K}{dT}\right)_P = \frac{\Delta H^\circ_{rxn}}{RT^2} \qquad (6.14)$$

If the heat of reaction is assumed independent of temperature over a particular range of temperature, integration and conversion to $\log_{10}$ form gives the approximate correlating equation:

$$\log_{10} K = A + B/T \qquad (6.15)$$

where T is the absolute temperature. Many chemical equilibrium curves are represented with reasonable accuracy by this equation. For example, for the gas-phase reaction of CO and H_2 to form methanol, over a temperature range of 273 to 773 K,

$$\log_{10} K = -12.275 + 4{,}938/T \qquad (6.16)$$

Typically, the methanol synthesis reaction is catalyzed by copper-zinc oxide, at a pressure of 100 bar and a temperature of 300°C. A large excess of hydrogen is used to help absorb the relatively high heat of reaction. At these conditions, $K = 0.0002202$ and $K_{\bar{\Phi}} = 0.61$. Therefore,

$$\frac{(y_{CH_3OH})}{(y_{CO})(y_{H_2})^2} = \frac{K}{K_\phi} P^2 = \frac{0.0002202}{0.61}(100)^2 = 3.61 \qquad (6.17)$$

If X is the equilibrium fractional conversion of the limiting reactant, CO, then using the same initial feed composition as before and the stoichiometry for the reaction, the equilibrium mole fractions are

$$y_{CH_3OH} = \frac{100\,X}{700 - 200\,X}$$

$$y_{CO} = \frac{100 - 100\,X}{700 - 200\,X}$$

$$y_{H_2} = \frac{600 - 200\,X}{700 - 200\,X}$$

Combining the above four equations to give a nonlinear equation in X and solving gives $X = 0.7087$.

The second method for computing chemical equilibrium is to apply the criterion that the total Gibbs energy, G, is a minimum at constant temperature and pressure. Alternatively, one could use the entropy, S, as a maximum or the Helmholtz energy, A, as a minimum, but the Gibbs energy is most widely applied. Two advantages of this second method are: (1) the avoidance of having to formulate stoichiometric equations (only the possible products need to be specified), and (2) the ease of formulation for multiple phases and simultaneous phase equilibrium. For a single phase, the total Gibbs energy at a specified T and P is given by

$$G = \sum_{i=1}^{C} N_i \overline{G}_i \qquad (6.18)$$

where N_i is the mole number of component i and $\overline{G}_i$ is the partial molar Gibbs energy of component i in the equilibrium mixture. The components are those in the feed plus those that may be produced by chemical reactions. The Gibbs energy is minimized with respect to N_i, which are constrained by atom balances. The method is readily extended to multiple phases.

It would seem that for simplicity or usefulness the second method would be preferred because when using this model, an independent set of chemical reactions need not be specified. Hence, the designer is not required to identify the reactions that take place. However, since most reactors are designed to emphasize desired reactions and curtail or exclude undesired reactions, the chemical reactions that take place in the reactor are usually known by the time the reactor is to be designed. Thus, the first model may be preferred, with the second model being useful for preliminary exploration of the thermodynamic possibilities. To make the second model more useful, restrictions should be placed on certain improbable reactions. If this is not done, the second method can produce results that are incorrect because they implicitly require reactions that are not kinetically feasible.

 For instructions on the use of the equilibrium constant and Gibbs reactor models in the process simulators, see the CD-ROM that accompanies this book (*ASPEN → Chemical Reactors → Equilibrium Reactors → REQUIL or RGIBBS* and *HYSYS → Chemical Reactors → Setting Up Reactors → Equilibrium* or *Gibbs*).

Kinetics

Fractional conversion and equilibrium reactor models are useful in the early stages of process design when conducting material and energy balance studies. However, eventually reactor systems must be configured and sized. This requires knowledge of reaction kinetics, which is obtained by conducting laboratory experiments. For homogeneous non-catalytic reactions, power-law expressions are commonly used for regression of laboratory kinetic data. These expressions are not always based on the stoichiometric equation because several elementary reaction steps may be involved, the sum of which is the stoichiometric equation, but one of which may control the overall reaction rate. Elementary reaction steps rarely involve more than two molecules. The general power-law kinetic equation is

$$-r_j = -\frac{dC_j}{dt} = k\prod_{i=1}^{C} C_i^{\alpha_i} \tag{6.19}$$

where $-r_j$ is the rate of disappearance of component j (in mol/time-volume), C_i is the concentration of component i (in mol/volume), t is time, k is the reaction rate coefficient, α_i is the order of reaction with respect to component i, and C is the number of components.

For gas-phase reactions, the partial pressure, P_i, is sometimes used in place of the concentration, C_i, in the kinetic equation. The reaction rate coefficient is a function of temperature as given by the empirical Arrhenius equation:

$$k = k_o \exp(-E/RT) \tag{6.20}$$

where k_o is the pre-exponential factor, and E is the activation energy.

For reactions that are catalyzed by solid porous catalyst particles, the sequence of elementary steps may include adsorption on the catalyst surface of one or more reactants and/or desorption of one or more products. In that case, a Langmuir–Hinshelwood (LH) kinetic equation is often found to fit the experimental kinetic data more accurately than the power-law expression of Eq. (6.19). The LH formulation is characterized by a denominator term that includes concentrations of certain reactants and/or products that are strongly adsorbed on the catalyst. The LH equation may also include a prefix, η, called an overall effectiveness factor that accounts for mass and heat transfer resistances, both external and internal, to the catalyst particles. As an example, laboratory kinetic data for the air-oxidation of SO_2 to SO_3 are fitted well by the following LH equation:

$$-r_{SO_2} = \frac{\eta k \left(P_{SO_2} P_{O_2} - \dfrac{P_{SO_3}^2}{K^2 P_{SO_2}} \right)}{[1 + K_1 P_{SO_2}^{1/2} + K_2 P_{SO_3}^{1/2}]} \tag{6.21}$$

where K is the chemical equilibrium constant and K_1 and K_2 are adsorption equilibrium constants.

Ideal Kinetic Reaction Models—CSTRs and PFRs

CSTR

The simplest kinetic reactor model is the CSTR (continuous-stirred-tank reactor), in which the contents are assumed to be perfectly mixed. Thus, the composition and the temperature are assumed to be uniform throughout the reactor volume and equal to the composition and temperature of the reactor effluent. However, the fluid elements do not all have the same residence time in the reactor. Rather, there is a residence-time distribution. It is not difficult to provide perfect mixing of the fluid contents of a vessel to approximate a CSTR model in a commercial reactor. A perfectly mixed reactor is used often for homogeneous liquid-phase reactions. The CSTR model is adequate for this case, provided that the reaction takes place under adiabatic or isothermal conditions. Although calculations only involve algebraic equations, they may be nonlinear. Accordingly, a possible complication that must be considered is the existence of multiple solutions, two or more of which may be stable, as shown in the next example.

EXAMPLE 6.1   *Adiabatic CSTR for Hydrolysis of Propylene Oxide*

Propylene glycol (PG) is produced from propylene oxide (PO) by liquid-phase hydrolysis with excess water under adiabatic and near-ambient conditions, in the presence of a small amount of soluble sulfuric acid as a homogeneous catalyst:

$$C_3H_6O + H_2O \rightarrow C_3H_8O_2 \tag{6.22}$$

Because the exothermic heat of reaction is appreciable, excess water is used. Furthermore, because PO is not completely soluble in water, methanol is added to the feed, which enters the reactor at 23.9°C, with the following flow rates:

Propylene oxide:	18.712 kmol/hr
Water, to be determined, within the range:	160 to 500 kmol/hr
Methanol:	32.73 kmol/hr

It is proposed to consider the use of an existing agitated reactor vessel, which can be operated adiabatically at 3 bar (to suppress vaporization), with a liquid volume of 1.1356 m³. The reaction occurs in a sequence of elementary steps, with the controlling step involving two molecules of PO. The power-law kinetic equation is:

$$-r_{PO} = 9.15 \times 10^{22} \exp(-1.556 \times 10^5/RT)\, C_{PO}^2 \tag{6.23}$$

where, C_{PO} is in kmol/m³, $R = 8.314$ kJ/kmol-K, and T is in K. Carry out a sensitivity analysis to investigate the effect of the water feed rate on the operating temperature and the PO conversion.

SOLUTION

As shown on the CD-ROM that accompanies this book (following the links: *HYSYS → Chemical Reactors → Setting Up Reactors → CSTR* or *ASPEN → Chemical Reactors → Kinetic Reactors → CSTRs → RCSTR*), analysis of this process shows the possibility of multiple steady states. For example, at a water flow rate of 400 kmol/hr, the following steady states are obtained: (1) conversion of 83% with an effluent temperature of 62°C, (2) conversion of 45% with an effluent temperature of 44°C, and (3) conversion of 3% with an effluent temperature of 25°C. The intermediate steady state at 45% conversion is unstable, while the other two steady states are stable. Furthermore, a controllability and resiliency (C&R) analysis on this process is carried out in Case Study 21.1, where a design involving a single CSTR is compared with one utilizing two CSTRs in series. ∎

PFR

More complex is the plug-flow tubular reactor (PFR or PFTR), in which the composition of the fluid, flowing as a plug, gradually changes down the length of the reactor, with no composition or temperature gradients in the radial direction. Furthermore, mass- and heat-transfer rates are negligible in the axial direction. Thus, the PFR is completely unmixed, with all fluid elements having the same residence time in the reactor. If the reactor operates under adiabatic or nonisothermal conditions, the temperature of the flowing fluid changes gradually down the length of the reactor.

All simulators provide one-dimensional, plug-flow models that neglect axial dispersion. Thus, there are no radial gradients of temperature, composition, or pressure; and mass diffusion and heat conduction do not occur in the axial direction. Operation of the reactor can be adiabatic, isothermal, or nonadiabatic, nonisothermal. For the latter, heat transfer to or from the reacting mixture occurs along the length of the reactor.

Consider the case of adiabatic operation with one chemical reaction. A mole balance for the limiting reactant, A, can be written as:

$$F_{A0} \frac{dX}{dV} = -r_A \{X, T\} \tag{6.24}$$

where F_{A0} is the molar flow rate of A entering the reactor, X is the fractional conversion of A, V is the reactor volume, and r_A is the rate of reaction of A written as a function of fractional conversion and temperature. Because the process simulators compute enthalpies referred to the elements, as described by Felder and Rousseau (2000), with values for the standard enthalpy of formation built into the component properties data bank, the heat of reaction is handled automatically, and the energy balance for adiabatic operation becomes simply:

$$H\{X, T\} = H\{X = 0, T = T_0\} \tag{6.25}$$

where H is the enthalpy flow rate of the reacting mixture in energy/unit time, and T_0 is the entering temperature. Equations (6.24) and (6.25) are solved by numerical integration. The following example illustrates the use of the simulator models for a PFR to size a plug-flow adiabatic reactor for the noncatalytic hydrodealkylation of toluene.

EXAMPLE 6.2   *Adiabatic PFR for Toluene Hydrodealkylation*

A hydrodealkylation reactor feed at 1,200°F and 494 psia consists of

Component	lbmol/hr
Hydrogen	2,049.1
Methane	3,020.8
Benzene	39.8
Toluene	362.0
Biphenyl	4.2
Total	5,475.9

These molar flow rates account for the small extent of reaction for the secondary reaction of benzene to biphenyl (2%), and ignore the negligible rate of the reverse reaction, leaving the main reaction to be considered:

$$H_2 + C_7H_8 \rightarrow CH_4 + C_6H_6 \tag{6.26}$$

Laboratory studies have shown that in the absence of a catalyst, this is a free-radical chain reaction that proceeds in three elementary steps:

$$\begin{aligned} H_2 &\rightleftharpoons 2H \cdot \\ H \cdot + C_6H_5CH_3 &\rightarrow C_6H_6 + CH_3 \cdot \\ CH_3 \cdot + H_2 &\rightarrow CH_4 + H \cdot \end{aligned} \tag{6.27}$$

Equilibrium is established rapidly in the first step to provide hydrogen free radicals. The sum of the next two steps is the stoichiometric equation. Step two is the slow or rate-controlling step. Thus, the overall reaction rate is not proportional to the product of the hydrogen and toluene concentrations as given by the law of mass action when applied to the stoichiometric reaction. Instead, the overall reaction rate is proportional to the product of the hydrogen free-radical and toluene concentrations as given in the second elementary step. For the above hydrodealkylation chain reaction, the power-law kinetic equation is derived as follows. Because the first elementary step approaches equilibrium:

$$K_1 = \frac{C_{H\cdot}^2}{C_{H_2}} \qquad (6.28)$$

Rearrangement gives

$$C_{H\cdot} = K_1^{1/2} C_{H_2}^{1/2} \qquad (6.29)$$

The power-law kinetic equation for the second elementary step, which determines the overall reaction rate, is

$$-\frac{dC_{\text{toluene}}}{dt} = k_2 C_{H\cdot} C_{\text{toluene}} \qquad (6.30)$$

Combining the last two equations,

$$-\frac{dC_{\text{toluene}}}{dt} = k_2 K_1^{1/2} C_{H_2}^{1/2} C_{\text{toluene}} \qquad (6.31)$$

This rate expression correlates well with laboratory kinetic data for temperatures in the range of 500–900°C and pressures from 1 to 250 atm, with $k_2 K_1^{1/2} = 6.3 \times 10^{10} \exp(-52,000/RT)$, concentrations in kmol/m^3, time in s, T in K, and $R = 1.987$ cal/mol-K. Use the PFR model in a process simulator to determine the length of a cylindrical plug flow reactor with a length-to-diameter ratio of six that yields a toluene conversion of 75%. Use the Peng-Robinson equation of state to estimate the thermophysical properties for this vapor-phase reaction.

SOLUTION

To see how an adiabatic PFR is designed to provide a 75% conversion of toluene, see the CD-ROM that accompanies this book. Follow the link *HYSYS → Chemical Reactors → Setting Up Reactors → PFR* for a solution obtained with HYSYS.Plant, and *ASPEN → Chemical Reactors → Kinetic Reactors → PFTRs → RPLUG* for a solution with ASPEN PLUS. Note that the results provided by these simulators are almost identical; the HYSYS.Plant result calls for a reactor volume of 3,690 ft^3 ($D = 9.2$ ft, $L = 55.3$ ft), while ASPEN PLUS gives a volume of 3,774 ft^3 ($D = 9.3$ ft, $L = 55.8$ ft). The main reason for the slight discrepancy is due to the neglected pressure drop in the HYSYS.Plant simulation (the ASPEN PLUS calculation assumes a pressure drop of 5 psia). ∎

At high flow rates (high Reynolds numbers) in a long tubular reactor, the PFR model is generally valid because turbulent flow may approximate plug flow without appreciable axial mass and heat transfer. At Reynolds numbers below 2,100, however, laminar flow persists and the PFR model is not valid because of the parabolic (nonplug) velocity profile. A partially mixed condition exists with a residence-time distribution for fluid elements. However, as shown on pp. 339–341 of *The Engineering of Chemical Reactions* by L.D. Schmidt (1998), the loss in conversion with laminar flow compared to plug flow is 12% at most. Simulation programs have both CSTR and PFR models, but not laminar flow models. Thus, for laminar-flow conditions, the PFR model can be used with a 15% safety factor applied.

For liquid-phase reactions, a single PFR or CSTR reactor is often used. For a single reaction at isothermal conditions, the volume of a PFR is smaller than that of a CSTR for the same conversion and temperature. However, for (1) autocatalytic reactions, where the reaction rate depends on the concentration of a product, or (2) autothermal reactions, where the feed is cold, but the reaction is highly exothermic, the volume of a CSTR can be smaller than a PFR, such that axial dispersion in a tubular reactor may actually be beneficial. In general, a

CSTR model is not used for a gas-phase reaction because of the difficulty in obtaining perfect mixing in the gas.

For noncatalytic homogeneous reactions, a tubular reactor is widely used because it can handle liquid or vapor feeds, with or without phase change in the reactor. The PFR model is usually adequate for the tubular reactor if the flow is turbulent and if it can be assumed that when a phase change occurs in the reactor, the reaction takes place predominantly in one of the two phases. The simplest thermal modes are isothermal and adiabatic. The nonadiabatic, nonisothermal mode is generally handled by a specified temperature profile or by heat transfer to or from some specified heat source or sink and a corresponding heat-transfer area and overall heat transfer coefficient. Either a fractional conversion of a limiting reactant or a reactor volume is specified. The calculations require the solution of ordinary differential equations.

For fixed-bed catalytic reactors, a PFR model with a pseudo-homogeneous kinetic equation is usually adequate and is referred to as a 1-D (one-dimensional) model. However, if the reactor is nonadiabatic with heat transfer to or from the wall, the PFR model is not usually adequate and a 2-D model, involving the solution of partial differential equations for variations in temperature and composition in both the axial and radial directions, is necessary. Simulators do not include 2-D models, but they can be generated by the user and inserted into the simulator.

Models for fluidized-bed catalytic reactors are the most complex and cannot be adequately modeled with either the CSTR or PFR models. Because some of the gas passing through the fluidized bed can bypass the suspended catalyst, the conversion in a fluidized bed can be less than that predicted by the CSTR model.

The CD-ROM that accompanies this book provides more complete coverage of the modeling of reactions and reactors using the process simulators. For ASPEN PLUS, follow the link: *ASPEN → Chemical Reactors → Overview*. For HYSYS.Plant, see: *HYSYS → Chemical Reactors → Overview*.

6.2 REACTOR DESIGN FOR COMPLEX CONFIGURATIONS

As discussed in Section 5.5, temperature control is an important consideration in reactor design. Adiabatic operation is always considered first because it provides the simplest and least-expensive reactor. However, when reactions are highly exothermic or endothermic, it is often desirable to exercise some control over the temperature. Methods for accomplishing this, as shown in Figure 6.1, include: (a) heat transfer to or from the reacting fluid, across a wall, to or from an external cooling or heating agent; (b) an inert or reactive heat carrier or diluent in the reacting fluid; (c) a series of reactor beds with a heat exchanger for cooling or heating between each pair of beds; and (d) cold-shot cooling (also called direct-contact quench) or hot-shot heating, where the combined feed is split into two or more parts, one of which enters at the reactor entrance while the remaining parts enter the reactor at other locations. The following are industrial examples of the application of these four methods. In considering these examples, a useful measure of the degree of exothermicity or endothermicity of a reaction is the adiabatic temperature rise (ATR) for complete reaction with reactants in the stoichiometric ratio.

An example of the industrial use of a heat-exchanger reactor in Figure 6.1a is in the manufacture of phthalic anhydride, produced by the oxidation of orthoxylene with air in the presence of vanadium pentoxide catalyst particles, as discussed by Rase (1977). The reaction, which is carried out at about 375°C and 1.2 atm, is highly exothermic with an ATR of about 1,170°C, even with nitrogen in the air providing some dilution. Adiabatic operation is not feasible. The reactor resembles a vertical shell-and-tube heat exchanger. Hundreds of long tubes of small diameter, inside the shell, are packed with catalyst particles and through which the reacting gas passes downwards. A heat-transfer medium consisting of a sodium nitrite–potassium nitrate fused salt circulates outside the tubes through the shell to remove

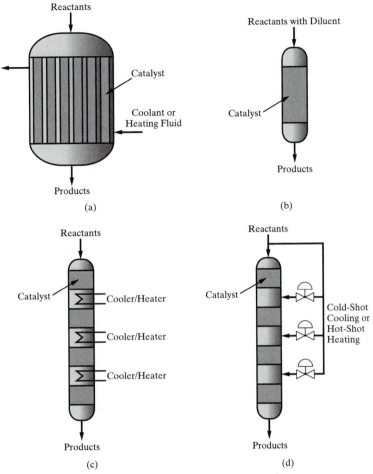

Figure 6.1 Reactors for handling large adiabatic temperature changes: (a) heat-exchanger reactor; (b) use of diluent; (c) external heat exchange; (d) hot/cold shot.

the heat of reaction. Water is ruled out as a heat-transfer medium in this case because the required water pressure would be very high. The heat transfer rate distribution is not adequate to maintain isothermal conditions for the reacting fluid, but its temperature changes by less than 40°C. In some applications, the arrangement involves catalyst-packed beds interspersed with tubes conveying a coolant (e.g., the TVA design in Figure 5.11).

Styrene is produced by the catalytic dehydrogenation of ethylbenzene at 1.2 atm and a temperature of about 575°C, as described by Smith (1981). The reaction is sufficiently endothermic, with an ATR of about −460°C, such that if the reactor were operated adiabatically with a feed of pure ethylbenzene, the temperature of the reacting fluid would decrease to such an extent that the reaction rate would be unduly compromised, resulting in a very large reactor volume. To maintain a reasonable temperature, a large amount of steam is added to the feed (typically with a molar ratio of steam to ethylbenzene equal to 20 : 1), which is preheated to 625°C before entering the reactor (Figure 6.1b). The steam is inert and is easily recovered from the reactor effluent by condensation. The presence of the steam reduces the reaction rate because the styrene concentration is reduced, but the reactor can be operated adiabatically in a simple manner.

Sulfur trioxide, which is used to make sulfuric acid, is produced by catalytic oxidation of sulfur dioxide in air with vanadium pentoxide catalyst particles at 1.2 atm and a temperature of about 450°C, as discussed by Rase (1977). Adiabatic operation is not feasible since the reaction is highly exothermic with an ATR of about 710°C, even with nitrogen in the air providing some dilution. Hence, the reactor system consists of four adiabatic reactor beds, of the same diameter but different height, in series, with a heat exchanger between each pair of beds, as shown in Figure 6.1c. The temperature rises adiabatically in each reactor bed, and the hot reactor effluent is cooled in the heat exchanger positioned before the next bed. When the ATR is higher, such as in the manufacture of ammonia from synthesis gas, as described by Rase (1977), the cold-shot design in Figure 6.1d is recommended.

For 1-D fixed-bed catalytic reactors, it is desirable to reduce the vessel volume to a minimum. As presented by Aris (1960), this is achieved by design. If z is the direction down the length of the reactor, the trajectory of the mass- and energy-balance equations for a single reaction in $X(z)$ and $T(z)$ space is adjusted to match the trajectory corresponding to the maximum reaction rate, (X^*, T^*)[curved line in Figure 6.2], as closely as possible. Thus, tube-cooled (or heated) reactors, cold-shot (or hot shot) converters, and multiple adiabatic beds with intercoolers (or interheaters) need to be carefully designed in such a way that $(X(z), T(z)) \approx (X^*, T^*)$.

As an example, consider an exothermic reversible reaction in a PFR. For this case, the rate of the reverse reaction increases more rapidly with increasing temperature than does the rate of the forward reaction. Also, the reverse reaction is slow and the forward reaction fast at low temperatures. Thus, for a maximum rate of reaction, the temperature should be high at low conversions and low at high conversions. This is shown in Figure 6.2, taken from Smith (1981), where the reaction rate for a sequence of fractional conversions, X, starting with $X_1 = 0$, is plotted against temperature, T. For each value of X, the reaction rate curve in Figure 6.2 shows a maximum value. A locus of maximum rates is shown, corresponding to the solid and dashed line passing through the points C and B, with the maximum reaction rate decreasing with increasing fractional conversion. At each conversion level, the desired temperature corresponds to the maximum reaction rate. In Figure 6.2, the feed enters at temperature, T_A, with a reaction rate at point A. Although the maximum reaction rate for X_1 is not shown, it is clear that T_A is not the temperature corresponding to the maximum rate. If the entering temperature cannot be increased, it is best to operate isothermally at T_A until the conversion at point C is reached, and then follow the optimal profile CB to the desired conversion. Suppose the reactor exit conversion is X_4. Then the desired reactor temperature trajectory is the solid line ACB, with reactor exit temperature, T_B. Corresponding to this trajectory, but not shown in Figure

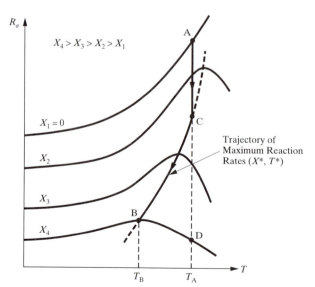

Figure 6.2 Temperature trajectories for an exothermic reversible reaction in a PFR (Smith, 1981).

6.2, is a heat-duty profile, which must be matched by heat exchange to achieve the optimal reaction rate trajectory. Alternatively, isothermal operation of the reactor at T_A corresponds to the trajectory ACD in Figure 6.2. In this case, the reaction rates are not at their maximum values except at point C, requiring a larger reactor volume. If instead of a PFR, a CSTR were used, the optimal temperature of operation for achieving conversion X_4 would be T_B, which corresponds to the maximum reaction rate for that conversion. By specifying a temperature profile for a PFR or an exit temperature for a CSTR, the optimal reactor volume can be determined together with the required corresponding heat-duty profile.

As discussed by Van Heerden (1953), the reactor feed temperature has an important effect on the stability of an autothermal reactor, that is, a reactor whose feed is preheated by its effluent. For a reversible exothermic reaction, as in ammonia synthesis, the heat generation rate varies nonlinearly with the reaction temperature, as shown by curve (a) in Figure 6.3. At low temperatures, the rate of heat generation is limited by the low rate of the forward reaction to ammonia. At very high temperatures, the rate of reaction is limited by equilibrium, so that again, low heat generation rates are to be expected. However, at some intermediate temperature, the reaction rate exhibits a maximum value. In contrast, because heat transfer by convection is dominant, the rate of heat removal is almost linear with the reaction temperature, with a slope dependent on the degree of heat exchange between the outlet and the inlet. Thus, the intersection of the heat removal line (b) and the heat generation line (a) sometimes leads to three possible operating conditions: (O) the non-reacting state, (I) the ignition point, and (S) the desired operating point. Both the non-reacting and the desired operating points are stable, since a small positive perturbation in the reactor temperature causes the heat removal rate to exceed the heat generation rate, decreasing the reactor temperature. Similarly, a small negative perturbation in the reactor temperature has the opposite effect, leading to a temperature rise. Using the same arguments, the ignition point is *unstable* because a small positive perturbation in temperature leads to a jump to the desired, stable operating point, whereas a negative perturbation leads to a so-called "blow-out," to the stable, non-reacting state. Note that a similar analysis for an adiabatic CSTR in Case Study 21.1 also detects the possibility of three steady states.

For tube-cooled converters, Van Heerden (1953), and for cold-shot converters, Stephens and Richards (1973) refer to the temperature difference between operating points I and S as the "stability margin." Clearly, operation at S with larger stability margins would be more robust to disturbances. Thus, a design with increased rate of heat transfer, indicated by the line (b′) in Figure 6.3, would clearly have a lower stability margin. In such cases, the proximity of the stable operating point to the unstable ignition point leads to an increased likelihood of loss of control in the face of process upsets. For example, ammonia synthesis catalyst undergoes deactivation, mainly by poisoning due to feed impurities, or by high temperature sintering, which reduces the catalyst surface area. Lewin and Lavie (1984) studied the effect of catalyst deactivation on the optimal operation of a tube-cooled ammonia converter, which can lead to loss of stability, since decreased catalyst activity leads to lower heat generation rates, as shown by line (a′) in Figure 6.3.

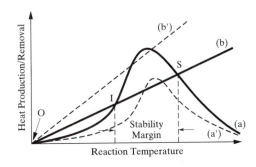

Figure 6.3 Multiple steady states in an autothermal reactor, with reaction rate limited by equilibrium: heat production rates for fully active (a) and deactivated catalyst (a′); heat removal rates for normal (b), and increased heat transfer (b′).

The following example illustrates how a cold-shot reactor is designed to maximize conversion while satisfying stability margins.

EXAMPLE 6.3   *Optimal Bypass Distribution in a Three-Bed, Cold-Shot, Ammonia Synthesis Converter*

A reactor for the synthesis of ammonia consists of three adiabatic beds, shown in Figure 6.4. As summarized in Table 6.1, the reactor feed consists of two sources, the first of which is a make-up feed stream of 20,000 kmol/hr at 25°C and 150 atmospheres containing mainly hydrogen and nitrogen in the stoichiometric molar ratio of 3:1. Since ammonia synthesis gas is produced from naphtha and air, it contains small concentrations of methane, from naphtha, and argon, from air. Both of these species reduce the partial pressures of the reagents, and thus affect the reaction rate. The second feed, which contains larger concentrations of the inert components, is a recycle stream of 40,000 kmol/hr at 25°C and 150 atmospheres consisting of unreacted synthesis gas, recovered after removing the ammonia product. The converter consists of three cylindrical, 2-m-diameter adiabatic beds, packed with catalyst for bed lengths of 1.5 m, 2 m, and 2.5 m, respectively. The reactor feed is split into three branches, with the first branch becoming the main feed entering the first bed after being preheated by the hot reactor effluent from the third bed. The second and third branches, with flow fractions ϕ_1 and ϕ_2, respectively, are controlled by adjusting valves V-1 and V-2, and provide cold-shot cooling at the first and second bed effluents, respectively. It is desired to optimize the allocation of the bypass fractions to maximize the conversion in the converter.

SOLUTION

Ammonia is synthesized in a reversible reaction, whose rate is correlated by the Tempkin equation (Tempkin and Pyzev, 1940), expressed in terms of the partial pressures, in atmospheres, of the reacting species:

$$R_a = 10^4 e^{-91,000/RT} [P_{N_2}]^{0.5} [P_{H_2}]^{1.5} - 1.3 \times 10^{10} e^{-140,000/RT} [P_{NH_3}] \tag{6.32}$$

where R_a is the rate of nitrogen disappearance in kmol/m^3−s, T is the temperature in K, P_i are the partial pressures of the reacting species in atm, and the activation energies for the forward and reverse reactions are in kJ/kmol. The species partial pressures can be expressed in terms of the ammonia mole fraction, x_{NH_3}, and the original feed composition:

$$P_{H_2} = \frac{F_o x_{f,H_2} - 1.5\xi}{F_o - \xi} P, \; P_{N_2} = \frac{F_o x_{f,N_2} - 0.5\xi}{F_o - \xi} P \text{ and}$$

$$P_{NH_3} = \frac{F_o x_{f,NH_3} + \xi}{F_o - \xi} P = x_{NH_3} P \tag{6.33}$$

where F_o is the total molar flow rate of the combined reactor feed, $\xi = F_o(x_{NH_3} - x_{f,NH_3})/(1 + x_{NH_3})$ is the molar conversion, P is the operating pressure, and $x_{f,i}$ is the feed mole fraction of species i. Consequently, the rate of reaction can be computed as a function of the temperature and x_{NH_3}, as shown in Figure 6.5a for an operating pressure of 150 atm. The ridge of maximum reaction rate in composition–temperature space defines an optimal decreasing temperature progression that is to be approximated by the appropriate design and operation of the converter.

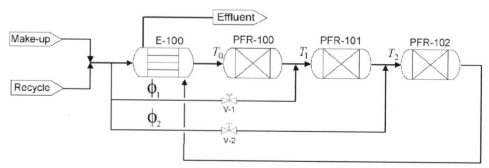

Figure 6.4 Cold-shot ammonia synthesis converter.

Table 6.1 Ammonia Converter—Make-up Feed and Recycle Streams

		Make-up stream		Recycle stream
Flow rate (kmol/hr)		20,000		40,000
Temperature (°C)			25	
Pressure (atm)			150	
Compositions (mol %):	H_2	72		61
	N_2	24		20
	NH_3	0		1.5
	CH_4	3		13
	Ar	1		4.5

The composition–temperature trajectory in the converter is plotted over contours of reaction rate in Figure 6.5b for suboptimal bypass fractions, $\underline{\phi} = [0.1, 0.1]^T$. Note that in the figure, the trajectories in the converter beds are solid lines, while those at the cold-shot mixing junctions are dotted lines. The temperature in Bed 1(PFR-100) rises to 415°C close to the equilibrium limit. The first cold shot cools the gas to 370°C. In Bed 2 (PFR-101), the temperature rises to 405°C, again close to the equilibrium limit. Before entering Bed 3, the final cold shot cools the gas to 360°C. The ammonia effluent concentration from the last bed is 12.7 mol%. Figure 6.5b also includes a dashed line for the optimal temperature progression when, instead of cold-shot cooling, heat is continuously removed while the reaction

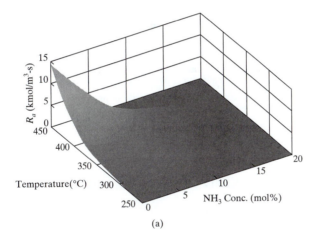

(a)

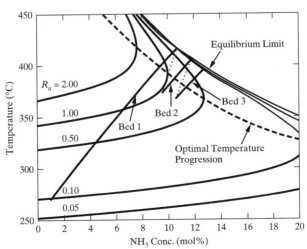

Figure 6.5 Composition–temperature space for ammonia synthesis converter: (a) reaction rate as a function of ammonia mole percent and temperature; (b) suboptimal cold-shot composition–temperature trajectory, plotted over reaction rate contours, with bypasses set to $\underline{\phi} = [0.1, 0.1]^T$.

proceeds in a single PFR. This line is the locus of maximum ammonia concentrations as a function of the reaction rate.

To maximize the conversion in the reactor, the following nonlinear program (NLP), of the type discussed in Chapter 18, is formulated:

$$\max_{\substack{\text{w.r.t.} \\ \phi_1, \phi_2}} \zeta \tag{6.34}$$

Subject to (s. t.)

$$\underline{f}(\underline{x}) = 0 \tag{6.35}$$
$$T_1 > 300°C \tag{6.36}$$
$$T_2 > 300°C \tag{6.37}$$
$$\phi_1 + \phi_2 \leq 0.6 \tag{6.38}$$

where Eq. (6.35) refers to the kinetics and material and energy balances for the converter in Figure 6.4, and Eqs. (6.36) and (6.37) define lower limits for the combined feed temperatures to the second and third beds. These minimum values are taken arbitrarily as 300°C, but are representative of minimum ignition temperatures. Note that in the first bed, where the rate of reaction, and with it the heat generation rate, is higher, the feed temperature is maintained constant at the lower value of 270°C, by appropriate design of the heat-exchanger, E-100. Finally, Eq. (6.38) ensures that a total maximum bypass of 60% is not exceeded, noting that this upper limit is arbitrary.

The NLP in Eqs. (6.34)–(6.38) is solved efficiently using successive quadratic programming (SQP), as described in Chapter 18. Figure 6.6a shows the optimal converged solution, which is obtained in four iterations. The final ammonia composition in the converter effluent is 15.9 mol%, obtained with optimal bypass fractions, $\phi_1 = 0.227$ and $\phi_2 = 0.240$. The composition–temperature trajectories for the optimal bypass distribution, shown in Figure 6.6b, confirm that the overall performance of the three beds is significantly improved through increased utilization of the second and third beds. These results can be reproduced with HYSYS.Plant, using the file NH3_CONVERTOR_OPT.hsc, and with ASPEN PLUS, using the file NH3_CONVERTOR.OPT.bkp. For full details, the reader is referred to the CD-ROM that accompanies this book, where this example is presented in multimedia tutorials under: *HYSYS → Tutorials → Reactor Design → Ammonia Converter Design*, and *ASPEN → Tutorials → Reactor Design Principles → Ammonia Converter Design*. Using simulators, complex reactor configurations are readily designed. ∎

6.3 REACTOR NETWORK DESIGN USING THE ATTAINABLE REGION

This section describes the use of the attainable region (AR), which defines the achievable compositions that may be obtained from a network of chemical reactors. This is analogous to the topic of feasible product compositions in distillation, presented in Section 7.5. The attainable region in composition space was introduced by Horn (1964), with more recent developments and extensions by Glasser and co-workers (Glasser et al. 1987; Hildebrandt et al., 1990).

Figure 6.7 illustrates the attainable region for van de Vusse kinetics (van de Vusse, 1964), based on the reactions:

$$A \underset{k_2}{\overset{k_1}{\rightleftharpoons}} B \overset{k_3}{\rightarrow} C$$

$$2A \overset{k_4}{\rightarrow} D \tag{6.39}$$

Reactions 1, 2, and 3 are first-order in A, B, and B, respectively, while reaction 4 is second-order in A. The rate constants at a particular temperature are: $k_1 = 0.01$ s^{-1}, $k_2 = 5$ s^{-1}, $k_3 = 10$ s^{-1}, and $k_4 = 100$ m^3/kmol·s^{-1}. The boundary of the attainable region, shown in Figure 6.7, is composed of arcs, each of which results from the application of a distinct reactor type, as described next.

For the case of van de Vusse kinetics with a feed of 1 kmol/m^3 of A, Figure 6.7 indicates that the AR boundary is composed of an arc representing a CSTR with bypass (curve C), a

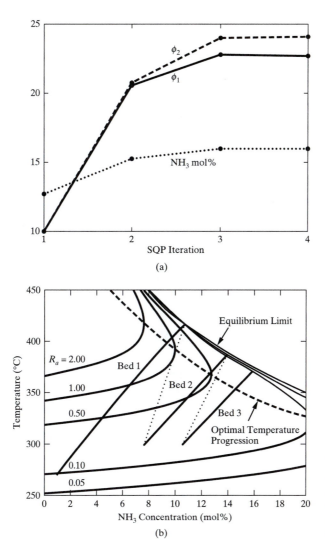

(a)

(b)

Figure 6.6 Optimal selection of bypass fractions for cold-shot ammonia converter: (a) convergence to the optimal solution; (b) optimal cold-shot profile in composition–temperature space with bypasses set to $\phi_2 = [0.227, 0.240]^T$.

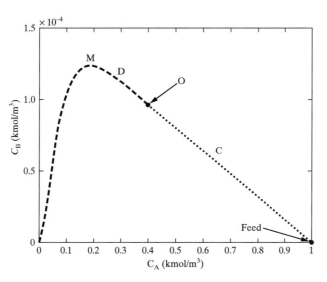

Figure 6.7 Attainable region for the van de Vusse reactions.

CSTR (point O), and a CSTR followed by a PFR (curve D). Within the region bounded by the three arcs and the horizontal base line ($C_B = 0$), product compositions can be achieved with some combination of these reactor configurations. The appropriate reactor configuration along the boundary of the attainable region depends on the desired effluent concentration of A. When $1 > C_A > 0.38$ kmol/m^3, a CSTR with bypass (curve C) provides the maximum concentration of B, while when $C_A < 0.38$ kmol/m^3, this is achieved using a CSTR (point O), followed by a PFR (Curve D). Note that the maximum achievable concentration, $C_B = 1.25 \times 10^{-4}$ kmol/m^3, is obtained using a CSTR followed by a PFR (at point M along curve D). Evidently, the attainable region provides helpful assistance in the design of optimal reactor networks. A procedure for the construction of attainable regions is discussed next.

Construction of the Attainable Region

A systematic method for the construction of the attainable region using CSTRs and PFRs, with or without mixing and bypass, for a system of chemical reactions, as presented by Hildebrandt and Biegler (1995), is demonstrated for van de Vusse kinetics:

Step 1: *Begin by constructing a trajectory for a PFR from the feed point, continuing to the complete conversion of A or chemical equilibrium.* In this case, the PFR trajectory is computed by solving simultaneously the kinetic equations for A and B:

$$\frac{dC_A}{d\tau} = -k_1 C_A + k_2 C_B - k_4 C_A^2 \qquad (6.40)$$

$$\frac{dC_B}{d\tau} = k_1 C_A - k_2 C_B - k_3 C_B \qquad (6.41)$$

where τ is the PFR residence time. Note that kinetic equations for C and D are not required for the construction of the attainable region in two-dimensional space because their compositions do not appear in Eqs. (6.40) and (6.41). The trajectory in $C_A - C_B$ space is plotted in Figure 6.8a as curve ABC. In this example, component A is completely converted.

Step 2: *When the PFR trajectory bounds a convex region, this constitutes a candidate attainable region. A convex region is one in which all straight lines drawn from one point on the boundary to any other point on the boundary lie wholly within the region or on the boundary. If not, the region is nonconvex. When the rate vectors, $[dC_A/d\tau, dC_B/d\tau]^T$ at concentrations outside of the candidate AR do not point back into it, the current limits are the boundary of the AR and the procedure terminates.* In this example, as seen in Figure 6.8a, the PFR trajectory is not convex from A to B, so proceed to the next step to determine if the attainable region can be extended beyond the curve ABC.

Step 3: *The PFR trajectory is expanded by linear arcs, representing mixing between the PFR effluent and the feed stream, extending the candidate attainable region. Note that a linear arc connecting two points on a composition trajectory is expressed by the equation:*

$$\underline{c}^* = \alpha \underline{c}_1 + (1 - \alpha)\underline{c}_2 \qquad (6.42)$$

where $\underline{c}_1$ and $\underline{c}_2$ are vectors for two streams in the composition space, $\underline{c}^*$ is the composition of the mixed stream, and α is the fraction of the stream with composition $\underline{c}_1$ in the mixed stream. The linear arcs are then tested to ensure that no rate vectors positioned on them point out of the AR. If there are such vectors, proceed to the next step, or return to step 2. As shown in Figure 6.8a, a linear arc, ADB, is added, extending the attainable region to ADBC. Since for this example, rate vectors computed along this arc are found to point out of the extended AR, proceed to the next step.

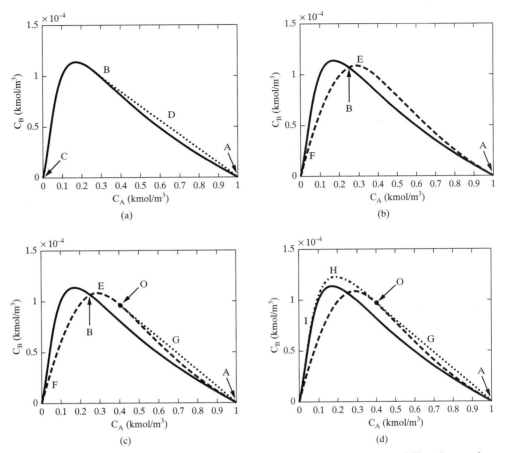

Figure 6.8 Construction of the attainable region for the van de Vusse reaction: (a) PFR trajectory from $\underline{C}(0) = [1, 0]^T$ (solid line), with mixing line (dotted line); (b) CSTR trajectory from $\underline{C}(0) = [1, 0]^T$ (dashed line); (c) addition of bypass to CSTR (dotted line); (d) addition of PFR in series with CSTR (dot-dashed line).

Step 4: *Since there are vectors pointing out of the convex hull, formed by the union between the PFR trajectory and linear mixing arcs, it is possible that a CSTR trajectory enlarges the attainable region. After placing the CSTR trajectory that extends the AR the most, additional linear arcs that represent the mixing of streams are placed to ensure that the AR remains convex.* The CSTR trajectory is computed by solving the CSTR form of the kinetic equations for A and B, given by Eqs. (6.40) and (6.41) as a function of the residence time, τ:

$$C_{A0} - C_A = \tau(k_1 C_A - k_2 C_B + k_4 C_A^2) \tag{6.43}$$

$$C_B = \tau(k_1 C_A - k_2 C_B - k_3 C_B) \tag{6.44}$$

For this example, the CSTR trajectory that extends the AR most is that computed from the feed point, at C_{A0}, the largest concentration of A. This is indicated as curve AEF in Figure 6.8b, which passes through point B. Since the union of the previous AR and the CSTR trajectory is not convex, a linear arc, AGO, is augmented as shown in Figure 6.8c. This arc represents a CSTR with a bypass stream.

Step 5: *A PFR trajectory is drawn from the position where the mixing line meets the CSTR trajectory. If this PFR trajectory is convex, it extends the previous AR to form an* *expanded candidate AR. Then return to Step 2. Otherwise, repeat the procedure from Step 3.* As shown in Figure 6.8d, the PFR trajectory, OHI, leads to a convex attainable region. The boundaries of the region are: (a) the linear arc, AGO, which represents a CSTR with bypass stream; (b) the point O, which represents a CSTR; and (c) the arc OHI, which represents a CSTR followed by a PFR in series. It is noted that the maximum composition of B is obtained at point H, using a CSTR followed by a PFR.

Clearly, the optimal reactor design minimizes the annualized cost, computed to account for the capital and operating costs, and not simply the design that maximizes the yield or selectivity. Nonetheless, the maximum attainable region identifies the entire space of feasible concentrations. The following example shows how the attainable region is used to select the most appropriate reactor network to maximize the yield of a desired product where a number of competing reactions occur.

EXAMPLE 6.4 *Reaction Network Synthesis for the Manufacture of Maleic Anhydride*

Maleic anhydride, $C_4H_2O_3$, is manufactured by the oxidation of benzene with excess air over vanadium pentoxide catalyst (Westerlink and Westerterp, 1988). The following reactions occur:

$$\text{Reaction 1:} \quad C_6H_6 + \tfrac{9}{2}O_2 \rightarrow C_4H_2O_3 + 2CO_2 + 2H_2O \tag{6.45}$$

$$\text{Reaction 2:} \quad C_4H_2O_3 + 3O_2 \rightarrow 4CO_2 + H_2O \tag{6.46}$$

$$\text{Reaction 3:} \quad C_6H_6 + \tfrac{15}{2}O_2 \rightarrow 6CO_2 + 3H_2O \tag{6.47}$$

Since air is supplied in excess, the reaction kinetics are approximated using first-order rate laws:

$$A \xrightarrow{\; r_1 \;} P \xrightarrow{\; r_2 \;} B$$
$$\quad \searrow^{r_3} \; C$$

$$r_1 = k_1 C_A, \; r_2 = k_2 C_P, \; \text{and} \; r_3 = k_3 C_A \tag{6.48}$$

In the above, A is benzene, P is maleic anhydride (the desired product), and B and C are the undesired byproducts (H_2O and CO_2). The rate coefficients for Eqs. (6.48) are (in m^3/kg catalyst·s)

$$\left. \begin{array}{l} k_1 = 4{,}280 \exp\left[-12{,}660/T(K)\right] \\ k_2 = 70{,}100 \exp\left[-15{,}000/T(K)\right] \\ k_3 = 26 \exp\left[-10{,}800/T(K)\right] \end{array} \right\} \tag{6.49}$$

Given that the available feed stream contains benzene at a concentration of 10 mol/m^3, with a volumetric flow rate, v_0, of 0.0025 m^3/s (the feed is largely air), a network of isothermal reactors is proposed to maximize the yield of maleic anhydride.

First, an appropriate reaction temperature is selected. Following Heuristic 7 in Chapter 5, Figure 6.9 shows the effect of temperature on the three rate coefficients, and indicates that in the range 366 $< T <$ 850 K, the rate coefficient, k_1, of the desired reaction to MA is larger than those of the competing reactions. An operating temperature at the upper end of this range is recommended, as the rate of reaction increases with temperature.

Since all of the reaction rate expressions involve only benzene and maleic anhydride, the system can be expressed in a two-dimensional composition space. For this system, the attainable region is straightforward to construct. This begins by tracing the composition space trajectory for a packed bed reactor (PBR), modeled as a PFR, which depends on the solution of the molar balances:

$$v_0 \frac{dC_A}{dW} = -k_1 C_A - k_3 C_A, \quad C_A(0) = C_{A0} = 10 \text{ mol/m}^3 \tag{6.50}$$

$$v_0 \frac{dC_P}{dW} = k_1 C_A - k_2 C_P, \quad C_P(0) = 0 \text{ mol/m}^3 \tag{6.51}$$

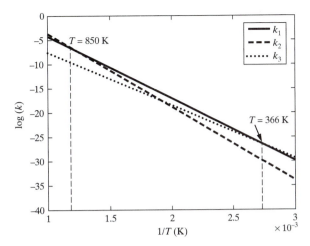

Figure 6.9 Influence of temperature on rate constants for MA manufacture.

where W is the kg of catalyst. In the above equations, the temperature-dependent rate constants are computed using Eqs. (6.49). Figure 6.10 presents solutions of these equations as trajectories in $C_A - C_P$ space for several operating temperatures. Since these trajectories are convex, and rate vectors computed along their boundaries are tangent to them, it is concluded that each trajectory bounds the attainable region for its corresponding temperature.

Evidently, a single plug-flow reactor (or packed-bed reactor in this case) provides the maximum production of maleic anhydride, with the required space velocity being that which brings the value of C_P to its maximum in Figure 6.10. At 800 K, it is determined that the maximum concentration of maleic anhydride is 3.8 mol/m³, requiring 4.5 kg of catalyst. At 600 K it is 5.3 mol/m³, but at this low temperature, 1,400 kg of catalyst is needed. A good compromise is to operate the PBR at an intermediate temperature, for example, 770 K, with a maximum concentration of maleic anhydride of 4.0 mol/m³, requiring 8 kg of catalyst.

Figure 6.11a shows composition profiles for all species as a function of bed length (proportional to the weight of catalyst), indicating that the optimal catalyst loading, where the concentration of maleic anhydride is a maximum, is about 8 kg for isothermal operation at 770 K. Figure 6.11b indicates that the yield (the ratio of the desired product rate and feed rate) under these conditions is 61%, while the selectivity is only about 10%. The selectivity (the ratio of the desired product and total products) for this reaction system is poor, due to the large amounts of CO_2 and H_2O produced, with the highest selectivity, achieved by repressing both of the undesired reactions, at 22%.   ■

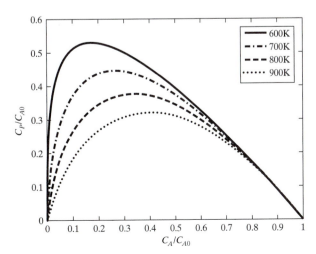

Figure 6.10 Attainable regions for MA manufacture at various temperatures.

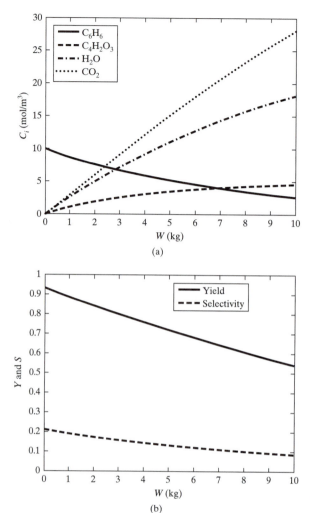

Figure 6.11 Composition profiles for MA manufacture in an isothermal PBR at 770 K: (a) composition profiles; (b) selectivity and yield.

Thus far, the attainable region has been shown for the analysis of systems with two key compositions to be tracked. In the following, the principle of reaction invariants is used to reduce the composition space in systems of larger dimension.

The Principle of Reaction Invariants

Because the attainable region depends on geometric constructions, it is effectively limited to the analysis of systems involving two independent species. However, as shown by Omtveit et al. (1994), systems involving higher dimensions can be analyzed using the two-dimensional AR approach by applying the principle of reaction invariants of Fjeld et al. (1974). The basic idea consists of imposing atom balances on the reacting species. These additional linear constraints impose a relationship between the reacting species, permitting the complete system to be projected onto a reduced space of independent species. The AR analysis above may be used when this reduced space is in two dimensions.

Let the reacting system consist of n_i moles of each species i, each containing a_{ij} atoms of element j. The molar changes in each of the species due to reaction are combined in the vector $\Delta \underline{n}$, and the coefficients a_{ij} form the atom matrix $\underline{\underline{A}}$, noting that since the number of gram-atoms for each element remain constant, $\underline{\underline{A}} \, \Delta \underline{n} = 0$. Partitioning $\Delta \underline{n}$ and $\underline{\underline{A}}$ into dependent, d, and independent, i, components:

$$\underline{\underline{A}} = [\underline{\underline{A}}_d \mid \underline{\underline{A}}_i] \tag{6.52}$$

$$\Delta \underline{n}^T = [\Delta \underline{n}_d^T \mid \Delta \underline{n}_i^T] \tag{6.53}$$

Assuming that $\underline{\underline{A}}_d$ is square and nonsingular, an expression for the changes in the number of moles of each dependent species is obtained by algebraic manipulation:

$$\Delta \underline{n}_d = -\underline{\underline{A}}_d^{-1} \underline{\underline{A}}_i \Delta \underline{n}_i \tag{6.54}$$

The dimension of i is equal to the number of species minus the number of elements (atoms) in the species. When this dimension is two or less, the principle of reaction invariants permits the application of the attainable region to complex reaction systems. This is illustrated in the following example, introduced by Omtveit et al. (1994).

EXAMPLE 6.5   *Attainable Region for Steam Reforming of Methane*

Construct the attainable region for the steam reforming of methane at 1,050 K, and use it to identify the networks that provide for the maximum composition and selectivity of CO.

SOLUTION

The following reactions, involving five species and three elements, dominate in the steam reforming of methane:

$$CH_4 + 2H_2O \rightleftharpoons CO_2 + 4H_2 \tag{6.55}$$

$$CH_4 + H_2O \rightleftharpoons CO + 3H_2 \tag{6.56}$$

$$CO + H_2O \rightleftharpoons CO_2 + H_2 \tag{6.57}$$

By evoking the principle of reaction invariants, the number of species that need to be tracked for this system is reduced to two so that the attainable region can be shown in two dimensions. Accordingly, the vector of molar changes is

$$\Delta \underline{n}^T = [\Delta \underline{n}_d^T \mid \Delta \underline{n}_i^T] = [\Delta n_{H_2}, \Delta n_{H_2O}, \Delta n_{CO_2} \mid \Delta n_{CH_4}, \Delta n_{CO}]^T \tag{6.58}$$

where methane and carbon monoxide have been selected as the independent components. The atom balances for the three elements, C, H, and O, are

$$\text{C balance:} \quad \Delta n_{CO_2} + \Delta n_{CH_4} + \Delta n_{CO} = 0$$
$$\text{H balance:} \quad 2\Delta n_{H_2} + 2\Delta n_{H_2O} + 4\Delta n_{CH_4} = 0$$
$$\text{O balance:} \quad \Delta n_{H_2O} + 2\Delta n_{CO_2} + \Delta n_{CO} = 0$$

The atom matrix $\underline{\underline{A}}$ with rows corresponding to C, H, and O, respectively, is

$$\underline{\underline{A}} = [\underline{\underline{A}}_d \mid \underline{\underline{A}}_i] = \begin{bmatrix} 0 & 0 & 1 & 1 & 1 \\ 2 & 2 & 0 & 4 & 0 \\ 0 & 1 & 2 & 0 & 1 \end{bmatrix} \tag{6.59}$$

The dependent molar changes, $\Delta \underline{n}_d$, are expressed in terms of the molar changes in methane and carbon monoxide, using Eq. (6.54):

$$\Delta \underline{n}_d = \begin{bmatrix} \Delta n_{H_2} \\ \Delta n_{H_2O} \\ \Delta n_{CO_2} \end{bmatrix} = -\underline{\underline{A}}_d^{-1} \underline{\underline{A}}_i \Delta \underline{n}_i = \begin{bmatrix} -4 & -1 \\ 2 & 1 \\ -1 & -1 \end{bmatrix} \begin{bmatrix} \Delta n_{CH_4} \\ \Delta n_{CO} \end{bmatrix} \tag{6.60}$$

For example, if $\Delta n_{CH_4} = -5$ mol and $\Delta n_{CO} = 3$ mol, Eq. (6.60) gives $\Delta n_{H_2} = 17$ mol, $\Delta n_{H_2O} = -7$ mol, and $\Delta n_{CO_2} = 2$ mol. The feed would have to contain more than 5 moles of methane and 7 moles of water.

Xu and Froment (1989) provide the kinetic expressions for the reversible reactions in Eqs. (6.55)–(6.57) in terms of the partial pressures of the participating species. Noting that the number of moles increases by two in each of the reactions in which methane is consumed, the total number of moles in the system is given by

$$n_T = [n_{H_2} + n_{H_2O} + n_{CO_2} + n_{CH_4} + n_{CO}]_0 - 2\Delta n_{CH_4} \tag{6.61}$$

The partial pressure of each of the five species is expressed as: $P\, n_i/n_T$, for i, where the number of moles of the dependent species, H_2, H_2O, and CO_2 are expressed in terms of the number of moles of CH_4 and CO using Eq. (6.60). This allows the construction of the attainable region for the steam reforming reactions at 1,050 K, which was computed by Omtveit et al. (1994) as follows:

Step 1: *Begin by constructing a trajectory for a PFR from the feed point, continuing to the complete conversion of methane or chemical equilibrium.* Here, the PFR trajectory is computed by solving the kinetic equations for the reactions of Eqs. (6.55)–(6.57) to give the mole numbers of CH_4 and CO. This leads to trajectory (1) in Figure 6.12, which tracks the compositions from the feed point, A, to chemical equilibrium at point B.

Step 2: *When the PFR trajectory bounds a convex region, this constitutes a candidate attainable region. When the rate vectors at concentrations outside of the candidate AR do not point back into it, the current limits are the boundary of the AR and the procedure terminates.* In Figure 6.12, the PFR trajectory is not convex, so proceed to the next step.

Step 3: *The PFR trajectory is expanded by linear arcs, representing mixing between the PFR effluent and the feed stream, extending the candidate attainable region.* Here, two linear arcs are introduced to form a convex hull, tangent to the PFR trajectory from below, connecting to the chemical equilibrium point B (line 2), and from the feed point to a point tangent to the PFR trajectory from above (line 3). In this example, line 2 constitutes the lower boundary of the attainable region. It is found that rate trajectories point out of the convex hull, so proceed to the next step.

Step 4: *Since there are vectors pointing out of the convex hull, formed by the union between the PFR trajectory and linear mixing arcs, a CSTR trajectory may enlarge the attainable region. After placing the CSTR trajectory that extends the AR the most, additional linear arcs that represent the mixing of streams are placed to ensure that the AR remains convex.* Here, the CSTR trajectory is computed by solving the molar balances for CH_4 and CO as the residence time, τ, varies. This gives trajectory (4), augmented by two linear arcs, connecting the feed point to a point tangent to the CSTR trajectory (line 5) at point C, and an additional line (6) connecting the CSTR to the PFR trajectories at two tangent points. This forms the new candidate attainable region, on which trajectories are identified that point outwards.

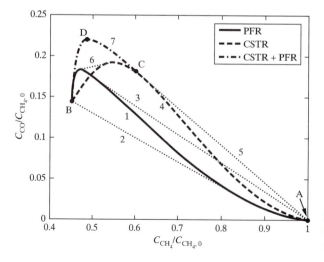

Figure 6.12 Development of the attainable region for steam-reforming reactions at $T = 1,050$ K.

Step 5: *A PFR trajectory is drawn from the position where the mixing line meets the CSTR trajectory. When this PFR trajectory is convex, it extends the previous AR to form an expanded candidate AR. Return to Step 2. Otherwise, repeat the procedure from Step 3.* As shown in Figure 6.12, the PFR trajectory (line 7) leads to a convex attainable region. The boundaries of the region are: (a) the linear arc (line 5) from A to C, which represents a CSTR with a bypass stream; (b) the point C, which represents a CSTR; and line 7 from C to B, which represents a CSTR followed by a PFR in series. Note that the maximum composition of CO is obtained at point D, using a CSTR and PFR in series. The maximum selectivity, defined by the ratio of CO/CH_4, is also achieved at point D, where the ratio is 0.47, as compared to point C, where the ratio is only 0.30. ∎

6.4 SUMMARY

This chapter has introduced the design of chemical reactors and reactor networks. Different methods of reaction temperature control have been presented, with emphasis on the use of multiple adiabatic reactors or beds, using cold shots or heat exchangers between reactors or beds. The attainable region is presented to define the reactor network that maximizes either the yield or the selectivity of a desired product, given the feed to the reactor. However, since reactor yield is often sacrificed in favor of selectivity, conversion is rarely complete, with unreacted species recycled. Thus, the optimal reactor feed conditions depend on the plant economics and the reactor network should be synthesized as part of an overall plant design, as described in Chapter 8. In addition, the reader is referred to Biegler et al. (1997) and Floudas (1995), for mixed-integer nonlinear programming (MINLP) approaches, which address this plantwide process optimization. As seen in Examples 10.15 and 10.16, MINLP approaches begin by constructing a superstructure of possible design configurations, with the optimal design being a subset of the superstructure. Note that many modern MINLP approaches for the design of plants involving reactor networks draw on the AR method for superstructure construction.

After completing this chapter and reviewing the CD-ROM that accompanies this book, the reader should

1. be able to use effectively ASPEN PLUS and/or HYSYS.Plant to model chemical reactors, implementing complex configurations involving tube cooling and cold shots.
2. have an appreciation for the complex configurations that are often used in commercial reactor designs, especially when it is required to handle highly exothermic or endothermic reactions.
3. be able to define the combination of CSTRs and/or PFRs that maximize the yield or selectivity of the desired reactor product for a particular feed composition, given the reaction kinetics, using attainable region analysis.

REFERENCES

Aris, R., "The Optimum Design of Adiabatic Reactors with Several Beds," *Chem. Eng. Sci.*, **12**, 243 (1960).

Biegler, L.T., I.E. Grossmann, and A.W. Westerberg, *Systematic Methods of Chemical Process Design*, Prentice-Hall, Upper Saddle River, New Jersey (1997).

Felder, R.M., and R.W. Rousseau, *Elementary Principles of Chemical Processes*, Third Edition, John Wiley & Sons, New York, pp. 451–452 (2000).

Fjeld, M., O.A. Asbjornsen, and H.J. Astrom, "Reaction Invariants and their Importance in the Analysis of Eigenvectors, Stability and Controllability of CSTRs," *Chem. Eng. Sci.*, **30**, 1917 (1974).

Floudas, C.A., *Nonlinear and Mixed-integer Optimization: Fundamentals and Applications*, Oxford University Press, Oxford (1995).

Fogler, H.S., *Elements of Chemical Reaction Engineering*, Third Edition, Prentice-Hall, Englewood Cliffs, New Jersey (1999).

Glasser, D., C. Crowe, and D.A. Hildebrandt, "A Geometric Approach to Steady Flow Reactors: The Attainable Region and Optimization in Concentration Space," *Ind. Eng. Chem. Res.*, **26**(9), 1803 (1987).

Hildebrandt, D., and L.T. Biegler, "Synthesis of Chemical Reactor Networks", *AIChE Symp. Ser.* No. 304, Vol. 91, 52 (1995).

Hildebrandt, D.A., D. Glasser, and C. Crowe, "The Geometry of the Attainable Region Generated by Reaction and Mixing with and without Constraints," *Ind. Eng. Chem. Res.*, **29**(1), 49 (1990).

Horn, F.J.M., "Attainable Regions in Chemical Reaction Technique," in the *Third European Symposium of Chemical Reaction Engineering*, Pergamon Press, London (1964).

Levenspiel, O., *Chemical Reaction Engineering*, 3rd ed., Wiley, New York (1999).

Lewin, D.R., and R. Lavie, "Optimal Operation of a Tube Cooled Ammonia Converter in the Face of Catalyst Bed Deactivation," *I. Chem. Eng. Symp. Ser.*, **87**, 393 (1984).

Omtveit, T., J. Tanskanen, and K.M. Lien, "Graphical Targeting Procedures for Reactor Systems," *Comput.Chem. Eng.*, **18**(S), S113 (1994).

Rase, H.F., *Chemical Reactor Design for Process Plants, Volume Two: Case Studies and Design Data*, Wiley-Interscience, New York (1977).

Schmidt, L.D., *The Engineering of Chemical Reactions*, Oxford University Press, Oxford (1998).

Smith, J.M., *Chemical Engineering Kinetics*, 3rd ed., McGraw-Hill, New York (1981).

Stephens, A.D., and R.J. Richards, "Steady State and Dynamic Analysis of an Ammonia Synthesis Plant," *Automatica*, **9**, 65 (1973).

Temkin, M., and V. Pyzhev, "Kinetics of Ammonia Synthesis on Promoted Iron Catalyst," *Acta Physicochim. U.R.S.S.*, **12**(3), 327 (1940).

Trambouze, P.J., and E.L. Piret, "Continuous Stirred Tank Reactors: Designs for Maximum Conversions of Raw Materials to Desired Product," *AIChE J.*, **5**, 384 (1959).

van de Vusse, J. G., "Plug Flow vs. Tank Reactor," *Chem. Eng. Sci.*, **19**, 994 (1964).

van Heerden, C., "Autothermic Processes—Properties and Reactor Design," *Ind. Eng. Chem.*, **45**(6), 1242 (1953).

Westerlink, E.J., and K.R. Westerterp, "Safe Design of Cooled Tubular Reactors for Exothermic Multiple Reactions: Multiple Reaction Networks," *Chem. Eng. Sci.*, **43**(5), 1051 (1988).

Xu, J., and G. Froment, "Methane Steam Reforming: Diffusional Limitations and Reactor Simulation," *AIChE J.*, **35**(1), 88 (1989).

EXERCISES

6.1 Carry out a modified design for the ammonia converter in Example 6.3, consisting of three diabatic reactor bed sections, each of 2 m diameter and 2 m length (note that the total bed length is the same as before). Assuming the same reactor inlet temperature of 270°C, compute the optimal heat duties and effluent temperatures for each bed, such that the effluent ammonia mole fraction for the reactor is maximized. Plot the temperature composition trajectory for the modified converter design and compare it to the three-bed cold-shot design of Example 6.3.

6.2 A system of three parallel reactions (Trambouze and Piret, 1959) is:

$$A \xrightarrow{k_1} B \quad A \xrightarrow{k_2} C \quad A \xrightarrow{k_3} D$$

where the reactions are zero-order, first-order, and second-order, respectively, with $k_1 = 0.025$ mol/L-min, $k_2 = 0.2$ min^{-1}, and $k_3 = 0.4$ L/mol-min, and the initial concentration of $C_A = 1$ mol/L. Use the attainable region algorithm to find the reactor network that maximizes the selectivity of C from A.

6.3 Repeat Exercise 6.2, taking the first two reactions as first order, and the last as second order, with $k_1 = 0.02$ min^{-1}, $k_2 = 0.2$ min^{-1}, and $k_3 = 2.0$ L/mol-min, and the initial concentration of $C_A = 1$ mol/L. Use the attainable region method to find the reactor network that maximizes the selectivity of C from A.

6.4 For the reaction system:

$$A + B \xrightarrow{r_1} C + A \xrightarrow{r_2} E$$
$$\quad \searrow r_3 \quad D + A \quad \nearrow r_4$$

where $r_1 = k_1 C_A^2$, $r_2 = k_2 C_C$, $r_3 = k_3 C_A$, and $r_4 = k_4 C_A$. The rate constants are $k_1 = 3$ m^3/kmol-min, $k_2 = 10$ min^{-1}, $k_3 = 0.5$ min^{-1}, $k_4 = 1.5$ min^{-1}, and the feed concentration of A is 1 kmol/m^3. Use the attainable region method to find the reactor network that maximizes the selectivity of C from A.

6.5 In Example 6.5, choose methane and hydrogen as independent components. Derive relationships for the mole numbers of the remaining components in terms of methane and hydrogen.

Chapter 7

Synthesis of Separation Trains

7.0 OBJECTIVES

Most chemical processes are dominated by the need to separate multicomponent chemical mixtures. In general, a number of separation steps must be employed, where each step separates between two components of the feed to that step. During process design, separation methods must be selected and sequenced for these steps. This chapter discusses some of the techniques for the synthesis of separation trains. More detailed treatments are given by Douglas (1995), Barnicki and Siirola (1997), and Doherty and Malone (2001).

After studying this chapter, the reader should

1. Be familiar with the more widely used industrial separation methods and their basis for separation.
2. Understand the concept of the separation factor and be able to select appropriate separation methods for vapor, liquid, and solid–fluid mixtures.
3. Understand how distillation columns are sequenced and how to apply heuristics to narrow the search for a near-optimal sequence.
4. Be able to apply algorithmic methods to determine an optimal sequence of distillation-type separations.
5. Be familiar with the difficulties in and techniques for determining feasible sequences when azeotropes can form.
6. Be able to determine feasible separation systems for gas mixtures and solid–fluid systems.

7.1 INTRODUCTION

Almost all chemical processes require the separation of mixtures of chemical species (components). In Section 5.4, three flowsheets (Figures 5.7, 5.8, and 5.9) are shown for processes involving a reactor followed by a separation system. A more general flowsheet for a process involving one reactor system is shown in Figure 7.1, where separation systems are shown before as well as after the reactor section. A *feed separation system* may be required to purify the reactor feed(s) by removing catalyst poisons and inert species, especially if present as a significant percentage of the feed. An *effluent separation system*, which follows the reactor system and is almost always required, recovers unconverted reactants (in gas, liquid, and/or solid phases) for recycle to the reactor system and separates and purifies products and byproducts. Where separations are too difficult, purge streams are used to prevent buildup of certain species in recycle streams. Processes that do not involve a reactor system also utilize

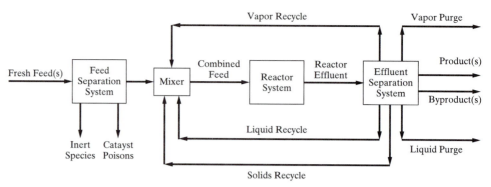

Figure 7.1 General flowsheet for a process with one reactor system.

separation operations if the feed is a mixture that requires separation. Frequently, the major investment and operating costs of a process will be those costs associated with the separation equipment, rather than with the chemical reactor(s).

Feed Separation System

As shown in Figure 7.1, the combined feed to a reactor section may consist of one or more feed streams and one or more recycle streams when conversion of reactants is incomplete. When a feed separation system is needed and more than one feed enters the process, it is usually preferable to provide separate separation operations for the individual feed streams before mixing them with each other and with any recycle streams. Some industrial examples of chemical processes that require a feed separation system are:

1. Production of polypropylene from a feed of propylene and propane. Propane, which is not involved in the propylene polymerization reaction, is removed from the propylene by distillation.
2. Production of acetaldehyde by the dehydrogenation of ethanol using a chromium-copper catalyst. If the feed is a dilute solution of ethanol in water, distillation is used to concentrate the ethanol to the near-azeotrope composition (89.4 mol% ethanol at 1 atm) before it enters the reactor.
3. Production of formaldehyde by air-oxidation of methanol using a silver catalyst. The entering air is scrubbed with aqueous sodium hydroxide to remove any SO_2 and CO_2, which are catalyst poisons.
4. Production of vinyl chloride by the gas-phase reaction of HCl and acetylene with a mercuric chloride catalyst. Small amounts of water are removed from both feed gases by adsorption to prevent corrosion of the reactor vessel and acetaldehyde formation.
5. Production of phosgene by the gas-phase reaction of CO and chlorine using an activated carbon catalyst. Both feed gases are treated to remove oxygen, which poisons the catalyst; sulfur compounds, which form sulfur chlorides; hydrogen, which reacts with both chlorine and phosgene to form HCl; and water and hydrocarbons, which also form HCl.

Phase Separation of Reactor Effluent

In Figure 7.1, the reactor effluent may be a heterogeneous (two or more phases) mixture, but most often is a homogeneous (single-phase) mixture. When the latter, it is often advantageous to change the temperature and/or (but less frequently) the pressure to obtain a partial separation of the components by forming a heterogeneous mixture of two or more phases. Following the change in temperature and/or pressure, phase equilibrium is rapidly attained, resulting in the following possible phase conditions of the reactor effluent, where two liquid phases may form (phase splitting) when both water and hydrocarbons are present in the reactor effluent:

Possible Phase Conditions of Reactor Effluent

Vapor	Liquid
Vapor and Liquid	Liquid 1 and Liquid 2
Vapor, Liquid 1, and Liquid 2	Liquid and Solids
Vapor and Solids	Liquid 1, Liquid 2, and Solids
Vapor, Liquid, and Solids	Solids
Vapor, Liquid 1, Liquid 2, and Solids	

With the reasonable assumption that the phases in a heterogeneous mixture are in phase (physical) equilibrium for a given reactor effluent composition at the temperature and pressure to which the effluent is brought, process simulators can readily estimate the amounts and compositions of the phases in equilibrium by an isothermal (two-phase)-flash calculation, provided that solids are not present. When the possibility of two liquid phases exists, it is necessary to employ a three-phase flash model, rather than the usual two-phase flash model. The three-phase model considers the possibility that a vapor phase may also be present, together with two liquid phases.

In the absence of solids, the resulting phases are separated, often by gravity, in a flash vessel for the V-L case, or in a decanter for the V-L1-L2 or L1-L2 cases. For the latter two cases, centrifugal force may be employed if gravity settling is too slow because of small liquid-density differences or high liquid viscosities. If solids are present with one or two liquid phases, it is not possible to separate completely the solids from the liquid phase(s). Instead, a centrifuge or filter is used to deliver a wet cake of solids that requires further processing to recover the liquid and dry the solids.

Several examples of phase-separation equipment are shown in Figure 7.2. Each exiting phase is either recycled to the reactor, purged from the system, or, most often, sent to separate vapor, liquid, or slurry separation systems, as shown in Figure 7.3. The effluents from these separation systems are products, which are sent to storage; byproducts, which also leave the process; *reactor-system recycle streams*, which are sent back to the reactor; or *separation-system recycle streams*, which are sent to one of the other separation systems. Purges and byproducts are either additional valuable products, which are sent to storage; fuel byproducts, which are sent to a fuel supply or storage system; and/or waste streams, which are sent to waste treatment, incineration, or landfill.

Consider the following examples of phase-equilibria calculations for industrial reactor effluents:

1. *Vapor-liquid case.* The reactor effluent for a toluene hydrodealkylation process, of the type discussed in Section 4.3, is a gas at 1,150°F and 520 psia. When brought to 100°F at say 500 psia by a series of heat exchangers, the result is a vapor phase in equilibrium with a single liquid phase. A two-phase flash calculation using the SRK equation of state gives the following results:

Reactor Effluent Phase Equilibrium for a
Toluene Hydrodealkylation Process

Component	Effluent (lbmol/hr)	Vapor (lbmol/hr)	Liquid (lbmol/hr)
H_2	1,292	1,290	2
CH_4	1,167	1,149	18
Benzene	280	16	264
Toluene	117	2	115
Biphenyl	3	0	3
Total	2,859	2,457	402

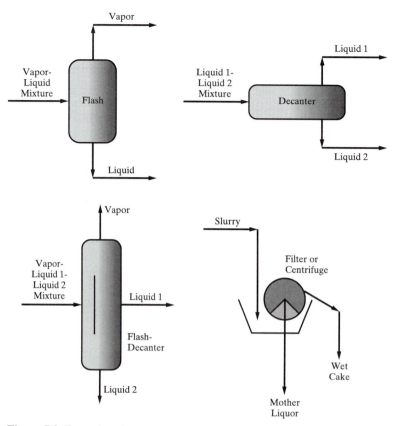

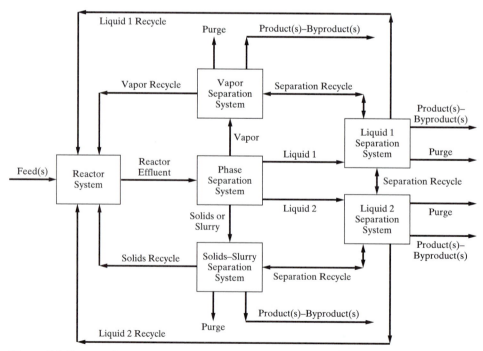

Figure 7.2 Examples of phase-separation devices.

Figure 7.3 Process flowsheet showing separate separation systems with reactor-system and separation-system recycles.

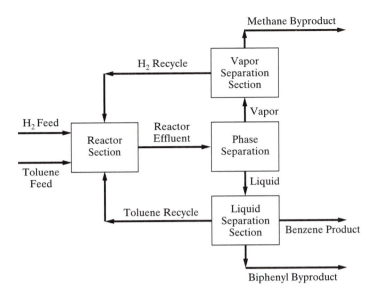

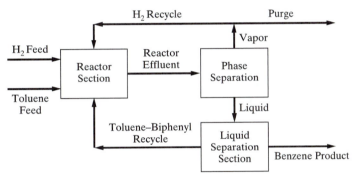

Figure 7.4 Alternative flowsheets for hydrodealkylation of toluene to benzene.

As seen, a reasonably good separation is made between the light gases, H_2 and CH_4, and the three less-volatile aromatic hydrocarbons. The vapor is sent to a vapor separation system to recover CH_4 as a byproduct and H_2 for recycle. The liquid is sent to a liquid separation system to recover benzene as the main product, toluene for recycle to the reactor, and biphenyl as a fuel byproduct. Alternatively, the vapor can be divided, without component separation, into a reactor recycle stream and a vapor purge stream to prevent buildup of CH_4, while the biphenyl can be separated with the toluene and recycled to extinction. These two alternatives are shown in Figure 7.4.

2. *Vapor-liquid 1-liquid 2 case.* The reactor effluent in a styrene production process, involving the reaction

$$\text{methanol} + \text{toluene} = \text{styrene} + \text{hydrogen} + \text{water}$$

with a side reaction, of the same reactants, that produces ethylbenzene and water, is a gas at 425°C and 330 kPa. When brought to 38°C at say 278 kPa by a series of heat exchangers, the result is a vapor phase in equilibrium with an organic-rich liquid phase and a water-rich liquid phase. A three-phase flash calculation using the NRTL method for estimating liquid-phase activity coefficients gives the following results:

Reactor Effluent Phase Equilibrium for a Styrene Process

Component	Effluent (kmol/hr)	Vapor (kmol/hr)	Liquid 1 (kmol/hr)	Liquid 2 (kmol/hr)
H_2	352.2	352.2	0.0	0.0
Methanol	107.3	9.9	31.0	66.4
Water	489.1	8.0	0.5	480.6
Toluene	107.3	1.7	105.5	0.1
Ethylbenzene	140.7	0.5	140.0	0.2
Styrene	352.2	2.0	350.1	0.1
Total	1,548.8	374.3	627.1	547.4

In this case, the vapor is 94 mol% H_2, for which a vapor separation section may not be needed. The organic-rich liquid phase (L1) is sent to a liquid separation section to recover a combined methanol and toluene stream for recycle to the reactor, ethylbenzene as a byproduct, and styrene as the main product. The water-rich liquid phase (L2) is sent to another liquid separation section to recover methanol for recycle to the reactor and water, which is sent to wastewater treatment to remove small quantities of soluble organic components. It is important to note that a two-phase flash calculation would produce erroneous results. If in doubt, perform a three-phase flash calculation, rather than a two-phase flash calculation.

3. *Vapor-solids case.* Phthalic anhydride is manufactured mainly by the vapor-phase partial oxidation of orthoxylene with excess air in a shell-and-tube fixed-bed reactor using vanadium pentoxide catalyst packed inside the tubes. Typically the reactor feed is very dilute in the orthoxylene, with only 1.27 moles of orthoxylene per 100 moles of air. The main reaction consumes 80% of the orthoxylene to produce phthalic anhydride and water. The remaining 20% of the orthoxylene is completely and unavoidably oxidized to CO_2 and water vapor. Typical reactor effluent conditions are 660 K and 25 psia. The reactions are exothermic, with heat removal in the reactor by molten salt of the eutectic mixture of sodium and potassium nitrites and nitrates, which recirculates between the shell-side of the reactor and a heat exchanger that produces steam. The reactor effluent is cooled, in a heat exchanger to produce steam from boiler feed water, to 180°C, which is safely above the primary dew point of 140°C, corresponding to condensation of liquid phthalic anhydride. The effluent then passes to one of two parallel desublimation condensers using cooling water, where the effluent is cooled to 70°C at 20 psia. Under these conditions, the phthalic anhydride desublimes on the outside of the extended-surface tubes of the heat exchanger as a solid because the temperature is well below its normal melting point of 131°C. The desublimation temperature of 70°C is safely above the secondary dew point of 36°C for the condensation of water. Two dew points can occur because water and phthalic anhydride are almost insoluble in each other. The water vapor will not begin to condense until its partial pressure in the vapor reaches its vapor pressure. At phase-equilibrium conditions of 70°C and 20 psia (1,034 torr), a two-phase flash calculation on the reactor effluent, gives the following results when the Clausius–Clapeyron vapor-pressure equation of Crooks and Feetham (1946) is used for solid phthalic anhydride:

$$Log_{10} P^s = 12.249 - 4,632/T$$

where vapor pressure, P^s, is in torr and temperature, T, is in K. This equation is valid for temperatures in the range of 30°C to the normal melting point of 131°C and predicts a vapor pressure of 0.000517 torr at 25°C, which is in good agreement with the often-quoted value of 0.000514 torr. Assuming that the solid phase is pure phthalic anhydride, its partial pressure in the equilibrium vapor phase is equal to its vapor pressure.

Reactor Effluent Phase Equilibrium for a Phthalic Anhydride Process
Basis: 100 Moles of Reactor Effluent

Component	Effluent (moles)	Vapor (moles)	Solids (moles)
N_2	77.70	77.70	0.00
O_2	15.05	15.05	0.00
Orthoxylene	0.00	0.00	0.00
CO_2	2.00	2.00	0.00
H_2O	4.25	4.25	0.00
Phthalic anhydride	1.00	0.005	0.995
Total	100.0	99.005	0.995

At these equilibrium conditions of 70°C and 1,034 torr, the partial pressure of phthalic anhydride in the vapor is $(0.005/99.005)1,034 = 0.05$ torr, which is equal to its vapor pressure. The partial pressure of water in the vapor is $(4.25/99.005)1,034 = 44.4$ torr, which is well below its vapor pressure of 234 torr at 70°C. Thus, water does not condense at these conditions. The amount of solids in the table above corresponds to a 99.5% desublimation of phthalic anhydride. At 85°C the percent desublimation is only 98%, while at 96.4°C it is only 95%. Thus, the recovery of phthalic anhydride from the reactor effluent is sensitive to the desublimation condenser temperature.

While one desublimation condenser is removing 99.5% of the phthalic anhydride from the effluent, phthalic anhydride in the other condenser is melted with hot water at 160°C, flowing inside the tubes, and sent to a liquid separation section for the removal of small amounts of any impurities. Thus, the reactor effluent gas is switched back and forth between the two parallel cooling-water condensers. The vapor leaving the desublimation condenser is sent to a vapor separation section.

4. Vapor-liquid-solids case. Magnesium sulfate as Epsom salts ($MgSO_4 \cdot 7H_2O$) is produced by the reaction of solid $Mg(OH)_2$ with an aqueous solution of sulfuric acid. A typical reactor effluent is a 10 wt% aqueous solution of $MgSO_4$ at 70°F and 14.7 psia. The effluent is concentrated to 37.75 wt% $MgSO_4$ in a double-effect evaporation system with forward feed, after which filtrate from a subsequent filtering operation is added. Crystallization is then carried out in a continuous adiabatic vacuum flash crystallizer, operating at 85.6°F and 0.577 psia to produce a vapor and a magma (slurry of liquid and solid). By making an adiabatic enthalpy balance that accounts for the heat of crystallization, heat of vaporization, and the activity coefficient of water in a sulfate solution; for a vapor phase of H_2O, an aqueous phase of dissolved sulfate using Figure 7.5 to obtain $MgSO_4$ solubility as a function of temperature, and a solid phase of hydrated magnesium sulfate crystals, the following phase equilibrium conditions are calculated:

Crystallizer Phase Equilibrium for a Magnesium Sulfate Process

Component	Effluent (lb/hr)	Vapor (lb/hr)	Liquid (lb/hr)	Solids (lb/hr)
H_2O	9,844	581	7,803	0
$MgSO_4$	4,480	0	3,086	0
$MgSO_4 \cdot 7H_2O$	0	0	0	2,854
Total	14,324	581	10,889	2,854

The vapor is condensed without further treatment. The magma of combined liquid and solids is sent to a slurry separation system to obtain a product of dry crystals of $MgSO_4 \cdot 7H_2O$.

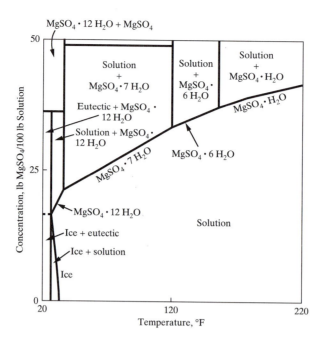

Figure 7.5 Phase diagram for the $MgSO_4 \cdot H_2O$ system.

Industrial Separation Operations

Following phase separation, the individual vapor, liquid, solids, and/or slurry streams are sent to individual separation systems, the most common of which is the liquid separation system. When the feed to a vapor or liquid separation system is a binary mixture, it may be possible to select a separation method that can accomplish the separation task in just one piece of equipment. In that case, the separation system is relatively simple. More commonly, however, the feed mixture involves more than two components. Although some progress is being made in devising multicomponent separation systems involving a single piece of equipment, most systems involve a number of units, in which the separations are sequenced, with each unit separating its feed stream into two effluent streams of different composition. The separation in each piece of equipment (unit) is made between two components designated as the key components for that particular separation unit. Each effluent is either a final product or a feed to another separation device. The synthesis of a multicomponent separation system can be very complex because it involves not only the selection of the separation method(s), but also the manner in which the pieces of separation equipment are sequenced. This chapter deals with both aspects of the synthesis problem.

As an example of the complexity of a multicomponent separation system, consider the synthesis of a separation system for the recovery of butenes from a C_4 concentrate from the catalytic dehydrogenation of n-butane. The specifications for the separation process are taken from Hendry and Hughes (1972), and are shown in Figure 7.6. The process feed, which contains propane, 1-butene, n-butane, trans-2-butene, cis-2-butene, and n-pentane, is to be separated into four fractions: (1) a propane-rich stream containing 99% of the propane in the feed, (2) an n-butane-rich stream containing 96% of the nC_4 in the feed, (3) a stream containing a mixture of the three butenes, at 95% recovery, and (4) an n-pentane-rich stream containing 98% of the nC_5 in the feed. The C_3 and nC_5 streams are final products, the nC_4 stream is recycled to the catalytic dehydrogenation reactor, and the butenes stream is sent to another dehydrogenation reactor to produce butadienes.

Many different types of separation devices and sequences thereof can accomplish the separations specified in Figure 7.6. In general, the process design engineer seeks the most eco-

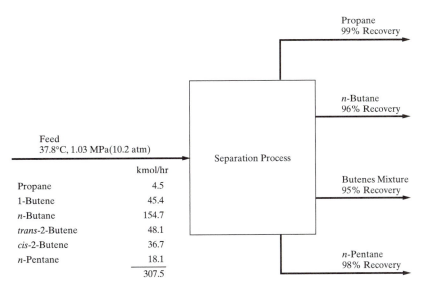

Figure 7.6 Specification for butenes recovery system.

nomical system. One such system, based on mature technology and the availability of inexpensive energy, is shown in Figure 7.7. The system involves two separation methods, distillation and extractive distillation. The process feed from the butane dehydrogenation unit is sent to a series of two distillation columns (1-butene columns, C-1A and C-1B), where the more volatile propane and 1-butene are removed as distillate and then separated in a second distillation column (depropanizer, C-2) into propane and 1-butene. Distillation unit C-1 consists of two columns because 150 trays are required, which is too many for a single column (since the tray spacing is typically 2 ft, giving a 300-ft high tower, while most towers do not exceed 200 ft for structural reasons). The bottoms from unit C-1A, which consists mainly of n-butane, the 2-butene isomers, and nC_5, is sent to another distillation column (deoiler, C-3), where nC_5 product is removed as bottoms. The distillate stream from unit C-3 cannot be separated into nC_4-rich and 2-butenes-rich streams economically by ordinary distillation because the relative volatility is only about 1.03. Instead, the process in Figure 7.7 uses extractive distillation with a solvent of 96% furfural in water, which enhances the relative volatility to about 1.17. The separation occurs in columns C-4A and C-4B, with nC_4 taken off as distillate. The bottoms is sent to a furfural stripper (C-5), where the solvent is recovered and recycled to unit C-4 and the 2-butenes are recovered as distillate. The 1-butene and 2-butenes streams are mixed and sent to a butenes dehydrogenation reactor. Although the process in Figure 7.7 is practical and economical, it does involve the separation of 1-butene from the 2-butenes. Perhaps another sequence could avoid this unnecessary separation.

The separation process of Figure 7.7 utilizes only distillation-type separation methods. These are usually the methods of choice for liquid or partially vaporized feeds unless the relative volatility between the two key components is less than 1.10 or extreme conditions of temperature and pressure are required. In those cases or for vapor, solid, or wet solid feeds, a number of other separation methods should be considered. These are listed in Table 7.1 in order of technical maturity as determined by Keller (1987), except for a few added separation methods not considered by Keller.

As noted in Table 7.1, the feed to a separation unit usually consists of a single vapor, liquid, or solid phase. If the feed is comprised of two or more coexisting phases, consideration should be given to separating the feed stream into two phases by some mechanical means, of the type shown in Figure 7.2, and then sending the separated phases to different separation units, each appropriate for the phase condition of the stream.

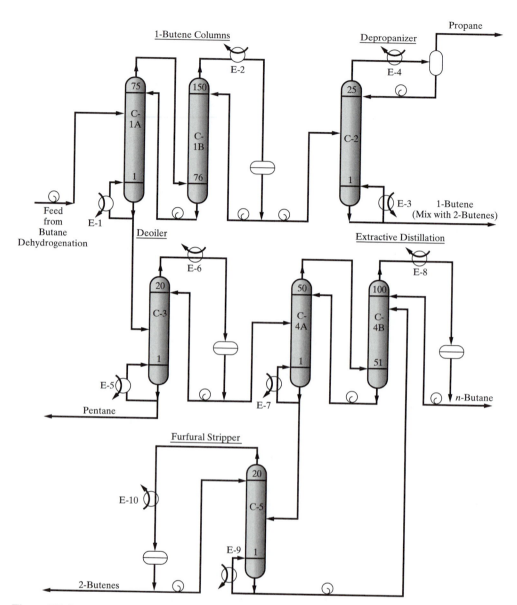

Figure 7.7 Process for butenes recovery: C = distillation column; E = heat exchanger.

The separation of a feed mixture into streams of differing chemical composition is achieved by forcing individual species into different spatial locations. This is accomplished by any one or a combination of four common industrial techniques: (1) the creation by heat transfer, shaft work, or pressure reduction of a second phase; (2) the introduction into the system of a second fluid phase; (3) the addition of a solid phase on which selective adsorption can occur; and (4) the placement of a selective membrane barrier. Unlike the mixing of chemical species, which is a spontaneous process, the separation of a mixture of chemicals requires an expenditure of some form of energy. In the first technique, no other chemicals are added to the feed mixture and the separation is achieved by an energy-separating agent (ESA), usually heat transfer, which causes the formation of a second phase. The components are separated by differences in volatility, thus causing each species to favor one phase over another. In the second technique, a second phase is added to the separation unit in the form

Table 7.1 Common Industrial Separation Methods

Separation Method	Phase Condition of Feed	Separating Agent(s)	Developed or Added Phase	Separation Property
Flash	L and/or V	Pressure reduction or heat transfer ESA	V or L	Volatility
Distillation (ordinary)	L and/or V	Heat transfer or shaft work ESA	V or L	Volatility
Gas absorption	V	Liquid absorbent MSA	L	Volatility
Stripping	L	Vapor stripping agent MSA	V	Volatility
Extractive distillation	L and/or V	Liquid solvent and heat transfer MSA	L and V	Volatility
Azeotropic distillation	L and/or V	Liquid entrainer and heat transfer MSA	L and V	Volatility
Liquid–liquid extraction	L	Liquid solvent MSA	Second L	Solubility
Crystallization	L	Heat transfer ESA	S	Solubility or melting point
Gas adsorption	V	Solid adsorbent MSA	S	Adsorbability
Liquid adsorption	L	Solid adsorbent MSA	S	Adsorbability
Membrane	L or V	Membrane ESA	Membrane	Permeability and/or solubility
Supercritical extraction	L or V	Supercritical solvent MSA	Supercritical fluid	Solubility
Leaching	S	Liquid solvent MSA	L	Solubility
Drying	S and L	Heat transfer ESA	V	Volatility
Desublimation	V	Heat transfer ESA	S	Volatility

of a solvent as a mass-separating agent (MSA) that selectively dissolves or alters the volatility of certain species of the mixture. A subsequent separation step is usually required to recover the solvent for recycle. The third technique involves the addition of solid particles that selectively adsorb certain species of the mixture. Subsequently, the particles must be treated by another separation method to recover the adsorbed species and regenerate the adsorbent for further use. Thus, the particles act as an MSA. The fourth technique imposes a barrier that allows the permeation of some species over others. A mechanical energy loss accompanies the permeation. Thus, this technique involves an ESA. For all four techniques, mass transfer controls the rate of migration of species from one phase to another. Except for the fourth technique, the extent of mass transfer is limited by thermodynamic equilibrium between the phases. In the case of membrane separations, the exiting phases do not approach equilibrium; rather the separation occurs strictly because of differences in the rates of permeation through the membrane.

7.2 CRITERIA FOR SELECTION OF SEPARATION METHODS

The development of a separation process requires the selection of (1) separation methods, (2) ESAs and/or MSAs, (3) separation equipment, (4) the optimal arrangement or sequencing of the equipment, and (5) the optimal operating conditions of temperature and pressure for the equipment.

When the process feed is a binary mixture and the task is to separate that mixture into two products, a single separation device may suffice if an ESA is used. If an MSA is necessary, an additional separation device will be required to recover the MSA for recycle. For a multicomponent feed that is to be separated into nearly pure components and/or one or more multicomponent products, more than one separation device is usually required. Not only must these devices be selected, an optimal arrangement of the devices must be sought. In devising such a separation sequence, it is preferable not to separate components that must be blended later to form desired multicomponent products. However, many exceptions exist to this rule. For example, in Figure 7.6, a six-component mixture is separated into four products, one of which contains 1-butene and *cis*- and *trans*-2-butene. However, the process in Figure 7.7 shows the separation of 1-butene from the 2-butenes and subsequent blending to obtain the desired olefin mixture. The unnecessary separation is carried out because the volatility of *n*-butane is intermediate between that of 1-butene and the two 2-butene isomers. The process shown in Figure 7.7 is the most economical one known. In a multicomponent separation process, each separation operation generally separates between two components, in which case the minimum number of operations is one less than the number of products. However, there are a growing number of exceptions to this rule, and cases are described later for which a single separation operation may produce only a partial separation.

Phase Condition of the Feed as a Criterion

When selecting a separation method from Table 7.1, the phase condition of the feed is considered first.

Vapor Feeds

If the feed is a vapor or is readily converted to a vapor, the following operations from Table 7.1 should be considered: (1) partial condensation (the opposite of a flash or partial vaporization), (2) distillation under cryogenic conditions, (3) gas absorption, (4) gas adsorption, (5) gas permeation with a membrane, and (6) desublimation.

Liquid Feeds

If the feed is a liquid or is readily converted to a liquid, a number of the operations in Table 7.1 may be applicable: (1) flash or partial vaporization, (2) (ordinary) distillation, (3) stripping, (4) extractive distillation, (5) azeotropic distillation, (6) liquid–liquid extraction, (7) crystallization, (8) liquid adsorption, (9) dialysis, reverse osmosis, ultrafiltration, and pervaporation with a membrane, and (10) supercritical extraction. A flash and the different types of distillation are also applicable for feeds consisting of combined liquid and vapor phases.

Slurries, Wet Cakes, and Dry Solids

Slurry feeds are generally separated first by filtration or centrifugation to obtain a wet cake, which is then separated into a vapor and a dry solid by drying. Feeds consisting of dry solids can be leached with a selective solvent to separate the components.

Separation Factor as a Criterion

The second consideration for the selection of a separation method is the *separation factor*, SF, that can be achieved by the particular separation method for the separation between two key components of the feed. This factor, for the separation of key component 1 from key component 2 between phases I and II, for a single stage of contacting, is defined as

$$SF = \frac{C_1^I/C_2^I}{C_1^{II}/C_2^{II}} \tag{7.1}$$

where C_j^i is a composition (expressed as mole fraction, mass fraction, or concentration) of component j in phase i. If phase I is to be rich in component 1 and phase II is to be rich in component 2, then SF must be large. The value of SF is limited by thermodynamic equilibrium, except for membrane separations that are controlled by relative rates of mass transfer through the membrane. For example, in the case of distillation, using mole fractions as the composition variable and letting phase I be the vapor and phase II be the liquid, the limiting value of SF is given in terms of vapor and liquid equilibrium ratios (K values) by

$$\text{SF} = \frac{y_1/y_2}{x_1/x_2} = \frac{y_1/x_1}{y_2/x_2} = \frac{K_1}{K_2} = \alpha_{1,2} \tag{7.2}$$

where α is the relative volatility. In general, components 1 and 2 are designated in such a manner that SF > 1.0. Consequently, the larger the value of SF, the more feasible is the particular separation operation. However, when seeking a desirable SF value, it is best to avoid extreme conditions of temperature that may require refrigeration or damage heat-sensitive materials; pressures that may require gas compression or vacuum; and MSA concentrations that may require expensive means to recover the MSA. In general, operations employing an ESA are economically feasible at a lower value of SF than are those employing an MSA. In particular, provided that vapor and liquid phases are readily formed, distillation should always be considered first as a possible separation operation if the feed is a liquid or partially vaporized.

When a multicomponent mixture forms nearly ideal liquid and vapor solutions, and the ideal gas law holds, the K values and relative volatility can be readily estimated from vapor pressure data. Such K values are referred to as ideal or Raoult's law K values. Then, the SF for vapor–liquid separation operations employing an ESA (partial evaporation, partial condensation, or distillation) is given by

$$\text{SF} = \alpha_{1,2} = \frac{P_1^s}{P_2^s} \tag{7.3}$$

where P_i^s is the vapor pressure of component i. When the components form moderately nonideal liquid solutions (hydrocarbon mixtures or homologous series of other organic compounds) and/or pressures are elevated, an equation-of-state, such as Soave–Redlich–Kwong (SRK) or Peng–Robinson (PR), may be necessary for the estimation of the separation factor, using

$$\text{SF} = \alpha_{1,2} = \frac{\overline{\phi}_1^L/\overline{\phi}_1^V}{\overline{\phi}_2^L/\overline{\phi}_2^V} \tag{7.4}$$

where $\overline{\phi}_i$ is the mixture fugacity coefficient of component i.

For vapor–liquid separation operations (e.g., azeotropic and extractive distillation) that use an MSA that causes the formation of a nonideal liquid solution, but operate at near-ambient pressure, expressions for the K values of the key components are based on a modified Raoult's law that incorporates liquid-phase activity coefficients. Thus, the separation factor is given by

$$\text{SF} = \alpha_{1,2} = \frac{\gamma_1^L P_1^s}{\gamma_2^L P_2^s} \tag{7.5}$$

where γ_i is the activity coefficient of component i, which is estimated from the Wilson, NRTL, UNIQUAC, or UNIFAC equations and is a strong function of mixture composition.

If an MSA is used to create two liquid phases, such as in liquid–liquid extraction, the SF is referred to as the relative selectivity, β:

$$\text{SF} = \beta_{1,2} = \frac{\gamma_1^{II}/\gamma_2^{II}}{\gamma_1^{I}/\gamma_2^{I}} \tag{7.6}$$

where phase II is usually the MSA-rich phase and component 1 is more selective for the MSA-rich phase than is component 2.

In general, MSAs for extractive distillation and liquid–liquid extraction are selected according to their ease of recovery for recycle and to achieve relatively large values of SF. Such MSAs are often polar organic compounds (e.g., furfural), used in the example earlier to separate *n*-butane from 2-butenes. In some cases, the MSA is selected in such a way that it forms one or more homogeneous or heterogeneous azeotropes with the components in the feed. For example, the addition of *n*-butyl acetate to a mixture of acetic acid and water results in a heterogeneous minimum-boiling azeotrope of the acetate with water. The azeotrope is taken overhead, the acetate and water layers are separated, and the acetate is recirculated.

Although the degree of separation that can be achieved for a given value of SF is almost always far below that required to attain necessary product purities, the application of efficient countercurrent-flow cascades of many contacting stages, as in distillation operations, can frequently achieve sharp separations. For example, consider a mixture of 60 mol% propylene and 40 mol% propane. It is desired to separate this mixture into two products at 290 psia, one containing 99 mol% propylene and the other 95 mol% propane. By material balance, the former product would constitute 58.5 mol% of the feed and the latter 41.5 mol%. From equilibrium thermodynamics, the relative volatility for this mixture is approximately 1.12. A single equilibrium vaporization at 290 psia to produce 58.5 mol% vapor results in products that are far short of the desired compositions: a vapor containing just 61.12 mol% propylene and a liquid containing just 51.36 mol% propane at 51.4°C. However, with a countercurrent cascade of such stages in a simple (single-feed, two-product) distillation column with reflux and boilup, the desired products can be achieved with 200 stages and a reflux ratio of 15.9.

Single-stage operations (e.g., partial vaporization or partial condensation with the use of an ESA) are utilized only if SF between the two key components is very large or only a rough or partial separation is needed. For example, if SF = 10,000, a mixture containing equimolar parts of components 1 and 2 could be partially vaporized to give a vapor containing 99 mol% of component 1 and a liquid containing 99 mol% of component 2. At low values of SF, lower than 1.10 but greater than 1.05, ordinary distillation may still be the most economical choice. However, an MSA may be able to enhance the value of SF for an alternative separation method to the degree that the method becomes more economical than ordinary distillation. As illustrated in Figure 7.8, from Souders (1964), extractive distillation or liquid–liquid extraction may be preferred if the SF can be suitably enhanced. If SF = 2 for ordinary distillation, it must be above 3.3 for extractive distillation to be an acceptable alternative, and above 18 for liquid–liquid extraction.

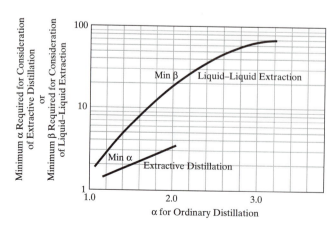

Figure 7.8 Relative selectivities for equal-cost separators (Souders, 1964).

Unless values of SF are about 10 or above, absorption and stripping operations cannot achieve sharp separation between two components. Nevertheless, these operations are used widely for preliminary or partial separations where the separation of one key component is sharp, but only a partial separation of the other key component is adequate. The degree of sharpness of separation is given by the recovery factor RF,

$$RF = \frac{n_i^I}{n_i^F} \tag{7.7}$$

where n is moles or mass, I is the product rich in i, and F is the feed.

The separation of a solid mixture may be necessary when one or more (but not all) of the components is (are) not readily melted, sublimed, or vaporized. Such operations may even be preferred when boiling points are close but melting points are far apart, as is the case with many isomeric pairs. The classic example is the separation of metaxylene from paraxylene, whose normal boiling points differ only by 0.8°C, but whose melting points differ by 64°C. With an SF of only 1.02, as determined from Eq. (7.2), ordinary distillation to produce relatively pure products from an equimolar mixture of the two isomers would require about 1,000 stages and a reflux ratio of more than 100. For the separation by crystallization, the SF is nearly infinity because essentially pure paraxylene is crystallized. However, the mother liquor contains at least 13 mol% paraxylene in metaxylene, corresponding to the limiting eutectic composition. When carefully carried out, crystallization can achieve products of very high purity.

The separation factor for adsorption depends on either differences in the rate of adsorption or adsorption equilibrium, with the latter being more common in industrial applications. For equilibrium adsorption, Eq. (7.1) applies, where the concentrations are those at equilibrium on the adsorbed layer within the pores of the adsorbent and in the bulk fluid external to the adsorbent particles. High selectivity for adsorbents is achieved either by sieving, as with molecular-sieve zeolites or carbon, or by large differences in adsorbability. For example, in the case of molecular-sieve zeolites, aperture sizes of 3, 4, 5, 8, and 10 Å are available. Thus, nitrogen molecules, with a kinetic diameter of about 3.6 Å, can be separated from ammonia, with a kinetic diameter of about 2.6 Å, using a zeolite with an aperture of 3 Å. Only the ammonia is adsorbed. Adsorbents of silica gel and activated alumina, having wide distributions of pore diameters in the range of 20 to 100 Å, are highly selective for water, while activated carbon with pore diameters in the same range is highly selective for organic compounds. When adsorption is conducted in fixed beds, essentially complete removal from the feed of those components with high selectivity can be achieved until breakthrough occurs. Before breakthrough, regeneration or removal of the adsorbent is required.

If only a small amount of one component is present in a mixture, changing the phase of the components in high concentrations should be avoided. In such a case, absorption, stripping, or selective adsorption best removes the minor component. Adsorption is particularly effective because of the high selectivity of adsorbents and is widely used for purification, where small amounts of a solute are removed from a liquid or vapor feed.

For membrane separation operations, SF may still be defined by Eq. (7.1). However, SF is governed by relative rates of mass transfer, in terms of permeabilities, rather than by equilibrium considerations. For the ideal case where the downstream concentration is negligible compared to the upstream concentration, the separation factor reduces to:

$$SF = \frac{P_{M_1}}{P_{M_2}} \tag{7.8}$$

where P_{M_i} is the permeability of species i. In most cases, the value of SF must be established experimentally. In general, membrane separation operations should be considered whenever adsorption methods are considered. Membranes are either porous or nonporous. If porous, the permeability is proportional to the diffusivity through the pore. If nonporous, the permeability is the product of the solubility of the molecule in the membrane and its diffusivity for travel through the membrane. An example of the use of membranes is gas permeation with non-porous hollow fibers to separate hydrogen, helium, carbon dioxide, and/or water vapor from gases containing oxygen, nitrogen, carbon monoxide, and/or light hydrocarbons. For a typical membrane, the SF between hydrogen and methane is 6. Because it is difficult to achieve large numbers of stages with membranes, an SF of this magnitude is not sufficient to achieve a sharp separation, but is widely used to make a partial separation. Sharp separations can be achieved by sieving, when the kinetic molecular diameters of the components to be separated differ widely, and membrane pore diameter lies between those kinetic diameters.

Supercritical extraction utilizes the solvent power of a gas at near-critical conditions. It is the preferred method for the removal of undesirable ingredients from foodstuffs with carbon dioxide. The separation factor, which is given by Eq. (7.1), is difficult to estimate from equations of state using Eq. (7.4) and is best determined by experiment. Equation (7.1) also applies for leaching (solid–liquid extraction), often using a highly selective solvent. As with supercritical extraction, the value of SF is best determined by experiment. Because mass transfer in a solid is very slow, it is important to preprocess the solid to drastically decrease the distance for diffusion. Typical methods involve making thin slices of the solid or pulverizing it. Desublimation is best applied when a sublimable component is to be removed from noncondensable components of a gas stream, corresponding to a very large separation factor.

Reason for the Separation as a Criterion

A final consideration in the selection of a separation method is the reason for the separation. Possible reasons are (1) purification of a species or group of species, (2) removal of undesirable constituents, and (3) recovery of constituents for subsequent processing or removal. In the case of purification, the use of an MSA method may avoid exposure with an ESA method to high temperatures that may cause decomposition. In some cases, removal of undesirable species together with a modest amount of desirable species may be economically acceptable. Likewise, in the recovery of constituents for recycle, a high degree of separation from the product(s) may not be necessary.

7.3 SELECTION OF EQUIPMENT

Only a very brief discussion of equipment for separation operations is presented here. Much more extensive presentations, including drawings and comparisons, are given in *Perry's Chemical Engineers' Handbook* (Perry and Green, 1997) and by Kister (1992), Walas (1988), Seader and Henley (1998), and in *Visual Encyclopedia of Chemical Engineering Equipment for MacIntosh and Windows 95/NT* by Montgomery, as described at www.engin.umich.edu/labs/mel/equipflyer/equip.html. In general, equipment selection is based on stage or mass-transfer efficiency, pilot-plant tests, scale-up feasibility, investment and operating cost, and ease of maintenance.

Absorption, Stripping, and Distillation

For absorption, stripping, and all types of distillation (i.e., vapor–liquid separation operations), either trayed or packed columns are used. Trayed columns are usually preferred for initial installations, particularly for columns 3 ft or more in diameter. However, packed columns should be given serious consideration for operation under vacuum or where a low-pressure drop is desired. Other applications favoring packed columns are corrosive systems, foaming systems, and cases where low liquid holdup is desired. Packing is also generally specified for revamps. Applications favoring trayed columns are feeds containing solids, high liquid-to-gas ratios, large-diameter columns, and where operation over a wide range of conditions is necessary. The three most commonly used tray types are sieve, valve, and bubble-cap. However, because of high cost, the latter is specified only when a large liquid holdup is required on the tray, for example, when conducting a chemical reaction simultaneously with distillation. Sieve trays are the least expensive and have the lowest pressure drop per tray, but they have the narrowest operating range (turndown ratio). Therefore, when flexibility is required, valve trays are a better choice. Many different types of packings are available. They are classified as random or structured. The latter are considerably more expensive than the former, but the latter have the lowest pressure drop, the highest efficiency, and the highest capacity compared to both random packings and trays. For that reason, structured packings are often considered for column revamps.

Liquid–Liquid Extraction

For liquid–liquid extraction, an even greater variety of equipment is available, including multiple mixer-settler units or single countercurrent-flow columns with or without mechanical agitation. Very compact, but expensive, centrifugal extractors are also available. When the equivalent of only a few theoretical stages is required, mixer-settler units may be the best choice because efficiencies approaching 100% are achievable in each unit. For a large number of stages, columns with mechanical agitation may be favored. Packed and perforated tray columns can be very inefficient and are not recommended for critical separations.

Membrane Separation

Most commercial membrane separations use natural or synthetic, glassy or rubbery polymers. To achieve high permeability and selectivity, nonporous materials are preferred, with thicknesses ranging from 0.1 to 1.0 micron, either as a surface layer or film onto or as part of much thicker asymmetric or composite membrane materials, which are fabricated primarily into spiral-wound and hollow-fiber-type modules to achieve a high ratio of membrane surface area to module volume.

Adsorption

For commercial applications, an adsorbent must be chosen carefully to give the required selectivity, capacity, stability, strength, and regenerability. The most commonly used adsorbents are activated carbon, molecular-sieve carbon, molecular-sieve zeolites, silica gel, and activated alumina. Of particular importance in the selection process is the adsorption isotherm for competing solutes when using a particular adsorbent. Most adsorption operations are conducted in a semicontinuous cyclic mode that includes a regeneration step. Batch slurry systems are favored for small-scale separations, whereas fixed-bed operations are preferred for large-scale separations. Quite elaborate cycles have been developed for the latter.

Leaching

Equipment for leaching operations is designed for either batchwise or continuous processing. For rapid leaching, it is best to reduce the size of the solids by grinding or slicing. The solids are contacted by the solvent using either percolation or immersion. A number of different patented devices are available.

Crystallization

Crystallization operations include the crystallization of an inorganic compound from an aqueous solution (solution crystallization) and the crystallization of an organic compound from a mixture of organic chemicals (melt crystallization). On a large scale, solution crystallization is frequently conducted continuously in a vacuum evaporating draft-tube baffled crystallizer to produce crystalline particles, whereas the falling-film crystallizer is used for melt crystallization to produce a dense layer of crystals.

Drying

A number of factors influence the selection of a dryer from the many different types available. These factors are dominated by the nature of the feed, whether it be granular solids, a paste, a slab, a film, a slurry, or a liquid. Other factors include the need for agitation, the type of heat source (convection, radiation, conduction, or microwave heating), and the degree to which the material must be dried. The most commonly employed continuous dryers include tunnel, belt, band, turbo-tray, rotary, steam-tube rotary, screw-conveyor, fluidized-bed, spouted-bed, pneumatic-conveyor, spray, and drum dryers.

7.4   SEQUENCING OF ORDINARY DISTILLATION COLUMNS FOR THE SEPARATION OF NEARLY IDEAL FLUID MIXTURES

Multicomponent mixtures are often separated into more than two products. Although one piece of equipment of complex design might be devised to produce all the desired products, a sequence of two-product separators is more common.

For nearly ideal feeds, such as hydrocarbon mixtures and mixtures of a homologous series, for example, alcohols, the most economical sequence will often include only ordinary distillation columns provided that the following conditions hold:

1. The relative volatility between the two selected key components for the separation in each column is > 1.05.
2. The reboiler duty is not excessive. An example of an excessive duty occurs in the distillation of a mixture with a low relative volatility between the two key components, where the light key component is water, which has a very high heat of vaporization.
3. The tower pressure does not cause the mixture to approach its critical temperature.
4. The overhead vapor can be at least partially condensed at the column pressure to provide reflux without excessive refrigeration requirements.
5. The bottoms temperature at the tower pressure is not so high that chemical decomposition occurs.
6. Azeotropes do not prevent the desired separation.
7. Column pressure drop is tolerable, particularly if operation is under vacuum.

Column Pressure and Type of Condenser

During the development of distillation sequences, it is necessary to make at least preliminary estimates of column operating pressures and condenser types (total or partial). The estimates are facilitated by the use of the algorithm in Figure 7.9, which is conservative. Assume that cooling water is available at 90°F, sufficient to cool and condense a vapor to 120°F. The bubble-point pressure is calculated at 120°F for an estimated distillate composition. If the computed pressure is less than 215 psia, use a total condenser unless a vapor distillate is required in which case use a partial condenser. If the pressure is less than 30 psia, set the condenser pressure to 30 psia and avoid near-vacuum operation. If the distillate bubble-point pressure is greater than 215 psia, but less than 365 psia, use a partial condenser. If it is greater than 365 psia, determine the dew-point pressure for the distillate as a vapor. If the pressure is greater than 365 psia, operate the condenser at 415 psia with a suitable refrigerant in place of cooling water. For the selected condenser pressure, add 10 psia to estimate the bottoms pressure,

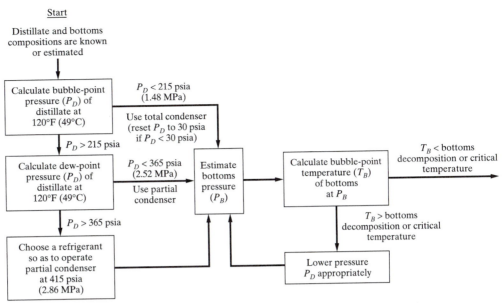

Figure 7.9 Algorithm for establishing distillation column pressure and condenser type.

and compute the bubble-point temperature for an estimated bottoms composition. If that temperature exceeds the decomposition or critical temperature of the bottoms, reduce the condenser pressure appropriately.

Number of Sequences of Ordinary Distillation Columns

Initial consideration is usually given to a sequence of ordinary distillation columns, where a single feed is sent to each column and the products from each column number just two, the distillate and the bottoms. For example, consider a mixture of benzene, toluene, and biphenyl. Because the normal boiling points of the three components (80.1, 110.8, and 254.9°C, respectively) are widely separated, the mixture can be conveniently separated into three nearly pure components by ordinary distillation. A common process for separating this mixture is the sequence of two ordinary distillation columns shown in Figure 7.10a. In the first column, the most volatile component, benzene, is taken overhead as a distillate final product. The bottoms is a mixture of toluene and biphenyl, which is sent to the second column for separation into the two other final products: a distillate of toluene and a bottoms of biphenyl, the least volatile component.

Even if a sequence of ordinary distillation columns is used, not all columns need give nearly pure products. For example, Figure 7.10b shows a distillation sequence for the separation of a mixture of ethylbenzene, p-xylene, m-xylene, and o-xylene into only three products: nearly pure ethylbenzene, a mixture of p- and m-xylene, and nearly pure o-xylene. The para and meta isomers are not separated because the normal boiling points of these two compounds differ by only 0.8°C, making separation by distillation impractical.

Note in Figure 7.10 that it takes a sequence of two ordinary distillation columns to separate a mixture into three products. Furthermore, other sequences can produce the same final products. For example, the separation of benzene, toluene, and biphenyl, shown in Figure 7.10a, can also be achieved by removing biphenyl as bottoms in the first column, followed by the separation of benzene and toluene in the second column. However, the separation of toluene from benzene and biphenyl by ordinary distillation in the first column is impossible,

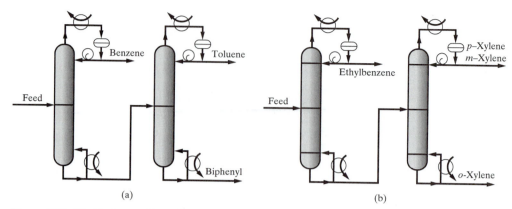

Figure 7.10 Distillation configurations for separation of ternary mixtures: (a) separation of a benzene-toluene-biphenyl mixture; (b) separation of xylene isomers.

because toluene is intermediate in volatility. Thus, the number of possible sequences is limited to two for this case of the separation of a ternary mixture into three nearly pure products.

Now consider the more general case of the synthesis of all possible ordinary distillation sequences for a multicomponent feed that is to be separated into P final products, which are nearly pure components and/or multicomponent mixtures. The components in the feed are ordered by volatility, with the first component being the most volatile. This order is almost always consistent with that for normal boiling point if the mixture forms nearly ideal liquid solutions, such that Eq. (7.3) applies. Assume that the order of volatility of the components does not change as the sequence proceeds. Furthermore, assume that any multicomponent products contain only components that are adjacent in volatility. For example, suppose that the previously cited mixture of benzene, toluene, and biphenyl is to be separated into toluene and a multicomponent product of benzene and biphenyl. With ordinary distillation, it would be necessary first to produce products of benzene, toluene, and biphenyl, and then blend the benzene and biphenyl.

An equation for the number of different sequences of ordinary distillation columns, N_s, to produce a number of products, P, can be developed in the following manner. For the first separator in the sequence, $P - 1$ separation points are possible. For example, if the desired products are A, B, C, D, and E in order of decreasing volatility, then the possible separation points are $5 - 1 = 4$, as follows: A–B, B–C, C–D, and D–E. Now let j be the number of final products that must be developed from the distillate of the first column. For example, if the separation point in the first column is C–D, then $j = 3$ (A, B, C). Then $P - j$ equals the number of final products that must be developed from the bottoms of the first column. If N_i is the number of different sequences for i final products, then, for a given separation point in the first column, the number of sequences is $N_j N_{P-j}$. But in the first separator, $P - 1$ different separation points are possible. Thus, the number of different sequences for P products is the following sum:

$$N_s = \sum_{j=1}^{P-1} N_j N_{P-j} = \frac{[2(P-1)]!}{P!(P-1)!}$$

(7.9)

Application of Eq. (7.9) gives results shown in Table 7.2 for sequences producing up to 10 products. As shown, the number of sequences grows rapidly as the number of final products increases.

Equation (7.9) gives five possible sequences of three columns for a four-component feed. These sequences are shown in Figure 7.11. The first, where all final products but one are distillates, is often referred to as the *direct sequence*, and is widely used in industry because

Table 7.2 Number of Possible Sequences for Separation by Ordinary Distillation

Number of Products, P	Number of Separators in the Sequence	Number of Different Sequences, N_s
2	1	1
3	2	2
4	3	5
5	4	14
6	5	42
7	6	132
8	7	429
9	8	1,430
10	9	4,862

distillate final products are more free of impurities such as objectionable high-boiling compounds and solids. If the purity of the final bottoms product (D) is critical, it may be produced as a distillate in an additional column called a *rerun* (or *finishing*) column. If all products except one are bottoms products, the sequence is referred to as the *indirect sequence*. This sequence is generally considered to be the least desirable sequence because of difficulties in achieving purity specifications for bottoms products. The other three sequences in Figure 7.11 produce two products as distillates and two products as bottoms. In all sequences except one, at least one final product is produced in each column.

EXAMPLE 7.1

Ordinary distillation is to be used to separate the ordered mixture C_2, $C_3^=$, C_3, $1-C_4^=$, nC_4 into the three products C_2; $(C_3^=, 1-C_4^=)$; (C_3, nC_4). Determine the number of possible sequences.

SOLUTION

Neither multicomponent product contains adjacent components in the ordered list. Therefore, the mixture must be completely separated with subsequent blending to produce the $(C_3^=, 1-C_4^=)$ and (C_3, nC_4) products. Thus, from Table 7.2 with P taken as 5, $N_s = 14$. ∎

Heuristics for Determining Favorable Sequences

When the number of products is three or four, designing and costing all possible sequences can best determine the most economical sequence. Often, however, unless the feed mixture has a wide distribution of component concentrations or a wide variation of relative volatilities for the possible separation points, the costs will not vary much and the sequence selection may be based on operation factors. In that case, the direct sequence is often the choice. Otherwise, a number of heuristics that have appeared in the literature, starting in 1947, have proved useful for reducing the number of sequences for detailed examination. The most useful of these heuristics are:

1. Remove thermally unstable, corrosive, or chemically reactive components early in the sequence.
2. Remove final products one by one as distillates (the direct sequence).
3. Sequence separation points to remove, early in the sequence, those components of greatest molar percentage in the feed.
4. Sequence separation points in the order of decreasing relative volatility so that the most difficult splits are made in the absence of the other components.
5. Sequence separation points to leave last those separations that give the highest-purity products.
6. Sequence separation points that favor near equimolar amounts of distillate and bottoms in each column.

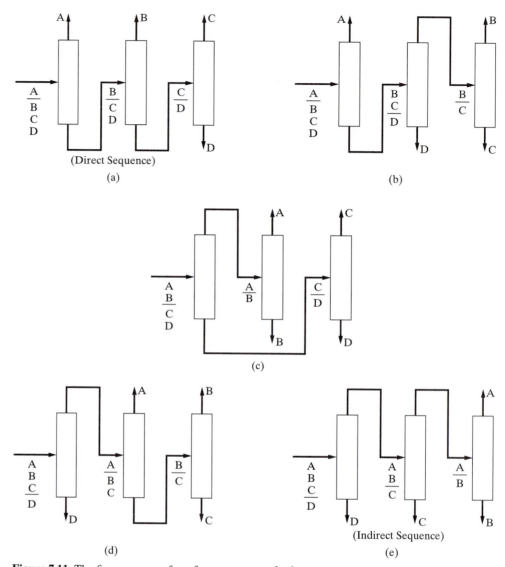

Figure 7.11 The five sequences for a four-component feed.

None of these heuristics require column design and costing. Unfortunately, however, these heuristics often conflict with each other. Thus, more than one sequence will be developed, and cost and other factors will need to be considered to develop an optimal final design. When energy costs are relatively high, the sixth heuristic often leads to the most economical sequence. Heuristics 2–6 are consistent with observations about the effect of the nonkey components on the separation of two key components. These nonkey components can increase the reflux and boilup requirements, which, in turn, increase column diameter and reboiler operating cost. These, and the number of trays, are the major factors affecting the investment and operating costs of a distillation operation.

EXAMPLE 7.2

Consider the separation problem shown in Figure 7.12a, except that separate isopentane and *n*-pentane products are also to be obtained with 98% recoveries. Use heuristics to determine a good sequence of ordinary distillation units.

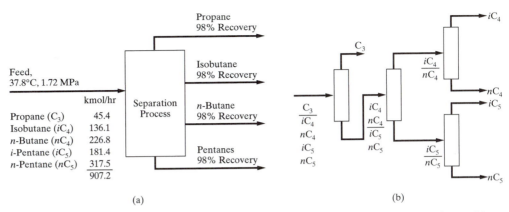

Figure 7.12 Synthesis problem and separation train for Example 7.2: (a) paraffin separation problem; (b) sequence developed from heuristics.

SOLUTION

Approximate relative volatilities for all adjacent pairs are

Component pair	Approximate α at 1 atm
C_3/iC_4	3.6
iC_4/nC_4	1.5
nC_4/iC_5	2.8
iC_5/nC_5	1.35

For this example, there are wide variations in both relative volatility and molar percentages in the process feed. The choice is Heuristic 4, which dominates over Heuristic 3 and leads to the sequence shown in Figure 7.12b, where the first split is between the pair with the highest relative volatility. This sequence also corresponds to the optimal arrangement. ∎

Marginal Vapor Rate Method

When application of the above heuristics for sequencing ordinary distillation columns is uncertain or conflicting results are obtained, it is preferable to employ sequencing methods that rely on column design and, in some cases, cost estimation. Exhaustive search to calculate the annualized cost of every sequence can determine the optimal sequence, provided that column-operating conditions are optimized, and may be justified for sequences involving just three or possibly four products. However, less rigorous methods are available that can produce good, although not always optimal, sequences. These methods, which attempt to reduce the search space, include those of Hendry and Hughes (1972), Rodrigo and Seader (1975), Gomez and Seader (1976), Seader and Westerberg (1977), and the marginal vapor rate (MV) method of Modi and Westerberg (1992). The latter method outperforms the other methods and can be applied without the necessity of complete column designs and calculations of costs.

For a given split between two key components, Modi and Westerberg (1992) consider the difference in costs between the separation in the absence of nonkey components and the separation in the presence of nonkey components, defining this difference as the marginal annualized cost (MAC). They show that a good approximation of MAC is the MV, which is the corresponding difference in molar vapor rate passing up the column. The sequence with the minimum sum of column MVs is selected. The good approximation is due to the fact that vapor rate is a good measure of cost because it is a major factor in determining column diameter as well as reboiler and condenser areas (thus, column and heat exchanger capital costs) and reboiler and condenser duties (thus, heat exchanger annual operating costs).

A convenient method for determining the molar vapor rate in an ordinary distillation column separating a nearly ideal system uses the Underwood equations to calculate the minimum reflux ratio, R_{min}. This is readily accomplished, as in the example below, with a process simulation program. The design reflux ratio is taken as $R = 1.2 R_{min}$. By material balance, the molar vapor rate, V, entering the condenser is given by $V = D(R + 1)$, where D is the molar distillate rate. Assuming that the feed to the column is a bubble-point liquid, the molar vapor rate through the column will be nearly constant at this value of V. In making the calculations of MV, the selection of product purities is not critical because the minimum reflux ratio is not sensitive to those purities. Thus, to simplify the material balance calculations, it is convenient to assume nearly perfect separations with the light and lighter-than-light key components leaving in the distillate and the heavy and heavier-than-heavy key components leaving in the bottoms. Column top and bottom pressures are estimated with Figure 7.9. The column feed pressure is taken as the average of the top and bottom pressures.

EXAMPLE 7.3

Use the marginal vapor rate (MV) method to determine a sequence for the separation of light hydrocarbons specified in Figure 7.12a, except: (1) remove the propane from the feed, (2) ignore the given temperature and pressure of the feed, and (3) strive for recoveries of 99.9% of the key components in each column. Use a process simulation program, with the Soave–Redlich–Kwong equation of state for K-values and enthalpies, to set top and bottom column pressures and estimate the reflux ratio with the Underwood equations.

SOLUTION

To produce four nearly pure products from the four-component feed, five sequences of three ordinary distillation columns each are shown in Figure 7.11. Let A = isobutane, B = n-butane, C = isopentane, and D = n-pentane. A total of 10 unique separations are embedded in Figure 7.11. These are listed in Table 7.3, together with the results of the calculations for the top column pressure, P_{top}, in kPa; the molar distillate rate, D, in kmol/hr; and the reflux ratio, R, using the shortcut (Fenske–Underwood–Gilliland or FUG) distillation model of the CHEMCAD process simulation program. This model applies the Underwood equations to estimate the minimum reflux ratio, as described by Seader and Henley (1998). Column feeds were computed as bubble-point liquids at $P_{top} + 35$ kPa. Also included in Table 7.3 are values of the column molar vapor rate, V, in kmol/hr and marginal vapor rate, MV, in kmol/hr.

From Table 7.3, the sum of the marginal vapor rates is calculated for each of the five sequences in Figure 7.11. The results are given in Table 7.4.

Table 7.4 shows that the preferred sequence is the one that performs the two most difficult separations, A/B and C/D, in the absence of nonkey components. These two separations are far more difficult than the separation, B/C. The direct sequence is the next best. ∎

Table 7.3 Calculations of Marginal Vapor Rate, MV

Separation	Column Top Pressure (kPa)	Distillate Rate, D (kmol/hr)	Reflux Ratio ($R = 1.2 R_{min}$)	Vapor Rate, $V = D(R + 1)$ (kmol/hr)	Marginal Vapor Rate (kmol/hr)
A/B	680	136.2	10.7	1,594	0
A/BC	680	136.2	11.9	1,757	163
A/BCD	680	136.2	13.2	1,934	340
B/C	490	226.8	2.06	694	0
AB/C	560	362.9	1.55	925	231
B/CD	490	226.8	3.06	921	227
AB/CD	560	362.9	2.11	1,129	435
C/D	210	181.5	13.5	2,632	0
BC/D	350	408.3	6.39	3,017	385
ABC/D	430	544.4	4.96	3,245	613

Table 7.4 Marginal Vapor Rates for the Five Possible Sequences

Sequence in Figure 7.11	Marginal Vapor Rate, MV (kmol/hr)
(a) Direct	567
(b)	725
(c)	435
(d)	776
(e) Indirect	844

Complex and Thermally Coupled Distillation Columns

Following the development of an optimal or near-optimal sequence of simple, two-product distillation columns, revised sequences involving complex, rather than simple, distillation columns should be considered. Some guidance is available from a study by Tedder and Rudd (1978a, b) of the separation of ternary mixtures (A, B, and C in order of decreasing volatility) in which eight alternative sequences of one to three columns were considered, seven of which are shown in Figure 7.13. The configurations include the direct and indirect sequences (I and II), two interlinked cases (III and IV), five cases that include the use of side streams (III, IV, V, VI, and VII), and one case (V) involving a column with two feeds. All columns in Cases I, II, V, VI, and VII have condensers and reboilers. In Cases III and IV, the first column has a condenser and reboiler. In Case III, the rectifier column has a condenser only, while the stripper in Case IV has a reboiler only. The interlinking streams that return from the second column to the first column thermally couple the columns in Cases III and IV.

As shown in Figure 7.14, optimal regions for the various configurations depend on the process feed composition and on an ease-of-separation index (ESI), which is defined as the relative volatility ratio, $\alpha_{A,B}/\alpha_{B,C}$. It is interesting to note that a ternary mixture is separated into three products with just one column in Cases VI and VII, but the reflux requirement is excessive unless the feed contains a large amount of B, the component of intermediate volatility, and little of the component that is removed from the same section of the column as B. Otherwise, if the feed is dominated by B but also contains appreciable amounts of A and C, the prefractionator case (V) is optimal. Perhaps the biggest surprise of the study is the superiority of distillation with a vapor sidestream rectifier, which is favored for a large region of the feed composition when ESI > 1.6. The results of Figure 7.14 can be extended to multicomponent separation problems involving more than three components, if difficult ternary separations are performed last.

Case V in Figure 7.13 consists of a prefractionator followed by a product column, from which all three final products are drawn. Each column is provided with its own condenser and reboiler. As shown in Figure 7.15, eliminating the condenser and reboiler in the prefractionator and providing, instead, reflux and boilup to that column from the product column can thermally couple this arrangement, which is referred to as a Petlyuk system after its chief developer and is described by Petlyuk et al. (1965). The prefractionator separates the ternary-mixture feed, ABC, into a top product containing A and B and a bottom product of B and C. Thus, component B is split between the top and bottom streams exiting from the prefractionator. The top product is sent to the upper section of the product column, while the bottom product is sent to the lower section. The upper section of the product column provides the reflux for the prefractionator, while the lower section provides the boilup. The product column separates its two feeds into a distillate of A, a sidestream of B, and a bottoms of C. Fidkowski and Krolikowski (1987) determined the minimum molar boilup vapor requirements for the Petlyuk system and the other two thermally coupled systems (III and IV) in Figure 7.13, assuming constant relative volatilities, constant molar overflow, and bubble-point liquid feed and products. They compared the requirements to those of the conventional direct and indirect sequences shown as Cases I and II in Figure 7.13 and proved

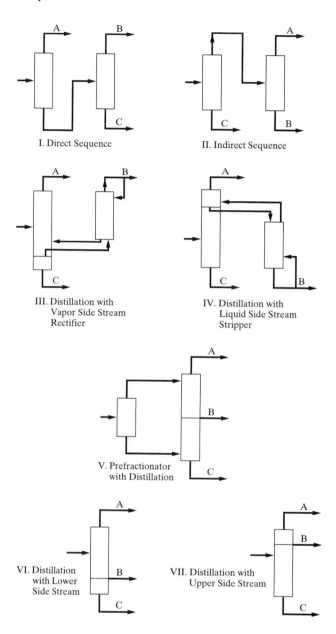

Figure 7.13 Configurations for ternary distillation.

I. Direct Sequence

II. Indirect Sequence

III. Distillation with Vapor Side Stream Rectifier

IV. Distillation with Liquid Side Stream Stripper

V. Prefractionator with Distillation

VI. Distillation with Lower Side Stream

VII. Distillation with Upper Side Stream

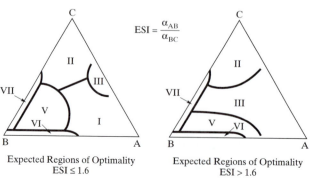

$$ESI = \frac{\alpha_{AB}}{\alpha_{BC}}$$

Expected Regions of Optimality
ESI ≤ 1.6

Expected Regions of Optimality
ESI > 1.6

Figure 7.14 Regions of optimality for ternary distillation configurations (Tedder and Rudd, 1978a, b).

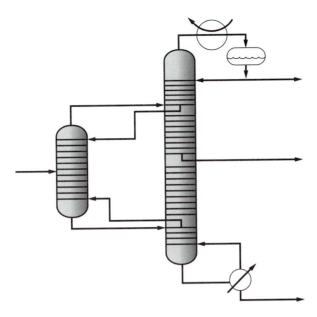

Figure 7.15 Thermally coupled Petlyuk system.

that for all combinations of feed flow rates of the components A, B, and C, as well as all values of relative volatilities, that: (1) the Petlyuk system has the lowest minimum molar boilup vapor requirements and (2) Cases III and IV in Figure 7.13 are equivalent and have lower minimum molar boilup vapor requirements than either the direct or indirect sequence.

Despite its lower vapor boilup requirements, no industrial installations of a two-column Petlyuk system have been reported. Two possible reasons for this, as noted by Agrawal and Fidkowski (1998), are: (1) an unfavorable thermodynamic efficiency when the three feed

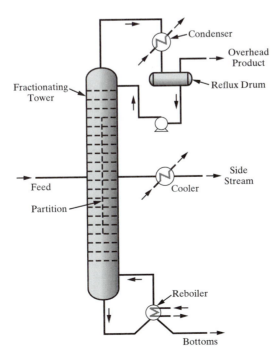

Figure 7.16 Dividing-wall (partition) column of Wright.

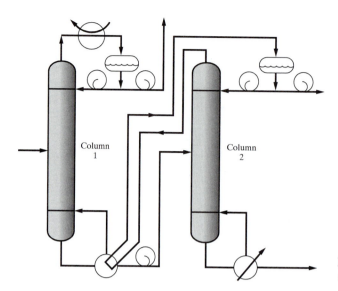

Figure 7.17 Heat-integrated direct sequence of two distillation columns.

components are not close-boiling because all of the reboiler heat must be supplied at the highest temperature and all of the condenser heat must be removed at the lowest temperature; and (2) the difficulty in controlling the fractions of vapor and liquid streams in the product column that are returned to the prefractionator as boilup and reflux, respectively. The Petlyuk system can be embodied into a single column, with a significantly reduced capital cost, by using a dividing-wall column (also called divided wall and column-in-column), a concept described in a patent by Wright (1949) and shown by his patent drawing in Figure 7.16. Because the dividing-wall column makes possible savings in both energy and capital, and because control difficulties appear to have been solved, it is attracting much attention. The first dividing-wall column was installed by BASF in 1985. A number of such columns using packing have been installed in the last 15 years and the first dividing-wall column using trays was recently announced. Agrawal and Fidkowski (1998) present other thermally fully coupled (FC) systems of distillation columns that retain the benefit of a minimum vapor requirement and afford easier control. Energy savings can also be achieved by heat-integrating the two columns in a direct sequence. In Figure 7.17, Column 2 is operated at a higher pressure than Column 1, such that the condenser duty of Column 2 can provide the reboiler duty of Column 1. Rev et al. (2001) show that heat-integrated systems are often superior in annualized cost to the Petlyuk system. For further discussion of heat-integrated distillation columns, see Sections 10.9, 20.1, and 21.3.

7.5   SEQUENCING OF OPERATIONS FOR THE SEPARATION OF NONIDEAL FLUID MIXTURES

When a multicomponent fluid mixture is nonideal, its separation by a sequence of ordinary distillation columns will not be technically and/or economically feasible if relative volatilities between key components drop below 1.05 and, particularly, if azeotropes are formed. For such mixtures, separation is most commonly achieved by sequences comprised of ordinary distillation columns, enhanced distillation columns, and/or liquid–liquid extraction equipment. Membrane and adsorption separations can also be incorporated into separation sequences, but their use is much less common. Enhanced distillation operations include extractive distillation, homogeneous azeotropic distillation, heterogeneous azeotropic distillation, pressure-swing distillation, and reactive distillation. These operations are considered in detail in *Perry's Chemical Engineers' Handbook* (Perry and Green, 1997) and by Seader and Henley (1998), Stichlmair and Fair (1998), and Doherty and Malone (2001). A design-oriented introduction to enhanced distillation is presented here.

In many processes involving oxygenated organic compounds, such as alcohols, ketones, ethers, and acids, often in the presence of water, distillation separations are complicated by the presence of azeotropes. Close-boiling mixtures of hydrocarbons (e.g., benzene and cyclohexane, whose normal boiling points only differ by 1.1°F) can also form azeotropes. For these and other mixtures, special attention must be given to the *distillation boundaries* in the composition space that confine the compositions for any one column to lie within a bounded region of the composition space. To introduce these boundaries, leading to approaches for the synthesis of separation trains, several concepts concerning azeotropes and *residue curves* and *distillation lines* are reviewed in the subsections that follow.

Azeotropy

Azeotrope is an ancient Greek word that is translated "to boil unchanged," meaning that the vapor emitted has the same composition as the liquid (Swietoslawski, 1963). When classifying the many azeotropic mixtures, it is helpful to examine their deviations from Raoult's law (Lecat, 1918).

When two or more fluid phases are in physical equilibrium, the chemical potential, fugacity, or activity of each species is the same in each phase. Thus, in terms of species mixture fugacities for a vapor phase in physical equilibrium with a single liquid phase,

$$\bar{f}_j^V = \bar{f}_j^L \qquad j = 1, \ldots, C \tag{7.10}$$

Substituting expressions for the mixture fugacities in terms of mole fractions, activity coefficients, and fugacity coefficients,

$$y_j \bar{\phi}_j^V P = x_j \gamma_j^L f_j^L \qquad j = 1, \ldots, C \tag{7.11}$$

where $\bar{\phi}$ is a mixture fugacity coefficient, γ is a mixture activity coefficient, and f is a pure-species fugacity.

For a binary mixture with an ideal liquid solution ($\gamma_j^L = 1$) and a vapor phase that forms an ideal gas solution and obeys the ideal gas law ($\bar{\phi}_j^V = 1$ and $f_j^L = P_j^s$), Eq. (7.11) reduces to the following two equations for the two components 1 and 2:

$$y_1 P = x_1 P_1^s \tag{7.12a}$$
$$y_2 P = x_2 P_2^s \tag{7.12b}$$

where P_j^s is the vapor pressure of species j.

Adding Eqs. (7.12a and 7.12b), noting that mole fractions must sum to one,

$$(y_1 + y_2) P = P = x_1 P_1^s + x_2 P_2^s = x_1 P_1^s + (1 - x_1) P_2^s = P_2^s + (P_1^s - P_2^s) x_1 \tag{7.13}$$

This linear relationship between the total pressure, P, and the mole fraction, x_1, of the most volatile species is a characteristic of Raoult's law, as shown in Figure 7.18a for the benzene-toluene mixture at 90°C. Note that the bubble-point curve (P–x) is linear between the vapor pressures of the pure species (at $x_1 = 0, 1$), and the dew-point curve (P–y_1) lies below it. When the (x_1, y_1) points are graphed at different pressures, the familiar vapor–liquid equilibrium curve is obtained, as shown in Figure 7.18b. Using McCabe–Thiele analysis, it is shown readily that for any feed composition, there are no limitations to the values of the mole fractions of the distillate and bottoms products from a distillation tower.

However, when the mixture forms a nonideal liquid phase and exhibits a positive deviation from Raoult's law ($\gamma_j^L > 1, j = 1, 2$), Eq. (7.13) becomes

$$P = x_1 \gamma_1^L P_1^s + (1 - x_1) \gamma_2^L P_2^s \tag{7.14}$$

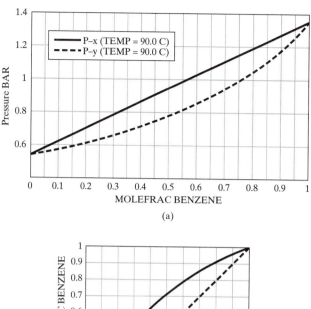

Figure 7.18 Phase diagrams for the benzene–toluene mixture at 90°C, calculated using ASPEN PLUS: (a) *P–x–y* diagram: (b) *x–y* diagram.

Furthermore, if the boiling points of the two components are close enough, the bubble- and dew-point curves may reach a maximum at the same composition, which by definition is the azeotropic point. Such a situation is illustrated in Figure 7.19a for the isopropyl ether (1)–isopropyl alcohol (2) binary at 70°C. Figure 7.19b shows the corresponding *x–y* diagram, and Figure 7.19c shows the bubble- and dew-point curves on a *T–x–y* diagram at 101 kPa. Note the *minimum-boiling azeotrope* at 66°C, where $x_1 = y_1 = 0.76$. Feed streams having lower isopropyl ether mole fractions cannot be purified beyond 0.76 in a distillation column, and streams having higher isopropyl ether mole fractions have distillate mole fractions that have a lower bound of 0.76. Consequently, the azeotropic composition is commonly referred to as a *distillation boundary*.

Similarly, when the mixture exhibits the less-common negative deviation from Raoult's law ($\gamma_j^L < 1, j = 1, 2$), both the bubble- and dew-point curves drop below the straight line that represents the bubble points for an ideal mixture, as anticipated by examination of Eq. (7.14). Furthermore, when the bubble- and dew-point curves have the same minimum, an azeotropic composition is defined, as shown in Figure 7.20a for the acetone–chloroform binary at 64.5°C, where $x_1 = y_1 = 0.35$. For this system, Figures 7.20b and 7.20c show the corresponding *x–y* diagram and *T–x–y* diagram at 101 kPa. On the latter diagram, the azeotropic point is at a maximum temperature, and consequently, the system is said to have a *maximum-boiling azeotrope*. In this case, feed streams having lower acetone mole fractions cannot be purified beyond 0.35 in the bottoms product of a distillation column, and streams having

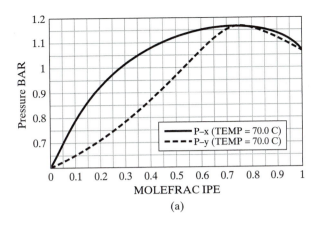

(a)

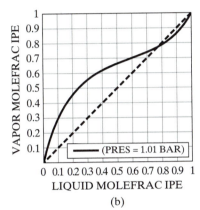

(b)

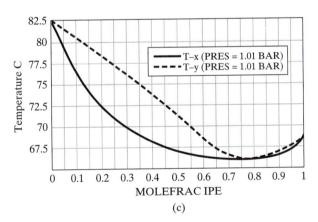

(c)

Figure 7.19 Phase diagrams for the isopropyl ether–isopropyl alcohol binary computed using ASPEN PLUS: (a) *P–x–y* diagram at 70°C; (b) *x–y* diagram at 101 kPa; (c) *T–x–y* diagram at 101 kPa.

higher acetone mole fractions have a lower bound of 0.35 in the acetone mole fraction of the bottoms product.

In summary, at a homogeneous azeotrope, $x_j = y_j$, $j = 1, \ldots, C$, the expression for the equilibrium constant, K_j for species j becomes unity. Based on the general phase equilibria Eq. (7.11), the criterion for azeotrope formation is:

$$K_j = \frac{y_j}{x_j} = \frac{\gamma_j^L f_j^L}{\bar{\phi}_j^V P} = 1 \qquad j = 1, \ldots, C \tag{7.15}$$

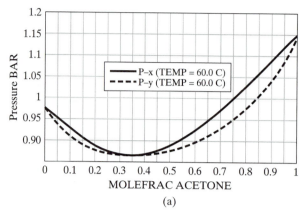

(a)

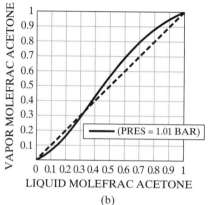

(b)

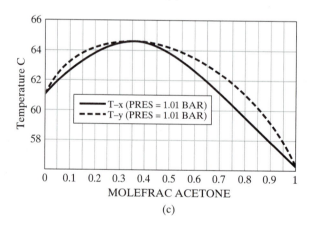

(c)

Figure 7.20 Phase diagrams for the acetone–chloroform binary computed using ASPEN PLUS: (a) *P–x–y* diagram at 60°C; (b) *x–y* diagram at 101 kPa; (c) *T–x–y* diagram at 101 kPa.

where the degree of nonideality is expressed by the deviations from unity of the activity coefficients and fugacities for the liquid phase and the fugacity coefficients for the vapor phase. At low pressure, $\bar{\phi}_j^V = 1$ and $f_j^L = P_j^s$ so that Eq. (7.15) reduces to

$$K_j = \frac{y_j}{x_j} = \gamma_j^L \frac{P_j^s}{P} = 1 \qquad j = 1, \ldots, C \tag{7.16}$$

Because the *K*-values for all of the species are unity at an azeotrope point, a simple distillation approaches this point, at which no further separation can occur. For this reason, an azeotrope is often called a *stationary* or *fixed* or *pinch point*.

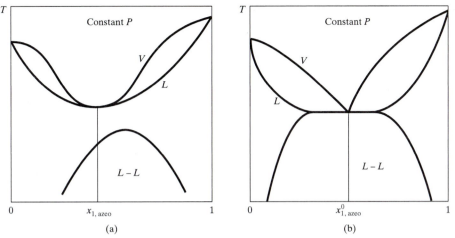

Figure 7.21 Binary phase diagram at a fixed pressure for: (a) homogeneous azeotrope; (b) heterogeneous azeotrope.

For a minimum-boiling azeotrope, when the deviations from Raoult's law are sufficiently large ($\gamma_j^L \gg 1.0$, usually > 7), splitting of the liquid phase into two liquid phases (phase splitting) may occur and a minimum-boiling, heterogeneous azeotrope may form that has a vapor phase in equilibrium with the two liquid phases. A heterogeneous azeotrope occurs when the vapor–liquid envelope overlaps with the liquid–liquid envelope, as illustrated in Figure 7.21b. For a homogeneous azeotrope, when $x_1 = x_{1,\text{azeo}} = y_1$, the mixture boils at this composition, as shown in Figure 7.21a; whereas for a heterogeneous azeotrope, when the overall liquid composition of the two liquid phases, $x_1 = x_{1,\text{azeo}}^0 = y_1$, the mixture boils at this overall composition, as illustrated in Figure 7.21b, but the three coexisting phases have distinct compositions.

Residue Curves

To understand better the properties of azeotropic mixtures that contain three chemical species, it helps to examine the properties of *residue curves* on a ternary diagram. A collection of residue curves, which is called a *residue curve map*, can be computed and drawn by any of the major simulation programs. Each residue curve is constructed by tracing the composition of the equilibrium liquid residue of a simple (Rayleigh batch) distillation in time, starting from a selected initial composition of the charge to the still, using the following numerical procedure.

Consider L moles of liquid with mole fractions x_j ($j = 1, \ldots, C$), in a simple distillation still at its bubble point, as illustrated in Figure 7.22. Note that the still contains no trays and

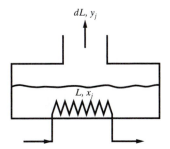

Figure 7.22 Simple distillation still.

no reflux is provided. As heating begins, a small portion of this liquid, ΔL moles, is vaporized. The instantaneous vapor phase has mole fractions y_j $(j = 1, \ldots, C)$, assumed to be in equilibrium with the remaining liquid. Since the residual liquid, $L - \Delta L$ moles, has mole fractions $x_j + \Delta x_j$, the mass balance for species j is given by

$$Lx_j = (\Delta L)y_j + (L - \Delta L)(x_j + \Delta x_j) \qquad j = 1, \ldots, C - 1 \tag{7.17}$$

In the limit, as $\Delta L \rightarrow 0$,

$$\frac{dx_j}{dL/L} = x_j - y_j = x_j \left(1 - K_j \{ T, P, \underline{x}, \underline{y} \} \right) \qquad j = 1, \ldots, C - 1 \tag{7.18}$$

and setting, $d\hat{t} = dL/L$,

$$\frac{dx_j}{d\hat{t}} = x_j - y_j \qquad j = 1, \ldots, C - 1 \tag{7.19}$$

where K_j is given by Eq. (7.15). In Eq. (7.19), $\hat{t}$ can be interpreted as the dimensionless time, with the solution defining a family of residue curves, as illustrated in Figure 7.23. Note that each residue curve is the locus of the compositions of the residual liquid in time, as vapor is boiled off from a simple distillation still. Often, an arrow is assigned in the direction of increasing time (and increasing temperature). Note that the residue curve map does not show the equilibrium vapor composition corresponding to each point on a residue curve.

Yet another important property is that the *fixed points* of the residue curves are points where the driving force for a change in the liquid composition is zero; that is, $dx/d\hat{t} = 0$. This condition is satisfied at the azeotropic points and the pure species vertices. For a ternary mixture with a single binary azeotrope, as in Figure 7.23, there are four fixed points on the triangular diagram: the binary azeotrope and the three vertices. Furthermore, the behavior of the residue curves in the vicinity of the fixed points depends on their stability. When all of the residue curves are directed by the arrows to the fixed point, it is referred to as a *stable node*, as illustrated in Figure 7.24a; when all are directed away, the fixed point is an *unstable node* (as in Figure 7.24b); and finally, when some of the residue curves are directed to and others are directed away from the fixed point, it is referred to as a *saddle point* (as in Figure 7.24c). Note that for a ternary system, the stability can be determined by calculating the eigenvalues of the Jacobian matrix of the nonlinear ordinary differential equations that comprise Eq. (7.19).

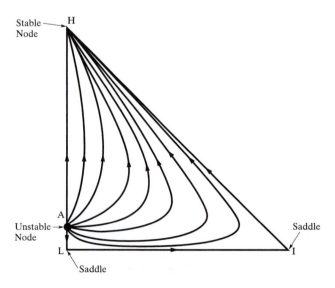

Figure 7.23 Residue curves of a ternary system with a minimum-boiling binary azeotrope.

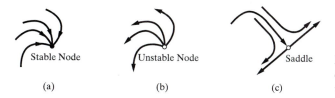

Figure 7.24 Stability of residue curves for a ternary system in the vicinity of a binary azeotrope.

As an example, consider the residue curve map for a ternary system with a minimum-boiling binary azeotrope of heavy (H) and light (L) species, as shown in Figure 7.23. There are four fixed points: one unstable node at the binary azeotrope (A), one stable node at the vertex for the heavy species (H), and two saddles at the vertices of the light (L) and intermediate (I) species.

It is of special note that the boiling points and the compositions of all azeotropes can be used to characterize residue curve maps. In fact, even without a simulation program to compute and draw the detailed diagrams, this information alone is sufficient to sketch the key characteristics of these diagrams using the following procedure. First, the boiling points of the pure species are entered at the vertices. Then the boiling points of the binary azeotropes are positioned at the azeotropic compositions along the edges, with the boiling points of any ternary azeotropes positioned at their compositions within the triangle. Arrows are assigned in the direction of increasing temperature in a simple distillation still. As examples, typical diagrams for mixtures involving binary and ternary azeotropes are illustrated in Figure 7.25. Figure 7.25a is for a simple system, without azeotropes, involving nitrogen, oxygen, and

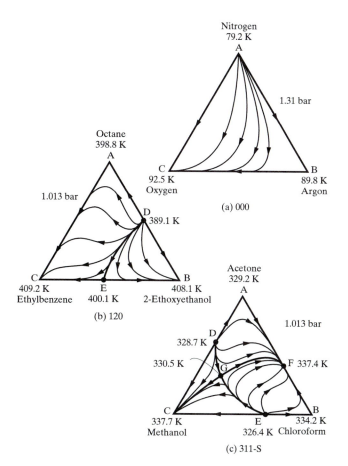

Figure 7.25 Maps of residue curves or distillation lines: (a) system without azeotropes; (b) system with two binary azeotropes; (c) system with binary and ternary azeotropes (Stichlmair et al., 1989).

argon. In this mixture, nitrogen is the lowest-boiling species (L), argon is the intermediate boiler (I), and oxygen is the highest-boiling species (H). Thus, along the oxygen–argon edge the arrow is pointing to the oxygen vertex, and on the remaining edges the arrows point away from the nitrogen vertex. Since these arrows point away at the nitrogen vertex, it is an unstable node, and all of the residue curves emanate from it. At the argon vertex, the arrows point to and away from it. Since the residue curves turn in the vicinity of this vertex, it is not a terminal point. Rather, it is referred to as a saddle point. All of the curves end at the oxygen vertex, which is a terminal point or stable node. For this ternary mixture, the map shows that pure argon, the intermediate boiler, cannot be obtained in a simple distillation.

Simple Distillation Boundaries

The graphical approach described here is effective in locating the starting and terminal points and the qualitative locations of the residue curves. As illustrated in Figures 7.25b and 7.25c, it works well for binary and ternary azeotropes that exhibit multiple starting and terminal points. In these cases, one or more *simple distillation boundaries* called *separatrices* (e.g., curved line DE in Figure 7.25b) divide these diagrams into regions with distinct pairs of starting and terminal points. For the separation of homogeneous mixtures by simple distillation, these separatrices cannot be crossed unless they are highly curved. A feed located in region ADECA in Figure 7.25b has a starting point approaching the composition of the binary azeotrope of octane and 2-ethoxyethanol and a terminal point approaching pure ethylbenzene, whereas a feed located in region DBED has a starting point approaching the same binary azeotrope but a terminal point approaching pure 2-ethoxyethanol. In this case, a pure octane product is not possible. Figure 7.25c is even more complex. It shows four distillation boundaries (curved lines GC, DG, GF, and EG), which divide the diagram into four distillation regions.

Distillation Towers

When tray towers are modeled assuming vapor–liquid equilibrium at each tray, the residue curves approximate the liquid composition profiles at *total reflux*. To show this, a species balance is performed for the top n trays, counting down the tower, as shown in Figure 7.26:

$$L_{n-1}\underline{x}_{n-1} + D\underline{x}_D = V_n\underline{y}_n \tag{7.20}$$

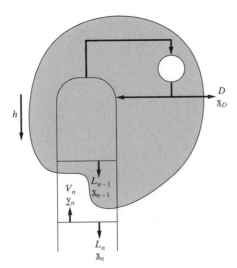

Figure 7.26 Schematic of rectifying section.

where D and $\underline{x}_D$ are the molar flow rate and vector of mole fractions of the distillate. Similarly, L_{n-1} and $\underline{x}_{n-1}$ are for the liquid leaving tray $n - 1$, and V_n and $\underline{y}_n$ are for the vapor leaving tray n. Defining h as the dimensionless distance from the top of the tower, a backward-difference approximation at tray $n + 1$ is

$$\left.\frac{d\underline{x}}{dh}\right|_{n+1} \cong \underline{x}_n - \underline{x}_{n-1} \tag{7.21}$$

Rearranging Eq. (7.20) and substituting in Eq. (7.21),

$$\left.\frac{d\underline{x}}{dh}\right|_{n+1} \cong \underline{x}_n - \frac{V_n}{L_{n-1}}\underline{y}_n + \frac{D}{L_{n-1}}\underline{y}_D \tag{7.22}$$

At total reflux, with $D = 0$ and $V_n = L_{n-1}$, Eq. (7.22) becomes

$$\left.\frac{d\underline{x}}{dh}\right|_{n+1} \cong \underline{x}_n - \underline{y}_n \tag{7.23}$$

Hence, Eq. (7.23) approximates the operating lines at total reflux and, because $\hat{t}$ and h are dimensionless variables and Eq. (7.19) is identical in form, the residue curves approximate the operating lines of a distillation tower operating at total reflux.

Distillation Lines

An exact representation of the operating line for a distillation tower at total reflux, also known as a *distillation line* [as defined by Zharov (1968) and Zharov and Serafimov (1975)], is shown in Figure 7.27. Note that, at total reflux,

$$\underline{x}_n = \underline{y}_{n+1} \qquad n = 0, 1, \dots \tag{7.24}$$

Furthermore, assuming operation in vapor–liquid equilibrium, the mole fractions on tray n, $\underline{x}_n$, and $\underline{y}_n$ lie at the ends of the equilibrium tie lines.

To appreciate better the differences between distillation lines and residue curves, consider the following observations. First, Eq. (7.19) requires the tie line vectors connecting liquid composition $\underline{x}$ and vapor composition $\underline{y}$, at equilibrium, to be tangent to the residue curves, as illustrated in Figure 7.28. Since these tie line vectors must also be chords of the distillation lines, the residue curves and the distillation lines must intersect at the liquid composition $\underline{x}$. Note that when the residue curve is linear (as for binary mixtures), the tie lines and the residue curve are collinear, and consequently, the distillation lines coincide with the residue curves.

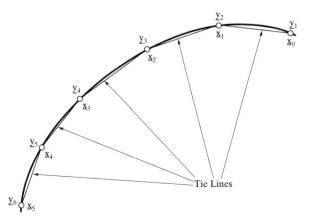

Figure 7.27 Distillation line and its tie lines.

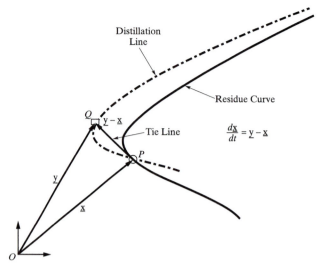

Figure 7.28 Residue curve and distillation line through P.

Figure 7.29a shows two distillation lines (δ_1 and δ_2) that intersect a residue curve at points A and B. As a consequence of Eq. (7.19), their corresponding vapor compositions at equilibrium, a and b, lie at the intersection of the tangents to the residue curves at A and B with the distillation lines δ_1 and δ_2. Clearly, the distillation lines do not coincide with the residue curves, an assumption that is commonly made but that may produce significant errors. In Figure 7.29b, a single distillation line connects the compositions on four adjacent trays (at C, D, E, F) and crosses four residue curves (ρ_C, ρ_D, ρ_E, ρ_F) at these points.

Note that distillation lines are generated by computer as easily as residue curves and, because they do not involve any approximations to the operating line at total reflux, are preferred for the analyses to be performed in the remainder of this section. However, simulation programs compute and plot only residue curves. It can be shown that distillation lines have the same properties as residue curves at fixed points, and hence, both families of curves are sketched similarly. Their differences are pronounced in regions that exhibit extensive curvature.

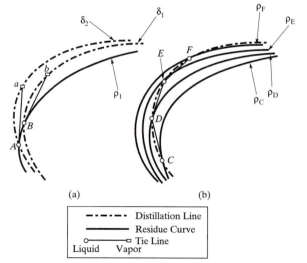

(a)   (b)

	Distillation Line
	Residue Curve
Liquid Vapor	Tie Line

Figure 7.29 Geometric relationship between distillation lines and residue curves.

Computing Azeotropes for Multicomponent Mixtures

Gmehling (1994) provides data on more than 15,000 binary azeotropes and 900 ternary azeotropes. Undoubtedly, many more ternary azeotropes exist, as well as untold numbers of azeotropes involving more than three components. When a process simulation program is used to compute a residue curve map for a ternary system at a specified pressure, compositions and temperatures of all azeotropes are automatically estimated. The results depend, of course, on the selected vapor pressure and liquid-phase activity coefficient correlations. For quaternary and higher systems, the arclength homotopy-continuation method of Fidkowski, Malone, and Doherty (1993) can be used for homogeneous systems to estimate all azeotropes. They find all roots to the following equations, which define a homogeneous azeotrope:

$$y_j - x_j = 0, \qquad j = 1, 2, \dots, C - 1 \tag{7.25}$$

$$y_j = \left(\frac{\gamma_j^L P_j^s}{\bar{\phi}_j^V P} \right) x_j, \qquad j = 1, 2, \dots, C \tag{7.26}$$

$$\sum_{j=1}^{C} x_j = 1 \tag{7.27}$$

$$\sum_{j=1}^{C} y_j = 1 \tag{7.28}$$

$$x_j \geq 0, \qquad j = 1, 2, \dots, C \tag{7.29}$$

To find the roots, they construct the following homotopy to replace Eqs. (7.25) and (7.26), based on gradually moving from an ideal K-value based on Raoult's law to the more rigorous expression of Eq. (7.26):

$$y_j - x_j = \left[(1 - t) + t \frac{\gamma_j^L}{\bar{\phi}_j^V} \right] \frac{P_j^s}{P} x_j = H(t, x_j) = 0, \qquad j = 1, 2, \dots, C \tag{7.30}$$

Initially, the homotopy parameter, t, is set to 0 and all values of x_j are set to 0 except for one, which is set to 1.0. Then t is gradually and systematically increased until a value of 1.0 is obtained. With each increase, the temperature and mole fractions are computed. If the resulting composition at $t = 1.0$ is not a pure component, it is an azeotrope. By starting from each pure component, all azeotropes are computed. The method of Fidkowski, Malone, and Doherty is included in many of the process simulation programs. Eckert and Kubicek (1997) extended the method of Fidkowski, Malone, and Doherty to the estimation of heterogeneous multicomponent azeotropes.

Distillation-Line Boundaries and Feasible Product Compositions

Of great practical interest is the effect of distillation boundaries on the operation of distillation towers. To summarize a growing body of literature, it is well established that the compositions of a distillation tower operating at total reflux cannot cross the distillation-line boundaries, except under unusual circumstances, where these boundaries exhibit a high degree of curvature. This provides the total-reflux bound on the possible (feasible) compositions for the distillate and bottoms products.

As shown in Figure 7.30a, at total reflux, x_B and y_D reside on a distillation line. Furthermore, these compositions lie collinear with the feed composition, x_F, on the overall material balance line. As the number of stages increases, the operating curve becomes more convex and in the limit approaches the two sides of the triangle that meet at the intermediate boiler. As an example, an operating line at total reflux (minimum stages) is the curve AFC in Figure 7.31a. At the other extreme, as the number of stages increases, the operating curve becomes more convex approaching ABC, where the number of stages approaches infinity (corresponding to minimum

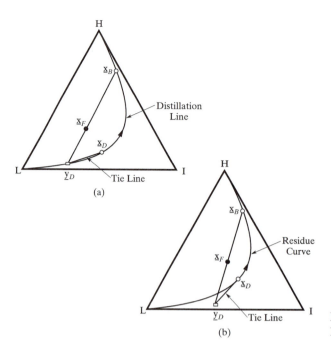

Figure 7.30 Overall mass balance line with a partial/total condenser.

reflux). Hence, the operating line for a distillation tower that operates within these limiting regimes lies within the region ABCFA in Figure 7.31a. Note that when a distillation tower operates with a partial condenser, as the feed and product streams are decreased toward total reflux, the last bubble of vapor distillate has the mole fractions, y_D, as shown in Figures 7.30a and 7.30b. Consequently, as total reflux is approached, the material balance line connecting the bottoms, feed, and distillate mole fractions is shown. Figure 7.30a shows the distillation line that passes through the x_B and y_D mole fractions, while Figure 7.30b shows the residue curve that passes through the x_D mole fractions, and approximately through the x_B mole fractions.

Two additional bounds in Figure 7.31a are obtained as follows. First, in the limit of a pure nitrogen distillate, the line AFE represents a limiting overall material balance for a feed composition at point F, with point E at the minimum concentration of oxygen in the bottoms product. Similarly, in the limit of a pure oxygen bottoms, the line CFD represents a limiting overall material balance, with point D at the minimum concentration of nitrogen in the distillate along the nitrogen–argon axis. Hence, the distillate composition is confined to the shaded region ADFA, and the bottoms product composition lies in the shaded region, CEFC. Operating lines that lie within the region ABCFA connect the distillate and bottoms product compositions in these shaded regions. At best, only one pure species can be obtained. In addition, only those species located at the end points of the distillation lines can be recovered in high purity, with one exception to be noted. Hence, the end points of the distillation lines determine the potential distillate and bottoms products for a given feed. This also applies to the complex mixtures in Figures 7.31b and 7.31c. Here, the location of the feed point determines the *distillation region* in which the potential distillate and bottoms product compositions lie. For example, in Figure 7.31b, for feed F, only pure 2-ethoxyethanol can be obtained. When the feed is moved to the left across the distillation-line boundary, pure ethylbenzene can be obtained. In Figure 7.31c, only methanol can be recovered in high purity for feeds in the region LTGCL. For a feed in the region EDTHGBE, no pure product is possible. Before attempting rigorous distillation calculations with a simulation program, it is essential to establish, with the aid of computer-generated residual curve maps, regions of product-composition feasibility such as shown in Figure 7.31. Otherwise it is possible to waste much time and effort in trying to converge distillation calculations when specified product compositions are impossible.

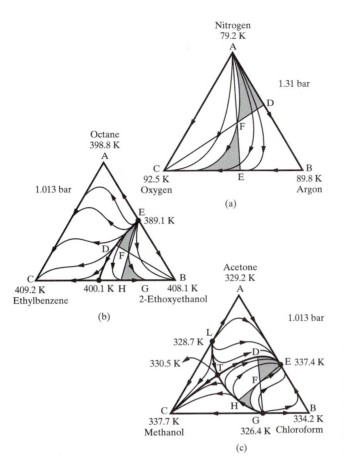

Figure 7.31 Regions of feasible distillate and bottoms product compositions (shaded) for a ternary mixture: (a) system without azeotropes; (b) system with two binary azeotropes; (c) system with binary and ternary azeotropes (Stichlmair et al., 1989).

Heterogeneous Distillation

In heterogeneous azeotropic distillation, an entrainer is utilized that concentrates in the overhead vapor and, when condensed, causes the formation of a second liquid phase that can be decanted and recirculated to the tower as reflux. The other liquid phase as well as the bottoms are the products from the distillation. This is possible when the entrainer forms a heterogeneous azeotrope with one or more of the species in the feed. Figure 7.32a shows one possible configuration, with an accompanying triangular diagram in Figure 7.32b for the dehydration of ethanol using toluene as an entrainer. In Column C-1, the feed is preconcentrated in ethanol. Column C-2 is the azeotropic tower. Unfortunately, both products, B1 and B2 are bottoms. Ethanol and water form a minimum-boiling azeotrope at 89 mol% ethanol and 1 atm, as shown in Figures 7.32c and 7.32d, which were prepared by ASPEN PLUS. Although toluene is the highest-boiling species, it is an appropriate entrainer because it forms minimum-boiling azeotropes with both water and ethanol. Hence, the arrows on the residue curves point toward both the ethanol and water vertices, allowing ethanol to be recovered in a high-purity bottoms product. Since toluene forms a ternary, minimum-boiling, heterogeneous azeotrope (point D2 in Figure 7.32b), the overhead vapor approaches this composition and condenses into two liquid phases, one rich in toluene (point S2 in Figure 7.32b) and the other rich in water (point S1 in Figure 7.32b), which are separated in the decanter. The former is recycled to the azeotropic tower, while the latter is recycled to the preconcentrator. All column product compositions are shown in Figure 7.32b. A binodal curve for the distillate temperature of the azeotropic tower is included in Figure 7.32b together with a tie line through the azeotropic composition of D2 to show the phase split of condensed overhead D2 into liquid phases S1 and S2.

When residue curve and distillation-line maps are constructed for heterogeneous systems, using process simulation programs, the composition spaces are also divided into regions with simple distillation boundaries. However, the residue curve and distillation-line maps for systems containing heterogeneous azeotropes are far more restricted. Their azeotropic points can only be minimum-boiling saddles or unstable nodes. More importantly, the compositions of the two liquid phases lie within different distillation regions. This unique property, which is not shared by homogeneous systems, enables the decanter to bridge the distillation regions. This is the key that permits the compositions of a single distillation column to cross from one distillation region into another, as illustrated in Figures 7.32a and 7.32b. In this system, for the dehydration of ethanol using toluene, the preconcentrator, C-1 with mixed feed, M1, removes water, B1, as the bottoms product. Its distillate, at D1, lies just to the right of the simple distillation boundary, K(D2)L, as shown in Figure 7.32b. The addition of entrainer, S2, to the distillate, D1, produces a C-2 feed stream, M2, that crosses this boundary into the distillation region just to the left of boundary K(D2) where high-purity ethanol, B2,

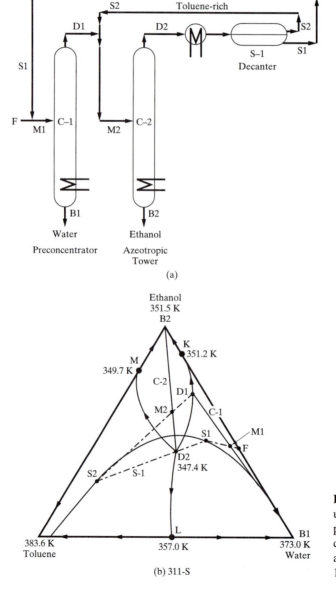

Figure 7.32 Dehydration of ethanol using toluene as an entrainer: (a) process flow diagram; (b) ternary composition diagram; (c) $T–x–y$ diagram at 1 atm; (d) $x–y$ diagram at 1 atm (Stichlmair et al., 1989).

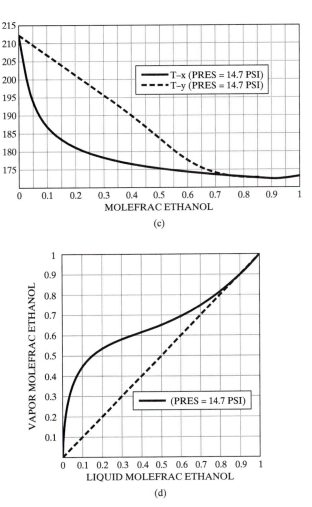

(c)

(d)

Figure 7.32 (*Continued*)

is obtained as the bottoms product of the azeotropic tower, C-2. Its overhead vapor, D2, is in the vicinity of the heterogeneous ternary azeotrope, and when condensed and subcooled forms two liquid phases that are decanted easily. The organic phase, at S2, lies in a different distillation region than the feed, M1, to column C-1. When combined with D1, the feed, M2, is on the other side of the simple distillation boundary, in the region M(D2)K(B2)M. The toluene-rich phase, S2, is recycled to column C-2, and the water-rich phase, S1, is combined with the fresh feed (F) to column C-1. The distillate and bottoms products of both towers and their overall mass balance lines are shown in Figure 7.32b.

The distillation sequence shown in Figure 7.32a is only one of several sequences, involving from two to four columns, that have been proposed and/or applied in industry for separating a mixture by employing heterogeneous azeotropic distillation. Most common is the three-column sequence from the study of Ryan and Doherty (1989), as shown in Figure 7.33a. When used to separate a mixture of ethanol and water using benzene as the entrainer, the three columns perform the separation in the following manner, where the material-balance lines for Columns 2 and 3 are shown in Figure 7.33b. The aqueous feed, F1, dilute in ethanol, is pre-concentrated in Column 1 to obtain a pure water bottoms, B1, and a distillate, D1, whose composition approaches that of the homogeneous minimum-boiling binary azeotrope. The distillate becomes the feed to Column 2, the azeotropic column, where nearly pure ethanol, B2, is removed as bottoms. The overhead vapor from Column 2, V2, is close to the composition of the heterogeneous ternary azeotrope of ethanol, water, and benzene. When condensed, it separates into two liquid phases in the decanter. Most of the organic-rich phase, L2, is returned to

(a) Column sequence for separation of ethanol and water with benzene

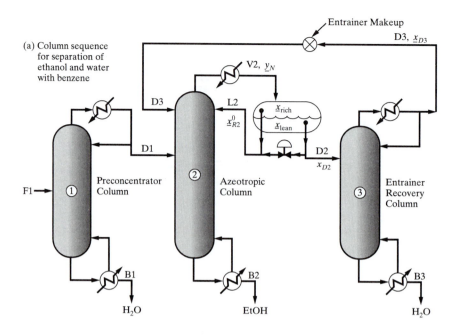

(b) Material balance lines

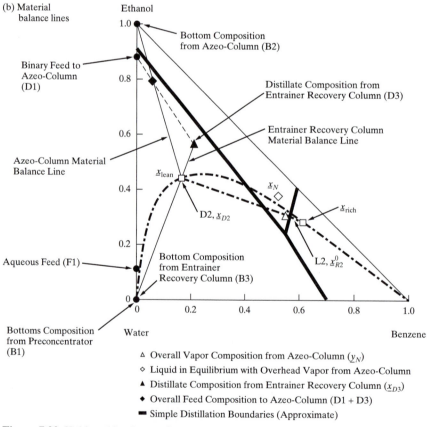

△ Overall Vapor Composition from Azeo-Column ($\underline{y}_N$)

◇ Liquid in Equilibrium with Overhead Vapor from Azeo-Column

▲ Distillate Composition from Entrainer Recovery Column ($\underline{x}_{D3}$)

◆ Overall Feed Composition to Azeo-Column (D1 + D3)

▬ Simple Distillation Boundaries (Approximate)

Figure 7.33 Kubierschky three-column system.

Column 2 as reflux. Most of the water-rich phase, D2, is sent to Column 3, the entrainer recovery column. Here, the distillate, D3, consisting mainly of ethanol, but with appreciable amounts of benzene and water, is recycled to the top of Column 2. The bottoms, B3, from Column 3 is nearly pure water. All columns operate at close to 1 atm pressure.

Multiple Steady States

The occurrence of multiple steady states in chemical reactors has been well recognized for at least 50 years. The most common example is an adiabatic CSTR, for which in some cases, for the same feed and reactor size, three possible products may be obtained, two of which are stable and one unstable, as shown in Case Study 21.1. The product obtained in actual operation depends upon the startup procedure for the reactor. Only in the last 25 years has the existence of multiple steady states in distillation towers been shown by calculations and verified by experimental data from tower operation. In particular, azeotropic distillation is particularly susceptible to multiple steady states. Disturbances during operation of an azeotropic tower can cause it to switch from one steady state to another as shown by Prokopakis and Seider (1983). Methods for computing multiple steady states for homogeneous and heterogeneous azeotropic distillation are presented in a number of publications. Kovach and Seider (1987) computed, by an arclength homotopy-continuation method, five steady states for the ethanol–benzene–water distillation. Bekiaris et al. (1993, 1996, 2000) studied multiple steady states for ternary homogeneous- and ternary heterogeneous-azeotropic distillation, respectively. Using the distillate flow rate as the bifurcation parameter, they found conditions of feed compositions and distillation-region boundaries for which multiple steady states can occur in columns operating at total reflux (infinite reflux ratio) with an infinite number of equilibrium stages (referred to as the $\infty-\infty$ case). They showed that their results have relevant implications for columns operating at finite reflux ratios with a finite number of stages. Vadapalli and Seader (2001) used ASPEN PLUS with an arclength continuation and bifurcation method to compute all stable and unstable steady states for azeotropic distillation under conditions of finite reflux ratio and finite number of equilibrium stages. Specifications for their heterogeneous azeotropic distillation example, involving the separation of an ethanol–water mixture using benzene, are shown in Figure 7.34a. The total feed rate to the column is 101.962 kmol/hr. The desired bottoms product is pure ethanol. Using the bottoms flow rate, as the bifurcation parameter, computed results for the mole fraction of ethanol in the bottoms are shown in Figure 7.34b as a function of the bifurcation parameter. In the range of bottoms flow rate from approximately 78 to 96 kmol/hr, three steady states exist, two stable and one unstable. For a bottoms rate equal to the flow rate of ethanol in the feed (89 kmol/hr), the best stable solution is an ethanol mole fraction of 0.98; the inferior stable solution is only 0.89. Figure 7.34b shows the computed points. In the continuation method, the results of one point are used as the initial guess for obtaining an adjacent point.

While heterogeneous azeotropic distillation towers are probably used more widely than their homogeneous counterparts, care must be taken in their design and operation. In addition to the possibility of multiple steady states, most azeotropic distillation towers involve sharp fronts as the temperatures and compositions shift abruptly from the vicinity of one fixed point to the vicinity of another. Furthermore, in heterogeneous distillations, sharp fronts often accompany the interface between trays having one and two liquid phases as well. Consequently, designers must select carefully the number of trays and the reflux rates to prevent these fronts from exiting the tower with an associated deterioration in the product quality. While these and other special properties of azeotropic towers (e.g., maximum reflux rates above which the separation deteriorates, and an insensitivity of the product compositions to the number of trays) are complicating factors, fortunately, they are usually less important when synthesizing separation trains, and consequently they are not discussed further here. For a review of the literature on this subject, see the article by Widagdo and Seider (1996).

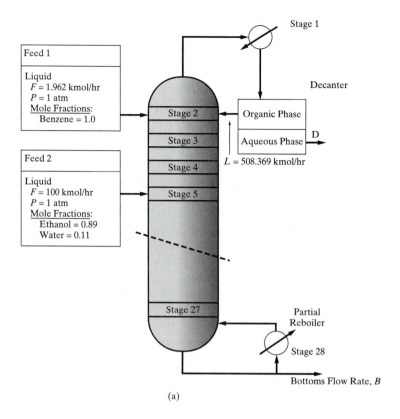

(a)

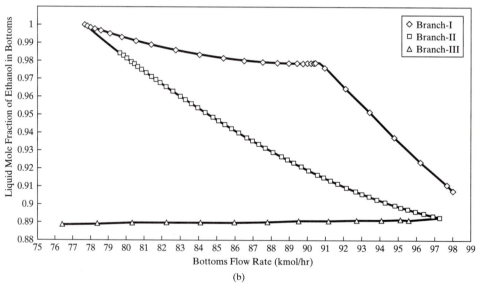

(b)

Figure 7.34 Heterogeneous azeotropic distillation: (a) specifications, (b) bifurcation diagram; branches I and III-stable, branch II-unstable.

Pressure-Swing Distillation

In some situations, azeotropic points are sensitive to moderate changes in pressure. When this is the case, pressure-swing distillation can be used in place of azeotropic distillation to permit the recovery of two nearly pure species that are separated by a distillation boundary. This section introduces pressure-swing distillation.

The effect of pressure on the temperature and composition of the ethanol–water and ethanol–benzene azeotropes, two minimum-boiling, binary azeotropes, is shown in Figure 7.35. For the first, as the pressure is decreased from 760 to 100 torr, the mole fraction of ethanol increases from 0.894 to 0.980. Although not shown, at a lower pressure, below 70 torr, the azeotrope disappears entirely. While the temperature changes are comparable for the ethanol–benzene azeotrope, the composition is far more sensitive. Many other binary azeotropes are pressure-sensitive, as discussed by Knapp and Doherty (1992), who list 36 systems taken from the compilation of azeotropic data by Horsley (1973).

An example of pressure-swing distillation, described by Van Winkle (1967), is provided for the mixture, A–B, having a minimum-boiling azeotrope, with the T–x–y curves at two pressures shown in Figure 7.36a. To take advantage of the decrease in the composition of A as the pressure decreases from P_2 to P_1, a sequence of two distillation towers is shown in

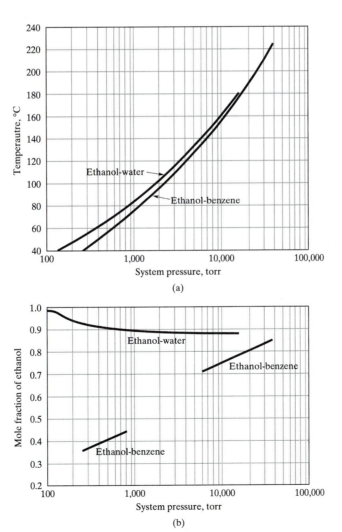

Figure 7.35 Effect of pressure on azeotrope conditions: (a) temperature of azeotrope; (b) composition of azeotrope.

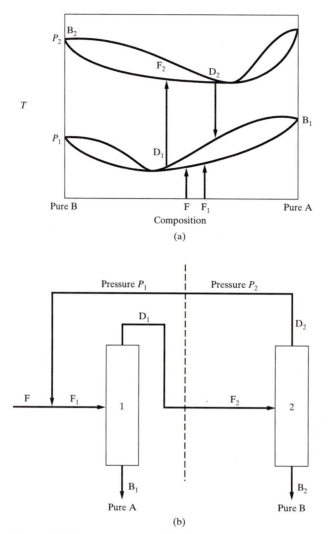

Figure 7.36 Pressure-swing distillation for the separation of a minimum-boiling azeotrope: (a) T–y–x curves at pressures P_1 and P_2 for minimum-boiling azeotrope; (b) distillation sequence for minimum-boiling azeotrope.

Figure 7.36b. The total feed to column 1, F_1, operating at the lower pressure, P_1, is the sum of the fresh feed, F, whose composition is richer in A than the azeotrope, and the distillate, D_2, whose composition is close to that of the azeotrope at P_2, and which is recycled from column 2 to column 1. The compositions of D_2, and consequently, F_1, are richer in A than the azeotropic composition at P_1. Hence, the bottoms product, B_1, that leaves column 1 is nearly pure A. Since the distillate, D_1, which is slightly richer in A than the azeotropic composition, is less rich in A than the azeotropic composition at P_2, when it is fed to column 2, the bottoms product, B_2, is nearly pure B. Yet another example is provided by Robinson and Gilliland (1950) for the dehydration of ethanol, where the fresh-feed composition is less rich in ethanol than the azeotrope. In this case, ethanol and water are removed as bottoms products also, but nearly pure B (water) is recovered from the first column and A (ethanol) is recovered from the second. Similar pressure-swing distillations are designed to separate maximum-boiling, binary azeotropes, which are less common.

When designing pressure-swing distillation sequences, the recycle ratio must be adjusted carefully. Note that it is closely related to the differences in the compositions of the azeotrope at P_1 and P_2. Horwitz (1997) illustrates this for the dehydration of ethylenediamine.

EXAMPLE 7.4

Consider the separation of 100 kmol/hr of an equimolar stream of tetrahydrofuran (THF) and water using pressure-swing distillation, as shown in Figure 7.37. The tower T1 operates at 1 bar, with the pressure of the tower T2 increased to 10 bar. As shown in the T–x–y diagrams in Figure 7.38, the binary azeotrope shifts from 19 mol% water at 1 bar to 33 mol% water at 10 bar. Assume that the bottoms product from T1 contains pure water and that from D2 contains pure THF. Also, assume that the distillates from T1 and T2 are at their azeotropic compositions. Determine the unknown flow rates of the product and internal streams. Note that data for the calculation of vapor–liquid equilibria are provided in Table 7.5.

SOLUTION

Since the bottoms products are pure, $B_1 = 50$ kmol/hr H_2O and $B_2 = 50$ kmol/hr THF. To determine the distillate flow rates, the following species balances apply.

$$H_2O \text{ balance on column T2:} \quad 0.19D_1 = 0.33D_2$$

$$THF \text{ balance on column T1:} \quad 0.81D_1 = 0.67D_2 + 50$$

Solving these two equations simultaneously, $D_1 = 117.9$ kmol/hr and $D_2 = 67.9$ kmol/hr. Exercise 7.19 examines the effect of pressure on the internal flow rates. ∎

Membranes, Adsorbers, and Auxiliary Separators

When operating homogeneous azeotropic distillation towers, a convenient vehicle for permitting the compositions to cross a distillation boundary is to introduce a membrane separator, adsorber, or other auxiliary separator. These are inserted either before or after the condenser of the distillation column and serve a similar role to the decanter in a heterogeneous azeotropic distillation tower, with the products having their compositions in adjacent distillation regions.

Reactive Distillation

Yet another important vehicle for crossing distillation boundaries is through the introduction of chemical reaction(s) on the trays of a distillation column. As discussed in Section 5.3, it is often advantageous to combine reaction and distillation operations so as to drive a reversible

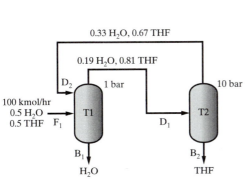

Figure 7.37 Pressure-swing distillation for dehydration of THF with stream compositions in mole fractions.

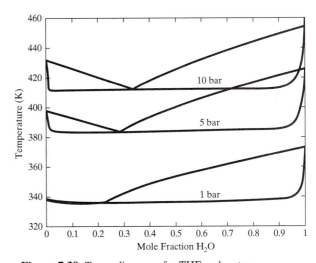

Figure 7.38 T–x–y diagrams for THF and water.

Table 7.5 Data for Vapor–Liquid Equilibria for THF–H_2O

Extended Antoine Coefficients		
	H_2O	THF
C^1	7.36	5.490
C^2	−7,258	−5,305
C^3	0.0	0.0
C^4	0.0	0.0
C^5	−7.304	−4.763
C^6	4.16530E-06	1.42910E-17
C^7	2.0	6.0

$$\ln P_i^s = C_i^1 + C_i^2/(T + C_i^3) + C_i^4 T + C_i^5 \ln T + C_i^6 T^{C_i^7}; \; P^s, \text{Pascal}$$

Wilson Interaction Coefficients					
A_{ij}	H_2O	THF	B_{ij}	H_2O	THF
H_2O	0.0	−23.709	H_2O	0.0	7,500
THF	−2.999	0.0	THF	−45.07	0.0

reaction(s) toward completion through the recovery of its products in the vapor and liquid streams that leave the trays. Somewhat less obvious, perhaps, is the effect the reaction(s) can have on repositioning or eliminating the distillation boundaries that otherwise complicate the recovery of nearly pure species. For this reason, the discussion that follows concentrates on the effect of a reaction on the residue curve maps. Several constructs must be introduced, however, to prepare for the main concepts.

For reactive systems, it is helpful to begin with a more rigorous definition of an azeotrope, that is, a mixture whose phases exhibit no changes in composition during vaporization or condensation. On this basis, for vapor and liquid phases, with $dx_j/dt = dy_j/dt = 0, j = 1, \ldots,$ C, in the presence of a homogeneous chemical reaction, $\Sigma_j \nu_j A_j = 0$, at equilibrium, the conditions for a *reactive azeotrope* can be derived (Barbosa and Doherty, 1988a) such that

$$\frac{y_j - x_j}{\nu_j - x_j \nu_T} = \frac{d\xi}{d\upsilon} = \kappa \qquad j = 1, \ldots, C \tag{7.31}$$

where ν_j is the stoichiometric coefficient of species j, $\nu_T = \Sigma_j \nu_j$, ξ is the extent of the reaction, υ is the moles of vapor, and κ is a constant. Furthermore, it can be shown that the mass balances for simple distillation in the presence of a chemical reaction can be written in terms of transformed variables (Barbosa and Doherty, 1988b):

$$\frac{dX_j}{d\tau} = X_j - Y_j \qquad j = 1, \ldots, C - 1; j \neq j' \tag{7.32a}$$

where

$$X_j = \frac{x_j/\nu_j - x_{j'}/\nu_{j'}}{\nu_{j'} - \nu_T x_{j'}} \tag{7.32b}$$

$$Y_j = \frac{y_j/\nu_j - y_{j'}/\nu_{j'}}{\nu_{j'} - \nu_T y_{j'}} \tag{7.32c}$$

$$\tau = \frac{H}{\upsilon}\left(\frac{\nu_{j'} - \nu_T y_{j'}}{\nu_{j'} - \nu_T x_{j'}}\right)t \tag{7.32d}$$

Here, H is the molar liquid holdup in the still, and j' denotes a reference species. Clearly, Eq. (7.32a) corresponds to the mass balances without chemical reaction [Eq. (7.19)]. By integration of the latter equation for a nonreactive mixture of isobutene, methanol, and methyl tertiary-butyl ether (MTBE), the residue curve map in Figure 7.39a is obtained. There are two minimum-boiling, binary azeotropes and a distillation boundary that separates two distillation regions.

When the chemical reaction is turned on and permitted to equilibrate, Eq. (7.32a) is integrated and at long times,

$$X_j = Y_j \qquad j = 1, \ldots, C \tag{7.33}$$

define the fixed point and are the conditions derived for a reactive azeotrope [Eq. (7.31)]. At shorter times, *reactive residue curves* are obtained, as shown in Figure 7.39d, where the effect of the chemical reaction can be seen. It is clear that the residue curves have been distorted significantly and pass through the reactive azeotrope, or so-called *equilibrium tangent pinch*. Furthermore, the distillation boundary has been eliminated completely. The reactive azeotrope of this mixture is shown clearly in an *X–Y* diagram (Figure 7.40), which is similar

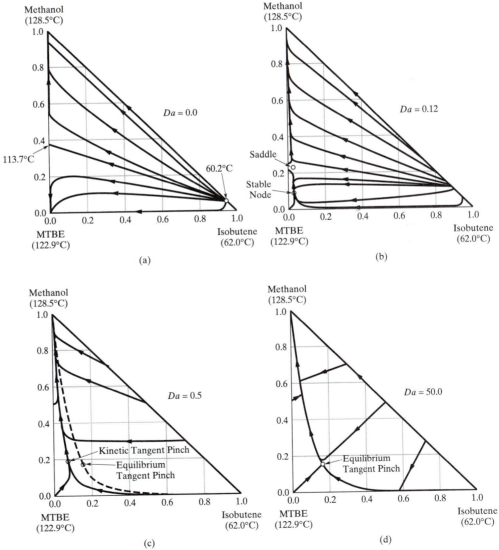

Figure 7.39 Residue curve maps for isobutene, methanol, and MTBE as a function of *Da* at 8 atm. (Reprinted from Venimadhavan et al., 1994).

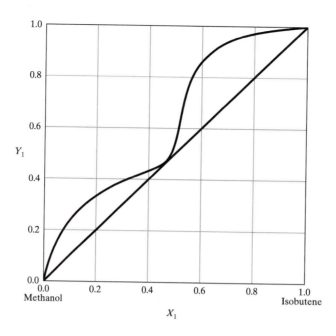

Figure 7.40 Transformed compositions for isobutene, methanol, and MTBE in chemical and phase equilibrium. (Reprinted from Doherty and Buzad, 1992).

to the x–y diagram when reaction does not occur. Finally, through the use of a kinetic model, involving a well-stirred reactor, it is possible to show the residue curves as a function of the residence time (that is, the Damkohler number, Da). Figures 7.39b and 7.39c show how the residue curves change as the residence time increases (Venimadhavan et al., 1994).

Separation Train Synthesis

Beginning with the need to separate a C-component mixture into several products, alternative sequences of two-product distillation towers are considered in this section. Although the synthesis strategies are not as well defined for highly nonideal and azeotropic mixtures, several steps are well recognized and are described next. It should be mentioned that these strategies continue to be developed, and variations are not uncommon.

1. *Identify the azeotropes.* Initially, it is very helpful to obtain estimates of the temperature, pressure, and composition of the binary, ternary, . . . , azeotropes associated with the C-component mixture. For all of the ternary submixtures, these can be determined, as described above, by preparing residue curve or distillation-line maps. When it is necessary to estimate the quaternary and higher-component azeotropes, as well as the binary and ternary azeotropes, the methods of Fidkowski et al. (1993) and Eckert and Kubicek (1997) are recommended. When the C-component mixture is the effluent from a chemical reactor, it may be helpful to include the reacting chemicals, that is, to locate any azeotropes involving these chemicals as well as the existence of reactive azeotropes. This information may show the potential for using reactive distillation operations as a vehicle for crossing distillation boundaries that complicate the recovery of nearly pure species.

2. *Identify alternative separators.* Given estimates for the azeotropes, the alternatives for the separators involving all C species are identified. These separate two species that may or may not involve a binary azeotrope. When no binary azeotrope is involved, a normal distillation tower may be adequate, unless the key components are close boiling. For close-boiling binary pairs, or binary pairs with an azeotrope separating the desired products, the design of an extractive distillation tower or an azeotropic distillation tower should be considered. The former is preferred when a suitable solvent is available.

3. ***Entrainer selection.*** Probably the most difficult decision in designing an azeotropic distillation tower involves the selection of the entrainer. This is complicated by the effect of the entrainer on the residue curves and distillation lines that result. In this regard, the selection of the entrainer for the separation of binary mixtures, alone, is a large combinatorial problem, complicated by the existence of 113 types of residue curve maps, involving different combinations of low- and high-boiling, binary and ternary azeotropes with associated distillation boundaries. This classification, which involves several indices that characterize the various kinds of azeotropes and vertices, was prepared by Matsuyama and Nishimura (1977) to aid in screening potential entrainers.

In view of the above, many factors need to be considered in selecting an entrainer, factors that can have a significant impact on the resulting separation train. Two of the more important guidelines are the following:

a. When designing homogeneous azeotropic distillation towers, select an entrainer that does not introduce a distillation boundary between the two species to be separated.
b. To cross a distillation boundary between two species to be separated, select an entrainer that induces liquid-phase splitting, as in heterogeneous azeotropic distillation.

The effects of these and other guidelines must be considered as each separator is designed and as the separation sequence evolves. More recently, Peterson and Partin (1997) showed that temperature sequences involving the boiling points of the pure species and the azeotrope temperatures can be used to effectively categorize many kinds of residue curve maps. This classification simplifies the search for an entrainer that has a desirable residue curve map, for example, that does not involve a distillation boundary.

4. ***Identify feasible distillate and bottoms-product compositions.*** When positioning a two-product separator, it is usually an objective to recover at least one nearly pure species, or at least to produce two products that are easier to separate into the desired products than the feed mixture. To accomplish this, it helps to know the range of feasible distillate and bottoms-product compositions. For a three-component feed stream, the feed composition can be positioned on a distillation-line map and the feasible compositions for the distillate and bottoms product identified using the methods described above in the subsection on distillation-line boundaries and feasible product compositions. For feed mixtures containing four or more species ($C > 3$), a common approach is to identify the three most important species that are associated with the separator being considered. Note, however, that the methods for identifying the feasible compositions assume that they are bounded by the distillation line, at total reflux, through the feed composition. For azeotropic distillations, however, it has been shown that the best separations may not be achieved at total reflux. Consequently, a procedure has been developed to locate the bounds at finite reflux. This involves complex graphics to construct the so-called *pinch-point trajectories*, which are beyond the scope of this presentation but are described in detail by Widagdo and Seider (1996). Because the composition bounds at finite reflux usually include the feasible region at total reflux, the latter usually leads to conservative designs.

Having determined the bounds on the feasible compositions, the first separator is positioned usually to recover one nearly pure species. At this point in the synthesis procedure, the separator can be completely designed (to determine number of trays, reflux ratio, installed and operating costs, etc.). Alternatively, the design calculations can be delayed until a sequence of separators is selected, with its product compositions positioned. In this case, Steps 2–4 are repeated for the mixture in the other product stream. Initially, the simplest separators are considered, that is, ordinary distillation, extractive distillation, and homogeneous azeotropic distillation. However, when distillation boundaries are encountered and cannot be eliminated through the choice of a suitable entrainer, more complex separators are considered, such as heterogeneous azeotropic distillation, pressure-swing distillation, the addition of membranes, adsorption, and auxiliary separators, and reactive distillation. Normally, a sequence is synthesized involving many two-product separators without chemical reaction. Subsequently, after the separators are designed completely, steps are taken to carry out task integration as described in Section 3.4. This involves the combination of two or more separators and seeking opportunities to combine the reaction and separation steps in reactive distillation towers. As an example, Siirola (1995) describes the development of a process for the manufacture of methyl acetate and the dehydration of acetic acid. Initially, a sequence was synthesized involving a reactor, an extractor, a decanter, and eight distillation columns, incorporating two mass separating agents. The flowsheet was reduced subsequently to four columns, using evolutionary strategies and task integration, before being reduced finally to just two columns, one involving reactive distillation.

As illustrated throughout this section, process simulators have extensive facilities for preparing phase-equilibrium diagrams (T–x–y, P–x–y, x–y, . . .), and residue curve maps and binodal curves for ternary systems. In addition, related but independent packages have been developed for the synthesis and evaluation of distillation trains involving azeotropic mixtures. These include SPLIT™ by Aspen Technology, Inc., and DISTIL™ by Hyprotech (now Aspen Technology, Inc., which contains MAYFLOWER developed by M.F. Doherty and M.F. Malone at the University of Massachusetts).

EXAMPLE 7.5 *Manufacture of Di-Tertiary-Butyl Peroxide*

This example involves the manufacture of 100 million pounds per year of di-tertiary-butyl peroxide (DTBP) by the catalytic reaction of tertiary-butyl hydroperoxide (TBHP) with excess tertiary-butyl alcohol (TBA) at 170°F and 15 psia according to the reaction

$$CH_3-\underset{\underset{CH_3}{|}}{\overset{\overset{CH_3}{|}}{C}}-OH + CH_3-\underset{\underset{CH_3}{|}}{\overset{\overset{CH_3}{|}}{C}}-OOH \rightarrow CH_3-\underset{\underset{CH_3}{|}}{\overset{\overset{CH_3}{|}}{C}}-OO\underset{\underset{CH_3}{|}}{\overset{\overset{CH_3}{|}}{C}}-CH_3 + H_2O$$

| Tertiary-butyl alcohol (TBA) | Tertiary-butyl hydroperoxide (TBHP) | DTBP |

Assume that the reactor effluent stream contains

	lbmol/hr	Mole Fraction
TBA	72.1	0.272
H₂O	105.6	0.398
DTBP	87.7	0.330
		1.000

and small quantities of isobutene, methanol, and acetone, which can be disregarded. A separation sequence is to be synthesized to produce 99.6 mol% pure DTBP, containing negligible water. It may be difficult to separate TBA and water. Therefore, rather than recovering and recycling the unreacted TBA, the conversion of TBA to isobutene and water in the separation sequence should be considered. In the catalytic reactor, the TBA dehydrates to isobutene, which is the actual molecule that reacts with TBHP to form DTBP. Thus, isobutene, instead of TBA, can be recycled to the catalytic reactor.

SOLUTION

A residue curve map at 15 psia, prepared using ASPEN PLUS (with the NRTL option set and proprietary interaction coefficients), is displayed in Figure 7.41a. There are three minimum-boiling binary azeotropes:

	T, °F	
DTBP–TBA	177	$x_{TBA} = 0.82$
TBA–H$_2$O	176	$x_{H_2O} = 0.38$
H$_2$O–DTBP	188	$x_{DTBP} = 0.47$

and the boiling points of the pure species are 181, 212, and 232°F, for TBA, H$_2$O, and DTBP, respectively. In addition, there is a minimum-boiling ternary azeotrope at $x_{TBA} = 0.44$, $x_{H_2O} = 0.33$, and $x_{DTBP} = 0.23$, and 174°F. Consequently, there are three distinct distillation regions, with the feed composition in a region that does not include the product vertex for DTBP.

To cross the distillation boundaries, it is possible to take advantage of the partial miscibility of the DTBP–H$_2$O system, as well as the disappearance of the ternary azeotrope at 250 psia as illustrated in Figure 7.41b. One possible design is shown in Figure 7.42, where the reactor effluent is in

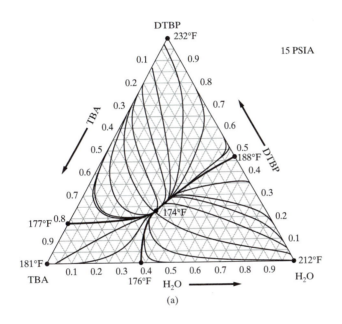

(a)

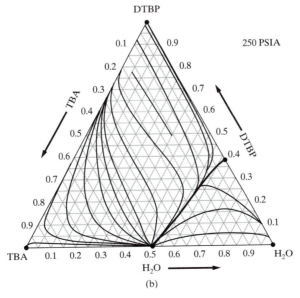

(b)

Figure 7.41 Residue curve map for the TBA–H$_2$O–DTBP system. (a) 15 psia; (b) 250 psia.

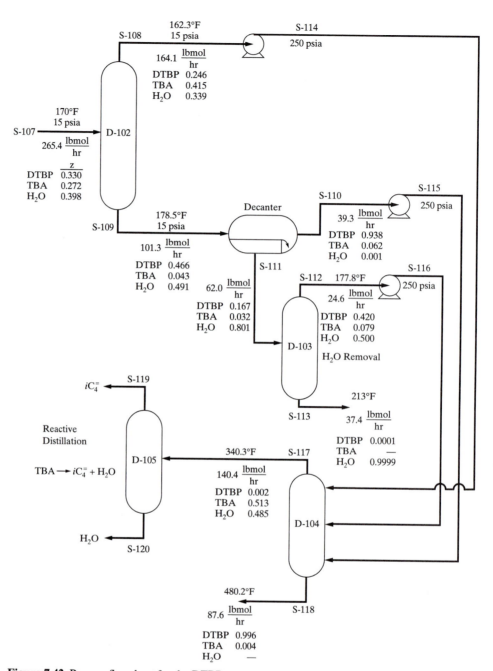

Figure 7.42 Process flowsheet for the DTBP process.

stream S-107. Column D-102 forms a distillate in stream S-108 whose composition is very close to the ternary azeotrope, and a bottoms product in stream S-109, as shown on the ternary diagram in Figure 7.43a. The latter stream, containing less than 5 mol% TBA, is split into two liquid phases in the decanter. The aqueous phase in stream S-111 enters the distillation tower, D-103, which forms nearly pure water in the bottoms product, stream S-113. The distillate from tower D-102, S-108, the organic phase from the decanter, S-110, and the distillate from tower D-103, S-112, are pumped to 250 psia and sent to the distillation tower, D-104, where they enter on stages that have comparable compositions. The compositions of the streams at elevated pressure, S-114, S-115, and S-116, and the mix point, M, are shown in Figure 7.43b. Note that at 250 psia, M lies in the distillation region that contains the

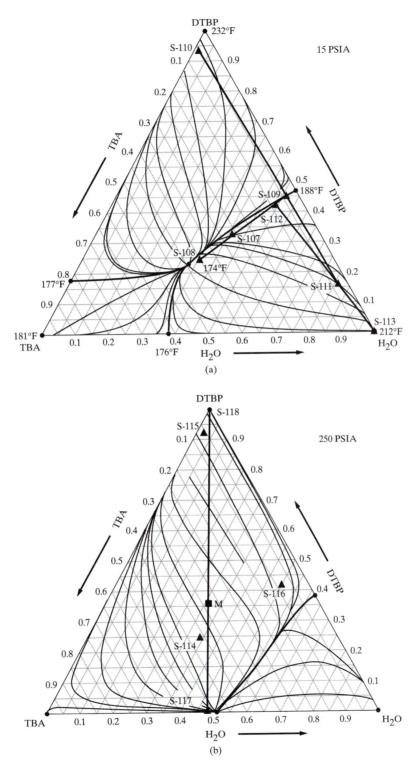

Figure 7.43 Distillation boundaries and material balance lines for the TBA–H₂O–DTBP system: (a) 15 psia; (b) 250 psia.

DTBP vertex. Consequently, tower D-104 produces nearly pure DTBP in the bottoms products, S-118, and its distillate, S-117, is sent to the reactive distillation tower, D-105, where the TBA is dehydrated according to the reaction

$$TBA \rightarrow i-\text{butene} + H_2O$$

with *i*-butene recovered in the distillate, S-119, which is recycled to the catalytic reactor, and water in the bottoms product, S-120. As seen in Figures 7.43a and 7.43b, the material balance lines associated with the distillation towers lie entirely within separate distillation regions. The process works effectively because of the phase split and because the distillation boundaries are repositioned at the elevated pressure. Note, however, that the material balance line for the tower, D-102, would preferably be positioned farther away from the distillation boundary to allow for inaccuracies in the calculation of the distillation boundary.

Since this design was completed, the potential for DTBP to decompose explosively at temperatures above 255°F was brought to our attention. At 250 psia, DTBP is present in the bottoms product of tower D-104 at 480.2°F. Given this crucial safety concern, a design team would seek clear experimental evidence. If positive, lower pressures would be explored, recognizing that the distillation boundaries are displaced less at lower pressures. ∎

For additional details of this process design, see the design report by Lee et al. (1995). Also, see Problem A-II.1.10 in the PDF, Design Problem Statements.pdf, on the multimedia CD-ROM that accompanies this textbook for the design problem statement that led to this design.

7.6 SEPARATION SYSTEMS FOR GAS MIXTURES

Sections 7.4 and 7.5 deal primarily with the synthesis of separation trains for liquid–mixture feeds. The primary separation techniques are ordinary and enhanced distillation. If the feed consists of a vapor mixture in equilibrium with a liquid mixture, the same techniques and synthesis procedures can often be employed. However, if the feed is a gas mixture and a wide gap in volatility exists between two groups of chemicals in the mixture, it is often preferable, as discussed in Section 7.1, to partially condense the mixture, separate the phases, and send the liquid and gas phases to separate separation systems as discussed by Douglas (1988) and shown in Figure 7.44. Note that if a liquid phase is produced in the gas separation system, it is routed to the liquid separation system and vice versa.

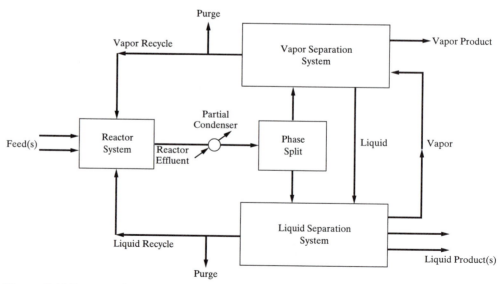

Figure 7.44 Process with vapor and liquid separation systems. (Modified and reprinted with permission from Douglas, 1988.)

In some cases, it has been found economical to use distillation to separate a gas mixture, with the large-scale separation of air by cryogenic distillation into nitrogen and oxygen being the most common example. However, the separation by distillation of many other gas mixtures, such as hydrogen from methane or hydrogen from nitrogen, is not practical because of the high cost of partially condensing the overhead vapor to obtain reflux. Instead, other separation methods, such as absorption, adsorption, or membrane permeation, are employed. In just the past 25 years, continuous adsorption and membrane processes have been developed for the separation of air that economically rival the cryogenic distillation process at low to moderate production levels.

Barnicki and Fair (1992) consider in detail the selection and sequencing of equipment for the separation of gas mixtures. Whereas ordinary distillation is the dominant method for the separation of liquid mixtures, no method is dominant for gas mixtures. The separation of gas mixtures is further complicated by the fact that whereas most liquid mixtures are separated into nearly pure components, the separation of gas mixtures falls into the following three categories: (1) sharp splits to produce nearly pure products, (2) enrichment to increase the concentration(s) of one or more species, for example, oxygen and nitrogen enrichment, and (3) purification to remove one or more low-concentration impurities. The first category is often referred to as *bulk separation*, the purpose of which is to produce high-purity products at high recovery. Separations in this category can be difficult to achieve for gas mixtures. The best choices are cryogenic distillation, absorption, and adsorption. By contrast, the second category achieves neither high purity nor high recovery and is ideally suited for any of the common separation methods for gas mixtures, including membrane separation by gas permeation. To produce high-purity products by purification, adsorption and absorption with chemical reaction are preferred.

The synthesis of a separation train for a gas mixture can be carried out by first determining the feasible separation methods, which depend on the separation categories and the separation factors, and then designing and costing systems involving these methods to determine the optimal train. The design of equipment for absorption, adsorption, distillation, and membrane separations is covered by Seader and Henley (1998). Besides the separation category and separation factor, the production scale of the process is a major factor in determining the optimal train because economies of scale are most pronounced for cryogenic distillation and absorption, and least pronounced for adsorption and membrane separations. For example, for the separation of air into nitrogen- and oxygen-enriched products, membrane separations are most economical at low production rates, adsorption at moderate rates, and cryogenic distillation at high rates.

Membrane Separation by Gas Permeation

In gas permeation, the gas mixture is compressed to a high pressure and brought into contact with a very thin membrane to produce two products: (1) a permeate that passes through the membrane and is discharged at a low pressure, and (2) a *retentate* that does not pass through the membrane and is maintained at close to the high pressure of the feed. The separation factor defined by Eqs. (7.2) and (7.8) can be applied to gas permeation, when the retentate-side pressure is much greater than the permeate-side pressure, if y is the mole fraction in the permeate and x is the mole fraction in the retentate. The relative volatility is replaced by the ratio of the membrane permeabilities for the two key components of the feed–gas mixture, sometimes called the permselectivity. Most commercial membranes for gas permeation are nonporous (dense) amorphous or semicrystalline polymers. To pass through such polymers, the gas molecules first dissolve in the polymer and then pass through it by diffusion. Thus, the permeability depends on both solubility and diffusivity in the particular polymer at the conditions of temperature and pressure. The permeability is the product of the solubility and diffusivity. Permeabilities are best determined by laboratory measurements. However, a predictive method given by Barnicki (1991) for a number of glassy and rubbery polymers, which depends on species van der Waals volume and critical temperature, can be applied in the absence of data. In general, gas permeation is commercially feasible when the ratio of permeabilities (permselectivity) for the two components is greater than 15. However, some processes that require only rough enrichments use membranes having permselectivities of only 5. Commercial applications include the recovery of carbon dioxide from hydrocarbons, the adjustment of the hydrogen-to-carbon monoxide ratio in synthesis gas, the recovery of hydrocarbons from hydrogen, and the separation of air into nitrogen- and oxygen-enriched streams.

Adsorption

Adsorption differs from the other techniques in that it is a cyclic operation with adsorption and desorption steps. However, adsorption is a very versatile separation technique. To be economical, the adsorbent must be regenerable. This requirement precludes the processing of gas mixtures that contain (1) high-boiling organic compounds because they are preferentially adsorbed and are difficult to remove during the regeneration part of the cycle, (2) lower-boiling organic compounds that may polymerize on the adsorbent surface, and (3) highly acidic or basic compounds that may react with the adsorbent surface. In some cases, such compounds can be removed from the gas mixture by guard beds or other methods prior to entry into the adsorption system.

Selectivity in adsorption is controlled by (1) molecular sieving or (2) adsorption equilibrium. When components differ significantly in molecular size and/or shape, as characterized by the kinetic diameter, zeolites and carbon molecular-sieve adsorbents can be used to advantage because of the strong selectivity achieved by molecular sieving. These adsorbents have very narrow pore-size distributions that prevent entry into the pore structure of molecules with a kinetic diameter greater than the nearly uniform pore aperture. Zeolites are readily available with nominal apertures in angstroms of 3, 4, 5, 8, and 10. Thus, for example, consider a gas mixture containing the following components, with corresponding kinetic diameters in angstroms in parentheses: nitrogen (<3), carbon dioxide (>3 and <4), and benzene (>7 and <8). The zeolite with a 3-Å aperture could selectively adsorb the nitrogen, leaving a mixture of carbon dioxide and benzene that could be separated with a zeolite of 4-Å aperture. Barnicki (1991) gives methods for estimating kinetic diameters. In effect, the separation factor for a properly selected sieving-type adsorbent is infinity.

Adsorbents made of activated alumina, activated carbon, and silica gel separate by differences in adsorption equilibria, which must be determined by experiment. Equilibrium-limited adsorption can be applied to all three categories of separation, but is usually not a favored method when the components to be selectively adsorbed constitute an appreciable fraction of the feed gas. Conversely, equilibrium-limited adsorption is ideal for the removal of small quantities of selectively adsorbed impurities. At a given temperature, the equilibrium loading of a given component, in mass of adsorbate per unit mass of adsorbent, depends on the component partial pressure and to a lesser extent on the partial pressures of the other components. For equilibrium-limited adsorption to be feasible, Barnicki and Fair (1992) suggest that the ratio of equilibrium loadings of the two key components be used as a separation factor. This ratio should be based on the partial pressures in the feed gas. A ratio of at least 2, and preferably much higher, makes equilibrium adsorption quite favorable. However, two other conditions must also be met: (1) the more highly adsorbed component should have a concentration in the feed of less than 10 mol% and (2) for an adsorption time of 2 hr, the required bed height should not exceed 20 ft. Equilibrium-limited adsorption is usually the best alternative for the removal of water and organic chemicals from mixtures with light gases, and should also be considered for enrichment applications.

Absorption

Absorption of components of a gas mixture into a solvent may take place by physical or chemical means. When no chemical reaction between the solute and absorbent occurs (physical absorption), the separation factor is given by Eq. (7.2). Thus, if component 1 is to be selectively absorbed, a small value of SF is desired. Alternatively, Barnicki and Fair (1992) suggest that consideration of physical absorption should be based on a selectivity, $S_{1,2}$, defined as the ratio of liquid-phase mole fractions of the two key components in the gas mixture. This selectivity can be estimated from the partial pressures of the two components in the gas feed and their K values for the given solvent. For components whose critical temperatures are greater than the system temperature,

$$S_{1,2} = \frac{x_1}{x_2} = \frac{\gamma_2^\infty p_1 P_2^s}{\gamma_1^\infty p_2 P_1^s} \qquad (7.34)$$

where γ^∞ is the liquid-phase activity coefficient at infinite dilution, p is partial pressure, and P^s is vapor pressure. For components, whose critical temperatures are less than the system temperature, the selectivity can be estimated from Henry's law constants:

$$S_{1,2} = \frac{x_1}{x_2} = \frac{H_2 p_1}{H_1 p_2} \qquad (7.35)$$

where $H = yP/x$. For enrichment, the selectivity should be 3 or greater; for a sharp separation, 4 or greater. The number of theoretical stages should be at least 5. For the removal of readily soluble organic compounds from light gases, Douglas (1988) recommends the use of 10 theoretical stages and a solvent molar flow rate, L, based on an absorption factor, A, for solute of 1.4, where

$$A = \frac{L}{KV} \qquad (7.36)$$

with V = gas molar flow rate. When the partial pressure, in the gas feed, of the component to be absorbed is very small and a high percentage of it is to be removed, physical absorption may not be favorable. Instead, particularly if the solute is an acid or base, chemical absorption may be attractive.

Partial Condensation and Cryogenic Distillation

The previously discussed separation techniques for gas mixtures all involve a mass separating agent. Alternatively, thermal means is employed with partial condensation and cryogenic distillation. Barnicki and Fair (1992) recommend that partial condensation be considered for enrichment when the relative volatility between the key components is 7. For large-scale (>10–20 tons/day of product gas) enrichment and sharp separations, cryogenic distillation is feasible when the relative volatility between the key components is greater than 2. However, if the feed gas contains components, such as carbon dioxide and water that can freeze at the distillation temperatures, those components must be removed first.

7.7 SEPARATION SEQUENCING FOR SOLID–FLUID SYSTEMS

The final product from many industrial chemical processes is a solid material. This is especially true for inorganic compounds, but is also common for a number of moderate- to high-molecular-weight organic compounds. Such processes involve the separation operations of leaching, evaporation, solution crystallization (solutes with high melting points that are crystallized from a solvent), melt crystallization (crystallization from a mixture of components with low to moderate melting points), precipitation (rapid crystallization from a solvent of nearly insoluble compounds that are usually formed by a chemical reaction), desublimation, and/or drying, as well as the phase-separation operations of filtration, centrifugation, and cyclone separation. In addition, because specifications for solid products may also include a particle-size distribution, size-increase and size-reduction operations may also be necessary. If particle shape is also a product specification, certain types of crystallizers and/or dryers may be dictated. Even when the final product is not a solid, solid–liquid or solid–gas separation operations may be involved. For example, liquid mixtures of *meta*- and *para*-xylene cannot be separated by distillation because their normal boiling points differ by only 0.8°C. Instead, they are separated industrially by melt crystallization, because their melting points differ by 64°C. Nevertheless, the final products are liquids. Another example is phthalic anhydride, which, although a solid at room temperature, is usually shipped in the molten state. It is produced by the air oxidation of napthalene or *ortho*-xylene. The separation of the anhydride from the reactor effluent gas mixture is accomplished by desublimation, followed by distillation to remove impurities and produce a melt.

A common flowsheet for the separation section of a process for the manufacture of inorganic salt crystals from their aqueous solution is shown in Figure 7.45. If the feed is aqueous MgSO$_4$, a typical process proceeds as follows. A 10 wt% sulfate feed is concentrated, without crystallization, to 30 wt% in a double-effect evaporation system. The concentrate is mixed with recycled mother liquors from the hydroclone and centrifuge before being fed to an evaporative vacuum crystallizer, which produces, by solution crystallization, a magma of 35 wt% crystals of MgSO$_4$·7H$_2$O, the stable hydrate at the temperature in the crystallizer. The magma is thickened to 50 wt% crystals in a hydroclone, and then sent to a centrifuge, which discharges a cake containing 35 wt% moisture. The cake is dried to 2 wt% moisture in a direct-heat rotary dryer. Approximately 99 wt% of the dried crystals are retained on a 100-mesh screen and 30 wt% are retained on a 20-mesh screen. The crystals are bagged for shipment. Rossiter (1986) presents a similar flowsheet for the separation of aqueous NaCl, where a fluidized-bed dryer replaces the rotary dryer in Figure 7.45.

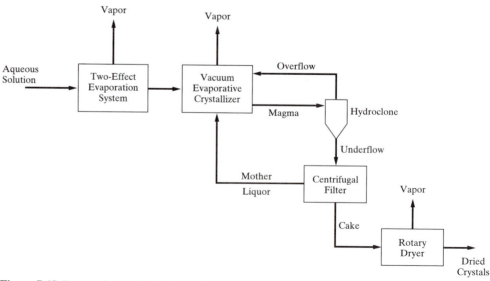

Figure 7.45 Process for producing inorganic salt crystals.

When a solid mixture of two components is to be separated, the process is more complicated. Such a process, using solution crystallization and shown in Figure 7.46, is considered by Rajagopal et al. (1988) for the production of crystalline potash (KCl) from sylvinite ore (mixture of 40 wt% KCl and 60 wt% NaCl). The separation scheme is feasible because KCl is less soluble than NaCl in water, and the solubility of KCl in water decreases with decreasing temperature whereas the reverse is true for NaCl. In the first step of the process, KCl is completely dissolved (leached) by a mixture of makeup water and filtrate from the second filter. The NaCl in the ore is not dissolved because conditions are selected so that the water in the dissolver is saturated with NaCl. Thus, a slurry of solid (undissolved) NaCl and aqueous KCl–NaCl leaves the dissolver. The slurry is filtered in a rotary vacuum filter, which sends the wet cake of NaCl to further processing and the mother liquor to an evaporative crystallizer. There, evaporation lowers the temperature below that in the dissolver, causing crystallization only of the KCl. The magma from the crystallizer is sent to a rotary vacuum filter, from which the mother liquor is recycled to the dissolver, and the filter cake is sent to a direct-heat rotary dryer to produce crystalline potash. Other sequences for multicomponent mixtures are considered by Rajagopal et al. (1991), Cisternas and Rudd (1993), and Dye and Ng (1995b).

In both of the processes just described, a crystallizer produces a solid and, following a solid–liquid phase separation, a dryer removes the moisture. In some cases, all three of these operations can be carried out in a single piece of equipment, a spray dryer or a drum dryer, but at the expense of increased utility cost because all of the solvent is evaporated. Such dryers are used extensively to produce dried milk and detergents. For these products, spray dryers are particularly desirable, because the drying process produces porous particles that are readily dissolved in water. Spray dryers can also handle slurries and pastes.

As discussed by Barnicki and Fair (1990), melt crystallization is an alternative to other separation techniques for liquid mixtures, including ordinary distillation, enhanced distillation, liquid–liquid extraction, adsorption, and membrane permeation. Melt crystallization should be considered only when ordinary distillation is not feasible, but may be an attractive alternative when the melting-point difference between the two key components exceeds 20°C and a eutectic is not formed. If a eutectic is formed, high recovery may not be possible, as discussed by King (1980). Methods for circumventing the eutectic limitation are discussed by Dye and Ng (1995a).

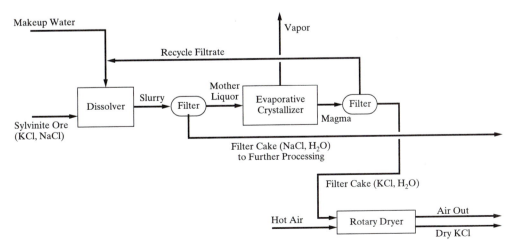

Figure 7.46 Process for separating a solid mixture.

7.8 SUMMARY

Having studied this chapter, the reader should

1. Know how each of the important industrial separation methods can be applied to the separation of multicomponent mixtures.
2. Know the importance of the separation factor.
3. Know how to determine near-optimal and optimal distillation sequences for nearly ideal systems.
4. Know how to develop separation sequences for nonideal systems that involve the formation of azeotropes.
5. Know how to develop a sequence for separating a gas mixture.
6. Know how to separate solid–fluid and multicomponent solid mixtures.

REFERENCES

Agrawal, R., and Z.T. Fidkowski, "More Operable Arrangements of Fully Thermally Coupled Distillation Systems," *AIChE J.*, **44**, 2565–2568 (1998).

Barbosa, D., and M.F. Doherty, "The Influence of Equilibrium Chemical Reactions on Vapor-Liquid Phase Diagrams," *Chem. Eng. Sci.*, **43**, 529 (1988a).

Barbosa, D., and M.F. Doherty, "The Simple Distillation of Homogeneous Reactive Mixtures," *Chem. Eng. Sci.*, **43**, 541 (1988b).

Barnicki, S.D., *Separation System Synthesis: A Knowledge-Based Approach*, Ph.D. dissertation, Dept. of Chemical Engineering, University of Texas at Austin (1991).

Barnicki, S.D., and J.R. Fair, "Separation System Synthesis: A Knowledge-Based Approach. 1. Liquid Mixture Separations," *Ind. Eng. Chem. Res.*, **29**, 421–432 (1990).

Barnicki, S.D., and J.R. Fair, "Separation System Synthesis: A Knowledge-Based Approach. 2. Gas/Vapor Mixtures," *Ind. Eng. Chem. Res.*, **31**, 1679–1694 (1992).

Barnicki, S.D., and J.J. Siirola, "Separations Process Synthesis," in J.I. Kroschwitz and M. Howe-Grant, Eds., *Kirk-Othmer Encyclopedia of Chemical Technology*, 4th ed., Vol. 21, pp. 923–962, Wiley, New York (1997).

Bekiaris, N., G.A. Meski, C.M. Radu, and M. Morari, "Multiple Steady States in Homogeneous Azeotropic Distillation," *Ind. Eng. Chem. Res.*, **32**, 2023–2038 (1993).

Bekiaris, N., G.A. Meski, C.M. Radu, and M. Morari, "Multiple Steady States in Heterogeneous Azeotropic Distillation," *Ind. Eng. Chem. Res.*, **35**, 207–227 (1996).

Bekiaris, N., T.E. Guttinger, and M. Morari, "Multiple Steady States in Distillation: Effect of VL(L)E Inaccuracies," *AIChE J.*, **46**, 5, 955–979 (2000).

Cisternas, L.A., and D.F. Rudd, "Process Designs for Fractional Crystallization from Solution," *Ind. Eng. Chem. Res.*, **32**, 1993–2005 (1993).

Crooks, D.A., and F.M. Feetham, *J. Chem. Soc.*, 899–901 (1946).

Doherty, M.F., and G. Buzad, "Reactive Distillation by Design," *Trans. IChemE*, **70**, 448–458 (1992).

Doherty, M.F., and M.F. Malone, *Conceptual Design of Distillation Systems*, McGraw-Hill, Boston (2001).

Douglas, J.M., *Conceptual Design of Chemical Processes*, McGraw-Hill, New York (1988).

Douglas, J.M., "Synthesis of Separation System Flowsheets," *AIChE J.*, **41**, 2522–2536 (1995).

Dye, S., and K.M. Ng, "Bypassing Eutectics with Extractive Crystallization: Design Alternatives and Tradeoffs," *AIChE J.*, **41**, 1456–1470 (1995a).

Dye, S., and K.M. Ng, "Fractional Crystallization: Design Alternatives and Tradeoffs," *AIChE J.*, **41**, 2427–2438 (1995b).

Eckert, E., and M. Kubicek, "Computing Heterogeneous Azeotropes in Multicomponent Mixtures," *Comput. Chem. Eng.*, **21**, 347–350 (1997).

Fidkowski, Z., and L. Krolikowski, "Minimum Energy Requirements of Thermally Coupled Distillation Systems," *AIChE J.*, **33**, 643–653 (1987).

Fidkowski, Z.T., M.F. Malone, and M.F. Doherty, "Computing Azeotropes in Multicomponent Mixtures," *Comput. Chem. Eng.*, **17**, 1141 (1993).

Gmehling, J., *Azeotropic Data*, VCH Publishers, Deerfield Beach, Florida (1994).

Gomez, A., and J.D. Seader, "Separation Sequence Synthesis by a Predictor Based Ordered Search," *AIChE J.*, **22**, 970–979 (1976).

Hendry, J.E., and R.R. Hughes, "Generating Separation Process Flowsheets," *Chem. Eng. Prog.*, **68**(6), 69 (1972).

Horsley, L.H., *Azeotropic Data—III, Advances in Chemistry Series No. 116*, American Chemical Society, Washington, DC (1973).

Horwitz, B.A., "Optimize Pressure-Sensitive Distillation," *Chem. Eng. Prog.*, **93**(4), 47 (1997).

Keller, G.M., "Separations: New Directions for an Old Field," *AIChE Monogr. Ser.* No. 17, **83**(1987).

King, C.J., *Separation Processes*, 2nd ed., McGraw-Hill, New York (1980).

Kister, H.Z., *Distillation Design*, McGraw-Hill, New York (1992).

Knapp, J.P., and M.F. Doherty, "A New Pressure-Swing Distillation Process for Separating Homogeneous Azeotropic Mixtures," *Ind. Eng. Chem. Res.*, **31**, 346 (1992).

Kovach III, J.W., and W.D. Seider, "Heterogeneous Azeotropic Distillation: Homotopy-Continuation Methods," *Comput. Chem. Eng.*, **11**, 593 (1987).

Lecat, M., *L'azcotropism: la tension de vapeur des melanges de liquides. 1. ptie., Donnes experimentales. Bibliographie*, A. Hoste, Gand, Belgium (1918).

Lee, K.-S., C. Levy, and N. Steckman, *Ditertiary-Butyl Peroxide Manufacture*, Towne Library, University of Pennsylvania, Philadelphia (1995).

Matsuyama, H., and H. Nishimura, "Topological and Thermodynamic Classification of Ternary Vapor–Liquid Equilibria," *J. Chem. Eng. Jpn.*, **10**(3), 181 (1977).

Modi, A.K., and A.W. Westerberg, "Distillation Column Sequencing Using Marginal Price," *Ind. Eng. Chem. Res.*, **31**, 839–848 (1992).

Perry, R.H., and D.W. Green, Ed., *Perry's Chemical Engineers' Handbook*, 7th ed., McGraw-Hill, New York (1997).

Peterson, E.J., and L.R. Partin, "Temperature Sequences for Categorizing All Ternary Distillation Boundary Maps," *Ind. Eng. Chem. Res.*, **36**, 1799–1811 (1997).

Petlyuk, F.B., V.M. Platonov, and D.M. Slavinskii, "Thermodynamically Optimal Method for Separating Multicomponent Mixtures," *Int. Chem. Eng.*, **5**, 555 (1965).

Prokopakis, G.J., and W.D. Seider, "Dynamic Simulation of Azeotropic Distillation Towers," *AIChE J.*, **29**, 1017–1029 (1983).

Rajagopal, S., K.M. Ng, and J.M. Douglas, "Design of Solids Processes: Production of Potash," *Ind. Eng. Chem. Res.*, **27**, 2071–2078 (1988).

Rajagopal, S., K.M. Ng, and J.M. Douglas, "Design and Economic Trade-offs of Extractive Crystallization Processes," *AIChE J.*, **37**, 437–447 (1991).

Rev, E., M. Emtir, Z. Szitkai, P. Mizsey, and Z. Fonyo, "Energy Savings of Integrated and Coupled Distillation Systems," *Comput. Chem. Eng.*, **25**, 119–140 (2001).

Robinson, C.S., and E.R. Gilliland, *Elements of Fractional Distillation*, McGraw-Hill, New York (1950).

Rodrigo, B.F.R., and J.D. Seader, "Synthesis of Separation Sequences by Ordered Branch Search," *AIChE J.*, **21**(5), 885 (1975).

Rossiter, A.P., "Design and Optimisation of Solids Processes. Part 3—Optimisation of a Crystalline Salt Plant Using a Novel Procedure," *Chem. Eng. Res. Des.*, **64**, 191–196 (1986).

Ryan, P.J., and M.F. Doherty, "Design/Optimization of Ternary Heterogeneous Azeotropic Distillation Sequences," *AIChE J.*, **35**, 1592–1601 (1989).

Seader, J.D., and E.J. Henley, *Separation Process Principles*, Wiley, New York (1998).

Seader, J.D., and A.W. Westerberg, "A Combined Heuristic and Evolutionary Strategy for Synthesis of Simple Separation Sequences," *AIChE J.*, **23**, 951 (1977).

Siirola, J.J., "An Industrial Perspective of Process Synthesis," in L.T. Biegler and M.F. Doherty, Eds., *Foundations of Computer-Aided Process Design*, *AIChE Symp. Ser.* No. 304, **91**, 222–233 (1995).

Souders, M., "The Countercurrent Separation Process," *Chem. Eng. Prog.*, **60**(2), 75–82 (1964).

Stichlmair, J.G., and J.R. Fair, *Distillation—Principles and Practice*, Wiley-VCH, New York (1998).

Stichlmair, J.G., J.R. Fair, and J.L. Bravo, "Separation of Azeotropic Mixtures via Enhanced Distillation," *Chem. Eng. Prog.*, **85**(63), 1 (1989).

Swietoslawski, W., *Azeotropy and Polyazeotropy*, Pergamon Press, New York (1963).

Tedder, D.W., and D.F. Rudd, "Parametric Studies in Industrial Distillation: I. Design Comparisons," *AIChE J.*, **24**, 303 (1978a).

Tedder, D.W., and D.F. Rudd, "Parametric Studies in Industrial Distillation: II. Heuristic Optimization," *AIChE J.*, **24**, 316 (1978b).

Vadapalli, A., and J.D. Seader, "A Generalized Framework for Computing Bifurcation Diagrams Using Process Simulation Programs," *Comput. Chem. Eng.*, **25**, 445–464 (2001).

Van Winkle, M., *Distillation*, McGraw-Hill, New York (1967).

Venimadhavan, G., G. Buzad, M.F. Doherty, and M.F. Malone, "Effect of Kinetics on Residue Curve Maps for Reactive Distillation," *AIChE J.*, **40**(11), 1814 (1994).

Walas, S.M., *Chemical Process Equipment—Selection and Design*, Butterworth, Boston (1988).

Widagdo, S., and W.D. Seider, "Azeotropic Distillation," *AIChE J.*, **42**(1), 96 (1996).

Wright, R.O., "Fractionation Apparatus," U.S. Patent 2,471,134, (May 24, 1949).

Zharov, V.T., "Phase Representations and Rectification of Multicomponent Solutions," *J. Appl. Chem. USSR*, **41**, 2530 (1968).

Zharov, V.T., and L.A. Serafimov, *Physicochemical Fundamentals of Distillations and Rectifications* (in Russian), Khimiya, Leningrad (1975).

EXERCISES

7.1 Stabilized effluent from a hydrogenation unit, as given below, is to be separated by ordinary distillation into five relatively pure products. Four distillation columns will be required. According to Eq. (7.9) and Table 7.2, these four columns can be arranged into 14 possible sequences. Draw sketches, as in Figure 7.11, for each of these sequences.

Component	Feed Flow Rate (lbmol/hr)	Approximate Relative Volatility Relative to C5
Propane (C3)	10.0	8.1
Butene-1 (B1)	100.0	3.7
n-Butene (NB)	341.0	3.1
Butene-2 isomers (B2)	187.0	2.7
n-Pentane (C5)	40.0	1.0

7.2 The feed to a separation process consists of the following species:

Species Number	Species
1	Ethane
2	Propane
3	Butene-1
4	*n*-Butane

It is desired to separate this mixture into essentially pure species. The use of two types of separators is to be explored:

1. Ordinary distillation

2. Extractive distillation with furfural (species 5)

The separation orderings according to relative volatility are

		Separator Type	
		1	**2**
Species number		1	1
		2	2
		3	4
		4	3
			5

Notice that the addition of furfural causes *n*-butane (4) to become more volatile than butene-1 (3). Determine the number of possible separation sequences.

7.3 Thermal cracking of naphtha yields a gas that is to be separated by a distillation train into the products indicated in Figure 7.47. If reasonably sharp separations are to be achieved, determine by the heuristics of Section 7.4 two good sequences.

7.4 Investigators at the University of Calforina at Berkeley have studied all 14 possible sequences for separating the following mixture at a flow rate of 200 lbmol/hr into its five components at about 98% purity each (D.L. Heaven, M.S. Thesis in Chemical Engineering, University of California, Berkeley, 1969).

Species	Symbol	Feed Mole Fraction	Approximate Volatility Relative to *n*-Pentane
Propane	A	0.05	8.1
Isobutane	B	0.15	4.3
n-Butane	C	0.25	3.1
Isopentane	D	0.20	1.25
n-Pentane	E	0.35	1.0
		1.00	

For each sequence, they determined the annual operating cost, including depreciation of the capital investment. Cost data for the best three sequences and the worst sequence are in Figure 7.48.

Explain in detail, as best you can, why the best sequences are best and the worst sequence is worst using the heuristics. Which heuristics appear to be the most important?

7.5 The effluent from a reactor contains a mixture of various chlorinated derivatives of the hydrocarbon RH_3, together with the hydrocarbon itself and HCl. Based on the following information and the heuristics of Section 7.4, devise the best two feasible separation sequences. Explain your reasoning. Note that HCl may be corrosive.

Species	lbmol/hr	α, Relative to RCl_3	Purity Desired
HCl	52	4.7	80%
RH_3	58	15.0	85%
RCl_3	16	1.0	98%
RH_2Cl	30	1.9	95%
$RHCl_2$	14	1.2	98%

7.6 The following stream at 100°F and 20 psia is to be separated into the four indicated products. Determine the best distillation sequence by the heuristics of Section 7.4. Compare your result with the result obtained by applying the Marginal Vapor Rate method.

	Feed (lbmol/hr)	Percent Recovery			
Species		Product 1	Product 2	Product 3	Product 4
Benzene	100	98			
Toluene	100		98		
Ethylbenzene	200			98	
p-Xylene	200			98	
m-Xylene	200			98	
o-Xylene	200				98

7.7 The following cost data, which include operating cost and depreciation of capital investment, pertain to Exercise 7.1. Determine by finding the total cost for each of the 14 possible sequences:
a. The best sequence
b. The second-best sequence
c. The worst sequence
Are the heuristics of Section 7.4 in agreement with the results based on costs?

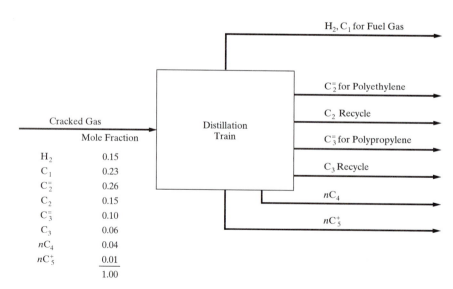

Figure 7.47 Thermal cracking of naphtha.

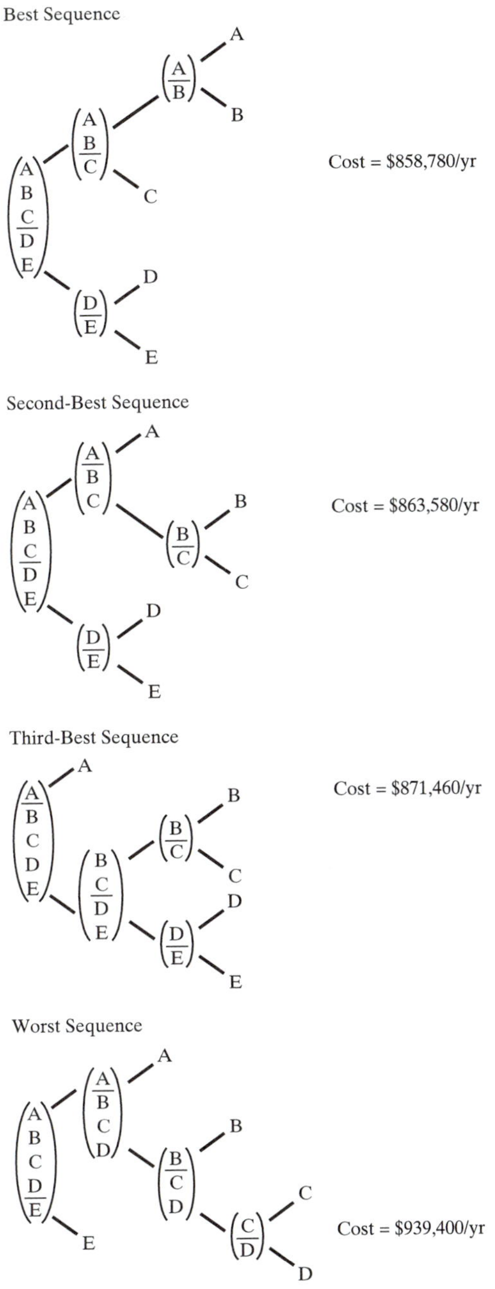

Best Sequence

Cost = $858,780/yr

Second-Best Sequence

Cost = $863,580/yr

Third-Best Sequence

Cost = $871,460/yr

Worst Sequence

Cost = $939,400/yr

Figure 7.48 Cost data for Exercise 7.4.

Split	Cost ($/yr)
C3/B1	15,000
B1/NB	190,000
NB/B2	420,000
B2/C5	32,000
C3, B1/NB	197,000
C3/B1, NB	59,000
B1, NB/B2	500,000
B1/NB, B2	247,000

Split	Cost ($/yr)
NB, B2/C5	64,000
NB/B2, C5	460,000
C3, B1, NB/B2	510,000
C3, B1/NB, B2	254,000
C3/B1, NB, B2	85,000
B1, NB, B2/C5	94,000
B1, NB/B2, C5	530,000
B1/NB, B2, C5	254,000
C3, B1, NB, B2/C5	95,000
C3, B1, NB/B2, C5	540,000
C3, B1/NB, B2, C5	261,000
C3/B1, NB, B2, C5	90,000

7.8 A hypothetical mixture of four species, A, B, C, and D, is to be separated into the four separate components. Two different separator types are being considered, neither of which requires a mass separating agent. The order of separation for each of the two types are

Separator Type I	Separator Type II
A	B
B	A
C	C
D	D

Annual cost data for all the possible splits are given below. Determine by considering each possible sequence:
a. The best sequence
b. The second-best sequence
c. The worst sequence
For each answer, draw a diagram of the separation train, being careful to label each separator as to whether it is type I or II.

Subgroup	Split	Type Separator	Annual Cost × $10,000
(A, B)	A/B	I	8
		II	15
(B, C)	B/C	I	23
		II	19
(C, D)	C/D	I	10
		II	18
(A, C)	A/C	I	20
		II	6
(A, B, C)	A/B, C	I	10
	B/A, C	II	25
	A, B/C	I	25
		II	20
(B, C, D)	B/C, D	I	27
		II	22
	B, C/D	I	12
		II	20
(A, C, D)	A/C, D	I	23
		II	10
	A, C/D	I	11
		II	20

(Continued)

Subgroup	Split	Type Separator	Annual Cost × $10,000
(A, B, C, D)	A/B, C, D	I	14
	B/A, C, D	II	20
	A, B/C, D	I	27
		II	25
	A, B, C/D	I	13
		II	21

7.9 The following stream at 100°F and 250 psia is to be separated into the four indicated products. Also given is the cost of each of the unique separators. Determine:
a. The best sequence
b. The second-best sequence

		Feed Rate	Percent Recovery			
Species	Symbol	(lbmol/hr)	Product 1	Product 2	Product 3	Product 4
Propane	A	100	98			
i-Butane	B	300		98		
n-Butane	C	500			98	
i-Pentane	D	400				98

Unique Separator	Cost ($/yr)
A/B	26,100
B/C	94,900
C/D	59,300
A/BC	39,500
AB/C	119,800
B/CD	112,600
BC/D	76,800
A/BCD	47,100
AB/CD	140,500
ABC/D	94,500

7.10 The following stream at 100°F and 300 psia is to be separated into four essentially pure products. Also given is the cost of each unique separator. Determine the best sequence.

Species	Symbol	Feed rate (lbmol/hr)
i-Butane	A	300
n-Butane	B	500
i-Pentane	C	400
n-Pentane	D	700

Unique Separator	Cost ($/yr)
A/B	94,900
B/C	59,300
C/D	169,200
A/BC	112,600
AB/C	76,800
B/CD	78,200
BC/D	185,300
A/BCD	133,400
AB/CD	94,400
ABC/D	241,800

7.11 Consider the problem of separation, by ordinary distillation, of propane A, isobutane, B, n-butane, C, isopentane, D, and n-pentane, E. Using the heuristics of Section 7.4, develop flowsheets for:
a. Equimolal feed with product streams A, (B, C) and (D, E) required
b. Feed consisting of A = 10, B = 10, C = 60, D = 10 and E = 20 (relative moles) with products A, B, C, D, and E

Component	Average Relative Volatility
A >	2.2
B >	1.44
C >	2.73
D >	1.25
E	

7.12 Derive the right-hand side of Eq. (7.9).

7.13 a. Consider binary mixtures of acetone and chloroform at 101 kPa, with vapor–liquid equilibria in Figure 7.20.

Using distillation, identify the maximum and minimum mole fractions of acetone in the product streams for feed streams containing:

1. 90 mol% acetone
2. 25 mol% acetone

What are the bubble-point temperatures of the associated distillate and bottoms products?

b. Repeat (a) for isopropyl ether (IPE) and isopropyl alcohol (IPA), using Figure 7.19, with mole fractions of IPE replacing those of acetone.

7.14 A multicomponent mixture is boiled in a flask at 1 atm. The vapors are condensed and recovered as a liquid product. It is desired to examine the mole fractions of the residual liquid in the flask as vaporization proceeds. Although sketches of the residue curve maps are called for in (b)–(d), a process simulator can be used to prepare the drawings accurately.
a. For a mixture of 60 mol% n-butane (1), and 40 mol% n-pentane (2), determine the residual mole fraction of n-butane after 10% of the liquid has vaporized.
b. Consider mixtures of n-butane (1), n-pentane (2), and n-hexane (3). For three typical feed compositions:

Component	Mole Fractions		
	I	II	III
1	0.7	0.15	0.15
2	0.15	0.7	0.15
3	0.15	0.15	0.7

sketch the residue curves (solutions of the ODEs—do not solve them analytically or numerically) on triangular graph paper. Use arrows to show the direction along the trajectories in time.

c. Repeat (b) for mixtures of acetone (1), chloroform (2), and benzene (3). Note that the acetone–chloroform binary exhibits a maximum-boiling azeotrope (64°C) at 35 mol% acetone, with no other azeotropes existing. Sketch any boundaries across which the residue curves cannot traverse.

d. Repeat (c) for mixtures of methyl acetate (1), methanol (2), and *n*-hexane (3). Note the existence of four azeotropes, where compositions are in mol%.

	T°C (1)
Methyl acetate (65%)–methanol (35%)	53.5
Methanol (51%)–*n*-hexane (49%)	50.0
Methyl acetate (60%)–*n*-hexane (40%)	51.8
Methyl acetate (31%)–*n*-hexane (40%)–methanol (29%)	49.0

7.15 Prepare residue curve maps using a process simulation program for the following mixtures at 1 atm. Identify any distillation boundaries.
a. Acetone, *n*-heptane, toluene
b. Methanol, ethanol, water
c. Acetone, chloroform, ethanol

7.16 For a mixture of 70 mol% chloroform, 15 mol% acetone, and 15 mol% ethanol at 1 atm, show on a residue curve map the feasible compositions of the distillate and bottoms product.

7.17 Consider the process for the dehydration of ethanol using toluene in Figure 7.32. Estimate the ratios of the flow rates in the following streams:
a. S1 and S2
b. S2 and D1
c. S1 and F
d. B1 and D1
e. B2 and D2

7.18 For the manufacture of di-tertiary-butyl peroxide in Example 7.5, synthesize an alternative process and show the flow rate and composition of each stream.

7.19 For the pressure-swing dehydration of THF, determine the internal flow rates when the high-pressure column is at 5 bar.

Chapter 8[1]

Reactor–Separator–Recycle Networks

8.0 OBJECTIVES

The presence of at least one chemical reactor and one or more separation sections for the separation of the effluent mixture leaving the reactor(s) characterizes many chemical processes. In almost all cases, one or more of the streams leaving the separation section(s) are recycled to the reactor. In Chapter 6, the design of reactors and reactor networks was considered without regard for the separation section(s) and possible recycle there from. Chapter 7 was concerned with the design of separation sections in the absence of any consideration of the reactor section. Chapter 5, which dealt with the synthesis of the entire process, included a few examples of the interaction between the reactor and separation sections. This chapter extends that introduction to give a detailed treatment of reactor–separator–recycle networks.

After studying this chapter, the reader should

1. Be able to determine the best location for the separation section, either before or after the reactor.
2. Understand the trade-offs between purge-to-recycle ratio, recycle ratio, and raw material loss, when dealing with inert or byproduct chemicals that are difficult to separate from the reactants.
3. Understand the need to determine the optimal reactor conversion, involving the trade-off between the cost of the reactor section and the cost of the separation section(s) in the presence of recycle, even when chemical equilibrium greatly favors the products of the reaction.
4. Understand the conditions under which recycle of byproducts to extinction can be employed to reduce waste and increase yield.
5. Be aware of the snowball effect in a reactor–separator–recycle network and the importance of designing an adequate control system, which is presented in Sections 20.3 (Example 20.11) and 21.5 (Case Study 21.3).

---

[1]Chapter 8 appears in the file, Chapter 8.pdf, on the CD-ROM that accompanies this text. Only the "Objectives" appear here.

Chapter 9[1]

 # Second-Law Analysis

9.0 OBJECTIVES

The first law of thermodynamics is widely used in design to make energy balances around equipment. Much less used are the entropy balances based on the second law of thermodynamics. Although the first law can determine energy transfer requirements in the form of heat and shaft work for specified changes to streams, it cannot even give a clue as to whether energy is being used efficiently. As shown in this chapter, calculations with the second law or a combined first and second law can determine energy efficiency. The calculations are tedious to do by hand, but are readily carried out with a process simulation program. When the efficiency of a process is found to be low, a better process should be sought. The average second-law efficiency for chemical plants is in the range of only 20–25%. Therefore, chemical engineers need to spend more effort in improving energy efficiency.

After studying this chapter, the reader should

1. Understand the limitations of the first law of thermodynamics.
2. Understand the usefulness of the second law and a combined statement of the first and second laws.
3. Be able to specify a system and surroundings for conducting a second-law analysis.
4. Be able to derive and apply a combined statement of the first and second laws for the determination of lost work or exergy.
5. Be able to determine the second-law efficiency of a process and pinpoint the major areas of inefficiency (lost work).
6. Understand the causes of lost work and how to remedy them.
7. Be able to use a process simulation program to perform a second-law analysis.

---

[1]Chapter 9 appears in the file, Chapter 9.pdf, on the CD-ROM that accompanies this text. Only the "Objectives" appear here.

Chapter 10

Heat and Power Integration

10.0 OBJECTIVES

This chapter introduces several *algorithmic* approaches that have been developed for process integration to satisfy the cooling, heating, and power demands of a process. After studying this material, the reader should

1. Be able to determine minimum energy requirement (MER) targets; that is, to compute the minimum usage of heating and cooling utilities when exchanging heat between the hot and cold streams in a process. Three methods are introduced: the temperature-interval (TI) method, a graphical approach, and the formulation and solution of a linear program (LP).

2. Be able to design a network to meet the MER targets; that is, to position heat exchangers in a network, assuming overall heat-transfer coefficients. Two methods are introduced: a unit-by-unit method beginning at the closest-approach temperature difference (the *pinch*), and the formulation and solution of a mixed-integer linear program (MILP).

3. Be able to reduce the number of heat exchangers in MER networks, by relaxing the MER target and *breaking the heat loops* (i.e., allowing heat to flow across the pinch), or alternatively, by employing *stream splitting*.

4. Be able to design a network when the minimum approach temperature is below a threshold value, at which either heating or cooling utility is used, but not both.

5. Be able to use the grand composite curve to assist in the selection and positioning of appropriate types of hot and cold utilities in the network.

6. Understand the importance of the specified minimum approach temperature difference on the design of a heat exchanger network (HEN).

7. Understand how to set up a superstructure for the design of a HEN that minimizes the annualized cost and how to formulate and solve its nonlinear program (NLP) using the General Algebraic Modeling System (GAMS).

8. Understand several approaches to designing energy-efficient distillation trains, including the adjustment of tower pressure, multiple-effect distillation, and heat pumping, vapor recompression, and reboiler flashing.

9. Understand the need to position *heat engines* to satisfy power demands of processes, and the need to position *heat pumps* to accomplish refrigeration to reduce power requirements. A methodology is introduced that does not require the usage of formal optimization methods.

10.1   INTRODUCTION

At the start of the task-integration step in process synthesis, the source and target temperatures, T^s and T^t, and power demands for pumping and compression of all streams are known. Heat and power integration seeks to utilize the energy in the high-temperature streams that need to be cooled and/or condensed to heat and/or vaporize the cold streams, and provide power to compressors from turbines and heat engines where possible. In most designs, it is common initially to disregard power demands in favor of designing an effective network of heat exchangers by heat integration, without using the energy of the high-temperature streams to produce power. To accomplish this, N_H hot process streams, with specified source and target temperatures $T^s_{h_i}$ and $T^t_{h_i}$, $i = 1, \ldots, N_H$, are cooled by N_C cold process streams, with specified source and target temperatures, $T^s_{c_j}$ and $T^t_{c_j}$, $j = 1, \ldots, N_C$, as shown schematically in Figure 10.1a. When either: (a) the sum of the heating requirements does not equal the sum of the cooling requirements; or (b) some source temperatures may not be sufficiently high or low to achieve some target temperatures through heat exchange; or (c) when other restrictions exist, as discussed in Section 10.2, it is always necessary to provide one or more auxiliary heat exchangers for heating or cooling through the use of utilities such as steam and cooling water. It is common to refer to the heat exchangers between the hot and cold process streams as comprising the *interior network*, and those between the hot or cold streams and the utilities as comprising the *auxiliary network*, as shown schematically in Figure 10.1b.

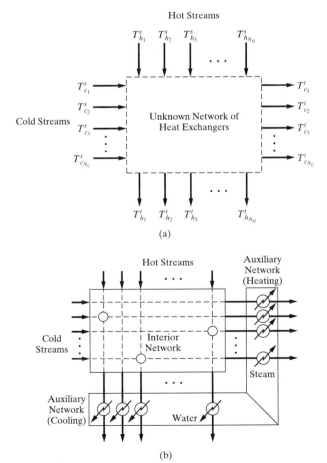

Figure 10.1 Heat integration schematics: (a) source and target temperatures for heat integration; (b) interior and auxiliary networks of heat exchangers.

When carrying out the design, given the states of the source and target streams (flow rates of the species, temperature, pressure, and phase), it is desired to synthesize the most economical network of heat exchangers. Several measures of economic goodness are possible, as discussed in Section 17.4. Usually, when generating and comparing alternative flowsheets, an approximate profitability measure is sufficient, such as the annualized cost:

$$C_A = i_m (C_{\mathrm{TCI}}) + C \tag{10.1}$$

where C_{TCI} is the total capital investment, as defined in Table 16.9, i_m is a reasonable return on investment annually (i.e., when $i_m = 0.33$, a chemical company charges itself annually for one-third of the cost of the capital invested), and C is the annual cost of sales, as defined in the *cost sheet* of Table 17.1. In Tables 16.9 and 17.1, many factors are involved, most of which are necessary for a detailed profitability analysis. However, to estimate an approximate profitability measure for the comparison of alternative flowsheets, it is adequate to approximate C_{TCI} as the sum of the purchase costs for each of the heat exchangers (without including installation costs and other capital investment costs). The purchase costs can be estimated based on the area for heat transfer, A, estimated from the heat transfer rate equation discussed in Section 13.2 [Eq. (13.7)]:

$$A = Q/(UF_T \Delta T_{\mathrm{LM}}) \tag{10.2}$$

where Q is the heat duty, U is the overall heat-transfer coefficient, F_T is the correction factor for a multiple-pass exchanger, and ΔT_{LM} is the log-mean temperature-driving force for countercurrent flow based on the approach-temperature differences at the two ends. Equation (10.2) must be used with care because of its restrictions, as discussed in Chapter 13. If both a phase change and a significant temperature change occur for one or both streams, U is not constant and a ΔT_{LM} is not appropriate. Furthermore, multiple-pass exchangers may be required, for which F_T is in the range 0.75–0.9. Nevertheless, to develop a reasonably optimal heat exchanger network, it is common to apply Eq. (10.2) with $F_T = 1.0$. It is adequate to approximate C as the annual cost of the utilities for heating and cooling, typically using steam and cooling water. In summary, with these approximations, Eq. (10.1) is rewritten as

$$C_A = i_m \left[\sum_i C_{P,I_i} + \sum_j C_{P,A_j} \right] + sF_s + (cw)F_{cw} \tag{10.3}$$

where C_{P,I_i} and C_{P,A_j} are the purchase costs of the heat exchangers in the interior and auxiliary networks, respectively, F_s is the annual flow rate of steam (e.g., in kilograms per year), s is the unit cost of steam (e.g., in dollars per kilogram), F_{cw} is the annual flow rate of cooling water, and cw is the unit cost of cooling water. Clearly, when other utilities, such as fuel, cool air, boiler feed water, and refrigerants are used, additional terms are needed.

Many approaches have been developed to optimize Eq. (10.3) and similar profitability measures, several of which are presented in this chapter. Probably the most widely used, an approach developed immediately after the OPEC oil embargo in 1973, which triggered a global energy crisis, utilizes a two-step procedure:

1. A network of heat exchangers is designed having the *minimum usage of utilities* (i.e., an MER network), usually requiring a large number of heat exchangers. However, when the cost of fuel is extremely high, as it was in the late 1970s, a nearly optimal design is obtained.

2. The number of heat exchangers is reduced toward the minimum, possibly at the expense of increasing the consumption of utilities.

Clearly, as step 2 is implemented, one heat exchanger at a time, capital costs are reduced due to the economy of scale in equations of the form:

$$C_P = KA^n \tag{10.4}$$

where K is a constant and n is less than unity, typically 0.6. As each heat exchanger is removed, with the total area for heat transfer approximately constant, the area of each of the remaining heat exchangers is increased, and because $n < 1$, the purchase cost per unit of area is decreased. In addition, as step 2 is implemented, the consumption of utilities is normally increased. At some point, the increased cost of utilities overrides the decreased cost of capital and C_A increases beyond the minimum. When the cost of fuel is high, the minimum C_A is not far from that for a network of heat exchangers using the minimum utilities. Finally, note that Eqs. (16.38)–(16.44) provide more accurate estimates than Eq. (10.4), but are not commonly used when comparing alternative heat exchanger networks during process synthesis.

The selection of the minimum approach temperature, ΔT_{min}, for the heat exchangers is a key design variable in the synthesis of heat exchanger networks (HENs), because of its impact on lost work associated with heat transfer. Consider the heat transfer between the high- and low-temperature reservoirs in Figure 10.2. Equation (9.27), which is the result of combining the first and second laws of thermodynamics for the general process in Figure 9.18, can be simplified to eliminate the term involving the flowing streams, the work term, and the term for unsteady operation, to give

$$LW = \left(1 - \frac{T_0}{T_1}\right)Q + \left(1 - \frac{T_0}{T_2}\right)(-Q) \tag{10.5a}$$

$$= Q\left(\frac{T_0}{T_2} - \frac{T_0}{T_1}\right) \tag{10.5b}$$

$$= QT_0\left(\frac{T_1 - T_2}{T_1 T_2}\right) \tag{10.5c}$$

$$= QT_0\frac{\Delta T}{T_1 T_2} \tag{10.5d}$$

where LW is the rate of lost work and T_0 is the absolute temperature of the environment. Note that a simpler notation suffices in this chapter, in which all analysis is in the steady state; hence, $LW \equiv L\dot{W}$, $Q \equiv \dot{Q}$, and $m \equiv \dot{m}$. It can be seen that, for a given rate of heat transfer and a given ΔT approach, the rate of lost work increases almost inversely with the decrease in the square of the absolute temperature level. Thus, as the temperature levels move lower into the cryogenic region, the approach temperature difference, ΔT, must decrease approximately as the square of the temperature level to maintain the same rate of lost work. This explains the need to use very small approach temperature differences, on the order of 1°C, in the cold boxes of cryogenic processes. If the approach temperature differences were not reduced, the large increases in the rate of lost work would sharply increase the energy requirements to operate these processes, especially the operating and installation costs for compressors.

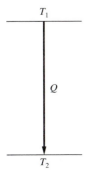

Figure 10.2 Heat exchange between two reservoirs.

10.2 MINIMUM UTILITY TARGETS

A principal objective in the synthesis of HENs is the efficient utilization of energy in the hot process streams to heat cold process streams. Thus, it is desirable to compute the maximum energy recovery (MER) before synthesizing the HEN; that is, to determine the minimum hot and cold utilities in the network, given the heating and cooling requirements of the process streams. This important first step is referred to as *MER targeting*, and is useful in that it determines the utility requirements for the most thermodynamically efficient network. To introduce this targeting step, the example provided by Linnhoff and Turner (1981) is presented here. This example involves only sensible heat. Later, examples are presented that also involve the latent heat of phase change and the heat of reaction, under either isothermal or nonisothermal conditions.

EXAMPLE 10.1

Two cold streams, C1 and C2, are to be heated and two hot streams, H1 and H2, are to be cooled without phase change. Their conditions and properties are as follows:

Stream	T^s (°F)	T^t (°F)	mC_p [Btu/(hr°F)]	$Q(10^4$ Btu/hr)
C1	120	235	20,000	230
C2	180	240	40,000	240
H1	260	160	30,000	300
H2	250	130	15,000	180

It is assumed that the *heat-capacity flow rate, $mC_P = C$*, which is the product of the specific heat and the mass flow rate, does not vary with temperature. As shown later, when it is necessary to account for a variation in the heat capacity with the temperature, a stream is *discretized* into several substreams, each involving a different segment of the temperature range and a different C (see Example 10.5). Design a HEN that uses the smallest amounts of heating and cooling utilities possible, such that the closest approach temperature differences never fall below a minimum value. For the temperature range in this example, a reasonable assumption is $\Delta T_{min} = 10$°F.

SOLUTION

For this system, a total of 480×10^4 Btu/hr must be removed from the two hot streams, but only a total of 470×10^4 Btu/hr can be consumed by the two cold streams. Hence, from the first law of thermodynamics, a minimum of 10×10^4 Btu/hr must be removed by a cold utility such as cooling water. As shown below, this is *not* the minimum utility usage, which, from the second law of thermodynamics, depends on ΔT_{min}. One possible HEN is shown in Figure 10.3, involving six heat exchangers and 57.5×10^4 and 67.5×10^4 Btu/hr of hot and cold utilities, respectively. Note that the *difference* between the hot and cold utility duties equals that given by the first law of thermodynamics. ∎

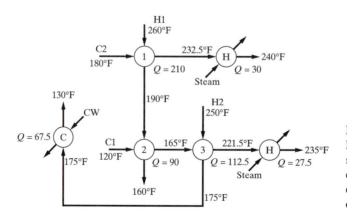

Figure 10.3 Proposed HEN for Example 10.1 with $\Delta T_{min} = 10$°F, showing interior heat exchangers (1–3) and auxiliary heat exchangers (H, C). Multiply heat duties, Q, by 10^4 Btu/hr.

Since the design of this HEN has neither considered the MER targets, nor utilized procedures for optimal HEN synthesis, its assessment focuses on two questions: (a) How do the heating and cooling utility duties of 57.5×10^4 and 67.5×10^4 Btu/hr compare with the MER targets? (b) Is it possible to synthesize a HEN with fewer heat exchangers, and if so, what are its utility requirements? In this section, the first of these two questions is addressed, with the second, which involves higher utility costs and decreased capital costs, postponed until later. Three methods are introduced to estimate MER targets: (1) the temperature-interval method, (2) a graphical method using *composite heating and cooling* curves, to be defined, and (3) the formulation and solution of a linear programming (LP) problem.

Temperature-Interval (TI) Method

The *temperature-interval method* was developed by Linnhoff and Flower (1978a, b) following the pioneering work of Hohmann (1971). The method is applied to the hot and cold streams introduced in Example 10.1, and as will be seen, a systematic procedure unfolds for determining the minimum utility requirements over all possible HENs, given just the heating and cooling requirements for the process streams and the minimum approach temperature in the heat exchangers, ΔT_{min}.

EXAMPLE 10.2   **(Example 10.1 Revisited)**

Returning to Example 10.1, the temperature-interval (TI) method is used for the calculation of MER targets for $\Delta T_{min} = 10°F$.

SOLUTION

The first step in the TI method is to adjust the source and target temperatures using ΔT_{min}. Somewhat arbitrarily, this is accomplished by reducing the temperatures of the hot streams by ΔT_{min}, while leaving the temperatures of the cold streams untouched as follows:

	T^s (°F)	T^t (°F)	Adjusted Temps T^s (°F)	Adjusted Temps T^t (°F)	
C1	120		120		T_5
		235		235	T_2
C2	180		180		T_3
		240		240	T_1
H1	260		250		T_0
		160		150	T_4
H2	250		240		T_1
		130		120	T_5

The adjustment of the hot stream temperatures by subtracting ΔT_{min} brings both the hot and cold streams to a common frame of reference so that when performing an energy balance involving hot and cold streams at the same temperature level, the calculation accounts for heat transfer with at least a driving force of ΔT_{min}. Next, the adjusted temperatures are rank ordered, beginning with T_0, the highest temperature. These are used to create a cascade of *temperature intervals* within which energy balances are carried out. As shown in Table 10.1 and Figure 10.4, each interval, i, displays the enthalpy difference, ΔH_i, between the energy to be removed from the hot streams and the energy to be taken up by the cold streams in that interval. For example, in interval 1, 240°F to 250°F ($\Delta T = 10°F$), only stream H1 is involved. Hence, the enthalpy difference is:

$$\Delta H_1 = \left(\sum C_h - \sum C_c\right)_1 (T_0 - T_1) = (3 \times 10^4) \times (250 - 240) = 30 \times 10^4 \text{ Btu/hr}$$

Table 10.1 Enthalpy Differences for Temperature Intervals

Interval, i	$T_{i-1} - T_i$, °F	$\Sigma C_h - \Sigma C_c$, 10^4 Btu/hr-°F	ΔH_i, 10^4 Btu/hr
1	$250 - 240 = 10$	3	30
2	$240 - 235 = 5$	$3 + 1.5 - 4 = 0.5$	2.5
3	$235 - 180 = 55$	$3 + 1.5 - 4 - 2 = -1.5$	-82.5
4	$180 - 150 = 30$	$3 + 1.5 - 2 = 2.5$	75
5	$150 - 120 = 30$	$1.5 - 2 = -0.5$	-15

It is noted that initially, no energy is assumed to enter this interval from a hot utility, such as steam at a higher temperature; that is, $Q_{steam} = 0$. Hence, 30×10^4 Btu/hr are available and flow down as a residual, R_1, into the next lowest interval 2; that is, $R_1 = 30 \times 10^4$ Btu/hr. Interval 2 involves streams H1, H2, and C2 between 235°F and 240°F ($\Delta T = 5$°F), and hence, the enthalpy difference is

$$\Delta H_2 = \left(\sum C_h - \sum C_c\right)_2 (T_1 - T_2) = [(3 + 1.5 - 4) \times 10^4] \times (240 - 235) = 2.5 \times 10^4 \text{ Btu/hr}$$

When this is added to the residual from interval 1, R_1, this makes the residual from interval 2, $R_2 = 32.5 \times 10^4$ Btu/hr. Note that no temperature violations of ΔT_{min} occur when the streams are matched in interval 2 because the hot stream temperatures are reduced by ΔT_{min}. Interval 3, 180°F to 235°F ($\Delta T = 55$°F), involves all four streams, and hence, the enthalpy difference is -82.5×10^4 Btu/hr as detailed in Table 10.1, making the residual from interval 3 equal to $(32.5 - 82.5) \times 10^4 = -50 \times 10^4$ Btu/hr. Similarly, the enthalpy differences in intervals 4 and 5 are 75×10^4 and -15×10^4 Btu/hr, respectively, with the residuals leaving these intervals being 25×10^4 and 10×10^4 Btu/hr. Note that for

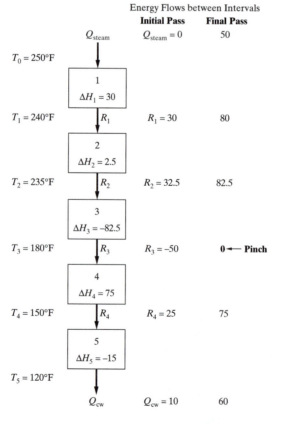

Figure 10.4 Cascade of temperature intervals, energy balances, and residuals; multiply ΔH_i and R_i by 10^4 Btu/hr.

Table 10.2   Interval Heat Loads (Multiply Q by 10^4 Btu/hr)

	Cooling			Heating		
Interval i	Temperature Range (°F)	Q	Cum. Q	Temperature Range (°F)	Q	Cum. Q
1	250–260	30	480.0	240–250	0	—
2	245–250	22.5	450.0	235–240	20	470.0
3	190–245	247.5	427.5	180–235	330	450.0
4	160–190	135	180.0	150–180	60	120.0
5	130–160	45	45.0	120–150	60	60.0

$Q_{steam} = 0$, the largest negative residual is from interval 3, $R_3 = -50 \times 10^4$ Btu/hr. Clearly, to satisfy the second law of thermodynamics, all negative residuals must be removed because heat cannot flow from a low- to a high-temperature interval. The only way to avoid negative residuals is to add energy at higher temperatures. This is achieved using low-pressure steam above 250°F. In Figure 10.4, note that when $Q_{steam} = 50 \times 10^4$ Btu/hr, R_1 becomes 80×10^4 Btu/hr, $R_2 = 82.5 \times 10^4$ Btu/hr, $R_3 = 0$, $R_4 = 75 \times 10^4$ Btu/hr, and $R_5 = Q_{cw} = 60 \times 10^4$ Btu/hr. Thus, $Q_{steam} = 50 \times 10^4$ Btu/hr is the smallest amount of energy that must be added above 180°F, and hence, it becomes a lower bound on the hot utility duty, just as $Q_{cw} = 60 \times 10^4$ Btu/hr becomes the lower bound on the cold utility duty. These are referred to as the *MER targets*; evidently, the HEN in Figure 10.3 exceeds these targets, by 7.5×10^4 Btu/hr each.

Note that $Q_{steam} - Q_{cw} = -10 \times 10^4$ Btu/hr, which is consistent with the first law. Furthermore, at minimum utilities, no energy flows between intervals 3 and 4. This is referred to as the *pinch*, with associated temperatures of 180°F for the cold streams and $180 + \Delta T_{min} = 180 + 10 = 190$°F for the hot streams. To maintain minimum utilities, it is recognized that *no energy is permitted to flow across the pinch*. If, as in the HEN in Figure 10.3, Q_{steam} were increased to 57.5×10^4 Btu/hr, R_3, the transfer of heat across the pinch, would be 7.5×10^4 Btu/hr, and Q_{cw} would increase to 67.5×10^4 Btu/hr.

Table 10.2 summarizes the cooling and heating loads in each interval, with the actual temperature ranges shown. Included are the cumulative loads starting from the lowest temperatures.

HEN synthesis is facilitated using the stream representation in Figure 10.5, in which arrows moving from left to right denote the hot streams, while arrows moving from right to left denote the cold streams. The arrows for the hot and cold streams either pass through or begin at the pinch temperatures.

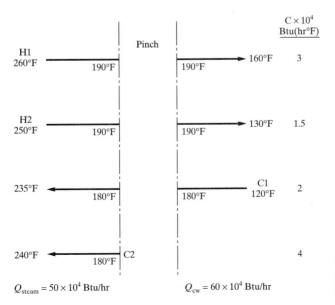

Figure 10.5 Pinch decomposition of the hot and cold streams for Example 10.2.

To maintain minimum utilities, two separate HENs *must* be designed, one on the hot side and one on the cold side of the pinch. Energy is added from hot utilities on the hot side of the pinch (50×10^4 Btu/hr), and energy is removed using cold utilities on the cold side of the pinch (60×10^4 Btu/hr), and no energy is permitted to flow across the pinch. If energy were exchanged between a hot stream on the hot side of the pinch and a cold stream on the cold side of the pinch, this energy would not be available to heat the cold streams on the hot side of the pinch and additional energy from the hot utilities would be required. Similarly, the cold stream on the cold side of the pinch would not have the ability to remove this energy from the hot streams on the cold side of the pinch and the same amount of additional energy would have to be removed from the cold streams on the cold side of the pinch using cold utilities. ∎

Composite Curve Method

The terminology, *pinch*, is understood more clearly in connection with a graphical display, introduced by Umeda et al. (1978), in which composite heating and cooling curves are positioned no closer than ΔT_{min}. As $\Delta T_{min} \rightarrow 0$, the curves *pinch* together and the area for heat exchange approaches infinity. In this respect, there is a close parallel to the graphical approach introduced by McCabe and Thiele (1925) in their classic method for the design of distillation towers to separate binary mixtures. It should be recalled that, on the McCabe–Thiele diagram, a *pinch* occurs when the operating lines intersect the equilibrium curve and the feed line. This occurs at the minimum reflux ratio, R_{min}, where an infinite number of stages accumulate in the vicinity of the pinch point, as shown in Figure 10.6. The parallel is illustrated further in Example 10.3.

EXAMPLE 10.3 *(Example 10.1 Revisited)*

In this example, the minimum utility requirements for a HEN involving the four streams in Example 10.1 are determined using the graphical approach by Umeda et al. (1978).

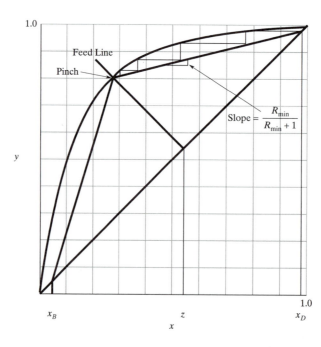

Figure 10.6 McCabe–Thiele diagram showing a pinch at minimum reflux in binary distillation.

SOLUTION   For each of the streams, the temperature, T, is graphed on the ordinate as a function of the enthalpy or heat transferred on the abscissa, with the slope being the inverse of the heat-capacity flow rate, C. When C is constant (i.e., not a function of T), the curves are straight lines. For the hot streams, they are *cooling* curves that begin at the highest temperature and finish at the lowest temperature after the energy has been removed. For the cold streams, they are *heating curves* that begin at the lowest temperature and finish at the highest temperature after heat has been added. In Figure 10.7a, the two heating and two cooling curves are displayed, with each of the lines positioned arbitrarily along the abscissa to avoid intersections and crowding.

To display the results of the TI method graphically, Table 10.2 is used to prepare *hot composite* and *cold composite* curves, which combine curves H1 and H2 in Figure 10.7a into one hot composite curve, and curves C1 and C2 into one cold composite curve. First, the hot composite curve is graphed starting with an enthalpy datum of 0 at 130°F, the lowest temperature of a hot stream. From Table 10.2, the hot composite enthalpies are

T (°F)	130	160	190	245	250	260
H (Btu/hr $\times 10^{-4}$)	0	45	180	427.5	450	480

These points form the hot composite curve in Figure 10.7b. As seen in the figure, the hot composite curve has a segment from stream H1 between 260 and 250°F. From 250 to 160°F, both streams H1 and

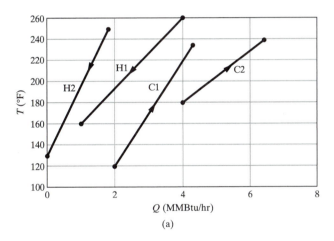

(a)

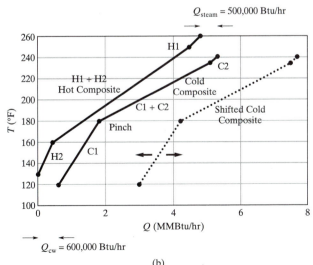

(b)

Figure 10.7 Graphical method to determine MER targets: (a) heating and cooling curves for the streams; (b) composite hot and cold curves.

H2 coexist, and hence, their cooling requirements are combined. Note that the combined heat-capacity flow rate is increased, and consequently, the slope of the hot composite curve is reduced. Finally, from 160 to 130°F, only stream H2 appears.

Next, the cold composite curve is graphed. For $\Delta T_{min} = 10$°F, the TI method determined a minimum cooling utility of 60×10^4 Btu/hr. Therefore, the graph begins with an enthalpy datum of that value. From Table 10.2, the cold composite enthalpies are

T (°F)	120	150	180	235	240
H (Btu/hr $\times 10^{-4}$)	60	120	180	510	530

These points form the cold composite curve in Figure 10.7b. From 120 to 180°F, only stream C1 appears in the cold composite curve. From 180°F to 235°F, the streams C1 and C2 coexist and their heating curves are combined. Finally, from 235 to 240°F, only stream C2 exists.

As shown by the solid lines in Figure 10.7b, the composite curves have a closest point of approach of $\Delta T_{min} = 10$°F at the point where stream C2 begins along the cold composite curve; that is, 180°F. The corresponding temperature on the hot composite curve is 190°F. Consequently, these two points provide the temperatures at the pinch. If ΔT_{min} is reduced to zero, the cold composite curve is shifted to the left until it touches the hot composite curve. As mentioned earlier, this corresponds to an infinite area for heat exchange.

In this example, with $\Delta T_{min} = 10$°F, heat in the segments of the hot composite curve is transferred vertically to heat the segments of the cold composite curve that lie below them. At the high-temperature ends, however, no segments of the hot composite curve lie vertically above the upper end of the cold composite curve. There, an additional 50×10^4 Btu/hr must be supplied from a hot utility, such as steam. This is consistent with the results using the temperature-interval method in Example 10.2. Similarly, at the low-temperature ends of the composite curves, no segments of the cold composite curve lie vertically below the lower end of the hot composite curve. Here, an additional 60×10^4 Btu/hr must be removed using a cold utility, such as cooling water, a result again consistent with the TI analysis.

Figure 10.7b also includes a dashed, cold composite curve, shifted to the right to give a ΔT_{min} of 65°F. Corresponding minimum utilities increase to $Q_{cw} = 300 \times 10^4$ Btu/hr and $Q_{steam} = 290 \times 10^4$ Btu/hr. Although not shown in Figure 10.7b, if the cold composite curve is shifted further to the right so that ΔT_{min} is increased to 140°F, all heat would have to be transferred from steam and to cooling water. ∎

Many additional observations are noteworthy in connection with the hot and cold composite curves. One is that the slopes of the composite curves *always* decrease at the inlet temperature of a stream and increase at the outlet temperature of a stream. It follows that points at which the slope decreases are candidate pinch points, and furthermore, when a pinch temperature exists, one of the inlet temperatures is *always* a pinch temperature. Hence, to locate a potential pinch temperature, one needs only to examine the inlet temperatures of the streams.

Yet another observation is that for some ΔT_{min} there are no pinch temperatures. In such cases, either hot or cold utilities (not both) are required in an amount equal to the difference between the total energy to be removed from the hot streams and that to be added to the cold streams. The ΔT_{min} at which the pinch disappears is referred to as the *threshold* ΔT_{min}, as discussed in Section 10.5.

Linear Programming Method

A closer examination of the temperature-interval (TI) method shows that the minimum hot and cold utilities can be calculated by creating and solving a linear programming (LP) problem, as discussed in Section 18.4. This approach is illustrated in the example that follows.

EXAMPLE 10.4

It is desired to determine the minimum hot and cold utilities for a HEN involving the four streams in Example 10.1 by creating and solving a linear programming problem, using the energy balance for each interval in the cascade of Figure 10.4.

SOLUTION

The LP is formulated:

$$\text{Minimize} \quad Q_{\text{steam}} \quad \textbf{(LP)}$$

With respect to (w.r.t):

$$Q_{\text{steam}}$$

Subject to (s. t.):

$$Q_{\text{steam}} - R_1 + 30 = 0 \quad \textbf{(LP.1)}$$
$$R_1 - R_2 + 2.5 = 0 \quad \textbf{(LP.2)}$$
$$R_2 - R_3 - 82.5 = 0 \quad \textbf{(LP.3)}$$
$$R_3 - R_4 + 75 = 0 \quad \textbf{(LP.4)}$$
$$R_4 - Q_{\text{cw}} - 15 = 0 \quad \textbf{(LP.5)}$$
$$Q_{\text{steam}}, Q_{\text{cw}}, R_1, R_2, R_3, R_4 \geq 0 \quad \textbf{(LP.6)}$$

Note that only Q_{steam} is needed in the objective function because when Q_{steam} is at its minimum, Q_{cw} is also at its minimum. Furthermore, the equality constraints are the energy balances for each of the temperature intervals in Figure 10.4. These must be satisfied at the solution of the LP.

Using the General Algebraic Modeling System (GAMS), the following linear programming problem is defined. Note that the solution is equivalent to that obtained using the TI method, as expected. For an introduction to GAMS, see the file GAMS.pdf on the CD-ROM that accompanies this text.

GAMS Program

```
VARIABLES
    Qs, Qcw, R1, R2, R3, R4
    Z     min. util.;
POSITIVE VARIABLE Qs, Qcw, R1, R2, R3, R4;

EQUATIONS
    COST     define objective function
    T1, T2, T3, T4, T5;

    COST ..    Z = E = Qs;
    T1 .. Qs - R1 + 30    =E= 0;
    T2 .. R1 - R2 + 2.5   =E= 0;
    T3 .. R2 - R3 - 82.5  =E= 0;
    T4 .. R3 - R4 + 75    =E= 0;
    T5 .. R4 - Qcw - 15   =E= 0;

MODEL HEAT/ALL/;
SOLVE HEAT USING LP MINIMIZING Z;
DISPLAY R1.L, R2.L, R3.L, R4.L, Qs.L, Qcw.L;
```

GAMS Solution

```
****  SOLVER STATUS   1 NORMAL COMPLETION
****  MODEL STATUS    1 OPTIMAL
****  OBJECTIVE VALUE    50.0000

    VARIABLE R1        =   80.000
    VARIABLE R2        =   82.500
    VARIABLE R3        =    0.000
    VARIABLE R4        =   75.000
    VARIABLE QS        =   50.000
    VARIABLE QCW       =   60.000
```

Note that the residual across the pinch temperatures, $R_3 = R_p$, is zero, as must be the case when the utilities are minimized. These results can be reproduced using the GAMS input files, CASC.1 or CASC.2, on the CD-ROM that accompanies this book. ∎

Thus far, only sensible heat changes have been considered. Furthermore, the specific heat or heat capacity has been assumed constant over the range between the source and target temperatures so that the stream heat-capacity flow rates are constant. However, in many processes, latent heat of phase change, heat of reaction, and heat of mixing may also be involved under isothermal or nonisothermal conditions. In addition, the specific heat may not be constant or sensible heat may be combined with latent heat, heat of reaction, or heat of mixing, such as for multicomponent mixtures passing through condensers, vaporizers, reboilers, and nonadiabatic reactors and mixers. In such cases, a fictitious heat-capacity flow rate can be used based on the change in enthalpy flow rate due to all applicable effects divided by a temperature range. In general, a plot of stream enthalpy flow rate as a function of temperature is curved, but can be discretized into straight-line segments. Particular attention should be paid to the accuracy of the discretization in the vicinity of the pinch temperatures. Note that a conservative approach is recommended, in which the linear approximations provide a bound for the cold stream temperature–enthalpy curves from above, and for the hot stream curves from below. This conservative approach ensures that the true temperature approach is greater than that computed in terms of the linear approximations. The following example shows how linear piece-wise approximations are used to generate stream data for a design problem involving both the vaporization and condensation of process streams.

EXAMPLE 10.5   *MER Targeting for a Process Exhibiting Phase Changes and Variable Heat Capacities*

Figure 10.8 shows a process for the manufacture of toluene by the dehydrogenation of *n*-heptane. Note that a furnace, E-100, heats the feed stream of pure *n*-heptane, S1, at 65°F, to the reactor feed, S2, at 800°F. Furthermore, the reactor effluent, S3, containing a multicomponent mixture of *n*-heptane, hydrogen, and toluene at 800°F is cooled to 65°F and fed to a separator as stream S4. The pressure is 1 atm throughout the process. It is planned to install a heat exchanger to heat the feed stream, S1, using the hot reactor effluent, S3, and thus reduce the required duty of the preheater, E-100. (a) Generate stream data using piece-wise linear approximations for the heating and cooling curves for the reactor feed and effluent streams. (b) Using the stream data, compute the MER targets for $\Delta T_{\min} = 10°F$.

SOLUTION

(a) Generation of stream data

 HYSYS.Plant is used to simulate the process in Figure 10.8, as shown in the multimedia CD-ROM accompanying this book (see *HYSYS* → *Tutorials* → *Heat Transfer* → *Toluene Manufacture*). The Peng–Robinson equation of state is used to estimate thermodynamic properties. Sensitivity analyses are performed, in which enthalpies for S2 and S4 are computed as a function of temperature, giving the temperature–enthalpy diagrams in Figures 10.9 and 10.10. In Figure 10.9, which is for pure *n*-heptane feed, only liquid sensible heat is involved from 65°F to 209°F, where isothermal vaporization occurs, as represented by the horizontal line. From there until 800°F, only vapor sensible heat is involved. In Figure 10.10, which is for the ternary reactor effluent, only vapor sensible heat is involved from 800°F to 183°F, which is the dew point. Then, condensation occurs, involving both latent heat and sensible heat, to the target temperature of 65°F. Shown in the diagrams are the piece-wise linear approximations, defined by critical coordinates on the original heating and cooling curves. The piece-wise linear approximations are defined in terms of temperature–enthalpy coordinates, (h_k, T_k) and (h_{k+1}, T_{k+1}) through which linear arcs are drawn to approximate the true heating or cooling curves. Each arc represents a new stream, with the source and target temperatures being the abscissa coordinates, T_k and T_{k+1}, and the heat-capacity flow rate is

$$C_k = \frac{h_{k+1} - h_k}{T_{k+1} - T_k} \qquad (10.6)$$

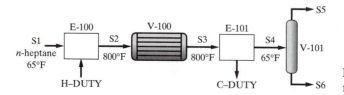

Figure 10.8 Process flow diagram for dehydrogenation of *n*-heptane.

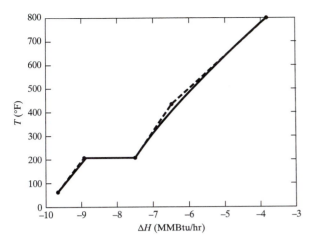

Figure 10.9 Temperature–enthalpy diagram for the cold stream, S2, showing simulation results (solid line) and piece-wise linear approximation (dashed line).

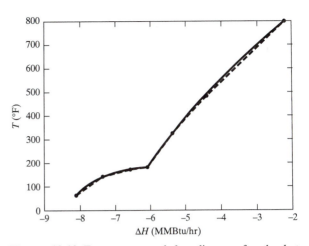

Figure 10.10 Temperature–enthalpy diagram for the hot stream, S4, showing simulation results (solid line) and piece-wise linear approximation (dashed line).

where h is the enthalpy flow rate and MM stands for million. This is the inverse of the slope of each linear segment in Figures 10.9 and 10.10. Reasonably accurate linear approximations are obtained using four segments for stream S2 and six for stream S4, whose coordinates are positioned to ensure accuracy in the vicinity of the pinch temperatures. The temperature coordinates are determined to the nearest degree. Thus, in Figure 10.9, the horizontal line for the vaporization of n-heptane at 209°F is taken to occur over a 1°F interval from 209 to 210°F, giving a fictitious heat-capacity flow rate, C, of 1.4282 MMBtu/hr/1°F = 1.4282 MMBtu/hr-°F.

Having determined the coordinate positions, the stream data for the four cold and six hot streams are computed directly, where the heat-capacity flow rate for the kth stream is given by Eq. (10.6), as shown in Table 10.3. Note the increased values of the heat-capacity flow rate in the region where streams exhibit phase change, and in particular, the large value for the cold stream (pure n-heptane), where vaporization is assumed to occur over 1°F.

Table 10.3   Stream Data for Example 10.5

(a) Cold Streams				
Stream	T^s (°F)	T^t (°F)	Duty (MMBtu/hr)	C (MMBtu/hr°F)
S2A	65	209	0.7446	0.5171×10^{-2}
S2B	209	210	1.4282	1.4282
S2C	210	435	1.0492	0.4663×10^{-2}
S2D	435	800	2.6446	0.7245×10^{-2}

(b) Hot Streams				
Stream	T^s (°F)	T^t (°F)	Duty (MMBtu/hr)	C (MMBtu/hr°F)
S4A	800	485	2.1745	0.6903×10^{-2}
S4B	485	326	0.9823	0.6178×10^{-2}
S4C	326	183	0.7304	0.5108×10^{-2}
S4D	183	172	0.4863	4.421×10^{-2}
S4E	172	143	0.7780	2.683×10^{-2}
S4F	143	65	0.7652	0.9810×10^{-2}

Table 10.4   Computation of MER Targets Using the TI Method

Interval	(°F) T_i	(°F) $T_{i-1} - T_i$	(MMBtu/hr) ΔH_i	(MMBtu/hr) Q ($Q_{H,min} = 0$)	(MMBtu/hr) Q ($Q_{H,min} = 1.4208$)	
T_0	800			0	1.4208	
T_1	790	10	−0.0725	−0.0725	1.3483	
T_2	475	315	−0.1077	−0.1802	1.2406	
T_3	435	40	−0.0427	−0.2229	1.1979	
T_4	316	119	0.1803	−0.0426	1.3782	
T_5	210	106	0.0472	0.0046	1.4254	
T_6	209	1	−1.4231	−1.4185	0.0023	
T_7	173	36	−0.0023	−1.4208	0.0000	← Pinch
T_8	162	11	0.4294	−0.9913	0.4294	
T_9	133	29	0.6281	−0.3633	1.0575	
T_{10}	65	68	0.3155	−0.0478	1.3729	
T_{11}	55	10	0.0981	0.0502	1.4710	

(b) Computing MER targets

The TI method is applied using the data in Table 10.3, with the hot temperatures reduced by ΔT_{min}. The results, summarized in Table 10.4, indicate that the cold pinch temperature is 173°F, giving the hot and cold utility targets of $Q_{H,min} = 1.421$ MMBtu/hr and $Q_{H,min} = 1.471$ MMBtu/hr. The location of the cold pinch temperature in the HYSYS.Plant simulation of the heat-integrated process designed for $\Delta T_{min} = 10°F$ is 172°F, with the heat duties of the preheater and cooler being 1.396 MMBtu/hr and 1.446 MMBtu/hr, respectively. These differences are the result of the linear approximations for the heating and cooling curves.   ∎

10.3   NETWORKS FOR MAXIMUM ENERGY RECOVERY

Having determined the minimum utilities for heating and cooling, it is common to design two networks of heat exchangers, one on the hot side and one on the cold side of the pinch, as shown in Figure 10.5. In this section, two methods are presented for this purpose. The first, introduced by Linnhoff and Hindmarsh (1983), places emphasis on positioning the heat exchangers by working out from the pinch. The second is an algorithmic strategy that utilizes a mixed-integer linear program (MILP), which was introduced by Papoulias and Grossmann (1983b) and is solved with GAMS.

Stream Matching at the Pinch

To explain the approach of Linnhoff and Hindmarsh (1983), it helps to refer to a diagram showing the *pinch decomposition* of the hot and cold streams, as shown in Figure 10.5 for the four streams in Example 10.1. Attention is focused at the pinch where the temperatures of the hot and cold streams are separated by ΔT_{min}. This, of course, is the location of the closest temperature approach.

Consider the schematic of a countercurrent heat exchanger in Figure 10.11. The hot stream, having a heat-capacity flow rate of C_h, enters at T_{h_i} and exits at T_{h_o}. It transfers heat,

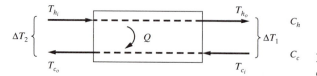

Figure 10.11 Schematic of a countercurrent heat exchanger.

Q, to the cold stream that has a heat-capacity flow rate of C_c, entering at T_{c_i} and exiting at T_{c_0}. On the cold end of the heat exchanger, where the temperatures of the hot and cold streams are the lowest, the approach temperature difference is ΔT_1. On the hot end, where the temperatures are the highest, the approach temperature difference is ΔT_2. Carrying out energy balances for the hot and cold streams:

$$Q = C_h(T_{h_i} - T_{h_0}) \quad \text{or} \quad T_{h_i} - T_{h_0} = \frac{Q}{C_h} \tag{10.7}$$

$$Q = C_c(T_{c_0} - T_{c_i}) \quad \text{or} \quad T_{c_0} - T_{c_i} = \frac{Q}{C_c} \tag{10.8}$$

and subtracting Eq. (10.8) from Eq. (10.7):

$$(T_{h_i} - T_{c_0}) - (T_{h_0} - T_{c_i}) = Q\left(\frac{1}{C_h} - \frac{1}{C_c}\right) \tag{10.9}$$

or:

$$\Delta T_2 - \Delta T_1 = \frac{Q(C_c - C_h)}{C_h C_c} \tag{10.10}$$

Following the approach introduced by Linnhoff and Hindmarsh (1983), the potential locations for the heat exchangers at the pinch are considered next. When a heat exchanger is positioned on the hot side of the pinch, which is considered first arbitrarily, $\Delta T_1 = \Delta T_{\min}$, and Eq. (10.10) becomes:

$$\Delta T_2 = \Delta T_{\min} + \frac{Q(C_c - C_h)}{C_h C_c} \tag{10.11}$$

Then, to assure that $\Delta T_2 \geq \Delta T_{\min}$, since $Q > 0$ and the heat-capacity flow rates are positive, it follows that $C_c \geq C_h$ is a necessary and sufficient condition. That is, for a match to be *feasible* at the pinch, on the hot side, $C_c \geq C_h$ must be satisfied. If two streams are matched at the pinch with $C_c < C_h$, the heat exchanger is *infeasible* because $\Delta T_2 < \Delta T_{\min}$.

When a heat exchanger is positioned on the cold side of the pinch, $\Delta T_2 = \Delta T_{\min}$, and Eq. (10.10) becomes:

$$\Delta T_1 = \Delta T_{\min} - \frac{Q(C_c - C_h)}{C_h C_c} \tag{10.12}$$

In this case, to assure that there are no approach temperature violations (i.e., $\Delta T_1 \geq \Delta T_{\min}$), it is necessary and sufficient that $C_h \geq C_c$. Note that this condition is just the reverse of that on the hot side of the pinch. These stream-matching rules are now applied to design a HEN for Example 10.1.

EXAMPLE 10.6   *(Example 10.1 Revisited)*

Design a HEN to meet the MER targets for Example 10.1: $Q_{\text{H,min}} = 50 \times 10^4$ Btu/hr and $Q_{\text{C,min}} = 60 \times 10^4$ Btu/hr, where $Q_{\text{H,min}}$ and $Q_{\text{C,min}}$ are the minimum hot and cold utility loads.

SOLUTION

As stated previously, when designing a HEN to meet MER targets, no heat is transferred across the pinch, and hence, two HENs are designed, one on the hot side and one on the cold side of the pinch, as shown in Figure 10.12. Arbitrarily, the HEN on the hot side of the pinch is designed first. At the pinch, an appropriate match with stream H1 is sought. Since $C_{\text{H1}} = 3 \times 10^4$ Btu/(hr°F), stream C2 must be selected, with $C_{\text{C2}} = 4 \times 10^4$ Btu/(hr°F), to ensure that $C_c \geq C_h$. Note that if C1 were selected, $C_{\text{C1}} < C_{\text{H1}}$, and an approach temperature violation occurs, $\Delta T_2 < \Delta T_{\min}$. Consequently, interior heat exchanger 1 is installed, with a heat duty equal to 210×10^4 Btu/hr, the entire cooling requirement of stream H1 on the hot side of the pinch. Similarly, because $C_{\text{C1}} \geq C_{\text{H2}}$, streams H2 and C1 are matched

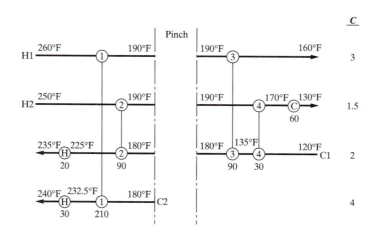

Figure 10.12 Interior heat exchangers (1–4) and auxiliary heat exchangers (H, C). Multiply heat duties by 10^4 Btu/hr and heat-capacity flow rates by 10^4 Btu/(hr°F).

on the hot side of the pinch, using interior heat exchanger 2, with a heat duty equal to 90×10^4 Btu/hr, the entire cooling requirement of H2 on the hot side of the pinch. Since these two heat exchangers bring streams C1 and C2 to 225 and 232.5°F, respectively, utility heaters (labeled 'H') are added, to complete the design on the hot side of the pinch, with a total duty of 50×10^4 Btu/hr, which matches the MER heating target. Note that each unit exchanges heat between two process streams in countercurrent flow, with the inlet and outlet temperatures for each stream shown on either side of circles, identified by the heat exchanger number, connected by a vertical line to represent the match, and the heat duty annotated below the circle associated with the cold stream.

On the cold side of the pinch, only streams H1 and C1 can be matched, since $C_{H1} \geq C_{C1}$. Note that $C_{H2} < C_{C1}$, and consequently, if streams H2 and C1 are matched, an approach temperature violation occurs, $\Delta T_1 < \Delta T_{\min}$. Thus, interior heat exchanger 3 is installed, with a heat duty equal to 90×10^4 Btu/hr, the entire cooling requirement of stream H1 on the cold side of the pinch. By energy balance, the temperature of stream C1 entering heat exchanger 3 is 135°F. This allows an additional internal heat exchanger to be positioned to pair streams H2 and C1, noting that the pairing rule $C_h \geq C_c$ applies only *at the pinch*. Heat exchanger 4 is installed with a heat duty equal to 30×10^4 Btu/hr, the remaining heating requirement of stream C1 on the cold side of the pinch. The HEN on the cold side of the pinch is completed by installing a cooler on stream H2 (labeled 'C') with a heat duty of 60×10^4 Btu/hr, which matches the MER cooling target. The final design, shown in Figure 10.12, meets the MER targets with a total number of seven heat exchangers. These are displayed in the flowsheet in Figure 10.13, which should be compared with the HEN in Figure 10.3. Note that the former meets the MER energy targets, while both the cold and hot utility targets are exceeded by 7.5×10^4 Btu/hr in the latter. In contrast, the latter involves only six units, compared to seven utilized in the MER design. Cost

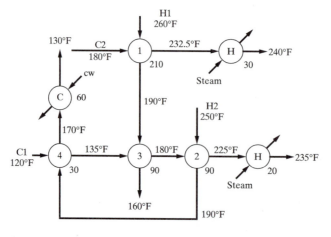

Figure 10.13 Flowsheet for HEN in Figure 10.12.

estimates are needed to select between these and other alternatives. As will be seen in later examples, the trade-off between capital and operating costs is at the heart of HEN synthesis. ∎

In summary, the HEN design procedure to meet MER targets consists of the following steps:

1. *MER Targeting:* The pinch temperatures are determined, together with minimum hot and cold utility targets, $Q_{H,min}$ and $Q_{C,min}$, respectively. Either the temperature-interval or composite curve methods are used, or a linear programming problem is formulated and solved.
2. The synthesis problem is decomposed at the pinch, yielding two independent HENs to be designed, using the representation shown in Figure 10.5. It is helpful to place the heat-capacity flow rates for each stream in a column to the right, for reference.
3. The HEN is designed on the hot side of the pinch, *starting at the pinch*, and working outwards. *At the pinch*, streams are paired such that $C_c \geq C_h$. In general, the heat duty of each interior heat exchanger is selected to be as large as possible, to reduce the total number of exchangers. In some cases (e.g., Example 10.7), duties are selected to retain sufficient temperature driving forces for additional matches. Finally, hot utilities are added to meet cold temperature targets (up to a total of $Q_{H,min}$). *Cold utilities are not used on the hot side of the pinch.*
4. The HEN is designed on the cold side of the pinch, *starting at the pinch*, and working outwards. *At the pinch*, streams are paired such that $C_h \geq C_c$. In general, the duty of each interior heat exchanger is selected to be as large as possible, to reduce the total number of exchangers. As mentioned above, it may be necessary to select duties to retain sufficient temperature driving forces for additional matches. Finally, cold utilities are added to meet cold temperature targets (up to a total of $Q_{C,min}$). *Hot utilities are not used on the cold side of the pinch.*

In this synthesis procedure, the designer positions the first heat exchangers *at the pinch*, where the approach temperature difference at one end of each heat exchanger is constrained at ΔT_{min}. Then, working outwards, the utility exchangers are positioned last. In cases that do not require stream splitting, to be considered in Section 10.4, this simple procedure is sufficient to guarantee compliance with the MER targets. However, it often leads to designs with a large number of heat exchangers, as illustrated in the following example.

EXAMPLE 10.7

Consider the design of a HEN for the four streams below in a problem presented by Linnhoff and Flower (1978a, b)

Stream	T^s (°C)	T^t (°C)	C (kW/°C)	Q (kW)
C1	60	180	3	360
C2	30	130	2.6	260
H1	180	40	2	280
H2	150	40	4	440

Let $\Delta T_{min} = 10°C$, with the following specifications:

Cooling water (cw): $T^s = 30°C$, $T^t \leq 80°C$, cost of cw = 0.00015 \$/kg
Steam (sat'd., s): $T = 258°C$, $\Delta H^v = 1,676$ kJ/kg, cost of s = 0.006 \$/kg

Overall heat-transfer coefficients:

$$U_{heater} = 1 \text{ kW/m}^2 \text{ °C}, U_{cooler} = U_{exch} = 0.75 \text{ kW/m}^2 \text{ °C}$$

Purchase cost of heat exchangers:

$$C_P = 3,000A^{0.5} \quad (\$, m^2)$$

Equipment operability = 8,500 hr/yr, return on investment = $i_m = 0.1$ [for Eq.(10.3)].

SOLUTION

First, the MER targets are computed using the TI method, as summarized in the cascade diagram of Figure 10.14, where the pinch temperatures are 140 and 150°C, and the minimum hot and cold utility duties are 60 kW and 160 kW, respectively.

Next, the MER synthesis procedure is used to design the HEN. Since only streams H1 and C1 appear on the hot side of the pinch, and $C_{H1} < C_{C1}$, a heat exchanger is installed between streams H1 and C1, as shown in Figure 10.15a, with a heat duty equal to 60 kW, the cooling demand of H1 on the hot side of the pinch. Finally, a 60-kW heater is installed to complete heating stream C1.

On the cold side, all four streams are present, but only streams H1, H2, and C1 exist at the pinch, since the target temperature of stream C2 is 130°C, below 140°C, the pinch temperature of the cold streams. Only the H2–C1 match is feasible since $C_{H2} > C_{C1}$, while $C_{H1} < C_{C1}$. Following the guidelines, one would be tempted to install an internal heat exchanger with a duty of 240 kW, as shown in Figure 10.15b, which would satisfy the entire energy requirement of stream C1 on the cold side of the pinch. Furthermore, the H1–C2 match is possible, even though $C_{H1} < C_{C2}$, because stream C2 doesn't reach the pinch. Thus, a heat exchanger can be installed between these streams, but since $\Delta T_1 < \Delta T_2$ and $\Delta T_1 \geq \Delta T_{min} = 10°C$, only 86.58 kW can be transferred, as shown in Figure 10.15b. Unfortunately, this requires that a portion of stream C1 be heated with hot utility (173.42 kW). The addition of heat from a hot utility on the cold side of the pinch requires that an equivalent amount of heat be removed from the hot streams using cooling water, thereby exceeding the minimum cold utility (160 kW).

Alternatively, a more careful design is performed in which heat exchangers are added with duties assigned such that sufficient temperature differences are retained for additional matches, as shown in Figure 10.15c. The first match, H2–C1, is assigned a heat duty of only 40 kW, so that the H2 effluent temperature is reduced to only 140°C. This allows it to be used to heat stream C2, with a heat duty of 120 kW, which reduces the effluent temperature of H2 to 110°C, allowing its subsequent use to heat stream C1, and so on. As seen in Figure 10.15c, this rather complicated design, involving six heat exchangers, meets the MER cooling target of 160 kW.

When combined with the network in Figure 10.15a, the resulting HEN involves six interior heat exchangers, an auxiliary heater, and an auxiliary cooler, as shown in Figure 10.15d. Note that this network utilizes the minimum hot and cold utilities, but involves eight heat exchangers, three above the minimum, as discussed in Section 10.4. The next step is to consider the possibilities for reducing the number of heat exchangers by removing the *heat loops*, or to introduce stream splitting, as demonstrated later in Section 10.4. ∎

Mixed-Integer Linear Programming

The *transshipment model* for heat transfer, introduced by Papoulias and Grossmann (1983b), provides a more systematic method for stream matching. In this model, the hot streams and hot utilities are viewed as the sources of energy that distribute among the temperature intervals, and that in turn become a source of energy for the sinks (or destinations), that is, the cold streams and cold utilities. The temperature levels of the sources and sinks establish the temperature intervals within which the sources can provide energy, and similarly, within which the sinks can receive energy. Energy can also be carried by the hot streams, as *residuals*, to the adjacent intervals at a lower temperature. Such a transshipment model is illustrated schematically in Figure 10.16a, in which hot streams, H1 and H2, potentially exchange energy with cold stream, C1, and cooling water, W, in interval k. As shown, energy from the hot streams in interval k, $Q_{H1,k}^H$ and $Q_{H2,k}^H$, combines with residual energy in the hot streams from interval $k - 1$, $R_{H1,k-1}$ and $R_{H2,k-1}$, to be transferred to cold stream C1 or to cooling water, W, or rejected to the next interval as residual energy, $R_{H1,k}$ and $R_{H2,k}$. Stated differently, Figure 10.16a shows a superstructure in which all possible exchanges of heat between the hot streams, cold streams, and cold utility are included. Using this superstructure to design a network with a low capital cost having the minimum utilities, Papoulias and Grossmann created a MILP that minimizes the number of matches (i.e., that minimizes the number of heat exchangers). For this MILP, Figure 10.16b defines the nomenclature, where Q_{ijk} is the heat exchanged between hot stream i and cold stream j in temperature interval k, Q_{ik}^H is the heat available from hot stream i in interval k, $R_{i,k-1}$ is the residual heat in hot stream i that is transferred from the adjacent hotter interval $k - 1$, $R_{i,k}$ is the residual heat in hot stream i that

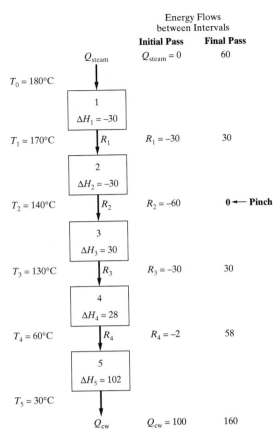

Energy Flows
between Intervals

Initial Pass	Final Pass

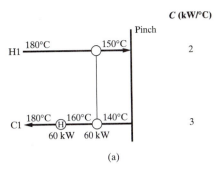

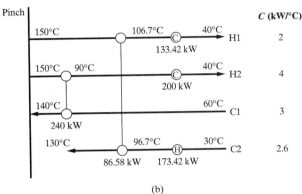

Figure 10.14 Cascade diagram for Example 10.7, showing temperature intervals, heat balances, and residuals, ΔH_i and R_i in kilowatts.

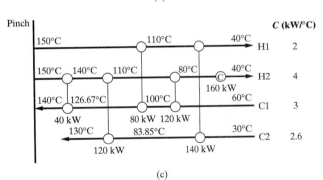

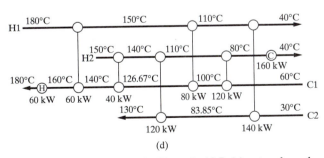

Figure 10.15 MER design for Example 10.7: (a) network on the hot side of the pinch; (b) network on the cold side of the pinch—additional utilities required; (c) network on the cold side of the pinch—minimum utilities; (d) combined network involving eight heat exchangers.

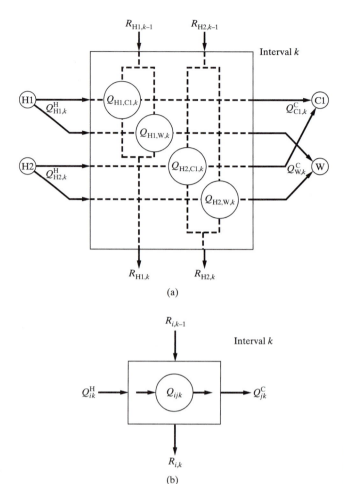

Figure 10.16 Transshipment model for stream matching: (a) super-structure; (b) nomenclature.

is transferred to the adjacent colder interval $k + 1$, and Q_{ik}^C is the heat that must be transferred to cold stream j in interval k. Then, using this nomenclature, the MILP takes the form:

$$\text{Minimize } z = \sum_i \sum_j w_{ij} y_{ij}$$
$$\text{w.r.t}$$

$$Q_{ijk}, \, y_{ij}$$

s.t.

$$R_{ik} - R_{i,k-1} + \sum_{j \in C_k} Q_{ijk} = Q_{ik}^H \quad i \in H_k \quad k = 1, \ldots, K \quad \textbf{(MILP.1)}$$

$$\sum_{i \in H_k} Q_{ijk} = Q_{ik}^C \quad j \in C_k \quad k = 1, \ldots, K \quad \textbf{(MILP.2)}$$

$$\sum_k Q_{ijk} - y_{ij} U_{ij} \leq 0 \quad i \in H \quad j \in C \quad \textbf{(MILP.3)}$$

$$R_{ik} \geq 0 \quad Q_{ijk} \geq 0 \quad y_{ij} \in 0, 1 \quad \textbf{(MILP.4)}$$

$$R_{i0} = R_{iK} = 0 \quad \textbf{(MILP.5)}$$

Here, y_{ij} is a binary variable that equals unity when a match exists between hot stream i and cold stream j, and is zero otherwise. The objective is to minimize the number of matches, and hence, the objective function sums over all of the possible matches, with the weighting coefficient, w_{ij}, increased as certain matches become less desirable. Constraints (MILP.1) and (MILP.2) are the energy balances for each of the K temperature intervals where i, the hot stream index, belongs to the set of hot streams in interval k, H_k, and j, the cold stream index, belongs to the set of cold streams in interval k, C_k.

Constraints (MILP.3) place bounds on the heat to be transferred when hot stream i and cold stream j are matched. Of course, when $y_{ij} = 0$, there is no match. However, when $y_{ij} = 1$, there is an upper bound on the heat that can be transferred between the two streams in all of the temperature intervals ($\Sigma\ Q_{ijk}$). This upper bound, U_{ij}, is the minimum of the heat that can be released by hot stream i [$Q_i = C_i(T_i^s - T_i^t)$] and that which can be taken up by cold stream j [$Q_j = C_j(T_j^t - T_j^s)$]; that is, $U_{ij} = \min \{Q_i, Q_j\}$. Note that H and C are the sets of the hot and cold streams, respectively.

Constraints (MILP.4) assure that all of the residuals and rates of heat transfer are greater than or equal to zero and define the binary variables. Finally, constraints (MILP.5) indicate that the residuals at the lower and upper bounds of the temperature intervals are zero, and hence, all streams that exchange energy must have their temperatures within the bounds of the temperature intervals, including the hot and cold utilities. This is one of the principal departures from the temperature intervals in the previous analysis (see Examples 10.2 and 10.6), where the hot and cold utilities are not included in the temperature intervals.

The next example is provided to illustrate the creation and solution of a typical MILP for MER design.

EXAMPLE 10.8

In this example, taken from Linnhoff and Flower (1978a, b), a network of heat exchangers is to be synthesized for two hot and two cold streams, with steam and cooling water as the utilities, as shown below. Note that the minimum utilities have been determined for $\Delta T_{\min} = 10°C$, using one of the methods in Section 10.2, and that the temperatures and heat duties of the utilities are given as well:

Stream	T^s (°C)	T^t (°C)	C(kW/°C)	Q(kW)
C1	60	160	7.62	762
C2	116	260	6.08	875.52
H1	160	93	8.79	588.93
H2	249	138	10.55	1171.05
S	270	270	—	127.68
W	38	82	5.685	250.14

The pinch temperatures are 249°C and 239°C. Note also that demineralized water would be required to achieve a target temperature of 82°C without scaling the heat exchanger tubes.

SOLUTION

First, the temperature intervals are identified, somewhat differently than when implementing the TI method to determine the minimum utilities, as shown in Figure 10.17. In this case, the *inlet* (or source) temperatures denote the bounds on the temperature intervals. These are circled in the figure. At each interval boundary, both the hot and cold stream temperatures, on the left and right, respectively, are shown, separated by $\Delta T_{\min} = 10°C$. The steam temperature is selected to be 10°C higher than the highest temperature of the cold streams, 260°C. On the left side of the temperature intervals, the hot streams are shown, so that it is clear within which intervals each hot stream appears. Similarly, on the right side, the cold streams are shown. Note that the target temperatures of the hot and cold streams could be used to define additional temperature intervals, but these do not affect the results of the MILP.

Using Figure 10.17, the streams in each interval are

Interval	Stream
1	S, C2
2	H2, C1, C2
3	H1, H2, C1, C2
4	H1, H2*, C1, W
5	H1*, H2*, W

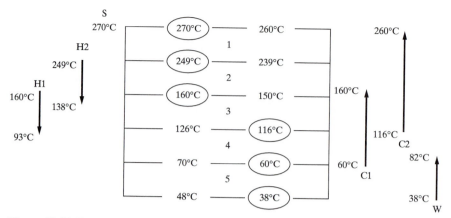

Figure 10.17 Temperature intervals for stream matching.

where the asterisks denote that the hot streams can exchange heat in these intervals through their residuals. This is the basis for the construction of the MILP that follows, for the case where all of the weighting coefficients are unity.

Minimize $\quad z = y_{S,C2} + y_{H1,C1} + y_{H1,C2} + y_{H1,W} + y_{H2,C1} + y_{H2,C2} + y_{H2,W}$
w. r. t.
$\quad y_{ij}$

s. t.

S:	$R_{S,1} + Q_{S,C2,1} = Q_{S,1}^H$
H2:	$R_{H2,2} + Q_{H2,C1,2} + Q_{H2,C2,2} = Q_{H2,2}^H$
	$R_{H2,3} - R_{H2,2} + Q_{H2,C1,3} + Q_{H2,C2,3} = Q_{H2,3}^H$
	$R_{H2,4} - R_{H2,3} + Q_{H2,C1,4} + Q_{H2,W,4} = Q_{H2,4}^H$
	$-R_{H2,4} + Q_{H2,W,5} = Q_{H2,5}^H$
H1:	$R_{H1,3} + Q_{H1,C1,3} + Q_{H1,C2,3} = Q_{H1,3}^H$
	$R_{H1,4} - R_{H1,3} + Q_{H1,C1,4} + Q_{H1,W,4} = Q_{H1,4}^H$
	$- R_{H1,4} + Q_{H1,W,5} = Q_{H1,5}^H$
C2:	$Q_{S,C2,1} = Q_{C2,1}^C$
	$Q_{H2,C2,2} = Q_{C2,2}^C$
	$Q_{H1,C2,3} + Q_{H2,C2,3} = Q_{C2,3}^C$
C1:	$Q_{H2,C1,2} = Q_{C1,2}^C$
	$Q_{H1,C1,3} + Q_{H2,C1,3} = Q_{C1,3}^C$
	$Q_{H1,C1,4} + Q_{H2,C1,4} = Q_{C1,4}^C$
W:	$Q_{H1,W,4} + Q_{H2,W,4} = Q_{W,4}^C$
	$Q_{H1,W,5} + Q_{H2,W,5} = Q_{W,5}^C$
S–C2:	$Q_{S,C2,1} - y_{S,C2}U_{S,C2} \leq 0$
H1–C1:	$Q_{H1,C1,3} + Q_{H1,C1,4} - y_{H1,C1}U_{H1,C1} \leq 0$
H1–C2:	$Q_{H1,C2,3} - y_{H1,C2}U_{H1,C2} \leq 0$
H1–W:	$Q_{H1,W,4} + Q_{H1,W,5} - y_{H1,W}U_{H1,W} \leq 0$
H2–C1:	$Q_{H2,C1,2} + Q_{H2,C1,3} + Q_{H2,C1,4} - y_{H2,C1}U_{H2,C1} \leq 0$
H2–C2:	$Q_{H2,C2,2} + Q_{H2,C2,3} - y_{H2,C2}U_{H2,C2} \leq 0$
H2–W:	$Q_{H2,W,4} + Q_{H2,W,5} - y_{H2,W}U_{H2,W} \leq 0$

From the energy balances for the streams in each interval:

$Q_{S,1}^H = 127.68$ kW	$Q_{C2,1}^C = 127.68$ kW
$Q_{H1,3}^H = 298.86$ kW	$Q_{C2,2}^C = 541.12$ kW
$Q_{H1,4}^H = 290.07$ kW	$Q_{C2,3}^C = 206.72$ kW
$Q_{H1,5}^H = 0$	$Q_{C1,2}^C = 76.2$ kW
$Q_{H2,2}^H = 938.95$ kW	$Q_{C1,3}^C = 259.08$ kW
$Q_{H2,3}^H = 232.1$ kW	$Q_{C1,4}^C = 426.72$ kW
$Q_{H2,4}^H = 0$	$Q_{W,4}^C = 125.07$ kW
$Q_{H2,5}^H = 0$	$Q_{W,5}^C = 125.07$ kW

Furthermore, the upper bounds, U_{ij}, for the potential matches are

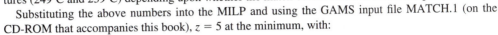

$$
\begin{array}{lll}
\text{S–C2:} & U_{S,C2} & = \min[127.68, 127.68] = 127.68 \\
\text{H1–C1:} & U_{H1,C1} & = \min[762, 588.93] = 588.93 \\
\text{HI–C2:} & U_{H1,C2} & = \min[875.52, 588.93] = 588.93 \\
\text{H1–W:} & U_{H1,W} & = \min[250.14, 588.93] = 250.14 \\
\text{H2–C1:} & U_{H2,C1} & = \min[1{,}171.05, 762] = 762 \\
\text{H2–C2:} & U_{H2,C2} & = \min[1{,}171.05, 747.84] = 747.84 \\
\text{H2–W:} & U_{H2,W} & = \min[1{,}171.05, 250.14] = 250.14
\end{array}
$$

When determining U_{ij}, to maintain the minimum utilities, heat cannot be exchanged across the pinch. Hence, the heat loads for each of the streams are for the segments above or below the pinch temperatures (249°C and 239°C) depending upon whether the match occurs above or below the pinch.

Substituting the above numbers into the MILP and using the GAMS input file MATCH.1 (on the CD-ROM that accompanies this book), $z = 5$ at the minimum, with:

$$ y_{S,C2} = y_{H1,C1} = y_{H1,W} = y_{H2,C1} = y_{H2,C2} = 1 $$

and:

$$
\begin{array}{lll}
\text{S–C2:} & Q_{S,C2,1} = 127.68 & \\
\text{H1–C1:} & Q_{H1,C1,3} = 259.08 & Q_{H1,C1,4} = 79.71 \\
\text{H1–W:} & Q_{H1,W,4} = 125.07 & Q_{H1,W,5} = 125.07 \\
\text{H2–C1:} & Q_{H2,C1,2} = 76.2 & Q_{H2,C1,3} = 0 \quad Q_{H2,C1,4} = 347.01 \\
\text{H2–C2:} & Q_{H2,C2,2} = 541.12 & Q_{H2,C2,3} = 206.72
\end{array}
$$

with:

$$
\begin{array}{llll}
\text{H1:} & R_{H1,3} = 39.78 & R_{H1,4} = 125.07 & \\
\text{H2:} & R_{H2,2} = 321.63 & R_{H2,3} = 347.01 & R_{H2,4} = 0
\end{array}
$$

As can be seen, these results satisfy all of the equality and inequality constraints. There are five matches ($z = 5$), but these translate into six heat exchangers, as shown in Figure 10.18. Streams H2 and C2 exchange heat in two adjacent intervals, 2 and 3, and hence, heat exchanger 1 is sufficient for this purpose. On the other hand, streams H2 and C1 exchange heat in intervals 2 and 4, with stream C1 exchanging heat with stream H1 in interval 3. Hence, two heat exchangers, 2 and 4, are installed, separated by heat exchanger 3.

In summary, six heat exchangers is the minimum for this network when it is required that the hot and cold utilities be minimized as well. As discussed in Section 10.4, the minimum number of heat exchangers for this system is five, which can be achieved either by *breaking heat loops*, usually at the price of exceeding the MER targets, or by *stream splitting*. ■

To complete this section, it is important to note that only an introduction to the application of MILP models for stream matching has been presented. A presentation that is far more

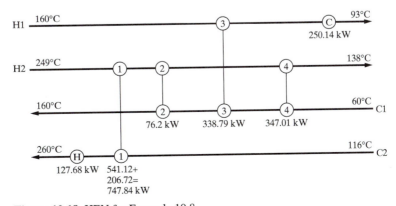

Figure 10.18 HEN for Example 10.8.

complete is provided by Floudas (1995) in *Nonlinear and Mixed-integer Optimization: Fundamentals and Applications*. The latter discusses the theory behind the MILP formulations and shows more advanced techniques for excluding specific matches, matching with multiple hot and cold utilities, and allowing matches between two hot or two cold streams, when this is desirable.

10.4 MINIMUM NUMBER OF HEAT EXCHANGERS

Having designed a HEN that meets the MER targets, it is common to reduce the number of heat exchangers toward the minimum while permitting the consumption of utilities to rise, particularly when small heat exchangers can be eliminated. In this way, lower annualized costs may be obtained, especially when the cost of fuel is low relative to the purchase cost of the heat exchangers. Alternatively, an attempt can be made to design the HEN to minimize the number of heat exchangers *and satisfy the MER target*, by invoking stream splitting.

Reducing the Number of Heat Exchangers—Breaking Heat Loops

Before proceeding, it is important to recognize that, as pointed out by Hohmann (1971), the minimum number of heat exchangers in a HEN is

$$N_{HX,min} = N_S + N_U - 1 \tag{10.13}$$

where N_S is the number of streams and N_U is the total number of distinct hot and cold utility sources. Thus, for hot utilities, fuel, steam at high pressure (HPS), intermediate pressure (IPS), and low pressure (LPS), and for cold utilities, boiler feed water (BFW), cooling water (CW), and refrigeration, each count as distinct utility sources. Hence, for the networks in Examples 10.7 and 10.8, which involve four streams and two utilities (steam and cooling water), $N_{HX,min} = 5$. Equation (10.13) indicates that the minimum number of heat exchangers, and with it the minimum capital cost of the HEN, increases with each distinct utility source utilized in a design. However, as shown in Section 10.8, the operating costs of a HEN are reduced by replacing a portion of the utility duty requirement provided by a costly utility (e.g., refrigeration) by one at a lower cost (e.g., cooling water).

In a more general result, Douglas (1988) shows that the minimum number of heat exchangers is also dependent on the number of *independent networks*, N_{NW}; that is, the number of sub-networks consisting of linked paths between the connected streams:

$$N_{HX,min} = N_S + N_U - N_{NW} \tag{10.14}$$

When all streams in a process are connected directly or indirectly by heat exchangers, as in Figure 10.19a, $N_{NW} = 1$, and Eq. (10.14) reverts to Eq. (10.13), with $N_{HX,min} = 4$, noting that the HEN in Figure 10.19a has seven heat exchangers, three in excess of the minimum possible. In contrast, Figure 10.19b illustrates a network of five streams and two utilities, comprised of two independent subsystems: one connecting streams H1, C1 and the cold utility, and the second connecting streams H2, C2, C3 and the hot utility. In this case, $N_{HX,min} = N_S + N_U - N_{NW} = 5 + 2 - 2 = 5$, which is the number of heat exchangers in the HEN shown.

In one of the first studies of the methods for heat integration, Hohmann (1971) showed that in a HEN with N_{HX} heat exchangers, $N_{HX} - N_{HX,min}$ independent *heat loops* exist, that is, subnetworks that exhibit cyclic heat flows between two or more streams. The simplest case of a heat loop is shown in Figure 10.20a, noting that streams H1 and C1 are matched twice, with heat exchangers 1 and 3. A similar heat loop is shown in Figure 10.20b, where heat flows between streams H1 and C2 in two matches, in heat exchangers 2 and 4. A more complex heat loop is shown in Figure 10.20c, in which heat flows between streams H1 and C1 in heat exchanger 1, between H1 and C2 in heat exchanger 2, and between C1 and C2 and the hot utility, forming a closed heat cycle. Note that if C1 and C2 were to be serviced by two

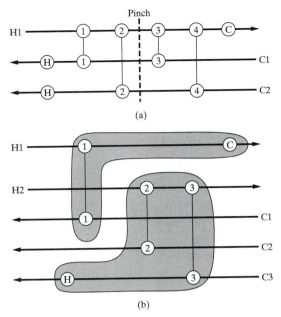

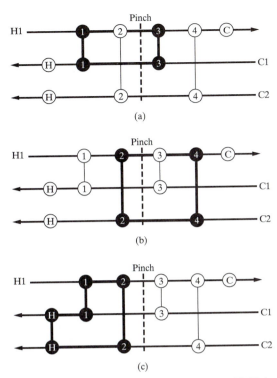

Figure 10.19 HENs involving: (a) one network; (b) two independent networks.

Figure 10.20 The three heat loops in the HEN in Figure 10.19a.

distinct sources of hot utility, there would be no heat loop in Figure 10.20c. Note also that the three heat loops identified in Figure 10.20 explain why the HEN has three heat exchangers more that $N_{HX,min}$.

Since a pinch exists in Figure 10.20, a HEN that meets the MER targets consists of two *independent* networks, one on each side of the pinch, so that: $N_{HX,min}^+ = N_{HX,min}^- = N_S + N_U - 1 = 3 + 1 - 1 = 3$, noting that the '+' and '−' superscripts indicate the hot and cold side of the pinch, respectively. Thus, the HEN in Figure 10.20 can be designed to meet the MER targets with just six heat exchangers, ($N_{HX,min}^{MER} = N_{HX,min}^+ + N_{HX,min}^- = 6$). The simplest change is to eliminate heat exchanger 1, transferring its duty to heat exchanger 2, decreasing the duty of the heater on stream C2 by this amount, and increasing the duty of the heater on stream C1 by this amount. This is referred to as *loop breaking*. Often, the smallest heat exchanger in the heat loop is eliminated, because the cost of the area saved by eliminating a small exchanger is usually more than the cost incurred in increasing the area of a large exchanger by the same amount.

As will be shown in Example 10.9, each heat loop is broken by removing a heat exchanger and adjusting the heat loads accordingly, which often leads to heat flow across the pinch, moving the HEN away from its MER targets.

EXAMPLE 10.9   *(Example 10.7 Revisited)*

Return to the HEN in Figure 10.15d, which was designed to meet its MER targets, and involves eight heat exchangers, three more than $N_{HX,min}$ given by Eq. (10.13). In this example, the heat loops are identified and removed, one by one, taking note of the impact on the capital cost, the cost of the utilities, and the annualized cost, C_A, in Eq. (10.3).

SOLUTION

Beginning with the HEN in Figure 10.15d, using the specifications at the start of Example 10.6, the heat transfer area for each heat exchanger is computed [Eq. (10.2)], the purchase costs are estimated using $C_P = 3,000A^{0.5}$, and the total purchase cost is computed to be $66,900. The annual cost of steam and cooling water is $10,960/yr, which combines with the purchase cost, multiplied by a return on investment ($i_m = 0.1$), to give $C_A = $17,650/yr.

To eliminate the first heat loop, in which heat is exchanged in two heat exchangers between streams H1 and C1, as shown in Figure 10.21a, the 80-kW exchanger is combined with the 60-kW exchanger, as shown in Figure 10.21b. This causes an approach-temperature violation ($110°C - 113.33°C = -13.33°C$), which must be eliminated by transferring less heat, x, in this heat exchanger. This amount of heat is computed so as to give $\Delta T_{min} = 10°C$ on the cold side; that is, the temperature of stream H1 is reduced to 123.33°C. Then, by heat balance,

$$140 - x = 2(180 - 123.33)$$

and the reduction in the heat load is $x = 26.66$ kW.

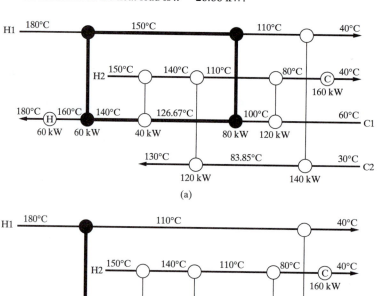

(a)

(b)

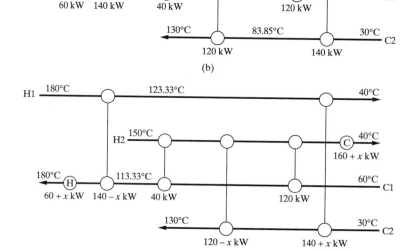

(c)

Figure 10.21 Breaking heat loops—Example 10.9: (a) eight heat exchangers, first heat loop; (b) seven heat exchangers, ΔT_{min} violation; (c) seven heat exchangers, shifting heat loads;

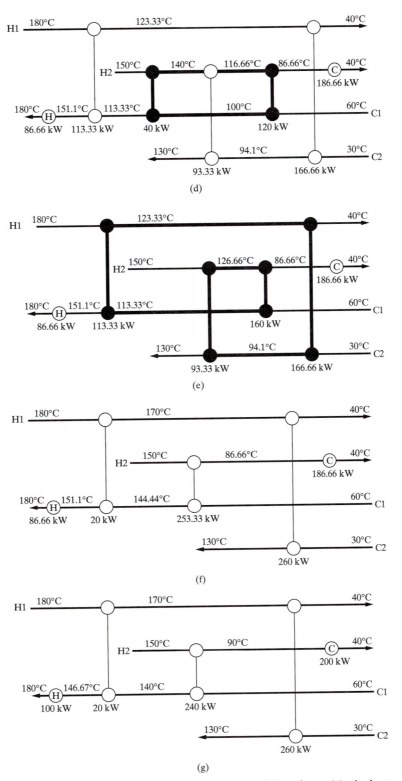

Figure 10.21 (d) seven heat exchangers, second heat loop; (e) six heat exchangers, third heat loop, (f) five heat exchangers, ΔT_{min} violation; (g) five heat exchangers.

Table 10.5   Cost Comparison for Example 10.9

Network	Utilities Cost ($/yr)	C_P, Purchase Cost ($)	C_A, Annualized Cost ($/yr)
Design for MER: 8 HXs	10,960	66,900	17,650
7 HXs	13,250	57,470	19,000
6 HXs	13,250	54,430	18,690
Design for $N_{HX,min}$: 5 HXs	15,340	45,930	19,930
Design for MER: stream-split design, 6 HXs[a]	10,960	54,890	16,450

[a]See Example 10.11 for details of the stream-split design to meet $N_{HX,min}^{MER}$

Furthermore, to account for this reduction, the steam consumption must be increased by x to heat stream C1 to 180°C and the loads of five other heat exchangers must be adjusted by the same amount, as shown in Figure 10.21c. This includes the same increase in the consumption of cooling water because x units of heat are transferred across the pinch. After the heat loads are adjusted, the resulting network is shown in Figure 10.21d. A glance at Table 10.5 shows the total purchase cost of the seven heat exchangers is reduced to $57,470, but the cost of utilities is increased to $13,250/yr, and hence, C_A is increased to $19,000/yr.

To eliminate the second heat loop, in which heat is exchanged in two heat exchangers between streams H2 and C1, as shown in Figure 10.21d, the 40-kW exchanger is combined with the 120-kW exchanger, as shown in Figure 10.21e. Normally, the heat exchanger with the smaller load is eliminated and its load transferred to the large one, unless the loads are comparable (as in the case of the first heat loop). In this case, since there are no temperature-approach violations, no adjustments in the loads of the heat exchangers are needed. Hence, as shown in Table 10.5, the cost of utilities is unchanged and the total purchase cost is reduced to $54,430, which reduces C_A to $18,690/yr.

The final heat loop has four heat exchangers involving four streams, as shown in Figure 10.21e. One of these can be eliminated by shifting the load of the smallest heat exchanger around the heat loop. The network in Figure 10.21f results, but it has a temperature approach violation ($150 - 144.44 = 5.56 \leq 10$°C). To eliminate this, the 253.33-kW heat load is reduced by y, where:

$$253.33 - y = 3(140 - 60)$$

to enable stream C1 to leave the heat exchanger at 140°C. This amount of heat ($y = 13.33$ kW) must be supplied and removed by additional steam and cooling water, respectively, as shown in Figure 10.21g, where the final HEN is displayed, with only five heat exchangers. In Table 10.5, for this network, the total purchase cost is reduced to $45,930, but the cost of utilities is increased to $15,340/yr, resulting in the largest C_A at $19,930. Hence, for this system, the HEN with eight heat exchangers has the lowest C_A when $\Delta T_{min} = 10$°C. Of course, the trade-off between the total purchase cost and the cost of utilities shifts as the cost of fuel decreases and the capital cost or the return on investment increases. Under different conditions, another configuration can have the lowest annualized cost. ∎

As seen in Table 10.5, an alternative design involving stream splitting gives the lowest annualized cost. Indeed, as will be shown in the next section, the use of stream splitting often enables the design of HENs that satisfy MER targets with the minimum number of heat exchangers.

Reducing the Number of Heat Exchangers—Stream Splitting

When designing a HEN to meet its MER targets, stream splitting *must* be employed if: (a) the number of hot streams *at the pinch*, on the cold side, is less than the number of cold streams; or (b) the number of cold streams *at the pinch*, on the hot side, is less than the number of hot streams. In this way, parallel pairings that fully exploit the temperature differences between energy sources and sinks are possible. Moreover, stream splitting helps to reduce the number of heat exchangers in a HEN without increasing the utility duties, which often occurs when heat loops are broken. The following example illustrates the use of stream splitting to achieve simultaneously the MER and the minimum units for a HEN involving two hot streams and one cold stream.

EXAMPLE 10.10

It is required to design a HEN, with a minimum number of heat exchangers that satisfy $\Delta T_{min} = 10°C$ and a hot utility MER target of 300 kW, for three streams on the hot side of the pinch:

Stream	T^s (°C)	T^t (°C)	C (kW/°C)	Q (kW)
H1	200	100	5	500
H2	150	100	4	200
C1	90	190	10	1,000

SOLUTION

Since no cold utility is allowed, Eq. (10.13) gives $N_{HX,min} = 3$. Furthermore, because the two hot streams need to be cooled entirely by the cold stream, to avoid an approach-temperature violation (i.e., $\Delta T_1 < \Delta T_{min}$), the cold stream must be split, as shown in Figure 10.22a. Note that the duties of all three heat exchangers have been set to satisfy the MER targets, but the portion of the heat-capacity flow rate assigned to the first stream, x, must be determined by solving the energy balances for the split streams:

$$x(T_1 - 90) = 500 \tag{10.15}$$

$$(10 - x)(T_2 - 90) = 200 \tag{10.16}$$

subject to the constraints:

$$200 - T_1 \geq \Delta T_{min} = 10°C \tag{10.17}$$

$$150 - T_2 \geq \Delta T_{min} = 10°C \tag{10.18}$$

To minimize lost work, as explained in Section 9.4, it is desirable to mix the split streams isothermally; that is, to select $T_1 = T_2 = 160°C$, with $x = 7.143$ kW/°C. Unfortunately, isothermal mixing is infeasible since inequality (10.18) limits T_2 to be less than or equal to 140°C. Arbitrarily, the equality is set

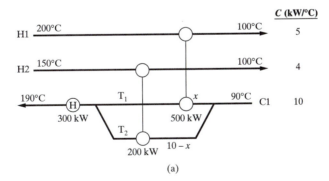

(a)

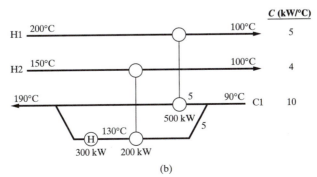

(b)

Figure 10.22 MER design for Example 10.10: (a) split determination; (b) improved design.

active; that is, $T_2 = 140°C$, giving $x = 6$ kW/°C and $T_1 = 173.33°C$. Note that by making a minor structural change in which the heater is moved upstream of the mixing junction, as in Figure 10.22b, the design is improved to provide isothermal mixing, as desired. ∎

As illustrated in Example 10.10 and introduced by Linnhoff and Hindmarsh (1983), generalized rules for stream splitting on both sides of the pinch to satisfy MER requirements are

Hot Side of Pinch

H-1. Let N_H and N_C be the number of hot and cold streams at the pinch. Noting that cold utilities cannot be used on the hot side of the pinch, if $N_H > N_C$, *a cold stream must be split*, since for feasibility, $N_H \leq N_C$.

H-2. On the hot side of the pinch, all feasible matches must ensure $C_h \leq C_c$. If this is not possible for every match, split one or more hot streams as necessary. If hot streams are split, return to step H-1 above.

Cold Side of Pinch

C-1. Let N_H and N_C be the number of hot and cold streams at the pinch. Noting that hot utilities cannot be used on the cold side of the pinch, if $N_C > N_H$, *a hot stream must be split*, since for feasibility, $N_C \leq N_H$.

C-2. On the cold side of the pinch, all feasible matches must ensure $C_c \leq C_h$. If this is not possible for every match, split one or more cold streams as necessary. If cold streams are split, return to step C-1 above.

As mentioned previously, stream splitting is also used to reduce the number of heat exchangers for HENs that satisfy the MER targets, as shown in the following example.

EXAMPLE 10.11   *(Example 10.9 Revisited)*

Return to Example 10.9 and recall that, as heat loops are broken to reduce the number of heat exchangers, heating and cooling utilities are normally increased. Here, stream splitting is used to reduce the number of heat exchangers while satisfying MER targets.

SOLUTION

Since an MER design implies separate HENs on both sides of the pinch, it is helpful to compute the minimum number of heat exchangers in each HEN. On the hot side of the pinch, only streams H1 and C1 exist, and so $N_{HX,min}^+ = 2$, while on the cold side of the pinch, all streams participate, and $N_{HX,min}^- = 4$. Thus, $N_{HX,min}^{MER} = N_{HX,min}^+ + N_{HX,min}^- = 6$. The HEN in Figure 10.15a has the minimum number of heat exchangers, while that in Figure 10.15c exceeds the minimum by two heat exchangers.

Figure 10.23 shows a possible design, in which stream H2 is split *at the pinch*, ensuring that $C_c \leq C_h$, with the largest portion of its heat-capacity flow rate, $4 - x$, paired with stream C1 using heat exchanger 1. The remaining branch is paired with a portion of stream C2, y, using heat exchanger 2. The remaining portion of stream C2, with a heat-capacity flow rate of $2.6 - y$, is paired with stream H1. To determine x and y, the energy balances associated with the stream splits are solved to ensure isothermal mixing, giving $x = 40/70 = 0.57$ kW/°C and $y = 40/100 = 0.4$ kW/°C. Note that $C_c \leq C_h$ must be satisfied only in match 1, because it is the only heat exchanger having both streams at the pinch. The overall HEN, a combination of Figures 10.15a and 10.23, satisfies the MER targets and has the minimum number of heat exchangers (six). Consequently, the cost of utilities is at the minimum, $10,960, with the total purchase cost, $54,890, providing an annualized cost of $16,450, the lowest in Table 10.5. ∎

While stream splitting provides these advantages, its use should be considered with caution because it reduces flexibility and complicates process operability. When possible, it is usually preferable to break heat loops without stream splitting, as shown later in Example 10.17.

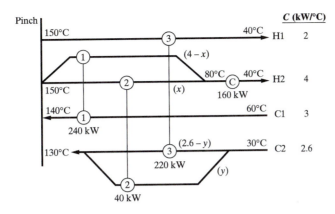

Figure 10.23 Stream splitting on the cold side of the pinch to achieve $N_{HX,\ min}^-$ for Example 10.11.

10.5   THRESHOLD APPROACH TEMPERATURE

In many cases, the selected ΔT_{min} is such that no pinch exist, and MER design calls for either hot or cold utility to be used, but not both. The critical ΔT_{min} below which no pinch exists is referred to as the *threshold approach temperature difference*, ΔT_{thres}. The following two examples illustrate how this arises and demonstrate how the guidelines presented previously are adapted for HEN design.

EXAMPLE 10.12

Compute MER targets for the following streams as a function of the minimum approach temperature:

Stream	T^s (°C)	T^t (°C)	C (kW/°C)	Q (kW)
H1	300	200	1.5	150
H2	300	250	2	100
C1	30	200	1.2	204

SOLUTION

For $\Delta T_{min} = 10°C$, no pinch exists, as shown in Figure 10.24a using the TI method. Note that 46 kW of cold utility are required. When the analysis is repeated as ΔT_{min} is varied, for $\Delta T_{min} > 100°C$, a pinch exists, as illustrated in Figure 10.24b for $\Delta T_{min} = 105°C$. Figure 10.25 shows the threshold, $\Delta T_{min} = \Delta T_{thres}$, at which the pinch appears. Since no heating utility is required for $\Delta T_{min} < \Delta T_{thres}$, where the cooling requirement is constant at 46 kW, no energy is saved while capital costs are increased as ΔT_{min} is decreased. The impact of ΔT_{min} on the economics of HENs is considered in Section 10.6. ∎

EXAMPLE 10.13

For the following streams:

Stream	T^s (°C)	T^t (°C)	C (kW/°C)	Q (kW)
H1	590	400	2.376	451.4
H2	471	200	1.577	427.4
H3	533	150	1.320	505.6
C1	200	400	1.600	320.0
C2	100	430	1.600	528.0
C3	300	400	4.128	412.8
C4	150	280	2.624	341.1

design a HEN at the threshold approach temperature difference, $\Delta T_{min} = \Delta T_{thres} = 50°C$, to meet its MER targets: $Q_{H,min} = 217.5$ kW and $Q_{C,min} = 0$ kW, as well as its $N_{HX,min}$ target of seven exchangers.

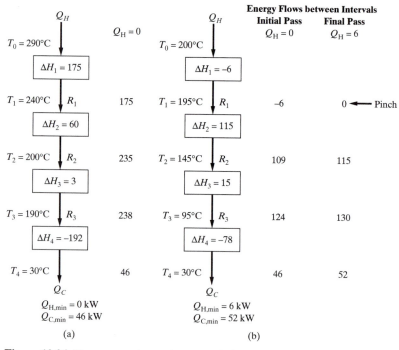

Figure 10.24 Temperature intervals, energy balances, and residuals for Example 10.12 at: (a) $\Delta T_{min} = 10°C$; (b) $\Delta T_{min} = 105°C$.

SOLUTION

Although no pinch exists, the MER design procedure is used, starting at the cold end, where matches are placed with $\Delta T_1 = \Delta T_{min} = 50°C$, and reserving the allocation of the utility heaters until last. Furthermore, matches at the limiting ΔT_{min} must have $C_h \leq C_c$, as occurs on the hot side of the pinch in the MER design procedure.

As seen in Figure 10.26, the first match is made bearing in mind that no cooling utility is allowed, and stream H3 must be cooled to 150°C. Since C2 is the only cold stream that can be used, internal heat exchanger 1 is installed, with a heat duty of 505.6 kW, which completes the cooling requirement of stream H3. Similarly, the second match is between streams H2 and C4, with a heat duty of 341.1 kW, which completes the heating requirement of stream C4. The remaining matches are far less constrained, and are positioned with relative ease. Note that the last units positioned are the auxiliary heaters, with a total duty equal to the MER target. Also, seven heat exchangers are utilized (five internal heat exchangers and two auxiliary heaters). Note that this network has no heat loops and stream splitting is not employed. ∎

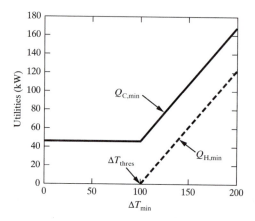

Figure 10.25 Minimum utility requirements as a function of ΔT_{min} for Example 10.12.

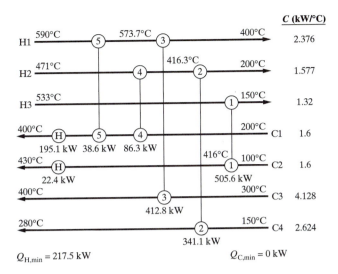

Figure 10.26 HEN with minimum number of exchangers to meet MER target at $\Delta T_{min} = \Delta T_{thres} = 50°C$ in Example 10.13.

10.6   OPTIMUM APPROACH TEMPERATURE

The importance of the minimum approach temperature, ΔT_{min}, has been emphasized in the previous sections. Clearly, as $\Delta T_{min} \to 0$, the true pinch is approached at which the area for heat transfer approaches infinity, while the utility requirements are reduced to the absolute minimum. At the other extreme, as $\Delta T_{min} \to \infty$, the heat transfer area approaches zero and the utility requirements are increased to the maximum, with no heat exchange between the hot and cold streams. The variations in heat transfer area and utility requirements with ΔT_{min} translate into the variations in capital and operating costs shown schematically in Figure 10.27. As discussed in the previous section, as ΔT_{min} decreases, the cost of utilities decreases linearly until a threshold temperature, ΔT_{thres}, is reached, below which the cost of utilities is not reduced. Thus, when $\Delta T_{min} \leq \Delta T_{thres}$, the trade-offs between the capital and utility costs do not apply.

In summary, when designing a HEN, it is important to consider the effect of ΔT_{min}. The next example illustrates this effect, while applying the techniques for stream matching described in Section 10.3.

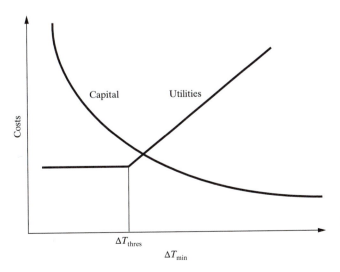

Figure 10.27 Trade-off between capital and utilities costs as a function of ΔT_{min}.

EXAMPLE 10.14

Consider the design of a HEN for four streams in a problem generated initially by Nishida et al. (1977):

Stream	T^s (K)	T^t (K)	C (kW/K)	Q (kW)
C1	269.3	488.7	36.93	8,102.44
H1	533.2	316.5	10.55	2,286.19
H2	494.3	383.2	26.38	2,930.82
H3	477.5	316.5	15.83	2,548.63

The following specifications apply:

> Cooling water (cw): $T^s = 310.9$ K, $T^t \leq 355.4$ K, cost of cw = 0.11023 $/1,000 kg
> Steam (sat'd., s): $T = 508.7$ K, $\Delta H^v = 1,785.2$ kJ/kg, cost of s = 2.2046 $/1,000 kg

Overall heat-transfer coefficients:

$$U_{\text{heater}} = 0.3505 \text{ kW/m}^2 \text{ K}, U_{\text{cooler}} = U_{\text{exch}} = 0.2629 \text{ kW/m}^2 \text{ K}$$

Purchase cost of heat exchangers:

$$C_P = 1,456.3A^{0.6} \text{ ($, m}^2\text{)}$$

Equipment operability = 8,500 hr/yr
Return on investment = $i_m = 0.1$ [for Eq.(10.3)]

SOLUTION

For this example, when $\Delta T_{\min} \geq \Delta T_{\text{thres}} = 25.833$ K, two pinches exist. This is because the heat-capacity flow rates of streams H1 and H2 sum to the heat-capacity flow rate of stream C1. Consequently, when $\Delta T_{\min} = 30$ K, at the first pinch the hot and cold stream temperatures are 494.3 K and 464.3 K, and at the second pinch the temperatures are 477.5 K and 447.5 K. The temperature interval between these temperatures involves just streams H1, H2, and C1, and hence, $\Delta T = 30$ K throughout this interval. Above the pinches, steam provides 490.7 kW and below the pinches, cooling water removes 153.89 kW. These minimum utility requirements, at $\Delta T_{\min} = 30$ K, are shown in Figure 10.28a.

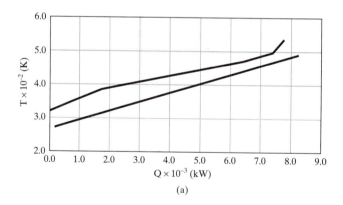

(a)

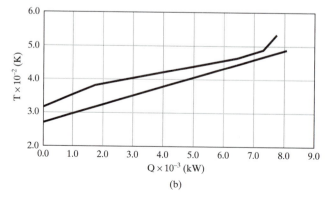

(b)

Figure 10.28 Composite heating and cooling curves for Example 10.14: (a) $\Delta T_{\min} = 30$ K; (b) $\Delta T_{\min} = \Delta T_{\text{thres}} = 25.833$ K.

When $\Delta T_{\min} = \Delta T_{\text{thres}} = 25.833$ K, the temperatures of the hot streams at the pinches are unchanged, but the temperatures of the cold streams are 468.5 K and 451.7 K. In this case, on the low-temperature side of the pinches, no cooling water is required, because the composite cold curve begins where the composite hot curve ends, as shown in Figure 10.28b.

Figure 10.29 shows the designs for three HENs, having minimum utilities, at three values of $\Delta T_{\min}$: 30, 25.833, and 16.9 K. For the first two, in Figures 10.29a and 10.29b, using the method of Linnhoff and Hindmarsh (1983), stream splitting is required below the first pinch. Stream C1 is split into two streams between the pinches because just streams H1 and H2 are present. Below the second pinch, stream C1 is split into three streams because all three hot streams are present. As shown in Table 10.6, both the purchase and utility costs are lower at ΔT_{thres}. The former is reduced because the cooling water exchanger is no longer needed. At $\Delta T_{\min} = 16.9$ K, the HEN is much simpler because no pinches exist, as shown in Figure 10.29c. Hence, the purchase cost is lower and the cost of utilities is equal to that at

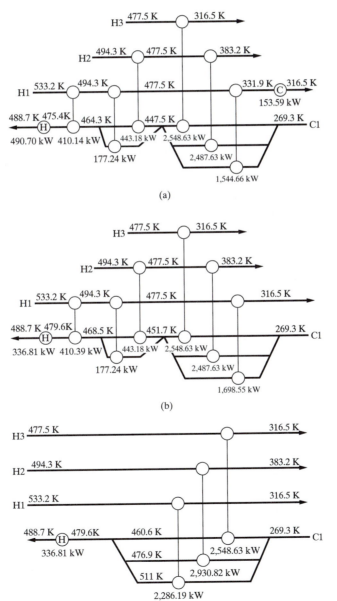

Figure 10.29 HENs for Example 10.14: (a) $\Delta T_{\min} = 30$ K; (b) $\Delta T_{\min} = \Delta T_{\text{thres}} = 25.833$ K; (c) $\Delta T_{\min} = 16.9$ K.

Table 10.6 Cost Comparison for Example 10.14

Network	Utilities Cost ($/yr)	C_P, Purchase Cost ($)	C_A, Annualized Cost ($/yr)
$\Delta T_{min} = 30$ K	27,050	82,740	35,320
$\Delta T_{min} = 25.833$ K	12,730	79,410	20,670
$\Delta T_{min} = 16.9$ K	12,730	68,440	19,580

ΔT_{thres}. When compared with the other two networks having minimum utilities, the latter has a lower annualized cost. For a more realistic comparison, the heat loops in Figures 10.29a and 10.29b should be broken to determine whether a lower annualized cost is possible. ∎

10.7   SUPERSTRUCTURES FOR MINIMIZATION OF ANNUAL COSTS

Thus far, emphasis has been placed on the two-step procedure introduced in Section 10.1, in which HENs are designed initially to have the minimum utilities, followed by a reduction in the number of heat exchangers. This strategy is particularly effective when the cost of fuel is high relative to the purchase costs of the heat exchangers. It is also relatively straightforward to implement compared with strategies that have been devised to minimize directly the annualized cost in Eq. (10.3). As shown in the previous sections, when solved formally as optimization problems, given ΔT_{min}, the determination of the minimum utilities involves the solution of a linear programming (LP) problem, and the determination of the minimum number of matches involves the solution of the transshipment problem in the form of a mixed-integer linear program (MILP). Furthermore, solutions of MILPs are straightforward using systems like GAMS, even for fairly large systems. Then, in the second step, the possibilities

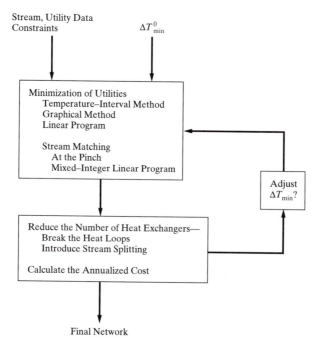

Figure 10.30 Two-step strategy for design of HENs.

of breaking the heat loops and stream splitting can be explored using the strategies in Section 10.4. This can be followed by the systematic adjustment of $\Delta T_{\min}$ to obtain HENs that have annualized costs closer to the global optimum. The overall strategy, which is illustrated in Figure 10.30, tends to be less effective when the cost of fuel is low relative to the purchase costs of the heat exchangers. In this case, networks having the minimum utilities and a large number of heat exchangers are far from optimal.

During the past two decades, perhaps because the cost of fuel has become less of a factor, much research has been directed toward the development of optimization formulations that locate the global optimum of C_A without decomposition into a two-step strategy. In their most general form, the formulations are mixed-integer nonlinear programs (MINLPs), involving nonlinear terms in the objective function (e.g., $A^{0.7}$) and in the constraints {e.g., $\Delta T_{\mathrm{LM}} = [(T_{h,i} - T_{c,o}) - (T_{h,o} - T_{c,i})]/\ln[(T_{h,i} - T_{c,o})/(T_{h,o} - T_{c,i})]$}. While the MINLPs can be formulated rather easily, unfortunately, the solution algorithms available today, in systems such as GAMS, are limited in their ability to obtain converged solutions. As computational speeds increase, along with the size of computer memories, algorithms that are more effective are being developed and more complex and larger mathematical programs are being solved.

It is beyond the scope of this text to cover in detail the latest approaches to formulating and solving MINLPs. Instead, following the approach of Floudas (1995), a superstructure is presented in Examples 10.15 and 10.16 to introduce several aspects of the approaches. Then, a brief review of the approaches is presented, with the Floudas text referred to for the details. MINLP formulations for HEN synthesis can also be solved reliably using stochastic optimization. Lewin et al. (1998) and Lewin (1998) present an approach using genetic algorithms for the synthesis of large-scale HENs.

EXAMPLE 10.15

In this example, a superstructure is to be created that embeds all of the alternative HENs involving one cold stream, C1, and three hot streams, H1, H2, and H3, as presented by Floudas (1995). The elements of the HENs include (1) heat exchangers, (2) stream mixers, and (3) stream splitters.

SOLUTION

Figure 10.31 shows all of the possible heat exchangers, mixers, and splitters, embedded within a single superstructure. A close examination of this superstructure identifies five alternative embedded substructures, including sequences in parallel, series, and combinations thereof, including the possibility of bypasses.

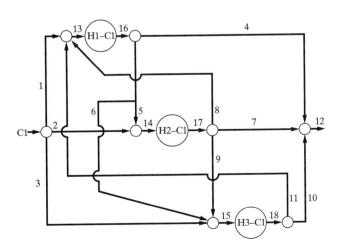

Figure 10.31 Superstructures for Example 10.15 (Reproduced from Floudas, 1995, with permission).

EXAMPLE 10.16

In this example, a superstructure is created for the potential HENs that involve hot stream, H1, and cold streams, C1 and C2, as specified below:

Stream	T^s (K)	T^t (K)	C (kW/K)	Q (kW)	h [kW/(m^{2}K)]
H1	440	350	22	1,980	2
C1	349	430	20	1,620	2
C2	320	368	7.5	360	0.6667

Note that there is no pinch and no hot or cold utilities are required. This example, taken from Floudas (1995), illustrates the formulation of the nonlinear program (NLP) to locate the HEN that minimizes the annualized cost. The purchase cost of a heat exchanger is given by $C_P = 13,000A^{0.6}$ ($, m^2), and the return on investment is 0.1.

SOLUTION

For this system, the superstructure that contains all possible HENs is shown in Figure 10.32. Annotated are the unknown variables, $F_1, \ldots, F_8, T_3, T_4, T_{56}, T_{78}, A_{H1,C1}$, and $A_{H1,C2}$. Note that in this formulation, the heat-capacity flow rates are denoted by F.

The NLP is formulated as follows:

$$\text{Minimize} \quad C_A = 1,300\, A_{H1,C1}^{0.6} + 1,300\, A_{H1,C2}^{0.6}$$
$$\text{w. r. t.}$$
$$F_i$$
$$\text{s. t.}$$

$$
\begin{aligned}
& F_1 + F_2 = 22 && \text{Mass balances for splitters} \\
& F_3 - F_5 - F_6 = 0 && \\
& F_4 - F_7 - F_8 = 0 && \\[4pt]
& F_3 - F_1 - F_8 = 0 && \text{Mass balances for mixers} \\
& F_4 - F_2 - F_6 = 0 && \\
& 440\,F_1 - F_8 T_{78} - F_3 T_3 = 0 && \text{Energy balances for mixers} \\
& 440\,F_2 - F_6 T_{56} - F_4 T_4 = 0 && \\[4pt]
& F_3 (T_3 - T_{56}) = 1,620 && \text{Energy balances for heat exchangers} \\
& F_4 (T_4 - T_{78}) = 360 && \\[4pt]
& T_3 - 430 \geq 10 && \text{Feasibility constraints} \\
& T_{56} - 349 \geq 10 && \\
& T_4 - 368 \geq 10 && \\
& T_{78} - 320 \geq 10 && \\[4pt]
& F_1, F_2, \ldots, F_8 \geq 0 && \text{Non-negativity constraints}
\end{aligned}
$$

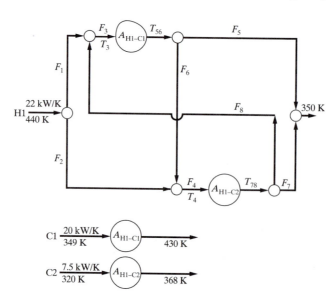

Figure 10.32 Superstructure for Example 10.16. (Reproduced from Floudas, 1995, with permission.)

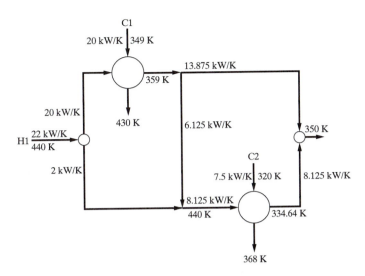

Figure 10.33 Optimal HEN for Example 10.16. (Reproduced from Floudas, 1995, with permission.)

where the areas in the objective function are defined by

$$A_{H1,C1} = \frac{1,620}{(1.0)(\Delta T_{LM})_{H1,C1}}, \quad A_{H1,C2} = \frac{360}{(0.5)(\Delta T_{LM})_{H1,C2}}$$

$$(\Delta T_{LM})_{H1,C1} = \frac{(T_3 - 430) - (T_{56} - 349)}{\ln\left[(T_3 - 430)/(T_{56} - 349)\right]}$$

$$(\Delta T_{LM})_{H1,C2} = \frac{(T_4 - 368) - (T_{78} - 320)}{\ln\left[(T_4 - 368)/(T_{78} - 320)\right]}$$

Note that there are eight heat-capacity flow rates, four temperatures, and nine equality constraints, and hence, three decision variables (e.g., F_1, F_6, and F_8). Also, ΔT_{min} is set at 10 K. In a more general formulation, it could be adjusted as an optimization variable.

As reported by Floudas (1995), the solution is:

$F_1 = 20$ kW/K	$F_2 = 2$ kW/K	$F_3 = 20$ kW/K
$F_4 = 8.125$ kW/K	$F_5 = 13.875$ kW/K	$F_6 = 6.125$ kW/K
$F_7 = 8.125$ kW/K	$F_8 = 0$	
$T_3 = 440$ K $T_4 = 378.9$ K	$T_{56} = 359$ K	$T_{78} = 334.64$ K
$A_{H1,C1} = 162$ m^2	$A_{H1,C2} = 56.717$ m^2	

with the corresponding HEN shown in Figure 10.33. It includes a splitter associated with the H1–C1 exchanger, a mixer associated with the H1–C2 exchanger, and a stream that connects a portion of the effluent from the former to the latter. Stream 2 bypasses the H1–C1 exchanger and stream 5 bypasses the H1–C2 exchanger.

See the CD-ROM that accompanies this book for the GAMS input file, COST.1, which obtains a solution using the Chen approximation to the log-mean temperature difference.   ∎

Solutions of the most general NLPs are complicated by the existence of nonconvex terms in the bilinear equality constraints. These lead most solvers to locate local optima rather than the global optimum. As global optimizers are being developed that are guaranteed to provide global convergence, improved superstructures and solution algorithms are evolving (Floudas, 1995).

10.8   MULTIPLE UTILITIES

Thus far, multiple sources of hot and cold utilities have not been considered. For example, steam is normally available at several pressures, with its cost a function of the pressure (and temperature) level, as discussed in Section 17.1 (Table 17.1). To reduce the cost of utilities, as well as the lost work, this section shows how to construct and use the *grand composite curve* (GCC) to reduce the temperature driving force in the auxiliary heat exchangers.

Designing HENs Assisted by the Grand Composite Curve

As described in Section 10.2, temperature intervals are identified and residuals computed to estimate the minimum heating and cooling utilities and to locate the pinch temperatures. For example, in the cascade diagram of Figure 10.34, computed for $\Delta T_{min} = 10°C$, the cold pinch temperature is 190°C, with the minimum hot and cold utility levels being 1,000 kW each. The residual enthalpies leaving each temperature interval indicate the excess heating or cooling capacity of the cascade above and including the temperature interval. This suggests the representation shown in Figure 10.35, referred to as the *grand composite curve*, in which the enthalpy residuals are displayed as a function of the interval temperatures. The enthalpy residuals corresponding to the highest and lowest interval temperatures are the minimum heating and cooling utility duties. Furthermore, as seen previously, the process is divided into a portion above the pinch (indicated by a solid circle) in which there is excess heating demand, and a portion below the pinch in which there is excess cooling demand.

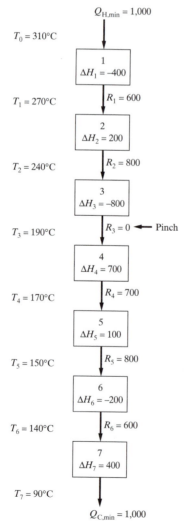

Figure 10.34 Temperature intervals, energy balances, and residuals; ΔH_i, R_i, Q_{Hmin}, and Q_{Cmin} in kilowatts.

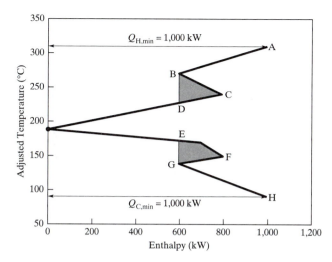

Figure 10.35 Grand composite curve for the temperature intervals in Figure 10.34.

In addition to this information, Figure 10.35 provides the following insights:

a. The local rise in the enthalpy residual along the arc BC, which indicates an increase in the residual heat available, can be used to supply the increased demand denoted by the decrease in the enthalpy residual along the arc CD. This is accomplished using internal heat exchange. Similarly, the heat demand along the arc FG can be supplied by internal heat exchange with the arc EF.

b. When 1,000 kW are provided at 330°C, a fired-heater is required that burns fuel. Alternatively, the GCC replotted in Figure 10.36 shows that up to 600 kW can be provided at 230°C using less expensive high-pressure steam at 400 psi, noting that the alternative utility temperature level and duty are limited by the intercept D. This leaves only 400 kW to be supplied in the fired-heater. Similarly, instead of removing 1,000 kW at 90°C using cooling water, up to 600 kW can be removed at 170°C by heating boiler feed water, leading to further savings in operating costs. As before, the temperature level and duty are limited by intercept E. Evidently, significant savings in operating costs are revealed by the GCC.

The grand composite curve helps to better utilize utility resources when designing HENs, as illustrated in the following example.

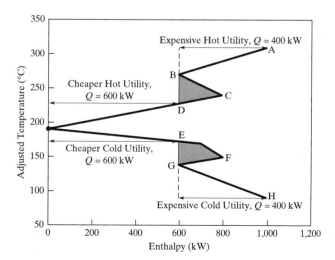

Figure 10.36 Grand composite curve for the temperature intervals in Figure 10.34, showing possible savings by utilizing HPS and BFW.

EXAMPLE 10.17

Consider the design of a HEN for the four streams below, with $\Delta T_{min} = 10°C$:

Stream	T^s (°C)	T^t (°C)	C (kW/°C)
H1	180	40	20
H2	160	40	40
C1	60	220	30
C2	30	180	22

To reduce operating costs, the design should consider alternative utility sources: high-pressure steam (HPS) and intermediate-pressure steam (IPS), boiler feed water (BFW), and cooling water.

The TI method is used to construct the grand composite curve shown in Figure 10.37a, which indicates that MER targets are $Q_{H,min} = 2,360$ kW and $Q_{H,min} = 1,860$ kW, with pinch temperatures at 150 and 160°C. Since stream H2 does not appear on the hot side of the pinch, the minimum number of heat exchangers that meets MER targets, $N_{HX,min}^{MER} = N_{HX,min}^+ + N_{HX,min}^- = 3 + 4 = 7$. A HEN designed for $N_{HX,min}^{MER} = 7$ is shown in Figure 10.38a, in which cooling water and HPS are used. A simpler design, ob-

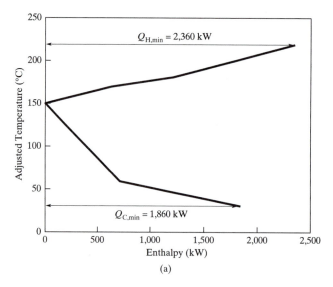

(a)

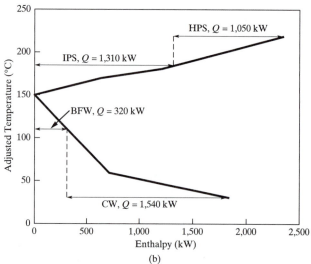

(b)

Figure 10.37 Grand composite curve for Example 10.17: (a) MER targets; (b) possible positioning of multiple utilities.

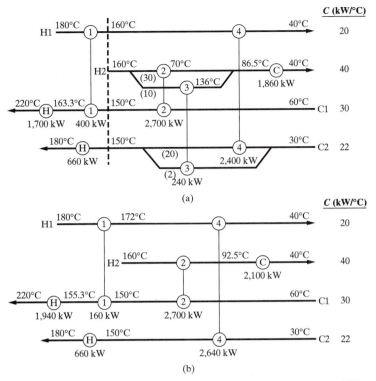

Figure 10.38 HENs for Example 10.17: (a) design to meet the $N_{HX,min}^{MER}$ target; (b) simplified design after breaking a heat loop.

tained by eliminating heat exchanger 3 and thus breaking one of the heat loops in Figure 10.38a, is shown in Figure 10.38b, noting that this involves 240 kW of additional heating and cooling utilities.

An alternative design is suggested in Figure 10.37b, in which a portion of the hot utility, 1,310 kW, is supplied as IPS at 195°C (at an adjusted temperature of 185°C). Furthermore, a portion of the cold utility duty, 320 kW, is used to generate steam from boiler feed water (BFW) at 110°C. These substitutions lead to the more complex design in Figure 10.39, involving three pinches: a process pinch at 150 and 160°C, and utility pinches at 110 and 120°C, and 185 and 195°C. The utility pinches arise because of the infinite heat-capacity flow rates associated with the utility streams. Note that the $N_{HX,min}^{MER}$ target of 12 heat exchangers is met by stream splitting and careful matching to permit a feasible design. Furthermore, the

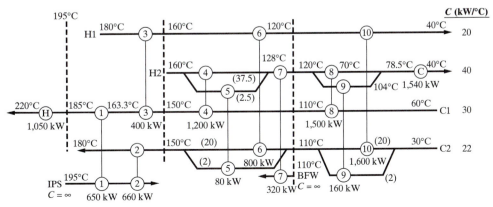

Figure 10.39 HEN for Example 10.17 utilizing cheaper utilities while meeting the $N_{HX,min}^{MER}$ target.

number of heat exchangers can be reduced by: (a) combining heat exchangers 6 and 10; (b) eliminating exchanger 1, reducing the capital costs at the expense of shifting 650 kW of hot utility duty from IPS to HPS; (c) eliminating heat loops, usually at the expense of additional utilities (see Exercise 10.20).   ■

10.9   HEAT-INTEGRATED DISTILLATION TRAINS

Although distillation is highly energy-intensive, having a low thermodynamic efficiency (less than 10% for a difficult separation, as shown in Example 9.4), it continues to be widely used for the separation of organic chemicals in large-scale chemical processes. As discussed in Chapter 7, the designer normally seeks to utilize more effective separation processes, but in many cases has no choice but to resort to distillation because it is more economical, especially in the manufacture of commodity chemicals.

In the previous sections of this chapter, methods have been discussed for exchanging heat between high-temperature sources and low-temperature sinks. When distillation operations are present in a process flowsheet, it is particularly important to consider the heating requirements in the reboilers and the cooling requirements in the condensers as HENs are designed. Furthermore, over the past two decades, several approaches have been suggested for the energy-efficient incorporation of distillation columns into a process flowsheet. This section is intended to present some of these approaches.

Impact of Operating Pressure

As discussed in Section 7.4, the column pressure of a distillation column is a key design variable because it determines the temperature levels in the reboiler and condenser, and consequently, the possible heating and cooling media. Earlier, when synthesizing the vinyl chloride process in Figure 3.8, it was noted that: " . . . heat is needed to drive the reboiler in the first distillation column at 93°C, but the heat of reaction (available from the direct chlorination of ethylene at 90°C) cannot be used for this purpose unless the temperature levels are adjusted." Furthermore, the question was raised: "How can this be accomplished?"

The principal vehicle for enabling these kinds of energy exchanges in most processes is through the adjustment of the pressure of the distillation towers, although in many cases it is possible to operate the reactor at a higher temperature. For the direct chlorination of ethylene in the liquid phase, it should be possible to increase the reactor temperature without increasing the rate of undesirable side reactions, and consequently, this alternative should be considered. The other alternative is to reduce the pressure of the distillation tower, thereby reducing the reboiler temperature and permitting the condensation of the dichloroethane product in the reboiler of the first tower. This reduces the usage of cooling water and steam, or eliminates them entirely, depending upon the cooling and heating demands. Also, as the distillation pressure is reduced, the separation is made easier and the number of stages is decreased. On the down side, however, the temperature of the condenser is reduced. If it is reduced below the temperature at which cooling water can be used, the cost of refrigeration becomes a significant cost factor (largely through an increased compression load). Furthermore, the integrated process is more difficult to control. In many cases, however, the combined savings in the utilities and the capital cost of the heat exchangers and the column exceeds the added refrigeration costs, and a reliable control system can be designed for the integrated process, as discussed in Chapters 20 and 21.

When working with the composite heating and cooling curves for a process, it helps to examine the heating and cooling requirements for a distillation column using a $T–Q$ diagram, as shown in Figure 10.40. This diagram shows the heat provided to the reboiler, Q_{reb}, at temperature, T_{reb}. Similarly, the diagram shows the heat removed from the condenser, Q_{cond}, at temperature, T_{cond}. The principal assumption in Figure 10.40 is that $Q_{reb} \approx Q_{cond}$. At first, this assumption may appear unjustified, but is reasonable for most distillation towers, especially

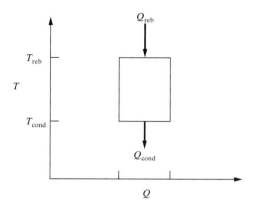

Figure 10.40 T–Q diagram for a distillation column.

when the feed and product streams are saturated liquids. To justify this, for the distillation tower in Figure 10.41, the energy balance in the steady state is

$$FH_F - DH_D - BH_B + Q_{reb} - Q_{cond} = 0 \qquad (10.19)$$

Then, for saturated liquids, where $T_{cond} < T_F < T_{reb}$, a reasonable approximation is:

$$FH_F - DH_D - BH_B \approx 0 \qquad (10.20)$$

and consequently,

$$Q_{reb} \approx Q_{cond} \qquad (10.21)$$

As shown in the multimedia CD-ROM that accompanies this textbook, the approximations in Eqs. (10.20) and (10.21) apply for the separation of propane from n-butane in a mixture of five normal paraffins, from ethane to n-hexane. Using both the RADFRAC subroutine in ASPEN PLUS (*SEPARATIONS → Distillation → MESH Equations → RADFRAC*) and the **Column** object in HYSYS.Plant (*SEPARATIONS → Distillation → Column Setup*), the approximation in Eq. (10.21) applies. Also, as shown in Dhole and Linnhoff (1993), the grand composite curve provides helpful insights when the column feeds are not saturated liquids.

Returning to Figure 10.40, it is important to note that the heat duty, $Q \approx Q_{reb} \approx Q_{cond}$, is related directly to the reflux ratio. When Q is reduced, because the cost of fuel is high, the

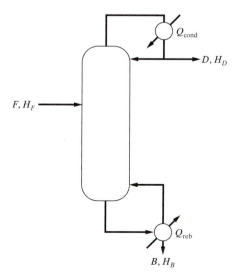

Figure 10.41 Schematic of a distillation tower.

number of trays (or the height of the packing) increases. Clearly, the trade-offs between operating and capital costs significantly influence the optimal design.

In one design strategy, to achieve a high degree of heat integration when the price of fuel is relatively high, the composite hot and cold curves are created without including the heat duties of the reboiler and condenser for a potential distillation tower. Then, when possible, the pressure level of the tower is adjusted to position its $T–Q$ rectangle to lie below the hot composite and above the cold composite curves, as shown in Figure 10.42a. In this way, heat can be transmitted from the hot process streams to the reboiler and from the condenser to the cold process streams without increasing the external utilities. Note, however, that unlike the example in Figure 10.42a, it is possible for the reboiler to accept energy from hot streams above the pinch and for the condenser to reject heat to the cold streams below the pinch. In this case, heat flows across the pinch, with the consumption of the hot and cold utilities increasing by Q. Clearly, when the cost of fuel is high, this design is not recommended.

Unfortunately, it can be difficult to position a tower as shown in Figure 10.42a unless the chemical species being separated are close-boiling (e.g., the split between propylene and propane in Example 9.4). Alternatively, the pressure of the tower can be adjusted and the tower positioned so that the reboiler receives its energy from a hot utility and the condenser rejects its energy to the cold process streams above the pinch, as shown in Figure 10.42b.

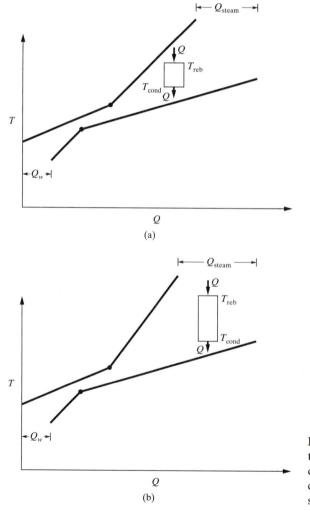

Figure 10.42 Positioning distillation towers between hot and cold composite curves: (a) exchange between hot and cold streams; (b) exchange with cold streams.

This reduces the load of the hot utilities required by the HEN for the remainder of the plant, while satisfying the heating demand of the reboiler. When the cost of fuel is high, this provides an attractive design.

Multiple-Effect Distillation

For separations where the T–Q rectangle for a distillation tower cannot be positioned under the hot composite and above the cold composite curves, as in Figure 10.42b, several possibilities exist for creating a more energy-efficient distillation operation when the price of fuel is high. One widely used configuration for distilling water is *multiple-effect distillation*, in which the feed stream is split into approximately equal portions and sent to the same number of separate distillation towers, each operating at a different pressure, as illustrated in Figure 10.43a for a cascade of two effects. The pressures in the towers decrease from the bottom to the top of the cascade so that the temperatures of the adjacent condensing vapors and boiling liquids differ by ΔT_{min}. In this way, heat from the condensing vapor in the tower below, at a pressure of P_2, is transferred to boil the bottoms liquid from the tower above, at the lower pressure of P_1. Note that the flow rates of the feed streams are adjusted to equate the duties of the adjacent condensing and boiling streams. In this way, the heat duties associated with condensation and boiling, Q_{effect}, are approximately equal to the heat duty for a single effect, Q, divided by the number of effects, N_{effect}; that is, $Q_{effect} \approx Q/N_{effect}$. When the price of fuel is high, this represents a substantial reduction in the operating costs. On the down side, however, the pressure level is increased in $N_{effect} - 1$ of the distillation towers, and multiple towers are needed. While pumping costs to increase the pressure of liquid feed streams are low, the tower walls are thicker, increasing the purchase costs, and the relative volatility is decreased at higher pressures, increasing the number of trays (or height of the packing) required to maintain the same reflux; that is, the same Q_{effect}. In many cases, these increases in

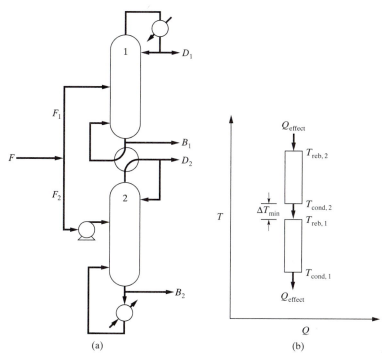

Figure 10.43 Two-effect distillation: (a) tower and heat exchanger configuration; (b) T–Q diagram.

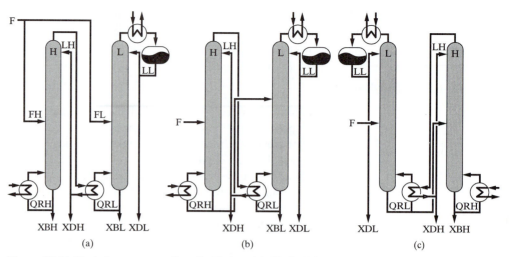

Figure 10.44 Variations on two-effect distillation: (a) FS; (b) LSF; (c) LSR.

the operating and purchase costs are small when compared with the large savings in the utilities for condensing and especially boiling.

The net effect of dividing the feed stream into N_{effect} nearly equal portions is to elongate the T–Q diagram as shown in Figure 10.43b. Note that the total area is not conserved because the pressure level in each effect determines the bubble-point temperatures of its condensing vapor and boiling liquid streams.

Two variations on multiple-effect distillation do not involve *feed-splitting* (FS), as described above, and shown in Figure 10.44a. For these configurations, as shown in Figures 10.44b and 10.44c, the entire feed stream is sent to the first tower. In the *light split/forward heat-integration* (LSF) configuration (Figure 10.44b), the feed is pumped and sent to the high-pressure column. About half of the light key component is removed in the distillate at high purity. The bottoms product, which contains the remainder of the light key component, is fed to the low-pressure column. In this case, the heat integration is in the direction of the mass flow. For the other variation. Figure 10.44c, referred to as the *light split/reverse heat-integration* (LSR) configuration, the feed is sent to the low-pressure column. Here, also, about half of the light key component is removed in the distillate, with the bottoms product pumped and sent to the high-pressure column. In this case, the heat integration is in the reverse direction of the mass flow. Note that these configurations are compared among themselves and with a single column in Example 21.7, where the dehydration of methanol is examined. First, the configurations are compared when operation is in the steady state. Then, the controllability and resiliency (C&R) of each configuration is assessed in response to typical disturbances, and verified by dynamic simulations of the FS and LSR configurations, using HYSYS.Plant, confirming the findings of the C&R analysis.

Heat Pumping, Vapor Recompression, and Reboiler Flashing

Other more sophisticated configurations, designed to increase the thermodynamic efficiency when the price of fuel is high, permit the vapor overhead to be condensed with the bottoms liquid from the *same* distillation column. Each of three configurations, *heat-pumping*, *vapor recompression*, and *reboiler flashing*, involves expensive compression of a vapor stream, as shown in Figure 10.45. The heat pump operates like a refrigeration cycle and requires an external fluid as the working medium. It pumps available heat from a low-temperature level up to a higher temperature level where it can be used more effectively. The other two configurations do not have external working fluids. Instead, they use the internal process fluids. To be

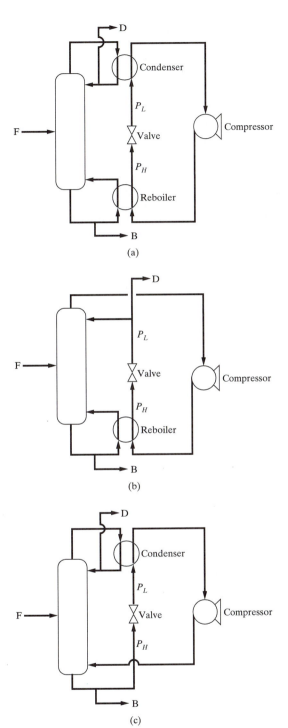

Figure 10.45 Distillation configurations involving compression: (a) heat pumping: (b) vapor recompression; (c) reboiler flashing.

effective, the savings in the cost of utilities and purchase costs for the heat exchangers must be greater than the increased utility and capital costs associated with the compressor.

For a detailed analysis of the reboiler-flashing configuration, which is usually the most financially attractive of the three configurations, the reader is referred to Example 9.4, in which the lost work and thermodynamic efficiency are computed for the separation of propylene and propane. Note that these configurations are most attractive for the separation of close-boiling mixtures because relatively small pressure changes are required, and consequently, the cost of compression is not too high.

10.10 HEAT ENGINES AND HEAT PUMPS

In the previous section, a refrigeration cycle, referred to as a *heat pump*, was introduced as one of the alternatives to permit the overhead vapor from a distillation tower to be condensed by vaporizing the bottoms liquid in a heat exchanger. This design is preferable when the savings in the cost of utilities for condensation and boiling and the purchase cost of the heat exchangers are greater than the costs of operation and installation of a compressor. It illustrates the use of *heat and power integration* to achieve a more profitable design.

The objective of this section is to consider the general role of heat engines, which convert heat to shaft work, and heat pumps in satisfying the heating, cooling, and power demands within a chemical process. In other words, after the source and target temperatures of the streams to be heated and cooled have been established, together with the power requirements for the pumps and compressors, heat exchangers, heat engines, and heat pumps are inserted to satisfy these demands in a profitable manner. The section begins with two examples that show typical processes for which these demands need to be satisfied. Then, some of the important considerations in positioning heat engines and heat pumps are considered.

EXAMPLE 10.18

This example involves the *ABCDE process* in Figure 10.46, which was created by Papoulias and Grossmann (1983c) to illustrate their approach to satisfying the heating, cooling, and power demands using a MILP. As can be seen, a vapor stream containing the species A, B, and C is fed to a two-stage compressor. It is combined with a recycle stream and sent to an exothermic reactor in which the species D and E are formed from A and B, with C being an inert. A flash vessel is used to concentrate A, B, and C in the vapor and D and E in the liquid. Water scrubs D and E from the vapor and the rich water is combined with the liquid from the flash vessel and sent to a distillation tower, which recovers nearly pure D in the distillate. The bottoms product is sent to a second distillation tower in which nearly pure

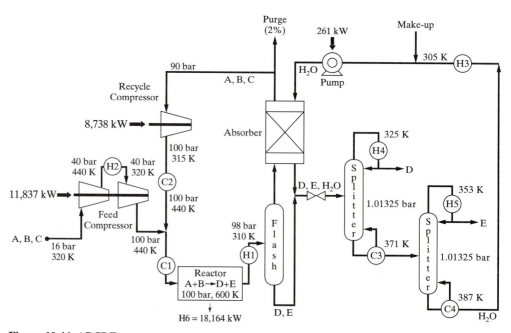

Figure 10.46 ABCDE process.

streams of E and water are produced. The water is combined with a make-up stream, pumped, and recycled to the absorber. Inert species C is removed in a small purge stream from the lean vapor from the absorber. The remainder is compressed and recycled. Note that the conditions shown in Figure 10.46 are typical of those produced by a process simulator during the fourth step in process synthesis in which the heating and cooling requirements are established, as well as the power demands for pumping and compression. The latter are annotated on the flowsheet and the former are as follows. Note that values of C (heat-capacity flow rates) for streams undergoing an isothermal phase change or an isothermal reaction are listed as infinite in the table.

Stream	T^s (K)	T^t (K)	C (kW/K)	Q (kW)
H1	600	310	901.0	261,300
H2	440	320	49.36	5,900
H3	387	310	36.44	2,800
H4	325	325	∞	65,300
H5	353	353	∞	171,000
H6	600	600	∞	18,200
C1	440	600	889.6	142,300
C2	315	440	840.5	105,100
C3	371	371	∞	119,200
C4	387	387	∞	90,800

SOLUTION

As can be seen, the heat recovered from the reactor, shown as stream H6, and the heat to be removed from the reactor effluent stream, H1, contain considerable energy at elevated temperatures. In synthesizing a profitable design, an important question concerns whether it is advantageous to utilize a heat engine(s) to generate the power required by the compressors and pump. This question will be addressed after several principles are established for the proper placement of heat engines.   ■

EXAMPLE 10.19

This example involves the ethylene plant in Figure 10.47, which was synthesized by Lincoff (1983) and used by Colmenares and Seider (1989b) to illustrate their method for designing cascade refrigeration systems. The feedstock to the process is a pyrolysis gas containing a mixture of water, hydrogen, methane, ethane, ethylene, propane, propylene, butadiene, butylenes, and steam-cracked naphtha (SCN). This is the quenched, gaseous product of a pyrolysis reactor, in which a mixture of paraffins and steam is cracked at 1,100 K. The pyrolysis gas enters a five-stage compression train at 333 K and 136.5 bar in which it is compressed to 350 bar. After each compression stage, the gas is cooled, condensed water is removed, and a vapor–liquid mixture is separated in a flash vessel. The liquid streams from the flash vessels are fed to a condensate splitter whose overhead vapors are recycled to the fourth stage of compression and the bottoms are sent to a depropanizer. The vapor feed from the fifth flash vessel is dried in a bed of zeolite molecular sieves and sent to a separation train, where low-temperature refrigeration is required to separate the light products (ethylene, propylene, etc.).

In the separation train, the gas stream is partially liquefied before entering the demethanizer at 320 bar. The overhead vapor, containing methane and hydrogen, is sent to a membrane separator in which these products are separated. The pressure of the bottoms product is reduced to 270 bar and fed to the deethanizer. In this column, the ethylene and ethane are removed in the distillate, whose pressure is reduced to 160 bar before the species are separated in the C-2 splitter. The bottoms product from the deethanizer, containing propylene, propane, and the heavier species, is throttled to 190 bar, mixed with the bottoms product from the condensate splitter, and fed to the depropanizer. The overhead product of the depropanizer is a mixture of propane and propylene and the bottoms product is throttled to 50 bar and sent to the debutanizer. In this column, the butylenes and butadiene are separated from the SCN.

The conditions shown in Figure 10.47 are typical of those produced by a process simulator during the fourth step in process synthesis, in which the heating and cooling requirements are established, as well as the power demands for the five-stage compression train. The former are tabulated below; the

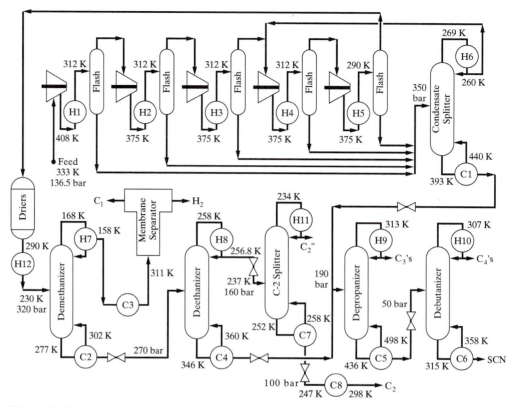

Figure 10.47 Ethylene process.

latter are omitted because they do not affect the positioning of the heat pumps. Values of C for streams H9, H10, and H11 are infinite because of isothermal phase change.

Stream	T^s (K)	T^t (K)	C (kW/K)	Q (kW)
H1	408	312	12.35	1,186
H2	375	312	7.397	466
H3	375	312	6.143	387
H4	375	312	6.032	380
H5	375	290	6.729	572
H6	269	260	2.222	20
H7	168	158	15.70	157
H8	258	256.8	317.5	381
H9	313	313	∞	224
H10	307	307	∞	141
H11	234	234	∞	1,081
H12	290	230	7.517	451
C1	393	440	2.362	111
C2	277	302	6.960	174
C3	158	311	1.360	208
C4	346	360	36.86	516
C5	436	498	7.226	448
C6	315	358	2.956	133
C7	252	256	280.0	1.120
C8	247	298	1.882	96

SOLUTION   As can be seen, refrigeration is needed to condense the overhead vapor streams at low temperatures, with the lowest temperatures in the condenser of the demethanizer. In synthesizing a profitable design, an important question concerns how to position the heat pumps (refrigeration cycles) to satisfy the heating and cooling demands. This question will be addressed after several principles are established for the proper placement of heat pumps.   ∎

Positioning Heat Engines and Heat Pumps

When processes have significant power demands, usually in compressor loads, it is normally sound practice to operate at or near the minimum utilities for heating and cooling. This is because the annualized cost is dominated often by the operating and capital costs associated with satisfying these power demands. Given the desirability of operating these processes at minimum utilities, Townsend and Linnhoff (1983a, b) make recommendations concerning the positioning of heat engines and heat pumps relative to the pinch, discussed next.

Figure 10.48 shows the temperature intervals for the streams to be heated and cooled in a chemical process, separated into two sections, above (a) and below (c) the pinch temperatures, T_p. Consider the three alternatives for positioning a typical heat engine, as shown in Figure 10.49. The latter is a closed cycle in which condensate, at T_1 and P_c, is pumped to T_2 and P_b, and sent to a boiler, where it leaves as a superheated vapor at T_3 and P_b. The boiler effluent is expanded across a turbine to T_4 and P_c, before it is condensed. The net heat energy, Q_b-Q_c, is converted to the net power, $W_{out}-W_{in}$, typically at a thermodynamic efficiency (see Section 9.7) of about 35%. Returning to Figure 10.48a, where the heat engine is positioned above the pinch, to satisfy the demand for hot utilities, Q^{HU}, Q_b is required by the boiler and the net power produced is W_{out}, neglecting the small power requirement of the pump. Hence, W_{out} is produced by adding the equivalent heat to Q^{HU}. In Figure 10.48c, where the heat engine is positioned below the pinch, the heat that would be rejected to cold utilities, Q^{CU}, is sent to the boiler. W_{out} is recovered from the turbine and the remainder is rejected to the cold utilities. The alternative to these two placements, in Figure 10.48b, has the heat engine accepting heat above the pinch and rejecting heat below the pinch. As shown, the

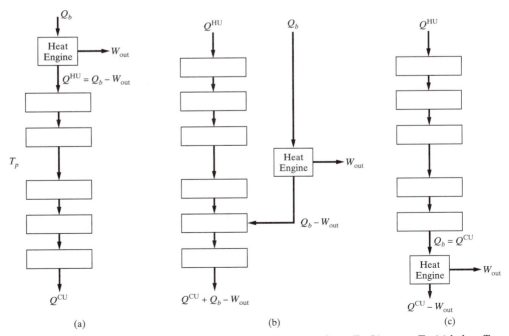

Figure 10.48 Alternatives for the placement of heat engines: (a) above T_p; (b) across T_p; (c) below T_p.

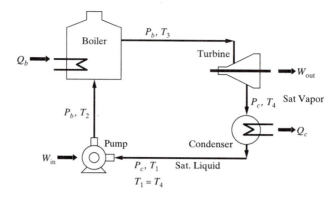

Figure 10.49 Heat engine.

total utilities above the pinch are incremented by Q_b, and below the pinch, the cold utilities are incremented by Q_b-W_{out}. Clearly, when the heat engine is positioned across the pinch, both the hot and cold utility loads are incremented. When the cost of fuel is high, this configuration is less profitable than the configurations with the heat engine entirely above or below the pinch.

Similarly, the three alternatives for positioning heat pumps relative to the pinch are shown in Figure 10.50. A typical heat pump is shown in Figure 10.51, where saturated vapor, at T_1 and P_b, is compressed to T_2 and P_c, and condensed, by rejecting its heat (often to the environment, but possibly to the boiler of another heat pump at lower temperature and pressure). The condenser effluent, at T_3 and P_c, is expanded across a valve to reduce its pressure and temperature to T_4 and P_b, while flashing some of the liquid. The remaining liquid is vaporized in the boiler. In Figure 10.50c, where the heat pump is positioned across the pinch, heat is removed from a temperature interval below the pinch and rejected to a temperature interval above the pinch, causing a reduction in both the hot and cold utility loads but at the expense of shaft work. Alternatively, when the heat pump is positioned above the pinch, as in

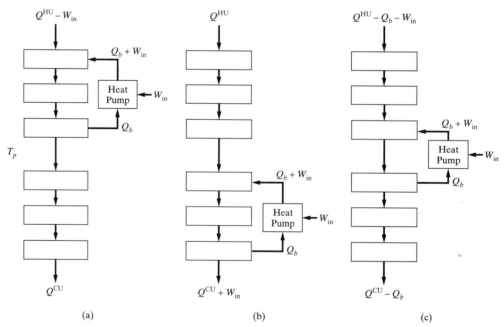

Figure 10.50 Alternatives for the placement of heat pumps: (a) above T_p; (b) below T_p; (c) across T_p.

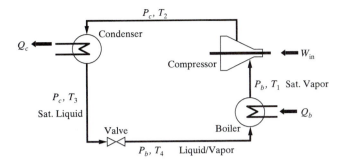

Figure 10.51 Heat pump.

Figure 10.50a, its compression load, W_{in}, reduces the hot utility load by this amount but does not reduce the cold utility load below the pinch. In this case, expensive power is converted directly to less valuable heat to reduce the hot utility load. Finally, when the heat pump is positioned below the pinch, as in Figure 10.50b, its compression load, W_{in}, increases the cold utility load by the same amount without affecting the hot utility load. Clearly, this is less desirable than when the heat pump is positioned across the pinch, where both the hot and cold utilities are decreased.

In summary, two heuristics result:

Townsend and Linnhoff Heuristics:

1. *When positioning heat engines, to reduce the total utilities, place them entirely* **above** *or* **below** *the pinch.*
2. *When positioning heat pumps, to reduce the total utilities, place them* **across** *the pinch.*

Optimal Design

Following the heuristics of Townsend and Linnhoff, optimal design strategies have been developed (Colmenares and Seider, 1989a, b). These involve the following steps:

1. Carry out the temperature-interval method to locate the pinch temperatures and the minimum hot and cold utility loads.
2. Lump the temperature intervals together above the pinch to create intervals having heat deficits; that is, intervals in which more heat is required to heat the cold streams than is available from the hot streams to be cooled. Similarly, lump the temperature intervals together below the pinch to create intervals having negative heat deficits; that is, intervals in which more heat must be removed from the hot streams than can be consumed by the cold streams.
3. Create a superstructure that includes the candidate heat engines and heat pumps. These can add heat to the lumped intervals above the pinch and remove heat from the lumped intervals below the pinch.
4. Formulate a NLP to minimize the total annualized cost of the heat engines and heat pumps. The design variables include the pressure levels and the flow rates of the working fluids in the heat engines and the heat pumps. The constraints include the heat balances for the condensers and boilers, the heat balances for the temperature intervals, an energy balance to satisfy the power demand, bounds on the heat removed or added to the temperature intervals, and bounds on the temperatures and pressures.
5. Solve the NLP using a solver such as MINOS within GAMS.

It should be pointed out that appropriate placement of heat pumps and heat engines should also account for the utility pinches, and not just the process pinch as stated above. However, even the simplified approach of Colmenares and Seider cannot be described in detail in the limited space available in this section. Instead, the highlights are summarized in the solutions to Examples 10.18 and 10.19 which are completed below.

EXAMPLE 10.20   *(Example 10.18 Revisited)*

SOLUTION (continued)   For $\Delta T_{min} = 10$ K, the pinch temperatures are 381 K and 371 K. After the temperature intervals are lumped together, the superstructure in Figure 10.52 is created. Note that only the temperature intervals above the pinch are included because heat pumps are not needed for refrigeration in the ABCDE process. Three candidate heat engines are included, one for each of the temperature intervals having a large heat deficit and one that provides power without providing heat to satisfy the heating requirements of the process streams. Also, three hot utilities are considered, provided by high-, medium-, and low-pressure steam. The lowest temperature interval is associated with the heating requirement in the reboiler of the first distillation column. Here, vaporization occurs at 371 K, and hence, $T^s = 371^-$ K and $T^t = 371^+$ K.

The annualized cost was minimized given the following data for the utilities:

Steam

	T (K)	P (bar)	$/ton	ΔH (kJ/kg)	$/kW
HPS	672.4	68.95	6.5	1,895.2	103.7
MPS	605.4	17.24	5.4	2,149.0	73.32
LPS	411.0	3.45	4.0	2,227.2	56.29

Cooling Water (cw)

$T^s = 300$ K, $T^t = 322$ K, cost of cw = $0.07/1,000 gal (= $6.011/kW)

Fuel (f)

$\Delta H^c = 43.95$ kJ/kg, cost of f = $143/ton (=$109.3/kW)

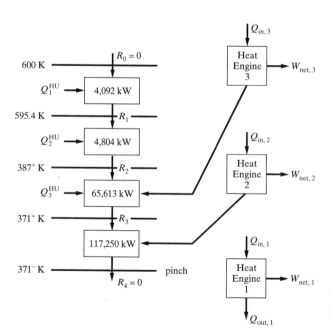

Figure 10.52 Superstructure for integration of ABCDE process with heat engines.

At the minimum, the annualized cost of the heat engines and utilities is $17,942,000/yr. Only Heat Engine 2 is utilized with $P_b = 53.6$ bar and $P_c = 1.26$ bar. The steam loads are $Q_1^{HU} = 4,092$ kW, $Q_2^{HU} = 4,804$ kW, and $Q_3^{HU} = 147,451$ kW. Note that the details of this solution, together with solutions for more general superstructures involving heat engines that exchange mass and energy, are presented by Colmenares and Seider (1989c). ∎

EXAMPLE 10.21   *(Example 10.19 Revisited)*

SOLUTION (continued)   For $\Delta T_{min} = 10$ K, the pinch temperatures are 408 K and 398 K. After the temperature intervals are lumped together, the superstructure in Figure 10.53 is created. Note that only the temperature intervals below the pinch are included, because the power demands for the five-stage compression system do not influence the design of the refrigeration system. As seen, the superstructure involves six potential heat pumps, each having a different working fluid, including commercial refrigerants R-13, R-22, and R-115. Ethane and ethylene are the candidates for removal of heat from the lowest temperature interval, with R-13 and propylene reserved for the intermediate temperature interval, and R-22 and R-115 reserved for the next highest temperature interval. Note that the highest temperature interval below the pinch can reject its heat to cooling water. It should also be noted that in the cascade structure, the heat pumps at the lower temperature levels can reject heat from their condensers to the boilers of heat pumps at the next higher temperature levels. The variables, Q_{ij}, denote the rate of heat transfer between heat pumps i and j. These variables, together with the pressure levels and the flow rates of the working fluids, are adjusted to minimize the annualized cost of the refrigeration system.

The annualized cost was minimized, including the installed cost of the compressors only, estimated using $C_{comp} = 1,925 W_{in}^{0.963}$, where the power required is in kilowatts. The cost of electricity was estimated to be \$0.04/kWh, and the cost of cooling water was estimated to be \$0.07/1,000 gal. At the minimum, computed by MINOS and shown in Figure 10.54, three heat pumps, involving ethane, propylene, and R-22, are in a cascade. All of the heat from the temperature interval below the pinch is rejected to cooling water, as well as the heat from the condenser of the R-22 heat pump. Note that no residual heat flows between the temperature intervals. For the complete details of this solution, showing how the temperature intervals are lumped and the NLP is formulated, see Colmenares and Seider (1989b). ∎

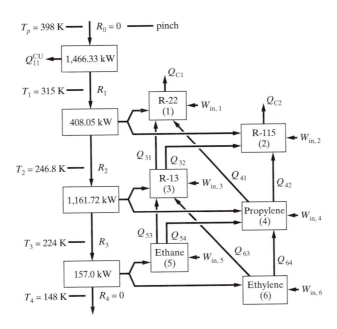

Figure 10.53 Superstructure for integration of the ethylene process with the cascade of heat pumps.

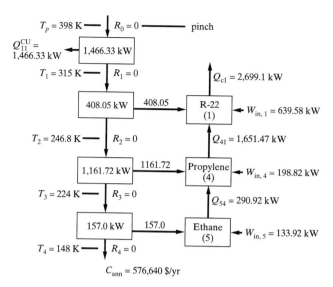

Figure 10.54 Cascade refrigeration system for the ethylene process at the minimum annualized cost.

10.11   SUMMARY

Having studied this chapter, and having solved many of the exercises, the reader should have learned how to achieve effective heat integration using several approaches. In one of the approaches, optimization problems are formulated and solved using GAMS: including linear programs (LPs), mixed-integer linear programs (MILPs), and nonlinear programs (NLPs). Although the chapter cannot provide a comprehensive treatment of the methods for achieving heat-integrated distillation trains and for satisfying power and refrigeration demands with heat engines and heat pumps, the introduction to these topics should enable a design team to include these features in its designs more systematically.

More specifically, the reader should

1. Be able to determine the minimum cooling and heating utilities (MER targets) for a network of heat exchangers using the temperature-interval (TI) method, the composite-curve method, or the formulation and solution of a linear program (LP).
2. Be able to design networks of heat exchangers on the hot and cold sides of the pinch, to meet the MER targets, using the heuristic method of Linnhoff and Hindmarsh (1983) or the transshipment model in a MILP.
3. Be able to reduce the number of heat exchangers toward the minimum by breaking the heat loops, and/or using stream splitting.
4. Be able to design a HEN when the minimum approach temperature difference is below ΔT_{thres}, at which no pinch occurs.
5. Understand the key role of the minimum approach temperature and appreciate the need to adjust it to achieve more optimal designs.
6. Be able to use the grand composite curve to consider the use of multiple hot and/or cold utilities and to find their optimal locations in the network.
7. Recognize the advantages and disadvantages of formulating superstructures for the design of HENs having the minimum annualized cost.
8. Be able to adjust the pressure in distillation columns to achieve heat integration and to consider the usage of multiple-effect distillation and the usage of compression to achieve designs that are more profitable.
9. Be able to insert turbines and heat engines, and compressors, refrigerators, and heat pumps, to satisfy both the heating and power demands for a process.

Heat Integration Software

Most of the methods introduced in this chapter, especially those for the design of HENs, are implemented in commercial software. Of special note, are ASPEN PINCH by Aspen Technology, Inc. (in the Aspen Engineering Suite), HX-NET by Hyprotech, HEXTRAN by Simulation Sciences, Inc., and TARGET by the Linnhoff–March Corp. As discussed in the chapter, many of the methods involve the solution of linear and nonlinear programs (LPs and NLPs), whose fundamentals are introduced in Chapter 18, and their mixed-integer counterparts (MILPs and MINLPs). While solutions using the General Algebraic Modeling System (GAMS) are presented herein and in the CD-ROM that accompanies this text, the former packages provide, in addition, excellent graphical user interfaces (GUIs) that simplify their usage.

REFERENCES

Colmenares, T.R., and W.D. Seider, "Synthesis of Utility Systems Integrated with Chemical Processes," *I&EC Res.*, **28**, 84 (1989a).

Colmenares, T.R., and W.D. Seider, "Synthesis of Cascade Refrigeration Systems Integrated with Chemical Processes," *Comput. Chem. Eng.*, **13**(3), 247–258 (1989b).

Colmenares, T.R., and W.D. Seider, "Heat and Power Integration of Chemical Processes," *AIChE J.*, **33**, 898 (1989c).

Dhole, V. R., and B. Linnhoff, "Distillation Column Targets," *Comput. Chem. Eng.*, **17**(5–6), 549 (1993).

Douglas, J.M. *Conceptual Design of Chemical Processes*, McGraw-Hill, New York (1988).

Floudas, C.A., *Nonlinear and Mixed-integer Optimization: Fundamentals and Applications*, Oxford University Press, Oxford (1995).

Hohmann, E.C., *Optimum Networks for Heat Exchange*, Ph.D. dissertation, University of Southern California, Los Angeles (1971).

Lewin, D.R., "A Generalized Method for HEN Synthesis Using Stochastic Optimization: (II) The Synthesis of Cost-Optimal Networks," *Comput. Chem. Eng.*, **22**(10), 1,387–1,405 (1998).

Lewin, D.R., H. Wang, and O. Shalev, "A Generalized Method for HEN Synthesis Using Stochastic Optimization: (I) General Framework and MER Optimal Synthesis," *Comput. Chem. Eng.*, **22**(10), 1,503–1,513 (1998).

Lincoff, A.M., *Separation System for Recovery of Ethylene and Light Products from a Naphtha-pyrolysis Gas Stream*, Process Design Case Study, CACHE Corp., Austin, Texas (1983).

Linnhoff, B., and J.R. Flower, "Synthesis of Heat Exchanger Networks: I. Systematic Generation of Energy Optimal Networks," *AIChE J.*, **24**, 633 (1978a).

Linnhoff, B., and J.R. Flower, "Synthesis of Heat Exchanger Networks: II. Evolutionary Generation of Networks with Various Criteria of Optimality," *AIChE J.*, **24**, 642 (1978b).

Linnhoff, B., and E. Hindmarsh, "The Pinch Design Method for Heat Exchanger Networks," *Chem. Eng. Sci.*, **38**, 745 (1983).

Linnhoff, B., and J.A. Turner, "Heat Exchanger Network Design: New Insights Yield Big Savings," *Chem. Eng.*, **77**, 56, November (1981).

Linnhoff, B., D.W. Townsend, D. Boland, G.E. Hewitt, B.E.A. Thomas, A.R. Guy, and R.H. Marsland, *A User Guide on Process Integration for the Efficient Use of Energy*, Revised 1st ed., The Institution of Chemical Engineers (IChemE), Rugby, England (1994).

McCabe, W., and E. Thiele, "Graphical Design of Fractionating Towers," *Ind. Eng. Chem.*, **17**, 605 (1925).

Nishida, N., Y.A. Liu, and L. Lapidus, "Studies in Chemical Process Design and Synthesis: III. A Simple and Practical Approach to the Optimal Synthesis of Heat Exchanger Networks," *AIChE J.*, **23**, 77 (1977).

Papoulias, S., and I.E. Grossmann, "A Structural Optimization Approach in Process Synthesis—I: Utility Systems," *Comput. Chem. Eng.*, **7**, 695 (1983a).

Papoulias, S., and I.E. Grossmann, "A Structural Optimization Approach in Process Synthesis—II: Heat Recovery Networks," *Comput. Chem. Eng.*, **7**, 707 (1983b).

Papoulias, S., and I.E. Grossmann, "A Structural Optimization Approach in Process Synthesis—III: Total Processing Systems," *Comput. Chem. Eng.*, **7**, 723 (1983c).

Townsend, D.W., and B. Linnhoff, "Heat and Power Networks in Process Design. 1. Criteria for Placement of Heat Engines and Heat Pumps in Process Networks," *AIChE J.*, **29**, 742 (1983a).

Townsend, D.W., and B. Linnhoff, "Heat and Power Networks in Process Design. II. Design Procedure for Equipment Selection and Process Matching," *AIChE J.*, **29**, 748 (1983b).

Umeda, T., J. Itoh, and K. Shiroko, "Heat Exchange System Synthesis," *Chem. Eng. Prog.*, **74**, 70, July (1978).

Yee, T.F., and I.E. Grossmann, "Simultaneous Optimization Models for Heat Integration—II: Heat Exchanger Network Synthesis," *Comput. Chem. Eng.*, **10**, 1,165 (1990).

EXERCISES

10.1 Four streams are to be cooled or heated:

Stream	T^s (°C)	T^t (°C)	C (kW/°C)
H1	180	60	3
H2	150	30	1
C1	30	135	2
C2	80	140	5

a. For $\Delta T_{min} = 10$°C, find the minimum heating and cooling utilities. What are the pinch temperatures?

b. Design a heat exchanger network for MER both on the hot and cold sides of the pinch.

10.2 a. For $\Delta T_{min} = 10$°C, find the minimum utility requirements for a network of heat exchangers involving the following streams:

Stream	T^s (°C)	T^t (°C)	C (kW/°C)
C1	60	180	3
C2	30	105	2.6
H1	180	40	2
H2	150	40	4

b. Repeat (a) for the following streams:

Stream	T^s (°C)	T^t (°C)	C (kW/°C)
C1	100	430	1.6
C2	180	350	3.27
C3	200	400	2.6
H1	440	150	2.8
H2	520	300	2.38
H3	390	150	3.36

c. For (a) and (b), design HENs that require the minimum utilities.

10.3 To exchange heat between four streams with $\Delta T_{min} = 20°C$, the HEN in Figure 10.55 is proposed. Determine if the network has the minimum utility requirements. If not, design a network with the minimum utility requirements. As an alternative, design a network with the minimum number of heat exchangers. Using the specifications in Example 10.7, which alternative is preferred?

10.4 Consider the network of heat exchangers in Figure 10.56:

a. Determine $N_{HX,min}$.

b. Identify the heat loop.

c. Show one way to break the heat loop using $\Delta T_{min} = 10°F$. For the resulting network, prepare a revised diagram, showing all temperatures and heat duties.

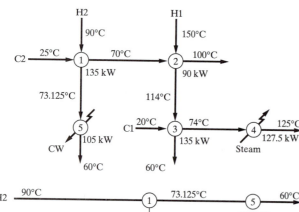

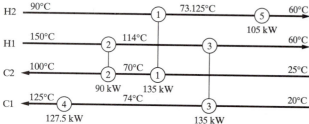

Figure 10.55 HEN for Exercise 10.3.

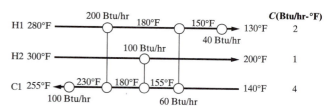

Figure 10.56 HEN for Exercise 10.4.

10.5 For the "pinch match" in Figure 10.57, show that to have a feasible match, that is $T_{h2} - T_{c1} \geq \Delta T_{min}$, the heat-capacity flow rate of the streams must satisfy $C_h \geq C_c$.

10.6 Consider the design of a network of heat exchangers that requires the minimum utilities for heating and cooling. Is it true that a pinch temperature can occur *only* at the inlet temperature of a hot or cold stream? *Hint:* Sketch typical composite hot and cold curves for two hot and two cold streams.

10.7 Consider Test Case No. 2 by Linnhoff and Flower (1978a):

Stream	T^s (°C)	T^t (°C)	C (kW/°C)
C1	60	180	3
C2	30	130	2.6
H1	180	40	2
H2	150	40	4

a. Use Figure 10.58 to find the minimum hot and cold utility loads when $\Delta T_{min} = 10°C$ and 50°C.

b. For $\Delta T_{min} = 50°C$, design a network of heat exchangers having minimum utilities.

10.8 In Example 10.6, a HEN is designed to meet the MER targets for the problem defined in Example 10.1:

a. Determine the number of heat loops in the design in Figures 10.12 and 10.13.

b. Systematically remove the loops, one at a time, and adjust the heat duties of the internal and auxiliary heat exchangers as needed.

c. How does your design for part (b) compare with that in Figure 10.3?

10.9 Design a HEN for the streams in Exercise 10.4 that meets the MER targets with the minimum number of heat exchangers.

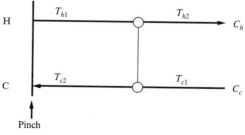

Figure 10.57 A heat exchanger positioned on the cold side of the pinch.

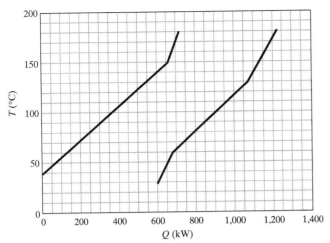

Figure 10.58 Composite hot and cold curves for Exercise 10.7.

10.10 A process has streams to be heated and cooled above its pinch temperatures, as illustrated in Figure 10.59. Complete a design that satisfies the MER targets with the minimum number of heat exchangers.

10.11. Design a HEN to meet the MER targets for $\Delta T_{min} = 10°C$ and $N_{HX,min}$ for a process involving five hot streams and one cold stream, as introduced by Yee and Grossmann (1990):

Stream	T^s (K)	T^t (K)	C (kW/K)
H1	500	320	6
H2	480	380	4
H3	460	360	6
H4	380	360	20
H5	380	320	12
C1	290	670	18

10.12 Consider a process with the following streams:

Stream	T^s (°C)	T^t (°C)	C (kW/°C)
H1	500	50	5
H2	400	100	4
H3	400	100	3
H4	200	50	2
C1	250	450	10
C2	30	430	6

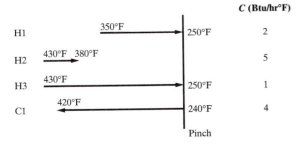

Figure 10.59 Streams for Exercise 10.10.

a. Determine MER targets for $\Delta T_{min} = 10°C$.

b. Design a HEN for MER using no more than 10 heat exchangers (including auxiliary heaters and coolers).

c. Add a stream to your HEN, without increasing the total number of exchangers. The data for the additional stream are

Stream	T^s (°C)	T^t (°C)	C (kW/°C)
C3	40	200	4

10.13 Consider a process with the following streams:

Stream	T^s (°C)	T^t (°C)	C (kW/°C)
H1	350	160	3.2
H2	400	100	3
H3	110	60	8
C1	50	250	4.5
C2	70	320	2
C3	100	300	3

When $\Delta T_{min} = 10°C$, the minimum utilities for heating and cooling are 237 kW and 145 kW, respectively, with pinch temperatures at 110°C and 100°C. Design a HEN that satisfies the MER targets and has the minimum number of heat exchangers, $N_{HX,min}^{MER}$. Show the heat duties and temperatures for each heat exchanger.

10.14 Consider the following heating and cooling demands:

Stream	T^s (°C)	T^t (°C)	C (kW/°C)
H1	525	300	2
H2	500	375	4
H3	475	300	3
C1	275	500	6

A HEN is to be designed with $\Delta T_{min} = 30°C$:

a. Find the MER targets.

b. Design a subnetwork of heat exchangers below the pinch that meets the MER targets.

10.15 Design a HEN with $N_{HX,min}^{MER}$ heat exchangers for Example 10.8. *Hint:* The solution requires stream splitting.

10.16 Consider a process with the following streams:

Stream	T^s (°F)	T^t (°F)	C ($\times 10^{-4}$ Btu/hr°F)
H1	480	250	2.0
H2	430	180	3.0
C1	100	400	2.5
C2	150	360	2.5
C3	200	400	2.5

a. Compute ΔT_{thres}, as well as the minimum external heating and cooling requirements as a function of ΔT_{min}.

b. Design a HEN to meet the MER targets with $N_{HX,min}^{MER}$ heat exchangers, for $\Delta T_{min} = 20°F$. Show the heat duties and temperatures for each heat exchanger.

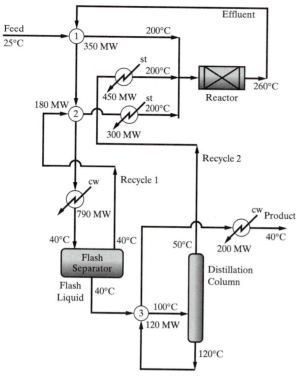

Figure 10.60 Process flowsheet for Exercise 10.17.

10.17 Consider the process flowsheet in Figure 10.60, where the duties required for each heat exchanger are in MW, and the source and target stream temperatures are

Process Stream	T^s (°C)	T^t (°C)
Feed	25	200
Effluent	260	40
Recycle 1	40	200
Flash liquid	40	100
Recycle 2	50	200
Product	120	40

a. The flowsheet calls for 990 MW to be removed by cooling water and 750 MW to be provided by steam. It is claimed that this design does not meet MER targets for $\Delta T_{min} = 10°C$. Verify or refute this claim.

b. If verified, design a HEN to meet MER targets for $\Delta T_{min} = 10°C$.

10.18 Design a heat exchanger network for MER, with at most 15 heat exchangers (including utility heaters) and $\Delta T_{min} = 10°C$, for the following streams:

Stream	T^s (°C)	T^t (°C)	C (kW/°C)
H1	140	50	10
H2	320	20	9
H3	370	20	8
C1	50	130	10
C2	130	430	8

Stream	T^s (°C)	T^t (°C)	C (kW/°C)
C3	100	300	6
C4	30	230	5
C5	30	130	4
C6	30	430	1

When MER targets are satisfied, the hot pinch temperature is 140°C, with $Q_{H,min} = 760$ kW and $Q_{C,min} = 960$ kW.

10.19 Design a heat exchanger network for MER, with at most 18 heat exchangers (including utility heaters) and $\Delta T_{min} = 10°C$, for the following streams:

Stream	T^s (°C)	T^t (°C)	C (kW/°C)
H1	400	20	10
H2	200	50	15
H3	350	230	5
H4	400	100	8
C1	80	450	7
C2	20	320	10
C3	50	450	5
C4	50	350	4
C5	100	500	1

When MER targets are satisfied, the hot pinch temperature is 200°C, with $Q_{H,min} = 1,170$ kW and $Q_{C,min} = 1,030$ kW.

10.20 In Example 10.17, HENs are designed for a process involving two hot and two cold streams. Note that three designs are proposed: (1) involving only HPS and cooling water that meets the $N_{HX,min}^{MER}$ target (shown in Figure 10.38a); (2) involving HPS and cooling water with no stream splitting and one less heat exchanger (shown in Figure 10.38b); (3) utilizing HPS and IPS, cooling water, and boiler feed water (shown in Figure 10.39). Which of these designs has the lowest annualized cost, given the following specifications:

Cooling water (CW): $T^s = 30°C$, $T^t \leq 80°C$,
cost of CW = 0.00015 $/Kg
Boiler feed water (BFW): $T = 110°C$, $\Delta H^v = 2,230$ kJ/kg,
revenue on BFW = 0.001 $/kg
Saturated IPS: $T = 195°C$, $\Delta H^v = 1,958$ kJ/kg,
cost of IPS = 0.003 $/kg
Saturated HPS: $T = 258°C$, $\Delta H^v = 1,676$ kJ/kg,
cost of HPS = 0.006 $/kg
Overall heat-transfer coefficients: $U_{heater} = U_{cooler} = U_{exch}$
$= 1$ kW/m²°C
Purchase cost of heat exchangers: $C_P = 3,000A^{0.5}$ ($, m²)
Return on investment, $i_m = 0.1$.

Bonus
Adapt the design in Figure 10.39 to produce a cheaper HEN.

10.21 *Flowsheet analysis and HEN synthesis problem.* A material balance has been completed for a process to manufacture styrene and an ethylbenzene byproduct from reactions involving methanol and toluene. See Figure 10.61 for a block flow diagram of the process with the results of material balance calculations. You are to develop an optimal heat exchanger network for this process. Note that:

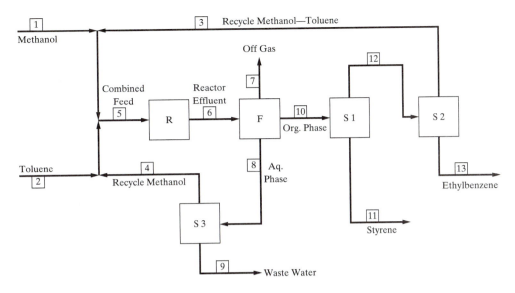

Figure 10.61 Styrene process.

Material Balance for Styrene Process										
Streams with flow rates in kmol/hr:										
Component	1	2	4	3	6	7	10	8	11	13
Hydrogen					352.2	352.2				
Methanol	493.4		37.0	66.0	107.3	4.3	66.0	37.0		
Water					489.1	7.9		481.2		
Toluene		491.9		104.5	107.3	1.5	105.8			1.3
Ethylbenzene				3.8	140.7	0.7	140.0			136.2
Styrene					352.2	1.6	350.6		346.7	3.9
Total	493.4	491.9	37.0	174.3	1,548.8	368.2	662.4	518.2	346.7	141.4

1. Stream 1 is fresh methanol feed, which enters at 25°C and 600 kPa.

2. Stream 2 is fresh toluene feed, which enters at 25°C and 600 kPa.

3. Stream 4 is methanol vapor recycle from the aqueous methanol recovery, S3, system. This stream returns at 113°C at saturation pressure.

4. Stream 3 is an organic recycle from the distillation section. This stream returns at 175°C and saturation pressure as a vapor.

5. Streams 1 and 2 must be brought to the bubble point in separate heat exchangers and then vaporized in separate heat exchangers. These streams can then be mixed as desired with streams 3 and/or 4 to obtain the combined feed to the reactor, which must enter the reactor at 500°C and 400 kPa. A furnace must be used to heat streams from 400 to 500°C.

6. Stream 6, the reactor effluent, leaves the adiabatic reactor as vapor at 425°C and 330 kPa and must be cooled and partially condensed at 278 kPa before entering the three-phase separator. During that condensation, primary and secondary dew points are observed, with the final effluent comprised of a vapor phase and two liquid phases. Be sure to determine a cooling curve for cooling and partially condensing the reactor effluent.

7. Column S1 has a reboiler duty of 2.83×10^7 kcal/hr to reboil the bottoms in the temperature range of 144–145°C.

8. Column S2 has a reboiler duty of 4.64×10^6 kcal/hr to reboil the bottoms in the temperature range of 153–154°C.

9. Column S1 needs a condenser duty of 2.59×10^7 kcal/hr to condense the overhead in the temperature range of 103–80°C.

10. Column S2 needs a condenser duty of 2.62×10^6 kcal/hr to condense the overhead in the temperature range of 108–89°C.

11. Stream 11, the liquid styrene product, needs to be cooled from 145 to 38°C before being sent to storage.

12. Stream 13, the liquid ethylbenzene byproduct, needs to be cooled from 153 to 38°C before being sent to storage.

Your tasks:

1. Solve the material and energy balances for the flowsheet in Figure 10.61 using a process simulator. Adjust the pressure drops in each equipment item to satisfy the pressure

specifications above. Two parallel reactions occur in the reactor, R, which may be modeled using a **Conversion Reactor** in HYSYS.Plant, with the conversion in each reaction specified to closely match the material balances in Figure 10.61. Furthermore, for simplicity, you may use **Component Splitters** for units S1, S2, and S3. Finally, note that it may be necessary to install additional equipment items between units F and S1 and between units S2 and the mixer for streams 1 and 3.

2. Using the solution of the material and energy balances in step 1, extract information necessary to define the HEN synthesis problem. Pay attention to possible phase changes in the streams.

3. Compute the pinch temperatures and MER targets for ΔT_{min} = 10°C.

4. Carry out an MER design to meet the targets in step 3. Avoid temperature-driving forces greater than 50°C when boiling a pure species or a mixture.

5. Refine your solution to eliminate heat loops, and minimize the annualized cost of the HEN. For the annualized cost, include estimates for:

a. the cost of all heat exchangers (both interior and utility exchangers). Estimate bare module costs, assume operation 330 days/year, and use a return on investment of 20%. Note that furnace costs are higher than heat exchanger costs.

b. the annual cost of utilities. Identify costs for refrigerant (if needed), cooling water, steam (at one pressure), and natural gas used in the furnace.

6. Repeat step 1 for the heat-integrated process. You may wish to fine-tune the design parameters using the simulator optimizer.

Submit a typed report, which:

- addresses the six tasks above
- describes your HEN and the steps involved in developing it
- provides a process flow diagram showing all of the heat exchangers
- provides a list of the heat exchangers with their duties in kcal/hr and log-mean temperature differences in °C
- summarizes the utility requirements (for fuel, steam, cooling water, and refrigeration) in kcal/hr.

Chapter **11**

Mass Integration

11.0 OBJECTIVES

This chapter extends the strategies for heat and power integration in Chapter 10 to apply for the mass integration of chemical processes during process synthesis. In Chapter 10, procedures for developing heat exchanger networks (HENs) were presented. In this chapter, analogous procedures for developing mass exchanger networks (MENs) are discussed. Mass exchangers use mass-separating agents (MSAs) to transfer solutes from solute-rich streams to solute-lean streams. Mass integration takes place after the demands for this transfer have been specified.

After studying this chapter, the reader should

1. Be able to compute the minimum usage of external mass-separating agents (MSAs) to determine the minimum operating cost (MOC) target. Two methods are introduced: the composition-interval (CI) method (analogous to the TI method for HENs) and a graphical method known as the composite-curve method (analogous to the use of temperature–composite curves in synthesizing HENs).
2. Be able to design a mass exchanger network (MEN) that meets the MOC targets. A method is introduced that inserts mass exchangers, one at a time, beginning at the closest approach mass-fraction difference, referred to as the *pinch*.
3. Be able to reduce the number of mass exchangers in MENs by relaxing the MOC target and *breaking mass loops* (i.e., allowing solute to be exchanged across the pinch).

11.1 INTRODUCTION

Almost all commercial operations for separating chemical mixtures utilize either an energy- (heat or shaft work) separating agent (ESA) as in distillation and certain high-pressure membrane separations, or an MSA, as in absorption, stripping, liquid–liquid extraction, solid–liquid extraction, adsorption, ion exchange, and membrane separations using a sweep fluid. With MSAs, solutes in so-called rich process streams are transferred into MSA streams, referred to as lean streams. The solute may then be removed from the MSA to permit its reuse. The network of equipment used to transfer solutes into MSAs is called a mass exchanger network (MEN). In general, it is assumed that the equipment used in the MEN employs countercurrent flow of the rich and lean streams. This is analogous to the assumption of the use of countercurrent flow heat exchangers in HENs. Major goals in the development of a MEN are to find and minimize the need for MSAs.

In Chapter 10, procedures for carrying out heat and power integration were discussed. These procedures are often implemented during the fourth step in process synthesis, where the differences in temperature, pressure, and phase are eliminated, when the source and target temperatures, T^s and T^t, for the streams to be heated and cooled, as well as power demands, have been specified. Emphasis is placed on determining the minimum hot and cold utility requirements [so-called MER (maximum energy recovery) targets for the HEN to be synthesized], and stream-matching when positioning heat exchangers in the HEN. These procedures are normally carried out before the detailed design of the individual heat exchangers and turbines, for which techniques are discussed in Chapters 13 and 15.

In this chapter, similar procedures are introduced for carrying out mass integration, which are often implemented during the third step in process synthesis, where the differences in composition are eliminated by introducing separation operations. These differences may be eliminated by the use of ESAs or MSAs. The use of ESAs are usually considered first. When not feasible, MSAs are used and mass integration becomes an important consideration. In this chapter, the separations are achieved by MSAs and the goal is to synthesize an efficient MEN. The development begins with specification of the source and target concentrations, c^s and c^t, of rich and lean streams. Emphasis is placed on determining the minimum amounts of MSAs to be introduced, and stream-matching when positioning separators in the MEN. These procedures are normally carried out before the detailed design of the individual separators, for which techniques are discussed in Chapter 14 and *Perry's Chemical Engineers' Handbook* (Perry and Green, 1997). While the first procedures for heat integration appeared in the early 1970s, parallel procedures for MENs were not introduced until two decades later by El-Halwagi and Manousiouthakis (1989). Major applications of MENs have been made in pollution abatement and waste minimization.

To define the MEN synthesis problem, N_R rich streams at mass flow rates, F_{R_i}, with specified source and target compositions, say mass fractions, y_i^s and y_i^t, $i = 1, \ldots, N_R$, have their solutes removed by N_L lean streams at mass flow rates, F_{L_i}, with specified source and target mass fractions, x_i^s and x_i^t, $i = 1, \ldots, N_L$, as shown schematically in Figure 11.1. Notice the similarity between Figure 11.1 for mass integration and Figure 10.1 for heat integration. The streams may be gas, liquid, or solid. Often, the lean streams are already present in the process flowsheet. These are referred to as the N_{LP} *process* mass-separating agents (MSAs). For example, monochlorobenzene (MCB) is the MSA used to separate HCl from benzene in the absorber of Figure 4.23. The remaining N_{LE} mass-separating agents are transferred to the process from *external* sources (like heating and cooling utilities in HEN synthesis). The external source might be steam when being used to strip volatile organic compounds (VOCs) from wastewater. However, note that the auxiliary network of Figure 10.1 for heat integration does not appear in Figure 11.1 for mass integration. To achieve sufficiently low concentrations, the lean stream concentrations must be sufficiently low or external mass-separating agents must be acquired. When it is desired to concentrate the lean streams above the inlet concentrations of the rich streams, more concentrated rich streams must be acquired, although this is not common and not necessary in waste-removal operations.

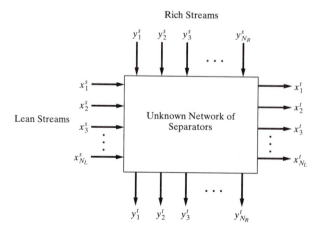

Figure 11.1 Mass integration schematic with source and target concentrations of rich and lean streams.

In aqueous waste-removal operations, the solute is often an undesirable species to be removed from wastewater. After solute recovery, the rich streams, with solute in low concentrations, are disposed of, returned to the environment, recycled, or reused. Clearly, when returned to the environment, these streams must meet the latest federal, state, or local regulations. When recycled or reused, solute concentrations must be sufficiently low to meet the requirements of sinks elsewhere in the process.

When carrying out the design, given the states of the source and target streams (flow rates and compositions of the solute), it is desired to synthesize the most economical network of mass exchangers. Several measures of economic goodness are possible, as discussed in Section 17.4. Usually, when generating and comparing alternative flowsheets, an approximate profitability measure is sufficient, such as the annualized cost:

$$C_A = i_m(C_{\text{TCI}}) + C \qquad (11.1)$$

where C_{TCI} is the total capital investment, as defined in Table 16.9, i_m is a reasonable return on investment annually (i.e., when $i_m = 0.33$, a chemical company charges itself annually for one-third of the cost of the capital invested), and C is the annual cost of sales, as defined in the *cost sheet* of Table 17.1. In Tables 16.9 and 17.1, many factors are involved, most of which are necessary for a detailed profitability analysis. However, to estimate an approximate profitability measure for the comparison of alternative flowsheets, it is adequate to approximate C_{TCI} as the sum of the purchase costs for each of the separators (without including installation costs and other capital investment costs). The purchase costs can be estimated based on the diameter, height, and weight of the process vessels, which are estimated using the procedures in Chapters 14 and 16. It is adequate to approximate C as the annual cost of the external mass-separating agents. In summary, with these approximations, Eq. (11.1) is rewritten as:

$$C_A = i_m\left(\sum_i C_{P_i}\right) + \sum_i e_i F_{E_i} \qquad (11.2)$$

where C_{P_i} is the purchase cost of separator i, F_{E_i} is the annual flow rate of external mass separating agent i (e.g., in kilograms per year), and e_i is the unit cost of external mass separating agent i (e.g., in dollars per kilogram). Clearly, when utilities, such as cooling water, air, steam, fuel, and refrigerants are used, additional terms are needed.

As in the synthesis of heat exchanger networks, two principal steps are typically carried out when synthesizing MENs:

1. A network of mass exchangers is designed having the *minimum amounts of mass-separating agents*, usually requiring a large number of mass exchangers. When the unit costs of the mass-separating agents are high, a nearly optimal design is obtained.
2. The number of mass exchangers is reduced toward the minimum, possibly at the expense of increasing the consumption of MSAs.

As step 2 is implemented, one mass exchanger at a time, capital costs are reduced due to the economy of scale in Eqs. (16.49) – (16.58). As each exchanger is removed, the diameters, heights, and weights of the exchangers are increased, and because the slope of the curves in Figure 16.13 are less than unity, the purchase costs per unit volume are decreased. Also, as step 2 is implemented, the consumption of MSAs is normally increased. At some point, the increased cost of MSAs overrides the decreased cost of capital and C_A increases beyond the minimum. When the costs of MSAs are high, the minimum C_A is not far from C_A for a network of mass exchangers using the minimum MSAs.

11.2 MINIMUM MASS-SEPARATING AGENT

When minimizing the utilities in heat integration, the approach temperature difference is the key specification. As $\Delta T_{\min}$ decreases, the utilities decrease, but the heat exchange area increases in inverse proportion. Similarly, when minimizing the flow rates of MSAs in mass integration, an approach composition difference must be specified. Here, it is convenient to specify the compositions of the rich and lean streams on the same scale. Commonly used compositions are mass fractions, mole fractions, and parts per million (ppm) on a volume basis for gases and a mass basis for liquids and solids.

Approach to Phase Equilibrium

Beginning arbitrarily with the rich phase, having solute mass fraction, y, the composition of the solute in the lean phase that approaches equilibrium with the rich phase is denoted, x^*. Here, the approach to phase equilibrium, $\Delta x_{\min}$, can be specified, where:

$$x^* = x + \Delta x_{\min} \tag{11.3}$$

Then, on an x–y diagram, in a dilute region where the equilibrium curve is linear, the x mass fraction is displaced to the left by $\Delta x_{\min}$, as shown in Figure 11.2. Let the equation of the equilibrium curve be a straight line given by:

$$y = mx^* + b \tag{11.4}$$

where m is the slope and b is the ordinate intercept. Substituting Eq. (11.3) in Eq. (11.4) and rearranging,

$$x = \frac{y - b}{m} - \Delta x_{\min} \tag{11.5}$$

Now, consider a countercurrent, direct-contact mass exchanger, such as a packed column. The packed height of the column is the product of the height of a transfer unit (HTU) and the number of transfer units (NTU). As $\Delta x_{\min}$ decreases, the NTU increases and, in turn, the height of the column and its capital cost increases. However, as will be shown, the amounts of the MSAs decrease.

Concentration–Interval (CI) Method

Consider the following example, which is similar to one introduced by El-Halwagi and Manousiouthakis (1989) in their pioneering paper. To determine the minimum flow rate of an MSA, the concentration–interval method is introduced first, after which, in the next subsection, the composite-curve method is introduced.

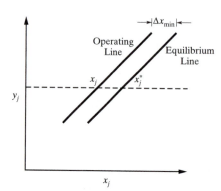

Figure 11.2 Equilibrium curve and approach composition difference.

EXAMPLE 11.1   *H₂S Removal from Sour Coke Oven Gas*

The process in Figure 11.3 is being designed to remove H_2S from sour coke oven gas (COG), which is a mixture of H_2, CH_4, CO, N_2, NH_3, CO_2, and H_2S. The removal is necessary because H_2S is corrosive and becomes the pollutant SO_2 when the gas is combusted. It is proposed to remove the H_2S and send it to a Claus unit to convert it to sulfur. However, because the conversion of the H_2S is incomplete, the tail gases must be recycled for H_2S removal. Distillation to remove the H_2S is not feasible, but absorption is feasible. Thus, it is proposed to design a MEN based on absorption. One possible MSA is aqueous ammonia, noting that ammonia is already present in the COG and that the flow rate and composition of the recycle stream are specified before the HEN is designed. An alternative MSA is chilled methanol, which is an external MSA. Both ammonia and chilled methanol are to be considered as possible absorbents for the removal of H_2S from the COG and the tail gas. As shown in Figure 11.3, the rich absorbent streams are regenerated by stripping to recover the acid gases, which are sent to the Claus unit.

To begin the development of the MEN, the sour COG and the tail gases are not mixed, and absorption can utilize ammonia, methanol, or both. Mass transfer in all mass exchangers is from the gas phase to the liquid phase.

The specifications for the rich and lean streams are as follows, where compositions, y for gases and x for liquids, are in mass fractions, F is the stream mass flow rate, and n is the mass flow rate of H_2S transferred to or from the stream:

Stream	y^s or x^s	y^t or x^t	F (kg/s)	n (kg/s)
R1 (COG)	0.0700	0.0005	0.9	0.06255
R2 (Tail gases)	0.0510	0.0003	0.1	0.00507
L1 (Aq. NH₃)	0.0008	0.0310	2.3	0.06946
L2 (Methanol)	0.0001	0.0035	Unlimited	Unlimited

Note that the flow rate of aqueous ammonia is limited, but chilled methanol is considered to be available in unlimited amounts. Note, also that the total amount of H_2S to be transferred to the absorbent(s) is $0.06255 + 0.00507 = 0.06762$ kg/s. This is less than the capacity of the aqueous ammonia. However, as in heat exchange, where a driving force is necessary to transfer the heat, mass exchange also requires a driving force and, at this point in the synthesis, it is not known whether sufficient mass-transfer driving forces exist to utilize the capacity of the aqueous ammonia. If not, then the use of chilled methanol must be considered.

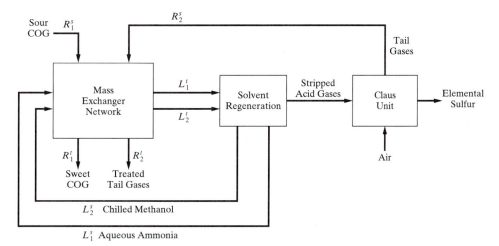

Figure 11.3 Process for recovery of H_2S.

All conditions in the above specifications table are considered to be dilute in the solute, H_2S. Therefore, stream flow rates are assumed constant and at the expected operating conditions of temperature and pressure, the following linear equilibrium equations apply:

$$\text{Aqueous ammonia (1)}, \quad y = m_1 x = 1.45x$$
$$\text{Chilled methanol (2)}, \quad y = m_2 x = 0.26x$$

For concentrated solutes, it is preferable to use solute-free flow rates and the mass ratios of solute to solute-free solvent.

At this stage in process synthesis, it is desired to determine, by the CI method, the minimum amount of chilled methanol required for a MEN involving these four streams, noting that it may be possible to eliminate the need for chilled methanol. In the solution below, the COG and lean gas streams are first matched with the aqueous ammonia stream. Alternatively, it may be preferable to consider first matches with the chilled methanol stream.

SOLUTION

The first step in the CI method is to rank-order the source and target mass fractions of streams R_1, R_2, and L_1, regardless of whether they are associated with the rich or lean phase. This includes computing the mass fractions in the corresponding phase, accounting for an approach to phase equilibrium using Eq. (11.5) with an assumed value of Δx_{min}, taken for this example as 0.0001. Thus, for the rich vapor streams, $x = y/1.45 - 0.0001$, and for the lean liquid streams, $y = 1.45(x + 0.0001)$. The results are given in Table 11.1, where specified values are in boldface and the two columns are rank-ordered, starting with the largest mass fraction at the top.

In the second step, the rank-ordered mass fractions are used to create a cascade of composition intervals, established in Table 11.2, within which mass balances are carried out. This is analogous to the cascade diagram in the TI method for HENs. As shown in Figure 11.4, each interval i displays the difference, Δn_i, between the mass to be removed from the rich streams and the mass to be taken up by the lean streams in the interval. For example, in interval 1 ($0.051 \leq y \leq 0.07$), only R1 is involved. Hence, as shown in Table 11.2, the difference, Δn_1, is $F_{R1} \times (0.07 - 0.051) = 0.0171$ kg/s. Since no excess mass, n_{Excess}, is assumed to enter this interval, 0.0171 kg/s are available and flow down into interval 2; that is, the residual from interval 1 is $R_1 = 0.0171$ kg/s. Interval 2 involves both rich streams, but the lean stream is not present. As shown in Table 11.2, its difference, Δn_2, is $(F_{R1} + F_{R2}) \times (0.051 - 0.0451) = 0.0059$ kg/s. When added to the residual from interval 1, R_1, the residual from interval 2, $R_2 = 0.0230$ kg/s. Interval 3 involves both rich streams ($0.001305 \leq y \leq 0.0451$) and stream L1 ($0.0008 \leq x \leq 0.0310$). Its difference, shown in Table 11.2, is negative, -0.025665 kg/s, which when added to R_2 gives $R_3 = -0.002665$ kg/s. In this interval, the lean stream requires $\Delta n_3 = 0.025665$ kg/s more solute than is available from the rich streams. The residual, R_2, provides 0.0230 kg/s, but this is insufficient, and consequently, the residual from interval 3 is negative; that is, $R_3 = -0.002665$ kg/s. Clearly, the H_2S solute cannot be transferred from the rich streams in interval 4, which are in a lower concentration range. The only way to avoid a negative residual is to add solute at a higher concentration. Note that the least amount of solute to be added is 0.002665 kg/s. Turning next to interval 4, only the rich streams are present ($0.0005 \leq y \leq 0.001305$) and $\Delta n_4 = 0.00080$ kg/s, which when added to R_3, gives $R_4 = -0.001865$ kg/s. Finally, only stream R2 is present in interval 5 ($0.0003 \leq y \leq 0.0005$) and $\Delta n_5 = 0.00002$ kg/s, which when added to R_4, gives $R_5 = -0.001845$ kg/s. Residuals R_3, R_4, and R_5 are negative and infeasible. Note that $n_{LE} = R_5$ is the minimum amount of lean external MSA (chilled methanol) required to remove the H_2S solute from streams R1 and R2. Clearly, it cannot be negative.

Table 11.1 Rank-Ordered Compositions Including the Approach to Phase Equilibrium

Rich Streams	Lean Streams
$y_0 = \mathbf{0.0700}$	$x_0 = 0.048276 - 0.0001 = 0.048176$
$y_1 = \mathbf{0.0510}$	$x_1 = 0.0035172 - 0.0001 = 0.035072$
$y_2 = 1.45(0.0310 + 0.0001) = 0.0451$	$x_2 = \mathbf{0.0310}$
$y_3 = 1.45(0.0008 + 0.0001) = 0.001305$	$x_3 = \mathbf{0.0008}$
$y_4 = \mathbf{0.0005}$	$x_4 = 0.000345 - 0.0001 = 0.000245$
$y_5 = \mathbf{0.0003}$	$x_5 = 0.000207 - 0.0001 = 0.000107$

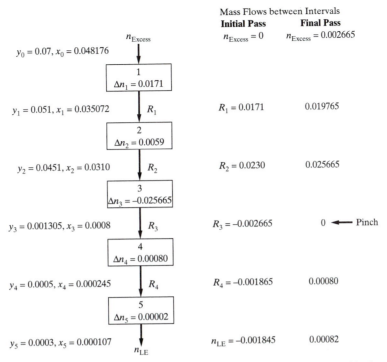

Figure 11.4 Cascade of composition intervals, mass balances, and residuals. R_i, n_{Excess}, and n_{LE} are in kilograms per second; x and y are mass fractions of the lean and rich phases.

Table 11.2 summarizes the solute loads to be removed and added in each interval, as well as the difference, or excess, and the residuals. These results constitute the "Initial Pass" in Figure 11.4.

Clearly, all negative residuals must be removed through the addition of solute at higher concentrations because the solute cannot be transferred from a low to a high concentration.[1] In this example, the largest negative residual is -0.002665 kg/s, and consequently, no negative residuals remain when $n_{Excess} = 0.002665$ kg/s is added at the top of the cascade, as shown in the "Final Pass" column of Figure 11.4. Alternatively, the internal MSA, aqueous ammonia, must be adjusted to reduce its consumption of solute, H_2S, by reducing its flow rate and/or its target mass fraction. By revising values of R_1 to R_4, a final residual of 0.00082 kg/s is found at the bottom of the cascade. This is the minimum amount of the solute that must be removed by the external MSA. To accomplish this,

Table 11.2 Internal Mass Loads

Interval	From Rich Streams (kg/s)	To Lean Streams (kg/s)	Excess (kg/s)	Residual, R (kg/s)
1	$(0.07 - 0.051) \times 0.9 = 0.0171$	—	0.0171	0.0171
2	$(0.051 - 0.0451) \times 0.9 +$ $(0.051 - 0.0451) \times 0.1 = 0.0059$	—	0.0059	0.0230
3	$(0.04505 - 0.001305) \times 0.9 +$ $(0.04505 - 0.001305) \times 0.1 = 0.04375$	$(0.031 - 0.0008) \times 2.3 = 0.06946$	-0.025665	-0.002665
4	$(0.001305 - 0.0005) \times 0.9 +$ $(0.001305 - 0.0005) \times 0.1 = 0.00080$	—	0.00080	-0.001865
5	$(0.0005 - 0.0003) \times 0.1 = 0.00002$	—	0.00002	-0.001845

---

[1] Actually, in a nonideal multicomponent system, as shown by Toor (1957), it is possible for mass transfer of a component to occur against a composition driving force because of cross-coupling effects.

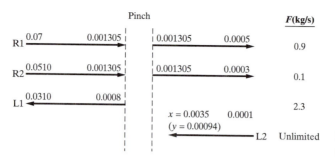

Figure 11.5 Pinch decomposition of the rich and lean streams.

$0.00082/(0.0035 - 0.0001) = 0.2412$ kg/s of chilled methanol is required. At the minimum usage of external MSA, no solute flows between intervals 3 and 4. This is referred to as the *pinch*, as shown in Figure 11.4, with associated mass fractions of 0.001305 for the rich streams and 0.0008 for the lean stream. Assuming that the mass fractions satisfy phase equilibrium, this is the location of the closest approach mass fraction, $\Delta x_{min} = 0.0001$. To maintain the minimum external MSA, *no solute is permitted to flow across the pinch*. Should additional solute, say 0.0002 kg/s, be added to the rich streams above the pinch, 0.0002 kg/s would be transferred across the pinch, and the amount of solute to be removed by the external MSA would be increased to $0.00082 + 0.0002 = 0.00102$ kg/s.

In Figure 11.5, which helps to define the design requirements when the external MSA is at a minimum, arrows moving from left to right denote the rich streams; lean streams are denoted by arrows moving from right to left. The arrows for the rich and lean streams either pass through or begin at the pinch mass fractions. Note that the smallest mass fraction of the rich streams that can enter a counter-current mass exchanger with the external MSA is, from Equation (11.5), 0.00094 [= 0.26 (0.0035 + 0.0001)]. It lies below the mass fraction of the rich streams at the pinch, 0.001305. To maintain the minimum external MSA, two separate MENs *must* be designed, one on the rich side and one on the lean side of the pinch mass fractions. Solute is not permitted to flow across the pinch. On the rich side of the pinch, the flow rate of the internal MSA has been reduced to permit all of the solute in the rich streams to be removed while meeting the target mass fraction of the internal MSA. On the lean side of the pinch, 0.00082 kg/s of H_2S solute are removed by the minimum amount of the external MSA. If solute from the rich streams on the rich side of the pinch were removed by a lean stream on the lean side of the pinch (in this case, the external MSA), solute would flow across the pinch and the amount of the external MSA would be increased above the minimum. ∎

In the next section, methods for inserting the mass exchangers (*stream matching*) are described. Before this, a graphical approach, the *composite-curve method* is discussed in the next subsection.

Composite-Curve Method

Similar to the discussion of heat integration in Section 10.2, the terminology *pinch* is understood more clearly in connection with a graphical display, as introduced by El-Halwagi and Manousiouthakis (1989) for mass integration, in which composite rich and lean curves are positioned no closer than the phase equilibrium departure plus Δx_{min}. As $\Delta x_{min} \to 0$, the curves pinch together toward the compositions at phase equilibrium and the area for mass transfer approaches infinity. The use of these curves is illustrated next in the composite-curve method.

EXAMPLE 11.2 **H_2S Removal from Sour Coke Oven Gas (Example 11.1 Revisited)**

In this example, the minimum external MSA requirement for a MEN involving the four streams in Example 11.1 is determined using the graphical approach of El-Halwagi and Manousiouthakis (1989).

SOLUTION

For each of the rich streams and the internal MSA lean stream (aqueous ammonia), y or x is graphed on the ordinate as a function of the mass of solute transferred on the abscissa, with the slope being the inverse of the mass flow rate, F (kg/s). For mixtures dilute in the solute, F is nearly constant, and consequently, the curves are approximately straight lines. For the rich streams, the curves begin at the highest mass fractions and finish at the lowest mass fraction after the solute has been removed. For the lean streams, they begin at the lowest mass fractions and finish at the highest mass fractions after the solute has been added. In Figure 11.6a, the three curves are displayed, with each of the lines positioned arbitrarily along the abscissa to avoid intersections and crowding.

To display the results of the CI method graphically, Table 11.2 is used to prepare the *rich composite curve*, which combines curves R1 and R2 in Figure 11.6a into one composite curve. Beginning with zero mass of solute at $y = 0.0003$, the lowest mass fraction of a rich stream, and using Table 11.2, the cumulative mass of solute removed at the interval mass fractions are

y	0.0003	0.0005	0.001305	0.0451	0.051	0.07
n_{cum}	0	0.00002	0.00082	0.04457	0.05047	0.06757

These points form the rich composite curve in Figures 11.6b and 11.6c. Note that the low concentration end is expanded in the *area of detail* into Figure 11.6c.

Normally, the lean composite curve is prepared in the same way. However, in this example, the curve is that for the one internal MSA in stream L1. It is simply copied from Figure 11.6a, but shifted

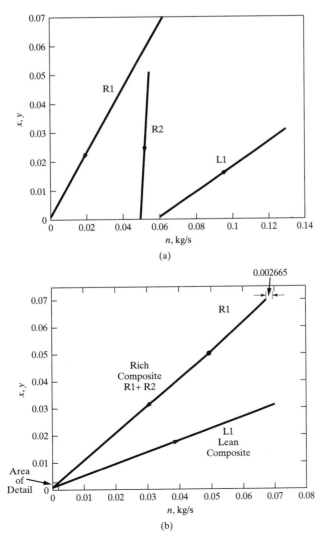

(a)

(b)

Figure 11.6 Graphical method to locate the minimum external MSA: (a) mass exchange curves for each stream, (b) and (c) composite rich and lean curves.

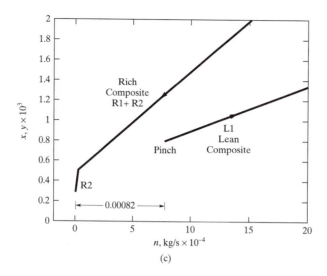

(c)

Figure 11.6 (*Continued*)

to the left so as to begin at $n = 0.00082$, the minimum amount of solute that must be removed by the external MSA, chilled methanol.

As shown in Figure 11.6c, the composite curves have a closest point of approach at the point where stream L1 begins along the lean composite curve; that is, $x = 0.0008$. The corresponding mass fraction on the rich composite curve is $y = 0.001305$, with a corresponding equilibrium mass fraction in the aqueous ammonia liquid phase of $0.001305/1.45 = 0.0009$. Thus, the approach is $0.0009 - 0.0008 = 0.0001$, which is Δx_{min}. Consequently, these two points provide the mass fractions at the pinch. If Δx_{min} is reduced to zero, the lean composite curve is shifted to the left until $y = 0.00116$ ($= 1.45 \times 0.0008$), the mass fraction at equilibrium with $x = 0.0008$. As mentioned earlier, this corresponds to an infinitely large mass exchanger.

In this example, with $\Delta x_{min} = 0.0001$, mass in the segments of the rich composite curve is transferred down vertically to the segments of the lean composite curve that lie below them. At the high-concentration ends, however, no segments on the rich composite curve lie vertically above the upper end of the lean composite curve. There, as computed previously and shown in the "Final Pass" of Figure 11.4, an additional 0.002665 kg/s of solute must be added. Alternatively, the flow rate of L1 can be reduced or its target mass fraction reduced from 0.0310 to eliminate the need for additional solute. This is consistent with the results using the composition-interval method in Example 11.1. Similarly, at the low-concentration ends of the composite curves, no segments of the lean composite curve lie vertically below the lower end of the rich composite curve. Here, 0.00082 kg/s must be removed by the external MSA, a result again consistent with the CI analysis. ■

As for heat integration in Section 10.2, many additional observations are noteworthy in connection with the rich and lean composite curves. One is that the slopes of the composite curves *always* decrease at the inlet concentration of a stream and increase at the outlet concentration of a stream. It follows that points at which the slope decreases are candidate pinch points, and furthermore, that one of the inlet concentrations is *always* a pinch concentration, when a pinch exists. Hence, to locate a potential pinch concentration, one needs only to examine the inlet concentrations of the streams.

Similarly, for some Δx_{min}, there are no pinch concentrations, which is analogous to HENs. In such cases, either excess internal MSAs exist or external MSAs are required, but not both. Their amounts equal the difference between the mass of solute to be removed from the rich streams and that required by the lean streams. In these cases, $\Delta x_{min} \leq \Delta x_{thres}$, where Δx_{thres} is the threshold approach concentration difference, above which a pinch exists. The threshold condition for HENs, which is similar to that for MENs, is discussed in Section 10.5.

In other cases, two or more pinch points arise. This occurs when the total flow rates of the rich and lean streams in a concentration interval are equal and the interval contains a pinch point.

11.3   MASS EXCHANGE NETWORKS FOR MINIMUM EXTERNAL MSA

Having determined the minimum flow rate of an external MSA (that is, the MOC target) using one of the two methods above, or using a linear program (similar to that for heat integration in Section 10.2), it is common to design two networks of mass exchangers, one on the rich side and one on the lean side of the pinch, as shown in Figure 11.5. In this section, a method, introduced by El-Halwagi and Manousiouthakis (1989) and similar to the method of Linnhoff and Hindmarsh (1983) for heat integration, is presented that places emphasis on positioning the mass exchangers (*stream matching*) by working out from the pinch.

Stream Matching at the Pinch

To explain the approach of El-Halwagi and Manousiouthakis (1989), it helps to refer to Figure 11.5, which shows the *pinch decomposition* of the rich and lean streams in Example 11.1. Attention is focused at the pinch where the mass fractions of the rich and lean streams are those determined as in Table 11.1 using Eq. (11.5) and the assumed value of Δx_{min}.

Consider the schematic of a countercurrent mass exchanger in Figure 11.7. The rich stream, having a flow rate of F_R, enters at y_i and exits at y_o. It transfers mass, n, to the lean stream that has a flow rate of F_L, enters at x_i and exits at x_o. Carrying out mass balances for the solute in the rich and lean streams:

$$n = F_R(y_i - y_o) \tag{11.6}$$

$$n = F_L(x_o - x_i) \tag{11.7}$$

$$F_R(y_i - y_o) = F_L(x_o - x_i) \tag{11.8}$$

When a mass exchanger is positioned on the rich side of the pinch, which is considered first arbitrarily, conditions at the lower end of the exchanger in Figure 11.7 become $y_o = y_{pinch}$, $x_i = x_{pinch}$, and Eq. (11.8) becomes:

$$F_R(y_i - y_{pinch}) = F_L(x_o - x_{pinch}) \tag{11.9}$$

Furthermore, accounting for the approach to phase equilibrium at the pinch, Eq. (11.5) becomes:

$$y_{pinch} = m(x_{pinch} + \Delta x_{min}) + b \tag{11.10}$$

At the upper end of the mass exchanger in Figure 11.7, it follows that the mass fractions at equilibrium must be separated by greater than Δx_{min}; that is,

$$y_i \geq m(x_o + \Delta x_{min}) + b \tag{11.11}$$

Substituting Eqs. (11.10) and (11.11) in Eq. (11.9):

$$F_R[m(x_o + \Delta x_{min}) + b - m(x_{pinch} + \Delta x_{min}) - b] \leq F_L(x_o - x_{pinch}) \tag{11.12}$$

Figure 11.7 Schematic of a countercurrent mass exchanger.

Rearranging, it follows that to install a mass exchanger at the pinch, on the rich side:

$$F_R \leq \frac{F_L}{m} \tag{11.13}$$

When a mass exchanger is positioned on the lean side of the pinch, $y_i = y_{pinch}$, $x_o = x_{pinch}$, and Eq. (11.8) becomes:

$$F_R(y_{pinch} - y_o) = F_L(x_{pinch} - x_i) \tag{11.14}$$

Substituting Eqs. (11.10) and

$$y_o \geq m(x_i + \Delta x_{min}) + b \tag{11.15}$$

in Eq. (11.14):

$$F_R[m(x_{pinch} + \Delta x_{min}) + b - m(x_i + \Delta x_{min}) - b] \geq F_L(x_{pinch} - x_i) \tag{11.16}$$

Rearranging, it follows that to install a mass exchanger at the pinch, on the lean side:

$$F_R \geq \frac{F_L}{m} \tag{11.17}$$

which is just the reverse of that on the rich side of the pinch. These two conditions are analogous to those used in HENs for working out from the pinch.

Stream Splitting at the Pinch

Similar to the design of HENs, when matching rich and lean streams at the pinch, on the rich side, it is necessary that the number of rich streams be less than or equal to the number of lean streams. When this is not the case, lean streams must be split until the number of rich and lean streams is equal. Also, on the lean side, it is necessary that the number of lean streams be less than or equal to the number of rich streams. Here, rich streams must be split until the number of rich and lean streams is equal. The development of MENs above and below the pinch, including the need for stream splitting, is illustrated in the next example.

EXAMPLE 11.3   *H₂S Removal from Sour Coke Oven Gas (Examples 11.1 and 11.2 Revisited)*

Returning to Figure 11.5, it is desired to synthesize a network of mass exchangers that utilizes the minimum external MSA. Two MENs are needed, one on the rich side and one on the lean side of the pinch.

SOLUTION

As discussed earlier, the capacity of stream L1 to remove solute exceeds the solute in steams R1 and R2 by 0.002665 kg/s. Rather than add solute to the rich streams, it is assumed that the flow rate of stream L1 can be reduced accordingly. Consequently, stream L1 is adjusted to remove 0.06946 − 0.002665 = 0.06680 kg/s of solute. Its adjusted flow rate is

$$F_{L1} = \frac{0.06680}{0.031 - 0.0008} = 2.2119 \text{ kg/s}$$

which is $2.3 - 2.2119 = 0.0881$ kg/s smaller.

Next, inequality (11.13) is checked for each potential match on the rich side of the pinch:

Match	$F_R \leq \frac{F_L}{m}$?
R1–L1	$0.9 \leq \dfrac{2.2119}{1.45} = 1.525$
R2–L1	$0.1 \leq \dfrac{2.2119}{1.45} = 1.525$

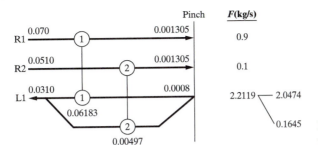

Figure 11.8 MEN on rich side of the pinch.

Both inequalities are satisfied, but stream L1 must be split to permit matches with both streams R1 and R2 at the pinch. This is accomplished as shown in Figure 11.8, based on the amounts of the solute to be removed above the pinch from R1 and R2. Using Table 11.2, $0.9(0.07 - 0.001305) = 0.06183$ kg/s of solute is to be removed from R1, and $0.1(0.051 - 0.001305) = 0.00497$ kg/s from R2. Therefore, stream L1 is split into the following portions to remove the entire amounts of solute from streams R1 $[0.06183/(0.0310 - 0.0008) = 2.0474$ kg/s] and R2 $[0.00497/(0.0310 - 0.0008) = 0.1645$ kg/s].

On the lean side of the pinch, stream L1 does not appear. Thus, the chilled methanol, stream L2, must be used. Its target mass fraction is $x = 0.0035$, which can be contacted by rich streams having mass fractions greater than or equal to $y = 0.00094$ [$= 0.26(0.0035 + 0.0001)$], which, as shown in Figure 11.5, is less than the mass fraction at the pinch ($y = 0.001305$). Thus, Eq. (11.17) does not apply. Two potential MENs are shown in Figure 11.9, one of which utilizes stream splitting. In both MENs, 0.000725 kg/s of solute must be removed from R1 and 0.000101 kg/s of solute from R2, as

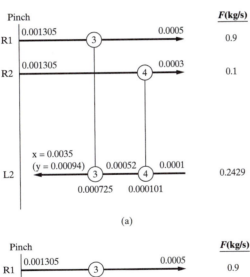

(a)

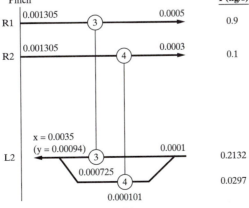

(b)

Figure 11.9 MENs on lean side of the pinch.

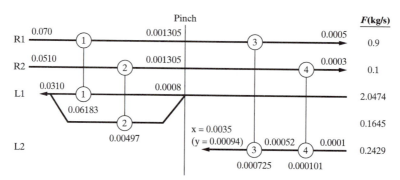

Figure 11.10 MEN having minimum number of mass exchangers, while satisfying the minimum usage of external MSA.

shown in Figure 11.9. In both cases, the equivalent value of y for the entering methanol $= 0.26(0.0001 + 0.0001) = 0.000052$. In both cases, the flow rate of methanol $= (0.000725 + 0.000101)/(0.0035 - 0.0001) = 0.2429$ kg/s, the minimum amount of external MEA.

Finally, the MEN on the rich side of the pinch in Figure 11.8 is combined in Figure 11.10 with the MEN in Figure 11.9a to give a MEN that utilizes the minimum external MSA. Alternatively, Figure 11.9b could be combined with Figure 11.8. The final selection would consider both capital and operating costs. ∎

11.4 MINIMUM NUMBER OF MASS EXCHANGERS

Having designed a MEN that meets the MOC target of minimum usage of an external MSA, it is common to consider the possible reduction in the number of mass exchangers toward the minimum while permitting the consumption of the external MSA to rise, particularly when small mass exchangers can be eliminated. In this way, lower annualized costs may be obtained, especially when the cost of external MSAs is low relative to the purchase cost of the mass exchangers.

Reducing the Number of Mass Exchangers—Breaking Mass Loops

By analogy to HENs, the minimum number of mass exchangers in a HEN is

$$N_{MX,min} = N_R + N_L - N_{NW} \tag{11.18}$$

where N_R and N_L are the number of rich and lean streams, and N_{NW} is the number of *independent networks*; that is, the number of subnetworks consisting of linked paths between the connected streams. Also, when the number of mass exchangers in a MEN exceeds the minimum, the difference, $N_{MX} - N_{MX,min}$, equals the number of independent *mass loops*. As the mass loops are broken, by combining mass exchangers, it is common for the amount of external MSA to increase. This is illustrated in the next example.

EXAMPLE 11.4 *H₂S Removal from Sour Coke Oven Gas (Example 11.3 Revisited)*

The MEN in Figure 11.10 contains the minimum number of mass exchangers with the minimum usage of external MSA, assumed equivalent to the minimum operating costs (MOC), given by:

$$N_{MX,min}^{MOC} = N_{MX,min}^{+} + N_{MX,min}^{-} \tag{11.19}$$

where $N_{MX,min}^{+}$ and $N_{MX,min}^{-}$ are the minimum number of mass exchangers on the rich and lean side of the pinch, keeping the external MSA at a minimum. For these streams, $N_{MX,min}^{+} = N_{MX,min}^{-} = 2 + 1 - 1 = 2$,

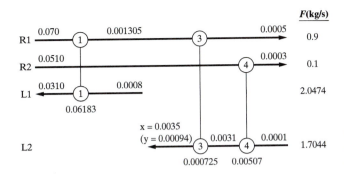

Figure 11.11 MEN having the minimum number of mass exchangers.

and consequently, $N_{\mathrm{MX,min}}^{\mathrm{MOC}} = 4$. Since, according to Eq. (11.18), treating the MEN as a whole, $N_{\mathrm{MX,min}} = 2 + 2 - 1 = 3$, one mass loop exists. To reduce to the minimum number of mass exchangers, this mass loop must be broken.

SOLUTION   Suppose the mass loop is broken by eliminating the smallest mass exchanger on the rich side of the pinch and shifting its mass load around the loop. The resulting MEN is shown in Figure 11.11, where the smallest mass exchanger is assumed to be the one with the smallest amount of solute transferred. Observe that the mass of the solute to be removed by the external MSA is increased in exchanger 4 by the amount removed by the internal MSA in mass exchanger 2, or 0.00497 kg/s, giving a new total for that exchanger of 0.00507 kg/s. This corresponds to a substantial increase in the flow rate of the external MSA to 1.7044 kg/s [= (0.000725 + 0.00507)/(0.0035 − 0.0001)] from 0.2429 kg/s.   ∎

11.5   ADVANCED TOPICS

Many of the concepts covered in Chapter 10 on heat and power integration apply in the design of MENs. These include the threshold $\Delta x_{\mathrm{thres}}$ and the optimal Δx_{min}, as well as strategies for mathematical programming, using MILP and MINLP formulations (El-Halwagi and Manousiouthakis, 1990; Papalexandri et al., 1994). For coverage of the former, see the book by El-Halwagi (1997). In addition, Hallale and Fraser (2000a,b) show how to extend the methods with simple approximations to obtain capital and operating cost estimates when calculating annualized costs.

Yet another topic involves the extension of the synthesis methods to processes with multiple solutes. Here, the impact of concentration on the slope of phase equilibrium Eq. (11.4) may become a factor for highly nonideal solutions at high concentrations. The analysis techniques presented herein can be extended when the slopes of the equilibrium curves can be approximated as constant, independent of mixture composition. Also, the analyses are simplified when the minimum external MSA for the principal solute is capable of removing the other solutes as well (El-Halwagi and Manousiouthakis, 1989).

Finally, it is possible to synthesize HENs and MENs simultaneously, for example, in the design of heat-integrated distillation networks (Bagajewicz and Manousiouthakis, 1992; Bagajewicz et al., 1998).

11.6   SUMMARY

Having studied this chapter, the reader should understand the parallels between heat and mass integration and be prepared to carry out analyses to identify the minimum external mass-separating agent. In addition, the reader should be able to position mass exchangers in a MEN beginning at the pinch and working outward. Lastly, the reader should have learned a strategy for stream splitting and for breaking mass loops while allowing solute to be exchanged across the pinch.

REFERENCES

Bagajewicz, M.J., and V. Manousiouthakis, "Mass/Heat-Exchange Network Representation of Distillation Networks," *AIChE J.*, **38**(11), 1769 (1992).

Bagajewicz, M.J., R. Pham, and V. Manousiouthakis, "On the State Space Approach to Mass/Heat Exchanger Network Design," *Chem. Eng. Sci.*, **53**(14), 2595–2621 (1998).

El-Halwagi, M.M., *Pollution Prevention Through Process Integration: Systematic Design Tools*, Academic Press, San Diego (1997).

El-Halwagi, M.M., and V. Manousiouthakis, "Synthesis of Mass Exchange Networks," *AIChE J.*, **35**(8), 1233–1244 (1989).

El-Halwagi, M.M., and V. Manousiouthakis, "Automatic Synthesis of Mass Exchange Networks with Single-component Targets," *Chem. Eng. Sci.* **45**, 2813–2831 (1990).

Hallale, N., and D.M. Fraser, "Capital and Total Cost Targets for Mass Exchange Networks, Part 1: Simple Capital Cost Models," *Comput. Chem. Eng.*, **23**, 1661–1679 (2000a).

Hallale, N., and D.M. Fraser, "Capital and Total Cost Targets for Mass Exchange Networks, Part 2: Detailed Capital Cost Models," *Comput. Chem. Eng.*, **23**, 1681–1699 (2000b).

Linnhoff, B., and E. Hindmarsh, "The Pinch Design Method for Heat Exchanger Networks," *Chem. Eng. Sci.*, **38**, 745 (1983).

Papalexandri, K.P., E.N. Pistikopoulos, and C.A. Floudas, "Mass Exchange Networks for Waste Minimization: A Simultaneous Approach," *Trans. Inst. Chem. Eng.*, **72**, 279–293 (1994).

Perry, R.H., and D.W. Green, Ed., *Perry's Chemical Engineer's Handbook*, 7th ed., McGraw-Hill, New York (1997).

Toor, H.L., "Diffusion in Three-component Gas Mixtures," *AIChE J.*, **3**, 198–207 (1957).

EXERCISES

11.1 A copolymerization plant uses benzene solvent. Benzene must be recovered from its gaseous waste stream. Two lean streams in the process, an additive stream and a catalytic solution, are potential process MSAs. Organic oil, which can be regenerated using flash separation, is the external MSA. The stream data are shown below:

Stream	F (kmol/s)	y^s or x^s	y^t or x^t
R1 (off-gas)	0.2	0.0020	0.0001
L1 (additives)	0.08	0.003	0.006
L2 (catalytic soln.)	0.05	0.002	0.004
L3 (organic oil)	Unlimited	0.0008	0.0100

In these concentration ranges, the following equilibrium equations apply.

Additives

$$y = 0.25x$$

Catalytic solution

$$y = 0.5x$$

Organic oil

$$y = 0.1x$$

a. Show how to utilize the process MSAs and minimize the amount of the external MSA required to remove benzene from the rich stream. Let $\Delta x_{min} = 0.0001$.

b. Design a MEN. Assuming that the operating cost of recirculating oil (including pumping, makeup, and regeneration) is $0.05/kg oil, calculate the annual cost of the oil.

11.2 An oil-recycling plant is shown in Figure 11.12 (El-Halwagi, 1997). Gas oil and lube oil streams are deashed and demineralized. Atmospheric distillation provides light gases, gas oil, and a heavy product, which is distilled under vacuum to produce lube oil. Subsequently, the gas oil is steam stripped to remove light and sulfur impurities. Similarly, the lube oil is dewaxed and deasphalted before it is steam stripped to remove light and sulfur impurities. The two wastewater streams contain phenol, a toxic pollutant that depletes oxygen, causes turbidity, and potentially causes objectionable taste and odor in fish and potable water. A MEN is to be designed involving the following internal streams:

Stream		y^s or x^s	y^t or x^t	F (kg/s)
R1	Condensate from Stripper 1	0.050	0.010	2
R2	Condensate from Stripper 2	0.030	0.006	1
L1	Gas oil	0.005	0.015	5
L2	Lube oil	0.010	0.030	3

Potential separations include solvent extraction using gas oil or lube oil. Note that phenol acts as an oxygen inhibitor, improves color stability, and reduces sediment formation in the oils. External MSAs include air, adsorption on activated carbon, and ion exchange on a polymeric resin. The following phase equilibrium data are provided for Eq. (11.4), with $b = 0$:

MSA	m
Gas oil	2.00
Lube oil	1.53
Air	0.04
Activated carbon	0.02
Ion exchange resin	0.09

Let $\Delta x_{min} = 0.001$ (kg phenol/kg MSA).

a. Locate the minimum amount of external MSA for a MEN. Use either the composition-interval method or the composite-curve method.

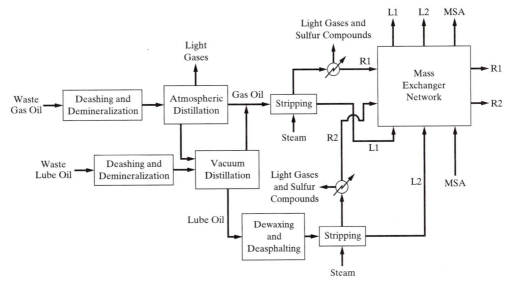

Figure 11.12 Oil-recycling plant.

b. Using activated carbon as the external MSA, design a MEN that has the minimum number of mass exchangers while utilizing the minimum amount of activated carbon.

c. Identify any mass loops in the solution to part b. Break these mass loops to design a MEN that has the minimum number of mass exchangers.

11.3 Ammonium nitrate is a fertilizer, also used in the production of explosives and other chemicals. It is commonly manufactured by neutralizing ammonia with nitric acid. Aqueous waste streams even containing low concentrations of ammonia and ammonium nitrate are toxic to aquatic life, also leading to eutrophication of lakes.

Two rich waste streams, R1 and R2, are produced: (1) wastewater containing ammonia, and (2) condensate, from the off-gas condenser downstream from the nitric-acid ammonia reactor, containing ammonium nitrate and ammonia. The treated condensate can be used as boiler feed water. Two potential recovery operations are air stripping and ion exchange. The table below provides data for the rich waste streams (R1 and R2) and the potential lean streams, L1 (containing air) and L2 (containing ion exchange resin).

where the flow rates are expressed on a solute-free basis and the compositions are in kg of solute per kg of solute-free solvent. In these concentration ranges, the following equilibrium equations apply.

NH₃

$$y_{air} = 0.788(x_{air} + 0.001) - 0.0002$$
$$y_{resin} = 0.11x_{resin} - 0.0006$$

NH₄NO₃

$$y_{air} = 0.98x_{air}$$
$$y_{resin} = 0.168x_{resin} - 0.0001$$

Let $\Delta x_{min} = 0.0001$.

a. Use NH_3 as the MSA. Find the minimum amount of NH_3 to achieve the separation.

b. Use NH_4NO_3 as the MSA. Find the minimum amount of NH_4NO_3 to achieve the separation.

c. For unit costs of air, 0.001 \$/kg, and ion exchange resin, 0.05 \$/kg, design MENs for parts a and b.

d. Repeat part c for the cheapest network using NH_3 and NH_4NO_3.

Stream	F (kg/s)	Species	y^s or x^s (kg/kg)	y^t or x^t (kg/kg)
R1 (wastewater)	2.6	NH_3	0.006	0.004
		NH_4NO_3	0.0	0.0
R2 (condensate)	0.8	NH_3	0.02	0.001
		NH_4NO_3	0.05	0.002
L1 (air)	Unlimited	NH_3	0.0	0.005
		NH_4NO_3	0.0	0.005
L2 (ion exchange resin)	Unlimited	NH_3	0.0	0.005
		NH_4NO_3	0.0	0.04

Chapter **12**

Optimal Design and Scheduling of Batch Processes

12.0 OBJECTIVES

This chapter introduces strategies for designing and scheduling batch processes. It begins with single equipment items, focusing on methods for achieving the optimal batch time and batch size. Then, reactor–separator processes are examined, where trade-offs exist between the reaction conversion, as it varies with reaction time, the cost of separation, which decreases with conversion, and the batch cycle time. Subsequently, methods of scheduling batch processes with recipes having numerous tasks and equipment items are considered. Initially, schedules are considered for plants involving a single product, produced in production trains that are repeated from batch to batch. The chapter concludes with strategies for designing multiproduct batch plants.

After studying this chapter, the reader should

1. Be knowledgeable about process units executed in batch mode and approaches for optimizing their designs and operations.
2. Know how to determine the optimal reaction time for a batch reactor–separator process.
3. Be able to schedule recipes for the production of a single chemical product.
4. Understand how to schedule batch plants for the production of multiple products.

12.1 INTRODUCTION

Continuous processes are dominant in the chemical process industries for the manufacture of commodity chemicals, plastics, petroleum products, paper, etc. When production rates are low, however—say, in the manufacture of specialty chemicals, pharmaceuticals, and electronic materials—it is difficult to justify the construction of a continuous plant comprised of small vessels and pipes. In these cases, it is common to design *batch processes* or *semicontinuous processes* that are hybrids between batch and continuous processes. The alternatives are illustrated schematically in Figure 12.1, with a continuous process shown in Figure 12.1a. In the batch process of Figure 12.1b, the chemicals are fed *before* (step 1) and the products are removed *after* (step 3) the processing (step 2) occurs. *Fed-batch processes* combine the first two steps with some or all chemicals being fed continuously during the processing. Then, when the processing is finished, the products are removed batchwise, as shown in

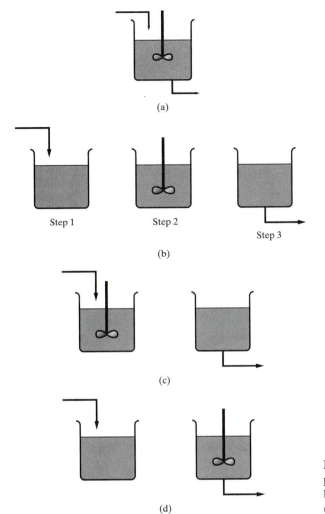

Figure 12.1 Continuous and batch processes: (a) continuous process; (b) batch process; (c) fed-batch process; (d) batch-product removal process.

Figure 12.1c. In *batch-product removal*, the chemicals are fed to the process before processing begins and steps 2 and 3 are combined; that is, the product is removed continuously as the processing occurs, as shown in Figure 12.1d. In effect, fed-batch and batch-product removal processes are semicontinuous processes.

The challenge in designing a batch, fed-batch, or batch-product removal process is in deciding on the size of the vessel and the processing time. This is complicated for the latter two processes where the flow rate and concentration of the feed stream or the flow rate of the product stream as a function of time strongly influence the performance of the process. Note that the determination of optimal operating profiles is referred to as the solution to the *optimal control problem*. This subject is introduced in the next section.

Batch and semicontinuous processes are utilized often when production rates are small, residence times are large, and product demand is intermittent, especially when the demand for a chemical is interspersed with the demand for one or more other products and the quantities needed and the timing of the orders are uncertain. Even when the demand is continuous and the production rates are sufficiently large to justify continuous processing, batch and semicontinuous processes are often designed to provide a reliable, though inefficient, route to the production of chemicals. For example, in the emulsion polymerization of resins, large batch reactors are installed often to avoid carrying out these highly exothermic reactions in continuous stirred-tank reactors. Note, however, that while operation at a low-conversion steady state is often less profitable than batch or semicontinuous processing, operation at an open-loop unstable steady state is often more profitable. Rather than install a control system to stabilize the operation, many companies prefer to operate in batch or semicontinuous mode. Similarly, design teams often opt for batch and semicontinuous processes when the chemicals are hazardous or toxic or when safety aspects are of great concern.

Because the designs for continuous and batch processes are usually very different, the choice of processing mode is made commonly during process synthesis, in the task integration step, as discussed in Section 3.4. At this stage, the decision to reject continuous processing is based upon rules of thumb, rather than a detailed comparison of the alternatives. Through process simulation, as discussed in Chapter 4, and the optimization methods presented in this chapter, more algorithmic methods are available for selecting from among the various batch and continuous processes.

Usually, for the production of small quantities of high-priced chemicals, such as in the manufacture of pharmaceuticals, foods, electronic materials, and specialty chemicals, batch, fed-batch, and batch-product removal processes are preferred. This is often the case in bioprocessing, for example, when drugs are synthesized in a series of chemical reactions, each having small yields, and requiring difficult separations to recover small amounts of product. This is also the case for banquet facilities in hotels, which prepare foods in batches, and for many unit operations in the manufacture of semiconductors. As discussed in Chapters 3 and 4, these processes usually involve a *recipe*, that is, a sequence of *tasks*, to be carried out in various items of equipment. In the latter sections of this chapter, variations on batch process schedules are discussed, as well as methods for optimizing the schedules.

12.2 DESIGN OF BATCH PROCESS UNITS

When designing a process unit to operate in batch mode, it is usually desired to determine the *batch time*, τ, and the *size factor*, S, which is usually expressed as the volume per unit mass of product, that maximize an objective like the amount of product. To accomplish this, a dynamic model of the process unit is formulated and the degrees of freedom adjusted, as illustrated in the examples that follow. As will be seen, there are many ways to formulate this *optimal control problem*. To simplify the discussion, models are presented and studied for various input profiles, to see how they affect the objectives. Emphasis is not placed on the formal methods of optimization.

Batch Processing

For conventional batch processing, with no material transfer to or from the batch, performance is often improved by adjusting the operating variables, such as temperature and agitation speed. Through these adjustments, reactor conversion is improved, thereby reducing the batch time to achieve the desired conversion. An example is presented next that shows how to achieve this objective by optimizing the temperature during batch processing.

EXAMPLE 12.1   *Exothermic Batch Reactor*

Consider a batch reactor to carry out the exothermic reversible reaction:

$$n_1 A \rightleftharpoons n_2 B$$

where the rate of consumption of A is:

$$r\{c_A, c_B, t\} = c_A^{n_1} k_1^{\circ} e^{\frac{-E_1}{RT}} - c_B^{n_2} k_2^{\circ} e^{\frac{-E_2}{RT}} \tag{12.1}$$

and where $E_1 < E_2$ for the exothermic reaction. The reaction is charged initially with A and B at concentrations, c_{A_o} and c_{B_o}. To achieve a specified fractional conversion of A, $X = (c_{A_o} - c_A)/c_{A_o}$, determine the profile of operating temperature in time that gives the minimum batch time. This example is based upon the development by Denn (1969).

SOLUTION

The minimum batch time, τ_{min}, is achieved by integrating the mass balances:

$$\frac{dc_A}{dt} = -r\{c_A, c_B, t\} \tag{12.2}$$

$$c_B\{t\} = c_{B_o} + \frac{n_1}{n_2}[c_{A_o} - c_A\{t\}] \tag{12.3}$$

while adjusting T at each point in time to give the maximum reaction rate.

The temperature at the maximum reaction rate is obtained by differentiation of Eq. (12.1) with respect to T:

$$\frac{dr}{dT} = 0 \tag{12.4}$$

Rearranging:

$$T_{opt} = \frac{E_2 - E_1}{R \ln \dfrac{c_B^{n_2} k_2^{\circ} E_2}{c_A^{n_1} k_1^{\circ} E_1}} \tag{12.5}$$

When an upper bound in temperature, T^U, is assigned, the typical solution profile is shown in Figure 12.2. Initially, when $T_{opt} > T^U$, the reactor temperature is adjusted to the upper bound, T^U. Then, as conversion increases, the reactor temperature decreases, leveling off to the equilibrium conversion. In practice, this optimal temperature trajectory is approached using feedback control, with the coolant flow rate adjusted to give temperature measurements that track the optimal temperature trajectory. ■

Fed-Batch Processing

Fermentation processes for the production of drugs are usually carried out in fed-batch reactors. In these reactors, it is desirable to find the best profile for feeding substrate into the fermenting broth, as illustrated in the next example.

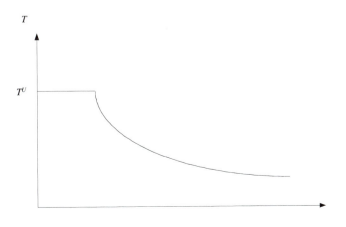

Figure 12.2 Temperature profile to minimize batch reactor time.

EXAMPLE 12.2   *Biosynthesis of Penicillin*

Consider the fed-batch reactor in Figure 12.3. Initially, the reactor is charged with an aqueous volume, $V\{0\}$, containing *E. coli* cells (referred to as biomass) in concentration, $X\{0\}$. Then, an aqueous solution of sucrose (referred to as the substrate; i.e., the substance being acted on) at a concentration, S_f (g/L), is fed to the reactor at a variable flow rate, $F\{t\}$ (g/hr). The reactor holdup, $V\{t\}$, contains *E. coli* cells in concentration, $X\{t\}$ (g/L), penicillin product in concentration, $P\{t\}$ (g/L), and sucrose in concentration, $S\{t\}$ (g/L). Using Monod kinetics, the specific growth rate of the cell mass (g cell growth/g cell) is

$$\mu = \mu_{max}\left(\frac{S}{K_x X + S}\right)$$

Lim and co-workers (1986) developed the following expressions for the specific rate of penicillin production (g penicillin/g cell):

$$\rho = \rho_{max}\left(\frac{S}{K_p + S(1 + S/K_{in})}\right)$$

and for the specific consumption rate of substrate (g substrate/g cell):

$$s = m_s\left(\frac{S}{K_m + S}\right)$$

Using mass balances for the cell mass, penicillin, and substrate, and the overall mass balance, the following rate equations can be derived (see Exercise 12.2):

$$\dot{X}\{t\} = \mu\{X, S\}X - \frac{X}{S_f V}F$$

$$\dot{P}\{t\} = \rho\{S\}X - K_{deg}P - \frac{P}{S_f V}F$$

$$\dot{S}\{t\} = -\mu\{X, S\}\frac{X}{Y_{X/S}} - \rho\{S\}\frac{X}{Y_{P/S}} - s\{S\}X + \left(1 - \frac{S}{S_f}\right)\frac{F}{V}$$

$$\dot{V}\{t\} = \frac{F}{S_f}$$

where $Y_{X/S}$ and $Y_{P/S}$ are the yield coefficients which relate the rate of substrate consumption to the rates of cell growth and penicillin production, respectively.

Using the feed concentration, $S_f = 500$ g S/L, with the kinetic parameters by Lim and co-workers (1986), $\mu_{max} = 0.11$ hr^{-1}, $K_x = 0.006$ g S/g X, $K_P = 0.0001$ g S/L, $\rho_{max} = 0.0055$ g P/(g X·hr), $K_{in} = 0.1$ g S/L, $K_{deg} = 0.01$ hr^{-1}, $m_s = 0.029$ g S/(g X·hr), $K_m = 0.0001$ g S/L, $Y_{X/S} = 0.47$ g X/g S, and $Y_{P/S} = 1.2$ g P/g S, for the initial conditions, $V\{0\} = 7$ L, $X\{0\} = 1.5$ g/L, and $P\{0\} = S\{0\} = 0$, and for the constraints,

$$0 \le X\{t\} \le 40$$
$$0 \le S\{t\} \le 100$$
$$0 \le V\{t\} \le 10$$
$$0 \le F\{t\} \le 50$$
$$72 \le X\{\tau\} \le 200$$

$F\{t\}, S_f$

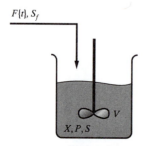

Figure 12.3 Batch penicillin reactor.

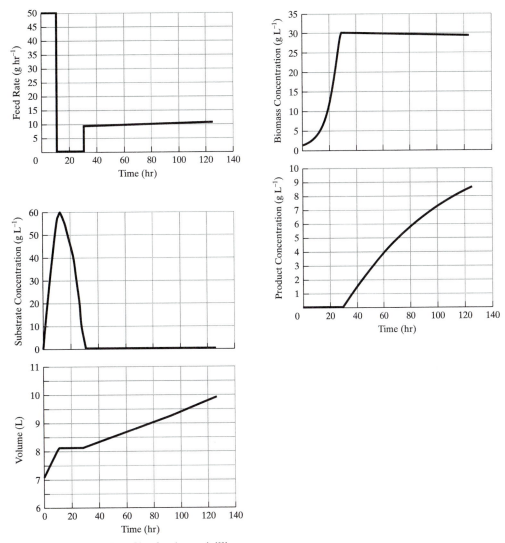

Figure 12.4 Optimal profiles for the penicillin reactor.

Cuthrell and Biegler (1989) maximize the production of penicillin, $P\{\tau\}V\{\tau\}$, where τ is the batch time, using variational calculus (Pontryagin maximum principle) to obtain the solution in Figure 12.4. As seen, at the optimum, the batch time is 124.9 hr, with the production of 87.05 g of penicillin. It is notable that the optimal feed flow rate is 50 g/hr for the first 11.21 hr, after which the feed stream is turned off until 28.79 hr, when it is held constant at 10 g/hr. This "on–off" control strategy is commonly referred to as *bang–bang control*. To confirm the cell mass, penicillin, and substrate concentration profiles, the differential equations can be integrated using a mathematical software package such as MATLAB. Furthermore, Cuthrell and Biegler (1989) show how to solve numerically for the optimal solution, which is often referred to as the "optimal control profile," using orthogonal collocation on finite elements, to discretize the differential equations, and successive quadratic programming (SQP). ∎

Batch-Product Removal

When distillations are carried out in batch mode, the still is charged with the feed mixture and the heat is turned on in the reboiler. The lightest species concentrate in the distillate, which is condensed and recovered in *batch-product removal* mode. As the light species is recovered, it is accompanied by increasing fractions of the heavier species unless a strategy is applied to maintain a high concentration of light species.

For multicomponent separations, to simplify operation it is often satisfactory to adjust the reflux rate once or twice while recovering each species. When the purity of a species, which is being collected in the product accumulator, drops below its specification, the contents of the product accumulator are dumped into its product receiver, and the reflux rate is adjusted. Stated differently, the reflux rate is increased as the difficulty of the separation between the light and heavy key components increases. This is illustrated in the next example.

EXAMPLE 12.3   *Batch Distillation*

A 100 lbmole mixture of methanol, water, and propylene glycol, with mole fractions, 0.33, 0.33, and 0.34, is separated using a 15-tray batch distillation operation, as shown in Figure 12.5. Assume operation at a nominal pressure of 1 atm, realizing that the pressure in the still will have to be somewhat higher than this to avoid a vacuum in the reflux accumulator. The tray and condenser liquid holdups are 0.1 ft^3/tray and 1.0 ft^3, respectively.

In an attempt to devise a satisfactory operating strategy, the following recipe (also called a campaign) was proposed:

Methanol Recovery

1. Bring the column to total reflux operation, with the distillate valve closed.
2. Using a reflux ratio of 3, send 5 lbmol/hr of distillate continuously to the product accumulator. Continue until the mole fraction of water in the instantaneous distillate reaches 0.001.
3. Bring the column to total reflux.
4. Using a reflux ratio of 5, send 2.5 lbmol/hr of distillate continuously to the product accumulator. Continue until the mole fraction of water in the instantaneous distillate reaches 0.001, at which point the distillate valve is closed. Dump the contents of the product accumulator into the methanol product receiver.

Propylene Glycol Recovery

5. The column is now at total reflux.
6. Using a reflux ratio of 3, send 20 lbmol/hr of distillate continuously to the product accumulator. Continue until the mole fraction of propylene glycol in the instantaneous distillate reaches 0.001. Dump the contents of the product accumulator into the water product receiver.
7. Pump the contents of the still pot into the propylene glycol product receiver.

To examine the performance of this recipe, it is helpful to use a batch distillation program in a process simulator, such as BATCHFRAC by Aspen Technology, Inc. (Boston et al., 1981). Then, a

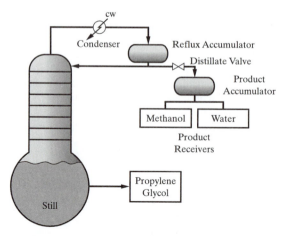

Figure 12.5 Batch distillation operation.

processing objective can be specified (such as minimum batch time, energy consumption, or reject chemicals, or some combination of these) and variations on the recipe can be explored in an effort to achieve more optimal operation, as in Exercise 12.3.

SOLUTION

Using the BATCHFRAC simulator, the results are as follows, where Distil. and Accum. refer to the instantaneous distillate and product accumulator, respectively:

After Step		Methanol	Water	Propylene Glycol	Total Amount (lbmol)	Step Time (hr)	Total Time (hr)
1	Charge	0.3300	0.3300	0.3400	100		0
2	Distil.	0.9990	0.0010	—	—	5.63	5.63
	Accum.	0.9999	0.0001	—	28.14		
	Still^x	0.0676	0.4592	0.4732	71.86		
3	Total reflux						
4*	Distil.	0.9990	0.0010	—	—	0.69	6.32
	Accum.	0.9999	0.0001	—	29.88		
	Still^x	0.0446	0.4705	0.4849	70.12		
5	Total reflux						
6*	Distil.	—	0.9990	0.0010	—	1.65	7.97
	Accum.^{x}	0.0947	0.9053	—	33.03		
	Still	—	0.0836	0.9164	37.09		

*Before the reflux accumulator is dumped.
^{x}Includes tray and condenser liquid holdups

As can be seen, the total batch time is nearly 8 hr, the amounts of 99.9 mol% methanol and 91.64 mol% propylene glycol products are 29.88 lbmol and 37.09 lbmol, respectively. Note that after step 4 the methanol product accumulator contains $0.999 \times 29.88 = 29.85$ lbmol methanol. The remainder, $33.33 - 29.85 = 3.48$ lbmol methanol is recovered initially in the water product accumulator during step 6. Hence, the water product accumulator contains a "slop cut" of water. Nearly all of the propylene glycol is recovered in the still. These results can be reproduced using the BATCHFRAC file, EXAM12–3.bkp, on the multimedia CD-ROM. ∎

12.3 DESIGN OF REACTOR–SEPARATOR PROCESSES

In this section, an approach to solving the optimal control problem is introduced for reactor–separator processes. The approach involves the simultaneous determination of the batch times and size factors for both of the process units. Furthermore, the interplay between the two units involves trade-offs between them that are adjusted in the optimization. It should be noted that simpler models, than in normal practice, are used here to demonstrate the concept and, in the first example, provide an analytical solution that is obtained with relative ease.

EXAMPLE 12.4

Consider the batch reactor–separator combination in Figure 12.6, initially presented by Rudd and Watson (1968). In the reactor, the isothermal, irreversible reaction, A → B, is carried out. The separator recovers the product B from unreacted A. The rate constant for the reaction is $k = 0.534$ day^{-1}. It is assumed that the cost of A is negligible compared to the value of the product, B, $C_B = \$2.00/lb$, that the operating costs of the reactor are proportional to the batch time (that is, $C_O = \alpha t$, where $\alpha = \$100/day$), and that the cost of separation per batch, C_S, is inversely proportional to the conversion of A in the reactor, X (that is, $C_S = KV/X$), where K is the proportionality constant and V is the holdup volume in the reactor. Find the batch time for the reactor when the gross profit in $/day is maximized. Let $c_{Ao} = 10$ lb/ft^3, $V = 100$ ft^3, and $K = 1.0$ \$/(ft^3·batch). Note that as the conversion of A to product B increases, the cost of separation decreases, as shown in Figure 12.7b.

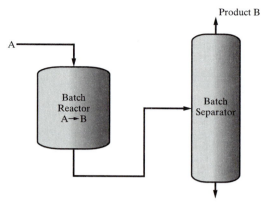

Figure 12.6 Reactor–separator process.

SOLUTION

For the first-order reaction, it can be shown (Exercise 12.4) that:

$$X = 1 - e^{-kt} \qquad (12.6)$$

as illustrated in Figure 12.7a, and

$$c_B = c_{Ao}X \qquad (12.7)$$

It is desired to maximize the gross profit, GP, in \$/day; that is,

$$\max_{t} \text{GP} = \frac{1}{t}[C_B V c_B\{t\} - C_O\{t\} - C_S\{X\{t\}\}] \qquad (12.8)$$

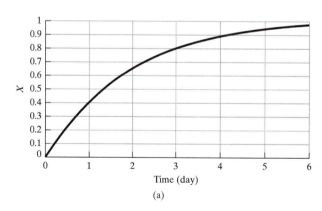

(a)

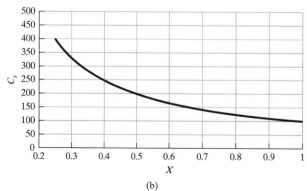

(b)

Figure 12.7 Conversion and cost of separation. (a) First-order response of conversion; (b) cost of separation as a function of conversion.

To locate the maximum, substitute the equations above and differentiate:

$$\frac{d\text{GP}}{dt} = 0$$

It can be shown (Exercise 12.4) that at the maximum, the reactor batch time is $\tau_{opt} = 1.35$ day, with a fractional conversion, x_{opt}, of 0.514. At longer times, while the revenues increase due to increased conversion and the separation costs decrease, the gross profit decreases due to the increase in the batch time in the denominator of Eq. (12.8). At shorter times, the revenues decrease due to smaller conversion and separation costs increase due to a more difficult separation, more than countering the decrease in the denominator and leading to smaller gross profit. At the maximum, $\text{GP}_{opt} = \$516.8/\text{day}$. Note that $\text{GP} < 0$ when $t < 0.507$ day.   ∎

EXAMPLE 12.5

Figure 12.8 shows an isothermal batch reactor, in which the irreversible reactions, $A \rightarrow B \rightarrow C$, take place, with B the desired product and C an undesired byproduct. The reactions are irreversible and first-order in the reactants. The reaction rate constants are $k_1 = 0.628 \text{ hr}^{-1}$ and $k_2 = 0.314 \text{ hr}^{-1}$. The reactor products are fed from an intermediate storage tank (not shown) to a batch distillation column, from which the most volatile species, A, is removed first in the distillate, followed by the product B, which is the intermediate boiler. For specified reactor volume, V_r, and column volume, V_c, it is desired to determine the batch times for the reactor and column that minimize the cost of producing the desired amount of product B, B_{tot} moles, in a single campaign. Following Barrera and Evans (1989), the analysis is simplified by assuming that the distillation column produces pure A until it is depleted from the still, followed by pure B. Furthermore, specifications include the total molar concentration in the reactor, C, the distillate flow rate, F_d, the time horizon within which the campaign must be completed, τ_{hor}, and the cleaning times between batches for the reactor and distillation column, t_{cr} and t_{cc}. In addition, several cost coefficients are specified, including the cost of fresh feed, P_A (\$/mol), recycling credit for A, P_{rA}, cost or credit for the byproduct C, P_C; the rental rates of the reactor, r_r (\$/hr), intermediate storage, r_s, and distillation column, r_c; the costs of cleaning the reactor, C_{clr} (\$/batch), and distillation column, C_{clc}; and the utility cost per mole of distillate, P_u. These specifications are given in Table 12.1.

SOLUTION

By integration of the kinetic rate equations, it can be shown (Exercise 12.5) that the mole fraction profiles in time are given by:

$$x_A = e^{-k_1 t_r} \tag{12.9}$$

$$x_B = \frac{k_1(e^{-k_1 t_r} - e^{-k_2 t_r})}{k_2 - k_1} \tag{12.10}$$

$$x_C = 1 - x_A - x_B \tag{12.11}$$

These profiles are graphed in Figure 12.9.

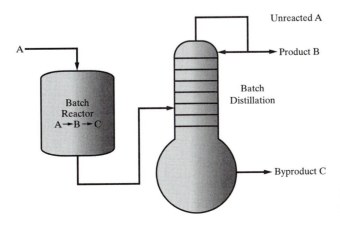

Figure   12.8 Reactor–distillation process.

Table 12.1 Specifications for Example 12.5

$V_r = 3{,}503$ L, $V_c = 784$ L
$B_{tot} = 800$ mol
$C = 0.5$ mol/L
$F_d = 206.2$ mol/hr
$\tau_{hor} \le 24$ hr
$t_{cr} = 0.8$ hr, $t_{cc} = 0.5$ hr
$P_A = 1$ \$/mol, $P_{rA} = 0.2$ \$/mol, $P_C = 0.4$ \$/mol
$r_r = 100$ \$/hr, $r_s = 20$ \$/hr, $r_c = 100$ \$/hr
$C_{clr} = 100$ \$/batch, $C_{clc} = 100$ \$/batch
$P_u = 0.45$ \$/mol

The minimization of the campaign costs is expressed as:

$$\min_{t_r, \tau_c} \phi = \phi_{rm} + \phi_{eq} + \phi_{cl} + \phi_u$$

where t_r is the reaction time (reactor operating time), τ_c is the batch time for the distillation column, and the campaign costs for the raw materials, equipment rental, cleaning, and utilities, are

$$\phi_{rm} = CV_r[P_A + P_{rA}x_A + P_C x_C]\tau_{tot}/\tau_r$$
$$\phi_{eq} = [r_r + r_s + r_c]\tau_{tot}$$
$$\phi_{cl} = [C_{clr}/\tau_r + C_{clc}/\tau_c]\tau_{tot}$$
$$\phi_u = P_u F_d t_c \tau_{tot}/\tau_c$$

Note that the quantity $P_u F_d$ is \$/hr charge for utilities, t_c/τ_c is the fraction of column batch time during which the column is operational (and using utilities), and τ_{tot} is the total batch time. These quantities are multiplied to give ϕ_u, the cost of utilities per batch. This assumes that the utilities are applied during a t_c/τ_c fraction of the reactor operation. The minimization is carried out subject to a mass bálance for the reactor:

$$B_{tot} = CV_r x_B \tau_{tot}/\tau_r,$$

an equation that assures periodicity in the storage tank (i.e., equal volumes processed per unit time):

$$\frac{V_r}{\tau_r} = \frac{V_c}{\tau_c},$$

a mass balance for the distillation column:

$$t_c = V_r C(1 - x_C)/F_d$$

and the total campaign time:

$$\tau_{tot} = \tau_r + \tau_c$$

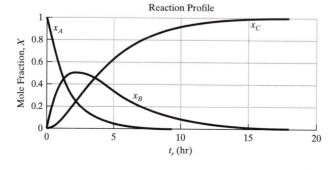

Figure 12.9 Mole fraction profiles for batch reactions: $k_1 = 0.628$ hr^{-1} and $k_2 = 0.314$ hr^{-1}.

where τ_r is the batch time for the reactor and t_c is the column operating time. In addition, as the optimization proceeds, it is necessary to satisfy the following inequality constraints:

$$\tau_{tot} \leq \tau_{hor}$$
$$\tau_r - (t_r + t_{cr}) \geq 0$$
$$\tau_c - (t_c + t_{cc}) \geq 0$$

For the specifications in Table 12.1, the following solution was obtained using successive quadratic programming (SQP) in GAMS (see Section 18.6). The reactor operates for $t_r = 4.44$ hr to produce a product containing 0.062, 0.373, and 0.565 mole fractions of A, B, and C. For periodic cycling, its batch time, $\tau_r = 18.74$ hr, which exceeds the total of the reactions and cleaning times. The distillation column operates for $t_c = 3.69$ hr, which together with its cleaning time, $t_{cc} = 0.5$ hr, gives a batch time for the column, $\tau_c = 4.19$ hr. The total time, or cycle time, is $\tau_{tot} = 22.93$ hr, which falls within the horizon time specified, $\tau_{hor} = 24$ hr. These times correspond to the minimum cost, $\phi = \$10,240$. By varying t_r, it can be shown that this is the minimum cost. ∎

12.4 DESIGN OF SINGLE PRODUCT PROCESSING SEQUENCES

Having examined small optimal control problems for batch process units in Section 12.2 and for reactor–separator sequences in Section 12.3, it should be clear that the determination of optimal batch times, given batch sizes expressed as batch volumes per unit mass of product, can be demanding computationally. Since most processes, in practice, have recipes with numerous tasks and a comparable number of processing units (e.g., the tPA process in Chapters 3 and 4), it is normally not practical to optimize the batch times for the individual processing units when preparing a schedule of tasks and equipment items for the manufacture of a product. Consequently, when preparing a schedule of tasks and equipment items, it is common to specify batch times for tasks to be performed in specific units, usually with batch sizes, and to optimize cycle times for a specific recipe. In some cases, using the rates of production and yields, the vessels are designed as well; that is, vessel sizes are determined to minimize the cost of the plant while determining the cycle times for a specific recipe. In this section, schedules are determined for the batch processes that involve only single products. In the next section, the methodology is extended for multi-product batch processes.

Batch process design begins with the specification of a *recipe* of *tasks* to produce a product. In continuous processing, each task is carried out in a specific equipment item, with one-to-one correspondence between them, shown on a flowsheet that remains fixed in time. Similarly, in batch processes, the tasks are assigned to equipment items, but over specific intervals of time, which vary with batch size, which is often determined by the available equipment sizes. For example, in the tPA process in Sections 3.4 and 4.5, given the rate of tPA production [50 pg tPA/(cell-day), where pg are picograms $= 10^{-12}$] and the cell concentration (between 0.225×10^6 and 3×10^6 cell/mL), the availability of a 5,000-L cultivator determines the 14-day batch time and the batch size (2.24 kg of tPA, produced in 4,000 L of medium, yielding 1.6 kg of final product) for the cultivator. As discussed in Section 3.4, process synthesis involves the creation of a sequence or flowsheet of operations, which can be referred to as a *recipe* of operations or tasks. During the *task integration* step, tasks are often combined to be carried out in a single equipment item; for example, heating and reaction in a pyrolysis furnace. Also, during this step, the decision to use continuous or batch processing is made. At this point, the available equipment sizes often determine the batch sizes and times.

Batch Cycle Times

When scheduling and designing batch processes, several formalisms are widely used, as reviewed by Reklaitis (1995). In this section, and those that follow, portions of the presentation are derived from his article.

In batch processes, it is common for a task to consist of a sequence of *steps* to be carried out in the same equipment unit. For example, Figure 12.10 shows a typical recipe with its tasks and steps. Note that each step involves a *batch time*, which is determined by the processing rates and the *batch size*; that is, the amount of the *final* product manufactured in one batch. Furthermore, a *production line* is a set of equipment items assigned to the tasks in a recipe to produce a product. When a production line is used to produce a sequence of identical batches, the *cycle time* is the time between the completions of batches. To better visualize the schedule of production, an equipment occupation diagram, known as a *Gantt chart*, is prepared, showing the periods of time during which each equipment item is utilized, as shown in Figure 12.11a. Note that the unit having the longest batch time (6 hr), U2, is the *bottleneck* unit, as it is always in operation. Note also that the second batch is begun in time to produce the feed to the unit, U2, when the latter becomes available after processing the first batch. In this diagram, the batches are transferred from unit-to-unit immediately (so-called *zero-wait* strategy, with no intermediate storage utilized). Clearly, the cycle time, 6 hr, is the batch time of U2.

In the schedule in Figure 12.11a, the serial process has a distinct task assigned to each equipment item. Often, to utilize the equipment more efficiently, it is possible to use an equipment item to carry out two or more tasks. Note that this may not be possible when manufacturing specialty chemicals that are very sensitive to contamination, as in the manufacture of pharmaceuticals. Returning to the schedule in Figure 12.11a, when the fourth task can be carried out in U1, this unit is better utilized and U4 can be released for production elsewhere in the batch plant, as shown in Figure 12.11b. Note that to achieve this schedule, without adding intermediate storage, it is necessary to retain the batch within U3 until U1 becomes available. Furthermore, to increase the efficiency of the schedule, that is, reduce the cycle time, it is common to add one or more units in parallel. When in phase, it is clear that the batch time for the unit is reduced to τ_j/n_j, where n_j is the number of units in parallel for task j. For example, when two U2 units, each half-size, are installed in parallel, the effective batch time for unit U2 is reduced to 3 hr, and the cycle time is reduced to 4 hr, with U3 the bottleneck unit. Alternatively, the parallel units can be sequenced out-of-phase, without altering their batch time, as shown in Figure 12.11c. In both cases, the U2 bottleneck is eliminated and the cycle time is reduced to 4 hr.

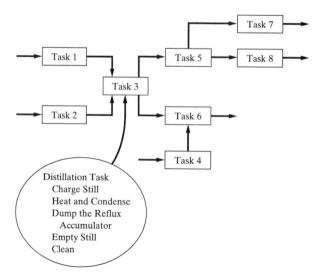

Figure 12.10 Recipes, tasks, and subtasks.

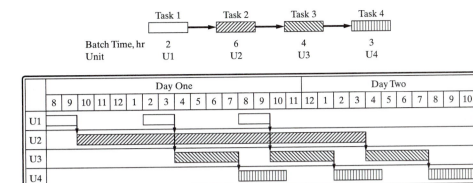

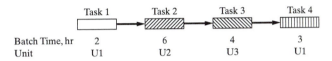

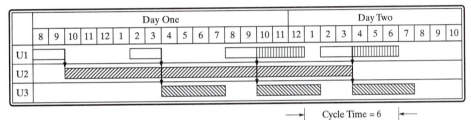

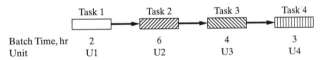

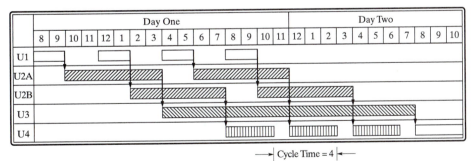

Figure 12.11 Serial recipe and Gantt charts.

Clearly, without parallel operation, the batch cycle time, CT, is the maximum of the batch times, $\tau_j, j = 1, \ldots, M$:

$$CT = \max_{j=1,\ldots,M} \tau_j \qquad (12.12)$$

where M is the number of unique equipment units. With n_j units in parallel and in phase, the cycle time is given by:

$$CT = \max_{j=1,\ldots,M} \frac{\tau_j}{n_j} \qquad (12.13)$$

Returning to the example, when two units U2 are installed in parallel to perform task 2:

$$CT = \max_{j=1,\ldots,M} \frac{\tau_j}{n_j} = \max\{2, \tfrac{6}{2}, 4, 3\} = 4 \text{ hr} \qquad (12.14)$$

Intermediate Storage

Thus far, two storage options have been illustrated. No storage is used in the schedules of Figures 12.11a and 12.11c, with the contents of each unit transferred immediately to the next unit, experiencing no delay after its task has been completed. As mentioned above, this is the so-called *zero-wait* (ZW) strategy. In the schedule of Figure 12.11b, U3 provides intermediate storage until U1 becomes available. Hence, a zero-wait strategy is implemented, with some intermediate storage when necessary. This is referred to as an *intermediate storage* (IS) strategy. The third strategy involves unlimited intermediate storage (UIS), sufficient to hold the contents of the products from a unit having a lengthy batch time, to be used repeatedly in a unit having half the batch time or less, as illustrated in Figure 12.12. Here, U1 is utilized at all times and the cycle time is reduced from 9 to 3 hr. To produce a specified amount of product, the batch size is reduced by a factor of one-third since the cycle time is divided by three.

Batch Size

It is convenient to define the size factor, S_j, for task j, as the capacity required per unit of product. Commonly, it is defined as the volume required to produce a unit mass of product. For example, for the third cultivator in the tPA process of Sections 3.4 and 4.5, 4,000 L of medium yields 2.24 kg of tPA, which eventually yields 1.6 kg of final tPA product. Consequently, its size factor is 4,000 L/ 1.6 kg = 2,500 L/kg tPA product. Size factors can be computed for each task in a recipe. Normally, equipment vessel sizes are selected that exceed batch volume by 10 to 20%. Clearly, the batch factor in volume/mass produced is determined by the rate of processing the batch (e.g., kg/hr) multiplied by the batch time (hr) and divided by the density of the batch (kg/L) and the mass of product produced (kg).

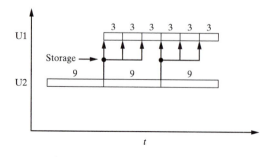

Figure 12.12 Gantt chart with unlimited intermediate storage (UIS).

12.5   DESIGN OF MULTIPRODUCT PROCESSING SEQUENCES

A *multiproduct batch plant* produces a set of products whose recipe structures are the same, or nearly identical. One example is a foundry that manufactures integrated circuit (IC) chips in which several different devices are produced simultaneously, each involving hundreds of tasks and utilizing several equipment items. In these plants, each product is produced in the same production line, with multiple processing tasks carried out using the same equipment items. The recipes are expressed in serial campaigns for each product. Figure 12.13 shows schedules in which a campaign of two batches to produce product A is followed by a campaign of two batches to produce product B. It should be noted, however, that because the tasks for products A and B differ in equipment utilized, the plant is not a multiproduct batch plant; instead, it is referred to as a *multipurpose batch plant*. Although the cycle times for both products are identical (4 hr), it is common for the product cycle times to be unequal. The use of alternating product cycles is a limitation that does not apply to *general multipurpose plants* in which there are no well-defined production lines and no cyclic patterns of batch completion, as shown in Figure 12.14. Such plants are more flexible and effective for a large number of products that are produced in small volumes, where their vessels are cleaned easily and the presence of trace contaminants in the products is not a concern. Their equipment items are utilized more completely, without the idle-time gaps in plants with cyclic campaigns for each product. Consequently, multiproduct batch plants are used for larger volume products having similar recipes, as is often the case for plants that produce a family of grades of a specific product.

Scheduling and Designing Multiproduct Plants

For an existing plant, the scheduling problem involves a specification of the: (1) product orders and recipes, (2) number and capacity of the equipment items, (3) a listing of the equipment items available for each task, (4) limitations on the shared resources (e.g., involving the usage of utilities and manpower), and (5) restrictions on the use of equipment due to operating or

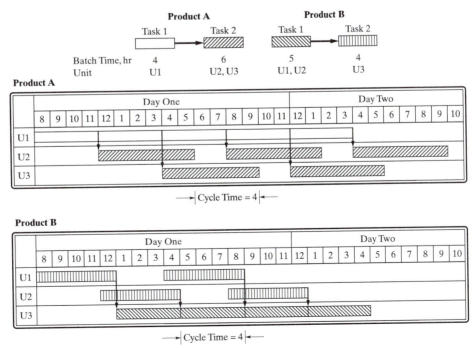

Figure 12.13 Gantt charts for a multipurpose plant.

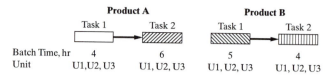

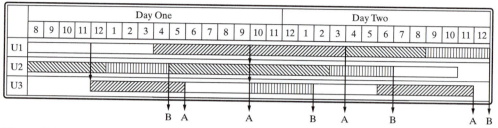

Figure 12.14 Gantt chart for a general multipurpose plant.

safety considerations. In solving the problem, that is, determining an optimal schedule, the order in which tasks use the equipment and resources is determined, with specific timings of the tasks provided, that optimize the plant performance (which can be specified in many ways, e.g., to maximize the gross profit).

When the plant does not exist, that is, when a new plant is to be designed, the product orders are usually not well defined. Otherwise, the specifications are identical. In fact, the design problem encompasses the scheduling problem in that its solution involves determining the number and capacity of the equipment items in addition to the optimal schedule. For the design problem, these are determined to optimize an objective that includes the investment costs of the equipment, such as the annualized cost. Because the product orders are not as well known during the design stage, it is common to solve the scheduling problem less rigorously.

As mentioned earlier, it is common to specify size factors and input/output ratios as known constants when defining recipes. Also, batch times for each task are often specified as constant, or as known functions of the batch size. These can be determined by optimizing the operation of each equipment item, as discussed in Section 12.2.

It is common to formulate the design problem for a multiproduct batch plant involving the processing of batch campaigns in series (i.e., one-at-a-time—commonly referred to as a *Flowshop plant*) as a mixed-integer nonlinear program (MINLP). Then, the formulation is simplified for solution using strategies that are beyond the scope of this book (Biegler et al., 1998). Herein, as an introduction, a typical formulation is presented without simplification. It begins with the objective, that is, to minimize the total investment cost, C:

$$\min C = \sum_{j=1}^{M} m_j a_j V_j^{\alpha_j}$$

where m_j is the number of out-of-phase units assigned to task j (integer variables), M is the number of tasks, and V_j is the size of the unit assigned to task j (usually in L; a_j and α_j are cost coefficients). This objective is minimized commonly subject to inequalities that involve the vessel size:

$$V_j \geq B_i S_{ij}$$

where B_i is the batch size of product i (i.e., the final product size, typically in kg), and S_{ij} is the size factor for task j in producing product i (typically in L/kg). This inequality insures that the unit size exceeds the smallest size required to produce all of the products. In addition, lower and upper bounds are specified on the equipment sizes in accordance with manufacturing limitations:

$$V_j^L \leq V_j \leq V_j^U$$

Inequalities are associated also with the cycle time and time horizon:

$$CT_i \geq \tau_{ij}/m_j$$

$$\sum_{i=1}^{N} \frac{Q_i}{B_i} CT_i \leq H$$

where CT_i is the cycle time for producing product i [which can be determined using Eqs. (12.12) and (12.13)], τ_{ij} is the batch time for task j in producing product i, Q_i is the annual demand for product i (typically, in kg/yr), and H is the production hours available annually.

12.6 SUMMARY

Initially, this chapter focuses on the optimal control of batch processing units, with emphasis on reducing the batch time and batch size. Then, the batch times for reactor–separator processes are optimized with emphasis on the interactions between the process units and the trade-offs in adjusting their batch times. Finally, the problem of determining operating schedules for single- and multiproduct batch plants, involving the possibility of intermediate storage and complex recipes with numerous tasks in numerous process units is examined.

REFERENCES

Barrera, M.D., and L.B. Evans, "Optimal Design and Operation of Batch Processes," *Chem. Eng. Comm.*, **82**, 45–66 (1989).

Biegler, L.T., I.E. Grossmann, and A.W. Westerberg, *Systematic Methods of Chemical Process Design*, Prentice-Hall, Englewood Cliffs, New Jersey (1998).

Boston, J.F., H.I. Britt, S. Jirapongphan, and V.B. Shah, "An Advanced System for Simulation of Batch Distillation Operations," in R.S.H. Mah and W.D. Seider, eds., *Foundations of Computer-aided Process Design*, Vol. 2, AIChE, New York (1981).

Cuthrell, J.E., and L.T. Biegler, "Simultaneous Optimization and Solution Methods for Batch Reactor Control Profiles," *Comput. Chem. Eng.*, **13**, 49–62 (1989).

Denn, M.M., *Optimization by Variational Methods*, McGraw-Hill, 1969.

Lim, H.C., Y.J. Tayeb, J.M. Modak, and P. Bonte, "Computational Algorithms for Optimal Feed Rates for a Class of Fed-batch Fermentations: Numerical Results for Penicillin and Cell Mass Production," *Biotechnol. Bioeng.*, **28**, 1408–1420 (1986).

Reklaitis, G.V., "Computer-aided Design and Operation of Batch Processes," *Chem. Eng. Ed.*, 76–85, Spring (1995).

Rudd, D.F., and C.C. Watson, *Strategy of Process Engineering*, Wiley, New York (1968).

EXERCISES

12.1 In Example 12.2, derive the mass balances for the cell mass, penicillin, and substrate, and the overall mass balance.

12.2 For the penicillin reactor in Example 12.2, using repeated simulations search for the optimal feed profile to maximize:

$$0.025P\{\tau\}V\{\tau\} - 1.68\tau - 0.00085 \int_0^\tau F\{t\}\, dt$$

where τ is the batch time. This objective function maximizes the penicillin produced while penalizing long batch times and the cost of the substrate feed stream. Indicate how the penalty terms affect the feed rate profile.

12.3 For the batch distillation column in Example 12.3, devise a recipe that will decrease the batch time without reducing the amount of product recovered. Estimate the increase in the utility usage.

12.4 In Example 12.4, derive Eqs. (12.6) and (12.7). Then, graph the gross profit as a function of the reactor batch time for various values of the rate constant, k, over the range 0.4–0.6 day^{-1}.

12.5 In Example 12.5, derive Eqs. (12.9)–(12.11).

12.6 For the reactor–distillation process in Example 12.5, recompute the solution when the reactor and column volumes are decreased by 20%.

12.7 Construct a Gantt chart for the general multipurpose plant in Figure 12.14, but with the unit assignments specified in Figure 12.13.

12.8 A batch process requires the following operations to be completed in sequence: 3 hr of mixing, 5 hr of heating, 4 hr of reaction, 7 hr of purification, and 2 hr of transfer.

a. When the five operations are carried out in vessels U1, U2, U3, U4, and U5, respectively, determine the cycle times, and construct Gantt charts corresponding to the zero-wait, intermediate storage, and unlimited intermediate storage inventory strategies.

b. When a new purification vessel U4A is purchased, so that two 7-hr purifications can take place in parallel, determine the system bottleneck using the intermediate storage, inventory strategy.

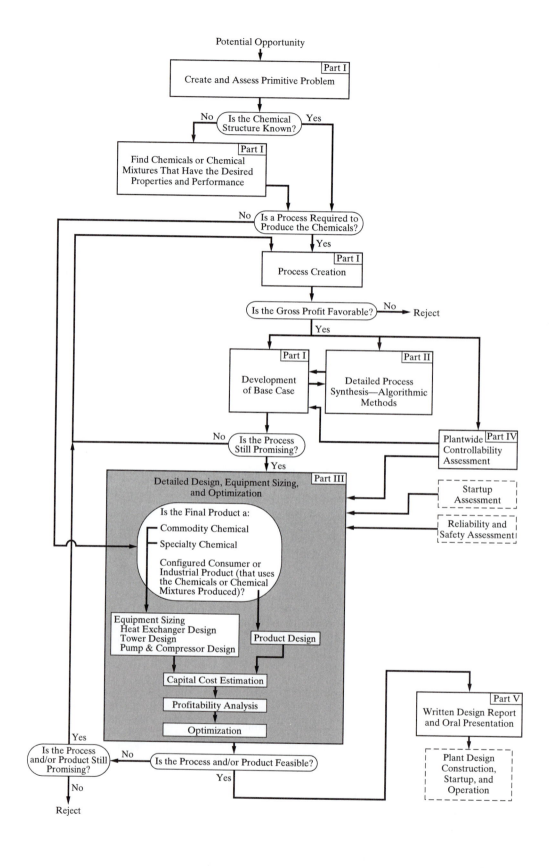

Part Three

DETAILED DESIGN, EQUIPMENT SIZING, AND OPTIMIZATION - CONFIGURED PRODUCT DESIGN

This part concentrates on the steps carried out approximately during process creation and more rigorously after a base-case design(s) has been developed, as discussed in Part One. It covers the steps of detailed design in (1) determining equipment sizes, (2) carrying out economic calculations, and (3) optimization. In all of the discussions, approximate and rigorous methods are presented. The approximate methods are used throughout the steps of process creation covered in Part One and the algorithmic methods for process synthesis covered in Part Two. They are referred to repeatedly in those parts and, consequently, this part of the book should be used long before Parts One and Two are completed.

In addition, Chapter 19 introduces the steps in configured product design. These are illustrated using several case studies.

Chapter 13, on the design of heat exchangers, contains some basic introductory material that may be helpful before or during the study of Chapter 10 on heat and power integration. As mentioned therein, the emphasis in that chapter is on the design of an energy-efficient heat exchanger network (HEN) and not on the details of the individual heat exchangers. This is the subject of Chapter 13. Students who have had a strong course in heat transfer are likely to find much of this material a review. They should, however, find the many recommendations for design helpful, including the selection of heating and cooling media, avoidance of temperature crossovers, and the frequent need for multiple shell and tube passes in heat exchangers.

Chapters 14 and 15 provide similar coverage on the details of the design of multistage and packed towers, and of pumps, compressors, and expanders, respectively. While some of this material may be a review for students, the recommendations for equipment design should be helpful.

Chapters 16 and 17, on cost accounting and capital cost estimation, and profitability analysis, respectively, provide a comprehensive treatment of these subjects. Both approximate and rigorous methods are presented. Equations are provided for estimating the purchase cost of a broad array of process equipment. Also, instruction is provided in the use of the Aspen Icarus Process Evaluator (IPE), together with the process simulators, for the estimation of purchase costs and the total permanent investment. In addition, an EXCEL spreadsheet is presented for calculation of profitability analyses.

Finally, the process simulators have the ability to optimize a process flowsheet by adjusting the continuous variables such as the purge/recycle ratio and the reflux ratio. Chapter 18 presents a general discussion of optimization methods, followed by two case studies.

To see how Part Three relates to the other four parts of this text and to the entire design process, see the figure on page 402.

Chapter 13

Heat Exchanger Design

13.0 OBJECTIVES

Storage tanks, reactors, and separation units in a chemical process are operated at specified temperatures, pressures, and phase conditions. In continuous processes, pressure conditions are established by valves and pumps for liquids, and valves, compressors, and turbines or expanders for gases. Valves are also used to partially or completely convert liquids to gases. Temperature and phase conditions are established mainly by heat exchangers, which are the subject of this chapter.

After studying this chapter, and the multimedia materials on heat exchangers on the CD-ROM that accompanies this book, the reader should

1. Understand how the temperature and phase condition of a stream can be changed by using a heat exchanger.
2. Be able to specify a heat exchanger when modeling just one side.
3. Be able to select heat transfer media for the other side of the exchanger.
4. Understand the importance of heating and cooling curves and how to generate them and use them to avoid crossover violations of the second law of thermodynamics.
5. Be familiar with the major types of heat exchange equipment and how they differ in flow directions of the two fluids exchanging heat, and the corresponding effect on the temperature-driving force for heat transfer.
6. Be able to specify a heat exchanger when modeling both sides.
7. Know how to estimate overall heat transfer coefficients, including the effect of fouling.
8. Understand the limitations of boiling heat transfer.
9. Be able to design a shell-and-tube heat exchanger with the help of a simulator.

13.1 INTRODUCTION

This chapter begins with consideration of the effects of changing temperature, pressure, and phase condition, for a single stream, on stream enthalpy and heat duty. Then heating and cooling media are discussed, and the temperature-driving force for effecting a desired change in stream conditions is considered. Selection of heat exchange equipment is followed by a discussion of methods of determining exchanger sizes from estimates of overall heat transfer coefficients. The chapter concludes with a comprehensive design problem for a shell-and-tube heat exchanger. In addition, the multimedia CD-ROM that accompanies this book shows how to model heat exchangers using ASPEN PLUS and HYSYS.Plant; see *ASPEN →* *Heat Exchangers* and *HYSYS → Heat Exchangers.*

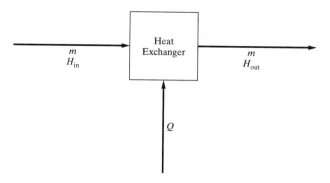

Figure 13.1 One-sided heat exchanger.

Heat Duty

In the early stages of process design, heating and cooling of liquids and vapors, partial and complete vaporization of liquids, partial and complete condensation of vapors, and sensible and latent heat changes for streams containing solids are treated without regard to (1) the source or sink of thermal energy transferred to or from the stream, (2) the rate at which the energy is transferred, or (3) the type and size of heat exchanger needed. Only the overall enthalpy change (*heat duty*) of the stream for the specified heat exchanger inlet and outlet conditions, and the variation of enthalpy with intermediate conditions in the exchanger, are of interest. The variation is represented most conveniently by *heating and cooling curves*. The heat duty and these curves are most easily obtained, especially for streams that are multicomponent mixtures undergoing phase change, with a steady-state process simulator. The calculations are not simple because effects of temperature, pressure, and composition on enthalpy are taken into account, and the phase condition is established by a phase equilibrium calculation.

Consider the heat exchanger in Figure 13.1. The continuous, steady-state heat duty is given by

$$Q = m(H_{out} - H_{in}) \tag{13.1}$$

where Q is the heat duty (rate of heat transfer), m is the flow rate of the stream (mass or molar), H_{in} is the enthalpy of the stream entering (per unit mass or mole), and H_{out} is the enthalpy of the stream leaving (per unit mass or mole). Simulation programs refer to this type of model as a *one-sided heat exchanger* because only one of the two streams exchanging heat is considered. The calculations are illustrated in the following example.

EXAMPLE 13.1

In Figure 3.7, the reactor effluent from the pyrolysis reactor consists of 58,300 lb/hr of HCl, 100,000 lb/hr of vinyl chloride, and 105,500 lb/hr of 1,2-dichloroethane at 500°C and 26 atm. Before entering a distillation section, this stream is cooled and condensed to 6°C at 12 atm. Assume that this is to be done in three steps: (1) cooling in heat exchanger 1 at 26 atm to the dew-point temperature, (2) adiabatic expansion across a valve to 12 atm, and (3) cooling and condensation in heat exchanger 2 at 12 atm to 6°C. Determine the heat duties and cooling curves for each heat exchanger. Note that the pressure drop in each of the two exchangers is neglected.

SOLUTION

This example was solved using ASPEN PLUS. The flowsheet is shown in Figure 13.2, where the HEATER subroutine (block) is used to model heat exchanger 1 (E-1) and heat exchanger 2 (E-2). The pressure is dropped using the VALVE subroutine (block) to model the valve (V-1). The Soave–Redlich–Kwong (SRK) equation of state is used to compute thermodynamic properties. The results of the simulation are included in Figure 13.2, where the heat duties, computed from Eq. (13.1) are shown to be 46,780,000 Btu/hr for E-1 and 53,000,000 Btu/hr for E-2. Stream conditions leaving E-1 are at the dew-point temperature of 157.6°C at 26 atm. The stream leaves valve V-1 as a vapor at

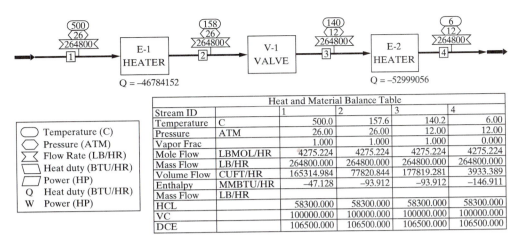

Heat and Material Balance Table					
Stream ID		1	2	3	4
Temperature	C	500.0	157.6	140.2	6.00
Pressure	ATM	26.00	26.00	12.00	12.00
Vapor Frac		1.000	1.000	1.000	0.000
Mole Flow	LBMOL/HR	4275.224	4275.224	4275.224	4275.224
Mass Flow	LB/HR	264800.000	264800.000	264800.000	264800.000
Volume Flow	CUFT/HR	165314.984	77820.844	177819.281	3933.389
Enthalpy	MMBTU/HR	–47.128	–93.912	–93.912	–146.911
Mass Flow	LB/HR				
HCL		58300.000	58300.000	58300.000	58300.000
VC		100000.000	100000.000	100000.000	100000.000
DCE		106500.000	106500.000	106500.000	106500.000

Figure 13.2 ASPEN PLUS flowsheet for Example 13.1

140.2°C and 12 atm. Thus, the adiabatic expansion lowers the temperature by 17.4°F. Stream conditions leaving E-2 are liquid at 6°C and 12 atm. The cooling curve for E-1 is given in Figure 13.3a. Vapor conditions persist throughout E-1; thus, the enthalpy change is all sensible heat. Because the vapor heat capacity varies only slightly with temperature, the graph of the temperature as a function of the enthalpy change is almost linear. The cooling curve for E-2 is given in Figure 13.3b. Entering E-2, the stream is slightly superheated at 140.2°F, with the dew point occurring at 126°C, as seen by the significant change in the slope of the curve in Figure 13.3b. Another significant change in slope occurs at 10°C, which is the bubble point. Between the dew point and the bubble point, both sensible and latent heat changes occur, with the curve deviating somewhat from a straight line. ∎

Heat Transfer Media

Heat is transferred to or from process streams using other process streams or *heat transfer media*. In a final process design, every effort is made to exchange heat between process streams and thereby minimize the use of heat transfer media (usually referred to as *utilities*). Inevitably, however, some use of media, mostly cooling water, steam, and the products of combustion, is necessary. When media must be used, the heat exchangers are called *utility exchangers*.

Heat transfer media are classified as *coolants* (heat sinks) when heat is transferred to them from process streams, and as *heat sources* when heat is transferred from them to process streams. Process design includes the selection of appropriate heat transfer media, data for which are listed in Table 13.1, where the media are ordered by temperature range of application.

The most common coolant, by far, is cooling water, which is circulated through a cooling tower. As indicated in Heuristic 27 of Chapter 5, the water typically enters the utility exchanger at 90°F and exits at no higher than 120°F. The cooling tower restores the cooling water temperature to 90°F by contacting the water with air, causing evaporation of a small amount of the water. The enthalpy of evaporation is supplied mainly from the water, causing it to cool. The evaporated water is replaced by treated water. With cooling water, process streams can be cooled and/or condensed to temperatures as low as about 100°F (depending on seasonal temperatures). When the plant is located near an ocean or river, that water is sometimes used for cooling without using a cooling tower. When water is scarce at the plant location, air is used for cooling, but air can only cool process streams economically to about 120°F.

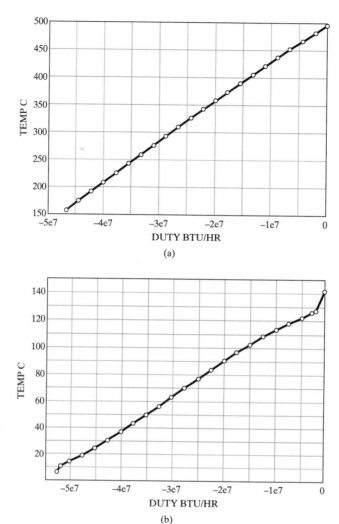

Figure 13.3 Cooling curves for Example 13.1: (a) exchanger E-1; (b) exchanger E-2.

When exchanger inlet temperatures of process streams to be cooled are higher than 250°F, consideration is given to transferring at least some of the heat to treated boiler feedwater to produce steam. The steam is produced at as high a pressure, and corresponding saturation temperature, as possible, subject to a reasonable temperature-driving force for heat transfer in the utility exchanger. For process design purposes, the boiler feedwater enters the utility exchanger as a saturated liquid at the selected pressure, and exits without temperature change as a saturated vapor. The steam is available for use elsewhere in the process. For process stream temperatures above the critical temperature of water, supercritical water is sometimes used as the coolant.

When process streams must be cooled below 100°F in utility exchangers, refrigerants are used, which are designated with an R number by the American Society of Heating, Refrigerating and Air-Conditioning Engineers (ASHRAE). When the process involves light hydrocarbons, the refrigerant can be one of the hydrocarbons, for example, propane (R-290). Otherwise, a commercial refrigerant, for example, R-717 (ammonia) or R-134a (tetrafluoroethane), is selected. A widely used refrigerant, R-12 (dichlorodifluoromethane), is being phased out because of the accepted hypothesis that chlorine and bromine, but not fluorine, atoms from halocarbons, when released to the air, deplete ozone in the atmosphere. Feasible

Table 13.1 Heat Transfer Media

Medium	Typical Temperature Range (°F)	Mode
Coolants:		
Ethylene	−150 to −100	Vaporizing
Propylene	−50 to 10	Vaporizing
Propane	−40 to 20	Vaporizing
Ammonia	−30 to 30	Vaporizing
Tetrafluoroethane	−15 to 60	Vaporizing
Chilled brine	0 to 60	Sensible
Chilled water	45 to 90	Sensible
Cooling water	90 to 120	Sensible
Boiler feedwater	220 to 450	Vaporizing
Heat sources:		
Hot water	100 to 200	Sensible
Steam	220 to 450	Condensing
Heating oils	30 to 600	Sensible
Dowtherm A	450 to 750	Condensing
Molten salts	300 to 1,100	Sensible
Molten metals	100 to 1,400	Sensible
Combustion gases	30 to 2,000	Sensible

refrigerants are included in Table 13.1 for a range of coolant temperatures. When the refrigerant is a pure compound, as it often is, the process design calculation assumes that the refrigerant enters the utility exchanger, at a specified pressure, as a saturated liquid and exits, without temperature change, as a saturated vapor. The refrigerant is circulated through a refrigeration cycle, often consisting of a compressor (to increase the pressure), a condenser (to condense the compressed vapor), a throttle valve (to reduce the pressure), and the utility exchanger (also called the refrigerant boiler), as discussed in Example 9.2. The refrigerant boiling temperature is chosen to avoid freezing the process stream at the wall of the exchanger, unless it is a crystallizer. When process streams are to be cooled to temperatures between 45 and 90°F, chilled water is often used as the coolant rather than a boiling refrigerant. Chilled aqueous brines can be used to temperatures as low as 0°F. Extensive information on refrigerants is given in the *ASHRAE Handbook*.

The most common heat source for heating and/or vaporizing process streams in a utility exchanger is steam, which is available in most chemical plants, from a boiler, at two, three, or more pressure levels. For example, the available levels might be 50, 150, and 450 psig, corresponding to saturation temperatures of 298, 366, and 459°F for a barometric pressure of 14.7 psia. For process design purposes, the steam enters the utility exchanger as a saturated vapor, and exits without pressure change as a saturated liquid (condensate), which is returned to the boiler. Uncondensed steam is prevented from leaving the utility exchanger by a steam trap.

Although condensing steam can be used as a heat source to temperatures as high as about 700°F (critical temperature = 705.4°F), steam pressures become very high at high temperatures (3,206 psia at the critical temperature). It is more common to use other media for temperatures above about 450°F. As listed in Table 13.1, these include the diphenyl (26.5 wt%)–diphenyloxide (73.5 wt%) eutectic (Dowtherm A) for temperatures from 450 to 750°F, and various heating oils, molten salts, and molten metals for higher temperatures. Alternatively, as indicated in Heuristic 25 of Chapter 5, a furnace (fired heater), burning gas, fuel oil, or coal is often used in place of a utility heat exchanger when the chemicals being heated are not subject to decomposition and heating is required above 750°F.

Temperature-Driving Force for Heat Transfer

When streams on both sides of a heat exchanger are considered in process design with a simulation program, a two-sided heat exchanger model is used. The model applies Eq. (13.1) to each side under conditions of equal heat transfer rates, assuming that the exchanger is well insulated such that heat losses are negligible. Thus, all of the heat released by one side is taken up by the other side. In addition, a transport equation is applied:

$$Q = UA \, \Delta T_m \qquad\qquad (13.2)$$

where U is the overall heat transfer coefficient, A is the area for heat transfer, and ΔT_m is the mean temperature-driving force for heat transfer.

The driving force is a critical component of Eq. (13.2). For a given heat exchange task, the rate of heat transfer, Q, is computed from Eq. (13.1). Depending on the geometry and extent of fouling of the heat exchanger, and the conditions of the streams passing through the exchanger, the overall coefficient, U, can be computed from correlations described later in this chapter. The mean driving force, ΔT_m, then determines the heat exchanger area, A. The driving force depends on the entering and exiting stream temperatures, the variation of enthalpy with temperature and pressure of each of the two streams as they pass through the exchanger (as given by the heating and cooling curves), and the stream flow patterns in the exchanger. The latter requires careful consideration.

Examples of a few standard flow patterns are shown in Figure 13.4. The standard and most efficient pattern is countercurrent flow of the two streams. For this case, reference temperature-driving forces are those at the two ends of the exchanger. At one end, ΔT is the difference between the temperatures of the entering hot stream and exiting cold stream. At the other end, ΔT is the difference between the temperatures of the exiting hot stream and the entering cold stream. The smallest of the two differences is called the *closest* or *minimum temperature approach*. It is common to specify the design of a two-sided heat exchanger in terms of inlet conditions for each stream, the pressure drop across the exchanger for each stream, and a minimum approach temperature that reflects economics, as shown in Section 10.6. The simulation program determines to which end of the exchanger the minimum applies, and then calculates the exiting stream temperatures and the heat duty.

The optimal minimum approach temperature is a function mainly of the temperature levels of the two streams as indicated in Heuristic 25 of Chapter 5, and the lost work analysis in

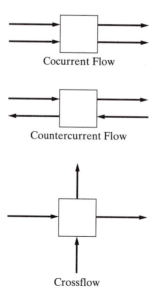

Cocurrent Flow

Countercurrent Flow

Crossflow

Figure 13.4 Standard flow patterns in heat exchangers.

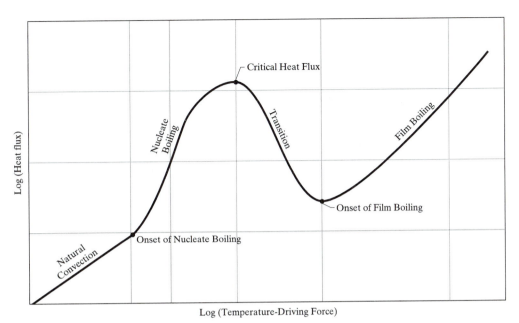

Figure 13.5 Modes in boiling heat transfer.

Section 10.1. At temperatures below ambient, it is less than 10°F and may be only 1–2°F at highly cryogenic conditions. At ambient temperature it is about 10°F. At temperatures above ambient, up to 300°F, it is about 20°F. At higher temperatures it may be 50°F. In a furnace, the flue gas temperature may be 250 to 350°F above the inlet process stream temperature. When one stream is boiled, a special consideration is necessary. Evaporation can take place in any of four different modes, as shown in Figure 13.5. At temperature-driving forces on the boiling side of less than about 10°F, natural convection is dominant and heat transfer rates are low. At driving forces between about 20 and 45°F, *nucleate boiling* occurs, with rapid heat transfer rates because of the turbulence generated by the bubbles. For driving forces above about 100°F, *film boiling* takes place and heat transfer rates are again low because the mechanism is conduction through the gas film. The region between about 50 and 100°F is in transition. Heat exchangers for vaporization and reboiling avoid film boiling and are designed for the nucleate boiling region to maximize heat transfer rates. A conservative rule of thumb is to employ Heuristic 28 of Chapter 5, which suggests using a mean overall temperature-driving force of 45°F. This driving force can be achieved by adjusting the pressure at which boiling takes place or the temperature of the heating medium.

EXAMPLE 13.2

Toluene is converted to benzene by hydrodealkylation. Typically, a 75% conversion is used in the reactor, which necessitates the recovery and recycle of unreacted toluene. In addition, a side reaction occurs that produces a small amount of a biphenyl byproduct, which is separated from the toluene. A hydrodealkylation process is being designed that includes a distillation column for separating toluene from biphenyl. The feed to the column is 3.4 lbmol/hr of benzene, 84.6 lbmol/hr of toluene, and 5.1 lbmol/hr of biphenyl at 264°F and 37.1 psia. The distillate is to contain 99.5% of the toluene and 2% of the biphenyl. If the column operates at a bottoms pressure of 38.2 psia, determine the bottoms temperature and select a suitable heat source for the reboiler. Steam is available at pressures of 60, 160, and 445 psig. The barometer reads 14 psia.

SOLUTION

Assume that no benzene is present in the bottoms because it has a much higher vapor pressure than toluene, and a sharp separation between toluene and biphenyl is specified. By material balance, the bottoms contains 0.423 lbmol/hr of toluene and 4.998 lbmol/hr of biphenyl. A bubble-point calculation for this composition at 38.2 psia, using ASPEN PLUS with the SRK equation of state for K values, gives a temperature of 510.5°F. The highest-pressure steam available is at 459 psia, with a saturation temperature of 458°F. Thus, steam cannot be used as the heat source for the reboiler. Instead, Dowtherm A is specified. It enters the exchanger as a saturated vapor and exits as a saturated liquid. To ensure nucleate boiling, the overall temperature-driving force for reboiling the biphenyl bottoms is taken as 45°F. Thus, the condensing temperature for the Dowtherm A is 555.5°F. From data supplied by Dow Chemical Co., the saturation pressure at this temperature is only 28.5 psia, and the heat of vaporization is 116 Btu/lb. If saturated steam at 555.5°F were available, the pressure would be 1,089 psia with a heat of vaporization of 633 Btu/lb. Thus, the use of Dowtherm A at high temperatures results in much lower pressures, but its low heat of vaporization requires a higher circulation rate. ∎

EXAMPLE 13.3	A mixture of 62.5 mol% ethylene and 37.5 mol% ethane is separated by distillation to obtain a vapor distillate of 99 mol% ethylene with 98% recovery of ethylene. When the pressure in the reflux drum is 200 psia, determine the distillate temperature and select a coolant for the condenser. What pressure is required to permit the use of cooling water in the condenser?

SOLUTION

Using the CHEMCAD simulator, the dew-point temperature for 99 mol% ethylene and 1 mol% ethane at 200 psia is −42°F. Assuming a minimum approach temperature of 5°F and a boiling refrigerant, the refrigerant temperature is −47°F. From Table 13.1, a suitable refrigerant is propylene, but ethylene, which is available at 99 mol% purity in the plant, is also a possibility, with a boiling pressure of 185 psia.

The critical temperatures of ethylene and ethane are 49 and 90°F, respectively, at critical pressures of 730 and 708 psia, respectively. The critical point for 99 mol% ethylene is approximately at 50°F and 729 psia. Therefore, it is not possible to use cooling water in the condenser because it can only achieve a condensing temperature of 100°F. ∎

When a process stream is both heated and vaporized, or both cooled and condensed, the minimum approach temperature can occur within the exchanger, away from either end. This can be determined from heating and cooling curves, as illustrated in the following example.

EXAMPLE 13.4	A mixture of 100 lbmol/hr of ethyl chloride and 10 lbmol/hr of ethanol at 200°F and 35 psia is cooled with 90 lbmol/hr of ethanol at 90°F and 100 psia in a countercurrent heat exchanger. Determine the stream outlet conditions and the heat duty for a minimum approach temperature of 10°F. Assume a pressure drop of 5 psi on the hot side and 10 psi on the cold side.

SOLUTION

The calculations are made with the CHEMCAD program using the HTXR model with the UNIFAC method for computing K values. The hot stream is found to enter the exchanger as superheated vapor and exit partially condensed. The cold stream is found to be a liquid throughout the exchanger. The *plot* menu is used to generate heating and cooling curves, which are shown in Figure 13.6a. It is seen that the minimum approach temperature of 10°F is placed by the HTXR model at the 200°F hot stream feed end to give a cold stream outlet temperature of 190°F. At the other end of the exchanger, the hot stream exits at 105.5°F, so the driving force at that end is $105.5 − 90 = 15.5°F$. The heat duty is 277,000 Btu/hr. However, Figure 13.6a shows a temperature crossover within the exchanger, which violates the second law of thermodynamics. This crossover is caused by the condensation of the hot stream, which begins at a dew-point temperature of approximately 120°F. This results in a sharp change in slope of the temperature–enthalpy curve for the hot stream. From 120°F to the exit temperature, the hot stream undergoes partial condensation to an exit condition of 93 mol% vapor. The HTXR model has an option that can be used to detect a crossover during execution. This option, which is suggested in Heuristic 29 of Chapter 5, is a zone analysis called "No of Zones." If, for example, 20 zones are specified, stream temperatures are computed at 19 intermediate points in the exchanger. From these temperatures, the intermediate temperature-driving forces for the heat exchanger are checked to determine if any are negative. If so, the HTXR model terminates, with a warning to the user.

When a crossover occurs, a trial-and-error procedure can be applied to place the minimum approach temperature within the exchanger. This involves increasing the specified minimum approach temperature, which, as mentioned above, is placed at one end or the other. For this example, the result is shown in Figure 13.6b, where it is seen that the minimum approach temperature occurs at the dew-point tem-

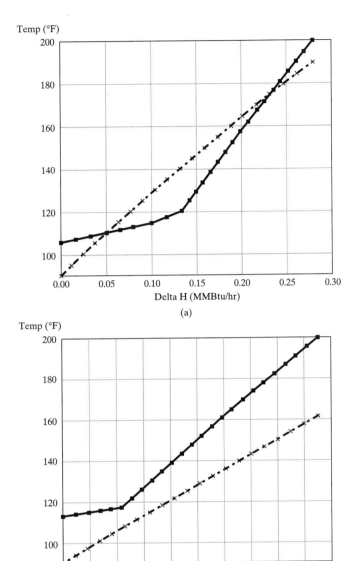

Figure 13.6 Heating and cooling curves for Example 13.4: (a) temperature crossover; (b) no temperature crossover.

perature of the hot stream. This is achieved by specifying a minimum approach temperature of 23°F, which is placed by the HTXR model at the hot stream exit end of the exchanger. Now the hot stream is cooled only to 113°F and the cold stream is heated only to 161°F. The heat duty is reduced to 190,000 Btu/hr. The hot stream exits with 97.8 mol% vapor. ∎

To see how ASPEN PLUS and HYSYS.Plant are used to model a heat exchanger in which both streams undergo phase changes, see *ASPEN PLUS → Tutorials → Heat Transfer → Toluene Manufacture* and *HYSYS → Tutorials → Heat Transfer → Toluene Manufacture* on the multimedia CD-ROM that accompanies this book.

Pressure Drop

The final design of a heat exchanger includes pressure-drop calculations on each side. For process design when using a simulation program, preliminary conservative estimates of pressure drops due to friction are as follows as suggested in Heuristic 31 of Chapter 5. An additional pressure change occurs if the exchanger is placed vertically, due to energy conversions between pressure head and potential energy.

	Pressure Drop	
Liquid streams with no phase change	5 to 9 psi	35 to 62 kPa
Vapor streams with no phase change	3 psi	21 kPa
Condensing streams	1.5 psi	10 kPa
Boiling streams	1.5 psi	10 kPa
Process stream passing through a furnace	20 psi	140 kPa

Methods for determining pressure drop when heat exchanger geometry is known are discussed in Section 13.3.

13.2 EQUIPMENT FOR HEAT EXCHANGE

As listed in Table 13.2, a wide variety of equipment is available for conducting heat exchange. Commercial units range in size from very small, *double-pipe heat exchangers*, with less than 1 ft² of heat transfer surface, to large, air-cooled units called *fin-fan heat exchangers* because they consist of tubes with external peripheral fins and fans to force air past the tubes. Finned area in a single unit is as large as 20,000 ft². By far the most common units are *shell-and-tube heat exchangers*, which come in a variety of configurations in sizes from 50 to 12,000 ft². For specialized applications, *compact heat exchangers* are challenging shell-and-tube units.

Double-Pipe Heat Exchangers

A typical double-pipe unit is shown in Figure 13.7a. In its simplest form, it consists of an inner straight pipe of circular cross section, concentric to and supported within an outer straight pipe by means of packing glands. One stream flows through the inner pipe, while the other stream flows countercurrently through the annular passage between the outer wall of the inner pipe and the inner wall of the outer pipe. When the inner pipe is 12-ft-long, $1\frac{1}{4}$-in., schedule 40 pipe, the heat transfer area from Table 13.3 is 5.22 ft² based on the outside wall of the inner pipe. When the inner pipe is 20-ft-long, 3-in., schedule 40 pipe, the heat transfer area is 18.34 ft². When more heat transfer area is needed, return bends and heads are used

Table 13.2 Heat Exchange Equipment

Double-pipe
Shell-and-tube
  Countercurrent flow
  Parallel (cocurrent) flow
  Crossflow
  1-2, 1-4, 1-6, 1-8
  2-4, 2-8
  3-6
  4-8
  6-12
Air-cooled (fin-fan)
Compact
  Plate-and-frame
  Spiral-plate
  Spiral-tube
  Plate-fin

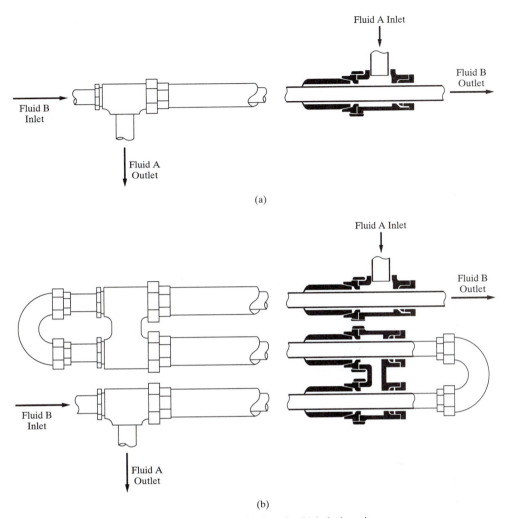

Figure 13.7 Double-pipe heat exchangers: (a) single unit; (b) hairpin unit.

with additional pipes to build a hairpin unit, as shown in Figure 13.7b. Hairpin units are available in sizes up to about 200 ft^2 of heat transfer area, and are competitive with shell-and-tube exchangers in the range of 100 to 200 ft^2. To prevent sagging of the inner pipe with a resulting distortion of the annular cross section, pipe length is limited to 20 ft. Therefore, a 200-ft^2 unit of 3-in.-diameter inner pipes requires 10 hairpin connections. When one stream is at high temperature and/or high pressure, and/or is corrosive, it is passed through the inner pipe. If the other stream is a gas, longitudinal fins can be added to the outside surface of the inner pipe to help balance the inner and outer heat transfer resistances. If crystallization occurs from a liquid stream flowing through the inner pipe, scrapers can be added inside that pipe to prevent buildup of crystals on the inner wall. Double-pipe exchangers are not recommended for use in boiling or vaporization services.

Shell-and-Tube Heat Exchangers

Heat transfer area per unit volume is greatly increased by placing a large number of small-diameter tubes inside a shell, that is, a pressure vessel. Shell-and-tube heat exchangers, whose design is standardized by the Tubular Exchanger Manufacturers Association (TEMA)

Table 13.3 Steel Pipe Data

Nominal Pipe Size (in.)	O.D. (in.)	Schedule No.	I.D. (in.)	Flow Area per Pipe (in.^{2})	Surface per Linear Foot (ft^2)		Weight per Linear Foot (lb steel)
					Outside	Inside	
$\frac{1}{8}$	0.405	40†	0.269	0.058	0.106	0.070	0.25
		80‡	0.215	0.036	0.106	0.056	0.32
$\frac{1}{4}$	0.540	40	0.364	0.104	0.141	0.095	0.43
		80	0.302	0.072	0.141	0.079	0.54
$\frac{3}{8}$	0.675	40	0.493	0.192	0.177	0.129	0.57
		80	0.423	0.141	0.177	0.111	0.74
$\frac{1}{2}$	0.840	40	0.622	0.304	0.220	0.163	0.85
		80	0.546	0.235	0.220	0.143	1.09
$\frac{3}{4}$	1.05	40	0.824	0.534	0.275	0.216	1.13
		80	0.742	0.432	0.275	0.194	1.48
1	1.32	40	1.049	0.864	0.344	0.274	1.68
		80	0.957	0.718	0.344	0.250	2.17
$1\frac{1}{4}$	1.66	40	1.380	1.50	0.435	0.362	2.28
		80	1.278	1.28	0.435	0.335	3.00
$1\frac{1}{2}$	1.90	40	1.610	2.04	0.498	0.422	2.72
		80	1.500	1.76	0.498	0.393	3.64
2	2.38	40	2.067	3.35	0.622	0.542	3.66
		80	1.939	2.95	0.622	0.508	5.03
$2\frac{1}{2}$	2.88	40	2.469	4.79	0.753	0.647	5.80
		80	2.323	4.23	0.753	0.609	7.67
3	3.50	40	3.068	7.38	0.917	0.804	7.58
		80	2.900	6.61	0.917	0.760	10.3
4	4.50	40	4.026	12.7	1.178	1.055	10.8
		80	3.826	11.5	1.178	1.002	15.0
6	6.625	40	6.065	28.9	1.734	1.590	19.0
		80	5.761	26.1	1.734	1.510	28.6
8	8.625	40	7.981	50.0	2.258	2.090	28.6
		80	7.625	45.7	2.258	2.000	43.4
10	10.75	40	10.02	78.8	2.814	2.62	40.5
		60	9.75	74.6	2.814	2.55	54.8
12	12.75	30	12.09	115	3.338	3.17	43.8
16	16.0	30	15.25	183	4.189	4.00	62.6
20	20.0	20	19.25	291	5.236	5.05	78.6
24	24.0	20	23.25	425	6.283	6.09	94.7

†Schedule 40 designates former "standard" pipe.

‡Schedule 80 designates former "extra-strong" pipe.

and has changed little in almost 70 years, is shown in one configuration in Figure 13.8a. Data for heat exchanger tubes are given in Table 13.4.

The following heuristic is useful in making preliminary calculations:

> ***Heuristic 54: For shell-and-tube heat exchangers, tubes are typically ³/₄-in. O.D., 16 ft long, and on 1-in. triangular spacing. A single-tube-pass shell of 1-ft inside diameter accommodates a tube outside area of approximately 300 ft²; 2-ft inside diameter, 1,330 ft², and 3-ft inside diameter, 3,200 ft².***

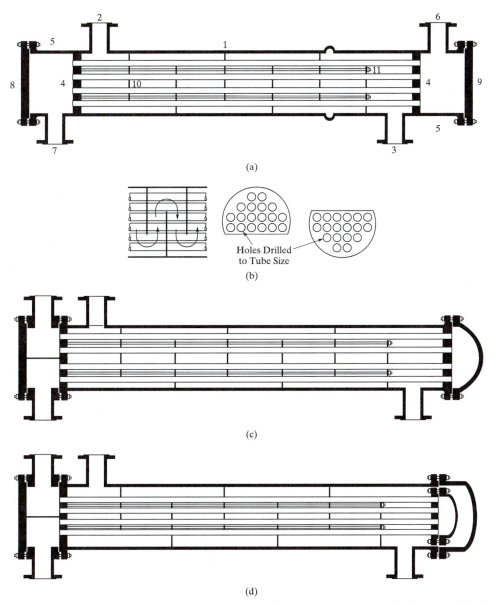

Figure 13.8 Shell-and-tube heat exchangers: (a) 1-1 fixed head; (b) segmental baffles; (c) 1-2 fixed head; (d) 1-2 floating head.

As a further example of this type of heat exchanger, a standard 37-in. I.D. shell can accommodate 1,074 $\frac{3}{4}$-in. O.D., 16 BWG (Birmingham wire gauge, which determines the tube wall thickness) tubes on a 1-in. triangular pitch (tube center-to-center distance). When the tubes are 20 ft long, the heat transfer area, based on the outside tube surface, is 4,224 ft^2. The inside volume of the shell is 149 ft^3, resulting in almost 30 ft^2 of heat transfer surface area per cubic foot of exchanger volume. A double-pipe heat exchanger consisting of a 1$\frac{1}{4}$-in., schedule 40 pipe inside a 2-in., schedule 40 pipe has only 1.17 ft^2 of heat transfer surface area per cubic foot of exchanger volume.

Many configurations of shell-and-tube heat exchangers are available, with Figure 13.8a being the simplest. It is a one-tube-pass, one-shell-pass, fixed (stationary)-head heat exchanger.

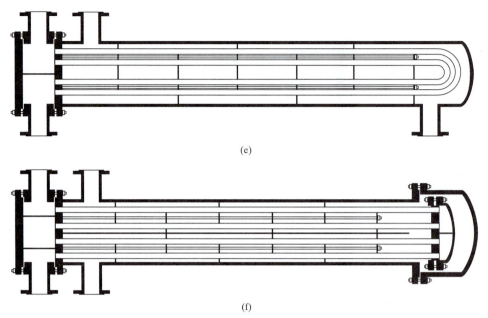

Figure 13.8 (e) 1-2 U-tube; (f) 2-4 floating head.

One stream (tube-side fluid) flows through the tubes; the other (shell-side fluid) flows through the shell, across the outside of the tubes. The exchanger consists of a shell (1), to which are attached an inlet nozzle (2) and an outlet nozzle (3) for the shell-side fluid. At either end of the shell are tube sheets (4), into which tubes are expanded to prevent leakage of streams between the tube side and the shell side. Attached to the tube sheets are channels (5) with inlet and outlet nozzles (6, 7) for the tube-side fluid. Attached to channels are covers (8, 9). To induce turbulence and increase the velocity of the shell-side fluid, transverse baffles (10), through which the tubes pass, are employed on the shell side. The baffles, shown in Figure 13.8b, cause the shell-side fluid to flow mainly at right angles to the axes of the tubes. Baffle spacing (*baffle pitch*) is fixed by baffle spacers (11), which consist of through-bolts screwed into the tube sheets and covered with pipes of length equal to the baffle spacing. Minimum spacing is 20% of the shell inside diameter; maximum is 100%. Various types of baffles are available, but the segmental is the most common, with a segment height equal to 75% of the shell inside diameter. This is often referred to as a *baffle cut* of 25%. Maximum baffle cut is 45%. It is not practical to fit the baffles snugly to the inside surface of the shell. Instead, there is a shell-to-baffle clearance, which depends on shell inside diameter. The diametric shell-to-baffle clearance (twice the clearance) varies from approximately $\frac{1}{8}$ to $\frac{3}{8}$ in. for shell inside diameters of 12 to 84 in.

Several different tube layout patterns are used, four of which are shown in Figure 13.9. Tube spacing is characterized by the *tube pitch*, which is the closest center-to-center distance between the adjacent tubes; or *tube clearance*, which is the shortest distance between two adjacent tube holes. The most common tube layouts are

Layout	Tube O.D. (in.)	Tube Pitch (in.)
Square	$\frac{3}{4}$	1
Square	1	$1\frac{1}{4}$
Triangular	$\frac{3}{4}$	$1\frac{5}{16}$
Triangular	$\frac{3}{4}$	1
Triangular	1	$1\frac{1}{4}$

Table 13.4 Heat Exchanger Tube Data

Tube O.D. (in.)	BWG	Wall Thickness (in.)	I.D. (in.)	Flow Area per Tube (in.^{2})	Surface per Linear Foot (ft^2) Outside	Inside	Weight per Linear Foot (lb steel)
$\frac{1}{2}$	12	0.109	0.282	0.0625	0.1309	0.0748	0.493
	14	0.083	0.334	0.0876	0.1309	0.0874	0.403
	16	0.065	0.370	0.1076	0.1309	0.0969	0.329
	18	0.049	0.402	0.127	0.1309	0.1052	0.258
	20	0.035	0.430	0.145	0.1309	0.1125	0.190
$\frac{3}{4}$	10	0.134	0.482	0.182	0.1963	0.1263	0.965
	11	0.120	0.510	0.204	0.1963	0.1335	0.884
	12	0.109	0.532	0.223	0.1963	0.1393	0.817
	13	0.095	0.560	0.247	0.1963	0.1466	0.727
	14	0.083	0.584	0.268	0.1963	0.1529	0.647
	15	0.072	0.606	0.289	0.1963	0.1587	0.571
	16	0.065	0.620	0.302	0.1963	0.1623	0.520
	17	0.058	0.634	0.314	0.1963	0.1660	0.469
	18	0.049	0.652	0.334	0.1963	0.1707	0.401
1	8	0.165	0.670	0.335	0.2618	0.1754	1.61
	9	0.148	0.704	0.389	0.2618	0.1843	1.47
	10	0.134	0.732	0.421	0.2618	0.1916	1.36
	11	0.120	0.760	0.455	0.2618	0.1990	1.23
	12	0.109	0.782	0.479	0.2618	0.2048	1.14
	13	0.095	0.810	0.515	0.2618	0.2121	1.00
	14	0.083	0.834	0.546	0.2618	0.2183	0.890
	15	0.072	0.856	0.576	0.2618	0.2241	0.781
	16	0.065	0.870	0.594	0.2618	0.2277	0.710
	17	0.058	0.884	0.613	0.2618	0.2314	0.639
	18	0.049	0.902	0.639	0.2618	0.2361	0.545
$1\frac{1}{4}$	8	0.165	0.920	0.665	0.3271	0.2409	2.09
	9	0.148	0.954	0.714	0.3271	0.2498	1.91
	10	0.134	0.982	0.757	0.3271	0.2572	1.75
	11	0.120	1.01	0.800	0.3271	0.2644	1.58
	12	0.109	1.03	0.836	0.3271	0.2701	1.45
	13	0.095	1.06	0.884	0.3271	0.2775	1.28
	14	0.083	1.08	0.923	0.3271	0.2839	1.13
	15	0.072	1.11	0.960	0.3271	0.2896	0.991
	16	0.065	1.12	0.985	0.3271	0.2932	0.900
	17	0.058	1.13	1.01	0.3271	0.2969	0.808
	18	0.049	1.15	1.04	0.3271	0.3015	0.688
$1\frac{1}{2}$	8	0.165	1.17	1.075	0.3925	0.3063	2.57
	9	0.148	1.20	1.14	0.3925	0.3152	2.34
	10	0.134	1.23	1.19	0.3925	0.3225	2.14
	11	0.120	1.26	1.25	0.3925	0.3299	1.98
	12	0.109	1.28	1.29	0.3925	0.3356	1.77
	13	0.095	1.31	1.35	0.3925	0.3430	1.56
	14	0.083	1.33	1.40	0.3925	0.3492	1.37
	15	0.072	1.36	1.44	0.3925	0.3555	1.20
	16	0.065	1.37	1.47	0.3925	0.3587	1.09
	17	0.058	1.38	1.50	0.3925	0.3623	0.978
	18	0.049	1.40	1.54	0.3925	0.3670	0.831

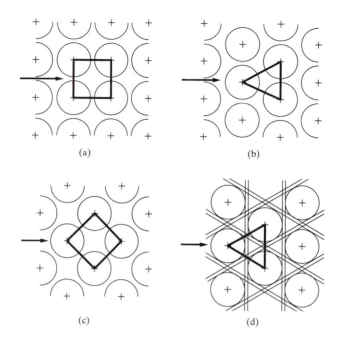

Figure 13.9 Tube layout patterns: (a) square pitch; (b) triangular pitch; (c) square pitch rotated; (d) triangular pitch with cleaning lanes.

It is not practical to fit tubes tightly to the baffles. Accordingly, some shell-side fluid leaks through the clearance between the tubes and the baffle holes. This leakage is in addition to the leakage through the clearance between the shell and the baffles. Although tubes can completely fill the shell, there must be a clearance between the outermost tubes and the shell. Typical clearance between the outer-tube limit (OTL) and the shell inside diameter is $\frac{1}{2}$ in. Common tube lengths are 8, 12, 16, and 20 ft.

The 1-1 fixed-head shell-and-tube heat exchanger of Figure 13.8a has several limitations:

1. The inside surfaces of the tubes can be cleaned, when necessary, by removing the end covers of the shell and reaming out the tubes, but the outside surfaces of the tubes cannot be cleaned because the tube bundle is fixed inside the shell.
2. If large temperature differences exist between the shell-side and tube-side fluids, differential expansion between the shell and tubes may exceed limits for bellows or expansion joints.
3. The velocity of the tube-side fluid may be too low to obtain a reasonable heat transfer coefficient.

These limitations are avoided by other configurations in Figure 13.8. The floating-head unit of Figure 13.8d eliminates the differential expansion problem. Also, the pull-through design permits removal of the tube bundle from the shell so that the outside surfaces of the tubes can be cleaned. The square-pitch tube layout is preferred for cleaning.

To increase the tube-side fluid velocity, a one-shell-pass, two-tube-pass (1-2) exchanger, shown in Figures 13.8c, 13.8d, and 13.8e is used, which are, respectively, fixed-head, floating-head, and U-tube units. A disadvantage of the U-tube unit is the inability to clean the insides of the tubes completely.

With the one-tube-pass exchangers of Figures 13.8a and 13.8b, efficient countercurrent flow between the tube-side and shell-side fluids is closely approximated. This is not the case with the 1-2 exchangers of Figures 13.8c, 13.8d, and 13.8e because of the reversal of the tube-side-fluid flow direction. The flow is countercurrent in one tube pass and cocurrent (parallel) in the other. As shown later in this section, this limits heat recovery because of the reduction in the mean temperature-driving force for heat transfer. Note that a video of an

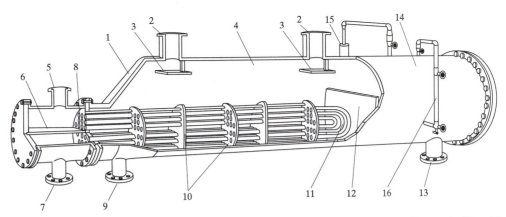

Figure 13.10 Kettle reboiler: (1) shell; (2) shell outlet nozzles (vapor); (3) entrainment baffles; (4) vapor-disengaging space; (5) channel inlet nozzle; (6) channel partition; (7) channel outlet nozzle; (8) tube sheet; (9) shell inlet nozzle; (10) tube support plates; (11) U-tube returns; (12) weir; (13) shell outlet nozzle (liquid); (14) liquid holdup (surge) section; (15) top of level—instrument housing (external displacer); (16) liquid level gauge.

 industrial 1-2 exchanger is provided on the CD-ROM that accompanies this text. See *ASPEN → Heat Exchangers → Introduction with Video* or *HYSYS → Heat Exchangers → Theory.*

The shell-side fluid velocity is increased and the exchanger heat recovery is improved with the two-shell-pass, four-tube-pass (2-4) configuration shown in Figure 13.8f, where a longitudinal baffle creates the two shell passes in a single shell. Alternatively, two exchangers in series, each with a single shell pass and two tube passes can be employed. Further improvements are achieved with 3-6 and 4-8 exchangers, but at the cost of more complexity in the exchanger design. Customarily, not more than two shell passes are provided in a single shell. Thus, a 3-6 pass exchanger would consist of three shells (exchangers) in series, each with two tube passes. When even higher tube-side velocities are desired, 1-4, 1-6, or 2-8 exchangers can be specified. Heat recovery for these various combinations of shell-and-tube passes is considered in detail later in this section.

The exchangers in Figure 13.8 are suitable for heating, cooling, condensation, and vaporization. However, a special design, the *kettle reboiler*, shown in Figure 13.10, is also in common use for vaporization or boiling. Compared to a 1-2 exchanger, the kettle reboiler has a weir to control the liquid level in the shell and a disengagement region in the space above the liquid level. In a typical service, steam is condensed inside the tubes and liquid is vaporized from the pool of liquid outside the tubes.

When employing a shell-and-tube heat exchanger, a decision must be made as to which fluid passes through the tubes (tube side) and which passes through the shell outside the tubes (shell side). The following heuristic is useful in making this decision:

> **Heuristic 55: *The tube side is for corrosive, fouling, scaling, hazardous, high-temperature, high-pressure, and more expensive fluids. The shell side is for more viscous, cleaner, lower-flow rate, evaporating, and condensing fluids.***

Air-Cooled Heat Exchangers

 When cooling water is scarce, air is used for cooling and condensing liquid streams in fin-fan heat exchangers. A common configuration is shown in Figure 13.11. See, also, a video of an industrial fin-fan cooler on the CD-ROM that accompanies this text. The liquid to be cooled

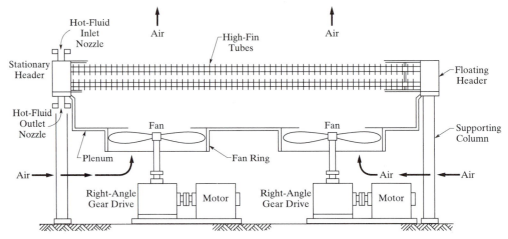

Figure 13.11 Fin-fan heat exchanger.

and/or condensed passes through the inside of the tubes. Peripheral fins on the outside of the tubes, across which the air flows, increase the outside heat transfer area and thereby lower the outside thermal resistance so that it approaches the tube inside resistance. The tubes are arranged in banks, with the air forced across the tubes in crossflow by one or more fans. No shell is needed, fouling on the outside of the tubes does not occur, and inside tube cleaning is readily accomplished. For initial design, the following heuristic is useful:

> ***Heuristic 56: For an air-cooled exchanger, the tubes are typically 0.75–1.00 in. in outside diameter. The ratio of fin surface area to tube outside bare area is large at 15–20. Fan power requirement is in the range of 2–5 Hp per million Btu/hr transferred, or about 20 Hp per 1,000 ft² of tube outside bare surface (fin-free) area. Minimum approach temperature is about 50°F, which is much higher than with water-cooled exchangers. Without the fins, overall heat transfer coefficients would be about 10 Btu/hr ft² °F. With the fins, U = 80–100 Btu/hr ft² °F, based on the tube outside, bare surface area.***

Design is usually based on an entering air temperature of 90°F (hot summer day), for which the process stream can be assumed to exit the air-cooled heat exchanger at 140°F. For air at 70°F, a process stream can be cooled typically to 120°F. Special design considerations may be required for the use of air coolers in the Middle East, where air temperatures may vary from 130°F during the day to 35°F at night. Overhead condensers sometimes combine an air cooler with a cooling-water condenser to reduce the cooling-water load.

Compact Heat Exchangers

Compact heat exchangers have been available for more than a century, but have been slow to replace shell-and-tube exchangers. This has been due to the standards established by TEMA for shell-and-tube exchangers and their applicability to high pressures and temperatures, and to streams containing particulate matter. Nevertheless, for nondemanding services, compact exchangers can offer significant economies and deserve consideration.

When the two fluids exchanging heat must be kept clean, *plate-and-frame heat exchangers* made of stainless steel are commonly used. A typical configuration, shown in Figure 13.12a, consists of a series of pressed corrugated plates on close spacing. Hot and cold fluids flow on opposite sides of a plate. Heat transfer coefficients are high because of the enhancement of

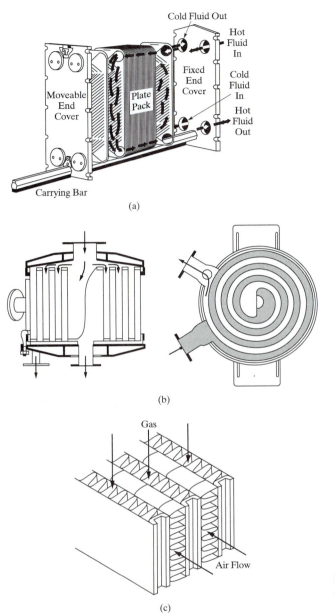

Figure 13.12 Compact heat exchangers: (a) plate-and-frame; (b) spiral plate; (c) plate fin.

turbulence by the corrugations. Fouling of the surfaces is low, and the heat transfer surfaces are readily cleaned. Because gasket seals are necessary in the grooves around the periphery of the plates to contain and direct the fluids, operating pressures and temperatures are limited to 300 psig and 400°F. Plate-and-frame units with as much as 16,000 ft^2 of heat transfer surface area are available. They are suitable only for heating and cooling with no phase change. They can be designed for very small minimum approach temperatures and are ideal for viscous, corrosive fluids. They are also well suited for high sanitation services, where in stainless-steel construction they may be 25–50% of the cost of a shell-and-tube unit.

Heat transfer coefficients can also be enhanced by using spiral flow passageways as in the *spiral-plate heat exchanger* shown in Figure 13.12b. This unit provides true counter-current flow. Typically, the hot fluid enters at the center of the spiral and flows outward, while the cold fluid enters at the periphery and flows inward. This unit is competitive with the shell-and-tube exchanger for heating and cooling of highly viscous, corrosive, fouling, and scaling fluids at ambient to moderate pressures. Units with up to 2,000 ft^2 of heat transfer surface area are available.

For operation at high pressures, a spiral of adjacent tubes can be used. One fluid flows through the tube coil, while the other fluid flows countercurrently in the spiral gap between turns of the coil. The shell side is readily cleaned, but the tube side is not. Sizes of the *spiral-tube heat exchanger* are limited to 500 ft^2 of heat transfer surface area.

When sensible heat is to be exchanged between two gases, extended heat transfer surface in the form of fins is desirable on both sides. This is accomplished by *plate-fin heat exchangers*, an example of which is shown in Figure 13.12c. These compact units achieve heat transfer surface areas of 350 ft^2/ft^3 of unit, which is much higher (up to 4 times) than for shell-and-tube heat exchangers. The fins consist of corrugated surfaces of 0.2- to 0.6-mm thickness and 3.8- to 11.8-mm height. Fin density is 230–700 fins/m. Plate-fin units can be designed for high pressures, and for countercurrent or crossflow. Two, three, or more streams can exchange heat in a single unit.

Furnaces

Furnaces (also called fired heaters) are often used to heat, vaporize, and/or react process streams at high temperatures and high flow rates. Heat duties of commercial units are in the range of 3 to 100 MW (10,000,000 to 340,000,000 Btu/hr). A number of different designs exist, using either rectangular and cylindrical steel chambers, lined with firebrick. The process fluid flows through tubes that are arranged in a so-called radiant section around the inside wall of the furnace enclosure. In this section, heat transfer to the outer surface of the tubes is predominantly by radiation from combustion gases resulting from burning of the furnace fuel with air. To recover as much energy as possible from the combustion gases, a so-called convection section, where the gases flow over a bank of extended-surface tubes, surmounts the radiant section. In this section, heat transfer from the gases to the tubes is predominantly by forced convection. In some cases, plain tubes are placed in the bottom part of the convection section to shield the extended-surface tubes from excessive radiation. Furnaces are purchased as package units (complete units ready for connection to other units), with preliminary estimates of purchase cost based on the heat duty. Typical designs are based on the following heuristic:

> **Heuristic 57: Typical heat fluxes in fired heaters are 12,000 Btu/hr-ft² in the radiant section and 4,000 Btu/hr-ft² in the convection section, with approximately equal heat duties in the two sections. Typical process liquid velocity in the tubes is 6 ft/s. Thermal efficiency for modern fired heaters is 80–90%, while older fired heaters may have thermal efficiencies of only 70–75%.**

As stated in Heuristic 30 of Chapter 5, stack gas (exit) temperatures are in the range 650 to 950°F. However, the flue gas must not be cooled below its dew point, called the *acid dew point*. Otherwise corrosion of the stack may occur.

Temperature-Driving Forces in Shell-and-Tube Heat Exchangers

The rate of heat transfer between two streams flowing through a heat exchanger is governed by Eq. (13.2). Except for a few simple, idealized cases, the mean temperature-driving force, ΔT_m, is a complicated function of the exchanger flow configuration and the thermodynamic and transport properties of the fluids. When a phase change occurs, an additional complication enters into its determination.

The simplest expression for ΔT_m is determined when the following assumptions hold:

1. Stream flows are at steady state.
2. Stream flows are countercurrent or cocurrent to each other.
3. The overall heat transfer coefficient is constant throughout the exchanger.
4. Each stream undergoes only sensible enthalpy changes (heating or cooling), with constant specific heat.
5. Heat losses are negligible.

For these assumptions, changes in the stream temperatures with distance through the exchanger, or with stream enthalpy, are linear, as shown in the heating and cooling curves of Figure 13.13a for countercurrent flow and Figure 13.13b for cocurrent flow. The ΔT_m is then a function only of the driving forces at the two ends of the exchanger, ΔT_1 and ΔT_2, in the form of a log mean:

$$\Delta T_{LM} = \frac{\Delta T_1 - \Delta T_2}{\ln(\Delta T_1/\Delta T_2)} \tag{13.3}$$

If one or both of the streams undergoes isothermal condensation or boiling, the specific heats are constant, and the above assumptions 1, 3, and 5 apply, the log-mean temperature difference applies to all heat exchanger configurations, including multiple tube- or shell-pass arrangements.

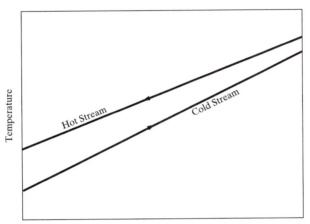

(a)

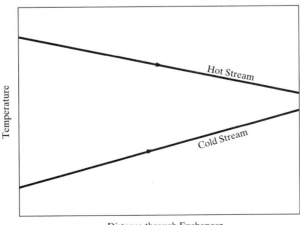

Distance through Exchanger

(b)

Figure **13.13** Ideal heating and cooling curves: (a) countercurrent flow; (b) cocurrent flow.

When shell-and-tube exchangers with multiple-tube passes, or multiple shell-and-tube passes are used, the flow directions of the two fluids are combinations of countercurrent and cocurrent flow. The resulting ΔT_m for given values of ΔT_1 and ΔT_2, based on countercurrent flow, is less than ΔT_{LM} given by Eq. (13.3). For assumptions 1, 3, 4, and 5 above, the true mean temperature-driving force for a 1-2 exchanger was derived by Nagle (1933) and Underwood (1934). The resulting equation is commonly expressed in terms of the ratio, F_T = correction factor = $\Delta T_m/\Delta T_{LM}$:

$$F_T = \frac{\sqrt{R^2 + 1}\ \ln[(1 - S)/(1 - RS)]}{(R - 1)\ln\left[\dfrac{[2 - S(R + 1 - \sqrt{R^2 + 1})]}{[2 - S(R + 1 + \sqrt{R^2 + 1})]}\right]} \tag{13.4}$$

where

$$R = \frac{T_{\text{hot in}} - T_{\text{hot out}}}{T_{\text{cold out}} - T_{\text{cold in}}} \tag{13.5}$$

$$S = \frac{T_{\text{cold out}} - T_{\text{cold in}}}{T_{\text{hot in}} - T_{\text{cold in}}} \tag{13.6}$$

The rate of heat transfer in multipass exchangers then becomes

$$Q = UAF_T\Delta T_{\text{LM for countercurrent flow}} \tag{13.7}$$

A graph of Eq. (13.4) appears in Figure 13.14 with F_T, as a function of S and R as a parameter. Values of F_T are always less than 1. In heat exchanger applications, it is desirable to have a value of F_T of 0.85 or higher. Values of less than 0.75 are generally unacceptable because below this value, the curves in Figure 13.14 turn sharply downward. Thus, small errors in R and S, or small deviations from the above assumptions can result in values of F_T much lower than anticipated. Values of F_T are not significantly decreased further by using exchangers with additional tube passes, such as 1-4, 1-6, or 1-8. At most, F_T for a 1-8 exchanger differs by less than 2% from that for a 1-2 exchanger.

When F_T is unsatisfactory, a multiple-shell-pass heat exchanger is used. The more shell passes, the higher is the value of F_T. For a given number of shell passes, the number of tube passes has very little effect on F_T. Charts for correction factors of multiple-shell-pass exchangers are given in Figure 13.15, from the work of Bowman et al. (1940). Crossflow

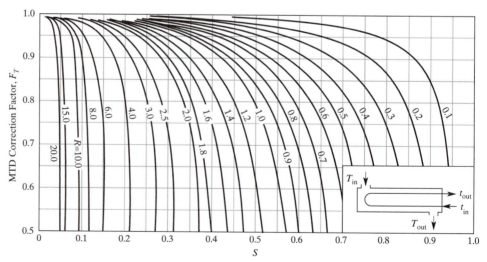

Figure 13.14 Temperature-driving-force correction factor for 1-2 shell-and-tube exchanger. [Adapted from Bowman et al., *Trans. ASME*, **62**, 283 (1940).]

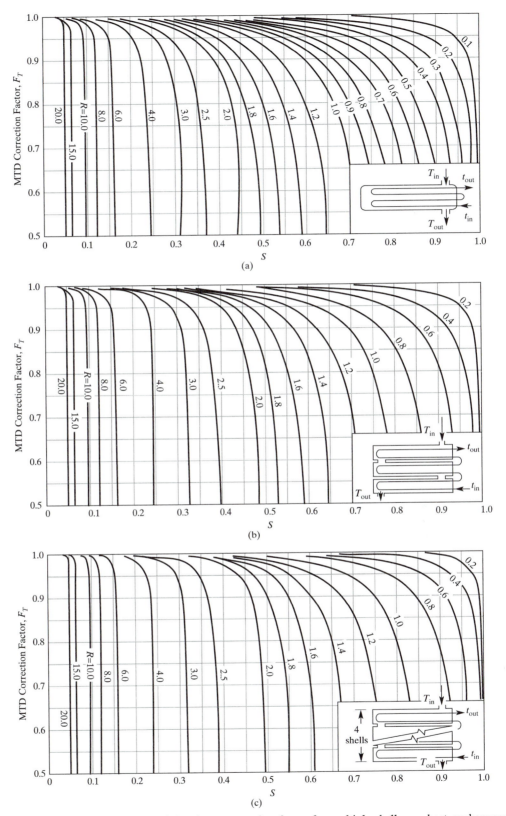

Figure 13.15 Temperature-driving-force correction factor for multiple-shell-pass heat exchangers: (a) 2-4 exchanger; (b) 3-6 exchanger; (c) 4-8 exchanger. [Adapted from Bowman et al., *Trans. ASME*, **62**, 283 (1940).]

exchangers are also less efficient than countercurrent exchangers. Charts of correction factors for crosscurrent flow are given in Figure 13.16. In Figures 13.14 to 13.16, the symbols T and t differentiate between shell- or tube-side fluids. Use of Figures 13.14 to 13.16 with Eqs. (13.5) to (13.7) is independent of whether the hot fluid flows on the shell or tube side. The use of the correction-factor charts is illustrated by the following example.

EXAMPLE 13.5

A hot stream is being cooled from 200°F to 140°F by a cold stream that enters the exchanger at 100°F and exits at 190°F. Determine the true mean temperature-driving force for multiple-tube-pass shell-and-tube exchangers.

SOLUTION

For countercurrent flow, the temperature-driving forces at the two ends of the exchanger are $200 - 190 = 10°F$ and $140 - 100 = 40°F$. The log-mean driving force, using Eq. (13.3), is

$$\Delta T_{LM} = \frac{40 - 10}{\ln(40/10)} = \frac{30}{1.386} = 21.6°\,F$$

For multiple-pass exchangers, using Eqs. (13.5) and (13.6),

$$R = \frac{200 - 140}{190 - 100} = \frac{60}{90} = 0.667 \quad \text{and} \quad S = \frac{190 - 100}{200 - 100} = \frac{90}{100} = 0.9$$

For a 1-2 exchanger, using Figure 13.14, the value of F_T cannot be read because it is less than 0.5. When it is computed from Eq. (13.4), the argument of the ln term in the denominator of Eq. (13.4) is negative. Thus, a real value of F_T cannot be computed. This indicates that a temperature crossover occurs in a 1-2 exchanger.

For a 2-4 exchanger, using Figure 13.15a, F_T is again less than 0.5. For a 3-6 exchanger, using Figure 13.15b, $F_T = 0.7$, which is risky. For a 4-8 exchanger, using Figure 13.15c, $F_T = 0.85$, which is satisfactory. The mean temperature-driving force is $F_T \Delta T_{LM} = 0.85(21.6) = 18.4°F$. ∎

13.3 HEAT TRANSFER COEFFICIENTS AND PRESSURE DROP

To determine the heat transfer area of a heat exchanger from Eq. (13.7), an overall heat transfer coefficient is required. It can be estimated from experience or from the sum of the individual thermal resistances. For double-pipe and shell-and-tube heat exchangers, the area for heat transfer increases across the pipe or tube wall from the inside to the outside surface. Accordingly, the overall heat transfer coefficient is based on the inner wall, i, the outer wall, o, or, much less frequently, a mean, m. The three coefficients are related by

$$\frac{1}{UA} = \frac{1}{U_o A_o} = \frac{1}{U_i A_i} = \frac{1}{U_m A_m} \tag{13.8}$$

When the outer wall is used, the area is A_o and

$$U_o = \frac{1}{R_{f,o} + \left(\frac{1}{h_o}\right) + \left(\frac{t_w A_o}{k_w A_m}\right) + \left(\frac{A_o}{h_i A_i}\right) + R_{f,i}\left(\frac{A_o}{A_i}\right)} \tag{13.9}$$

where $R_{f,o}$ is the outside fouling factor, $R_{f,i}$ is the inside fouling factor, h is the individual heat transfer coefficient, k_w is the thermal conductivity of the cylindrical wall, t_w is the thickness of the cylindrical wall,

$$A_o = \pi D_o L \qquad A_i = \pi D_i L \qquad A_m = \frac{\pi L(D_o - D_i)}{\ln(D_o/D_i)}$$

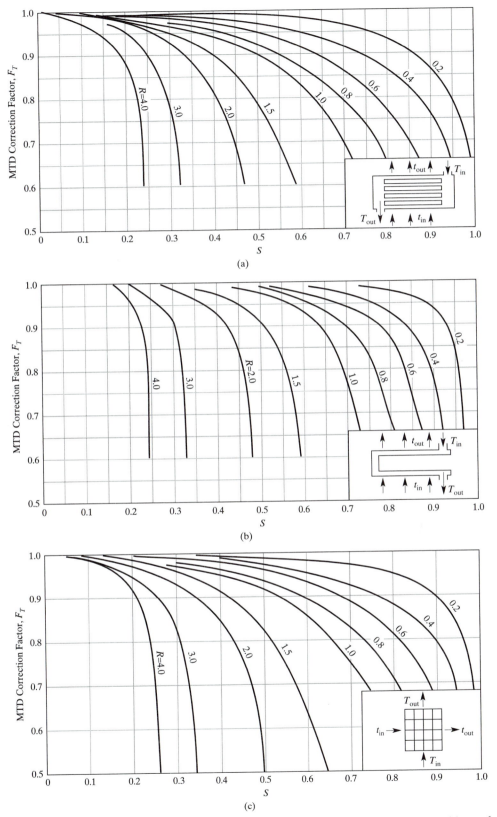

Figure 13.16 Temperature-driving-force correction factor for crossflow heat exchangers: (a) one shell pass, one or more parallel rows of tubes; (b) two shell passes, two rows of tubes (for more than two passes, use $F_T = 1$); (c) one shell pass, one tube pass, both fluids unmixed. [Adapted from Bowman et al., *Trans. ASME*, **62**, 283 (1940).]

D is the tube or pipe diameter, and L is the tube or pipe length. When the inner wall is used, the area is A_i and

$$U_i = \frac{1}{R_{f,o}\left(\dfrac{A_i}{A_o}\right) + \left(\dfrac{1}{h_o}\right)\left(\dfrac{A_i}{A_o}\right) + \left(\dfrac{t_w A_i}{k_w A_m}\right) + \left(\dfrac{1}{h_i}\right) + R_{f,i}} \tag{13.10}$$

Estimation of Overall Heat Transfer Coefficients

For preliminary design, the heat transfer area is computed from Eq. (13.7) using a rough estimate of the overall heat transfer coefficient, U, based on the service. Because the values are rough, the basis for the area is of no concern. Typical values of U for shell-and-tube heat exchangers are given in Table 13.5. The values include a fouling-factor contribution referred to as *total dirt*, equal to $R_{f,o} + R_{f,i}$. For example, for gasoline on the shell side and water in the tubes, U is given as 60–100 Btu/°F-ft²-hr with total dirt of 0.003 (hr-ft²-°F)/Btu. The U in Table 13.5 can be referred to as U_{dirty}. Thus, $1/U_{\text{dirty}} = 0.010$–$0.017$ (hr-ft²-°F)/Btu. For a clean exchanger, $1/U_{\text{clean}} = 1/U_{\text{dirty}} - (R_{f,o} + R_{f,i}) = 0.007$–$0.014$ (hr-ft²-°F)/Btu or $U_{\text{clean}} = 70$–140 Btu/°F-ft²-hr.

EXAMPLE 13.6

A mixture of 60 mol% propylene and 40 mol% propane at a flow rate of 600 lbmol/hr is distilled at 300 psia to produce a distillate of 99 mol% propylene and a bottoms of 95 mol% propane. The bottoms temperature is 138°F and the heat duty of the reboiler, Q, is 33,700,000 Btu/hr. When waste heat, consisting of saturated steam at 220°F, is used as the heating medium in the reboiler, estimate the area of a shell-and-tube reboiler.

SOLUTION

Assume that the bottoms is on the shell side and steam is inside the tubes. Because the bottoms is almost pure, assume that it vaporizes at 138°F, whereas the steam condenses at 220°F. Therefore, $\Delta T_{\text{LM}} = \Delta T_m = 220 - 138 = 82°F$. From Table 13.5, under vaporizers, with propane on the shell side and steam condensing on the tube side, $U = 200$–300 Btu/°F-ft²-hr. Note that this includes a fouling resistance of 0.0015 (hr-ft²-°F)/Btu. The correction factor, F_T, is 1, regardless of the number of passes or flow directions, because at least one fluid is at a constant temperature in the exchanger. From Eq. (13.7), using 200 Btu/°F-ft²-hr for U,

$$A = \frac{Q}{U F_T \Delta T_{\text{LM}}} = \frac{33,700,000}{(200)(1.0)(82)} = 2,050 \text{ ft}^2$$

The heat flux in the reboiler is

$$\frac{Q}{A} = \frac{33,700,000}{2,050} = 16,400 \text{ Btu/ft}^2\text{-hr}$$

Note that ΔT_m greatly exceeds the maximum value of 45°F suggested earlier for reboilers. However, that value pertains to just the portion of the ΔT on the boiling side of the exchanger. In this example, when the total driving force of 82°F is divided among the five resistances, it is possible that the maximum value might not be exceeded. Alternative limits on reboilers for the vaporization of organic chemicals are maximum heat fluxes of 12,000 Btu/ft²-hr for natural circulation and 20,000 Btu/ft²-hr for forced circulation. Therefore, with a heat flux of 16,400 Btu/ft²-hr, a kettle reboiler should not be specified. Instead, a pump-through reboiler should be used to pump the bottoms through the shell side of the reboiler. Alternatively, the heating steam temperature could be reduced. However, this would result in vacuum steam, which is very undesirable because air that leaks into the steam can interfere with condensation. ∎

Estimation of Individual Heat Transfer Coefficients and Frictional Pressure Drop

An enormous amount of experimental work on convective heat transfer and skin-friction pressure drop has been reported during the 1900s. This has been accompanied by theoretical developments. For laminar flow, heat transfer coefficients and friction factors for well-

Table 13.5 Typical Overall Heat Transfer Coefficients for Shell-and-Tube Heat Exchangers [$U = Btu/(°F\text{-}ft^2\text{-}hr)$]

Shell Side	Tube Side	Design U	Includes Total Dirt
Liquid–liquid media			
Aroclor 1248	Jet fuels	100–150	0.0015
Cutback asphalt	Water	10–20	0.01
Demineralized water	Water	300–500	0.001
Ethanol amine (MEA or DEA) 10–25% solutions	Water or DEA, or MEA solutions	140–200	0.003
Fuel oil	Water	15–25	0.007
Fuel oil	Oil	10–15	0.008
Gasoline	Water	60–100	0.003
Heavy oils	Heavy oils	10–40	0.004
Heavy oils	Water	15–50	0.005
Hydrogen-rich reformer stream	Hydrogen-rich reformer stream	90–120	0.002
Kerosene or gas oil	Water	25–50	0.005
Kerosene or gas oil	Oil	20–35	0.005
Kerosene or jet fuels	Trichlorethylene	40–50	0.0015
Jacket water	Water	230–300	0.002
Lube oil (low viscosity)	Water	25–50	0.002
Lube oil (high viscosity)	Water	40–80	0.003
Lube oil	Oil	11–20	0.006
Naphtha	Water	50–70	0.005
Naphtha	Oil	25–35	0.005
Organic solvents	Water	50–150	0.003
Organic solvents	Brine	35–90	0.003
Organic solvents	Organic solvents	20–60	0.002
Tall oil derivatives, vegetable oil, etc.	Water	20–50	0.004
Water	Caustic soda solutions (10–30%)	100–250	0.003
Water	Water	200–250	0.003
Wax distillate	Water	15–25	0.005
Wax distillate	Oil	13–23	0.005
Condensing vapor–liquid media			
Alcohol vapor	Water	100–200	0.002
Asphalt (450°F)	Dowtherm vapor	40–60	0.006
Dowtherm vapor	Tall oil and derivatives	60–80	0.004
Condensing vapor–liquid media (continued)			
Dowtherm vapor	Dowtherm liquid	80–120	0.0015
Gas-plant tar	Steam	40–50	0.0055
High-boiling hydrocarbons V	Water	20–50	0.003
Low-boiling hydrocarbons A	Water	80–200	0.003
Hydrocarbon vapors (partial condenser)	Oil	25–40	0.004
Organic solvents A	Water	100–200	0.003
Organic solvents high NC, A	Water or brine	20–60	0.003
Organic solvents low NC, V	Water or brine	50–120	0.003
Kerosene	Water	30–65	0.004
Kerosene	Oil	20–30	0.005
Naphtha	Water	50–75	0.005
Naphtha	Oil	20–30	0.005
Stabilizer reflux vapors	Oil	80–120	0.003
Steam	Feed water	400–1,000	0.0005
Steam	No. 6 fuel oil	15–25	0.0055
Steam	No. 2 fuel oil	60–90	0.0025
Sulfur dioxide	Water	150–200	0.003
Tall oil derivatives, vegetable oils (vapor)	Water	20–50	0.004
Water	Aromatic vapor-stream azeotrope	40–80	0.005
Gas–liquid media			
Air, N2, etc. (compressed)	Water or brine	40–80	0.005
Air, N2, etc., A	Water or brine	10–50	0.005
Water or brine	Air, N2 (compressed)	20–40	0.005
Water	Air, N2, etc., A	5–20	0.005
Water	Hydrogen containing natural-gas mixtures	80–125	0.003
Vaporizers			
Anhydrous ammonia	Steam condensing	150–300	0.0015
Chlorine	Steam condensing	150–300	0.0015
Chlorine	Light heat-transfer oil	40–60	0.0015
Propane, butane, etc.	Steam condensing	200–300	0.0015
Water	Steam condensing	250–400	0.0015

NC = noncondensable gas present.

V = vacuum.

A = atmospheric pressure.

Dirt (or fouling factor) units are (hr-ft²-°F/Btu).

To convert British thermal units per hour-square foot-degrees Fahrenheit to joules per square meter-second-Kelvin, multiply by 5.6783; to convert hours per square foot-degree Fahrenheit-British thermal units to square meters per second-Kelvin-joules, multiply by 0.1761.

Source: From R.H. Perry and D.W. Green, ed., *Perry's Chemical Engineer's Handbook*, 7th ed., McGraw-Hill, New York (1997).

defined, simple geometries can be accurately predicted from theory. For turbulent flow, both theoretical equations and empirical correlations of data are available. No attempt is made in the brief space permitted here to present recommended methods for predicting convective heat transfer coefficients and friction factors for the wide variety of commercial heat exchanger geometries. Instead, the reader is referred to the *Handbook of Heat Exchanger Design*, edited by G.F. Hewitt (1992), which provides a comprehensive coverage by experts in the field. A brief discussion is given here of turbulent-flow convective heat transfer and skin friction without phase change. In general, turbulent flow is preferred in heat exchangers because of the higher heat transfer coefficients that can be achieved.

Turbulent Flow in Straight, Smooth Ducts, Pipes, and Tubes of Circular Cross Section

In double-pipe and shell-and-tube heat exchangers, fluids flow through straight, smooth pipes and tubes of circular cross section. Many correlations have been published for the prediction of the inside-wall, convective heat transfer coefficient, h_i, when no phase change occurs. For turbulent flow, with Reynolds numbers, $N_{Re} = D_i G / \mu$, greater than 10,000, three empirical correlations have been widely quoted and applied. The first is the Dittus–Boelter equation (Dittus and Boelter, 1930) for liquids and gases in fully developed flow ($D_i/L <$ 60), and with Prandtl numbers, $N_{Pr} = C_p \mu / k$, between 0.7 and 100:

$$N_{Nu} = \frac{h_i D_i}{k_b} = 0.023 \left(\frac{D_i G}{\mu_b} \right)^{0.8} \left(\frac{C_{p_b} \mu_b}{k_b} \right)^n \tag{13.11}$$

where D_i is the inside duct, pipe, or tube diameter, G is the fluid mass velocity (flow rate/cross-sectional area for flow), k is the fluid thermal conductivity, C_p is the fluid specific heat, μ is the fluid viscosity, subscript b refers to average bulk fluid conditions, and exponent $n = 0.4$ for heating the fluid and 0.3 for cooling.

The Colburn equation (Colburn, 1931) also applies to liquids and gases and is almost identical to the Dittus–Boelter equation, but is usually displayed in a *j*-factor form in terms of a Stanton number, $N_{St} = h_i/GC_p$. It is considered valid to a Prandtl number of 160:

$$\frac{h_i}{GC_{p_b}} \left(\frac{C_{p_i} \mu_f}{k_f} \right)^{2/3} = 0.023 \left(\frac{D_i G}{\mu_f} \right)^{-0.2} \tag{13.12}$$

where the subscript *f* refers to a film temperature midway between the wall and bulk condition.

The Sieder–Tate equation (Sieder and Tate, 1936) is specifically for liquids, especially for viscous liquids where the viscosities at the wall and in the bulk may be considerably different. It is claimed to be valid for very high Prandtl numbers. In Nusselt number form, it is

$$N_{Nu} = \frac{h_i D_i}{k_b} = 0.027 \left(\frac{D_i G}{\mu_b} \right)^{0.8} \left(\frac{C_{p_b} \mu_b}{k_b} \right)^{1/3} \left(\frac{\mu_b}{\mu_w} \right)^{0.14} \tag{13.13}$$

where the subscript *w* refers to the temperature at the wall.

In Section 2.5.1 of Hewitt (1992), prepared by Gnielinski, a more accurate and more widely applicable correlation is given, which accounts for tube diameter-to-tube length ratio for $0 < D_i/L < 1$, and is applicable to wide ranges of Reynolds and Prandtl numbers of 2,300 to 1,000,000 and 0.6 to 2,000, respectively. The correlation has a semitheoretical basis in the Prandtl analogy to skin friction in terms of the Darcy friction factor, f_D:

$$N_{Nu} = \frac{h_i D_i}{k_b} = \frac{(f_D/8)(N_{Re} - 1,000)N_{Pr}}{1 + 12.7 \sqrt{f_D/8}(N_{Pr}^{2/3} - 1)} \left[1 + \left(\frac{D_i}{L} \right)^{2/3} \right] \tag{13.14}$$

where

$$f_D = (1.82 \log_{10} N_{Re} - 1.64)^{-2} \tag{13.15}$$

The Darcy friction factor is related to the Fanning friction factor by $f_D = 4f$. The application of Eq. (13.14) is made easy because all properties are evaluated at the bulk fluid conditions. However, for viscous liquids, the right-hand side is multiplied by a correction factor K, where

$$K = \left(\frac{N_{Pr_b}}{N_{Pr_w}}\right)^{0.11} \tag{13.16}$$

For gases being heated, a different correction factor is employed:

$$K = \left(\frac{T_b}{T_w}\right)^{0.45} \tag{13.17}$$

where T is absolute temperature. The Gnielinski equations are preferred for computer calculations in heat exchanger design programs.

The pressure drop for the flow of a liquid or gas under isothermal conditions without phase change through a straight circular tube or pipe of constant cross-sectional area is given by either the Darcy or Fanning equation:

$$-\Delta P = \frac{f_D G^2 L}{2g_c \rho D_i} = \frac{2fG^2 L}{g_c \rho D_i} \tag{13.18}$$

where:

$-\Delta P = P_{in} - P_{out} =$ pressure drop
$L =$ length of tube or pipe
$g_c =$ conversion factor $= 32.17$ ft-lbm/lbf-s$^2 = 1$ in SI units

For turbulent flow at $N_{Re} > 10{,}000$ with a smooth wall, f_D is given by Eq. (13.15) or a Fanning friction factor chart can be used to obtain f.

Equation (13.18) accounts for only skin friction at the inside wall of the tube or pipe. Pressure drop also occurs as the fluid enters (by contraction) or leaves (by expansion) the tube or pipe from or to, respectively, the header, and as the fluid reverses flow direction in exchangers with multiple-tube passes. In addition, pressure drop occurs as the fluid enters the exchanger from a nozzle and passes out through a nozzle. For nonisothermal flow in a multitube-pass exchanger, Eq. (13.18) is modified to:

$$-\Delta P_i = K_P \frac{N_P f_D G^2 L}{2g_c \rho D_i \phi} = K_P \frac{2N_P fG^2 L}{g_c \rho D_i \phi} \tag{13.19}$$

where:

$K_P =$ correction factor for contraction, expansion, and reversal losses
$N_P =$ number of tube passes
$\phi =$ correction factor for nonisothermal turbulent flow $= 1.02(\mu_b/\mu_w)^{0.14}$, where subscript w refers to the average inside wall temperature

A reasonable value for K_P is 1.2. If the exchanger is vertical and flow is upward, the outlet pressure is further reduced by the height of the heat exchanger times the fluid density. If the flow is downward, the outlet pressure is increased by the same amount.

Turbulent Flow in the Annular Region between Straight, Smooth, Concentric Pipes of Circular Cross Section

In double-pipe heat exchangers, one fluid flows through the annular region between the inner and outer pipes. To predict the heat transfer coefficient at the outside of the inner pipe, Eqs. (13.14) and (13.15), with the K corrections, can be used by (1) replacing D_i by $D_2 - D_1$, where D_2 is the inside diameter of the outer pipe and D_1 is the outer diameter of the inner pipe. Then the following correction is made:

$$\frac{N_{Nu,\,annulus}}{N_{Nu,\,tube}} = 0.86 \left(\frac{D_1}{D_2}\right)^{-0.16} \tag{13.20}$$

When the flow is through the annulus of a double-pipe heat exchanger, Eqs. (13.15) and (13.19) can be used to estimate the frictional pressure drop, provided that the inside diameter, D_i, of the tube or pipe is replaced by the hydraulic diameter, D_H, which is defined as 4 times the channel cross-sectional area divided by the wetted perimeter. For an annulus, $D_H = D_2 - D_1$.

Turbulent Flow on the Shell Side of Shell-and-Tube Heat Exchangers

Accurate predictions of the shell-side heat transfer coefficient and pressure drop are difficult because of the complex geometry and resulting flow patterns. A number of correlations are available, none of which is as accurate as those above for the tube side. All are based on crossflow past an ideal tube bank, either staggered (triangular pitch pattern) or inline (square pitch pattern). Corrections are made for flow distortion due to baffles, leakage, and bypassing. From 1950 to 1963, values of h_o, the shell-side, convective heat transfer coefficient, were most usually predicted by the correlations of Donohue (1949) and Kern (1950), which are suitable for hand calculations. Both of these correlations are of the general Nusselt number form

$$N_{Nu} = \frac{h_o D}{k_b} = C \left(\frac{DG}{\mu_b}\right)^n \left(\frac{C_{P_b}\mu_b}{k_b}\right)^{1/3} \left(\frac{\mu_b}{\mu_w}\right)^{0.14} \tag{13.21}$$

The two correlations differ in how D and G are defined, and how C and n are determined. For D, Donohue uses the tube outside diameter, whereas Kern uses the hydraulic diameter. For the mass velocity, G, Donohue uses a geometric mean of (1) the mass velocity in the free area of the baffle window, parallel with the tubes, and (2) the mass velocity normal to the tubes for the row closest to the centerline of the exchanger; Kern uses just the latter mass velocity. Donohue uses $n = 0.6$ and $C = 0.2$; Kern uses 0.55 and 0.36, respectively. Kern's correlation is valid for N_{Re} from 2,000 to 1,000,000. Donohue's correlation is considered to be conservative.

For flow of a gas or liquid across the tubes on the shell side of a shell-and-tube heat exchanger, a preliminary estimate of the shell-side pressure drop can be made by the method of Grimison (1937). The pressure drop is given by a modified Fanning equation:

$$-\Delta P_t = K_S \frac{2N_R f' G_S^2}{g_c \rho \phi} \tag{13.22}$$

where K_S is a correction factor for friction due to inlet and outlet nozzles and to the presence of shell-side baffles that cause reversal of the flow direction, recrossing of tubes, and variation in cross-sectional area for flow. K_S may be taken as approximately 1.10 times (1 + number of baffles). N_R is the number of tube rows across which the shell fluid flows, which equals the total number of tubes at the center plane minus the number of tube rows that pass through the cut portions of the baffles. For 25% cut baffles, N_R may be taken as 50% of the number of tubes at the center plane. For example, if the inside shell diameter is 25 in., the tube outside diameter is 0.75 in., and the tube clearance is 0.25 in. (pitch = 1 in.), the number of tubes in the row at the center plane is 25. With 25% cut baffles, $N_R = 0.5 \times 25 \cong 13$. G_S is the fluid mass velocity based on the flow area at the center plane, which equals the distance between baffles times the tube clearance times the number of tubes at the center plane. f' is the modified friction factor such that:

$$f' = b \left(\frac{D_o G_S}{\mu_b} \right)^{-0.15} \tag{13.23}$$

where b for triangular pitch (staggered tubes) is

$$b = 0.23 + \frac{0.11}{(x_T - 1)^{1.08}} \tag{13.24}$$

and for tubes in line, for example, square pitch, b is

$$b = 0.044 + \frac{0.08 x_L}{(x_T - 1)^{0.43 + 1.13/x_L}} \tag{13.25}$$

Here, x_T is the ratio of the pitch transverse to flow-to-tube outside diameter, and x_L is the ratio of pitch parallel to flow-to-tube outside diameter. For square pitch, $x_T = x_L$.

In 1963, Bell and co-workers at the University of Delaware published a comprehensive method for predicting the shell-side pressure drop and convective heat transfer coefficient. This method is often referred to as the Bell–Delaware method, and is described in detail in Section 11 of *Perry's Chemical Engineers' Handbook* (1997). Experts in Hewitt (1992) consider it to be the best method available. To use the method, geometric and construction details of the exchanger must be known. The calculations are best carried out with a computer. The method considers the effects of tube layout, bypassing, tube-to-baffle leakage, shell-to-baffle leakage, baffle cut, baffle spacing, and adverse temperature gradients. These effects are applied as corrections to an equation of the form of Eq. (13.21). However, the exponent n on the Reynolds number depends on the Reynolds number.

When making estimates of the heat transfer coefficients and pressure drop for shell-and-tube heat exchangers, using the methods discussed previously or the more accurate methods in *Perry's Chemical Engineers' Handbook* (Perry and Green, 1997), tubesheet layouts must be known as a function of shell-and-tube diameters. Typical layouts are given in Table 13.6 for shell diameters ranging from 8 to 37 in., and for $\frac{3}{4}$- and 1-in. O.D. tubes.

Table 13.6 Tube Sheet Layouts

Shell I.D., in.	One-Pass Square Pitch	One-Pass Triangular Pitch	Two-Pass Square Pitch	Two-Pass Triangular Pitch	Four-Pass Square Pitch	Four-Pass Triangular Pitch
\multicolumn{7}{c}{$\frac{3}{4}$-in.-O.D. Tubes on 1-in. Pitch}
8	32	37	26	30	20	24
12	81	92	76	82	68	76
$15\frac{1}{4}$	137	151	124	138	116	122
$21\frac{1}{4}$	277	316	270	302	246	278
25	413	470	394	452	370	422
31	657	745	640	728	600	678
37	934	1,074	914	1,044	886	1,012
\multicolumn{7}{c}{1-in.-O.D. Tubes on $1\frac{1}{4}$-in. Pitch}
8	21	21	16	16	14	16
12	48	55	45	52	40	48
$15\frac{1}{4}$	81	91	76	86	68	80
$21\frac{1}{4}$	177	199	166	188	158	170
25	260	294	252	282	238	256
31	406	472	398	454	380	430
37	596	674	574	664	562	632

Heat Transfer Coefficients for Laminar-Flow, Condensation, Boiling, and Compact Heat Exchangers

Correlations are available for predicting pressure drops and convective heat transfer coefficients for laminar flow inside and outside of ducts, tubes, and pipes; for pipes with longitudinal and peripheral fins; for condensation and boiling; and for several different geometries used in compact heat exchangers. No attempt is made to discuss or summarize these correlations here. They are presented by Hewitt (1992).

13.4   DESIGN OF SHELL-AND-TUBE HEAT EXCHANGERS

The design of a shell-and-tube heat exchanger is an iterative process because heat transfer coefficients and pressure drop depend on many geometric factors, including shell and tube diameters, tube length, tube layout, baffle type and spacing, and the numbers of tube and shell passes, all of which are initially unknown and are determined as part of the design process.

A procedure for an iterative design calculation is as follows, where it is assumed that the inlet conditions (temperature, pressure, composition, flow rate, and phase condition) are known for the two streams entering the heat exchanger and that an exit temperature or some equivalent specification is given for one of the two streams. If a heating or cooling utility is to be used for one of the two streams, it is selected from Table 13.1, together with its entering and leaving temperatures. A decision is made as to which stream will flow on the tube side and which will flow on the shell side. Shell-and-tube side pressure drops are estimated using the values suggested at the end of Section 13.1. With this information, an overall energy balance is used, as discussed in Section 13.1, to calculate the heat duty and the remaining exiting conditions for the two streams. If a heating or cooling utility is to be used, its flow rate is calculated from an energy balance.

A one-tube-pass, one-shell-pass, countercurrent-flow exchanger is assumed. A check is made to make sure that the second law of thermodynamics is not violated and that a reasonable temperature driving force exists at the two ends of the exchanger, as discussed in Section 13.1. If a phase change occurs on either side of the exchanger, a heating and/or cooling curve is calculated as discussed in Section 13.1, and a check is made to make sure that a temperature crossover is not computed within the exchanger.

A preliminary estimate of the heat exchanger area is made by using Table 13.5 to estimate first the overall heat transfer coefficient and then using the heating and/or cooling curves or Eq. (13.3) to compute the mean driving force for heat transfer, followed by Eq. (13.7) to estimate the heat exchanger area, with $F_T = 1$. If the area is greater than 8,000 ft^2, multiple exchangers of the same area are used in parallel. For example, if an area of 15,000 ft^2 is estimated, then two exchangers of 7,500 ft^2 each are used.

From the estimated heat transfer area, preliminary estimates are made of the exchanger geometry. A tube-side velocity in the range of 1 to 10 ft/s is selected, with a typical value being 4 ft/s. The total inside-tube cross-sectional area is then computed from the continuity equation. A tube size is selected, for example, $\frac{3}{4}$-in. O.D., 14 BWG, which, from Table 13.4, has an inside diameter of 0.584 in. and an inside flow area, based on the inside cross-sectional area, of 0.268 in.^{2} From this, the number of tubes per pass per exchanger is calculated. A tube length is selected, for example, 16 ft, and the number of tube passes per exchanger is calculated. The tube-side velocity and tube length are adjusted, if necessary, to obtain an integer number for the number of tube passes.

If more than one tube pass is necessary, the log-mean temperature-driving force is corrected, using Figures 13.14 through 13.16. This may require using more than one shell pass, as discussed in Section 13.2 and illustrated in Example 13.5. A tube-sheet layout is then selected from Table 13.6, and a baffle design and spacing is selected for the shell side. This completes a preliminary design of the heat exchanger.

A revised design is made next by using the geometry of the preliminary design to estimate an overall heat transfer coefficient from calculated individual heat transfer coefficients and estimated fouling factors, as well as pressure drops, using the methods discussed in Section 13.3. Then the entire procedure for sizing the heat exchanger is iterated until changes to the design between iterations are within some tolerance.

The previous procedure is tedious if done by hand calculations. Therefore, it is more convenient to conduct the design with available computer programs. For example, the HEATX subroutine of the ASPEN PLUS simulator computes heat transfer coefficients, pressure drops, and outlet conditions for a shell-and-tube heat exchanger of known geometry, as illustrated in Example 13.7. It can be used by trial-and-error with the iterative procedure to design an exchanger.

EXAMPLE 13.7

An existing 2-8 shell-and-tube heat exchanger in a single shell (equivalent to two shells in series with 4 tube passes in each shell) is to be used to transfer heat to a toluene feed stream from a styrene product stream. The toluene enters the exchanger on the tube side at a flow rate of 125,000 lb/hr at 100°F and 90 psia. The styrene enters on the shell side at a flow rate of 150,000 lb/hr at 300°F and 50 psia. The exchanger shell and tubes are carbon steel. The shell has an inside diameter of 39 in. and contains 1,024 $\frac{3}{4}$-in., 14 BWG, 16-ft-long tubes on a 1-in. square pitch. Thirty-eight segmental baffles are used with a baffle cut of 25%. Shell inlet and outlet nozzles are 2.5-in., schedule 40 pipe, and tube-side inlet and outlet nozzles are 4-in., schedule 40 pipe. Fouling factors are estimated to be 0.002 (hr-ft²-°F)/Btu on each side. Determine the exit temperatures of the two streams, the heat duty, and the pressure drops.

SOLUTION

The HEATX subroutine (block) of the ASPEN PLUS simulator is used to make the calculations. It has built-in correlations of the type described above for estimating shell-side and tube-side heat transfer coefficients and pressure drops. The following results are obtained (both streams are liquid):

Toluene exit temperature = 257.4°F
Styrene exit temperature = 175.9°F
Tube-side tube pressure drop = 3.59 psi
Tube-side nozzle pressure drop = 0.56 psi
Toluene exit pressure = 85.85 psia
Shell-side baffled pressure drop = 4.57 psia
Shell-side nozzle pressure drop = 4.92 psia
Styrene exit pressure = 40.52 psia
Heat transfer area (tube outside) = 3,217 ft²
Heat duty = 8,775,000 Btu/hr
Estimated overall heat-transfer coefficient, U_o, clean = 101.6 Btu/(hr-ft²-°R)
Estimated overall heat-transfer coefficient, U_o, dirty = 69.4 Btu/(hr-ft²-°R)
Log-mean temperature difference based on countercurrent flow = 57.6°F
Correction factor for 2-8 exchanger, F_T = 0.682
Maximum velocity in the tubes = 2.90 ft/s
Maximum Reynolds number in the tubes = 34,000
Maximum crossflow velocity in the shell = 2.36 ft/s
Maximum crossflow Reynolds number in the shell = 32,400
Flow regime on tube and shell sides = turbulent

Note that the file EXAM13-7.bkp on the CD-ROM can be used to reproduce these results. ∎

Of even greater utility are the B-JAC programs of Aspen Technology, Inc., which are a suite of three programs: (1) HETRAN for the detailed thermal design, rating, and simulation of shell-and-tube heat exchangers, including sensible heating and cooling, condensation, and vaporization; (2) AEROTRAN for the detailed design, rating, and simulation of air-cooled heat exchangers; and (3) TEAMS for the mechanical design of shell-and-tube heat exchangers, using the pressure vessel code. Of particular importance is HETRAN, which can determine the optimal geometry for a shell-and-tube heat exchanger. This program evaluates all possible baffle and shell- and tube-pass arrangements, and seeks the exchanger with the smallest shell diameter, shortest tube length, minimum reasonable baffle spacing, and maximum reasonable number of tube passes, subject to allowable shell- and tube-side pressure drops. The result is a complete TEMA (Tubular Exchanger Manufacturers Association) specification sheet.

EXAMPLE 13.8

Design a new shell-and-tube heat exchanger for the conditions of Example 13.7, but with maximum shell-side and tube-side pressure drops of 10 psi each.

SOLUTION

In this case, it is convenient to use a heat-exchanger design program such as HETRAN in B-JAC. For this example, the inlet conditions for the toluene and styrene streams are taken from Example 13.7. That example used physical properties of ASPEN PLUS. In this example, the physical property correlations provided with the B-JAC programs were used. The results showed less than a 5% difference. The current versions of ASPEN PLUS and B-JAC now show no difference. To provide the best comparison with Example 13.7, the same two exit temperatures computed in Example 13.7 (165.2°F for styrene and 268.7°F for toluene) were specified. To maintain an energy balance, the toluene flow rate was increased by 4.5%. The computed heat duty was 9,970,000 Btu/hr compared to 9,472,000 Btu/hr in Example 13.7. HETRAN considered 17 designs, with up to three shell passes in series and total tube passes ranging from two to eight. Maximum tube length was limited to 20 ft. Most designs resulted in pressure drops that exceeded the 10-psi maximum. The recommended design was a 3-12 exchanger with three exchangers in series, each with one shell pass and 4 tube passes. Tubes of 0.75-inch O.D., 0.065-inch thickness, 16-ft long, and on 0.9375-inch triangular spacing were selected. Other results are as follows in the same order as Example 13.7:

Toluene exit temperature = 268.7°F
Styrene exit temperature = 165.2°F
Tube-side tube pressure drop = 5.37 psi
Tube-side nozzle pressure drop = 1.02 psi
Toluene exit pressure = 83.61 psia
Shell-side baffled pressure drop = 8.02 psi
Shell-side nozzle pressure drop = 1.16 psi
Styrene exit pressure = 40.82 psia
Heat transfer area (tube outside) = 3,663.2 ft^2 or 1,221.1 ft^2 in each of the three shells
Heat duty = 9,970,000 Btu/hr
Estimated heat-transfer coefficient on the tube side = 304 Btu/hr-ft^2-°F
Estimated heat-transfer coefficient on the shell side = 344 Btu/hr-ft^2-°F
Estimated overall heat-transfer coefficient, clean = 140 Btu/hr-ft^2-°F
Estimated overall heat-transfer coefficient, fouled = 86.6 Btu/hr-ft^2-°F
Log-mean temperature difference based on countercurrent flow = 46.4°F
Correction factor for 3-12 exchanger, $F_T = 0.75$
Velocity in the tubes = 3.49 ft/s
Nominal Reynolds number in the tubes = 44,000
Velocity in the shell = 1.67 ft/s
Nominal Reynolds number in the shell = 29,000
Flow regime on tube and shell sides = turbulent

Additional results were 20 baffles in each shell on an 8.5-inch spacing and with a 25% baffle cut, 392 tubes in each shell for a total of 1,176 tubes, and a shell inside diameter of 21.25 inches. The setting plan and tube layout for each of the three shells in series is shown in Figure 13.17, while the heat exchanger specification sheet is shown in Figure 13.18. In Figure 13.17, the dimensions are in inches and the line marked "O" is a reference line for dimensions to the left and right of it. Note that the file EXAM13-8.bjt on the CD-ROM can be used to reproduce these results. ∎

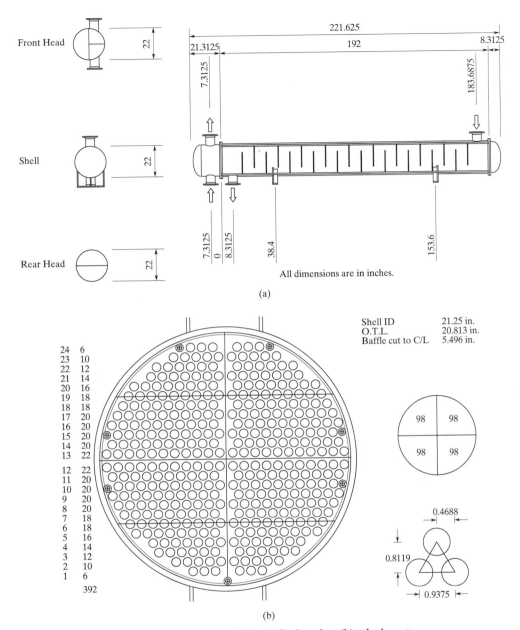

Figure 13.17 Heat exchanger for Example 13.8: (a) Setting plan, (b) tube layout.

13.5 SUMMARY

Having studied this chapter, the reader should

1. Know how the temperature and phase condition of a stream can be changed by using a heat exchanger.
2. Be able to specify and use a simulation program to calculate a heat exchanger when modeling just one side.
3. Be able to select heat transfer media for the other side of the exchanger.
4. Know the importance of heating and cooling curves and how to generate them with a simulation program, and use them to avoid crossover violations of the second law of thermodynamics.

1	Company:							
2	Location:							
3	Service of Unit:		Our reference:					
4	Item No..		Your reference:					
5	Date:	Rev No.:	Job No.:					
6	Size	21–192	in	Type BEM	hor	Connected in	1 parallel	3 series
7	Surf./unit(eff.)	3663.2	ft2	Shells/unit 3		Surf/shell (eff.)	1221.1	ft2
8				**PERFORMANCE OF ONE UNIT**				

9	Fluid allocation			Shell Side		Tube Side	
10	Fluid name						
11	Fluid quantity, Total		lb/h	150000		130714	
12	Vapor (In/Out)		lb/h				
13	Liquid		lb/h	150000	150000	130714	130714
14	Noncondensable		lb/h				
15							
16	Temperature (In/Out)		F	300	165.2	100	268.7
17	Dew / Bubble point		F				
18	Density		lb/ft3	48.617	53.007	53.284	47.245
19	Viscosity		cp	0.214	0.381	0.478	0.217
20	Molecular wt, Vap						
21	Moleculat wt, NC						
22	Specific heat		BTU/(lb*F)	0.5491	0.447	0.4234	0.4855
23	Thermal conductivity		BTU/(ft*h*F)	0.066	0.074	0.077	0.061
24	Latent heat		BTU/lb				
25	Inlet pressure (abs)		psi	50		90	
26	Velocity		ft/s	1.67		3.49	
27	Pressure drop, allow./calc.		psi	10	9.177	10	6.392
28	Fouling resist. (min)		ft2*h*F/BTU	0.002		0.002	
29	Heat exchanged	9969642	BTU/h	MTD corrected		34.83	F
30	Transfer rate, Service	78.14	Dirty 86.58	Clean	140.24	BTU/(h*ft2*F)	
31			**CONSTRUCTION OF ONE SHELL**				Sketch

32			Shell Side	Tube Side		
33	Design /Test pressure	psi	75/ /Code	90/ /Code		
34	Design temperature	F	360	330		
35	Number passes per shell		1	4		
36	Corrosion allowance	in	0.0625	0.0625		
37	Connections	In	6 / 150 ANSI	6 /150 ANSI		
38	Size/rating	Out	6 / 150 ANSI	6 / 150 ANSI		
39	in/	Intermediate	/ 150 ANSI	/ 150 ANSI		
40	Tube No.	OD 0.75	Tks-avg 0.065	in Length 16	ft Pitch 0.9375	in
41	Tube type		Material CS		Tube pattern 30	
42	Shell CS ID OD22	in	Shell cover			
43	Channel or bonnet CS		Channel cover			
44	Tubesheet-stationary CS		Tubesheet-floating			
45	Floating head cover		Impingement protection None			
46	Baffle-crossing CS	Type single seg	Cut(%d) 24 hor Spacing: c/c 8.5			in
47	Baffle-long	Seal type		Inlet 14.4375		in
48	Supports-tube	U-bend		Type		
49	Bypass seal		Tube-tubesheet joint groove/expand			
50	Expansion joint		Type			
51	RhoV2-Inlet nozzle 927	Bundle entrance 489		Bundle exit 448		lb/(ft*s2)
52	Gaskets - Shell side		Tube Side			
53	Floating head					
54	Code requirements	ASME Code Sec VIII Div 1		TEMA class B		
55	Weight/Shell	5400.9	Filled with water 7927	Bundle 3600.5		lb
56	Remarks					
57						
58						

Figure 13.18 Heat exchanger specification sheet for Example 13.8.

5. Know the major types of heat exchange equipment and how they differ in flow directions of the two fluids exchanging heat, and how to determine the corrected temperature driving force for heat transfer.

6. Know how to specify a heat exchanger when modeling both sides with a simulation program.

7. Know how to estimate overall heat transfer coefficients, including the effect of fouling.

8. Know the limitations of boiling heat transfer.

9. Be able to design a shell-and-tube heat exchanger with the help of a simulator.

REFERENCES

Bowman, R.A., A.C. Mueller, and W.M. Nagle, "Mean Temperature Difference in Design," *Trans. ASME*, **62**, 283–293 (1940).

Colburn, A.P., *Trans. AIChE*, **29**, 166 (1931).

Dittus, F.W., and L.M.K. Boelter, *Univ. Calif. (Berkeley) Pub. Eng.*, **2**, 443 (1930).

Donohue, D.A., *Ind. Eng. Chem.*, **41**, 2,499 (1949).

Grimison, E.D., *Trans. ASME*, **59**, 583 (1937).

Hewitt, G.F., Ed., *Handbook of Heat Exchanger Design*, Begell House, New York (1992).

Kern, D.Q., *Process Heat Transfer*, McGraw-Hill, New York (1950).

Nagle, W.M., "Mean Temperature Differences in Multipass Heat Exchangers," *Ind. Eng. Chem.*, **25**, 604–609 (1933).

Perry, R.H., and D.W. Green, *Perry's Chemical Engineers' Handbook*, 7th ed., McGraw-Hill, New York (1997).

Sieder, E.N., and G.E. Tate, "Heat Transfer and Pressure Drops of Liquids in Tubes," *Ind. Eng. Chem.*, **28**, 1,429–1,436 (1936).

Underwood, A.J.V., "The Calculation of the Mean Temperature Difference in Multipass Heat Exchangers," *J. Inst. Petroleum Technol.*, **20**, 145–158 (1934).

EXERCISES

13.1 In Example 13.7, an existing exchanger is used to transfer sensible heat between toluene and styrene streams. A minimum approach temperature of 31.3°F is achieved. Design a new shell-and-tube heat exchanger for a 10°F minimum approach temperature.

13.2 A heat exchange system is needed to cool 60,000 lb/hr of acetone at 250°F and 150 psia to 100°F. The cooling can be achieved by exchanging heat with 185,000 lb/hr of acetic acid, which is available at 90°F and 75 psia, and needs to be heated. Four 1-2 shell-and-tube heat exchangers are available. Each has an inside shell diameter of 21.25 in. and contains 270 $\frac{3}{4}$-in.-O.D., 14 BWG, 16-ft-long carbon steel tubes in a square layout on a 1-in. pitch. Segmental baffles with a 25% baffle cut are spaced 5 in. apart. Determine whether one or more of these exchangers can accomplish the task. Note that if two, three, or four of the exchangers are connected in series, they will be equivalent to one 2-4, 3-6, or 4-8 exchanger, respectively. If the exchangers are not adequate, design a new exchanger or exchanger system that is adequate. Assume a combined fouling factor of 0.004 (hr-ft²-°F)/Btu.

13.3 A trim heater is to be designed to heat 116,000 lb/hr of 57 wt% ethane, 25 wt% propane, and 18 wt% *n*-butane from 80 to 96°F. The stream will enter the exchanger at 520 psia and must not reach the bubble point in the exchanger. The stream will be heated with gasoline, which will enter at 240°F and 95 psia, with a flow rate of 34,000 lb/hr. Standard practice of the company is

to use 1-2 shell-and-tube heat exchangers with $\frac{3}{4}$-in., 16 BWG carbon steel tubes, 20 ft long, 1-in. square pitch. Tube count depends on shell diameter, with the following diameters available:

Shell I.D. (in.)	Tube Count
10	52
12	78
13.25	96
15.25	136
17.25	176
19.25	224

The gasoline will flow on the shell side. Assume a combined fouling factor of 0.002 (hr-ft²-°F)/Btu. Design a suitable heat exchange system, assuming a 25% overdesign factor.

13.4 Design a shell-and-tube heat exchanger to cool 60,000 lb/hr of 42° API (American Petroleum Institute) kerosene from 400 to 225°F by heating a 35° API distillate from 100 to 200°F under the following specifications. Allow a 10-psi pressure drop for each stream and a combined fouling factor of 0.004 (hr-ft²-°F)/Btu. Neglect the tube-wall resistance. Use $\frac{3}{4}$-in., 16 BWG tubing, O.D. = 0.75 in., I.D. = 0.620 in., flow area/tube = 0.302 in.², surface/linear foot = 0.1963 ft² outside and 0.1623 ft² inside. Use 1-in. square pitch. Place kerosene on the shell side. If necessary, change the configuration to keep the tube lengths below 20 ft and the pressure drops below 10 psi.

DATA

	42°API		35°API	
	400°F	**225°F**	**100°F**	**200°F**
C_p, Btu/lb°F	0.67	0.56	0.53	0.47
μ, cP	0.20	0.60	1.3	3.4
k, Btu/hr-ft-°F	0.074	0.078	0.076	0.078
sp. gr.	0.685	0.75	0.798	0.836

13.5 Hot water at 100,000 lb/hr and 160°F is cooled with 200,000 lb/hr of cold water at 90°F, which is heated to 120°F in a countercurrent shell-and-tube heat exchanger. The exchanger has 20-ft steel tubes with 0.75-in. O.D. and 0.62-in. I.D. The tubes are on a 1-in. square pitch. The thermal conductivity of copper is 25.9 Btu/(ft-hr-°F). The mean heat transfer coefficients are estimated as $h_i = 200$ Btu/(ft²-hr-°F) and $h_o = 200$ Btu/(ft²-hr-°F). Estimate:

a. The area for heat transfer

b. The diameter of the shell

13.6 A horizontal 1-4 heat exchanger is used to heat gas oil with saturated steam. Assume that $h_o = 1,000$ Btu/(ft²-hr-°F) for condensing steam and the fouling factor = 0.004 ft²-hr-°F/Btu [1 bbl (barrel) = 42 gal].

a. For a tube-side velocity of 6 ft/s, determine the number and length of tubes and the shell diameter.

b. Determine the tube-side pressure drop.

	Shell Side		Tube Side	
	Inlet	**Outlet**	**Inlet**	**Outlet**
Fluid	Saturated Steam		Gas oil	
Flow rate (bbl/hr)			1,200	
Temperature (°F)			60	150
Pressure (psig)	50	50	60	
Viscosity (cP)			5.0	1.8
Sp. gr.			0.840	0.810
Thermal conductivity [Btu/(ft-hr-°F)]			0.078	0.083
Heat capacity (Btu/lb-°F)			0.480	0.461

The tubes are 1-in. O.D. by 16 BWG on a 1.25-in. square pitch.

13.7 An alternative heating medium for Exercise 13.6 is a distillate:

	Shell Side	
	Inlet	**Outlet**
Fluid	35° API distillate	
Flow rate	—	—
Temperature (°F)	250	150
Pressure (psig)	80	
Viscosity (cP)	1.3	3.4
Sp. gr.	0.798	0.836
Thermal conductivity (Btu-hr-ft-°F)	0.076	0.078
Heat capacity (Btu/lb-°F)	0.53	0.47

Determine the tube-side velocity, number and length of tubes, and shell diameter for a 1-6 shell-and-tube heat exchanger using the 1-in.-O.D. by 16 BWG tubes on a 1.25-in. square pitch. Design to avoid pressure drops greater than 10 psia. If necessary, change the configuration to keep the tube length below 20 ft.

13.8 Ethylene glycol at 100,000 lb/hr enters the shell of a 1-6 shell-and-tube heat exchanger at 250°F and is cooled to 130°F with cooling water heated from 90 to 120°F. Assume that the mean overall heat transfer coefficient (based on the inside area of the tubes) is 100 Btu/(ft²-hr-°F) and the tube-side velocity is 5 ft/s. Use $\frac{3}{4}$-in., 16 BWG tubing (O.D. = 0.75 in., I.D. = 0.62 in.) arranged on a 1-in. square pitch.

a. Calculate the number of tubes, length of the tubes, and tube-side heat transfer coefficient.

b. Calculate the shell-side heat transfer coefficient to give an overall heat transfer coefficient of 100 Btu/(ft²-hr-°F).

DATA

	Ethylene Glycol 190°F	**Water 105°F**
C_p, Btu/lb-°F	0.65	1.0
μ, cP	3.6	0.67
k, Btu/hr-ft-°F	0.154	0.363
Sp. gr.	1.110	1.0

Chapter **14**

Separation Tower Design

14.0 OBJECTIVES

The most commonly used separation method in industrial chemical processes is distillation, including enhanced distillation (extractive, azeotropic, and reactive), which is carried out in towers of cylindrical shape containing either plates or packing for contacting the vapor flowing up the tower with the liquid flowing down. The process design of such towers consists of a number of calculations, which are described and illustrated in this chapter. Most of these calculations are readily made with a simulator. The same calculations apply to any multistage separation involving mass transfer between vapor and liquid phases, including absorption and stripping.

After studying this chapter and the materials on distillation on the multimedia CD-ROM that accompanies this book, the reader should

1. Be able to determine the tower operating conditions of pressure and temperature and the type of condenser to use.
2. Be able to determine the number of equilibrium stages and reflux required.
3. Be able to select an appropriate contacting method (plates or packing).
4. Be able to determine the number of actual plates or packing height required, together with feed and product locations.
5. Be able to determine the tower diameter.
6. Be able to determine other factors that may influence tower operation.

14.1 OPERATING CONDITIONS

Multistage towers for separations involving mass transfer between vapor and liquid phases can operate anywhere within the two-phase region, but proximity to the critical point should be avoided. Typical operating pressures for distillation range from 1 to 415 psia. For temperature-sensitive materials, vacuum distillation is very common, with pressures as low as 5 mm Hg. Except for low-boiling components and cases where a vapor distillate is desired, a total condenser is used. Before determining a feasible and, hopefully, a near-optimal operating pressure, a preliminary material balance must be made to estimate the distillate and bottoms product compositions. As a starting point for establishing a reasonable operating pressure and a type of condenser, the graphical algorithm in Figure 7.9 can be applied in the following manner, noting that it is based on the use of cooling water that enters the condenser at 90°F and exits at 120° F. The pressure at the exit of the condenser (or in the reflux drum), P_D, is determined so as to permit condensation with cooling water, if possible. This pressure is computed as the bubble-point pressure at 120°F. If this pressure is less than 215 psia, a total condenser is used. However, if the pressure is less than 30 psia, set the condenser outlet pressure at 20 to 30 psia to avoid vacuum operation. If the pressure at 120°F is greater than 215 psia, calculate the dew-point pressure of the distillate at 120°F. If that pressure is less than 365 psia, use a partial condenser; if it is greater than 365 psia, select a refrigerant that gives a minimum approach temperature of 5 to 10°F, in place of cooling water, for the partial condenser, such that the distillate dew-point pressure does not exceed 415 psia. Up to this point, the tower operating pressure has been determined by the composition of the distillate. Conditions based on the composition of the bottoms product must now be checked. Using the determined condenser outlet pressure, assume a condenser pressure drop in the range of 0–2 psia. Assume a tower pressure drop of from 5 to 10 psia. This will give a pressure at the bottom of the column, P_B, in the range of 5–12 psia higher than the condenser-outlet pressure. Almost all reboilers are partial reboilers that produce a bottoms product at or close to the bubble point. Therefore, determine the bottoms temperature, T_B, by a bubble-point calculation based on the estimated bottoms composition and the bottoms pressure. If this exceeds the decomposition, polymerization, or critical temperature of the bottoms, then compute a bottoms pressure based on a bottoms temperature safely below the limiting temperature. Then using the assumed pressure drops, calculate a new condenser outlet pressure and corresponding temperature. This may require a change in the coolant used in the condenser and the type of condenser. Also, the new condenser outlet pressure may be less than about 15 psia, in which case a vacuum system for the tower will be necessary, as discussed later in Section 16.5.

In some distillations, the overhead vapor may contain components covering a wide range of volatility. For example, the overhead vapor from a vacuum tower will contain air from leakage into the tower, mixed with other components that could be condensed with cooling water or a modest refrigerant. In other cases, the overhead vapor may contain hydrogen and other light gases mixed with easily condensable components. In those cases, neither a total or partial condenser is used. Instead, the condenser is designed to produce both a vapor distillate and a liquid distillate. The latter has the same composition as the reflux. For vacuum operation, the vapor distillate is sent to a vacuum pump. To determine the pressure, P_D, compositions of the vapor distillate and the liquid distillate are calculated for a series of pressures at a temperature of 120°F, for cooling water, or at a lower temperature if a refrigerant is necessary to recover a higher percentage of the less-volatile components in the liquid distillate. When using a refrigerated condenser, one should always consider placing a water-cooled partial condenser ahead of it. From the results of the calculations, a reasonable pressure is selected.

For extractive and azeotropic distillation, the condenser outlet pressure is usually near ambient pressure, in the range of 20–30 psia, and a total condenser is used. An exception is azeotropic distillation when a low molecular-weight entrainer is used that necessitates a higher pressure. For reactive distillation, the pressure must be sufficiently high to give corresponding temperatures in the range of reasonable reaction rates.

Absorbers and strippers usually involve components that cover a very wide range of volatility. For example, an absorber might have a feed gas that contains methane, while the absorbent might be an oil of 150 molecular weight. For these two separation operations, which frequently do not utilize either a condenser or reboiler, the tower operating pressure cannot be determined from bubble- and/or dew-point calculations because they can be extremely sensitive to assumed product vapor and/or liquid compositions. Instead, the following rules may apply:

Absorption favors high pressures and low temperatures. Therefore, cool the feed gas and absorbent with cooling water or a refrigerant. If the internal temperature rise in an absorption column is large, interstage coolers can be added. However, because of the high cost of gas compression, it may not be economical to increase the pressure of the feed gas. But do not decrease the pressure of the feed gas.

Stripping favors low pressures and high temperatures. Therefore, heat the liquid feed and stripping agent and lower the pressure to near ambient, but not to a vacuum.

14.2 FENSKE-UNDERWOOD-GILLILAND (FUG) SHORTCUT METHOD FOR ORDINARY DISTILLATION

For ordinary distillation of a single feed to give only distillate and bottoms products, the FUG method, which is included in the library of equipment models of all simulators, is useful for making an initial estimate of the reflux ratio, the number of equilibrium stages, and the location of the feed stage. The method is quite accurate for ideal mixtures of a narrow-boiling range. However, for nonideal mixtures, particularly those that form azeotropes, and for wide-boiling feeds, the FUG method can be quite inaccurate. Therefore, before applying the method, the vapor–liquid equilibrium of the feed should be carefully examined for the magnitude of liquid-phase activity coefficients and the possibility of azeotropes over the range of possible compositions. Note that for nonideal mixtures, especially, design engineers often skip this approximate method, in preference to running a few iterations using a rigorous model, as discussed in Section 14.4. Often, reasonable guesses can be provided for the number of stages and reflux ratio to achieve a satisfactory simulation that can be fine tuned to satisfy product specifications.

The FUG method, which applies to binary and multicomponent feeds, is described in detail by Seader and Henley (1998) and in *Perry's Chemical Engineers' Handbook* (1997). Only the procedure is discussed here. The method involves five steps based on the desired separation of two key components in the feed. It includes an estimation of the separation of the nonkey components.

Step 1: *Estimation by the Fenske equation of the minimum number of equilibrium stages, N_{min} (corresponding to total reflux or infinite reflux ratio), needed to separate the two key components.* The Fenske equation is simple and readily applied, even by hand. It involves only one assumption, that of an average relative volatility, $\alpha_{LK,HK}$, between the two key components, throughout the tower. This may be the geometric average of the distillate and bottoms, or the geometric average of the feed, distillate, and bottoms. The Fenske equation may be written as follows:

$$N_{min} = \frac{\log\left[\left(\dfrac{d_{LK}}{b_{LK}}\right)\left(\dfrac{b_{HK}}{d_{HK}}\right)\right]}{\log(\alpha_{LK,HK})} \tag{14.1}$$

where d is a component flow rate in the distillate and b is a component flow rate in the bottoms product.

Step 2: *Estimation by the Fenske equation [Eq. (14.1)] of the distribution, d/b, of the nonkey components between distillate and bottoms at total reflux using the value of N_{min} computed in Step 1, the b/d ratio for the heavy key, and the relative volatility between the nonkey and the heavy key, $\alpha_{NK,HK}$.* Although this estimate is for total reflux conditions, it is a surprisingly good estimate for the distribution of the nonkey components at finite reflux conditions for nearly ideal mixtures.

Step 3: *Estimation by the Underwood equations of the minimum reflux ratio, R_{min} (corresponding to an infinite number of equilibrium stages), needed to separate the two key components.* This calculation is complicated because it involves the solution of nonlinear equations and requires a calculation of the distribution of the nonkey components at minimum reflux even though that distribution is not used for any other purpose. The application of the Underwood equations involves two serious assumptions: (1) The molar liquid flow rate is constant throughout the rectifying section, and (2) the relative volatility is constant in the pinch region. When these assumptions are not valid, the estimated minimum reflux ratio can be less than the true value, making the method nonconservative. More details of the use of the Underwood equations are given by Seader and Henley (1998).

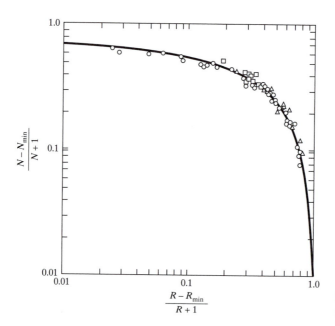

Figure 14.1 Gilliland correlation for ordinary distillation.

Step 4: *Estimation by the Gilliland correlation of the actual number of equilibrium stages, N, for a specified ratio of actual reflux ratio, R, to minimum reflux ratio, R_{min}.* The Gilliland correlation, which is shown in Figure 14.1, has no theoretical foundation, but is an empirical fit of many rigorous binary and multicomponent calculations when plotted as $(N - N_{min})/(N + 1)$ as a function of $(R - R_{min})/(R + 1)$. The accuracy of the Gilliland method is limited because it ignores the effect of the feed condition (from subcooled to superheated), and can be badly in error when stripping is much more important in the separation than rectification. For optimal design, the recommended value of R/R_{min} to use with the Gilliland method is typically in the range of 1.1 to 1.5, with the lower value for difficult separations requiring more than 100 equilibrium stages and the higher value for easy separations of less than 10 equilibrium stages. At $R/R_{min} = 1.3$, N/N_{min} is often equal to approximately 2.

Step 5: *Estimation of the feed-stage location by the Fenske equation.* The calculation is made with Eq. (14.1) by applying it to the section of stages between the feed composition and the distillate composition to obtain the minimum number of rectification stages, $N_{R,min}$, and then to the section of stages between the feed and bottoms product to obtain the minimum number of stripping stages, $N_{S,min}$. The ratio of $N_{R,min}$ to $N_{S,min}$ is assumed to be the same as the ratio of N_R to N_S at finite reflux conditions. Alternatively, the empirical, but often more accurate, Kirkbride equation can be applied.

14.3 KREMSER SHORTCUT METHOD FOR ABSORPTION AND STRIPPING

For adiabatic absorbers and strippers with one feed, one absorbent or stripping agent, and two products, a simple and useful shortcut method for estimating the minimum absorbent or stripping agent flow rate is the Kremser method. It applies in the limit of an infinite number of equilibrium stages for the specified absorption or stripping of one component, the key component, from the feed. It also applies for a finite number of equilibrium stages, *N*. Although the method is not included in the library of equipment models of most simulators, it is quite straightforward to apply the Kremser method using hand calculations or a spreadsheet. The derivation of the equations is presented in detail by Seader and Henley (1998) and in *Perry's Chemical Engineers' Handbook* (1997).

The separation factor in the Kremser method is an effective absorption factor, A_e, for absorption and a stripping factor, S_e, for stripping, rather than a relative volatility as in the FUG method for distillation. These two factors, which are different for each component, are defined by:

$$A_e = L/KV \tag{14.2}$$

$$S_e = KV/L \tag{14.3}$$

The total molar liquid rate down the tower, L, the total molar vapor rate up the tower, V, and the K-value all vary from the top stage to the bottom stage of the tower. However, sufficiently good estimates by the Kremser method can be achieved by using average values based on the flow rates and temperatures of the two streams entering the tower.

For an absorber, the design basis is the tower pressure; the flow rate, composition, temperature, and pressure of the entering vapor feed; the composition, temperature, and pressure of the absorbent; and the fraction to be absorbed of one key component. The minimum molar absorbent flow rate is estimated from:

$$L_{\min} = K_K V_{\text{in}} (1 - \phi_{A_K}) \tag{14.4}$$

where K_K is the K-value of the key component computed at the average temperature and pressure of the two entering streams and $(1 - \phi_{A_K})$ is the fraction of the key component in the feed gas that is to be absorbed. Typically, the operating absorbent rate is 1.5 times the minimum value. Then, the following equation, due to Kremser and shown in Figure 14.2, is used

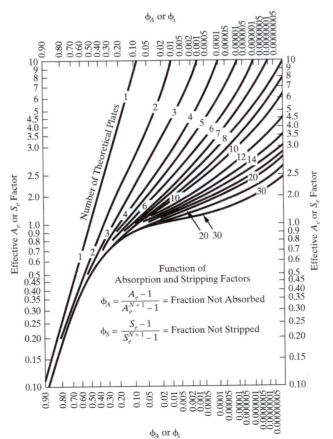

Figure 14.2 Plot of the Kremser equation for absorbers and strippers.

to compute the number of equilibrium stages required. This equation assumes that the absorbent does not contain the key component.

$$\phi_{A_K} = \frac{A_{e_K} - 1}{A_{e_K}^{N+1} - 1} \tag{14.5}$$

With the value of N computed for the key component, Eq. (14.5) is then used to compute the values of ϕ_A for the other components in the feed gas using their absorption factors. From this, a material balance around the tower can be completed.

For a stripper, the design basis is the tower pressure; the flow rate, composition, temperature, and pressure of the entering liquid feed; the composition, temperature, and pressure of the stripping agent; and the fraction of a key component to be stripped. The minimum molar stripping agent flow rate is estimated from:

$$V_{\text{min}} = \frac{L_{\text{in}}}{K_K}(1 - \phi_{S_K}) \tag{14.6}$$

where K_K is the K-value of the key component computed at the average temperature and pressure of the two entering streams and $(1 - \phi_{S_K})$ is the fraction of the key component in the feed gas that is to be stripped. Typically, the stripping agent rate is 1.5 times the minimum value. Then, the following equation is used to compute the number of equilibrium stages required. This equation assumes that the stripping agent does not contain the key component.

$$\phi_{S_K} = \frac{S_{e_K} - 1}{S_{e_K}^{N+1} - 1} \tag{14.7}$$

With the value of N computed for the key component, Eq. (14.7) is then used to compute the values of ϕ_S for the other components in the feed liquid using their stripping factors. From this, a material balance around the tower can be completed.

EXAMPLE 14.1

The feed gas to an absorber at 105°F and 400 psia contains 150 kmol/hr of methane, 350 kmol/hr of ethane, 250 kmol/hr of propane, and 50 kmol/hr of n-butane. The absorber is to absorb 90% of the n-butane with an oil at 90°F and 50 psia. Estimate, with the Kremser equation, the number of stages required and the amounts absorbed of the other three components in the feed gas.

SOLUTION

Set the absorber pressure at the feed gas pressure of 400 psia and neglect the pressure drop in the absorber. Use a pump to increase the pressure of the absorbent to 400 psia. The entering vapor rate is $V = 150 + 350 + 250 + 50 = 800$ kmol/hr. The average temperature of the two entering streams is $(105 + 90)/2 = 97.5°F$. The K-value for the key component, n-butane, at 400 psia and 97.5°F, is 0.22 by the SRK equation of state. Using Eq. (14.4) with $(1 - \phi_{A_K}) = 0.90$, the minimum absorbent rate is

$$L_{\text{min}} = 0.22(800)(0.90) = 158 \text{ kmol/hr}$$

Select an operating absorbent flow rate of $L = 1.5\,L_{\text{min}} = 1.5(158) = 237$ kmol/hr. The absorption factor for n-butane, from Eq. (14.2) $= A_{e_K} = 237/[0.22(800)] = 1.35$. This is close to 1.40, which is often quoted as the optimal value of the absorption factor. Equation (14.5), which is nonlinear in N, is now applied with $(1 - \phi_{A_K}) = 0.90$, which gives $\phi_{A_K} = 0.10$.

$$0.10 = \frac{1.35 - 1}{1.35^{N+1} - 1}$$

Solving, $N = 4$ equilibrium stages. The result for this exercise is very useful as a first approximation for a rigorous equilibrium-stage method using a simulator, as described in the next section. ∎

14.4 RIGOROUS MULTICOMPONENT, MULTI-EQUILIBRIUM-STAGE METHODS WITH A SIMULATOR

Almost all multistage, multicomponent, vapor–liquid separation towers, whether plate or packed, are routinely designed with simulators. The calculations are usually based on the assumption of equilibrium stages, but more realistic mass-transfer models are also available (e.g., see Chapter 12 of Seader and Henley, 1998). The equilibrium–stage calculations apply component mole balances, enthalpy balances, and vapor–liquid phase equilibrium at each stage, and utilize any of a number of reasonably rigorous thermodynamic correlations based on equations of state or liquid-phase activity coefficients to estimate K-values and enthalpies. The resulting large set of equations is nonlinear and is solved iteratively for stage-wise profiles of vapor flows and compositions, liquid flows and compositions, and temperatures, from a set of starting guesses by either an *inside-out method* or a *Newton method*, both of which are described in some detail by Seader and Henley (1998) and in *Perry's Chemical Engineers' Handbook* (1997). The inside-out method is fast and is the most widely used, but Newton's method is sometimes preferred for highly nonideal systems. However, convergence of the solution of the nonlinear equations is not guaranteed for either method. When a method fails to converge within the default number of iterations (usually 20): (1) more iterations can be specified, (2) a damping factor can be applied to limit the changes made by the method to the guesses of the unknowns between iterations to prevent wild swings, and/or (3) the initial guesses of the unknowns can be changed. In this way, most problems, unless infeasibly specified, can be converged. Infeasible specifications include those where an inadvertent attempt is made to violate the order of volatility of the components.

Equilibrium-stage methods are usually adequate for nearly ideal distillation systems when coupled with calculations of plate efficiency to estimate actual trays or, in the case of packed towers, when HETS (height equivalent of a theoretical stage) or HETP (height equivalent to a theoretical plate) values are known from experience or from experiment to enable the estimation of packed height. For absorbers, strippers, and nonideal distillation systems, mass-transfer models are preferred, but their use requires a value for the tower diameter and a tray layout or type and size of packing. Even when mass-transfer models are preferred, initial calculations are usually made with equilibrium-stage models. Also, note that data for reliable mass-transfer coefficients is often difficult to obtain.

Both of the equilibrium-stage methods can handle almost any tower configuration, including multiple feeds, vapor and liquid sidestreams, and interheaters and intercoolers. Some of these methods can also handle pumparounds (liquid side draws returned to the column at a higher tray after heat exchange with other streams), bypasses, two liquid phases, chemical reaction, interlinked towers, and specified plate efficiencies. Thus, these models can be applied to ordinary and complex distillation, extractive distillation, homogeneous azeotropic distillation, heterogeneous azeotropic distillation, reactive distillation, absorption, stripping, reboiled stripping, and reboiled absorption.

When using an equilibrium-stage model, the following must be specified: (1) all stage pressures; (2) type of condenser (total, partial, or mixed) and type of reboiler; (3) all tower feed streams and feed stage locations, including total feed flow rate, composition, temperature, and pressure; (4) and number of equilibrium stages. In addition, stage locations for sidestreams, intercoolers, and interheaters are necessary. From a degrees-of-freedom analysis, as discussed by Seader and Henley (1998), in *Perry's Chemical Engineers' Handbook* (1997), and in Section 4.2, this leaves one additional specification for each stream leaving the tower and each intermediate heat exchanger. In addition, some models require the user to provide initial guesses of vapor and liquid flow rates at the top of the tower and stage temperatures at the top and bottom of the tower.

For ordinary distillation of nearly ideal systems, the FUG method, described in Section 14.2, provides an excellent starting point because it estimates the number of equilibrium stages, the feed stage location, and the reflux ratio. The latter can be used for the degree of freedom for the distillate product. For the degree of freedom of the bottoms product, a preferred initial specification is the bottoms flow rate, because it almost always results in a converged solution. However, these two specifications may not give the desired split of the two key components. If not, the calculation is repeated by specifying the desired heavy-key flow rate or mole fraction in the distillate and the desired light-key flow rate or mole fraction in the bottoms product, using the results of the previous calculation as an initial approximation of the solution. The reflux ratio and bottoms flow rate now become initial guesses that are varied to achieve the desired split of the two key components.

If convergence for the desired split is not achieved, then estimates of the reflux ratio and/or the bottoms product flow rate may have to be revised to achieve convergence when specifying the desired split of the two key components. It is usually not difficult to judge the direction in which these estimates should be revised. Rarely does the number of equilibrium stages have to be increased or decreased. However, the degree of separation as high purities are approached is more sensitive to the number of stages than to the reflux ratio. Finally, it is useful to vary the feed stage location to determine its optimal value, which corresponds to the lowest necessary reflux ratio.

For converged calculations, simulators can provide tables and graphs of temperature, vapor and liquid flow rates, and vapor and liquid compositions as a function of stage number. These profiles should be examined closely to detect the existence of any pinch points where little or no change occurs over a section of stages. If a pinch point is found, say over a region of 4 stages, then the number of stages in that section of the column can probably be reduced by 4 without changing the degree of separation. This should be confirmed by calculations.

For simple absorbers and strippers, the Kremser method described in Section 14.3 can be used to obtain an initial approximation to the number of equilibrium stages and the flow rate of the absorbent or stripping agent. Then, with the rigorous method, the latter can be varied to achieve the desired separation of the key component for a fixed number of stages.

When the FUG method is not valid for obtaining initial estimates for use with the rigorous methods, the following procedure may be useful. It focuses on an attempt to at least estimate the number of equilibrium stages required for each section of stages bounded by feeds and/or products. These estimates are provided by the Fenske equation, applied to key-component concentrations at either end of the section, where the computed N_{min} is multiplied by 2 to approximate the necessary N. This is illustrated in the following example.

EXAMPLE 14.2

A distillation column for the separation between propane and *n*-butane is to have the following two feeds:

	Upper Feed	Lower Feed
Temperature, °F	170	230
Pressure, psia	245	245
Component feed rates, lbmol/hr:		
Ethane	2.5	0.5
Propane	14.0	6.0
n-Butane	10.0	18.0
n-Pentane	5.0	30.0
n-Hexane	0.5	4.5

Use the Fenske equation to estimate the number of stages that should be placed between the two feeds.

SOLUTION First compute the relative volatility between propane and n-butane at 245 psia and the average temperature of the two feeds of $(170 + 230)/2 = 200°F$. The respective average K-values by the SRK equation of state are 1.76 and 0.84, giving $\alpha_{LK,HK} = 1.76/0.84 = 2.10$. Applying the Fenske equation [Eq. (14.1)] between the two feeds, using the key component feed flow rates, gives:

$$N_{min} = \frac{\log\left[\left(\dfrac{14}{6}\right)\left(\dfrac{18}{10}\right)\right]}{\log(2.10)} = \frac{0.623}{0.322} = 1.93$$

Therefore, $N = 2(1.93) = 3.86$. If this is rounded up to a value of 4, then four equilibrium stages should be placed between the two feed stages. ■

Rigorous calculations for extractive distillation are usually readily converged once the user determines which components the solvent forces to the bottom of the tower. The Fenske equation can be applied, in a manner similar to that in Example 14.2, to determine at which stage down from the top to bring in the solvent so as to minimize its loss to the distillate. Rigorous calculations for azeotropic distillation are another matter. Before even attempting a rigorous calculation, a triangular residue-curve map, which can be drawn by the simulators, should be used to determine feasible entrainer flow rates and product compositions, as described in Section 7.5. In addition, for heterogeneous azeotropic distillation, a triangular liquid–liquid phase equilibrium diagram should be used to determine preliminary values for the flows and compositions of the phase split that occurs in the overhead decanter. Failure to make these preliminary studies can result in much time and effort spent in trying to converge an infeasible tower specification. Most difficult of all are reactive distillation calculations. Again, preliminary calculations are necessary, including (1) independent reactor calculations, with a CSTR model, to determine an operating temperature range that gives reasonable reaction rates, and (2) flash calculations to determine component volatilities of reaction mixtures.

14.5 PLATE EFFICIENCY AND HETP

If a mass-transfer (rate-based) model is used, of the type described by Seader and Henley (1998), the stages will be actual trays or packed height in the case of packings. If an equilibrium-stage model is used, plate efficiencies for tray towers or HETP values for packed towers must be estimated to convert equilibrium stages to actual trays or to packed height. One of the major factors that influences mass transfer is the viscosity of the liquid phase. In distillation, liquid viscosities are generally low, often in the range of 0.1 to 0.2 cP, and overall plate efficiencies, E_o, are relatively high, in the range of 50 to 100%. Because of a liquid crossflow effect in large-diameter distillation towers, efficiencies even higher than 100% have been measured. Liquid viscosity in absorbers and some strippers is often in the range of 0.2 to 2.0 cP, and overall plate efficiencies are in the range of 10 to 50%. Very approximate estimates that are sometimes used are 70% for distillation, 50% for strippers, and 30% for absorbers. The number of actual plates required is

$$N_{actual} = N_{equilibrium}/E_o \tag{14.8}$$

A better estimate of overall plate efficiency can be made with the Lockett and Leggett version of the empirical O'Connell correlation, as shown in Figure 14.3. In this plot, the overall plate efficiency depends on the product of the average liquid-phase viscosity in cP and a dimensionless volatility factor. For distillation, the volatility factor is the average relative volatility between the light and heavy key components, $\alpha_{LK,HK}$. For absorbers and strippers, the volatility factor is 10 times the average K-value of the key component. If an even better estimate of the plate efficiency is desired, and in particular one that depends on plate location and component, a semitheoretical method developed by Chan and Fair (1984a), based on the

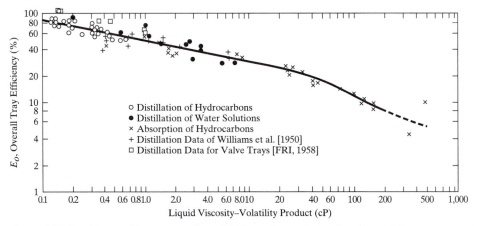

Figure 14.3 Lockhart and Leggett version of O'Connell correlation for plate efficiency.

definition of the Murphree vapor phase efficiency, can be applied, as discussed by Seader and Henley (1998).

For packed columns, HETP values are usually used to convert equilibrium stages to packed height even though the alternative concept of HTU (height of a transfer unit) together with NTU (number of transfer units) is on a more firm theoretical foundation. Values of HETP are generally derived from experimental data for a particular type and size of packing, and are often available from packing vendors. Typically cited, in the absence of data, is an HETP of 2 ft for modern random packings, and 1 ft for structured packings. However, Kister (1992) suggests the following, where D_P is the nominal diameter of random packings and a is the specific surface area of structured packings:

1. For modern random packings with low-viscosity liquids:

$$\text{HETP, ft} = 1.5(D_p, \text{in.})$$

2. For structured packings at low-to-moderate pressures and low-viscosity liquids:

$$\text{HETP, ft} = 100/a, \text{ft}^2/\text{ft}^3 + 0.333$$

3. For absorption with a viscous liquid:

$$\text{HETP} = 5 \text{ to } 6 \text{ ft}$$

4. For Vacuum service:

$$\text{HETP, ft} = 1.5(D_p, \text{in.}) + 0.50$$

5. For high-pressure service with structured packings:

$$\text{HETP, ft} > 100/a, \text{ft}^2/\text{ft}^3 + 0.333$$

6. For small-diameter towers less than 2 ft in diameter:

$$\text{HETP, ft} = \text{tower diameter in feet, but not less than 1 ft.}$$

The packed height is given by:

$$\text{Packed height} = N_{\text{equilibrium}} (\text{HETP}) \tag{14.9}$$

If a more accurate estimate of packed height is desired, correlations of experimental mass-transfer coefficients or heights of transfer units should be used for the particular packing selected. Some of these correlations are provided in simulators and the method of calculation is given in detail by Seader and Henley (1998).

14.6 TOWER DIAMETER

The tower diameter depends on the vapor and liquid flow rates and their properties up and down the tower. The tower diameter is computed to avoid flooding, where the liquid begins to fill the tower and leave with the vapor because it cannot flow downward at the required rate.

Tray Towers

For a given vapor flow rate in a tray tower, *downcomer flooding* occurs when the liquid rate is increased to the point where the liquid froth in the downcomer backs up to the tray above. This type of flooding is not common, because most tray towers have downcomers with an adequate cross-sectional area for liquid flow. A common rule is to compute the height of clear liquid in the downcomer. At low to moderate pressures, if the height is less than 50% of the tray spacing, it is unlikely that downcomer flooding will occur. However, at high pressures, this value may drop to 20–30%. Another rule is to provide a downcomer cross-sectional area of at least 10–20% of the total tower cross-sectional area, with the larger percentage pertaining to high pressure.

More commonly, the diameter of a tray tower is determined to avoid *entrainment flooding*. For a given liquid rate, as the vapor rate is increased, more and more liquid droplets are carried by the vapor to the tray above. Flooding occurs when the liquid entrainment by the vapor is so excessive that column operation becomes unstable.

The tower inside cross-sectional area, A_T, is computed at a fraction, f (typically 0.75 to 0.85), of the vapor flooding velocity, U_f, from the continuity equation for one-dimensional steady flow, applied to the vapor flowing up to the next tray through area $(A_T - A_d)$:

$$\dot{m}_V = G = (fU_f)(A_T - A_d)\rho_G \qquad (14.10)$$

where G = mass flow rate of vapor, A_d = downcomer area, and ρ_G = vapor density. Substituting $A_T = \pi(D_T)^2/4$ for a circular cross-section into Eq. (14.10) and solving for the tower inside diameter, D_T, gives

$$D_T = \left[\frac{4G}{(fU_f)\pi\left(1 - \dfrac{A_d}{A_T}\right)\rho_G} \right]^{1/2} \qquad (14.11)$$

The flooding velocity is computed from an empirical capacity parameter, C, based on a force balance on a suspended liquid droplet:

$$U_f = C\left(\frac{\rho_L - \rho_G}{\rho_G}\right)^{1/2} \qquad (14.12)$$

The capacity parameter is given by:

$$C = C_{SB}F_{ST}F_F F_{HA} \qquad (14.13)$$

The parameter, C_{SB}, for towers with perforated (sieve) plates is given by the correlation of Fair (1961), based on data from commercial-size towers, covering tray spacings from 6 to 36 in. A revision by Fair of the original correlation, shown in Figure 14.4, applies to all common crossflow plates (sieve, valve, and bubble-cap), with tray spacing, T_S, in mm, from 150 to 900 and C_{SB} in m/s. The abscissa in Figure 14.4 is a flow ratio parameter, $F_{LG} = (L/G)(\rho_G/\rho_L)^{1/2}$, where both the liquid rate, L, and vapor rate, G, are mass flow rates. The surface tension factor, F_{ST}, is equal to $(\sigma/20)^{0.20}$, where the surface tension, σ, is in dyne/cm. The foaming factor, F_F, is 1 for non-foaming systems, typical of distillation, and 0.5 to 0.75 for foaming systems, typical of absorption with heavy oils. The hole-area factor, F_{HA}, is 1 for

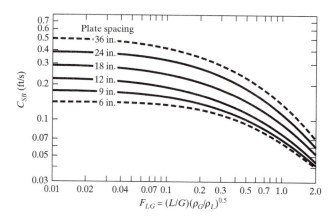

Figure 14.4 Flooding correlation for sieve, valve, and bubble-cap trays.

valve and bubble-cap trays. For sieve trays, it is 1 for $(A_h/A_a) \geq 0.10$, and $[5(A_h/A_a) + 0.5]$ for $0.06 \leq (A_h/A_a) \leq 1.0$, where A_h is the total hole area on a tray and A_a is the active area of the tray $= (A_T - 2A_d)$ where bubbling occurs.

In Eq. (14.11), the ratio (A_d/A_T) may be estimated by

$$\frac{A_d}{A_T} = \begin{cases} 0.1, & F_{LG} \leq 0.1 \\ 0.1 + \dfrac{(F_{LG} - 0.1)}{9}, & 0.1 \leq F_{LG} \leq 1.0 \\ 0.2, & F_{LG} \geq 1.0 \end{cases}$$

Example 14.3 below illustrates the calculation of the tower diameter for a sieve tray.

Packed Towers

If a packed tower is irrigated from a good distributor by a downflow of liquid, the liquid flows over the packing surface and a volumetric holdup of liquid in the tower is observed. As vapor is passed up the tower at low flow rates, countercurrent to the liquid, little or no drag is exerted by the vapor on the liquid and the liquid holdup is unchanged. The liquid has no difficulty leaving the tower as fast as it enters. However, if the gas flow rate is increased, a point is eventually reached where, because of drag, the liquid holdup begins to increase significantly with increasing vapor rate. This is called the loading point. Further increases in the vapor rate eventually reach the point where liquid begins to fill the tower, causing a rapid increase in pressure drop. The flooding point can be defined as the point where the pressure drop rapidly increases with a simultaneous decrease in mass-transfer efficiency. Typically, the flooding point is accompanied by a pressure head of approximately 2 in. of water/ft of packing. For a given liquid flow rate, the loading gas flow rate, which is typically 70% of the flooding gas flow rate, is often used to compute the tower inside diameter.

The diameter of a packed tower is calculated from an estimated flooding velocity with a continuity equation similar to Eq. (14.11) for tray towers:

$$D_T = \left[\frac{4G}{(fU_f)\pi\rho_G} \right]^{1/2} \tag{14.14}$$

For towers with random packing, the generalized correlation of Leva (1992) gives reasonable estimates of the flooding velocity in terms of a packing factor, F_P, which depends on the type and size of packing, and the same flow ratio parameter, F_{LG}, used for tray towers. The Leva flooding correlation fits the following equation:

$$Y = \exp[-3.7121 - 1.0371(\ln F_{LG}) - 0.1501(\ln F_{LG})^2 - 0.007544(\ln F_{LG})^3] \tag{14.15}$$

Table 14.1 Packing Factors for Calculating Flooding Velocity

Type Packing	Material	Nominal Diameter, D_P (in.)	Packing Factor, F_P (ft^2/ft^3)
Raschig rings	Ceramic	1.0	157
		2.0	58
		3.0	33
Raschig rings	Metal	1.0	165
		2.0	71
		3.0	40
Intalox saddles	Ceramic	1.0	92
		2.0	30
		3.0	15
Intalox saddles	Plastic	1.0	36
		2.0	25
Pall rings	Metal	1.0	56
		1.5	29
		2.0	27
		3.5	16
Pall rings	Plastic	1.0	53
		2.0	25
		3.5	15

where:

$$Y = \frac{U_f^2 F_P}{g}\left(\frac{\rho_G}{\rho_{H_2O_{(L)}}}\right)f\{\rho_L\}f\{\mu_L\} \tag{14.16}$$

The flooding velocity factor Y is dimensionless, with U_f in ft/s, F_P in ft^2/ft^3, and $g = 32.2$ ft/s^2. Values of F_P for several representative packings are listed in Table 14.1. Equation (14.15) is valid for $Y = 0.01$ to 10.

The density function is given by:

$$f\{\rho_L\} = -0.8787 + 2.6776\left(\frac{\rho_{H_2O_{(L)}}}{\rho_L}\right) - 0.6313\left(\frac{\rho_{H_2O_{(L)}}}{\rho_L}\right)^2 \tag{14.17}$$

for density ratios from 0.65 to 1.4.

For random packings of 1 in. or greater nominal diameter, the viscosity function is

$$f\{\mu_L\} = 0.96\,\mu_L^{0.19} \tag{14.18}$$

for liquid viscosities from 0.3 cP to 20 cP.

For a value of F_{LG}, Y is computed from Eq. (14.15), U_f is then computed from Eq. (14.16) for a given packing type and size, with F_P from Table 14.1 and using Eqs. (14.17) and (14.18). Then for $f = 0.7$, the tower diameter is computed from Eq. (14.14). The tower inside diameter should be at least 10 times the nominal packing diameter and preferably closer to 30 times.

The determination of the flooding velocity in structured packings is best carried out by using interpolation of the flooding and pressure drop charts for individual structured packings in Chapter 10 of Kister (1992).

14.7 PRESSURE DROP AND WEEPING

In general, pressure drop per unit height is least for towers with structured packings and greatest with tray towers, with randomly packed towers in between. For sieve trays, the components of the pressure drop are (1) pressure drop through the holes in the tray, which depends

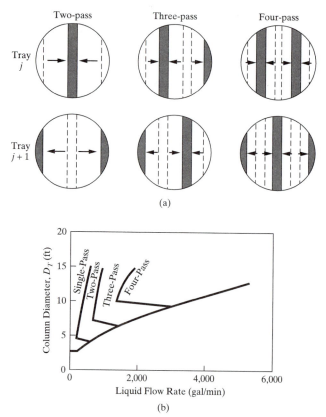

(a)

(b)

Figure 14.5 Selection of multipass trays. (a) Multipass trays: (1) two-pass; (2) three-pass; (3) four-pass. (b) Flow pass correlation. (Derived from *Koch Flexitray Design Manual, Bulletin 960*, Koch Engineering Co., Inc., Wichita, Kansas, 1960.)

on the hole diameter, hole area, and vapor volumetric flow rate; (2) pressure drop due to surface tension; and (3) the head of equivalent clear liquid on the tray, which depends on the weir height, weir length, and froth density. Detailed methods of calculation of tray pressure drop are presented by Kister (1992), Seader and Henley (1998), and in *Perry's Chemical Engineers' Handbook* (1997). Most simulators perform this calculation. However, the user should minimize the hydraulic gradient of the liquid flowing across the tray before requesting the calculation, by considering the number of liquid passes to use. Columns of diameter larger than 4 ft and operating with liquid rates greater than 500 gal/min frequently employ multipass trays to increase weir length and shorten the liquid flow path across the tray. Figure 14.5 shows three multipass arrangements and a correlation for selecting the number of passes to use. For preliminary design, a pressure drop of 0.10 psi/tray can be assumed for columns operating at ambient pressure or higher. For vacuum operation, trays should be designed so as not to exceed 0.05 psi/tray or packing should be considered as a substitute for trays to give an even lower pressure drop. Methods for estimating pressure drop in packed towers are found in Kister (1992), Seader and Henley (1998), and in *Perry's Chemical Engineers' Handbook* (1997) and are performed by simulators.

For sieve trays, the possibility of weeping of liquid through the holes in the trays should be checked, particularly when the vapor flow rate is considerably below the flooding point. Methods for checking this are used by the simulators. Note that, in general, weeping rates as high as 10% do not effect the tray efficiency, primarily because the weeping liquid is in contact with the vapor as it falls to the tray below.

EXAMPLE 14.3

Several alternative distillation sequences are being examined for the separation of a mixture of light hydrocarbons. The sequences are to be compared on the basis of annualized cost, discussed below in

Chapter 17 and given by Eq. (17.10). This requires estimates of the total capital cost and the annual operating cost of the columns, trays, condensers, reboilers, and reflux accumulators. To estimate these costs, equipment sizes must be determined. In this example, calculations of the height and diameter are illustrated for one column in one of the sequences.

The column to be sized is a deisobutanizer with a saturated liquid feed of 500 lbmol/hr of isobutane and 500 lbmol/hr of n-butane. The distillate is to be 99 mol% isobutane and the bottoms 99 mol% n-butane. The column shell is carbon steel, with carbon-steel sieve trays on 24-in. spacing. The trays have 0.25-in. diameter holes with a hole area to active area ratio of 0.1. The weir height is 2 in.

SOLUTION

Following the procedure outlined above, the following results are obtained:

1. Using a simulator, with the Soave–Redlich–Kwong (SRK) equation of state for thermodynamic properties, a bubble-point pressure of 98 psia is computed at 120°F for the distillate composition. Therefore, from Figure 7.9, a total condenser should be used with cooling water. Assuming a pressure drop of 2 psia across the condenser, the pressure at the top of the column is 100 psia. Assuming a 10-psi drop across the tower, the tower bottoms pressure is 110 psia. This gives a bubble-point bottoms temperature of 152°F, which is far below the decomposition temperature of n-butane. The assumed tower pressure drop is checked by a simulator after the column diameter is determined.

2. Using the Fenske–Underwood–Gilliland shortcut model, with a process simulator, for a reflux-to-minimum reflux ratio of 1.10 (because this is a difficult separation with a relative volatility predicted by the SRK equation of state of approximately 1.30), gives 36.4 minimum stages, a minimum reflux ratio of 6.6, 85.6 equilibrium stages at a reflux ratio of 7.25, and a feed-stage location of 43 stages from the top (approximately at the middle stage). Using these results as a first approximation, a rigorous equilibrium-stage calculation for 84 equilibrium stages in the column, an equilibrium-stage reboiler, and a total condenser (86 stages in all), with a feed stage at the middle gives a reflux ratio of 7.38 (only 2% greater than the FUG value) to achieve the specified distillate and bottoms purities. Thus, for this nearly ideal system, the FUG method is in close agreement with a rigorous method. The computed condenser duty is 31,600,000 Btu/hr and the reboiler duty is 31,700,000 Btu/hr.

3. Use Figure 14.3 to estimate the plate efficiency for average conditions in the tower. Using a simulator, the estimated average liquid viscosity = 0.12 cP, while the average relative volatility = 1.30. Using the product of these two factors = 0.12(1.3) = 0.156, Figure 14.3 predicts E_O = 0.80. Therefore, the number of actual trays = 84/0.80 = 105, with the partial reboiler counted as an additional stage.

4. For a 24-in. tray spacing, allowing a 10-ft-high liquid bottoms storage (sump) below the bottom tray, and a 4-ft disengagement height above the top tray, the tower height is 222 ft (tangent-to-tangent, i.e., not including the top and bottom tower heads).

5. Assume that the tower diameter will be determined from the entrainment flooding velocity rather than by downcomer flooding. The clear liquid height in the downcomer is one of a number of items computed by a simulator when a tray design is specified. That height should be checked to determine if it is less than 50% of the tray spacing. If not, to prevent downcomer flooding, the downcomer cross-sectional area should be increased. For conditions at the top stage of the column, a process simulation program gives the following results.

Liquid phase:

Surface tension = 7.1 dyne/cm

Flow rate = 215,000 lb/hr

Density = 32.4 lb/ft^3 or 4.33 lb/gal

Molecular weight = 58.12

Vapor phase:

Flow rate = 244,000 lb/hr

Density = 1.095 lb/ft^3

Molecular weight = 58.12

The flow ratio parameter = F_{LG} = (215,000/244,000)(1.095/32.4)$^{0.5}$ = 0.162
From Figure 14.4 for 24-in. (approximately 600-mm tray spacing), C_{SB} = 0.09 m/s

The surface-tension factor = F_{ST} = (7.1/20)$^{0.2}$ = 0.81. Assume F_F = 1. Also, F_{HA} = 1. Therefore, from Eq. (14.13), C = 0.09(0.81)(1)(1) = 0.073 m/s. From Eq. (14.12), U_f = 0.073[(32.4 − 1.095)/1.095]$^{0.5}$ = 0.390 m/s = 4,610 ft/hr. Assume operation on the top tray of 80% of flooding (f = 0.80). To determine the ratio, A_d/A_T, 0.1 + (F_{LG} − 0.1)/9 = 0.1 + 0.062/9 = 0.107 = A_d/A_T. From Eq. (14.11),

$$D_T = \left[\frac{4(244,000)}{0.80(4,610)(3.14)(1 - 0.107)(1.095)}\right]^{1/2} = 9.3 \text{ ft}$$

For this large a tower diameter, the need for a multipass tray needs to be considered, using Figure 14.5. The volumetric liquid flow rate = (215,000/60)/4.33 = 828 gpm. For this diameter and liquid flow rate, a three-pass tray is indicated. For a one-pass tray, a simulator gives a tower diameter of 9.5 ft, when the diameter is restricted to increments of 0.5 ft. For a three-pass tray, the tower diameter remains at 9.5 ft.

Other calculations from a simulator for both single-pass and three-pass trays are as follows:

	Single-Pass Sieve Tray	Three-Pass Sieve Tray
Weir length, ft	7.3	23.3
Flow path length, ft	6.1	2.2
Active area, ft^2	70.9	70.9
Weeping tendency	Barely	No
Pressure drop, psi	0.067	0.056
Downcomer backup, ft	0.70	0.54
Downcomer area/Tower area	0.122	0.122

Both the single-pass and three-pass trays have the same ratio of downcomer area to tower area, which is only 14% greater than the assumed value of 0.107. The much shorter flow path length of the three-pass tray reduces the hydraulic gradient so that a more uniform vapor distribution over the tray active area is achieved. The weeping tendency is not a problem with either tray. The total pressure drop for the 105 trays is 7.0 psi for the single-pass tray and 5.9 psi for the three-pass tray compared to the assumed 10-psi drop. The downcomer backups, which are based on clear liquid, are safely below a possible problem of downcomer flooding provided that the volume fraction of vapor in the downcomer froth is not much greater than the commonly assumed value of 0.50. ∎

14.8 SUMMARY

After studying this chapter and completing a few exercises, the reader should have learned to

1. Select an appropriate operating pressure for a multistage tower and a condenser type for distillation.
2. Determine the number of equilibrium stages required for a separation and a reasonable reflux ratio for distillation.
3. Determine whether trays, packing, or both should be considered.
4. Determine the number of actual trays or packing height required.
5. Estimate the tower diameter.
6. Consider other factors for successful tower operation.

REFERENCES

Chan, H., and J.R. Fair, "Prediction of Point Efficiencies on Sieve Trays. 1. Binary Systems," *Ind. Eng. Chem. Process Des. Dev.*, **23**, 814–819 (1984a).

Chan, H., and J.R. Fair, "Prediction of Point Efficiencies on Sieve Trays. 1. Multicomponent Systems," *Ind. Eng. Chem. Process Des. Dev.*, **23**, 820–827 (1984b).

Fair, J.R., *Petro./Chem. Eng.*, **33**, 211–218 (Sept., 1961).

FRI (Fractionation Research Institute) report of Sept. 3, 1958, *Glitsch Ballast Tray*, published as Bulletin No. 159 of Fritz W. Glitsch and Sons, Inc., Dallas, Texas (1958).

Kister, H.Z., *Distillation Design*, McGraw-Hill, New York (1992).

Leva, M., "Reconsider Packed-Tower Pressure-Drop Correlations," *Chem. Eng. Prog.*, **88** (1), 65–72 (1992).

Perry, R.H. and D.W. Green, Eds., *Perry's Chemical Engineers' Handbook*, 7th ed., McGraw-Hill, New York (1997).

Seader, J.D. and E.J. Henley, *Separation Process Principles*, John Wiley & Sons, New York (1998).

Williams, G.C., E.K. Stigger, and J.H.Nichols, "A Correlation of Plate Efficiencies in Fractionating Columns," *Chem. Eng. Prog.*, **46**, 1, 7–16 (1950).

EXERCISES

14.1 In Example 14.1, an absorber with an absorbent rate of 237 kmol/hr and 4 equilibrium stages absorbs 90% of the entering *n*-butane. Repeat the calculations for:

a. 474 kmol/hr of absorbent (twice the flow) and 4 equilibrium stages.

b. 8 equilibrium stages (twice the stages) and 237 kmol/hr of absorbent.

Which case results in the most absorption of *n*-butane? Is this result confirmed by the trends of the curves in the Kremser plot of Figure 14.2.

14.2 The feed to a distillation tower consists of 14.3 kmol/hr of methanol, 105.3 kmol/hr of toluene, 136.2 kmol/hr of ethylbenzene, and 350.6 kmol/hr of styrene. The bottoms product is to contain 0.1 kmol/hr of ethylbenzene and 346.2 kmol/hr of styrene. Determine a suitable operating pressure at the top of the tower noting that the bottoms temperature is limited to 145°C to prevent the polymerization of styrene.

14.3 A mixture of benzene and monochlorobenzene is to be separated into almost pure products by distillation. Determine an appropriate operating pressure at the top of the tower.

14.4 In a reboiled absorber, operating as a deethanizer at 400 psia to separate a light hydrocarbon feed, conditions at the bottom tray are

Liquid phase:

Molar flow = 1,366 lbmol/hr
MW = 91.7
Density = 36.2 lb/ft^3
Surface tension = 10.6 dyne/cm

Vapor phase:

Molar flow = 735.2 lbmol/hr
MW = 41.2
Density = 2.83 lb/ft^3

If sieve trays are used with hole area of 10% and a 24-in. tray spacing, determine the tower diameter. Assume 80% of flooding and a foaming factor of 0.75.

14.5 A distillation tower with sieve trays is to separate benzene from monochlorobenzene. Conditions at a plate near the bottom of the column are

Vapor phase:

Mass flow rate = 24,850 lb/hr
Density = 0.356 lb/ft^3

Liquid phase:

Mass flow rate = 41,850 lb/hr
Density = 59.9 lb/ft^3
Surface tension = 24 dyne/cm

Determine a reasonable tower diameter.

14.6 Water is to be used to absorb acetone from a dilute mixture with air in a tower packed with 3.5-in. metal Pall rings. Average conditions in the tower are

Temperature = 25°C
Pressure = 110 kPa

Liquid phase:

Water = 1,930 kmol/hr
Acetone = 5 kmol/hr
Density = 62.4 lb/ft^3
Surface tension = 75 dyne/cm

Vapor phase:

Air = 680 kmol/hr
Water = 13 kmol/hr
Acetone = 5 kmol/hr

Determine the column diameter for operation at 70% of flooding.

Chapter **15**

Pumps, Compressors, and Expanders

15.0 OBJECTIVES

This chapter presents brief descriptions and some theoretical background of the most widely used pumps for liquids, and compressors and expanders for gases, all of which are modeled in simulators. Heuristics for the application of these devices during the synthesis of a chemical process are presented in Chapter 5. Further information on their selection and capital cost estimation is covered in Chapter 16. More comprehensive coverage of the many types of pumps, compressors, and expanders available is presented in Sandler and Luckiewicz (1987) and in *Perry's Chemical Engineers' Handbook* (1997). After studying this chapter and the materials on pumps, compressors, and turbines on the multimedia CD-ROM that accompanies this book, the reader should be able to explain how the more common types of pumps, compressors, and expanders work and how a simulator computes their power input or output.

15.1 PUMPS

The main purpose of a pump is to provide the energy needed to move a liquid from one location to another. The net result of the pumping action may be to increase the elevation, velocity, and/or pressure of the liquid. However, in most process applications, pumps are designed to increase the pressure of the liquid. In that case, the power required is

$$\dot{W} = F\upsilon\,(\Delta P), \tag{15.1}$$

where F is the molar flow rate, υ is the molar volume, and P is pressure. Because the liquid molar volume is usually much smaller than that of a gas, pumps require relatively little power compared to gas compressors for the same molar flow rate and increase in pressure. Therefore, when a vapor stream is produced from a liquid stream with increased pressure and temperature, it is generally more economical to increase the pressure while the stream is a liquid. Except for very large changes in pressure, the temperature of the liquid being pumped increases only slightly.

The main methods used to move a liquid are centrifugal force, displacement, gravity, electromagnetic force, and transfer of momentum from another fluid, with the first two methods being the most common for chemical processes. Pumps that use centrifugal force are sometimes referred to as kinetic pumps, but more commonly as *centrifugal pumps*. Displacement of one part of a fluid with another part takes place in so-called *positive-displacement pumps*, whose action is either reciprocating or rotary. The use of electromagnetic force is limited to fluids that can conduct electricity. Jet pumps, either eductors or injectors, are simple devices that transfer momentum from one fluid to another. Their application is also limited because the motive and pumped fluids contact each other and may mix together, and the efficiency of transfer is very low.

The two most important characteristics of a pumping operation are the *capacity* and the *head*. The capacity refers to the flow rate of the fluid being pumped. It may be stated as a mass flow rate, a molar flow rate, or a volumetric flow rate. Most common is the volumetric flow rate, Q, in units of either m^3/hr or gal/min (gpm). The head, or pump head, H, refers to the increase in total head across the pump from the suction, s, to the discharge, d, where the head is the sum of the velocity head, static head, and pressure head. Thus,

$$H, \text{ pump head} = \left(\frac{V_d^2}{2g} + z_d + \frac{P_d}{\rho_d g}\right) - \left(\frac{V_s^2}{2g} + z_s + \frac{P_s}{\rho_s g}\right) \tag{15.2}$$

where V is the average velocity of the liquid, z is the elevation, P is the pressure of the liquid, g is the gravitational acceleration (32.2 ft/s^2, 9.81 m/s^2) and ρ is the liquid density. The head is expressed in units of ft or m of liquid. The required pump head or pressure increase is determined by an energy balance as discussed with a heuristic and example in Section 5.7.

Centrifugal Pumps

As shown in Figure 15.1, a centrifugal pump consists of an impeller, mounted on a shaft and containing a number of blades, rotating within a stationary casing that is provided with an inlet and outlet for the liquid being pumped. Power, usually from an electric motor, rotates the shaft, which rotates the impeller. The rotating blades reduce the pressure at the inlet or eye of the impeller, causing liquid to enter the impeller from the suction of the pump. This liquid is forced outward along the blades to the blade tips at an increasing tangential velocity. At this point the liquid has acquired an increased velocity head from the power input to the pump. The velocity head is then reduced and converted to a pressure head as the liquid passes into the annular (volute) chamber within the casing and beyond the blades, and thence to the pump outlet or discharge.

When a centrifugal pump is installed in a pumping system and operated at a particular rotational rate, N, (usually 1,750 to 3,450 rpm), the flow rate can be varied by changing the opening on a valve located in the pump discharge line. The variation of H with Q defines a unique *characteristic curve* for the particular pump operating at N with a fluid of a particular viscosity. Each make and model of a centrifugal pump is supplied by the manufacturer with a

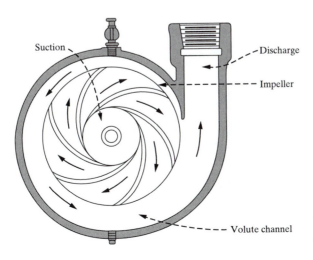

Figure 15.1 Schematic of centrifugal pump.

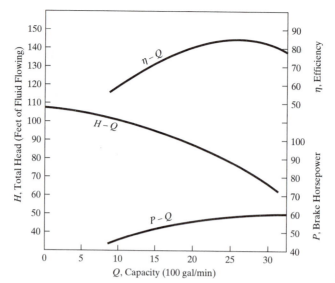

Figure 15.2 Characteristic curves for a centrifugal pump.

characteristic curve determined by the manufacturer when pumping water. Corrections are necessary when other fluids are pumped. Corresponding to the variation of H with Q, curves representing the effect of Q on the brake horsepower, P, and the pump efficiency, η, are shown in Figure 15.2. Typically, the pump head decreases with increasing flow rate, while the brake horsepower increases with increasing flow rate. The pump efficiency passes through a maximum. The pump will only operate at points on the characteristic curve. Therefore, for a particular pumping task, the required head-volumetric flow rate point must lie somewhat below the characteristic curve. The difference between the two heads, (pump head − required head), can be throttled across a control valve in the discharge line. Ideally, a centrifugal pump should be selected so that the operating point is located on the characteristic curve at the point of maximum efficiency.

For a given centrifugal pump, the characteristic curve moves upward with increasing rate of rotation, N, as shown in Figure 15.3. Similarly, for a pump of a particular design, the characteristic curve moves upward with increasing impeller diameter, D, as shown in Figure

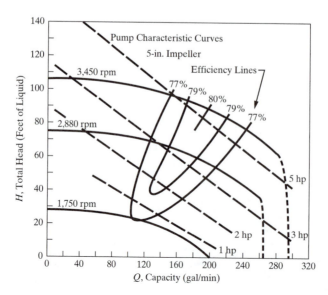

Figure 15.3 Effect of rate of rotation on characteristic curves.

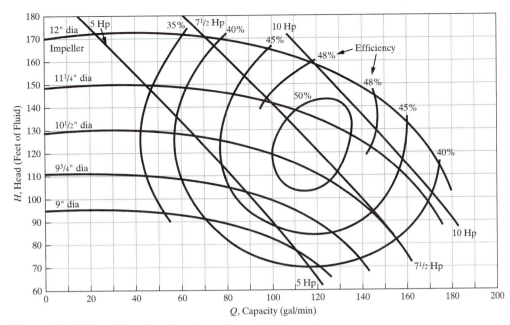

Figure 15.4 Effect of impeller diameter on characteristic curves.

15.4. When a characteristic curve for just one rotation rate and/or impeller diameter is available and an approximate characteristic curve is desired for another rotation rate and/or impeller diameter, the affinity laws for a centrifugal pump can be applied:

$$Q_2 = Q_1 \left(\frac{N_2}{N_1} \right) \tag{15.3}$$

$$H_2 = H_1 \left(\frac{N_2}{N_1} \right)^2 \tag{15.4}$$

$$Q_2 = Q_1 \left(\frac{D_2}{D_1} \right) \tag{15.5}$$

$$H_2 = H_1 \left(\frac{D_2}{D_1} \right)^2 \tag{15.6}$$

More difficult is the correction for viscosity. In general, increasing viscosity for a fixed capacity, Q, decreases the pump head and the pump efficiency, and increases the brake horsepower. Typical effects of viscosity are shown in Figure 15.5. As seen, the effect of viscosity can be substantial.

Because centrifugal pumps operate at high rates of rotation, the imparted high liquid velocities can lower the local pressure. If that pressure falls below the vapor pressure of the liquid, vaporization will produce bubbles that may collapse violently against surfaces where a higher pressure exists. This phenomenon is called *cavitation* and must be avoided. Otherwise, besides a lowering of efficiency and flow rate, the pump may be damaged. The tendency for cavitation is measured by a quantity, peculiar to each pump and available from the manufacturer, called *required NPSH* (net positive-suction head), expressed as a head. It is typically in the range of 2–10 ft of head. The *available NPSH* is defined as the difference between the liquid pressure at the pump inlet and the vapor pressure of the liquid, expressed as a head. To avoid cavitation, the available NPSH must be greater than the manufacturer's value for the required NPSH. An example of the application of the NPSH is given later in Example 16.5.

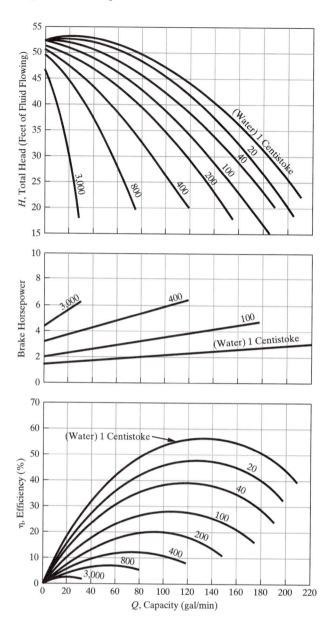

Figure 15.5 Effect of viscosity on characteristic curves.

Centrifugal pumps are limited by the rate of rotation of the impeller to the pump head they can achieve in a single stage. A typical maximum head for a single stage is 500 ft. By going to multiple stages, heads as high as at least 3,200 ft can be achieved.

Positive-Displacement Pumps

Positive-displacement pumps, either reciprocating or gear, are essentially metering pumps, designed to deliver a volumetric flow rate, Q, that is independent of the required pump head, H. Thus, the characteristic curve of a positive displacement pump, if it can be called that, is a vertical line on a plot of Q as a function of H. The pump head is limited only by the Hp of the driver, the strength of the pump, and/or possible leakage through clearances between moving pistons, plungers, gears, or screws, and stationary cylinders or casings. Unlike centrifugal pumps, where the flow rate can be changed (while staying on the characteristic curve) by adjusting a valve on the discharge line, the flow rate of a positive-displacement pump must be changed by a bypass or with a speed changer on the motor. The efficiency of positive-displacement pumps is greater than for centrifugal pumps because less friction occurs in the former, and cavitation is not usually a concern with positive-displacement pumps.

The three main classes of reciprocating pumps are piston, plunger, and diaphragm, which are shown schematically in Figure 15.6. They all contain valves on the inlet and outlet. During suction, a chamber is filled with liquid, with the inlet valve open and the outlet valve closed. During discharge of the liquid from the chamber, the inlet valve is closed and outlet valve opened. This type of action causes pressure pulsations, which causes a fluctuating flow rate and discharge pressure. These fluctuations can be reduced by employing a gas-charged

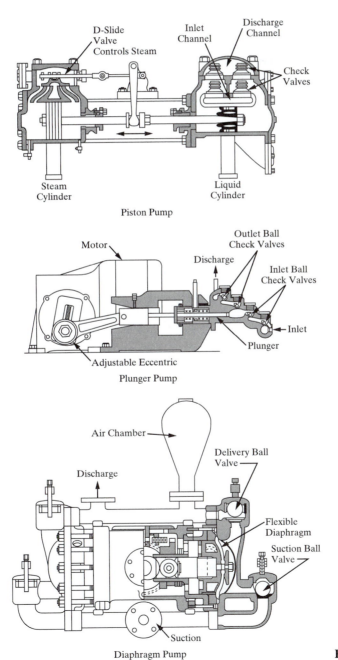

Figure 15.6 Reciprocating pumps.

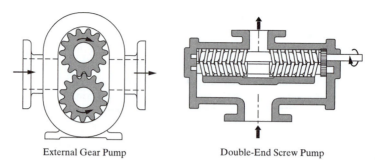

External Gear Pump   Double-End Screw Pump

Figure 15.7 Rotary pumps.

surge chamber in the discharge line and/or by using multiple cylinders in parallel. In addition, if pistons are used, the pump can be double-acting, with chambers on either side of the piston. With a plunger, only a single-action is used. Reciprocating pumps with a flexible diaphragm of metal, rubber, or plastic eliminate packing and seals, making them useful for hazardous or toxic liquids.

Rotary pumps include gear pumps and screw pumps, which are shown schematically in Figure 15.7. These must be designed to tight tolerances to avoid binding and excessive wear. They are best suited for liquids of high viscosity. Flow rates are more steady than for reciprocating pumps but less steady than for centrifugal pumps.

Pump Models in Simulators

The pump models in process simulators do not differentiate between centrifugal pumps and positive-displacement pumps when calculating theoretical power requirement from the product of the capacity and required head [Eq. (15.1)]. In some cases the models do utilize built-in efficiency equations, which differentiate between the two types of pumps, when computing the brake horsepower. Most models calculate a discharge temperature, which is based on the small variation of density with temperature and an assumption that all of the pump inefficiency produces friction that causes an increase in the liquid temperature. In most cases, the temperature and enthalpy changes of the liquid across the pump are small. Simulators do not provide built-in characteristic curves to help select a suitable centrifugal pump nor do they consider multiple stages or cylinders. Pump subroutines are discussed further on the CD-ROM, accompanied by a video of an industrial-scale centrifugal pump. (*ASPEN → Pumps, Compressors & Expanders → Pumps* and *HYSYS → Pumps, Compressors & Expanders → Pumps*)

EXAMPLE 15.1

In a toluene hydrodealkylation process, 25,000 lb/hr of toluene feed is pumped from 75°F and 30 psia to 570 psia. Use a process simulator to compute the capacity in gpm, the pump head in feet of toluene, the exit temperature, and brake horsepower (BHp) for:

a. A pump efficiency of 100%.
b. A pump efficiency of 75%.

SOLUTION

Using the SRK equation of state for thermodynamic properties, the following results are obtained

	Pump Efficiency = 100%	Pump Efficiency = 75%
Capacity, gpm	57.3	57.3
Pump head, ft of toluene	1,440	1,440
Outlet temperature, °F	75.78	77.37
Brake horsepower, BHp	18.2	24.3

The temperature rise is very small even for the 75% efficiency case. The pump head is well above the limit of 500 ft for a single-stage centrifugal pump. Therefore, a multistage centrifugal pump would be required.   ■

15.2   COMPRESSORS AND EXPANDERS

Gas compressors (including fans and blowers), unlike pumps, are designed to increase the velocity and/or pressure of gases rather than liquids. In fact, small amounts of liquid can cause significant amounts of degradation to the compressor blades, and, consequently, most compressor systems are designed to prevent liquid from entering the compressor and to avoid condensation in the compressor. The main methods used to move a gas are centrifugal force, displacement, and transfer of momentum. There are no sharp boundaries among fans, blowers, or compressors, but one convenient classification is based on discharge pressure or compression ratio. By this classification, a fan mainly increases the kinetic energy of the gas with a discharge pressure of no more than 110% of the suction pressure. A blower increases the pressure head more than the velocity head, with a compression ratio of not more than 2. A compressor increases the velocity head very little, with a compression ratio of greater than 2.

Centrifugal Compressors

Centrifugal fans, blowers, and compressors are widely used in chemical processes because they produce a continuous flow, are relatively small, and are free of vibration. Because gases are compressible, the temperature difference between the compressed gas and the feed gas is significant at even moderate compression ratios and may limit the compression ratio possible in a single stage. However, the need for multiple stages in centrifugal compressors is usually dictated instead by impellor rotation-rate limitations, which limit the compression ratio that can be achieved.

Like pumps, the feed (stream 1) to a centrifugal compressor, at its suction pressure, enters the eye of the impeller unit, as shown in Figure 15.8. The compressed gas leaves as stream 2.

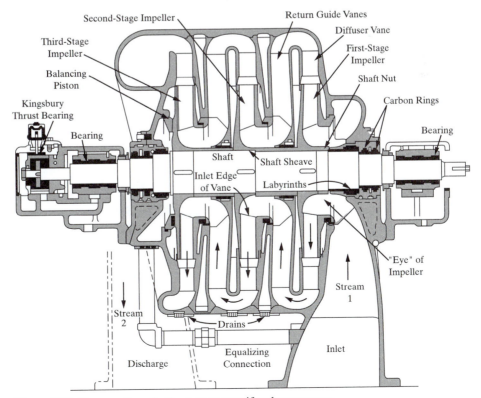

Figure 15.8 Cross section of a three-stage centrifugal compressor.

A large amount of power input, in comparison with pumps, is required to increase the pressure of a gas, primarily because of the large molar volume of a gas. Although compressors are much larger than pumps, they can be well insulated so that heat losses are negligible in comparison with their power requirements. Accordingly, adiabatic operation is usually assumed. The characteristic curves for centrifugal compressors are similar to those for a centrifugal pump, as shown in Figures 15.2–15.4, except that the coordinates may be static pressure (in place of head) and actual ft³/min (ACFM) at inlet conditions (in place of gpm). Also, for some impeller designs, as ACFM is increased from zero, the static pressure first decreases, goes through a minimum, rises to a maximum and then drops sharply. As with a centrifugal pump, a centrifugal fan, blower, or compressor should be selected for operation at the point of maximum efficiency on the characteristic curve.

Positive-Displacement Compressors

Positive-displacement fans, blowers, and compressors are similar in action to positive-displacement pumps, and include reciprocating compressors, two- or three-lobe blowers, and screw compressors. However, with gases, the almost vertical characteristic curves, bend to the left more than for liquids because of the greater tendency for slip.

Reciprocating compressors use pistons with either single- or double-action. As discussed in Section 16.5, compression ratios in a single stage are limited to a discharge temperature of 400°F. This corresponds to compression ratios about 2.5–6 as the specific heat ratio of the gas decreases from 1.67 (monotomic gas) to 1.30 (methane). Compression ratios of even 8 are possible with high molecular-weight gases. If higher compression ratios are needed, a multistage reciprocating compressor is used with intercooling, usually by water. See, for example, the video of a two-stage reciprocating compressor with an intercooler on the CD-ROM (*ASPEN → Pumps, Compressors & Expanders → Compressors & Expanders* and *HYSYS → Pumps, Compressors & Expanders → Compressors & Expanders*). Reciprocating compressors must be protected by knock-out drums to prevent the entry of liquid.

A lobed blower, shown in Figure 15.9, is similar to a gear pump. Both two- and three-lobe units are common. They are limited to low capacity and low heads because shaft deflection must be kept small to maintain clearance between the rotating lobes and the casing. If higher compression ratios are required, multiple stages can be used. A screw compressor, as shown in Figure 15.10, with two screws, male and female, that rotate at speeds typical of centrifugal pumps, can operate at higher capacities to give higher compression ratios that may be limited by temperature. If so, higher compression ratios can be achieved with multiple stages separated by intercoolers. Screw compressors can run dry or can be flooded with oil.

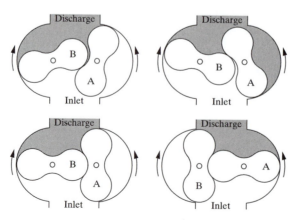

Figure 15.9 Lobed blower.

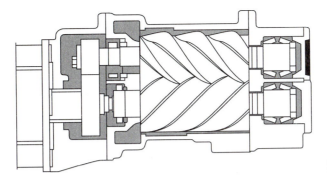

Figure 15.10 Screw compressor.

Expanders

Expanders (also called turboexpanders and expansion turbines) are often used, in place of valves, to recover power from a gas when its pressure must be decreased. At the same time, the temperature of the gas is reduced, and often the chilling of the gas is more important than the power recovery. Most common is the radial-flow turbine, as shown in Figure 15.11, which resembles a centrifugal pump, and can handle inlet pressures to 3,000 psi and temperatures to 1,000°F. With an impeller tip speed of 1,000 ft/s, a single stage of expansion can reduce the enthalpy of the gas by as much as 50 Btu/lb (116 kJ/kg). When calculations show that condensation may occur during the expansion, the expander must be designed to avoid erosion of the impeller. Expanders are widely used at cryogenic conditions. Although power can also be recovered by decreasing the pressure of a liquid with a turbine, it is usually not economical to do so.

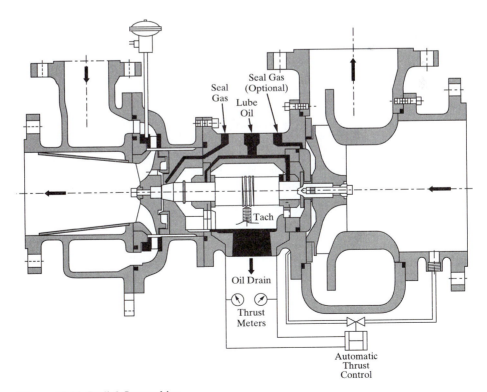

Figure 15.11 Radial-flow turbine.

Compressor and Expander Models in Simulators

Either of two methods can be used to take into account efficiency when calculating power requirements for compressors, whether they are centrifugal, reciprocating, or screw. One method is the *polytropic method*, based on the expression, PV^n = constant during compression, where V is the gas volume and n is the polytropic coefficient, which lies between 1 and the specific heat ratio.

Since the advent of simulation programs that routinely calculate entropy, the second method, called the *isentropic method*, has become preferred, because it has a sound theoretical basis. The theoretical horsepower delivered to the gas is computed for a reversible, adiabatic (isentropic) compression from inlet, 1, to outlet, 2. The entropy balance in terms of the molar entropy, s, is

$$s\{T_1, P_1\} = s\{T_{2,\text{isentropic}}, P_2\} \tag{15.7}$$

Since T_1, P_1, and P_2 are known, Eq. (15.7) is solved iteratively for $T_{2,\text{isentropic}}$. With T_2 known, the exit enthalpy can be computed. Then the first law of thermodynamics for an adiabatic compression of molar gas flow, F, assuming no change in potential or kinetic energy of the gas and written in terms of molar enthalpy, h, can be applied to calculate the theoretical or isentropic power:

$$\dot{W}_{\text{isentropic}} = F(h_{2,\text{isentropic}} - h_1) \tag{15.8}$$

The excess power required, because of inefficiency of the compressor, is the difference between the brake power, $\dot{W}_{\text{brake}}$, and the isentropic power, $\dot{W}_{\text{isentropic}}$. These two powers define an isentropic efficiency, with the assumption that the excess power increases the enthalpy to an actual value, h_2:

$$\eta_s = \frac{\dot{W}_{\text{isentropic}}}{\dot{W}_{\text{brake}}} = \frac{h_{2,\text{isentropic}} - h_1}{h_2 - h_1} \tag{15.9}$$

The actual temperature of the discharged compressed gas, T_2, is then computed iteratively from the actual enthalpy, h_2. The actual temperature T_2 can be significantly higher than the isentropic temperature, $T_{2,\text{isentropic}}$.

The isentropic method is also applied to an expander. Eq. (15.7) is used for calculating the isentropic exit temperature, but taking into account possible condensation of the gas. Like the exit pressure, the exit temperature will be less than the inlet value. Then, the exit isentropic enthalpy is computed, from which Eq. (15.8) is used to calculate the power recovered, which will be a negative value. The effect of the expander efficiency is just the opposite of the compressor efficiency, as indicated by a revision of Eq. (15.9) for applicability to expanders:

$$\eta_s = \frac{\dot{W}_{\text{brake}}}{\dot{W}_{\text{isentropic}}} = \frac{h_1 - h_2}{h_1 - h_{2,\text{isentropic}}} \tag{15.10}$$

Because of inefficiency, the brake horsepower recovered is less than the isentropic horsepower and the exit temperature is higher than the isentropic exit temperature. Thus, inefficiency will reduce the tendency for condensation to occur.

EXAMPLE 15.2

A natural gas stream of 5,000 kmol/hr at 25°C and 1,500 kPa contains 90% methane, 7% ethane, 3% propane. Currently this gas is expanded adiabatically across a valve to 300 kPa. Use a process simulator to determine the exit temperature and recovered power if the valve is replaced with:

    a. an isentropic expansion turbine, and
    b. an expansion turbine with an isentropic efficiency of 75%.

SOLUTION   Using the SRK equation of state for thermodynamic properties, the following results are obtained.

	Valve	Isentropic Expander	Expander
Isentropic efficiency, η_s	—	1.00	0.75
Exit temperature, °C	18.5	−69.7	−47.1
Power recovered, kW	0	4,480	3,360
Power recovered, BHp	0	6,010	4,510

The results show that not only does the expander recover a significant amount of power, but it is also very effective in reducing the temperature, compared to the valve. However, the actual exit temperature is almost 20°C higher than the isentropic value. In all cases, no condensation is found to occur, since the dew point of the exit gas at 300 kPa is computed to be −83.2°C   ∎

15.3 SUMMARY

Having studied this chapter, the reader should

1. Be able to explain how the more common types of pumps, compressors, and expanders work.
2. Understand the types of calculations made by a simulator for pumps, compressors, and expanders.

REFERENCES

Sandler, H.J., and E.T. Luckiewicz, *Practical Process Engineering*, McGraw-Hill, New York (1987).

Perry, R.H., and D.W. Green, Eds., *Perry's Chemical Engineers' Handbook*, 7th ed., McGraw-Hill, New York (1997).

EXERCISES

15.1 Liquid oxygen is stored in a tank at −298°F and 35 psia. It is to be pumped at 100 lb/s to a pressure of 300 psia. The liquid oxygen level in the tank is 10 ft above the pump, and friction and acceleration losses from the tank to the pump suction are negligible. If the pump efficiency is 80%, calculate the BHp, the oxygen discharge temperature, and the available NPSH, using a simulator to make the calculations.

15.2 Use a simulator to design a compression system with intercoolers to compress 600 lb/hr of a mixture of 95 mol% hydrogen and 5 mol% methane at 75°F and 20 psia to a pressure of 600 psia, if the maximum exit temperature from a compressor stage is 400°F and compressor efficiency is 80%. Assume gas outlet temperatures from the intercoolers at 120°F. For each compressor stage, compute the BHp. For each intercooler, compute the heat duty in Btu/hr.

15.3 Superheated steam, available at 800 psia and 600°F, is to be expanded to a pressure of 150 psia at the rate of 100,000 lb/hr. Calculate, with a simulator, the exit temperature, phase condition, and Hp recovered for:

a. an adiabatic valve,

b. an isentropic expansion turbine, and

c. an expansion turbine with an isentropic efficiency of 75%.

15.4 Propane gas at 300 psia and 150°F is sent to an expansion turbine with an efficiency of 80%. What is the lowest outlet pressure that can be achieved without condensing any of the propane?

Chapter **16**

Cost Accounting and Capital Cost Estimation

16.0 OBJECTIVES

Throughout the design process, as discussed in the previous chapters, estimates of the cost of equipment and other costs related to the capital investment play a crucial role in selecting from among the design alternatives. This chapter presents the various methods in common use for making preliminary estimates of capital costs of ventures for new processing plants or revamps of existing plants and should be studied in connection with the other chapters as needed. Some readers may prefer to study Sections 16.1 to 16.4 even before working with the previous chapters, especially when studying the techniques for process synthesis that require estimates of capital costs. In Chapter 17, capital cost estimates prepared according to the information presented in this chapter are combined with process operating costs and other expenses to determine the profitability of a proposed venture. However, even though a venture is predicted to be profitable, the financial condition of the company exploring the venture may not be sufficient to justify a decision to proceed with the venture, or competing ventures may be more attractive. In the former case, it is important to understand measures used to determine the financial condition of a company. These measures are intimately tied to the field of cost accounting, which is the subject of the first section of this chapter.

After studying this chapter, the reader should

1. Be able to assess the financial condition of a company from its annual report.
2. Be able to use the equations provided to estimate the purchase cost of representative types of process equipment.
3. Be able to estimate the cost of installation of the equipment units, including the cost of materials, labor, and indirect costs.
4. Be able to estimate the total capital investment for the process, including the costs of spare equipment, storage tanks, surge vessels, site preparation and service facilities, allocated costs for utilities and related facilities, contingencies, land, royalties, start-up, and working capital. Understand the need to reestimate the working capital after preparing the cost sheet, as discussed in Chapter 17.

16.1 ACCOUNTING

Accounting is the systematic recording, reporting, and analysis of financial transactions of a business. Accounting is necessary and valuable to a company because it:

1. Provides a record of property owned, debts owed, and money invested.
2. Provides a basis for the preparation of a report that gives the financial status of the company.
3. Gives assistance and direction to those managing the affairs of the company.
4. Provides a basis for stockholders and others to evaluate management of the company.

Debits and Credits

By tradition, since the fifteenth century, the recording of financial transactions by accountants is carried out by the *double-entry method* of *debits* and *credits*. Surprisingly, debits are increases (not decreases) in assets, where an *asset* is anything of economic value that is owned by the company. Credits are just the opposite; that is, they are decreases in assets. One possible explanation for the definitions of debits and credits is that the giver receives a credit while the receiver receives a debit. By custom, all transactions are initially recorded chronologically, in terms of debits and credits, in a *journal*, where for every debit, there is a credit of equal amount. The debits and credits apply to different accounts (cash, land, equipment, bank account, etc.), which are maintained in separate *ledgers* for each account. Entries in the journal are posted to the ledgers, usually with the debit entry going to one ledger and the corresponding credit entry going to a different ledger. Although the journal might seem superfluous, it serves two useful purposes besides being a chronological record: (1) It reduces the possibility of error because for each transaction the debit and corresponding credit are recorded together, and (2) if desired, a detailed explanation for the transaction is easily entered into the journal. In both the journal and the separate account ledgers, debits are entered to the left of the credits. At any point in time, the sum of debit entries for all ledgers must equal the sum of credit entries for all ledgers. Although it is not necessary for engineers to be accountants, it is important for engineers to understand what accountants do and why they do it.

Tables 16.1 and 16.2 show an example of double-entry bookkeeping, with a journal and ledgers, for the following transactions. The company purchases a heat exchanger for

Table 16.1 Typical Journal Page

JOURNAL					Page 43
2002				Debit ($)	Credit ($)
June	3	Heat Exchanger	15	80,450	
		Cash	11		80,450
		Purchase of a heat exchanger for ammonia plant			
	4	Cash	11	125,000	
		Ammonia product	12		125,000
		Sales of product from ammonia plant to ABC			
	6	Land	20	265,000	
		Cash	11		265,000
		Purchase of land in Iowa from XYZ			

Table 16.2 Typical Ledger Page

2002				BANK ACCOUNT, LEDGER 11 Debit ($)		2002				Page 5 Credit ($)
June	1	Balance forward	42	500,000		June				
	4	Sales	43	125,000			3	Purchase	43	80,450
							6	Purchase	43	265,000

$80,450, paid for by a check. The next day the company sells $125,000 of ammonia product, with payment by check. Two more days later, the company purchases land for $265,000, paid for by check. In all cases, the checks are handled with the same bank account. Note that four separate accounts are involved. Suppose they have been assigned the following account numbers: Bank Account, 11; Plant Equipment, 15; Sales of Products, 12; and Land, 20. The journal page (say page 43) is shown in Table 16.1, where the ledger account numbers are added in the column to the left of the debits column when the journal entries are posted to the ledgers. Postings to the ledger page (page 5 of this ledger) for the Bank Account are shown in Table 16.2. Included to the left of the debits and credits columns are the corresponding pages in the journal, in this case just page 43. This ledger page is for the month of June, for which the initial balance in the bank account, page 42, was an amount of $500,000. At the end of June 6, the bank account balance was $500,000 + 125,000 - 80,450 - 265,000 = $279,550$. Balances may be entered into the ledger at the end of each month.

The Annual Report

Although not legally required to do so, every publicly held company publishes an impressive-looking annual report at the end of each calendar year or fiscal year. This report, which is written primarily for stockholders, but is available to other interested parties, usually provides the following useful information:

1. Nature of the company's business.
2. Summary of important events and new developments of the year.
3. New acquisitions and formation of partnerships.
4. Plans for the near future.
5. Summary of concerns that might influence the company's business.
6. Collection of financial statements, including:
   a. Balance Sheet
   b. Income Statement
   c. Cash Flow Statement

The current financial condition of a company can be assessed by an analysis of the financial statements in the annual report. The latest annual reports of more than 3,600 U.S. companies can be obtained free of charge from PRARS (The Public Register's Annual Report Service) from their Web site at www.prars.com.

The Balance Sheet

The *balance sheet*, also called the *consolidated balance sheet* or *statement of consolidated financial condition*, is a quantitative summary of a company's financial condition at a specific point in time (at the end of the calendar or fiscal year), including *assets*, *liabilities*, and *net worth* (share owners' equity, stockholders' equity, or proprietorship). *Equity* means ownership, generally in the form of common stock or as a holding company. The balance sheet is prepared from balances in the ledger accounts. The overall entries in the balance sheet must conform to the fundamental accounting equation:

$$\text{Assets} = \text{Liabilities} + \text{Net Worth} \qquad \textbf{(16.1)}$$

For publicly held companies, the net worth is the stockholders' equity. Bankers and other grantors of credit to companies are concerned with the margin of security for their loans. The balance sheet provides them with two important measures: (1) The assets owned by the company, and (2) the liabilities owned by the company. A representative consolidated balance sheet for a large fictitious corporation, U.S. Chemicals, is given in Table 16.3 for the calendar year ending December 31, 2002. In the United States, a corporation is the most common form of business organization, one that is chartered by a state and given many legal rights as

Table 16.3 Consolidated Balance Sheet for U.S. Chemicals in Millions of Dollars as of 31 December 2002

ASSETS			
Current Assets			
Cash and cash equivalents	107		
Marketable securities	45		
Accounts receivable	2,692		
Inventories:			
Finished products and work in progress	1,420		
Materials and supplies	312		
Deferred income tax assets	54		
Total current assets		4,630	
Investments			
In nonconsolidated affiliates	544		
Other	1,476		
Total investments		2,020	
Property			
Land	200		
Buildings	2,190		
Plant machinery and equipment	7,684		
Office equipment	645		
Computer software	242		
Less accumulated depreciation	(6,006)		
Net property		4,955	
Other Assets			
Goodwill	952		
Other intangible assets	1,654		
Total other assets		2,606	
TOTAL ASSETS			14,211

(Continued)

Table 16.3 (*Continued*)

LIABILITIES			
Current Liabilities			
Short-term debt (payable within one year)	150		
Accounts payable	2,773		
Income taxes payable	130		
Deferred income tax liabilities	21		
Dividends payable	104		
Accrued current liabilities	975		
Total current liabilities		4,153	
Long-Term Debt		3,943	
Other Noncurrent Liabilities			
Pension and other postretirement benefits	892		
Reserve for discontinued operations	78		
Other noncurrent obligations	784		
Total other noncurrent liabilities		1,754	
TOTAL LIABILITIES			9,850
STOCKHOLDERS' EQUITY			
Common stock (authorized 2,000,000,000 shares at $1.00 par value; 1,000,000,000 issued)	1,000		
Capital in excess of par value of common stock	4,230		
Retained earnings	2,559		
Less treasury stock at cost, 300,000,000 shares	(3,428)		
NET STOCKHOLDERS' EQUITY			4,361
TOTAL LIABILITIES + STOCKHOLDERS' EQUITY			14,211

an entity separate from its owners. It is characterized by the limited liability of its owners, the issuance of shares of easily transferable stock, and existence as a going concern. The balance sheet is divided, according to Eq. (16.1), into three sections, Assets, Liabilities, and, in this case, Shareholders' Equity (in place of net worth). Each section is divided into accounts, where the entries are the balances in the ledger accounts as of the date of the balance statement. All numbers in the table represent millions of U.S. dollars.

As shown in Table 16.3, assets for this corporation are divided into Current Assets, Investments, Property, and Other Assets. *Current assets* are items of economic value that could be converted to cash in less than one year, including cash and cash equivalents, marketable securities, accounts receivable, inventories, prepaid expenses, and deferred income taxes. The current assets total $4,630,000,000. *Investments* pertain to investments in companies in which ownership interest by U.S. Chemicals is 50% or less, but where U.S. Chemicals exercises significant influence over operating and financial policies. *Property* constitutes fixed assets, including land, buildings, machinery, equipment, and software, and is listed at its *book value*, which is the original cost (the so-called basis), corrected for accumulated depreciation. *Depreciation* is the allocation of the cost of an asset over a period of time for accounting and tax purposes. It accounts for a decline in the value of a property due to general wear and tear or obsolescence. Property still in use remains on the books and the balance sheet even after it is completely depreciated. *Goodwill* is an intangible asset that provides to a company a competitive advantage, such as a well-known and accepted brand name, reputation, or high employee morale. In addition to goodwill, other intangible assets may be listed, such as patents and trademarks. The total assets is shown as $14,211,000,000.

The second part of the balance sheet in Table 16.3 lists the liabilities and stockholders' equity. *Current liabilities* include all payments that must be made by the company within one year. The total for U.S. Chemicals is $4,153,000,000. *Long-term debts*, often in the form of bonds, are due after more than one year from the date of the balance sheet. They total $3,943,000,000. *Other noncurrent liabilities* total $1,754,000,000 and include pension and other postretirement benefits as well as reserves for any company operations that are discontinued. Total liabilities are $9,850,000,000. We note that liabilities are less than assets by $4,361,000,000. Thus, by Eq. (16.1), this difference must be the stockholders' equity. This equity includes the par value of issued common stock, which totals $1,000,000,000. The par value of a share of stock is an arbitrary amount that has no relationship to the market value of the stock, but is used to determine the amount credited to the stock account. If the stock is issued for more than its par value, the excess is credited to the account shown as *capital in excess of par value*. In Table 16.3, the par value is $1.00 per share but the stock was issued at $4.23 per share. Companies frequently repurchase shares of their common stock, resulting in a reduction of stockholders' equity. Because the shares are placed in a treasury, the transaction appears as *treasury stock at cost*. In Table 16.3, that amount is $3,428,000,000. The other account under stockholders' equity is *retained earnings*, which is the accumulated retained earnings that is increased each year by net income. The amount of this entry must be such that Eq. (16.1) is satisfied. This is seen to be the case in Table 16.3, where the net stockholders' equity is $4,361,000,000, giving total liabilities plus stockholders' equity as $14,211,000,000, which is equal to total assets.

The Income Statement

An annual report also contains *the income statement*, also called the *statement of consolidated income* (loss) or *consolidated statement of income*, which is an accounting of sales, expenses, and net profit (same as net earnings and net income) for a given period. In the annual report, the period is for one calendar or fiscal year. However, many companies also issue quarterly statements. Bankers, other grantors of credit, and investors and speculators pay close attention to the income statement because it provides the net profit of the company, which is an indication of the ability of the company to pay its debts and grow. *Net profit* is defined as *revenues* (sales) minus cost of sales, operating expenses, and taxes, over a given period of time; with *gross profit* (gross earnings or gross income) being revenues minus just cost of sales and operating expenses (i.e., profit before taxes).

A representative consolidated income statement for a large fictitious corporation, U.S. Chemicals, is given in Table 16.4 for the calendar year 2002. *Net sales* is gross sales minus returns, discounts, and allowances. The *cost of goods sold* (cost of sales) is the cost of purchasing the necessary *raw materials* to produce the goods plus the *cost of manufacturing* the finished products. Operating expenses are expenses other than those of manufacture and include: research and development expenses; selling, general, and administrative expenses; insurance and finance company operations; and amortization and adjustments of goodwill. *Amortization* is the gradual elimination of a liability, such as a mortgage, in regular payments over a specified period of time, where the payments are sufficient to cover both principal and interest. Income from operations equals gross profit minus operating expenses. From this, interest expenses are subtracted to give gross income (sometimes called net profit before

Table 16.4 Consolidated Income Statement for U.S. Chemicals
in Millions of Dollars for the Calendar Year 2002

Net sales	11,504	
Cost of goods sold	9,131	
GROSS PROFIT		2,373
OPERATING EXPENSES		
Research and development expenses	446	
Selling, general, and administrative expenses	439	
Insurance and finance company operations	34	
Amortization and adjustments of goodwill	64	
TOTAL OPERATING EXPENSES		983
INCOME FROM OPERATIONS		1,390
Interest expense	185	
GROSS INCOME		1,205
Provision for income taxes	402	
NET INCOME		803

income taxes). Interest expenses pertain to interest payments to bond holders and interest on loans. Subtraction of income taxes gives net income. Table 16.4 shows that from an annual net sales of $11,504,000,000 the net profit is $803,000,000 or 6.98%.

The Cash Flow Statement

The *cash flow statement*, also called the *consolidated statement of cash flow* or *statement of consolidated cash flow* is a summary of the cash flow of a company over a given period of time. The *cash flow* equals cash receipts minus cash payments over a given period of time or equivalently, net profit plus amounts charged off for depreciation, depletion, and amortization. These latter three items are added back because they do not represent any cash transactions. *Depletion*, which is similar to depreciation, accounts for the exhaustion of natural resources such as oil, timber, and minerals. The cash flow statement is a measure of a company's financial health and, in recent years, has become a very important feature of the annual report.

A representative consolidated cash flow statement for a fictitious company, Chicago Chemicals, is given in Table 16.5 for the calendar year 2002. The statement is divided into three parts: operating activities, investing activities, and financing activities. Cash flows are either into or out of the company. In this statement, cash flows out of the company are stated in parentheses. Under operating activities, to the net income available for holders of common stock is added depreciation, depletion, amortization, and provision for deferred income tax; subtracted is a net loss on sales of property. Since the cash is not yet in hand, receivables and inventory are subtracted, but accounts payable (not yet paid) is added. The resulting cash flow into the company for operating activities for the year 2002 is $4,202,000,000. Under investing activities, capital expenditures by the company are subtracted from the sum of sales of property and sales of investments to give a cash flow of $1,225,000,000 into the company. Under financing activities, cash flows out of the company due to payments of long-term debt, purchases of treasury stock, and dividends paid to stockholders are partially offset by proceeds to the company from sales of preferred stock to give a cash flow out of the company of $404,000,000. For the combined three activities, the cash flow into the company is $5,023,000,000.

Table 16.5 Consolidated Cash Flow Statement for Chicago Chemicals in Millions of Dollars for the Calendar Year 2002

OPERATING ACTIVITIES		
Net income available for common stockholders	3,151	
Adjustments to reconcile net income to net cash:		
Depreciation	675	
Depletion	383	
Amortization	486	
Provision for deferred income tax	125	
Net gain (loss) on sales of property	(103)	
Changes in assets and liabilities involving cash:		
Accounts and notes receivable	(441)	
Inventories	(389)	
Accounts payable	315	
CASH PROVIDED BY OPERATING ACTIVITIES		4,202
INVESTING ACTIVITIES		
Capital expenditures	(1,227)	
Sales of property	231	
Sales (purchases) of investments	2,221	
CASH PROVIDED IN INVESTING ACTIVITIES		1,225
FINANCING ACTIVITIES		
Payments on long-term debt	(524)	
Purchases of treasury stock	(15)	
Proceeds from sales of preferred stock	620	
Dividends paid to stockholders	(485)	
CASH PROVIDED (USED) IN FINANCING ACTIVITIES		(404)
INCREASE (DECREASE) IN CASH AND CASH EQUIV.		5,023

Financial Ratio Analysis

The analysis of the performance and financial condition of a company is carried out by computing several ratios from information given in its annual report. Such analysis must be done carefully because seemingly good performance might be due more to such factors as inflation and reduction of inventory than to improvements in company operations.

Current Ratio

The *current ratio* is defined as current assets divided by current liabilities. It is an indication of the ability of a company to meet short-term debt obligations. The higher the current ratio, the more liquid the company is. However, too high a ratio may indicate that the company is not putting its cash or equivalent cash to good use. A reasonable ratio is two, but it is better to compare current ratios of companies in a similar business. From Table 16.3, the current assets ratio of U.S. Chemicals is 4,630/4,153 = 1.11, which is a low value. At the end of the year 2000, Monsanto Company had a much better current ratio of 1.80.

Acid-Test Ratio

The *acid-test ratio*, also called the *quick ratio*, is a modification of the current ratio with the aim of obtaining a better measure of the liquidity of a company. In place of current assets, only assets readily convertible to cash, called *quick assets*, are used. Thus, it is defined as the ratio of current assets minus inventory to current liabilities. Marketable securities, accounts receivable, and deferred income tax assets are considered to be part of quick assets. From Table 16.3, the quick assets for U.S. Chemicals, in millions of dollars, is $4,630 - 1,420 - 312 = 2,898$. This gives an acid-test ratio of $2,898/4,153 = 0.70$, which is not a desirable ratio, since it is less than one. At the end of the year 2000, Monsanto Company had a much better acid-test ratio of 1.35.

Equity Ratio

The *equity ratio* is defined as the ratio of stockholders' equity to total assets. It measures the long-term financial strength of a company. From Table 16.3, the equity ratio for U.S. Chemicals is $4,361/14,211 = 0.31$, which, again, is too low a value. This ratio should be about 0.50. If the equity ratio is too high, the company may be curtailing its growth. At the end of the year 2000, Monsanto Company had an equity ratio of 0.63.

The above three financial ratios use only data from the balance sheet. We next consider two ratios that use both the balance sheet and the income statement, followed by two ratios that use data from the income statement only. These ratios are particularly susceptible to economic conditions, which can, sometimes, change quickly from year to year. In the United States, the year 2000 started with economic prosperity, but a downturn occurred in early 2001, leading to a recession that continued to the end of that year.

Return on Total Assets (ROA)

One measure of how a company uses its assets is the *return on total assets*, which is defined as the ratio of income before interest and taxes to total assets. Using data in Tables 16.3 and 16.4, this ratio for U.S. Chemicals is $1,390/14,211 = 0.098$ or 9.8%, which is higher than the 4.84% achieved by Monsanto Company in the year 2000.

Return on Equity (ROE)

A more widely quoted return measure is the *return on stockholders' equity*, which measures the ability of a company to reinvest its earnings to generate additional earnings. It is identical to ROA except that the divisor in the ratio is common stockholders' equity instead of total assets. For U.S. Chemicals, the ROA from data in Tables 16.3 and 16.4 is $1,390/4,361 = 0.319$ or 31.9%, which, again, is higher than the 7.72% achieved by Monsanto Company in the year 2000. In the third quarter of the year 2001, the average ROE for 900 companies in the United States was 7.5%, with the 25 largest publicly held chemical companies averaging 3.0%.

Operating Margin

The operating margin is defined as the ratio of income from operations (called revenues) to net sales. For some industrial groups, it is highly susceptible to general economic conditions. For U.S. Chemicals, Table 16.4 gives an operating margin of $1,390/11,504 = 12.1\%$, which is somewhat higher than the 10.3% achieved by Monsanto Company in the year 2000.

Profit Margin

The *profit margin*, also called the *net profit ratio*, is defined as the ratio of net income after taxes to net sales. It is more widely quoted than the operating margin and is also more susceptible to general economic conditions. When used over a period of years, it is a very useful measure of the growth of a company. Using the data of Table 16.4, the profit margin for U.S. Chemicals was 803/11,504 = 0.070 or 7.0%. In the third quarter of the recession year 2001, the average operating margin for 900 companies in the United States was only 3.0%, with the 25 largest publicly held chemical companies averaging a dismal 0.9%. A year earlier, when general economic conditions were much more favorable, these two values were higher, at 6.8% and 6.6%, respectively. Companies in the medical products group are less subject to general economic conditions and had the highest average operating margin in the United States, 15.5% in the years 2000 and 2001. For some companies, operating margins can be quite high. Microsoft, a large and very successful software company in the United States achieved 44.8% in the third quarter of the year 2000, but fell to 20.9%, still a highly desirable value, in the third quarter of the recession year, 2001.

Cost Accounting

Cost accounting is a branch of accounting that deals specifically with the identification, recording, tracking, and control of costs. Accountants allocate these costs among: (1) *direct costs*, both labor and materials, (2) *indirect costs* or *overhead*, and (3) other miscellaneous expenses. Direct costs are those costs directly attributable to a project such as the construction of a new plant or the operation of an existing plant. Indirect costs or overhead are costs that are generally shared among several projects and are allocated to the individual projects by a formula or some other means. Miscellaneous expenses include administration, distribution and selling, research, engineering, and development. Direct costs can be more accurately identified, measured, and controlled; and are generally the largest fraction of the total cost.

Cost accounting is of great interest in plant construction and plant operation. It is also of importance when making an economic evaluation of a process design to determine whether a plant should be built. By studying company cost accounting records for existing manufacturing plants, chemical engineers preparing economic evaluations of proposed new processing plants or revamps of existing processes are less likely to omit or neglect costs that may have a significant influence on estimated process profitability measures. Large companies that engage in a number of plant construction and plant operation projects use cost accounting records to make comparisons of costs. These records are invaluable to company process design engineers when preparing, for new projects, estimates of investment and operating costs.

Cost accounting for direct costs is accomplished in terms of unit cost and quantity. The product of these two is the cost. For example, an existing process may have used 11.2 million kg/yr of a raw material with an average unit cost of $0.52/kg. The cost is then $5,824,000/yr. At the beginning of the year, a standard cost and a standard quantity are established for the year, for example, 10,500,000 kg and $0.51/kg. The budgeted cost for the year is 10,500,000 × 0.51 = $5,355,000. The differences between the actual and standard unit costs and quantities are called variances. Accountants can set variance flags to warn process managers of possible cost overruns. In the example, the quantity variance is 11.2 − 10.5 or 0.7 million kg/yr. This is a % variance of 0.7/10.5 × 100% = 6.7%. The cost variance is $0.01/kg or a % variance of 0.01/0.51 × 100% = 2.0%.

A *variance in quantity* may reveal the extent of waste. For example, suppose a plant is scheduled to produce 22,700,000 kg for the coming year. Design calculations indicate that

1.2 kg of raw material is required to produce each kilogram of product. The actual production rate for the year is 21,800,000 kg. The design rate for the raw material at the actual production rate is 1.2 × 21,800,000 = 26,160,000 kg/yr. However, the accounting records show for the raw material a beginning inventory of 1,070,000 kg, an ending inventory of 1,120,000 kg, and purchase of 26,980,000 kg. Thus, the amount of raw material used is 26,980,000 + 1,070,000 − 1,120,000 = 26,930,000 kg. If the design calculations are accepted as the basis, the waste is 26.93 − 26.16 = 0.77 million kg. At $0.52/kg, this represents a loss for the year of $0.52 × 770,000 = $400,400, a significant amount of money and an incentive to find the reasons for the waste and eliminate it. Similar calculations can be made for utility usage by expressing both the design and the actual values for the quantities of utilities on a per kilogram production of product basis.

The analysis of *variance in cost* is complicated when the price of the material, whether it be the raw material or finished product, varies during the year. This is because of the need for the company to maintain inventories of raw materials and finished products. Two methods of costing in common use are: (1) *first-in, first-out*, and (2) *last-in, first-out*. In the first method, abbreviated to *fifo*, the cost of the oldest material in the inventory is used first. In the second method, abbreviated to *lifo*, the cost of the most recent material in the inventory is used first; that is, the older material is kept in the inventory. Companies may also use any of several average costing methods. To illustrate the possible significance to the company of choosing between the fifo and lifo methods, consider the following example. At the beginning of the year 2001, ABC Oil Producing Company has an inventory of 100,000 barrels (bbl) of crude oil with a unit cost of $25/barrel. During the first quarter of 2001, purchases of crude oil are made at three different prices as follows:

Month	Barrels Purchased	Cost ($/bbl)
January	80,000	25
February	100,000	30
March	150,000	15

At the end of March, the inventory is 75,000 barrels. Determine the cost of the barrels sold during the quarter. The total barrels sold during the quarter is 100,000 − 75,000 + 80,000 + 100,000 + 150,000 = 355,000 barrels.

Fifo method:

Cost of barrels sold by first-in, first-out is

$$100,000 \times 25 = \$2,500,000$$
$$80,000 \times 25 = 2,000,000$$
$$100,000 \times 30 = 3,000,000$$
$$75,000 \times 15 = 1,125,000$$

Total barrels sold = 355,000

Total cost of barrels sold = $8,625,000

Lifo method:

Cost of barrels sold by last-in, first-out is

$$150,000 \times 15 = \$2,250,000$$
$$100,000 \times 30 = 3,000,000$$
$$80,000 \times 25 = 2,000,000$$
$$25,000 \times 25 = 625,000$$

Total barrels sold = 355,000

Total cost of barrels sold = $7,875,000

The total cost of barrels sold by the lifo method is $750,000 less than that of the fifo method. The unit costs are $24.30/bbl for the fifo method and $22.18/bbl for the lifo method.

16.2   COST INDEXES AND CAPITAL INVESTMENT FOR COMMODITY CHEMICALS

In all stages of the design process, estimates of both the total capital investment (TCI) and the annual cost of manufacture (COM) are crucial for the evaluation of processing alternatives. As discussed in Chapters 5, 6, and 7, many heuristics have been developed to create process flowsheets that reduce costs and increase the profitability of the processes being designed. In Chapters 7 and 10, approximate measures are used, such as the annualized cost (involving both the capital investment and the annual manufacturing cost), for the comparison of alternative process flowsheets. In some cases, when the manufacturing costs, especially the cost of fuel, are high, it is possible to compare the alternatives on the basis of the lost work or thermodynamic efficiency. This is the subject of Chapter 9, in which several considerations are presented for adjusting the minimum approach temperature in heat exchangers, replacing valves with turbines, and reducing pressure drops in pipelines.

In this chapter and the next, commonly used methods are developed for assessing the profitability of a process design. This chapter focuses on the so-called *direct permanent* (capital) *investment*, C_{DPI}, that is, the estimation of the purchase cost of required equipment and the cost of its installation in a potential chemical process. To this is added a contingency, the cost of land, any applicable royalties, and the cost of starting up the plant, to give the *total permanent investment*, C_{TPI}. The *contingency* accounts for uncertainty in the estimate and the possibility of not accounting for all of the costs involved. *Royalties* are payments made for the use of property, especially a patent, copyrighted work, franchise, or natural resource through its use. In Chapter 17, the *annual manufacturing costs*, together with the general annual expenses such as administration and marketing, which are listed in a *Cost Sheet*, are considered. These costs are the basis for an estimate of the *working capital* needed to compute the *total capital investment* for a chemical process. Then, together with depreciation and tax schedules, cash flows are computed that lead to profitability measures such as the *investor's rate of return* (IRR), also known as the *discounted cash flow rate of return* (DCFRR).

Cost Indexes

As discussed in Section 16.5, the purchase cost of processing equipment is generally obtained from charts, equations, or quotes from vendors. However, costs are not static. Because of inflation, they generally increase with time. Thus, charts and equations apply to a particular date, usually month and year, or to an average for a particular year. Quotes from vendors are often applicable only for a month or two. An estimate of the purchase cost at a later date is obtained by multiplying the cost from an earlier date by the ratio of a *cost index*, I at that later date to a base cost index, I_{base}, that corresponds to the date that applies to the purchase cost.

$$Cost = Base\ Cost\left(\frac{I}{I_{base}}\right) \tag{16.2}$$

The indexes most commonly considered by chemical engineers are

1. **The Chemical Engineering (CE) Plant Cost Index**
   It is published in each monthly issue of the magazine *Chemical Engineering* with $I = 100$ for 1958. A complete description of the index appears in *Chemical Engineering*, **70**(4), 143 (February 18, 1963).
2. **The Marshall & Swift (MS) Equipment Cost Index**
   It is published in each monthly issue of the magazine *Chemical Engineering* with $I = 100$ for 1926. A complete description of the index appears in *Chemical Engineering*, **54**(11), 124 (November 1947).

3. *The Nelson–Farrar (NF) Refinery Construction Cost Index*

It is published in the first issue each month of the magazine *Oil & Gas Journal* with $I = 100$ for 1946. A complete description of the index appears in the magazine, *Oil & Gas Journal*, **63**(14), 185 (1965).

4. *The Engineering News-Record (ENR) Construction Cost Index*

It is published each week in the magazine *Engineering News-Record* and in each monthly issue of *Chemical Engineering*, with $I = 100$ for 1967. A complete description of the index appears in *Engineering News-Record*, **143**(9), 398 (1949).

The CE and NF indexes pertain to the entire processing plant, taking into account labor and materials to fabricate the equipment, deliver it, and install it. However, the NF index is restricted to the petroleum industry, while the CE index pertains to an average of all chemical processing industries. The ENR index, which is a more general index that pertains to the average of all industrial construction, is a composite of the costs of structural steel, lumber, concrete, and labor. The MS index pertains to an all-industry average equipment purchase cost. However, it is accompanied by a more useful process industries average equipment cost index, averaged mainly for the chemicals, petroleum products, paper, and rubber industries. The CE and NF indexes also provide cost indexes for the purchase cost only of several categories of processing equipment, including heat exchangers, pumps and compressors, and machinery.

Figure 16.1 compares, on a semilogarithmic plot, the values of the CE Plant Cost Index, MS Process Industries Average Cost Index, NF Refinery Cost Index, and ENR Construction Cost index for the period of 1975 to 2000. The same values are tabulated in Table 16.6. It can be seen that costs increased at a more rapid annual rate from 1975 to 1981 than from 1981 to 2000. In the latter 19-year period, the cost indexes have increased by factors of 1.327 for CE, 1.480 for MS, 1.707 for NF, and 1.755 for ENR. These factors correspond to the following respective increases per year: 1.50%, 2.08%, 2.85%, and 3.00%. Included in Figure 16.1 and Table 16.6 are values for the U.S. Consumer Price Index (CPI), published by the

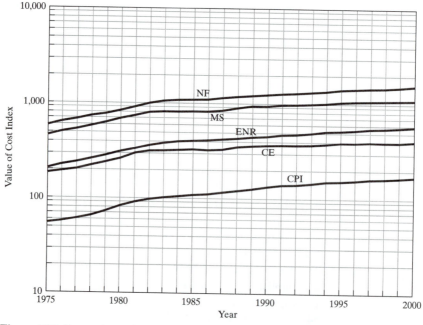

Figure 16.1 Comparison of annual average cost indexes.

Table 16.6 Comparison of Cost Indexes

Year	CE	MS	NF	ENR	CPI
1975	182	452	576	207	53.8
1976	192	479	616	224	56.9
1977	204	514	653	241	60.6
1978	219	552	701	259	65.2
1979	239	607	757	281	72.6
1980	261	675	823	303	82.4
1981	297	745	904	330	90.9
1982	314	774	977	357	96.5
1983	317	786	1,026	380	99.6
1984	323	806	1,061	387	103.9
1985	325	813	1,074	392	107.6
1986	318	817	1,090	401	109.6
1987	324	830	1,122	412	113.6
1988	343	870	1,165	422	118.3
1989	355	914	1,196	429	124.0
1990	358	935	1,226	441	130.7
1991	361	952	1,253	450	136.2
1992	358	960	1,277	464	140.3
1993	359	975	1,311	485	144.5
1994	368	1,000	1,350	503	148.2
1995	381	1,037	1,392	509	152.4
1996	382	1,051	1,419	524	156.9
1997	387	1,068	1,449	542	160.5
1998	390	1,075	1,478	551	163.0
1999	391	1,083	1,497	564	166.6
2000	394	1,103	1,543	579	172.3
2001	395	1,110	1,578	594	177.2

federal government and used to measure the rate of inflation, with a basis of 10.0 for the year 1914. In the 19-year period from 1981 to 2000, the CPI increased by a factor of 1.894, giving an average consumer inflation rate of 3.42% per year.

Commodity Chemicals

Manufactured chemicals can be classified as either: (1) primary chemicals, usually referred to as *commodity chemicals* or *bulk chemicals*, and (2) secondary chemicals, which include *fine chemicals* and *specialty chemicals*. Commodity chemicals have a known chemical structure, are most often produced by continuous processing, and are characterized by high production rates (typically more than 10 million lb/yr), high sales volumes, low selling prices, and global competition. Secondary chemicals are most often produced by batch processing, and are characterized by low production rates (typically less than 1 million lb/yr), low sales volume, and high selling prices. Fine chemicals are defined by their chemical structure and are, thus, distinguished from specialty chemicals, which are identified by their performance rather than their chemical content. Specialty chemicals are incorporated into *performance products* so as to achieve certain results. Specialty chemicals often have trademarked names and their chemical composition is often proprietary.

Some of the high-volume commodity chemicals are listed in Table 16.7, including U.S. production rates in the year 2000 from the June 25, 2001 issue of *Chemical & Engineering News*, typical sales price in the year 2001 from the *Chemical Market Reporter*, and required raw materials. Note that 11 commodity chemicals are produced at total rates of more than 10

Table 16.7 Major U.S. Commodity Chemicals

Chemical	U.S. Production in 2000 (Millions of Pounds)	Typical Price in 2001 ($/lb)	Typical Raw Materials Required
Ethylene	55,397	0.24	Crude oil fractions
Sulfuric acid	43,639	0.01	Sulfur dioxide, oxygen, water
Propylene	31,825	0.15	Crude oil fractions
Ethylene dichloride	21,850	0.12	Ethylene, chlorine
Ammonia	16,550	0.09	Nitrogen, hydrogen
Urea	15,322	0.05	Carbon dioxide, ammonia
Chlorine	13,225	0.10	Sodium chloride, water
Ethylbenzene	13,156	0.25	Benzene, ethylene
Phosphoric acid	12,895	0.34	Phosphorus, oxygen, water
Sodium hydroxide	12,102	0.03	Sodium chloride, water
Styrene	11,916	0.25	Ethylbenzene
Nitric acid	8,797	0.11	Ammonia, oxygen
Ethylene oxide	8,526	0.45	Ethylene, oxygen
Cumene	8,247	0.26	Benzene, propylene
Ammonium nitrate	8,244	0.08	Ammonia, nitric acid

billion lb/yr. Typical large plants produce 1 billion lb/yr, equivalent to about 125,000 lb/hr. A large petroleum refinery, producing many products, feeds 100,000 barrels of crude oil per day, equivalent to about 1,300,000 lb/hr. Prices for the commodity chemicals in Table 16.7 range from as low as $0.01/lb for sulfuric acid to $0.45/lb for ethylene oxide. These prices may be compared to regular-grade gasoline at the pump, before state and federal taxes are added, at $0.60/gal, equivalent to about $0.10/lb. Most fine and specialty chemicals cost much more than commodity chemicals. For example, cetyl (palmityl) alcohol in crystalline flake form, which is used as a textile conditioner, an emulsifier, and a component of cosmetics, costs about $16 for 1 lb. Many pharmaceuticals cost significantly more. For example, tPA, a drug to degrade blood clots, costs as much as $2,000 for a 100 mg dose.

Economy of Scale and the Six-Tenths Factor

When demand is high for commodity chemicals, advantage can be taken of the *economy of scale*. This principle holds as long as each major piece of equipment in the plant can be made larger as the production rate is increased. This makes possible a single-train plant, with no or few pieces of equipment duplicated. However, when the equipment size exceeds the maximum size that can be fabricated and shipped, then equipment must be duplicated and the economy of scale is lost because two or more trains of equipment are required. The economy of scale is embedded in the following relationship, which correlates the variation of cost with capacity:

$$\frac{Cost_2}{Cost_1} = \left(\frac{Capacity_2}{Capacity_1}\right)^m \tag{16.3}$$

This relationship has been found to give reasonable results for individual pieces of equipment and for entire plants. Although, as shown by Williams (1947a,b), the exponent, m, may vary from 0.48 to 0.87 for equipment and from 0.38 to 0.90 for plants, the average value is close to 0.60. Accordingly, Eq. (16.3) is referred to as the "six-tenths rule." Thus, if the capacity is doubled, the 0.6 exponent gives only a 52% increase in cost. Equation (16.3) is used in conjunction with Eq. (16.2) to take cost data from an earlier year at a certain capacity and estimate the current cost at a different capacity. As an example, suppose the total depreciable capital investment for a plant to produce 1,250 tonnes/day (1 tonne = 1,000 kg) of ammonia was $140 million in 1990. In the year 2000, the estimated investment for a 2,500 tonnes/day plant is as follows, where the CE index in Table 16.6 is used:

$$\text{Estimated investment, millions of U.S. \$} = 140\left(\frac{2,500}{1,250}\right)^{0.6}\left(\frac{394}{358}\right) = 140(1.52)(1.10) = 234$$

As discussed below in Section 16.7, the Aspen Engineering Suite provides methods more accurate than the six-tenths factor method of Equation (16.3) for determining the effect of scale on capital cost. The Aspen methods apply engineering-based scale-up rules to each piece of process equipment and to buildings, site development, and other items of capital cost.

Typical Plant Capacities and Capital Investments for Commodity Chemicals

Because of the economy of scale, one might ask: How large are the capacities of the plants used to produce commodity chemicals and what are the corresponding capital investments? Haselbarth (1967) presented investment and plant capacity data for 60 types of chemical plants. This was followed by a more extensive compilation by Guthrie (1970, 1974) for 54 chemical processes. Unfortunately, because competition for the manufacture of commodity chemicals has become so keen in recent years, companies are now reluctant to divulge investment figures for new plants. However, Table 16.8 presents recent data for the 15 commodity chemicals in Table 16.7. Plant production rates are large, 0.360 to 4.0 billion lb/yr. Corresponding total depreciable capital investments are also large, ranging from 20 million to 400 million U.S. dollars in 1995.

Note that in several cases, the plants produce combined products. Both ethylene and propylene are produced from a naphtha cut obtained from the fractionation of crude oil. A combined ammonia–urea fertilizer plant is common. The electrolysis of a brine solution produces both chlorine and sodium hydroxide. Recent literature data are usually given for plant capacities in tonnes per year (1 tonne = 1,000 kg) or tons per day (1 ton = 2,000 lb) of product, but the capacity data in Table 16.8 are given in pounds of product per year. Also included in the table is the value of C_b for use in the following modification of Eq. (16.3):

$$C_{TDC} = C_b\left(\frac{\text{pounds/year}}{\text{production rate in Table 16.8}}\right)^{0.6}\left(\frac{I}{I_b}\right) \tag{16.4}$$

Table 16.8 Representative Plant Capacity and Capital Investment for Some Commodity Chemicals

Commodity Chemical(s)	Production Rate(s) (Millions of Pounds/Year)	Capital Investment Factor [C_b in Eq. (16.4) for 1995]
Ethylene and propylene	1,200 and 600	$300,000,000
Sulfuric acid	4,000	$30,000,000
Ethylene dichloride	1,000	$80,000,000
Ammonia and urea	400 and 1,500	$400,000,000
Chlorine and sodium hydroxide	360 and 400	$80,000,000
Ethylbenzene	2,800	$80,000,000
Phosphoric acid	3,200	$50,000,000
Styrene	2,500	$200,000,000
Nitric acid	1,400	$50,000,000
Ethylene oxide	600	$80,000,000
Cumene	600	$30,000,000
Ammonium nitrate	800	$20,000,000

where C_b is the total depreciable capital (TDC) investment, for the production rate given in the table. The data in Table 16.8 are indexed to the year 1995, when, according to Table 16.6, the Chemical Engineering Plant Cost Index was 381. Thus, for the year 2000, the right side of Eq. (16.4) includes a cost index ratio of 394/381 = 1.034.

EXAMPLE 16.1

Estimate the total depreciable capital investment in the year 2000 for a plant to produce 90 ton/day of chlorine and 100 ton/day of sodium hydroxide. Assume that the plant will operate continuously for 330 days of the 365-day year.

SOLUTION

The production rate can be based upon either the chlorine or the sodium hydroxide since both are produced in the same plant. The annual production rate of, say, chlorine in lb/year = 90(2,000)(330) = 59,400,000 lb/year. In Table 16.8, C_b = $80,000,000 for a chlorine production rate of 360,000,000 lb/yr. Using Eq. (16.4),

$$C_{TDC} = \$80 \text{ million} \left(\frac{59.4}{360}\right)^{0.6} \times 1.034 = \$28 \text{ million}$$

∎

16.3 CAPITAL INVESTMENT COSTS

The total capital investment (TCI) of a chemical plant is a one-time expense for the design, construction, and start-up of a new plant or a revamp of an existing plant. It is analogous to the purchase price of a new house, where the price includes purchase of the land, building-permit fees, excavation of the land, improvements to the land to provide utilities and access, preparation of architectural and construction drawings, construction of the house, landscaping, and contractor's fee. For convenience in cost accounting and cost estimation, Busche (1995) divides the TCI into the components listed in Table 16.9.

Before Table 16.9 is discussed, it is important to make a few distinctions. A new chemical processing plant may be an addition to an existing *integrated complex*, such as the addition

Table 16.9 Components of Total Capital Investment (TCI)

Total bare-module costs for fabricated equipment	C_{FE}					
Total bare-module costs for process machinery	C_{PM}					
Total bare-module costs for spares	C_{spare}					
Total bare-module costs for storage and surge tanks	$C_{storage}$					
Total cost for initial catalyst charges	$C_{catalyst}$					
Total bare-module investment, TBM		C_{TBM}				
Cost of site preparation		C_{site}				
Cost of service facilities		C_{serv}				
Allocated costs for utility plants and related facilities		C_{alloc}				
Total of direct permanent investment, DPI			C_{DPI}			
Cost of contingencies and contractor's fee			C_{cont}			
Total depreciable capital, TDC				C_{TDC}		
Cost of land				C_{land}		
Cost of royalties				C_{royal}		
Cost of plant startup				$C_{startup}$		
Total permanent investment, TPI					C_{TPI}	
Working capital					C_{WC}	
Total capital investment, TCI						C_{TCI}

of a polyethylene plant to a petroleum refinery that produces ethylene as one of its products; or it may be a *grass-roots plant*, with no other chemical plants nearby. In both cases, a new plant requires *auxiliary facilities*, including utilities, such as steam, cooling water, and electricity; and other services, such as waste treatment and railroad facilities. A grass-roots plant may also require other new facilities, such as a cafeteria and a maintenance shop. In the integrated complex, the auxiliary facilities may be shared among the various plants in the complex. For either an integrated complex or a grass-roots plant, it is customary to separate the processing equipment directly associated with the manufacturing process from the auxiliary facilities by an imaginary fence that defines so-called *battery limits*, with the chemical processing plant inside the limits in an *on-site* area. The utilities and other services are outside the battery limits and are referred to as *offsite facilities*. Figure 16.2 shows typical offsite auxiliary facilities that might be associated with a grass-roots plant. Depending on the extent of the offsite facilities, they can be a significant fraction of the total capital investment.

Table 16.9 begins with the sums of so-called bare-module costs for fabricated process equipment and process machinery. These refer to the on-site part of the plant, which can be divided into modules, each of which contains a processing unit, which may be a piece of *fabricated process equipment*, such as a heat exchanger, vessel, or distillation column; or an item of *process machinery*, such as a pump, compressor, or centrifuge. Fabricated equipment is custom designed, usually according to the pressure vessel code and other design standards, for any size and shape that can be shipped. Process machinery is selected from a vendor-supplied list of standard sizes and often includes a driver, such as an electric motor. A module contains not only the piece of equipment or machinery, but also all other materials for installing it (setting it and connecting it to equipment in other modules), including the piping to and from other modules; the concrete (or other) foundation; ladders and other steel supporting structures; the instruments, controllers, lighting, and electrical wiring; insulation; and painting. Also, depending on plant location and size, some equipment may be housed in process buildings or shelters.

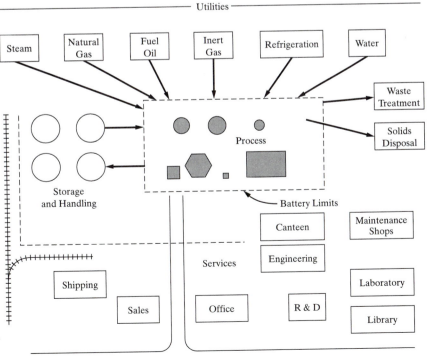

Figure 16.2 Plant services outside of process battery limits (courtesy of C.A. Miller).

Table 16.10 Example of Installation Costs for a Heat Exchanger to Give the Bare-Module and Total-Module Costs

	Cost ($)	Total Costs ($)	Fraction of f.o.b. Purchase Cost (C_P)
Direct module expenses			
Equipment purchase price, f.o.b., C_P		10,000	1.00 C_P
Field materials used for installation			
Piping	4,560		
Concrete	510		
Steel	310		
Instruments and controllers	1,020		
Electrical	200		
Insulation	490		
Paint	50		
Total of direct field materials, C_M		7,140	$C_M = 0.714\ C_P$
Direct field labor for installation			
Material erection	5,540		
Equipment setting	760		
Total of direct field labor, C_L		6,300	$C_L = 0.63\ C_P$
Indirect module expenses			
Freight, insurance, taxes, C_{FIT}	800		$C_{FIT} = 0.08\ C_P$
Construction overhead, C_O	5,710		$C_O = 0.571 C_P$
Contractor engineering expenses, C_E	2,960		$C_E = 0.296\ C_P$
Total indirect expenses, C_{IE}		9,470	$C_{IE} = 0.947\ C_P$
Bare-module cost, C_{BM}		32,910	$C_{BM} = 3.291\ C_P$
			$F_{BM} = 3.291$

Given the purchase cost of a process unit, the installed cost is obtained by adding the cost of installation. It is common to estimate the cost of installation using *factored-cost methods* based on the f.o.b. purchase cost of the process equipment. For each piece of equipment, Guthrie (1969, 1974) provides factors to estimate the direct costs of materials and labor, as well as indirect costs, involved in the installation procedure. When these costs are added to the purchase cost, Guthrie calls the result the *bare-module cost* instead of the installed cost. As an illustration, see Table 16.10, in which the installation costs for a heat exchanger are given as a fraction of the f.o.b. purchase cost, C_P, where f.o.b. means free on board (i.e., the purchase cost does not include the delivery cost to the plant site). Although the f.o.b. purchase cost is $10,000, the bare-module cost is $32,910. The components of the installed cost are as follows.

Direct Materials and Labor (M&L)

The costs of materials, C_M, for installation include the costs of concrete for the foundations, steel for structural support, piping from and to the other modules, instruments and controllers, lighting and electrical materials, insulation, and paint. Piping costs can be very substantial. Guthrie (1969, 1974) indicates that the cost of piping for a heat exchanger is typically 45.6% of the f.o.b. purchase cost, while the total cost of materials for installation is estimated at 71.4% of the f.o.b. purchase cost, as shown in Table 16.10. Hence, for a $10,000 heat exchanger, the cost of materials for installation is $7,140.

Similarly, Guthrie provides a field labor factor of 0.63 for installation; that is, the cost of labor for installation of the heat exchanger in Table 16.10, C_L, is 63% of the f.o.b. purchase cost or $6,300. The field labor cost accounts for the setting in place of the exchanger, installation of the associated piping, and all other labor costs utilizing the field materials. Combined with the cost of materials, the total cost of direct materials and labor for installation of the heat exchanger is $13,440, corresponding to a combined *materials and labor factor* of 1.344.

When using a factored-cost method like that of Guthrie, it is important to check the materials and labor factors for a specific job. Piping costs are usually underestimated and, in some cases, rival the purchase cost of the equipment. In this respect, separation towers often have the highest piping costs. Instruments and controllers can also be very expensive (with factors from 0.1 to 0.75) when they include analyzers, distributed control systems, and so on. The larger factor applies to a high degree of instrumentation and control of small equipment.

Other considerations involve materials of construction and the design pressure. As discussed later in this chapter, equations provide estimates of purchase costs based on the use of carbon steel at low-to-moderate pressures, with multiplication factors to estimate the purchase costs for more expensive materials and higher pressures. Often the materials and labor factors are applied incorrectly to the resulting purchase costs. More specifically, the concrete foundation for a titanium vessel is no more expensive than for a carbon steel vessel and is, therefore, a much smaller percentage of the vessel cost. Instrument and electrical costs are also a smaller percentage of the vessel cost. An example of the correct application of materials and labor factors is given below in Example 16.4.

The above discussion supposes that the vendor assembles all fabricated process equipment before shipping it to the plant for installation. In some cases, equipment cannot be shipped to the plant site in one piece and pre-installation field assembly will be required. Examples are furnaces and very large distillation columns and other vessels, which cannot be trucked, barged, or sent by rail in one piece to the plant site. Large columns may be fabricated in sections in the shops of the vendor and transported to the plant site where the sections are welded in a horizontal orientation before the column is erected to a vertical position. In this chapter, the purchase cost of field-assembled equipment includes the cost of pre-installation assembly at the plant site. Field-assembly costs are usually included in the purchase-cost quote from a vendor.

Indirect Costs

Other costs, such as the costs of freight to deliver the equipment to the plant site, with associated insurance and taxes, are considered to be *indirect expenses*, C_{FIT}. As shown in Table 16.10, these are estimated to be approximately 8% of C_P; that is, $800 for the heat exchanger. These are accompanied by construction overhead, C_O, which includes fringe benefits for the workers (health insurance, vacation pay, sick leave, etc.), so-called *burden* (Social Security taxes, unemployment insurance, etc.), and salaries, fringe benefits, and burden for the supervisory personnel. The construction overhead also includes the costs of temporary buildings, roads, parking areas, cranes and machinery (purchased or rented), job site cleanup, security costs, and so on. These costs are estimated at approximately 57.1% of the f.o.b. purchase cost of the equipment, or $5,710 for a $10,000 heat exchanger. Contractor engineering expenses, C_E, are also included in the indirect expense category. This covers the costs of engineering, including salaries for project and process engineers, designers, procurement expenses, home office expenses, and so on. They are estimated to be 29.6% of C_P, that is, $2,960 for a $10,000 heat exchanger.

Summing the indirect module expenses ($9,470), and combining the result with the cost of materials and labor, the *bare-module cost* for the $10,000 heat exchanger, C_{BM}, is $32,910, and, hence, the *bare-module factor*, F_{BM}, is 3.291.

Guthrie also computes and lists factors for starting with the f.o.b purchase cost and arriving at the bare-module cost. For the heat exchanger example of Table 16.10, the factors are derived as follows:

Start with the equipment purchase price, f.o.b. of $10,000.

The direct field materials, including the equipment, total (10,000 + 7,140) = $17,140.

Guthrie defines the direct materials factor as (17,140)/10,000 = 1.714.

The direct field labor totals $6,300.

Guthrie defines the direct labor factor as (6,300)/10,000 = 0.630.

The combined direct field materials and labor totals $23,440.

Guthrie defines the combined direct field materials and labor factor as (23,440) 10,000 = 2.344.

The indirect module expenses total $9,470.

Guthrie defines the indirect factor as (23,440 + 9,470)/23,440 = 1.404

The total bare module factor is 2.344 times 1.404 = 3.291, the same as above.

Bare-module factors vary among the various types of fabricated equipment and process machinery, decreasing somewhat with increasing size. The extent of this variation for ordinary materials of construction and low-to-moderate pressures can be seen in Table 16.11, which is taken from Guthrie (1974), based on single units of smaller size, where the factors are as much as 10% lower for multiple units of the same type. For the solids-handling equipment, the indirect factor is taken from Guthrie (1969) as 1.29. The bare-module factors vary from a value of 1.39 for crushers to separate solid particles by size to 4.16 for vertical pressure vessels, which are widely used for distillation, absorption, stripping, and flash drums.

Table 16.11 Bare-Module Factors of Guthrie (1974) for Ordinary Materials of Construction and Low-to-Moderate Pressures

	Bare-module Factor (F_{BM})
Furnaces and direct-fired heaters, Shop-fabricated	2.19
Furnaces and direct fired heaters, Field-fabricated	1.86
Shell-and-tube heat exchangers	3.17
Double-pipe heat exchangers	1.80
Fin-tube air coolers	2.17
Vertical pressure vessels	4.16
Horizontal pressure vessels	3.05
Pumps and drivers	3.30
Gas compressors and drivers	2.15
Centrifuges	2.03
Horizontal conveyors	1.61
Bucket conveyors	1.74
Crushers	1.39
Mills	2.30
Crystallizers	2.06
Dryers	2.06
Evaporators	2.45
Filters	2.32
Flakers	2.05
Screens	1.73

All of the equipment for handling solids and fluid–solids mixtures have factors less than 2.45. In Table 16.9, the sum of the bare-module costs for all items of fabricated equipment is C_{FE}, while the sum of the bare-module costs for all items of process machinery is C_{PM}.

Other Investment Costs

In addition to the bare-module costs associated with the process equipment in a flowsheet, capital costs are incurred for spare items of equipment, C_{spare}, for storage and surge tanks, $C_{storage}$, and for initial charges of catalyst, $C_{catalyst}$. As shown in Table 16.9, these costs are added to the bare-module costs for the on-site equipment to give the *total bare-module investment*, C_{TBM}. Other investment costs include site preparation or development, C_{site}, service facilities (e.g., utility lines and industrial buildings), C_{serv}, and allocated costs to purchase or upgrade the utility plants and other off-site facilities (e.g., for steam and electricity generation and waste disposal), C_{alloc}, shown in Figure 16.2. These are added to C_{TBM} to give the *direct permanent investment*, C_{DPI}. After adding funds (typically 18% of C_{DPI}) to cover contingencies and a contractor fee, the *total depreciable capital*, C_{TDC}, is obtained. Depreciation is very important to companies that use equipment to manufacture goods because it permits companies to reduce their taxes. As will be discussed in detail in the next chapter, depreciation is the allocation of the cost of an asset, such as a processing plant, over a period of time for accounting and tax purposes. Depreciation accounts for the decline in the value of equipment due to general wear and tear. To the C_{TDC} is added the investment in non-depreciable items, including the cost of land, the cost of royalties for the use of processes patented by others, and costs for plant startup, to give the *total permanent investment*, C_{TPI}, also referred to as *total fixed capital*. After the working capital is added, the *total capital investment*, C_{TCI} is obtained.

The working capital is described in detail in the next chapter. It includes the initial investment in temporary and consumable materials, as well as cash for initial payments of salaries and other operating expenses, prior to the receipt of payments for plant products. These other costs, which represent a significant fraction of the total capital investment, are considered next.

Spares, Storage Tanks, Surge Vessels, and Catalyst Costs

In addition to the bare-module costs for each process unit in the flowsheet as discussed above, it is often recommended to provide funds for *spares*, C_{spare}, especially for liquid pumps, to permit uninterrupted operation when a process unit becomes inoperable. Pumps are relatively inexpensive but require frequent maintenance to prevent leaks. Funds are also provided for storage units and surge tanks, $C_{storage}$, to provide improved control and intermediate storage between sections of the plant, so that one section can continue operation when an adjoining section is down or operating under capacity. The amount of storage depends on the anticipated periods of downtime and the importance of maintaining steady, uninterrupted operation. Finally, it is common to include the cost of the initial charge of catalyst, $C_{catalyst}$, which is a sizable investment cost in some plants. These costs are included in C_{TBM}, as indicated in Table 16.9.

Site Preparation

Site preparation typically involves making land surveys, dewatering and drainage, surface clearing, rock blasting, excavation, grading, piling; and addition of fencing, roads, sidewalks, railroad sidings, sewer lines, fire protection facilities, and landscaping. Costs for site preparation and development, C_{site}, can be quite substantial for grass-roots plants, in the range of 10–20% of the total bare-module cost of the equipment. For an addition to an existing integrated complex, the cost may only be in the range of 4–6% of the total bare-module cost of the equipment.

Table 16.12 Allocated Capital Investment
Costs for Utility Plants and Related Facilities

	Capital Cost Rate
Steam	$50/(lb/hr)
Electricity	$203/kW
Cooling water	$58/gpm
Process water	$347/gpm
Refrigeration	$1,330/ton
Liquid waste disposal	$3/1,000 gpy

gpy = gallons per year

Service Facilities

Costs for service facilities, C_{serv}, include utility lines, control rooms, laboratories for feed and product testing, maintenance shops, and other buildings. Service facility costs can be substantial for a grass-roots plant when administrative offices, medical facilities, cafeterias, garages, and warehouses are needed.

Allocated Costs for Utilities and Related Facilities

Allocated costs, C_{alloc}, are included to provide or upgrade off-site utility plants (steam, electricity, cooling water, process water, boiler feed water, refrigeration, inert gas, fuels, etc.) and related facilities for liquid waste disposal, solids waste disposal, off-gas treatment, and wastewater treatment. Some typical capital investment costs for utility plants, estimated by Busche (1995), are shown in Table 16.12. Cogeneration plants can provide both steam and electricity by burning a fuel. When utilities, such as electricity, are purchased from vendors at so many cents per kilowatt-hour, that cost includes the vendor investment cost. Thus, a capital cost for the plant is then not included in the capital cost estimate.

Contingencies and Contractor's Fee

Contingencies are unanticipated costs incurred during the construction of a plant. To account for the cost of contingencies, it is common to set aside 15% of the direct permanent investment, C_{DPI}, which is comprised of the components in Table 16.9. In addition, Guthrie (1969) adds a contractor fee of 3% of the direct permanent investment. When this total of 18%, designated C_{cont}, is added, the total depreciable capital, C_{TDC}, is obtained.

The cost of contingencies varies considerably, with 15% being a useful estimate when the design team is unable to make a better estimate. With processes for which the company has considerable experience, the cost of contingency is much lower than when a plant is being designed to produce a new chemical, just discovered by a research group. When deciding on a contingency cost, the design team should address the following three groups of questions:

1. How well is the process known? Has the process been demonstrated commercially, in a pilot plant, or in a laboratory? How long has the process run? Are the corrosion rates associated with the equipment well established? Has a demonstration of the process included all of the recycle streams?
2. How complete is the design? Has a simulation model been prepared? Is the detailed design complete? How much is known about the plant site?
3. How accurate are the estimates? Is the equipment conventional, or are there new and complex equipment items with which the company has little experience and cost history? In most cases, a 15% contingency is low, except when the experience factor is

very great. Typical designs that have not entered the final design stage in Figure 1.2, which are representative of most designs by student groups at universities, are likely to require 35% for contingency costs. When the chemistry is new and not well understood, 100% might be more realistic.

Land

The cost of land, C_{land}, is nondepreciable, since land rarely decreases in value, and in the absence of data can be taken as 2% of the total depreciable capital, C_{TDC}.

Royalties

When a company desires to use a process that is covered by patents owned by another company, a license may sometimes be negotiated. The license fee may be a one-time fee, in which case that fee is included in the capital investment as a one-time royalty or paid-up license, C_{royal}. A more common arrangement is to pay an initial license fee, included in the capital investment, and an annual royalty based on the amount or dollar value of product sold. The amount of the annual royalty depends on the uniqueness of the process and the chemical being produced, with a range of 1–5% of product sales. In the absence of data, an initial royalty fee of 2% of C_{TDC} may be assumed together with an annual royalty of 3% of product sales.

Startup

The cost of plant startup, $C_{startup}$, is typically estimated as 10% of C_{TDC}. However, according to Feldman (1969), if the process and equipment are well known to skilled operators and the new process is not dependent on the operation of another plant, the startup cost may be as low as 2% of C_{TDC}. At the other extreme, if the process and the equipment are radically new, and the new process is dependent on another plant, the startup cost may be as high as 30% of C_{TDC}. In this latter case, it may be necessary to modify the process and add more equipment. With respect to startup, it is important that the process design include additional equipment, such as heat exchangers, to achieve the startup. This is particularly important for processes involving significant recycle streams and/or a high degree of energy integration. Startup time depends on the same factors as startup costs and is generally taken as a percentage of construction time, varying from 10 to 40%.

Some company accountants may prefer to divide plant startup costs into two categories: (1) those costs incurred by the contractor in checking equipment performance, calibrating controllers and other plant equipment, and commissioning the plant, and (2) those costs incurred by plant operating personnel when starting up and shutting down the plant. The former costs are included in the capital cost while the latter are considered operating costs.

Investment Site Factors

In many companies, *investment site factors*, F_{ISF}, are used to multiply the total permanent investment, C_{TPI}, to account for different costs in different localities, based on the availability of labor, the efficiency of the workforce, local rules and customs, union status, and other items. Typical factors in recent use by one of the major chemical companies are provided in Table 16.13, where a plant in the U.S. Gulf Coast area is given a base factor of 1.0. The factors range from 0.85 in India to 1.25 in the U.S. West Coast area. The corrected total permanent investment is computed as:

$$C_{TPI_{corrected}} = F_{ISF} C_{TPI}$$

Table 16.13 Typical
Investment Site Factors, F_{ISF}

U.S. Gulf Coast	1.00
U.S. Southwest	0.95
U.S. Northeast	1.10
U.S. Midwest	1.15
U.S. West Coast	1.25
Western Europe	1.20
Mexico	0.95
Japan	1.15
Pacific Rim	1.00
India	0.85

Working Capital

Working capital funds, C_{WC}, are needed to cover operating costs required for the early operation of the plant, including the cost of the inventory and funds to cover accounts receivable. Because they involve the costs of the raw materials and the values of the intermediates, products, and byproducts, the working capital is normally estimated in connection with the calculation of the operating "Cost Sheet", which is presented in Table 17.1 and discussed in Section 17.3. Note that funds are usually allocated for a spare charge of catalyst, often kept in a warehouse, as a backup in case an operating problem causes the catalyst to become ineffective.

Example of an Estimate of Capital Investment

An example of an estimate of the total capital investment for a processing plant is given in Tables 16.14 and 16.15 for an ammonia plant producing 1 billion lb/yr. The costs are for the year 2000 at a U.S. Midwest location. The plant is part of an integrated complex. The process involves a variety of equipment, including gas compressors, pumps, heat exchangers, a catalytic reactor, a distillation column, an absorber, a flash drum, a gas adsorber, and gas permeation membrane separators. The material of construction is almost exclusively carbon steel.

Table 16.14 Capital Cost Estimate of Bare-Module
Equipment Cost for an Ammonia Plant—Costs in Millions
of U.S. Dollars (Year 2000)

	C_P	F_{BM}	C_{BM}
Fabricated equipment			
Heat exchangers	5.25	3.3	17.33
Flash drum	0.01	4.3	0.04
Distillation column	0.07	4.3	0.30
Adsorbers	0.18	4.3	0.77
Absorber	0.20	4.3	0.86
Membrane separators	3.56	3.2	11.39
Reactor	0.34	4.3	1.46
Process machinery			
Gas compressors	21.84	3.5	76.44
Pumps	0.07	3.4	0.24
Total bare-module cost for on-site equipment			108.83

Table 16.15 Total Capital Investment for an Ammonia Plant—Costs in Millions of U.S. Dollars (Year 2000)

Total bare-module cost for on-site equipment	108.83				
Cost for spares	0.52				
Cost for storage and surge tanks	0.45				
Cost for initial catalyst charge	0.50				
Total bare-module investment		110.30			
Cost of site preparation		3.31			
Cost of service facilities		1.65			
Allocated costs for utility plants and related facilities		15.45			
Direct permanent investment			130.71		
Cost of contingencies and contractor's fee			23.53		
Total depreciable capital				154.24	
Cost of land				3.08	
Cost of plant startup				12.32	
Total permanent investment					169.64
Working capital					10.09
Total capital investment					179.73

For the ammonia process, which operates at high pressure (200 atm), mostly in the gas phase, the total f.o.b. purchase cost of the on-site process equipment is $31,520,000. Installation costs boost this amount by a factor of 3.453 to a total bare-module cost of $108,830,000. As seen in Table 16.14, this cost is dominated by the gas compressors, with significant contributions from the heat exchangers and the membrane separators. Surprisingly, the reactor cost is a small fraction of the total cost. This is often the case for chemical plants. The reactor may not cost much, but it is the heart of the process and it better produce the desired results.

Table 16.15 continues the cost estimate to obtain the total capital investment. After all other investment costs are added to the total bare-module cost of the on-site equipment, the total capital investment becomes $179,730,000. The total permanent investment is $169,640,000, which is a factor of 5.38 times the total f.o.b purchase cost of the on-site process equipment. The start-up cost here is taken as 8% of C_{TDC}.

16.4   ESTIMATION OF THE TOTAL CAPITAL INVESTMENT

As a project for manufacturing a new or existing chemical by a new process progresses from laboratory research through pilot-plant development to a decision for plant construction, a number of process-design studies of increasing complexity may be made, accompanied at each step by capital-cost estimates of increasing levels of accuracy as follows:

1. *Order-of-magnitude estimate* based on bench-scale laboratory data sufficient to determine the type of equipment and its arrangement to convert the feedstock(s) to product(s).
2. *Study estimate* based on a preliminary process design.
3. *Preliminary estimate* based on detailed process design studies leading to an optimized process design.
4. *Definitive estimate* based on a detailed plant design, including detailed drawings and cost estimates, sufficient to apply cost accounting.

If the process is well known, having been verified by one or more commercial operating plants, only estimate levels 3 and 4 are necessary. Methods for making capital investments at the first three levels are discussed next. This chapter is concluded with an example of a definitive estimate using the Aspen Icarus Process Evaluator (IPE), which is part of the Aspen Engineering Suite that includes ASPEN PLUS.

Currently, Aspen Icarus software systems are the only commercially available systems that are recognized as the standard of the industry for estimating process costs. These systems, which are discussed in Sections 16.7 and 17.8, have more than 30 years of field-testing on commercial plants and are in use worldwide by owner companies and engineering design and construction firms. In our experience, application of the Aspen Icarus Process Simulator (IPE) is rather quickly understood and applied by chemical engineering students and practitioners after having studied the simpler, but less accurate costing methods presented here in Sections 16.1 to 16.6.

Method 1. Order-of-Magnitude Estimate (Based on the Method of Hill, 1956)

This estimation method can be applied rapidly and is useful in determining whether a new process is worth pursuing, especially when there are competing routes. The method is particularly useful for low-pressure petrochemical plants, where it has an accuracy of approximately ±50%. For moderate-to-high pressure processes, the actual cost may be as much as twice the estimate. To produce the estimate, only two things are needed, a production rate in pounds per year and a flowsheet showing the gas compressors, reactors, and separation equipment required. Heat exchangers and liquid pumps are not considered in making the estimate. Also, it is not necessary to compute a mass and energy balance or to design or size the equipment, but it is important that the process has been sufficiently studied that the flowsheet is complete with all the major pieces of gas movement, reactors, and separation equipment and their required materials of construction. Another important factor in making the estimate is the design pressure of each major piece of equipment if it is greater than 100 psi. The method proceeds as follows based on a year 2000 Marshall and Swift Process Industries Average Cost Index of 1,103, a base production rate of 10,000,000 lb/yr for the main product, carbon steel construction, and a design pressure of less than 100 psi.

Step 1: Establish the production rate of the main product in pounds per year. Compute a production rate factor, F_{PR}, using the six-tenths rule:

$$F_{PR} = \left(\frac{\text{Main product flow rate, lb/yr}}{10,000,000}\right)^{0.6} \tag{16.5}$$

Step 2: Using a process flowsheet, calculate from the following equation a module cost, C_M, for purchasing, delivering, and setting in place each major piece of equipment, including gas compressors and blowers (but not low-compression ratio recycle compressors and blowers); reactors; separators such as distillation columns, absorbers, strippers, adsorbers, membrane units, extractors, electrostatic precipitators, crystallizers, and evaporators; but not heat exchangers, flash and reflux drums, and liquid pumps:

$$C_M = F_{PR} F_M \left(\frac{\text{design pressure, psia, if} > 100 \text{ psi}}{100}\right)^{0.25} (\$130,000) \tag{16.6}$$

where F_M is a material factor, as follows:

Material	F_M
Carbon steel	1.0
Copper	1.2
Stainless steel	2.0
Nickel alloy	2.5
Titanium clad	3.0

Step 3: Sum the values of C_M, multiply the sum by a factor, F_{PI}, to account for piping, instrumentation and automatic controls, and indirect costs; and update with the current MS cost index, giving the total bare-module investment, C_{TBM}:

$$C_{TBM} = F_{PI}\left(\frac{MS\ index}{1,103}\right)\sum C_M \tag{16.7}$$

where the factor, F_{PI}, depends on the whether the plant processes solids, fluids, or a mixture of the two, as follows:

Type of Process	F_{PI}
Solids handling	1.85
Solids–fluids handling	2.00
Fluids handling	2.15

Step 4: To obtain the direct permanent investment, C_{DPI}, multiply C_{TBM} by the following factors to account for site preparation, service facilities, utility plants, and related facilities:

$$C_{DPI} = (1 + F_1 + F_2)C_{TBM} \tag{16.8}$$

where the factors F_1 and F_2 are

	F_1
Outdoor construction	0.15
Mixed indoor and outdoor construction	0.40
Indoor construction	0.80

	F_2
Minor additions to existing facilities	0.10
Major additions to existing facilities	0.30
Grass-roots plant	0.80

Outdoor construction is common except where winters are very severe and/or solids handling is critical.

Step 5: Obtain the total permanent investment and the total capital investment by the following equations, where a large contingency of 40% is used because of the approximate nature of the capital-cost estimate, and the costs of land, royalties, and plant startup are assumed to add an additional 10%. Working capital is taken as 15% of the total permanent investment.

$$C_{TPI} = 1.50C_{DPI}$$
$$C_{TCI} = 1.15C_{TPI}$$

EXAMPLE 16.2

Make an order-of-magnitude estimate of the total capital investment, as of the year 2001 (MS = 1,110), to produce benzene according to the toluene hydrodealkylation process shown in Figure 5.13. Assume an overall conversion of toluene to benzene of 95% and 330 days of operation per year. Also, assume the makeup gas enters at the desired pressure and a clay adsorption treater must be added to the flow sheet after the stabilizer. The treater removes contaminants that would prevent the benzene product from meeting specifications. In addition, in order for the reactor to handle the high temperature, it must have a brick lining on the inside, so take a material factor of $F_M = 1.5$. Otherwise, all major equipment is constructed of carbon steel. The plant will be constructed outdoors with major additions to existing facilities.

SOLUTION

Step 1: The plant will operate 330(24) = 7,920 hr/yr. For a toluene feed rate of 274.2 lbmol/hr, the annual benzene production rate is 0.95(274.2)(78.11)(7,920) = 161,000,000 lb/yr. Thus,

$$F_{PR} = \left(\frac{161,000,000}{10,000,000}\right)^{0.6} = 5.3$$

Step 2: The flowsheet includes one reactor (with $F_M = 1.5$) operating at 570 psia, three distillation columns operating at pressures less than 100 psia, one compressor operating at 570 psia, and one adsorption tower, assumed to operate at less than 100 psia. Therefore, the sum of the C_M values is

$$\sum C_M = 5.3\left[1.5\left(\frac{570}{100}\right)^{0.25} + 3 + 1\left(\frac{570}{100}\right)^{0.25} + 1\right](\$130,000) = \$5,400,000$$

Step 3: From Eq. (16.7), the total bare-module investment for a fluids handling process is

$$C_{TBM} = 2.15\left(\frac{1,110}{1,103}\right)(\$5,400,000) = \$11,700,000$$

Steps 4 and 5:

$$C_{DPI} = (1 + 0.15 + 0.30)(\$11,700,000) = \$17,000,000$$
$$C_{TPI} = 1.50(\$17,000,000) = \$25,500,000$$
$$C_{TCI} = 1.15(\$25,500,000) = \$29,300,000$$

■

Method 2. Study Estimate (Based on the Overall Factor Method of Lang, 1947a,b, and 1948)

In a series of three articles from 1947 to 1948, Lang developed a method for estimating the capital cost of a chemical plant using overall factors that multiply estimates of the delivered cost of the major items of process equipment. This method requires a process design, complete with a mass and energy balance, and equipment sizing. In addition, materials of construction for the major items of equipment, including the heat exchangers and pumps, must be known. Considerably more time is required for making a study estimate than for the preceding order-of-magnitude estimate. But, the accuracy is improved to ±35%. To apply the method, the f.o.b. purchase cost of each piece of major equipment must be estimated. F.o.b purchase costs of a wide range of chemical processing equipment are given in the next section of this chapter. The Lang method proceeds by steps as follows:

Step 1: From the process design, prepare an equipment list, giving the equipment title, label, size, material of construction, design temperature, and design pressure.

Step 2: Using the data in Step 1 with f.o.b. equipment cost data, add to the equipment list the cost and the corresponding cost index of the cost data. Update the cost data to the current cost index, sum the updated purchase costs to obtain the total f.o.b. purchase cost, C_P, and multiply by 1.05 to account for delivery of the equipment to the plant site. Then, multiply the result by an appropriate Lang factor, f_L, to obtain the total permanent investment (fixed capital investment), C_{TPI} (i.e., without the working capital), or the total capital investment, C_{TCI} (i.e., including an estimate of the working capital at 15% of the total capital investment or 17.6% of the total permanent investment).

$$C_{TPI} = 1.05 f_{L_{TPI}} \sum_i \left(\frac{I_i}{I_{b_i}}\right) C_{P_i} \tag{16.9}$$

$$C_{TCI} = 1.05 f_{L_{TCI}} \sum_i \left(\frac{I_i}{I_{b_i}}\right) C_{P_i} \tag{16.10}$$

The original Lang factor, based on capital costs for 14 different chemical plants, was found to depend on the extent to which the plant processes solids or fluids. Lang's factors, which at that time did not account for working capital, are given in the second column of Table 16.16.

Table 16.16 Original and Recommended Lang Factors

	Original Lang Factors, *Not* Including Working Capital	$f_{L_{TPI}}$ Recommended Lang Factors of Peters and Timmerhaus, *Not* Including Working Capital	$f_{L_{TCI}}$ Recommended Lang Factors of Peters and Timmerhaus, Including Working Capital
Solids processing plant	3.10	3.9	4.8
Solids–fluids processing plant	3.63	4.1	4.9
Fluids processing plant	4.74	4.6	5.7

A more detailed development of Lang factors, based on an analysis of 156 capital-cost estimates, was published by the editors of *Chemical Engineering* magazine in the September 30, 1963 issue on pages 120 and 122, as "Cost File 81." A further refinement, carried out by Peters and Timmerhaus (1968), resulted in the most widely accepted values of the Lang factor, which are included in Table 16.16 and are the factors recommended here for use in Eqs. (16.9) and (16.10). The detailed breakdown of costs by Peters and Timmerhaus is given in Table 16.17, which assumes that major plant additions are made to an existing site, but that additional land is purchased. The numbers in the table are based on a value of 100 for the total delivered cost of the process equipment. Here, the delivered cost is estimated as 1.05 times the

Table 16.17 Breakdown of Lang Factors by Peters and Timmerhaus (1968)

	Percent of Delivered-Equipment Cost for		
	Solids Processing Plant	Solids–fluids Processing Plant	Fluids Processing Plant
Delivered cost of process equipment	100	100	100
Installation	45	39	47
Instrumentation and controls	9	13	18
Piping	16	31	66
Electrical	10	10	11
Buildings (including services)	25	29	18
Yard improvements	13	10	10
Service facilities	40	55	70
Land	6	6	6
Total direct plant cost	264	293	346
Engineering and supervision	33	32	33
Construction expenses	39	34	41
Total and indirect plant costs	336	359	420
Contractor's fee	17	18	21
Contingency	34	36	42
Fixed capital investment	387	413	483
Lang factor, $f_{L_{TPI}}$, for use in Eq. (16.9)	3.9	4.1	4.8
Working capital	68	74	86
Total capital investment	455	487	569
Lang factor, $f_{L_{TCI}}$, for use in Eq. (16.10)	4.6	4.9	5.7

f.o.b. purchase cost. The Lang factors apply to total permanent investments of up to approximately $100 million U.S. dollars. Note that a contractor's fee of 5% and a contingency of only 10% is assumed. The latter value seems low when considering the accuracy of the estimate. If Eq. (16.9) is used, a detailed estimate of the working capital should be made according to the method presented in Section 17.3. No provision is made in the Lang-factor estimates for spares, storage and surge tanks, initial catalyst charge, royalties, or plant startup. However, these additional items can be added when desired. The fixed capital investment in Table 16.17 is the same as the total permanent investment in Table 16.9.

EXAMPLE 16.3

Use the Lang-factor method to estimate the total capital investment, as of the year 2001 (MS = 1,110), to produce cyclohexane according to the benzene hydrogenation process shown in Figure 9.24. However, the makeup H_2 feed is not available at 335 psia, but at 75 psia. Therefore, a feed-gas compressor, K2, has been added. Also two heat exchangers have been added. Reactor effluent stream S7 now enters new exchanger H2, which cools the effluent to 260°F by producing 10 psig steam from boiler feed water. The effluent leaves H2 as stream S7A and enters new exchanger H3, where it is heat exchanged with the feed benzene, heating the benzene to 235°F while being cooled to 201°F and leaving as stream S7B, which now enters existing exchanger H1. The process design has been completed, with the equipment sizes in the "Equipment List" of Table 16.18. Also included in Table 16.18 are estimates of the f.o.b. purchase costs in the year 1977 (MS index of 514). All equipment is fabricated from carbon steel.

SOLUTION

Referring to Table 16.18, the total f.o.b. purchase cost corresponding to an MS index of 514 is $176,900. However, if we provide spares for the two pumps, the total becomes $178,600. From Eq. (16.10), using a Lang factor for fluids processing of 5.7 from Table 16.16 and an updated MS index of 1,110, the estimated total capital investment is

$$C_{TCI} = 1.05(5.7)\left(\frac{1,110}{514}\right)\$178,600 = \$2,308,000$$

For this example, the order-of-magnitude estimate of Method 1 gives $3,800,000. ∎

Table 16.18 Equipment List, Including Purchase Costs, for Cyclohexane Process

Equipment Name	Equipment Label	Size	Design Temperature (°F)	Design Pressure (psia)	C_P, f.o.b. Purchase Cost (MS Index = 514)
Recycle compressor	K1	3 Hp	120	350	2,000
Feed-gas compressor	K2	296 Hp	450	350	80,000
Benzene feed pump	P1	4 Hp	120	350	1,200
Recycle pump	P2	1 Hp	120	350	500
Cooler	H1	210 ft²	210	300	4,000
Effluent-BFW HX	H2	120 ft²	400	320	2,500
Effluent-benzene HX	H3	160 ft²	270	310	3,200
High-pressure flash	F1	2 ft diam. × 8 ft height	120	300	5,000
Low-pressure flash	F2	2 ft diam. × 8 ft height	120	20	3,500
Reactor	R1	8 ft diam. × 30 ft height	400	330	75,000

Method 3. Preliminary Estimate (Based on the Individual Factors Method of Guthrie, 1969, 1974)

This method is best carried out after an optimal process design has been developed, complete with a mass and energy balance, equipment sizing, selection of materials of construction, and development of a process control configuration as incorporated into a P&ID. More time is required for making a preliminary estimate than for the preceding study estimate, but the accuracy is improved to perhaps ±20%. To apply the method, the f.o.b. purchase cost of each piece of major equipment must be estimated, as was the case with the Lang method. However, instead of using an overall Lang factor to account for installation of the equipment and other capital costs, individual factors for each type of equipment, given in Table 16.14 above, are used, as developed first by Hand (1958) and later in much more detail by Guthrie (1969, 1974), who introduced the bare-module concept. Furthermore, the Guthrie method takes into account the fact that the installation cost of equipment made of stainless steel or other expensive materials and/or for operation at a high pressure is not as large a factor of the purchased cost as with carbon steel and/or at near ambient pressure because with stainless steel and/or at high pressure, the foundation, supporting structures and ladders, electrical, insulation, and paint will be the same cost as when carbon steel is used and/or the design pressure is near ambient. The only difference is in the cost of the attached piping, because it is stainless steel, for example, instead of carbon steel, and may have to be thicker for high pressure. This is illustrated by an example below. F.o.b purchase costs of a wide range of chemical processing equipment are given in the next section. The Guthrie method involves the summation of estimates of module costs for the four different modules listed below. To this summation is added a contingency and contractor fee, in terms of a factor, to obtain the total permanent investment. An appropriate estimate of the working capital is added to obtain the total capital investment. Thus, the components of the total permanent investment are accounted for in a manner somewhat different from Table 16.9, but the overall result is the same. The equation for the total capital investment by the Guthrie method is

$$C_{TCI} = C_{TPI} + C_{WC} = 1.18(C_{TBM} + C_{site} + C_{buildings} + C_{offsite\ facilities}) + C_{WC} \quad \textbf{(16.11)}$$

Equation (16.11) does not account for royalties or plant startup. These additional costs should be added if they are known or can be estimated

The total bare-module cost, C_{TBM}, refers to the summation of bare-module costs for all items of process equipment, including fabricated equipment, process machinery, spares, storage tanks, and surge tanks. The initial charge of catalyst is included with the corresponding catalytic reactor cost. As shown in the heat-exchanger example of Table 16.10, the bare-module cost is based on the f.o.b. equipment purchase cost, to which is factored in direct field materials and labor, and indirect expenses such as freight, insurance, taxes, overhead, and engineering.

Site development costs, C_{site}, are discussed above in the section on capital investment costs. In lieu of a detailed estimate, which is not normally prepared at this stage of cost estimation, a value of 10–20% of C_{TBM} may be assigned for a grass-roots plant and 4–6% for an addition to an integrated complex.

Building costs, $C_{buildings}$, are also discussed above in the section on capital investment costs. In the Guthrie method, buildings include process buildings and non-process buildings. Again, a detailed estimate is not generally made at this stage of cost estimation. Instead, an approximate estimate is sufficient, but must consider whether some or all the process equipment must be housed in buildings because of weather or other conditions, and whether a grass-roots location or an addition to an integrated complex is being considered. If the equipment is housed, the cost of process buildings may be estimated at 10% of C_{TBM}. If a grass-roots plant is being considered, the non-process buildings may be estimated at 20% of C_{TBM}. If the process is to be an addition to an integrated complex, the non-process buildings may be estimated at 5% of C_{TBM}.

Offsite facilities include utility plants, when the company provides its own utilities, pollution control, ponds, waste treatment, offsite tankage, and receiving and shipping facilities. The utility plants may be estimated with the help of Table 16.12. To this may be added 5% of C_{TBM} to cover other facilities.

The factor, 1.18, in Eq. (16.11) covers a contingency of 15% and a contractor fee of 3%. As with the Lang-factor method, the working capital can be estimated at 15% of the total capital investment, which is equivalent to 17.6% of the total permanent investment, or it can be estimated in detail by the method in Section 17.3.

The Guthrie method proceeds by steps as follows:

Step 1: From the process design, prepare an equipment list, giving the equipment title, label, size, material of construction, design temperature, and design pressure.

Step 2: Using the data in Step 1 with f.o.b. equipment purchase cost data, add to the equipment list the cost, C_{P_b}, and the corresponding cost index, I_b, of the cost data. In the Guthrie method, the f.o.b. purchase cost is a base cost corresponding to a near-ambient design pressure, carbon steel as the material of construction, and a base design.

Step 3: Update the cost data to the current cost index. For each piece of equipment, determine the bare-module cost, using bare-module factors, F_{BM}, from Table 16.11, being careful to determine it properly when the material of construction is not carbon steel and/or the pressure is not near ambient, as given by Eq. (16.12) and illustrated by the following example, before moving to Step 4. As discussed earlier, the bare module cost accounts for delivery, insurance, taxes, and direct materials and labor for installation.

$$C_{BM} = C_{P_b}\left(\frac{I}{I_b}\right)[F_{BM} + (F_d F_p F_m - 1)] \tag{16.12}$$

where:

$$F_{BM} = \text{bare-module factor}$$
$$F_d = \text{equipment design factor}$$
$$F_p = \text{pressure factor}$$
$$F_m = \text{material factor}$$

Step 4: Obtain the total bare-module cost, C_{TBM}, by summing the bare-module costs of the process equipment.

Step 5: Using Eq. (16.11), estimate the total permanent investment. Add to this an estimate of the working capital to obtain the total capital investment.

EXAMPLE 16.4

The base f.o.b. purchase cost for a fabricated vertical pressure vessel, 6 ft in inside diameter and 100 ft in height (tangent-to-tangent) made of carbon steel for a design pressure of not greater than 50 psig is given as $102,000 as of 1995 (CE index = 381). Calculate the bare-module cost for the year 2000 (CE index = 394) if the vessel is made of 316 clad stainless steel for a design pressure of 200 psig. For these conditions, Guthrie (1974) gives $F_{BM} = 4.16$, $F_d = 1$, $F_p = 1.55$, $F_m = 2.60$.

SOLUTION

Using Eq. (16.12),

$$C_{BM} = \$102,000\left(\frac{394}{381}\right)[4.16 + (1 \times 1.55 \times 2.60 - 1)] = \$758,000 \qquad \blacksquare$$

EXAMPLE 16.5

The total bare-module cost for a process to produce 40,000,000 lb/yr of butyl alcohols by the catalytic hydration of butylenes is $10,200,000, indexed to the year 2002. Estimate the total capital investment. The process will be an addition to an existing integrated complex and no process buildings will be required. Offsite utility plants have been estimated at $1,200,000 and the working capital has been estimated at $1,350,000.

SOLUTION

$$C_{TBM} = \$10,200,000$$

Estimates of the other terms in Eq. (16.11) are as follows:

$$C_{site} = 0.05 C_{TBM} = 0.05(10,200,000) = \$510,000$$
$$C_{buildings} = 0.05 C_{TBM} = 0.05(10,200,000) = \$510,000$$
$$C_{offsite\ facilities} = 1,200,000 + 0.05 C_{TBM} = 1,200,000 + 0.05(10,200,000) = \$1,710,000$$
$$C_{TPI} = 10,200,000 + 510,000 + 510,000 + 1,710,000 = \$12,930,000$$
$$C_{WC} = \$1,350,000$$
$$C_{TCI} = 12,930,000 + 1,350,000 = \$14,280,000$$

For a grass-roots plant, an additional 25% of C_{TBM} is added for site development and buildings. This amounts to $2,550,000, giving a total capital investment of $16,830,000.   ■

16.5   PURCHASE COSTS OF THE MOST WIDELY USED PROCESS EQUIPMENT

The Lang and Guthrie methods for estimating the total capital investment require f.o.b. purchase costs for all major items of process equipment. Since 1949, a number of literature articles and book chapters have presented such data. Some of the more widely used sources of equipment cost data are given in Table 16.19. Included is the cost index of the cost data. Typically, equipment cost data are presented in the form of graphs and/or equations of f.o.b. purchase cost as a function of one or more equipment size factors. Graphs show clearly the effect of the size factors on the cost and may be quickly read; however, equations are more consistent, especially compared to graphs using logarithmic coordinates. Furthermore, equations are easily incorporated into computer programs. In this section, equations and graphs are presented for f.o.b. purchase costs of the most widely used chemical processing equipment: pumps, electric motors, fans, blowers, compressors, shell-and-tube and double-pipe heat exchangers, general-purpose fired heaters (furnaces), pressure vessels and towers, trays, and packings. Then, in Section 16.6, equations alone are presented for a wide variety of other

Table 16.19 Sources of Purchase Costs of Process Equipment

Author	Reference	Cost Index
Chilton, E.H.	*Chemical Engineering*, **56**(6), 97–106, June 1949	
Walas, S.M., and Spangler, C.D.	*Chemical Engineering*, **67**(6) 173–176, March 21, 1960	MS = 234.3
Bauman, H.C.	"Fundamentals of Cost Engineering in the Chemical Industry," Reinhold (1964)	MS = 237.3
Mills, H.E.	*Chemical Engineering*, **71**(6), 133, March 16, 1964	MS = 238.8
Guthrie, K.M.	*Chemical Engineering*, **76**(6), 114–142, March 24, 1969	MS = 273.1
Guthrie, K.M.	"Process Plant Estimating Evaluation and Control," Craftsman Book (1974)	MS = 303.3
Woods, D.R.	"Financial Decision Making in the Process Industry," Prentice-Hall (1975)	MS = 300
Pikulik, A., and Diaz, H.E.	*Chemical Engineering*, **84**(21), 107–122, October 10, 1977	MS = 460
Hall, R.S., Matley, J., and McNaughton, K.J.	*Chemical Engineering*, **89**(7), 80–116, April 5, 1982	CE = 305
Ulrich, G.D.	"A Guide to Chemical Engineering Process Design and Economics," Wiley (1984)	CE = 315
Walas, S.M.	"Chemical Process Equipment," Butterworth (1988)	CE = 325
Peters, M.S., and Timmerhaus, K.D.	"Plant Design and Economics for Chemical Engineers," 4th ed., McGraw-Hill (1991)	CE = 356

chemical processing equipment. The equipment cost equations should be used, even when one of the graphs below might apply.[1]

The form of the equations is a modification of the equation, $C_P = A(\text{size factor, } S)^b$ (where A and b are constants), obtained by taking the natural logarithm of both sides, adding additional higher-order terms, as with a polynomial, and solving for C_P to obtain

$$C_P = \exp\left\{A_0 + A_1[\ln(S)] + A_2[\ln(S)]^2 + \cdots\right\}$$

The equations are usually based on the more common materials of construction, such as carbon steel. For other materials, multiplying factors are provided. Assistance in choosing the materials of construction is given in Appendix III.

As discussed by Woods (1975) and Walas (1988), when cost data are assembled from vendor quotes, they exhibit scatter due to differing qualities of equipment fabrication, design differences, market conditions, vendor profit, and other considerations. Accordingly, the accuracy of published equipment cost data, such as referenced in Table 16.19, may be no better than ±25%. More accurate estimates can be obtained from computing systems, such as the Aspen Icarus Process Evaluator (IPE) in the Aspen Engineering Suite discussed in Section 16.7, which account for the details of equipment design, fabrication, and materials and labor requirements for installation, as well as related costs for site preparation, service facilities, indirect expenses, etc. Cost systems like IPE are rapidly gaining popularity in the chemical industry through the medium of personal computers. Although the equipment purchase-cost equations presented in Sections 16.5 and 16.6 may be convenient to use, the Aspen IPE system is preferred because it is more accurate, more consistent, and is periodically updated. Final equipment cost estimates can only be obtained by bids from the equipment manufacturers. These require special requests that are sometimes costly to prepare and, for that reason, are often not obtained until a decision has been made to construct the plant and a final detailed capital-cost estimate is needed for making an appropriation request. For some proprietary equipment systems, even a preliminary cost estimate might have to be requested from a vendor.

Pumps and Electric Motors

Pumps are used widely in chemical processing plants to move liquids through piping systems from one piece of equipment to another. The three most commonly used pumps are radial centrifugal, piston or plunger reciprocating, and external rotary gear, as discussed in some detail in Section 15.1. Of these three, the radial centrifugal pump (referred to here as just the centrifugal pump) is selected for industrial service approximately 90% of the time because it is

1. Relatively inexpensive to purchase and install.
2. Operated at high speed so that it can be driven directly with an electric motor.
3. Relatively simple in construction with no closely fitting parts that might wear, with a resulting low maintenance cost.
4. Available from a large number of vendors, many of whom comply with industry standards, such as those of American Petroleum Institute (API), American Society of Mechanical Engineers (ASME), and the International Organization for Standardization (ISO).

---

[1]In developing the cost equations for the purchase cost of processing equipment, presented in this chapter, available literature sources as far back as 1960 were consulted. After determining a suitable equipment size factor, all or much of the cost data were plotted. When a wide spread in the data was evident, which was not uncommon, an attempt was made to assess the validity of the data by comparison with costs of similar equipment. When the validity could not be determined, the data were averaged. In some cases, especially where available data were sparse, cost data were obtained from vendors of the equipment. It must be understood that the only accurate cost data are bids from a vendor and bids from different vendors can sometimes differ significantly.

5. Available in a wide range of materials of construction.
6. Applicable over a wide range of volumetric flow rate and temperature.
7. Applicable by staging for the achievement of heads up to 3,200 ft.
8. Capable of pumping a liquid at a smooth flow rate with a constant discharge pressure.
9. Capable of handling slurries.
10. Easy to operate, with control of the flow rate by a valve on the discharge line.
11. Free from damage if the valve on the discharge line is inadvertently or purposely closed.

However, the centrifugal pump has the following limitations:

1. It cannot efficiently pump liquids with a kinematic viscosity greater than100 centistokes.
2. The maximum efficiency of a particular pump is limited to a narrow range of its characteristic curve (flow rate versus head).
3. For most models, it cannot produce heads greater than 3,200 ft.
4. For most models, the volumetric flow rate must be greater than 10 gpm.
5. Most models are subject to air binding and must be primed.
6. Because of potential cavitation and the NPSH limitation, most models cannot pump liquids that are close to their bubble point.
7. Spares are normally specified because seals to prevent leakage may require attention more often than scheduled maintenance.

Pump Selection

A *centrifugal pump*, of the common radial type, should be given first consideration when the pumping requirements fall in the following ranges:

1. Volumetric flow rate from 10 gpm (0.000631 m^3/s) to 5,000 gpm (0.3155 m^3/s).
2. Head from 50 ft (15.24 m) to 3,200 ft (975.4 m).
3. Kinematic viscosity less than 100 centistokes (0.0001 m^2/s).
4. Available NPSH greater than 5 ft (1.52 m).

When one or more of these requirements is outside of the above ranges, a suitable centrifugal pump may still be available. This is particularly the case for the volumetric flow rate, where centrifugal pumps may be found for capacities up to 15,000 gpm or more. Alternatively, two or more centrifugal pumps may be placed in parallel. However, when one or more of the other three requirements cannot be met, one of the following two types of pumps should be considered for the service.

External rotary gear pumps are particularly suitable for moderate-to-very high-kinematic viscosity liquids in the range of 100 to 500,000 centistokes for flow capacities to at least 1,500 gpm (0.252 m^3/s) and heads to at least 3,000 ft (914.4 m). They are moderately priced, but usually more expensive than radial centrifugal pumps. They cannot be used with liquids containing particles and do not give as smooth a flow rate as do radial centrifugal pumps.

Reciprocating pumps of the plunger type can achieve the highest heads, up to at least 20,000 ft and flow rates to at least 500 gpm, at a maximum Hp of 200, using 3 to 5 cylinders to reduce flow pulsations. The piston type can achieve heads to at least 5,000 ft and flow rates to at least 100 gpm, at a maximum Hp of 90, using 2 to 3 cylinders. For a given horsepower, the highest heads are only achieved with the lowest flow rates and vice versa. Reciprocating pumps can handle liquids of moderate kinematic viscosity, up to 100,000 centistokes. Reciprocating pumps are more expensive than gear pumps and deliver a pulsating, rather than a smooth, steady flow, which is reduced but not eliminated when multiple cylinders are used.

Many other specialized pumps are available for services that cannot be handled by radial centrifugal, external rotary gear, and plunger and piston reciprocating pumps. Specialized pumps are not considered here. Purchase-cost data are given next only for the three most commonly used types of pumps.

Pump and Motor Purchase Costs

Centrifugal Pumps

Much purchase-cost data for centrifugal pumps, of the most common radial type, have been published. The cost most often includes the pump, a base plate, and a direct-drive coupling. In some cases, an electric motor drive is included. There is no general agreement on the equipment size factor to be used for correlating purchase costs. Most common are (1) brake horsepower and (2) the product of the capacity and the head (or pressure increase). Here, the cost correlation method used is that of the Monsanto Company in their FLOWTRAN simulation program, which was subsequently adopted by Corripio, Chrien, and Evans (1982a). Their size factor, S, which recognizes the fact that a given centrifugal pump can operate over a range of flow rate and head combinations, is

$$S = Q(H)^{0.5} \tag{16.13}$$

where Q is the flow rate through the pump in gallons per minute and H is the pump head in feet of fluid flowing (pressure rise/liquid density). The pump purchase cost is correlated with the maximum value of S that the pump can handle.

In addition to the size factor, the purchase cost of a centrifugal pump depends on its rate of rotation (usually in the range of 1,800 to 3,600 rpm), the number of impellers (usually in the range of 1 to 4) in series (called stages) to reach the desired head, the orientation of the splitting of the bolted-together pump case (HSC, horizontal split case, or VSC, vertical split case), and the material of construction. Typical ranges of flow rate and head, and maximum horsepower of the electric motor used to drive the pump, taken from Corripio et al. (1982a), are given in Table 16.20. From their cost data, indexed to mid-2000 (CE = 394), the f.o.b. purchase cost of a single-stage centrifugal pump with VSC construction of cast iron and operating at 3,600 rpm (referred to here as the base cost, C_B), is plotted in Figure 16.3. The cost includes the base plate and driver coupling but not the electric motor. The cost curve in Figure 16.3 is given by the equation:

$$C_B = \exp\left\{9.2951 - 0.6019[\ln(S)] + 0.0519\,[\ln(S)]^2\right\} \tag{16.14}$$

For other type centrifugal pumps and other materials of construction, the f.o.b. purchase cost is given by:

$$C_P = F_T F_M C_B \tag{16.15}$$

where F_M is a material factor given in Table 16.21 and F_T is a pump-type factor included in Table 16.20.

Table 16.20 Typical Types of Radial Centrifugal Pumps and F_T Factors

No. of Stages	Shaft rpm	Case-Split Orientation	Flow Rate Range (gpm)	Pump Head Range (ft)	Maximum Motor Hp	Type Factor [F_T in Eq. (16.15)]
1	3,600	VSC	50–900	50–400	75	1.00
1	1,800	VSC	50–3,500	50–200	200	1.50
1	3,600	HSC	100–1,500	100–450	150	1.70
1	1,800	HSC	250–5,000	50–500	250	2.00
2	3,600	HSC	50–1,100	300–1,100	250	2.70
2[+]	3,600	HSC	100–1,500	650–3,200	1,450	8.90

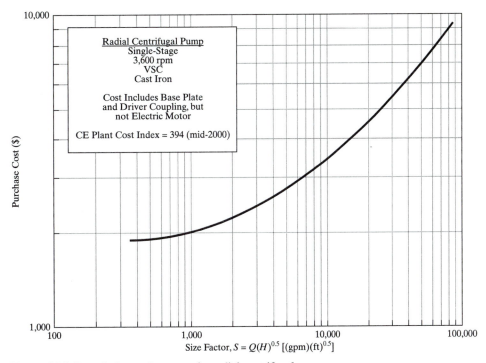

Figure 16.3 Base f.o.b. purchase cost for radial centrifugal pumps.

Electric Motors

A centrifugal pump is usually driven by an electric motor whose cost is added to the pump cost from Eq. (16.15). The size parameter for the motor is its power consumption, P_C, which is determined from the theoretical horsepower of the pump, P_T, its fractional efficiency, η_P, and the fractional efficiency of the electric motor, η_M, by the equation:

$$P_C = \frac{P_T}{\eta_P \eta_M} = \frac{P_B}{\eta_M} = \frac{QH\rho}{33,000\eta_P\eta_M} \qquad (16.16)$$

where, as previously, Q is the flow rate through the pump in gallons per minute, H is the pump head in feet of fluid flowing, and P_B is the pump brake horsepower, with ρ equal to the liquid density in pounds per gallon. Corripio et al. (1982a) give the following equations for

Table 16.21 Materials of Construction
Factors, F_M, for Centrifugal Pumps

Material of Construction	Material Factor [F_M, in Eq. (16.15)]
Cast iron	1.00
Ductile iron	1.15
Cast steel	1.35
Bronze	1.90
Stainless steel	2.00
Hastelloy C	2.95
Monel	3.30
Nickel	3.50
Titanium	9.70

estimating η_P as a function of the volumetric flow rate, and η_M as a function of pump brake horsepower:

$$\eta_P = -0.316 + 0.24015(\ln Q) - 0.01199(\ln Q)^2 \qquad (16.17)$$

for Q in the range of 50 to 5,000 gpm.

$$\eta_M = 0.80 + 0.0319(\ln P_B) - 0.00182(\ln P_B)^2 \qquad (16.18)$$

for P_B in the range of 1 to 1,500 Hp.

The f.o.b. purchase cost of an electric motor depends on its power consumption, P_C, the rotation rate of its shaft, and the type of motor enclosure. Cost data are given here for two common motor speeds (3,600 rpm and 1,800 rpm) and three common types of motor enclosures:

1. *Open, drip-proof enclosure*, which is designed to prevent the entrance of liquid and dirt particles, but not airborne moisture, dust, and corrosive fumes, into the internal working parts of the motor.
2. *Totally enclosed, fan-cooled* (TEFC) enclosure, which prevents any air from getting inside, thus protecting against moisture, dust, dirt, and corrosive vapors.
3. *Explosion-proof enclosure*, which protects the motor against explosion hazards from combustible gases, liquids, and dust, by pressurizing the enclosure with a safe gas.

From the electric motor cost correlations of Corripio et al. (1982a), indexed to mid-2000 (CE = 394), the f.o.b. purchase cost of an electric motor operating at 3,600 rpm, with an open, drip-proof enclosure (referred to here as the base cost, C_B), is plotted in Figure 16.4 as a function of the horsepower consumption, P_C. The cost curve is given by the equation:

$$C_B = \exp\{5.4866 + 0.13141[\ln(P_C)] + 0.053255\,[\ln(P_C)]^2$$
$$+\ 0.028628\,[\ln(P_C)]^3 - 0.0035549[\ln(P_C)]^4\} \qquad (16.19)$$

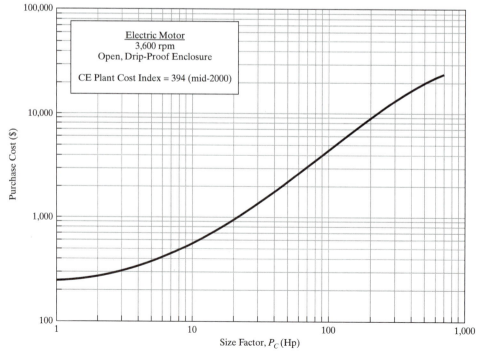

Figure 16.4 Base f.o.b. purchase cost for electric motors.

Table 16.22 F_T Factors in Eq. (16.20) and Ranges for Electric Motors

Type Motor Enclosure	3,600 rpm	1,800 rpm
Open, drip-proof enclosure		
1 to 700 Hp	1.0	0.90
Totally enclosed, fan-cooled		
1 to 250 Hp	1.4	1.3
Explosion-proof enclosure		
1 to 250 Hp	1.8	1.7

which applies over the range of 1 to 700 Hp. For other motor speeds and type enclosures, the f.o.b. purchase cost is given by:

$$C_P = F_T C_B \tag{16.20}$$

where F_T is a motor-type factor given in Table 16.22, applicable within a range of electric motor power consumption, P_C, from 1 to 1,500 Hp.

External Gear Pumps

Purchase-cost data for external gear pumps are not as widely available as for radial centrifugal pumps. The cost most often includes the gear pump, a base plate, and a driver coupling. In some cases an electric motor drive is included. There is no general agreement on what equipment size factor to use for correlating purchase costs. Most common are (1) brake horsepower and (2) flow capacity. Here, the cost correlation method used is in terms of flow capacity, Q, in gallons per minute as used by Walas (1988).

In addition to the size factor, the purchase cost of a gear pump depends on the material of construction. Although gear pumps can be designed to operate over a wide range of flow rate and discharge pressure, typical ranges are 10 to 1,500 gpm and up to 200 psia for high viscosity fluids. Typical pump efficiencies are 80% for low-kinematic viscosity liquids (20 centistokes) and 50% for high-kinematic viscosity liquids (500 centistokes). The f.o.b. purchase cost of an external gear pump of cast iron construction and for a cost index for mid-2000 (CE = 394) (referred to here as the base cost, C_B) is plotted in Figure 16.5. The cost includes the base plate and driver coupling, but not the electric motor. The cost curve in Figure 16.5 is given in terms of Q by the equation:

$$C_B = \exp\{7.2744 + 0.1986[\ln(Q)] + 0.0291[\ln(Q)]^2\} \tag{16.21}$$

which is applicable over a range of 10 to 900 gpm. For other materials of construction, the f.o.b. purchase cost is given by:

$$C_P = F_M C_B \tag{16.22}$$

where F_M is a material factor given above in Table 16.21. The power requirement for the electric motor to drive the pump depends on the head, H, and the flow rate, Q, as given by Eq. (16.16).

Reciprocating Plunger Pumps

Although piston pumps are common, the plunger type is the best choice for the most demanding applications and is available for a wider range of flow rate. Purchase-cost data for reciprocating plunger pumps are not as widely available as for radial centrifugal pumps. The cost most often includes the pump and a driver coupling for a motor or a V-belt drive. In

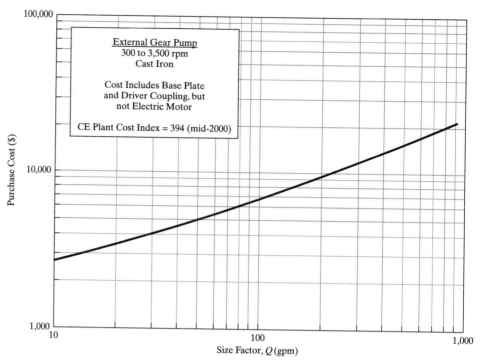

Figure 16.5 Base f.o.b. purchase cost for external gear pumps.

most cases an electric motor drive is not included. There is no general agreement on what equipment size factor to use for correlating purchase costs. Most common are (1) brake horsepower and (2) flow capacity. The cost of most models is based on the brake horsepower. By changing the plunger and cylinder diameter, a reciprocating pump of a specified horsepower can operate over a 10-fold range of flow rate and head. Here, the cost correlation method used is in terms of brake horsepower, P_B, as given by Eq. (16.16), where the pump efficiency, η_P, is typically 0.90 (90%).

In addition to the size factor, the purchase cost of a reciprocating plunger pump depends on the material of construction. The f.o.b. purchase cost of a reciprocating plunger pump of ductile iron construction and a cost index for mid-2000 (CE = 394) (referred to here as the base cost, C_B), is plotted in Figure 16.6. The cost includes a V-belt drive, but not the electric motor. The cost curve in Figure 16.6 is given by the equation:

$$C_B = \exp\left\{7.3883 + 0.26986[\ln(P_B)] + 0.06718[\ln(P_B)]^2\right\} \qquad (16.23)$$

which is applicable over the range of 1 to 200 BHp. For other materials of construction, the f.o.b. purchase cost is given by Eq. (16.15), where F_M is a material factor, as follows:

Ductile iron	$F_M = 1.00$
Ni$-$Al$-$Bronze	$F_M = 1.15$
Carbon steel	$F_M = 1.50$
Stainless steel	$F_M = 2.20$

EXAMPLE 16.5

In Chapter 3, a vinyl chloride process is synthesized, with a detailed process flow diagram shown in Figure 3.19. In that process, Reactor Pump P-100 takes stream 4 (a mixture of streams 3 and 16) of 263,800 lb/hr of 1,2-dichloroethane at 90°C and 1.5 atm and delivers it to an evaporator operating at a much higher pressure of 26 atm. Select a suitable pump and electric motor and estimate the f.o.b. purchase cost at a CE index of 400.

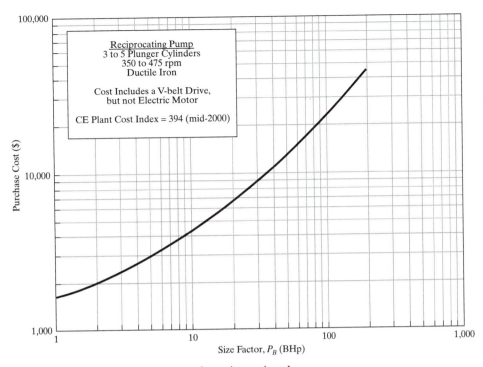

Figure 16.6 Base f.o.b. purchase cost for reciprocating plunger pumps.

SOLUTION

A process simulation program is used to obtain the density, viscosity, and vapor pressure of the feed at 90°C and 1.5 atm. The density is 9.54 lb/gal (71.4 lb/ft^3 or 1.14 g/cm^3), the viscosity is 0.37 cP, and the vapor pressure is 1.212 atm.

The feed volumetric flow rate is $263,800/[(60)(9.54)] = Q = 461$ gpm.

The pressure increase across the pump is $26 - 1.5 = 24.5$ atm.

The pump head is $(24.5)(14.696)(144)/71.4 = H = 726$ ft.

The kinematic viscosity is $0.37/1.14 = 0.33$ centistokes, which is quite low.

Choose a radial centrifugal pump.

However, it is necessary to first check the available NPSH, as discussed in Section 15.2.

$$\text{NPSH}_A = \frac{\text{Suction pressure} - \text{Vapor pressure}}{\text{Liquid density}} = \frac{(1.5 - 1.212)(14.696)(144)}{71.4} = 8.54 \text{ ft}$$

Assume we can purchase a radial centrifugal pump with a required NPSH > 5.

From Eq. (16.13), the centrifugal pump size parameter is

$$S = Q(H)^{0.5} = 461(726)^{0.5} = 12,420 \text{ (gpm)(ft)}^{0.5} \text{ and } \ln(S) = 9.427$$

From Eq. (16.14), the base pump purchase cost at a CE cost index of 394 is

$$C_B = \exp\left\{9.2951 - 0.6019[\ln(12,420)] + 0.0519[\ln(12,420)]^2\right\} = \$3,760$$

From Table 16.20, for the given flow rate and relatively high head, choose a 2-stage, 3,600 rpm, HSC centrifugal pump, with $F_T = 2.70$.

From Table 16.21, choose cast steel, with $F_M = 1.35$ because of the relatively high discharge pressure.

From Eq. (16.15) and correcting for a CE cost index of 400,

$$C_P = (2.70)(1.35)(400/394)(\$3,760) = \$13,910$$

From Eq. (16.17), the pump efficiency for $Q = 461$ gpm and $\ln(Q) = 6.13$ is

$$\eta_P = -0.316 + 0.24015(6.13) - 0.01199(6.13)^2 = 0.706$$

From Eq. (16.16), the pump brake horsepower, P_B, is

$$P_B = \frac{QH\rho}{33,000\eta_P} = \frac{(461)(726)(9.54)}{33,000(0.706)} = 137 \text{ BHp}$$

From Eq. (16.18), the motor efficiency for $\ln(P_B) = 4.92$ is

$$\eta_M = 0.80 + 0.0319(4.92) - 0.00182(4.92)^2 = 0.913$$

From Eq. (16.16), the power consumption of the motor is

$$P_C = \frac{P_B}{\eta_M} = \frac{137}{0.913} = 150 \text{ Hp}$$

From Eq. (16.19), the base cost of the motor for $\ln(P_C) = 5.01$ and a CE index of 394 is

$$C_B = \exp\{5.4866 + 0.13141[\ln(5.01)] + 0.053255[\ln(5.01)]^2$$
$$+ 0.028628[\ln(5.01)]^3 - 0.0035549[\ln(5.01)]^4\} = \$6,920$$

Because of the possible flammability hazard of 1,2-dichloroethane, specify an explosion-proof electric motor to drive the pump. From Table 16.22, for 3,600 rpm, $F_T = 1.8$ and using Eq. (16.20), but with added updating of the CE cost index to 400,

$$C_P = F_T C_B = (1.80)(\$6,920)\left(\frac{400}{394}\right) = \$12,650$$

Total cost of centrifugal pump and motor = \$13,910 + \$12,650 = \$26,560. Consideration should be given to purchasing a spare pump and motor. ∎

Fans, Blowers, and Compressors

When energy input is required to move a gas through various pipelines or ducts in a chemical processing plant, a fan, blower, or compressor is used. As discussed in Section 15.3, the power input to the gas mover increases the total head of the gas, which, ignoring a change in potential energy of the gas due to change in elevation above sea level, includes the velocity (dynamic) head and the pressure (static) head. According to Papanastasiou (1994), the gas mover is defined as: (1) a *fan*, if almost all of the energy input increases the velocity head: (2) a *blower*, if the energy input increases both the velocity head and the pressure head; and (3) a *compressor*, if almost all of the energy input increases the pressure head. However, that definition is not widely accepted. In practice, one vendor may refer to a particular gas mover as a fan, while another vendor may refer to it as a blower. The same situation applies to blowers and low-compression-ratio compressors. Here, we classify a fan as a gas mover that is generally limited to near-ambient suction pressures and pressure increases of less than 10%. Blowers can operate at any suction pressure, with compression ratios of up to 2. Thus, the main purpose of a fan is to move large quantities of gas with an increase in pressure head of up to 40 in. of H_2O head, while a blower can take a gas at 1 atm and deliver it at up to 2 atm. For larger compression ratios, a compressor is generally specified.

Fans are widely used for high-flow, low-pressure-increase applications such as heating and ventilating systems; air supply to cooling towers, low-pressure-drop dryers, and finned-tube air coolers; and removal of fumes, flue gas, and gas from a baghouse. Blowers are used for supplying combustion air to boilers and fired heaters, air to strippers, purge gas for regeneration of fixed-bed adsorbers, air to dryers with more pressure drop than a fan can handle, and for pneumatic conveying of particles. Compressors are widely used with a variety of gases and gas mixtures to increase their pressure to required levels for chemical reaction and separation.

Because a gas is compressible with a density much lower than that of a liquid, the temperature of a gas rises when compressed and this temperature rise is a limiting factor in determining the permissible compression ratio for a single-stage gas compressor. As discussed in Section 15.3, frictional dissipation in the gas mover causes a further rise in the gas temperature. Although some specialized compressors permit a gas discharge temperature of up to 600°F, the more widely used compressors are limited to discharge temperatures in the range of 375 to 400°F. For a diatomic gas or gas mixture (e.g., air) with a specific heat ratio of 1.4, a 400°F limit corresponds to a maximum single-stage compression ratio of 3.75, after taking into account an assumed compressor isentropic efficiency of 85%. This limiting compression ratio decreases to 2.50 for a monatomic gas with a higher specific heat ratio of 1.67, but increases to 6.0 for a gas with a lower specific heat ratio of 1.30 (e.g., methane). For gases with a specific heat ratio less than 1.30, even higher compression ratios may be possible, but most compressors are limited to a compression ratio of 8.0. If a higher compression ratio is required, the compression is accomplished in stages that are separated by heat exchangers that cool the gas to about 100°F before entering the next stage. Because the cooling may cause some condensation, a vapor–liquid separator, which is usually a vertical vessel containing a demister pad to coalesce small liquid droplets and is called a *knock-out drum*, must be provided after each intercooler to remove liquid from the gas prior to its entry into the next stage of compression. Because most compressors cannot tolerate any liquid in the gas, compressor inlet and outlet phase conditions should always be checked. As discussed below in the subsection "Compressors," the maximum compression ratio for a centrifugal compressor may be limited by the maximum velocity at the blade tips rather than by the exit temperature.

Fans

Most fans are of the centrifugal or axial-flow type. *Centrifugal fans* achieve the highest discharge pressures, while *axial-flow fans* provide the highest flow rates. Although forward-curved and airfoil blade designs are available for centrifugal fans, the two most popular are the *backward-curved blade* and the *straight-radial blade*. The former is the cheapest for a given flow capacity and the most efficient, but the discharge pressure decreases rapidly from its maximum value as the flow rate is increased. It is only suitable for air and clean gases. The straight-radial centrifugal fan is less efficient, but is suitable for dust-laden gases and maintains the discharge pressure, up to a compression ratio of 1.2, over a wider range of flow rate. However, at this level of pressure increase, a fan may be called a blower. Axial-flow fans come in two main types: *vane-axial* (compression ratio to 1.04) and *tube-axial* (compression ratio to 1.025). Less efficient are propeller fans. Typical operating ranges of centrifugal and axial-flow fans are given in Table 16.23.

The equipment size factor for a fan is the actual cubic feet per minute, ACFM, entering the fan. Fans are usually driven by an electric motor with either a direct drive or a belt. Base

Table 16.23 Typical Operating Ranges of Fans

Fan Type	Flow Rate (ACFM)[a]	Total Head (in. H$_2$O)
Centrifugal backward curved	1,000–100,000	1–40
Centrifugal straight radial	1,000–20,000	1–30
Vane axial	1,000–800,000	0.02–16
Tube axial	2,000–800,000	0.00–10

[a]ACFM = actual cubic feet per minute.

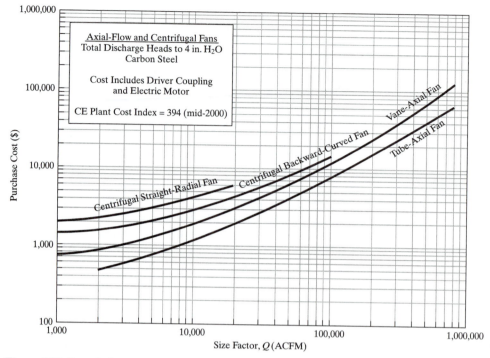

Figure 16.7 Base f.o.b. purchase costs for axial-flow and centrifugal fans.

f.o.b. purchase costs, C_B, for the four most common types of fans, averaged from several published sources listed in Table 16.19 and vendor quotes, are plotted in Figure 16.7 as a function of Q in ACFM at a cost index for mid-2000 (CE = 394). The base cost, which includes an electric motor drive, is for carbon-steel construction and total discharge heads to 4 in. H_2O. For other materials of construction and higher discharge heads, the f.o.b. purchase cost is given by

$$C_P = F_H F_M C_B \qquad (16.24)$$

where the following values of F_M apply to other materials of construction:

$$\begin{array}{ll} \text{Fiberglass} & F_M = 1.8 \\ \text{Stainless steel} & F_M = 2.5 \\ \text{Nickel alloy} & F_M = 5.0 \end{array}$$

The head factor, F_H, for total heads greater than 4 in. H_2O is given in Table 16.24.

The base cost curves in Figure 16.7 for a CE cost index of 394 are given by the following equations, with Q in actual cubic feet per minute (ACFM) of gas entering the fan.

Table 16.24 Head Factor, F_H, for Fans in Eq. (16.24)

Head (in. H_2O)	Centrifugal Backward Curved	Centrifugal Straight Radial	Vane Axial	Tube Axial
5–8	1.15	1.15	1.15	1.15
9–15	1.30	1.30	1.30	
16–30	1.45	1.45		
31–40	1.55			

Centrifugal backward-curved fan:

$$C_B = \exp\left\{10.8375 - 1.12906[\ln(Q)] + 0.08860[\ln(Q)]^2\right\} \qquad \textbf{(16.25)}$$

Centrifugal straight-radial fan:

$$C_B = \exp\left\{11.9296 - 1.31363[\ln(Q)] + 0.09974[\ln(Q)]^2\right\} \qquad \textbf{(16.26)}$$

Vane-axial fan:

$$C_B = \exp\left\{9.2847 - 0.97566[\ln(Q)] + 0.08532[\ln(Q)]^2\right\} \qquad \textbf{(16.27)}$$

Tube-axial fan:

$$C_B = \exp\left\{5.89085 - 0.40254[\ln(Q)] + 0.05787[\ln(Q)]^2\right\} \qquad \textbf{(16.28)}$$

The brake horsepower for a fan may be computed in any of three ways, depending on whether the total change in head is mostly dynamic, static, or a mixture of the two. The corresponding nominal fan efficiency, η_F, is 40% for mostly a dynamic change, 60% for mostly a static change, and 70% for a mixture of the two. The power consumption is given by the following equation, which is similar to Eq. (16.16) and where the electric motor efficiency, η_M, can be taken as 90%:

$$P_C = \frac{P_B}{\eta_M} = \frac{QH_t}{6{,}350\eta_F\eta_M} \qquad \textbf{(16.29)}$$

where Q = gas inlet flow rate, cubic feet per minute and H_t = change in total head, inches of water.

EXAMPLE 16.6

A flue gas at 200°F and 740 torr, with an average molecular weight of 31.3 is to be discharged at a rate of 12,000 standard cubic feet per minute at 60°F and 1 atm (SCFM) to a pressure of 768 torr in a duct where the velocity, V, will be 150 ft/s. Calculate the actual inlet flow rate in cubic feet per minute and the power consumption. Select a suitable fan and estimate the purchase cost (CE = 410).

SOLUTION

The actual fan inlet flow rate is $12{,}000\left(\dfrac{660}{520}\right)\left(\dfrac{740}{760}\right) = 14{,}830$ ft^3/min.

Assume the inlet velocity is zero.

The increase in dynamic head is $\dfrac{V^2}{2g_c} = \dfrac{150^2}{2(32.2)} = 349$ ft-lbf/lbm of gas.

From the ideal gas law, the average density of the gas in passing through the fan, assuming no change in temperature, is 0.0644 lb/ft^3.

The increase in pressure head $= \dfrac{\Delta P}{\rho} = \dfrac{\dfrac{(768 - 740)}{760}(14.7)(144)}{0.0644} = 1{,}211$ ft-lbf/lbm of gas. The total

change in head is $349 + 1{,}211 = 1{,}560$ ft-lbf/lbm of gas $= 19.3$ in. of H$_2$O. Because static head is predominant, let $\eta_F = 0.60$. From Eq. (16.29),

$$P_C = \frac{14{,}830(19.3)}{6{,}350(0.60)(0.90)} = 83.4 \text{ Hp}$$

Because the head is greater than 16 in. of H$_2$O and the flue gas is not likely to be clean, using Table 16.23, select a centrifugal fan with a straight-radial impeller. From Table 16.24, the head factor, F_H, is 1.45. From Eqs. (16.24) and (16.26), correcting for the cost index, the purchase cost, including the motor, is

$$C_B = 1.45\left(\frac{410}{394}\right)\exp\left\{11.9296 - 1.31363[\ln(14{,}830)] + 0.09974[\ln(14{,}830)]^2\right\} = \$7{,}510 \qquad \blacksquare$$

Blowers

Most blowers, with compression ratios up to two, are of the multistage centrifugal (often called turboblower) type or the rotary positive-displacement type. Axial-flow units can also be used, but must be multistaged. The centrifugal units are similar to centrifugal fans, with the same type of blades, but operate at higher speeds and are built to withstand higher discharge pressures. The most common rotary blower is the straight-two-lobe (Roots) blower, developed by the Roots brothers in 1854, or a modification with three straight lobes. Typical operating ranges are 100 to 50,000 ICFM for centrifugal blowers and 20 to 50,000 ICFM for rotary straight-lobe blowers with two lobes, where ICFM equals the cubic feet per minute at inlet conditions. Typical mechanical efficiencies, η_B, are 70–80% for centrifugal blowers and 50–70% for straight-lobe blowers. However, with straight-lobe blowers, the higher efficiency is only achieved for compression ratios from 1.2 to 1.3; from 1.3 to 2.0, the efficiency falls off rapidly. The centrifugal blower delivers a smooth flow rate, but as discussed in Section 15.2, the straight-lobe units deliver a somewhat pulsing flow. As with pumps, rotary blowers deliver a fixed volumetric flow rate with varying inlet and outlet pressures, while the volumetric throughput of centrifugal blowers varies with changes in inlet or discharge pressures. Both types of blowers have found a wide range of applications, but centrifugal blowers are more common in chemical processing plants and are widely used to supply air to strippers, dryers, and combustion devices. Rotary blowers are useful for pneumatic conveying and are well suited to other applications when a fixed volumetric flow rate is essential.

The equipment size factor for a blower is the brake horsepower, P_B, which is computed from the inlet volumetric flow rate, Q_I, in cubic feet per minute and pressures in lbf/in² at the inlet, P_I, and outlet, P_O, by the following equation, which assumes the ideal gas law and a constant specific heat ratio, k:

$$P_B = 0.00436\left(\frac{k}{k-1}\right)\frac{Q_I P_I}{\eta_B}\left[\left(\frac{P_O}{P_I}\right)^{\frac{k-1}{k}} - 1\right] \tag{16.30}$$

Blowers are usually driven by an electric motor with a direct drive for the centrifugal type and a belt or chain drive for the rotary type. Except for very large units, centrifugal blowers must be staged to achieve pressures greater than 40 in. of H_2O gauge. Base f.o.b. purchase costs, C_B, for the two major types of blowers, based on data from Garrett (1989) and recent data from vendors, are plotted in Figure 16.8 as a function of P_C in horsepower for a cost index in mid-2000 (CE = 394). The base cost, which includes an electric motor drive, is for a cast iron construction housing, with compression ratios up to 2. The centrifugal blower uses sheet metal blades. For other materials of construction, the f.o.b. purchase cost is given by:

$$C_P = F_M C_B \tag{16.31}$$

where the metal material factors given above for fans can be used. In addition, centrifugal blowers are available with cast aluminum blades with $F_M = 0.60$.

The base blower purchase cost curves in Figure 16.8, for a CE cost index of 394, are given by the following equations, with P_C in horsepower:

Centrifugal (turbo) blower:

$$C_B = \exp\{6.6547 + 0.7900[\ln(P_C)]\} \tag{16.32}$$

Rotary straight-lobe blower:

$$C_B = \exp\{7.35356 + 0.79320[\ln(P_C)] - 0.012900[\ln(P_C)]^2\} \tag{16.33}$$

EXAMPLE 16.7

Air, available at 70°F and 14.5 psia, is to be supplied by a blower at 6,400 ACFM to a column with 15 plates to strip VOCs (volatile organic compounds) from 1,000 gpm of wastewater. The column is 6 ft in diameter. The pressure drop through the inlet line, the column, and the outlet line has been estimated

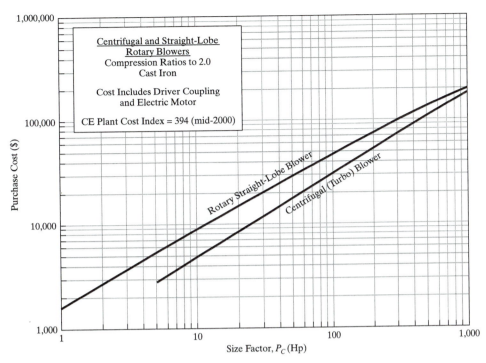

Figure 16.8 Base f.o.b. purchase costs for centrifugal and straight-lobe blowers.

to be 3 psi. The gas exiting from the column is to be sent to the next unit at a pressure of 18 psia. Select and size a blower, calculate the required power consumption, and estimate the f.o.b. purchase cost for a CE cost index of 400.

SOLUTION

The total pressure increase required across the blower is $18.0 - 14.5 + 3.0 = 6.5$ psi. At the blower inlet, the pressure is 14.5 psia, giving an air density of 0.074 lbm/ft³. This gives an inlet pressure head of $(14.5)(144)/0.074 = 28,200$ ft-lbf/lbm. At the blower exit, the pressure is $18.0 + 3.0 = 21.0$ psia, giving an air density of approximately 0.095 lbm/ft³ and an outlet pressure head of $(21)(144)/0.095 = 31,800$ ft-lbf/lbm. This gives a change in pressure head of $31,800 - 28,200 = 3,600$ ft-lbf/lbm. For the increase in kinetic energy head, assume the blower inlet velocity is zero and the blower discharge air velocity in the exit line is 75 ft/s. The increase in kinetic energy is $(75)^2/[2(32.2)] = 87.3$ ft-lbf/lbm. In this example, the change in kinetic-energy head is only about 2.5% of the total increase in head. Neglecting the increase in kinetic-energy head, the blower brake horsepower from Eq. (16.30), using $k = 1.4$ and $\eta_B = 0.75$ for a centrifugal blower, is:

$$P_B = 0.00436 \left(\frac{1.4}{1.4 - 1} \right) \frac{6,400(14.5)}{0.75} \left[\left(\frac{21}{14.5} \right)^{\frac{1.4-1}{1.4}} - 1 \right] = 211 \text{ BHp}$$

Using Eq. (16.18), the motor efficiency $= \eta_M = 0.92$. From Eq. (16.16), the consumed power for a centrifugal blower is $211/0.92 = 229$ Hp. A straight-lobe blower with $\eta_B = 0.65$ requires a consumed power of 265 Hp. For this application, either the centrifugal blower or straight-lobe blower is suitable, but with a compression ratio of $21/14.5 = 1.45$, the centrifugal blower is more efficient and is more widely used for this application. Using Eq. (16.32), with a CE cost index of 400, the estimated f.o.b. purchase cost of the centrifugal blower of iron and steel construction, which would be a multistage unit, including an electric motor with a direct drive is

$$C_B = \frac{400}{394} \exp \left[6.6547 + 0.7900[\ln(229)] \right] = \$57,700$$

A centrifugal blower with aluminum blades, with $F_M = 0.60$, is also suitable and reduces the f.o.b. purchase cost to $0.60(\$57,700) = \$34,600$. ∎

Compressors

Compressors are used widely to move gases for compression ratios greater than 2. As discussed in Section 15.3, the major types are the trunk-piston and crosshead reciprocating compressors, diaphragm compressor, centrifugal compressor, axial compressor, and the screw, sliding-vane, and liquid-ring (piston) rotary compressors. Of these, the most commonly used in chemical processing plants are the: (1) double-acting crosshead reciprocating compressor, (2) multistage centrifugal compressor, and (3) rotary twin-screw compressor. These are referred to here as simply reciprocating, centrifugal, and screw compressors.

Reciprocating compressors can handle the widest range of pressure, from vacuum to 100,000 psig, but the narrowest range of flow rates, from 5 to 7,000 ACFM, with horsepowers up to 20,000 per machine. By using many stages, centrifugal compressors can deliver pressures up to 5,000 psig for the largest flow rates, from 1,000 to 150,000 ACFM, with horsepowers to 2,000 per machine. Screw compressors have the smallest pressure range, up to 400 psig, for flow rates from 800 to 20,000 ACFM, with horsepowers to 6,000 per machine.

Because reciprocating and screw compressors are of the positive-displacement type, they are designed for a particular flow rate, with their discharge pressure set by the downstream system provided that the power input to the compressor is sufficient. The maximum compression ratio per stage is set by a limiting temperature rise of the gas being compressed, as discussed at the beginning of this subsection. Compared to the screw compressor, the reciprocating compressor is more efficient (80–90% compared to 75–85%), more expensive, larger in size, somewhat more flexible in operation, accompanied in operation by large shaking forces that require a large foundation and more maintenance, are less noisy, and do not deliver as smooth a flow rate. Reciprocating compressors cannot tolerate the presence of liquid or solid particles in the feed gas, and consequently, must be protected by a knock-out drum.

The centrifugal compressor has become exceedingly popular in the last few decades because it is easily controlled, delivers a smooth flow rate (which, however, is dependent on the required discharge pressure), has small foundations and low maintenance, and can handle large flow rates and fairly high pressures. However, it is less efficient (70–75%) and more expensive than a screw compressor for the same application. The velocity of the blade tips sets the maximum compression ratio per machine stage. This limitation almost always translates into multiple stages in a single machine for compression ratios greater than 2, even though a limiting temperature is not achieved. Single machines may have as many as 10 stages. Further compression after reaching the temperature limit requires an intercooler followed by another machine.

Process simulation programs are preferred to compute the theoretical and brake horsepower requirements, as well as the exit temperature, of a compressor because the ideal gas law is not usually applicable for pressures above two atmospheres. However, Eq. (16.30) can be used to obtain a preliminary estimate of the brake horsepower. An estimate of the exit temperature, including the effect of compressor efficiency, η_c, can be made with the following modification of the equation for the isentropic exit temperature:

$$T_O = T_I + \frac{T_I\left[\left(\frac{P_O}{P_I}\right)^{\frac{k-1}{k}} - 1\right]}{\eta_C} \qquad \textbf{(16.34)}$$

Compressors may be driven by electric motors, steam turbines, or gas turbines, but the former is the most common driver and the gas turbine is the least common. All drivers are available up to at least 20,000 Hp. For applications below about 200 Hp, electric motors are used almost exclusively. Most efficient is the electric motor; least efficient is the gas turbine. Efficiencies of all three drivers increase with horsepower. At 1,000 Hp, typical efficiencies are

95%, 65%, and 35%, respectively, for the electric motor, steam turbine, and gas turbine. Therefore, unless excess steam or low-cost combustion gas is available, the electric motor is the driver of choice over the entire horsepower range. An exception is sometimes made for centrifugal compressors, where the steam turbine is an ideal driver because the speeds of the two devices can be matched. See Section 16.6 and Table 16.32 for cost equations for steam and gas turbines.

Base f.o.b. purchase costs, C_B, for the three major types of compressors, based on data from Garrett (1989) and Walas (1988) in Table 16.19, are plotted in Figure 16.9 as a function of consumed power, P_C, in horsepower for a cost index in mid-2000 (CE = 394). The base cost, which includes an electric motor drive, is for cast iron or carbon-steel construction. For other drives and materials of construction, the f.o.b. purchase cost is given by:

$$C_P = F_D F_M C_B \qquad (16.35)$$

where, in place of an electric motor drive, $F_D = 1.15$ for a steam turbine drive and 1.25 for a gas turbine drive, and $F_M = 2.5$ for stainless steel and 5.0 for nickel alloy.

The base compressor purchase cost curves in Figure 16.9, for a CE cost index of 394, are given by the following equations, with P_C in horsepower:

Centrifugal Compressor:

$$C_B = \exp\{7.2223 + 0.80[\ln(P_C)]\} \qquad (16.36)$$

Reciprocating Compressor:

$$C_B = \exp\{7.6084 + 0.80[\ln(P_C)]\} \qquad (16.37)$$

Screw Compressor:

$$C_B = \exp\{7.7661 + 0.7243[\ln(P_C)]\} \qquad (16.38)$$

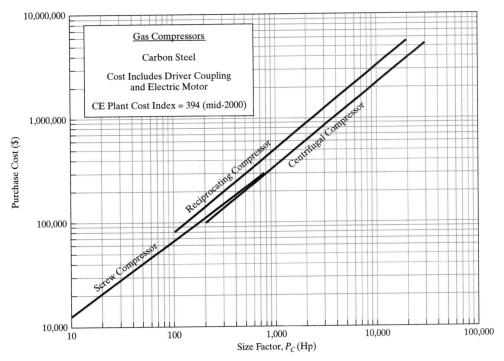

Figure 16.9 Base f.o.b. purchase costs for centrifugal, reciprocating, and screw compressors.

EXAMPLE 16.8

In an ammonia process, where hydrogen and nitrogen are combined at high temperature and high pressure in a catalytic reactor, a multistage, gas compression system with intercoolers is needed to compress the feed gas to the reactor pressure. For one of the compression stages, the feed gas, at 320 K, 30 bar, and 6,815 kmol/hr, has a composition in mol% of 72.21 H_2, 27.13 N_2, 0.61 CH_4, and 0.05 Ar. It is to be compressed to 70 bar in an uncooled, adiabatic compressor. Size and select the compressor and the drive and estimate the f.o.b. purchase cost for a CE index of 400.

SOLUTION

At the high-pressure conditions, the ideal gas law does not apply and, therefore, it is preferred to size the compressor with a process simulation program. Using the SRK equation of state, the results give an inlet volumetric flow rate of 6,120 m^3/hr or 3,602 cfm. This is in the range of all three compressor types. However, the discharge pressure of 70 bar or 1,016 psi is beyond the range of the screw compressor. Select a centrifugal compressor of steel construction, with an assumed isentropic efficiency of 75%. This gives a discharge temperature of 439 K or 331°F, which is below the suggested 400°F discharge limit for a compressor. The corresponding theoretical kilowatts is 4,940 or 6,630 THp (theoretical horsepower). With a 75% compressor efficiency, the brake horsepower is 8,840. Assume a 95% efficiency for the electric motor. This gives P_C = 9,300 Hp. From Eq. (16.36), the f.o.b. purchase cost for CE = 400 is

$$C_B = \left(\frac{400}{394}\right) \exp\left[7.2223 + 0.80[\ln(9,300)]\right] = \$2,079,000$$

An alternative that might be considered is the use of two identical centrifugal compressors, each delivering 50% of the required flow, in place of the single compressor.   ∎

Heat Exchangers

As discussed in Chapter 13, a wide variety of heat exchangers is available for heating, cooling, condensing, and vaporizing process streams, particularly liquids and gases. The most important types are shell-and-tube, double-pipe, air-cooled fin-fan, and compact heat exchangers, including plate-and-frame, spiral-plate, spiral-tube, and plate-fin types. For most applications in chemical processing plants, shell-and-tube heat exchangers, which are governed by TEMA (Tubular Exchanger Manufacturers Association) standards and ASME (American Society of Mechanical Engineers) pressure-vessel code, as well as other standards are selected. However, for heat-exchanger areas less than 200 ft^2, double-pipe heat exchangers are often selected, and when streams are cooled by air, fin-fan units are common. Compact heat exchangers are usually reserved for non-demanding applications. In this section, graphs and equations are presented only for shell-and-tube and double-pipe heat exchangers. Equations for air-cooled fin-fan and compact heat exchangers that are described in Section 13.2 are presented below in Section 16.6 and Table 16 32.

Shell-and-Tube Heat Exchangers

These exchangers cover a wide range of geometrical variables, including tube diameter, wall thickness, length, spacing, and arrangement; baffle type and spacing; numbers of tube and shell passes; and fixed-head, floating-head, U-tube, and kettle designs. However, most published purchase-cost data are correlated in terms of heat-exchange surface area (usually based on the outside surface area of the tubes) for a base-case design, with correction factors only for pressure and materials for the shell and tubes. In some cases, corrections for tube length are given. Here, the following cost correlations are based on several of the references in Table 16.19. The base cost curves in Figure 16.10 for a CE cost index of 394 are given by the following equations, with tube outside surface area, A, in square feet, ranging from 150 to 12,000 ft^2. These base-case exchangers include $\frac{3}{4}$-in. or 1-in. O.D., 16 BWG (Birmingham Wire Gage) carbon-steel tubes, 20 ft long, on square or triangular pitch in a carbon-steel shell for use with shell-side pressures up to 100 psig.

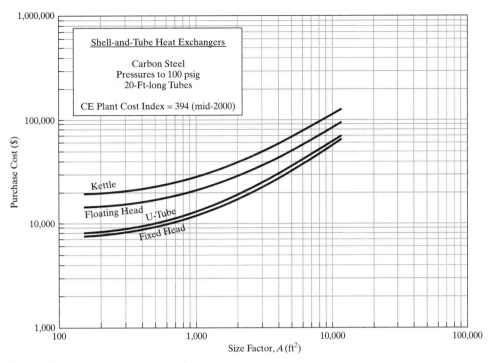

Figure 16.10 Base f.o.b. purchase costs for shell-and-tube heat exchangers.

Floating head:

$$C_B = \exp\left\{11.667 - 0.8709[\ln(A)] + 0.09005[\ln(A)]^2\right\} \tag{16.39}$$

Fixed head:

$$C_B = \exp\left\{11.0545 - 0.9228[\ln(A)] + 0.09861[\ln(A)]^2\right\} \tag{16.40}$$

U-tube:

$$C_B = \exp\left\{11.147 - 0.9186[\ln(A)] + 0.09790[\ln(A)]^2\right\} \tag{16.41}$$

Kettle vaporizer:

$$C_B = \exp\left\{11.967 - 0.8709[\ln(A)] + 0.09005[\ln(A)]^2\right\} \tag{16.42}$$

The f.o.b. purchase cost for each of these four types of heat exchangers is determined from

$$C_P = F_P F_M F_L C_B \tag{16.43}$$

where F_M is a material factor for various combinations of tube and shell materials, as given in Table 16.25 as a function of the surface area, A, in square feet according to the equation:

$$F_M = a + \left(\frac{A}{100}\right)^b \tag{16.44}$$

The factor F_L is a tube-length correction as follows:

Tube length (ft)	F_L
8	1.25
12	1.12
16	1.05
20	1.00

Table 16.25 Materials of Construction Factors, F_M, for Shell-and-Tube Heat Exchangers

Materials of Construction Shell/Tube	a in Eq. (16.44)	b in Eq. (16.44)
Carbon steel/carbon steel	0.00	0.00
Carbon steel/brass	1.08	0.05
Carbon steel/stainless steel	1.75	0.13
Carbon steel/Monel	2.1	0.13
Carbon steel/titanium	5.2	0.16
Carbon steel/Cr–Mo steel	1.55	0.05
Cr–Mo steel/Cr–Mo steel	1.70	0.07
Stainless steel/stainless steel	2.70	0.07
Monel/Monel	3.3	0.08
Titanium/titanium	9.6	0.06

The pressure factor, F_P, is based on the shell-side pressure, P, in psig and is given by the following equation, which is applicable from 100 to 2,000 psig:

$$F_P = 0.9803 + 0.018\left(\frac{P}{100}\right) + 0.0017\left(\frac{P}{100}\right)^2 \qquad \textbf{(16.45)}$$

Double-Pipe Heat Exchangers

For heat-exchange surface areas of less than 200 ft^2 and as low as 2 ft^2, double-pipe heat exchangers are often selected over shell-and-tube heat exchangers. The area, A, is usually based on the outside surface area of the inner pipe. The cost correlation here is based on the average of several of the references in Table 16.19. The base cost curve in Figure 16.11 for a

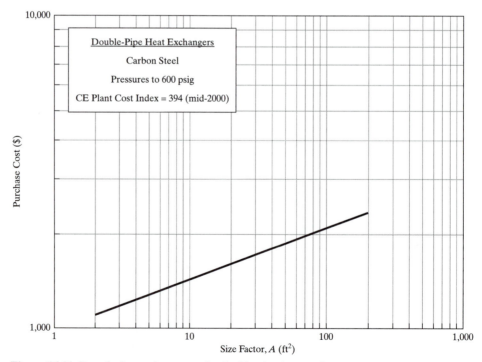

Figure 16.11 Base f.o.b. purchase costs for double-pipe heat exchangers.

CE cost index of 394 is for carbon-steel construction for pressures to 600 psig, with the area in square feet. The correlating equation is

$$C_B = \exp\{7.1248 + 0.16[\ln(A)]\} \tag{16.46}$$

The f.o.b. purchase cost is determined from

$$C_P = F_P F_M C_B \tag{16.47}$$

where the material factor, F_M, is 2.0 for an outer pipe of carbon steel and an inner pipe of stainless steel. If both pipes are stainless steel, the factor is 3.0. The pressure factor, F_P, for the range of pressure, P, from 600 to 3,000 psig is given by:

$$F_P = 0.8510 + 0.1292\left(\frac{P}{600}\right) + 0.0198\left(\frac{P}{600}\right)^2 \tag{16.48}$$

EXAMPLE 16.9

In Section 4.3, a toluene hydrodealkylation process is synthesized. Material and energy balance calculations on that process give a combined feed to the hydrodealkylation reactor of 5,802 lbmol/hr, containing mainly 35 vol% hydrogen, 58 vol% methane, and 7 vol% toluene, at 127.6°F and 569 psia. This stream is heated to 1,000°F in a heat exchanger by 6,010 lbmol/hr of quenched reactor effluent (which also contains a significant percentage of hydrogen), entering the exchanger at 1,150°F and 494 psia, and exiting at 364.2°F and 489 psia. The calculated heat duty, Q, is 69,360,000 Btu/hr. Estimate the area of the heat exchanger and the f.o.b. purchase cost at a CE index of 394.

SOLUTION

The combined feed enters the heat exchanger with a molar vapor fraction of 93.1% and leaves as a superheated vapor. The quenched effluent enters and exits as a superheated vapor. Thus, some vaporization occurs on the combined feed side. A zone analysis for a countercurrent heat exchanger gives a mean temperature driving force, ΔT_M, of 190.4°F, compared to end driving forces of 150°F and 236.6°F. Assuming an overall heat-transfer coefficient, U, of 50 Btu/hr-ft^2-°F, the heat exchanger area is

$$A = \frac{Q}{U(\Delta T_M)} = \frac{69,360,000}{50(190.4)} = 7,290 \text{ ft}^2$$

For these size and temperature-difference conditions, select a floating-head shell-and-tube heat exchanger with 20-ft-long tubes. Pressures on both the shell and tube sides are in the range of 500 to 600 psig. Select a design pressure of 700 psig. Because temperatures are as high as 1,000 to 1,150°F, carbon steel cannot be used as the material of construction for either the shell or tubes. Because the hydrogen content of both streams is significant, Cr–Mo alloy steel, which is often used in the temperature range of this exchanger, is not suitable either, and stainless steel must be selected. From Eq (16.39), the base purchase cost at a CE index of 394 is

$$C_B = \exp\{11.667 - 0.8709[\ln(7,290)] + 0.09005[\ln(7,290)]^2\} = \$62,600$$

From Eq. (16.44) and Table 16.25, for stainless-steel construction, $F_M = 2.70 + (7,290/100)^{0.07} = 4.05$. For a pressure of 700 psig, using Eq. (16.45):

$$F_P = 0.9803 + 0.018\left(\frac{700}{100}\right) + 0.0017\left(\frac{700}{100}\right)^2 = 1.19$$

From Eq. (16.43), the f.o.b. purchase cost is

$$C_P = 1.19(4.05)(1.0)(62,600) = \$301,700 \qquad \blacksquare$$

Fired Heaters

Indirect-fired heaters of the box type, also called fired heaters, process heaters, and furnaces, are commonly used to heat and/or vaporize non-reacting process streams at elevated temperatures beyond where steam is normally employed. The fuel for combustion is either gas or

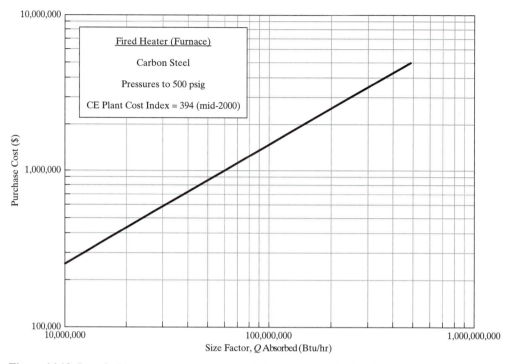

Figure 16.12 Base f.o.b. purchase costs for indirect-fired heaters of the box type.

fuel oil. As discussed in Section 13.2, heat duties of fired heaters are in the range of 10 to 340 million Btu/hr (3,000 to 100,000 kJ/s or 3 to 100 MW). Typically, fired heaters are complete package units with standard horizontal tubes of carbon steel, adequate for temperatures to 1,100°F and pressures to 500 psig. For higher temperatures and/or pressures, other materials of construction may be needed and tubes must have an increased wall thickness. Thermal efficiencies range for 70 to 90%, with the higher value corresponding to units designed for energy conservation. The base cost depends on the heat duty, Q, absorbed by the process stream in Btu/hr. The cost correlation, shown in Figure 16.12, is based on the average of several of the references in Table 16.19. For CE = 394, the base cost is

$$C_B = \exp\{0.08505 + 0.766[\ln(Q)]\} \tag{16.49}$$

The f.o.b. purchase cost is determined from:

$$C_P = F_P F_M C_B \tag{16.50}$$

where the material factor, F_M, is 1.4 for tubes of Cr–Mo alloy steel and 1.7 for stainless steel. The pressure factor, F_P, for the range of pressure, P, from 500 to 3,000 psig is given by:

$$F_P = 0.986 - 0.0035\left(\frac{P}{500}\right) + 0.0175\left(\frac{P}{500}\right)^2 \tag{16.51}$$

Fired heaters for specific purposes are discussed below in Section 16.6.

EXAMPLE 16.10

After being heated to 1,000°F and before entering the reactor, the combined feed in Example 16.9 is heated further to 1,200°F in a fired heater. Determine the f.o.b. purchase cost of a fired heater at a CE index of 394.

SOLUTION

The calculated absorbed heat duty, Q, is 18,390,000 Btu/hr. Assume a design pressure for the tubes of 700 psig. Because of the significant hydrogen concentration in the combined feed, stainless steel tubes are required. From Eq. (16.49), the base cost is

$$C_B = \exp\{0.08505 + 0.766[\ln(18,390,000)]\} = \$399,600$$

From stainless steel tubes, $F_M = 1.7$. For a pressure of 700 psig, using Eq. (16.51):

$$F_P = 0.986 - 0.0035\left(\frac{700}{500}\right) + 0.0175\left(\frac{700}{500}\right)^2 = 1.015$$

From Eq. (16.50), the f.o.b. purchase cost for a CE index = 394 is

$$C_P = 1.015(1.7)(399,600) = \$689,500 \qquad \blacksquare$$

Pressure Vessels and Towers for Distillation, Absorption, and Stripping

Pressure vessels containing little or no internals (largely empty) are widely used in chemical processing plants. Applications include reflux drums, flash drums, knock-out drums, settlers, chemical reactors, mixing vessels, vessels for fixed-bed adsorption, and storage drums. These vessels are usually cylindrical in shape, with an inside diameter, D_i, and consist of a cylindrical shell of length, L (often referred to as the *tangent-to-tangent length*), to which are welded two ellipsoidal or torispherical (dished) heads at opposite ends. In addition, the vessel includes nozzles for entering and exiting streams, manholes for internal access, connections for relief valves and instruments, skirts or saddles for support depending on whether the vessel is oriented horizontally or vertically, and platforms and ladders. Shell and head thicknesses are usually determined from the ASME Boiler and Pressure Vessel Code and may include allowance for corrosion, vacuum operation, wind loading, and earthquake.

Because many factors can affect the purchase cost of a pressure vessel, it is not surprising that a wide selection of size factors has been used to estimate the purchase cost; however, all methods differentiate between vertical and horizontal orientation of the vessel. The simplest methods base the cost on the inside diameter and tangent-to-tangent length of the shell, with a correction for design pressure. The most elaborate method is based on a complete design of the pressure vessel to obtain the vessel weight, and a sizing and count of the nozzles and manholes. Here, the method of Mulet, Corripio, and Evans (1981a,b) is employed, which is a method of intermediate complexity, based on the weight of the shell and two 2:1 elliptical heads. The f.o.b. purchase cost, which is for carbon-steel construction and includes an allowance for platforms, ladders, and a nominal number of nozzles and manholes, is given by

$$C_P = F_M C_V + C_{\text{PL}} \qquad (16.52)$$

The f.o.b. purchase cost, at a CE index = 394, of the empty vessel, C_V, but including nozzles, manholes, and supports, based on the weight in pounds of the shell and the two heads, W, depends on orientation, as shown in Figure 16.13. The correlating equations are

Horizontal Vessels for $1,000 < W < 920,000$ lb:

$$C_V = \exp\{8.717 - 0.2330[\ln(W)] + 0.04333[\ln(W)]^2\} \qquad (16.53)$$

Vertical Vessels for $4,200 < W < 1,000,000$ lb:

$$C_V = \exp\{6.775 + 0.18255[\ln(W)] + 0.02297[\ln(W)]^2\} \qquad (16.54)$$

The added cost, C_{PL}, for platforms and ladders depends on the vessel inside diameter, D_i, in feet and, for a vertical vessel, on the tangent-to-tangent length of the shell, L, in feet, and is given by:

Horizontal Vessels for $3 < D_i < 12$ ft:

$$C_{\text{PL}} = 1,580 \, (D_i)^{0.20294} \qquad (16.55)$$

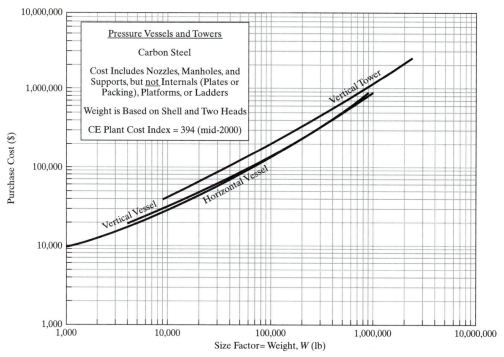

Figure 16.13 Base f.o.b. purchase costs for pressure vessels and towers.

Vertical Vessels for $3 < D_i < 21$ ft and $12 < L < 40$ ft:

$$C_{PL} = 285.1 \, (D_i)^{0.73960} \, (L)^{0.70684} \tag{16.56}$$

In Figure 16.13, it is seen that for a given shell weight, vertical vessels cost more than horizontal vessels up to 500,000 lb.

Towers are vertical pressure vessels for various separation operations, including distillation, absorption, and stripping. They contain plates and/or packing plus additional nozzles and manholes, and internals for multiple feed entries and management of bottoms liquid and its withdrawal. Figure 16.13 includes a curve for the f.o.b. purchase cost in U.S. dollars, as of a CE index = 394, of vertical towers, C_T, including nozzles, manholes, a skirt, and internals (but not plates and/or packing), based on the weight in pounds of the shell and the two heads, W. The correlating equation is

Towers for $9,000 < W < 2,500,000$ lb

$$C_V = \exp\{7.0374 + 0.18255[\ln(W)] + 0.02297[\ln(W)]^2\} \tag{16.57}$$

The added cost, C_{PL}, for platforms and ladders for towers depends on the tower inside diameter, D_i, in feet and on the tangent-to-tangent length of the shell, L, in feet, and is given by:

Towers for $3 < D_i < 24$ ft and $27 < L < 170$ ft:

$$C_{PL} = 237.1 \, (D_i)^{0.63316} \, (L)^{0.80161} \tag{16.58}$$

The weight, W, in the cost correlations for a pressure vessel or tower depends on the wall thicknesses of the shell and the two heads. Although the thickness of the heads may be required to be somewhat thicker than the shell, particularly at a high pressures, it is sufficient for cost estimation purposes to assume head thicknesses equal to the shell thickness, t_S. Then, with 2:1 elliptical heads, the weight of the shell and the two heads is approximately:

$$W = \pi(D_i + t_S)(L + 0.8D_i)t_S\rho \tag{16.59}$$

where the term in L accounts for the cylinder, the term in 0.8 D_i accounts for the two heads, and ρ is the density of the carbon steel, which can be taken as 490 lb/ft^3 or 0.284 lb/in^3.

In the absence of corrosion, wind, and earthquake considerations and for internal pressures greater than the external pressure (i.e., excluding vacuum operation), the cylindrical shell wall thickness is computed from the ASME pressure-vessel code formula:

$$t_P = \frac{P_d D_i}{2SE - 1.2P_d} \tag{16.60}$$

where t_P = wall thickness in inches to withstand the internal pressure, P_d = internal design gauge pressure in psig, D_i = inside shell diameter in inches, S = maximum allowable stress of the shell material at the design temperature in pounds per square inch, and E = fractional weld efficiency.

Sandler and Luckiewicz (1987) recommend that the design pressure, P_d in psig, be greater than the operating pressure, P_o. The following recommendations are similar to theirs. For operating pressures between 0 and 5 psig, use a design pressure of 10 psig. In the range of operating pressure from 10 psig to 1,000 psig, use the following equation:

$$P_d = \exp\{0.60608 + 0.91615[\ln(P_o)] + 0.0015655[\ln(P_o)]^2\} \tag{16.61}$$

For operating pressures greater than 1,000 psig, use a design pressure equal to 1.1 times the operating pressure. However, safety considerations may dictate an even larger difference between design pressure and operating pressure, especially when runaway reactions are possible.

The maximum allowable stress, S, in Eq. (16.60), depends on the design temperature and the material of construction. The design temperature may be taken as the operating temperature plus 50°F. However, again, safety considerations may dictate an even larger difference. At a given temperature, different steel compositions have different values for the maximum allowable stress. In the design temperature range of −20°F to 650°F, in a non-corrosive environment that is free of hydrogen, a commonly used carbon steel is SA-285, grade C with a maximum allowable stress of 13,750 psi. In the temperature range from 650°F to 900°F, in a non-corrosive environment including the presence of hydrogen, a commonly used low-alloy (1% Cr and 0.5% Mo) steel is SA-387B. Its maximum allowable stress in its recommended temperature range is as follows:

Temperature (°F)	Maximum Allowable Stress (psi)
−20 to 650	15,000
700	15,000
750	15,000
800	14,750
850	14,200
900	13,100

The weld efficiency, E, in Eq. (16.60), accounts mainly for the integrity of the weld for the longitudinal seam. For carbon steel up to 1.25 in. in thickness, only a 10% spot X-ray check of the weld is necessary and a value of 0.85 for E should be used. For larger wall thicknesses, a 100% X-ray check is required, giving a value of 1.0 for E.

At low pressures, wall thicknesses calculated from Eq. (16.60) may be too small to give sufficient rigidity to vessels. Accordingly, the following minimum wall thicknesses should be used:

Vessel Inside Diameter (ft)	Minimum Wall Thickness (in.)
Up to 4	1/4
4–6	5/16
6–8	3/8
8–10	7/16
10–12	1/2

Equation (16.60) does not account for wind and earthquake, and is not applicable to vessels under vacuum. Mulet et al. (1981a) present a method for determining average wall thickness, t_V, of a vertical vessel or tower to withstand, not only the internal design pressure, but also a substantial wind load based on a wind velocity of 140 miles/hr, which is assumed to be sufficient to handle additionally an earthquake. A simple, but somewhat less accurate implementation of that method, is given by the following equation, which depends on the ratio of the tangent-to-tangent vessel height, L, and the inside diameter of the vertical vessel, D_i, as well as the internal design pressure, P_d, in psig. The equation applies for the range of $10 > (L/D_i)^2/P_d > 1.34$. Below a value of 1.34, there is no effect of wind load and, therefore, $t_V = t_P$ from Eq. (16.60).

$$t_V = t_P \left[0.75 + 0.22E \frac{(L/D_i)^2}{P_d} \right] \tag{16.62}$$

Equation (16.60) does not apply to vacuum vessels for which the internal pressure is less than the external pressure. Such vessels must be sufficiently thick to withstand a collapsing pressure or they must be provided with stiffening rings around their outer periphery. For the former alternative, Mulet et. al. (1981a) present a method for computing the necessary wall thickness, t_E, based mainly on the vessel length-to-outside diameter ratio and the modulus of elasticity, E_M, of the metal wall. A simple approximation of their method, which is applicable to $t_E/D_o < 0.05$, is given by the following equation, where D_o = outside diameter:

$$t_E = 1.3D_o \left(\frac{P_d L}{E_M D_o} \right)^{0.4} \tag{16.63}$$

However, to the value of t_E, the following correction, t_{EC}, must be added:

$$t_{EC} = L(0.18D_i - 2.2) \times 10^{-5} - 0.19 \tag{16.64}$$

where all variables are in inches. The total thickness for a vacuum vessel is

$$t_V = t_E + t_{EC} \tag{16.65}$$

The modulus of elasticity, E_M, depends on temperature, with the following values for carbon steel and low-alloy steel:

E_M, Modulus of elasticity, psi (multiply values by 10^6)

Temperature (°F)	Carbon Steel	Low-alloy Steel
−20	30.2	30.2
200	29.5	29.5
400	28.3	28.6
650	26.0	27.0
700	—	26.6
800	—	25.7
900	—	24.5

Even for noncorrosive conditions, a corrosion allowance, t_C, of $\frac{1}{8}$ in. should be added to t_V to give the value of t_S to be used in Eq. (16.59) for vessel weight. In addition, it is important

Table 16.26 Materials-of-Construction
Factors, F_M, for Pressure Vessels

Material of Construction	Material Factor [F_M, in Eq. (16.52)]
Carbon steel	1.0
Low-alloy steel	1.2
Stainless steel 304	1.7
Stainless steel 316	2.1
Carpenter 20CB-3	3.2
Nickel-200	5.4
Monel-400	3.6
Inconel-600	3.9
Incoloy-825	3.7
Titanium	7.7

to note that vessels are generally fabricated from metal plate, whose thicknesses can be assumed to come in the following increments:

$\frac{1}{16}$-in. increments for $\frac{3}{16}$ to $\frac{1}{2}$ in. inclusive

$\frac{1}{8}$-in. increments for $\frac{5}{8}$ to 2 in. inclusive

$\frac{1}{4}$-in. increments for $2\frac{1}{4}$ to 3 in. inclusive

The final vessel thickness is obtained by rounding up to the next increment.

The material-of-construction factor for pressure vessels and towers, F_M, is given in Table 16.26.

Before presenting cost data for plates and packings in distillation, absorption, and stripping towers, the following example is presented to estimate the purchase cost of an adiabatic, homogeneous, gas-phase reactor.

EXAMPLE 16.11

An adiabatic reactor consists of a cylindrical vessel with elliptical heads, with an inside diameter of 6.5 ft (78 in.) and a tangent-to-tangent length of 40 ft (480 in.). Gas enters the reactor at a pressure of 484 psia and 800°F. Exit conditions are 482 psia and 850°F. The vessel will be oriented in a horizontal position. Estimate the vessel thickness in inches, weight in pounds, and purchase cost in dollars for a CE cost index of 400. The vessel contains no internals and the gas is non-corrosive. The barometric pressure at the plant site is 14 psia.

SOLUTION

The operating pressure, based on the higher pressure, is $(484 - 14) = 470$ psig. Using Eq. (16.61), the design pressure is

$$P_d = \exp\{0.60608 + 0.91615[\ln(470)] + 0.0015655[\ln(470)]^2\} = 546 \text{ psig}$$

The higher operating temperature is 850°F. Take a design temperature of 50°F higher, or 900°F. A suitable material of construction is low-alloy steel. From a table above, its maximum allowable stress, S, is 13,100 psi. Assume that the wall thickness will be greater than 1.25 in., giving a weld efficiency, E, of 1.0. From the pressure-vessel code formula of Eq. (16.60),

$$t_P = \frac{(546)(78)}{2(13,100)(1.0) - 1.2(546)} = 1.667 \text{ in.}$$

which is greater than the assumed 1.25 in. and also greater than the minimum value of $\frac{3}{8}$ in. required for rigidity. Because the orientation of the vessel is horizontal, the vessel is not subject to wind load or earthquake considerations. Adding a corrosion allowance of $\frac{1}{8}$ in. gives a total thickness of 1.792 in. Since this is in the range of $\frac{5}{8}$ to 2 in., the steel plate comes in increments of $\frac{1}{8}$ in. Since 1.792 in. is greater than $1\frac{6}{8}$ in., specify a plate thickness, t_S, of $1\frac{7}{8}$ in. or 1.875 in. for use in Eq. (16.59), to give a vessel weight of:

$$W = (3.14)(78 + 1.875)[480 + 0.8(78)](1.875)(0.284) = 72,500 \text{ lb}$$

The purchase cost of the vessel given by Eq. (16.53) is

$$C_V = \exp\left\{8.717 - 0.2330[\ln(72,500)] + 0.04333[\ln(72,500)]^2\right\} = \$102,400$$

From Table 16.26, the material factor is 1.20. Thus, $F_M C_V = \$122,900$.
   To this is added the cost of the platforms and ladders from Eq. (16.55):

$$C_{PL} = 1,580\,(78)^{0.20294} = \$3,800$$

Using Eq. (16.52), the purchase cost at a CE index of 400 is

$$C_P = \left(\frac{400}{394}\right)(122,900 + 3,800) = \$128,600$$ ∎

Plates

Vertical towers for absorption, distillation, and stripping utilize trays (plates) and/or packing. Mulet et al. (1981b) present a method for estimating the purchase cost of trays installed in a vertical tower. This cost is added to Eq. (16.52) to obtain the total purchase cost. The cost for the installed trays, C_T, all with downcomers, is given by

$$C_T = N_T F_{NT} F_{TT} F_{TM} C_{BT} \tag{16.66}$$

Here, the base cost, C_{BT}, is for sieve trays at a CE cost index of 394, where the inside diameter of the tower is in feet and the equation is valid for 2 to 16 ft.

$$C_{BT} = 369\,\exp(0.1739\,D_i) \tag{16.67}$$

In Eq. (16.66), N_T = the number of trays. If that number is greater than 20, the factor $F_{NT} = 1$. If $N_T < 20$, the factor is greater than one, as given by

$$F_{NT} = \frac{2.25}{1.0414^{NT}} \tag{16.68}$$

the factor F_{TT} accounts for the type of tray:

Tray Type	F_{TT}
Sieve	1.0
Valve	1.18
Bubble cap	1.87

The factor F_{TM}, which depends on column diameter in feet, corrects for the material of construction:

Material of Construction	F_{TM}
Carbon steel	1.0
303 Stainless steel	$1.189 + 0.0577\,D_i$
316 Stainless steel	$1.401 + 0.0724\,D_i$
Carpenter 20CB-3	$1.525 + 0.0788\,D_i$
Monel	$2.306 + 0.1120\,D_i$

EXAMPLE 16.12

A distillation column is to be used to separate isobutane from n-butane. The column, which is equipped with 100 sieve trays, has an inside diameter of 10 ft (120 in.) and a tangent-to-tangent length of 212 ft (2,544 in.). Operating conditions are 110 psia and 150°F at the bottom of the tower and 100 psia and

120°F at the top. The material of construction is carbon steel. The barometric pressure at the plant location is 14.5 psia. Estimate the purchase cost of the distillation column at a CE index of 410.

SOLUTION

The operating pressure, based on the higher pressure, is $(110 - 14.5) = 95.5$ psig. Using Eq. (16.61), the design pressure is

$$P_d = \exp\{0.60608 + 0.91615[\ln(95.5)] + 0.0015655[\ln(95.5)]^2\} = 123 \text{ psig}$$

The higher operating temperature is 150°F. Take a design temperature of 50°F higher or 200°F. For carbon steel at this temperature, the maximum allowable stress, S, is 15,000 psi. Assume that the wall thickness will be less than 1.25 in., giving a weld efficiency, E, of 0.85. From the pressure-vessel code formula, Eq. (16.60),

$$t_P = \frac{(123)(120)}{2(15,000)(0.85) - 1.2(123)} = 0.582 \text{ in.}$$

which is greater than the minimum value of 7/16 (= 0.4375) in. required for rigidity. Because orientation of the vessel is vertical and quite tall, it may be subject to wind load or earthquake considerations. For this column, $(L/D_i)^2/P_d = (212/10)^2/123 = 3.65$. Since this is greater than 1.34, the vessel thickness for wind load is calculated from Eq. (16.62):

$$t_V = 0.625[0.75 + 0.22(0.85)(3.65)] = 0.625(1.43) = 0.895 \text{ in.}$$

Adding a corrosion allowance of $\frac{1}{8}$ in. gives a total thickness of 1.020 in. Since this is in the range of $\frac{5}{8}$ to 2 in., the steel plate comes in increments of $\frac{1}{8}$ in. Therefore, specify a plate thickness, t_S, of $1\frac{1}{8}$ in. or 1.125 in. for use in Eq. (16.59), to give a vessel weight of:

$$W = (3.14)(120 + 1.125)[2,544 + 0.8(120)](1.125)(0.284) = 320,800 \text{ lb}$$

The purchase cost at the vertical tower given by Eq. (16.57) is

$$C_V = \exp\{7.0374 - 0.18255[\ln(320,800)] + 0.02297[\ln(320,800)]^2\} = \$462,400$$

From Table 16.26, the material factor is 1.00. Thus, $F_M C_V = \$462,400$. To this is added the cost of the platforms and ladders from Eq. (16.58), which, however, is applied outside of its range of 27 to 170 ft for the tangent-to-tangent length of the shell.

$$C_{PL} = 237.1 \, (10)^{0.63316} \, (212)^{0.80161} = \$74,600$$

The purchase cost at the CE index of 400 for just the tower, platforms, and ladders is

$$C_P = \left(\frac{400}{394}\right)(462,400 + 74,600) = \$558,800$$

To this must be added the cost of 100 sieve trays. From Eq. (16.67),

$$C_{BT} = 369 \exp [0.1739(10)] = \$2,100 \text{ per tray}$$

Using Eq. (16.66), and upgrading the cost index to 410,

$$C_T = N_T F_{NT} F_{TT} F_{TM} C_{BT} = 100(1.0)(1.0)(1.0)(2,100)\left(\frac{410}{394}\right) = \$218,500$$

The total purchase cost of the distillation column is

$$\$558,800 + \$218,500 = \$777,300$$ ∎

Packings

Packings for towers are classified as dumped (random) or structured. When packings are used, the total purchase cost of the packed tower becomes

$$C_P = F_M C_V + C_{PL} + V_P \, C_{PK} + C_{DR} \tag{16.69}$$

Table 16.27 Installed Costs of Some Dumped Packings

Size	Installed Cost ($/ft^3)				
	1 in.	1.5 in.	2 in.	3 in.	4 in.
Berl saddles					
Ceramic	38	29	22		
Raschig rings					
Carbon steel	43	32	27	21	
Stainless steel	142	110	87	50	
Ceramic	21	17	15	12	
Intalox saddles					
Ceramic	27	22	19	15	
Polypropylene	29		18	9	
Pall rings					
Carbon steel	39	29	25		
Stainless steel	133	102	87		
Polypropylene	29	21	17	13	
Cascade mini-rings					
Stainless steel	106		75	55	41
Ceramic	71		55	44	
Polypropylene	71		55	44	
Tellerettes					
Polyethylene	60				

where F_M for the vessel is given in Table 16.26, C_V for a vertical tower is given by Eq. (16.57), C_{PL} for the platforms and ladders is given by Eq. (16.58), V_P is the volume of the packing in cubic feet, C_{PK} is the installed cost of the packing in dollars per cubic foot and C_{DR} is the installed cost of high-performance liquid distributors and redistributors required for obtaining satisfactory performance with packings.

Installed costs for dumped packings are given by several of the references in Table 16.19. Table 16.27, which was developed by taking averages of those costs, indexed to CE = 394, and adding some additional values from vendors, includes costs for six different dumped packings and five different materials.

Compared to dumped packings and trays, structured packings offer reduced pressure drop, higher stage efficiency in terms of reduced HETS (height equivalent to a theoretical stage), and capacity in terms of reduced diameter. However, they are more expensive in dollars per cubic foot of packing and are not normally available in carbon steel. They are most often installed when revamping existing towers to reduce pressure drop, increase capacity, and/or increase purity of products. Accordingly, installed costs of structured packings are not widely available. In the absence of a vendor quote and for a very approximate estimate, the installed cost of structured packing of the corrugated-sheet type in stainless steel can be taken as $200/ft^3.

Installed-cost data for high-performance liquid distributors and redistributors are also not widely available. Distributors should be placed at every feed point and, conservatively, redistributors should be used every 20 ft. In the absence of a vendor quote and for a very approximate estimate, the installed cost of a liquid distributor can be taken as $100/ft^2 of column cross-sectional area.

EXAMPLE 16.13

A distillation column has two sections. The one above the feed is 14 ft in inside diameter with a 20-ft height, 15 ft of which is packed with structured packing of the corrugated-sheet type. The bottom section is 16 ft in diameter with a 70-ft height, 60 ft of which is packed with 4-in. Cascade mini-rings.

The column is made of carbon steel, but both packings are of stainless steel. The column will operate under vacuum with conditions of 55 kPa and 60°C at the top and 60 kPa and 125°C at the bottom. A total of four liquid distributors or redistributors will be used. Estimate the f.o.b. purchase cost of the column, including installed packings, distributors, and redistributors, for a CE cost index of 410. The barometric pressure is 100 kPa.

SOLUTION

Take a design temperature or 50°F higher than the highest temperature of 125°C (257°F), or 307°F. For a vacuum vessel, use Eq. (16.63) to estimate the wall thickness. The maximum pressure difference between the inside and outside of the vessel is $100 - 55 = 45$ kPa or 6.5 psig. Use this as the design pressure, P_d. From a table above, the modulus of elasticity for carbon steel at 307°F is 28.9×10^6 psi. For the top section, $D_i = 14$ ft or 168 in. Assume for this large a diameter that $D_i = D_o$ But for L, use the total length of $(20 + 70) = 90$ ft. Then the wall thickness for the top section is

$$t_E = 1.3(168)\left(\frac{6.5(90)}{28.9 \times 10^6(14)}\right)^{0.4} = 1.01 \text{ in.}$$

This is well within the limit of applicability of $t_E/D_o < 0.05$. To this must be added the correction of Eq. (16.64):

$$t_{EC} = (90)(12)[0.18(168) - 2.2] \times 10^{-5} - 0.19 = 0.11 \text{ in.}$$

The total thickness for the top section, from Eq. (16.65), is $t_V = 1.01 + 0.11 = 1.12$ in. Adding a $\frac{1}{8}$-in. corrosion allowance gives 1.245 in. and, therefore, use a plate thickness including the next $\frac{1}{8}$-in. increment or 1.25 in. Similar calculations for the bottom section give a wall thickness of 1.375 in.

The weights of the two sections are estimated from Eq. (16.59), but with only one head per section. However, as recommended by Mulet et al. (1981a) for a two-diameter tower, the weight of each section is based on the total length. The base cost for the two sections is then calculated from:

$$C_V = \frac{L_1 C_{V1} + L_2 C_{V2}}{L_1 + L_2} \tag{16.70}$$

For the top section,

$$W_1 = 3.14(168 + 1.25)[1,080 + 0.8(168)](1.25)(0.284) = 229,100 \text{ lb}$$

A similar calculation for the bottom section gives $W_2 = 319,300$ lb. Base purchase costs for a tower section are given by Eq. (16.57). For the top section,

$$C_{V1} = \exp\{7.0374 + 0.18255[\ln(229,100)] + 0.02297[\ln(229,100)]^2\} = \$358,400$$

A similar calculation for the bottom section gives $C_{V2} = \$460,800$. Using Eq. (16.70), the purchase cost of the entire empty tower is

$$C_V = \frac{20(358,400) + 70(460,800)}{20 + 70} = \$438,000$$

To this is added the costs of the platforms and ladders for the tower. The concept of Eq. (16.70) is again applied after substituting C_{PL} for C_V for each section. From Eq. (16.58) for the top section, again using the total length of the tower:

$$C_{PL1} = 237.1(14)^{0.63316} (90)^{0.80161} = \$46,500$$

A similar calculation for the bottom section gives $C_{PL2} = \$50,600$. Using the form of Eq. (16.70), the total cost of the platforms and ladders is

$$C_{PL} = \frac{20(46,500) + 70(50,600)}{20 + 70} = \$49,700$$

The structured packing occupies a volume, V_{P1}, of $3.14(14)^2(15)/4 = 2,310$ ft³. The estimated installed cost, $C_{PK,1}$, is \$200/ft³. The random packing occupies a volume. V_{P2}, of $3.14(16)^2(60)/4 = 12,060$ ft³. From Table 16.27, the installed cost of 4-in. Cascade mini-rings in stainless steel = $C_{PK,2} = \$41/$ft³.

For the four distributors, assume one has a diameter of 14 ft, while the other three have a diameter of 16 ft. This corresponds to areas of 154 ft^2 and 201 ft^2, respectively. At a cost of $100/ft^2, the total installed cost of the four distributors or redistributors is

$$C_{\text{DR}} = 154(100) + 3(201)(100) = \$75,700$$

From Equation (16.67), after including the CE cost index ratio, the f.o.b. purchase cost of the vacuum tower, complete with packings, distributors and redistributors, platforms, and ladders is

$$C_P = (410/394)[(1.0)(438,000) + 49,700 + [2,310(200) + 12,060(41)] + 75,700]$$
$$= (410/394)[438,000 + 49,700 + 956,500 + 75,700] = \$1,582,000$$

Note that the packings are a large fraction of the total cost of the tower.   ∎

16.6   PURCHASE COSTS OF OTHER CHEMICAL PROCESSING EQUIPMENT

In this section, equations are presented for the estimation of the f.o.b. purchase cost of chemical processing equipment not covered in Section 16.5. In each equipment category, so many different designs are available that it is not possible to consider them all. Instead, an attempt has been made to include only the most widely used designs for which, in some cases, heuristics are included for estimating equipment sizes. This should be sufficient for preliminary estimates of capital cost. For final plant design, vendors of the different types of equipment should be consulted for assistance in selecting and sizing the most suitable design and to obtain more accurate estimates of purchase cost so as to achieve the most operable and economical process. The purchase-cost equations for the equipment in this section, which are based on a size factor, S, valid for a stated range and an average cost index for the year 2000 (CE = 394), are listed in Table 16.32, which appears at the end of this chapter. For the most part, the purchase-cost equations were developed from the sources of cost data given in Table 16.19 and the data at the Internet site, *www.matche.com/EquipCost*. When the pressure and material of construction are not mentioned in Table 16.32, low-to-moderate pressures and conventional materials of construction such as carbon steel may be assumed. In lieu of data for other materials of construction, the purchase cost for another material may be estimated by multiplying the cost obtained from the equation in Table 16.32 by one of the following factors:

Material	Factor
Carbon steel	1.0
Copper	1.2
Stainless steel	2.0
Nickel	2.5
Monel	2.7
Titanium-clad	3.0
Titanium	6.0

Table 16.32 is accompanied by the following brief equipment descriptions, which include, in some cases, heuristics for estimating the size factor when design data are not readily available. More equipment descriptions and detailed methods for determining size for most of the types of equipment described below may be found in *Perry's Chemical Engineers' Handbook* (1997).

Adsorption Equipment

Adsorption from liquids is carried out in stirred vessels (*slurry adsorption*) or in fixed beds while *fixed-bed adsorption* is used for gases. For slurry adsorption, which is usually conducted batchwise, the purchase cost includes the vessel, a motor-driven agitator, and the adsorbent particles. The size of the vessel and the amount of adsorbent required depend on the amount of solute to be adsorbed from the feed, adsorption equilibrium, solids content of the slurry in the vessel, and the desired time of treatment. A reasonable slurry composition is 5 vol% solids. An agitator sized at 10 Hp/1,000 gal being stirred is generally sufficient to keep the solid particles in suspension. Costs of adsorbents and motor-driven agitators are included in Table 16.32.

For fixed-bed adsorption of gases, a reasonable superficial gas velocity through the bed is 100 ft/min, while for liquids, 1 ft/min. Typical bed heights are 1 to 3 times the bed diameter. A conservative equilibrium adsorbent loading is 10 lb of adsorbate per 100 lb of adsorbent. Breakthrough times should account for mass-transfer resistance effects by adding 2 ft to the bed height calculated on the basis of equilibrium loading.

Agitators (Propellers and Turbines)

Motor-driven propellers and turbines are the most widely used devices for agitation in vessels. Propellers are small in diameter and use high rates of rotation by direct coupling, with motors running typically at 1,150 or 1,750 rpm. They are often used to agitate large liquid storage tanks by mounting several of them sideways at locations around the periphery of the tank. Turbines are larger in diameter, typically one-third of the vessel diameter, and, by using speed reducers with electric motors, rotate at from 37 to 320 rpm. Turbines, which are available in several designs, are more versatile than propellers and are usually the preferred type of agitator for applications involving mixing of miscible and immiscible liquids in reactors, mixing of immiscible liquids in liquid–liquid extraction vessels, suspension of fine adsorbent particles in slurry adsorption, enhancement of heat transfer to or from a liquid in a jacketed tank, and dispersion of a gas into a liquid in a tank. Typical horsepower requirements, for turbines, based on the fluid volume in the vessel or tank, are

Application	Hp/1,000 Gallons
Blending miscible liquids	0.5
Homogeneous liquid reaction	1.5
Reaction with heat transfer	3
Liquid–liquid reaction or extraction	5
Gas dispersion in a liquid	10
Suspension of solid particles	10

Autoclaves

An autoclave is predominantly a vertical, cylindrical, stirred-tank reactor. It can be operated continuously or batchwise over a wide range of production rates, temperatures, and pressures. The stirring is achieved by internal agitators (turbines or propellers) or by forced circulation through the vessel with an external pump. However, the contents of small autoclaves are stirred by rocking, shaking, or tumbling the vessel. Most autoclaves are provided with a means of transferring heat to or from the contents of the vessel. That means may be a jacket, internal coils or tubes, an external pump–heat exchanger combination, an external reflux condenser when vapors are evolved, an electrically heated mantel, or direct firing by partial submergence of the autoclave in a furnace. In Table 16.32, equations for estimating f.o.b. purchase costs are listed for autoclaves made of steel, stainless steel, and glass-lined steel. These autoclaves are provided with turbine agitators and heat-transfer jackets.

Crystallizers

Most industrial crystallization operations are *solution crystallization* involving the crystallization of inorganic compounds from aqueous solutions. Only the inorganic compound crystallizes. However, a growing number of applications are being made for *melt crystallization*, which involves a mixture of two or more organic components whose freezing points are not far removed from each other. In that case, impure crystals (solid solutions) may be obtained that require repeated melting and freezing steps to obtain pure crystals of the component with the highest freezing temperature. Only solution crystallization is considered here.

Solution crystallization occurs from a supersaturated aqueous solution, which is achieved from the feed by cooling, evaporation, or a combination of cooling and evaporation. The application of cooling crystallization is limited because for many dissolved inorganic compounds, the decrease in solubility with decreasing temperature is not sufficient to make the method practical. Therefore, evaporative crystallizers are more common. Table 16.32 contains f.o.b. purchase costs for four types of crystallizers. The continuous jacketed scraped-wall crystallizer is based on length, which can be estimated by a heat-transfer rate using a scraped-surface cooling area of 3 ft^2 per foot of length and an overall heat transfer coefficient of 20 Btu/hr-ft^2-°F. The heat transfer rate is obtained by an energy balance that accounts for both sensible heat and the heat of crystallization. The purchase costs of continuous forced-circulation evaporative crystallizers or the popular continuous draft-tube baffled (DTB) crystallizers are based on the rate of production of crystals in tons (2,000 lbs) per day. The purchase cost of batch evaporative crystallizers, which usually operate under vacuum, depends on the vessel size.

Drives Other than Electric Motors

When the required shaft horsepower for power input to an item of process equipment is less than 100 Hp, an electric motor is usually the selected drive. For higher horsepower input, consideration is given to combustion gas turbines, steam turbines, and internal combustion gas engines. However, except for remote, mobile, or special situations, steam turbines are the most common alternative to electric motors. Furthermore, steam turbines are considerably more efficient, 50–80%, than gas turbines or engines, which have efficiencies of only 30–40%. Equations for f.o.b. purchase costs of steam and gas turbine drives are included in Table 16.32 as a function of shaft horsepower.

Dryers

No single drying device can handle efficiently the wide variety of feed materials, which includes granular solids, pastes, slabs, films, slurries, and liquid. Accordingly, many different types of commercial dryers have been developed for both continuous and batchwise operation. Batch dryers include tray and agitated types. Continuous dryers include tunnel, belt, tray, direct and indirect rotary, screw conveyor, fluidized-bed, spouted-bed, pneumatic-conveyor, spray, drum, infrared, dielectric, microwave, and freeze types. The selection and sizing of dryers often involves testing on a pilot-plant level. The f.o.b. purchase costs for several of the more widely used dryers are included in Table 16.32. Different size factors are used depending on the type of dryer. In the *batch compartment dryer*, the feed is placed in stacked trays over which hot air passes. Trays typically measure 30 × 30 × 3 in. Typical drying time is a few hours. The cost depends on the tray surface area.

Two types of *rotary dryers* are available for continuous drying. In the direct-heat type, longitudinal flights, which extend inward radially from the inner periphery of the slightly inclined rotating dryer cylinder, lift and shower the granular solids through hot air flowing through the cylinder. The inclination of the cylinder causes the solids to flow from the feed end to the discharge end of the cylinder. Moisture evaporation rates are generally in the range of 0.3–3 lb/hr-ft^3 of dryer volume depending on the amount of free moisture and the desired moisture content of the product. Direct-heat rotary dryers vary in size from 1-ft diameter by 4-ft long to 20-ft diameter by 150-ft long. The cost depends on the peripheral area of the shell. Typical length-to-diameter ratios vary from 4 for small dryers to 8 for large dryers. In the indirect-heat type, the material is dried by contact with the outer surface area of tubes, arranged in one or two circular rows around the inner periphery of the rotating shell. The purchase cost depends on the outside surface area of the tubes, which carry condensing steam. Typical heat fluxes range from 600 to 2,000 Btu/hr-ft^2. Indirect-heat rotary dryers vary in size from 3 ft in diameter by 15 ft long to 15 ft in diameter by 80 ft long.

Drum dryers take a solution or thin slurry and spread it as a thin film over a rotating drum heated internally by condensing steam to produce a flaked product. Typical moisture evaporation rates are 3–6 lb/hr-ft^2. One or two drums (side-by-side) may be used. Drum dimensions range from 1 ft in diameter by 1.5 ft long to 5 ft in diameter by 12 ft long. The cost of drum dryers depends on the surface area of the drum.

Spray dryers produce small porous particles, such as powdered milk and laundry detergent, from a liquid solution by evaporation of the volatile component of the feed, with the purchase cost correlated with the evaporation rate. Size and cost data for other dryers as well as considerations in dryer selection are found in Section 12 of *Perry's Chemical Engineers' Handbook* (1997).

Dust Collectors

The removal of dust particles, typically 1 to 1,000 microns in diameter, from gas streams (also called gas cleaning) is accomplished on an industrial scale by four main types of equipment: (1) *bag filters*, (2) *cyclones* using centrifugal force, (3) *electrostatic precipitators*, and (4) *venturi scrubbers* using washing with a liquid. Reasons for dust collection include air-pollution control, elimination of safety and health hazards, recovery of a valuable product, improvement in the quality of other products, and reduction of equipment maintenance. Typical ranges of particle size that can be efficiently removed by each of the four methods are as follows:

Equipment	Dust Particle Range (microns)
Bag filters	0.1 to 50
Cyclones	10 to 1,000
Electrostatic precipitators	0.01 to 10
Venturi scrubbers	0.1 to 100

Bag filters, also referred to as fabric filters and baghouses, use natural or synthetic woven fabrics or felts. Typically, the openings in the filter fabric are larger than the particles to be removed. Therefore an initiation period must take place during which particles build up on the fabric surface by impingement. After this occurs, the collected particles themselves act as a precoat to become the actual filter medium for subsequent dust removal. Bag filters are automatically cleaned periodically by shaking, using reverse flow of clean gas, or by pulsing with clean gas. Dust collection efficiencies approach 100%.

The theory of particle removal by cyclones is well developed and equipment dimension ratios are well established, making it possible to design cyclones to remove particles of a known size range with reasonable estimates of particle collection efficiency as a function of particle size. However, the efficiency of cyclones is greater for the larger particles. Smaller particles can only be removed with cyclones of small diameter. Pressure drops in cyclones can be substantial and often limit the smallest particle size that can be removed. Multiple cyclones are sometimes used in series or parallel.

Electrostatic precipitators operate in a fashion that is just the opposite of a cyclone, removing the smallest particles at the highest collection efficiencies. They can remove the smallest dust particles as well as even smaller particles in the fume and smoke ranges down to 0.01 μm. In the precipitator, the particles are charged by passing the gas between two electrodes charged to a potential difference of tens of thousands of volts. This requires that the gas contain ionizable components such as carbon dioxide, carbon monoxide, sulfur dioxide, and water vapor. If not, water vapor can be added. The ions attach to the particles, carrying them to a collecting electrode. The particles are continuously or periodically removed from the electrode by rapping the collecting electrode.

The venturi scrubber usually consists of a venturi contactor followed by a cyclone separator. The particle-laden gas flows downward to the approach section of the venturi, where water is injected tangentially to flood the wall. After achieving intimate contact of the particles with the liquid in the throat of the venturi, the stream then turns and enters the cyclone where clean gas leaves at the top and a slurry leaves at the bottom. Venturi scrubbers are particularly effective at high concentrations of particles.

Equations for estimating the f.o.b. purchase costs of dust collectors are given in Table 16.32, where the size factor is the actual gas flow rate. All collectors are of mild steel construction. For relatively large dust particles, cyclones are adequate and are the least expensive alternative. For gases containing a wide range of particle sizes, a cyclone followed by an electrostatic precipitator is a common combination.

Evaporators

Aqueous solutions of inorganic salts and bases are concentrated, without crystallization, in evaporators. Because the volatility of water is much higher than the dissolved inorganic salts and bases, only water is evaporated. Evaporators are also used with aqueous solutions of organic compounds that have little volatility. Most popular is the *long-tube vertical (rising-film) evaporator* with tubes from 12 to 35 ft long with boiling inside the tubes. For viscous solutions, a pump is added to give a *forced-circulation evaporator*. Less efficient, but less expensive is the *horizontal-tube evaporator* where boiling occurs outside the tubes. For temperature-sensitive applications, a (long-tube vertical) *falling-film evaporator* is used, typically with a small overall temperature driving force of less than 15°F. Many evaporators operate under vacuum and are frequently multistaged to reduce the cost of the heating steam. The f.o.b. purchase costs for evaporators are included in Table 16.32, in terms of the heat transfer area, for carbon-steel construction and pressures to 10 atmospheres. Typical ranges of overall heat-transfer coefficients, U, which increase with the boiling temperature, are as follows for a boiling-point range of 120 to 220°F:

Evaporator Type	U (Btu/hr-ft^2-°F)
Horizontal tube	80–400
Long-tube vertical	150–650
Forced circulation	450–650
Falling film	350–750

Fired Heaters for Specific Purposes

Fired heaters are available for providing heat transfer media of the types described in Section 13.2, including hot water, steam, mineral oils, silicon oils, chlorinated diphenyls (Dowtherm A), and molten (fused) salts. Fired heaters are also used as reactors, such as reformers in petroleum refineries and for pyrolysis of organic chemicals. Purchase costs for these specific types of fired heaters, based on heat duty, are presented in Table 16.32.

Liquid–Liquid Extractors

A wide variety of commercial equipment is available for carrying out separations by liquid–liquid extraction. The most efficient are those that provide mechanical agitation of the liquid phases. When the number of equilibrium stages is small, for example, five or less, and floor space is available but headroom is at a premium, a battery of *mixer-settler vessels* may be the best choice. For preliminary cost estimates, and in lieu of mass-transfer data, the mixing vessels can be pressure vessels that have a height-to-diameter ratio of 1 and provide 5 min or less residence time, depending on the liquid viscosity and the interfacial tension. The mixers are equipped with vertical side baffles and a flat-blade turbine agitator that delivers 4 Hp/1,000 gal. The settlers can be horizontal vessels with inlet baffles and a length-to-diameter ratio of 4. The capacity of the settler can be determined based on 5 gal/min of feed/ft^2 of phase disengaging area (diameter $\times$ length), provided that the specific-gravity difference between the two liquid phases is greater than 0.10. Each mixing vessel in the battery will approximate an equilibrium stage. Purchase costs of mixers and settlers can be estimated from the costs for pressure vessels and agitators discussed above.

When headroom is available, a number of column designs, with mechanical agitation from impellers on a vertical shaft, can be used. Typical of these is the *rotating-disk contactor (RDC)*, which has a maximum diameter of 25 ft and maximum total liquid throughput of 120 ft^3 of liquid/hr-ft^2 of column cross-sectional area. Typical HETP (height equivalent to a theoretical stage) values range from 2 to 4 ft depending on column diameter and interfacial tension. The f.o.b. purchase cost for an RDC unit is included in Table 16.32, where the size factor is the product of the column diameter raised to a 1.5 exponent and the column height.

EXAMPLE 16.14

An aqueous feed of 30,260 lb/hr of 22 wt% acetic acid is contacted with 71,100 lb/hr of a solvent of 96.5 wt% ethyl acetate at 100°F and 25 psia to extract 99.8% of the acetic acid. A process simulation program computes 6 equilibrium stages for the separation. The densities of the two entering liquid phases are 62.4 lb/ft^3 for the feed and 55.0 lb/ft^3 for the solvent. Estimate the size and f.o.b. purchase cost of an RDC liquid–liquid extraction column for a CE cost index of 410.

SOLUTION

The volumetric flow rate of the feed = 30,260/62.4 = 485 ft^3/hr

The volumetric flow rate of the solvent = 71,100/55.0 = 1,293 ft^3/hr

The total volumetric flow rate through the column = 485 + 1,293 = 1,778 ft^3/hr

For a maximum throughput of 120 ft^3/hr-ft^2, cited above, the minimum cross-sectional area for flow = 1,778/120 = 14.8 ft^2.

Assume a throughput of 60% of the maximum value.

Actual cross-sectional area for flow = 14.8/0.6 = 24.7 ft^2

Column diameter = $[24.7(4)/3.14]^{0.5}$ = 5.6 ft. Specify a diameter, D, of 5.5 ft.

Assume an HETP of 4 ft. This gives a total stage height of 4(6) = 24 ft. Add an additional 3 ft at the top and 3 ft at the bottom to give a total tangent-to-tangent height, H, of 24 + 3 + 3 = 30 ft.

From Table 16.32, the size factor = $S = H(D)^{1.5} = 30(5.5)^{1.5} = 387$ ft$^{2.5}$

For carbon steel at a CE index of 394, the f.o.b. purchase cost is

$$C_P = 250(387)^{0.84} = \$37,300$$

Because the feed contains water and acetic acid, assume stainless-steel construction with a material factor of 2.0 and correct for the cost index. This gives an estimated f.o.b. purchase cost of

$$C_P = 37,300(2.0)(410/394) = \$77,600$$

Membrane Separations

Commercial membrane separation processes include *reverse osmosis, gas permeation, dialysis, electrodialysis, pervaporation, ultrafiltration*, and *microfiltration*. Membranes are mainly synthetic or natural polymers in the form of sheets that are spiral wound or hollow fibers that are bundled together. Reverse osmosis, operating at a feed pressure of 1,000 psia, produces water of 99.95% purity from seawater (3.5 wt% dissolved salts) at a 45% recovery, or with a feed pressure of 250 psia from brackish water (less than 0.5 wt% dissolved salts). Bare-module costs of reverse osmosis plants based on purified water rate in gallons per day are included in Table 16.32. Other membrane separation costs in Table 16.32 are f.o.b. purchase costs.

Gas permeation is used to separate gas mixtures, for example, hydrogen from methane. High pressures on the order of 500 psia are used to force the molecules through a dense polymer membrane, which is packaged in pressure-vessel modules, each containing up to 4,000 ft^2 of membrane surface area. Membrane modules cost approximately \$35/ft^2 of membrane surface area. Multiple modules are arranged in parallel to achieve the desired total membrane area.

Pervaporation is used to separate water–organic and organic–organic mixtures that form azeotropes and may be difficult to separate by enhanced distillation. Typical membrane modules cost \$30/ft^2 of membrane surface area.

Ultrafiltration uses a microporous polymer membrane, which allows water and molecules of less than some cut-off molecular weight to pass through, depending on the pore diameter, while retaining larger molecules. A typical membrane module may contain 30 ft^2 of membrane surface area at a cost of from 8 to 20 \$/ft^2 of surface area.

Mixers for Powders, Pastes, and Doughs

A wide variety of designs is available for mixing powders or pastes, polymers, and doughs of high viscosity. Among the more widely used designs are *ribbon and tumbler mixers* for dry powders, and *kneaders and mullers* for pastes and doughs. Equations for f.o.b. purchase costs of these devices are included in Table 16.32 in terms of the volumetric size. All designs operate batchwise; some can operate continuously. Typical residence times for mixing are less than 5 min. A comprehensive treatment of this type of mixing is given in *Perry's Chemical Engineers' Handbook* (1997).

Power Recovery

Valves are often used to reduce the pressure of a gas or liquid process stream. By replacing the valve with a turbine, called an *expander, turboexpander, or expansion turbine* in the case of a gas and a *liquid expander* or *radial-inflow, power-recovery turbine* in the case of a liquid, power can be recovered for use elsewhere. Power recovery from gases is far more common than from liquids because for a given change in pressure and mass flow rate, far more power can be recovered from a gas than from a liquid because of the lower density of the gas. Equations for f.o.b. purchase costs of power recovery devices are included in Table 16.32 in terms of horsepower that can be extracted. Typical efficiencies are 75–85% for gases and 50–60% for liquids. Condensation of gases in expanders up to 20% can be tolerated, but vapor evolution from liquid expansion requires a special design. Whenever more than 100 Hp for a gas and more than 150 Hp for a liquid can be extracted, a power recovery device should be considered.

Screens

Solid particles are separated according to particle size by screening. Ideally, particles of size smaller than the opening of the screen surface, called *undersize* or *fines*, pass through, while larger particles, called *oversize* or *tails*, do not. However, a perfect separation is not made. If the undersize is the desired product, the screen efficiency is the mass ratio of undersize obtained from the screen to the undersize in the feed. A typical efficiency might be 75%. For particles with sizes greater than 2 in, a vibrating, inclined *grizzly*, which consists of parallel bars of fixed spacing are commonly used. The opening between adjacent parallel bars may be from 2 to 12 in. Inclined *vibrating screens* are used to separate particles smaller than 2 in. Standard screen sizes of the U.S. Sieve Series are used. The screens consist of woven wire with square apertures (openings). Screen sizes are quoted in millimeters above 8 mm and in *mesh* for 8 mm and lower. Mesh refers to the number of openings per inch. However, because wire diameter is not constant, the actual opening size cannot be easily calculated from the mesh. Instead, a table, such as found in *Perry's Chemical Engineers' Handbook*, must be used to obtain the opening size. For example, a 20-mesh (No. 20) screen has square openings of 0.841 mm. Purchase costs of grizzlies and vibrating screens, included in Table 16.32, depend on the screen surface area. Screen capacities are directly proportional to screen surface area and approximately proportional to the opening between bars or the screen aperture. However, capacity for vibrating screens drops off dramatically for particle sizes below that for 140 mesh (0.105 mm). Typical capacities for grizzly screens range from 1 to 4 tons of feed/ft^2 of screen surface-hr-inch of opening between bars, with the lower value for coal and the higher value for gravel. For vibrating screens, typical capacities range from 0.2 to 0.8 ton/ft^2-hr-mm of aperture. Vibrating screens are available in single-, double-, and triple-deck machines, where the screen surface area refers to the total area in square feet of all screens in the deck. For the separation of very small particles of less than 0.1 mm in diameter, air classifiers are used, which can be costed as a cyclone separator.

Size Enlargement

Solid products are frequently produced in preferred shapes, such as tablets, rods, sheets, etc. Such shapes are produced by a variety of size enlargement or agglomeration equipment by pressure compaction, as by a *pellet mill*, *pug mill extruder*, *roll-type press*, *screw extruder*, or *tableting press*; or by tumbling compaction in *disk or pan agglomerators*. Equations for f.o.b. purchase costs of these devices are included in Table 16.32 in terms of the feed rate. A comprehensive treatment of size enlargement is given in *Perry's Chemical Engineers' Handbook* (1997).

Size Reduction Equipment

The size of solid particles can be reduced by one or more of the following actions: (1) impact by a single rigid force, (2) compression between two rigid forces, (3) shear, and (4) attrition between particles or a particle and a wall. The energy required to reduce the size of particles is far more than that theoretically required to increase the surface area of the particles, with the excess energy causing an increase in the temperature of the particles and the surroundings. A wide variety of equipment is available for particle size reduction, but except for special cases, most crushing and grinding tasks can be accomplished with the types of equipment listed in Table 16.28, which is taken from Walas (1988) and is organized by feed particle size. Included in the table are representative solids feed rates and power consumption. In general, primary and secondary crushers, including *jaw crushers, gyratory crushers*, and *cone crushers*, are used with feed particle sizes greater than about 2 in. They accomplish a size (diameter) reduction ratio of approximately 8. For feeds of smaller particle sizes, grinding is used with *hammer mills* and *ball mills*, but at lower capacities, especially with hammer mills. With these two mills, the size reduction ratio is much larger, 100 to 400, to achieve particle sizes in the 0.01 to 0.1 mm range. For the production of even smaller particles, in the range of 3 to 50 μm, a *jet mill* (pulverizer), which uses gas jets, can be used. The f.o.b. purchase costs of the crushers and grinders in Table 16.28, including electric motor drives, are listed in Table 16.32.

Table 16.28 Operating Ranges of Widely Used Crushing and Grinding Equipment[a]

Equipment	Feed size (mm)	Product Size (mm)	Size Reduction Ratio	Capacity [ton/hr (1 ton = 2,000 lb)]	Power (Hp)
Gyratory crushers	200–2,000	25–250	8	100–500	135–940
Jaw crushers	100–1,000	25–100	8	10–1,000	7–270
Cone crushers	50–300	5–50	8	10–1,000	27–335
Hammer mills	5–30	0.01–0.1	400	0.1–5	1.3–135
Ball mills	1–10	0.01–0.1	100	10–300	65–6,700
Jet mills	1–10	0.003–0.05	300	0.1–2	2.7–135

[a]Reprinted with permission from Walas (1988).

Solid–Liquid Separation Equipment (Thickeners, Clarifiers, Filters, Centrifuges, and Expression)

Slurries of solid particles and liquid are separated into more concentrated slurries or wet cakes and relatively clear liquid (overflow) by means of gravity, pressure, vacuum, or centrifugal force. The solid particles and/or the clear liquid may be the product of value. Continuous *clarifiers* remove small concentrations of solid particles by settling (sedimentation) to produce a clear liquid. Continuous *thickeners* are similar in design to clarifiers, but the emphasis is on producing a more concentrated slurry that can be fed to a filter or centrifuge to produce a wet cake. With thickeners, the solid particles are usually more valuable. Clarifiers and thickeners operate continuously, most often with gravity causing the solid particles to settle to the bottom of the equipment, where a rake is used to remove concentrated slurry. Typically, the feed to a thickener is 1–30 wt% solids, while the underflow product is 10–70 wt% solids. For a clarifier, the wt% solids in the feed is usually less than 1%. The f.o.b. purchase costs for thickeners and clarifiers listed in Table 16.32 are based on a size factor of the settling area. Most thickeners and clarifiers are circular, with a diameter from 10 to 400 ft. Below 100 ft, construction is usually of carbon steel, while concrete is used for diameters greater than 100 ft. The required settling area is best determined by scaling up laboratory sedimentation experiments. In the absence of such data, very preliminary cost estimates can be made by estimating a settling area based on the equation: Settling area, ft^2 = C_1 (tons solids/day), where C_1 typically is in the range of 2–50, with the lower value applying to large particles of high density and the higher value applying to fine particles of low density. The settling area for a clarifier can be estimated by the equation: Settling area, ft^2 = C_2 (gpm of overflow liquid), where C_2 is typically 0.5 to 1.5. Power requirements for thickeners and clarifiers are relatively low because of the low rotation rate of the rake. For example, a 200-ft-diameter unit only requires 16 Hp.

Continuous wet classifiers, of the *rake* and *spiral* type, and *hydroclones* (hydrocyclones) can also produce concentrated slurries, but the overflow liquid is not clear, containing the smaller particles. A separation of particles by size and density occurs. Hydroclones are inexpensive, typically 6 in. in diameter or less, and a single unit can handle only low flow rates. For high flow rates, parallel units are employed in a *multiclone*. Table 16.32 contains f.o.b. purchase costs for wet classifiers and hydroclones. The size factor is the solids flow rate for classifiers and the liquid flow rate for hydroclones.

Wet cakes can be produced by sending concentrated slurries to filters operating under pressure or vacuum. Liquid passes through a porous barrier (filter media), which retains most of the solid particles to form a wet cake. Filters are designed to operate either continuously (e.g., the *continuous rotary-drum vacuum filter* or the *rotary pan filter*) or batchwise (e.g., the *plate-and-frame filter* or the *pressure leaf filter*). Table 16.32 contains f.o.b. purchase costs for most of the widely used filters. All costs are based on filtering area, which must be determined by laboratory experiments with a handheld vacuum or pressure leaf filter. For preliminary cost estimates, in the absence of such tests, select a continuous rotary-drum vacuum filter operating at a rotation rate of 20 rev/hr, with a filtering area estimated for fine particles (e.g., those produced by a precipitation reaction to produce relatively insoluble inorganic particles) at a filtrate rate of 1,500 lb/day-ft^2 (vacuum of 18–25 in. of Hg) and, for coarse solids (e.g., those produced by crystallization) at a filtrate rate of 6,000 lb/day-ft^2 (vacuum of 2–6 in. of Hg).

An alternative to a filter is a centrifuge, where the removal of some of the liquid in the feed is accomplished under a high centrifugal force, up to 50,000 times the gravitational force on the earth, by sedimentation with a solid bowl or by filtration with a perforated bowl, either continuously or batchwise. Table 16.32 contains f.o.b. purchase costs for many of the more widely used centrifuges in 304 stainless steel. Included are two manual *batch centrifugal filters*, two *automatic-batch centrifugal filters*, and two *continuous centrifuges*. The size factor for the latter two centrifuges is the tons per hour of solid particles, while the size factor for the other four centrifuges is the bowl diameter. With auto-batch operation, from 1 to 24 ton/hr of solids can be processed. For manual batch operation, a cycle time of 2 min for coarse particles and 30 min for fine particles is representative. From 0.15 to 5 ton/hr of solids can be processed in a 40-in-diameter manual batch centrifuge.

Some wet solids consist of fibrous pulps or other compressible materials, from which liquid cannot be removed by settling, filtering, or centrifuging. Instead *expression* is used with *screw presses* or *roll presses*. Table 16.32 contains f.o.b. purchase costs for these two presses in stainless steel.

EXAMPLE 16.15

Consider a process similar to that of Figure 7.45 for the continuous production of crystals of MgSO$_4$·7H$_2$O from an aqueous solution of 10 wt% MgSO$_4$. However, replace the hydroclone-centrifugal filter combination with a rotary-drum vacuum filter. Thus, in Figure 7.45, the magma goes directly to the filter, which produces filtrate (mother liquor) that is recycled to the crystallizer and a wet cake that is sent to the dryer. The crystallizer operates adiabatically. Therefore, the filtrate is heated with an external heat exchanger before being recycled to the crystallizer. For the production of 3,504 lb/hr of MgSO$_4$·7H$_2$O, containing 1.5 wt% moisture, from a feed of a 10 wt% aqueous solution of MgSO$_4$ at 14.7 psia and 70°F, estimate the f.o.b. purchase costs of all major items of equipment at a CE cost index of 394, using the following results from material and energy balances, obtained with a process simulation program, and preliminary equipment sizing:

Feed to the process:

H_2O = 15,174 lb/hr

MgSO$_4$ = 1,686 lb/hr

Evaporator Effect 1 (long-tube vertical):

Operating pressure = 7.51 psia

Evaporation rate = 6,197 lb/hr

U = 450 Btu/hr-ft^2-°F

ΔT = 30°F

Heat duty = 7,895,000 Btu/hr

Area, A = 585 ft^2

Evaporator Effect 2 (long-tube vertical):

Operating pressure = 2.20 psia

Evaporation rate = 6,197 lb/hr

U = 250 Btu/hr-ft^2-°F

ΔT = 40°F

Heat duty = 5,855,000 Btu/hr

Area, A = 585 ft^2

Feed to crystallizer from Evaporator Effect 2:

H_2O = 2,780 lb/hr

MgSO$_4$ = 1,686 lb/hr

Recycle filtrate (mother liquor) to crystallizer from external heater:

H_2O = 7,046 lb/hr

$MgSO_4$ = 2,792 lb/hr

Temperature after being heated = 180°F

Crystallizer (continuous adiabatic draft-tube baffled):

Operating pressure = 0.577 psia

Temperature = 85.6°F

Solubility of $MgSO_4$ at these conditions = 28.3 wt% $MgSO_4$

H_2O evaporation rate = 583 lb/hr

Magma flow rate to filter:

H_2O = 7,783 lb/hr

Dissolved $MgSO_4$ = 3,084 lb/hr

$MgSO_4 \cdot 7H_2O$ crystals produced = 2,854 lb/hr or 34.2 ton/day

Filter (continuous rotary vacuum):

Cake, assumed to contain 73.5 wt% solids = 3,883 lb/hr

Total filtrate = 9,838 lb/hr = 236,100 lb/day

Assumed filtrate flux = 5,000 lb/day-ft^2 because crystals are fairly coarse

Filter area = 236,100/5,000 = 47 ft^2

Dryer (continuous direct-heat rotary)

Production rate of crystals with 1.5 wt% moisture = 3,504 lb/hr

Moisture evaporation rate by contact with hot air = 379 lb/hr

Moisture in crystals = 52 lb/hr

Outlet temperature of crystals = 113°F

Assumed volumetric moisture evaporation rate = 2 lb/hr/ft^3 of dryer volume because crystal moisture will be mainly free and final moisture content is not particularly low.

Dryer volume = 379/2 = 190 ft^3

Dryer dimensions: 3.5 ft diameter by 20 ft long

Peripheral area = 220 ft^2

SOLUTION

The process system consists of two long-tube vertical evaporators, a draft-tube baffled crystallizer, a rotary-drum vacuum filter, and a direct-heat rotary dryer. Also, pumps are needed to move the solution from evaporator 1 to evaporator 2, to recycle the filtrate from the filter to the crystallizer, and to move the magma from the crystallizer to the filter; and a heat exchanger is needed to heat the recycle filtrate. However, the purchase costs for the three pumps and the heat exchanger are not considered here because examples for these types of equipment are presented in Section 16.5. For the equipment considered here, assume fabrication from stainless steel, with a material factor of 2 for the ratio of stainless steel cost to carbon steel cost. For the process, using the following size factors and the equations in Table 16.32, the estimated f.o.b. equipment purchase costs at a CE index of 394 are included in the following table.

Equipment	Type	Size Factor	f.o.b. Purchase Cost, in Carbon Steel	f.o.b. Purchase Cost, in Stainless Steel
Evaporator 1	Long-tube vertical	A = 585 ft^2	$150,000	$300,000
Evaporator 2	Long-tube vertical	A = 585 ft^2	$150,000	$300,000
Crystallizer	Draft-tube baffled	W = 34.2 ton/day	$205,000	$410,000
Filter	Rotary vacuum	A = 47 ft^2	$101,000	$202,000
Dryer	Direct-heat rotary	A = 220 ft^2	—	$155,000	■

Solids Handling Systems

The handling of bulk solids in a chemical process is considerably more complex than the handling of liquids and gases, and requires much more attention by operators. A typical handling system may include a storage bin, hopper, or silo; a feeder; and a conveyor and/or elevator to send the bulk solids to a piece of processing equipment. The selection of equipment for handling bulk solids depends strongly on the nature of the solids. The classification scheme of the FMC Corporation, as presented in Section 21 of *Perry's Chemical Engineers' Handbook* (1997), is particularly useful. In that scheme, bulk solids are classified by size, flowability, abrasiveness, and other special characteristics. For example, fine sodium chloride is classified as being particles between 100 mesh and $\frac{1}{8}$ in. diameter, free flowing with an angle of repose of 30 to 45°, mildly abrasive, mildly corrosive, and hygroscopic. Titanium dioxide particles, widely used as a pigment for whiteness, are particularly difficult to handle because of their small particle size (minus 325 mesh), irregular shape, and stickiness. In general, increased moisture content, increased temperature, decreased particle size, and increased time of storage at rest cause increased cohesiveness of the particles and decreased flowability.

A typical storage vessel for bulk solids, called a *bin*, consists of an upper section with vertical walls and a lower section with at least one sloping side, referred to as a hopper. The upper part of the bin is square or circular in cross section, while the hopper is frequently conical in shape. Below the hopper is a chute, gate, or a rotary star valve. The design of a bin is best accomplished by a method devised by Andrew W. Jenike in the 1960s and described in *Perry's Chemical Engineers' Handbook* (1997). This method differentiates between the more desirable uniform mass flow with all particles moving downward and the less desirable funnel flow where all particles are not in motion and solids flow downward only in a channel in the center of the vessel. The elimination of bridging and assistance in obtaining uniform flow across the cross-section of the bin can often be achieved by using a vibrating hopper. Typically, bin storage is provided for 8 hrs of operation. Most bins are constructed of carbon steel with or without rubber lining, fiberglass, or stainless steel. An equation for estimating the f.o.b. purchase cost of carbon steel bins is included in Table 16.32.

Bins may discharge bulk solids directly into a piece of processing equipment. Alternatively, the solids may be dropped onto a conveyor or into a feeder. Feeders are classified as volumetric or gravimetric. Volumetric feeders, which are the most common, discharge a volume of material per unit time, while more-expensive gravimetric feeders weigh the solids being discharged. If the bulk density of the solids is reasonably constant, volumetric feeders can confine mass flow rates to within a range of 5%. Volumetric feeders include *belts* (aprons), *rotary valves, screws, tables*, and *vibratory feeders*. Screw feeders are best for sticky materials, but belt and vibratory feeders also work in some cases. Gravimetric feeders work only with free-flowing solids and include weigh belts, loss-in-weight systems, and gain-in-weight systems. Purchase-cost equations for belt, screw, and vibratory feeders are included in Table 16.32.

Bulk solids are moved to other locations by conveyors (usually horizontal) or elevators (usually vertical). A wide range of conveyors is available, but the most common are the belt, screw, and vibratory conveyors. *Belt conveyors* can move coarse, corrosive, and abrasive particles, of 100 mesh to several inches in size, for distances up to 1,000 ft, with a modest degree of inclination but at temperatures normally limited to 150°F. Typical belt widths range from 14 in. to 60 in., with belt speeds ranging from 100 to 600 ft/min. Typical heights of bulk solids on the belt range from 1 in. for the narrowest belt to 6 in. for the widest belt. Typical volumetric flow capacities range from 660 ft³/hr for a 14-in.-wide belt moving at 100 ft/min to 86,000 ft³/hr for a 60-in.-wide belt moving at 600 ft/min.

Screw conveyors, consisting of a screw mounted in a trough, are widely used. They can move particles of any size up to a few inches, in any straight direction, for distances up to

300 ft horizontally and up to 30 ft vertically, and at temperatures up to 900°F. Screw conveyors can be fitted with special screws for sticky materials, can be sealed to keep in dust and keep out moisture, and can be jacketed for cooling and heating. The screw, which ranges from 6 to 20 in. in diameter, typically rotates at from 25 to 100 rpm. Volumetric flow capacities, when the trough is 30% full and the screw is rotating at 50 rpm, range from 75 ft³/hr for a 6-in.-diameter screw to 3,000 ft³/hr for a 20-in. diameter screw.

Vibratory conveyors are limited to straight distances, usually horizontal, up to 100 ft, but are not suited for fine particles less than 100 mesh in size. Solids must be free flowing, but temperatures up to 250°F can be handled. Widths range from 1 to 36 in. with pan heights to at least 5 in. Experiments are usually necessary to properly size a vibratory conveyor. For an 18-in.-wide conveyor with a 5-in. pan height and a 20-ft length, a typical mass flow capacity is 70,000 lb/hr or 700 ft³/hr for solid particles having a bulk density of 100 lb/ft³.

A *bucket elevator*, consisting of an endless chain of buckets, is best for moving solids vertically. The elevator loads at one level and discharges at another. Vertical distances of more than 3,000 ft have been spanned, but commonly available elevators are limited to 150 ft. Most often, discharge is by centrifugal force from buckets moving at speeds up to 1,200 ft/min. However, for materials that are sticky or that tend to pack, discharge is by gravity by completely inverting the buckets, which travel at lower speeds, up to 400 ft/min. Typical buckets range in width from 6 to 20 in., with bucket volumes from 0.06 to 0.6 ft³ at bucket spacings from 1 to 1.5 ft. For a bucket speed of 150 ft/min, typical volumetric capacities range from 300 to 7,500 ft³/hr.

Equations for estimating the purchase costs for the above four conveyor and elevator systems are included in Table 16.32. Belt and vibratory conveyors are priced by the width and the length of the conveyor. Screw conveyors are priced by the diameter of the screw and the length of the conveyor. Bucket elevators are priced by the width of the bucket and the elevated height. The costs do not include the electric motor and drive system. The required horsepower input of the electric motor drive, which depends on the mass flow rate, m, and the length of the conveyor, L, may be estimated from the equations given in Table 16.29, taken from Ulrich (1984). To the equations, additional power must be added for elevating the material by a height, h. The screw conveyor requires the largest amount of power, while the belt conveyor requires the least.

EXAMPLE 16.16

Flakes of phthalic anhydride with a bulk density of 30 lb/ft³ are to be conveyed a horizontal distance of 40 ft from a bin to a packaging facility at the rate of 1,200 ft³/hr. Size and cost a bin and conveying system as of a CE index of 394.

Table 16.29 Power Requirements of Mechanical Conveyors[a]

Conveyor Type	Power Equation[b]
Belt	$P = 0.00058 \, (m)^{0.82} L$
Screw	$P = 0.0146 \, (m)^{0.85} L$
Vibratory	$P = 0.0046 \, (m)^{0.72} L$
Bucket	$P = 0.020 \, m(L)^{0.63}$

[a]Reproduced with permission from Ulrich (1984)

[b]Units: P = Hp, m = lb/s, L = ft. For an elevation change, h, in ft, add or subtract $P = 0.00182 \, mh$.

SOLUTION

Assume a bin storage time of 8 hr. Therefore, the bulk solids volume is $8(1,200) = 9,600$ ft^3. Assume an outage (gas space above the bulk solids) of 20%. Thus, the bin volume above the hopper $= 9,600/(1 - 0.20) = 12,000$ ft^3. Neglecting the volume of the hopper below the bin and assuming a cylindrical bin with a height equal to 150% of the diameter, the hopper dimensions are 22 ft in diameter by 33 ft high. From Table 16.32, the purchase cost of the bin in carbon steel is $450(12,000)^{0.46} = \$34,000$. Because flakes may tend to mat and interlock, consideration should be given to the addition of a vibrator to the hopper.

A screw conveyor is a reasonable choice to transport the flakes and it can also act as a feeder to remove the flakes from the hopper. From the above discussion, for a trough running 30% full at 50 rpm, a 6-in. screw can transport 75 ft^3/hr while a 20-in. screw can transport 3,000 ft^3/hr. Assume that the flow rate is proportional to the screw diameter raised to the exponent, x. Then, the exponent is computed as 3. Therefore, the required screw diameter is computed as 15 in. From the equation in Table 16.32, for a length of 40 ft, the purchase cost of the screw conveyor is $55.6(15)(40)^{0.59} = \$7,400$. The cost of the motor and a belt drive must be added. From Table 16.29, for a mass flow rate, m, equal to $(1,200/3,600)30 = 10$ lb/s and with no elevation change, the electric motor Hp is $0.0146(10)^{0.85}40 = 4.13$. Assume a 5-Hp motor. From Table 16.22, select a totally enclosed, fan-cooled motor rotating at 1,800 rpm. Thus, F_T in Eq. (16.20) is 1.3. From Eq. (16.19), for $P_C = 5$ Hp, the base cost is $C_B =$ approximately \$380, giving a purchase cost of $1.3(380) = \$500$. This gives a total of \$7,900 for the conveyor with motor or a total of \$41,900 for the bin and conveyor. Add 10% to vibrate the hopper, cover the conveyor, and add a belt drive to the motor. This gives a total purchase cost for the system of \$46,100. ∎

Bulk solids may also be conveyed pneumatically as a dilute suspension in a gas, often air, through a piping system over distances of up to several hundred feet. Materials ranging in size from fine powders to $\frac{1}{4}$-in.-diameter pellets and in bulk densities up to 200 lb/ft^3 have been routinely conveyed in this manner. A pneumatic conveying system usually includes a blower to move the gas, a rotary air lock valve to control the rate of addition of the bulk solids to the gas, a piping system, and a cyclone to separate the solids from the gas at the discharge point. The pressure in the piping system may be below or above ambient pressure. Air velocities required to convey the solids typically range from 50 ft/s for low-bulk-density solids to 200 ft/s for high-bulk-density solids. The purchase cost of a *pneumatic conveying* system depends mainly on the bulk solids flow rate and the equivalent length (pipe plus fittings) of the piping system, but the particle size and bulk density of the solids are also factors. Table 16.32 includes an equation for estimating the purchase cost of a system to convey solids having a bulk density of 30 lb/ft^3. The power requirement, P, in horsepower depends mainly on the solids flow rate, m, in pounds per second as estimated by:

$$P = 10 \, m^{0.95} \tag{16.71}$$

Storage Tanks and Vessels

Storage tanks and vessels are used to store process liquid and gas feeds and products, as well as to provide intermediate storage between sections of a plant operating continuously or between operations for a batch or semicontinuous process. For liquid storage at pressures less than approximately 3 psig, so-called atmospheric tanks are used. These tanks may be open (no roof), cone roof, or floating-roof types. *Open tanks*, which may be made of fiberglass, are commonly used only for water and some aqueous solutions because they are subject to moisture, weather, and atmospheric pollution. *Cone roof* (or *other fixed-roof*) *tanks* require a vent system to prevent pressure changes due to changes in temperature and during changes in liquid level during filling or emptying. When the vapor pressure of the liquid over the expected range of storage temperature causes a significant rate of evaporation, a *floating-roof* (or *other variable volume*) *tank* should be used. Such tanks do not vent. Current EPA regulations dictate the use of a floating-roof tank when, at the maximum atmospheric temperature at the plant site, the vapor pressure of the liquid is greater than 3.9 psia for storage of less than 40,000 gal or greater than 0.75 psia for storage of more than 40,000 gal.

Storage of liquid feeds and products is usually provided off-site with residence times varying from one week to one month, depending on the frequency of delivery and distribution. The capacity of atmospheric liquid storage tanks should be at least 1.5 times the size of the transportation equipment, typically 4,000 to 7,500 gal for tank trucks and up to 34,500 gal for tank cars. A shipment by barge may be as large as 420,000 gal. The maximum size for a single cone-roof or floating-roof tank is approximately 20,000,000 gal, which corresponds to a diameter of about 300 ft and a height of about 50 ft. Storage of liquid feeds, products, and intermediates may also be provided onsite in so-called surge tanks or day tanks, which provide residence times of 10 min to one day. Equations for estimating the f.o.b. purchase costs of open, cone-roof, and floating-roof tanks are included in Table 16.32.

For liquid stored at pressures greater than 3 psig or under vacuum, spherical or horizontal (or vertical) cylindrical (bullet) pressure vessels are used. Vertical vessels are not normally used for volumes greater than 1,000 gal. Horizontal pressure vessels for storage are at least as large as 350,000 gal. Spherical pressure vessels are also common, with more than 5,000 having been constructed worldwide. For liquid storage, spheres as large as 94 ft in diameter (3,000,000 gal) have been installed. The design and costing of cylindrical pressure vessels is considered in detail in Section 16.5. Purchase costs are plotted in Figure 16.13. For spherical pressure vessels, Eq. (16.60) for cylindrical pressure vessels is revised to:

$$t_P = \frac{P_d D_i}{4SE - 0.4P_d} \qquad (16.72)$$

Equations for estimating the f.o.b. purchase costs of spherical pressure vessels, based on just the vessel volume, are included in Table 16.32 for two different pressure ranges.

Pressure vessels are also used for the storage of gases at pressures greater than 3 psig. For pressures between 0 and 3 psig, a *gas holder*, similar to a floating-roof tank for liquids, is used. An equation for estimating the f.o.b. purchase cost of a gas holder is included in Table 16.32.

Vacuum Systems

In some chemical processes, operations are conducted at pressures less than ambient. To achieve such pressures, vacuum systems are required. Pressures below ambient are commonly divided into four vacuum regions:

Vacuum Region	Pressure Range (torr)
Rough	760 to 1
Medium	1 to 0.001
High	0.001 to 10^{-7}
Ultrahigh	10^{-7} and below

Of greatest interest to chemical processing is the rough region, which covers most polymer reactors, vacuum distillation columns, vacuum stripping columns, pervaporation membrane separations, vacuum-swing adsorbers, and vacuum crystallizers, evaporators, filters, and dryers.

In the rough region, the available vacuum systems include: (1) one-, two-, and three-stage *ejectors* driven with steam and with or without interstage surface or barometric (direct-contact) condensers, (2) one- or two-stage *liquid-ring pumps* using oil or water as the sealant, and (3) *dry vacuum pumps* including rotary lobe, claw, and screw compressors. Although the first two systems have been the most widely used, dry vacuum pumps are gaining attention because they are more efficient and do not require a working fluid such as steam, water, or oil, which can contribute to air pollution. Typical flow capacities and lower limits of suction

Table 16.30 Lower Limits of Suction Pressure and Capacities of Vacuum Systems[a]

System Type	Lower Limit of Suction Pressure (torr)	Volumetric Flow Range at Suction Conditions (ft³/min)
Steam-jet ejectors		10–1,000,000
One-stage	100	
Two-stage	15	
Three-stage	2	
Liquid-ring pumps		3–18,000
One-stage water sealed	50	
Two-stage water sealed	25	
Oil-sealed	10	
Dry vacuum pumps		
Three-stage rotary-lobe	1.5	60–240
Three-stage claw	0.3	60–270
Screw compressor	0.1	50–1,400

[a]Reprinted with permission from Ryans and Bays (2001).

pressure for process applications of these three types of vacuum systems are given in Table 16.30, taken from Ryans and Bays (2001). This table is useful in making a preliminary selection of candidate vacuum systems based on the flow rate and pressure at suction conditions.

For batch processes where a vessel is being evacuated, the flow rate to be handled by the vacuum system depends on the selected time period for evacuation and on the contents of the vessel. When the flow contains condensables, a precondenser upstream of the vacuum system should be considered so as to reduce the flow rate to the vacuum pump. For continuous processes, the flow rate to be handled by the vacuum system is usually based on an estimate of air leakage into the equipment operating under vacuum. Air leakage occurs at gasketed joints, porous welds, and cracks and fissures in vessel walls. A simple, but often adequate estimate can be made based on the equipment volume and operating pressure with the following equation, derived from recommendations of the Heat Exchange Institute:

$$W = 5 + \{0.0298 + 0.03088[\ln(P)] - 0.0005733[\ln(P)]^2\}V^{0.66} \qquad \textbf{(16.73)}$$

where W is the air leakage rate in lb/hr, P is the absolute operating pressure in torr assuming a barometric pressure of 760 torr, and V is the vessel volume in ft³. For many pieces of equipment operating under a vacuum, the air leakage leaving the equipment will be accompanied by volatile process components. To partially recover these components and reduce the load on the vacuum pump, the exiting gas should first pass through a precondenser before proceeding to the vacuum system. The flow rates of process components still in the gas leaving the precondenser with the air can be determined by a flash calculation as illustrated in the example below.

Note, in Table 16.30, that steam-jet ejector systems can handle a very wide range of conditions. They have no moving parts and are inexpensive to maintain, but are very inefficient because of the high usage of motive steam. The maximum compression ratio per stage is approximately 7.5. The required motive steam rate for each stage depends on the ratio of suction pressure-to-discharge pressure, steam pressure, temperature, gas properties, and ejector nozzle-to-throat ratio. A reasonably conservative range for the total motive steam requirement for all stages, when using 100-psig steam to evacuate mostly air, is 5–10 lb of steam per pound of gas being pumped. A detailed procedure for designing an ejector vacuum system is given by Sandler and Luckiewicz (1987).

Liquid-ring pumps are limited to a suction pressure of 10 torr with a moderate capacity and efficiency (25–50%). Dry vacuum pumps can achieve very low pressures at higher efficiencies,

Table 16.31 Multiplying Factors for Steam-Jet Ejector Vacuum Systems

Items	Cost Multiplying Factors
1 Stage	1.0
2 Stages	1.8
3 Stages	2.1
1 Surface condenser	1.6
2 Surface condensers	2.3
1 Barometric condenser	1.7
2 Barometric condensers	1.9
Carbon steel	1.0
Stainless steel	2.0
Hastelloy	3.0

but only for low capacities. Since vacuum pumps are actually gas compressors, a tendency exists for the gas temperature to increase in an amount corresponding to the compression ratio. However, this temperature rise is greatly minimized or eliminated in ejector systems and with the liquid-ring pump because of the addition of another fluid. The temperature rise must not be overlooked with dry vacuum pumps.

The f.o.b. purchase costs for vacuum systems are included in Table 16.32. The equation for the one-stage ejector system in carbon steel is based on indexed data from Pikulik and Diaz (1977). Use the multiplying factors in Table 16.31 to add stages and interstage condensers, and change materials of construction. For the other vacuum systems, f.o.b. purchase costs in Table 16.32 were taken from Ryans and Bays (2001).

EXAMPLE 16.17

A vacuum distillation column produces an overhead vapor of 1,365 kmol/hr of ethylbenzene and 63 kmol/hr of styrene at 30 kPa. The volume of the column, vapor line, condenser, and reflux drum is 50,000 ft^3. The overhead vapor is sent to a condenser where most of the vapor is condensed. The remaining vapor at 50°C and 25 kPa is sent to a vacuum system. Determine the air leakage rate in the distillation operation and the flow rate to the vacuum system. Select an appropriate vacuum system and determine its f.o.b. purchase cost at a CE cost index of 400.

SOLUTION

The amount of air leakage, W, is estimated from Eq. (16.73). Using a pressure of 25 kPa = 188 torr:

$$W = 5 + [0.0298 + 0.03088[\ln(188)] - 0.0005733[\ln(188)]^2]V^{0.66} = 227 \text{ lb/hr}$$

This is equivalent to 103 kg/hr or 3.6 kmol/hr. Adding this to the overhead vapor and performing a flash calculation at 50°C and 25 kPa (188 torr) gives a vapor leaving the reflux drum and entering the vacuum system of 3.6 kmol/hr of air and 0.7 kmol/hr of ethylbenzene. The volumetric flow rate to the vacuum system is 272 ft^3/min. The flow rate in pounds per hour is 394. From Table 16.30, applicable vacuum systems are a single-stage steam-jet ejector, a single-stage liquid-ring pump, and a screw compressor. The three-stage claw unit is just out of the range of the volumetric flow rate.

From Table 16.32, the size factor for the ejector = S = 394/188 = 2.1. From the cost equation in Table 16.32, the f.o.b. purchase cost of the ejector in carbon steel and for a cost index of 400 is

$$C_P = (400/394)(1,330)(2.1)^{0.41} = \$1,830$$

The estimated 100-psig steam consumption is 10(394) = 3,940 lb/hr. Assuming a steam cost of $4.00/1,000 lb from Table 17.1 and operation for 8,000 hr/yr, the annual steam cost is 3.94(4.00)(8,000) = \$126,100/yr, which is far more than the cost of the ejector.

Next, consider the liquid-ring pump. From Table 16.32, with a size factor of 272 ft^3/min, in stainless steel and at CE = 400, the f.o.b. purchase cost is

$$C_P = (400/394)(6,500)(272)^{0.35} = \$47,000$$

Table 16.32 Purchase Costs (f.o.b.) of Other Chemical Processing Equipment, CE Index = 394.
Equations for pumps, compressors, motors, heat exchangers, and pressure vessels are in Section 16.5

Equipment Type	Size Factor (S)	Range of S	f.o.b. Purchase Cost Equation ($)	Notes
Adsorbents				
Activated alumina	Bulk volume, ft^3		$C_P = 35\,S$	
Activated carbon	Bulk volume, ft^3		$C_P = 25\,S$	
Silica gel	Bulk volume, ft^3		$C_P = 90\,S$	
Molecular sieves	Bulk volume, ft^3		$C_P = 60\,S$	
Agitators				Includes motor and shaft
Propeller, open tank	Motor Hp	1–8 Hp	$C_P = 1{,}810\,S^{0.34}$	Direct coupling to motor
Propeller, closed vessel	Motor Hp	1–8 Hp	$C_P = 2{,}600\,S^{0.17}$	Direct coupling to motor, pressures to 150 psig
Turbine, open tank	Motor Hp	2–60 Hp	$C_P = 2{,}590\,S^{0.54}$	Includes speed reducer
Turbine, closed vessel	Motor Hp	2–60 Hp	$C_P = 2{,}850\,S^{0.57}$	Includes speed reducer, pressures to 150 psig
Autoclaves				Includes turbine agitator and heat-transfer jacket
Steel	Vessel volume, gal	30–8,000 gal	$C_P = 825\,S^{0.52}$	Pressures to 300 psig
Stainless steel	Vessel volume, gal	30–2,000 gal	$C_P = 1{,}560\,S^{0.58}$	Pressures to 300 psig
Glass-lined	Vessel volume, gal	30–4,000 gal	$C_P = 1{,}450\,S^{0.54}$	Pressures to 100 psig
Crystallizers				
Continuous cooling				
Jacketed scraped wall	Length, L, ft	15–200 ft	$C_P = 11{,}400\,L^{0.67}$	Stainless steel
Continuous evaporative				
Forced circulation	Tons crystals/day, W	10–1,000 ton/day	$C_P = 27{,}500\,W^{0.56}$	Carbon steel
Draft-tube baffled	Tons crystals/day, W	10–250 ton/day	$C_P = 22{,}200\,W^{0.63}$	Carbon steel
Batch evaporative	Volume, V, ft^3	50–1,000 ft^3	$C_P = 32{,}200\,V^{0.41}$	Stainless steel
Drives other than electric motors				
Steam turbines (noncondensing)	Shaft power, P, Hp	250–10,000 Hp	$C_P = 7{,}400\,P^{0.41}$	Carbon steel
Steam turbines (condensing)	Shaft power, P, Hp	250–10,000 Hp	$C_P = 20{,}000\,P^{0.41}$	Carbon steel
Gas turbines	Shaft power, P, Hp	100–10,000 Hp	$C_P = 2{,}000\,P^{0.76}$	Carbon steel
Internal combustion engines	Shaft power, P, Hp	100–4,000 Hp	$C_P = 1{,}100\,P^{0.75}$	Carbon steel
Dryers				
Batch tray	Tray area, A, ft^2	20–200 ft^2	$C_P = 3{,}500\,A^{0.38}$	Stainless steel
Direct-heat rotary	Drum peripheral area, A, ft^2	200–3,000 ft^2	$C_P = \exp\{10.158 - 0.1003[\ln(A)] + 0.04303[\ln(A)]^2\}$	Stainless steel
Indirect-heat steam-tube rotary	Heat-transfer area, A, ft^2	100–1,400 ft^2	$C_P = 1{,}200\,A^{0.92}$	Stainless steel
Drum	Heat-transfer area, A, ft^2	60–480 ft^2	$C_P = 25{,}000\,A^{0.38}$	Stainless steel
Spray	Evaporation rate, W, lb/hr	30–3,000 lb/hr	$C_P = \exp\{8.0556 + 0.8526[\ln(W)] - 0.0229[\ln(W)]^2\}$	Stainless steel
Dust collectors				
Bag filters	Gas flow rate, actual ft^3/min	5,000–2,000,000	$C_P = \exp\{10.020 - 0.4381[\ln(S)] + 0.05563[\ln(S)]^2\}$	Carbon steel
Cyclones	Gas flow rate, actual ft^3/min	200–100,000	$C_P = \exp\{8.9845 - 0.7892[\ln(S)] + 0.08487[\ln(S)]^2\}$	Carbon steel
Electrostatic precipitators	Gas flow rate, actual ft^3/min	10,000–2,000,000	$C_P = \exp\{11.442 - 0.5300[\ln(S)] + 0.05454[\ln(S)]^2\}$	Carbon steel
Venturi scrubbers	Gas flow rate, actual ft^3/min	2,000–20,000	$C_P = \exp\{9.3773 - 0.3281[\ln(S)] + 0.0500[\ln(S)]^2\}$	Carbon steel
Evaporators				
Horizontal tube	Heat-transfer area, A, ft^2	100–8,000 ft^2	$C_P = 3{,}200\,A^{0.53}$	Carbon steel
Long-tube vertical (rising film)	Heat-transfer area, A, ft^2	100–8,000 ft^2	$C_P = 4{,}500\,A^{0.55}$	Carbon steel
Forced circulation	Heat-transfer area, A, ft^2	150–8,000 ft^2	$C_P = \exp\{8.0604 + 0.5329[\ln(A)] - 0.000196[\ln(A)]^2\}$	Carbon steel
Falling film	Heat-transfer area, A, ft^2	150–4,000 ft^2	$C_P = 10{,}800\,A^{0.55}$	Stainless steel tubes, carbon steel shell

(Continued)

Table 16.32 (*Continued*)

Equipment Type	Size Factor (S)	Range of S	f.o.b. Purchase Cost Equation ($)	Notes
Fired heaters for specific purposes				
Reformer	Heat absorbed, Q, Btu/hr	10–500 million Btu/hr	$C_P = 0.677\, Q^{0.81}$	Carbon steel, pressure to 10 atm
Pyrolysis	Heat absorbed, Q, Btu/hr	10–500 million Btu/hr	$C_P = 0.512\, Q^{0.81}$	Carbon steel, pressure to 10 atm
Hot water	Heat absorbed, Q, Btu/hr	0.5–70 million Btu/hr	$C_P = \exp\{9.3548 - 0.3769[\ln(Q)] + 0.03434[\ln(Q)]^2\}$	
Molten salt, mineral and silicon oils	Heat absorbed, Q, Btu/hr	0.5–70 million Btu/hr	$C_P = 9.71\, Q^{0.64}$	
Dowtherm A	Heat absorbed, Q, Btu/hr	0.5–70 million Btu/hr	$C_P = 9.83\, Q^{0.65}$	
Steam boiler	Heat absorbed, Q, Btu/hr	0.5–70 million Btu/hr	$C_P = 0.289\, Q^{0.77}$	Carbon steel, pressure to 20 atm
Heat exchangers, other				
Air-cooled fin-fan	Bare-tube heat-transfer area, A, ft^2	40–150,000 ft^2	$C_P = 1,970\, A^{0.40}$	Carbon steel
Compact units				
Plate-and-frame	Heat-transfer area, A, ft^2	150–15,000 ft^2	$C_P = 7,000\, A^{0.42}$	Stainless steel
Spiral plate	Heat-transfer area, A, ft^2	20–2,000 ft^2	$C_P = 4,900\, A^{0.42}$	Stainless steel
Spiral tube	Heat-transfer area, A, ft^2	1–500 ft^2	$C_P = \exp\{7.8375 + 0.4343[\ln(A)] + 0.03812[\ln(A)]^2\}$	Stainless steel
Liquid–liquid extractors				
Rotating disk contactors (RDC)	$S = $ (Height, H, ft)(Diameter, D, ft)$^{1.5}$	3–2,000 ft$^{2.5}$	$C_P = 250\, S^{0.84}$	Carbon steel
Membrane separations				
Reverse osmosis, seawater	Purified water, Q, gal/day	2–50 million gal/day	$C_{BM} = \exp\{0.8020[\ln(Q)] + 0.01775[\ln(Q)]^2\}$	Bare-module cost
Reverse osmosis, brackish water	Purified water, Q, gal/day	0.2–14 million gal/day	$C_{BM} = 2.1\, Q$	Bare-module cost
Gas permeation	Membrane surface area, A, ft^2	—	$C_P = 35\, A$	Membrane module
Pervaporation	Membrane surface area, A, ft^2	—	$C_P = 30\, A$	Membrane module
Ultrafiltration	Membrane surface area, A, ft^2	—	$C_P = 8\, A$ to 20 A	Membrane cartridge
Mixers for powders, pastes, polymers and doughs				
Kneaders, tilting double arm	Volume, V, ft^3	10–56 ft^3	$C_P = 1,400\, V^{0.58}$	Carbon steel
Kneaders, sigma double arm	Volume, V, ft^3	20–380 ft^3	$C_P = 1,300\, V^{0.60}$	Carbon steel
Muller	Volume, V, ft^3	10–380 ft^3	$C_P = 11,000\, V^{0.56}$	Carbon steel
Ribbon	Volume, V, ft^3	25–320 ft^3	$C_P = 1,700\, V^{0.60}$	Carbon steel
Tumblers, double cone	Volume, V, ft^3	50–270 ft^3	$C_P = 2,700\, V^{0.42}$	Carbon steel
Tumblers, twin shell	Volume, V, ft^3	35–330 ft^3	$C_P = 1,200\, V^{0.60}$	Carbon steel
Power recovery turbines				
Gas expanders (pressure discharge)	Power extracted, P, Hp	20–5,000 Hp	$C_P = 420\, P^{0.81}$	Carbon steel
Gas expanders (vacuum discharge)	Power extracted, P, Hp	200–8,000 Hp	$C_P = 940\, P^{0.81}$	Carbon steel
Liquid expanders	Power extracted, P, Hp	150–2,000 Hp	$C_P = 1,100\, P^{0.70}$	Carbon steel
Screens				
Vibrating grizzlies	Surface area, A, ft^2	6–40 ft^2	$C_P = 4,600\, A^{0.34}$	Carbon steel
Vibrating screens, 1 deck	Surface area, A, ft^2	32–60 ft^2	$C_P = 1,100\, A^{0.71}$	Carbon steel
Vibrating screens, 2 decks	Surface area, A, ft^2	32–192 ft^2	$C_P = 970\, A^{0.78}$	Carbon steel
Vibrating screens, 3 decks	Surface area, A, ft^2	48–192 ft^2	$C_P = 700\, A^{0.91}$	Carbon steel
Size enlargement				
Disk agglomerators	Feed rate, F, lb/hr	800–80,000 lb/hr	$C_P = \exp\{10.4947 - 0.4915[\ln(F)] + 0.03648[\ln(F)]^2\}$	Carbon steel
Drum agglomerators	Feed rate, F, lb/hr	800–240,000 lb/hr	$C_P = \exp\{11.1885 - 0.5981[\ln(F)] + 0.04451[\ln(F)]^2\}$	Carbon steel
Pellet mills	Feed rate, F, lb/hr	800–80,000 lb/hr	$C_P = 5,500\, F^{0.11}$	Carbon steel
Pug mill extruders	Feed rate, F, lb/hr	80–40,000 lb/hr	$C_P = \exp\{9.2486 - 0.01453[\ln(F)] + 0.01019[\ln(F)]^2\}$	Carbon steel
Screw extruders	Feed rate, F, lb/hr	8–800 lb/hr	$C_P = \exp\{10.5546 + 0.02099[\ln(F)]^2\}$	Carbon steel
Roll-type presses	Feed rate, F, lb/hr	8,000–140,000 lb/hr	$C_P = 91\, F^{0.59}$	Carbon steel
Tableting presses	Feed rate, F, lb/hr	800–8,000 lb/hr	$C_P = \exp\{8.9188 + 0.1050[\ln(F)] + 0.01885[\ln(F)]^2\}$	Carbon steel

(*Continued*)

Table 16.32 (Continued)

Equipment Type	Size Factor (S)	Range of S	f.o.b. Purchase Cost Equation ($)	Notes
Size reduction equipment				
Gyratory crushers	Feed rate, W, ton/hr	25–1,200 ton/hr	$C_P = 8,300\,W^{0.60}$	Includes motor and drive
Jaw crushers	Feed rate, W, ton/hr	10–200 ton/hr	$C_P = 1,800\,W^{0.89}$	Includes motor and drive
Cone crushers	Feed rate, W, ton/hr	20–300 ton/hr	$C_P = 1,400\,W^{1.05}$	Includes motor and drive
Hammer mills	Feed rate, W, ton/hr	2–200 ton/hr	$C_P = 3,000\,W^{0.78}$	Includes motor and drive
Ball mills	Feed rate, W, ton/hr	1–30 ton/hr	$C_P = 45,000\,W^{0.69}$	Includes motor and drive
Jet mills	Feed rate, W, ton/hr	1–5 ton/hr	$C_P = 27,000\,W^{0.39}$	Includes motor and drive
Solid–liquid separators				
Thickener, steel	Settling area, A, ft²	80–8,000 ft²	$C_P = 2,650\,A^{0.58}$	Carbon steel
Thickener, concrete	Settling area, A, ft²	8,000–125,000 ft²	$C_P = 1,900\,A^{0.58}$	Concrete
Clarifier, steel	Settling area, A, ft²	80–8,000 ft²	$C_P = 2,400\,A^{0.58}$	Carbon steel
Clarifier, concrete	Settling area, A, ft²	8,000–125,000 ft²	$C_P = 1,700\,A^{0.58}$	Concrete
Filters				
Plate-and-frame	Filtering area, A, ft²	130–800 ft²	$C_P = 3,800\,A^{0.52}$	Carbon steel
Pressure leaf	Filtering area, A, ft²	30–2,500 ft²	$C_P = 960\,A^{0.71}$	Carbon steel
Rotary-drum vacuum	Filtering area, A, ft²	10–800 ft²	$C_P = \exp\{11.432 - 0.1905[\ln(A)] + 0.0554[\ln(A)]^2\}$	Carbon steel
Rotary pan	Filtering area, A, ft²	100–1,100 ft²	$C_P = 19,500\,A^{0.48}$	Carbon steel
Wet classifiers (rake and spiral)	Solids feed rate, F, lb/hr	8,000–800,000 lb/hr	$C_P = 0.013\,F^{1.33}$	Carbon steel
Hydroclones	Liquid feed rate, Q, gal/min	8–1,200 gal/min	$C_P = 190\,Q^{0.50}$	Carbon steel
Centrifuges				
Batch top-drive vertical basket	Bowl diameter, D, in.	20–43 in.	$C_P = 1,600\,D^{0.95}$	Stainless steel
Batch bottom-drive vertical basket	Bowl diameter, D, in.	20–43 in.	$C_P = 680\,D^{1.00}$	Stainless steel
Vertical auto-batch	Bowl diameter, D, in.	20–70 in.	$C_P = 4,300\,D^{0.94}$	Stainless steel
Horizontal auto-batch	Bowl diameter, D, in.	20–43 in.	$C_P = 1,700\,D^{1.11}$	Stainless steel
Continuous reciprocating pusher	Tons solids/hr, S	1–20 tons solids/hr	$C_P = 120,000\,S^{0.30}$	Stainless steel
Continuous scroll solid bowl	Tons solids/hr, S	2–40 tons solids/hr	$C_P = 47,000\,S^{0.50}$	Stainless steel
Expression				
Screw presses	Wet solids flow rate, F, lb/hr	150–12,000 lb/hr	$C_P = \exp\{10.7951 - 0.3380[\ln(F)] + 0.05853[\ln(F)]^2\}$	Stainless steel
Roll presses	Wet solids flow rate, F, lb/hr	150–12,000 lb/hr	$C_P = \exp\{10.6167 - 0.4467[\ln(F)] + 0.06136[\ln(F)]^2\}$	Stainless steel
Solids handling systems				
Bins	Volume, ft³	10–100,000 ft³	$C_P = 450\,S^{0.46}$	Carbon steel at atmospheric pressure
Feeders				
Belt	Volumetric flow rate, ft³/hr	120–500 ft³/hr	$C_P = 565\,S^{0.38}$	Includes motor and belt drive
Screw	Volumetric flow rate, ft³/hr	400–10,000 ft³/hr	$C_P = 760\,S^{0.22}$	Includes motor and belt drive
Vibratory	Volumetric flow rate, ft³/hr	40–900 ft³/hr	$C_P = 32.4\,S^{0.90}$	Includes motor and belt drive
Conveyors				
Belt	width, W, in. Length, L, ft	14–60 in. up to 300 ft	$C_P = 16.9\,WL$	Does not include motor or drive
Screw	Diameter, D, in. Length, L, ft	6–20 in. up to 300 ft	$C_P = 55.6\,DL^{0.59}$	Does not include motor, drive, lid, or jacket
Vibratory	Width, W, in. Length, L, ft	12–36 in. up to 100 ft	$C_P = 64.3\,W^{0.57}L^{0.87}$	Does not include motor or drive
Bucket elevators	Bucket width, W, in. Height, L, ft	6–20 in. 15–150 ft	$C_P = 480\,W^{0.5}L^{0.57}$	Does not include motor or drive
Pneumatic conveyors	Solids flow rate, m, lb/s Equivalent length, L, feet	3–30 lb/s 30–600 ft	$C_P = 12,000\,m^{0.63}L^{0.20}$	Includes blower, motor, piping, rotary valve, and cyclone

(Continued)

Table 16.32 (*Continued*)

Equipment Type	Size Factor (S)	Range of S	f.o.b. Purchase Cost Equation ($)		Notes
Storage tanks					
Open	Volume, V, gal	1,000–30,000 gal	$C_P = 14\,V^{0.72}$		Fiberglass
Cone roof	Volume, V, gal	10,000–1,000,000 gal	$C_P = 210\,V^{0.51}$		Carbon steel, pressure to 3 psig
Floating roof	Volume, V, gal	30,000–1,000,000 gal	$C_P = 375\,V^{0.51}$		Carbon steel, pressure to 3 psig
Spherical, 0–30 psig	Volume, V, gal	10,000–1,000,000 gal	$C_P = 47\,V^{0.72}$		Carbon steel
Spherical, 30–200 psig	Volume, V, gal	10,000–750,000 gal	$C_P = 37\,V^{0.78}$		Carbon steel
Gas holders	Volume, V, ft^3	4,000–400,000 ft^3	$C_P = 2,500\,V^{0.43}$		Carbon steel, pressure to 3 psig
Vacuum systems					
One-stage jet ejector	(lb/hr)/(suction pressure, torr)	0.1–100 lb/hr-torr	$C_P = 1,330\,S^{0.41}$		See Table 16.31 for multistage units and condensers
Liquid-ring pumps	Flow at suction, ft^3/min	50–350 ft^3/min	$C_P = 6,500\,S^{0.35}$		Stainless steel with sealant recirculation
Three-stage lobe	Flow at suction, ft^3/min	60–240 ft^3/min	$C_P = 5,610\,S^{0.41}$		Includes intercoolers
Three-stage claw	Flow at suction, ft^3/min	60–270 ft^3/min	$C_P = 6,800\,S^{0.36}$		Includes intercoolers
Screw compressors	Flow at suction, ft^3/min	50–350 ft^3/min	$C_P = 7,560\,S^{0.38}$		With protective controls
Wastewater treatment					
Primary	Wastewater rate, Q, gal/min	75–75,000 gal/min	$C_{BM} = 11,700\,Q^{0.64}$		Bare-module cost
Primary + Secondary	Wastewater rate, Q, gal/min	75–75,000 gal/min	$C_{BM} = 34,000\,Q^{0.64}$		Bare-module cost
Primary + Secondary + Tertiary	Wastewater rate, Q, gal/min	75–75,000 gal/min	$C_{BM} = 69,000\,Q^{0.64}$		Bare-module cost

At an overall efficiency of 40% for compression to 100 kPa, the calculated horsepower input is 16.8 or 12.6 kW. Assuming an electricity cost of $0.04 kW/hr from Table 17.1 for 8,000 hr/yr, the annual electricity cost is 12.6(0.04)(8,000) = $4,000/yr, which is much less than for the ejector system.

These two vacuum systems can be compared on an annualized basis as discussed in Section 17.4, but it seems clear that the higher cost of the liquid-ring pump is more than offset by the much higher utility cost to operate the ejector system. The screw compressor is also a candidate, but its purchase cost, $65,000, is significantly higher and the annual electricity cost, at an overall efficiency of 70%, is only about $1,700/yr less than for the liquid-ring pump. ∎

Wastewater Treatment

Wastewater can contain inorganic and organic compounds in soluble, colloidal, insoluble liquid, and solid particulate forms. Before wastewater can be sent to a sewer or converted to drinking water, process water, boiler-feed water, or cooling water, it must be treated to remove certain impurities. Such treatment may consist of as many as three major treatment steps: primary, secondary, and tertiary. Primary treatment involves physical separation operations such as screening to remove large solids and sedimentation to remove smaller particulate matter, which settles to the bottom, and insoluble organic liquid, which floats to the top and is skimmed. Secondary treatment removes dissolved organic compounds by biological degradation with aerobic or anaerobic microorganisms in a recirculating activated sludge. This may produce settleable solids, which are removed by filtration. Removal of nitrogen and phosphorus nutrients, residual organic compounds, and dissolved inorganic compounds is accomplished in a tertiary treatment, which involves such operations as carbon adsorption, demineralization, and reverse osmosis. The water may also be disinfected with chlorine, ozone, or ultraviolet light. Equations for typical investment costs for wastewater treatment are included in Table 16.32. These are bare-module costs, rather than f.o.b. purchase costs.

16.7 EQUIPMENT SIZING AND CAPITAL COST ESTIMATION USING THE ASPEN ICARUS PROCESS EVALUATOR (IPE)

This section is provided in the file Section 16.7.pdf on the CD-ROM that accompanies this book. It introduces Aspen IPE and shows how to estimate equipment sizes, purchase costs, installation costs, and the total permanent investment. Two examples are provided:

1. *Depropanizer distillation tower*. This tower is presented on the CD-ROM (either HYSYS → Separations → Distillation or ASPEN PLUS → Separations → Distillation. Reference is made to the design procedure on the CD-ROM, which is carried out prior to the estimation of equipment sizes and costs. Beginning with the ASPEN PLUS file, RADFRAC.bkp, Aspen IPE maps the distillation unit (that is, estimates equipment sizes for the column, condenser, reflux accumulator, condenser pump, reboiler, and reboiler pump) and estimates its purchase and installation costs.

2. *Monochlorobenzene separation process*. This process was introduced in Section 4.4, with simulation results using ASPEN PLUS provided on the CD-ROM (ASPEN PLUS → Principles of Flowsheet Simulation → Interpretation of Input and Output → Sample Problem). Beginning with the file, MCB.bkp, the equipment sizes, purchase costs, and installation costs are estimated using Aspen IPE.

16.8 SUMMARY

Having completed this chapter, and some of the associated exercises, the reader should

1. Be able to assess the financial condition of a company by applying financial ratio analysis to data given in its annual report.

2. Be able to estimate the purchase costs of equipment items using the equations provided together with cost indexes to update those costs.

3. Be able to estimate each of the other costs included in the capital cost of a plant and apply the concept of the bare-module cost.

4. Be able to estimate the total capital investment of a plant by three methods of increasing complexity that range from order-of-magnitude to preliminary estimates.

5. Be prepared to use the Aspen Icarus Process Evaluator (IPE) system provided by Aspen Technology, Inc. to prepare a more definitive estimate of capital cost.

REFERENCES

Busche, R.M., *Venture Analysis: A Framework for Venture Planning—Course Notes*, Bio-en-gene-er Associates, Wilmington, Delaware (1995).

Corripio, A.B., K.S. Chrien, and L.B. Evans, "Estimate Costs of Centrifugal Pumps and Electric Motors," *Chem. Eng.*, **89**, 115–118, February 22 (1982a).

Corripio, A.B., K.S. Chrien, and L.B. Evans, "Estimate Costs of Heat Exchangers and Storage Tanks via Correlations," *Chem. Eng.*, **89**, 125–127, January 25 (1982b).

Feldman, R.P., "Economics of Plant Startups," *Chem. Eng.*, **76**, 87–90, November 3 (1969).

Garrett, D.E., *Chemical Engineering Economics*, Van Nostrand-Rineholt, New York (1989).

Guthrie, K.M., "Data and Techniques for Preliminary Capital Cost Estimating," *Chem. Eng.*, **76**, 114–142, March 24 (1969).

Guthrie, K.M., "Capital and Operating Costs for 54 Chemical Processes," *Chem. Eng.*, **77**(13), 140 (1970).

Guthrie, K.M., *Process Plant Estimating, Evaluation, and Control*, Craftsman, Solano Beach, California (1974).

Hand, W.E., "From Flow Sheet to Cost Estimate," *Petroleum Refiner*, **37**(9), 331–334, September (1958).

Haselbarth, J.E., "Updated Investment Costs for 60 Types of Chemical Plants," *Chem. Eng.*, **74**(25), 214–215, December 4 (1967).

Hill, R.D., "What Petrochemical Plants Cost," *Petroleum Refiner*, **35**(8), 106–110, August (1956).

Lang, H.J., "Engineering Approach to Preliminary Cost Estimates," *Chem. Eng.*, **54**(9), 130–133 (1947b).

Lang, H.J., "Cost Relationship in Preliminary Cost Estimation," *Chem. Eng.*, **54**(10), 117–121 (1947b).

Lang, H.J., "Simplified Approach to Preliminary Cost Estimates," *Chem. Eng.*, **55**(6), 112–113 (1948).

Mulet, A., A.B. Corripio, and L.B. Evans, "Estimate Costs of Pressure Vessels via Correlations," *Chem. Eng.*, **88**(20) 145–150, October 5 (1981a).

Mulet, A., A.B. Corripio, and L.B. Evans, "Estimate Costs of Distillation and Absorption Towers via Correlations," *Chem. Eng.*, **88**(26) 77–82, December 28 (1981b).

Papanastasiou, T.C., *Applied Fluid Mechanics*, Prentice-Hall, Englewood Cliffs, New Jersey (1994).

Perry, R.H., and D.W. Green, Eds., *Perry's Chemical Engineers' Handbook*, 7th ed., McGraw-Hill, New York (1997).

Peters, M.S., and K.D. Timmerhaus, *Plant Design and Economics for Chemical Engineers*, 2nd ed., McGraw-Hill, New York (1968).

Pikulik, A., and H.E. Diaz, *Chemical Engineering*, **84**(21), 107–116 October 10 (1977).

Ryans, J., and J. Bays, "Run Clean with Dry Vacuum Pumps," *Chem. Eng. Prog.*, **97**(10) 32–41 (2001).

Sandler, H.J., and E.T. Luckiewicz, *Practical Process Engineering*, McGraw-Hill, New York (1987).

Seader, J.D., and E.J. Henley, *Separation Process Principles*, Wiley, New York (1998).

Ulrich, G.D., *A Guide to Chemical Engineering Process Design and Economics*, John Wiley & Sons, Inc., New York (1984).

Walas, S.M., *Chemical Process Equipment*, Butterworth, London (1988).

Williams, R., "Standardizing Cost Data on Process Equipment," *Chem. Eng.*, **54**(6), 102, June (1947a).

Williams, R., "Six-Tenths Factor Aids in Approximating Costs," *Chem. Eng.*, **54**(12), 124–125, December (1947b).

Woods, D.R., *Financial Decision Making in the Process Industry*, Prentice-Hall, Englewood Cliffs, NJ (1975).

EXERCISES

16.1 On the Internet, find a recent annual report for the company, Merck & Co., Inc. As of August 2002, the 1998 report was available in PDF format. Based on information in that report, determine the following:

a. The nature of the business of Merck.

b. The new developments by Merck.

c. The new acquisitions or partnerships, if any.

d. Stated concerns of the company.

e. A financial ratio analysis, including:

1. Current ratio.

2. Acid-test ratio.

3. Equity ratio.

4. Return on total assets.

5. Return on equity.

6. Operating margin.

7. Profit margin.

f. Your stock purchase recommendation, including reasons for buying or selling the stock.

16.2 At the beginning of the year 2001, Company XYZ had an inventory of 8,000 widgets with a unit cost of $6.00. During that year, the following purchases of widgets were made:

Month of Purchase	Number of Units	Unit Cost ($)
February	10,000	$7.00
May	5,000	$8.00
June	15,000	$9.00
August	25,000	$10.00
November	20,000	$10.50

At the end of 2001, the number of units in the inventory is 2,900. Use both the fifo and lifo methods to determine the cost of goods sold for 2001.

16.3 Based on the following data for the reactors, compressors/ expanders, and distillation columns of a plant to produce 1,500 metric ton/day of methanol, with an operating factor of 0.95, estimate by Method 1 (order-of-magnitude estimate) the total capital investment. The two methanol reactors are of the shell-and-tube type. The plant will be constructed outdoors and is a major addition to existing facilities.

Equipment	Size	Material	Pressure (kPa)	Temperature (°C)
Steam reformer	620 million Btu/hr	316 ss	2,000	350
2 Methanol reactors	Each with 4,000 1.5-in. o.d. tubes by 30-ft long on 2.25-in. triangular pitch	cs shell 316 ss tubes	6,000	320
Reformed gas centrifugal compressor	16,000 kW	cs	6,000	200
Recycle gas centrifugal compressor	5,000 kW	cs	6,000	200
Tail-gas expander	4,500 kW	cs	6,000	200
Light ends tower	3-ft diameter 60 sieve trays	ss	500	200
Finishing tower	18-ft diameter 80 sieve trays	ss	200	200

ss = stainless steel

16.4 The feed to a sieve-tray distillation column operating at 1 atm is 700 lbmol/hr of 45 mol% benzene and 55 mol% toluene at 1 atm and its bubble-point temperature of 201°F. The distillate contains 92 mol% benzene and boils at 179°F. The bottoms product contains 95 mol% toluene and boils at 227°F. The column has 23 trays spaced 18 in. apart, and its reflux ratio is 1.25. Column pressure drop is neglected. Tray efficiency is 80%. Estimate the total bare-module cost of the column, condenser, reflux accumulator, combined reflux and distillate pump, reboiler, and reboiler pump. Also, estimate the total permanent investment. Results should be computed using (1) the equations in Chapter 16, and (2) Aspen IPE (Icarus Process Evaluator). Compare the results.

Data

Molal heat of vaporization of distillate = 13,700 Btu/lbmol
Molal heat capacity of distillate = 40 Btu/lbmol-°F
Overall U of condenser = 100 Btu/hr-ft^2°F
Cooling water from 90°F to 120°F
Heat flux for reboiler = 12,000 Btu/hr-ft^2
Saturated steam at 60 psia
Reflux accumulator residence time = 5 min at half full; L/D = 4
Pump heads = 50 psia; suction pressure = 1 atm, efficiency = 1

Calculate the flooding velocity of the vapor using the procedure in Example 14.3. Use 85% of the flooding velocity to determine the column diameter.

Notes

The file, BENTOLDIST.bkp, is included on the CD-ROM that accompanies this book. It contains the simulation results using the RADFRAC subroutine in ASPEN PLUS. This file should be used to prepare the report file for Aspen IPE. Note that the simulation was carried out using 20 stages (18 trays plus the condenser and reboiler). When using Aspen IPE, set the tray efficiency to 0.8 and Aspen IPE will adjust the number of trays to 23.

Since Aspen IPE does not size and cost a bottoms pump, a centrifugal pump should be added.

Aspen IPE estimates the physical properties and heat-transfer coefficients. Do not adjust these.

In Aspen IPE, reset the temperatures of cooling water (90 and 120°F) and add a utility for 60 psia steam. Use the steam tables to obtain the physical properties.

Use a kettle reboiler with a floating head.

Aspen IPE sizes the tower using a 24 in. tray spacing as the default. After sizing (mapping) is complete, adjust the tray spacing to 18 in. Note that the height of the tower must be adjusted accordingly.

Note that Aspen IPE estimates the costs of *Direct Material and Manpower* for each equipment item. These are also referred to as the costs of direct materials and labor, $C_{DML} = C_P + C_M + C_L$.

Include your hand calculations using the equations and the methods in Chapters 14 and 16. (Note that these methods are identical to those in Examples 14.3 and 16.12.)

Do not submit the entire Aspen IPE *Capital Estimate Report*. Instead, prepare a table showing a comparison of the equipment sizes and purchase costs. When using the methods in Chapter 16, show the bare module cost. For Aspen IPE, show the direct cost of materials and labor. It is sufficient to take the numbers from Aspen IPE. For both methods, show the calcula-

tions leading to the total permanent investment. Discuss the results.

16.5 Figure 16.14 shows a system designed to recover argon from the purge stream in an ammonia synthesis plant. Estimate the total bare-module cost associated with the addition of this argon recovery system to an existing plant. Assuming no allocated costs for utilities and related facilities, estimate the direct permanent investment, the total depreciable capital, and the total

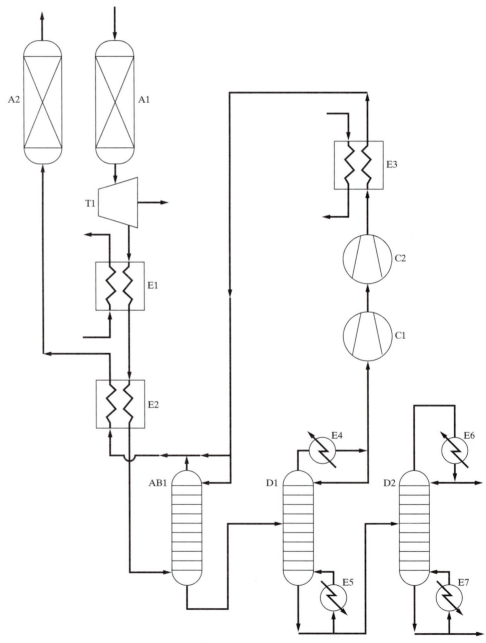

Figure 16.14 Argon recovery process.

permanent investment for the process. Include only the equipment shown in the flowsheet and specified below.

Equipment Specifications

Molecular sieve adsorbers (ignore packing):

	A1	A2
Diameter (ft)	10	10
Height (ft)	7	7
Pressure (psia)	2,000	1,000
Material	s.s.	s.s.

Absorber and distillation columns:

	AB1	D1	D2
Material	s.s.	s.s.	s.s.
Diameter (ft)	2	3.5	1.7
Height (ft)	45	50	15
Pressure (psia)	1,000	70	14.7
No. of trays	40	45	12
Tray spacing (ft)	1	1	1
Tray type	Sieve	Sieve	Sieve

Heat exchangers, reboilers, and condensers:

	E1	E2	E3	E4	E5	E6	E7
Area (ft^2)	993	401	891	2,423	336	2,828	55
Pressure (psia)	1,000	1,000	1,000	70	70	14.7	14.7

E1, E2, E3, E4, and E6 are shell-and-tube, floating-head, stainless steel heat exchangers.
E5 and E7 are stainless steel kettle reboilers.

Compressors:

	C1	C2
Type	Centrifugal	Centrifugal
Material	s.s	s.s
Efficiency (%)	80	80
Hydraulic horsepower (theoretical work)	122.4	130.4

Turbine (T1):

Type	Axial gas turbine
Material	s.s
Theoretical power	228.6 Hp
Efficiency (%)	90

16.6 The following are the major units in a chemical plant. Evaluate the bare-module cost for each unit and for the entire process. Assuming no allocated costs for utilities and related facilities, estimate the direct permanent investment, the total depreciable capital, and the total permanent investment for the process.

a. Two cast-steel centrifugal pumps (one a standby), each to handle 200 gal/min and producing a 200 psia head. The suction pressure is 25 psia and the temperature is ambient.

b. A process heater with heat duty of 20,000,000 Btu/hr. The tube material is carbon steel. The pressure is 225 psia.

c. A distillation column with 5 ft diameter, 25 sieve plates, and 2 ft tray spacing. The pressure is 200 psia and the material of construction is 316 stainless steel.

d. A distillation column with 2 ft diameter, 15 sieve plates, and 2 ft tray spacing. The pressure is 200 psia and the material of construction is 316 stainless steel.

e. A shell-and-tube heat exchanger with 3,200 ft^2 transfer surface, floating heads, a carbon steel shell, and stainless steel tubes at 200 psia.

f. A shell-and-tube heat exchanger with 7,800 ft^2 transfer surface, floating head, and stainless steel shell and tubes at 200 psia.

16.7 Determine the total bare-module cost for the flowsheet in Figure 16.15 at ambient temperature and pressure.

Design Specifications

Pumps: reciprocating, motor driven, 25 psia head
Heat exchangers: floating-head, $\Delta T_{LM} = 30°F$, $U = 15$ Btu/hr-ft^2-°F
Reslurry vessel and crystallizer: vertical, with $H/D = 2$
   All equipment: carbon steel

16.8 A chemical plant contains the following equipment:

Two gas-fired heaters, each with a heat duty of 20,000,000 Btu/hr. The tubes are carbon steel and the heaters operate at 225 psia.
Three distillation columns with 8 ft diameter, 25 sieve trays, and 2-ft tray spacing, constructed using solid 316 stainless steel and operating at 200 psia.
Evaluate the total bare-module cost for the equipment and for the entire plant. Assuming no allocated costs for utilities and related facilities, estimate the direct permanent investment, the total depreciable capital, and the total permanent investment for the process.

16.9 Estimate the total bare-module cost for installation of nine 600-Hp centrifugal compressors of carbon steel with explosion-proof electric motors, a stainless steel direct-heat rotary dryer of 6-ft diameter by 30-ft long and a continuous stainless steel scroll solid-bowl centrifuge processing 20 ton/hr of solids.

16.10 A plant contains

2 centrifugal compressors of carbon steel and 500 kW rating, with explosion-proof electric motors
1 jaw crusher at 10 kg/s capacity
3 floating-head shell-and-tube heat exchangers of stainless steel, rated at 400 m^2 and 100 barg on the tube side

Calculate the total bare-module cost.

16.11 Consider a 1-2, shell-and-tube heat exchanger:

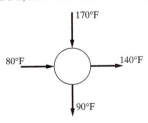

The cold stream has a heat-capacity flow rate $C = 40,000$ Btu/hr-°F. Its heat transfer coefficients are $h_i = h_o = 50$ Btu/(ft^2-hr-°F). For a stainless steel heat exchanger with a floating head, built to withstand pressures up to 100 barg, estimate the bare-module cost.

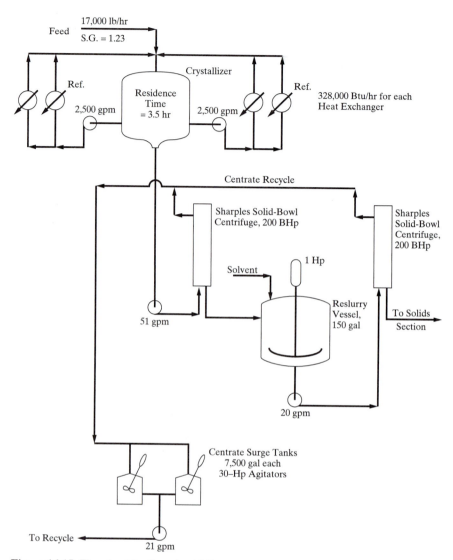

17,000 lb/hr
Feed
S.G. = 1.23

Crystallizer

Ref.

Residence
Time
= 3.5 hr

2,500 gpm

2,500 gpm

Ref.

328,000 Btu/hr for each
Heat Exchanger

Centrate Recycle

Sharples Solid-Bowl
Centrifuge, 200 BHp

Sharples
Solid-Bowl
Centrifuge,
200 BHp

1 Hp

Solvent

51 gpm

Reslurry
Vessel,
150 gal

To Solids
Section

20 gpm

Centrate Surge Tanks
7,500 gal each
30–Hp Agitators

To Recycle

21 gpm

Figure 16.15 Flowsheet for Exercise 16.7.

16.12 A chemical plant contains

3 drum dryers of nickel alloy, each containing 540 ft^2

40 kettle reboilers, with a carbon steel shell and copper tubes, at 1,450 psia, each containing 325 ft^2

Calculate the total bare-module cost.

16.13 Toluene Hydrodealkylation Process—Capital Cost Estimation. See Exercise 17.21 for a complete economic analysis, including equipment sizing, cost estimation, and calculation of the total capital investment.

Chapter 17

Annual Costs, Earnings, and Profitability Analysis

17.0 OBJECTIVES

Like the capital cost estimates of Chapter 16, methods presented in this chapter for the estimation of annual costs, annual earnings, and profitability measures play a crucial role throughout the design process in helping the design team to select the best design alternatives. The methods presented are those in common use and should be studied in connection with other chapters in this book as needed. In many cases, readers may prefer to study Sections 17.1 to 17.7 even before reading other chapters, especially when studying the techniques for process synthesis that require estimates of capital costs, annual costs, and annual earnings, followed by the calculation of profitability measures.

After studying this chapter, the reader should

1. Be able to estimate annual costs using a standard cost sheet and estimate the annual cash flows and the working capital. The latter completes the estimation of the total capital investment, C_{TCI}, in Table 16.9.
2. Be able to compute approximate profitability measures, such as return on investment (ROI), payback period (PBP), venture profit (VP), and annualized cost (C_A). These measures provide a snapshot view of the economic goodness, usually in the third year of operation. They do not include the time value of money, that is, compound interest.
3. Be able to compute the present worth and future worth of single payments and annuities and the capitalized-cost perpetuity. These measures are often used to compare proposals for the purchase of two competitive equipment items.
4. Be able to compute cash flows and depreciation, and use them to project the net present value and investor's rate of return (IRR) (also known as the discounted cash-flow rate of return, DCFRR), two measures that account for projections of revenues and costs over the life of the proposed process, and the time value of money.
5. Be able to use Aspen IPE in the Aspen Engineering Suite and an economics spreadsheet to carry out a profitability analysis for a potential process.

17.1 INTRODUCTION

Having completed an estimate for the total permanent investment, C_{TPI}, in Table 16.9, of a proposed plant, it remains to estimate the total annual sales revenue, S, the total annual production cost, C, and the annual pre-tax and after-tax earnings. This includes the development of the so-called *Cost Sheet*. Then the working capital can be estimated and added to the total permanent investment to give the total capital investment for the plant, as shown in Table 16.9. These provide the ingredients for an approximate measure of economic goodness, called the *return on investment*, defined by

$$\text{Return on Investment} = \frac{\text{annual earnings}}{\text{capital investment}} \qquad (17.1)$$

which is generally stated as a percentage per year. This definition includes a number of alternatives, depending on whether the annual earnings are before or after taxes and whether the capital investment includes land and working capital. The most common alternative for return on investment is based on the annual earnings after taxes and the total capital investment. This alternative is referred to here as ROI. A new commercial venture must compete with the *commercial interest rate* (or *cost of capital*), i, which is the annual rate at which money is returned to investors for the use of their capital, say in the purchase of high-grade bonds. The commercial interest rate is considered to be essentially without risk. Investments in chemical processing plants always entail *risk*. Therefore, to be attractive, an investment in a venture involving a new or revamped chemical processing plant must have an ROI greater than i. The greater the risk of the venture, the greater must be the difference between a financially attractive ROI and i. Establishment of the degree of risk involves answers to the following questions:

1. Is a new chemical to be produced? If so, are uses for it established and is there a sure market for it at its projected price?
2. Is an already commercially available chemical to be produced? If so, is the new plant going to utilize new technology that is predicted to reduce investment and/or operating costs? If so, how certain is the new technology? Does the technology involve potential environmental, safety, and/or control issues? Does the technology involve uncertainties with respect to materials of construction? If the new plant is going to use established, mature technology, is future demand for the chemical predicted to be greater than the current supply?
3. For the new plant, is the availability of the feedstocks (raw materials) ensured at a known price, or are the feedstocks controlled and/or produced by the company installing the new process?

As an example, suppose the current commercial interest rate is 10%. The proposed venture involves the manufacture of a new chemical at conditions of high temperature and pressure using new technology. The uses of the new chemical have been established and buyers have signed contracts to purchase the new chemical at an agreed upon price. The feedstocks are produced by and available from the company that will produce the new chemical. This degree of risk might be considered moderate, requiring an (ROI − i) of 15% or an ROI of at least 25%. A high-risk venture might require an (ROI − i) of 50%.

This chapter begins with the methods for estimating the remaining elements of the approximate ROI measure, as well as other comparable measures such as the *venture profit* (VP), *payback period* (PBP), and *annualized cost* (C_A), which are utilized often to compare alternatives during the early stages of process design, particularly during process synthesis. Rigorous profitability measures, which involve consideration of the *time value of money* and estimates of the *cash flows* throughout the life of the proposed process, are applied before making a final decision on a project when a company must assess carefully how it expends its limited capital. These measures, which include the *net present value* (NPV) and the *investor's rate of return* (IRR) (also referred to as the *discounted cash-flow rate of return*, DCFRR), incorporate one of a number of equipment depreciation schedules, based on U.S. tax laws, and account for the time value of money over the life of the process. These rigorous measures permit the design team to account for anticipated changes as well—for example, the need to replace the catalyst charge every 4 years, or the recognition that in 7 years the company patent will expire and the selling price will be reduced, and so on. To initiate the discussion of these rigorous measures, several subjects that involve the time value of money are discussed, including compound interest, annuities, and perpetuities such as capitalized costs. The effects of depreciation, inflation, depletion, and salvage value at the end of the life of a processing plant are also discussed.

The chapter concludes with a discussion of the use of the Aspen Icarus Process Evaluator (IPE) in the Aspen Engineering Suite and an economics spreadsheet to calculate profitability measures. As described in the previous chapter, Aspen IPE can be used in connection with a simulation program, such as ASPEN PLUS, CHEMCAD, HYSYS.Plant, or PRO/II, and it can be used independently.

17.2 ANNUAL SALES REVENUES, PRODUCTION COSTS, AND THE COST SHEET

Many continuing costs are associated with the operation of a chemical plant. These are included in the cost sheet shown in Table 17.1, which is patterned after one prepared by Busche (1995) and includes representative unit costs (typical factors) that can be used for early estimates when more exact costs are not available.

Sales Revenue

Before estimating the annual costs listed on the Cost Sheet, the total annual sales revenue, S, should be estimated. If S is not greater than the costs of the feedstock(s), there is no need to consider the process further. Typically, this calculation is made early, during the preliminary process synthesis step, as discussed in Section 3.4 and shown in Table 3.3 for the development of a vinyl chloride process, when five different reaction paths are being considered. As the process design proceeds, this calculation needs to be repeated after conversions and yields are better established by process design. The total sales revenue is based on the unit selling price(s) and on the quantity of product(s) produced for sale. If the process produces more than one main product, such as in Table 16.8, where plants produce ethylene–propylene, ammonia–urea, and chlorine–sodium hydroxide, the total sales revenue can include both products as co-products. Otherwise, additional products can be considered byproducts, for which an annual credit can be taken toward the cost of manufacture. Other possible credits include (1) gas, liquid, or solid effluents that can be used for fuel, (2) steam produced from boiler-feed water, and (3) electrical energy produced from a gas expander (turbine). If the streams with fuel value, steam, and/or electricity are used within the process, then the credit will be automatically accounted for. Otherwise, if to be used elsewhere, a transfer cost can be assigned to determine the credit against the cost of manufacture. The quantity of product(s) is obtained from the process design material balance and the estimated plant-operating factor or annual hours of plant operation.

Feedstocks

A major consideration in determining the cost of manufacture are the costs of the feedstocks, which may be natural resources such as petroleum, commodity chemicals such as chlorine, or fine chemicals. In the production of commodity chemicals, feedstock costs can be a significant contribution to the cost of manufacture, often in the range of 40 to 60% and even

Table 17.1 Cost Sheet Outline[a]

Cost factor	Typical factor in American engineering units	Typical factor in SI units
Feedstocks (raw materials)		
Utilities		
Steam, 450 psig	$5.50/1,000 lb	$12.10/1,000 kg
Steam, 150 psig	$4.00/1,000 lb	$8.80/1,000 kg
Steam, 50 psig	$2.50/1,000 lb	$5.50/1,000 kg
Electricity	$0.040/kW-hr	$0.040/kW-hr
Cooling water (CW)	$0.050/1,000 gal	$0.013/$m^3$
Process water	$0.50/1,000 gal	$0.13/$m^3$
Boiler feed water (BFW)	$1.50/1,000 gal	$0.40/$m^3$
Refrigeration, $-150°F$	$3.20/ton-day	$10.50/GJ
Refrigeration, $-90°F$	$2.60/ton-day	$8.55/GJ
Refrigeration, $-30°F$	$2.00/ton-day	$6.60/GJ
Refrigeration, $10°F$	$1.40/ton-day	$4.60/GJ
Chilled water, $40°F$	$1.00/ton-day	$3.30/GJ
Natural gas	$2.70/1,000 SCF	$0.113/SCM
Fuel oil	$0.75/gal	$200/$m^3$
Coal	$40/ton	$44/1,000 kg
Wastewater treatment	$0.10/lb organic removed	$0.22/kg organic removed
Landfill	$0.06/dry lb	$0.13/dry kg
Operations (labor-related) (O)		
(See Table 17.3)		
Direct wages and benefits (DW&B)	$30/operator-hr	$30/operator-hr
Direct salaries and benefits	15% of DW&B	15% of DW&B
Operating supplies and services	6% of DW&B	6% of DW&B
Technical assistance to manufacturing	$52,000/(operator/shift)-yr	$52,000/(operator/shift)-yr
Control laboratory	$57,000/(operator/shift)-yr	$57,000/(operator/shift)-yr
Maintenance (M)		
Wages and benefits (MW&B)		
Fluid handling process	3.5% of C_{TDC}	3.5% of C_{TDC}
Solid–fluid handling process	4.5% of C_{TDC}	4.5% of C_{TDC}
Solid handling process	5.0% of C_{TDC}	5.0% of C_{TDC}
Salaries and benefits	25% of MW&B	25% of MW&B
Materials and services	100% of MW&B	100% of MW&B
Maintenance overhead	5% of MW&B	5% of MW&B
Operating overhead		
General plant overhead	7.1% of M&O-SW&B	7.1% of M&O-SW&B
Mechanical department services	2.4% of M&O-SW&B	2.4% of M&O-SW&B
Employee relations department	5.9% of M&O-SW&B	5.9% of M&O-SW&B
Business services	7.4% of M&O-SW&B	7.4% of M&O-SW&B
Property taxes and insurance	2% of C_{TDC}	2% of C_{TDC}
Depreciation (see also Section 17.6)		
Direct plant	8% of $(C_{TDC} - 1.18C_{alloc})$	8% of $(C_{TDC} - 1.18C_{alloc})$
Allocated plant	6% of $1.18C_{alloc}$	6% of $1.18C_{alloc}$
COST OF MANUFACTURE (COM)	Sum of above	Sum of above
General Expenses		
Selling (or transfer) expense	3% (1%) of sales	3% (1%) of sales
Direct research	4.8% of sales	4.8% of sales
Allocated research	0.5% of sales	0.5% of sales
Administrative expense	2.0% of sales	2.0% of sales
Management incentive compensation	1.25% of sales	1.25% of sales
TOTAL GENERAL EXPENSES (GE)		
TOTAL PRODUCTION COST (C)	COM + GE	COM + GE

[a]DW&B = direct wages and benefits; MW&B = maintenance wages and benefits; M&O-SW&B = maintenance and operations salary, wages, and benefits. See Table 16.9 for C_{TDC} and C_{alloc}. 1 ton of refrigeration = 12,000 Btu/hr
Source: Busche (1995) with modifications.

higher. The required quantity of feedstock is obtained from the process design material balance and the estimated plant-operating factor. For example, in the earlier discussion in Chapter 3 of the production of vinyl chloride at the rate of 100,000 lb/hr, the process design material balance gives a chlorine feedstock flow rate of 113,400 lb/hr. If the plant-operating factor is based on 330 days/yr (operating factor = 330/365 = 0.904), the annual chlorine flow rate is 113,400(24)(365)(0.904) = 898.02 million lb/yr. If the delivered purchase cost of the chlorine is $0.11/lb, the annual chlorine feedstock cost is $98,782,000/yr. Similarly, the annual flow rate of ethylene, the other feedstock, is 44,900 lb/hr or 355.56 million lb/yr. At $0.18/lb, the annual ethylene feedstock cost is $64,001,000/yr. The total annual feedstock cost is $162,783,000. The process produces 100,000 lb/hr of vinyl chloride or 791.90 million lb/yr. At a selling price of $0.22/lb, the annual sales revenue is $174,218,000, which is greater than the total annual feedstock costs. The process also produces a byproduct of gaseous HCl, which, if it could be sold, could enhance the potential earnings of the process.

The feedstocks may be purchased from suppliers, or the company itself may produce one or more of the feedstocks or have control over a required natural resource. If a feedstock is to be purchased, availability from more than one supplier can keep the cost down. However, a long-range contract can ensure the availability of the feedstock. In the absence of price quotations from prospective suppliers of feedstocks, especially for early cost evaluations where the raw materials are commodity chemicals, the weekly newspaper *Chemical Market Reporter*, can be consulted, where most of the costs are for tank-car quantities. The prices quoted are representative of prices in the United States, but they are not location specific and may require an added delivery cost. Furthermore, the prices do not reflect the discounts that usually accompany long-term contracts. Also, larger production rates can increase the supply to such an extent that the price is reduced for a given demand. For specialty chemicals, such as pharmaceuticals, prices can be obtained from their manufacturers. If the company manufactures a feedstock or controls it, a *transfer price* must be assigned. The transfer price is an agreed-upon price between the company division that supplies the feedstock and the division that manufactures the products from the feedstock. The transfer price may be: (1) the market price, or (2) a price negotiated between the two divisions, recognizing that the price influences the selling division's revenue and the buying division's cost.

Utilities

Except for certain processes involving inexpensive feedstocks, such as the manufacture of oxygen, nitrogen, and argon from air or the production of hydrogen and oxygen from water, the annual cost of utilities, while much smaller than the feedstock costs, is not an insignificant contribution to the selling price of the product(s), often in the range of 5 to 10%. As listed in the Cost Sheet of Table 17.1, utilities include steam for heating at two or more pressure levels, electricity, cooling water, process water, demineralized boiler feed water, refrigeration at different temperature levels, fuels such as natural gas, wastewater treatment, waste disposal, and landfill. Often, the largest utility cost is that of steam.

The company may purchase utilities from a public or private utility company or build their own utility plants. Credit can be taken for any utilities, for example, fuel, steam, and electricity, produced by the process. Purchased utility costs are based on consumption. For company-owned utilities, both capital costs and operating costs apply. A cogeneration unit using a fuel can supply electricity accompanied by low-, medium-, and high-pressure steam. For early estimates, the purchase of utilities can be assumed, using the unit costs in Table 17.1.

Steam

Steam has many potential uses in a process, both as a process fluid and as a utility. In the former category, it may be used as a feedstock, as an inert diluent in a reactor to absorb heat of reaction, as a direct heating agent, and as a stripping agent in absorbers and adsorbers. As a utility, it can be used in place of electricity to drive pumps and compressors, and in ejectors to produce a vacuum. Steam is used in heat exchangers to heat liquids and gases and vaporize liquids. Typical pressure levels for steam are 50, 150, and 450 psig, with high-pressure steam more costly than steam at the lower pressures. In computing steam utility requirements, credit is only taken for the latent heat of vaporization. No credit is taken for sensible heat of the steam even though the steam may arrive at the heat exchanger superheated and leave as subcooled condensate. This is illustrated in the following example.

EXAMPLE 17.1

A kettle reboiler is used to evaporate toluene at 375°F with a heat duty of 3,000,000 Btu/hr. Steam is available at 50, 150, and 450 psig. Determine the steam pressure level to use, the steam flow rate in lb/hr and lb/yr, and estimate the annual steam cost if the plant-operating factor is 0.90.

SOLUTION

Assuming a barometric pressure of 14 psia, the saturation temperature of 150-psig steam is 365°F. Since this is lower than 375°F, 450-psig steam must be used. However, to avoid film boiling, as discussed in Chapter 13, use an overall temperature driving force of 45°F, which sets the steam condensing temperature at 375 + 45 = 420°F, corresponding to a saturated steam pressure of 309 psia. With a valve, assumed to operate adiabatically, the pressure of the high-pressure steam is reduced from 464 psia to 309 psia, producing a superheated vapor. However, the steam is assumed to enter the heat exchanger as a saturated vapor at 420°F and 309 psia and leave as a saturated liquid at 420°F. At these conditions, the latent heat of vaporization of steam is 806 Btu/lb. By an energy balance, the hourly steam requirement is 3,000,000/806 = 3,722 lb/hr or 3,722(24)(365)(0.9) = 29,350,000 lb/yr. From Table 17.1, the cost of 450-psig steam is $5.50/1,000 lb. Therefore, the annual steam cost = 29,350,000(5.50)/1,000 = $161,400/yr. ∎

Electricity

The operation of many items of processing equipment requires energy input in the form of a drive or motor. This equipment includes pumps, compressors, blowers, fans, agitators and mixers, feeders, conveyors, elevators, crushers, grinders, mills, scraped-wall crystallizers, agitated-film evaporators, agitated and centrifugal liquid–liquid extractors, centrifuges and rotary vacuum filters, rotary dryers and kilns, and spray and drum dryers. Although many types of drives are available, including air expanders, combustion gas turbines, internal combustion engines, and steam turbines, the most common drives are electric motors because they are very efficient (>90%); very reliable; readily available in a wide range of wattages (Hp), shaft speeds, and designs; long lasting; and offer convenience, small footprint, favorable cost, and ease of maintenance. Electric motors are almost always used for power up to 200 Hp and are even available for applications requiring in excess of 10,000 Hp.

Alternating, rather than direct, current is used almost exclusively in electric motors. In the United States of America the alternating current cycles from positive to negative and back to positive 60 times a second, referred to as 60 hertz (Hz). This rate works well with electric clocks, since there are 60 seconds in a minute. In Europe, for some reason, 50 Hz is used. Alternating-current electricity originates at offsite private, public, governmental, or company facilities, generally at 18, 22, or 24 kilovolts (kV). For transmission to the plant onsite location, transformers step up the voltage to as high as 700 kV and then step it down to user voltages that are mainly in the range of 120 to 600 volts, but may be as high as 13,800 volts for motors of very high horsepower. Of the many electric motor designs, the most common is the three-phase, alternating-current, constant-speed, squirrel-cage induction motor. Induction motors of 60 Hz are capable of being operated satisfactorily on 50 Hz circuits if their voltage and horsepower ratings are reduced by a factor of 50/60. Motor enclosures may be explosion-proof (for locations near combustible fluids and dust), open drip-proof (to prevent entrance of liquid drips and dirt particles, but not vapor, dust, or fumes), or totally enclosed.

For a given amount of energy transfer, the cost of electricity usually is greater than the cost of steam, as illustrated in the following example.

EXAMPLE 17.2	An electric motor is to be used to drive a compressor of 1,119 brake horsepower (BHp). The efficiency of the motor is 95%. Therefore, the electrical input to the motor must be 1,119(0.7457)/0.95 = 878 kW. This is equivalent to 3,000,000 Btu/hr, which is the basis for the previous example. Calculate the kW-hr required per year for the motor if the plant-operating factor is 0.9, and calculate the cost of electricity per year.
SOLUTION	The plant will operate (365)(24)(0.9) = 7,884 hr/yr. Therefore, the motor requires 878(7,884) = 6,922,000 kW-hr/yr. From Table 17.1, the cost of the electricity is 6,922,000(0.04) = $276,900/yr. ∎

Cooling Water

Cooling water is used to cool liquids and gases and condense vapors. Typically cooling water circulates between a cooling tower and process heat exchangers by means of a pump. For preliminary design purposes, it can be assumed that the cooling water enters a heat exchanger at 90°F and exits at 120°F. In the cooling tower, direct contact of downward flowing water with air, forced upward by a fan, causes the water temperature to approach within about 5°F of the wet-bulb temperature of the air. Approximately eighty percent of the reduction of the temperature of the cooling water is accomplished by evaporation of a small amount of the cooling water, with the balance caused by the transfer of heat from the cooling water to the surrounding air. In addition to the evaporation in the cooling tower, cooling water is also lost by drift (entrained water droplets in the cooling tower air discharge) and blowdown (deliberate purging of untreated cooling water to prevent buildup and subsequent precipitation) of dissolved salts in the cooling water. Typically, makeup cooling water is 1.5 to 3% of the circulating cooling water rate. Alternatives to cooling towers include spray ponds and cooling ponds. Warm water spread out over a large area of impervious ground in an open pond can cool by evaporation, convection, and radiation. The rate of cooling can be increased by recirculating the water through spray nozzles, much like a fountain.

When a plant is located near a river or large body of water, cooling water can be drawn off at one location, pumped to the heat exchangers, and discharged downstream in a river or to another location in a lake, bay, or ocean. It is customary to filter this water, but not treat it to remove salts and similar impurities.

In general, the cost of cooling water is much less than the cost of steam for a given heat exchanger duty, as illustrated in the following example.

EXAMPLE 17.3	Cooling water is used in the overhead condenser of a distillation column, with a heat duty of 3,000,000 Btu/hr. Determine the gallons per minute (gpm) of cooling water required and the annual cost of the cooling water, if the plant has an operating factor of 0.90.
SOLUTION	Assume the cooling water enters the condenser at 90°F and exits at 120°F. Water has a specific heat of 1 Btu/lb-°F and a density of 8.33 lb/gal. Therefore, by an energy balance, the condenser requires 3,000,000/[(1)(120 − 90)] = 100,000 lb/hr or 100,000/[(60)(8.33)] = 200 gpm. The total gallons per year = 200(60)(24)(365)(0.9) = 94,600,000 gal. From Table 17.1, the cost of cooling water = $0.05/1,000 gal. Therefore the annual cost = 0.05(94,600,000)/(1,000) = $4,730/yr. ∎

Process Water and Boiler Feed Water

Water is needed for many purposes in a chemical processing plant, including cooling water (discussed above), boiler feed water, and process water, which is water that enters directly into the process, rather than being used indirectly as a heat-transfer agent. Process water must be purified to the extent necessary to avoid introduction of any undesirable chemicals into the process that could poison catalysts, foul equipment, and/or introduce impurities into products. Boiler feed water (BFW) is used to produce steam in offsite boiler or cogeneration facilities. BFW can also be used as a cooling agent in a process, in place of cooling water, when the temperature of a process stream to be cooled exceeds approximately 300°F. In the heat exchanger, the BFW is vaporized to steam, which may find use elsewhere in the process. Whether BFW is used onsite or offsite, it must be demineralized before use to avoid fouling of heat-exchanger tubes. Sources of water include municipal water, well water, river water, lake water, ocean water, brackish water, treated wastewater, and condensate. The cost of process water, given in Table 17.1 as $0.50/1,000 gal, corresponds to only a moderate degree of pretreatment. The annual cost of process water, when needed, is usually very small compared to other feedstocks. When BFW is used to produce steam in a process heat exchanger, the cost of the BFW is partially offset by the value of the steam produced from it. This value is taken as a credit. In Table 17.1, the cost of BFW is given as $1.50/1,000 gal, which accounts for the credit. Extensive treatment of water containing large amounts of impurities can raise the cost to as much as $5.00/1,000 gal. Sterilized water for the manufacture of pharmaceuticals can cost as much as $450/1,000 gal. The use of BFW in a process heat exchanger is illustrated in the following example.

EXAMPLE 17.4

A process for hydrogenating benzene to cyclohexane, described in Chapter 8, includes a well-mixed reactor, where an exothermic reaction occurs at 392°F and 315 psia. A total of 4,704,200 Btu/hr of heat must be transferred out of the reactor. Although this heat could be transferred into cooling water, the temperature in the reactor is sufficiently high to consider transferring the heat, by means of a heat exchanger, to boiler feed water to produce steam. Determine the pressure level of the steam that could be produced, the pounds per hour of BFW required, and the annual cost of the BFW. The plant-operating factor is 0.9.

SOLUTION

To ensure nucleate boiling of the BFW, assume an overall driving force of 45°F, as discussed in Chapter 13. Thus, the BFW will be converted to steam at $(392 - 45) = 347°F$. This corresponds to a saturation pressure of 130 psia. Assume the BFW enters the heat exchanger as liquid at 90°F and exits as saturated vapor at 347°F. The change in enthalpy = 1,134 Btu/lb. Therefore, by an energy balance, the steam produced from the BFW = 4,704,200/1,134 = 4,148 lb/hr. The annual cost of the BFW at $1.50/1,000 gal with water at a density of 8.33 lb/gal is $4,148(24)(365)(0.9)(1.50)/[(8.33)(1,000)] = $5,889/yr. ∎

Refrigeration

The two most common coolants are cooling water and air. In general, cooling water from a cooling tower or pond can be used to cool a process stream to 100°F, while air, which is used in desert locations or where water is in short supply, can cool to only 120°F. To cool and/or condense process streams to temperatures below 100°F, chilled water, chilled brine, or a refrigerant is necessary, with the latter category being the most common. Several refrigerants are listed in Table 13.1. Prior to 1995, when the U.S. Clean Air Act Amendments of 1990 went into effect, two of the most popular refrigerants were CFC Freon R-12 (dichlorodifluoromethane) and HCFC Freon R-22 (chlorodifluoromethane). The chlorine atoms in these refrigerants were found to be released in the stratosphere, causing a depletion of the ozone layer. Since 1995, production of these two refrigerants and other chlorofluorocarbons has ceased or has been curtailed, as discussed in Section 2.3. A common replacement for R-12 is 1,1,1,2-tetrafluoroethane (R-134a, which is the same as HFC-134a), which according to the EPA will not propagate a flame under normal conditions in open air, shows no evidence of toxicity below 400 ppm, and does not deplete the ozone layer. This refrigerant, as well as ammonia, and several light hydrocarbons, cools by transferring heat from the process stream in a heat exchanger where the refrigerant is evaporated. A typical propane refrigeration system is shown in Figure 9.20, where the propane is circulated by a compressor through a condenser, a valve or turbine, and an evaporator where the cooling takes place. The temperature driving force in the refrigerant evaporator is typically in the range of 2 to 10°F. Thus, with R-134a, which from Table 13.1 has an operating range of −15 to 60°F, a process stream can be cooled down to within 2°F of −15°F or −13°F. For lower temperatures, ammonia or the light hydrocarbons can be used. For example, an ethylene refrigerant can cool a stream down to as low as −148°F. The lower-limit temperature of operation of the refrigerant corresponds approximately to its normal boiling point. Lower temperatures would require an undesirable vacuum on the refrigerant side of the evaporator.

Typically, a chemical plant must provide its own refrigeration, either offsite, or more typically onsite. Petroleum refineries use light-hydrocarbon refrigerants, while other plants may consider ammonia and R-134a, unless very low temperatures ($< −30°F$) are required, where cascade refrigeration systems are often used, as discussed in Examples 10.19 and 10.21. These systems are often included with the equipment in the process flowsheet.

The temperature level is the key factor in determining the cost of refrigeration. For moderate temperatures, estimates of annual operating cost are based on a ton-day of refrigeration, where a ton is defined as the heat removal to freeze 1 ton (2,000 lb) per day of water at 32°F, which corresponds to 12,000 Btu/hr. Calculation of the annual refrigeration cost is illustrated in the following example. For a given energy transfer rate, the cost of a moderate level of refrigeration compares to the cost of steam for heating.

EXAMPLE 17.5

A process stream in a petroleum refinery is to be partially condensed and cooled to 10°F, with a cooling duty of 3,000,000 Btu/hr. Select a suitable refrigerant, and calculate the tons of refrigeration required and the annual operating cost if the plant-operating factor is 0.9.

SOLUTION

From Table 13.1, a suitable refrigerant for a petroleum refinery is propane, since it has an evaporation range of −40 to 20°F. The propane evaporation temperature would be approximately 5°F. The tons of refrigeration = 3,000,000/12,000 = 250 tons. The annual refrigeration is 250(365)(0.9) = 82,130 ton-day. From Table 17.1, for this temperature level, the annual operating cost = 82,130(2.00) = $164,300/yr. Most of this cost is the cost of electricity to drive the propane compressor. ∎

Fuels

Various fuels may be combusted in a chemical process to provide heat or work. Besides the fuels needed for the off-site facilities, such as boilers, electrical power generation, and cogeneration, fuels such as coal, natural gas, manufactured gas, and/or fuel oil may be needed for high-temperature heating in furnaces and fired heaters. Also, fuels may be used to drive pumps and compressors. Typically, the fuel, whether it be solid, liquid, or gas, is burned completely with an excess amount of air. To determine the amount of fuel required, the heating value (heat of combustion) of the fuel must be known. Two heating values are in common use, the *higher heating value* (also called the gross heating value), HHV, and the *lower heating value* (also called the net heating value), LHV. The heating value is the total heat evolved by complete combustion of a fuel with dry air when the fuel and air are at 60°F before combustion and all of the flue gas (product from combustion) is brought to 60°F after combustion. If the water vapor in the cooled flue gas is not condensed, the total heat is the LHV. If the water vapor is condensed, additional heat is evolved, giving the HHV. Some typical heating values for common fuels are given in Table 17.2. A manufactured gas is not listed; typically, it contains mainly H_2, CO, CH_4, and N_2 over wide ranges of composition. Note that heating values for solid and liquid fuels are usually quoted on a mass basis, while gaseous fuels are on a volume basis, usually standard cubic foot (SCF) at 1 atm and 60°F. For a given heat transfer rate to a process stream being heated and/or vaporized in a fired

Table 17.2 Typical Heating Values of Fuels

Fuel	HHV	LHV
Pennsylvania anthracite coal	13,500 Btu/lb	
Illinois bituminous coal	12,500 Btu/lb	
Wyoming subbituminous coal	9,500 Btu/lb	
North Dakota lignite coal	7,200 Btu/lb	
No. 2 fuel oil (33° API)	139,000 Btu/gal	131,000 Btu/gal
No. 4 fuel oil (23.2° API)	145,000 Btu/gal	137,000 Btu/gal
Low-sulfur No. 6 fuel oil (12.6° API)	153,000 Btu/gal	145,000 Btu/gal
Methyl alcohol	9,550 Btu/lb	
Ethyl alcohol	12,780 Btu/lb	
Benzene	17,986 Btu/lb	17,259 Btu/lb
Hydrogen	322 Btu/SCF	272 Btu/SCF
Carbon monoxide	321 Btu/SCF	321 Btu/SCF
Methane	1,012 Btu/SCF	907 Btu/SCF
Ethane	1,786 Btu/SCF	1,601 Btu/SCF
Propane	2,522 Btu/SCF	2,312 Btu/SCF
Natural gas (85–95 vol% methane)	1,020–1,090 Btu/SCF	920–990 Btu/SCF

heater, the amount of fuel required is greater than that based on its HHV because of heat losses, a flue gas temperature much greater than 60°F, and the presence of water as vapor in the flue gas. The ratio of the amount of fuel based on the HHV to the actual amount is the fired heater thermal efficiency, which may range from 50 to 80%. Typical fuel costs are included in Table 17.1. The calculation of the fuel requirement for a fired heater is illustrated in the following example.

EXAMPLE 17.6

A fired heater is to be used to heat and vaporize, from 1,000 to 1,200°F, the feed to a reactor. The heat duty is 3,000,000 Btu/hr. The fuel is natural gas with an HHV of 1,050 Btu/SCF. The thermal efficiency is 70%. If the plant-operating factor is 0.9, compute the SCF/hr and SCF/yr of natural gas required and the annual fuel cost.

SOLUTION

For an efficiency of 70%, the heat evolved from combustion of the fuel is 3,000,000/0.7 = 4,286,000 Btu/hr. The natural gas must be supplied at a rate of 4,286,000/1,050 = 4,082 SCF/hr or a rate of 4,082(24)(365)(0.9) = 32,180,000 SCF/yr. From Table 17.1, the cost of natural gas is $2.70/1,000 SCF. Therefore, the annual cost is 32,180,000(2.70)/1,000 = $86,900/yr. ∎

Waste Treatment

Most chemical processes produce waste streams: gaseous (with or without particles), liquid (with or without particles, dissolved gases, and dissolved solids), solids (wet or dry), and slurries. In some cases, valuable by-products can be removed from waste streams by additional processing. However, when this is not economical, federal regulations require that waste streams be treated to remove pollutants before being sent to the surrounding air, a sewer, a pond, a nearby river, a lake, an ocean, or a landfill.

Air-Pollution Abatement

Waste gases may contain particulates and/or gaseous pollutants, inorganic or organic. Additional equipment must be added to the process to remove these pollutants. If that equipment requires utilities, their costs must be added to the other utility costs. The removal of particles is usually accomplished with cyclone collectors, wet scrubbers, electrostatic precipitators, and fabric-filter systems. Inorganic gaseous pollutants, such as ammonia; chlorine and fluorine; oxides of sulfur, carbon, and nitrogen; hydrogen sulfide, chloride, fluoride, and cyanide; and organic gaseous pollutants such as hydrocarbons and oxygenated organic compounds can be removed by absorption, adsorption, condensation, and/or combustion.

Wastewater Treatment

When water is fed into a process and/or the process produces water, wastewater is usually one of the process effluents. This wastewater must be treated for the removal of pollutants before being discharged to a sewer, pond, or body of water. In the United States of America, the treatment is regulated by the U.S. Water Pollution Act of 1972. The treatments necessary depend on the nature of the foreign material, whether it is suspended or dissolved in the water. When private or municipal sewage treatment plants are nearby, the wastewater can be sent directly to those plants. However, pretreatment may be required to neutralize the water, remove large solids, and remove grease and oil. If the treatment facilities are located onsite or offsite, equipment for several treatments should be considered. *Primary treatment*, using gravity sedimentation or clarification, is used to remove suspended solids. *Secondary treatment* adds aerobic biological organisms (those requiring molecular oxygen for metabolism) in the form of a sludge to cause oxidation of the dissolved biodegradable organic compounds to carbon dioxide, water, sulfates, etc. Excess flocculated suspension (activated sludge) is removed from the water by clarification or air flotation. A measure of the biodegradability is the biochemical oxygen demand (BOD), which is the amount of oxygen required in parts per million of water by mass, in milligrams per liter of water. *Tertiary treatment* adds one or more additional chemical treatments to remove acids, alkalies, colloidal matter, color, odor, metals, and other pollutants not removed in earlier steps. Of most concern in chemical processing plants is the removal of dissolved organic compounds, particularly those that are carcinogenic, such as benzene. The following example illustrates the calculation of the cost of removal of dissolved organic compounds by biodegradation.

EXAMPLE 17.7

A wastewater stream of 500 gpm at 70°F contains 150 mg/L of benzene that is to be removed by biodegradation. If 99.9% of the benzene is removed, determine the amount of benzene removed per year and the operating cost of removal using a cost from Table 17.1 of $0.10/lb benzene and a plant-operating factor of 0.9.

SOLUTION

First, determine if all of the benzene is dissolved in the wastewater. The solubility of benzene in water at 70°F is a mole fraction of 0.00040. The mole fraction of benzene (MW = 78) in the wastewater (55.5 mol/L) is (0.150/78)/55.5 = 0.000035. Therefore, all of the benzene is dissolved. The flow rate of 500 gpm is equivalent to 500(3.785)(60) = 113,600 L/hr. The pounds of benzene removed per year = (0.150/454)(113,600)(24)(365)(0.9) = 296,000 lb/yr. The cost of benzene removal = 0.10(296,000) = $29,600/yr. ∎

Solid Wastes

According to U.S. federal regulations, solid wastes must be classified as hazardous or non-hazardous. Hazardous wastes, due to their ignitability, corrosivity, reactivity, and/or toxicity, pose a substantial threat to human, plant, or animal life and must be treated onsite or near-site by physical, chemical, thermal, or biological means before being put in containers and removed. Nonhazardous solid wastes may be placed in containers and removed to a landfill or, in some cases, incinerated. The annual cost of solid waste treatment and disposal varies widely. Typical costs are $0.02/lb of waste for nonhazardous dry or wet solids and $0.08/lb for hazardous dry or wet solids.

Labor-Related Operations, O

One of the most difficult annual costs to estimate is direct wages and benefits (DW&B) for operating the plant. It and the other annual costs that are proportional to it are often an important fraction of the cost of manufacture. Table 17.1 lists the labor-related charges associated with operations. These include *direct wages and benefits* (DW&B), calculated from an hourly rate for the operators of a proposed plant. To estimate all labor-related operations, it is necessary to estimate the number of operators for the plant per shift and to account for three shifts daily. Typically, each shift operator works 40 hr per week, and, hence, for each operator required during a 7(24) = 168-hr week, 4.2 shifts must be covered. In practice, due to illness, vacations, holidays, training, special assignments, overtime during startups, etc., it is common to provide for 5 shifts for each operator required.

Estimates of the number of plant operators needed per shift are based on the type and arrangement of the equipment, the multiplicity of units, the amount of instrumentation and control for the process, whether solids are handled, whether the process is continuous or batchwise or includes semicontinuous operations, and company policy in establishing labor requirements, particularly as it relates to operator unions. For preliminary estimates of the number of operators required per shift, the process may be divided into sections as discussed in Chapter 7 and shown in Figures 7.1 and 7.3. These sections may include: (1) feed preparation system using separation steps, (2) reactor system, (3) vapor recovery system, (4) liquid-separation system, (5) solids separation and purification system, and (6) pollution abatement system. When a process includes two or more reactor systems and/or two or more liquid-separation systems, each is counted separately. As given in Table 17.3, for a continuously operating, automatically controlled, fluids-processing plant with a low-to-medium capacity of 10 to 100 ton/day of product, one operator/shift is assigned to each section. For solids–fluids processing and solids processing, the number of operators per shift is increased as noted in Table 17.3. For large capacities, for example, 1,000 ton/day of product, the number of operators/shift in Table 17.3 are doubled for each section. Batch and semicontinuous processing also require more operators than a continuous process, as indicated in Table 17.3. A process should always have at least two operators present per shift. Each shift operator is paid for 40 hr/week and 52 weeks/yr or a total of 2,080 hr/yr. The annual cost of direct wages and benefits (DW&B) is obtained from:

$$\text{DW\&B, \$/yr} = (\text{operators/shift})(5 \text{ shifts})(2{,}080 \text{ hr/yr-operator})(\$/\text{hr}) \qquad \textbf{(17.2)}$$

Table 17.3 Direct Operating Labor Requirements
for Chemical Processing Plants
Basis: Plant with Automatic Controls
and 10–100 Ton/Day of Product

Type of process	Number of operators per process section[a]
Continuous operation	
Fluids processing	1
Solids–fluids processing	2
Solids processing	3
Batch or Semibatch operation	
Fluids processing	2
Solids–fluids processing	3
Solids processing	4

[a]Note: For large continuous-flow processes (e.g., 1,000 ton/day of product), multiply the number of operators by 2.

where the $/hr covers wages and benefits, and depends on locality and whether operators are unionized. In Table 17.1, a figure of $30/hr is typical in the United States.

To obtain the total annual labor-related operations cost, O, direct salaries and benefits for supervisory and engineering personnel at 15% of DW&B and operating supplies and services at 6% of DW&B are added to DW&B. In addition, $52,000/(operator/shift)-yr for technical assistance to manufacturing and $57,000/(operator/shift)-yr for control laboratory are added. An estimate of the total annual cost of labor-related operations is illustrated in the following example.

EXAMPLE 17.8

The vinyl chloride process discussed in Sections 3.4 and 3.5 and shown in Figures 3.8 and 3.19 produces 100,000 lb/hr of vinyl chloride or 1,200 ton/day. Estimate the annual cost of labor-related operations, O.

SOLUTION

This is a continuous fluids process of large capacity. Assume it is automatically controlled. From the block flow diagram, the process is comprised of two reactor sections and one liquid separation section. Therefore, from Table 17.3, three operators per shift are required for a moderate-capacity plant. However, this is a large-capacity plant, requiring twice that number or 6 operators per shift and five shifts or a total of 30 shift operators. Also, a large-capacity plant requires one labor-yr each for technical assistance and control laboratory. Using Eq. (17.2), the annual costs are

$$\text{Annual DW\&B} = (30 \text{ operators})(2{,}080 \text{ hr/yr})(\$30.00/\text{hr}) = \$1{,}872{,}000$$

Using Table 17.1, the other annual labor-related operation costs are

$$\text{Direct salaries and benefits} = 0.15(\$1{,}872{,}000) = \$280{,}800$$
$$\text{Operating supplies and services} = 0.06(\$1{,}872{,}000) = \$112{,}300$$
$$\text{Technical assistance to manufacturing} = \$52{,}000(5) = \$260{,}000$$
$$\text{Control laboratory} = \$57{,}000(5) = \$285{,}000$$

The total labor-related operations annual cost, O, is

$$O = \$1{,}872{,}000 + \$280{,}800 + \$112{,}300 + \$260{,}000 + \$285{,}000 = \$2{,}810{,}100/\text{yr} \quad \blacksquare$$

Maintenance, M

A second category of labor-related costs is associated with the maintenance of a proposed plant. Processing equipment must be kept in acceptable working order, with repairs and replacement of parts made as needed. Annual maintenance costs, M, are sometimes greater than the cost of labor-related operations, O. Included in Table 17.1, under annual maintenance costs, M, is the main item, maintenance wages and benefits (MW&B), which is estimated as a fraction of the total depreciable capital, C_{TDC}, depending on whether the process handles fluids, solids, or a combination of fluids and solids. The range is from a low of 3.5% for fluids to 5.0% for solids, with 4.5% for solids–fluids processing. Salaries and benefits for the engineers and supervisory personnel are estimated at 25% of MW&B. Materials and services for maintenance are estimated at 100% of MW&B, while maintenance overhead is estimated at 5% of MW&B. Thus, the total annual cost of maintenance varies from 8.05 to 11.5% of C_{TDC}. Maintenance costs can be controlled by selecting the proper materials of construction for the processing equipment, sparing pumps, avoiding high rotation speeds of shafts, restricting the highest fouling streams to the tube side of heat exchangers, selecting long-life catalysts for reactors, scheduling routine maintenance, and practicing preventative maintenance based on experience, supplier information, and record-keeping. Routine maintenance includes cleaning of heat exchanger tubing and lubrication and replacement of packing and mechanical seals in pumps, compressors, blowers, and agitators. A main goal should be to provide most of the maintenance during scheduled plant shutdowns, which might be during a two- or three-week period each year.

EXAMPLE 17.9

The total depreciable capital investment, C_{TDC}, for a plant to produce 300,000 tons per year of cumene is estimated to be $31,000,000. The process only involves fluids processing. Estimate the annual plant maintenance cost, M.

SOLUTION

Using Table 17.1, the annual maintenance costs are

$$\text{Wages at benefits (MW\&B) at 3.5\% of } C_{TDC} = \$1,085,000$$
$$\text{Salaries and benefits at 25\% of MW\&B} = 271,000$$
$$\text{Materials and services at 100\% of MW\&B} = 1,085,000$$
$$\text{Maintenance overhead at 5\% of MW\&B} = 54,000$$

The total annual maintenance cost, M, is $2,495,000/yr.   ■

Operating Overhead

To this point in Table 17.1, all costs have been directly related to plant operation. However, a company always incurs many other expenses, which while not directly related to plant operation can be estimated as a fraction of the combined salary, wages, and benefits for maintenance and labor-related operations, referred to here as M&O-SW&B. Overhead expenses include the costs of providing the following services: cafeteria; employment and personnel; fire protection, inspection, and safety; first aid and medical; industrial relations; janitorial; purchasing, receiving, and warehousing; automotive and other transportation; and recreation. In Table 17.1, overhead costs are divided into four categories: general plant overhead, provision for the services of the mechanical department and for the employee relations department, as well as business services, with the total annual operating overhead cost equal to the sum of these four categories or $(7.1 + 2.4 + 5.9 + 7.4) = 22.8\%$ of M&O-SW&B.

EXAMPLE 17.10

Estimate the annual cost of operating overhead for the cumene plant of Example 17.9, assuming that the cost of labor-related operations is the same as in Example 17.8.

SOLUTION

The previous two examples provide the following wages, salaries, and benefits per year for labor-related operations and maintenance:

$$\text{Direct wages and benefits (DW\&B)} = \$1,872,000$$
$$\text{Direct salaries and benefits} = 280,800$$
$$\text{Maintenance wages and benefits (MW\&B)} = 1,085,000$$
$$\text{Maintenance salaries and benefits} = 271,000$$

The total annual M&O-SW&B is the sum, which equals $3,508,800/yr.
The total annual operating overhead cost is 22.8% of M&O-SW&B or

$$0.228(3,508,800) = \$800,000/yr$$ ■

Property Taxes and Insurance

Annual property taxes are assessed by the local municipality as a percentage of the total depreciable capital, C_{TDC}, with a range from 1% for plants located in sparsely populated areas to 3% when located in heavily populated areas. Property taxes are not related to federal income taxes levied by the Internal Revenue Service and considered below. Liability insurance costs depend on the pressure and temperature levels of plant operation and on whether flammable, explosive, or toxic chemicals are involved. The annual cost of insurance is also estimated as a percentage of the total depreciable capital, C_{TDC}, with a range of 0.5 to 1.5%. In the absence of data, annual property taxes and insurance may be estimated at 2% of C_{TDC}, as given in Table 17.1. This corresponds to a process of low risk located away from a heavily populated area.

Depreciation, D

The subject of depreciation is a complex subject that is often confusing because depreciation has several definitions and applications. Most commonly, it is simply a measure of the decrease in value of something over time. Some companies use depreciation as a means to set aside a fund to replace a plant when it is no longer operable. In its most complex application, depreciation is an annual allowance, whose calculation is controlled by the U.S. federal government when determining federal income tax. The larger the depreciation in a given year, the smaller the federal income tax and the greater the net profit. This is considered in detail below in the discussion of cash flow in Section 17.6.

For use with approximate profitability measures, as applied here to the preliminary calculation of the annual manufacturing cost, depreciation, D, is estimated as a constant percentage of the total depreciable capital, C_{TDC}. This type of depreciation is referred to as *straight-line (SL) depreciation*. Although it has been customary to take that percentage as 10% for each of 10 yr, here, in Table 17.1, the direct plant (onsite) depreciation is taken as 8% of $(C_{TDC} - 1.18C_{alloc})$ (equivalent to a plant life of about 12 yr), while the allocated plant (offsite) depreciation is taken as 6% of the contribution of the allocated costs for utilities and related facilities to the total depreciable capital, C_{TDC}, $1.18C_{alloc}$ (equivalent to a life of about 16 yr), where the 1.18 factor accounts for the share of the contingency and contractor's fee, C_{cont}.

Cost of Manufacture, COM

The total annual cost of manufacture, COM, as shown in Table 17.1, is the sum of (1) *direct manufacturing costs*: feedstocks, utilities, labor-related operations, and maintenance; (2) *operating overhead*; and (3) *fixed costs*: property taxes, insurance, and depreciation.

Total Production Cost, C

The total annual production cost equals the sum of the cost of manufacture and general expenses,

$$C = COM + \text{general expenses} \qquad (17.3)$$

General expenses refer to activities that are conducted by the central operations of a company, perhaps at the corporate headquarters, and are financed from profits made by the company from their operating plants. In Table 17.1, general expenses comprise five categories: selling (or transfer) expense, research (direct and allocated), administrative expense, and management incentive compensation. The *selling expense* covers all the costs involved in selling the products, including expenses of the sales office, advertising, traveling salesmen, containers and shipping, commissions, and technical sales service. The research expense covers both research and development costs for new products and new manufacturing methods for existing products. Administrative expense covers those top-management and general administrative activities that are not direct manufacturing costs. In Table 17.1, all general expenses are estimated as a percentage of the total sales revenue. The total general expenses range from 9.55 to 11.55% of S.

EXAMPLE 17.11

For the MCB separation process shown in Figure 4.23, estimate the annual production cost, C, where products will be used in-house, and the total annual sales, S. Basis:

Continuous plant operation	330 day/yr or 7,920 hr/yr
Feedstock	9,117 lb/hr @ $0.30/lb
MCB product	5,572 lb/hr @ $0.52/lb
Benzene co-product	3,133 lb/hr @ $0.15/lb
HCl gas co-product	412 lb/hr @ $0.05/lb
Total bare-module costs, C_{TBM}	$1,711,000
Cost of site preparation and service facilities, $C_{site} + C_{serv}$	$142,000
Allocated costs for utilities and related facilities, C_{alloc}	$74,000
Cost of contingencies @18% of C_{DPI}	
50-psig steam	1,258 lb/hr @ $2.50/1,000 lb
150-psig steam	3,480 lb/hr @ $4.00/1,000 lb
Electricity	9.60 kW @ $0.04/kW-hr
Cooling water	182 gpm @ $0.050/1,000 gal
Operators	two/shift

SOLUTION

The total depreciable capital, using Table 16.9, is computed as follows:

$$C_{DPI} = \$1,711,000 + \$142,000 + \$74,000 = \$1,927,000$$
$$C_{cont} = 0.18\ C_{DPI} = 0.18(1,927,000) = \$346,900$$
$$C_{TDC} = C_{DPI} + C_{cont} = \$1,927,000 + \$346,900 = \$2,273,900$$

For this moderate-size plant, with one section, use 2 operators/shift. Using Table 17.1 with the above data, the following annual costs are computed

Cost factor	Annual cost
Feedstocks (raw materials)	$21,662,000
Utilities	
Steam, 150 psig	110,300
Steam, 50 psig	24,900
Electricity	3,000
Cooling water (CW)	4,300
Total utilities	$142,500
Operations (O)	
Direct wages and benefits (DW&B)	624,000
Direct salaries and benefits	93,600
Operating supplies and services	37,400
Technical assistance to manufacturing	104,000
Control laboratory	114,000
Total labor-related operations	$973,000
Maintenance (M)	
Wages and benefits (MW&B)	79,600
Salaries and benefits	19,900
Materials and services	79,600
Maintenance overhead	4,000
Total maintenance	$183,100

Cost factor	Annual cost
Total of M&O-SW&B	817,100
Operating overhead	
General plant overhead	58,000
Mechanical department services	19,600
Employee relations department	48,200
Business services	60,500
Total operating overhead	$186,300
Property taxes and insurance	$45,500
Depreciation (D)	
Direct plant	174,900
Allocated plant	5,200
Total depreciation	$180,100
COST OF MANUFACTURE (COM)	$23,372,500
General Expenses	
Transfer expense	268,300
Direct research	1,288,000
Allocated research	134,200
Administrative expense	536,700
Management incentive compensation	335,400
TOTAL GENERAL EXPENSES (GE)	$2,562,600
TOTAL PRODUCT COST (C)	$25,935,100
Sales	
Monochlorobenzene product	22,947,700
Benzene co-product	3,722,000
HCl co-product	163,200
TOTAL SALES, S	$26,832,900

■

Pre-tax (Gross) Earnings and After-tax (Net) Earnings (Profit)

The annual *pre-tax earnings or profit*, also called the *gross earnings or profit*, is the difference between the annual sales revenue and the annual product cost:

$$\text{Gross earnings or profit} = S - C \qquad (17.4)$$

The annual *after-tax earnings or profit*, also called the *net earnings or profit*, is the gross earnings minus U.S. federal and state income taxes on the gross earnings. Since at least the year 1913, U.S. corporations have been subject to federal income tax on gross earnings. During the period from 1913 to 2001, the federal corporation income tax has been as low as 1% and as high as 52%. During World War II and the Korean War, an additional excise tax brought the total tax to as high as 80%. Since 1988, the federal income tax rate for corporate annual gross earnings greater than $335,000 has been 34%. State income tax rates vary greatly. In the absence of data, a combined federal and state income tax rate, t, of $34 + 3 = 37\%$ is used here. Thus,

$$\text{Net earnings or profit} = (1 - t) \text{ gross earnings} = 0.63(S - C) \qquad (17.5)$$

EXAMPLE 17.12

For the data and results of Example 17.11, calculate the annual gross earnings and annual net earnings.

SOLUTION   From Eq. (17.4),

Annual gross earnings or profit = S − C = $26,832,900 − $25,935,100 = $897,800/yr

From Eq. (17.5),

Annual net earnings or profit = 0.63(897,800) = $565,600/yr

Of the costs in the cost sheet of Table 17.1, only the costs for labor-related operations, maintenance, operations overhead, property taxes and insurance, and depreciation are considered to be *fixed costs*, which do not vary with the production rate of the plant. Fixed costs are contrasted with the costs of feedstocks, utilities, and general expenses, which are referred to as *variable costs*, because they vary directly with the production rate. For a large plant, with a large total capital investment and significant economies of scale, the profitability can be sharply increased by substantial savings in utilities such as steam. A smaller plant, in contrast, has a larger fraction of its costs in the investment and the fixed costs of operation, and hence the same percentage decrease in the utilization of steam results in a smaller increase in its profitability. This is one of the reasons why small plants usually have difficulty competing with larger plants in the chemical industry. The reader should keep this in mind while the profitability measures in Sections 17.4 and 17.7 are studied. Before leaving Table 17.1, the reader is reminded that the prices listed there refer to the year 2000 and should be adjusted in subsequent years, possibly escalated by the rate of inflation. ∎

17.3   WORKING CAPITAL AND TOTAL CAPITAL INVESTMENT

To complete the estimation of the total capital investment, a more accurate estimate of working capital is needed to replace the 15% of total capital investment used in conjunction with Eq. (16.10). In general, working capital is funds, in addition to fixed capital and start-up funds, needed by a company to meet its obligations until payments are received from others for goods they have received. Accountants define working capital as current assets minus current liabilities, where current assets consist of cash reserves, inventories, and accounts receivable, while current liabilities include accounts payable. It is fairly standard to provide working capital for a one-month period of plant operation, because those buying the product are usually given 30 days to make their payments, while the company has 30 days to pay for raw materials. Inventories of products may be much less than 30 days. Here, 7 days is assumed. Working capital is fully recoverable and, therefore, is not depreciated. If we apply the definition of working capital to the operation of a chemical plant, working capital is

$$C_{WC} = \text{cash reserves} + \text{inventory} + \text{accounts receivable} - \text{accounts payable} \quad \textbf{(17.6)}$$

with the following basis for calculation, which follows general accounting practices:

1. 30 days of cash reserves for raw materials, utilities, operations, maintenance, operating overhead, property taxes, insurance, and depreciation. This amounts to 8.33% of the annual cost of manufacture, COM (assuming 30 days is $\frac{1}{12}$ of a year).
2. 7 days of inventories of liquid and solid (but not gas) products at their sales price, which assumes that these products are shipped out once each week, while gas products are not stored, but are pipelined. This amounts to 1.92% of the annual sales of liquid and solid products.
3. 30 days of accounts receivable for product at the sales price. This amounts to 8.33% of the annual sales of all products.
4. 30 days of accounts payable by the company for feedstocks at the purchase price. This amounts to 8.33% of the annual feedstock costs.

EXAMPLE 17.13   For the MCB plant considered in Example 17.11, estimate the working capital and compute the total capital investment if land cost and royalty costs are zero, but the startup cost is taken as 2% of C_{TDC}.

SOLUTION   The data required, obtained from Example 17.11, are

$$C_{TDC} = \$2,273,900$$
$$COM = \$23,372,500/yr$$
Sales for MCB and benzene = $\$26,669,700/yr$
Sales for MCB, benzene, and HCl = $\$26,832,900/yr$
Cost of feedstock = $\$21,662,000/yr$

Therefore,

Cash reserves = $0.0833(23,372,500) = \$1,946,900$
Inventories = $0.0192(26,669,700) = \$512,100$
Accounts receivable = $0.0833(26,832,900) = \$2,235,200$
Accounts payable = $0.0833(21,662,000) = \$1,804,400$

From Eq. (17.6),

Working capital, $C_{WC} = \$1,946,900 + \$512,100 + \$2,235,200 - \$1,804,400 = \$2,889,800$, which is much greater than the total depreciable capital

The startup cost = $C_{start} = 0.02(2,273,900) = \$45,500$

The total capital investment = $C_{TCI} = C_{TDC} + C_{start} + C_{WC} = \$2,273,900 + \$45,500 + \$2,889,800 = \$5,209,200$

Note that for this case, the working capital is 55.5% of the total capital investment and much more than the commonly used approximate estimate of 15% of total capital investment. In this example, it appears that working capital is more a function of annual sales (perhaps 10%) than of total depreciable capital. ∎

17.4 APPROXIMATE PROFITABILITY MEASURES

To be a worthwhile investment, a venture for the installation of a new chemical plant or a re-vamp of an existing plant must be profitable. However, it is not sufficient that a venture make a large net profit. That profit over the life of the venture must be more than the original capital investment for the venture. The greater the excess of profits over investment, the more attractive is the venture. To compare alternative ventures that vie for capital investment, a number of profitability measures have been developed. They are all based on the estimates of capital investment and annual earnings that have been presented in Chapter 16 and the previous sections of this chapter. The simpler, approximate measures discussed in this section and summarized in Table 17.4, ignore the effect of inflation or so-called time value of money and use simple straight-line depreciation. Therefore, they are only useful in the early stages of project evaluation. The rigorous measures that account for the time value of money and faster depreciation are considered in the three subsequent sections and must be considered before a final decision is made on whether to proceed with a new venture.

Return on Investment (ROI)

This profitability measure, introduced earlier as Eq. (17.1) is also called rate of return on investment (ROROI), simple rate of return (ROR), return on original investment, engineer's method, and operator's method. ROI is the annual interest rate made by the profits on the original investment. ROI provides a snapshot view of the profitability of the plant, normally using estimates of the elements of the investment, in Table 16.9, and the pre-tax or after-tax earnings in, say, the third year of operation and assuming that they remain unchanged during the life of the process. For ROI, and all of the approximate profitability measures of this section, the production cost is computed using straight-line depreciation, and, after some

Table 17.4 Approximate Profitability Measures
Time Value of Money Is Ignored and Straight-Line Depreciation Is Used
(Details Presented in Section 17.4)

Approximate profitability measure	Formula[a]
Return on investment (ROI)	$\text{ROI} = \dfrac{\text{net earnings}}{\text{total capital investment}} = \dfrac{(1-t)(S-C)}{C_{\text{TCI}}}$
Payback period (PBP)	$\text{PBP} = \dfrac{C_{\text{TDC}}}{(1-t)(S-C)+D}$
Venture profit (VP)	$\text{VP} = (1-t)(S-C) - i_m(C_{\text{TCI}})$
Annualized cost (AC)	$\text{AC} = C_A = C + i_m(C_{\text{TCI}})$

[a] i_m = reasonable return on investment; t = sum of U.S. federal and state income tax rates; C = annual production cost; D = annual depreciation; S = annual sales revenues; C_{TCI} = total capital investment; C_{TDC} = total depreciable capital.

startup period, the plant is assumed to operate each year at full capacity (or at some percentage of full capacity) for the same number of days per year. As was stated earlier, many definitions of ROI have been suggested and used. Here, the most common definition is applied

$$\text{ROI} = \frac{\text{net earnings}}{\text{total capital investment}} = \frac{(1-t)(S-C)}{C_{\text{TCI}}} \qquad (17.7)$$

The calculation of ROI is readily made and the concept is easy to understand. However, as stated above, the definition of ROI involves many assumptions. Furthermore, ROI does not consider the size of the venture. Would a large company favor many small projects over a few large projects, when the small projects have just slightly more favorable values of ROI?

Payback Period (PBP)

The *payback period* is the time required for the annual earnings to equal the original investment. Payback period is also called payout time, payout period, payoff period, and cash recovery period. Because it is simple and even more understandable than ROI, PBP is widely used in early evaluations to compare alternatives. Like ROI, the payback period in years has several definitions, but the following is used here. This definition is not consistent with the definition of ROI in Eq. (17.7), because only the depreciable capital is used and the annual depreciation, D, is added back to the net earnings because that depreciation is retained by the company.

$$\text{PBP} = \frac{C_{\text{TDC}}}{(1-t)(S-C)+D} = \frac{C_{\text{TDC}}}{\text{net earnings} + \text{annual depreciation}} = \frac{C_{\text{TDC}}}{\text{cash flow}} \qquad (17.8)$$

High-risk ventures should have payback periods of less than 2 yr. In these times of rapid progress in technology, most companies will not consider a project with a PBP of more than 4 yr. PBP is especially useful for simple equipment replacement problems. For example, should an old, inefficient pump be replaced with a new, energy-efficient model. This decision is clear if the PBP is less than 1 yr. PBP should never be used for final decisions on large projects because it gives no consideration to the period of plant operation after the payback period.

EXAMPLE 17.14

A process, projected to have a total depreciable capital, C_{TDC}, of $90 million, with no allocated costs for off-site utilities, is to be installed over a 3-yr period (1997–1999). Just prior to startup, $40 million of working capital is required. At 90% of production capacity (projected for the third and subsequent operating years), sales revenues, S, are projected to be $150 million/yr and the total annual production cost, excluding depreciation, is projected to be $100 million/yr. Also, the plant is projected to operate at 0.5 of 90% and 0.75 of 90% of capacity during the first and second operating years. Thus, during those years, S = $75 million/yr and $113 million/yr, respectively. Take straight-line depreciation at 8%/yr. Using the third operating year as a basis, compute

    a. return on investment (ROI)
    b. payback period (PBP)

SOLUTION

$$\text{Depreciation} = 0.08(\$90,000,000) = \$7,200,000/\text{yr}$$
$$\text{Total production cost} = \$100,000,000 + \$7,200,000 = \$107,200,000/\text{yr}$$
$$\text{Pre-tax earnings} = \$150,000,000 - \$107,200,000 = \$42,800,000/\text{yr}$$
$$\text{Income taxes} = 0.37(\$42,800,000) = \$15,800,000/\text{yr}$$
$$\text{After-tax earnings} = \$42,800,000 - \$15,800,000 = \$27,000,000/\text{yr}$$
$$C_{\text{TCI}} = \$90,000,000 + \$40,000,000 = \$130,000,000$$

a. From Eq. (17.7),

$$\text{ROI} = \frac{\$27,000,000}{\$130,000,000} = 0.208 \text{ or } 20.8\%$$

b. From Eq. (17.8),

$$\text{PBP} = \frac{\$90,000,000}{\$27,000,000 + \$7,200,000} = 2.63 \text{ yr}$$

In this example, values of both ROI and PBP are sufficient to merit some interest in the project, but they are not sufficient to attract a high degree of interest unless the process is of very low risk and only less-profitable ventures are under consideration. ∎

Venture Profit (VP)

An approximate measure of the profitability of a potential process that does take into account the size of the project is *venture profit*. It is used often for preliminary estimates when comparing alternative flowsheets during the process synthesis stage of process design. VP is the annual net earnings in excess of a minimum acceptable return on investment, i_{min}. Thus,

$$\text{VP} = (1 - t)(S - C) - i_{\text{min}}C_{\text{TCI}} = \text{net earnings} - i_{\text{min}}C_{\text{TCI}} \qquad \textbf{(17.9)}$$

Sometimes, for crude comparisons of flowsheets with different arrangements of process units, the total capital investment in Eq. (17.9) is estimated as the sum of the bare-module costs, or even the sum of the purchase costs; and annual production cost, C, includes only the cost of the raw materials, the utilities, and the labor-related operations. The return on investment, i_{min}, is that desired by the company. Here, we take $i_{\text{min}} = 0.20$ (20%).

EXAMPLE 17.15

For the MCB process considered in Examples 17.11, 17.12, and 17.13, calculate

    a. return on investment (ROI)
    b. payback period (PBP)
    c. venture profit (VP)

SOLUTION

From the previous examples,

$$C_{\text{TCI}} = \$5,209,200$$
$$C_{\text{TDC}} = \$2,273,900$$
$$\text{Net earnings} = \$565,600/\text{yr}$$
$$\text{Depreciation} = \$181,900/\text{yr}$$

a. From Eq. (17.7),

$$\text{ROI} = \frac{\$565,600}{\$5,209,200} = 0.109 \text{ or } 10.9\%$$

b. From Eq. (17.8),

$$\text{PBP} = \frac{\$2,273,900}{\$565,600 + \$181,900} = 3.04 \text{ yr}$$

c. From Eq. (17.9),

$$\text{VP} = \$565,600 - 0.20(\$5,209,200) = -\$476,200/\text{yr}$$

These results are conflicting with respect to the profitability of the MCB process. The PBP is reasonably good, although less than 2 yr would be preferred because the process is low-risk. The ROI is too low for the proposed plant to be given serious consideration, except when going interest rates are very low. The VP is negative for $i_{min} = 0.20$, eliminating the plant from further consideration. Even if i_{min} were lowered to 0.10, there is no incentive to build the plant. The discrepancy among the three profitability measures is caused mainly by the large magnitude of the working capital compared to the depreciable capital.   ∎

Annualized Cost (C_A)

A measure of economic goodness, which does not involve sales revenues for products and is also used for preliminary estimates when comparing alternative flowsheets during process synthesis, is the *annualized cost*. It is the sum of the production cost and a reasonable return on the original capital investment where, again, the reasonable return on investment, i_{min}, is taken here as 0.2. Thus,

$$C_A = C + i_{min}(C_{TCI}) \tag{17.10}$$

This criterion is also useful for comparing alternative items of equipment in a process or alternative replacements for existing equipment.

EXAMPLE 17.16

Several alternative distillation sequences are being examined for the separation of a mixture of light hydrocarbons. The sequences are to be compared on the basis of annualized cost, given by Eq. (17.10). However, for the total capital investment, only the bare module costs of the columns, trays, condensers, reboilers, and reflux accumulators will be summed. For the total annual production cost, C, only the annual utility costs for the condenser cooling water and reboiler steam will be summed. For one of the columns, design calculations have been completed and the costs have been computed, with the results given below. The column is a deisobutanizer with a saturated liquid feed of 500 lbmol/hr of isobutane and 500 lbmol/hr of n-butane. The distillate is 99 mol% isobutane and the bottoms is 99 mol% n-butane. The column shell is carbon steel, with carbon steel sieve trays on 24-in. spacing. The trays have 0.25-in.-diameter holes with a hole area of 10%. The weir height is 2 in. The column pressure is set at 100 psia at the top so that cooling water can be used in the total condenser, while the bottoms pressure is 110 psia. Calculations give 100 trays, at a reflux ratio of 7.4. This corresponds to a condenser duty of 33,600,000 Btu/hr and a reboiler duty of 33,800,000 Btu/hr. For 24-in. tray spacing, allowing a 10-ft-high bottoms sump below the bottom tray, and a 4-ft disengagement height above the top tray, the column height is 212 ft (tangent to tangent). Based on entrainment flooding, the column diameter is determined to be constant at 9.5 ft.

The bare-module cost of the tower vessel is estimated to be $1,480,000 and the accompanying tray cost is $120,000, giving a total bare-module cost for the column of $1,600,000. The bare-module costs for the column auxiliaries are computed to be

2 Condensers in parallel	$378,000
Reboiler	94,000
Reflux drum	94,000
Reflux pump + a spare	60,000

The total bare-module cost for the column and its auxiliaries = $2,226,000

The annual heating steam cost for the reboiler is computed = $728,000/yr

The annual cooling water cost for the two condensers = $84,000/yr

The annual electricity cost for the reflux pump = $38,000/yr

The total utility cost = $850,000/yr

Compute the annualized cost.

SOLUTION

For purposes of comparison of alternatives, the bare-module cost of the distillation column and its aux-iliary equipment replaces C_{TCI}. The annual utility cost replaces the total annual production cost.
From Eq. (17.10), C_A = $850,000 + 0.20($2,226,000) = $1,295,200/yr.   ∎

Product Selling Price for Profitability

In some cases, especially when a new chemical is to be produced, the selling price may not be known or easily established. Rather than guess a selling price, a desired return on invest-ment (say, 20%) can be assumed and Eq. (17.7) can then be used to back-calculate the sell-ing price necessary to achieve this objective. Another useful procedure is to set the venture profit to zero and use Eq. (17.9) to back-calculate a minimum selling price. More elaborate methods for determining a selling price are implemented using the rigorous profitability measures in Section 17.6 that account for the time value of money.

EXAMPLE 17.17

In Example 17.15, approximate profitability measures, when applied to the MCB plant, are not favorable. However, one of the chemicals produced, MCB, is given a selling price of $0.52/lb, which is not well established by current competition.

a. Use the ROI measure of Eq. (17.7) to estimate a selling price for a 20% return on investment.
b. Use the VP measure of Eq. (17.9) to estimate a minimum selling price.

SOLUTION

a. From Example 17.11, C = $25,935,100/yr and the total annual sales of all three products is $26,832,900/yr. This includes 44,130,200 lb/yr of MCB at $0.52/yr. Thus, if we let x = the sell-ing price of MCB, the total annual sales of all three products, in terms of x, becomes

$$S = \$26,832,900 - 44,130,200(0.52 - x) = \$3,885,200 + 44,130,200\,x$$

From Example 17.13, C_{TCI} = $5,209,200
Substitution into Eq. (17.7) gives

$$\text{ROI} = 0.20 = \frac{(1 - 0.37)(3,885,200 + 44,130,200\,x - 25,935,100)}{5,209,200}$$

Solving this equation, x = selling price of MCB = $0.537/lb, which is slightly higher than the price of $0.52. Clearly, the ROI is very sensitive to the selling price of MCB.

b. Substitution into Eq. 17.9, with VP = 0, gives

$$VP = 0 = (1 - 0.37)[(3,885,200 + 44,130,200\,x) - 25,935,100] - 0.20(5,209,200)$$

Solving this equation, x = selling price of MCB = $0.537/lb, which is the same result as in part (a). This is not surprising because Eqs. (17.7) and (17.9) are identical when VP is set to zero and ROI = i_{min}.   ∎

17.5   TIME VALUE OF MONEY

All of the profitability measures discussed so far give only a snapshot view at a given point in time. The total annual sales revenues, S, and the total annual production cost, C, are estimated at critical points, normally for the third operating year. Furthermore, a simple depreciation

Table 17.5 Time Value of Money Interest Formulas—Single Payments
(Details Presented in Section 17.5)

Interest Type	Formula[a]
Amount of simple interest	$I_S = F - P = niP$
Single-payment simple-amount factor	$\dfrac{F}{P} = 1 + ni$
Single-payment simple-present worth factor	$\dfrac{P}{F} = \dfrac{1}{1 + ni}$
Compound interest	
Amount of compound interest	$I_C = F - P = P[(1 + i)^n - 1]$
Single-payment compound-amount factor	$\dfrac{F}{P} = (1 + i)^n$
Single-payment present-worth factor	$\dfrac{P}{F} = \dfrac{1}{(1 + i)^n}$
Nominal interest rate per year	$r = im$
Effective discrete compound interest	
Effective discrete annual compound interest rate	$i_{\text{eff}} = (1 + i)^m - 1 = \left(1 + \dfrac{r}{m}\right)^m - 1$
Amount of discrete compound interest	$I_C = F - P = P[(1 + i_{\text{eff}})^{n_y} - 1]$
Discrete single-payment compound-amount factor	$\dfrac{F}{P} = (1 + i_{\text{eff}})^{n_y}$
Discrete single-payment present-worth factor	$\dfrac{P}{F} = \dfrac{1}{(1 + i_{\text{eff}})^{n_y}}$
Effective continuous compound interest	
Effective continuous annual compound interest rate	$i_{\text{eff}} = e^r - 1 = e^{im} - 1$
Amount of continuous compound interest	$I_C = F - P = P[(1 + i_{\text{eff}})^{n_y} - 1]$
Continuous single-payment compound-amount factor	$\dfrac{F}{P} = (1 + i_{\text{eff}})^{n_y} = e^{rn_y}$
Continuous single-payment present-worth factor	$\dfrac{P}{F} = \dfrac{1}{(1 + i_{\text{eff}})^{n_y}} = e^{-rn_y}$

[a]i = interest rate per period; m = number of periods per year; r = nominal interest rate per year; n_y = number of years; i_{eff} = effective annual compound interest rate; n = number of interest periods.

schedule, typically straight-line depreciation, is used. As mentioned earlier, company resources are often sufficiently limited so as not to justify a more careful examination of the revenues and costs over the life of a proposed plant at the early stages of consideration. However, because of the compounding effect of interest and inflation, it eventually becomes important to account for the time value of money and to charge for depreciation in accordance with the schedule required by the U.S. Internal Revenue Service since 1986 (modified in 1988). In the next section, the methods of calculating cash flows for each year in the life of a proposed plant project are presented, with which rigorous profitability measures can be computed in Section 17.7. Before doing this, however, it is necessary, in this section, to examine how interest is compounded and to discuss annuities and perpetuities. A number of useful formulas are derived and/or presented for single-payment interest in this section. They are summarized in Table 17.5.

Compound Interest

The time value of money recognizes the fact that an amount of money at the current time, referred to as *present amount, present sum, present value,* or *present worth* and given the symbol, P, may not be the same at a future date. Instead, if that money is invested at an *interest rate*, i, and the interest is added to P, the amount of money at the future date will be a *future amount, future value*, or *future worth*, given here the symbol F. The *interest*, which is the compensation for the use of the money or capital over a period of time, is the difference between F and P. The concept of interest is complicated because: (1) the *interest period* is not necessarily 1 yr, (2) interest may be simple or compound, and (3) compounding may be discrete or continuous.

Let us call the starting present worth or present sum the capital or principal, P. Simple interest over several interest time periods is calculated only on P. No interest is calculated on interest accrued in previous interest periods. Thus, the total amount of simple interest for n interest periods, where i is the simple interest rate per period, is

$$\text{Simple interest} = I_S = F - P = niP \qquad \textbf{(17.11)}$$

Simple interest is rarely used. It has been largely replaced by compound interest, which is calculated at each period on the principal plus the accumulated interest. The interest rate, i, is now referred to as the compound interest rate per period. The effect of compounding is shown in Table 17.6, where the future worth, F, of the principal, P, is calculated for n periods. Beginning at the start of the first period with principal (present worth) P, the interest accumulated during the first period is Pi, which when added to P gives the future worth at the end of the first period as $F = P + Pi = P(1 + i)^1$. After each period, the power to which $(1 + i)$ is raised increases, and consequently, after n compound-interest periods, the principal has grown to

$$F = P(1 + i)^n \qquad \textbf{(17.12)}$$

With compound interest, the total amount of interest after n periods is

$$\text{Compound interest} = I_C = F - P = P[(1 + i)^n - 1] \qquad \textbf{(17.13)}$$

The factor $(1 + i)^n$ in Eqs. (17.12) and (17.13) is commonly referred to as the *discrete single-payment compound-amount factor*. As shown in Eq. (17.12), when this factor is multiplied by P, we obtain the future worth, F, after n periods with interest rate per period, i. If Eq. (17.12) is solved for P, we obtain

$$P = F\left[\frac{1}{(1 + i)^n}\right] \qquad \textbf{(17.14)}$$

where the factor $[1/(1 + i)^n]$ is the discrete single-payment present-worth factor. When applied in this manner, this factor is a *discount factor* because the present worth is less than (is discounted from) the future worth.

Table 17.6 Compound Interest

No. of periods	Capital at start of period	Interest paid during period	F = Future worth at end of period
1	P	Pi	$P + Pi = P(1 + i)$
2	$P(1 + i)$	$P(1 + i)i$	$P(1 + i)^2$
3	$P(1 + i)^2$	$P(1 + i)^2 i$	$P(1 + i)^3$
.	.	.	.
.	.	.	.
.	.	.	.
n	$P(1 + i)^{n-1}$	$P(1 + i)^{n-1} i$	$P(1 + i)^n$

<table>
<tr><td>**EXAMPLE 17.18**</td><td>Determine the interest rate per year required to double $10,000 in 10 yr if the interest rate is

a. simple
b. compound</td></tr>
</table>

SOLUTION

$$P = \$10,000, \qquad F = 2(10,000) = \$20,000, \qquad n = 10 \text{ yr}$$
$$F - P = \$20,000 - \$10,000 = \$10,000$$

a. From Eq. (17.11), $\$10,000 = niP = 10\, i\, (\$10,000)$

Solving, $i = 0.10$ or 10%

b. From Eq. (17.13), $\$10,000 = P[(1 + i)^n - 1] = \$10,000[(1 + i)^{10} - 1]$

Solving, $i = 0.0718$ or 7.18%

Thus, money can double in 10 yr with an interest of just over 7% compounded annually.   ■

As seen in Example 17.18, there is a significant difference between simple interest and compound interest. Looking at this example from another perspective, if the compound interest rate for 10 yr were 10%, from Eq. (17.13), the future worth would be $25,937, compared to $20,000 for simple interest. When investing money, one should always seek compound interest so that interest is obtained on the interest.

Nominal and Effective Interest Rates

The interest period can be a day, week, month, year, etc. However, it is commonly defined in fractions of a year, for example, 1 yr, $\frac{1}{2}$ yr, . . ., $1/m$ yr, where m is the number of periods per year. When the interest period is not 1 yr, it is common to use the concepts of *nominal interest rate* and *effective interest rate* for compound interest, both based on 1 yr. The use of these two concepts permits the calculations to be carried out on an annual basis.

Given the value of m, the number of times per year to calculate interest at i, the interest rate per period of $1/m$ yr (m times per year), the *nominal interest rate* per year, r, is

$$r = im \tag{17.15}$$

If the interest rate is 3%/quarter, then with four quarters per year, the nominal interest rate, r, is $0.03(4) = 0.12$ or 12%/yr. In the case of simple interest (no compounding), $1,000 at the beginning of a year would yield $1,000(1.12) = \$1,200$. But, more commonly, nominal interest rates are stated on an annual basis with a compounding period, for example, 12% compounded quarterly.

To handle compound interest when the interest period is some fraction of a year, $(1/m)$, an effective interest rate per year, i_{eff}, is defined by

$$F_{\text{end of 1 yr}} = P(1 + i_{\text{eff}}) \tag{17.16}$$

Based on i, the actual interest rate per $1/m$ yr, we can also write

$$F_{\text{end of 1 yr}} = P(1 + i)^m = P\left(1 + \frac{r}{m}\right)^m \tag{17.17}$$

Equating Eqs. (17.16) and (17.17) and solving for i_{eff} gives

$$i_{\text{eff}} = (1 + i)^m - 1 = \left(1 + \frac{r}{m}\right)^m - 1 \tag{17.18}$$

<table>
<tr><td>**EXAMPLE 17.19**</td><td>An interest rate is reported as 3% compounded quarterly. Determine the nominal and effective interest rates per year.</td></tr>
</table>

SOLUTION $i = 3\%$/quarter of a year, $m = 4$ times per year

The nominal interest rate per year, from Eq. (17.15), is

$$r = 0.03(4) = 0.12 \text{ or } 12\%/\text{yr compounded quarterly}$$

From Eq. (17.18), the effective interest rate per year is

$$i_{\text{eff}} = \left(1 + \frac{0.12}{4}\right)^4 - 1 = 0.1255 \text{ or } 12.55\%, \text{ which is larger than the nominal rate.}\quad\blacksquare$$

Continuous Compounding of Interest

In the limit, as the number of periods per year approaches infinity, that is, as $m \to \infty$, *continuous compounding* occurs and i_{eff} tends to a maximum value for a given value of i. Equation (17.18) becomes

$$i_{\text{eff,cont}} = \lim_{m\to\infty}\left(1 + \frac{r}{m}\right)^m - 1 = \lim_{m\to\infty}\left(1 + \frac{1}{(m/r)}\right)^{(m/r)r} - 1$$

Since

$$\lim_{x\to\infty}\left(1 + \frac{1}{x}\right)^x = e = 2.71828\ldots$$

$$\text{and } \lim_{x\to\infty}\left(1 + \frac{1}{x}\right)^{xr} = e^r$$

Therefore,

$$i_{\text{eff}} = e^r - 1 \qquad\qquad (17.19)$$

where r is now the nominal annual interest rate compounded continuously, while i_{eff} is the effective annual interest rate compounded continuously. If $r = 10\%$ per year, from Eq. (17.18), $i_{\text{eff}} = \exp^{0.1} - 1 = 0.10517$ or 10.517%.

With continuous compounding, Eq. (17.12) for the future worth in terms of i_{eff} and the number of years, n_y, becomes

$$F = P(1 + i_{\text{eff}})^{n_y} = Pe^{rn_y} \qquad\qquad (17.20)$$

With continuous compound interest, the total amount of interest after n_y years is

$$\text{Continuous compound interest} = F - P = P[(1 + i_{\text{eff}})^{n_y} - 1] \qquad (17.21)$$

The factor $(1 + i_{\text{eff}})^{n_y}$ in Eqs. (17.12) and (17.13), which from Eq. (17.19) equals e^{rn_y}, is commonly referred to as the *continuous single-payment compound-amount factor.*

EXAMPLE 17.20

If it is assumed that $200,000 will be needed for a 4-yr college education starting 10 yr from now, how much must be invested today at a 6% nominal annual interest rate compounded (a) continuously, and (b) twice annually?

SOLUTION $F = \$200,000$, $r = 0.06$,

a. For continuous compounding, from Eq. (17.19), $i_{\text{eff}} = e^{0.06} - 1 = 0.06184$
From Eq. (17.19), with $n_y = 10$ yr,

$$P = \frac{F}{(1 + i_{\text{eff}})^{n_y}} = \frac{\$200,000}{(1 + 0.06184)^{10}} = \$109,760$$

b. Eq. (17.17) gives the future worth for m periods per year of compounding at the end of the first year. For n_y years, that equation becomes

$$F = P\left(1 + \frac{r}{m}\right)^{mn_y} \qquad\qquad (17.22)$$

For compounding twice annually ($m = 2$),

$$P = \frac{F}{\left(1 + \dfrac{r}{m}\right)^{mn_y}} = \frac{\$200{,}000}{\left(1 + \dfrac{0.06}{2}\right)^{2(10)}} = \$110{,}740$$

Note that the minimum capital is obtained when the interest is compounded continuously, with a difference of $980, between it and the result for semi-annual compounding. ∎

Annuities

Early in this section, only two sums of money were considered, one at the beginning, called present worth, P, and one at the end, called future worth, F. One of these was referred to as the single payment. The two were related by equations involving the interest rate/period and the number of periods that interest was applied. The use of compound interest to determine sums earlier in time (e.g., present worth) that are equivalent to a later, larger sum (e.g., future worth) was referred to as *discounting*. Factors such as $1/(1 + i)^n$ are called *discount factors*. The concepts in the previous section can be extended to a very common situation, called the *annuity*, where instead of a single payment, a series of equal payments is made at equal time intervals. Annuities also involve discounting and discount factors.

Everyday applications of annuities include house, automobile, and other loan payments (installments), where the total amount paid back over the loan period includes not only the principal (original amount of the loan), but also interest, sometimes in substantial amounts. Those saving for retirement put payments into an annuity over a period of years, with interest added to their payments. Upon retirement, retirees receive periodic payments over a specified period of years, with the unpaid amount at any period still accumulating interest. Periodic payments are also made to life insurance policies. Other kinds of annuities are created for corporations to accumulate capital, perhaps for building a new chemical processing plant.

In this section, so-called *ordinary annuities* are defined, in which the payments are made at the end of each of n interest periods and interest, i, is compounded per period. The annuity begins at the start of the first period and finishes at the end of the last period with the duration referred to as the *annuity term*. At the close of the last period, the future worth, F, of all of the payments made is known as the *amount of the annuity*. A number of formulas are derived or presented below in this and the subsequent subsection of Section 17.5. For convenience, they are summarized in Table 17.7. Less common than ordinary annuities and not discussed here are *annuity due*, in which payments are made at the beginning of the period, and the *de-*

Table 17.7 Time Value of Money
Annuity Factors—Uniform-Series Payments—Compound Interest
(Details Presented in Section 17.5)[a]

Discrete or continuous factor	Periodic interest A, end of year, discrete factor	Continuous interest A, end of year, continuous factor	Continuous interest A, continuous factor
Uniform-series sinking-fund deposit factor	$\dfrac{A}{F} = \dfrac{i}{(1 + i)^n - 1}$	$\dfrac{\overline{A}}{F} = \dfrac{e^r - 1}{e^{rn_y} - 1}$	$\dfrac{\overline{A}}{F} = \dfrac{r}{e^{rn_y} - 1}$
Uniform-series compound-amount factor	$\dfrac{F}{A} = \dfrac{(1 + i)^n - 1}{i}$	$\dfrac{F}{\overline{A}} = \dfrac{e^{rn_y - 1}e^r - 1}{}$	$\dfrac{F}{\overline{A}} = \dfrac{e^{rn_y} - 1}{r}$
Uniform-series capital-recovery factor	$\dfrac{A}{P} = \dfrac{i(1 + i)^n}{(1 + i)^n - 1}$	$\dfrac{\overline{A}}{P} = \dfrac{e^r - 1}{1 - e^{-rn_y}}$	$\dfrac{\overline{A}}{P} = \dfrac{r}{1 - e^{-rn_y}}$
Uniform-series present-worth factor	$\dfrac{P}{A} = \dfrac{(1 + i)^n - 1}{i(1 + i)^n}$	$\dfrac{P}{\overline{A}} = \dfrac{1 - e^{-rn_y}}{e^r - 1}$	$\dfrac{P}{\overline{A}} = \dfrac{1 - e^{-rn_y}}{r}$

[a]i = periodic interest rate; A = payment per interest period; n = number of interest periods; $\overline{A}$ = total annual payments per year; r = nominal annual interest rate.

ferred annuity, in which the first payment is delayed to a specified date. A *perpetuity* is another form of annuity that continues payments forever.

Discrete Compounding

To determine F when discrete uniform payments of A each are made at the end of each of the n discrete interest periods, the future worth of all the accumulated amounts, payments and interest, are summed to give the amount of the annuity. Thus, starting with the first payment at the end of first period and finishing with the last payment at the end of the last period,

$$F = A(1 + i)^{n-1} + A(1 + i)^{n-2} + \ldots + A(1 + i) + A \tag{17.23}$$

Note that because the first payment is made at the end of the first period, it is compounded over the remaining $(n - 1)$ periods. Also, the last payment is made at the end of the last period, and consequently it is not compounded. Because as n becomes large, Eq. (17.23) becomes cumbersome to evaluate, it is useful to simplify the equation. This is accomplished by multiplying both sides of Eq. (17.23) by $(1 + i)$ to give

$$F(1 + i) = A(1 + i)^n + A(1 + i)^{n-1} + \ldots + A(1 + i)^2 + A(1 + i) \tag{17.24}$$

Then, if Eq. (17.23) is subtracted from Eq. (17.24), we obtain

$$Fi = A(1 + i)^n - A \tag{17.25}$$

which, when rearranged, gives

$$F = A\left[\frac{(1 + i)^n - 1}{i}\right] \tag{17.26}$$

where the factor $[(1 + i)^n - 1]/i$ is referred to as the *discrete uniform-series compound-amount factor*. If Eq. (17.26) is solved for A, we obtain

$$A = F\left[\frac{i}{(1 + i)^n - 1}\right] \tag{17.27}$$

where the factor $i/[(1 + i)^n - 1]$ is referred to as the *discrete uniform-series sinking-fund deposit factor*. A sinking fund consists of periodic deposits that accumulate with interest up to a maturity date. In the past, some companies have used a sinking fund as a depreciation allowance to recover an original capital investment.

Sometimes, periodic payments, A, are made two or more times per year and interest is also compounded the same number of times per year, that is, m times each year. In that case, it is convenient to express the annual total of all annuity payments by the variable $\bar{A}$. Then, the payment per period, A, is simply $\bar{A}/m$. Since $i = r/m$ and $n = mn_y$, Eq. (17.26) can be rewritten as

$$F = \frac{\bar{A}}{m}\left[\frac{\left(1 + \dfrac{r}{m}\right)^{mn_y} - 1}{r/m}\right] = \bar{A}\left[\frac{\left(1 + \dfrac{r}{m}\right)^{mn_y} - 1}{r}\right] \tag{17.28}$$

If, in the more general case, equal payments are made p times per year, while interest is compounded m times per year, then, according to Bauman (1964), the future worth becomes

$$F = \hat{A}\left[\frac{\left(1 + \dfrac{r}{m}\right)^{mn_y} - 1}{\left(1 + \dfrac{r}{m}\right)^{m/p} - 1}\right] \tag{17.29}$$

where, $\hat{A}$ is the amount of each payment and $\bar{A} = p\hat{A}$. Equation (17.29) is not included in Table 17.7, but is considered in Example 17.21.

Continuous Compounding

For continuous compounding of interest with continuous payments, as $m \to \infty$, Eq. (17.26) can be expressed as follows, with the limit obtained, as before from the derivation of Eq. (17.19):

$$F = \lim_{m \to \infty} \bar{A}\left[\frac{(1 + r/m)^{(m/r)(rn_y)} - 1}{r}\right] = \bar{A}\left(\frac{e^{rn_y} - 1}{r}\right) \tag{17.30}$$

The factor, $\left(\dfrac{e^{rn_y} - 1}{r}\right)$, is referred to as the *continuous uniform-series compound-amount factor*. Equation (17.30) seems hypothetical because, although interest can be credited continuously, payments cannot be made continuously.

More practical is continuous compounding of interest, but equal discrete payments at p times per year and totaling $\bar{A}$ each year, giving the limit of Eq. (17.29) as $m \to \infty$ as

$$F = \lim_{m \to \infty} \hat{A}\left[\frac{(1 + r/m)^{(m/r)(rn_y)} - 1}{\left(1 + \dfrac{r}{m}\right)^{(m/r)(r/p)} - 1}\right] = \hat{A}\left(\frac{e^{rn_y} - 1}{e^{(r/p)} - 1}\right) \tag{17.31}$$

Equation (17.31) is not included in Table 17.6, but the case for just one payment per year, $p = 1$, with continuous compounding is included.

EXAMPLE 17.21

For the college education savings plan considered in Example 17.20, which is estimated to require $200,000 10 yr from now, calculate the total of the payments made each year to an annuity at a 6% nominal interest rate for the following conditions:

  a. Interest compounded continuously and payments continuous.
  b. Interest compounded continuously but payments quarterly.
  c. Interest compounded continuously but payments annually.
  d. Interest compounded quarterly and payments quarterly.
  e. Interest compounded semi-annually and payments semi-annually.
  f. Interest compounded quarterly and payments monthly.

For the lowest and highest payments, on an annual basis, compute the total amount of payments.

SOLUTION

For this example, $r = 0.06$, $F = \$200,000$, $n_y = 10$ yr
  Use the $\hat{A}/F$ or $\bar{A}/F$ uniform-series sinking-fund deposit factors.

  a. Equation (17.30) applies:

$$\bar{A} = F\frac{r}{e^{rn_y} - 1} = \$200,000\frac{0.06}{e^{0.06(10)} - 1} = \$14,596/\text{yr}$$

  b. Equation (17.31) applies with $p = 4$ payment/yr and $A = \$/\text{payment}$. $\bar{A} = p\hat{A} = 4\hat{A}$

$$\hat{A} = F\left(\frac{e^{(r/p)} - 1}{e^{rn_y} - 1}\right) = \$200,000\left(\frac{e^{0.06/4} - 1}{e^{0.06(10)} - 1}\right) = \$3,677/\text{payment}$$

    Therefore, $\bar{A} = 4(\$3,677) = \$14,706/\text{yr}$
  c. Equation (17.31) applies with $p = 1$ payment/yr. Therefore, $\bar{A} = \hat{A}$.

$$\bar{A} = \hat{A} = F\left(\frac{e^{(r/p)} - 1}{e^{rn_y} - 1}\right) = \$200,000\left(\frac{e^{0.06/1} - 1}{e^{0.06(10)} - 1}\right) = \$15,043/\text{yr}$$

d. Equation (17.28) applies with $m = p = 4$ payment/yr.

$$\bar{A} = F \left[\frac{r}{\left(1 + \frac{r}{m}\right)^{mn_y} - 1} \right] = \$200,000 \left[\frac{0.06}{\left(1 + \frac{0.06}{4}\right)^{4(10)} - 1} \right] = \$14,742/\text{yr}$$

e. Equation (17.28) applies with $m = p = 2$ payment/yr.

$$\bar{A} = F \left[\frac{r}{\left(1 + \frac{r}{m}\right)^{mn_y} - 1} \right] = \$200,000 \left[\frac{0.06}{\left(1 + \frac{0.06}{2}\right)^{2(10)} - 1} \right] = \$14,886/\text{yr}$$

f. Equation (17.29) applies with $m = 4$ payment/yr and $p = 12$ payment/yr. $\bar{A} = p\hat{A}$.

$$\hat{A} = F \left[\frac{\left(1 + \frac{r}{m}\right)^{m/p} - 1}{\left(1 + \frac{r}{m}\right)^{mn_y} - 1} \right] = \$200,000 \left[\frac{\left(1 + \frac{0.06}{4}\right)^{4/12} - 1}{\left(1 + \frac{0.06}{4}\right)^{4(10)} - 1} \right] = \$1,222/\text{payment}$$

$$\bar{A} = p\hat{A} = 12(\$1,222) = \$14,669/\text{yr}$$

Because the nominal interest rate is relatively low, the differences between the answers are not large, ranging from a low of \$14,596/yr for continuous compounding of interest and continuous payments to a high of \$15,043/yr for annual payments with interest compounded continuously. Thus, the total amount of payments over the 10 yr of payments ranges from \$145,960 to \$150,430. The annual payment is even higher for discrete annual payments with interest compounded annually: \$15,174/yr or a total of \$151,740 for 10 yr.

EXAMPLE 17.22

An engineer begins employment at the age of 25 and plans to invest enough money to have \$1,000,000 at a retirement age of 65. Assume that payments to the retirement fund will be made each month and that the money will receive interest at 8% compounded quarterly. Calculate the amount of each payment and the total amount of the payments made during the 40-yr savings period.

SOLUTION

Equation (17.29) applies with $F = \$1,000,000$, $r = 0.08$, $m = 4$ times/yr, and $p = 12$ times/yr.

$$\hat{A} = F \left[\frac{\left(1 + \frac{r}{m}\right)^{m/p} - 1}{\left(1 + \frac{r}{m}\right)^{mn_y} - 1} \right] = \$1,000,000 \left[\frac{\left(1 + \frac{0.08}{4}\right)^{4/12} - 1}{\left(1 + \frac{0.08}{4}\right)^{4(40)} - 1} \right] = \$290.85/\text{month}$$

For the $12(40) = 480$ payments, the total amount of payments is only $480(\$291) = \$139,680$. The growth of the future worth is exponential as shown in the following table, where an additional 10 yr of payments is added, giving a future worth of \$2,261,096.

End of Year	Future Worth (\$)	Total Payments (\$)
10	53,054	34,902
20	170,200	69,805
30	428,862	104,707
40	1,000,000	139,680
50	2,261,096	174,582

This example shows clearly the remarkable power of compound interest. ∎

Present Worth of an Annuity

The *present worth of an annuity*, P, is the amount of money at the present time that if invested at a compound interest rate will yield the amount of the annuity, F, at a future time. This is useful for determining the periodic payments that can be made over a specified number of years in the future from an annuity.

Annuity equations relating F and the periodic payments, A, are converted to equations relating P to A by combining them with Eq. (17.12) for discrete interest or Eq. (17.20) for continuous interest. This is often referred to as discounting the amount of the annuity to determine its present worth. In Table 17.7, under periodic interest, the discrete uniform-series sinking-fund deposit factor becomes the discrete uniform-series capital-recovery factor in the following manner:

$$P = \frac{F}{(1 + i)^n} = A\left[\frac{(1 + i)^n - 1}{i(1 + i)^n}\right] \tag{17.32}$$

Similarly, the continuous uniform-series capital-recovery factor with payments, A, at the end of each year is obtained:

$$P = \frac{F}{e^{rn_y}} = \frac{A}{e^{rn_y}}\left[\frac{e^{rn_y} - 1}{e^r - 1}\right] = A\left[\frac{1 - e^{-rn_y}}{e^r - 1}\right] \tag{17.33}$$

When comparing two annuities involving many payments into the future, it can be very helpful to discount all of the payments to their present worth. This gives the principal required at the current time, invested at the current interest rate, to enable the payments to be made at the end of each annuity period. While the annuity is making payments, interest continues to be paid on the remaining balance. At the end of the term of the annuity, the balance is zero.

EXAMPLE 17.23

Upon retirement at the age of 65, an employee has a retirement fund of $1,000,000. If this fund is invested at 8% compounded quarterly, how much can be paid to the retiree at the end of each month, if the fund is to diminish to zero at the end of 20 yr when the retiree would be 85?

SOLUTION

$$r = 0.08,\ m = 4,\ p = 12,\ \text{and}\ P = \$1,000,000$$

Since the period of compounding and the payment period are different, none of the equations in Table 17.7 apply. Instead, use the following extension of Eq. (17.14),

$$P = \frac{F}{(1 + i)^n} = \frac{F}{\left(1 + \dfrac{r}{m}\right)^{mn_y}} \tag{17.34}$$

with Eq. (17.29) to give

$$\hat{A} = P\left[\frac{\left(1 + \dfrac{r}{m}\right)^{m/p} - 1}{1 - \left(1 + \dfrac{r}{m}\right)^{-mn_y}}\right]$$

Thus,

$$\hat{A} = \$1,000,000\left[\frac{\left(1 + \dfrac{0.08}{4}\right)^{4/12} - 1}{1 - \left(1 + \dfrac{0.08}{4}\right)^{-4(20)}}\right] = \$8,332/\text{month or } \$99,979/\text{yr}$$

It is interesting to note that if the monthly payments are reduced to $6,536, or $19,608 quarterly, then at the end of the first quarter, 2% interest will be paid on the balance of ($1,000,000 − $19,608 = $980,392), giving $19,608, which is the same as the amount paid out during the quarter. Thus, the amount of the annuity will remain at $1,000,000 and payments can continue forever. ∎

Comparing Alternative Equipment Purchases

It is often desirable to compare the purchases of two or more alternative items of equipment, each having a different installed cost and estimated performance life, maintenance cost, and salvage value. The two main methods for comparison, *present worth* and *capitalized cost*, are covered in this subsection. At the outset, it is important to recognize that these measures are examined, often on an ad-hoc basis, primarily for the purchase of an equipment item after the plant has been designed. During the comparison of alternative plant designs or major retrofits, when it is important to account for sales revenues, the calculation of *cash flows* is recommended, as described in the next section, for use in computing the *net present value* (NPV) or the *investor's rate of return* (IRR or DCFRR). Note, however, that the present worth and NPV are identical when there are no revenues, for example, when comparing alternative methods for treating a waste stream. In general, when making comparisons of alternatives, it is not necessary to consider so-called *sunk costs*, which are costs that occurred in the past, but have no effect on current or future decisions.

Present Worth

In the present worth technique, all of the costs and revenues are discounted to calculate the *present worth* of each alternative. Note that it is crucial to compare the alternatives over the same time period. This approach is illustrated in Example 17.24, in which diagrams show the projected costs and the recovery of the salvage values in time.

EXAMPLE 17.24

Two alternative pumps, A (carbon steel) and B (aluminum), have different installed and maintenance costs, salvage values, and anticipated service lives, as indicated below. It is desired to select one of the pumps on the basis of present worth when the effective interest rate is 10%.

	A	B
Installed cost	$18,000	$25,000
Uniform end-of-year maintenance	$4,000	$3,000
Salvage value	$500	$1,500
Service life	2 yr	3 yr

SOLUTION

Six years is the shortest time period for which the two pumps can be compared on a common basis because six is the smallest number divisible by both two and three. Thus, pump A is replaced twice and salvaged three times over 6 yr, during which time pump B is replaced once and salvaged twice. For pump A, the costs and salvage values are shown on the following diagram, in which the installed and maintenance costs are represented by downward vectors (i.e., negative, compared to zero along the horizontal axis) and the salvage values are represented by upward vectors (i.e., positive). Notice that at the end of the second, fourth, and sixth years, the maintenance costs appear even though the pump is being salvaged and replaced. These are charges that have accumulated over the prior year.

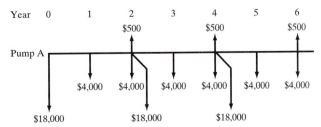

When discounting the costs and salvage values, the maintenance costs can be treated as an annuity, computed using Eq. (17.32) with $i = 0.10$ and $n = 6$, because the cost is periodic and constant at

$4,000/yr. Also, the salvage value can be credited against the purchase cost, giving $18,000 - $500 = $17,500, because they both occur at the end of the second and fourth years. Thus,

$$P_A = -\$18,000 - \$4,000 \left[\frac{(1 + 0.1)^6 - 1}{0.1(1 + 0.1)^6} \right] - \frac{\$17,500}{(1 + 0.1)^2} - \frac{\$17,500}{(1 + 0.1)^4} + \frac{\$500}{(1 + 0.1)^6} = -\$61,554$$

The corresponding diagram for pump B is

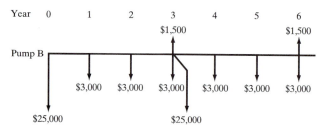

and the discounted costs and salvage values are

$$P_B = -\$25,000 - \$3,000 \left[\frac{(1 + 0.1)^6 - 1}{0.1(1 + 0.1)^6} \right] - \frac{\$23,500}{(1 + 0.1)^3} + \frac{\$1,500}{(1 + 0.1)^6} = -\$54,875$$

Although pump B has the higher installation cost, it is selected because its present worth is lower than pump A. ∎

Capitalized Costs and Perpetuities

Another method for comparing alternatives, which leads to conclusions identical to those of present worth, is to compute *capitalized costs*. This involves the creation of a *perpetuity*, in which periodic replacements continue indefinitely for each alternative. The capitalized cost, K, is defined as the original cost, C_I, plus the present value of the perpetuity for an infinite number of replacements made every n_R years. When a replacement is made, it is common to assign a salvage value, S_{equip}. Thus, if inflation of costs is ignored, the replacement cost is constant at $C_R = C_I - S_{equip}$. Note that better estimates for the replacement costs, taking into account inflation, likely market conditions, and similar factors, are not normally justified for comparisons involving perpetuities. To account for such factors, the cash flow analysis in the next section is preferred. Assuming a nominal interest rate, r, compounded m times per year, and using the general form of the discount factor in Eq. (17.34), the investment must provide a future worth, F, every n_y years, sufficient to pay for the replacement of the equipment item and replace the principal, P, so that it can be reinvested for another n_y years. Thus,

$$F_{n_y} = C_R + P = P\left(1 + \frac{r}{m}\right)^{mn_y} \tag{17.35}$$

Rearranging Eq. (17.35), the present value of the perpetuity is

$$P = \frac{C_R}{\left(1 + \dfrac{r}{m}\right)^{mn_y} - 1} \tag{17.36}$$

From the definition of the capitalized cost,

$$K = C_I + P = C_I + \frac{C_R}{\left(1 + \dfrac{r}{m}\right)^{mn_y} - 1} \tag{17.37}$$

When comparing alternatives, the equipment item having the lowest capitalized cost is selected.

EXAMPLE 17.25

Select one of the two pumps in Example 17.24 on the basis of capitalized costs.

SOLUTION

To use capitalized costs when annual operating costs are also required (in this case for maintenance), it is appropriate to discount the operating cost payments to present worth, using the annuity equation [Eq. (17.32)], and to add this to the equipment installed cost. For pump A, the initial adjusted installed cost is

$$C_{I_A} = \$18,000 + \$4,000\left[\frac{(1 + 0.1)^2 - 1}{0.1(1 + 0.1)^2}\right] = \$18,000 + \$6,940 = \$24,940$$

and the capitalized cost, using Eq. (17.37) with $r = 0.1$, $m = 1$, and $n_y = 2$, is

$$K_A = \$24,940 + \frac{\$24,940 - \$500}{(1 + 0.1)^2 - 1} = \$24,940 + \$116,380 = \$141,320$$

Turning to pump B,

$$C_{I_B} = \$25,000 + \$3,000\left[\frac{(1 + 0.1)^3 - 1}{0.1(1 + 0.1)^3}\right] = \$25,000 + \$7,460 = \$32,460$$

and

$$K_B = \$32,460 + \frac{\$32,460 - \$1,500}{(1 + 0.1)^3 - 1} = \$32,460 + \$93,540 = \$126,000$$

Indeed, pump B has the lower capitalized cost, a result consistent with the comparison in Example 17.24 of the two present worths. In fact, P_A/P_B in Example 17.24 equals K_A/K_B in Example 17.25.   ∎

17.6   CASH FLOW AND DEPRECIATION

The approximate profitability measures in Section 17.4, although often used to select the most promising flowsheets during process synthesis (when the details of the process units are deemphasized relative to the arrangements and sequences of the units in the flowsheet), are inadequate to enable management to make a final decision regarding the financial feasibility of a potential process. Dr. Robert M. Busche, President of Bio-en-gene-er Associates and a long-time engineer and venture analyst at DuPont, often reminds students that chemical companies like DuPont were awash in cash assets during the period following World War II, throughout the 1950s and 1960s. At that time, the principal challenges were to develop new products and commercialize them quickly. Approximate measures of financial goodness, like the return on investment (ROI), were used routinely to decide whether a process was sufficiently promising to fund. In the 1970s and 1980s, however, cash assets became less plentiful and the competition among potential processes for the limited resources of a company became much stiffer. To arrive at decisions, management required more accurate assessments of the financial aspects of potential processes, and consequently, companies began to require cash flow estimates for each year of operation of the most promising processes. This change in perspective was signaled in a pioneering paper by Souders (1966). Note, however, that the ROI, as well as the PBP, continue to be computed, primarily because they are easy to calculate. They permit a quick comparison of investments, with relatively few calculations, and are especially useful when the costs and revenues are not anticipated to change significantly over the life of the project.

As was discussed in Chapter 16, *cash flow* for a company has become an important financial factor. Cash flow is defined as the net passage of money into or out of a company due to an investment. It may be positive (into) or negative (out of). For investment evaluation, investments are considered a negative cash flow, while after-tax profits plus depreciation are positive cash flows. During the years of plant construction, the cash flow, CF, for a particular year, is

$$CF = -fC_{TDC} - C_{WC} - C_{land} \qquad (17.38)$$

where f is the fraction of the total depreciable capital, C_{TDC}, expended that year, and C_{WC} and C_{land} are the working capital and the cost of land that are expended, if any, during that year of construction.

To estimate the cash flow for a particular year of plant operation, the pretax earnings are computed from Eq. (17.4) and the after-tax earnings from Eq. (17.5). However, in the calculation of the production cost, a more elaborate depreciation schedule, discussed in the next section, replaces the straight-line depreciation used in Section 17.2. The cash flow from plant operations is the after-tax earnings plus the depreciation:

$$CF = (1 - t)(S - C) + D = 0.63(S - C) + D \qquad (17.39)$$

During the first year of operation, startup costs may occur. During all years of operation, there may be royalty costs. At the conclusion of plant operations, there may be a salvage value for used equipment, S_{equip}.

The annual cash flow, CF, for any year of the project, including the construction phase and possible salvage at the end of operation, is

$$CF = (1 - t)(S - C) + D - fC_{TDC} - C_{WC} - C_{land} - C_{startup} - C_{royal} + S_{equip} \qquad (17.40)$$

In Eq. (17.40), the depreciable capital is normally expended before operation of the plant begins, and the working capital is usually expended in the year preceding the beginning of operation. The working capital is recovered during the last year of operation as a negative entry (to give a positive cash flow) in Eq. (17.40). A common convention is that all cash transactions take place at the end of the year. It is also common to project cash flows for new products over 10 plus years, typically 15 yr, whereas cash flows are projected for established products for 20 yr or more. Another convention recommended by Busche (1995) is to design the plant on a capacity basis to operate during 330 days (7,920 hr) per year, with 35 days for shutdowns due to maintenance needs and malfunctions. This corresponds to an operating factor of 0.9041. Some companies prefer to round this figure to 8,000 hr/yr. In addition, Busche recommends that the total production cost, C (computed using the cost sheet of Table 17.1), and sales, S, be computed for production at less than 100% of capacity during the first and, perhaps, second year of operation, while the plant is being started up and any design flaws are being remedied.

Depreciation

Depreciation is the reduction in value of an asset. Recall that in the cost sheet of Table 17.1 that a company is allowed to treat depreciation as a cost of production, thereby reducing its income tax liability, even though depreciation is not an actual cash flow out of the company. When calculating approximate profitability measures, such as the return on investment (ROI), it is common to calculate the cost of sales using straight-line (SL) depreciation, as in Table 17.1. Since the approximate profitability measures give just a snapshot view of the economic goodness of a proposed project, usually projected for the third year of operation, no other method of accounting for the depreciation of the total depreciable capital is justified.

Two profitability measures, discussed below, that provide more rigor, net present value (NPV) and investor's rate of return (IRR), involve the discounting of cash flows to present worth, as discussed in the next section. These measures increase in magnitude when a larger fraction of the total depreciation is taken in the early years of operation, when the plant is probably operating below capacity and the discount factors are low. For these reasons, it is advantageous for a company to rapidly depreciate its capital investment early in the life of a process instead of using straight-line depreciation. Depreciation methods that favor the earlier years include the *declining-balance method* (DB), the *double declining-balance method* (DDB), and the *sum-of-the-years digits method* (SYD). More recently, in 1981, the U.S. federal income tax regulations provided an *Accelerated Cost Recovery System* (ACRS) for early depreciation. A *Modified Accelerated Cost Recovery System* (MACRS) went into effect in 1987. The ACRS and MACRS methods combine aspects of the DB or DDB methods with the SL method. These five methods are discussed next and compared to straight-line depreciation. Another depreciation method that is sometimes referred to is the sinking-fund method. However, it decelerates, rather than accelerates, depreciation and, therefore, is not of interest to most industries and is not considered here. It is important to note that a company may use two or more different depreciation methods, most commonly: (1) one method for *book depreciation* for internal financial accounting and (2) one method for *tax depreciation* that follows government regulations.

All of the depreciation methods to be discussed are based on the asset *book value*, which at any year is defined as the original cost of the asset (e.g., the depreciable capital investment) minus the sum of the depreciation charges made to the asset up to that year. This is in contrast to the *market value*, which is the price that could be obtained for the asset if it were placed for sale in the open market, and the *replacement value*, which is the cost of replacing the asset. The book value is the value shown on the accounting records. The book value decreases each year until it reaches a salvage value, at which time it is completely depreciated. The number of years, n, over which an asset can be depreciated is usually related to an estimate of the useful life of the asset, which is discussed below.

Declining-Balance (DB) and Double Declining-Balance (DDB) Methods

The declining-balance method is also referred to as the fixed percentage or the uniform percentage method because the amount of depreciation each year is a fixed fraction, d, of the book value of the depreciable asset. Let B = the original cost of the asset, which is usually called the *basis*; t = years of service of the asset; and BV_t = book value at the end of year t. Then, the amount of annual depreciation, D_t, for the year t is given by

$$D_t = BV_{t-1} - BV_t = dBV_{t-1} \qquad \textbf{(17.41)}$$

Consequently, after $t - 1$ yr, the book value is

$$BV_{t-1} = B(1 - d)^{t-1} \qquad \textbf{(17.42)}$$

Combining Eqs. (17.41) and (17.42) gives

$$D_t = dB(1 - d)^{t-1} \qquad \textbf{(17.43)}$$

Limits are placed on the value of d, allowing it to range only from $1/n$ to $2/n$, with $1.5/n$ (150% declining balance) and $2/n$ (200% or double declining balance) being common values. With the declining-balance methods, a salvage value is not used. However, the book value, which never reaches zero because it is only decreased each year by a fixed fraction, is not permitted to drop below the estimated salvage value. To force the book value to the salvage value at the end of year n, it is considered desirable to use the *combination method*, which involves switching from the declining-balance method to the straight-line method part way through the service life. Another scheme is to back-calculate the value of d that will give a book value equal to the salvage value at year $t = n$. This value of d is obtained from Eq. (17.42) by setting $(t - 1) = n$. Thus,

$$S_{\text{equip}} = B(1 - d)^n \qquad \textbf{(17.44)}$$

Solving Eq. (17.44) for d gives

$$d = 1 - \left(\frac{S_{\text{equip}}}{B}\right)^{1/n} \qquad \textbf{(17.45)}$$

However, if the computed $d > 0.2$, it is out of the accepted range and cannot be used. In that case, the only alternative declining-balance method is the combination method, as illustrated in the following example.

EXAMPLE 17.26

A new instrument is purchased for the control laboratory of a plant at a cost of $200,000. It is estimated to have a 10-yr useful life with a salvage value of $30,000. Estimate the amount of depreciation each year by the following methods:

a. Straight-line depreciation over 10 yr based on $200,000 − $30,000 = $170,000.
b. Declining-balance depreciation with $d = 1/n$.
c. 150% declining-balance depreciation.
d. Double declining-balance depreciation.
e. Combination method of double declining-balance depreciation switching to straight-line depreciation after 5 yr.

SOLUTION

a. The amount of depreciation each year is constant at $170,000/10 = $17,000
From Eq. (17.45), for declining-balance methods,

$$d = 1 - \left(\frac{S_{equip}}{B}\right)^{1/n} = 1 - \left(\frac{\$30,000}{\$200,000}\right)^{1/10} = 0.173$$

Therefore, the 100% and 150% declining-balance methods will not be attractive, since this $d > 0.10$ and 0.15. The 200% declining-balance method and the combination method will lead to good results.

b. The depreciation each year is computed from Eq. (17.41) with B = $200,000 and $d = 1/10 = 0.10$. See the table below.
c. Use Eq. (17.41) with B = $200,000 and $d = 1.5/10 = 0.15$. See the table below.
d. Use Eq. (17.41) with B = $200,000 and $d = 2/10 = 0.20$. See the table below.
e. Use Eq. (17.41) with B = $200,000 and $d = 1.5/10 = 0.15$ for the first five years and then subtract the salvage value from the book value and continue with straight-line depreciation. See the table below.

The results of the calculations, which are readily carried out on a spreadsheet, are as follows:
For Parts (a), (b), and (c):

End of Year	Straight-Line Depreciation		Declining Balance ($d = 0.1$)		Declining Balance ($d = 0.15$)	
	D ($/yr)	BV ($)	D ($/yr)	BV ($)	D ($/yr)	BV ($)
0		200,000		200,000		200,000
1	17,000	183,000	20,000	180,000	30,000	170,000
2	17,000	166,000	18,000	162,000	25,500	144,500
3	17,000	149,000	16,200	145,800	21,675	122,825
4	17,000	132,000	14,580	131,220	18,424	104,401
5	17,000	115,000	13,122	118,098	15,660	88,741
6	17,000	98,000	11,810	106,288	13,311	75,430
7	17,000	81,000	10,629	95,659	11,314	64,115
8	17,000	64,000	9,566	86,093	9,617	54,498
9	17,000	47,000	8,609	77,484	8,175	46,323
10	17,000	30,000	7,748	69,736	6,949	39,375

Note in the above table for Parts (b) and (c) that the salvage value of $30,000 is not reached by year 10. Thus, these are not good methods to apply.

For Parts (d) and (e):

End of Year	Declining Balance (d = 0.20)		Combination Method	
	D ($/yr)	BV ($)	D ($/yr)	BV ($)
0		200,000		200,000
1	40,000	160,000	40,000	160,000
2	32,000	128,000	32,000	128,000
3	25,600	102,400	25,600	102,400
4	20,480	81,920	20,480	81,920
5	16,384	65,536	16,384	65,536
6	13,107	52,429	7,107	58,429
7	10,486	41,943	7,107	51,322
8	8,389	33,554	7,107	44,214
9	3,554	30,000	7,107	37,107
10	0	30,000	7,107	30,000

For Part (d), the double declining-balance method, with $d = 0.2 > 0.173$ above, the salvage value is reached by the book value before 10 yr. As shown in the table, it is reached in year 9, so that no depreciation is taken in year 10. In Part (e), the combination method switches from the double declining-balance method to the straight-line method in year 6, such that the book value becomes the salvage value in year 10. For these two methods, the depreciation is greatly accelerated over the straight-line method in the first 4 yr. ∎

Sum-of-the-Years-Digits Method (SYD)

This is a classic depreciation acceleration method that has the advantage of being able to handle a salvage value, including zero. Its disadvantage is that the depreciation acceleration is less than the double declining-balance method. Its name is derived from the use of the sum of the digits from 1 to n, the number of years of useful life of the asset. This sum in compact form is given by

$$\text{SUM} = \sum_{j=1}^{n} j = \frac{n(n + 1)}{2} \tag{17.46}$$

Thus, for $n = 10$ yr, SUM $= 10(10 + 1)/2 = 55$. The annual depreciation is

$$D_t = \frac{\text{depreciable years remaining}}{\text{SUM}} (B - S_{equip}) \tag{17.47}$$

Thus, if $n = 10$ yr, the fraction depreciated the first year is $10/55 = 0.1818$, which is almost twice that of the straight-line method. For the next year, the fraction is $9/55 = 0.1636$. In year six, the fraction is $5/55 = 0.0909$, which is now less than straight-line depreciation of 0.10. If Example 17.26 is applied to the SYD method, the following results are obtained, which are compared to the DDB method:

End of Year	Sum-of-the-Years Digits		Declining Balance ($d = 0.20$)	
	D ($/yr)	BV ($)	D ($/yr)	BV ($)
0		200,000		200,000
1	30,909	169,091	40,000	160,000
2	27,818	141,273	32,000	128,000
3	24,727	116,545	25,600	102,400
4	21,636	94,909	20,480	81,920
5	18,545	76,364	16,384	65,536
6	15,455	60,909	13,107	52,429
7	12,364	48,545	10,486	41,943
8	9,273	39,273	8,389	33,554
9	6,182	33,091	3,554	30,000
10	3,091	30,000	0	30,000

In the first 3 yr, the DDB method accelerates the depreciation much more than the SYD method, although the depreciation in the first year of the SYD method, $30,909, is considerably higher than the $17,000 of the straight-line method.

ACRS and MACRS Methods for Tax Depreciation

From 1982 to 1986, the U.S. federal income-tax regulations required companies to use the Accelerated Cost Recovery System (ACRS) to depreciate property when computing federal income tax. In 1987, the Modified Accelerated Cost Recovery System (MACRS) replaced that system. Both systems are based on the declining-balance method with a switch to the straight-line method when it offers a faster depreciation write-off. However, both methods assume that assets are placed in service at the midpoint of the tax year. Therefore, for both methods, only 50% of the DB depreciation is allowed in the first year. Another departure occurs for the MACRS method, wherein the depreciation is continued for 1 yr beyond the life, but only 50% of the straight-line depreciation is taken in that final year. For both methods, the service life (called *class life*) is fixed by regulations for from 3 to 15 yr, and to 20 yr and even longer (in the case of some structures) for the MACRS method. The depreciation calculations are best carried out using the U.S. tables. The MACRS depreciation table is shown in Table 17.8 for class lives of 5, 7, 10, and 15 yr. The selection of class life is also regulated by the U.S. federal government, which offers two options: (1) the General Depreciation System (GDS) and (2) the Alternative Depreciation System (ADS). The GDS allows a more desirable shorter class life and is the preferred choice. However, the ADS is sometimes used by new businesses that do not need the tax benefit of accelerated depreciation. Table 17.9 gives the GDS class life for a number of different kinds of assets. For most new chemical-plant projects, a class life of 5, 7, or 10 yr is used. For these three class lives, Table 17.8 shows that depreciation begins with the double declining-balance method. For example, for a class life of 10 yr, depreciation in the first year is 50% of $2/n = 2/10 = 0.20$, which gives the 10% shown in the table. When 10% of the basis, B, is subtracted from B, the book value is 90% of the basis. In year 2, the DDB depreciation is 20% of the 90% or 18%, which is the value shown in the table. Also, for a same class life of 10 yr, the table shows a switch to straight-line depreciation of 6.55% in year 7, because the calculated DDB depreciation would be lower at 5.90%.

Table 17.8 MACRS Tax-Basis Depreciation

Percent of total depreciable capital (C_{TDC})
for class life of:

Year	5 Yr	7 Yr	10 Yr	15 Yr
1	20.00	14.29	10.00	5.00
2	32.00	24.49	18.00	9.50
3	19.20	17.49	14.40	8.55
4	11.52	12.49	11.52	7.70
5	11.52	8.93	9.22	6.93
6	5.76	8.92	7.37	6.23
7	100.00	8.93	6.55	5.90
8		4.46	6.55	5.90
9		100.00	6.56	5.91
10			6.55	5.90
11			3.28	5.91
12			100.00	5.90
13				5.91
14				5.90
15				5.91
16				2.95
				100.00

EXAMPLE 17.27

In Example 17.14, the total depreciable capital of a new plant is projected to be $90 million. Compute the annual depreciation by the MACRS method for class lives of 5, 7, and 10 yr and the income taxes saved because of depreciation during an 11-year period for a combined federal and state income tax rate of 37%. Assume no salvage value.

SOLUTION

The basis for depreciation is $90,000,000. The amount of depreciation for each year is the product of the basis and the fractional percentage depreciation from Table 17.8. The savings in income tax each

Table 17.9 GDS Class Life for Use with the MACRS Depreciation Method

Type of asset	GDS class life (years)
Special manufacturing and handling devices, e.g., tractors	3
Autos, trucks, buses; cargo containers, computers and peripherals; copy and duplicating equipment; some manufacturing equipment	5
Railroad cars, engines, tracks; agricultural machinery; office furniture; petroleum and natural gas equipment and some other manufacturing equipment; all other business assets not listed in another class	7
Equipment for water transportation, petroleum refining, agriculture product processing, durable-goods manufacturing, and shipbuilding	10
Land improvements, docks, roads, drainage, bridges, pipelines, landscaping, nuclear-power production, and telephone distribution	15
Farm buildings, telephone switching buildings, power production equipment, municipal sewers, and water utilities	20
Residential rental property, including mobile homes	27.5
Nonresidential real property attached to the land	39

year is 37% of the amount of depreciation. The calculations are readily made with a spreadsheet, which gives the following results:

Year	Class Life = 5 yr		Class Life = 7 yr		Class Life = 10 yr	
	D ($/yr)	Taxes Saved ($/yr)	D ($/yr)	Taxes Saved ($/yr)	D ($/yr)	Taxes Saved ($/yr)
1	18,000,000	6,660,000	12,861,000	4,758,570	9,000,000	3,330,000
2	28,800,000	10,656,000	22,041,000	8,155,170	16,200,000	5,994,000
3	17,280,000	6,393,600	15,741,000	5,824,170	12,960,000	4,795,200
4	10,368,000	3,836,160	11,241,000	4,159,170	10,368,000	3,836,160
5	10,368,000	3,836,160	8,037,000	2,973,690	8,298,000	3,070,260
6	5,184,000	1,918,080	8,028,000	2,970,360	6,633,000	2,454,210
7	0	0	8,037,000	2,973,690	5,895,000	2,181,150
8	0	0	4,014,000	1,485,180	5,895,000	2,181,150
9	0	0	0	0	5,904,000	2,184,480
10	0	0	0	0	5,895,000	2,181,150
11	0	0	0	0	2,952,000	1,092,240
Total $	90,000,000	33,300,000	90,000,000	33,300,000	90,000,000	33,300,000

These results show for the three cases the same total depreciation of $90,000,000, which equals the basis, and the same total income tax savings of $33,300,000 because of depreciation. However, when the present values of the tax savings for each year are computed from Eq. (17.14) and summed for each of the three cases, the results are different, with the shorter class life favored, as shown below for a nominal interest rate of 10% compounded annually.

Class Life (yr)	Present Value of Income Tax Savings
5	$25,750,000
7	$24,024,000
10	$21,783,000

The class life of 5 yr is superior to 7 yr, and even more so to 10 yr. ■

Before leaving this complex topic, it is important to emphasize that depreciation does not involve a transfer of cash; it is just an accounting artifact. In the calculation of cash flows, it is needed to calculate the taxable earnings, from which income tax is computed [Income tax $= 0.37(S - C)$]. Then, as shown in Eq. (17.39), depreciation is added back to the after-tax earnings to obtain cash flows. From an income-tax standpoint, depreciation should be taken as rapidly as the law permits.

Depletion

Whereas depreciation applies to assets that can be replaced, *depletion* applies to natural resources, which when removed for processing disappear forever, or are only renewed by nature over a period of many years. Depletion is applicable to fisheries, forests, mineral deposits, natural gas wells, oil deposits, orchards, quarries, vineyards, etc. The U.S. federal government permits those using natural resources a depletion allowance, which acts like depreciation as an expense against sales revenue. Two methods are used to calculate the annual depletion allowance: *cost* (or factor) *depletion* and *percentage depletion*.

Table 17.10 Allowable Percentages of Gross Income for
Depletion of Natural Resources When Using Percentage Depletion

Natural resource	Percentage of gross income
Gravel, peat, sand, and some stones	5
Coal, lignite, and sodium chloride	10
Most other minerals and metal ores	14
Copper, gold, iron ore, and silver	15
Oil and gas wells (only for small producers)	15
Lead, nickel, sulfur, uranium, and zinc	22

Cost depletion

This method is based on the usage of the resource each year, starting with an estimate of the amount of resource that can be removed (recovered) and its cost. Since it may be difficult to make an initial estimate of the amount of recoverable resource, the estimate can be revised at a later date. To use this method, a cost depletion factor, p_t, for the year t is defined

$$p_t = \frac{\text{first cost of the resource}}{\text{estimated units of recoverable resource}} \qquad (17.48)$$

where the units are barrels for oil, tons for ore, standard cubic feet for gas, board feet for lumber, etc. The depletion charge for year t is the product of the cost depletion factor and the recovered number of units in year t. The total depletion charge cannot exceed the first cost of the resource.

Percentage Depletion

For the natural resources listed below, a special consideration is given. A constant percentage of the sales revenue (referred to as the gross income) from the resource may be depleted provided that it does not exceed 50% of the taxable income (before the depletion allowance) of the company. Total depletion charges are allowed to exceed the first cost of the resource. The allowable percentage depends on the type of resource, as given in Table 17.10. However, for oil and gas, only small producers are allowed to use percentage depletion. When percentage depletion is applicable, cost depletion is also computed and the method giving the largest annual depletion charge is used.

EXAMPLE 17.28

A mining property, containing an estimated 900,000 tons of lead and zinc ore, is purchased for $4,500,000. In the first year of operation, 100,000 tons of the ore is mined and sold for $20/ton. The expenses that year are $1,200,000. Calculate the net profit and cash flow by (a) cost depletion and (b) percentage depletion.

SOLUTION

The sales revenue (gross income) = 20(100,000) = $2,000,000

a. From Eq. (17.48), p_t = $4,500,000/900,000 = $5/ton
Depletion charge = 5(100,000) = $500,000
Profit before taxes = $2,000,000 − $1,200,000 − $500,000 = $300,000
Income tax = 0.37(300,000) = $111,000
Net profit (after tax) = $300,000 − $111,000 = $189,000
Cash flow = net profit + depletion charge = $189,000 + $500,000 = $689,000
b. From Table 17.10, the allowable % of gross income for depletion = 22%
Depletion allowance = 0.22($2,000,000) = $440,000
Taxable income before depletion = $2,000,000 − $1,200,000 = $800,000

The percentage depletion allowance of $440,000 exceeds 50% of the taxable income before the depletion allowance. Therefore, the depletion allowance can only be $0.50(\$800,000) = \$400,000$.

Profit before taxes = $800,000 - $400,000 = $400,000

Income tax = $0.37(\$400,000) = \$148,000$

Net profit (after tax) = $400,000 - $148,000 = $252,000

Cash flow = $252,000 + $400,000 = $652,000

In this example, cost depletion is better than percentage depletion. ■

17.7 RIGOROUS PROFITABILITY MEASURES

The two principal profitability measures that involve the time-value-of-money in terms of discounted cash flows are the net present value or worth (NPV) and the investor's rate of return (IRR), which is also referred to as the discounted cash flow rate of return (DCFRR). These measures are anomalous in that, when used to compare alternative processes, they often give different results. This has led to substantial disagreement within the finance community (Brealey and Myers, 1984).

When using NPV and IRR, the discounting is normally made with Eq. (17.34) using a nominal interest rate, r, that is compounded annually ($m = 1$), with n_y starting from the beginning of the first year of plant construction. It is possible to account for *investment creep* in the projection of the cash flows. This usually arises through small annual increases in the investment, of the order of 1–1.5%, due to small projects to install additional equipment as the capacity of the process is increased or process modifications are needed. The additional investment is depreciated on the same schedule as the original investment. The calculations are more complex, but are readily made when calculating cash flows.

When carrying out a rigorous profitability analysis, some design teams adopt the convention of reducing the process yield by a small amount, such as 2%, to account for the loss of raw materials and products during startups, shutdowns, and periods when there are malfunctions. Often, raw materials are vented or flared during startup. In other cases, one part of the plant shuts down while the remainder continues to operate, with small amounts of intermediate products vented when they are nontoxic and not easily stored, until the idle portion of the plant is restarted.

Finally, when calculating the cash flow for the last year of operation, it is common to take credit for the working capital investment. Some companies also take credit for a projection of the salvage value of the plant, assuming that it is dismantled and sold at the end of its useful life. Because salvage values are difficult to estimate, and in some cases distort the NPV and IRR, many companies prefer to be conservative and assume a zero salvage value.

The NPV method is simpler to implement than the IRR method and is well defined, whereas the IRR is not defined in all situations. Because the latter involves an iterative computation of the net present value, the simpler-to-calculate NPV is discussed first.

Net Present Value (NPV)

To evaluate the *net present value* (NPV) of a proposed plant, its cash flows are computed for each year of the projected life of the plant, including construction and startup phases. Then, given the interest rate specified by company management (typically 15%), each cash flow is discounted to its present worth. The sum of all the discounted cash flows is the net present value. The NPV provides a quantitative measure for comparing the capital required for competing processes in current terms. However, the result is usually quite sensitive to the assumed interest rate, with proposed processes changing favor as the interest rate varies. An illustration of the calculation of NPV is given below in Example 17.29, following a discussion of the IRR method.

Investor's Rate of Return (IRR or DCFRR)

The *investor's rate of return* (IRR), also called the *discounted cash flow rate of return* (DCFRR), is the interest rate that gives a net present value of zero. Since the net present value is a complex nonlinear function of the interest rate, an iterative procedure (easily accomplished using a spreadsheet) is required to solve

$$\text{NPV}\{r\} = 0 \qquad \text{for } r \qquad\qquad (17.49)$$

When comparing proposed processes, the largest IRR is the most desirable. Note, however, that sometimes the process having the largest IRR has the smallest NPV. In many cases, especially when the alternatives have widely disparate investments, both the NPV and the IRR are effective measures. This is especially true when the alternatives are comparable in one measure but are very different in the other. The following example computes both the NPV and the IRR.

EXAMPLE 17.29

(Example 17.14 Revisited)

For the process considered in Example 17.14, but with MACRS depreciation for a 5-yr class life as determined in Example 17.27, calculate, over an estimated life of 15 yr, including years 1997–1999 when the plant is being constructed (a) the NPV for a nominal interest rate of 15% compounded annually and (b) the nominal interest rate for the IRR method (i.e., for NPV = 0). For the first 2 yr of plant operation, when at 45 and 67.5% of capacity, the cost of production, exclusive of depreciation, is $55 million and $78 million, respectively.

SOLUTION

a. The cash flows are shown in the following table in millions of dollars per year. Note that the total depreciable capital of $90 million is divided into three equal parts for the first 3 yr. The working capital of $40 million appears in the third year. Plant startup costs in the years 2000 and 2001 are not included and no salvage is taken at the end of the project. In the year 1999, the investment in millions is $-\$30 + -\$40 = -\$70$. The discount factor is $1/(1 + 0.15)^2 = 0.7561$. Therefore, the PV is $0.7561(-70) = -52.93$ or $-\$52.93$ million. Instead of showing negative signs in the table, negative values are enclosed in parentheses. When added to the $-\$56.09$ million for the cumulative PV for 1998, a cumulative PV of $-\$109.02$ million is obtained for 1999. In the first year of plant operation, 2000, sales revenue is $75 million, MACRS depreciation is $18 million, and production cost exclusive of depreciation is $55 million. Therefore, pre-tax earnings is $75 - 55 - 18 = \$2$ million. The combined federal and state income tax is $0.37(2) = \$0.74$ million. This gives an after-tax or net earnings of $2 - 0.74 = \$1.26$ million. The cash flow for year 2000 is $1.26 +$ the 18 depreciation allowance $= \$19.26$ million. The discount factor for that year is $1/1.15^3 = 0.6575$. Therefore, the PV for 2000 is $0.6575(19.26)$ or $\$12.67$ million. When added to the cumulative PV of $-\$109.02$ million, a cumulative PV of $-\$96.35$ is obtained. The calculations for the remaining years of the project are carried out in a similar manner, most conveniently with a spreadsheet. The final NPV at the end of year 2011 is $25.28 million. Notice that the cumulative PV remains negative until the year 2008. This is equivalent to a payback time of more than 8 yr from the start of plant operation. This is very different from the 2.63-yr payback period computed in Example 17.14, where both the time-value-of-money and the first 2 yr of operation at reduced capacity were ignored.

b. The IRR (or DCFRR) is obtained iteratively by conducting the same calculations as in part (a), but with different selected values for the nominal interest rate until an NPV of zero is obtained. Since an interest rate of 15% in part (a) produced a positive NPV, we know that the interest rate for a zero NPV must be greater than 15%. In fact, an IRR of 19% to the nearest integer is obtained.

Calculation of Cash Flows (in Millions of Dollars/Year)

Year	Investment fC_{TDC}	C_{WC}	D	$C_{Excl.\ Dep.}$	S	Net Earn	Cash Flow	Cum. PV @ 15%	IRR
1997	(30.00)						(30.00)	(30.00)	
1998	(30.00)						(30.00)	(56.09)	
1999	(30.00)	(40.00)					(70.00)	(109.02)	
2000			18.00	55.00	75.00	1.26	19.26	(96.35)	
2001			28.80	78.00	113.00	3.91	32.71	(77.65)	
2002			17.28	100.00	150.00	20.61	37.89	(58.81)	
2003			10.37	100.00	150.00	24.97	35.34	(43.53)	
2004			10.37	100.00	150.00	24.97	35.34	(30.25)	
2005			5.18	100.00	150.00	28.23	33.42	(19.32)	
2006				100.00	150.00	31.50	31.50	(10.37)	
2007				100.00	150.00	31.50	31.50	(2.58)	
2008				100.00	150.00	31.50	31.50	4.17	
2009				100.00	150.00	31.50	31.50	10.06	
2010				100.00	150.00	31.50	31.50	15.18	
2011		40.00		100.00	150.00	31.50	71.50	25.28	19%

Net earnings = $(S - C_{Excl.\ Dep.} - D) \times (1.0 - \text{income tax rate})$

Annual cash flow = $C = (\text{net earnings} + D) - fC_{TDC} - C_{WC}$

It is of interest to examine the annual cash flows on nondiscounted and discounted bases, as shown in the bar graphs below. The first graph is for the former. For discounted cash flows, the second graph is for an interest rate of 15%, while the third graph is for the IRR of 19%. Note that discounted cash flows during the time of plant operation are much smaller than those for the nondiscounted cash flows in the first graph.

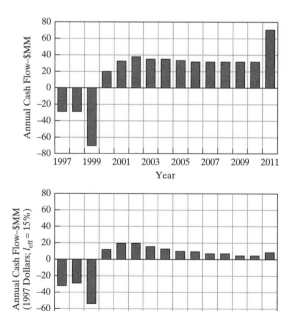

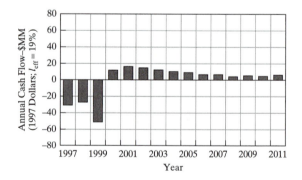

Finally, when calculating annual discounted cash flows, it is not difficult to account for inflation in estimating revenues and costs, when the inflation factors are known. Inflation is considered in the following subsection, but was not included in this example so as to enable the reader to trace the calculations of the cash flows more easily.   ■

Inflation

Inflation is the change in the value of a currency over time. Most often, the change is a loss in value. The effect of inflation on a profitability analysis is difficult to include because the future inflation rate is not known and there is no general agreement on how the effect of inflation should be incorporated into present worth calculations. Some argue that revenues and costs increase in the same proportion to the inflation rate factor, making it unnecessary to consider inflation when using a rigorous profitability measure. However, this ignores the fact that depreciation allowances are not adjusted for inflation and, therefore, if revenues and costs increase by the same percentage, the gross earnings increase, making it necessary to pay more income tax, as shown in the following example.

EXAMPLE 17.30

Consider the years 2002 and 2003 of the results in the table of Example 17.29. In 2002 and 2003, income tax in millions of dollars is, respectively,

$$0.37(150.00 - 100.00 - 17.28) = 12.11$$
$$0.37(150.00 - 100.00 - 10.37) = 14.66, \text{ with a cash flow of } 37.89$$

Assume that the results of 2002 are unchanged, but that the sales revenue and production cost, excluding depreciation, both increase by 8% due to inflation. Now the income tax in millions of dollars for 2003 is

$$0.37[150(1.08) - 100(1.08) - 10.37)] = 16.07, \text{ a } 9.6\% \text{ increase in income tax.}$$

The cash flow for 2003 is now 37.93, an increase of about 0.1% over the 37.89 million dollars of the first case of no inflation in 2003. Based on the effect of inflation on the value of currency, the company has fallen behind. Thus, it would appear that it is important to make some correction for inflation, especially if the inflation rate is high.   ■

The historic effect of inflation on costs was seen in Chapter 16 in Table 16.6, where four cost indexes and the consumer price index (CPI) were compared. From that table, the following average annual inflation rates, shown in Table 17.11, are obtained for the 10-yr periods of 1980–1990 and 1990–2000. Also included in the table below are the average annual hourly wage increases. In the period of 1980–1990, the average hourly labor wage rate in the United States increased from \$6.66 to \$10.01, while for 1990–2000 the increase was from \$10.01 to \$13.67.

The data in Table 17.11 show that recent attempts in the United States to control inflation have met with some success. In the most recent 10-yr period, the cost of chemical plants

Table 17.11   Average Annual Inflation Rates of Cost Indices and Hourly Labor Wages Rates Are Percent Increase/Year

Period	CE	MS	ENR	CPI	Hourly labor wages
1980–1990	3.21	3.31	3.82	4.72	4.60
1990–2000	0.97	1.67	2.76	2.80	2.80

(CE index) and chemical plant equipment (MS index) have increased at less than 2%/yr. The cost of all construction (ENR index) has risen at less than 3%/yr. The consumer price index (CPI), which rose 4.72%/yr from 1980 to 1990, rose only 2.80% from 1990 to 2000. The increase in hourly labor wages for the 20-yr period was almost identical to that of the CPI.

It is also of interest to consider the effect of inflation on the prices of commodity chemicals, such as those in Table 16.7. In the past 50 yr, the prices of the inorganic chemicals in that list have changed at the most by a factor of two. This represents an average annual inflation rate of only 1.4%. However, in the case of petrochemicals, their prices, like the price of gasoline, are tied to the price of oil, which has fluctuated greatly since the 1950s. For example, the price of ethylene was $0.05/lb in 1963, $0.24/lb in 1990, and was still $0.24/lb in 2001, making it impossible to predict the effect of inflation unless the future price of oil can be predicted. In addition, for a new chemical plant, raw material prices and product prices are often negotiated with contracts for at least a few years.

For the purpose of comparing alternative processes with rigorous profitability measures and in the absence of future inflation rates, the following recent inflation rates can be used

Cost of raw materials and price of products	1.5%/yr
Cost of utilities	1.5%/yr
Cost of processing equipment	1.5%/yr
Cost of hourly labor	3.0%/yr

In effect, these rates are 1.5% for material and 3.0% for labor.

Another aspect of inflation is its effect on purchasing power, both for consumers and companies. Equation (17.12) gives the future worth of a present amount of money, if it earns compound interest.

$$F = P(1 + i)^n$$

However, if inflation occurs, the purchasing power of that future sum will not be the same as the present sum. To account for a constant rate of inflation, Eq. (17.12) is modified to give the future worth in terms of the purchasing power of present dollars. Let the nominal interest rate compounded annually $= r = i$, the number of years $= n$, the annual inflation rate $= f$, and the future worth in present purchasing power $= F_{PP,0}$. Then,

$$F_{PP,0} = \frac{F}{(1 + f)^n} = P\left[\frac{(1 + i)^n}{(1 + f)^n}\right] \tag{17.50}$$

EXAMPLE 17.31

A present sum of money of $10,000 is invested for 10 yr at an interest rate of 7% compounded annually. During that 10-yr period, the inflation rate is constant at 3%. Compute the future worth, F, and the future worth in terms of purchasing power at the beginning of the investment, $F_{PP,0}$.

SOLUTION

$$P = \$10,000, i = 0.07, f = 0.03$$

From Eq. (17.12),

$$F = \$10,000(1 + 0.07)^{10} = \$19,672$$

Thus, when inflation is not taken into account, the future worth after 10 yr is almost twice as much as the present amount.

From Eq. (17.50),

$$F_{PP,0} = \frac{\$19,672}{(1 + 0.03)^{10}} = \frac{\$19,672}{1.3439} = \$14,638$$

When inflation is taken into account, the future purchasing power is only about 50% (rather than 100%) more than the present amount.   ■

17.8 PROFITABILITY ANALYSIS SPREADSHEET

 This section is provided on the CD-ROM that accompanies this book. It shows how to use purchase and installation cost estimates from Aspen IPE and other sources, together with an economics spreadsheet by Holger Nickisch (2003) to estimate profitability measures for the monochlorobenzene separation process, which was introduced in Section 4.4. In Section 16.7, Aspen IPE was used to estimate the total permanent investment for this process. The economics spreadsheet, Profitability Analysis − 1.0.xls, is on the CD-ROM that accompanies this book.

17.9 SUMMARY

Having studied this chapter, and completed several of the exercises, the reader should have learned to

1. Estimate annual costs using a standard cost sheet like that in Table 17.1.
2. Estimate annual cash flows, working capital, and the total capital investment in Table 16.9.
3. Compute approximate profitability measures, ROI, PBP, VP, and annualized cost.
4. Compute present worth and future worth of single payments and annuities.
5. Compute profitability measures that account for the time value of money, including net present value, IRR, and DCFRR.
6. Use the Aspen IPE in the Aspen Engineering Suite and an economics spreadsheet to perform a profitability analysis.

REFERENCES

Bauman, H.C., *Fundamentals of Cost Engineering in the Chemical Industry*, Reinhold Publishing Co., New York (1964).

Brealey, R., and S. Myers, *Principles of Corporate Finance*, McGraw-Hill, New York (1984).

Busche, R.M., *Venture Analysis: A Framework for Venture Planning*, Course Notes, Bio-en-gene-er Associates, Wilmington, Delaware (1995).

Nickisch, H., *Profitability Analysis Spreadsheet*, Univ. of Pennsylvania, Philadelphia (2003).

Souders, M., *Engineering Economy, Chem. Eng. Prog.*, **62** (3), 79–81 (1966).

EXERCISES

17.1 In the design of a chemical plant, the following costs and revenues (in the third year of production) are projected

Total depreciable capital,
excluding allocated power	$10,000,000
Allocated power utility	$ 2,000,000
Working capital	$ 500,000
Annual sales	$ 8,000,000/yr
Annual cost of sales excluding depreciation	$ 1,500,000/yr

Assume the costs of land, royalties, and startup are zero.

Determine

a. The return on investment (ROI)

b. The payback period (PBP)

17.2 For Exercise 17.1, the return on investment desired by the chemical company is 20%. Determine the venture profit.

17.3 It is desired to have $9,000 available 12 yr from now. If $5,000 is available for investment at the present time, what discrete annual rate of compound interest on the investment would be necessary to yield the desired amount?

17.4 What will be the total amount available 10 yr from now if $2,000 is deposited at the present time with nominal interest at the rate of 6% compounded semi-annually?

17.5 An original loan of $2,000 was made at 6% simple interest per year for 4 yr. At the end of this time, no interest had been paid and the loan was extended for 6 yr more at a new effective compound interest rate of 8%/yr. What is the total amount owed at the end of the 10 yr if no intermediate payments are made?

17.6 A concern borrows $50,000 at an annual effective compound interest rate of 10%. The concern wishes to pay off the debt in 5 yr by making equal payments at the end of each year. How much will each payment have to be?

17.7 How many years are required for money to double when invested at a nominal interest rate of 14% compounded semi-annually? Determine the shortest time in years, allowing any number of compounding periods.

17.8 A person at age 30 is planning for retirement at age 60. He projects that he will need $100,000 a year until age 80. Determine the uniform annual contribution (by him and his company) to provide these funds. Assume that the effective interest rate is 8%/yr and the rate of inflation is zero.

17.9 A heat exchanger has been designed for use in a chemical process. A standard type of heat exchanger with a negligible scrap value costs $4,000 and will have a useful life of 6 yr. Another proposed heat exchanger of equivalent design capacity costs $6,800 but will have a useful life of 10 yr and a scrap value of $800. Assuming an effective compound interest rate of 8%/yr, determine which heat exchanger is cheaper by comparing the capitalized costs.

17.10 Two machines, each with a service life of 5 yr, have the following cost comparison. If the effective interest rate is 10%/yr, which machine is more economical?

	A	B
First cost	$25,000	$15,000
Uniform end-of-year maintenance	2,000	4,000
Overhaul, end of third year		3,500
Salvage value	3,000	
Benefit from quality as a uniform end-of-year amount	500	

17.11 Two pumps are being considered for purchase:

	A	B
Initial cost	$8,450	$10,000
Salvage value	1,500	4,000

Determine the service life, n_y, at which the two pumps are competitive. The annual effective interest rate is 10%.

17.12 Two heat exchangers are being considered for installation in a chemical plant. It is projected that:

	A	B
Installed cost	$70,000	$80,000
Uniform end-of-year maintenance	?	$4,000
Salvage value	$7,000	$8,000
Service life	8 yr	7 yr

For an effective interest rate of 10%, determine the uniform end-of-year maintenance for heat exchanger A at which the two are competitive.

17.13 Consider the following two alternatives for the installation of a pump:

	A	B
Installed cost	$30,000	$16,000
Uniform end-of-year maintenance	$ 1,600	$ 2,400
Salvage value	?	$ 1,000
Life	5 yr	3 yr

The effective annual interest rate is 10%. Determine the salvage value for pump A when the two pumps are competitive.

17.14 You are offered two options to finance a compressor, with a nominal interest rate at 7.25% compounded monthly.

a. You pay $590 per month for 134 months.

b. You pay $545 per month for 151 months.

Compare these with the alternative of $590 per month for 151 months (at a nominal interest rate of 8.75% compounded monthly).

Compared to the alternative, it is claimed that Option a saves $10,030 and Option b saves $6,795. Do you agree? Justify your response.

(Hint: Use present-value analysis.)

17.15 Two pumps are under consideration:

	A	B
Installed cost	$10,000	$18,000
Service life	1 yr	2 yr

Determine the interest rate at which the two pumps are competitive.

17.16 A chemical plant is to be constructed in 1997 with operation scheduled to begin in 1998. In 2000, the plant is projected to operate at 90% of capacity, with

Sales	$10 MM
Cost of sales (Excl. Dep.)	$ 5 MM

a. Calculate the return on investment (ROI) in 2000 given that the total depreciable capital is $18 MM and the working capital is $2 MM. Assume straight-line depreciation at 8% per year.

b. Calculate the cash flow in 2000 and discount it to present value assuming an effective interest rate of 15%. Use the MACRS depreciation schedule for a class life of 5 yr.

17.17 A proposed chemical plant has the following projected costs and revenues in millions of dollars:

	Investment	Working Capital	Cost of Sales (Excl. Dep.)	Sales
1993	(40.0)	(4.0)		
1994			4.0	10.0
1995			5.6	14.0
1996			7.0	17.5
1997			8.0	20.0
1998			9.0	22.5
1999			9.6	24.0
2000		4.0	10.0	25.0

Using an MACRS depreciation schedule having a class life of 5 yr.

a. Compute the cash flows.

b. At an effective interest rate of 20%, determine the net present value.

c. Is the investor's rate of return less than or greater than 20%? Explain.

d. Compute the investor's rate of return.

17.18 An engineer in charge of the design of a plant must choose either a batch or a continuous system. The batch system offers a lower initial outlay but, owing to high labor requirements, exhibits a higher operating cost. The cash flows relevant to this problem have been estimated as follows:

	Year		Investor's Rate of Return	Net Present Worth at 10%
	0	**1–10**		
Batch system	−$20,000	$5,600	25%	$14,400
Continuous system	− 30,000	7,650	22	17,000

Check the values given for the investor's rate of return and net present worth. If the company requires a minimum rate of return of 10%, which system should be chosen?

17.19 An oil company is offered a lease of a group of oil wells on which the primary reserves are close to exhaustion. The major condition of the purchase is that the oil company must agree to undertake a water-flood project at the end of 5 yr to make possible secondary recovery. No immediate payment by the oil company is required. The relevant cash flows have been estimated as follows:

	Year			Investor's Rate of Return	Net Present Worth at 10%
0	**1–4**	**5**	**6–20**		
0	$50,000	−$650,000	$100,000	?	$227,000

Should the lease-and-flood arrangement be accepted? How should this proposal be presented to the company board of directors who understand and make it a policy to evaluate by the investor's rate of return?

17.20 *Sequencing of two distillation columns.* The demand for styrene monomer continues to increase. Other companies produce styrene by alkylating benzene with ethylene to ethylbenzene, followed by dehydrogenation to styrene. Our chemists have developed a new reaction path to styrene that involves other chemicals that appear to be available from our own supplies at a relatively low cost. These chemicals are toluene and methanol. We have been asked to prepare a preliminary design and economic evaluation for this new route to determine if it merits further consideration. If so, we will consider entering the styrene manufacturing business.

The new process can be broken down into four sections: (1) the reactor section, where toluene is partially reacted with methanol to produce styrene, water, and hydrogen, with an unfortunate side reaction that produces ethylbenzene and water; (2) an aqueous stream separation system; (3) an organic stream separation system; and (4) a vapor separation system. Fortunately, we have a potential buyer for the ethylbenzene byproduct. We are assigning you the design and economic calculations of just the organic stream separation system. Others are preparing the designs for the other three sections.

The chemical engineer working on the reactor section has already calculated the following reactor effluent:

Component	kmol/hr
Hydrogen	352
Methanol	107
Water	484
Toluene	107
Ethylbenzene	137
Styrene	352

This effluent is cooled to 38°C and enters a flash-decanter vessel at 278 kPa. Three phases leave that vessel. The vapor phase (hydrogen rich) is sent to the vapor separation system. The aqueous phase (mostly water, with some methanol) is sent to the aqueous stream separation system. The organic-rich phase is sent to the organic stream separation system, which you will design. To obtain the composition of the feed to your section, use a simulator with the UNIFAC method to perform a three-phase flash for the above conditions. If the resulting organic liquid stream contains small amounts of hydrogen and water, assume they can be completely removed at no cost before your stream enters your separation section.

Your separation system must produce the following streams with two distillation operations in series:

A methanol–toluene-rich distillate stream for recycle back to the reactor. This stream should not contain more than 5 wt% of combined ethylbenzene and styrene.

An ethylbenzene byproduct stream containing 0.8 wt% max. toluene and 3.9 wt% max. styrene.

A styrene product stream containing 300 ppm (by wt.) max. ethylbenzene.

We have a serious problem with styrene. If any stream contains more than 50 wt% styrene, the temperature of the stream must not exceed 145°C. Otherwise, the styrene will polymerize. This must be carefully considered when establishing the operation conditions for the two distillation operations. You may have to operate one or both columns under vacuum. This will require you to estimate the amount of air that leaks into the vacuum columns and select and cost vacuum systems. In designing the distillation system, you are to consider the direct sequence and the indirect sequence.

Please submit a report on the two designs and cost estimates (fixed capital and utility operating costs only). For the capital cost of each of the two alternative sequences, sum the purchase costs of the distillation columns, heat exchangers, and any vacuum equipment. Multiply that cost by the appropriate Lang factor. To annualize the capital cost, multiply by 0.333. Add to this annualized cost the annual utility cost for steam and cooling water. Call this the total annualized cost for the alternative.

Use a simulator to do as many of the calculations as possible, including the very important column pressure-drop calculations (because of the need for vacuum in one or more columns). Assume a condenser pressure drop of 5 kPa and no pressure drop across the reboiler. You may select column internals from the following list:

Sieve trays on 18-in. spacing with overall tray efficiency of 75%.
Pall rings random packing with HETP = 24 in.
Mellapak structured packing with HETP = 12 in.

Submit your results as a short report complete with an introduction to the problem, a process description, process flow diagram, discussion of equipment operating conditions and how you arrived at them, a material-balance table, cost tables, conclusions, and your recommendations. Make it clear which alternative you favor and whether it might offer some technical challenges if it is selected for final design.

17.21 *Toluene hydrodealkylation process–economic analysis using ASPEN PLUS, Aspen IPE and the economics spreadsheet.* This assignment begins with a completed simulation of the toluene hydrodealkylation process in Figure 17.1 and involves the completion of an economic evaluation. Note that the simulation results for this process were developed in Chapter 4 and can be reproduced using the HDA.bkp file on the CD-ROM that accompanies this book. Aspen IPE will be used for equipment-cost and capital-cost estimation. The economics spreadsheet, Profitability Analysis—1.0.xls, on the CD-ROM that accompanies this book will be used for profitability analysis. This spreadsheet should enable you to complete an economic analysis using the specifications recommended for capital investment costs (Section 16.3, Tables 16.9 and 16.12) and for the cost sheet (Section 17.2, Table 17.1), and using the approximate profitability measures in Section 17.4 and those involving cash flows in Section 17.6. In most cases, only the nondefault Aspen IPE entries are provided in the problem statement. This, together with the description in Section 16.7, should enable you to understand the items on the Aspen IPE input forms.

The information you will use to complete the economic analysis is as follows.

Cost Options

The project startup date is January 2003 and the project duration is 1 yr. For simplicity, the effect of inflation is disregarded in this assignment.

Equipment

Unless otherwise stated, the default equipment types and materials of construction are acceptable.

Tanks

Toluene storage tank

Create a tank to store a 2-day supply of toluene.

Benzene storage tank

Create a tank to store a 2-day supply of the benzene product.

Stream 11 storage tank

Create a tank to store a 1-day supply of the intermediate stream 11.

Flash vessel F1

Size a vertical flash vessel to have a 10-minute liquid retention time.

Reactor R1

The elevated temperature (1,268°F) and pressure (494 psia) present special complications when designing and sizing the hydrodealkylation reactor. This is a large-diameter vessel that is lined with refractory brick to insulate the steel alloy that comprises the retaining wall. Initially, use Aspen IPE to design a vessel that has a 10-ft diameter and 60-ft tangent-to-tangent length. To estimate the thickness of the refractory-brick and the temperature at the brick-steel interface, a heat balance is necessary. In practice, the brick thickness is adjusted to give an interface temperature of 900°F. When using Aspen IPE, select refractory-lined carbon steel. Using the pull-down menu, select a 9-inch layer of 90% alumina fire brick, symbol 9FB90. Also, consider the possibility of using two or three smaller-diameter vessels in parallel. Finally, compare the costs of horizontal and vertical vessels.

Before mapping the reactor model in the ASPEN PLUS simulation, it is necessary to replace the RSTOIC subrou-

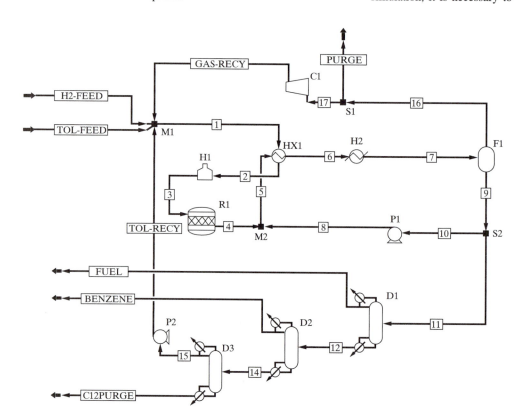

Figure 17.1 ASPEN PLUS simulation flowsheet for the toluene hydrodealkylation process.

tine with the RPLUG subroutine in ASPEN PLUS. Note, however, that the kinetics of the side reaction cannot be modeled using the RPLUG subroutine. Since the conversion of this reaction is small, its kinetics can be neglected in the reactor design. Rather, it is sufficient to account for this reaction using a dummy reactor unit, R1D, modeled with the RSTOIC subroutine, which follows the reactor unit, R1, modeled with the RPLUG subroutine. The dummy unit, R1D, cannot be sized by Aspen IPE.

Heat Exchangers

H2
 Use cooling water.
D1, D2, and D3 condensers
 Use cooling water.
D1, D2, and D3 reboilers
 Use steam.

Distillation Towers

In the ASPEN PLUS simulation, for each distillation column, use the RADFRAC subroutine in place of the DISTL subroutine. As a result, the reflux ratios change somewhat to achieve the same product specifications using the same number of trays.

In Aspen IPE, set the tray efficiency of the three columns to 90%.

The reflux accumulators should be horizontal vessels with a liquid holdup time of 10 min.

Pumps

Toluene pump
 Create a pump to bring the toluene feed from atmospheric pressure to 569 psia.
Benzene pump
 Increase the pressure of the benzene product stream by 25 psia.
D1 condenser pump
 Increase the pressure of the D1 reflux by 25 psia.
D1 reboiler pump
 Increase the pressure of the D1 reboiler pump by 25 psia.
D2 condenser pump
 Increase the pressure of the D2 reflux by 25 psia.
D2 reboiler pump
 Increase the pressure of the D2 reboiler pump by 25 psia.
D3 condenser pump
 Increase the pressure of the D3 reflux by 25 psia.
D3 reboiler pump
 Increase the pressure of the D3 reboiler pump by 25 psia.

Other Equipment

Compressor C1
 Use a centrifugal compressor.
 Use a motor drive and an electrical utility.
Fired heater H1
 In the ASPEN PLUS simulation, the HEATER subroutine was used. By default, units modeled with the HEATER subroutine are mapped as heat exchangers by IPE. Instead, this unit must be mapped as a furnace. Furthermore, before estimating the cost of the furnace, the default material, carbon steel, must be replaced by a material that can withstand temperatures at 1,200°F in the furnace. Note that Incoloy 1800 is selected from among the materials available in Aspen IPE that can withstand this temperature. To replace carbon steel, in row nine of the window that displays the equipment sizes for the furnace, depress *material selection*, which opens the pull-down menu on the right. Then, select the last entry, *TUBE MATERIAL.* This produces a second pull-down menu, from which the tube-material,"1800," is selected. These changes must be saved before leaving this window.

Labor Costs

Use a wage rate of $30/hr.

Materials

For the feed, product, and byproduct streams, H2-FEED, TOL-FEED, BENZENE, PURGE, FUEL, and C12PURGE, the following prices are typical:

Hydrogen feed	$0.80/lb
Toluene feed	$0.50/gal
Benzene product	$1.75/gal
FUEL, PURGE	$2.60/MM Btu heating value
C12PURGE	0

The prices are for toluene and benzene at 1 atm and 75°F. You can use ASPEN PLUS to estimate the densities needed to obtain the prices on a mass basis. (You can also use ASPEN PLUS to estimate the heating value of a stream, rather than calculate it independently. Note that the simulation results are arranged to report the heating values in the stream section of the report file.)

Utilities

For this process, cooling water, steam, electricity, and fuel are purchased.

Cooling water: $0.05/1,000 gal. Use inlet and outlet temperatures of 90°F and 120°F.

Steam: Use prices in Table 17.1.

Electricity: $0.04/kW-hr.

Fuel (to heater H1): assume the heater is gas-fired at a cost of $2.60/MM Btu.

Operating Costs

The plant is anticipated to operate 330 days a year.

Average hourly wage rate is $30/hr with two operators per shift.

Profitability Analysis

Project a 15-yr life for the plant.

Assume a 15% interest rate to calculate the net present value.

For cash flow analysis, use the 5-yr MACRS depreciation schedule. Estimate the effective tax rate to be 37% with no investment tax credit.

Assume a production schedule such that the plant operates at 50% of full scale (90% of capacity) in the first year, 75% of full scale in the second year, and 100% of full scale thereafter.

Provide for the cost of startup at 20% of the total materials and labor cost of the process units.

Provide for working capital to cover 2 days of raw material inventory, 2 days of in-process chemicals, 2 days of finished-product inventory, and 30 days of accounts receivable.

Chapter **18**

Optimization of Process Flowsheets

18.0 OBJECTIVES

This chapter begins with a brief discussion of the fundamental principles of optimization, which are presented in much more detail by Edgar et al. (2001), Reklaitis et al. (1983), and Beveridge and Schechter (1970). These principles are then applied to the use of flowsheet simulators to optimize the most promising flowsheets during process design. Two process examples are presented. The first involves maximizing the venture profit (VP) of a process for the production of ethyl chloride. In the second example, the separation afforded by a distillation column with sidestreams is maximized, typical of an optimization carried out during the development of a base-case design.

After studying this chapter and the multimedia CD-ROM that accompanies this book, the reader should

1. Understand the fundamentals of optimization.
2. Be able to formulate a nonlinear optimization problem (nonlinear program, NLP) to maximize or minimize an objective function by adjusting continuous decision variables in the model of the process.
3. Understand the nature of algorithms that optimize the process while simultaneously converging the recycle loops and design specifications associated with the process simulation.
4. Begin to understand the advantages and disadvantages of converting design specifications, associated with a simulation model, to equality constraints in the NLP.
5. Be able to use process simulators to solve the NLP.

18.1 INTRODUCTION

From a mathematical point of view, chemical engineers deal with three types of problems when solving equations. The first type, which is the subject of most undergraduate text books in chemical engineering and has been the main subject of this book to this point, is the *completely specified* case where the number of equations, $N_{\text{Equations}}$, to be solved is equal to the number of variables, $N_{\text{Variables}}$, to be determined. The equations to be solved are usually at least partially nonlinear and the challenge is to find a method that will solve them. This case is complicated by the fact that more than one practical solution may exist, a challenge that has been largely ignored in process simulators. The second type is common in experimental work and process operations, where $N_{\text{Variables}} < N_{\text{Equations}}$. This is the *overspecified* case, which is commonly referred to as the reconciliation (or data reconciliation and rectification) problem. An example of this case is a piece of equipment operating under steady-state conditions with all or some of the variables, for example, component flow rates, measured. Because the measurements are subject to error, they do not satisfy material-balance equations. The task with this case is to determine the most likely values for the variables. This case is not covered in this book, but is discussed by Mah (1990). The third case is the subject of this chapter, *optimization*, where the problem is *underspecified* such that $N_{\text{Variables}} > N_{\text{Equations}}$. The problem is to select from the set of variables, $\underline{x}$, a subset of size

$$N_{\text{Variables}} - N_{\text{Equations}} = N_D \qquad (18.1)$$

called the *decision variables*, $\underline{d}$, and iteratively adjust the decision variables to achieve the optimal solution to a specified objective.

Having studied the steps in process creation, and having solved problems involving the synthesis of alternative process flowsheets presented in previous chapters, it is appropriate to question whether a given flowsheet can and should be optimized by adjusting its key equipment parameters to, for example, increase some measure of its economic attractiveness. This optimization step, shown in connection with the detailed design stage in Figure 1.2, is applied to the most promising process flowsheets after a base case has been evaluated, where the equipment has been sized, the capital and operating costs have been estimated, and a profitability analysis has been completed. In many cases, however, optimizations are performed much earlier in process design using more approximate cost and profitability measures, as shown in Examples 18.4 and 18.5.

As introduced in Section 10.7, formal methods of optimization can be utilized to optimize a superstructure of process units with streams that can be turned on and off using binary (integer) variables. In principle, the *mixed-integer formulation* (involving both continuous and integer variables) of the optimization problem permits the optimizer to select simultaneously the best flowsheet and then optimize it with respect to its continuous variables, such as pressure levels, reflux ratios, residence times, and split fractions. In practice, however, most design problems are not solved using superstructures and mixed-integer optimization algorithms. Rather, as described throughout the earlier chapters, heuristics, many of which are presented in Chapter 5, together with simulation and algorithmic methods are utilized to build and analyze *synthesis trees*. Although substructures, such as networks of heat exchangers, can be optimized conveniently using mixed-integer methods, it is impractical, except for simple processes, to attempt the optimization of entire process flowsheets in this manner. Accordingly, this chapter is restricted to the case of optimization problems involving continuous variables, of either the LP (linear programming) type or the NLP (nonlinear programming) type. Thus, MILP (mixed-integer linear programs) and MINLP (mixed-integer nonlinear programs) are only briefly mentioned.

Emphasis is placed in this chapter on the usage of process simulators to carry out the optimization simultaneously with converging the recycle loops and/or decision variables. To do the optimization efficiently, simulators use one of three methods: (1) *successive linear programming* (SLP), (2) *successive quadratic programming* (SQP), and (3) *generalized reduced gradient* (GRG). Emphasis in this chapter is placed on SQP, used by ASPEN PLUS and HYSYS.Plant. GRG, which is used by CHEMCAD, is not discussed here, but is covered by Edgar et al. (2001).

18.2 GENERAL FORMULATION OF THE OPTIMIZATION PROBLEM

The formulation of an optimization problem involves:

1. A set of $N_{\text{variables}}$ variables, $\underline{x}$.
2. The selection of a set of appropriate *decision variables*, $\underline{d}$, from the set $\underline{x}$.

3. A measure of goodness called an *objective function*, $f\{x\}$. **(18.2)**

4. A set of $N_{\text{Equations}}$ equality constraints, $\underline{c}\{x\} = 0$. **(18.3)**

5. A set of N_{Inequal} inequality constraints, $\underline{g}\{x\} \leq 0$. **(18.4)**

6. Lower and upper bounds on some or all of the variables, $\underline{x}^L \leq \underline{x} \leq \underline{x}^U$. **(18.5)**

A general optimization problem is stated as follows:

$$\text{Minimize (or maximize)} \quad f\{\underline{x}\}, \text{ the objective function} \quad \textbf{(NLP1)}$$
$$\text{with respect to (w.r.t.)}$$
$$\underline{d},$$
$$\text{the design variables}$$

$$\text{subject to (s. t.):}$$
$$\underline{c}\lfloor x \rfloor = 0, \text{ the equality constraints}$$
$$\underline{g}\lfloor x \rfloor \leq 0, \text{ the inequality constraints}$$
$$\underline{x}^L \leq \underline{x} \leq \underline{x}^U, \text{ the lower and upper bounds}$$

Objective Function and Decision Variables

Candidates for the objective function (measure of goodness), $f\{\underline{x}\}$, are often the profitability measures of Chapter 17, beginning with the approximate measures, such as the return on investment (ROI), the venture profit (VP), the payback period (PBP), and the annualized cost (C_A). For more thorough analyses, the rigorous measures that involve the time value of money and cash flows are used. These include the net present value (NPV) and the investor's rate of return (IRR). Other objective functions may involve measures that are related to costs or may involve safety, control, or pollution aspects. A multiobjective function, consisting of two or more measures, each with a weighting coefficient, may also be employed. In any case, the objective function is a function of all of the decision variables, $\underline{d}$, and may also be a function of some or all of the other variables of the set, $\underline{x}$. Depending upon the nature of the objective function, it is minimized or maximized analytically or numerically by adjusting the values of the decision variables until the optimal solution is reached. When possible, it is best to select decision variables for which significant trade-offs are anticipated in the objective function. It is common to begin an optimization study with an approximate measure of goodness and switch to more rigorous measures when the design continues to be promising as the model of the flowsheet is refined.

Equality Constraints

The largest fraction of equality constraints, $c\{\underline{x}\}$, are the modeling equations (usually algebraic) associated with processing equipment. For example, a distillation column with equilibrium stages may be modeled with hundreds of material balance, energy balance, and phase-equilibria equations in terms of the set of variables, $\underline{x}$. The NLP problem often involves hundreds and even thousands of equations. However, in the implementation of most simulators, these equations are solved for each process unit, given equipment parameters and stream variables (usually for the feed streams), using subroutines in a program library. Hence, when using simulators, the equations for the process units are not shown explicitly in the statement of the nonlinear programming problem. Given assumed values for the decision variables, the simulators call upon these subroutines to solve the appropriate equations and obtain the unknowns that are needed to perform the optimization. However, certain specifications, especially those that involve more than one process unit, may require the user to formulate an equality constraint.

Inequality Constraints

Inequality constraints, $g\{x\}$, are expressions that involve any or all of the set of variables, x, and are used to bound the feasible region of operation. For example, when operating a centrifugal pump, the head developed decreases with increasing flow rate according to a pump characteristic curve. Hence, if the flow rate is varied when optimizing the process, care must be taken to make sure that the required pressure increase (head) does not exceed that available from the pump. The expression might be of the form,

$$\text{Pump head} \le a - b \text{ (Flow rate)} - c \text{ (Flow rate)}^2$$

Similar kinds of constraints involve the reflux ratio in distillation, which must exceed the minimum value for the required separation. If the distillation tower pressure is adjusted, the minimum reflux ratio will change and the actual ratio must be maintained above the minimum value. Even when optimization is not performed, the decision variable values must be selected to avoid violating the inequality constraints. In some cases, the violations can be detected when examining the simulation results. In other cases, the unit subroutines are unable to solve the equations as, for example, when the reflux ratio is adjusted to a value below the minimum value for a specified split of the key components.

Lower and Upper Bounds

Some inequality constraints simply place lower and upper bounds, $x^L \le x \le x^U$, on any or all of the variables, x. Others permit the specification of just a lower bound or just an upper bound, for example, a lower bound on the fractional recovery of a species in a product stream. Sometimes the upper and lower bounds are included with the inequality constraints, but here, they are considered separately.

18.3 CLASSIFICATION OF OPTIMIZATION PROBLEMS

The combination of the equality constraints, inequality constraints, and lower and upper bounds defines a *feasible region*. A *feasible solution* is one that satisfies the equality constraints, the inequality constraints, and the upper and lower bounds for a feasible set of decision variables. If the solution also minimizes (or maximizes) the objective function, it is a *local optimal solution*. Sometimes, other local optimal solutions exist in the feasible region, with one or more being a *global optimal solution*.

Many numerical methods have been devised for solving optimization problems. The choice of method depends upon the nature of the formulation of the problem. Therefore, it is useful to classify optimization problems with respect to certain categories.

When the objective function, equality constraints, and inequality constraints are linear with respect to the variables, x, the problem is referred to as a *linear programming* (LP) problem. If the objective function, any of the equality constraints, and/or any of the inequality constraints are nonlinear with respect to the variables, the problem is referred to as a *nonlinear programming* (NLP) problem.

The simplest optimization problems are those without equality constraints, inequality constraints, and lower and upper bounds. They are referred to as *unconstrained optimization*. Otherwise, if one or more constraints apply, the problem is one in *constrained optimization*.

For just a single decision variable (*unidimensional*), the feasible region and optimal solution(s) are conveniently displayed on a plot of $f\{x\}$ against x. This display and solution is easily carried out with a spreadsheet. The types of optimal solutions obtained depend both on the nature of the objective function and of the constraints. Figure 18.1 shows two different linear cases, where the only variable is designated x for the abscissa. An unconstrained *linear*

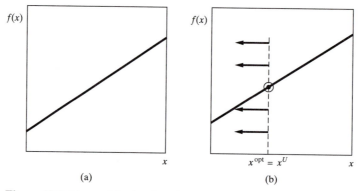

Figure 18.1 Linear objective function: (a) unconstrained; (b) subject to linear inequality constraint, $x \leq \bar{x}$.

objective function is shown in Figure 18.1a. In the absence of inequality constraints, no solution exists, unless one at $\pm\infty$ is meaningful.

In Figure 18.1b, a linear objective function is again shown, but an upper bound is placed on x. Therefore, the feasible region is to the left of this upper bound, as indicated by the arrows pointing to the left. A maximum objective function is desired. Thus, the optimal solution is at the bound. For linear objective functions, it can be shown that optimal solutions always occur at intersections between the objective function and the inequality constraints and/or lower and upper bounds.

Two cases of a nonlinear objective function that passes through a maximum are shown in Figure 18.2, with both constraints being a simple bound. The objective function is to be maximized. In Figure 18.2a, the constraint is a lower limit on x. Since the constraint is at a value of x less than the value at the maximum value of the objective function, the optimal value of x is that corresponding to the maximum value of the objective function and the constraint is referred to as a *slack constraint*.

In Figure 18.2b, an upper bound is placed on x, and that bound is below the value at the maximum value of the objective function. Therefore, the optimal value of x is its upper bound, and the constraint is referred to as a *binding constraint*.

For two decision variables, a common display for the feasible region and optimal solution(s) is a plot of one decision variable against the other, with contours of the objective

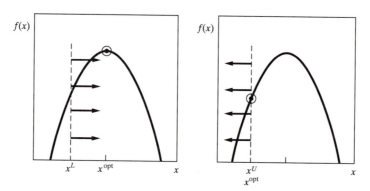

Figure 18.2 Nonlinear objective function: (a) subject to a slack constraint; (b) subject to a binding constraint.

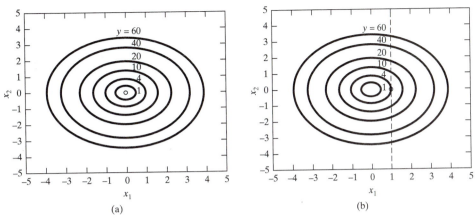

Figure 18.3 Two-variable optimization: (a) unconstrained; (b) constrained.

function, as shown in Figure 18.3. The optimization problem is to minimize the objective function:

$$f\{x\} = y = 4x_1^2 + 5x_2^2$$

Two solutions are shown in Figure 18.3. The first is for the unconstrained case, where the entire region is feasible. The solution, shown in Figure 18.3a, is at $x_2 = 0$ and $x_1 = 0$, where $y = 0$. The second is a constrained case with the inequality constraint:

$$g\{x\} = 1 - x_1 \le 0$$

This is a linear constraint that is easily converted to $x_1 \ge 1$. Now the feasible region is situated to the right of the constraint, as shown in Figure 18.3b. The optimal point for the unconstrained case is not located in the feasible region. Now the minimum value of the objective function occurs at $x_2 = 0$ and $x_1 = 1$, where $y = 4$. For more than two decision variables, a graphical representation is not usually attempted.

When one optimal solution exists in the feasible region, the objective function is *unimodal*. When two optimal solutions exist, it is bimodal; if more than two, it is *multimodal*. LP problems are unimodal unless the constraints are inconsistent, such that no feasible region exists. The solutions in Figures 18.1 to 18.3 are unimodal. A two-dimensional, multimodal case is shown in Figure 18.4, taken from Reklaitis et al. (1983) and called the Himmelblau problem. This is an unconstrained problem with the objective function:

$$f\{x_1,x_2\} = y = (x_1^2 + x_2 - 11)^2 + (x_1 + x_2^2 - 7)^2 \tag{18.6}$$

The contours of the objective function shown in Figure 18.4 range from 5 to 140. It can be shown that a point $y = 0$ exists at the center of each of the four closed contours identified as the global minima.

In some cases, using calculus to form derivatives, nonlinear optimization problems can be solved analytically. For a one-dimensional problem, it is only necessary to differentiate the objective function with respect to the decision variable, set that derivative to zero, and solve for the decision variable. When more than one solution exists, each solution is obtained and examined. For n decision variables, take n partial derivatives of the objective function, one each with respect to each decision variable; set each derivative to zero; and solve the resulting set of n equations for all solutions of the decision variables. Examine each solution. The following example illustrates the use of calculus to differentiate the Himmelblau function of Figure 18.4.

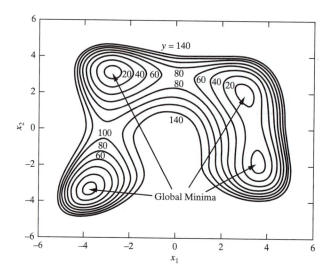

Figure 18.4 Example of a multimodal problem—Himmelblau's function.

EXAMPLE 18.1   *Determining Extrema for the Himmelblau Function*

Use calculus to determine all minimum and maximum values of the unconstrained Himmelblau function given by Eq. (18.6).

SOLUTION

The two partial derivatives are

$$\frac{\partial y}{\partial x_1} = F_1 = 4x_1(2x_1^2 + x_2 - 11) + 2(x_1 + x_2^2 - 7) = 0 \tag{18.7}$$

$$\frac{\partial y}{\partial x_2} = F_2 = 2(x_1^2 + x_2 - 11) + 4x_2(x_1 + x_2^2 - 7) = 0 \tag{18.8}$$

Both of the equations are third-degree polynomials. By the Theorem of Bezout, as discussed by Morgan (1987), the maximum number of solutions to a set of polynomial equations is the product of the highest degrees of the equations, which is called the total degree of the set of functions, and is equal to nine for the set consisting of Eqs. (18.7) and (18.8). Using a solver for a set of polynomial equations, nine solutions are found, as listed in Table 18.1.

The nine solutions, referred to as *stationary points*, correspond well with the contours of constant values of the objective function plotted on Figure 18.4. The type of solution is also listed in Table 18.1. There is one maximum at an objective function, $y = 181.62$. Larger values of the objective function

Table 18.1 Solutions to Himmelblau's Function

x_1	x_2	$f\{x_1, x_2\}$	F_1	F_2	Solution type
−0.2709	−0.9230	181.62	0.0023	−0.0004	Local maximum
−0.1279	−1.9538	178.34	−0.0023	−0.0022	Saddle point
3.5844	−1.8481	0	−0.0028	0.0006	Global minimum
3.3852	0.0739	13.31	0.0051	0.0000	Saddle point
3.0000	2.0000	0	0.0000	0.0000	Global minimum
0.0867	2.8843	67.72	−0.0001	0.0036	Saddle point
−2.8051	3.1313	0	0.0012	−0.0010	Global minimum
−3.0730	−0.0814	104.02	0.0024	0.0015	Saddle point
−3.7793	−3.2832	0	0.0016	−0.0015	Global minimum

occur as the two decision variables are increased to infinity. Four minima occur at an objective function of zero, which is the global minimum. There are four saddle points. Edgar et al. (2001) give necessary and sufficient conditions for determining whether a stationary point is a local maximum, local minimum, or saddle point. The former two types of points are referred to as *extrema*. For a maximum, any combination of small changes in x can only decrease the objective function. For a minimum, the opposite is true. For a saddle point, from which small changes in x are made, some directions will increase and others decrease the objective function. ∎

Much more common in applications to problems in chemical processing is the use of numerical methods for either nonlinear or linear problems. These methods, which are covered in the following sections of this chapter, are mostly *search* methods that start from an assumed solution for $\underline{d}$ and then move $\underline{d}$ in a series of iterations, by some strategy, to reduce (increase) the objective function to achieve a minimum (maximum).

Additional complications can be present in optimization problems. For example, the objective function and/or one or more equality constraints may be *discontinuous*. This might occur, for example, when steam is available to a process at two or three pressure levels, with a different cost at each level. When steam is used where pressure is a variable, the steam cost could change abruptly at a certain value of the pressure, causing a discontinuity in the objective function.

Another complication arises when one or more of the decision variables is an integer, rather than continuous. The most common case is when that integer is binary with just two values, 1 or 0. This gives rise to mixed-integer linear or nonlinear programming (MILP and MINLP) formulations. Although not covered in this chapter, examples of mixed-integer applications are presented in Chapter 10. MILP and MINLP arise in process optimization from the need to deal with binary as well as continuous decision variables, the former being a convenient way of representing alternative locations of a given equipment item in a flowsheet. MILP formulations are appropriate when both the objective function and the constraints are linear, such as in Example 10.8, which shows how a MILP is formulated and solved for design of a heat exchanger network (HEN) having the minimum energy requirement (MER). More commonly, both the objective function and the constraints are nonlinear, leading to a MINLP formulation. Example 10.16 shows how a MINLP is set up and solved to minimize the annual cost of a HEN. The *superstructure* for the MINLP, which incorporates all possible heat exchanger locations and flow configurations in the HEN, using binary decision variables, is described in Example 10.15. For comprehensive coverage of MILP and MINLP formulations in process design, the reader is referred to Floudas (1995) and Biegler et al. (1997).

18.4 LINEAR PROGRAMMING (LP)

Although LP problems are not common when optimizing a process design, they are common in many other applications of chemical engineering. Furthermore, a numerical solution of an NLP problem is sometimes achieved by approximating the nonlinear functions with linear functions, at each step of the iterative procedure, using a method called *successive linear programming* (SLP). Therefore, it is useful to have a basic understanding of LP methods.

Some of the common applications of LP methods are for: (1) assignment, (2) blending, (3) distribution, (4) determining network flows, (5) scheduling, (6) transportation, and (7) scheduling traveling salesmen. Example 10.4 demonstrates how an LP is used to determine the minimum hot and cold utilities for a HEN. For small problems that can be reduced to two decision variables, a graphical solution is instructive. The graphical solution method involves: (a) a definition of the decision variables; (b) formulation of the objective function; (c) formulation of the model; (d) reduction of the number of decision variables using equality constraints, applying Eq. (18.1); and (e) solution of the LP graphically, when the resulting number of decision variables is less than three. The following blending example, with three decision variables, although solved graphically, illustrates some of the characteristics of all LP problems.

EXAMPLE 18.2   *Beer Supply Problem*

During the 2002 Winter Olympics in Salt Lake City, Utah, a local microbrewery received a rush order for 100 gals of beer containing 4.0 vol% alcohol. Although no 4% beer was in stock, large quantities of Beer A with 4.5% alcohol at a price of $6.40/gal and Beer B with 3.7% alcohol priced at $5.00/gal were available, as well as water suitable for adding to the blend at no cost. The brewery manager wanted to use at least 10 gal of Beer A. Neglecting any volume change due to mixing, determine the gallons each of Beer A, Beer B, and water that should be blended together to produce the desired order at the minimum cost.

SOLUTION

Let V_A = gallons of Beer A, V_B = gallons of Beer B, and V_W = gallons of water. The optimization problem is stated as follows

$$\text{Minimize} \qquad \text{Cost, \$} = 6.40V_A + 5.00V_B + 0.00V_W \qquad (18.9)$$

w. r. t.

s. t.

V_A, V_B, V_W

$$0.045V_A + 0.037V_B + 0.00V_W = 0.04(100) = 4.00 \qquad (18.10)$$

$$V_A + V_B + V_W = 100 \qquad (18.11)$$

$$10 \le V_A, 0 \le V_B, 0 \le V_W$$

The problem consists of three variables, two equality constraints, and three lower bounds. The problem can be reduced to two decision variables by solving Eq. (18.11) for V_B,

$$V_B = 100 - V_A - V_W \qquad (18.12)$$

and substituting it into Eqs. (18.9) and (18.10) and to give the following restatement of the problem:

$$\text{Minimize} \qquad \text{Cost, \$} = 1.40V_A - 5.00V_W + 500 \qquad (18.13)$$

w. r. t.

s. t.

V_A, V_W

$$0.008V_A + 0.037V_W = 0.3 \qquad (18.14)$$

$$V_B = 100 - V_A - V_W \qquad (18.15)$$

$$10 \le V_A, 0 \le V_B, 0 \le V_W$$

Now, the optimal volume of Beer B need only be calculated from Eq. (18.15), after the optimal volumes of Beer A and water have been determined. Since the objective function, the equality constraints, and lower and upper bounds are all linear, this constitutes an LP problem. With just two decision variables, the problem can be shown graphically on a plot of V_A against V_W, as in Figure 18.5. The plot includes not only the lower bounds on the volumes of Beer A and water, but also the equality constraint, Eq. (18.14). The optimal solution to an LP problem occurs at a vertex of the set of constraints. Note that Eq. (18.14) can be rearranged to give

$$V_A = 4.625 V_W + 37.5 \qquad (18.16)$$

Thus, at the lower limit of V_W, $V_A = 37.5$. Thus, one vertex is at $V_A = 37.5$ and $V_W = 0$. The other vertex is at the intersection of two lower bounds ($V_A = 10$, $V_W = 0$). One might argue that another vertex exists at an upper bound on the volume of Beer A, corresponding to an upper bound on the volume of water. This occurs where the blend contains only Beer A and water ($V_A = 88.89$, $V_W = 11.11$). Now, evaluate the cost at each of these three vertices. The results are

V_A (gal)	V_B (gal)	V_W (gal)	% Alcohol	Cost ($)
37.50	62.50	0.00	4.00	552.50
10.0	90.00	0.00	3.78	514.00
88.89	0.00	11.11	4.00	568.90

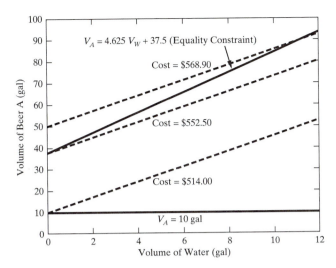

Figure 18.5 Constraints and costs for Example 18.2.

The optimal solution is the first one in the table because it is the lowest cost that satisfies the equality constraint that fixes the percentage of alcohol at 4.0. The second result need not have been calculated because it does not satisfy that constraint, giving only a 3.78% alcohol content. The solution must lie along the equality constraint line, Eq. (18.16), which was obtained from Eq. (18.14). It is perhaps surprising that the optimal blend involves no addition of water. An acceptable blend that includes water is 5 gal water, 60.63 gal Beer A, and 34.37 gal Beer B, but at a cost of $559.88. Figure 18.5 includes three lines of constant cost.

Note that the two equality constraints, Eqs. (18.10) and (18.11), permit two of the decision variables to be eliminated, leaving just one decision variable, say V_A. Consequently, the objective function, Eq. (18.9), can be minimized easily using a univariable search. If the problem statement is altered to place a *lower bound* on the beer concentration of 4%, Eq. (18.10) becomes an inequality constraint:

$$0.045\, V_A + 0.037\, V_B \geq 4.00$$

Then, two decision variables result and the graphical approach in Figure 18.5 is altered with the equality constraint replaced by the inequality constraint:

$$V_A \leq 4.625\, V_W + 37.5$$

At the minimum cost, the solution remains unchanged. ■

For large LP problems, which may involve more than 10,000 decision variables, one of two methods is applied. The first method, developed by Dantzig (1949) in the 1940s, is referred to as the *simplex method*. It is an iterative method that begins with initial values for the design variables (iterates) that satisfy the constraints at one of the vertices, and for which the objective function is computed. The optimal solution must be at this or another vertex. Therefore, subsequent iterations generate, in a systematic procedure, a sequence of iterates that move from one vertex of the feasible region to an adjacent vertex, each time finding an improved value for the objective function, until the vertex corresponding to the optimal solution is found. As described by Edgar et al. (2001), LP problems can be solved by the simplex method with the linear model of the Solver routine of the Microsoft Excel spreadsheet. An LP solver is also available in MATLAB and GAMS.

When the number of iterations required by the simplex method is found to increase exponentially with the number of decision variables, the second method, which is also iterative and was developed in the 1980s, may be more efficient because it may require far fewer iterations, although each iteration requires more calculations. This method, introduced by Karmarkar in

1984 and described in detail by Vanderbei (1999), is called the *interior-point method*. It differs from the simplex method in that all iterates are not required to satisfy the constraints and, therefore, need not be located on vertices. This allows the iterates to be points interior to the feasible region, which on successive iterations can cut clear across the feasible region, so as to locate the optimal point more quickly. Software for both LP methods is widely available from the Internet and is included in libraries of mathematical software.

18.5 NONLINEAR PROGRAMMING (NLP) WITH A SINGLE VARIABLE

Nonlinear optimization problems in just a single decision variable frequently arise in chemical engineering applications. If the objective function is unconstrained, the optimal solution(s) can often be obtained analytically using derivatives from calculus. When constrained, numerical methods are frequently necessary. Some applications that have appeared often in chemical engineering textbooks include the following, many of which involve a balance between capital and operating costs:

1. Optimal thickness of insulation for a pipe carrying steam, for which the insulation cost is balanced against the heat loss causing steam condensation.
2. Optimal reflux ratio for a distillation column, for which the capital cost of the column and heat exchangers is balanced against the utility costs for cooling water in the condenser and steam in the reboiler.
3. Optimal absorbent (stripping agent or extraction solvent) flow rate in an absorber (stripper or liquid–liquid extractor), for which the number of stages is balanced against the column diameter and absorbent (stripping agent or extraction solvent) cost.
4. Optimal pipe diameter for a flowing liquid, for which the capital cost of the pipe and the pump is balanced against the operating cost of the pump.
5. Optimal length (or height) and diameter of a cylindrical pressure vessel of a given volume to minimize the capital cost.
6. Optimal number of stages in a multieffect evaporation system, for which the capital cost is balanced against the cost of heating steam.
7. Optimal interstage pressures of a multistage gas compression system with intercoolers, for which the total power requirement is to be a minimum.
8. Optimal cooling water outlet temperature from a heat exchanger, for which the capital cost of the heat exchanger is balanced against the cost of the cooling water.
9. Optimal filter cake thickness in a batch filter, for which the rate of filtration is balanced against the cost of removing the cake.
10. Optimal number of CSTRs in series, for which the capital cost is to be minimized.

Solutions to optimization problems such as these have led to many of the heuristics presented in Chapter 5.

When the NLP problem consists of only one decision variable (or can be reduced to one), with lower and upper bounds, the optimal solution can be found readily with a spreadsheet, or by one of several structured and efficient search methods, including *region elimination*, *derivative based*, and *point estimation*, as described in detail by Reklaitis et al. (1983). Of the search methods, the golden-section method (involving region elimination) is reasonably efficient, reliable, easily implemented, and widely used. Therefore, it is described and illustrated by example here.

Golden-Section Search

The golden-section search method determines the optimal solution to a bounded objective function that is one-dimensional and unimodal. However, the function need not be continu-

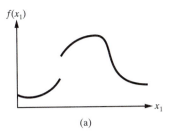 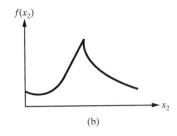

Figure 18.6 Objective function with discontinuities in (a) the function and (b) the derivative of the function.

ous in either the function or its derivative. Thus, the method can solve functions like those shown in Figure 18.6.

Referring to Figure 18.7, let a and b equal the lower and upper bounds of x, the only decision variable. It is not necessary to compute the objective function at these two bounds. The distance between the two points, $b - a = L^{(1)}$. The strategy employed in the golden-section search begins by locating two points in x that are symmetrically placed within the interval from a to b by means of a factor, τ. Thus, if the point farthest from a is located at $(a + \tau L^{(1)})$, then the other point is located at $(b - \tau L^{(1)})$, which is equal to $[a + (1 - \tau)L^{(1)}]$. The objective function is computed for each of the first two points. It is desired to eliminate one of the two points and the region between it and its closest bound. Next, a new point is tested, positioned symmetrical to the remaining point within the new interval. This enables the value of τ to be determined. Suppose a minimum in the objective function is sought, and let the point closest to a have the lowest value of the objective function. Then, because of the assumption of unimodality, the optimal value of x cannot lie to the right of the point closest to b. Therefore, the region between the point closest to b and the point b is eliminated from consideration, leaving a shorter search region interval of length $L^{(2)} = \tau L^{(1)}$. Where should the new point be placed? Note that the remaining point, which was located at $[a + (1 - \tau)L^{(1)}]$ is now located on the new interval at $(a + \tau L^{(2)})$, which is the same as $(a + \tau^2 L^{(1)})$. Hence, $(1 - \tau) = \tau^2$, whose positive solution is $\tau = 0.61803$.

Using this value of τ, subsequent steps in the golden-section method add new points symmetrical to the remaining point, calculate the objective function for the new point, and eliminate a point and the region between it and the closest bound of the remaining region. How many steps are required? Since each step reduces the search space by a factor $\tau = 0.61803$, the fraction of the search interval remaining after M steps is τ^M, requiring the computation of

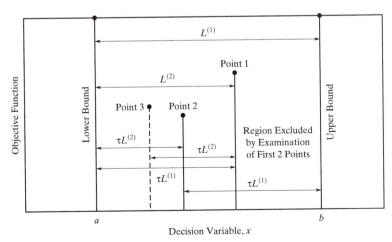

Figure 18.7 Development of the golden-section method.

$M + 1$ objective functions. Thus in 10 steps, the optimal solution is located in an interval that is less than 1% of the distance between the lower and upper bounds. In 20 steps, that interval is reduced to less than 0.01% of that distance.

Rudd and Watson (1968) point out that the ratio of 0.618 was known in ancient times as the "golden section." Greek temples were designed with this ratio because it was most pleasing to the eye. The ancient Badge of the Pythagoreans, a five-pointed star, consists of five isosceles triangles, whose bases are the sides of a pentagon. If the length of each side of the pentagon is 1, the length of each of the two long sides of the triangles is $1 + \tau$.

EXAMPLE 18.3   *Design of Heat Exchanger to Minimize Annual Costs*

In a petroleum refinery, 80,000 lb/hr of a light gas oil at 440°F from a sidecut stripper of a crude distillation tower is currently being cooled with cooling water before being sent to storage. The heat lost could be used to help preheat 500,000 lb/hr of the crude oil, which is available at 240°F and is being heated by other means at a cost of $3.00/million Btu. The plant operates 8,200 hr/yr. Based on the following data, determine what should be done, if anything, when a reasonable return on investment is $i_m = 0.20$. The savings in cooling water cost can be assumed negligible.

Data

Average specific heat of light gas oil = 0.50 Btu/lb-°F

Average specific heat of crude oil = 0.45 Btu/lb-°F

Use a floating-head, shell-and-tube heat exchanger for areas greater than 200 ft^2

For areas greater than 12,000 ft^2, use parallel units

Delivered cost of a heat exchanger = 1.05 times Eq. (16.39) = $1.05C_P\{A\}$

Bare-module factor for a heat exchanger, $F_{BM} = 3.17$ from Table 16.11

Add 5% for site preparation and 18% for contingency and contractor's fee

SOLUTION

For an objective function, use annualized cost, given by Eq. (17.10). Thus, if Q is the duty, in Btu/hr, of the light gas oil-to-crude oil heat exchanger of area, A, in ft^2, the annualized cost, to be minimized, is given by

$$C_A = C + i_m(C_{TCI}) = -\frac{8,200(3.00)Q}{1,000,000} + 0.20(1.05)(1.05)(1.18)(3.17)C_P\{A\}$$

$$= -0.0246\,Q + 0.8248\,C_P\{A\} \tag{18.17}$$

Thus, to be attractive, the annualized cost must be negative so that the absolute value of the savings in the annual cost of heating the crude oil (a negative quantity) is greater than the annualized cost of the heat exchanger installation. The more negative C_A is, the better.

From Table 13.5, for gas oil-to-oil, the overall heat-transfer coefficient, U, is 20 to 35 Btu/hr-°F-ft^2. Since the gas oil is a light gas oil, use $U = 35$ Btu/hr-°F-ft^2. For a mean temperature-driving force, use 0.7 times the log mean = $0.7\,\Delta T_{LM}$

The equality constraints are

1. Energy balances:

$$Q = 80,000(0.50)(440 - T_{LGO,out}) \tag{18.18a}$$

$$Q = 500,000(0.45)(T_{CO,out} - 240) \tag{18.18b}$$

2. Heat-transfer rate:

$$Q = 0.7\,UA\Delta T_{LM} = 0.7(35)A\Delta T_{LM} \tag{18.19}$$

3. Definition of log-mean temperature-driving force:

$$\Delta T_{LM} = \frac{(440 - T_{CO,out}) - (T_{LGO,out} - 240)}{\ln\left(\dfrac{440 - T_{CO,out}}{T_{LGO,out} - 240}\right)} \tag{18.20}$$

Thus, four equations relate five independent variables: $T_{LGO,out}$, $T_{CO,out}$, ΔT_{LM}, Q, and A, yielding one decision variable, which must be selected from the five variables. The best choice is the exit temperature of the light gas oil, $T_{LGO,out}$ because it is easily bounded and permits the remaining four variables to be calculated sequentially using the four equality constraints. An upper bound on its value is for no heat exchange, where $T_{LGO,out} = 440°F$ and $C_A = \$0$. A lower limit assumes infinite heat exchange area, where $T_{LGO,out} = 240°F$, the inlet temperature of the crude oil, because the light gas oil has a much lower flow rate than the crude oil, and C_A is infinite.

The optimization problem is one-dimensional with a nonlinear objective function, which may be discontinuous, depending on the heat-exchanger area. The single decision variable is bounded. Therefore, the golden-section search is suitable for determining the optimal solution. The calculations can be carried out conveniently in the following manner for each selection of $T_{LGO,out}$:

1. Calculate Q using Eq. (18.18a).
2. Calculate $T_{CO,out}$ using Eq. (18.18b).
3. Calculate ΔT_{LM} using Eq. (18.20).
4. Calculate A using Eq. (18.19).
5. Calculate C_A using Eq. (18.17).

To begin the steps in the golden-section search, note that the interval that bounds the decision variable is $440 - 240 = 200°F$. With $\tau = 0.61803$, the first two points are located at $T_{LGO,out} = [240 + 0.61803(200)] = 363.606°F$ and $[240 + (1 - 0.61803)(200)] = 316.394°F$.

Table 18.2 gives the results of the golden-section search, indicating that it is attractive to install the heat exchanger. The optimal exit temperature of light gas oil from the heat exchanger is approximately 250°F, giving an annualized cost of approximately $\$-144,200$ or a savings of $\$144,200$ per year. The final interval for search is less than 1°F. More steps could reduce this interval further, but the area of the heat exchanger would change less than 2%. The crude oil outlet temperature from the heat exchanger is approximately 274°F, so it is heated up only 34°F, compared to a decrease of 190°F in temperature of the light gas oil. The optimal minimum approach temperature is approximately $(250 - 240) = 10°F$. The heat exchangers in the table are all within the range of 200 to 12,000 ft^2 in area so that a single shell-and-tube heat exchanger is sufficient, giving a smooth curve for the objective function. That function is plotted in Figure 18.8, where it is observed that the optimum is not particularly sharp. Consequently, other factors, such as operability and reliability, might enter into a final decision on the size of the heat exchanger. ■

Table 18.2 Golden-Section Search Results for Example 18.3

Point	$T_{LGO,out}$ (°F)	$T_{CO,out}$ (°F)	ΔT_{LM} (°F)	$Q (\times 10^6$ Btu/hr)	A (ft^2)	C_A $(\times 10^3$ \$)
1	316.39	261.97	120.12	4.94	1,680	-100.2
2	363.61	253.58	152.87	3.06	815.8	-59.1
3	287.21	267.16	96.80	6.11	2,577	-123.7
4	269.18	270.37	79.80	6.83	3,495	-136.4
5	258.03	272.35	67.10	7.28	4,428	-142.4
6	251.15	273.57	57.44	7.55	5,368	-144.2
7	246.89	274.33	49.93	7.72	6,314	-143.4
8	253.77	273.10	61.38	7.45	4,954	-143.8
9	249.52	273.86	54.77	7.62	5,678	-144.2
10	252.15	273.39	58.99	7.51	5,199	-144.1
11	250.52	273.68	56.53	7.58	5,482	-144.2
12	250.14	273.75	55.81	7.59	5,554	-144.2
13	250.76	273.64	56.82	7.57	5,437	-144.2

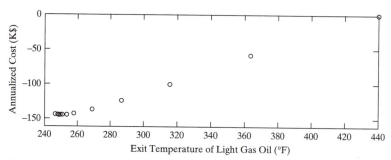

Figure 18.8 Golden-section search results for Example 18.3.

18.6 CONDITIONS FOR NONLINEAR PROGRAMMING (NLP) BY GRADIENT METHODS WITH TWO OR MORE DECISION VARIABLES

Optimization problems encountered in process design are often nonlinear programming (NLP) problems with two or more decision variables. Accordingly, much effort has been expended by researchers in the development of efficient search methods for finding an optimal solution. The remainder of this chapter deals with some of these methods, particularly those that are implemented in simulators.

The formulation of the NLP for application to large process design problems begins with the steady-state simulation of the process flowsheet for a *nominal* set of specifications or decision variables. As described in Section 4.2, during the creation of the simulation model (involving the material and energy balances, kinetic equations, etc., for the process units), a degrees-of-freedom analysis is performed. For the simulation model, the number of variables, $N_{Variables}$, normally exceeds the number of equations, $N_{Equations}$, with the difference between them referred to as the number of degrees of freedom or decision variables, N_D. The best procedure is to carry out a base-case simulation, where the specifications for the decision variables are set using heuristics, such as those of Chapter 5. Then, gradually, as experience is gained by carrying out several simulations, the values of the decision variables are adjusted to better achieve the design objectives. In addition, the process units are simulated with more accurate models, the thermodynamic and transport properties are tuned, often using experimental and pilot-plant data, and profitability measures are computed. Having completed these steps, and having gained a good appreciation of the operation of the process and some indication of the key optimization trade-offs, the engineer is well prepared to formulate a detailed NLP for solution by a simulator.

General Formulation

The general formulation of the NLP is given above at the beginning of Section 18.2. The NLP is usually solved using gradient-based methods. By using numerical partial derivatives of the objective function with respect to the decision variables, gradient-based methods are faster than nongradient methods, with the advantage increasing with the number of decision variables. The use of these methods require conditions for optimality, referred to as stationary or stationarity conditions. These conditions involve slack variables and Lagrangian functions. These are described next, used with *successive quadratic programming* (SQP), which is a widely used gradient method in simulators. Readers who prefer to skip these details should follow the text from Section 18.7.

Stationarity Conditions

To obtain the stationarity conditions, the Lagrangian is formed and differentiated, with the details of this procedure described by McMillan (1970) and by Beveridge and Schecter (1970). The development begins by converting the inequality constraints to equality constraints through the addition of *slack variables*, z_i^2, $i = 1, \ldots, N_{\text{Inequal}}$, such that constraints (18.4) become

$$g_i\{\underline{x}\} + z_i^2 = 0 \qquad i = 1, \ldots, N_{\text{Inequal}} \tag{18.21}$$

where z_i^2 takes up the slackness when $g_i\{\underline{x}\} < 0$. Then, the *unconstrained objective function*, or Lagrangian, is formed:

$$L\{\underline{x}, \underline{\pi}, \underline{\lambda}, \underline{z}\} = f\{\underline{x}\} + \underline{\pi}^{\text{T}} \underline{c}\{\underline{x}\} + \underline{\lambda}^{\text{T}}[\underline{g}\{\underline{x}\} + \underline{z}^2] = 0 \tag{18.22}$$

where $\underline{\pi}$ and $\underline{\lambda}$ are vectors of the Lagrange and Kuhn–Tucker multipliers. At the minimum,

$$\nabla L = 0, \tag{18.23}$$

which can be expanded to give the *stationarity conditions* or the Karush–Kuhn–Tucker (KKT) conditions:

$$\nabla_x L = \nabla_x f\{\underline{x}\} + \underline{\pi}^{\text{T}} \nabla_x \underline{c}\{\underline{x}\} + \underline{\lambda}^{\text{T}} \nabla_x \underline{g}\{\underline{x}\} = 0 \tag{18.24}$$

$$\nabla_\pi L = \underline{c}\{\underline{x}\} = 0 \tag{18.25}$$

$$\nabla_{z_i} L = g_i \lambda_i = 0, \quad i = 1, \ldots, m \tag{18.26}$$

$$\underline{\lambda} \geq 0 \tag{18.27}$$

Note that $\nabla_\lambda L = 0$ gives Eqs. (18.21), which are the definitions of the slack variables and need not be expressed in the KKT conditions. Note also that $\nabla_{z_i} L = 2\lambda_i z_i = 0$, and, using Eqs. (18.21), Eqs. (18.26) result. These are the so-called *complementary slackness* equations. For constraint i, either the residual of the constraint is zero, $g_i = 0$, or the Kuhn–Tucker multiplier is zero, $\lambda_i = 0$, or both are zero; that is, when the constraint is *inactive* ($g_i > 0$), the Kuhn–Tucker multiplier is zero, and when the Kuhn–Tucker multiplier is greater than zero, the constraint must be *active* ($g_i = 0$). Stated differently, there is slackness in either the constraint or the Kuhn–Tucker multiplier. Finally, it is noted that $\nabla_x \underline{c}\{\underline{x}\}$ is the Jacobian matrix of the equality constraints, $\underline{\underline{J}}\{\underline{x}\}$, and $\nabla_x \underline{g}\{\underline{x}\}$ is the Jacobian matrix of the inequality constraints, $\underline{\underline{K}}\{\underline{x}\}$.

Solution of the Stationarity Equations

The KKT conditions are a set of $N_{\text{Variables}} + N_{\text{Equations}} + N_{\text{Inequal}}$ nonlinear equations (NLEs) in $N_{\text{Variables}} + N_{\text{Equations}} + N_{\text{Inequal}}$ unknowns that can be solved, in principle, using an algorithm for the solution of NLEs such as the Newton–Raphson method, which for the equations

$$\underline{F}\{\underline{X}\} = 0 \tag{18.28}$$

takes the form

$$\underline{\Delta X}^{(k)} = -\underline{\underline{J}}\{\underline{X}^{(k)}\}\underline{F}\{\underline{X}^{(k)}\} \tag{18.29}$$

$$\underline{X}^{(k+1)} = \underline{X}^{(k)} + \underline{\Delta X}^{(k)} \tag{18.30}$$

Here, $\underline{X}^{(k)}$ is the vector of guessed values at the kth iteration (the initial guesses when $k = 0$), $\underline{F}\{\underline{X}^{(k)}\}$ is the vector of residuals at $\underline{X}^{(k)}$, $\underline{\underline{J}}\{\underline{X}^{(k)}\}$ is the Jacobian at $\underline{X}^{(k)}$, $\underline{\Delta X}^{(k)}$ is the vector of corrections computed using the Newton–Raphson linearization, and $\underline{X}^{(k+1)}$ is the vector of unknowns after the kth iteration.

When the Newton–Raphson method is applied to solve the KKT Eqs. (18.24)–(18.26), $\underline{X}^{T} = [\underline{x}^{T}\ \underline{\pi}^{T},\ \underline{\lambda}^{T}]$, and Eqs. (18.29) and (18.30) can be rewritten in terms of these variables. This was accomplished by Jirapongphan (1980) who showed that beginning with the vector of guesses, $\underline{X}^{(k)}$, one iteration of the Newton–Raphson method is equivalent to solving the following quadratic program (QP):

Minimize

w. r. t.

$\underline{\Delta x}^{(k+1)},\ \underline{\Delta \pi}^{(k+1)},\ \underline{\Delta \lambda}^{(k+1)}$

$$\nabla_{x}f\{\underline{x}^{(k)}\}^{T}\ \underline{\Delta x}^{(k+1)} + (1/2)\underline{\Delta x}^{(k+1)^{T}}\ \nabla_{xx}^{2}L\{\underline{x}^{(k)},\ \underline{\pi}^{(k)},\ \underline{\lambda}^{(k)}\}\underline{\Delta x}^{(k+1)} \quad \textbf{(QP)}$$

s. t.

$$\underline{\underline{J}}\{\underline{x}^{(k)}\}\underline{\Delta x}^{(k+1)} + \underline{c}\{\underline{x}^{(k)}\} = 0 \tag{18.31}$$

$$\underline{\underline{K}}\{\underline{x}^{(k)}\}\underline{\Delta x}^{(k+1)} + \underline{g}\{\underline{x}^{(k)}\} = 0 \tag{18.32}$$

Algorithms for the solution of quadratic programs, such as the Wolfe (1959) algorithm, are very reliable and readily available. Hence, these have been used in preference to the implementation of the Newton–Raphson method. For each iteration, the quadratic objective function is minimized subject to linearized equality and inequality constraints. Clearly, the most computationally expensive step in carrying out an iteration is in the evaluation of the Laplacian of the Lagrangian, $\nabla_{xx}^{2}L\{\underline{x}^{(k)},\ \underline{\pi}^{(k)},\ \underline{\lambda}^{(k)}\}$ which is also the Hessian matrix of the Lagrangian; that is, the matrix of second derivatives with respect to $\underline{x}^{(k)},\ \underline{\pi}^{(k)},\ \underline{\lambda}^{(k)}$.

To circumvent this calculation, Powell (1977) used the Broyden, Fletcher, Goldfarb, Shanno (BFGS) quasi-Newton method to approximate $\nabla_{xx}^{2}L\{\underline{x}^{(k)},\ \underline{\pi}^{(k)},\ \underline{\lambda}^{(k)}\}$. This saves considerable computation time and is the basis of Powell's SQP method.

A key problem that arises in the implementation of Powell's algorithm is due to the linearization that produces a quadratic objective function and linear constraints, which often lead to infeasible solution vectors, $\underline{X}^{T} = [\underline{x}^{T},\ \underline{\pi}^{T},\ \underline{\lambda}^{T}]$. This problem manifests itself in solutions, $\underline{x}^{(k+1)}$, that violate the inequality constraints, as well as multipliers that are often driven to zero prematurely. Assuming that the initial guesses do not violate the inequality constraints, Han (1977) designed a unidirectional search in the direction of $\underline{\Delta X}^{(k)}$ that is designed to reduce the step taken so as to keep the solution vector within the feasible space.

For more complete presentations of the SQP algorithm, the reader is referred to the textbooks by Reklaitis et al. (1983), Biegler et al. (1997), and Edgar et al. (2001).

18.7 OPTIMIZATION ALGORITHM

The most straightforward way to improve the objective function is by *repeated simulation*. In this procedure, the designer selects values of the decision variables and completes a simulation. Then, usually using a systematic strategy, the decision variables are adjusted and the simulation is repeated, for example, using *sensitivity analysis* in the process simulators, in which simulation results are recomputed as a decision variable is adjusted using uniform increments between bounds specified by the user. However, sensitivity analyses can be very time-consuming and can generate large files of information, much of which is associated with suboptimal processes. Alternatively, the designer can select a formal optimization algorithm built into the simulator to adjust the decision variables, a strategy that is usually more efficient. However, when many recycle loops are present, the simulation calculations can be very time-consuming, with on the order of 20 or more iterations required to achieve an optimum and, for each of these iterations, as many as 20 iterations to converge each of the recycle and control loops (involving design specifications). To overcome these inefficiencies, especially for integrated flowsheets, the latest strategies adjust the decision variables and the tear variables to converge the recycle loops simultaneously (Lang and Biegler, 1987). In these strategies, the optimization algorithm does not converge the recycle loops for each set

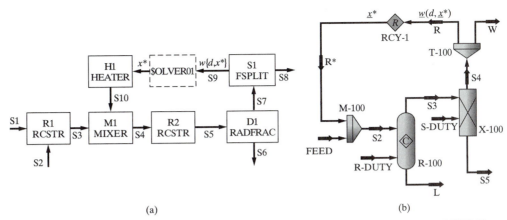

Figure 18.9 Process flowsheet with material recycle and tear stream: (a) ASPEN PLUS; (b) HYSYS.Plant.

of decision variables. Instead, it performs just one pass through the recycle loops before adjusting the decision variables, and consequently, the strategies are referred to as *infeasible path algorithms* because the solution for that path is not converged. In most cases, the infeasible path strategy is successful in ultimately converging the recycle loops to a feasible design while optimizing the process.

To clarify the infeasible path strategy, consider the simulation flowsheets in Figure 18.9. In Figure 18.9a, the ASPEN PLUS simulation flowsheet, in which a continuous-stirred-tank reactor (CSTR), followed by a recycle loop involving another CSTR, a distillation column, a purge splitter, and a heater, is optimized. In this case, the recycle convergence unit, $OLVER01, is positioned arbitrarily so as to tear stream S9, with x^* being the vector of guessed values for the tear variables and $w\{d, x^*\}$ being the vector of tear variables after one pass through the recycle loop. In Figure 18.9b, the HYSYS.Plant PDF, which consists of a recycle loop involving a **conversion reactor**, a **component splitter**, and a **tee**, is optimized. Note that in HYSYS.Plant, a recycle convergence unit is explicitly positioned to tear stream R, with $\underline{x}^*$ being the vector of guessed values for the tear variables and $\underline{w}\{d, \underline{x}^*\}$ being the vector of tear variables after one pass through the recycle loop. Note that w is used because f is reserved for the objective function, as shown in the revised NLP that follows:

$$\text{Minimize} \qquad f\{\underline{x}\} \qquad\qquad \textbf{(NLP2)}$$
$$\text{w. r. t.}$$
$$\underline{d}$$
$$\text{s. t.}$$

$$\underline{h}\{\underline{x}^*\} = \underline{x}^* - \underline{w}\{\underline{x}^*\} = 0 \qquad \text{(tear equations)}$$
$$\underline{c}\{\underline{x}\} = 0$$
$$\underline{g}\{\underline{x}\} \le 0$$
$$\underline{x}^L \le \underline{x} \le \underline{x}^U$$

Here, the equality constraints are augmented by the tear equations, $\underline{h}\{\underline{x}^*\} = 0$, which must be satisfied as well at the minimum of $f\{\underline{x}\}$. For this and similar flowsheets, the decision variables include the residence times in the reactors, the reflux ratio of the distillation tower, and the purge/recycle ratio. In one-dimensional space (i.e., with one decision variable), as $\underline{d}$ varies, the objective function can be displayed as shown in Figure 18.10a. Clearly, the optimizer seeks to locate the minimum efficiently, a task that is complicated when multiple minima exist and it is desired to locate the *global* minimum.

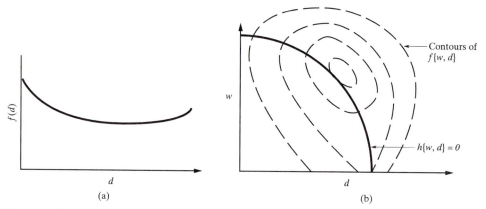

Figure 18.10 Optimization of a process with recycle: (a) objective function; (b) level contours and tear equation.

As the decision variables are adjusted by the optimizer, the values of the tear variables and the objective function change, and it is helpful to show these functionalities in a more explicit form of the NLP:

$$\text{Minimize} \qquad f\{\underline{x}, \underline{d}\} \qquad\qquad\qquad \textbf{(NLP3)}$$

$$\text{w. r. t.}$$
$$\underline{d}$$

$$\text{s. t.}$$

$$\underline{h}\{\underline{x}^*, \underline{d}\} = \underline{x}^* - \underline{w}\{\underline{x}^*, \underline{d}\} = 0$$
$$\underline{c}\{\underline{x}, \underline{d}\} = 0$$
$$\underline{g}\{\underline{x}, \underline{d}\} \leq 0$$
$$\underline{x}^L \leq \underline{x} \leq \underline{x}^U$$

Here, $\underline{x}$ is the vector of process variables excluding the decision variables and $\underline{x}^*$ are guessed values for the tear variables (which equal $\underline{w}$ at the minimum). As will be seen later, it helps to show the variations of the objective function, f, and the tear functions, h, in the schematic diagram of Figure 18.10b. In this diagram, level contours are displayed as a function of just one decision variable and one tear variable, and the locus of points is displayed at which the tear equation is satisfied as d varies along the abscissa. Clearly, the minimum of f must lie on the latter curve, with the unconstrained minimum of $f\{w, d\}$ being infeasible.

Repeated Simulation

The repeated simulation approach is illustrated in Figure 18.11a. Beginning with an initial guess for the decision and tear variables, in the small box, a simulation is completed in which the recycle loop is converged; that is, the tear equation is satisfied. Then, d is adjusted, somewhat arbitrarily, by Δ_d and the simulation is repeated using the previous solution for w as the initial guess. This strategy is repeated until convergence to the minimum is achieved.

Infeasible Path Approach

In the infeasible path approach, as illustrated in Figure 18.11b, both d and w are adjusted simultaneously by the optimizer (with $w \rightarrow x^*$ for the next iteration), usually using the SQP algorithm. This algorithm involves just one pass through the flowsheet per iteration, so the tear equations are normally not satisfied until the optimum is located. As will be seen in the ex-

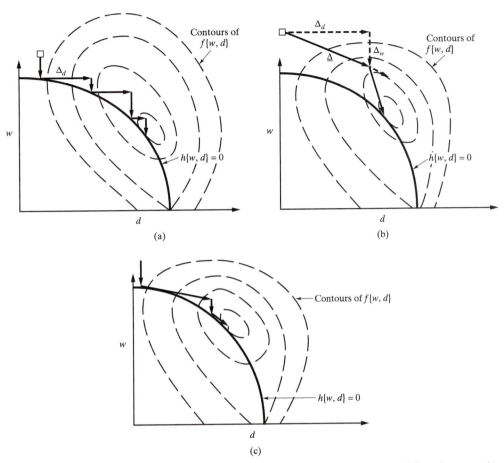

Figure 18.11 Optimization of a process with recycle: (a) repeated simulation (feasible path approach); (b) infeasible path approach; (c) compromise approach.

amples that follow, convergence is usually achieved in a few iterations. Figure 18.11b shows a schematic of $\underline{\Delta}$, the vector of the changes in d, Δ_d, and w, Δ_w, as computed by the SQP algorithm.

Compromise Approach

In a compromise approach, which is often necessary to achieve convergence, after the SQP step is taken (i.e., $\underline{\Delta}$), the tear equations, $\underline{h}\{\underline{x}^*\} = 0$, are converged as shown in Figure 18.11c. Often the tear equations are converged loosely using a convergence method, typically Wegstein's method, with a maximum number of iterations assigned, typically three or four. This compromise approach is often utilized when convergence cannot be achieved after several attempts using the infeasible path approach.

Practical Aspects of Flowsheet Optimization

In most flowsheet simulations, design specifications (or control loops) are included. The iterative calculations to achieve convergence of these specifications are embedded often within the recycle loops and are converged during each pass through the recycle loops. In other cases, these specifications are implemented as outer loops, with the recycle loops converged entirely during each iteration of the outer loop. Yet another alternative is to converge these loops simultaneously, usually using just one pass through the unit subroutines in the recycle and design specification loops.

When solving a NLP, to optimize a flowsheet, still another alternative exists. In many cases, it is preferable to incorporate the design specifications as equality constraints, $\underline{c}\{\underline{x}, \underline{d}\} = 0$, as shown in NLP3. Then, it is necessary to remove these design specifications when adding the optimization convergence unit. The latter usually replaces the recycle convergence units in the simulation flowsheet.

Before implementing an infeasible path optimization, it is very helpful to carry out preliminary searches by varying the key decision variables, somewhat randomly, to gain insights into the key trade-offs. For these searches, it is probably best not to use optimization algorithms that require derivatives, or approximations to them, such as SQP. A common approach is to use the sensitivity analysis facilities of the process simulators referred to earlier.

As a final caution, be sure not to use a gradient-based approach, such as SQP, when selecting a discrete decision variable such as the number of trays in a distillation tower. In these cases, it is meaningless to estimate partial derivatives in expressions for the gradient of the objective function or its Hessian matrix, because the decision variables are restricted to integer quantities. Similar problems arise when there are discrete changes in the installation costs of equipment, which can arise when a single unit is replaced with two or more units. This often occurs when size variables exceed upper bounds embedded in the subroutines for the calculation of equipment sizes and costs. These kinds of discrete changes are more difficult to detect when sizes and costs are computed for many process units in a complex profitability analysis. For this reason, it is often recommended to carry out the optimization initially using a simpler objective function that does not involve such discontinuities. Then, after the optimum is computed, more rigorous measures can be computed and further optimized, using simpler methods (that do not involve derivatives), in the vicinity of the optimum.

18.8 FLOWSHEET OPTIMIZATIONS—CASE STUDIES

In this section, two case studies are presented. The first is a relatively simple example involving one decision variable and one constraint, in which the venture profit for a process to manufacture ethyl chloride is maximized. In the second example, it is desired to optimize the operation of a multidraw distillation column, in which a mixture of normal paraffins is separated into four product streams, two of which are sidestreams. This involves four decision variables and a number of constraints. Although neither of the examples involves the optimization of accurate measures of plant profitability, detailed costing could be included.

EXAMPLE 18.4 *Maximizing the Venture Profit in Ethyl Chloride Manufacture*

In this example, the venture profit of the ethyl chloride process in Figure 18.12, introduced in the multimedia CD-ROM that accompanies this text (see *HYSYS → Tutorials → Material and Energy Balances → Ethyl Chloride Manufacture* and *ASPEN → Tutorials → Material and Energy Balances → Ethyl Chloride Manufacture*) is maximized by adjusting the purge (*W*) flow rate. To estimate the venture profit, the following information is supplied

Installed cost of equipment	$500 \left(\dfrac{330 \times 24}{1,000}(F_R) \right)^{0.6}$ monetary units
Cost of ethylene	1.5×10^{-3} monetary units/kg
Cost of HCl	1.0×10^{-3} monetary units/kg
Revenue for ethyl chloride	2.5×10^{-3} monetary units/kg

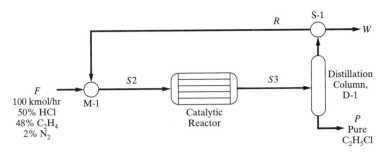

Figure 18.12 Process for the production of ethyl chloride.

where F_R is the reactor feed rate (in kg/hr). The venture profit [Eq. (17.9)] is formulated (in monetary units) assuming a 10% return on investment (ROI) and 330 operating day/yr:

$$\text{VP} = 330 \times 24 \times 10^{-3}[2.5P - (1.5x_{Et} + x_{HCl})F] - 0.1\left[500\left(\frac{330 \times 24}{1,000}(F_R)\right)^{0.6}\right] \quad \textbf{(18.33)}$$

where x_{Et} and x_{HCl} are the mass fractions of ethylene and HCl in the feed stream, respectively, and F and P are the feed and product flow rates (in kg/hr). The NLP is

Minimize $\qquad$ VP $\qquad\qquad\qquad\qquad\qquad\qquad\qquad\qquad$ **(18.34)**

w. r. t.

$\qquad W$

s. t.

$\qquad\qquad \underline{c}\{\underline{x}\} = 0 \qquad\qquad$ (material equations) $\qquad\qquad$ **(18.35)**

$\qquad\qquad R < 300 \text{ kg/hr} \qquad\qquad\qquad\qquad\qquad\qquad\qquad$ **(18.36)**

SOLUTION

As shown in the multimedia CD-ROM that accompanies this book (see *HYSYS → Principles of Flowsheet Simulation → Getting Started in HYSYS → Advanced features → Optimization*) and *ASPEN → Principles of Flowheet Simulation → Optimization* the VP is optimized with relative ease. As usual, the optimization is initialized from a feasible solution, for $W = 5$ kmol/hr. The SQP method is used, requiring four iterations of the SQP method, with 13 evaluations of the material and energy balances. The HYSYS spreadsheet is used to compute the VP based on flowsheet information, and the HYSYS optimizer is then invoked to enter the NLP objective function and constraints, the decision variables and numerical method parameters. The unconstrained solution [i.e., neglecting Eq. (18.36) in the NLP] gives the global maximum VP of 4,730 units, obtained with a value of $W = 7.27$ kmol/hr. Augmenting the NLP with the inequality constraint in Eq. (18.36) gives VP = 4,450 units, obtained with a value of $W = 8.96$ kmol/hr, for which R is at its upper bound of 300 kg/hr. Use the file ETHYLCHLORIDE_OPT.hsc on the multimedia CD-ROM to reproduce the results presented above. ∎

EXAMPLE 18.5 *Optimization of a Distillation Tower with Sidedraws*

In this example, the distillation tower in Figure 18.13 is optimized. A feed containing the normal paraffins from nC_5 to nC_9 is fed to the 25-stage tower (including the condenser and reboiler) on stage 15, counting upward from the partial-reboiler stage. The objective is to adjust the operating conditions so as to achieve a distillate (D) concentrated in nC_5, a sidedraw at stage 20 ($S1$) concentrated in nC_6, a second sidedraw at stage 10 ($S2$) concentrated in nC_7 and nC_8, and a bottoms product (B) concentrated in nC_9. No costing is involved. The operating conditions to be adjusted are the reflux ratio and flow rates

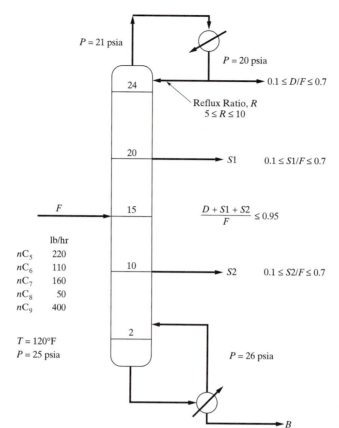

$P = 21$ psia

$P = 20$ psia

$0.1 \leq D/F \leq 0.7$

24

Reflux Ratio, R
$5 \leq R \leq 10$

20

S1 $0.1 \leq S1/F \leq 0.7$

F 15 $\dfrac{D + S1 + S2}{F} \leq 0.95$

lb/hr	
nC_5	220
nC_6	110
nC_7	160
nC_8	50
nC_9	400

10   S2 $0.1 \leq S2/F \leq 0.7$

$T = 120°F$
$P = 25$ psia

2

$P = 26$ psia

B

Figure 18.13 Distillation tower with sidedraws.

of the distillate and two sidedraws. This is accomplished by formulating a NLP in which the feed stage and the stages of the sidedraws are held fixed during the optimization:

Minimize $\qquad D_{C5} + S1_{C6} + S2_{C7} + S2_{C8} + B_{C9}$ $\qquad$ **(18.37)**

w. r. t.

$R, D, S1, S2$

s. t.

$$5 \leq R \leq 10 \qquad\qquad\qquad \textbf{(18.38)}$$
$$0.1 \leq D/F \leq 0.7 \qquad\qquad \textbf{(18.39)}$$
$$0.1 \leq S1/F \leq 0.7 \qquad\qquad \textbf{(18.40)}$$
$$0.1 \leq S2/F \leq 0.7 \qquad\qquad \textbf{(18.41)}$$
$$(D + S1 + S2)/F \leq 0.95 \qquad \textbf{(18.42)}$$

where R is the reflux ratio; $F, D, S1, S2,$ and B are the molar flow rates of the feed, distillate, two sidedraws, and bottoms product streams; and the subscript denotes the molar flow rate of a specific chemical species in that stream. Note that the product streams are withdrawn as saturated liquids at the low pressures shown. The liquid and vapor phases are assumed to be ideal and at equilibrium on the stages of the tower. All of the decision variables have inequality constraints as indicated by Eqs. (18.38)–(18.42). This example does not involve recycle convergence or any user-supplied equality constraints.

SOLUTION

As shown in the multimedia CD-ROM that accompanies this book (see *HYSYS → Tutorials → Process Design → Multi-draw Tower Optimization* and *ASPEN → Tutorials → Process Design → Multi-draw Tower Optimization*), the solution of this NLP requires care, since the gradient-based SQP method is sen-

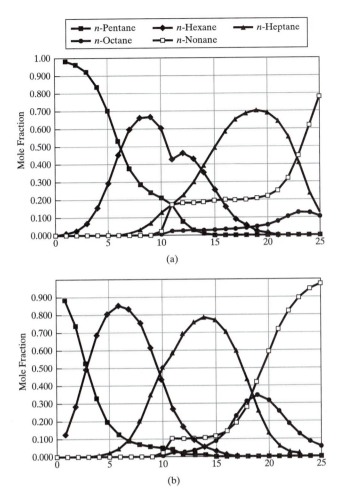

(a)

(b)

Figure 18.14 Composition profiles in multi-draw column as a function of the tray number, counting from the top: (a) initial design; (b) optimized design.

sitive to the numerical estimates of the derivatives in the Jacobian matrices $\underline{\underline{J}}\{\underline{x}\}$ and $\underline{\underline{K}}\{\underline{x}\}$. The optimization is initialized at a feasible solution: $R = 5$, $D = S1 = S2 = 2$ lbmol/hr, for which the composition profiles in the column are given in Figure 18.14a and the product recoveries summarized in Table 18.3.

Note that the peaks in the composition profiles are not at the draw stream locations, which explains the rather poor product recoveries, especially for C_8. After optimization, the decision variables are: $R = 10$, $D = 3.35$ lbmol/hr, $S1 = 1$ lbmol/hr, and $S2 = 2.16$ lbmol/hr, giving the composition profiles in Figure 18.14b, which show a marked improvement in positioning the composition peaks. This explains the significant improvement in the product recoveries shown in Table 18.3. Use the files MULTIDRAW_OPT.hsc and MULTIDRAW_OPT.bkp on the multimedia CD-ROM. ∎

Table 18.3 Product Recoveries in the Multidraw Column

	Percentage molar recovery					Objective function
	D-C_5	$S1$-C_6	$S2$-C_7	$S2$-C_8	B-C_9	
Initial design	65	72	74	18	87	6.87
Optimized design	96	65	96	68	91	8.44

18.6 SUMMARY

Having studied this chapter and the multimedia CD-ROM that accompanies this book, the reader should:

1. Understand the fundamentals of optimization concerning the use of analytical or numerical methods.
2. Be able to solve LP problems in one or two decision variables.
3. Be able to create a nonlinear program (NLP) to optimize a process using equality and inequality constraints.
4. Be able to use the golden-section search to solve a constrained NLP problem in one decision variable.
5. Recognize the advantages of calculating the objective function and constraints for at least a base case of the decision variables before using an optimization algorithm.
6. Understand the advantages of performing optimization and converging recycle calculations and design specifications simultaneously, as implemented using an *infeasible path* optimization algorithm.
7. Be able to optimize a process using ASPEN PLUS and HYSYS.Plant beginning with the results of a steady-state simulation.

REFERENCES

Beveridge, G.S.G., and R.S. Schecter, *Optimization: Theory and Practice*, McGraw-Hill, New York (1970).

Biegler, L.T., I.E. Grossmann, and A.W. Westerberg, *Systematic Methods of Chemical Process Design*, Prentice-Hall, Englewood Cliffs, New Jersey (1997).

Dantzig, G.B., "Programming of Independent Activities, II, Mathematical Model," *Econometrica*, **17**, 200–211 (1949).

Edgar, T.F., D.M. Himmelblau, and L.S. Lasdon, *Optimization of Chemical Processes*, 2nd ed., McGraw-Hill, New York (2001).

Floudas, C.A., *Nonlinear and Mixed-integer Optimization: Fundamentals and Applications*, Oxford University Press, Oxford (1995).

Han, S.-P., "A Globally Convergent Method for Nonlinear Programming," *J. Optimization Appl.*, **22**, 297 (1977).

Jirapongphan, S., *Simultaneous Modular Convergence Concept in Process Flowsheet Optimization*, Sc.D. Thesis, M.I.T., Cambridge, Massachusetts (1980).

Lang, Y.-D., and L.T. Biegler, "A Unified Algorithm for Flowsheet Optimization," *Comput. Chem. Eng.*, **11**, 143 (1987).

Mah, R.S.H., *Chemical Process Structures and Information Flows*, Butterworth, Boston (1990).

McMillan, Jr., C., *Mathematical Programming: An Introduction to Design and Application of Optimal Decision Machines*, Wiley, New York (1970).

Morgan, A., *Solving Polynomial Systems Using Continuation for Engineering and Scientific Problems*, Prentice-Hall, Englewood Cliffs, New Jersey (1987).

Powell, M.J.D., *A Fast Algorithm for Nonlinearly Constrained Optimization Calculations*, AERE Harwell, England (1977).

Reklaitis, G.V., A. Ravindran, and K.M. Ragsdell, *Engineering Optimization—Methods and Applications*, Wiley-Interscience, New York (1983).

Rudd, D.F., and C.C. Watson, *Strategy of Process Engineering*, John Wiley & Sons, New York (1968).

Seader, J.D., W.D. Seider, and A.C. Pauls, *FLOWTRAN Simulation—An Introduction*, 3rd ed., CACHE, Austin, Texas (1987).

Vanderbei, R.J., *Linear Programming Foundations and Extensions*, Kluwer Academic Publishers, Norwell, Massachusetts (1999).

Wolfe, P., "The Simplex Method of Quadratic Programming," *Econometrica*, **27**, 382–398 (1959).

EXERCISES

18.1 *Scheduling of batch distillation.* A batch distillation facility has a bank of columns of Type 1 and another bank of Type 2. Type 1 columns are available for processing 6,000 hr/week, while Type 2 columns are available 10,000 hr/week. It is desired to use these columns to manufacture two different slates of products, A and B. Distillation time to produce 100 gal of product A is 2 hr in Type 1 columns and 1 hr in Type 2 columns. Distillation time to produce 100 gal of product B is 1 hr in Type 1 columns and 4 hr in Type 2 columns. The net profit is $5.00/gal for product A and $3.50/gal for product B. Use an LP with a graph to determine the production schedule that maximizes the net profit in $/week.

18.2 *Analytical optimization.* Determine using calculus all maxima, minima, and saddle points for the following unconstrained two-dimensional objective functions:

a. $f\{x_1,x_2\} = 2x_1^3 + 4x_1x_2^2 - 10x_1x_2 + x_2^2$

b. $f\{x_1,x_2\} = 1,000x_1 + 4 \times 10^9 x_1^{-1}x_2^{-1} + 2.5 \times 10^5 x_2$

c. $f\{x_1,x_2\} = (1 - x_1)^2 + 100(x_2 - x_1^2)^2$

18.3 *Golden-section search.* Use 10 steps of the golden-section search method to find the optimal dimensions for the cylindrical reactor vessel in Example 16.11. In that example, the dimensions of the vessel are given as the inside diameter, $D = 6.5$ ft, and tangent-to-tangent length, $L = 40$ ft. These dimensions are not critical as long as the volume is maintained. Determine the optimal diameter and length, if the permissible range of the aspect ratio, L/D, is 1 to 50.

18.4 *Optimization of minimum temperature approach in a heat exchanger.* As shown in the ASPEN PLUS simulation flowsheet in Figure 18.15, liquid toluene is to be heated from 100 to 350°F, while liquid styrene is to be cooled from 300 to 100°F. Auxiliary heat exchangers E2 and E3, which use steam and cooling water, respectively, are provided to meet the target temperatures when they cannot be achieved by heat exchanger E1. The process is to be optimized with respect to the minimum temperature of approach in E1, which is to be within 1°F and 50°F. The temperature of stream S3 is constrained to be less than or equal to 200°F, and the temperature of stream S4 is constrained to be less than or equal to 300°F. The annualized cost is to be minimized with the return on investment, r, equal to 0.5. All of the necessary data are included in Figure 18.15.

18.5 *Separation of propylene and propane by high-pressure distillation.* A process to separate propylene and propane, to produce 99 mol% propylene and 95 mol% propane, is shown in Figure 18.16. Because of the high product purities and the low relative volatility, 200 stages may be required. Assuming 100% tray efficiency and a tray spacing of 24 in., two towers are installed because a single tower would be too tall. The distillate is vapor at 280 psia and a pressure increase of 0.1 psi is assumed on each tray, with a 0.2-psi increase in the condenser. Note that the stage numbers and reflux ratio are for a nominal design. In this exercise, a suitable objective function is to be selected and the number of stages and reflux ratio are to be adjusted to find

the optimum. Pay close attention to the determination of the proper feed stage to avoid pinch or near-pinch conditions.

18.6 *Petlyuk columns.* A process design for the disproportionation of toluene to benzene and the xylene isomers is being completed. Your assistance is needed on the design of the liquid separation section. It has been established that the feed to this section is at 100°F and 50 psia with the following flow rates in lbmol/hr:

Benzene	16.3
Toluene	70.9
p-Xylene	4.0
m-Xylene	7.5
o-Xylene	3.5

From this feed, it is desired to produce 99.5 mol% benzene, 98 mol% toluene for recycle, and 99 mol% mixed xylenes, by distillation in two columns. The assigned plant operators have informed us that they prefer the direct sequence of two columns. However, because of the high percentage of toluene in the feed, a thermally coupled system, shown in Figure 18.17, known as Petlyuk towers after the Russian inventor, may be a less expensive alternative. Please prepare process designs for these two alternatives, together with estimates of capital and operating costs, and indicate whether the Petlyuk towers are attractive. The plant-operating factor is assumed to be 95%. The following data are provided for the design of the towers:

1. Determine optimal feed preheat using the bottoms product where applicable.

2. Determine optimal reflux ratios.

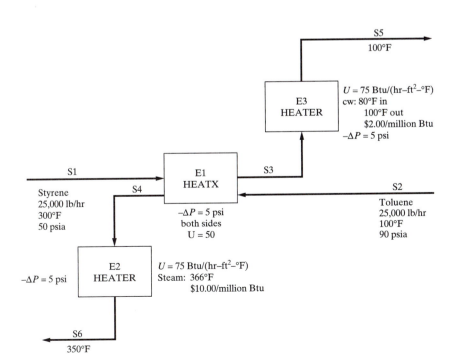

Figure 18.15 Heat exchange network.

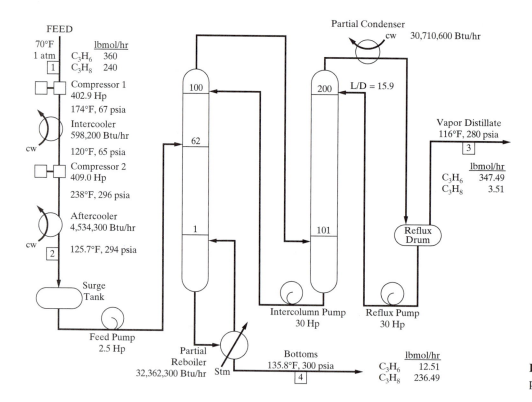

FEED
70°F
1 atm
     lbmol/hr
    C_3H_6 360
    C_3H_8 240

Compressor 1
402.9 Hp
174°F, 67 psia

Intercooler
598,200 Btu/hr
120°F, 65 psia

Compressor 2
409.0 Hp
238°F, 296 psia

Aftercooler
4,534,300 Btu/hr
125.7°F, 294 psia

Surge Tank

Feed Pump
2.5 Hp

Partial Reboiler
32,362,300 Btu/hr Stm

Intercolumn Pump
30 Hp

Reflux Pump
30 Hp

Partial Condenser cw 30,710,600 Btu/hr

L/D = 15.9

Vapor Distillate
116°F, 280 psia
     lbmol/hr
    C_3H_6 347.49
    C_3H_8 3.51

Reflux Drum

Bottoms
135.8°F, 300 psia
     lbmol/hr
    C_3H_6 12.51
    C_3H_8 236.49

Figure 18.16 Propylene–propane distillation tower.

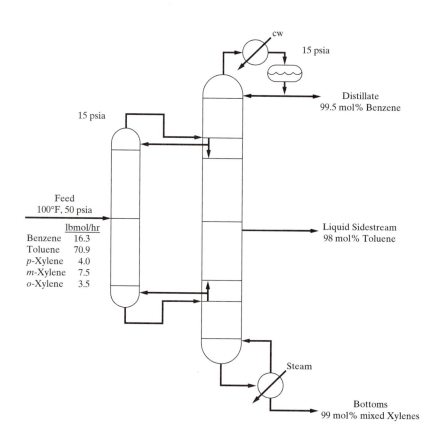

15 psia

Distillate
99.5 mol% Benzene

15 psia

Feed
100°F, 50 psia
     lbmol/hr
Benzene   16.3
Toluene   70.9
p-Xylene   4.0
m-Xylene   7.5
o-Xylene   3.5

Liquid Sidestream
98 mol% Toluene

Steam

Bottoms
99 mol% mixed Xylenes

Figure 18.17 Petlyuk towers.

3. Set a pressure of 15 psia at the top of each column, assuming no pressure drop through the condenser. Determine column bottom pressure drop from tray pressure drops.

4. Use standard sieve trays. Overall tray efficiency is estimated to be 90% for operation at 85% of flooding.

5. Determine column diameters to the next increment of 0.5 ft for each section and swedge the columns, when section diameters differ by more than 1.0 ft.

6. Use reflux subcooled to 120°F from each condenser.

7. Standard materials of construction, for example, carbon steel, can be used.

8. Minimum shell thickness for columns and vessels is as follows:

    0.25 in. for diameters less than 4 ft

    $\frac{5}{16}$ in. for diameters from 4 to 5.5 ft

    $\frac{3}{8}$ in. for diameters from 6 to 7.5 ft

    $\frac{7}{16}$ in. for diameters from 8 to 11.5 ft

    0.5 in. for diameters from 10 to 12 ft

9. Provide horizontal reflux drums that can hold liquid reflux and distillate for 5 min at half full.

10. Include all necessary liquid pumps with spares.

Chapter **19**

Product Design

19.0 OBJECTIVES

This chapter focuses on the design of configured industrial and consumer products, which were introduced in Section 1.2 in connection with Table 1.1. Having specified desired properties and performance, and having identified chemicals to achieve them, and possibly having designed a process to produce these chemicals, the design team concentrates on the three-dimensional configuration of the product and the methods for its manufacture.

An objective of this chapter is to introduce several approaches for the design of configured industrial and consumer chemical products using case studies. First, the steps in product design are reexamined, with emphasis on those steps specific to the design of configured industrial and consumer products. To obtain the desired functionality, in some cases, the rates of mass and heat transfer or the distribution of stresses due to loadings become important. For these designs, it is often helpful to carry out mass, heat, and/or force balances using finite-element analysis and computer packages such as FEMLAB and the Finite-element Toolbox in MATLAB. Through an example involving the chemical vapor deposition of polycrystalline silicon, an objective is to introduce the usage of these packages.

This chapter concentrates on the steps in designing the product, not on the manufacturing process, which often involves parts making (including die cutting and molding), parts assembly and integration, and finishing. Since design of the manufacturing process is usually carried out by mechanical, civil, and electrical engineering members of the design team, this aspect of product design is not covered in this book. An exception arises in processes to deposit thin films and coatings, for example, on electronic materials. Here, chemical engineers focus on the design of the equipment for chemical vapor deposition, as discussed in Section 19.6.

After studying this chapter, the reader should

1. Be knowledgeable about many of the steps in designing configured industrial and consumer products.
2. Be familiar with the design of a representative array of nine chemical products. These products include small hemodialysis devices, solar desalinators, hand warmers, fuel cells to power automobiles, and thin silicon films that coat microelectronic devices.
3. Understand aspects of theoretical analyses in product design, using material and energy balances, transport phenomena, and reaction kinetics to size products and obtain cost estimates.

19.1 STEPS IN DESIGNING INDUSTRIAL AND CONSUMER PRODUCTS

Returning to the steps in product and process design introduced in Section 1.2 and shown in Figure 1.2 in this section, the typical sequences design teams follow to design configured industrial and consumer products are reviewed. Given a potential opportunity, either suggested to the design team or devised by it, the team engages in creating and assessing the primitive problem. This often involves an extensive idea-generation phase, in which brainstorming is undertaken to generate a long list of ideas, but without careful analysis. Gradually, as potential customers are interviewed, as the literature is examined (especially the patent literature to locate competing products), and as technical feasibility, marketing, and business studies are undertaken, the list of ideas is winnowed to a few that deserve detailed design work. Often the technical feasibility study involves the specification of properties and performance for the desired chemicals, and a search for molecules to satisfy these specifications, as discussed in Chapter 2 on molecular structure design. After the molecules and mixtures have been selected (e.g., a polymer membrane for a hollow fiber, a surface coating, or a polymer for a fluid container), the team often concentrates on the synthesis of a process to produce the chemicals. Whether or not a process design is undertaken, when designing configured industrial and consumer products, three-dimensional design of the product is often necessary. Because this is usually a key aspect of configured product design, this chapter has been positioned in Part Three on *Detailed Design*, along with Chapters 13–15 on the detailed design of heat exchangers, distillation towers, pumps, and compressors.

As mentioned in the above "Objectives" section of this chapter, the design team also designs the manufacturing process, involving parts making and parts assembly. Since other engineers normally carry out this work, design of the manufacturing process for the product is not covered in this chapter. Note, however, that for the most promising designs, it is usually important to build prototypes and test them on a small scale. Only after the product passes these initial tests do the design engineers focus on the process to manufacture the product in commercial quantities. This scale-up often involves pilot-plant testing, as well as consumer testing, before a decision is made to enter commercial production. Here, the methods of quality control summarized in Table 1.1 become important. For industrial products, functional quality criteria include optical performance, weatherability, mechanical strength, and printability. For consumer products, in addition to these, there are use criteria including durability, life cycle, ease of use, and effectiveness. In recent years, designers have been following many of the steps recommended to achieve *six-sigma performance*, as discussed briefly in Section 1.2 (see Rath and Strong's *Design for Six Sigma Pocket Guide*, 2002) and more completely at the end of this section. In addition, environmental and safety criteria are addressed throughout the design process, as discussed in Sections 1.3 and 1.4.

In some cases, the design of processing equipment is the key to product design, for example, in the chemical vapor deposition of polysilicon for integrated circuit manufacture, as discussed in Section 19.6.

When evaluating the product and methods for its manufacture, possibly including the design of a process to produce the desired chemicals, the methods of capital-cost estimation, profitability analysis, and optimization, discussed previously in Chapters 16–18, are often utilized. However, for products that can be sold at a high price, due to large demand, lack of competition, and limited production, detailed profitability analysis and optimization are less important. For these products, it is most important to reduce the design and manufacturing time to capture the market before competitive products are developed.

In the sections that follow, the designs of several chemical products are presented with a discussion of aspects of the product design sequence. Since the entire design sequence is usually not available for most of these products, many of which have been patented, it is not possible, or even desirable in the space available, to attempt the presentation of step-by-step design sequences. Instead, some of the key design procedures and decisions are examined, with references provided to patents and other sources of information.

Before proceeding to the first product design, in Section 19.2, an introduction to the use of six-sigma methodology for the design of high-quality products is presented.

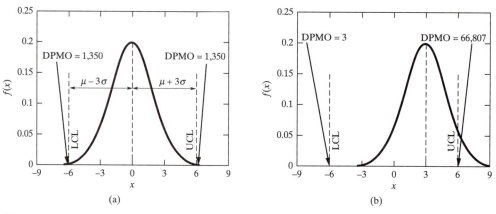

Figure 19.1 Distribution of product quality at 3σ, with $\sigma = 2$: (a) normal operation at $\mu = 0$; (b) abnormal operation shifted to $\mu + 1.5\sigma$.

Six Sigma in Product Design and Manufacturing

Definitions

Six sigma (6σ) is a structured methodology for eliminating defects and hence improving product quality in manufacturing and services. The methodology aims at identifying and reducing the variance in product quality, and involves a combination of statistical quality control, data analysis methods, and the training of personnel. The term six sigma defines a desired level of quality: 3.4 defects per million opportunities (DPMO). The symbol σ (sigma) is the standard deviation of the value of a quality variable, a measure of its variance, which is assumed to be normally distributed. Figure 19.1a shows such a distribution with $\sigma = 2$. Note that the distribution is normalized such that the total area under the curve is unity, with a probability density function given by

$$f(x) = \frac{1}{\sigma\sqrt{2\pi}} \exp\left[-\frac{1}{2}\left(\frac{x-\mu}{\sigma}\right)^2\right] \tag{19.1}$$

where $f(x)$ is the probability of the quality at a value of x, and μ is the average value of x. Assuming that operation at 3σ on either side of μ is considered normal, this defines the upper control limit (UCL) at $\mu + 3\sigma$, and the lower control limit (LCL) at $\mu - 3\sigma$. As shown in the figure, the number of defects per million opportunities above the UCL is

$$\text{DPMO} = 10^6 \int_{\mu+3\sigma}^{\infty} f(x)dx = \frac{1}{2} 10^6 \left(1 - \int_{\mu-3\sigma}^{\mu+3\sigma} f(x)dx\right) = 1,350 \tag{19.2}$$

This means that 1,350 DPMO can be expected in a normal sample above the UCL and the same number below the LCL. It is important, however, that the manufacturing process be insensitive to process drifts. In accepted six-sigma methodology, a worst-case shift of 1.5σ in the distribution of quality is assumed, to a new average value of $\mu + 1.5\sigma$, as shown in Figure 19.1b. For operation at 3σ, the expected DPMO above the UCL are 66,807, and below the LCL are 3. This gives a total expected DPMO of 66,810 — a significant deterioration in quality. In contrast, suppose that the variance can be reduced to $\sigma = 1$. Assuming operation at 6σ either side of the average value of the distribution, $\mu = 0$, this defines the UCL at $\mu + 6\sigma$ and the LCL at $\mu - 6\sigma$, as shown in Figure 19.2a, with 1 defect per billion opportunities on either side of the acceptance limits, which are insignificant defect levels. The improvement in performance is apparent when considering a shift of 1.5σ as before; for 6σ operation, the DPMO (above the UCL) increases to only 3.4, as shown in Figure 19.2b.

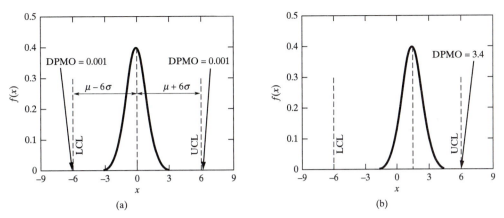

Figure 19.2 Distribution of product quality at 6σ, with σ = 1: (a) normal operation at μ = 0; (b) abnormal operation shifted to μ + 1.5σ.

Cost of Defects

Table 19.1 and Figure 19.3 present the effect of the sigma level on DPMO, assuming a 1.5σ shift in mean, as in Figures 19.1b and 19.2b. Note that the table shows the total DPMO, above the UCL and below the LCL.

The expected number of defects presented in Table 19.1 and Figure 19.3 apply to a single manufacturing step. Usually, device manufacture involves a number of steps. For n steps, and assuming that all defective device components are removed from the production sequence at the step where they occur, the overall defect-free throughput yield is

$$\text{TY} = \prod_{i=1}^{n} \left(1 - \frac{\text{DPMO}_i}{10^6}\right) \tag{19.3}$$

where DPMO_i are the expected number of defects per million opportunities in step i. If the DPMO is identical in each step, Eq. (19.3) reduces to

$$\text{TY} = \left(1 - \frac{\text{DPMO}}{10^6}\right)^n \tag{19.4}$$

The fraction of the production capacity lost due to defects is $1 - \text{TY}$. For example, consider the manufacture of a device involving 40 steps, each of which operates at 4σ. From Table 19.1, the expected DPMO are 6,210 per step, so TY = $(1 - 0.00621)^{40} = 0.779$. Thus, 22%

Table 19.1 Sigma Level on Expected DPMO with 1.5σ Shift in Mean, μ

Sigma Level	Expected DPMO
1.0	697,672
2.0	308,770
3.0	66,810
3.5	22,750
4.0	6,210
4.5	1,350
5.0	233
5.5	32
6.0	3.4

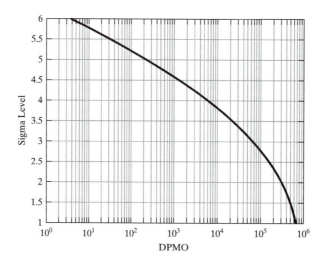

Figure 19.3 The relationship between DPMO and the sigma level.

of production capacity is lost due to defects, rendering the overall manufacturing operation a 2.3σ process (see Table 19.1 and Figure 19.3). In contrast, if each of the 40 steps operate at 6σ, $TY = (1 - 3.4/10^6)^{40} = 0.99986$, corresponding to about one faulty device for every 10,000 produced, and in this case, the overall operation is a 5.2σ process.

In the preceding discussion, it has been assumed that defective devices are eliminated in production, leaving only the impact on reduced throughput yield. In the likely event that a fraction of the defects are undiscovered and lead to shipped defective devices, the impact on sales resulting from customer dissatisfaction could be much greater. Noting that many manufacturing operations involve hundreds of steps (e.g., integrated-circuit chip manufacturing), it is clear that high levels of reliability, as expressed by low DPMO values, are generally required to ensure profitable manufacture. This is the driving force behind the extensive proliferation of six-sigma methodology (Wheeler, 2002).

Methods to Monitor and Reduce Variance

As described in detail by Rath and Strong (2000), an iterative five-step procedure is followed to progressively improve product quality. The five steps are (a) Define, (b) Measure, (c) Analyze, (d) Improve, and (e) Control, referred to by the acronym, DMAIC:

a. Define: First, a clear statement is made defining the intended improvement. Next, the project team is selected, and the responsibilities of each team member assigned. To assist in project management, a chart is prepared showing the suppliers, inputs, process, outputs, and customers (referred to by the acronym, SIPOC). A simplified block diagram usually accompanies a SIPOC, showing the principal steps in the process (usually 4–7 steps). At this stage, the main focus is on customer concerns, which are used to define critical-to-quality (CTQ) output variables. As an example, suppose the company, ACME Tubes, Inc., manufactures PVC tubing by extrusion of PVC melt. A SIPOC describing its operations is presented in Figure 19.4. The quality of the PVC tubing, measured in terms of its impact strength, is considered to be the principal CTQ, and customer specifications define the LCL and UCL.

b. Measure: The CTQ variables are monitored to check their compliance with the LCLs and UCLs. Most commonly, univariate statistical process control (SPC) techniques, such as the Shewart chart, are utilized (see Chapter 28 in Ogunnaike and Ray, 1994). The data for the critical quality variables are analyzed and used to compute the DPMO. This enables the sigma level of the process to be assessed using Table 19.1 or Figure

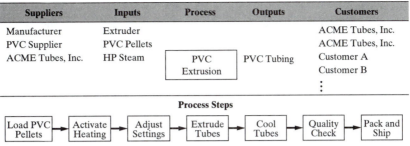

Figure 19.4 SIPOC for PVC tubing extrusion by ACME Tubes, Inc.

19.3. Continuing the PVC extrusion example, suppose this analysis indicates operation at 3σ, with a target to attain 5σ performance.

c. Analyze: When the sigma level is below its target, steps are taken to increase it, starting by defining the most significant causes for the excessive variability. This is assisted by a systematic analysis of the sequence of steps in the manufacturing process and the interactions between them. Using this analysis, the common root cause of the variance is identified. Continuing the PVC extrusion example, note that several factors contribute to an excessively high variance in product quality, among them, the variance in the purity of the PVC pellets, the variance in the fraction of volatiles in the pellets, and the variance in the operating temperature of the steam heater. Clearly, all of these factors interact, but suppose that after analysis it is determined that the variance in the operating temperature has the greatest impact on quality.

d. Improve: Having identified the common root cause of variance, it is eliminated or attenuated by redesign of the manufacturing process or by employing process control, or a combination of the two. Continuing the PVC tubing example, one possible solution would be to redesign the steam heater. Chapter 21 presents several examples in which process redesign can improve the controllability and resiliency of a process, and hence, reduce the variance in controlled output variables. Alternatively, a feedback controller could be installed, which manipulates the steam valve to enable tighter control of the operating temperature. In this way, the variance in the temperature is transferred to that of the mass flow rate of steam.

e. Control: After implementing steps to reduce the variance in the CTQ variable, this is evaluated and maintained. Thus, steps (b) to (e) in the DMAIC procedure are repeated to continuously improve process quality. Note that achieving 6σ performance is rarely the goal, and seldom achieved.

Six Sigma for Design

The DMAIC procedure is combined with ideas specific to product design to create a methodology that assists in applying the six-sigma approach to product design. Again, a five-step procedure is recommended

Step 1. Define project. In this step, the market opportunities are identified, a design team is assigned, and resources are allocated. Typically, the project timeline is summarized in a Gantt chart (see Section 12.4).

Step 2. Identify requirements. As in DMAIC, the requirements of the product are defined in terms of the needs of customers.

Step 3. Select concept. Innovative concepts for the new design are generated, first by brainstorming. These are evaluated, with the best selected for further development.

Step 4. Develop design. Often several teams work in parallel to develop and test competing designs, making modifications as necessary. The goal of this step is to prepare a detailed design, together with a plan for its management, manufacture, and quality assurance.

Step 5. Implement design. The detailed designs in Step 4 are critically tested. The most promising design is pilot-tested and if successful, proceeds to full-scale implementation.

The reader is referred to Rath and Strong (2002) for more details.

19.2 HEMODIALYSIS DEVICE

Hemodialysis is one of two types of artificial dialysis treatments (the other being peritoneal dialysis) that replaces the function of the kidneys, which is to regulate the composition of the bloodstream by removing waste products and excess fluids from the bloodstream while maintaining the proper chemical balance of the blood, which consists of red cells, white cells, platelets, and the accompanying plasma. The plasma, which accounts for about 55 vol% of the blood, consists of about 92% water with the balance being inorganic salts, organic chemicals, and dissolved gases. A typical male adult has a blood volume of 5 L with a resting cardiac output of 5 L/min and a systemic blood pressure of 120/80 mmHg. The most common application of hemodialysis is for patients with temporary or permanent kidney failure. It is the only treatment available for patients with end-stage renal disease (ESRD), whose kidneys are no longer capable of their function. This treatment, which is required three times per week for an average of 3–4 hr per dialysis, was performed on more than 200,000 patients in the United States in 1996.

A very common commercial device for hemodialysis is the C-DAK 4000 artificial kidney of Althin CD Medical, Inc. (acquired by Baxter International, Inc. in March, 2000). This disposable, sterilized membrane module, shown in Figure 19.5, resembles a shell-and-tube heat exchanger. The tubes, which number 10,000, are hollow fibers, 200 microns i.d. by 10 microns wall thickness by 22 cm long, made of hydrophilic microporous cellulose acetate of 15 to 100 Å pore diameter. Alternatively, fibers of polycarbonate, polysulfone, and other poly-

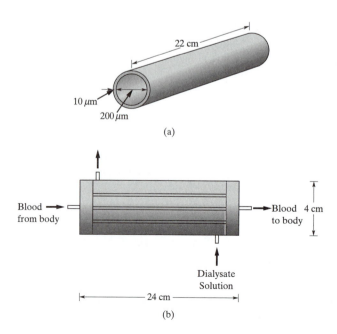

Figure 19.5 Hemodialysis device. (a) single tube; (b) complete module.

mers are used. The shell, made of acrylonitrile-butadiene-styrene (ABS) plastic with inlet and outlet side ports of polycarbonate, is 24 cm long by 4 cm in outside diameter with centered ports in the heads (caps) at either end of the shell for delivering blood flow into and out of the hemodialyzer. The fibers are potted at each end into polyurethane, sliced at each end to open the fibers, and sealed into the shell heads (caps) to prevent leakage between the tube side and the shell side. The total membrane area, based on the inside area of the hollow fibers, is 1.38 m^2. The packing density is 4,670 m^2 of membrane area per m^3 of membrane module volume. In 1992, 60 million of these units, which weigh less than 100 g each, were sold at 5 to 6 U.S. dollars each. Based on total membrane area used and dollar value, artificial kidneys are the single largest application of membranes.

A hemodialysis treatment with a C-DAK artificial kidney involves the insertion of two needles into the patient's vein, with attachment to plastic tubes to carry the patient's blood to and from the artificial kidney. The blood flows through the hollow fibers at a flow rate monitored by and controlled at 200 ml/min by a dialysis machine. A sweep dialysate solution of water, glucose, and salt passes countercurrently to the blood through the shell side of the artificial kidney at a rate of 500 ml/min. The pressure difference across the membrane from the tube side to the shell side is approximately equal to the diastolic pressure of the patient (e.g., 120 to 150 mmHg), since the dialysate is pulled through the module with suction. The pressure drop on the tube side is 30 mmHg, while that on the shell side is 12 mmHg. The pressure difference across the membrane causes excess fluids in the bloodstream to pass to the dialysate. Concentration differences between the blood and the dialysate cause urea (CH_4ON_2), uric acid ($C_5H_4O_3N_4$), creatinine ($C_4H_7ON_3$), phosphates, and other low-molecular-weight metabolites to transfer by diffusion from the blood to the dialysate; and glucose and salts to transfer by diffusion from the dialysate to the blood. When the kidneys of an adult function normally, the urea content of the blood is maintained in the range of 10–20 mg of urea nitrogen per 100 ml (1 deciliter, dl).

For the design of the C-DAK 4000 artificial kidney, and the many similar hemodialysis devices (Daugirdas and Ing, 1988), rates of permeation of the species through the candidate membranes are necessary. Estimates for the permeability of pure species in a microporous membrane can be made from the molecular diffusivity, and pore diameter, porosity, and tortuosity of the membrane (Seader and Henley, 1998), as shown in Example 19.1. For this reason, considerable laboratory experimentation is required when selecting membranes in the molecular structure design step.

Having selected the candidate polymer membranes, which must be available in hollow-fiber bundles, the design team must configure the dialysis device for selected flow rates of blood and dialysate. In practice, blood flow rates range from 100 to 400 ml/min, while dialysate flow rates range from 200 to 800 ml/min. Decisions regarding the fiber inside diameter, wall thickness, and length, and the number of fibers influence the pressure for the fluids on each side, the surface area for mass transfer, and the rates of mass transfer. The following example discusses the approach that might be taken to develop a preliminary design of a hemodialysis device of the hollow-fiber type shown in Figure 19.5.

EXAMPLE 19.1

Develop a design procedure for a hollow-fiber hemodialysis device of the type shown in Figure 19.5. Base the design on a blood flow rate of 200 ml/min and a dialysate flow rate of 500 ml/min. Assume the design will be controlled by mass transfer of one of the blood plasma components to be removed, for example, urea. The blood will flow through the hollow fibers, while the dialysate will flow past the outside surface of the fibers in a direction countercurrent to the flow of the blood plasma. A typical patient will require hemodialysis when the blood reaches a urea nitrogen level (BUN) of 100 mg/dl. A target for the hemodialysis device is to reduce the BUN to 30 mg/dl within 4 hr.

SOLUTION

Several key steps in the design procedure are presented next, beginning with the estimation of the mass-transfer coefficient for transport of urea across the membrane. Next, pressure drops are estimated both in and outside of the hollow fibers. Then, a mass-transfer model is solved for the concentration of urea in the bloodstream as a function of time, which assists in sizing the dialysis device.

1. Overall mass-transfer coefficient for urea:

The rate of mass transfer of urea from the blood plasma, through the membrane, and to the dialysate is given by

$$n = K_i A_i \Delta C_{LM} \tag{19.5}$$

where: n = rate of mass transfer of urea nitrogen, mg/min

K_i = overall mass-transfer coefficient based on the inside area, cm/min

A_i = mass-transfer area based on the inside area of the hollow fibers, cm^2

ΔC_{LM} = log-mean urea-nitrogen concentration difference for mass transfer

The overall mass-transfer coefficient, which must consider the resistances of the blood plasma, the membrane, and the dialysate, is given by

$$\frac{1}{K_i} = \frac{1}{k_b} + \frac{l_M}{P_M}\frac{A_i}{A_m} + \frac{1}{k_d}\frac{A_i}{A_o} = \frac{1}{k_b} + \frac{l_M}{P_M}\frac{D_i}{D_m} + \frac{1}{k_d}\frac{D_i}{D_o} \tag{19.6}$$

where: k_b = mass-transfer coefficient on the inside where the blood flows, cm/min

k_d = mass-transfer coefficient on the outside where the dialysate flows, cm/min

A_o = mass transfer area based on the outside area of the hollow fibers, cm^2

A_m = arithmetic mean of A_i and A_o, cm^2

l_M = membrane wall thickness, cm

P_M = membrane permeability, cm^2/min

The mass-transfer coefficients, k_b and k_d, can be estimated by analogy from available dimensionless empirical correlations for heat transfer, which are taken from Knudsen and Katz (1958). These analogous correlations depend on the flow regime and relate the Sherwood number to the Reynolds and Schmidt numbers.

For the flow of blood plasma inside the hollow fibers, the flow regime will be laminar because of the very small fiber diameter and the need to avoid high flow velocities that would stress the blood cells to destruction. For example, for the C-DAK artificial kidney described above, 200 ml/min (200 cm^3/min), Q_b, of blood flows through 10,000 fibers, each of 200 microns (0.02 cm) inside diameter, D_i, 20 microns wall thickness, t_w, rather than the 10 micron thickness shown in Figure 19.5a, and 22 cm in length, L. The cross-sectional area for flow in each fiber is $3.14(0.02)^2/4 = 0.000314$ cm^2. The total flow area, S_i, for 10,000 fibers is $10,000(0.000314) = 3.14$ cm^2. The average blood velocity, V_b, is $(200)/3.14 = 63.7$ cm/min = 1.062 cm/s. Blood has a density, ρ_b, of 1.06 g/cm^3 and a viscosity, μ_b, of approximately 0.014 g/cm-s. This gives a Reynolds number for blood flow through the fibers, $N_{Re_b} = D_i V_b \rho_b / \mu_b = 0.02(1.062)(1.06)/0.014 = 1.608$, which is in the laminar flow regime. Because the fully developed parabolic velocity profile for laminar flow is obtained by a fiber length, $L = 0.05 D_i N_{Re_b}$, the velocity profile will be developed in less than one fiber diameter. Thus, for mass transfer of urea through the blood plasma, the important criterion is the Peclet number for mass transfer, N_{Pe_M}, which is the product of the Reynolds number and the Schmidt number, $N_{Sc_b} = \mu_b/\rho_b D_{urea}$ or

$$N_{Pe_M} = \frac{D_i V_b}{D_{urea}} \tag{19.7}$$

where D_{urea} = molecular diffusivity of urea in blood plasma, cm^2/s. The diffusivity for urea in blood plasma at a body temperature of 37°C is 0.8×10^{-5} cm^2/s. Thus, the Peclet number = $0.02(1.062)/(0.8 \times 10^{-5}) = 2,650$. The Sherwood number depends on the ratio of N_{Pe_M} to (L/D_i), which equals $2,650/(22/0.02) = 2.41$. For this condition, the Sherwood number, $N_{Sh} = D_i k_b / D_{urea}$, is approximately constant at a value of 4.364. Thus, $k_b = 4.364(0.8 \times 10^{-5})/0.02 = 0.00175$ cm/s.

Estimation of the mass-transfer coefficient in the dialysate outside the fibers is considerably more difficult because of the complex geometry. In a shell-and-tube heat exchanger, baffles are used to help direct the shell-side fluid to flow back and forth in directions largely normal to the tube length. It would be extremely difficult to include such baffles in a hollow-fiber hemodialysis unit. Instead, as shown in Figure 19.5, the dialysate leaves the inlet port at one end to enter the shell normal to the fibers, turns 90° and flows parallel to the fibers, and then turns 90° to enter the exit port at the other end. Presumably, the dialysate flow is largely parallel to the fibers along their length. For the example being considered here, assume the inside diameter of the shell is 3.8 cm, giving a cross-sectional area of $3.14(3.8)^2/4 = 11.34$ cm^2. The 10,000 fibers, with an outside diameter of $200 + 40 = 240$ microns or 0.024 cm, occupy a cross-sectional area of $10,000(3.14)(0.024)^2/4 = 4.52$ cm^2. Thus, the area for flow of the dialysate is $11.34 - 4.52 = 6.82$ cm^2. As an approximation, estimate the shell-side mass-transfer coefficient, k_d, from a circular tube inside-flow correlation, as for k_b, by replacing tube inside diameter by an equivalent diameter equal to $4r_H$, where r_H is the hydraulic radius, which is equal to the cross-sectional area for flow divided by the wetted perimeter. For 10,000 fibers of 0.024 cm diameter on a square pitch, the wetted perimeter is $3.14(0.024)(10,000/4) = 188.4$ cm. This gives a hydraulic radius of $6.82/188.4 = 0.0362$ cm. The equivalent diameter $= 4(0.0362) = 0.145$ cm. Assume dialysate properties at 37°C of $\rho_d = 1.05$ g/cm^3 and $\mu_d = 0.007$ g/cm-s. Take a dialysate flow rate of 500 cm^3/min. The average velocity of the dialysate, V_d, is $(500/60)/6.82 = 1.22$ cm/s. The dialysate Reynolds number, $N_{Re_d} = 4r_H V_d \rho_d/\mu_d$, is $0.145(1.22)(1.05)/0.007 = 26.5$, which is much higher than for the tube side, but is still in the laminar-flow region. The estimated entry length, using the shell-side version of the tube-side equation given above, is $0.05(0.145)(26.5) = 0.192$ cm, making it possible to again assume fully developed flow. The molecular diffusivity of urea through the dialysate solution is 1.8×10^{-5} cm^2/s. The Peclet number, using a shell-side version of the equation given above, is $N_{Pe_M} = (0.145)(1.22)/(1.8 \times 10^{-5}) = 9,830$. The ratio of N_{Pe_M} to $(L/4r_H)$, which equals $9,830/(22/0.145) = 64.8$. For this condition, the Sherwood number, $N_{Sh} = 4r_H k_d/D_{urea}$, is approximately 7. Thus, $k_d = 7(1.8 \times 10^{-5})/0.145 = 0.00087$ cm/s.

For a microporous membrane, the membrane permeability is the effective diffusivity for urea. For the microporous membrane of 20 microns (0.002 cm) wall thickness, assume a porosity, ε, of 0.25 and a tortuosity, τ, of 1.5. The effective diffusivity or permeability in the membrane is given by $(D_{urea})_{eff} = P_M = \varepsilon(D_{urea})_d/\tau = 0.25(1.8 \times 10^{-5})/1.5 = 3 \times 10^{-6}$ cm^2/s. The permeance, $P_M/l_M = 3 \times 10^{-6}/0.002 = 0.0015$ cm/s.

The overall mass-transfer coefficient based on the inside area of the hollow fibers is obtained from Eq. (19.6):

$$\frac{1}{K_i} = \frac{1}{0.00175} + \frac{1}{0.0015}\left(\frac{0.020}{0.022}\right) + \frac{1}{0.00087}\left(\frac{0.020}{0.024}\right) = 571 + 606 + 958 = 2,135 \text{ s/cm}$$

This result shows that 27% of the mass-transfer resistance is on the blood side, 28% is in the membrane, and 45% is on the dialysate side. These percentages change as the fiber geometry is changed. Taking the reciprocal, the overall mass-transfer coefficient is $K_i = 0.000468$ cm/s.

2. Pressure drop for blood flow through the hollow fibers:

Because the blood flow through the hollow fibers of the hemodialyzer is laminar, the pressure drop is computed from the Hagen–Poiseuille equation:

$$-\Delta P_b = \frac{32\mu_b L V_b}{D_i^2} = \frac{32(0.014)(22)(1.062)}{0.02^2} = 26,170 \text{ g/cm-s}^2 \text{ or } 2.617 \text{ kPa or } 19.6 \text{ mmHg}$$

This pressure drop compares well with the pressure drop of 30 mmHg cited above for the commercial C-DAK 4000 artificial kidney.

3. Pressure drop for dialysate flow past the hollow fibers:

The flow on the shell side is also laminar. Using the hydraulic-radius concept, an estimate can be made of the pressure drop by replacing D_i in the above Hagen–Poiseuille equation with $4r_H$. Thus, the pressure drop is

$$-\Delta P_d = \frac{32\mu_d L V_d}{(4r_H)^2} = \frac{32(0.007)(22)(1.22)}{(0.145)^2} = 286 \text{ g/cm-s}^2 \text{ or } 0.0286 \text{ kPa or } 0.2 \text{ mmHg}$$

This is far less than the 12 mmHg quoted above for the C-DAK 4000. This difference could be due to the entering and exiting flow of dialysate normal to the fibers, which would increase the pressure drop considerably. A better estimate could be made with a computational fluid dynamics (CFD) program.

This pressure-drop calculation also sheds some doubt on the above calculation of the mass-transfer co-efficient on the shell side, which could also be greater than that calculated by assuming flow parallel to the length of the fibers.

4. Urea mass transfer and time required to reduce its concentration in the blood:

A compartment model, shown in Figure 19.6 has been successful in following changes in solute concentrations with time for a patient undergoing hemodialysis. The model consists of three perfectly mixed compartments and one membrane separator. Streams that are assumed to have negligible volume connect the compartments and separator. The upper compartment, of volume, V_P, represents a patient's body fluid, other than blood. A solute, such as urea, is transferred to the body fluid at the constant mass rate, G. The second compartment of volume, V_B, represents the patient's blood, which circulates between these two compartments at a volumetric rate, Q_P. Below the second compartment is the hemodialysis unit, which transfers solutes, such as urea, across hollow-fiber membranes to a dialysate. Blood circulates at a volumetric rate, Q_B, between the hemodialyzer and the second compartment. Dialysate circulates at a volumetric rate, Q_D, between the bottom compartment (dialysate holding tank) and the hemodialyzer, through which it flows countercurrently to the blood flow. From the dialysate holding tank, a constant volumetric flow rate of waste dialysate, Q_W, is withdrawn. An equal volumetric flow rate of fresh dialysate of zero waste solute concentrations, for example, urea, is added to the circulating dialysate before it enters the hemodialyzer. Also indicated in Figure 19.6 are symbols, C_j, for solute (for example, urea) concentrations in the various streams in units of mass/unit volume. Because of the assumption of perfectly mixed compartments, concentrations of solutes in the three compartments are equal to the solute concentrations in the streams leaving the corresponding compartments.

Model equations for a system similar to that of Figure 19.6 were developed and solved by Spaeth (1970), whose equations are applied here. Because solute concentrations change with time, the following solute mass balance equations apply to the three compartments:

$$V_P \frac{dC_P}{dt} = G - Q_P(C_P - C_B) \tag{19.8}$$

$$V_B \frac{dC_B}{dt} = Q_P(C_P - C_B) - Q_B(C_B - C_{B,\text{out}}) \tag{19.9}$$

$$V_D \frac{dC_W}{dt} = Q_D(C_{D,\text{out}} - C_W) \tag{19.10}$$

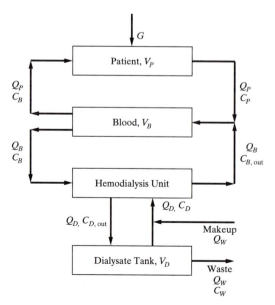

Figure 19.6 Hemodialysis model.

Equations for the rate of mass transfer of a solute across the walls of the hollow fibers in the membrane unit and a solute mass balance are as follows, where countercurrent flow is assumed in the hemodialyzer, with a corresponding log-mean concentration driving force:

$$Q_B(C_B - C_{B,\text{out}}) = K_i A_i \Delta C_{\text{LM}} = K_i A_i \left[\frac{(C_B - C_{D,\text{out}}) - (C_{B,\text{out}} - C_D)}{\ln \dfrac{(C_B - C_{D,\text{out}})}{(C_{B,\text{out}} - C_D)}} \right] \tag{19.11}$$

$$Q_B(C_B - C_{B,\text{out}}) = Q_D(C_{D,\text{out}} - C_D) \tag{19.12}$$

Finally, a solute mass balance around the mixing point, assuming that the makeup dialysate contains no solute, gives

$$Q_D C_D = (Q_D - Q_W)C_W \tag{19.13}$$

Equations (19.8) to (19.13) constitute six equations in the six variables, C_P, C_B, C_D, C_W, $C_{B,\text{out}}$, and $C_{D,\text{out}}$, all of which vary with time. The six equations can be reduced to the following three ordinary differential equations in the three variables, C_P, C_B, and C_D:

$$\frac{dC_P}{dt} = \frac{G}{V_P} - \left(\frac{Q_P}{V_P}\right)(C_P - C_B) \tag{19.14}$$

$$\frac{dC_B}{dt} = \left(\frac{Q_P}{V_B}\right)(C_P - C_B) - \left(\frac{Q_B}{V_B}\right)E(C_B - C_D) \tag{19.15}$$

$$\frac{dC_D}{dt} = \left[E\left(\frac{Q_B}{V_D}\right)\left(1 - \frac{Q_W}{Q_D}\right) \right]C_B - \left[\left(\frac{Q_W}{V_D}\right) + E\left(\frac{Q_B}{V_D}\right)\left(1 - \frac{Q_W}{Q_D}\right) \right]C_D \tag{19.16}$$

where the parameter, E, is defined by

$$E \equiv \frac{C_B - C_{B,\text{out}}}{C_B - C_D} \tag{19.17}$$

and is computed from

$$E = \frac{1 - \exp\left[\dfrac{K_i A_i}{Q_B}\left(1 - \dfrac{Q_B}{Q_D}\right)\right]}{\dfrac{Q_B}{Q_D} - \exp\left[\dfrac{K_i A_i}{Q_B}\left(1 - \dfrac{Q_B}{Q_D}\right)\right]} \tag{19.18}$$

which is derived from Eqs. (19.11) and (19.12).

Equations similar to Eqs. (19.14) to (19.16), but for a two-compartment model, are solved analytically by Bird et al. (2002). However, a numerical solution, suitable for a spreadsheet is used here, beginning with conditions at time, $t = 0$, for the three solute concentrations. Once, C_P, C_B, and C_D are obtained as functions of time, the other three concentrations, C_W, $C_{B,\text{out}}$, and $C_{D,\text{out}}$ are computed as functions of time from Eqs. (19.13), (19.17), and (19.12), respectively.

As an example of the application of the above equations to urea, the following values are used:

$$V_P = 40{,}000 \text{ cm}^3, \ V_B = 5{,}000 \text{ cm}^3, \text{ and } V_D = 1{,}000 \text{ cm}^3$$
$$Q_P = 5{,}000 \text{ cm}^3/\text{min}, \ Q_B = 200 \text{ cm}^3/\text{min}, \ Q_D = 500 \text{ cm}^3/\text{min}, \text{ and } Q_W = 250 \text{ cm}^3/\text{min}$$
$$G = 5 \text{ mg/min}, \ K_i = 0.000468 \text{ cm/s (from above)} = 0.0281 \text{ cm/min},$$
$$\text{and } A_i = 13{,}800 \text{ cm}^2$$

Initial conditions for the three concentrations are

$$C_P = 1 \text{ mg/cm}^3, \ C_B = 1 \text{ mg/cm}^3, \text{ and } C_D = 0 \text{ mg/cm}^3$$

The numerical solution gives the following results over a 6-hr period, where all three concentrations are given in mg urea/cm^3.

Time (hr)	C_P	C_B	C_D
0	1.00	1.00	0.00
1	0.86	0.84	0.20
2	0.74	0.73	0.18
3	0.64	0.63	0.15
4	0.55	0.54	0.13
5	0.48	0.47	0.11
6	0.42	0.41	0.10

Thus, in 4 hr, which is a typical treatment time, the urea concentrate has been reduced by 45%. It is left as an exercise at the end of this chapter for a study of the effect on the rate of urea removal of changing the hemodialyzer geometry, the blood and dialysate flow rates to the dialyzer, the rate of waste withdrawal, the volume of the dialysate tank, and the sensitivity of the rate of urea mass transfer to the mass-transfer coefficient. In particular, the above estimate of the coefficient on the shell side may be low because the entry to and exit from the hemodialyzer of the dialysate is normal to, rather than parallel to the fibers. This should enhance the shell-side coefficient. ∎

Example 19.1 concentrates on the three-dimensional design and performance of a hollow-fiber dialysis device. It is important to recognize that this detailed design work is undertaken midway through the design process, after the idea for this product has received positive responses from potential customers and initial technical feasibility, market, and business studies have been promising, as discussed in Sections 1.2 and 19.1. As the design calculations proceed, a prototype unit would be built, and subjected to consumer testing before a decision is made to enter commercial production; that is, to construct a full-scale manufacturing facility for the product.

19.3 SOLAR DESALINATION UNIT

Throughout the world, there are many arid and semiarid locations where annual rainfall is less than 10 in./yr and 25 in./yr, respectively. Since the mid-nineteenth century, designers have developed solar desalination processes capable of producing potable water, which is defined as containing less than 500 ppm of dissolved salt. This section is based upon a product-design problem developed by Dally and Regan (1997) in which a small desalination unit is to be designed, capable of producing on the order of several ounces per hour of potable water from a solution containing 2 lb/ft^3 of table salt (NaCl). A basin design is suggested that uses batch distillation, as illustrated in Figure 19.7. The sun's rays pass through a transparent glass cover and impinge on a water pool in a basin, which is insulated from the ground (or supporting surface) and colored black (to absorb best the sun's radiation). The glass cover is a radiation shield, which serves to reduce heat losses due to reflections from the water surface. As the basin absorbs energy, its temperature increases, causing the rate of evaporation to increase. Evaporated water is condensed on the cool cover, which must be positioned at a sufficiently large angle with the ground to permit water droplets to run down into the troughs

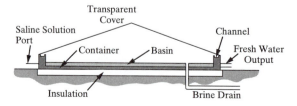

Figure 19.7 Schematic of a basin solar desalination unit. Reprinted with permission from Dally and Regan (1997).

that surround the cylindrical basin. Dally and Regan (1997) list many large commercial installations worldwide with outputs ranging from 140 to 6,900 gal/day, the largest being in Patmos, Greece, with a basin size of 93,000 ft³.

The effectiveness of the solar desalination unit depends on the amount of solar radiation available, which in turn depends on its location on the earth's surface and the time of year. The following example shows how to estimate the heat fluxes and the rate of water production.

EXAMPLE 19.2

Using the specifications of Dally and Regan (1997), design a basin solar desalination unit that produces at least one ounce per hour of potable water at a location close to the Equator where the total diurnal radiant flux is 2,200 Btu/(ft²-day) year-round, with a maximum flux of 260 Btu/(ft²-hr) at 11:00 a.m. The entire unit should be housed in an empty box of the size used to ship 10 reams of 8.5-in. by 11-in. paper for copy machines (17 in. × 22 in. × 10 in.). The possibility of using concentrating mirrors should be considered.

SOLUTION

In carrying out the design, a key question concerns the rate of evaporation of water. Since horizontal collectors are normally used, it is recommended that such a design be considered before attempting the installation of a tracking mechanism to increase the fraction of the radiant flux that impinges on the collector, or lenses or reflecting mirrors to increase the heat flux that impinges on the collector. Tracking mechanisms are designed to manipulate the orientation of the collector to more closely approach orthogonality to the sun's rays, thereby increasing the fraction of the heat flux that impinges on the collector. Since these are normally expensive to install and require substantial energy to operate, they are normally not used. Instead, collectors are often fixed in position, facing south, tilted to an angle of 20° plus the latitude of the collector in degrees. Well-designed lenses can increase the heat flux by a factor of 400, but these are normally too expensive to install, especially for the larger desalination units. The Fresnel lens, which has a poorly defined focal point but is relatively inexpensive to manufacture, is effective, especially for a collecting area less than 2–3 ft² (Dally and Regan, 1997). The alternative of flat booster mirrors is often considered, as these can double the heat flux that impinges on the collector. Finally, in recent years, it has been possible to coat metal films on plastic sheets (e.g., aluminized Mylar). These sheets are configured into conical mirrors that can increase the heat flux by a factor of 15. In summary, a design will be prepared initially assuming a horizontal configuration. If the desalination unit is sufficiently small, the impact of the impinging flux on the size and cost will be examined. If the desalination unit is too large, methods of concentrating the heat flux will be considered.

Potential Design
Consider the potential design in Figure 19.8, which shows the key heat fluxes that determine the mass flux of water produced; that is, the rate of water production per unit area, lb/(hr-ft²).

Energy Balances
Energy balances are created to solve for the mass flux of water produced. Then, given the need to produce 1 oz/hr = 0.0625 lb/hr, the collector area can be estimated.

Radiation Balance
As light impinges on a surface, the fraction absorbed, α, plus the fraction transmitted, τ, plus the fraction reflected, r, sum to unity; that is

$$\alpha + \tau + r = 1 \tag{19.19}$$

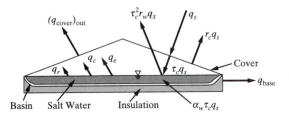

Figure 19.8 Potential desalination unit design showing principal heat fluxes. Reprinted with permission from Dally and Regan (1997).

Energy Balance on the Desalination Unit

In this balance, the heat accumulated equals the difference between the heat in and the heat out:

$$c\frac{dT_w}{dt} = q_{in} - q_{out} \tag{19.20}$$

where $c = 16$ Btu/(ft^2-°F) is the heat capacity flux, assuming a shallow basin, T_w is the water temperature, and the heat fluxes are in Btu/(ft^2-hr). The heat flux in is

$$q_{in} = (\alpha_c + \alpha_w\tau_c)q_s \tag{19.21}$$

where q_s is the radiation flux impinging on the desalination unit, α_w is the fraction absorbed by the water, and α_c and τ_c are the fractions absorbed and transmitted by the glass cover. The first term on the right involves the fraction absorbed by the glass cover, and the second term involves the fraction transmitted through the cover that is absorbed by the water pool. The heat flux out is

$$q_{out} = (\tau_c^2 r_w q_s + K_b(T_w - T_g) + q_{cover,out} \tag{19.22}$$

The first term on the right involves the heat flux reflected from the water and retransmitted through the glass cover. Note that the transmission fraction is squared because the light is transmitted twice through the cover. The second term involves the rate of conduction to the ground, where $K_b = 1$ Btu/(hr-ft^2-°F) is the conductivity flux, assuming the basin is insulated from the ground, and T_g is the ground temperature. The last term involves the heat flux that leaves the glass cover:

$$q_{cover,out} = 0.9\sigma[(T_c + 460)^4 - (T_a + 460)^4] + h_a(T_c - T_a) \tag{19.23}$$

where the first term on the right is the radiation heat flux from the glass cover to the surrounding air, T_a is the air temperature in °F, $\sigma = 0.174 \times 10^{-8}$ Btu/(hr-ft^2-°R^4) is the Stefan–Boltzmann constant, and the emissivity of a gray body is 0.9. The second term represents the rate of convective heat transfer from the glass cover to the surrounding air, with the heat-transfer coefficient, h_a, a function of the wind velocity, v_{wind} [2.6, 4.1, and 7.2 Btu/(hr-ft^2-°F) at 5, 10, and 20 mph].

Energy Balance on the Glass Cover

In this balance, the glass cover is assumed to have negligible heat capacity, and hence:

$$q_{cover,out} = q_{cover,in} \tag{19.24}$$

The heat flux into the glass cover is

$$q_{cover,in} = q_r + q_c + q_e + \alpha_c q_s \tag{19.25}$$

where the heat flux by radiation from the water pool is

$$q_r = 0.9\sigma[(T_w + 460)^4 - (T_c + 460)^4] \tag{19.26}$$

and the heat flux by convection from the water pool is

$$q_c = 0.128\left[(T_w - T_c) + \frac{P_w^s - P_c^s}{39 - P_w^s}(T_w + 460)\right]^{1/3}(T_w - T_c) \tag{19.27}$$

where P_w^s and P_c^s are the vapor pressures of water at the water pool and the glass cover, respectively, in psia. The heat flux due to condensation of water vapor at the cover surface is

$$q_e = 0.0254\left[(T_w - T_c) + \frac{P_w^s - P_c^s}{39 - P_w^s}(T_w + 460)\right]^{1/3}(P_w^s - P_c^s)\Delta H^v \tag{19.28}$$

where ΔH^v is the latent heat of vaporization of water, which can be taken as 1,025 Btu/lb (corresponding to 121°F).

Solve Equations

To determine the rate of evaporation,

$$\frac{dm_w}{dt} = q_e/\Delta H^v \tag{19.29}$$

it is first necessary to solve Eq. (19.24) for T_c. Then, Eq. (19.20) is solved for the rate of change of T_w, and using a finite-difference approximation, ΔT_w is estimated over the time interval, Δt. Clearly, these steps require that the initial condition be specified, T_w^0.

Specifications

The equations were solved for the following specifications:

$$T_a = 80°F$$
$$T_g = 70°F$$
$$\alpha_c = 0.1$$
$$\tau_c = 0.8$$
$$r_c = 0.1$$
$$\alpha_w = 0.9$$
$$r_w = 0.1$$
$$T_w^0 = 110°F$$
$$v_{wind} = 5 \text{ mph}$$

Results

The temperature of the glass cover, $T_c = 98°F$, the heat flux into or out of the glass cover, $q_{cover} = 73$ Btu/(hr-ft^2), the change in the water temperature from 11:00 A.M. to 12:00 noon, $\Delta T_w = 6.3°F$, and most importantly, the rate of evaporation, $dm_w/dt = 0.0273$ lb/(hr-ft^2). Hence, during this hour, the area required is

$$\frac{0.0625 \text{ lb/hr}}{0.0273 \text{ lb/(hr-ft}^2)} = 2.29 \text{ ft}^2$$

A basin with this area would nearly fit into the box provided. However, it is important to recognize that 0.0273 lb/(hr-ft^2) is the largest flux experienced during a day. On average, the flux could be approximately one-third of this value. Hence, to achieve the specified daily evaporation rate, either more area is required (not possible in this case) or a lens or mirror system is necessary. Such a design is considered in Exercise 19.6.

Costs for this laboratory-scale unit would be estimated initially for the construction of a prototype, probably on the order of $300. Then *economies-of-scale* sharply reduce the unit cost in mass production, perhaps by a factor of five.   ∎

As the design of this solar desalinator proceeds, before the engineering calculations in Example 19.2 are undertaken, it is important to have positive responses from potential customers and to have completed promising marketing and business studies. In some arid countries, bottled water is available at sufficiently low prices to discourage consumers from investing in a solar desalinator, with the need to collect salt water and regularly load the desalinator. Where bottled water is not available, or too expensive for regular use, consumers are likely to respond favorably to a low-cost product.

19.4   HAND WARMER

For therapeutic, health, and fitness purposes, until recently, hot-water containers to heat body parts and ice bags to cool body parts dominated the market. Now, a number of new products have been developed, principally thanks to advances in the ease of manufacture of leak-proof polymer containers. These include two hand warmers carried in the palm: (1) the GRABBER MYCOAL Hand Warmer made in Japan and distributed by GRABBER Warmers of Grand Rapids, Michigan, and (2) the Zap Pak Heat Pack made and distributed by Prism Enterprises in San Antonio, Texas. The former produces heat by a chemical reaction, but can only be used once. The latter produces heat by crystallization and can be regenerated.

GRABBER MYCOAL Hand Warmer

Patent citations on this product date back to 1924, but the commercial product was not introduced until 1978, when it was initially marketed in Japan. Import to the United States began in 1980 with wide distribution in recent years. The ingredients of the product are moisture and fine particles of iron, cellulose, vermiculite, activated carbon, and salt for a total weight of approximately 16.8 g. These are contained in a 1.05-g inner 3-in. × 2-in. polypropylene pouch, which is contained in an outer sealed plastic package. When the outer package is opened and the pouch removed, a slow exothermic chemical reaction occurs, wherein oxygen from the surrounding air permeates through the polypropylene pouch and oxidizes the iron to iron oxide (rust) by the reaction:

$$4Fe_{(s)} + 3O_{2(g)} \rightarrow 2Fe_2O_{3(s)}$$

which has a standard enthalpy of reaction of -824.2 kJ/mol of the oxide.

The salt acts as a catalyst, the carbon helps disperse the heat of reaction, the vermiculite acts as an insulator to slow heat loss and control the temperature, and the cellulose is added as a filler. The heat production lasts up to 7 hr, during which the temperature of the pouch ranges from 104 to 156°F, with an average temperature of 135°F. Because the reaction is irreversible, this hand warmer is not reusable.

Zap Pak Heat Pack

This product consists of a 4.5-in. × 3-in. flexible plastic pouch (a convenient size for holding with one hand or between two hands), containing approximately 50 g of a salt dissolved in 50 g of water. Also submerged within the liquid in the pouch is a thin metal disk of 0.625-in. diameter. Possible pouch materials are rubber, polyvinyl alcohol, polyethylene, or a vinyl-coated fabric, with clear vinyl of 0.13-mm thickness favored. Possible metal disk activators include stainless steel, a beryllium–copper alloy, and a phosphor–bronze alloy, with stainless steel favored. When the disk is flexed between the thumb and the fingers, the salt begins to crystallize, with a heat release sufficient to produce, within 5 s, a temperature of about 130°F. Note that this flexing probably causes very small particles, which act as nuclei for crystallization, to be dislodged. Heat is released to the hands holding the heat pack over a period of up to 30 min. This hand warmer can be recharged in a microwave oven or by placing it on a cloth in boiling water for 10 min or until the salt dissolves. In this manner, the hand warmer can be used over and over again. U.S. patents on the concept of using the heat of crystallization for a hand warmer date back to at least 1921. Many of the necessary features of the current product are discussed in a 1940 patent by Donald F. Othmer (U.S. Patent 2,220,777, dated November 5, 1940). However, a significant problem that had to be solved to make this hand warmer reliable was the activator for initiating the crystallization process. A significant advance was reported in U.S. Patent 4,077,390 of March 7, 1978. The Heat Pak cites protection under U.S. patent 4,872,442, dated October 10, 1989.

Consider the problem of designing a hand warmer based on the crystallization concept. During the idea-generation stage, a design team must have set objectives to locate a salt that is soluble in water and will crystallize, giving off its heat of crystallization at a comfortable temperature. Most importantly, it must be possible to significantly supercool the solution such that no crystals are present at room temperature prior to the activation of crystallization, even though the concentration of the salt is significantly higher than its equilibrium solubility at room temperature. It must also be possible, after the hand warmer is used, to redissolve the salt at a temperature that will not destroy the pouch. As evidenced by the large number of patents, a reliable method of activating the crystallization process must be found.

To begin the product design process, the team would probably seek to identify a salt with the following properties: (1) capable of remaining dissolved in a supersaturated solution at temperatures far below its equilibrium crystallization temperature, (2) capable of crystallization at a moderate temperature, such as 125°F, (3) having a large latent heat of crystallization, and (4) capable of being regenerated, that is, easily dissolved for reuse. Techniques for carrying out such a search, using experimental data, are illustrated in Example 19.3.

After selecting a salt, the problem of reliably activating the crystallization process must be solved. Then, a polymeric container would be designed, which is sufficiently malleable to adjust to the contours of two cupped hands and capable of withstanding agitation without leaking when transported. Prototype products would be created in the laboratory in sufficient quantity for market testing before the design of a manufacturing facility would be undertaken. Also, a patent search would be necessary to confirm the absence of patent infringements. Patents should be sought to protect any novel aspects of the product.

EXAMPLE 19.3

For a hand warmer, select a suitable salt and estimate by an energy balance a salt and water mixture that could achieve a temperature rise of 20°C under adiabatic conditions. An article by de Nevers (1991) discusses the hand warmer and the energy balance.

SOLUTION

The U.S. patent literature, which is readily accessible on the Internet at www.uspto.gov, suggests several possible salts, including sodium acetate, sodium thiosulfate, calcium nitrate, and lead acetate. These are all hydrated at ambient temperatures with high solubility and relatively high heats of crystallization, as shown by the following data taken from Mullin (1993):

Salt	Heat of Crystallization at ~20°C (kcal/mol)	Equilibrium Solubility at 20°C (g Anhyd./100 g Water)	Equilibrium Solubility at 40°C (g Anhyd./100 g Water)
$NaC_2H_3O_2 \cdot 3H_2O$	−4.7	46.5	65.5
$Na_2S_2O_3 \cdot 5H_2O$	−11.4	70	103
$Ca(NO_3)_2 \cdot 4H_2O$	−8.0	129	196
$Pb(C_2H_3O_2)_2 \cdot 3H_2O$	−5.9	44.1	116

Of great importance for a hand warmer is the degree to which solutions of these salts can exhibit supersaturation. A generic solubility–supersolubility diagram is shown in Figure 19.9, taken from Mullin (1993), where salt concentration in the overall mixture (liquid plus solid if any) is plotted against temperature. Three regions are shown: (1) a *stable region* lying below the equilibrium solubility curve, where the salt is soluble and crystallization is impossible; (2) a *metastable region* lying between the dashed supersolubility curve and the equilibrium solubility curve, where spontaneous crystallization is improbable, but crystallization can be initiated on a crystal seed; and (3) a *labile region*, above the supersolubility curve, where spontaneous crystallization is probable but not inevitable. The dashed supersolubility curve is not well defined. The supersaturation ratio, S, is defined as the ratio of the achievable supersaturation solubility to the equilibrium solubility at a given temperature, with values ranging from 1 to 2. No data for S are readily available for the above four salts, although many investigators have reported that supersaturated solutions of sodium acetate are readily achieved. Assume that all four salts

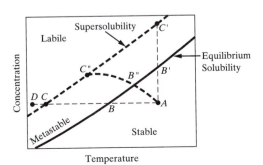

Figure 19.9 Generic solubility–supersolubility diagram. Reprinted with permission from Mullin (1993).

can achieve an S of 2. Determine, for sodium acetate, the feasibility of reaching a temperature of 40°C starting from a temperature of 20°C. Similar calculations for the other three salts for 40°C and, for all four salts, 50°C, and a salt selection, is left as an exercise at the end of this chapter.

Assume that upon the completion of crystallization at 40°C, we have a solution containing 100 g of water (MW = 18) and the equilibrium solubility of sodium acetate (MW = 82), or 65.5 g of sodium acetate from the above table. Let the solid crystals at 40°C be x g of solid crystalline trihydrate of sodium acetate (MW = 136). For a thermodynamic path, take crystallization at 20°C, but with the 40°C solubility, followed by heating the crystalline solids and the solution to 40°C. Therefore, specific heats are needed for the solid and the solution. In the absence of data for the trihydrate of sodium acetate, Kopp's rule (Felder and Rousseau, 2000) is applied with atomic heat capacities in cal/mol-°C of: 1.8 for C, 2.3 for H, 4.0 for O, 9.8 for water of crystallization, and 6.2 for all other atoms in the four salts. Thus, for $NaC_2H_3O_2 \cdot 3H_2O$,

$$C_p = 6.2 + 2(1.8) + 3(2.3) + 2(4.0) + 3(9.8) = 54.1 \text{ cal/mol-°C or 0.398 cal/g-°C}$$

In the absence of data for sodium acetate dissolved in water, Wenner's values for atomic heat capacities of dissolved solids, in cal/mol-°C, can be applied, with: 2.8 for C, 4.3 for H, 6.0 for O, and 8 for all other atoms in the four salts. Thus, for $NaC_2H_3O_2$,

$$C_p = 8 + 2(2.8) + 3(4.3) + 2(6.0) = 38.5 \text{ cal/mol-°C or 0.470 cal/g-°C}$$

For liquid water, $C_p = 1.0$ cal/g-°C. From the above table, the heat of crystallization of the trihydrate of sodium acetate is −4,700 cal/mol or −34.6 cal/g. For the 20°C temperature rise, the adiabatic energy balance is given by

$$34.6\,x = [0.398x + 1.0(100) + 0.470(65.5)](40 - 20)$$

Solving, $x = 98.2$ g of the solid trihydrate, containing $(82/136)(98.2) = 59.2$ g of sodium acetate and 39.0 g of water of crystallization. Thus, the original mixture before crystallization at supersaturation contains $(100 + 39) = 139$ g of water and $(65.5 + 59.2) = 124.7$ g of sodium acetate. Now check the supersaturation ratio, S, using an equilibrium solubility from the above table of 46.5 g/100 g water:

$$S = \frac{124.7/139}{46.5/100} = 1.93$$

which is slightly less than the limiting value of 2 and would need experimental verification. ■

19.5 MULTILAYER POLYMER MIRRORS

In an article on "Giant Birefringent Optics in Multilayer Polymer Mirrors" (Weber et al., 2000), the authors describe how to combine thin polymer films, each having a different refractive index, to achieve prescribed transmission, refraction, and reflection of light in specific wavelengths. Their article shows how these properties vary with the thickness and refractive index of the individual films. Given the ability to create multilayer stacks of polymer films, the product design problem becomes one of finding the proper combination of layers to achieve desired properties. This problem has been solved in many ways by engineers at the 3M Corporation (Petit, 2000). The result is a host of new products that manipulate light in helpful ways. These include thin films that: (1) coat automobile windows, allowing visible light to pass through, while reflecting infrared light waves, (2) protect rare paintings, permitting visible light to pass through, while reflecting ultraviolet light, and (3) that serve as mirrors in the backing for the new flat-panel displays and the screens of laptop computers, thereby increasing their brightness.

When designing a new product, a design team begins by specifying the desired optical properties. Then, it designs a stack of on the order of 100–1,000 layers of polymer films. Next, it designs extrusion devices to properly blend the polymer, and feed-block-splitters that divide the flow into 100–1,000 thin films. The effluent films from these splitters are stacked at high extrusion speeds.

A key to creating successful designs are the estimates of the fraction of light transmitted and reflected when thin polymer films are combined. Estimation methods are used to design stacks of films, which are subsequently produced in small quantities and distributed to customers for product testing and marketing studies.

19.6   CVD OF POLYSILICON IN IC MANUFACTURE

Integrated circuits (ICs) are manufactured by fabricating microscopic structures on a substrate of silicon. For example, an IC transistor involves the generation of a topography in which a gate straddles a doped region of the substrate, in between the source and drain. Together, this topography performs the function of a switch. Here, the gate is a portion of the overall structure that is made from amorphous silicon (referred to as *polysilicon*), set down onto the original silicon substrate, often by the rapid-thermal, chemical, vapor-deposition (RTCVD) process, involving parallel electrodes, at low pressure. In one process, a 500-Å film of amorphous silicon is deposited on an 8-cm wafer, which sits on top of the lower electrode (Armaou and Christofides, 1999; Christofides, 2001), as shown in Figure 19.10. The reactor is fed with 10% SiH_4 (silane) in He at 1 torr through a showerhead. An RF power source, at 13.56 MHz frequency, is used to generate a plasma (chemically reactive mixture of ions, electrons, and radicals) at 500 K, which is transported by convection and diffusion to the surface of the wafer where they react and deposit amorphous silicon. In product design, the objective is to examine this and other configurations for this reactor, to assure that a uniform thin film, containing few impurities, is obtained. The design team adjusts these specifications iteratively until satisfactory performance is achieved. Some combination of theoretical and experimental verification is used.

For theoretical analysis, Armaou and Christofides (1999) provide the following kinetic model for the chemical reactions.

Initial Dissociation

Initially, SiH_4 dissociates due to electron impact to form silylene radical, SiH_2, silyl radical, SiH_3, and atomic hydrogen:

$$e^- + SiH_4 \rightarrow SiH_2 + 2H + e^- \tag{R1}$$

$$e^- + SiH_4 \rightarrow SiH_3 + H + e^- \tag{R2}$$

Then atomic hydrogen reacts with SiH_4:

$$H + SiH_4 \rightarrow SiH_3 + H_2 \tag{R3}$$

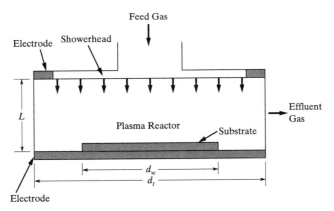

Figure 19.10 Cylindrical showerhead, electrode, plasma-enhanced, CVD reactor.

Silyl radicals diffuse toward the wafer surface where recombination reactions occur

$$2SiH_3 \rightarrow SiH_4 + SiH_2 \qquad\qquad (R4)$$

$$SiH_4 + SiH_2 \rightarrow Si_2H_6 \qquad\qquad (R5)$$

The intrinsic rates of consumption in mol/(cm^3-min) of the four species are

$$r_{SiH_4} = -k_1 n_e c_{SiH_4} - k_2 n_e c_{SiH_4} - k_3 c_{SiH_4} c_H + k_4 c_{SiH_3}^2 - k_5 c_{SiH_4} c_{SiH_2} \qquad (19.30)$$

$$r_{SiH_2} = k_1 n_e c_{SiH_4} + k_4 c_{SiH_3}^2 - k_5 c_{SiH_4} c_{SiH_2} \qquad (19.31)$$

$$r_{SiH_3} = k_2 n_e c_{SiH_4} + k_3 c_{SiH_4} c_H - k_4 c_{SiH_3}^2 \qquad (19.32)$$

$$r_H = 2k_1 n_e c_{SiH_4} + k_2 n_e c_{SiH_4} - k_3 c_{SiH_4} c_H \qquad (19.33)$$

where c is the concentration in mol/cm^3 and n_e is the electron density given by

$$n_e\{r, z\} = n_e^{max} J_0\left(2.405\frac{r}{r_t}\right) \sin\left(\frac{\pi z}{L}\right) \qquad (19.34)$$

Here, n_e^{max} is the maximum electron density in the reactor, r is the radial position in the reactor, r_t is the radius of the reactor cylinder, z is the axial coordinate, L is the height of the reactor (distance between the two electrodes), and J_0 is the zero-order Bessel function of the first kind. Clearly, the electron density is a maximum at the center of the reactor ($r = 0$, $z = L/2$). The rate constants are

Rate Constant	Units
$k_1 = 1.870 \times 10^{-11}$	s^{-1}cm^3
$k_2 = 1.590 \times 10^{-10}$	s^{-1}cm^3
$k_3 = 1.325 \times 10^{12}$	s^{-1}mol^{-1}cm^3
$k_4 = 9.033 \times 10^{13}$	s^{-1}mol^{-1}cm^3
$k_5 = 2.830 \times 10^{13}$	s^{-1}mol^{-1}cm^3

To examine the performance of this rapid deposition process, a design team would create a partial differential equation (PDE) model involving momentum, heat, and mass balances, as illustrated in the next example.

EXAMPLE 19.4

Using the model of Armaou and Christofides (1999), the plasma is assumed to be a continuum, with physical properties of the gas (independent of position and time, negligible volume change of the reacting gases, and velocity and concentration fields symmetric about the reactor centerline (azimuthal symmetry). They consider the design of a reactor having $L = 3.6$ cm, $r_t = 8.0$ cm, with a radius of holes in the showerhead, $r_h = 0.1$ cm, $N_h = 350$ holes in the showerhead, and a wafer having $r_w = 4$ cm. Silane is fed to the reactor at 50 cm^3/min, uniformly distributed among the showerhead holes, and the maximum electron density is 2.0×10^{10} cm^{-3}. At 1 torr and 500 K, the viscosity and density of the gas are 1.832×10^{-7} kg/(s-cm) and 1.030×10^{-9} kg/cm^3. The silicon density is 8.292×10^{-10} mol/(Å-cm^2). These design specifications can be varied to produce more uniform films, to adjust the deposition time, etc.

To determine the thickness of the silicon film as a function of radius, both the Navier–Stokes equations and the chemical species balances are written and solved. Then, the volumetric flow rate of the feed stream, and its distribution through the showerhead holes, is adjusted to obtain a more uniform film.

Navier–Stokes Equations

$$\frac{\partial v_r}{\partial t} + v_r \frac{\partial v_r}{\partial r} + v_z \frac{\partial v_r}{\partial z} = -\frac{1}{\rho}\frac{\partial P}{\partial r} + v\left(\frac{\partial^2 v_r}{\partial r^2} + \frac{1}{r}\frac{\partial v_r}{\partial r} - \frac{v_r}{r^2} + \frac{\partial^2 v_r}{\partial z^2}\right) \qquad (19.35)$$

$$\frac{\partial v_z}{\partial t} + v_r \frac{\partial v_z}{\partial r} + v_z \frac{\partial v_z}{\partial z} = -\frac{1}{\rho}\frac{\partial P}{\partial z} + v\left(\frac{\partial^2 v_z}{\partial r^2} + \frac{1}{r}\frac{\partial v_z}{\partial r} + \frac{\partial^2 v_z}{\partial z^2}\right) \qquad (19.36)$$

$$\frac{\partial v_r}{\partial r} + \frac{v_r}{r} + \frac{\partial v_z}{\partial z} = 0 \qquad (19.37)$$

Boundary conditions:

$$v_r\{r,0\} = 0, \qquad v_z\{r,0\} = 0, \tag{19.38}$$

$$v_r\{r,L\} = 0, \qquad v_z\{r,L\} = -v_w \tag{19.39}$$

where v_r is the velocity in the radial direction, v_z is the velocity in the axial direction, $\rho = 1.030 \times 10^{-9}$ kg/cm^3 is the density of the plasma, P is the pressure, $\nu = \mu/\rho$ is the kinematic viscosity of the plasma, $\mu = 1.832 \times 10^{-7}$ kg/(s-cm) is the viscosity of the plasma, and v_w is the velocity of the feed gas entering the reactor. When the velocity of the feed gas is uniform, it is $v_w = Q\{T, P\}/(N_h \pi r_h^2)$, where $Q\{T, P\}$ is the volumetric flow rate of the feed gas.

Species Mass Balances

$$\frac{\partial c_i}{\partial t} = D_i \left(\frac{\partial^2 c_i}{\partial r^2} + \frac{1}{r} \frac{\partial c_i}{\partial r} + \frac{\partial^2 c_i}{\partial z^2} \right) - v_r \frac{\partial c_i}{\partial r} - v_z \frac{\partial c_i}{\partial z}$$
$$+ r_i\{n_e, c_1, c_2, c_3, c_4\}, \qquad i = 1, \dots, 4 \tag{19.40}$$

where c_i is the concentration of species i, with: $1 = \mathrm{SiH_4}$, $2 = \mathrm{SiH_2}$, $3 = \mathrm{SiH_3}$, and $4 = \mathrm{H}$, and D_i is the diffusion coefficient of species i.

Boundary conditions:

$$\frac{\partial c_i}{\partial r}\{t,0,z\} = 0, \qquad\qquad \frac{\partial c_i}{\partial r}\{t,r_t,z\} = 0 \qquad i = 1, \dots, 4 \tag{19.41}$$

$$c_i\{t,r,L\} = c_{w,i}\{t,r\} \qquad\qquad i = 1, \dots, 4 \tag{19.42}$$

$$\frac{\partial c_1}{\partial z}\{t,r,0\} = -\gamma_3 \frac{D_3}{D_1} \frac{\partial c_3}{\partial z}\{t,r,0\} \tag{19.43}$$

$$c_2\{t,r,0\} = 0, \qquad\qquad c_3\{t,r,0\} = 0 \qquad 0 \le r \le r_w \tag{19.44}$$

$$\frac{\partial c_4}{\partial z}\{t,r,0\} = 0 \qquad\qquad 0 \le r \le r_w \tag{19.45}$$

$$\frac{\partial c_i}{\partial z}\{t,r,0\} = 0 \qquad\qquad i = 1, \dots, 4 \qquad r_w \le r \le r_t \tag{19.46}$$

Initial conditions:

$$c_i\{0,r,z\} = c_i^0\{r,z\}, \qquad i = 1, \dots, 4 \tag{19.47}$$

Here, $c_{w,i}\{t,r\}$ is the concentration of species i in the feed, and $c_i^0\{r,z\}$ is the initial concentration of species i in the reactor. Boundary condition (19.43) accounts for the production of SiH$_4$ by consumption of SiH$_3$ on the wafer surface, where γ_3 is the fraction of SiH$_3$ that reacts to form SiH$_4$ on the wafer surface. Boundary conditions (19.44) account for the instantaneous consumption of SiH$_2$ and SiH$_3$ on the wafer surface, boundary condition (19.45) accounts for zero mass transfer of H at the wafer surface, and boundary conditions (19.46) are due to the lack of reaction of the four species outside of the wafer surface. Finally, the physicochemical properties are

Properties	SiH$_4$	SiH$_3$	SiH$_2$	H
D_i, cm^{2}s^{-1}	285.05	321.86	319.16	2,107.75
γ_i	—	—	0.6	—
s_i	0	1.0	0.4	0

Deposition Rate of Silicon
The deposition rate of amorphous silicon is

$$r_{\mathrm{dep}}\{t,r\} = \frac{1}{\rho_{\mathrm{Si}}} \left[\sum_{i=1}^{4} s_i D_i \frac{\partial c_i}{\partial z}\{t,r,0\} \right] \tag{19.48}$$

where ρ_{Si} is the density of amorphous Si, and s_i is the fraction of the flux of species i toward the surface that leads to the deposition of amorphous silicon.

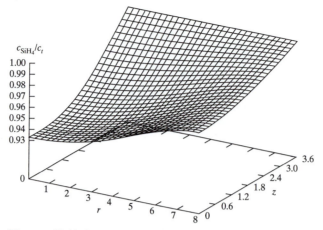

Figure 19.11 Steady-state concentration profile of SiH₄. Reprinted with permission from Armaou and Christofides (1999).

SOLUTION

A solution can be obtained using a finite-element package, such as FEMLAB. For this example, the results computed by Armaou and Christofides (1999) are presented. Initially, the reactor contains silane at the feed composition.

Figure 19.11 shows that the steady-state concentration of silane, where $r = z = 0$ is at the center of the surface of the wafer, is lowest toward the center of the reactor because the dissociation reactions of silane, R1 and R2, have the highest rates corresponding to the highest electron densities.

Figures 19.12 and 19.13 show that the steady-state concentrations of SiH_3 are much higher than the concentrations of SiH_2. The low concentrations suggest that amorphous silicon is produced mainly from SiH_3. Both concentrations are highest at the center of the reactor where the electron density is highest.

Consequently, the rate of deposition and the thickness of amorphous silicon are non-uniform, as shown in Figures 19.14 and 19.15.

To obtain a more uniform film, the design team could consider design changes and/or a control strategy. One approach is that described by Armaou and Christofides (1999), who adjust the concentration of silane as a function of the radial position in the showerhead. Using three PI (proportional-integral)-controllers to attain a near-uniform Si deposition rate at radial positions across the wafer, they show how to obtain a nearly uniform film. ∎

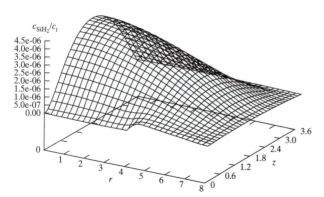

Figure 19.12 Steady-state concentration profile of SiH₂. Reprinted with permission from Armaou and Christofides (1999).

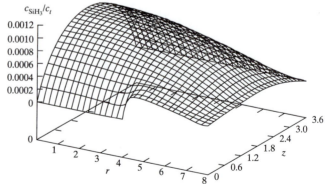

Figure 19.13 Steady-state concentration profile of SiH₃. Reprinted with permission from Armaou and Christofides (1999).

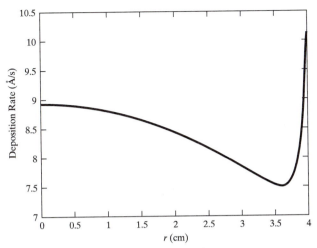

Figure 19.14 Steady-state deposition rate of amorphous silicon. Reprinted with permission from Armaou and Christofides (1999).

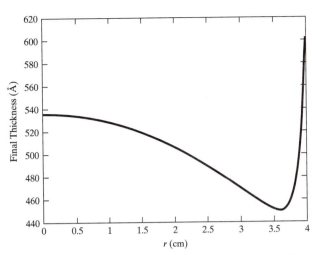

Figure 19.15 Final thickness of amorphous silicon film. Reprinted with permission from Armaou and Christofides (1999).

19.7   GERM-KILLING SURFACES

Devices that are used interchangeably by many persons, especially by children, such as telephones, doorknobs, toys, computer keyboards, and supermarket grocery carts, often contain bacteria inadvertently deposited through sneezes and coughs. Since these devices are rarely cleaned, bacteria can be transmitted among individuals by simple contact. For this reason, chemists have been developing a polymer coating capable of killing bacteria (Tiller et al., 2001). In laboratory tests, a coating of N-hexylated poly(4-vinylpyridine), known as hexyl-PVP, has been shown to kill up to 99% of common bacteria, such as staphylococcus, pseudomonas, and *E. coli*, all common disease-causing organisms. This polymer has a permanent positive charge, which has been shown to destroy the bacteria cell walls and membranes. While these authors have patented this invention, to our knowledge, no attempt has been made to design and commercialize a new product.

Given the laboratory findings, a fine "potential opportunity" is provided for a product design team. The team would examine the availability of hexyl-PVP and consider the possibility of designing a process to produce it. In addition, it would consider methods of packaging the product for application to the various devices mentioned above. One possibility would be to spray the polymer, possibly in a solvent, to coat the surfaces. This would require that a suitable solvent be identified, probably using the procedures discussed in Chapter 2 on molecular structure design. Alternatively, the polymer might be applied with a moist applicator or painted using a small brush, possibly attached to the bottle cap of a small jar that contains the liquid polymer.

19.8   INSECT REPELLING WRIST BAND

The problem with insects, especially in warm summer-like environments, has drawn the attention of product designers for many years. Many products are available, marketed especially to hikers, campers, and backpackers. While effective to varying extents, these products often have to be applied to the skin as liquids and have a distinctive odor, often not appealing to the person being protected.

A product, in the form of a wristband, called *Bug Off!* has been developed by the Stinger Division of Kaz, Inc., New York, NY. For 180 hr, this product gradually emits an aroma that repels insects. When not in use, the wristband is stored in a plastic container designed to prevent leakage of the active agent.

In designing this product, a design team sought a nontoxic chemical species or mixture capable of repelling insects. Since quantitative measures of insect reaction to insecticides are probably not available, much experimental work was probably necessary to carry out the molecular structure design. Then, given the chemical species or mixture, a delivery mechanism was designed. In this case, the chemicals are stored in a polymer spiral that wraps around the wrist. To assure their availability over 180 hr, the polymer material had to be selected with the proper permeability. Since permeabilities of large molecules through polymer members are difficult to estimate, considerable experimentation was probably necessary in the laboratory. When the product appeared to be viable, prototypes were probably prepared for testing and market analysis, prior to the design of a manufacturing facility.

19.9 AUTOMOTIVE FUEL CELL

With increasingly strict air standards and the desire to convert fuel to power more efficiently, considerable attention is being given to the possibility of using fuel cells to power automobiles. Low sulfur fuel, such as gasoline and methanol, would be produced in refineries. Each automobile would carry a small chemical plant to convert gasoline or methanol to hydrogen, which would be consumed with oxygen from air in a fuel cell.

The product to be designed involves the chemical plant and the fuel cell, as suggested by Ernst and co-workers (1999) in a problem statement distributed to chemical engineering students worldwide. In the example that follows, aspects of the fuel cell design are considered. The process to convert fuel to hydrogen can be designed using the strategies presented throughout this book, noting that laboratory-scale equipment is necessary.

EXAMPLE 19.5

A fuel cell is to be designed that provides all of the power necessary for a vehicle during all phases of driving, except during startup where additional energy is supplied. The latter will have been generated during operation and stored in batteries, or compressed hydrogen gas can be stored. This aspect of the design is not to be considered.

Vehicle Performance

Figure 19.16 shows a schematic of a vehicle with the positions of the fuel tank, the system for hydrogen generation and purification, and the fuel cell, shown approximately.

The maximum motive power, P_{motive}^{max}, required, which is the power to enable the vehicle to accelerate and cruise, is estimated to be

$$P_{motive}^{max} = 48m + 3,000$$

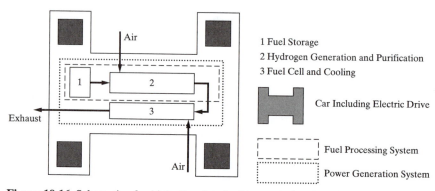

Figure 19.16 Schematic of vehicle. Reprinted with permission from Ernst et al. (1999).

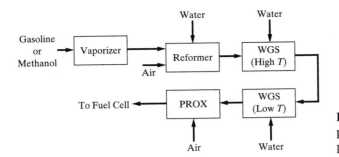

Figure 19.17 Schematic of fuel processing system. Reprinted with permission from Ernst et al. (1999).

where m is the mass of the vehicle in kilograms, excluding the fuel processing system, and the power is in Watts. For this design, let $m = 1,250$ kg. In addition, for air conditioning, lighting, windshield wipers, etc., 7 kW of auxiliary power are estimated to be required continuously; not including power required by the fuel processing system. Hence, the power demand is the sum of the motive, auxiliary, and fuel processing power. Furthermore, the cruising range of the vehicle is 600 km and the average motive power demand is assumed to be 35% of the maximum. The vehicle is anticipated to be driven 20,000 km/yr and to have a useful life of 5 yr, with scrap value approximated as 30% of its initial cost.

Fuels
Either low sulfur gasoline, approximated as $C_{7.14}H_{14.25}$, or methanol, are available, each assumed to be $1.00/gal.

Fuel Processing System
Ernst et al. (1999) suggest that the fuel processing system be comprised of a steam-reforming reactor, water–gas shift (WGS) reactors (at high and low temperatures), and a preferential oxidation (PROX) reactor to oxidize carbon monoxide, as shown in Figure 19.17. Kinetic data are provided for the reformer and the two shift reactors.

Fuel Cell Performance
A proton exchange membrane (PEM) design is to be utilized. A stack of fuel cells will be provided, each of which oxidizes hydrogen at its platinum–ruthenium anode:

$$H_2 = 2H^+ + 2e^-$$

The protons are transported through the PEM to the cathode, where they react with oxygen from air to form water:

$$\tfrac{1}{2}O_2 + 2H^+ + 2e^- = H_2O$$

The individual cells, each of which has a potential less than 1 volt, are connected in series.

Figure 19.18 shows the cell potential as a function of the current density and Figure 19.19 shows the power density as a function of the current density, both at 70°C and at 1 and 3 atm.

Oxygen is assumed to be in 100% excess. The key decision is the current density at which the fuel cell is operated. At higher current densities, smaller area electrodes are required, but the hydrogen consumption is increased. The efficiency of the fuel cell is defined as the power output divided by the heating value of its hydrogen fuel.

The pressures on the anode and cathode sides must be equivalent. Furthermore, between 60 and 90°C, the current density is a linear function of the temperature:

$$\frac{i}{A} = 2.5\,T \tag{19.49}$$

where the current density is in mA/cm² and temperature, T, is in °C.

The platinum–ruthenium cathode tolerates concentrations of carbon monoxide less than 100 ppm and the temperature must be less than 85°C. Not all of the hydrogen is consumed in the anode, with a minimum of 8 vol% on a dry basis permitted in the effluent gas. After this hydrogen is burned, the effluent gases are cooled.

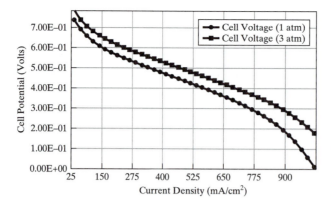

Figure 19.18 Fuel cell potential as a function of the current density at 70°C, 1-3 atm. Reprinted with permission from Ernst et al. (1999).

Sizing and Cost Data

For the sizing and estimation of purchase costs, when designing laboratory-scale equipment, data are provided by Ernst et al. (1999). Note that such data are based upon similar laboratory-scale equipment.

Unit	Weight	Cost ($)
Fuel cell	0.5 lb/kW	27/ft^2
Reformer (methanol)	800 kg/m^3 catalyst	35/kg catalyst
Reformer (gasoline)	800 kg/m^3 catalyst	53/kg catalyst
WGS (high T)	750 kg/m^3 catalyst	14/kg catalyst
WGS (low T)	820 kg/m^3 catalyst	22/kg catalyst
PROX	800 kg/m^3 catalyst	150/kg catalyst
Compressor/expander	9 lb[a]	600[a]
Pump (any fluid)	5 lb	150
Heat exchangers (gas/liquid)	200 ft^2/ft^3, 6.3 lb/ft^3	1.80/lb
Heat exchangers (gas/gas)	400 ft^2/ft^3, 12.6 lb/ft^3	1.80/lb
Reactor vessels	0.05 lb/ft^2	1.80/lb
Piping	0.04 lb/ft	1.80/lb

[a]Cost/weight scaling of compressor/expander $= (P/18)^{0.6}$ with pressure in psig.

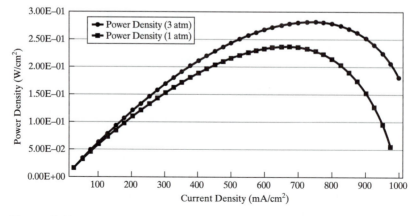

Figure 19.19 Fuel cell power density as a function of the current density at 70°C, 1-3 atm. Reprinted with permission from Ernst et al. (1999).

SOLUTION   The solution of this example is too extensive to provide herein. It is best undertaken in Exercise 19.8. In one solution, obtained by students at the Georgia Institute of Technology, using gasoline as the fuel, the entire vehicle weighs 1,750 kg and carries 74 L of fuel. Its fuel cell operates at 3 bar, provides $P^{max} = 111$ kW, has an efficiency of 43%, current density = 473 mA/cm^2, and area = 450,000 cm^2. Its overall system efficiency is 8.10 km/L, overall fuel efficiency is 35.4%, and the total cost of the power train is $24,700, far in excess of the cost of a conventional gasoline engine. Note that the cost of the fuel cell is $13,300, with a mass of 25.1 kg. ∎

19.10 ENVIRONMENTALLY SAFE REFRIGERANTS

Prior to the 1990s, fully halogenated chlorofluorocarbons (CFCs) were widely used as: (1) propelling agents in spray bottles, (2) blowing agents to make plastic foams, (3) cooling media in household refrigerators and automobile air conditioners, and (4) cooling media in industrial refrigeration cycles. The most popular CFC was dichlorodifluoromethane, CCl_2F_2, commonly known as refrigerant CFC-12 or R-12. In 1974, Mario J. Molina and F. Sherwood Rowland published a paper (*Nature*, **249**, 810–812), "Stratospheric Sink for Chlorofluoromethanes: Chlorine Atom-Catalyzed Destruction of Ozone," in which they warned that CFCs were depleting the ozone layer, which exists from 12 to 30 miles above the surface of the earth and protects us from harmful ultraviolet radiation. In this stratospheric layer, where the concentration of ozone is approximately 10 ppm by volume, CFCs undergo photodissociation that releases chlorine atoms. These react with ozone by the reactions:

$$Cl + O_3 \rightarrow ClO + O_2$$
$$\underline{ClO + O \rightarrow O_2 + Cl}$$
$$O_3 + O \rightarrow 2\,O_2$$

It has been estimated that one chlorine atom can destroy 10,000 molecules of ozone. Bromine also reacts with ozone, but fluorine does not. In 1995, Molina and Rowland, together with Paul Crutzen, who showed in 1970 that nitrogen oxides also reacted with ozone, received the $1 million Nobel Prize in Chemistry for their discovery of the depletion of the ozone layer, which the Nobel committee termed the Achilles heel of the universe.

In 1987, the Montreal Protocol, an international agreement, established a worldwide production phaseout schedule for CFCs. By the mid-1990s, a significant reduction in the use of CFCs had occurred, with a concurrent reduction in ozone depletion and a rise in the use of hydrochlorofluorocarbons, HCFCs and hydrofluorocarbons, HFCs, as replacement refrigerants. However, HCFCs are only a short-term solution and they too are scheduled for phaseout, by 2020 or 2030. Today, refrigerants, are compared on the basis of the following factors: (1) the ozone depletion potential, ODP, (2) the global warming potential, GWP, (3) toxicity, (4) flammability, (5) availability, (6) cost, (7) pressure of the refrigeration cycle, (8) density, and (9) theoretical refrigeration cycle efficiency. The ODP compares compounds to trichlorofluoromethane, CCl_3F (R-11), which along with R-12 has the highest ozone depletion potential, ODP = 1. Other CFCs have values of ODP greater than 0.5. HCFCs have ODP values ranging from 0.02 to 0.11, while HFCs, as well as commercial refrigerants such as hydrocarbons and ammonia have zero ODP. Accordingly, because some HFCs are nontoxic and nonflammable, they have been the choice for replacement of CFCs in household refrigerators and automobile air conditioners. The selection of a replacement for R-12 is considered in the following example and in Exercise 19.9.

EXAMPLE 19.6   The condenser of a distillation column uses chilled water instead of cooling water. The chilled water, at a flow rate of 120 gpm, enters at 50°F and exits at 60°F. This corresponds to 50 tons of refrigeration (1 ton of refrigeration = 200 Btu/min). The chilled water is obtained from a refrigeration cycle similar to that shown in Figure 9.13 However, instead of a propane refrigerant, a Freon, R-12 (dichlorodifluoromethane), has been used. In the refrigeration cycle, the chilled water enters the refrigerant evaporator at 60°F and leaves at 50°F. R-12 leaves the refrigerant evaporator as a saturated vapor at 40°F. R-12 then enters a screw compressor, where it is compressed to a pressure corresponding to a saturation temperature of 105°F. The compressed R-12 is a superheated vapor, which passes to the refrigerant condenser, where it is condensed to a saturated liquid by cooling water, which enters the condenser at 85°F and leaves at 95°F. To complete the cycle, R-12 then passes through a valve, which drops the pressure to a value corresponding to a saturation temperature of 40°F, at which point, the R-12 is partially vaporized. The overall heat transfer coefficients of the refrigerant evaporator and refrigerant condenser are both 50 Btu/hr-ft²-°F. The overall efficiency of the compressor is 80%.

The R-12 must be replaced because it can release chlorine atoms, which deplete ozone in the stratosphere. Accordingly, a new refrigerant must be selected. Hopefully, one can be found that will not require any major changes to the current equipment or its operation. Possible candidates are fluorinated paraffin hydrocarbons of the general formulas: CH_xF_y and $C_2H_xF_y$. However, it should be noted that the compressor lubricant used with R-12 is a soluble mineral oil. In general, mineral oil is not soluble in HFCs, but a synthetic oil known as polyolester is soluble. The physical properties of R-12 and all fluorinated methanes and ethanes (including isomers) are readily obtained from the chemical data bank of a process simulation program.

The problem is to select a suitable new refrigerant. To accomplish this, the following steps are to be taken using a process simulation program.

1. For the current refrigeration system, calculate the R-12 flow rate, all R-12 stream temperatures and pressures in the cycle, heat-exchanger duties and sizes, and compressor brake horsepower. Neglect pipeline and heat-exchanger pressure drops. For the refrigerant condenser, where the refrigerant enters as a superheated vapor, a zone analysis is necessary to compute a correct temperature-driving force for heat transfer, because the sensible heat at the higher temperatures is quickly transferred followed by latent heat transfer at a constant temperature.

2. Examine the vapor pressures, toxicities, and flammability characteristics of all HFC refrigerant candidates and select one that appears to be in the range that will permit use of the current equipment.

3. For that HFC refrigerant, perform the refrigeration cycle material balances, energy balances, and equipment sizing to determine if that refrigerant is a suitable match.

SOLUTION

Only the solution to the first part is given here. The other two parts are the subject of Exercise 19.9 at the end of this chapter.

Using a simulation program, start with the R-12 stream leaving the refrigerant evaporator, where the temperature is 40°F and the corresponding saturation pressure (vapor pressure) is determined to be 51.5 psia. At 105°F, which is the temperature of R-12 leaving the refrigerant condenser, the corresponding saturation pressure is determined to be 139.7 psia. A compressor calculation with a compression ratio of (139.7/51.5) and a mechanical efficiency of 80% gives a compressor exit temperature of 136°F as a superheated vapor. When condensed and passed through a valve to drop the pressure back to 51.5 psia before being sent to the refrigerant evaporator, the R-12 becomes 23% vaporized.

The duty of the refrigerant evaporator is computed from the 50 tons of refrigeration as 50(200)(60) = 600,000 Btu/hr. The log-mean temperature-driving force for the refrigerant evaporator is computed to be 14.4°F. With an overall heat-transfer coefficient of 50 Btu/hr-ft²-°F, the heat-transfer area is 600,000/[(50)(14.4)] = 833 ft².

The computed duty of the refrigerant condenser is 714,600 Btu/hr. Using a zone analysis with an overall heat-transfer coefficient of 50 Btu/hr-ft²-°F, the heat-transfer area is computed to be 924 ft². If a zone analysis is not used, an erroneous area of 488 ft² is computed.

The theoretical horsepower of the compressor is computed to be 36.1. With a mechanical efficiency of 80%, the brake horsepower is 45.1. ■

19.11 SUMMARY

Having studied this chapter, the reader should

1. Be well acquainted with many of the steps in designing configured industrial and consumer products. These include
   a. creating and assessing the primitive problem, involving idea generation, customer interviews, technical feasibility, marketing, and business studies.

b. setting product performance specifications and locating molecules to satisfy these specifications.

c. designing a process to produce the chemicals, when desirable.

d. three-dimensional design of the configured industrial or consumer product.

e. building a prototype product and consumer testing.

f. designing the manufacturing facility to produce the configured product—normally done by mechanical, civil, and electrical engineering members of the design team.

g. designing to satisfy quality control specifications; for example, six-sigma specifications.

h. designing environmentally friendly and safe products and processes.

2. Through study of one or more of the nine chemical products, close examination of the examples, and solution of the associated exercises, the student should be very knowledgeable concerning the key technical considerations in the design of configured industrial and consumer products. These include

a. product performance specifications and molecular structure design.

b. three-dimensional product design, including analysis and optimization using momentum, heat, and mass balances, and chemical kinetics.

c. product quality testing.

REFERENCES

Armaou, A., and P.D. Christofides, "Plasma Enhanced Chemical Vapor Deposition: Modeling and Control," *Chem. Eng. Sci.*, **54**, 3305–3314 (1999).

Bird, R.B., W.E. Stewart, and E.N. Lightfoot, *Transport Phenomena*, 2nd ed., Wiley, New York (2002).

Christofides, P.D., *Nonlinear and Robust Control of PDE Systems: Methods and Applications to Transport-Reaction Processes*, Birkhaeuser, Boston (2001).

Coe, J.T., *Unlikely Victory: How General Electric Succeeded in the Chemical Industry*, AIChE, New York (2000).

Cussler, E.L., and G.D. Moggridge, *Chemical Product Design*, Cambridge University Press, Cambridge (2001).

Dally, J.W., and T.M. Regan, *Introduction to Engineering Design: Book 1—Solar Desalination*, James W. Dally Associates, Knoxville, Tennessee (1997).

Daugirdas, J.T., and T.S. Ing, Eds., *Handbook of Dialysis*, Little, Brown & Co., Boston (1988).

de Nevers, N., "A Simple Heat of Crystallization Experiment," *Chem. Eng. Ed.*, **25**(3), 154–156 (1991).

Ernst, W.R., M.J. Realff, and J. Winnick, "Process Synthesis and Design of the Power Generation System for Automobiles: A Fuel Cell Approach," *2000 AIChE National Student Design Competition*, AIChE, New York (1999).

Felder, R.M., and R.W. Rousseau, *Elementary Principles of Chemical Processing*, 2nd ed., John Wiley & Sons, New York (2000).

Gundling, E., *The 3M Way to Innovation: Balancing People and Profit*, Kodansha, Int'l., New York (2000).

Knudsen, J.G., and D.L. Katz, *Fluid Dynamics and Heat Transfer*, McGraw-Hill, New York (1958).

Lachman-Shalem, S., B. Grosman, and D.R. Lewin, "Nonlinear Modeling and Multivariable Control of Photolithography," *IEEE Trans. Semiconductor Manufacturing*, **15**(3), 310–322 (2002).

"Membrane Technology," *Kirk-Othmer Encyclopedia of Chemical Technology*, Vol. 16, 4th ed., Wiley-Interscience, New York (1995).

Molina, M.J., and F.S. Rowland, "Stratospheric Sink for Chlorofluoromethanes: Chlorine Atom-Catalyzed Destruction of Ozone," *Nature*, **249**, 810–812 (1974).

Mullin, J.W., *Crystallization*, 3rd ed., Butterworth-Heinemann, Oxford (1993).

Ogunnaike, B.A., and W.H. Ray, *Process Dynamics, Modeling and Control*, Oxford University Press, New York (1994).

Petit, C.W., "A Discovery Makes the Light Fantastic," *U.S. News & World Report*, June 19 (2000).

Pisano, G.P., *The Development Factory: Unlocking the Potential of Process Innovation*, Harvard Business School Press, Boston (1997).

Rath and Strong, *Six Sigma Pocket Guide*, Rath and Strong Management Consultants/AON Management Consulting, Lexington, Massachusetts (2000).

Rath and Strong, *Design for Six Sigma Pocket Guide*, Rath and Strong Management Consultants/AON Management Consulting, Lexington, Massachusetts (2002).

Seader, J.D., and E.J. Henley, *Separation Process Principles*, John Wiley & Sons, New York (1998).

Spaeth, E.E., *Washington University Case Study 9, Analysis and Optimization of an Artificial Kidney System*, Department of Chemical Engineering, Washington University, St. Louis, Missouri (1970).

Tiller, J.C., C.-J. Liao, K. Lewis, and A.M. Klibanov, "Designing Surfaces that Kill Bacteria on Contact," *Proc. Natl. Acad. Sci. U.S.A.*, **98**(11), 5,981–5,985 (2001).

Ulrich, K.T., and S.D. Eppinger, *Product Design and Development*, 2nd ed., McGraw-Hill, New York (2000).

Weber, M.F., C.A. Stover, L.R. Gilbert, T.J. Nevitt, and A.J. Ouderkirk, "Giant Birefringent Optics in Multilayer Polymer Mirrors," *Science*, **287**, March 31 (2000).

Wheeler, J.M., "Getting Started: Six-Sigma Control of Chemical Operations," *Chem. Eng. Prog.*, 76–81, June (2002).

EXERCISES

19.1 Espresso coffee is prepared in a machine that injects high-pressure steam through a cake of ground coffee. In a conventional machine, the user manually loads ground coffee into a metal filter cup, locks the cup under the steam head, and then opens the steam heater.

a. Prepare a SIPOC for the preparation of coffee using an espresso machine.

b. Identify all of the sources of variance in the quality of the coffee produced using the above machine.

c. For a novel espresso machine, the user provides a vacuum-sealed container of ground coffee with a built-in filter. On insertion into the machine, the container is perforated and used to make a single cup of espresso coffee. Which of the sources of variance identified in (b) are addressed by this design? Can you suggest additional improvements?

19.2 a. Prepare a SIPOC for the manufacture of the hemodialysis device in Example 19.1.

b. Identify all of the sources of variance in the urea concentration in treated blood after a 4-hr treatment. Suggest improvements in the design to increase the sigma level of the product.

19.3 The primary product in a distillation column violates its specifications during 5 hr monthly, on average.

a. Determine the sigma level.

b. Determine the sigma level when improved process control reduces the specification violations to 0.5 hr monthly, on average.

c. A crude-oil distillation unit that has frequent changes in feedstock is expected to have a lower sigma level than one that relies on a single feedstock. Explain why.

19.4 An important process in integrated circuit manufacture is photolithography, where a circuit pattern is transferred from a *mask* onto a photosensitive polymer (the PR), ultimately replicating that pattern onto the surface of a silicon wafer (Lachman-Shalem et al., 2002). A typical photolithography process consists of seven consecutive steps: spin coat of the PR, prebake, chill, expose, postexposure bake (PEB), chill, and development. The overall objective is to produce printed lines with accurate and consistent width (referred to as the *critical dimension*, CD). Table 19.2 shows the steps required, together with the sigma level of each step.

a. Compute the sigma level of the complete process.

b. You are required to reduce the variance in this process by process improvements. Given the limited engineering time available, you can allocate only three instances of process improvements, each of which will increase the sigma level of the selected step by 0.5. Allocate process improvements optimally to maximize the increase in sigma level for the overall process.

19.5 Consider the hemodialysis device in Example 19.1.

a. Examine the effect on the rate of urea removal of changing the hemodialyzer geometry, the blood and dialysate flow rates to the dialyzer, the rate of waste withdrawal, the volume of the dialysate tank, and the sensitivity of the rate of urea mass transfer to the mass-transfer coefficient. In particular, the above estimate of the coefficient on the shell side may be low because the entry to and exit from the hemodialyzer of the dialysate is normal to, rather than parallel to the fibers. This should enhance the shell-side coefficient.

b. Modify the design such that the urea concentration can be reduced to 30% of the initial concentration within 4 hr.

19.6 For the solar desalination unit in Example 19.2, consider the use of a conical mirror that multiplies the incident heat flux by a factor of three. Recompute the temperature of the glass cover, the heat flux to the glass cover, the evaporation rate of water, and the collector area. Would this area be sufficiently small to satisfy the specifications? Furthermore, estimate the cost of producing one unit, as well the cost of producing 100,000 units annually.

19.7 Repeat Example 19.3 for a temperature of 40°C (a 20°C rise), using the data for the other three potential salts: sodium thiosulfate, calcium nitrate, and lead acetate. Repeat the calculations for all four salts for 50°C (a 30°C rise), noting that this will require finding solubility data for that temperature or extrapolating the given solubility data. Select, from the results for the four salts, the best candidate for a hand warmer and recommend the steps to be taken to prove the hand warmer and bring it to market.

19.8 Obtain a solution to Example 19.5.

Table 19.2 Steps in Photolithography and Their Sigma Levels

Step	Subprocess	Function	Sigma level
1	PR coating	Coats wafer with a thick layer of PR	3.5
2	Prebake	Hardens the PR before exposure in the stepper	4.5
3	Chill 1	Cools the wafer after prebake	5
4	Stepper	Exposes the PR through a negative of the pattern to be reproduced	4
5	PEB	Fully hardens the PR after exposure	4.5
6	Chill 2	Cools the wafer after PEB	5
7	Development	Develops the image imprinted on the PR	3

19.9 Complete the solution to Example 19.6 by

a. Examining the vapor pressures, toxicities, and flammability characteristics of all HFC refrigerant candidates and selecting one that appears to be in the range that will permit use of the current equipment.

b. For that HFC refrigerant, perform the refrigeration cycle material balances, energy balances, and equipment sizing to determine if that refrigerant is a suitable match for the existing equipment. You can assume that the equipment for the R-12 refrigerant is overdesigned by 10%.

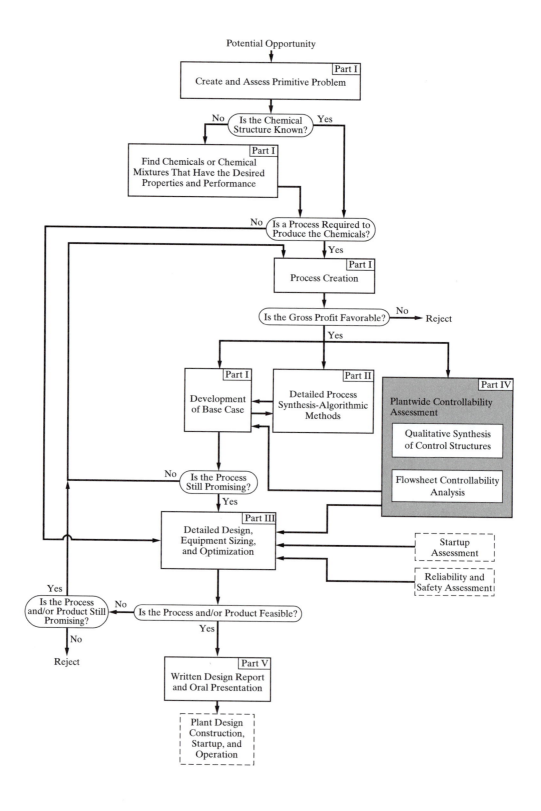

Part Four

PLANTWIDE CONTROLLABILITY ASSESSMENT

In Parts One through Three the emphasis is on the performance of potential processes, with little, if any, attention to the difficulties in meeting the operating specifications. In Part Four, the importance of assessing the degree to which a process is easy to control at the desired specifications, in the face of disturbances, is examined. An assessment of the controllability of the process is initiated after the detailed process flow diagram has been completed. Part Four begins with the *qualitative* synthesis of control structures for the entire process, as discussed in Chapter 20. Then, as shown in Chapter 21, *quantitative* analysis is used to determine the degree to which a process is easily controlled and is inherently resilient to disturbances. These methods permit alternative processes to be screened based on controllability and resiliency with relatively little effort, and for the most promising processes they identify appropriate control structures. The analysis is verified using dynamic simulation to check that potential processes perform adequately in the face of plant upsets and disturbances.

To see how Part Four relates to the other four parts of this text and to the entire design process, see the figure at the left.

Chapter 20

The Interaction
of Process Design
and Process Control

20.0 OBJECTIVES

In this chapter, and in Chapter 21, the importance of considering controllability and operability issues early in the design process is demonstrated by showing how controllability considerations can help to differentiate between processes that are easy rather than difficult to control. Chapter 20 also recommends a methodology to initiate the design of attractive plantwide control systems.

After reading this chapter, the student should

1. Be able to identify potential control problems in a process flowsheet.
2. Be able to classify and select controlled and manipulated variables for a plantwide control system.
3. Be able to perform a conceptual synthesis of plantwide control structures (pairings) based on degrees-of-freedom analysis and qualitative guidelines.

20.1 INTRODUCTION

The design of a continuous chemical process is usually carried out at steady state for a given operating range, it being assumed that a control system can be designed to maintain the process at the desired operating level and within the design constraints. However, unfavorable process static and dynamic characteristics could limit the effectiveness of the control system, leading to a process that is unable to meet its design specifications. A related issue is that, usually, alternative designs are judged based on economics alone, without taking controllability and resiliency into account. This may lead to the elimination of easily controlled, but slightly less economical, alternatives in favor of slightly more economical designs that may be extremely difficult to control. It is becoming increasingly evident that design based on steady-state economics alone is risky, because the resulting plants are often difficult to control (i.e., inflexible, with poor disturbance-rejection properties), resulting in off-specification product, excessive use of fuel, and associated profitability losses.

Consequently, there is a growing recognition of the need to consider the *controllability and resiliency* (C&R) of a chemical process during its design. *Controllability* can be defined as the ease with which a continuous plant can be held at a specific steady state. An associated concept is *switchability*, which measures the ease with which the process can be moved from one desired stationary point to another. *Resiliency* measures the degree to which a processing system can meet its design objectives despite external disturbances and uncertainties in its design parameters. Clearly, it would be greatly advantageous to be able to predict how well a given flowsheet meets these dynamic performance requirements as early as possible in the design process.

Table 20.1 summarizes the four main stages in the design of a chemical process. In the conceptual and preliminary stages, a large number of alternative process flowsheets, in the steady state (SS), are generated. Subsequent stages involve more detailed analysis in the steady state, followed by the testing of the dynamic (Dyn) performance of the controlled flowsheets. Here, considerably more engineering effort is expended than in the preliminary stages. Therefore, far fewer designs are considered, with many of the initial flowsheets having been eliminated from further consideration by screening in the preliminary stages.

The need to account for the controllability of competing flowsheets in the early design stages is an indication that simple screening measures, using the limited information available, should be employed to select from among the flowsheets. Here, if high fidelity, closed-loop, dynamic modeling were used, the engineering effort and time for development and analysis would *slow* the design process significantly. The right-hand-side columns in Table 20.1 show that the shortcut C&R tools provide a bridge between steady-state simulation for process design and the rigorous dynamic simulation required to verify switchability and other attributes of the closed-loop dynamics of the final design. For the former, ASPEN PLUS, PRO/II, HYSYS.Plant, and CHEMCAD are commonly used steady-state simulation packages, whereas dynamic simulation is often carried out using the ASPEN DYNAMICS and HYSYS.Plant simulators.

In the following examples, the impact of design decisions on controllability and resiliency is introduced for four processes. Chapter 21 expands upon this introduction and shows how to eliminate the less desirable alternatives and validate the performance of the most promising designs.

Table 20.1 Process Design Stages, Issues, and Tools

Design Stage (see Figure 1.2)	Issues	What Gets Fixed	Tools		
			SS	C&R	Dyn
1. Process creation	Selecting between alternative material pathways and flowsheets	Material pathways			
2. Development of base-case design	Feasibility studies based on fixed material pathways Unit operations selection Heat integration superstructure	Flowsheet structure			
3. Detailed design	Optimization of key process variables Analysis of process sensitivity to disturbances and uncertainties	Optimal flowsheet parameters			
4. Plantwide controllability assessment	Flowsheet controllability Dynamic response of the process to disturbances Selection of the control system structure and its parameters	Control structure and its parameters			

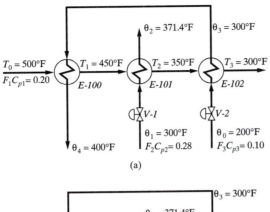

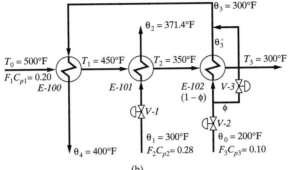

Figure 20.1 Heat-exchanger network: (a) Original configuration; (b) modification with bypass.

EXAMPLE 20.1 *Heat-Exchanger Networks*

The network shown in Figure 20.1a, which was introduced by McAvoy (1983), cools hot stream 1 from 500 to 300°F using cold streams 2 and 3 having feed temperatures of 300 and 200°F and corresponding target temperatures of 371.4 and 400°F, respectively, with the heat capacity flow rates in MMBtu/hr-°F. Furthermore, the feed rate and temperature of the hot stream are considered as disturbances.

As shown in Figure 20.1a, two of the target temperatures can be controlled by manipulating the flow rates of the two cold streams. This means that one of the target temperatures is left uncontrolled in the face of disturbances in the hot stream. An alternative design, involving a bypass around exchanger E-102, is illustrated in Figure 20.1b. As shown, this simple modification allows all three target temperatures to be regulated. Since the selection of the appropriate bypass flow fraction, ϕ, and of the most effective control configuration is not trivial, controllability analysis should be carried out on the alternative networks and their candidate control structures. This will assist in selecting one of the two designs as shown in Chapter 21. ∎

EXAMPLE 20.2 *Heat-Integrated Distillation Columns*

The production of methanol is carried out in a moderate-pressure synthesis loop by the direct hydrogenation of carbon dioxide,

$$CO_2 + 3H_2 \rightleftharpoons CH_3OH + H_2O \tag{20.1}$$

which generates a liquid product that contains a binary mixture of methanol and water in approximately equal proportions. To provide commercial methanol that is nearly free of water, dehydration is achieved commonly by distillation. To reduce the sizable energy costs, three double-effect,

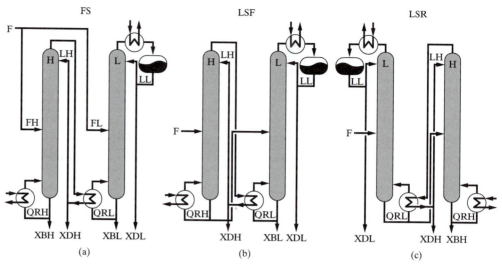

Figure 20.2 Three heat-integrated alternatives to a single distillation column.

heat-integrated configurations, shown in Figure 20.2, are considered commonly as alternatives to a single distillation column (SC):

FS (*Feed Split*). The feed is split nearly equally ($F_H \approx F_L$) between two columns to achieve optimal operation. The overhead vapor product of the high-pressure column supplies the heat required in the low-pressure column.

LSF (*Light-Split/Forward Heat Integration*). The entire feed is fed to the high-pressure column. About half of the methanol product is removed in the distillate from the high-pressure column, and the bottoms product is fed into the low-pressure column. In this configuration, heat integration is in the same direction as the mass flow.

LSR (*Light-Split/Reverse Heat Integration*). The entire feed is fed to the low-pressure column, with the bottoms product from the low-pressure column fed into the high-pressure column. Here, heat integration is in the direction opposite to that of the mass flow.

As discussed under "Multiple-Effect Distillation" in Section 10.9, these configurations reduce the energy costs by using the heat of condensation of the overhead stream from the high-pressure column (H) to supply the heat of vaporization of the boilup in the low-pressure column (L). Although they are more economical, assuming steady-state operation, they are more difficult to control because the configurations (1) are more interactive, and (2) have one less manipulated variable for process control, since the reboiler duty in the low-pressure column can no longer be manipulated independently.

To show the energy savings, the four flowsheets were simulated on the basis of an equimolar feed of 2,700 kmol/hr, producing 96 mol% methanol in the distillate and 4 mol% methanol in the bottoms product, assuming 75% tray efficiency and no heat loss to the surroundings, and using UNIFAC to estimate the liquid-phase activity coefficients. The total energy requirements for the four alternatives were computed as follows:

SC	2.12×10^7 kcal/hr	LSR	1.23×10^7 kcal/hr
LSF	1.33×10^7 kcal/hr	FS	1.23×10^7 kcal/hr

Clearly, the LSR and FS configurations save the most energy, although the energy consumption in the LSF configuration is only 8% higher. Based on steady-state economics alone, one of these three configurations would be selected. However, disturbance resiliency analysis (Chiang and Luyben, 1988; Weitz and Lewin, 1996) shows that either the LSR or LSF configurations are preferred for disturbance rejection, providing performance only slightly worse than that of a single column, SC. The FS configuration, on the other hand, does considerably worse. Chapter 21 shows how to obtain this information when selecting from among these alternatives. ∎

Figure 20.3 Two configurations for an exothermic reactor requiring feed preheating: (a) Reactor with independent preheat; (b) Heat-integrated reactor.

EXAMPLE 20.3   *Heat Recovery from an Exothermic Reactor*

Often, the heat from an exothermic reactor is used to preheat the reactor feed, thus saving energy, as discussed in Section 5.5. Figure 20.3b shows a configuration using a feed/product heat exchanger that is preferred commonly to the configuration with independent preheat in Figure 20.3a. However, the heat-integrated configuration shares the same disadvantages as the heat-integrated distillation systems (one less manipulated variable and possibly unfavorable dynamic interactions). In particular, the feed-effluent heat exchanger introduces positive feedback and the possibility of thermal runaway.   ■

EXAMPLE 20.4   *Reactor–Flash–Recycle System*

Reactor design for complete conversion may be impossible thermodynamically or undesirable because of reduced yields when byproducts are formed. In such cases, an economic alternative is to design a combined reactor–separator–recycle system, as illustrated in the simple example in Figure 20.4 and discussed in Chapter 8. Here, the reaction A → B is carried out in a CSTR, whose liquid feed is a stream containing pure A. In the event that B is sufficiently more volatile than A, the separation can be performed using a flash vessel and unreacted A is recycled to the reactor. As will be seen in the plantwide control examples at the end of this chapter and in the quantitative analysis in Chapter 21, the presence of the recycle complicates control of the process and requires special attention.   ■

To enable the evaluation of the controllability and resiliency of alternative process configurations, it is important to consider two aspects of the design of plantwide control systems:

1. The classification and selection of controlled and manipulated variables
2. The qualitative synthesis of plantwide control structures based on degrees-of-freedom analysis and qualitative guidelines. These are examined in the next two sections.

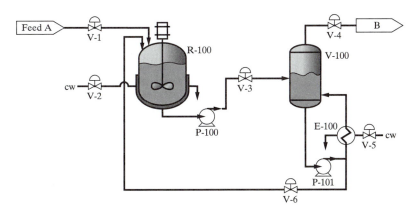

Figure 20.4 Reactor–flash–recycle system for the production of B.

20.2 CONTROL SYSTEM CONFIGURATION

The design of a control system for a chemical plant is guided by the objective to maximize profits by transforming raw materials into useful products while satisfying product specifications, safety and operational constraints, and environmental regulations. All four constraints require special consideration.

1. *Product specifications.* To satisfy customer expectations, it is important that product quality and rate meet specifications. This has been the driving force for the implementation of on-line, optimal process control in the chemical industry.
2. *Safety.* The plant must be operated safely to protect the well-being of plant personnel and nearby communities. As an example, a typical safety-driven constraint requires that the temperature and pressure of a steel vessel not exceed upper limits dictated by metallurgy. For other examples, see Section 1.4.
3. *Operational constraints.* These constraints are related to the desired operating range of the process units. Thus, upper bounds on the vapor velocity avoid flooding in distillation columns, and upper bounds on the reactor temperatures prevent degradation of the catalyst or the onset of undesirable side reactions.
4. *Environmental regulations.* These require that processing plants comply with constraints on the air and water quality as well as waste disposal. Many examples are discussed in Section 1.3.

Classification of Process Variables

When designing a plantwide control system, it is common to view the process in terms of its input and output variables. These variables include flow rates of streams entering and leaving process equipment, and temperatures, pressures, and compositions in entering and leaving streams and/or within equipment.

Process *output variables* are those that give information about the state of the process. They are usually associated with streams leaving the process or with measurements inside a process vessel. When designing a control system, output variables are usually referred to as *controlled variables*, which are measured (on-line or off-line).

Process *input variable*s are independent variables that affect the output variables of a process. They can be subdivided into two subgroups: (1) *manipulated variables* (also called *control variables*), which can be adjusted freely by an operator or a control mechanism, and (2) *disturbance variables* (also called *externally defined variables*), which are subject to the external environment and thus cannot be controlled. These variables are associated typically with the inlet and outlet streams. In a control system, manipulated variables cause changes to controlled variables.

There are three main reasons why it may be impossible to control all of the output variables of a process.

1. It may not be possible to measure on-line all of the output variables, especially compositions. Even when it is possible, it may be too expensive to do so.
2. By a degree-of-freedom analysis, described below, there may not be enough manipulated variables available to control all of the output variables.
3. Potential control loops may be impractical because of slow dynamics, low sensitivity to the manipulated variables, or interactions with other control loops.

Qualitative criteria have been suggested by Newell and Lee (1988) to guide the selection of controlled and manipulated variables, suitable for an initial analysis in the design of a plantwide control system. These guidelines, which are presented next, are driven by the plant and control objectives and should not be applied without due consideration. When two guidelines conflict, the most important of the two should be adopted. Following presentation of the guidelines, examples of the selection of variables are given.

Selection of Controlled (Output) Variables

Guideline 1. *Select output variables that are not self-regulating.* A *self-regulating process* is one that is described by a state equation of the form $\dot{x} = f\{x, u\}$, where x is an output variable and u is an input variable. A change in u will result in the process moving to a new steady state. A *non-self-regulating* process is described by $\dot{x} = f\{u\}$. As a result, changes in the input variable, u, affect the process output as a pure integrator. When the process is unstable in the open loop (i.e., in the absence of feedback control), a change in the input variable causes the system to go unstable. Clearly, *non-self-regulating* process output variables must be selected as controlled variables.

Guideline 2. *Choose output variables that would exceed the equipment and operating constraints without control.* Clearly, when safety or operational constraints are imposed, it is important to measure and control these output variables to comply with the constraints.

Guideline 3. *Select output variables that are a direct measure of the product quality or that strongly affect it.* Examples of variables that are a direct measure of the product quality are the composition and refractive index, whereas those that strongly affect it are the temperature and pressure. This guideline helps the control system to ensure that the product specifications are regulated and met.

Guideline 4. *Choose output variables that exhibit significant interactions with other output variables.* Plantwide control must handle the potential interactions in the process. Improved closed-loop performance is achieved by stabilizing output variables that interact significantly with each other.

Guideline 5. *Choose output variables that have favorable static and dynamic responses to the available manipulated variables.* All other things being equal, this guideline should be applied.

Selection of Manipulated Variables

Guideline 6. *Select manipulated variables that significantly affect the controlled variables.* For each control loop, select an input variable with as large a steady-state gain as possible and sufficient range to adjust the controlled variable. For example, when a distillation column operates with a large reflux ratio, that is, values greater than four (Luyben et al., 1999), it is much easier to control the level in the reflux drum using the reflux flow rate rather than the distillate flow rate.

Guideline 7. *Select manipulated variables that rapidly affect the controlled variables.* This precludes the selection of inputs that affect the outputs with large delays or time constants.

Guideline 8. *Select manipulated variables that affect the controlled variables directly rather than indirectly.* For example, when appropriate for the design of an exothermic reactor, it is preferable to inject a coolant directly rather than use a cooling jacket.

Guideline 9. *Avoid recycling disturbances.* It is better to eliminate the effect of disturbances by allowing them to leave the process in an exiting stream rather than having them propagate through the process by the manipulation of a feed or recycle stream.

Selection of Measured Variables

Both input and output variables may be measured variables, with on-line measurement preferred to off-line measurement. Seborg et al. (1989) discuss the importance of measurements in control, and provide three guidelines for the selection of variables to be measured and the location of the measurements.

Guideline 10. *Reliable, accurate measurements are essential for good control.* Examples of poorly designed measurements include orifices with insufficient straight piping, and/or saturated liquids that flash in the orifice.

Guideline 11. *Select measurement points that are sufficiently sensitive.* Consider, for example, the indirect control of the product compositions from a distillation column by the regulation of a temperature near the end of the column. In high-purity distillation columns, where the terminal temperature profiles are almost flat, it is preferable to move the temperature measurement point closer to the feed tray.

Guideline 12. *Select measurement points that minimize time delays and time constants.* Large time delays and dynamic lags in the process limit the achievable closed-loop performance. These should be reduced, whenever possible, in the process design and the selection of measurements.

Degrees-of-Freedom Analysis

Before selecting the controlled and manipulated variables for a control system, one must determine the number of manipulated variables permissible. As discussed under "Degrees of Freedom" in Section 4.2, the number of manipulated variables cannot exceed the number of degrees of freedom, which are determined using a process model according to

$$N_D = N_{\text{Variables}} - N_{\text{Equations}} \tag{20.2}$$

where N_D is the number of degrees of freedom, $N_{\text{Variables}}$ is the number of process variables, and $N_{\text{Equations}}$ is the number of independent equations that describe the process. However, the number of manipulated variables is generally less than the number of degrees of freedom, since one or more variables may be externally defined (disturbed); that is, $N_D = N_{\text{Manipulated}} + N_{\text{Externally Defined}}$. Consequently, the number of manipulated variables can be expressed in terms of the number of externally defined variables:

$$N_{\text{Manipulated}} = N_{\text{Variables}} - N_{\text{Externally Defined}} - N_{\text{Equations}} \tag{20.3}$$

The number of manipulated variables equals the number of controlled variables that can be regulated. When a manipulated variable is paired with a regulated output variable, its degree of freedom is transferred to the output's setpoint, which becomes the new independent variable.

Next, degrees-of-freedom analyses are carried out and their implications for control system design are considered for heat-exchanger networks, jacketed stirred-tank reactors, a utility system, a flash vessel, and a distillation column.

EXAMPLE 20.5 | *Control Configurations for Heat-Exchanger Networks (Example 20.1 Revisited)*

Referring to Figure 20.1a, the process can be described in terms of 15 variables: F_1, F_2, F_3, T_0, T_1, T_2, T_3, θ_0, θ_1, θ_2, θ_3, θ_4, Q_1, Q_2, and Q_3. Of these, assume that four variables can be considered to be externally defined: F_1, T_0, θ_0, and θ_1. A steady-state model for the process consists of three equations for each heat exchanger. For example, for the first heat exchanger, the following equations apply

$$Q_1 = F_1 C_{p1}(T_0 - T_1) \tag{20.4}$$

$$Q_1 = F_3 C_{p3}(\theta_4 - \theta_3) \tag{20.5}$$

$$Q_1 = U_1 A_1 \frac{(T_0 - \theta_4) - (T_1 - \theta_3)}{\ln\left\{\dfrac{T_0 - \theta_4}{T_1 - \theta_3}\right\}} \tag{20.6}$$

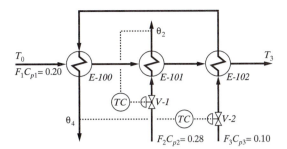

Figure 20.5 Control system for original heat-exchanger network.

In these equations, Q_i, U_i, and A_i are the heat duty, heat transfer coefficient, and heat transfer area for heat exchanger i. Values for the latter two are assumed known, so they are not process variables. Similar equations are written for the other two heat exchangers, making a total of nine equations. Consequently, the number of manipulated variables is computed as $N_{\text{Manipulated}} = N_{\text{Variables}} - N_{\text{Externally Defined}} - N_{\text{Equations}} = 15 - 4 - 9 = 2$.

Thus, two variables can be manipulated. Two candidates are the flow rates of the two cold streams: F_2 and F_3. Ideally, for the selection of the controlled variables, it would be desirable to regulate all three target temperatures: T_3, θ_2, and θ_4. However, with only two manipulated variables, only two controlled variables can be selected. The guidelines presented above are insufficient to select which two of the three should be picked, since all three provide a direct measure of the product quality (Guideline 3), and there are clearly significant interactions among all three variables (Guideline 4). Without quantitative analysis, one cannot gauge which of the three have the most favorable static and dynamic responses to the manipulated variables. If only T_3, θ_2, and θ_4 are considered as potential controlled variables, three possible control systems should be investigated. As an illustration, Figure 20.5 shows one possible configuration of two control loops. One loop adjusts the flow rate, F_2, to control the temperature, θ_2, while the other loop adjusts F_3 to control θ_4. An alternative configuration with reversed pairings (i.e., $\theta_4 - F_2$, $\theta_2 - F_3$) is unstable, as shown in Case Study 21.2.

The design in Figure 20.1b, involving a bypass on exchanger E-102, permits the regulation of all three target temperatures. In this case, the number of variables is increased by two (the bypass flow fraction, ϕ, and the temperature, θ_3'), giving a total of 17 variables, with the same four disturbance variables. For constant heat capacities and no phase change, the process is modeled by one additional energy balance for the mixer,

$$\theta_3 = (1 - \phi)\,\theta_0 + \phi\theta_3' \tag{20.7}$$

where ϕ is the E-102 bypass flow fraction and θ_3' is the temperature leaving heat exchanger E-102. Since $N_{\text{Manipulated}} = N_{\text{Variables}} - N_{\text{Externally Defined}} - N_{\text{Equations}} = 17 - 4 - 10 = 3$, three variables can be manipulated, namely, F_2, F_3, and ϕ. The flow rate of the second cold stream, F_3, affects two of the three heat exchangers, whereas F_2 affects only the second one directly, and ϕ affects T_3 directly (Guidelines 6, 7, and 8). The control structure shown in Figure 20.6 is the most resilient and controllable regulatory structure, as is demonstrated in Case Study 21.2 using quantitative analysis. ∎

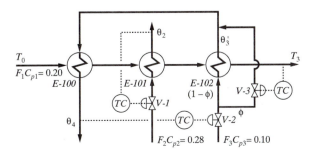

Figure 20.6 Control system for the modified heat-exchanger network.

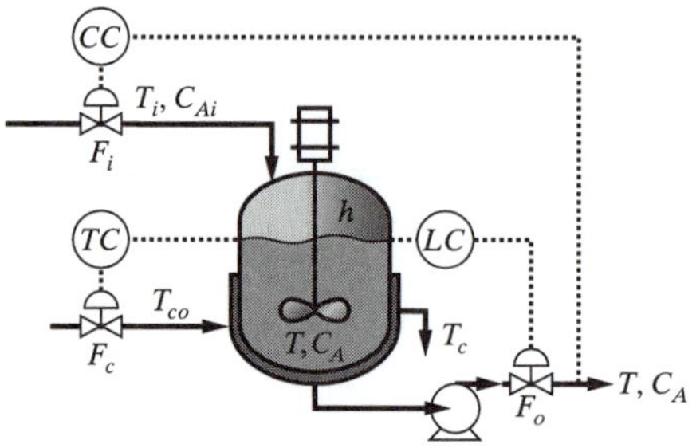

Figure 20.7 Control system for a jacketed CSTR.

EXAMPLE 20.6   *Control Configuration for a Jacketed CSTR*

Consider the control of a jacketed, continuous, stirred-tank reactor (CSTR) in which the exothermic reaction A → B is carried out. This system can be described by 10 variables, as shown in Figure 20.7: h, T, C_A, C_{Ai}, T_i, F_i, F_o, F_c, T_c, and T_{co}, three of which are considered to be externally defined : C_{Ai}, T_i, and T_{co}. Its model involves four equations, assuming constant fluid density.

1. Overall mass balance:

$$A\frac{dh}{dt} = F_i - F_o \tag{20.8}$$

2. Mass balance on component A:

$$A\frac{d}{dt}(hC_A) = F_iC_{Ai} - F_oC_A - Ah \cdot r\{C_A,T\} \tag{20.9}$$

3. Energy balance on the reacting mixture:

$$A\rho c_p \frac{d}{dt}(h \cdot T) = F_i\rho c_p T_i - F_o\rho c_p T$$
$$+ Ah \cdot r\{C_A,T\}(-\Delta H) - UA_s(T - T_c) \tag{20.10}$$

4. Energy balance on the jacket coolant:

$$V_c\rho_c c_{p_c}\frac{dT_c}{dt} = F_c c_{p_c} T_{co} - F_c c_{p_c} T_c + UA_s(T - T_c) \tag{20.11}$$

where A is the cross-sectional area of the vessel, h is the liquid level in the reactor, A_s is the area for heat transfer, U is the overall heat transfer coefficient, C_{Ai} and C_A are the inlet and reactor concentrations of A, T_i, and T are the inlet and reactor temperatures, F_i and F_o are the inlet and outlet volumetric flow rates, ρ is the fluid density, F_c is the coolant mass flow rate, ρ_c is the coolant density, T_{co} and T_c are the inlet coolant and jacket temperatures, V_c is the volume of the cooling jacket, r is the intrinsic rate of reaction, ΔH is the heat of reaction, and c_p and c_{pc} are the specific heats of the reacting mixture and coolant, respectively.

Here, the number of variables that can be manipulated independently is $N_{\text{Manipulated}} = N_{\text{Variables}} - N_{\text{Externally Defined}} - N_{\text{Equations}} = 10 - 3 - 4 = 3$.

Selection of controlled variables. C_A should be selected because it affects the product quality directly (Guideline 3). T should be selected because it must be regulated properly to avoid safety problems (Guideline 2) and because it interacts with C_A (Guideline 4). Finally, h must be selected as a controlled output because it is non-self-regulating (Guideline 1).

Selection of manipulated variables. The volumetric feed flow rate, F_i, should be selected because it directly and rapidly affects the conversion (Guidelines 6, 7, and 8). Using the same reasoning, F_c is selected to control the reactor temperature, T, and the flow rate of the reactor effluent, F_o, is selected to control h. This configuration, which is shown in Figure 20.7, should be compared with other pairings using the quantitative analysis presented in Chapter 21, it being noted that there are several opportunities for improvement. ∎

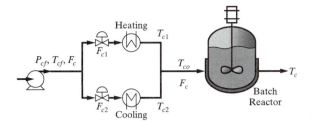

Figure 20.8 Utilities subsystem for a batch chemical reactor.

EXAMPLE 20.7 **Control Configuration for a Utilities Subsystem**

Often, the contents of a chemical batch reactor are heated initially to achieve ignition, and then cooled to remove the heat generated in reaction. In such cases, it is common to install a jacket supplied with both cooling and heating utility streams, as shown in Figure 20.8. The utilities subsystem involves eight variables: P_{cf}, T_{cf}, F_{c1}, T_{c1}, F_{c2}, T_{c2}, F_c, and T_{co}. Of these, two are externally defined and constitute disturbance variables: P_{cf} and T_{cf}. Four material and energy balances relate the subsystem variables: (1) an energy balance for the cooling branch, (2) an energy balance for the heating branch, (3) an energy balance for the mixing junction, and (4) a mass balance for the mixing junction. Hence, the number of variables to be manipulated independently is $N_{\text{Manipulated}} = N_{\text{Variables}} - N_{\text{Externally Defined}} - N_{\text{Equations}} = 8 - 2 - 4 = 2$. This is also the number of subsystem variables that can be controlled independently.

Selection of Controlled Variables. The guidelines in Section 20.2 are not helpful because no output variable has a direct effect on the product quality, all are self-regulating, and none are directly associated with equipment or operating constraints. Nonetheless, F_c and T_{co} are obvious choices for the controlled variables because the objective of this subsystem is to control the temperature and flow rate of the utility stream fed to the reactor jacket.

Selection of Manipulated Variables. The two obvious candidates are F_{c1} and F_{c2}, since both affect the two outputs directly and rapidly (Guidelines 7 and 8). However, linear and nonlinear combinations of these flow rates are also possible. As shown in Example 21.4, a quantitative analysis is needed to make the best selection. ∎

EXAMPLE 20.8 **Control Configuration for a Flash Drum**

The flash drum in Figure 20.9 illustrates a situation where a stream containing a binary mixture of two components, A and B, is flashed through a valve and separated in a flash drum into an overhead vapor stream and a residual liquid product stream. The liquid in the drum is cooled by external heat exchange with liquid recycle. This process is modeled with 11 variables: F_i, T, C_A, F_W, P_f, h, T_f, F_V, y_A, F_L, and x_A. Two variables are considered to be externally defined, T and C_A. The model involves five equations: a

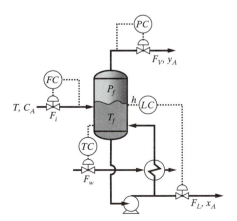

Figure 20.9 Control configuration for a flash drum.

total mass balance, a mass balance for component A, an overall energy balance, and a vapor–liquid equilibrium equation for each component. Thus, the number of variables to be manipulated independently is $N_{\text{Manipulated}} = N_{\text{Variables}} - N_{\text{Externally Defined}} - N_{\text{Equations}} = 11 - 2 - 5 = 4$.

Selection of Controlled Variables. P_f is selected because of the potential safety problems (Guideline 2) and because it affects the product concentrations (Guideline 3). T_f should be selected because it directly affects the product quality (Guideline 3). The liquid height in the drum, h, must be selected because it is non-self-regulating (Guideline 1), and F_i is selected because it controls the product flow rate directly, one of the overall control objectives (Guideline 3). Note that all of these outputs exhibit significant interaction.

Selection of Manipulated Variables. F_i is adjusted to achieve its setpoint (Guideline 8). F_V has a rapid, direct effect on the vessel pressure, P_f, and almost no effect on any other output (Guidelines 7 and 8). For similar reasons, F_L is selected to control the liquid level, h. F_W is selected because it directly controls the flash temperature, T_f (Guideline 8).   ∎

EXAMPLE 20.9   *Control Configuration for a Binary Distillation Column*

This analysis, for the distillation operation in Figure 20.10, is based on the following assumptions and specifications: (a) constant relative volatility, (b) saturated liquid distillate, (c) negligible vapor holdup in the column, (d) constant tray pressure drops (Luyben, 1990), and (e) negligible heat losses except for the condenser and reboiler. When using N_T trays, the column is modeled with $4N_T + 13$ variables:

Vapor and liquid compositions on each tray:	$2N_T$
Tray liquid flow rates and holdups:	$2N_T$
Reflux drum holdup and composition:	2
Reflux and distillate flow rates:	2
Column sump vapor and liquid compositions:	2
Column sump liquid holdup:	1
Bottoms product and reboiler vapor flow rates:	2
Feed flow rate and composition:	2
Condenser pressure:	1
Condenser duty:	1

The system is described by $4N_T + 6$ equations:

Species mass balances (trays, sump, reflux drum):	$N_T + 2$
Total mass balances (trays, sump, reflux drum):	$N_T + 2$
Vapor–liquid equilibrium (trays, sump):	$N_T + 1$
Tray hydraulics for tray holdup:	N_T
Total vapor dynamics:	1

Assuming that the feed flow rate and composition are externally defined, the number of variables to be manipulated independently is $N_{\text{Manipulated}} = N_{\text{Variables}} - N_{\text{Externally Defined}} - N_{\text{Equations}} = 4N_T + 13 - 2 - (4N_T + 6) = 5$.

Selection of Controlled Variables. The condenser pressure, P_D, should be regulated, since it strongly affects the product compositions (Guidelines 3 and 4). The reflux drum and sump liquid inventory levels, L_D and L_R, need to be regulated, since they are not self-regulating (Guideline 1). This leaves two additional variables that can be regulated. When distillate and bottoms streams are product streams, their compositions, x_D and x_B, are often selected as controlled variables (Guideline 3). However, as pointed out by Luyben et al. (1999), in plantwide control, it is often sufficient to control one composition; for example, when the composition of a recycle stream is not regulated. Since significant delay times are often associated with composition measurements, tray temperatures (which are measured without delay times) are often used to infer compositions (Guideline 12). In this regard, temperatures must be measured on trays that are sensitive to column upsets (Guideline 11).

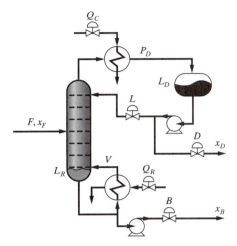

Figure 20.10 A distillation column with two liquid products.

Selection of Manipulated Variables. As shown in Figure 20.10, by labels on the valves, the five manipulated variables are the flow rates of the reflux, distillate and bottoms streams L, D, and B, and the cooling and heating duties, Q_C and Q_R. It is most common to control P_D in columns with a liquid overhead product by manipulating Q_C. Since it is impossible to control the product compositions while setting the product flow rates, the latter are commonly used to regulate the liquid inventory levels (i.e., the liquid levels in the reflux drum and the sump). This leaves two variables to control the product compositions. Figure 20.11a shows the so-called LV-configuration, where L controls x_D, Q_R (which is closely related to V) controls x_B, D controls L_D, B controls L_R, and Q_C controls P_D. In columns operating with large reflux ratios, it is advisable to use L to control L_D (Guideline 6). Then, D regulates x_D, in the so-called DV-configuration shown in Figure 20.11b. Alternative configurations involve ratios of manipulated variables, intended to decouple the control loops by reducing the interaction between them (Shinskey, 1984; Luyben et al., 1999). The dynamic performance of a column control system should be verified using the quantitative methods described in Chapter 21.   ■

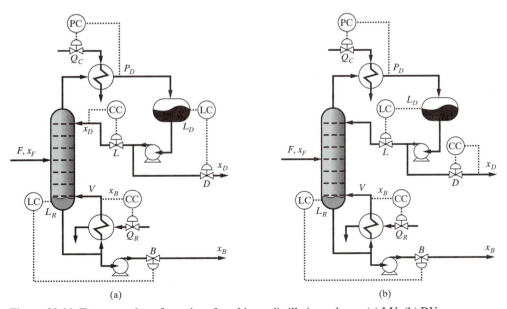

Figure 20.11 Two control configurations for a binary distillation column: (a) LV; (b) DV.

20.3   QUALITATIVE PLANTWIDE CONTROL SYSTEM SYNTHESIS

As pointed out by Luyben et al. (1999), the design of a plantwide control system should be driven by the objectives of the overall process rather than by considerations of the individual processing units as in the preceding section. Their strategy for control system design utilizes the available degrees of freedom to achieve these objectives in order of importance, by adopting a "top-down" approach, in common with successful programming practice. Alternatively, in a simpler "bottom-up" approach (Stephanopoulos, 1984), the process is divided into subsystems, with each subsystem often comprised of several process units that share a common processing goal. Then, a control system is formulated for each subsystem, relying on the qualitative guidelines in Section 20.2 or the quantitative analysis to be described in Chapter 21. Finally, an integrated system is synthesized by eliminating possible conflicts among the subsystems. The main disadvantage of this bottom-up approach is that good solutions at the subsystem level may not satisfy the process objectives. This can occur when manipulated variables are assigned to meet the control objectives of a subsystem, leaving less attractive inputs to satisfy those of the overall process. Furthermore, interactions between subsystems, such as those resulting from heat integration and material recycle, are not addressed in this decomposition approach, which may lead to unworkable solutions, as will be shown.

The qualitative design procedure for plantwide control by Luyben et al. (1999) consists of the following steps.

Step 1. Establish the control objectives. As mentioned above, these are related closely to the process objectives. For example, one may wish to impose a given production rate, while ensuring that the products satisfy the quality specified by the market, and guaranteeing that the process meets environmental and safety constraints.

Step 2. Determine the control degrees of freedom. In practice, the degrees-of-freedom analysis in Section 20.2 may be too cumbersome for the synthesis of plantwide control systems. In a more direct approach, the number of control valves in the flowsheet equals the degrees of freedom (Luyben et al., 1999). As the valves are positioned on the flowsheet, care must be taken to avoid the control of a flow rate by more than one valve. In cases where the degrees of freedom are insufficient to meet all of the control objectives, it may be necessary to add control valves, for example, by adding bypass lines around heat exchangers, as shown in Example 20.5, or by adding trim-utility heat exchangers.

Step 3. Establish the energy management system. In this step, control loops are positioned to regulate exothermic and endothermic reactors at desired temperatures. In addition, temperature controllers are positioned to ensure that disturbances are removed from the process through utility streams rather than recycled by heat-integrated process units.

Step 4. Set the production rate. This is accomplished by placing a flow control loop on the principal feed stream (referred to as *feed flow control*), or on the principal product stream (referred to as *on-demand product flow*), noting that these two options lead to very different plantwide control configurations. Alternatively, the production rate is controlled by regulating the reactor operating conditions; for example, temperature and feed composition.

Step 5. Control the product quality and handle safety, environmental, and operational constraints. Having regulated the production rate and the effect of temperature disturbances, secondary objectives to regulate product quality and satisfy safety, environmental, and operational constraints and process constraints are addressed in this step.

Step 6. Fix a flow rate in every recycle loop and control vapor and liquid inventories (vessel pressures and levels). Process unit inventories, such as liquid holdups and vessel pressures (measures of vapor holdups), are relatively easy to control. While vessel holdups are usually non-self-regulating (Guideline 1), the *dynamic performance* of their controllers is less important. In fact, level controllers are usually detuned to allow the vessel accumulations to dampen disturbances in the same way that shock absorbers cushion an automobile, as demonstrated in Chapter 21. Less obvious is the need to handle plantwide holdups in recycle loops. As will be shown qualitatively in several examples that follow, and quantitatively in Chapter 21, failure to impose flow control on each recycle stream can result in the loss of control of the process.

Step 7. Check component balances. In this step, control loops are installed to prevent the accumulation of individual chemical species in the process. Without control, chemical species often build up, especially in material recycle loops.

Step 8. Control the individual process units. At this point, the remaining degrees of freedom are assigned to ensure that adequate local control is provided in each process unit. Note that this comes *after* the main plantwide control issues have been handled.

Step 9. Optimize economics and improve dynamic controllability. When control valves remain to be assigned, they are utilized to improve the dynamic and economic performance of the process.

Next, the above procedure, which focuses the control system design on meeting the process objectives, is demonstrated on three processes of increasing complexity: (a) an acyclic process; (b) the reactor–flash–recycle process in Example 20.4; and (c) the vinyl chloride process in Sections 3.4 and 3.5. The last two examples feature recycle loops.

EXAMPLE 20.10   *Plantwide Control System Configuration for an Acyclic Process*

The chemical process shown in Figure 20.12 is based on an example by Stephanopoulos (1984). It consists of a CSTR in which the species A reacts to form B in an exothermic reaction. The reactor effluent is fed to a flash vessel, where the heavier product B is concentrated in the liquid stream, and unreacted A is discarded in the vapor stream. A preheater recovers heat from the hot reactor effluent, with a so-called "trim heater" installed to ensure that the liquid reactor feed is at the desired temperature. To ensure that the reactor temperature remains on target, the CSTR is equipped with a jacket, fed with cooling water to attenuate the heat released. Seven flow-control valves are shown. An eighth valve could be placed on the liquid recycle to V-100, but is not.

Applying the nine-step design procedure for plantwide control by Luyben and co-workers (1999), using, where possible, the guidelines of Section 20.2:

Step 1. *Set objectives*. The control objectives for this process are as follows:

 1. Maintain the production rate of component B at a specified level.
 2. Keep the conversion of the plant at its highest permissible value.
 3. Achieve constant composition in the liquid effluent from the flash drum.

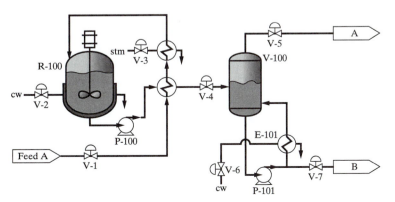

Figure 20.12 Process flowsheet for the acyclic process.

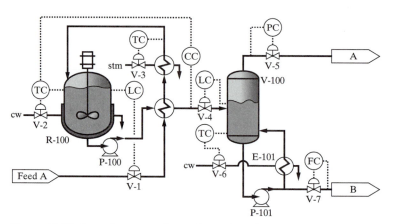

Figure 20.13 Control structure for on-demand product flow.

The structure of the plantwide control system depends on the primary control objective: that is, to maintain a desired production rate. The two possible interpretations of this goal are (a) ensure a desired flow rate of the product stream by flow control using valve V-7, which leads to the "on-demand product flow" configuration shown in Figure 20.13, or (b) ensure a desired production level by "feed flow rate control" using valve V-1, which leads to the control configuration shown in Figure 20.14. The *on-demand product flow* configuration is considered first.

Step 2. *Define control degrees of freedom.* As shown in Figure 20.12, the process has seven degrees of freedom for manipulated variables, with three valves controlling the flow rates of the utility streams (V-2, V-3, and V-6), one controlling the feed flow rate (V-1), two controlling the product flows (V-5 and V-7), and one controlling the reactor effluent flow rate (V-4). Having decided to design an *on-demand product flow* configuration, the valve controlling the B product flow rate (V-7) is reserved for independent flow control (i.e., it directly controls the flow rate). This leads to the control system shown in Figure 20.13.

Step 3. *Establish energy management system.* The critical energy management for the CSTR is handled next, since loss of control of the reactor would have serious plantwide consequences. Using the guidelines for controlled and manipulated variable selection, the reactor feed and effluent temperatures are identified as critical for safety (Guideline 2) and quality assurance (Guideline 3). The obvious choices for valves to control these two temperatures are V-2, the jacket coolant valve, and V-3, the steam valve for the trim heater, both of which have a direct effect (Guidelines 6 and 7). These are assigned to temperature control loops.

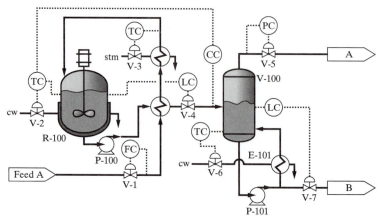

Figure 20.14 Control structure for fixed fresh feed.

Step 4. *Set the production rate.* As mentioned above, the B product valve, V-7, is assigned to a flow controller, whose setpoint directly regulates the production rate.

Step 5. *Control product quality and meet safety, environmental, and operational constraints.* The product quality is controlled by maintaining the operating temperature and pressure in the flash vessel at setpoints (Guideline 3). The former is regulated by adjusting the coolant water flow rate through V-6, while the latter is controlled by adjusting the flow rate through overhead valve, V-5. These valves are selected because of their rapid and direct effect on the outputs (Guidelines 6, 7, and 8). In addition, these two control loops satisfy the third control objective; that is, to provide tight product-quality control.

Step 6. *Fix recycle flow rates and vapor and liquid inventories.* The liquid inventories in the flash vessel and reactor are non-self-regulating, and therefore, need to be controlled (Guideline 1). Since the liquid product valve from the flash vessel has been assigned to control the product flow rate, the inventory control must be in the reverse direction to the process flow. Thus, the reactor effluent valve, V-4, controls the flash vessel liquid level, and the feed valve, V-1, controls the reactor liquid level. Both of these valves have rapid, direct effects on the liquid holdups (Guidelines 6, 7, and 8). The vapor product valve, V-5, which has been assigned to control the pressure in V-100, thereby controls the vapor inventory.

Step 7. *Check component balances.* With the controllers assigned above, A and B cannot build up in the process, and consequently, this step is not needed.

Step 8. *Control the individual process units.* Since all of the control valves have been assigned, no additional control loops can be designed for the process units.

Step 9. *Optimize economics and improve dynamic controllability.* While a temperature control system for the CSTR is in place, its setpoint needs to be established. To meet the second control objective, which seeks to maximize conversion, a cascade controller is installed in which the setpoint of the reactor temperature controller (TC on V-2) is adjusted to control the concentration of B (CC) in the reactor effluent. If the reaction is irreversible, conversion is maximized by operating the reactor at the highest possible temperature, making this controller unnecessary.

This completes the control system design for the *on-demand product flow* configuration in Figure 20.13. The performance of the control system needs to be verified by using controllability and resiliency assessment and by applying dynamic simulation, as described in Chapter 21. As an alternative to the *on-demand product flow* configuration, the production level can be maintained by fixing the feed flow rate, which leads to the control system shown in Figure 20.14. This control configuration is derived using the same procedure as with Figure 20.13, with the only difference being in Step 6, where the liquid levels are controlled in the direction of the process flow. For the fixed-feed configuration, reaction kinetics may dictate that the reactor holdup needs to be manipulated in concert with throughput changes. In this case, it may be necessary to coordinate the reactor level set point with the feed flow rate.   ■

Processes that involve significant heat integration and/or material recycle present more challenging plantwide control problems. The next two examples involve material recycle loops.

EXAMPLE 20.11   *Plantwide Control System Configuration for Reactor–Flash–Recycle Process (Example 20.4 Revisited)*

For the reactor–flash–recycle process introduced in Example 20.4, Figure 20.15 shows a control system with the control objectives to:

1. Maintain the production rate of component B at a specified level.
2. Keep the conversion of the plant at its highest permissible value.

This controller configuration results from using a unit-by-unit design approach, with each vessel inventory controlled by manipulation of its liquid effluent flow. Although the control pairings are acceptable for each process unit in isolation, the overall control system does not establish flow control of the

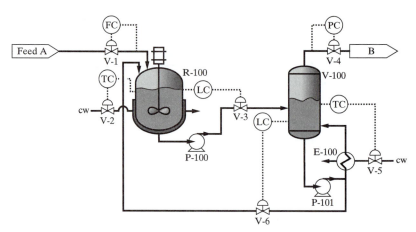

Figure 20.15 Control structure based on unit-by-unit design approach.

recycle stream. Consequently, a change in the desired feed rate, keeping the reactor inventory constant with level control, causes an excessive increase in the reactor effluent flow, which is transferred rapidly to the recycle flow by the flash level controller. This undesirable positive-feedback is referred to as the "snowball effect" by Luyben and co-workers (1999), and is the consequence of not ensuring flow control of the recycle stream.

Since Luyben identified the snowball effect (Luyben, 1994), the sensitivity of reactor–separator–recycle processes to external disturbances has been the subject of several studies (e.g., Wu and Yu, 1996; Skogestad, 2002). Recent work by Bildea and co-workers (Bildea et al., 2000 and Kiss et al., 2002) has shown that a critical reaction rate can be defined for each reactor–separator–recycle process using the Damköhler number, Da (dimensionless rate of reaction, proportional to the reaction rate constant and the reactor hold-up). When the Damköhler number is below a critical value, Bildea et al. show that the conventional unit-by-unit approach in Figure 20.15 leads to the loss of control. Furthermore, they show that controllability problems associated with exothermic CSTRs and PFRs are resolved often by controlling the total flow rate of the reactor feed stream.

The extent of the snowball effect is shown next by analysis of the controlled process in Figure 20.15. The combined feed of pure A and recycle is partially converted to B in reactor, R-100, by the isothermal, liquid-phase, irreversible reaction, A $\rightarrow$ B, which has first-order kinetics. The reactor effluent is flashed across valve, V-3, to yield a vapor product stream, assumed to be pure B, and a liquid product stream, assumed to be pure A. The liquid stream is recycled to the reactor where it is mixed with fresh A to give the combined feed stream. The control configuration consists of six control loops: (1) production rate, F_0, controlled using the valve, V-1, on the fresh A feed stream, (2) temperature control using valve, V-2, to ensure isothermal operation of R-100, (3) level control in R-100 using valve, V-3, (4) level control in V-100 using valve, V-6, (5) pressure control in V-100 using the vapor product valve, V-4, and (6) temperature control in V-100 (controlling product quality) using the coolant valve, V-5.

What happens when the fresh feed flow rate changes or is disturbed? Note, that even though the flow controller fixes the fresh feed rate in Figure 20.15, it can still be disturbed. The equations that apply are:

Combined molar feed to the CSTR:	$F_0 + B$
Molar material balance around the flash vessel:	$F_0 + B = D + B$
Overall material balance:	$F_0 = D$

where F_0, D, and B, are the molar flow rates of the feed, distillate, and bottoms streams. Finally, the rate of consumption of A in the reactor is:

$$r_A = kc_A \tag{20.12}$$

where r_A is the intrinsic rate of reaction, k is the first-order rate constant, and c_A is the molar concentration of A in the reactor effluent. Using c_{total} and x_A, the total molar concentration and the mole fraction of A in the reactor effluent, Eq. (20.12) becomes:

$$r_A = kx_A c_{total} \tag{20.13}$$

The molar flow rate of B in the reactor effluent is:

$$(1 - x_A)(F_0 + B) = kx_A c_{total} V_R \tag{20.14}$$

where V_R is the volume of the reactor holdup. Then, substituting $c_{total} V_R = n_T$:

$$(1 - x_A)(F_0 + B) = kx_A n_T \tag{20.15}$$

where n_T is the total molar holdup in the reactor. Rearranging Eq. (20.15) for the bottoms flow rate (recycled to the reactor), B:

$$B = \frac{x_A(F_0 + kn_T) - F_0}{1 - x_A} \tag{20.16}$$

With the reactor temperature and holdup fixed, a change to the fresh feed flow rate by a disturbance causes the mole fraction of A in the reactor effluent to change. So, to obtain the effect of a change in F_0 on B, x_A must be eliminated from Eq. (20.16). For a perfect separation, an overall balance on the disappearance of A gives:

$$F_0 = kx_A n_T \tag{20.17}$$

Rearranging Eq. (20.17) for x_A and substituting in Eq. (20.16) gives:

$$B = \frac{F_0^2}{kn_T - F_0} \tag{20.18}$$

Equation (20.18) shows clearly that the numerator increases as the square of F_0 while the denominator decreases with increasing F_0. Thus, B increases by more than the square of F_0! As an example, consider increasing F_0 from 50 to 150, with $kn_T = 200$. Then, Eq. (20.18) gives the distillate recycle rate, B:

F_0	B
50	16.7
75	45
100	100
125	208
150	450

Thus, when the feed rate is tripled from 50 to 150, the distillate rate increases by a factor of $450/16.7 = 27$. This result assumes fixed kn_T. A more general result relies on reformulating Eq. (20.18) in terms of the Damköhler number, $Da = kn_T/F_0$, giving:

$$B = \frac{F_0}{Da - 1} \tag{20.19}$$

Equation (20.19) shows that for values of Da much larger than unity, no snowball effect is expected. The snowball effect occurs as Da approaches a critical value of 1, but is eliminated by controlling the recycle flow rate, as shown next.

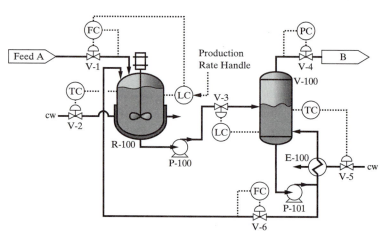

Figure 20.16 Workable plantwide control structure.

To generate a workable plantwide control system, as shown in Figure 20.16, the design procedure for plantwide control by Luyben and co-workers (1999) is applied

Step 1. *Set objectives.* To achieve the primary control objective, the production level is maintained by flow control of the feed stream using valve V-1.

Step 2. *Define control degrees of freedom.* As shown in Figure 20.4, the process has six degrees of freedom with two valves controlling the flow rates of the utility streams (V-2 and V-5), one controlling the feed flow rate (V-1), one controlling the product flow (V-4), one controlling the reactor effluent flow rate (V-3), and one controlling the recycle flow rate (V-6). Having chosen *constant feed flow* in Step 1, the feed valve (V-1) is reserved for independent flow control.

Step 3. *Establish energy management system.* The reactor temperature, which affects the process yield and stability (Guidelines 2 and 3), is controlled by adjusting the coolant flow rate, using valve V-2.

Step 4. *Set the production rate.* As stated previously, the feed valve, V-1, is assigned to a flow controller, whose setpoint regulates the production rate.

Step 5. *Control product quality and meet safety, environmental, and operational constraints.* A conventional pressure and temperature control system is set up for the flash vessel, as in the previous example.

Step 6. *Fix recycle flow rates and vapor and liquid inventories.* To eliminate the snowball effect, the recycle flow rate must be controlled by installing a flow controller, either on the reactor effluent or on the flash liquid effluent. As shown in Figure 20.16, the second option forces the reactor effluent valve, V-3, to control the flash vessel liquid inventory, in the absence of other alternatives. Then, to regulate the reactor inventory, a cascade control system is designed, in which the reactor level controller (LC) adjusts the setpoint of the feed flow controller (FC on V-1). This does not conflict with the objective to set the production rate by fixing the feed flow rate because, in stable operation, the reactor level and feed flow rate vary proportionally, through higher conversion in the CSTR. The vapor product valve, V-4, which has been assigned to control the pressure, thereby controls the vapor inventory.

Steps 7 and 8. *Check component balances and control individual process units.* As in Example 20.10, the controllers assigned thus far prevent the build up of A and B in the process. Furthermore, no valves are available to improve control of either process unit.

Step 9. *Optimize economics and improve dynamic controllability.* To maximize conversion, a cascade controller is installed as in the previous example, in which the setpoint of the reactor temperature controller (TC on V-2) is adjusted to control the concentration of B in the reactor effluent. Again, for an irreversible reaction, it is enough to operate the reactor at the highest possible temperature. ∎

EXAMPLE 20.12   *Plantwide Control System Configuration for the Vinyl Chloride Process*

For the vinyl chloride process synthesized in Section 3.4, a preliminary design of its plantwide control system helps to assess the ease of maintaining the desired production level. As shown in Figures 20.17 and 20.18, this is achieved following the design procedure of Luyben and co-workers (1999):

Step 1. *Set objectives.* Note that nearly 100% conversion is achieved in the dichloroethane reactor (R-100). Assuming that the conversion in the pyrolysis furnace (F-100) cannot be altered, the production level must be maintained by flow control of the ethylene feed flow rate using valve V-1.

Step 2. *Determine the control degrees of freedom.* Twenty control valves have been positioned in the PFD, as shown in Figure 20.17.

Step 3. *Establish energy management system.* The coolant valve, V-3, in the overhead condenser of the exothermic dichloroethane reactor, R-100, is used for temperature control. The yield in the pyrolysis furnace, F-100, is controlled by maintaining the outlet temperature at 500°C using the fuel gas valve, V-6. To attenuate the effect of temperature disturbances, the flow rates of the utility streams are adjusted to regulate effluent temperatures in the evaporator, E-101 (using V-5), the quench tank, V-100 (using the cooler E-102 and manipulating V-7), the partial condenser, E-103 (using V-8), and the recycle cooler (using V-20). All of these valves act rapidly and directly on the controlled outputs (Guidelines 6, 7, and 8). Note that the temperature control loops using utility exchangers ensure that temperature disturbances are not recycled (Guideline 9).

Step 4. *Set the production rate.* As stated previously, the feed valve, V-1, is assigned to a flow controller whose setpoint regulates the production rate.

Step 5. *Control product quality and meet safety, environmental, and operational constraints.* The overhead product compositions in both distillation columns are regulated by adjusting the reflux flow rates using valves V-11 and V-16, both of which provide fast, direct control action (Guidelines 6, 7, and 8). The bottoms product compositions are controlled using the reboiler steam valves, V-13 and V-18. Thus, the control systems for both columns are in the LV configuration. The process design calls for pressure reduction at the feed of each distillation column, using valves V-9 and V-14. However, these valves are not appropriate for the control of the column pressures, which are regulated in Step 6 by manipulation of downstream, rather than upstream valves. Thus, V-9 is maintained constant at its design position, while V-14 is utilized for inventory control in the sump of column T-100 from Step 6.

Step 6. *Fix recycle flow rates and vapor and liquid inventories.* The recycle flow rate must be held constant by flow control. However, since two-point composition control has been installed in T-101, it is not possible to also fix its bottoms flow rate by manipulating V-19. Thus, a flow controller is installed to fix the combined recycle and feed flow rates using V-4. In addition, liquid inventory control must be installed for the reactor R-100, as well as for the sumps and reflux drums of the two columns, with vapor inventory control needed for the two columns. Having assigned V-4 for recycle flow rate control, a level controller for R-100 is cascaded with the ethylene flow controller, making the level setpoint the production handle, as in Example 20.11 (Figure 20.16). Inventory control of T-101 is assigned first. With the total recycle flow rate under control, the bottoms flow rate in T-101, adjusted by valve V-19, is used for sump level control. The liquid level in reflux drum V-102 is controlled by manipulating the distillate valve, V-17. Inventory control for T-101 is completed by controlling the overhead pressure using the coolant valve, V-15. Turning to the HCl column, T-100, the bottoms product valve, V-14, is assigned to control the sump level. Since the overhead product is vapor, the condenser pressure is regulated using the distillate valve, V-12. This frees the condenser coolant valve, V-10, to regulate the reflux drum liquid level.

Steps 7 and 8. *Check component balances and control individual unit operations.* At this point, all but one of the valves (V-2) has been assigned. To ensure a stoichiometric ratio of reagents entering reactor R-100, the chlorine feed is adjusted to ensure complete conversion of the ethylene, using a composition controller on the reactor effluent.

Step 9. *Optimize economics and improve dynamic controllability.* As in the previous example, to improve the range of production levels that can be tolerated, the setpoint of the recycle flow controller is set in proportion to the feed flow rate, suitably lagged for synchronization with the propagation rate through the process. For clarity, this is not shown in Figure 20.18.

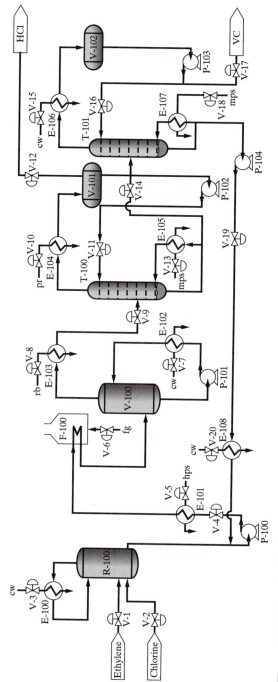

Figure 20.17 Control valve placement for the vinyl chloride process.

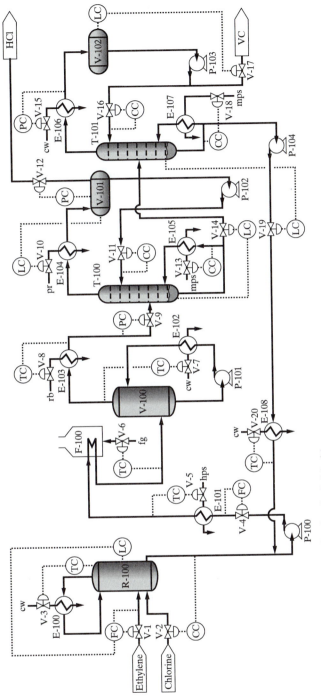

Figure 20.18 Control system for the vinyl chloride process.

The complete control system is shown in Figure 20.18. Many of the qualitative decisions need to be checked by quantitative analysis or by simulation. For example, the interaction between the control systems of the two columns may require careful controller tuning. These refinements are discussed in Chapter 21.   ■

20.4   SUMMARY

This chapter has introduced the importance of considering plantwide control early in the design process. A qualitative control synthesis method, combining the approaches suggested by Newell and Lee (1988) and Luyben and co-workers (1999), was presented to show how to generate alternative control configurations. The limitations of this qualitative approach have been highlighted and the need for the quantitative approach presented in Chapter 21, which involves analysis and dynamic simulation, has been established.

REFERENCES

Al-Arfaj, M.A., and W.L. Luyben, "Control of Ethylene Glycol Reactive Distillation Column," *AICHE J.*, **48**, 905–908 (2002).

Bildea, C.S., A.C. Dimian, and P.D. Iedema, "Nonlinear Behavior of Reactor–Separator–Recycle Systems," *Comput. Chem. Eng.*, **24**, 2–7, 209–215 (2000).

Chiang, T., and W.L. Luyben, "Comparison of the Dynamic Performances of Three Heat-Integrated Distillation Configurations," *Ind. Eng. Chem. Res.*, **27**, 99–104 (1988).

Kiss, A.A., C.S. Bildea, A.C. Dimian, and P.D. Iedema, "State Multiplicity in CSTR-Separator-Recycle Systems," *Chem. Eng. Sci.*, **57**, 4, 535–546 (2002).

Luyben, W. L., *Process Modeling, Simulation, and Control for Chemical Engineers*, 2nd ed., McGraw-Hill, New York (1990).

Luyben. W.L., "Snowball Effects in Reactor/Separator Processes with Recycle," *Ind. Eng. Chem. Res.*, **33**, 299–305 (1994).

Luyben, W.L., B.D. Tyreus, and M.L. Luyben, *Plantwide Process Control*, McGraw-Hill, New York (1999).

McAvoy, T.J., *Interaction Analysis*, Instrument Society of America, Research Triangle Park, North Carolina (1983).

Newell, R.B., and P.L. Lee, *Applied Process Control*, Prentice-Hall of Australia, Brookvale, NSW (1988).

Seborg, D.E., T.F. Edgar, and D.A. Mellichamp, *Process Dynamics and Control*, Chapter 28, Wiley, New York (1989).

Shinskey, F.G., *Distillation Control*, 2nd ed., McGraw-Hill, New York (1984).

Skogestad, S. "Plantwide Control: Towards a Systematic Procedure," in Grievink, J., and J. van der Schijndel (eds.). *Computer Aided Chemical Engineering–10, European Symposium on Computer Aided Process Engineering–12*, Elsevier (2002).

Stephanopoulos, G., *Chemical Process Control*, Chapter 23, Prentice-Hall, Englewood Cliffs, New Jersey (1984).

Weitz, O., and D.R. Lewin, "Dynamic Controllability and Resiliency Diagnosis Using Steady State Process Flowsheet Data," *Comput. Chem. Eng.*, **20**(4), 325–336 (1996).

Wu, K.L., and C.C. Yu, "Reactor/Separator Process with Recycle–1. Candidate Control Structures for Operability," *Comput. Chem. Eng.*, **20**, 11, 1,291–1,316 (1996).

EXERCISES

20.1 Perform a degrees-of-freedom analysis for the noninteracting exothermic reactor shown in Figure 20.3a. Suggest an appropriate control structure. Carry out the same exercise for the heat-integrated reactor shown in Figure 20.3b. Compare the results.

20.2 Consider the mixing vessel shown in Figure 20.19. The feed stream flow rate, F_1, and composition, C_1, are considered to be disturbance variables. The feed is mixed with a control stream of flow rate, F_2, and constant known composition C_2. To ensure a product of constant composition, it is also possible to manipulate the flow rate, F_3, of the product stream. Perform a degrees-of-freedom analysis and suggest alternative control system configurations. Note that unsteady-state balances are required.

20.3 Consider the FS two-column configuration for the separation of methanol and water in Figure 20.2 and (a) determine the number of degrees of freedom for the overall system, (b) determine the number of controlled and manipulated variables, and (c) select a workable control configuration using qualitative arguments.

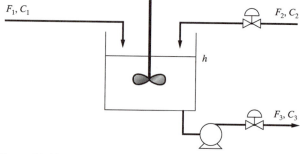

Figure 20.19 Mixing vessel.

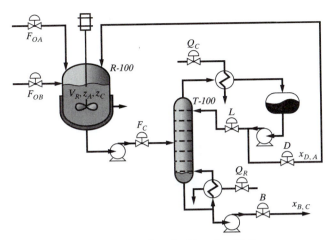

Figure 20.20 Process flowsheet for Exercise 20.6.

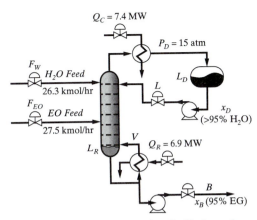

Figure 20.21 Ethylene glycol reactive distillation column.

20.4 Repeat Exercise 20.3 for the LSF configuration in Figure 20.2.

20.5 A control system is suggested for the exothermic reactor in Figure 20.7. Suggest alternative configurations and compare them with the original configuration.

20.6 Figure 20.20 shows a process for the isothermal production of C from A and B (A + B → C). The two reagents are fed to a CSTR, R-100, where complete conversion of B is assumed. The reactor effluent stream, consisting of C and unreacted A is separated in a distillation column, T-100, where the more volatile A is withdrawn in the distillate and recycled, and the product C is withdrawn in the bottoms stream. Your task is to devise a conceptual plantwide control system for the process. *Hint:* It may be helpful to reposition the feed stream of A.

20.7 Figure 20.21 shows the flowsheet for a reactive distillation column for the production of ethylene glycol (EG) from ethylene oxide (EO) and water (Al-Arfaj and Luyben, 2002):

$$EO + H_2O \rightarrow EG$$

Note that the reaction proceeds to 100% conversion in the column, with part of the EG undergoing a secondary (undesired) reaction to diethylene glycol (DEG):

$$EO + EG \rightarrow DEG$$

For this reason, the EO is fed to the column in slight excess. The EG product is withdrawn as the bottoms stream, and almost pure water concentrates at the top of the column. Your task is to *use the procedure of Luyben and co-workers, showing all steps, to* devise a conceptual plantwide control system for the process with the following objectives:

a. Control production rate

b. Ensure the EG product is at the required concentration

20.8 Figure 20.22 shows the monochlorobenzene separation process introduced in Section 4.4. The process involves a flash vessel, V-100, an absorption column, T-100, a distillation column, T-101, a reflux drum, V-101, and three utility heat exchangers. As shown in Figure 4.23, most of the HCl is removed at high purity in the vapor effluent of T-100. However, in contrast with the design shown in Chapter 4, that in Figure 20.22 does not include a "treater" to remove the residual HCl; instead, this is purged in a small vapor overhead product stream in T-101. The benzene and monochlorobenzene are obtained at high purity as distillate and bottoms liquid products in T-101. Note that the 12 available control valves are identified. Your task is to design a conceptual control system to ensure that the process provides stable production at a desired level, while meeting quality specifications.

20.9 How would the control configuration for the vinyl chloride process in Figure 20.18 change if the primary control objective is to provide on-demand, vinyl chloride product?

20.10 Figure 20.23 shows a heat-integrated process for the manufacture of vinyl chloride. As discussed in Section 3.4, this heat integration is possible when the rate of carbon deposition is sufficiently low. This design sharply reduces the utilization of utilities in Figure 20.17, without requiring additional heat exchangers. Design a conceptual control system for the same control objectives in Example 20.12. *Hint:* To provide sufficient degrees of freedom, it may be necessary to add heat exchanger bypasses and/or trim-utility exchangers.

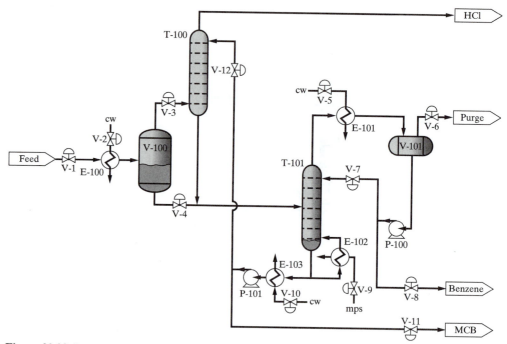

Figure 20.22 Process flowsheet for the MCB separation process.

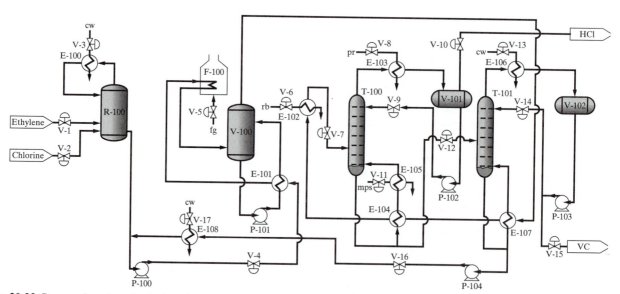

Figure 20.23 Process flowsheet for the heat-integrated vinyl chloride process.

Chapter 21

Flowsheet Controllability Analysis

21.0 OBJECTIVES

This chapter introduces quantitative measures for controllability assessment to be used when developing a base-case design and in the detailed design stage (Stages 2 and 3, Table 20.1; see also Figure 1.2) and highlights how their integration into the design process can help to generate improved flowsheets that satisfy control performance criteria. At this point, the process creation stage (Stage 1, Table 20.1) has been completed and several promising process flowsheets exist. As they are evaluated, the *control* objectives are considered as constraints, including

- Adequate disturbance resiliency, that is, the ability to reject disturbances quickly enough to meet specifications
- Insensitivity to model uncertainty, that is, the ability to control easily, and to provide adequate closed-loop performance, with relative insensitivity to model inaccuracies.

An approach is introduced to screen the potential designs *as early as possible*, to identify the most promising designs for rigorous testing in stage 4, in which plantwide controllability assessment is completed. As demonstrated in this chapter, it is important to verify the approximate analysis using rigorous dynamic simulation. The CD-ROM that accompanies this textbook: *Using Process Simulators in the Chemical Engineering Curriculum*, Version 2.0 (Lewin et al., 2003), provides multimedia instruction on the use of HYSYS.Plant for dynamic simulation. ASPEN DYNAMICS and CHEMCAD can be used also for dynamic simulation.

It is assumed that the reader is familiar with the basic concepts of linear systems theory. This material is covered typically in an introductory course on process dynamics and control at the undergraduate level. The subjects in that course that are prerequisite to understanding the concepts in this chapter are

1. Basic linear matrix theory, linearization, complex numbers, and Laplace and Fourier transforms. Note that Section 21.1 provides some of this background material.
2. Pole and zero positions in the complex plane, and their impact on the time-domain response of linear systems.
3. Linear stability theory and the impact of feedback.
4. Tuning of single-input, single-output (SISO) controllers (P, PI, and PID controllers). Note that Section 21.4 provides instruction on model-based PI-controller tuning.

Key concepts relating to linear process models are reviewed in the first section of the chapter. For deeper coverage, the reader is referred to the following undergraduate-level texts:

Bequette, B.W., *Process Control: Modeling, Design, and Simulation*, Prentice-Hall, Englewood Cliffs, New Jersey (2003).

Luyben, W.L., *Process Modeling, Simulation and Control for Chemical Engineers*, 2nd ed., McGraw-Hill, New York (1990).

Ogunnaike, B.A., and W.H. Ray, *Process Dynamics, Modeling and Control*, Oxford University Press, New York (1994).

Seborg, D.E., T.F. Edgar, and D.A. Mellichamp, *Process Dynamics and Control*, Wiley, New York (1989).

Stephanopoulos, G., *Chemical Process Control,* Prentice-Hall, Englewood Cliffs, New Jersey (1984).

This chapter

1. Explains how to generate linear process models in their standard forms.
2. Defines quantitative measures that are used to analyze the controllability and resiliency (C&R) of process flowsheets, and shows how to implement them using MATLAB.
3. Describes a method to carry out C&R analysis using the results of steady-state process simulations.
4. Shows how to use quantitative analysis with steady-state and dynamic relative gain arrays (RGA and DRGA) to select control loop pairings reliably and to use the IMC model-based approach to provide preliminary tuning of single-loop PI controllers.
5. Analyzes, in Section 21.5, selected case studies in Chapter 20 to demonstrate the utility of the quantitative methods. For completeness, these analyses are verified with dynamic simulations using single-loop PI controllers.

After reading this chapter, the student should

1. Be able to compute the frequency-dependent process transfer functions $\{\underline{\underline{P}}, \underline{\underline{P}}_d\}$ using MATLAB, given a linear model in one of its standard forms.
2. Be able to generate the C&R measures: relative-gain array (RGA), and disturbance cost (DC), given the matrices $\underline{\underline{P}}\{s\}$ and $\underline{\underline{P}}_d\{s\}$, describing the effects of the manipulated variables and disturbances on the process outputs, using MATLAB.
3. Be able to select the appropriate pairings for a decentralized control system for the process using the static and dynamic RGAs and appropriate resiliency measures, and provide preliminary tuning using IMC-PI tuning rules.
4. Be able to perform C&R analysis to select between alternative process configurations, given the results of process simulations using linearized models.

21.1 GENERATION OF LINEAR MODELS IN STANDARD FORMS

In this chapter, several methods are described to assist the designer in rejecting designs that do not provide acceptable closed-loop performance, using models linearized about a steady state. These are generated by expressing the open-loop response of the process outputs, $\underline{y}\{s\}$, in terms of the variations of the inputs, $\underline{u}\{s\}$, and disturbances, $\underline{d}\{s\}$:

$$\underline{y}\{s\} = \underline{\underline{P}}\{s\}\,\underline{u}\{s\} + \underline{\underline{P}}_d\{s\}\,\underline{d}\{s\} \tag{21.1}$$

The procedure for deriving the linear state–space model and the input–output transfer function model in Eq. (21.1) involves the following steps:

Step 1: The nonlinear state and output equations are derived from the material and energy balances that model the process. These are expressed in the form:

$$\frac{d\underline{x}}{dt} = \underline{f}\{\underline{x}, \underline{u}, \underline{d}\}$$

$$\underline{y} = \underline{g}\{\underline{x}, \underline{u}, \underline{d}\} \tag{21.2}$$

where $\underline{x}$ is a vector of n_x state variables, $\underline{y}$ is a vector of n_y output (measured) variables, $\underline{u}$ is a vector of n_u manipulated variables, $\underline{d}$ is a vector of n_d disturbances, and $\underline{f}$ and $\underline{g}$ are vectors of n_x and n_y nonlinear functions, respectively.

Step 2: The state and output equations are solved at a stationary (steady) state that is defined either in terms of the desired state variable values or those of the input variables:

$$0 = \underline{f}\{\underline{x}_*, \underline{u}_*, \underline{d}_*\}$$
$$\underline{y} = \underline{g}\{\underline{x}_*, \underline{u}_*, \underline{d}_*\} \qquad (21.3)$$

where the stationary point is at $\underline{x} = \underline{x}_*$, $\underline{u} = \underline{u}_*$, and $\underline{d} = \underline{d}_*$. The solution of Eq. (21.3) requires that the degrees of freedom of the process be resolved through the specification of $n_u + n_d$ values.

Step 3: The equations are linearized in the vicinity of the desired stationary point by a Taylor series expansion of Eq. (21.2):

$$\frac{d\underline{x}}{dt} \cong \underline{f}\{\underline{x}_*, \underline{u}_*, \underline{d}_*\} + \underline{\underline{A}}(\underline{x} - \underline{x}_*) + \underline{\underline{B}}_U(\underline{u} - \underline{u}_*) + \underline{\underline{B}}_D(\underline{d} - \underline{d}_*) + \text{h.o.t.}$$
$$\underline{y} \cong \underline{g}\{\underline{x}_*, \underline{u}_*, \underline{d}_*\} + \underline{\underline{C}}(\underline{x} - \underline{x}_*) + \underline{\underline{D}}_U(\underline{u} - \underline{u}_*) + \underline{\underline{D}}_D(\underline{d} - \underline{d}_*) + \text{h.o.t.} \qquad (21.4)$$

Note that the linear approximation is obtained by ignoring the higher order terms (h.o.t.) of the Taylor series expansion. The matrices $\underline{\underline{A}}$, $\underline{\underline{B}}_U$, $\underline{\underline{B}}_D$, $\underline{\underline{C}}$, $\underline{\underline{D}}_U$, and $\underline{\underline{D}}_D$ and are the Jacobian matrices of appropriate dimension evaluated at the stationary point, defined as follows:

$$\underline{\underline{A}} = \{a_{i,j}\} \equiv \left.\frac{\partial f_i}{\partial x_j}\right|_{\underline{x}_*, \underline{u}_*, \underline{d}_*} \qquad \underline{\underline{C}} = \{c_{i,j}\} \equiv \left.\frac{\partial g_i}{\partial x_j}\right|_{\underline{x}_*, \underline{u}_*, \underline{d}_*}$$

$$\underline{\underline{B}}_U = \{b_{U,i,j}\} \equiv \left.\frac{\partial f_i}{\partial u_j}\right|_{\underline{x}_*, \underline{u}_*, \underline{d}_*} \qquad \underline{\underline{D}}_U = \{d_{U,i,j}\} \equiv \left.\frac{\partial g_i}{\partial u_j}\right|_{\underline{x}_*, \underline{u}_*, \underline{d}_*}$$

$$\underline{\underline{B}}_D = \{b_{D,i,j}\} \equiv \left.\frac{\partial f_i}{\partial d_j}\right|_{\underline{x}_*, \underline{u}_*, \underline{d}_*} \qquad \underline{\underline{D}}_D = \{d_{D,i,j}\} \equiv \left.\frac{\partial g_i}{\partial d_j}\right|_{\underline{x}_*, \underline{u}_*, \underline{d}_*}$$

Step 4: The linearized equations are formulated in terms of perturbation variables that express the deviation from the stationary point (or steady state): $\hat{\underline{x}} = \underline{x} - \underline{x}_*$, $\hat{\underline{y}} = \underline{y} - \underline{y}_*$, $\hat{\underline{u}} = \underline{u} - \underline{u}_*$, and $\hat{\underline{d}} = \underline{d} - \underline{d}_*$. Substituting the perturbation variables into Eqs. (21.4) and ignoring higher-order terms:

$$\frac{d\hat{\underline{x}}}{dt} \cong \underline{\underline{A}}\,\hat{\underline{x}} + \underline{\underline{B}}_U\,\hat{\underline{u}} + \underline{\underline{B}}_D\,\hat{\underline{d}}$$
$$\hat{\underline{y}} \cong \underline{\underline{C}}\,\hat{\underline{x}} + \underline{\underline{D}}_U\,\hat{\underline{u}} + \underline{\underline{D}}_D\,\hat{\underline{d}} \qquad (21.5)$$

Eqs. (21.5) constitute the *linear state–space representation* of the system.

Step 5: The linearized equations are transformed into the Laplace domain:

$$\hat{\underline{y}}\{s\} = \underline{\underline{P}}\{s\}\hat{\underline{u}}\{s\} + \underline{\underline{P}}_d\{s\}\hat{\underline{d}}\{s\} \qquad (21.1)$$

where $\underline{\underline{P}}\{s\} = \underline{\underline{C}}(s \cdot \underline{\underline{I}} - \underline{\underline{A}})^{-1}\underline{\underline{B}}_U + \underline{\underline{D}}_U$ and $\underline{\underline{P}}_d\{s\} = \underline{\underline{C}}(s \cdot \underline{\underline{I}} - \underline{\underline{A}})^{-1}\underline{\underline{B}}_D + \underline{\underline{D}}_D$ are matrices of the appropriate dimension. Equation (21.1) constitutes the *input–output transfer function representation* of the linear system.

As an example, the procedure for generating linear models in standard form is demonstrated for an exothermic reactor, whose complete analysis is presented in Case Study 21.1 of Section 21.5.

EXAMPLE 21.1 **Standard Linear Models for an Exothermic Reactor**

A continuous-stirred-tank reactor (CSTR) for the production of propylene glycol is analyzed in Case Study 21.1, in Section 21.4 below. Approximate linear models for the reactor are generated using the five-step procedure as follows.

Step 1: *Define the state and output equations.* The hydrolysis of propylene oxide (PO) to propylene glycol is an exothermic reaction catalyzed by H_2SO_4:

$$CH_2\!-\!O\!-\!CH\!-\!CH_3 + H_2O \rightarrow CH_2OH\!-\!CH_2OH\!-\!CH_3$$

When water is supplied in excess, the reaction is second order with respect to the propylene oxide concentration and zero order with respect to the water concentration. Its rate constant exhibits an Arrhenius dependence on temperature, with $k_0 = 3.294 \times 10^{26}$ m³/(kmol-hr) and $E = 1.556 \times 10^5$ kJ/kmol. Furthermore, it is customary to dilute the PO feed with methanol (MeOH), while the H_2SO_4 catalyst enters the reactor with the feed. Operating conditions are sought for carrying out this liquid-phase reaction in a 47-ft³ CSTR, with the liquid holdup at 85% of its total volume (1.135 m³). The liquid feeds are fed at 23.9°C, with one consisting of 18.712 kmol/hr of PO and 32.73 kmol/hr of MeOH. The water feed rate is from 160 to 500 kmol/hr (2.84–8.88 m³/hr), selected to moderate the reactor temperature. To reduce the risk of vaporization, the reactor is operated at a pressure of 3 bar. Under these conditions, the transients for the PO concentration, C_{PO} (kmol/m³), and temperature, T (°C), are determined by solving the following species and enthalpy balances:

$$\frac{dC_{PO}}{dt} = \frac{\mathfrak{I}_{PO,in}}{V} - \frac{C_{PO}(q_0 + q_w)}{V} - k\{T\}C_{PO}^2 \tag{21.6}$$

$$\frac{dT}{dt} = \frac{1}{c_P}k\{T\}C_{PO}^2(-\Delta H) - \frac{(q_0 + q_w)(T - T_0)}{V} \tag{21.7}$$

where, $k\{T\} = k_o e^{-E/R(T+273.2)}$ m³/(kmol-hr), $R = 8.314$ kJ/kmol-K, the molar flow rate of PO in the feed, $\mathfrak{I}_{PO,in} = 18.712$ kmol/hr, $V = 1.135$ m³, $\Delta H = -9 \times 10^4$ kJ/kmol, the organic volumetric feed rate, $q_0 = 2.556$ m³/hr, the water volumetric feed rate is q_w, $T_0 = 23.9$°C, and $c_P = 3,558$ kJ/m³-°C. Implicit in the assumption of perfect level control is the pairing between the effluent volumetric flow rate, F, and the liquid level, L. This leaves the temperature, T, as the output, to be controlled by the water feed rate, q_w, as the manipulated variable. The disturbances to the process are the organic volumetric feed rate, q_0, and the feed temperature, T_0. Thus, $\underline{x} = [C_{PO}, T]^T$, $\underline{y} = [T]$, $\underline{u} = [q_w]$, and $\underline{d} = [\mathfrak{I}_{PO,in}, T_o]^T$.

Step 2: *Solve at the steady state.* The state equations are solved at the steady state. The degrees of freedom are resolved by fixing all of the input variable values and solving the two equations for the two unknown state variables.

$$\frac{\mathfrak{I}_{PO,in}}{V} - \frac{C_{PO}(q_0 + q_w)}{V} - k\{T\}C_{PO}^2 = 0 \tag{21.8}$$

$$\frac{1}{c_P}k\{T\}C_{PO}^2(-\Delta H) - \frac{(q_0 + q_w)(T - T_0)}{V} = 0 \tag{21.9}$$

Taking $\underline{u}_* = q_{w*} = 5.325$ m³/hr and $\underline{d}_* = [\mathfrak{I}_{PO,in*}, T_o]^T = [18.712, 23.9]^T$ and solving Eqs. (21.8) and (21.9), gives $\underline{x}_* = [0.06, 82.4]^T$. Note that the fractional conversion of PO is $X = 1 - C_{PO}/C_{PO,in} = 1 - 0.06/2.374 = 0.975$, where $C_{PO,in} = \mathfrak{I}_{PO,in}/(q_0 + q_w)$. This solution is obtained analytically, graphically (see Case Study 21.1), or using a numerical method (e.g., the Newton–Raphson method).

Steps 3 and 4: *Linearize in the vicinity of the steady state.* The Jacobian matrices for the linearized approximation are

$$A = \frac{1}{60}\begin{bmatrix} -\dfrac{(q_0 + q_{w*})}{V} - 2k\{T_*\}C_{PO*} & -\dfrac{\partial k\{T_*\}}{\partial T}C_{PO*}^2 \\[2ex] \dfrac{(-\Delta H)}{c_P}2k\{T_*\}C_{PO*} & -\dfrac{(q_0 + q_{w*})}{V} + \dfrac{\partial k\{T_*\}}{\partial T}\dfrac{(-\Delta H)C_{PO*}^2}{c_P} \end{bmatrix} \tag{21.10}$$

$$\underline{\underline{B}}_U = \frac{1}{60}\begin{bmatrix} -\dfrac{C_{PO}*}{V} \\[2mm] -\dfrac{(T_* - T_{0}*)}{V} \end{bmatrix}, \qquad \underline{\underline{B}}_D = \frac{1}{60}\begin{bmatrix} \dfrac{1}{V} & 0 \\[2mm] 0 & \dfrac{(q_0 + q_w*)}{V} \end{bmatrix} \qquad \textbf{(21.11)}$$

$$\underline{C} = \begin{bmatrix} 0 & 1 \end{bmatrix}, \underline{\underline{D}}_U = [0], \underline{\underline{D}}_D = \begin{bmatrix} 0 & 0 \end{bmatrix} \qquad \textbf{(21.12)}$$

The division by 60 in each matrix is to express time in minutes instead of hours. Note also that all of the variables are expressed in physical units. Substituting numerical values into Eqs. (21.10) and (21.11):

$$\underline{\underline{A}} = \begin{bmatrix} -9.203 & -0.0396 \\ 229.8 & 0.8870 \end{bmatrix},$$

$$\underline{\underline{B}}_U = \begin{bmatrix} -0.0009 \\ -0.8596 \end{bmatrix}, \qquad \underline{\underline{B}}_D = \begin{bmatrix} 0.0147 & 0 \\ 0 & 0.1157 \end{bmatrix}$$

These matrices are scaled by assuming that all outputs and manipulated variables are nominally at 50% of their ranges, and the disturbance variable values are constrained to vary in the range $\underline{\Delta d} = [\pm 50\%, \pm 5°C]^{T}$. Thus:

$$\underline{\underline{A}}_s = \underline{\underline{S}}_x^{-1}\, \underline{\underline{A}}\, \underline{\underline{S}}_x = \begin{bmatrix} -9.203 & -55.43 \\ 0.1644 & 0.8870 \end{bmatrix}, \qquad \underline{\underline{S}}_x = \begin{bmatrix} C_{PO}* & 0 \\ 0 & T_* \end{bmatrix}$$

$$\underline{\underline{B}}_{U,s} = \underline{\underline{S}}_x^{-1}\, \underline{\underline{B}}_U\, \underline{S}_u = \begin{bmatrix} -0.0782 \\ -0.0555 \end{bmatrix}, \qquad \underline{S}_u = q_w* \qquad \textbf{(21.13)}$$

$$\underline{\underline{B}}_{D,s} = \underline{\underline{S}}_x^{-1}\, \underline{\underline{B}}_D\, \underline{\underline{S}}_d = \begin{bmatrix} 2.330 & 0 \\ 0 & 0.0070 \end{bmatrix}, \qquad \underline{\underline{S}}_d = \begin{bmatrix} 0.5\tilde{\Im}_{PO,in}* & 0 \\ 0 & 5 \end{bmatrix}$$

These matrices relate the input (manipulated and disturbance) variables to the output (controlled) variables, with all of the variables scaled and in perturbation variable form.

Step 5: *Generate transfer functions.* These are computed using the scaled matrices in Eq. (21.13), for example

$$\underline{\underline{P}}\{s\} = \underline{C}(s \cdot \underline{I} - \underline{\underline{A}}_s)^{-1}\, \underline{\underline{B}}_{U,s} + \underline{\underline{D}}_{U,s} = \begin{bmatrix} 0 & 1 \end{bmatrix} \begin{bmatrix} s + 9.203 & 55.43 \\ -0.1644 & s - 0.8870 \end{bmatrix}^{-1} \begin{bmatrix} -0.0782 \\ -0.0555 \end{bmatrix}$$

Hence,

$$\underline{\underline{P}}\{s\} = \frac{-0.0555s - 0.524}{s^2 + 8.32s + 0.949} = \frac{-0.552(0.106s + 1)}{(0.122s + 1)(8.64s + 1)} \qquad \textbf{(21.14)}$$

Note that $\underline{\underline{P}}\{s\}$ is a scalar transfer function, since it relates perturbations in the single manipulated variable, $\hat{q}_w$, to those in the single process output variable, $\hat{T}$. Note that the process zero almost cancels the fast process pole, meaning that the response to the manipulated variable is effectively that of a first-order lag, with a time constant of approximately 9 min.

A similar computation yields the transfer function matrix, $\underline{\underline{P}}_d\{s\}$:

$$\underline{\underline{P}}_d\{s\} = \begin{bmatrix} 0 & 1 \end{bmatrix} \begin{bmatrix} s + 9.203 & 55.43 \\ -0.1644 & s - 0.8870 \end{bmatrix}^{-1} \begin{bmatrix} 2.330 & 0 \\ 0 & 0.0070 \end{bmatrix}$$

$$= \begin{bmatrix} \dfrac{0.410}{(0.122s + 1)(8.64s + 1)} & \dfrac{0.068(0.109s + 1)}{(0.122s + 1)(8.64s + 1)} \end{bmatrix} \qquad \textbf{(21.15)}$$

The columns of $\underline{\underline{P}}_d\{s\}$ define the responses of $\hat{T}$ to $\tilde{\Im}_{PO,in}$ and $\hat{T}_o$, respectively. Note that the temperature response to step changes in $\tilde{\Im}_{PO,in}$ is of second order, while its response to changes in the feed temperature is effectively of first order.

The reader is referred to the CD-ROM that accompanies this text for useful MATLAB functions and scripts for the generation of linear models in their standard forms. ∎

21.2 QUANTITATIVE MEASURES FOR CONTROLLABILITY AND RESILIENCY

The quantitative assessment of the controllability and resiliency of chemical processes has generated considerable interest. The term *resiliency* was introduced by Morari (1983), who also pioneered qualitative measures for its assessment. Furthermore, Perkins (1989) presented an approach for the simultaneous design of processes and their control systems that addresses plantwide controllability directly.

All of the linear C&R measures use the approximations, $\underline{\underline{P}}\{s\}$ and $\underline{\underline{P}}_d\{s\}$, which describe the effects of the control variables and disturbances, respectively, on the process outputs. A commonly used controllability measure is the *relative-gain array* (RGA; Bristol, 1966), which relies only on $\underline{\underline{P}}\{s\}$. The *disturbance condition number* (DCN; Skogestad and Morari, 1987) and the *disturbance cost* (DC; Lewin, 1996) are resiliency measures that require a disturbance model, $\underline{\underline{P}}_d\{s\}$, in addition to $\underline{\underline{P}}\{s\}$. These C&R measures are especially useful in Stages 2 and 3 of the design process (see Table 20.1) because they do not assume a controller structure or a specific controller design and tuning.

It is assumed that each input variable is nominally at the midpoint of its range and is expressed in perturbation variable form, and scaled by dividing by its nominal value. For example, if F_i is an inlet flow rate, nominally at 500 lbmol/hr, its operating range is $0 \leq F_i \leq$ 1,000 lbmol/hr, in perturbation variable form, $-500 \leq F_i \leq 500$, and in scaled form, $-1 \leq F_i \leq 1$. Thus, $\underline{\underline{P}}\{s\}$ and $\underline{\underline{P}}_d\{s\}$ are scaled by multiplying the gains in each column by the nominal value of the appropriate input variable. As a result, all of the scaled inputs vary over the same range $[-1, 1]$. Note, however, that the RGA is scale independent, whereas the DC is input scale dependent.

Steady-state RGA (Bristol, 1966)

Figure 21.1 shows the block diagram for a multiple-input, multiple-output (MIMO) process to be controlled by two single-loop controllers. Having closed one of the loops $(y_1 - u_1)$, the controller in the second loop, which manipulates u_2 based on the feedback of y_2, must be tuned. A desirable feature of the process, as seen by this controller, is to have the effective process gain remain invariant, regardless of the action of the other control loop.

When the controller c_1 is put into manual operation, that is, when it is turned off, the process gain as seen by controller c_2 is

$$\left.\frac{y_2}{u_2}\right|_{c_1,\text{OL}} = p_{22} \tag{21.16}$$

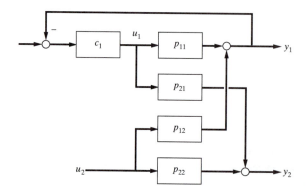

Figure 21.1 MIMO process with one control loop.

where u_2 and y_2 are the deviations of the input and output from their nominal values in the steady state. On the other hand, when c_1 is put into automatic operation, the process gain as seen by controller c_2 is

$$\left.\frac{y_2}{u_2}\right|_{c_1,\text{CL}} = p_{22} - p_{12}\frac{c_1}{1 + p_{11}c_1}p_{21} \tag{21.17}$$

In general, for MIMO systems, a useful measure is the ratio

$$\frac{\text{process gain as seen by a given controller with all other loops open}}{\text{process gain as seen by a given controller with all other loops closed}}$$

When this ratio is close to unity, the given controller is relatively insensitive to interaction. Computing this ratio for the MIMO process in Figure 21.1:

$$\frac{\left.\dfrac{y_2}{u_2}\right|_{c_1,\text{OL}}}{\left.\dfrac{y_2}{u_2}\right|_{c_1,\text{CL}}} = \frac{p_{22}}{p_{22} - p_{12}\dfrac{c_1}{1 + p_{11}c_1}p_{21}} \tag{21.18}$$

When the top loop is closed-loop stable, and when c_1 has integral action,

$$\lim_{s \to 0}\frac{c_1}{1 + p_{11}c_1} = \frac{1}{p_{11}}$$

Therefore, the ratio at steady state is

$$\lim_{s \to 0}\frac{\left.\dfrac{y_2}{u_2}\right|_{c_1,\text{OL}}}{\left.\dfrac{y_2}{u_2}\right|_{c_1,\text{CL}}} = \lim_{s \to 0}\frac{p_{22}}{p_{22} - p_{12}\dfrac{c_1}{1 + p_{11}c_1}p_{21}} = \frac{p_{11}p_{22}}{p_{11}p_{22} - p_{12}p_{21}} \tag{21.19}$$

Similarly,

$$\lim_{s \to 0}\frac{\left.\dfrac{y_2}{u_1}\right|_{c_1,\text{OL}}}{\left.\dfrac{y_2}{u_1}\right|_{c_1,\text{CL}}} = \frac{-p_{12}p_{21}}{p_{11}p_{22} - p_{12}p_{21}} \tag{21.20}$$

Thus, for a two-input, two-output process, the RGA is defined as

$$\underline{\underline{\Lambda}} = \begin{bmatrix} \lambda_{11} & \lambda_{12} \\ \lambda_{21} & \lambda_{22} \end{bmatrix} = \begin{bmatrix} \dfrac{\left.\dfrac{y_1}{u_1}\right|_{c_2,\text{OL}}}{\left.\dfrac{y_1}{u_1}\right|_{c_2,\text{CL}}} & \dfrac{\left.\dfrac{y_1}{u_2}\right|_{c_2,\text{OL}}}{\left.\dfrac{y_1}{u_2}\right|_{c_2,\text{CL}}} \\[4mm] \dfrac{\left.\dfrac{y_2}{u_1}\right|_{c_1,\text{OL}}}{\left.\dfrac{y_2}{u_1}\right|_{c_1,\text{CL}}} & \dfrac{\left.\dfrac{y_2}{u_2}\right|_{c_1,\text{OL}}}{\left.\dfrac{y_2}{u_2}\right|_{c_1,\text{CL}}} \end{bmatrix} = \begin{bmatrix} p_{11}p_{22} & -p_{12}p_{21} \\ -p_{12}p_{21} & p_{11}p_{22} \end{bmatrix} \cdot det(\underline{\underline{P}})^{-1} \tag{21.21}$$

In general, the RGA can be computed using

$$\underline{\underline{\Lambda}} = \underline{\underline{P}} \otimes (\underline{\underline{P}}^{-1})^{\text{T}} \tag{21.22}$$

where $\otimes$ denotes the element-by-element (Schur) product.

Theorem (2 × 2 Systems Only)

If λ_{11} ($= \lambda_{22}$) is positive, there exists a pair of single-input, single-output (SISO) controllers, c_1 and c_2, with integral action for the loops $u_1 - y_1$ and $u_2 - y_2$ such that the loops are *stable by themselves and together*. If λ_{11} is negative, there are no controllers that can guarantee stability *by themselves and together*. In other words, to guarantee closed-loop stability with either of the two SISO controllers in automatic or manual mode, the controllers should be paired such that the RGA elements corresponding to the pairings are positive. Negative RGA elements are an indication of the presence of destabilizing positive feedback due to unfavorable process interactions. Similarly, excessively large RGA elements are related to poorly conditioned processes; those in which the effective process gain may be orders of magnitude different, depending on the input direction.

For systems of higher rank, a sufficient condition for the stabilizability of a decentralized control system is the selection of pairings such that $\lambda_{ij} > 0$, and hence the RGA provides a useful screening tool. The *decentralized integral controllability* (DIC) conditions (see Morari and Zafiriou, 1989, pp. 359–367) provide additional sufficient conditions for the stability of higher-order systems, which depend only on the steady-state gain matrix, $\underline{\underline{P}}\{0\}$.

Properties of the Steady-state RGA

The following properties are especially noteworthy when working with the RGA:

1. $\sum_i \lambda_{ij} = \sum_j \lambda_{ij} = 1$ (rows and columns sum to unity)
2. If $\underline{\underline{P}}$ is triangular (lower or upper), $\underline{\underline{\Lambda}} = \underline{\underline{I}}$

$$\text{e.g., } \underline{\underline{P}} = \begin{bmatrix} -3 & 2 & 3 \\ 0 & 3 & 5 \\ 0 & 0 & 1 \end{bmatrix} \Rightarrow \underline{\underline{\Lambda}} = \begin{bmatrix} 1 & 0 & 0 \\ 0 & 1 & 0 \\ 0 & 0 & 1 \end{bmatrix}$$

In such systems, the process interaction is in one direction only, and therefore precludes the possibility of the occurrence of destabilizing feedback.

3. For 2 × 2 systems only:
   If $\underline{\underline{P}}$ has an odd number of positive elements, $0 \le \lambda_{ij} \le 1$
   If $\underline{\underline{P}}$ has an even number of positive elements, either $\lambda_{ij} < 0$ or $\lambda_{ij} > 1$

Dynamic RGA (McAvoy, 1983)

Considering the same MIMO process in Figure 21.1, y_2 is expressed in terms of u_1 and u_2:

$$y_2 = p_{21}u_1 + p_{22}u_2 \tag{21.23}$$

When c_1 is in manual operation, $u_1 = 0$ and

$$\left. \frac{y_2}{u_2} \right|_{c_1,\text{OL}} = p_{22}$$

as for the steady-state analysis. When c_1 is in automatic operation and it is assumed that the first loop can be designed to give perfect control (i.e., the first loop's output is assumed to be held at its setpoint),

$$y_1 = p_{11}u_1 + p_{12}u_2 = 0 \Rightarrow u_1 = -\frac{p_{12}}{p_{11}}u_2 \tag{21.24}$$

Substituting for u_1 in Eq. (21.23),

$$\left. \frac{y_2}{u_2} \right|_{c_1,\text{CL}} = p_{22} - \frac{p_{12}p_{21}}{p_{11}} \tag{21.25}$$

Hence, the dynamic RGA (DRGA) has precisely the same form as the steady-state array. Note that the dynamic RGA assumes perfect control, which may not be an appropriate assumption, especially at high frequencies. The computation of the DRGA requires care since it involves complex algebra. Because columns and rows sum to unity only at the steady state, the DRGA should be computed using

$$DRGA_{ij}\{\omega\} = sign(\lambda_{ij}\{0\}) \cdot \left|\lambda_{ij}\{j\omega\}\right| \qquad (21.26)$$

with $\lambda_{ij}\{j\omega\}$ computed conveniently for 2×2 systems using Eqs. (21.19) and (21.20), or using Eq. (21.22) in general.

An accepted rule of thumb is to avoid pairings between variables with negative RGA elements and to select those with values close to unity, as illustrated in the following example.

EXAMPLE 21.2   *LV Control of a Binary Distillation Column*

Figure 21.2 shows the LV configuration for the two-point composition control of a binary distillation column discussed in Example 20.9. After assigning manipulated variables to regulate the vapor and liquid inventories, the boilup rate, V, and the reflux flow rate, L, remain available to control the distillate and bottoms product compositions, x_D and x_B, respectively. To assess the controllability and resiliency of this configuration, the disturbances are taken to be the feed composition, x_F, and the flow rate, F. The column dynamics are approximated by a linear model in transfer function form (Sandelin et al., 1990):

$$\begin{bmatrix} x_D \\ x_B \end{bmatrix} = \begin{bmatrix} \dfrac{-0.045}{8.1s + 1} e^{-0.5s} & \dfrac{0.048}{11s + 1} e^{-0.5s} \\ \dfrac{-0.23}{8.1s + 1} e^{-1.5s} & \dfrac{0.55}{10s + 1} e^{-0.5s} \end{bmatrix} \begin{bmatrix} L \\ V \end{bmatrix} + \begin{bmatrix} \dfrac{-0.001}{10s + 1} e^{-s} & \dfrac{0.004}{8.5s + 1} e^{-s} \\ \dfrac{-0.16}{5.5s + 1} e^{-s} & \dfrac{-0.65}{9.2s + 1} e^{-s} \end{bmatrix} \begin{bmatrix} F \\ x_F \end{bmatrix} \qquad (21.27)$$

To complete the process model definition, it is noted that the process input ranges are as follows: $L = 60 \pm 60$ kmol/hr, $V = 72 \pm 72$ kmol/hr, $F = F_{nom} \pm 20$ kmol/hr, $x_F = x_{F,nom} \pm 6\%$. In Eq. (21.27), the gain coefficients are in the appropriate units, and time is in minutes.

The qualitative guidelines presented in Chapter 20 are not sufficient to decide how to pair the two manipulated variables with the two outputs. Without analysis, it is not clear whether this pairing should be diagonal (i.e., $\{x_D\text{–}L, x_B\text{–}V\}$ as shown in Figure 21.2) or off-diagonal (i.e., $\{x_D\text{–}V, x_B\text{–}L\}$). However, using Eq. (21.19), λ_{11} in the RGA is

$$\lambda_{11} = \frac{p_{11}p_{22}}{p_{11}p_{22} - p_{12}p_{21}} = 1.8 \qquad (21.28)$$

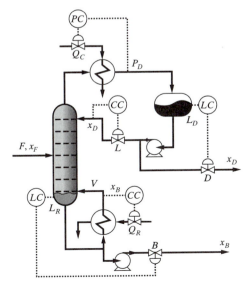

Figure 21.2 Control of a binary distillation column using the LV configuration.

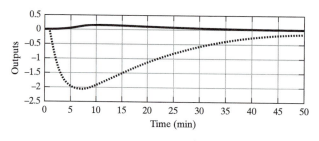

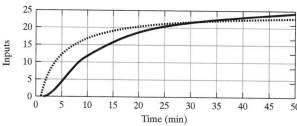

Figure 21.3 Closed-loop response of the LV configuration for binary distillation to the worst-case disturbance, $\underline{d} = [20, \ 6]^T$, with the decentralized PI control—Outputs: x_D (solid line), x_B (dashed line); Inputs: L (solid line), V (dashed line).

Using the property that the RGA rows and columns add to unity,

$$\underline{\underline{\Lambda}} = \begin{bmatrix} 1.8 & -0.8 \\ -0.8 & 1.8 \end{bmatrix}$$

and consequently, diagonal pairing is recommended, with the off-diagonal pairing resulting in stability problems, either when both of the controllers are on automatic or when one of the controllers is switched to manual operation. Although stable, significant interactions are anticipated when both loops are closed, because of the large RGA element.

To verify this, Figure 21.3 shows the closed-loop response for the process, diagonally paired with IMC-tuned PI controllers (x_D–L loop: $K_c = -50$, $\tau_I = 8$ min; x_B–V loop: $K_c = 5$, $\tau_I = 10$ min). For the IMC-PI tuning rules, the reader is referred to Section 21.4. The simulation is computed for the worst-case disturbance, $\underline{d} = [20, 6]^T$, identified using the disturbance cost analysis, to be discussed shortly. As expected, the response is stable but shows significant interactions, with the bottoms composition affected more significantly. The reader can reproduce the above results using the interactive C&R Tutorial CRGUI, which is implemented using MATLAB, on the CD-ROM that accompanies this text. In the main menu, opt for the "Binary Column."

In some cases, the RGA elements vary significantly with the frequency, which may indicate bandwidth limitations on the diagonal dominance of the process. For this example, Figure 21.4 shows λ_{11}

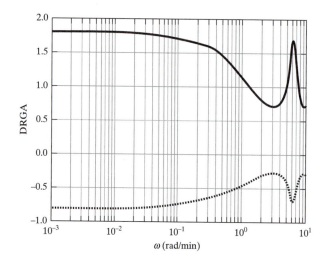

Figure 21.4 Dynamic RGA for the diagonal (solid line) and off-diagonal (dotted line) pairings for Example 21.2.

and λ_{12} as a function of the frequency. Although there is considerable variation at high frequencies, the diagonal dominance holds for the entire frequency range of interest. In this case, the RGA and DRGA give the same pairing recommendations. For some processes, however, the information furnished by the dynamic RGA can be crucial for the correct pairing selection. The following example provides one such case. ■

EXAMPLE 21.3   *Importance of the Dynamic RGA*

Consider the process:

$$\begin{bmatrix} y_1 \\ y_2 \end{bmatrix} = \begin{bmatrix} \dfrac{2.5}{(15s + 1)(2s + 1)}\,e^{-5s} & \dfrac{5}{4s + 1} \\ \dfrac{1}{3s + 1} & \dfrac{-4}{20s + 1}\,e^{-5s} \end{bmatrix} \begin{bmatrix} u_1 \\ u_2 \end{bmatrix} + \begin{bmatrix} \dfrac{-4}{10s + 1}\,e^{-2s} & \dfrac{3}{10s + 1}\,e^{-5s} \\ \dfrac{-1}{5s + 1}\,e^{-2s} & \dfrac{-2}{10s + 1}\,e^{-5s} \end{bmatrix} \begin{bmatrix} d_1 \\ d_2 \end{bmatrix} \quad (21.29)$$

Here the process inputs are limited to the ranges: $u_1 = 60 \pm 60$, $u_2 = 50 \pm 50$, $d_1 = d_{1,\mathrm{nom}} \pm 20$, and $d_2 = d_{2,\mathrm{nom}} \pm 5$, and time is in minutes.

For this system, $\lambda_{11} = \frac{2}{3}$ in the steady-state RGA, suggesting that the variables be paired diagonally. In the dynamic RGA, however, the diagonal dominance deteriorates at moderate frequencies, as shown in Figure 21.5. In fact, the process is *off-diagonally* dominant in the frequency range of interest. The open-loop time constants are on the order of 10 min, and hence, frequencies in the range $0.1 < \omega < 1$ rad/min are of particular interest. For this system, the off-diagonal pairing (i.e., y_1–u_2 and y_2–u_1) is preferred, contrary to the pairing suggested by the steady-state RGA. To verify the analysis in Figure 21.5, the two pairings are simulated using IMC-PI tuning rules (see Section 21.4). For the diagonal pairing, the controllers are tuned: y_1–u_1 loop, $K_c = 0.6$, $\tau_I = 15$ min; y_2–u_2 loop, $K_c = -0.37$, $\tau_I = 20$ min. In contrast, for the off-diagonal pairing, the controller tuning parameters are $K_c = 10$, $\tau_I = 3$ min for the y_2–u_1 loop and $K_c = 2$, $\tau_I = 4$ min for the y_1–u_2 loop. Note that the controller gains for the diagonal pairing are an order of magnitude lower than for the off-diagonal pairing, a reflection of the bandwidth limitations imposed by the delays on the diagonal elements of the process transfer-function matrix. For a unit-step increase in the y_1 setpoint, Figure 21.6 shows that although the diagonal controller is bandwidth limited, the tuning for the off-diagonal configuration can be arbitrarily aggressive, only restricted by the actuator constraints. The reader can test the effect of changing pairings using the interactive C&R Tutorial CRGUI, which is implemented using MATLAB, on the CD-ROM that accompanies this text. In the main menu, opt for the "Mystery Process." ■

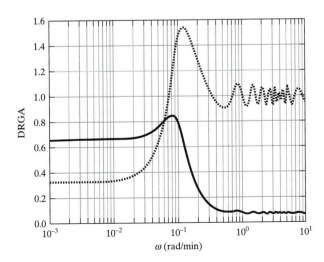

Figure 21.5 Dynamic RGA for the diagonal (solid line) and off-diagonal (dotted line) pairings for Example 21.3.

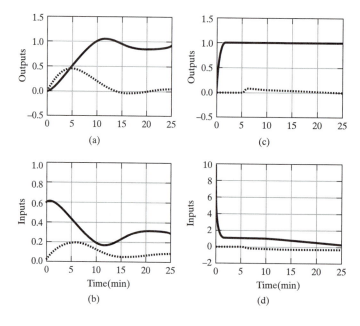

Figure 21.6 Closed loop response for Example 21.3 with PI control for a setpoint change in y_1 using: (a, b) diagonal pairing; (c, d) off-diagonal pairing. Shown on the top row are outputs: y_1 (solid line), y_2 (dashed line), and on the bottom row, inputs: u_1 (solid line), u_2 (dashed line).

EXAMPLE 21.4 *Control Configuration for a Utilities Subsystem (Example 20.7 Revisited)*

The analysis of the utilities subsystem in Figure 20.8 is based on the steady-state material and energy balances:

$$F_c = F_{c1} + F_{c2} \tag{21.30}$$

$$T_{co} = \frac{F_{c1}T_{c1} + F_{c2}T_{c2}}{F_{c1} + F_{c2}} \tag{21.31}$$

where the controlled variables are F_c and T_{co}. Selecting F_{c1} and F_{c2} as the manipulated variables, the steady-state gain matrix is computed by partial differentiation of Eqs. (21.30) and (21.31):

$$\begin{bmatrix} \Delta F_c \\ \Delta T_{co} \end{bmatrix} = \begin{bmatrix} 1 & 1 \\ \dfrac{F_{c2}(T_{c1} - T_{c2})}{(F_{c1} + F_{c2})^2} & -\dfrac{F_{c1}(T_{c1} - T_{c2})}{(F_{c1} + F_{c2})^2} \end{bmatrix} \begin{bmatrix} \Delta F_{c1} \\ \Delta F_{c2} \end{bmatrix} \tag{21.32}$$

The λ_{11} element of the RGA is computed using Eq. (21.19):

$$\lambda_{11} = \frac{p_{11}p_{22}}{p_{11}p_{22} - p_{12}p_{21}} = \frac{F_{c1}}{F_{c1} + F_{c2}} = x \tag{21.33}$$

where $0 \leq x \leq 1$. Since the RGA rows and columns add to unity, the RGA matrix is

$$\underline{\underline{\Lambda}} = \begin{bmatrix} x & 1 - x \\ 1 - x & x \end{bmatrix} \tag{21.34}$$

The recommended pairings depend on the operating conditions of the coolant subsystem, with significant interactions normally occurring between the control loops. To avoid this, the manipulated variables are defined as $\phi = F_{c1} + F_{c2}$ and $\mu = F_{c1}/(F_{c1} + F_{c2})$, transforming Eqs. (21.30) and (21.31) to

$$F_c = \phi \tag{21.35}$$

$$T_{co} = \mu \cdot T_{c1} + (1 - \mu)T_{c2} \tag{21.36}$$

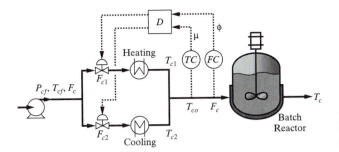

Figure 21.7 An attractive control configuration for the utilities subsystem.

After partial differentiation,

$$
\begin{bmatrix} \Delta F_c \\ \Delta T_{co} \end{bmatrix} = \begin{bmatrix} 1 & 0 \\ 0 & T_{c1} - T_{c2} \end{bmatrix} \begin{bmatrix} \Delta \phi \\ \Delta \mu \end{bmatrix}
\tag{21.37}
$$

This is a *decoupled* system that requires diagonal pairings, because $\underline{\underline{\Lambda}} = \underline{\underline{I}}$.

These pairings, shown in Figure 21.7, are intuitively correct in that the total flow rate is controlled by the sum of the two coolant streams, and the coolant temperature is controlled by the fraction of the coolant flowing through the heating system. Note that the temperature and flow controllers manipulate the variables μ and ϕ, respectively, which are processed by a decoupler, D, to generate corrections to the two flow rates, F_{c1} and F_{c2}, according to: ∎

$$
\left. \begin{array}{l} F_{c1} = \phi \cdot \mu \\ F_{c2} = \phi(1 - \mu) \end{array} \right\}
\tag{21.38}
$$

EXAMPLE 21.5   ***Control Configuration for a Debottlenecked Distillation Column***

Often, process design modifications can lead to potential control problems, as demonstrated by McAvoy (1983) for a distillation column in which the reboiler capacity is doubled by the addition of an identical reboiler in parallel with the original one (i.e., the column is *debottlenecked*), as shown in Figure 21.8.

A MIMO control system must be configured for the retrofitted column. To compute the RGA, a linearized model, in the steady state, relates the changes in the designated outputs, T, L_1, and L_2, to those of the manipulated variables, Q_1, Q_2, and B:

$$
\begin{bmatrix} \Delta T \\ \Delta L_1 \\ \Delta L_2 \end{bmatrix} = \begin{bmatrix} a_{11} & a_{12} & a_{13} \\ a_{21} & a_{22} & a_{23} \\ a_{31} & a_{32} & a_{33} \end{bmatrix} \begin{bmatrix} \Delta Q_1 \\ \Delta Q_2 \\ \Delta B \end{bmatrix}
\tag{21.39}
$$

Since B does not affect T directly, $\Delta T/\Delta B = 0$. Furthermore, by symmetry,

$$
a_{11} = a_{12} = \frac{\Delta T}{\Delta Q_1} = \frac{\Delta T}{\Delta Q_2} \qquad a_{23} = a_{33} = \frac{\Delta L_1}{\Delta B} = \frac{\Delta L_2}{\Delta B}
$$

$$
a_{21} = a_{32} = \frac{\Delta L_1}{\Delta Q_1} = \frac{\Delta L_2}{\Delta Q_2} \qquad a_{22} = a_{31} = \frac{\Delta L_1}{\Delta Q_2} = \frac{\Delta L_2}{\Delta Q_1}
$$

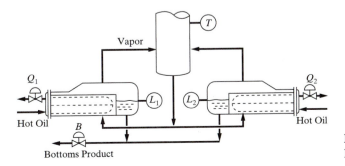

Figure 21.8 Debottlenecked distillation column.

Hence, Eq. (21.39) becomes

$$
\begin{bmatrix} \Delta T \\ \Delta L_1 \\ \Delta L_2 \end{bmatrix} = \begin{bmatrix} a_{11} & a_{11} & 0 \\ a_{21} & \beta a_{21} & a_{23} \\ \beta a_{21} & a_{21} & a_{23} \end{bmatrix} \begin{bmatrix} \Delta Q_1 \\ \Delta Q_2 \\ \Delta B \end{bmatrix}
\tag{21.40}
$$

where

$$
\beta = \frac{a_{22}}{a_{21}} = \frac{\text{effect of } Q_1 \text{ on } L_2}{\text{effect of } Q_1 \text{ on } L_1} < 1
$$

Using Eq. (21.22),

$$
\underline{\underline{\Lambda}} = \underline{\underline{P}} \otimes (\underline{\underline{P}}^{-1})^{\mathrm{T}} = \begin{bmatrix} 0.5 & 0.5 & 0 \\ \dfrac{0.5}{1 - \beta} & \dfrac{-0.5\beta}{1 - \beta} & 0.5 \\ \dfrac{-0.5\beta}{1 - \beta} & \dfrac{0.5}{1 - \beta} & 0.5 \end{bmatrix}
\tag{21.41}
$$

Note that $\beta < 1$ and is close to unity. Assuming $\beta = 0.95$, the RGA becomes

$$
\Lambda = \begin{bmatrix} 0.5 & 0.5 & 0 \\ 10 & -9.5 & 0.5 \\ -9.5 & 10 & 0.5 \end{bmatrix}
\tag{21.42}
$$

To ensure no loss of stability, pairings on negative RGA coefficients are avoided. Thus, only two possible pairings remain to be considered: $[Q_2\text{–}T, Q_1\text{–}L_1, B\text{–}L_2]$ and $[Q_1\text{–}T, B\text{–}L_1, Q_2\text{–}L_2]$. Neither alternative gives good performance since, in each case, one loop has a relative gain of 10, implying the need for severely detuned controllers. Clearly, the pairing selection for both of these controller configurations is limited by the available outputs and manipulated variables, and does not exploit the *symmetry* in the process design. This drawback can be avoided by selecting *other* manipulated and controlled variables. Here, it is desired to control the *total holdup* ($\psi = \Delta L_1 + \Delta L_2$), and the best manipulated variable to do this is intuitively the bottoms flow rate, B. Thus, the vector of manipulated variables is redefined as $\underline{u} = [\phi \ \ \Gamma \ \ \Delta B]^{\mathrm{T}}$, where $\phi = \Delta Q_1 - \Delta Q_2$ and $\Gamma = \Delta Q_1 + \Delta Q_2$, and the vector of controlled variables is redefined as $\underline{y} = [\Delta T \ \ \Omega \ \ \psi]^{\mathrm{T}}$, where $\Omega = \Delta L_1 - \Delta L_2$. The linear model, expressed in terms of these new variables, becomes

$$
\begin{bmatrix} \Omega \\ \Delta T \\ \psi \end{bmatrix} = \begin{bmatrix} (1 - \beta)a_{21} & 0 & 0 \\ 0 & a_{11} & 0 \\ 0 & (1 + \beta)a_{21} & 2a_{23} \end{bmatrix} \begin{bmatrix} \phi \\ \Gamma \\ \Delta B \end{bmatrix}
\tag{21.43}
$$

Note that this is a lower-triangular matrix, with the corresponding RGA:

$$
\underline{\underline{\Lambda}} = \begin{bmatrix} 1 & 0 & 0 \\ 0 & 1 & 0 \\ 0 & 0 & 1 \end{bmatrix}
$$

This result suggests the following pairings: (1) $\Omega\text{–}\phi$ (the imbalance in the holdups controlled by the imbalance in the heat duties of the two reboilers), (2) $\Delta T\text{–}\Gamma$ (the reboiler temperature controlled by the total heat duty), and (3) $\psi\text{–}\Delta B$ (the total holdup controlled by the bottoms flow rate). These control loops largely respond independently of each other (there is a small one-way interaction between the second and third loops), and are referred to as *decoupled*. ∎

The RGA as a Measure of Process Sensitivity to Uncertainty

Thus far, the RGA has been used to measure the process interactions and to aid in selecting the pairings for *decentralized* controller configurations. It is noteworthy that the magnitudes of the RGA elements are an indication of the degree of the process sensitivity to uncertainty. This is illustrated using a hypothetical process model,

$$
\underline{\underline{P}} = \begin{bmatrix} K(1 - \varepsilon) & 1 \\ 1 & K \end{bmatrix}
\tag{21.44}
$$

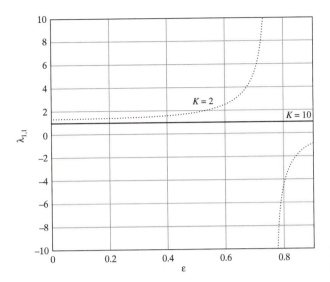

Figure 21.9 Effect of uncertainty on the RGA for $K = 2$ and $K = 10$.

in which the p_{11} coefficient is subject to a fractional uncertainty, ε. This uncertainty can significantly affect the RGA, depending on the value of K, as shown in Figure 21.9 where, $\lambda_{11} = K^2(1 - \varepsilon)/[K^2(1 - \varepsilon) - 1]$ is displayed as a function of ε for two values of K. For $K = 10$, the process is strongly diagonally dominant and hardly affected by the uncertainty (λ_{11} is close to unity). On the other hand, for $K = 2$, $\lambda_{11} = 1.33$ when $\varepsilon = 0$, indicating that the process has significant interactions. Furthermore, $\underline{\underline{P}}$ becomes singular at $\varepsilon = 0.75$ and the recommended pairings are switched, implying that a multivariable control system is unreliable at this level of uncertainty.

In summary, processes with RGA coefficients close to unity are relatively insensitive to uncertainties in the process model. Conversely, processes with large RGA coefficients tend to exhibit a high degree of sensitivity to model uncertainties.

Using the Disturbance Cost to Assess Resiliency to Disturbances

The design of process controllers for open-loop stable systems is motivated principally by the need to impart disturbance resiliency properties to processing operations. In other words, it is intended to maintain the outputs of multivariable processes at their setpoints despite external disturbances and uncertainties in the process model. The degree to which this requirement is satisfied is referred to as *resiliency*.

Given the process model of Eq. (21.1) and assuming perfect control, the action required to completely reject the disturbance, $\underline{d}$, is

$$\underline{u}\{s\} = -\underline{\underline{P}}^{-1}\{s\}\underline{d}'\{s\}, \text{ where } \underline{d}'\{s\} = \underline{\underline{P}}_d\{s\}\underline{d}\{s\} \tag{21.45}$$

By computing the norm of the actuator response, $\| \underline{u} \|$, as a function of the disturbance direction, the relative cost of rejecting a particular disturbance, $\underline{d}$, is computed as a function of its direction. One quantitative measure of the control effort to reject a given disturbance vector is the Euclidean norm:

$$\| \underline{u}\{s\} \|_2 = \| \underline{\underline{P}}^{-1}\{s\}\underline{\underline{P}}_d\{s\}\underline{d}\{s\} \|_2 \tag{21.46}$$

it being noted that the infinity norm provides an alternative resiliency measure. Parseval's theorem provides the direct translation of the 2-norm, in the frequency domain, to the total control action in the time domain. This norm, $\| \underline{u} \|_2$, is the *disturbance cost* (DC; Lewin, 1996). Often, it is more helpful to compute DC values for each manipulated variable separately. Since $\| \underline{u} \|_2$

is a frequency-dependent measure, it can be displayed as a function of frequency and the direction of $\underline{d}\{s\}$, to show the effect of two disturbances d_1 and d_2, where the disturbance direction is the angle of the disturbance vector with respect to the abscissa, that is, $\arg\{\underline{d}\}$. Contour maps of DC are displayed as a function of the disturbance direction and frequency. Since the DC is based on the assumption of perfect control, the results are independent of controller tuning or sophistication. For this reason, the DC is helpful for screening alternative flowsheets in Stages 2 and 3 of the design process, before it is practical to consider the details of the individual controllers. Even though perfect control is assumed, the values of the steady-state DC indicate:

1. *The Settling Time for Disturbance Rejection.* Note that disturbance directions for which the steady-state DC is high are those for which disturbance recovery is sluggish, regardless of the sophistication of the controller.

2. *The Limitations Due to Actuator Constraints.* Disturbance directions for which the steady-state DC exceeds the actuator constraints are those in which offset is incurred because of actuator saturation. Assuming that the process model has been scaled such that inputs are constrained to lie within $|\underline{u}| \le 1$, steady-state DC values above unity indicate that the actuator constraints are exceeded, and hence, such flowsheets should be avoided or modified to ensure adequate regulation.

Furthermore, by observing the DC variation at higher frequencies (e.g., at the closed-loop bandwidth specified), the disturbance directions are identified for which the high-frequency modes are attenuated with difficulty or not at all.

The next example shows the utility of the disturbance cost for predicting the ease of rejecting disturbances, as applied to the operation of a distillation tower.

EXAMPLE 21.6   *Resiliency Analysis of the Shell Process*

To test alternative control strategies, Prett and Morari (1986) provide a linearized model, referred to as the Shell process, of a distillation tower to separate crude oil into fractions in a refinery. Part of the model describes the dynamics of the two top compositions as a function of the manipulated variables (the two top draw rates) and two key disturbances (the heat removal loads in pump-around streams used to remove heat and create intermediate reflux). For this example, it is sufficient to examine the matrices specific to the nominal model:

$$\underline{\underline{P}}\{s\} = \begin{bmatrix} \dfrac{4.05}{50s+1}e^{-27s} & \dfrac{1.77}{60s+1}e^{-28s} \\ \dfrac{5.39}{50s+1}e^{-18s} & \dfrac{5.72}{60s+1}e^{-14s} \end{bmatrix} \qquad \underline{\underline{P}}_d\{s\} = \begin{bmatrix} \dfrac{1.2}{45s+1}e^{-27s} & \dfrac{1.44}{40s+1}e^{-27s} \\ \dfrac{1.52}{25s+1}e^{-15s} & \dfrac{1.83}{20s+1}e^{-15s} \end{bmatrix} \quad \textbf{(21.47)}$$

The time units in this model are minutes, and both manipulated variables and disturbances are in the range ± 0.5. After scaling, the inputs (both disturbances and manipulated variables) are in the range ± 1.

First, the disturbance cost at steady state is computed for various disturbance vectors. For $\underline{d} = [1,\,-1]^T$,

$$\arg(\underline{d}) = -45°: \; \| \underline{u}\{0\} \|_2 = \left\| \begin{bmatrix} 4.05 & 1.77 \\ 5.39 & 5.72 \end{bmatrix}^{-1} \begin{bmatrix} 1.2 & 1.44 \\ 1.52 & 1.83 \end{bmatrix} \begin{bmatrix} 1 \\ -1 \end{bmatrix} \right\|_2 = 0.0606 \quad \textbf{(21.48)}$$

and for three other disturbance directions,

$$\arg(\underline{d}) = 0°: \; \| \underline{u}\{0\} \|_2 = \left\| \begin{bmatrix} 4.05 & 1.77 \\ 5.39 & 5.72 \end{bmatrix}^{-1} \begin{bmatrix} 1.2 & 1.44 \\ 1.52 & 1.83 \end{bmatrix} \begin{bmatrix} 1 \\ 0 \end{bmatrix} \right\|_2 = 0.3072$$

$$\arg(\underline{d}) = 45°: \; \| \underline{u}\{0\} \|_2 = \left\| \begin{bmatrix} 4.05 & 1.77 \\ 5.39 & 5.72 \end{bmatrix}^{-1} \begin{bmatrix} 1.2 & 1.44 \\ 1.52 & 1.83 \end{bmatrix} \begin{bmatrix} 1 \\ 1 \end{bmatrix} \right\|_2 = 0.6748 \quad \textbf{(21.49)}$$

$$\arg(\underline{d}) = 90°: \; \| \underline{u}\{0\} \|_2 = \left\| \begin{bmatrix} 4.05 & 1.77 \\ 5.39 & 5.72 \end{bmatrix}^{-1} \begin{bmatrix} 1.2 & 1.44 \\ 1.52 & 1.83 \end{bmatrix} \begin{bmatrix} 0 \\ 1 \end{bmatrix} \right\|_2 = 0.3676$$

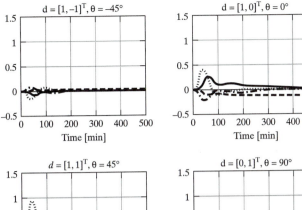

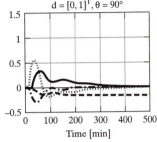

Figure 21.10 Closed-loop response of the Shell process to different disturbance directions: Solid line = y_1, dotted line = y_2, dashed line = u_1, dashed-dotted line = u_2.

Clearly, the worst disturbance to reject is $\underline{d} = [1, 1]^T$ (45° or $-135°$), whereas $\underline{d} = [1, -1]^T$ ($-45°$ or 135°) is the easiest to overcome.

These observations are verified by closed-loop simulations. The RGA for the matrix $\underline{P}\{0\}$, with $\lambda_{11} = 1.7$, indicates that the control loops can operate in a stable fashion only by pairing the inputs and outputs diagonally. Thus, diagonally paired PI controllers are tuned according to the improved IMC-based tuning rules (see Section 21.4), with $K_{C1} = 0.29$, $\tau_{I1} = 64$ min, $K_{C2} = 0.42$ and $\tau_{I2} = 67$ min. As shown in Figure 21.10, simulations confirm that disturbances acting in opposite directions, $\underline{d} = [1, -1]^T$, are the easiest to reject, whereas those in the same direction, $\underline{d} = [1, 1]^T$, are the most difficult.

To identify potential bandwidth limitations, the DC values are computed as a function of the frequency. For two disturbances, contours of DC values, computed separately for each manipulated variable, are displayed as a function of the disturbance direction (in degrees) and frequency, as shown in Figure 21.11. This confirms that the worst disturbance direction is $\underline{d} = [1, 1]^T$ (i.e., 45°), where the two manipulated variables are highest, with u_1 having relatively high steady-state values, but lower than unity, while u_2 has low values at steady state, but values that exceed unity at frequencies greater than $10^{-1.5} = 0.03$ rad/min. Consequently, the fast modes in disturbance vectors entering in this direction (45°) *are not attenuated, even if perfect control were possible.* Thus, the fastest settling time possible for this disturbance is approximately five times the inverse of the bandwidth limit, where DC = 1, that is, 150 min. This analysis is corroborated by the responses in Figure 21.10, which indicate that the most severe bandwidth limitations are exhibited for disturbances aligned at 45°, but with perfect steady-state disturbance rejection, with most of the static effects eliminated using u_1. Clearly, the response obtained with the decentralized PI control system for $\underline{d} = [1, 1]^T$ is considerably more sluggish, since delay times impose additional stability limitations, as discussed in Section 21.4. In contrast, disturbances in

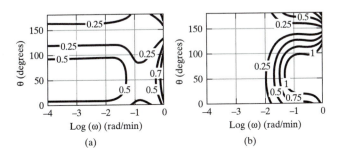

Figure 21.11 DC contour map for the Shell process: (a) u_1; (b) u_2.

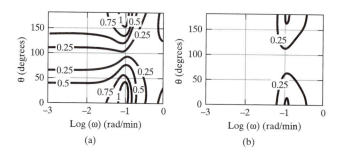

Figure 21.12 DC contour map for Example 21.7: (a) u_1; (b) u_2.

the direction, $\underline{d} = [1, -1]^T$, (i.e., $-45°$ or $135°$), are more easily rejected, as the DC for each manipulated variable remains low over the entire frequency range. Again, this is corroborated by the responses in Figure 21.10. The reader can test the effect of disturbance direction in this example using the interactive C&R Tutorial CRGUI, which is implemented using MATLAB, on the CD-ROM that accompanies this text. In the main menu, opt for the "Shell Process."   ∎

EXAMPLE 21.7   *Using DC to Improve Process Resiliency (Example 21.3 Revisited)*

Returning to the process in Example 21.3, and noting that DRGA analysis leads to the recommendation that the variables be paired off-diagonally (i.e., y_2–u_1 and y_1–u_2), the resiliency of the controlled system to disturbances is examined. DC contour maps for each of the manipulated variables are presented in Figure 21.12, where it is noted that the worst disturbance is $\underline{d} = [20, 0]^T$ (i.e., a disturbance direction of $0°$). Furthermore, in this direction, u_1 is bandwidth-limited, with saturation occurring at a frequency of approximately $10^{-1} = 0.1$ rad/min; that is, with a characteristic time of 10 min. Consequently, the settling time in response to such a disturbance is expected to be greater than 50 min. In contrast, the second input, u_2, has no bandwidth limitations.

The predictions afforded by the DC contour maps in Figure 21.12 are confirmed by simulation for $\underline{d} = [20, 0]^T$. As seen in Figures 21.13a and b, u_1 saturates at its upper bound (with manipulated variable bounds set at $|u_1| \leq 60$ and $|u_2| \leq 50$). This bandwidth limitation is the reason for the sluggish process recovery, confirming the DC analysis that anticipates saturation in u_1, a *design* problem that arises because the u_1 range is too small to provide adequate dynamic resiliency. Through redesign, the

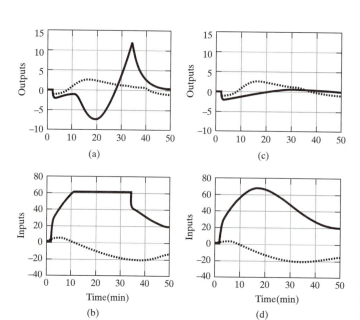

Figure 21.13 Closed-loop response with off-diagonal PI control to the disturbance $\underline{d} = [20, 0]^T$ for Example 21.7, with: (a, b) original bounds on u_1; (c, d) enlarged bounds on u_1 (to ±70). Shown on the top row are outputs: y_1 (solid line), y_2 (dashed line), and on the bottom row, inputs: u_1 (solid line), u_2 (dashed line).

range is increased to $|u_1| \leq 70$, and the performance is improved significantly, as shown in Figures 21.13c and d. The reader can investigate the effect of manipulated variable constraints in this example using the interactive C&R Tutorial CRGUI, which is implemented using MATLAB on the CD-ROM that accompanies this text. In the main menu, opt for the "Mystery Process." ∎

21.3   TOWARD AUTOMATED FLOWSHEET C&R DIAGNOSIS

This section describes a procedure for assessing the controllability and resiliency of a process flowsheet that relies on heuristics to create a linearized dynamic model of the process using the results of steady-state simulations. The derived model is used to test the flowsheet C&R, using the measures introduced in Section 21.2. As a tutorial exercise, the procedure is applied to screen the designs for the heat-integrated distillation columns in Example 20.2. Subsequently, in Section 21.5, case studies are presented for three additional processes. For each case, the results of the approximate linear analysis are compared with the results of closed-loop simulations using nonlinear dynamic models. As will be shown, the overall approach is very promising as a shortcut diagnostic and screening tool, which can be expected to be integrated into commercial simulation software.

Shortcut C&R Diagnosis

As discussed above, both steady-state and dynamic C&R analyses provide useful information for flowsheet assessment. Clearly, the second alternative provides more information and is more reliable. On the other hand, steady-state analysis requires much less work, and is often adequate for screening purposes. Consequently, both approaches are considered in this section.

The following steps are involved in steady-state analysis:

1. After the flowsheet is synthesized, the control structure is considered, first by selecting the process outputs to be controlled, $\underline{y}\{t\}$, the manipulated variables, $\underline{u}\{t\}$, and the disturbance variables, $\underline{d}\{t\}$. These are related by Eq. (21.1).
2. Steady-state simulation of the flowsheet is carried out using a process simulator.
3. Steady-state gains for the overall transfer functions, $\underline{\underline{P}}\{0\}$ and $\underline{\underline{P}}_d\{0\}$, are computed by perturbing each input, one at a time.
4. Steady-state C&R measures are computed using $\underline{\underline{P}}\{0\}$ and $\underline{\underline{P}}_d\{0\}$.

For the dynamic C&R analysis (Weitz and Lewin, 1996), the steps in the algorithm are as follows:

1. Step 1 of the steady-state algorithm.
2. Step 2 of the steady-state algorithm.
3. The flowsheet is decomposed into *component parts*. These are MIMO subsections of the flowsheet that are approximated by matrices of low-order transfer functions (usually first order with dead time). This decomposition permits process units to be modeled in sufficient detail, allowing inverse response and overshoot phenomena to be represented.
4. Steady-state gains for the component parts are computed by perturbation of each input, one at a time.
5. Time constants and delay times are estimated assuming perfect mixing or plug flow, as appropriate, with the flow rates at steady state. At this point, transfer function matrices are defined for each component part.
6. The transfer function matrices, $\underline{\underline{P}}\{s\}$ and $\underline{\underline{P}}_d\{s\}$, are generated for the complete flowsheet. This involves computing the frequency response of each component part, and recombining the component parts, as dictated by the plant topology.
7. The frequency-dependent C&R measures are computed using the approximate linear model, $\underline{\underline{P}}\{j\omega\}$ and $\underline{\underline{P}}_d\{j\omega\}$.

Many packages are available for steady-state simulation, as discussed in Chapter 4. To manipulate the linearized models in the Laplace, frequency, and time domains, MATLAB and SIMULINK are used commonly, and example scripts are introduced in Section 21.6. The most recent commercial packages permit steady-state and dynamic simulations. These include HYSYS.Plant, CHEMCAD, and ASPEN DYNAMICS, with the former used in this section and in Section 21.5.

All of the steps in the two algorithms are implemented using an array of computer packages that are becoming more integrated. Note that steps 4 and 5 deserve special attention, as they are the basis for the approximate models generated for the dynamic C&R analysis.

Generating Low-Order Dynamic Models

The linearized model for each component part, to be completed in steps 4 and 5 of the dynamic C&R analysis, has the form

$$\underline{y}^c\{s\} = \underline{\underline{K}}^c \otimes \underline{\underline{\Psi}}^c\{s\}\underline{u}^c\{s\} \tag{21.50}$$

where $\underline{u}^c\{s\}$ and $\underline{y}^c\{s\}$ are m-dimensional input and n-dimensional output vectors, in complex space, $\underline{\underline{K}}^c$ is a matrix of steady-state gains in $n \times m$ real space, and $\underline{\underline{\Psi}}^c\{s\}$ is a matrix describing the dynamics in $n \times m$ complex space (each element of which is typically a delayed, low-order transfer function with dead time). The term $\otimes$ denotes the Schur (or element-by-element) product. Each distillation column is characterized by a single time constant. For heat exchangers, separate time constants are associated with the tube- and shell-side fluids. In the subsections that follow, it is shown that the gains, time constants, and dead times can be estimated almost entirely using the results of steady-state simulations.

Steady-state Gain Matrix, $\underline{\underline{K}}^c$

The steady-state gains between the inputs and outputs for each component part are generated using the following procedure:

1. The material and energy balances, in the steady state, are solved for the complete flowsheet at the nominal operating point.
2. Small positive and negative perturbations are introduced for each input of each component part, one at a time, and the changes in the outputs are computed.
3. The steady-state gains for each component part are computed using finite differences: $K_{ij}^c = \Delta y_i^c / \Delta u_j^c$, where the perturbation, Δu_j^c, is sufficiently small to avoid precision losses.

Dynamics Matrix, $\underline{\underline{\Psi}}^c\{s\}$

In this section, an approach is suggested for estimating the time constants and delay times for distillation columns and heat exchangers.

Distillation Columns

Time Constants.   Following Skogestad (1987), the dominant time constant is estimated as

$$\tau = \tau_I + \tau_C + \tau_R \tag{21.51}$$

where τ_C and τ_R are the time constants (in minutes) associated with the condenser and reboiler, respectively, and τ_I is the time constant (in minutes) for the column, estimated according to

$$\tau_I = \sum_{i=1}^{N} \frac{M_i}{L_i} \tag{21.52}$$

where M_i is the volumetric holdup (m³) on tray i, L_i is the liquid flow rate (m³/min) from tray i, and N is the number of trays. The liquid holdup is expressed as

$$M_i = A_c(h_w + h_{ow}) = \frac{\pi D_c^2}{4}(h_w + h_{ow}) \tag{21.53}$$

where D_c is the column diameter (m), and h_w and h_{ow} are the weir height and fluid height above the weir (m), respectively. The latter can be expressed in terms of the weir length, l_w (m), using the Francis weir equation:

$$h_{ow} = \left(\frac{L_i}{111\, l_w}\right)^{2/3} \tag{21.54}$$

Delay Times.   When the internal liquid flow rate in the column changes, a delay time is associated with the change in the fluid holdup above the weir. For a single tray, this is estimated by considering the time taken for h_{ow} to stabilize after a change in the liquid flow rate (Shinskey, 1984):

$$\theta = \frac{dM_{ow}/dt}{dL_i/dt} = A_c\frac{dh_{ow}}{dL_i} = \frac{\pi D_c^2}{666\, l_w\, h_{ow}^{0.5}} \tag{21.55}$$

Thus, the overall delay experienced by the bottoms product after changes in the flow rates, temperatures, or compositions of the feed or reflux depends on the number of trays involved. In contrast, it is noted that the distillate composition responds *immediately* to a change in the reflux flow rate, but experiences a considerable delay after changes in the feed concentration or temperature. For the latter, the delay time is estimated as the sum of the residence times on all trays between the feed and the top tray, since such a change is assumed to propagate by affecting the entire tray holdup rather than just the over-weir fluid.

Typical Design Parameters.   The following heuristics are in common use: $\tau_C = \tau_R = 0.5\tau_I$, $l_w = 0.65D_c$, and $h_w = 2$ in.

Heat Exchangers

It is assumed that time delays associated with heat exchangers in the major processing units, such as the condensers and reboilers in distillation columns, are negligible. When heat exchangers are not included in the major processing units, they are modeled as first-order lags associated with single shell and single tube passes.

Time Constants.   These are estimated for the tube- and shell-side fluids using $\tau_T = V_T/q_T$ and $\tau_S = V_S/q_S$. The volumes of the fluid holdups in the tubes and shell, V_T and V_S, are estimated using the heat transfer area, the average fluid velocity in the tubes, v, and the tube and shell diameters. The volumetric flow rates, q_T and q_S, are estimated by the process simulator.

EXAMPLE 21.8   ***C&R Analysis for Heat-Integrated Distillation Columns (Example 20.2 Revisited)***

Dynamic C&R analysis is applied to screen the heat-integrated distillation configurations for the dehydration of methanol in Figure 20.2 of Example 20.2. Of the three heat-integrated designs, the FS and LSR configurations provide the maximum energy savings. Clearly, the most controllable and resilient of the two should be selected based on C&R screening. Note that Chiang and Luyben (1988) prepared nonlinear dynamic models of the three heat-integrated configurations. They carried out C&R analysis, using the RGA and minimum singular values, based on linear approximations to their dynamic models. Although their findings using linear analysis were inconclusive, they showed the FS configuration to be far less desirable using closed-loop simulations with their nonlinear models.

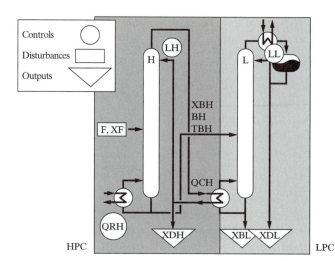

Figure 21.14 Component parts for the LSF configuration.

In the following, each step of dynamic C&R analysis is described as it is applied to the LSF configuration.

Step 1: *Selection of the outputs, manipulated variables, and disturbances.* As shown in Figure 21.14, for the LSF configuration, the process outputs are the mole fractions of methanol in the three product streams (x_{DH}, x_{DL}, and x_{BL}). The process inputs are the control variables (L_H, L_L, and Q_{RH}), and the disturbances are F and x_F.

Step 2: *Steady-state simulation.* The simulation was carried out using the PRO/II simulator and assuming that there is no pressure drop in the columns, no heat losses to the surroundings, and tray efficiencies of 75%. The thermodynamic properties were computed using the UNIFAC option. These conditions were used by Chiang and Luyben (1988), with the exception that they accounted for heat losses. The results for the four flowsheets are in Table 21.1. Note that the energy requirements for the LSR and FS configurations are the lowest (0.205×10^6 kcal/min), followed by the LSF configuration (0.222×10^6 kcal/min).

Table 21.1 Results from the Steady-State Simulation Using PRO/II of Simulation Sciences for the Heat-Integrated Configurations for the Dehydration of Methanol, Compared with a Single Column

Variable	SC COL1	FS COL1	FS COL2	LSF COL1	LSF COL2	LSR COL1	LSR COL2
F, feed flow (kmol/min)	45.00	22.04	22.96	45.00	33.95	45.00	32.96
x_F, feed mole frac. (CH_3OH)	0.50	0.50	0.50	0.50	0.35	0.50	0.33
D, distillate flow (kmol/min)	22.50	11.02	11.48	11.05	11.45	12.04	10.46
x_D, distillate mole frac. (CH_3OH)	0.96	0.96	0.96	0.96	0.96	0.96	0.96
B, bottoms flow (kmol/min)	22.50	11.02	11.48	33.95	22.50	32.96	22.50
x_B, bottoms mole frac. (CH_3OH)	0.04	0.04	0.04	0.35	0.04	0.33	0.04
N, number of trays	13	16	13	16	13	13	16
N_F, feed tray (1 ≡ top)	9	12	9	13	11	11	12
R, reflux ratio	0.82	1.12	0.82	1.06	1.10	0.75	1.15
P, working pressure (mmHg)	760	3,900	760	3,900	760	760	3,900
Q_R, reboiler duty (10^6 kcal/min)	0.353	0.205	0.180	0.222	0.175	0.180	0.205
Q_C, condenser duty (10^6 kcal/min)	0.348	0.180	0.178	0.175	0.205	0.179	0.180
T_R, reboiler temperature (°C)	93.7	146.3	93.7	126.5	93.7	77.2	127.2
T_C, condenser temperature (°C)	65.1	113.4	65.1	113.4	65.1	65.1	95.9
D_C, column diameter (m)	3.2	1.3	2.3	2.0	2.4	2.3	2.0

Step 3: *Decomposition into component parts.* It has been demonstrated that a first-order lag is a reasonable approximation for the dynamics of a distillation column (Skogestad, 1987). Thus, the LSF configuration is decomposed into two component parts, one for each column. Four intermediate variables are identified to model the information transfer between the component parts: x_{BH}, B_H, T_{BH}, and Q_{CH} ($= -Q_{RL}$). Note that both T_{BH} and Q_{CH} are needed for the energy balance in the reboiler because partial vaporization occurs.

The control variables, in perturbation variable form, are scaled between zero and their nominal values, and the disturbances are scaled using bounds 20% above and below their nominal values. The outputs are scaled to provide a reasonable match with the steady-state gains computed by Chiang and Luyben (1988), it being noted that the output scaling does not affect the RGA or the DC values.

Steps 4 and 5: *Computing $\underline{\underline{K}}^c$ and $\underline{\underline{\psi}}^c(s)$.* These are computed following the procedure in the section on "Generating Low-Order Dynamic Models," which gives linearized models for the high-pressure column in the LSF configuration:

$$\begin{bmatrix} x_{DH} \\ \hdashline x_{BH} \\ T_{BH} \\ B_H \\ Q_{CH} \end{bmatrix} = \frac{1}{13s+1} \begin{bmatrix} 0.017 & -1.109 & 0.001 & 0.090e^{-6.4s} \\ 0.011e^{-1.3s} & -1.859 & 0.006e^{-0.1s} & 1.296e^{-0.1s} \\ -0.33e^{-1.3s} & 59.0 & -0.2e^{-0.1s} & -41.05e^{-0.1s} \\ 0.916e^{-1.3s} & -123.7 & 1.127e^{-0.1s} & -0.02e^{-0.1s} \\ 0.00004e^{-1.3s} & -0.994 & 0.001e^{-0.1s} & 0.003e^{-0.1s} \end{bmatrix} \begin{bmatrix} L_H \\ Q_{RH} \\ \hdashline F \\ x_F \end{bmatrix} \quad \textbf{(21.56)}$$

and for the low-pressure column:

$$\begin{bmatrix} x_{BL} \\ x_{DL} \end{bmatrix} = \frac{1}{17s+1} \begin{bmatrix} 0.792e^{-0.1s} & -0.029e^{-0.1s} & 0.007e^{-0.1s} & -2.161 & 0.012e^{-1.4s} \\ 0.790e^{-8.5s} & -0.051 & 0.003 & -3.291 & 0.038 \end{bmatrix} \begin{bmatrix} x_{BH} \\ T_{BH} \\ B_H \\ Q_{RL} \\ \hdashline L_L \end{bmatrix} \quad \textbf{(21.57)}$$

Step 6: *Generation of transfer function matrices.* The linear approximation for the flowsheet dynamics is obtained by recombining the models for the component parts. Note that Figure 21.15 shows schematically how the linearized models for the component parts are linked in each of the configurations. The inputs and outputs associated with each of the configurations are represented by the terminal junctions to the left and right. Thus, for example, the FS configuration has four manipulated and two disturbance variables (six inputs in all), and four output variables. The blocks marked HPC and LPC represent the component parts for the high- and low-pressure distillation columns, respectively. The arcs represent the flow of information (intermediate variables) to and from the component parts.

The recombination to form overall transfer functions is accomplished by algebraic manipulation. For the LSF configuration, Eqs. (21.56) and (21.57) are rewritten in block-matrix form:

$$\begin{bmatrix} x_{DH} \\ \hdashline x_{BH} \\ T_{BH} \\ B_H \\ Q_{CH} \end{bmatrix} = \begin{bmatrix} \underline{\underline{P}}_H(1,1) & \vdots & \underline{\underline{P}}_H(1,2) \\ \hdashline \underline{\underline{P}}_H(2,1) & \vdots & \underline{\underline{P}}_H(2,2) \end{bmatrix} \begin{bmatrix} L_H \\ Q_{RH} \\ \hdashline F \\ x_F \end{bmatrix} \quad \textbf{(21.58)}$$

$$\begin{bmatrix} x_{BL} \\ x_{DL} \end{bmatrix} = [\underline{\underline{P}}_L(1,1) \vdots \underline{\underline{P}}_L(1,2)] \begin{bmatrix} x_{BH} \\ T_{BH} \\ B_H \\ Q_{RL} \\ \hdashline L_L \end{bmatrix} \quad \textbf{(21.59)}$$

where the matrix blocks contain elements from the transfer function matrices in Eqs. (21.56) and (21.57); for example,

$$\underline{\underline{P}}_H(1,1) = \frac{1}{13s+1} [0.017 \ -1.109]$$

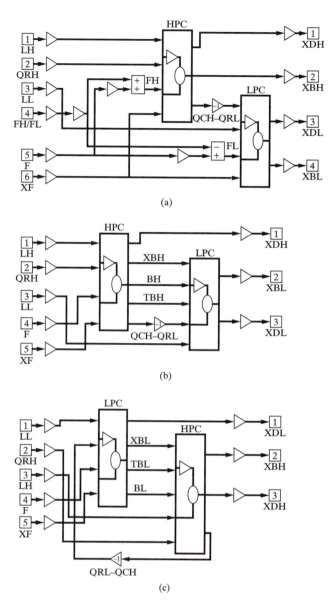

Figure 21.15 Information flows between the component parts of the heat-integrated distillation configurations: (a) FS; (b) LSF: (c) LSR.

Next, by algebraic manipulation, the vector of internal variables, namely $[x_{BH} \, T_{BH} \, B_H \, Q_{CH}]^T$, is eliminated, leading to

$$[x_{DH}] = [\underline{P_H}(1,1) \; \vdots \; 0] \begin{bmatrix} L_H \\ \underline{Q_{RH}} \\ L_L \end{bmatrix} + [\underline{P_H}(1,2)] \begin{bmatrix} F \\ x_F \end{bmatrix} \qquad (21.60)$$

$$\begin{bmatrix} x_{BL} \\ x_{DL} \end{bmatrix} = [\underline{P_L}(1,1) \cdot \underline{P_H}(2,1) \; \vdots \; \underline{P_L}(1,2)] \begin{bmatrix} L_H \\ \underline{Q_{RH}} \\ L_L \end{bmatrix} + [\underline{P_L}(1,1) \cdot \underline{P_H}(2,2)] \begin{bmatrix} F \\ x_F \end{bmatrix} \qquad (21.61)$$

Note that Eqs (21.60) and (21.61) are in the standard transfer function form of Eq. (21.1). Similar manipulations are used for the other two configurations. The LSR configuration involves the most complicated manipulations, since it involves the feedback of information.

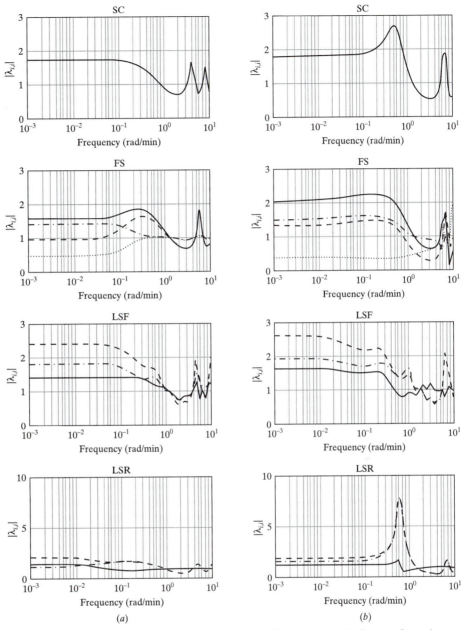

Figure 21.16 Diagonal RGA elements as a function of frequency for the four configurations to dehydrate methanol: (a) procedure in Section 21.3; (b) Chiang and Luyben (1988).

These models are compared with those derived by Chiang and Luyben (1988), who fitted linear transfer functions to the transient open-loop responses that were computed using their nonlinear model. In Figure 21.16, the diagonal RGA matrix coefficients for all four configurations are plotted against the frequency; the values reported by Chiang and Luyben appear on the right, while those computed using the linear models derived using the C&R analysis appear on the left. As shown, the results are in close agreement. The resonant peaks computed by Chiang and Luyben are the result of differences in the time constants and delay times in the transfer function elements. Furthermore, the relative gains, computed using the procedure in this section, do not vary significantly with frequency. Hence, diagonal pairings are preferred for the decentralized control system.

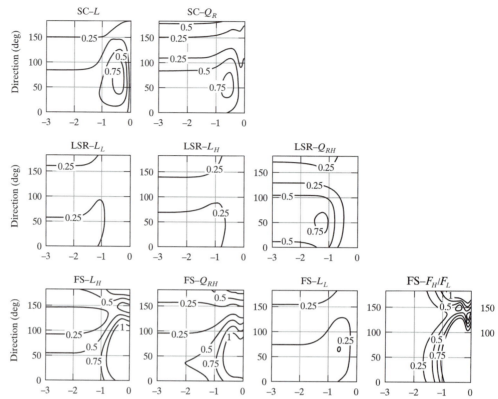

Figure 21.17 DC contour maps for the SC, FS, and LSR configurations to dehydrate methanol. The bounds on the disturbances are ±20% from their nominal values. The DC contour maps for each manipulated variable are computed separately, with bold solid lines indicating DC = 1. See Figure 21.39 for the DC contour maps for the LSF configuration.

Step 7: *Computation of C&R measures.* Figure 21.17 shows the DC contour maps computed for each of the manipulated variables associated with the configurations SC, LSR, and FS, where the ordinate is the direction of the disturbance $[F, x_F]^T$, and the abscissa is the log 10 of the frequency. Since DC values in excess of unity correspond to saturated manipulated variables, it is apparent that the disturbances are rejected adequately by all of the designs at the steady state (i.e., when $\omega = 0$). However, for a wide range of disturbance directions, the FS configuration has disturbance costs in excess of unity at frequencies beyond 0.1 rad/min in three of the manipulated variables (L_H, Q_{RH}, and F_H/F_L). Thus, disturbance rejection is expected to be very sluggish for this configuration. The other two configurations have low disturbance costs and are expected to reject these disturbances nearly as well as a single column. Thus, the FS configuration should be rejected and the LSR configuration selected, because its energy requirements are lower than for the LSF and SC configurations.

To confirm these results, nonlinear dynamic simulations are carried out using HYSYS. Three configurations are simulated: (1) the single column (SC) in the LV configuration; (2) the FS configuration with the pairing: x_{DH}–L_H, x_{BH}–Q_{RH}, x_{DL}–L_L, and x_{BL}–F_H/F_L; (3) the LSR configuration with the pairing: x_{DH}–L_H, x_{BH}–Q_{RH}, and x_{DL}–L_L. These pairings are selected on the basis of the RGA. The control loops are tuned using the IMC-PI tuning rules (see Section 21.4), with tuning parameters summarized in Table 21.2. Note that the nominal values of the manipulated variables are at the midpoint of their ranges.

Figures 21.18, 21.19, and 21.20 show the responses for the three configurations, subjected to the worst-case disturbance in which the feed flow rate and composition simultaneously undergo positive step changes to their design limits. As shown in Figure 21.18, the SC configuration is returned to its setpoints in approximately 100 min, with T_{10} most affected. Note that this response is qualitatively similar to that of the linear approximation shown in Figure 21.3. The simulation can be reproduced using the METH_SC.hsc file on the CD-ROM that accompanies this book.

Table 21.2 Tuning Parameters for the SC, LSR, and FS Configurations

Loop	SC[a]	LSR	FS
x_{DL}–L_L	$K_c = 29$; $\tau_I = 10$ min	$K_c = 1$; $\tau_I = 10$ min	$K_c = 1$; $\tau_I = 5$ min
x_{BL}–Q_{RL}	$K_c = 6$; $\tau_I = 10$ min		
x_{BL}–F_H/F_L[b]			$K_c = 0.5$; $\tau_I = 30$ min
x_{DH}–L_H		$K_c = 0.15$; $\tau_I = 5$ min	$K_c = 1$; $\tau_I = 5$ min
x_{BH}–Q_{RH}		$K_c = 1$; $\tau_I = 10$ min	$K_c = 0.1$; $\tau_I = 10$ min

[a]For the SC configuration, the temperatures on trays 2 and 10 are regulated instead of the compositions.

[b]The x_{BL} composition controller is the master of a lower-level flow controller to regulate F_H/F_L.

As shown in Figure 21.19, the response of the FS configuration to the same disturbance is very sluggish, settling in about 100 min, and exhibiting severe undershoots in two of the four mole fractions: x_{BH} (by 20 mol%) and x_{BL} (by 10 mol%). This verifies the predictions of the DC contour maps in Figure 21.17, which anticipate significant bandwidth limitations. These simulation results can be reproduced using the METH_FS.hsc file, also on the CD-ROM.

In contrast, the response of the LSR configuration to the same disturbance, shown in Figure 21.20, settles in about half the time, with significantly less undershoot in x_{DL}. This is because the controllers are significantly less bandwidth-limited, as predicted by the DC contour maps in Figure 21.17. Furthermore, the settling time of the LSR configuration is comparable to that for the single column, as predicted by the DC analysis. These simulation results can be reproduced using the METH_LSR.hsc file on the CD-ROM. ∎

It should be emphasized that the DC contour maps are based on the assumption of perfect control, assuming that there are no stability limitations to increases in the controller gain. In practice, when single-loop controllers are implemented, the controller gains are limited, as in this example, by process interactions or by single-loop stability limitations such as delay times. As a result, the bandwidth limitations are usually underestimated by the DC contour maps, but usually not sufficiently to affect their diagnoses when used to screen alternative designs. Note that the prediction that the FS configuration provides significantly worse disturbance rejection compared with that of the LSR configuration has been verified by simulation. Clearly, the LSR design is preferable based on energy efficiency and controllability.

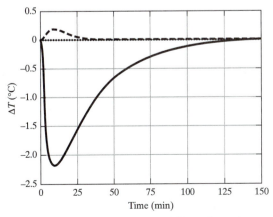

Figure 21.18 Response of the SC configuration to simultaneous disturbances in F (from 2,700 to 3,000 kmol/hr) and x_F (from 0.5 to 0.6 methanol mol fraction): T_2 (dashed line), T_{10} (solid line), setpoints (dotted lines).

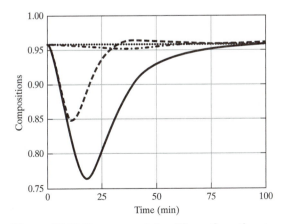

Figure 21.19 Response of the FS configuration to simultaneous disturbances in F (from 2,700 to 3,000 kmol/hr) and x_F (from 0.5 to 0.6 methanol mol fraction): x_{BH} (solid line), x_{BL} (dashed line), x_{DL} (dash-dotted line), and setpoints (dotted line).

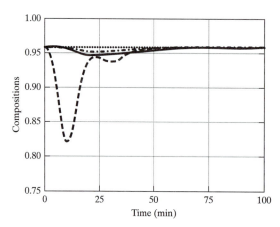

Figure 21.20 Response of the LSR configuration to simultaneous disturbances in F (from 2,700 to 3,000 kmol/hr) and x_F (from 0.5 to 0.6 methanol mol fraction): x_{DH} (dashed line), x_{BH} (solid line), x_{DL} (dash-dotted line), and setpoints (dotted line).

This approach has been used successfully for screening more complex heat-integrated flowsheets (Weitz, 1994), exothermic reactors (Naot and Lewin, 1995), and polymerization reactors (Lewin and Bogle, 1996). In all cases, the projections were confirmed using rigorous dynamic models. To further illustrate this screening technique, Section 21.5 provides three case studies, involving exothermic reactors in series, heat-exchanger networks, and a recycle process.

21.4 CONTROLLER LOOP DEFINITION AND TUNING

Since the regulatory loops use PI controllers for verification of the C&R analysis, a brief summary of their configuration and tuning is provided in this section.

Definition of PID Control Loop

This involves specifying

1. The *process variable* to be controlled, *PV*; that is, any stream- or operation-related variable in the flowsheet (e.g., pressure, temperature, liquid level, species mass or mole fraction, mass or molar flow rate). In addition, the minimum and maximum values of the *PV* are used to express the *PV* as a percentage of its full range:

$$PV(\%) = \left(\frac{PV - PV_{\min}}{PV_{\max} - PV_{\min}} \right) \times 100 \qquad \textbf{(21.62)}$$

2. The *controller output*, *OP*, to be manipulated by the controller, as a percentage of its full range. This variable is usually either a stream flow rate or the rate of heat transfer of an energy stream. Generally, its minimum value is specified as zero and its maximum is taken as twice its nominal value. Note that occasionally the nominal value is not positioned midway between the minimum and maximum values (e.g., when the nominal flow rate of a bypass stream lies near its maximum or minimum flow rate).

3. The *controller action*, either *direct* or *reverse acting*, which defines the direction of its effect. For a direct-acting controller, when the *PV* rises above the setpoint (*SP*), the *OP increases*, and vice versa. In this case, the static process gain is negative, as illustrated for a level controller in Figure 21.21a. Here, the liquid level is the *PV*, the flow rate of the effluent stream, Q_o, is the *OP*, and the controller action is set to *Direct*. In contrast, for a reverse-acting controller, when the *PV* rises above the *SP*, the *OP decreases*, and vice versa. In these cases, the static process gain is positive, as illus-

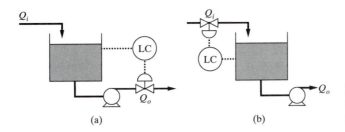

Figure 21.21 Level-control configurations for a surge tank: (a) direct acting; (b) reverse acting.

trated in Figure 21.21b, which shows a different controller configuration for the surge tank. Here, the flow rate of the feed stream, Q_i, is the *OP*, and the controller action is set to *Reverse*.

4. The *tuning parameters*. For a PID (Proportional-Integral-Derivative) controller, the output, $OP(t)$, is a function of the tracking error, $E\{t\}$:

$$OP\{t\} = OP_{SS} + K_C\left(E\{t\} + \frac{1}{\tau_i}\int_0^t E\{\theta\}d\theta + \tau_d\frac{dE\{t\}}{dt}\right) \qquad \textbf{(21.63)}$$

where OP_{SS} is the bias, or controller output at zero error, and K_C, τ_i, and τ_d are the proportional gain, integral time constant (or reset time), and derivative time constant (or rate time) of the controller. The tracking error at time t, $E(t)$, is the difference between the set point and the process variable:

$$E\{t\} = SP\{t\} - PV\{t\} \qquad \textbf{(21.64)}$$

As mentioned above, *SP*, *PV*, and *OP* are expressed as percentages of their full ranges. Consequently, the controller gain, K_C, is dimensionless, and represents the percentage change in *OP* for a one-percent change in *PV*. In the absence of other information, *factory settings* are used: $K_C = 1$, $\tau_i = 10$ min, and $\tau_d = 0$. These are tuned for improved performance, as discussed in the next section.

Controller Tuning

The PID controller is the most commonly used feedback controller in industry, with three tunable parameters as stated previously. The integral component ensures that the tracking error, $E\{t\}$, is asymptotically reduced to zero, whereas the derivative component imparts a predictive capability, potentially enhancing the performance. Despite its apparent simplicity, the subject of PID controller tuning has been discussed in several textbooks and thousands of research papers since the landmark work of Ziegler and Nichols (1942). In practice, despite these developments, most PID controllers are tuned as PI controllers for several reasons.

1. The improved performance attainable using the derivative term is often not required. In many cases, the derivative action causes the controller to respond more nervously, especially when step changes are imposed on the setpoints. Furthermore, as pointed out by Luyben and co-workers (1999), it is often sufficient to tune level controllers as proportional-only regulators (i.e., $1/\tau_i = 0$ and $\tau_d = 0$).
2. The derivative component amplifies measurement noise. It is highly recommended that derivative action be applied only to filtered feedback signals.
3. Using model-based tuning methods, as discussed in the section on "Model-Based PI Controller Tuning," higher-order models are needed to tune a PID controller. In many cases, the additional engineering effort is not justified.

4. Many PID controller loops remain on their factory settings long after plant startup. When the controller action is in the right direction and the *PV* range is defined wisely, these settings often give adequate performance.

For tuning, either on-line methods, implemented with controllers on-line, or model-based methods, which rely on process models, are utilized. The main advantages of the on-line methods are that tuning occurs under closed-loop control and a process model is not required. It should be recognized, however, that they provide initial controller settings that are usually improved iteratively during operation. Furthermore, typical on-line tuning rules apply strictly for a single control loop. For a multivariable control system, detuning is often necessary to prevent process interactions from introducing instability. Because no guidelines are available to modify the parameters of the on-line tuned controllers when this occurs, the discussion below focuses on model-based PI tuning rules. For more details on on-line tuning, the reader is referred to Seborg et al. (1989) and Luyben (1990).

Model-Based PI-Controller Tuning

All control systems are implicitly or explicitly model based. They can be as simple as PI or PID controllers, which are implicitly based on first- or second-order lag models of the process, or as sophisticated as a set of differential-algebraic equations (DAEs) that model the process and are solved in a nonlinear predictive-control algorithm. However, models approximate the true process dynamics and, when they involve nominal parameters with lower and upper bounds, are said to exhibit *parametric uncertainty*. Through the *internal model control* (IMC) theory (Morari and Zafiriou, 1989), the relationships among model reduction, model uncertainty, and closed-loop performance are well established. Both the selection of the nominal model order and its parameters (e.g., a first-order lag) and the model uncertainties (parameter ranges) limit the achievable closed-loop performance. Although space is not available to discuss these concepts further, the IMC tuning rules and their advantages are introduced and applied in the remainder of this section. The reader is referred to the text by Ogunnaike and Ray (1994) for a full exposition of model-based control.

The IMC structure, illustrated in Figure 21.22a, includes the process, $p\{s\}$, the process model, $\tilde{p}\{s\}$, and the IMC controller, $q\{s\}$. This structure is equivalent to the classic feedback structure, shown in Figure 21.22b, in which $c\{s\}$ is the feedback controller. It is convenient to carry out design using the IMC structure, and then implement the control system using the classic feedback structure, with $c\{s\}$ computed using the equation

$$c\{s\} = (1 - \tilde{p}\{s\}q\{s\})^{-1} q\{s\} \tag{21.65}$$

The order of the process model determines the order of the controller, and therefore has an impact on the achievable performance of the control system. Thus, a PID controller, which is of second order, is generated on the basis of a model, $\tilde{p}\{s\}$, of the same order. To design $c(s)$ as a PI controller, the process model is limited to a first-order transfer function: $\tilde{p}\{s\} =$

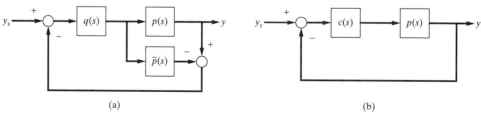

(a)   (b)

Figure 21.22 (a) IMC and (b) classic control structures.

$K_p (\tau s + 1)^{-1}$, in which case the IMC controller becomes $q\{s\} = (\tau s + 1)(\lambda s + 1)^{-1} K_p^{-1}$, where λ is the time constant of the IMC filter. Using Eq. (21.65), the classic feedback controller is

$$c\{s\} = (1 - \tilde{p}\{s\}q\{s\})^{-1} q\{s\} = (\tau s + 1)(K_p \lambda s)^{-1} \equiv K_C \left(1 + \frac{1}{\tau_i s} \right) \qquad \textbf{(21.66)}$$

By equivalence of the terms, the IMC-PI tuning rules are

$$\tau_i = \tau \text{ and } K_C = \tau_i / K_p \lambda \qquad \textbf{(21.67)}$$

For processes exhibiting time delay, for example, $\tilde{p}\{s\} = K_p e^{-\theta s} (\tau s + 1)^{-1}$, Rivera et al. (1986) recommend the IMC tuning rules:

$$\tau_i = \tau + \frac{\theta}{2} \text{ and } K_C = \tau_i / K_p \lambda \qquad \textbf{(21.68)}$$

IMC-based tuning parameters have been computed for a variety of open-loop stable transfer functions by Rivera et al. (1986), and for open-loop unstable transfer functions by Rotstein and Lewin (1991). The main advantage of the IMC tuning rules is that the time constant of the IMC filter is the *only tunable parameter*. This is of great practical importance because it provides guidance for detuning the controllers in the face of model uncertainty and multivariable system interactions. The value of λ should be set initially to either the desired closed-loop time constant (about one-fifth of the desired settling time), or to twice the process delay time. Increased robustness (less oscillations) is attained by increasing its value, whereas more aggressive control action results from decreasing its value.

EXAMPLE 21.9   *Tuning PI Control Loops for a Binary Distillation Column (Example 21.8 Revisited)*

As discussed in Example 21.8, this column, for the dehydration of methanol, is simulated using HYSYS.Plant. To assist in tuning the PI controllers, each control loop is placed in manual operation and a step change in the manipulated variable is applied. For a step change from 50 to 60% of the maximum reflux flow, the temperature of the second tray (used to regulate the distillate composition) is reduced from 66.7 to 66°C, with a settling time of about 50 min. Since the PV range is defined from 25 to 125°C, the 10% increase in OP causes a PV change of:

$$\Delta PV = \left(\frac{PV - PV_{min}}{PV_{max} - PV_{min}} \right) \times 100 = \frac{66 - 66.7}{125 - 25} \times 100 = -0.7\%$$

Hence, the dimensionless process gain is $K_p = -0.7/10 = -0.07$, with *direct action* control needed due to the negative process gain. Furthermore, the open-loop process time constant is approximately 10 min (assuming the settling time is on the order of five time constants). Thus, the IMC-based PI tuning parameters for this loop, computed using Eq. (21.67), are

$$\tau_i = \tau = 10 \text{ min, and } K_C = \frac{\tau_i}{K_p \lambda} = \frac{143}{\lambda}$$

The value of λ is tunable, allowing the designer to trade-off between robustness and performance. With $\lambda = 5$ min, $K_C = 29$. A similar approach for the bottom loop leads to the PI tuning: $K_C = 6$, $\tau_i = 10$ min. ∎

EXAMPLE 21.10   *Tuning PI Control Loops for the Shell Process (Example 21.6 Revisited)*

In Example 21.6, RGA analysis for the Shell Process indicates the need for diagonal pairing. For the loop, $y_1 – u_1$, the open-loop transfer function is $\tilde{p}\{s\} = 4.05e^{-27s} (50s + 1)^{-1}$. Using Eq. (21.68), with

$\lambda = 2\theta$, the PI tuning parameters are $K_{C1} = 0.29$, $\tau_{i1} = 64$ min. Similarly, for the second loop, y_2–u_2, the PI tuning parameters are $K_{C2} = 0.42$, $\tau_{i2} = 67$ min. As shown in Figure 21.10, these settings give adequate closed-loop response.   ∎

For more details on the implementation and tuning of PI controllers using HYSYS.Plant, the reader is referred to the multimedia CD-ROM that accompanies this text (*HYSYS → Dynamic Simulation → Tuning PI Controllers*). In the following case studies, C&R analysis is demonstrated, with results verified using dynamic simulations of the PI-controlled processes.

21.5   CASE STUDIES

The case studies presented in this section demonstrate the application of C&R analysis for screening the designs of three chemical processes that are representative of those encountered during process synthesis. The case studies also show how the results using steady-state and dynamic C&R analyses compare.

CASE STUDY 21.1   ***Exothermic Reactor Design for the Production of Propylene Glycol (Example 21.1 Revisited)***

Returning to Step 1 of Example 21.1, the CSTR to hydrolyze propylene oxide to propylene glycol is considered further. Its material and energy balances are in Eqs. (21.6) and (21.7), with variables defined and specifications provided. Beginning with these dynamic balances, the steady-state behavior patterns for this reactor are examined next.

Steady-State Solution

Equation (21.8) provides an expression for the PO concentration in the reactor in terms of the reactor temperature:

$$C_{PO}\{T\} = \frac{\sqrt{1 + 4k\{T\}C_{PO,in}\tau} - 1}{2k\{T\}\tau} \tag{21.69}$$

where $C_{PO,in} = \Im_{PO,in}/(q_0 + q_w)$ and the residence time, $\tau = V/(q_0 + q_w)$. Note that the fractional conversion of PO is $X = 1 - C_{PO}/C_{PO,in}$. Substituting for PO concentration, Eq. (21.9) provides expressions for the heat generation and removal rates in terms of the reactor temperature:

$$H_{GEN} = \frac{1}{c_P}k\{T\}\,C_{PO}^2\{T\}(-\Delta H), \quad H_{REM} = \frac{(T - T_0)}{\tau} \tag{21.70}$$

These are monotonically increasing functions of the reactor temperature. H_{GEN} is small at low temperatures, where the reaction rates are small, rising exponentially with temperature to a plateau limited by the complete conversion of PO. In contrast, H_{REM} varies linearly with the reactor temperature. Figure 21.23 shows these rates as a function of T for $q_w = 5.325$ m³/hr. Note that a steady state occurs at the intersection of the two curves, at $T = 82.4°C$, $X = 0.97$. For small positive perturbations in temperature in the vicinity of $T = 82.4°C$, the heat removal rate is higher than the heat generation rate, whereas the opposite is true for small negative perturbations. Such imbalances result in the temperature returning to its operating point, which is referred to as *stable in the open loop* or *open-loop stable*.

As q_w increases, τ decreases, shifting H_{REM} to the left, while increasing its gradient. At $q_w = 8$m³/hr, three intersections with the H_{GEN} curve occur, corresponding to three steady-state solutions, as shown in Figure 21.24. The upper and lower intersections at 62 and 25°C are stable in the open loop. However, for small positive perturbations in temperature in the vicinity of $T = 44°C$, the heat removal rate is *lower* than the heat generation rate. This imbalance causes the reactor to move to the upper (stable) operating point. Similarly, small negative perturbations in temperature cause the reactor to move to the lower (stable) operating point. Therefore, the intermediate operating point is referred to as *unstable in the open loop* or *open-loop unstable*. This operating point can only be maintained by installing a feedback control system, whose gain must be high enough to ensure closed-loop stability. The multiplicity

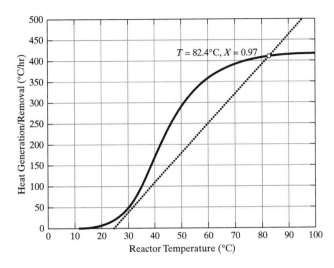

Figure 21.23 Solution diagram for PO hydration in a CSTR with $q_w = 5.325$ m³/hr: H_{GEN} = solid line, H_{REM} = dotted line.

of steady states exhibited by the CSTR may lead to hysteresis phenomena when operating the reactor. This is illustrated using ASPEN PLUS (*ASPEN → Chemical Reactors → Kinetic Reactors → Continuous Stirred Tank Reactors (CSTRs) → RCSTR*) and HYSYS.Plant (*HYSYS → Chemical Reactors → Setting Up Reactors → CSTR*) on the multimedia CD-ROM that accompanies this text.

Returning to the steady state for $q_w = 5.325$ m³/hr, a conversion in excess of 95% is obtained with a 47-ft³ CSTR operating at 85% capacity, which is denoted as the nominal operating point. Assuming an aspect ratio of $L/D = 2$, the diameter for a single reactor is $D_1 = 3.1$ ft. Consider an alternative design composed of two CSTRs in series, as shown in Figure 21.27b. Assuming perfect level control by manipulating the effluent streams, two output variables, T_1 and T_2, are controlled by two manipulated variables, q_{w1} and q_{w2} (the water feed rates). Assuming operation at approximately the same temperature in each CSTR (about 80°C), two 14-ft³ CSTRs in series provide the same conversion as for the base-case design involving a single reactor. Assuming the reactors are operated at 85% capacity and designed with the same aspect ratio ($L/D = 2$), the diameters are $D_2 = 2.08$ ft. Because of the small vessel volumes in both of the alternative designs, it is not possible to estimate their costs using Eq. (16.53).

C&R Diagnosis

Based on safety considerations, the design involving two CSTRs is preferred, since it involves a smaller inventory of the highly reactive mixture. It remains to examine the C&R measures for these

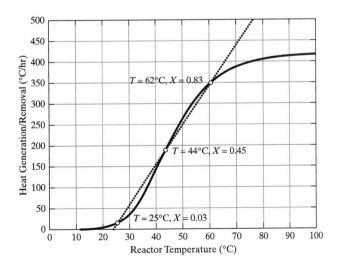

Figure 21.24 Solution diagram for PO hydration in a CSTR with $q_w = 8$ m³/hr: H_{GEN} = solid line, H_{REM} = dotted line.

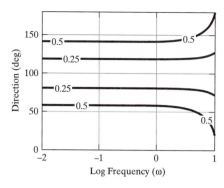

Figure 21.25 DC contour map for PO hydration in a single CSTR.

two designs. Beginning with a single CSTR, Eqs. (21.6) and (21.7) are linearized, as shown in Section 21.1, in the vicinity of the steady state to give $P\{s\}$ and $P_d\{s\}$ in Eq. (21.1):

$$P\{s\} = \underline{C}(s\underline{I} - \underline{\underline{A}})^{-1} \underline{\underline{B}}_U \text{ and } P_d\{s\} = \underline{C}(s\underline{I} - \underline{\underline{A}})^{-1} \underline{\underline{B}}_D \qquad \textbf{(21.71)}$$

where the scaled state–space matrices are

$$\underline{\underline{A}} = \begin{bmatrix} -9.203 & -55.43 \\ 0.1644 & 0.8870 \end{bmatrix} \qquad \underline{\underline{B}}_U = \begin{bmatrix} -0.0782 \\ -0.0555 \end{bmatrix}$$

$$\underline{\underline{B}}_D = \begin{bmatrix} 2.330 & 0 \\ 0 & 0.0070 \end{bmatrix} \qquad \underline{C} = \begin{bmatrix} 0 & 1 \end{bmatrix}$$

These Jacobian matrices can also be computed by making small positive and negative perturbations in each input variable, one at a time. The perturbation magnitude is reduced until the magnitude of the resulting change in the outputs is insensitive to the direction of the input perturbation. The matrices are input scaled, assuming that the manipulated variables are nominally at 50% of their full range, PO feed rate disturbances (i.e., production rate changes) are limited to $\pm 50\%$ of the nominal value, and feed temperature disturbances are limited to $\pm 5°C$. The time in Eq. (21.71) is in minutes.

Using the linear approximation, the DC contour map in Figure 21.25 identifies the worst disturbance as $\underline{\Delta d} = [+50\%, 0°C]^T$, that is, throughput changes. Even with this disturbance, the linear approximation indicates that there are no limitations to perfect disturbance rejection, since all of the DC values up to a frequency of 10 rad/min lie well below unity.

A similar analysis for the two CSTRs in series yields the DC contour maps in Figure 21.26. While the first reactor has disturbance rejection comparable to that for the single reactor, the control variable in the second reactor is saturated at steady state for the worst disturbance direction (feed rate change alone). Hence, even for perfect control in the second reactor, the temperature setpoint cannot be maintained for this disturbance. However, this is mitigated by the fact that the conversion in the second reactor is small, and hence, the offset in T_2 is expected to be small.

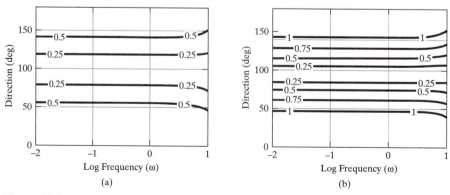

Figure 21.26 DC contour maps for PO hydration in two CSTRs in series: (a) DC contour map for q_{w1}; (b) DC contour map for q_{w2}.

Table 21.3 Controller Tuning Parameters for the Two
Reactor Configurations

Single CSTR configuration (see Figure 21.27a)

Loop	PV Range	K_c	τ_I (min)	Action
FC-1	0–100 kmol/hr	0.1	1	Reverse
TC-1	20–120°C	1.5	4	Direct
LC-1	0–100%	1	10	Direct

Two-CSTR configuration (see Figure 21.27b)

Loop	PV Range	K_c	τ_I (min)	Action
FC-1	0–100 kmol/hr	0.1	1	Reverse
TC-1	20–120°C	0.5	2	Direct
TC-2	20–120°C	3	4	Direct
LC-1	0–100%	1	5	Direct
LC-2	0–100%	1	5	Direct

It should be noted that these results assume that perfect control is achievable. Thus, the true response is expected to be worse than that predicted based on the linear analysis, especially when single-loop PI controllers are implemented. Based on the C&R analysis, two CSTRs are anticipated to provide the same disturbance rejection as obtained with a single CSTR. This suggests that the former should be selected, since it is the safest design. To confirm this conclusion, dynamic simulation of both systems is carried out using HYSYS.Plant. Table 21.3 summarizes the PI tuning parameters, with the control configurations shown in Figure 21.27. The IMC-PI tuning parameters, computed using the approach de-

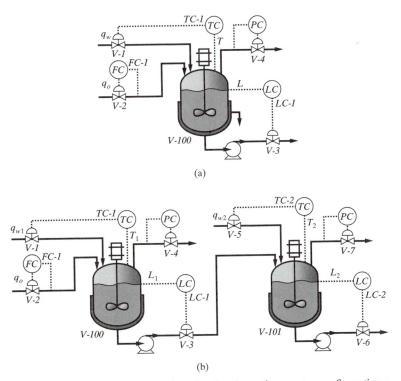

(a)

(b)

Figure 21.27 Control configurations for the alternative reactor configurations:
(a) single CSTR; (b) two CSTRs in series.

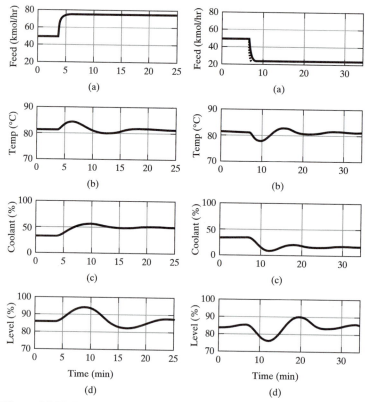

Figure 21.28 Response of the single CSTR to 50% positive and negative changes in throughput: (a) feed rate and setpoint; (b) reactor temperature; (c) water feed rate (%); (d) reactor level.

scribed in Section 21.4, are detuned for the level control loops, and ensure relatively tight control on the reactor temperatures. Perfect pressure control is assumed, by adjusting the small flow rates in the vapor vents (3 bar for the first vessel, and 2 bar for the second). The pressure control loops are not simulated explicitly.

Because the DC analysis indicates that the most challenging regulatory control is associated with throughput changes, the organic feed rate is adjusted, and closed-loop simulations are performed to check the effects of positive and negative changes in q_0. The response of the single CSTR is shown in Figure 21.28, and indicates that both positive and negative throughput changes of 50% are easily handled, with the reactor temperature returned to its setpoint in about 15 min, while the coolant flow rate, q_w, remains within its constraints, as predicted by the DC analysis. The response of the liquid level is more sluggish by design. The responses of the two-CSTR system compare well with that for the single CSTR, as shown in Figure 21.29. Its response to a positive feed rate change is rapid, with no evidence of saturation on either manipulated variable. In contrast, the coolant flow rate to the second reactor, q_{w2}, saturates in response to a negative change in the throughput, as predicted by the DC analysis. However, as seen in Figure 21.29, this does not significantly affect T_2.

For more details, the reader is referred to the section covering dynamic simulation with HYSYS.Plant on the multimedia CD-ROM that accompanies this text (*HYSYS → Dynamic Simulation → Overview*). Using the HYSYS.Plant files, CSTR_1.hsc and CSTR_2.hsc, the results in Figures 21.28 and 21.29 can be reproduced.   ■

In summary, it would appear that C&R analysis, though indicating potential saturation problems associated with the more complex two-CSTR system, shows that the dynamic responses of the two systems are approximately the same. This suggests that the two-CSTR system should be adopted, since it is the safest. However, designs involving smaller holdups

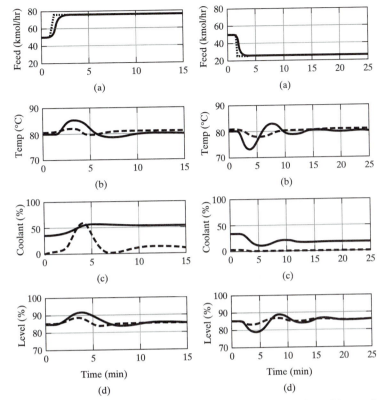

Figure 21.29 Response of the two-CSTR system to 50% positive and negative changes in throughput: (a) feed flow rate and setpoint; (b) reactor temperatures (T_1, solid; T_2, dashed); (c) water feed flow rate (q_{w1}, solid; q_{w2}, dashed, in %); (d) holdups (L_1, solid; L_2, dashed).

are often less resilient, especially in the face of disturbances that manifest themselves rapidly. This example has focused on the importance of performing C&R analysis in reactor design. For more examples, the reader is encouraged to study the following publications:

1. Luyben et al. (1999): Chapter 4 discusses the design of control systems for reactors in general. The design of heat-integrated reactor systems is discussed in Chapter 5.
2. Shinskey (1988): Chapter 10 discusses reactor control in industrial practice.
3. Lewin and Bogle (1996): This paper concerns the optimal operation and controllability of a continuous polymerization reactor.
4. Russo and Bequette (1998): This paper discusses the multiplicity of steady states associated with jacketed polymerization reactors.

CASE STUDY 21.2   *Two Alternative Heat-Exchanger Networks (Examples 20.1 and 20.5 Revisited)*

Here, the two alternative heat-exchanger networks (HENs) in Examples 20.1 and 20.5 are screened using C&R analysis. More specifically, the two designs are required to be resilient to $\pm 5\%$ changes in F_1 and $\pm 5°F$ in T_0. As discussed in Chapter 20, it is often necessary to augment the process degrees of freedom to meet control objectives, either by addition of trim utility exchangers or by adding bypasses, as is the case here. The focus of this study is in the use of resiliency analysis to select the design configuration and to adjust its nominal operating conditions.

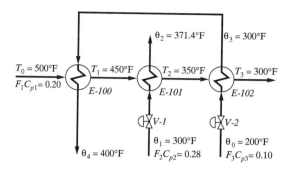

Figure 21.30 Heat-exchanger network without bypass.

Original HEN (No Bypass)

In the network shown in Figure 21.30, only two of the target temperatures, θ_2 and θ_4, are controlled by manipulation of the flow rates of the two cold streams, leaving the third target temperature, T_3, uncontrolled. The energy balances for this system involve 15 variables: F_1, F_2, F_3, T_0, T_1, T_2, T_3, θ_0, θ_1, θ_2, θ_3, θ_4, Q_1, Q_2, and Q_3, two of which, θ_0 and θ_1, are considered to be fixed, and two, F_1 and T_0, are considered to be external disturbances.

Three energy balances apply for each heat exchanger. For the first heat exchanger, E-100, they are

$$f_1\{\underline{x}\} = Q_1 - F_1 C_{p1}(T_0 - T_1) = 0 \tag{21.72}$$

$$f_2\{\underline{x}\} = Q_1 - F_3 C_{p3}(\theta_4 - \theta_3) = 0 \tag{21.73}$$

$$f_3\{\underline{x}\} = Q_1 - U_1 A_1 \frac{(T_0 - \theta_4) - (T_1 - \theta_3)}{\ln[(T_0 - \theta_4)/(T_1 - \theta_3)]} = 0 \tag{21.74}$$

For E-101, the equations are

$$f_4\{\underline{x}\} = Q_2 - F_1 C_{p1}(T_1 - T_2) = 0 \tag{21.75}$$

$$f_5\{\underline{x}\} = Q_2 - F_2 C_{p2}(\theta_2 - \theta_1) = 0 \tag{21.76}$$

$$f_6\{\underline{x}\} = Q_2 - U_2 A_2 \frac{(T_1 - \theta_2) - (T_2 - \theta_1)}{\ln[(T_1 - \theta_2)/(T_2 - \theta_1)]} = 0 \tag{21.77}$$

Finally, for E-102

$$f_7\{\underline{x}\} = Q_3 - F_1 C_{p1}(T_2 - T_3) = 0 \tag{21.78}$$

$$f_8\{\underline{x}\} = Q_3 - F_3 C_{p3}(\theta_3 - \theta_0) = 0 \tag{21.79}$$

$$f_9\{\underline{x}\} = Q_3 - U_3 A_3 \frac{(T_2 - \theta_3) - (T_3 - \theta_0)}{\ln[(T_2 - \theta_3)/(T_3 - \theta_0)]} = 0 \tag{21.80}$$

where U_i and A_i are the heat transfer coefficients and heat transfer areas for heat exchanger i, respectively, such that: $U_1 A_1 = 0.0811$ MMBtu/hr-°F, $U_2 A_2 = 0.3162$ MMBtu/hr-°F, and $U_3 A_3 = 0.1386$ MMBtu/hr-°F. The number of independent manipulated variables is $N_{\text{Manipulated}} = N_{\text{Variables}} - N_{\text{Externally Defined}} - N_{\text{Equations}} = 15 - 4 - 9 = 2$, and the pairings can be selected using the RGA. To accomplish this, a linearized model is generated using the following procedures.

1. The nonlinear state equations, $\underline{f}\{\underline{x}\} = 0$, in Eqs. (21.72)–(21.80) are solved for the nominal values of the manipulated variables, $\underline{u} = [F_2, F_3]^T$, disturbances, $\underline{d} = [F_1, T_0]^T$, and constants θ_0 and θ_1, to determine 9 state variables: $\underline{x} = [T_1, T_2, T_3, \theta_2, \theta_3, \theta_4, Q_1, Q_2, Q_3]^T$. This is accomplished using an appropriate numerical method (e.g., the Newton–Raphson method).

2. The output vector, $\underline{y} = [\theta_2, \theta_4]^T$, is recomputed for small positive and negative perturbations of magnitude Δu_i to each manipulated variable, u_i, one at a time, with the results stored in the vectors $\mathbf{y}_{p,i}$ and $\mathbf{y}_{n,i}$, respectively. Then, column i of the steady-state gain matrix, $P\{0,\}$ is computed: $p_{ji}\{0\} = \Delta u_i^{\max} \cdot (y_{p,i,j} - y_{n,i,j})/\Delta u_i$, $j = 1, \ldots, 3$. Note that a factor of $\Delta u_i^{\max}$ scales the input variables such that $|u_i| \le 1$.

3. The output vector is recomputed for small positive and negative perturbations of magnitude Δd_i to each disturbance variable, d_i, one at a time, with the results stored in the vectors $\mathbf{y}_{p,i}$ and $\mathbf{y}_{n,i}$,

respectively. Then, column i of the steady-state gain matrix, $\underline{\underline{P}}_d\{0\}$, is computed: $pd_{ji}\{0\} = \Delta d_i^{\max} \cdot (y_{p,i,j} - y_{n,i,j})/\Delta d_i, j = 1, \ldots, 3$. The disturbance gain matrix is scaled arbitrarily relative to the inputs using the scaling, $\underline{\Delta d}^{\max} = [5\%, 5°F]^T$.

Since the nominal values of the manipulated variables are $\underline{u} = [F_2, F_3]^T = [1.00, 1.00]^T$, the maximum perturbations are $\underline{\Delta u}^{\max} = [1.00, 1.00]^T$. The resulting linearized model is

$$
\begin{bmatrix} \Delta\theta_2 \\ \Delta\theta_4 \\ \Delta T_3 \end{bmatrix} = \underbrace{\begin{bmatrix} -58.7 & -73.3 \\ -7.14 & -112 \\ \hdashline -14.3 & -41.6 \end{bmatrix}}_{\begin{bmatrix} \underline{\underline{P}}_1\{0\} \\ \hdashline \underline{\underline{P}}_2\{0\} \end{bmatrix}} \cdot \begin{bmatrix} \Delta F_2 \\ \Delta F_3 \end{bmatrix} + \underbrace{\begin{bmatrix} 2.83 & 1.89 \\ 2.23 & 2.94 \\ 4.92 & 0.833 \end{bmatrix}}_{\begin{bmatrix} \underline{\underline{P}}_{d_1}\{0\} \\ \hdashline \underline{\underline{P}}_{d_2}\{0\} \end{bmatrix}} \cdot \begin{bmatrix} \Delta F_1 \\ \Delta T_0 \end{bmatrix}
$$

(21.81)

Note that the gains in Eq. (21.81) are presented as the change in °F in response to a full-scale change of each input. Thus, for example, the linear model predicts a 4.92°F increase in T_3 in response to a 5% increase in F_1. Using Eq. (21.22), the steady-state RGA is

$$
\underline{\underline{\Lambda}} = \underline{\underline{P}}_1\{0\} \otimes (\underline{\underline{P}}_1^{-1}\{0\})^T = \begin{bmatrix} 1.09 & -0.09 \\ -0.09 & 1.09 \end{bmatrix}
$$

(21.82)

The RGA indicates that the diagonal pairing shown in Figure 20.5, θ_2–F_2 and θ_4–F_3, provides responses that are almost perfectly decoupled. These are recommended, while the off-diagonal pairing has stability problems. This result is consistent with $\underline{\underline{P}}_1(0)$, which is *diagonally* dominant. Next, the resiliency of the HEN is examined by computing the DC at steady-state for disturbances of $\pm5\%$ in F_1 and $\pm5°F$ in T_0:

$$
\begin{bmatrix} \Delta F_2\{0\} \\ \Delta F_3\{0\} \end{bmatrix} = -\underline{\underline{P}}_1^{-1}\{0\} \cdot \underline{\underline{P}}_{d_1}\{0\} \cdot \begin{bmatrix} \Delta F_1 \\ \Delta T_0 \end{bmatrix}, \qquad DC = \left\| \begin{bmatrix} \Delta F_2\{0\} \\ \Delta F_3\{0\} \end{bmatrix} \right\|_2
$$

(21.83)

The values of the two manipulated variables, computed to completely reject the effect of the disturbances on θ_2 and θ_4, lead to changes in the value of T_3, computed by substituting Eq. (21.83) into Eq. (21.81):

$$
\Delta T_3\{0\} = (\underline{\underline{P}}_{d_2}\{0\} - \underline{\underline{P}}_2\{0\}\, \underline{\underline{P}}_1^{-1}\{0\} \cdot \underline{\underline{P}}_{d_1}\{0\}) \begin{bmatrix} \Delta F_1 \\ \Delta T_0 \end{bmatrix}
$$

(21.84)

Table 21.4 shows the changes in the control variables, ΔF_2 and ΔF_3 (assuming perfect control), the disturbance cost, and the resulting change in T_3, computed using Eq. (21.84) for four disturbance vectors. The results indicate that perfect disturbance rejection is achieved for θ_2 and θ_4 with negligible control effort. However, the uncontrolled temperature, T_3, is significantly perturbed, with the worst-case disturbance being ΔF_1 and ΔT_0 in opposite directions. Variations of $\pm5\%$ in F_1 and $\pm5°F$ in T_0 lead to variations of approximately $\pm4°F$ in T_3.

Modified HEN (with Bypass)

The PFD for the modified HEN, including a bypass around E-102 to eliminate the offsets in the third target temperature, is reproduced in Figure 21.31. Resiliency analysis is used to determine the required bypass fraction. The energy balances involve 17 variables: $F_1, F_2, F_3, T_0, T_1, T_2, T_3, \theta_0, \theta_1, \theta_2, \theta_3, \theta_3', \theta_4,$ $Q_1, Q_2, Q_3,$ and ϕ, two of which, θ_0 and θ_1, are assumed to be fixed, and two, F_1 and T_0, are considered

Table 21.4 Input Changes and Disturbance Cost for the HEN without Bypasses

ΔF_1	ΔT_0	ΔF_2	ΔF_3	DC $= \| \underline{u} \|_2$	ΔT_3
+5%	0	0.0253	0.0184	0.0313	3.79
+5%	+5°F	0.0246	0.0447	0.0511	3.59
0	+5°F	−0.0007	0.0264	0.0264	−0.20
−5%	+5°F	−0.0261	0.0080	0.0273	−4.00

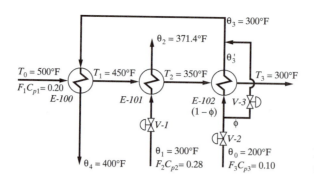

Figure 21.31 Modified heat-exchanger network.

to be external disturbances. The first six equations (21.72)–(21.77), for the HEN without bypasses apply. For heat exchanger E-102 and its bypass, the material and energy balances are

$$f_7\{\underline{x}\} = Q_3 - F_1 C_{p1}(T_2 - T_3) = 0 \tag{21.85}$$

$$f_8\{\underline{x}\} = Q_3 - F_3 C_{p3}(\theta_3' - \theta_0) = 0 \tag{21.86}$$

$$f_9\{\underline{x}\} = Q_3 - K_3 U_3 A_3 \frac{(T_2 - \theta_3') - (T_3 - \theta_0)}{\ln[(T_2 - \theta_3')/(T_3 - \theta_0)]} = 0 \tag{21.87}$$

$$f_{10}\{\underline{x}\} = (1 - \phi)\theta_0 + \phi\theta_3' - \theta_3 = 0 \tag{21.88}$$

In Eq. (21.87), the product $U_3 A_3$ is identical to that for the network without bypasses ($= 0.1386$ MMBtu/hr°F). As the bypass fraction, ϕ, increases, K_3 increases beyond unity, corresponding to an increase in the heat-transfer area. The number of independent manipulated variables is $N_{\text{Manipulated}} = N_{\text{Variables}} - N_{\text{Externally Defined}} - N_{\text{Equations}} = 17 - 4 - 10 = 3$. This leaves F_2, F_3, and ϕ as the manipulated variables, which are paired with the controlled variables, θ_2, θ_4, and T_3.

A linearized model is generated and used to assist in the selection of an appropriate bypass fraction, ϕ. The procedure followed for the HEN without bypasses is used, parameterized by values of ϕ. Since the nominal values of the manipulated variables are $\underline{u} = [F_2, F_3, \phi]^T = [1, 1, \phi]^T$, the maximum perturbations are $\Delta \underline{u}^{\max} = [1, 1, \phi]^T$. For example, for $\phi = 0.1$, the linearized model is

$$\begin{bmatrix} \Delta\theta_2 \\ \Delta\theta_4 \\ \Delta T_3 \end{bmatrix} = \underbrace{\begin{bmatrix} -58.7 & -72.3 & -0.068 \\ -7.15 & -108 & -0.285 \\ -14.3 & -44.9 & 0.237 \end{bmatrix}}_{\underline{\underline{P}}\{0\}} \cdot \begin{bmatrix} \Delta F_2 \\ \Delta F_3 \\ \Delta\phi \end{bmatrix} + \underbrace{\begin{bmatrix} 2.80 & 1.89 \\ 2.20 & 2.94 \\ 4.95 & 0.88 \end{bmatrix}}_{\underline{\underline{P}}_d\{0\}} \cdot \begin{bmatrix} \Delta F_1 \\ \Delta T_0 \end{bmatrix} \tag{21.89}$$

Using $\underline{P}(0)$ in Eq. (21.89), the steady-state RGA is computed using Eq. (21.22):

$$\underline{\underline{\Lambda}} = \underline{\underline{P}}\{0\} \otimes (\underline{\underline{P}}^{-1}\{0\})^T = \begin{bmatrix} 1.17 & -0.22 & 0.04 \\ -0.07 & 0.84 & 0.23 \\ -0.10 & 0.38 & 0.72 \end{bmatrix} \tag{21.90}$$

Hence, the diagonal pairing shown in Figure 20.6 is preferred; that is, θ_2–F_2, θ_4–F_3, and T_3–ϕ, with significant interactions between the second and third loops anticipated.

The impact of the bypass fraction on the resiliency of the HEN is examined next. The manipulated variable values and the disturbance cost are computed for disturbances of $\pm5\%$ in F_1 and $\pm5°F$ in T_0. Table 21.5 shows the changes in the control variables, ΔF_2, ΔF_3, and $\Delta\phi$ (assuming perfect control), and the disturbance cost, for four disturbance vectors, $\underline{d} = [F_1 + \Delta F_1, T_0 + \Delta T_0]^T$. Note that for the worst-case disturbance ($\Delta F_1 = -5\%$ and $\Delta T_0 = +5°F$), the scaled change in the bypass fraction is $\Delta\phi = 12.3$, which far exceeds unity. To avoid this, the nominal bypass fraction is increased further to account for the expected disturbance levels, noting that heat exchanger E-102 must be resized.

With the nominal bypass fractional flow increased to $\phi = 0.25$, the linearized model is recomputed

$$\begin{bmatrix} \Delta\theta_2 \\ \Delta\theta_4 \\ \Delta T_3 \end{bmatrix} = \underbrace{\begin{bmatrix} -58.7 & -69.8 & -0.720 \\ -7.15 & -97.1 & -3.02 \\ -14.3 & -53.7 & 2.52 \end{bmatrix}}_{\underline{\underline{P}}\{0\}} \cdot \begin{bmatrix} \Delta F_2 \\ \Delta F_3 \\ \Delta\phi \end{bmatrix} + \underbrace{\begin{bmatrix} 2.80 & 1.89 \\ 2.10 & 2.94 \\ 5.03 & 0.88 \end{bmatrix}}_{\underline{\underline{P}}_d\{0\}} \cdot \begin{bmatrix} \Delta F_1 \\ \Delta T_0 \end{bmatrix} \tag{21.91}$$

Table 21.5 Input Changes and Disturbance Cost for the HEN with $\phi = 0.1$

ΔF_1	ΔT_0	ΔF_2	ΔF_3	$\Delta \phi$	$DC = \| \underline{u} \|_2$
+5%	0	−0.0010	0.051	−11.4	11.4
+5%	+5°F	−0.0003	0.075	−10.3	10.3
0	+5°F	0.0007	0.025	0.98	0.98
−5%	+5°F	0.0017	−0.026	12.3	12.3

In this case, the steady-state RGA is

$$\underline{\underline{\Lambda}} = \underline{\underline{P}}\{0\} \otimes (\underline{\underline{P}}^{-1}\{0\})^{\mathrm{T}} = \begin{bmatrix} 1.17 & -0.21 & 0.04 \\ -0.07 & 0.75 & 0.32 \\ -0.10 & 0.46 & 0.64 \end{bmatrix} \tag{21.92}$$

This RGA is similar to that obtained with $\phi = 0.1$, again indicating a diagonal pairing, as shown in Figure 20.6. Next, the resiliency is tested, with the results reported in Table 21.6. Note that when $\phi = 0.25$, the disturbance rejection is nearly acceptable, with $DC_{\max} = 1.1$, only slightly above unity.

Clearly, the resiliency of the HEN increases with the nominal bypass fraction, but at the cost of increased heat-transfer area. Table 21.7 shows the trade-off between resiliency and heat-transfer area. Note that while only 12% additional heat-exchange area is required for $\phi = 0.1$, the resiliency is inadequate. In contrast, when $\phi = 0.30$, the resiliency is satisfactory (with DC significantly lower than unity), but the heat-transfer area is doubled. A good compromise is to select $\phi = 0.25$, which approximates the desired resiliency, while requiring only 55% more heat-exchange area.

The C&R analysis in the steady state predicts the superior performance of the modified HEN, which allows all three target temperatures to be controlled at their setpoints in the face of disturbances in the feed flow rate and temperature of the hot stream. More specifically, the steady-state RGA indicates that a decentralized control system can be configured for the modified HEN in which $\theta_2 − F_2$, $\theta_4 − F_3$ and $T_3 − \phi$ are paired, and in which the first loop is almost perfectly decoupled, with moderate coupling between the other two loops. Finally, aided by DC analysis, the nominal bypass fraction is selected to be 0.25, providing the best trade-off between increased plant costs and adequate resiliency.

Given the design decision to use $\phi = 0.25$, based on the steady-state C&R analysis, verification is performed by dynamic simulations with HYSYS.Plant. The hot stream of n-octane at 2,350 lbmol/hr is cooled from 500 to 300°F using n-decane as the coolant, with $F_2 = 3,070$ lbmol/hr and $F_3 = 1,200$ lbmol/hr. Note that these species and flow rates are chosen to match the heat-capacity flow rates defined by McAvoy (1983), with F_1 slightly increased to avoid temperature crossovers in the heat exchangers due to temperature variations in the heat capacities. Additional details of the HYSYS.Plant simulation are

a. The tubes and shells for the heat exchangers provide 2-min residence times.
b. The feed pressures of all three streams are set at 250 psi, with nominal pressure drops of 5 psi defined for the tubes, shells, and for the bypass valve, V-3. Subsequently, these pressure drops are computed based on the equipment and valve sizing and the pressure-flow relationships.
c. The bypass valve, V-3, is sized carefully, ensuring that the nominal bypass fraction is 0.25, with the nominal valve position being 50% open (selecting a linear characteristic curve).
d. IMC-PI tuning parameters are presented in Table 21.8.

Table 21.6 Input Changes and Disturbance Cost for the HEN with $\phi = 0.25$

ΔF_1	ΔT_0	ΔF_2	ΔF_3	$\Delta \phi$	$DC = \| \underline{u} \|_2$
+5%	0	−0.0010	0.051	−0.93	0.93
+5%	+5°F	−0.0003	0.075	0.75	0.75
0	+5°F	0.0007	0.025	0.18	0.18
−5%	+5°F	0.0017	−0.026	1.11	1.11

Table 21.7 Trade-off Between the Heat-Exchanger Area and Bypass Fraction

ϕ	DC = $\|\underline{u}\|_2$	K_3
0.10	12.3	1.12
0.15	4.63	1.21
0.20	2.16	1.33
0.25	1.11	1.55
0.30	0.58	2.05

Table 21.8 IMC-PI Tuning Parameters for the Alternative HENs

HEN without bypass (Figures 21.30 and 20.5)				
Loop	PV Range (°F)	K_c	τ_I (min)	Action
θ_2–F_2	300–500	2	1.5	Direct
θ_4–F_3	300–500	1.5	2.5	Direct
HEN with bypass (Figures 21.31 and 20.6)				
Loop	PV Range (°F)	K_c	τ_I (min)	Action
θ_2–F_2	300–500	2	1	Direct
θ_4–F_3	300–500	1	2	Direct
T_3–ϕ	300–500	1	1	Reverse

The regulatory responses of the two configurations are discussed next. Figure 21.32 shows that, as predicted by the DC analysis, even the worst-case disturbance has little effect on the two controlled variables, whose control loops are decoupled, as indicated by the RGA analysis. Moreover, the uncontrolled output, T_3, exhibits offsets of about ± 4.5°F, which compare well with the value of ± 4°F predicated by the linear DC analysis. In comparison, Figure 21.31 shows that, for the HEN with bypass, the response also corroborates the results of the linear DC analysis. Most importantly, the design with $\phi = 0.25$ rejects the worst-case disturbance with no saturation, indicating that the DC analysis is slightly conservative. In addition, the first control loop (θ_2–F_2) is perfectly decoupled, with slight interactions seen in the other two loops, again as predicted by the static RGA analysis. For more details, the reader

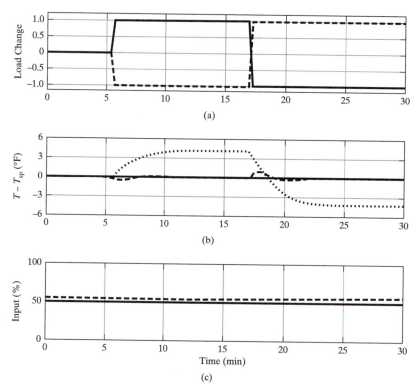

Figure 21.32 Response of HEN without bypass to the worst-case disturbances: (a) normalized changes in F_1 (solid) and T_0 (dashed); (b) tracking errors (θ_2, solid; θ_4, dashed; T_3, dotted); (c) manipulated variables (F_2, solid; F_3, dashed).

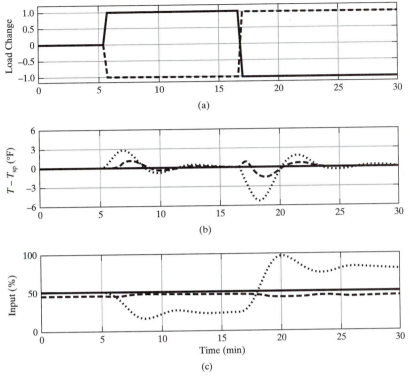

Figure 21.33 Response of HEN with bypass to the worst-case disturbances: (a) normalized changes in F_1 (solid) and T_0 (dashed); (b) tracking errors (θ_2, solid; θ_4, dashed; T_3, dotted); (c) manipulated variables (F_2, solid; F_3, dashed; ϕ, dotted).

is referred to the section covering dynamic simulation using HYSYS.Plant on the multimedia CD-ROM that accompanies this text (*HYSYS → Dynamic Simulation → Overview*), where the HYSYS.Plant files, HEN_1.hsc and HEN_2.hsc, are provided to enable the reproduction of the results in Figures 21.32 and 21.33. ∎

While steady-state C&R analysis often provides a good assessment of the controllability and resiliency, dynamic analysis should be considered when the steady-state analysis is inconclusive. The latter methods are discussed by Wolff et al. (1991) and Mathisen et al. (1993).

CASE STUDY 21.3   *Interaction of Design and Control in the MCB Separation Process*

Denn and Lavie (1982) show that recycles increase the process response time and static gain. Furthermore, when the recycle loop contains a time delay, resonant peaks comparable in magnitude to the steady-state gain may result. Since these phenomena are potentially destabilizing, control systems for recycle processes should be designed carefully. In this regard, control systems for recycle processes are designed using the nine-step design procedure of Luyben and co-workers (1999), presented in Section 20.3, with particular emphasis on the need to impose flow control on each recycle stream.

Process Description

Figure 21.34 shows the monochlorobenzene separation process introduced in Section 4.4. The process involves a flash vessel, V-100, an absorption column, T-100, a distillation column, T-101, a reflux drum, V-101, and three utility heat exchangers. As shown in Figure 4.23, most of HCl is removed at high purity (96 mol% by design) in the vapor effluent of T-100. However, rather than use the "treater" to remove the residual HCl, it is removed in the small vapor purge from T-101. The benzene by-product and monochlorobenzene product are obtained at high purity as distillate (99 mol% benzene) and

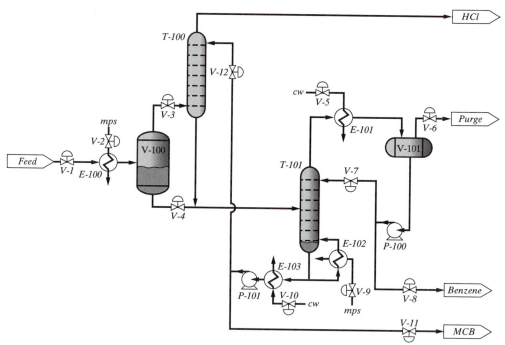

Figure 21.34 Flowsheet for the MCB separation process.

bottoms products (98 mol% MCB) from T-101. It is required to design a control system to ensure that the process meets its quality specifications in the face of an increase in the throughput demand (treated as a disturbance) and a feed composition change, as shown in Table 21.9. A preliminary control system configuration is proposed, and then refined and checked using C&R analysis. Finally, the performance of the control system is verified using dynamic simulation.

Preliminary Control System Configuration

The nine-step control-design procedure of Luyben and co-workers is applied to design the preliminary control structure in Figure 21.35.

Step 1: *Set objectives.* To achieve the primary control objective, the production level is maintained by flow control of the feed stream using valve V-1.

Step 2: *Define control degrees of freedom.* As shown in Figure 21.34, the process has 12 degrees of freedom with four valves controlling the flow rates of the utility streams (V-2, V-5, V-9, and V-10), one controlling the feed flow rate (V-1), three controlling product stream flow rates (V-6, V-8, and V-11), and the four remaining valves controlling internal process flow rates. Having chosen *constant feed flow* in Step 1, the feed valve (V-1) is reserved for independent flow control.

Step 3: *Establish energy management system.* The steam valve, V-2, is used to control the flash feed temperature. Furthermore, the temperature of the recycle and bottoms product streams is controlled by adjusting the coolant valve, V-10.

Table 21.9 Process Disturbances

Species	Nominal	d_1	d_2
	Molar flow rates (kmol/hr)		
HCl	10	15	15
Benzene	40	60	50
MCB	50	75	35
Total	100	150	100

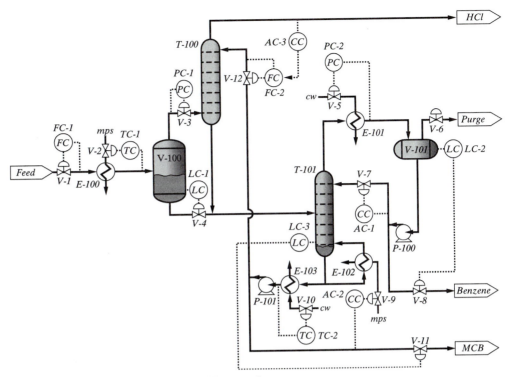

Figure 21.35 Control system for the MCB separation process.

Step 4. *Set the production rate.* As stated previously, the feed valve, V-1, is assigned to a flow controller, whose setpoint regulates the production rate.

Step 5. *Control product quality and meet safety, environmental, and operational constraints.* The pressure in V-100 is controlled by adjusting the flow rate of the vapor stream using valve V-3. Pressure regulation in T-101 is carried out by adjusting V-5, the coolant valve to the condenser E-101. Since both of the products from T-101 are required to meet specifications, the LV configuration is implemented, noting that the reflux ratio in the column is less than five. Thus, the reflux valve, V-7, is adjusted to control the distillate composition, and the reboiler steam valve, V-9, is used to regulate the bottoms composition.

Step 6. *Fix recycle flow rates and vapor and liquid inventories.* Valve V-12 is the obvious choice for recycle flow control. The liquid inventories in the flash drum, the reflux drum, and the column sump are regulated using the valves V-4, V-8, and V-11, respectively. Note that the purge stream that removes the residual HCl from the column overhead is less than 1 mol% of the feed by design. Thus, the valve V-6 is designed to be fixed at 50% open and left uncontrolled. The purge is sent to HCl product storage (not shown). The vapor inventories in both V-100 and T-101 are regulated using the pressure controllers discussed above.

Steps 7 and 8. *Check component balances and control individual process units.* The HCl is removed from the process, mostly in the T-100 overhead stream, with small amounts removed in the purge stream. Benzene and MCB are mostly removed in the distillate and bottoms streams from T-101, respectively, and in small amounts in the purge. No chemicals build up in the process.

Step 9. *Optimize economics and improve dynamic controllability.* It is noted that all of the control valves have been assigned, but the HCl product composition is still uncontrolled. To correct this, a cascade controller is installed to regulate the HCl composition. This controller can adjust the setpoint of either (a) the recycle flow controller, FC-2, or (b) the recycle temperature controller, TC-2. Figure 21.35 shows the first alternative, in which the liquid feed rate to the absorber is adjusted to control the mass transfer of the benzene from the vapor stream. Clearly, quantitative analysis is required to select the most appropriate configuration, as will be discussed next.

Control System Refinement Using C&R Analysis

To improve the control system in Figure 21.35, controllability and resiliency analysis has two roles. These involve the use of: (1) the RGA to aid in selecting the appropriate pairing between the controlled outputs and manipulated variables when interactions are anticipated, and (2) the DC to assist in checking that the operating ranges of the manipulated variables are sufficient to ensure adequate disturbance rejection. To provide data for these two analytical methods, a dynamic simulation of the MCB separation process is developed using HYSYS.Plant.

The equipment items are sized as follows:

a. The flash vessel, V-100, condenser, V-101, and reboiler, E-102, are installed assuming a liquid residence time of at least 10 min, computed on the basis of the liquid-feed flow rate at steady state. Since the liquid feed to V-100 is nominally at 137 ft³/hr, the vessel volume is $2 \times 10 \times 137/60 = 45.7$ ft³, which is rounded up to 50 ft³. Similar calculations give volumes of 120 ft³ for V-101 and 240 ft³ for E-102.

b. The absorption column, T-100, is a 10-stage packed bed with a diameter of 1.5 ft, computed with the assistance of the HYSYS.Plant column sizing utility.

c. The distillation column, T-101, has 10 valve-trays with a diameter of 2.5 ft, also computed with the assistance of the HYSYS.Plant column sizing utility.

d. Only the process stream is modeled in the two heat exchangers, assuming that the heat duty is the manipulated variable. Thus, E-100 is a **heater**, with a volume of 20 ft³ and E-103 is a **cooler**, with a volume of 50 ft³. Pressure drops are computed as a function of the flow rate.

e. Each valve is set at 50% open, sized on the basis of nominal flow rates, and then assigned to follow a pressure drop–flow rate relationship. When a valve is selected for control, it is assigned to a controller, which manipulates the percentage of the valve opening. Valve V-6 is maintained at 50% open, as mentioned above.

Several of the control loops in Figure 21.35 are provided for inventory control, in three level-control loops and two pressure-control loops. Note, however, that the pressure in V-100 is assumed to be constant and loop PC-1 is not simulated by HYSYS.Plant. In contrast, pressure control is crucial to maintain stable internal flows in the column. Finally, because the feed flow rate and temperature controllers are decoupled from the rest of the process, they are not included in the C&R analysis. Consequently, the interactions to be analyzed involve the four valves: V-7, V-9, V-10, and V-12, and four controlled variables: $x_{D,2}$, $x_{B,3}$, $x_{A,1}$ (mole fractions of benzene in the distillate, MCB in the bottoms, and HCl in the absorber overhead stream, respectively) and T_R, the recycle temperature. Note that to improve the dynamic performance, the temperature of tray 4 is controlled rather than the distillate benzene mole fraction.

Interaction analysis is performed using the steady-state RGA. To generate information to compute the RGA, the control loops are placed in "manual" mode, and the process is simulated to "line-out" the outputs at steady-state values. Then, one-by-one, step changes in the four valve positions are imposed, and new steady-state values of the outputs are recorded. Note that, for consistency, the step directions are chosen such that AC-1 changes in the same direction. The results are recorded in Table 21.10. Thus, for example, a 0.5% increase in the position of the reflux valve, V-7, leads to a decrease of 4.5°F in the temperature on tray 4.

Dimensionless static gains are computed, normalized by the full range of each variable. Thus, for example, the gain that relates the variation of $x_{D,2}$ to the change in R is

$$p_{11}\{0\} = \frac{\Delta x_{D,2}}{\Delta R} = \frac{-4.5/200}{0.5/100} = -4.5 \tag{21.93}$$

In this way, the other 15 static gains are computed, giving the steady-state gain matrix:

$$\begin{bmatrix} \Delta x_{D,2} \\ \Delta x_{B,3} \\ \Delta x_{A,1} \\ \Delta T_R \end{bmatrix} = \begin{bmatrix} -4.50 & 4.80 & -1.40 & -2.00 \\ -8.76 & 11.2 & -1.60 & -2.16 \\ -0.24 & 0.28 & -0.56 & 0.44 \\ -3.30 & 4.40 & 1.10 & -2.2 \end{bmatrix} \begin{bmatrix} \Delta R \\ \Delta Q_R \\ \Delta F_R \\ \Delta Q_C \end{bmatrix} \tag{21.94}$$

Table 21.10 Simulation Results for RGA Calculations

	R (V-7)	$x_{D,2}$ (AC-1)	$x_{B,3}$ (AC-2)	$x_{A,1}$ (AC-3)	T_R (TC-2)
Range	0–100%	100–300°F	0.5–1.0	0.5–1.0	50–250°F
Before	43.0%	226.3°F	0.9857	0.9596	121.2°F
After	43.5%	221.8°F	0.9638	0.9590	117.9°F
Change	+0.5%	−4.5°F	−0.0219	−0.0006	−3.3°F
	Q_R (V-9)	$x_{D,2}$ (AC-1)	$x_{B,3}$ (AC-2)	$x_{A,1}$ (AC-3)	T_R (TC-2)
Before	45.4%	226.3°F	0.9857	0.9596	121.2°F
After	44.9%	221.5°F	0.9576	0.9589	116.8°F
Change	−0.5%	−4.8°F	−0.0281	−0.0007	−4.4°F
	F_R (FC-2)	$x_{D,2}$ (AC-1)	$x_{B,3}$ (AC-2)	$x_{A,1}$ (AC-3)	T_R (TC-2)
Before	45.0%	226.3°F	0.9857	0.9596	121.2°F
After	45.5%	224.9°F	0.9817	0.9582	122.1°F
Change	+0.5%	−1.4°F	−0.0040	−0.0014	+0.9°F
	Q_C (V-10)	$x_{D,2}$ (AC-1)	$x_{B,3}$ (AC-2)	$x_{A,1}$ (AC-3)	T_R (TC-2)
Before	76.0%	226.3°F	0.9857	0.9596	121.2°F
After	76.5%	224.3°F	0.9803	0.9605	119.0°F
Change	+0.5%	−2.0°F	−0.0054	+0.0009	−2.2°F

Using Eq. (21.22), the RGA is computed

$$\underline{\underline{\Lambda}} = \begin{bmatrix} 15.8 & -11.7 & -1.35 & -1.73 \\ -100 & 84.2 & 7.44 & 9.63 \\ 15.7 & -11.8 & -15.1 & 12.1 \\ 69.8 & -59.8 & 9.98 & -19.0 \end{bmatrix} \tag{21.95}$$

While the RGA indicates that the pairings: $x_{D,2}$–R, $x_{B,3}$–Q_R, $x_{A,1}$–Q_C, and T_R–F_R, provide stable responses, the large coefficients are indicative of large interactions among the control loops, and of significant sensitivity to model uncertainty, which is often related to process nonlinearities.

On this basis, a simpler control structure is suggested, in which F_R is maintained constant, giving the steady-state gain relationship:

$$\begin{bmatrix} \Delta x_{D,2} \\ \Delta x_{B,3} \\ \Delta x_{A,1} \end{bmatrix} = \begin{bmatrix} -4.50 & 4.80 & -2.00 \\ -8.76 & 11.2 & -2.16 \\ -0.24 & 0.28 & 0.44 \end{bmatrix} \begin{bmatrix} \Delta R \\ \Delta Q_R \\ \Delta Q_C \end{bmatrix} \tag{21.96}$$

In this case, the RGA is

$$\underline{\underline{\Lambda}} = \begin{bmatrix} 5.225 & -4.475 & 0.250 \\ -4.841 & 5.874 & -0.033 \\ 0.616 & -0.399 & 0.784 \end{bmatrix} \tag{21.97}$$

Consequently, improved performance is anticipated with the diagonal pairings: $x_{D,2}$–R, $x_{B,3}$–Q_R, $x_{A,1}$–Q_C, as shown in Figure 21.36. Note that the third loop is almost decoupled, with strong interactions remaining in the two distillation-column loops. The large RGA elements associated with the LV configuration are significantly larger than those anticipated for a stand-alone column. See the RGA for the SC configuration in Figure 21.16. This is due to the additional positive feedback contributed by the material recycle stream.

Next, the DC is computed for the load changes and disturbances in Table 21.9. Two scenarios are considered: (1) d_1, a 50% increase in throughput, and (2) d_2, a composition disturbance in which the

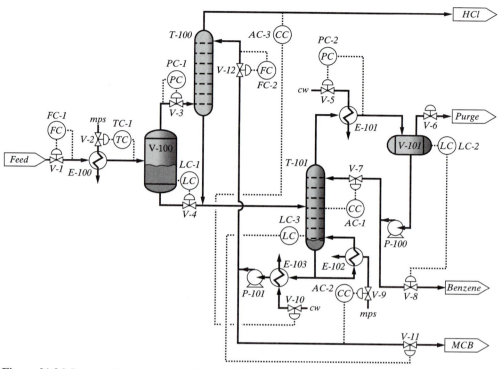

Figure 21.36 Improved control system for the MCB separation process.

three species flow rates are changed, but the total flow rate is unchanged. Table 21.11 shows the effect of each perturbation on the four outputs, indicating that the second disturbance has the greatest effect on the top composition in T-101. The effect of the two disturbances on the three outputs of the control system in Figure 21.36, expressed in scaled perturbation variable form, are

$$\text{For disturbance 1: } \underline{\underline{P_d}} \cdot \underline{d_1}\{0\} = \begin{bmatrix} -0.0195 \\ -0.1420 \\ -0.4796 \end{bmatrix} \tag{21.98}$$

$$\text{For disturbance 2: } \underline{\underline{P_d}} \cdot \underline{d_2}\{0\} = \begin{bmatrix} 0.1835 \\ 0.0398 \\ 0.0336 \end{bmatrix} \tag{21.99}$$

Note that the scaled perturbation variables are computed by dividing the changes in the output variables in Table 21.11 by their full-scale ranges. This allows the steady-state DC to be computed directly:
For disturbance 1: $DC\{0\} = -[\underline{\underline{P}}\{0\}]^{-1}\,\underline{\underline{P_d}}\,\underline{d_1}\{0\}$

$$= \begin{bmatrix} -4.50 & 4.80 & -2.00 \\ -8.76 & 11.2 & -2.16 \\ -0.24 & 0.28 & 0.44 \end{bmatrix}^{-1} \begin{bmatrix} -0.0195 \\ -0.1420 \\ -0.4796 \end{bmatrix} = \begin{bmatrix} -1.2559 \\ -0.7941 \\ 0.9103 \end{bmatrix} \tag{21.100}$$

For disturbance 2: $DC\{0\} = -[\underline{\underline{P}}\{0\}]^{-1}\,\underline{\underline{P_d}}\,\underline{d_1}\{0\}$

$$= \begin{bmatrix} -4.50 & 4.80 & -2.00 \\ -8.76 & 11.2 & -2.16 \\ -0.24 & 0.28 & 0.44 \end{bmatrix}^{-1} \begin{bmatrix} 0.1835 \\ 0.0398 \\ 0.0336 \end{bmatrix} = \begin{bmatrix} 0.2999 \\ 0.2208 \\ -0.0533 \end{bmatrix} \tag{21.101}$$

The linear analysis suggests that the effect of the first disturbance cannot be rejected completely, because the first control variable, R, saturates since the magnitude of its DC exceeds unity. In contrast, the linear DC analysis predicts that the second disturbance is rejected relatively easily.

Table 21.11 Data for Gains in DC Calculations

(a) Disturbance 1: 50% Increase in Throughput

	$x_{D,2}$ (AC-1)	$x_{B,3}$ (AC-2)	$x_{A,1}$ (AC-3)	T_R (TC-2)
Range	100–300°F	0.5–1.0	0.5–1.0	50–250°F
Before	226.3°F	0.9857	0.9596	121.2°F
After	222.4°F	0.9148	0.7200	141.1°F
Change	−3.9°F	−0.0709	−0.2396	+19.9°F

(b) Disturbance 2: Composition Change

	$x_{D,2}$ (AC-1)	$x_{B,3}$ (AC-2)	$x_{A,1}$ (AC-3)	T_R (TC-2)
Before	226.3°F	0.9857	0.9596	121.2°F
After	263.0°F	0.9976	0.9764	90.8°F
Change	36.7°F	0.0199	0.0168	−30.4°F

Dynamic simulation using HYSYS.Plant is used to verify the predictions of the linear C&R analysis. The control loops shown in Figure 21.36 are all PI controllers, tuned using the IMC-PI rules, given in Table 21.12. Note that the level controllers are loosely tuned, as in Case Study 21.1. In contrast, the distillation column pressure controller, PC-2, is tuned to ensure tight pressure control. The gains on the three composition controllers, AC-1, AC-2, and AC-3, are tuned to prevent the strong interaction between them from causing instability, while providing acceptable regulatory performance.

The simulations shown in Figures 21.37 and 21.38 show that:

a. The 3 × 3 control system, paired as suggested by the RGA, provides stable performance for both disturbances.
b. Both of the disturbances involve step changes in the molar feed rates of the three species. Note that the control system manipulates the product flow rates, while ensuring that the product compositions stay on specification, by the action of the level controllers (see Figures 21.37d and 21.38d).
c. The effects of both of the disturbances on the purities of the three products are rejected successfully, despite the prediction of the linear DC analysis (see Figures 21.37b and 21.38b). The control action perturbations required to reject the first disturbance are greater than for the second one. This is qualitatively in agreement with the DC analysis (see Figures 21.37c and 21.38c).

The results in Figures 21.37 and 21.38 can be reproduced using the HYSYS.Plant files, MCB_1.hsc and MCB_2.hsc, on the multimedia CD-ROM that accompanies this text.

Table 21.12 IMC-PI Tuning Parameters for the MCB Separation Process in Figure 21.36

Loop	PV Range	Setpoint	K_c	τ_i	Action
TC-1	150–350°F	270°F	3	2 min	Reverse
FC-2	0–200 lbmol/hr	90 lbmol/hr	1.4	0.5 min	Reverse
AC-1	200–300°F	226.3°F	5	25 min	Direct
AC-2	0.50–1.00 MCB	0.98 MCB	12	10 min	Reverse
AC-3	0.50–1.00 HCl	0.97 HCl	12	20 min	Reverse
PC-2	15–40 psia	26 psia	3	0.5 min	Direct
LC-1	0–100%	50%	2	30 min	Direct
LC-2	0–100%	50%	2	30 min	Direct
LC-3	0–100%	50%	2	30 min	Direct

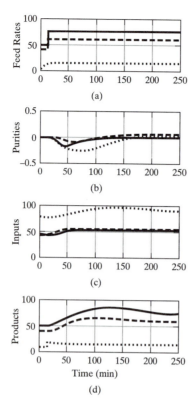

Figure 21.37 Response of the MCB separation process to a 50% increase in throughput (d_1): (a) molar feed rates in kmol/hr (solid, MCB; dashed, benzene; dotted, HCl); (b) changes in product mole percent (solid, MCB; dashed, benzene; dotted, HCl); (c) manipulated variables [solid, V-9 (Q_R); dashed, V-7 (R); dotted, V-10 (Q_C)]; (d) product flow rates in kmol/hr (solid, MCB; dashed, benzene; dotted, HCl).

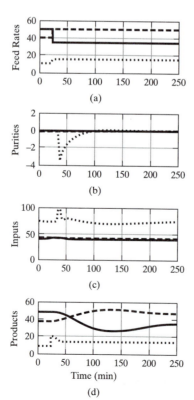

Figure 21.38 Response of the MCB separation process to composition change disturbance (d_2): (a) molar feed rates in kmol/hr; (b) changes in product mole percent; (c) manipulated variables; (d) product flow rates in kmol/hr. The variable keys are defined in Figure 21.37.

This case study has shown the advantages of employing C&R analysis to assist in the design of a plantwide control system using the procedure of Luyben and co-workers (1999). The control–configuration pairing is determined using the steady-state RGA. The disturbance rejection afforded by the process is predicted incorrectly by the linear DC analysis. For this process, nonlinear controllability and resiliency analysis is needed, using methods currently being developed (e.g., Seferlis and Grievink, 1999; Solovyev and Lewin, 2001). ∎

21.6   MATLAB FOR C&R ANALYSIS

MATLAB and SIMULINK are invaluable tools for the frequency- and time-domain calculations required for C&R analysis. In this section, several examples are carried out using MATLAB, it being assumed that the reader is familiar with the MATLAB syntax. The reader is referred to Bequette (1998) for details of MATLAB usage in dynamical analysis and control, and to the multimedia CD-ROM that accompanies this text for sources of these and other useful MATLAB functions and scripts for C&R analysis. In particular, the interactive C&R Tutorial CRGUI can be used to test three example linear processes for controllability and resiliency and simulate their closed-loop response under single-loop PI control.

EXAMPLE 21.11   *Computing the Dynamic RGA*

For the system given in Example 21.3, the MATLAB script that generates the dynamic RGA in Figure 21.5 is

```
%Example 21.3
%This script file computes the dynamic RGA for Example 21.3

%Define a vector of frequency values on a log scale
wmin = −3; wmax = 1;nw = 30*fix(wmax-wmin);
w = logspace (wmin,wmax, nw); s = i*w;

% Data for process model
kp = [2.5 5;1 −4];              %process gain matrix
tp1 = [15 4;3 20];              %process time constant
tp2 = [2 0;0 0];               %process time constant
thp = [5 0;0 5];               %process delay

% Compute the frequency response for each element of Pij
p11 = kp(1,1)./(tp1(1,1)*s + 1)./(tp2(1,1)*s + 1).*exp(−thp(1,1)*s);
p12 = kp(1,2)./(tp1(1,2)*s + 1)./(tp2(1,2)*s + 1).*exp(−thp(1, 2)*s);
p21 = kp(2,1)./(tp1(2,1)*s + 1)./(tp2(2,1)*s + 1).*exp(−thp(2, 1)*s);
p22 = kp(2,2)./(tp1(2,2)*s + 1)./(tp2(2,2)*s + 1).*exp(−thp(2,2)*s);

% Compute lambda (1,1) and lambda (1,2) as functions of frequency.
L11 = p11.*p22./(p11.*p22 − p12.*p21);
lam11 = sign(real(L11(1))).*abs(L11);
L12 = −p12.*p21./(p11.*p22 − p12.*p21);
lam12 = sign(real(L12(1))).*abs(L12);

% Plot the results
figure
semilogx (w,lam11,'−k', w,lam12,':k','Line Width',2)
xlabel('\omega [rad/min]', 'FontName','Times','FontSize',14)
ylabel('DRGA','FontName','Times','FontSize',14)
```

As discussed in Example 21.3, the steady-state RGA suggests diagonal pairings. However, the dynamic RGA implies that these pairings are unstable for frequencies higher than about 0.5 rad/min. Thus, antidiagonal pairings should be used. ∎

EXAMPLE 21.12   *Computing Disturbance Cost Maps*

Consider the component parts in the LSF configuration represented by Eqs. (21.56) and (21.57). In this example, the elements of the transfer function matrices are entered into MATLAB and used to compute the DC contour maps for this configuration. $\underline{\underline{P}}\{j\omega\}$ and $\underline{\underline{P}}_d\{j\omega\}$ are computed for each frequency, and used to compute DC for all of the disturbance directions. By looping over all frequencies, the entire DC map is calculated and repeated for each manipulated variable separately. Note that, as mentioned in Example 21.8, the inputs are nominally at 50% of the full range. Here, the nominal inputs are taken as $L_H = L_L = 11$ kmol/min, $Q_{RH} = 0.222 \times 10^6$ kcal/min, and the maximum disturbance magnitudes are taken as $|F| = 18$ kmol/min and $|x_F| = 0.2$ ($\pm 20\%$ of the full range).

```
% LFS:        This script computes P(s) and Pd(s) for the LSF configuration, given the
%             transfer-function matrices for the two component parts. It then uses the
%             matrices to compute DC contours

%       Definition of frequency and direction vectors.
n = 41; i = sqrt(−1); wmin = −3; wmax = 0; dw = (wmax − wmin)/(n − 1);
tmin = 0; tmax = 180; dt = (tmax − tmin)/(n − 1);
w = logspace(wmin,wmax,n);              % Frequency vector [rad/min]
ome = wmin:dw:wmax;                     % Frequency vector in log scale.
phi = tmin:dt:tmax;                     % Direction vector [degrees]
a = pi*phi/180:                         % Direction vector [radians]
s = w*i;                                % Vector complex s
tt = exp(i*a);                          % Computing the direction in radian coordinates
dd(1:n, 1:2) = [real(tt'), imag(−tt')]; % tt in cartesian coordinates
z = zeros(1:n,1:n);                     % matrix for storing computed DC values.

% Gains and delay items for the high pressure column:
KH = [0.017 − 1.109 0.001 0.090; 0.011 −1.859 0.006 1.296;
−0.33 59.0 −0.2 −41.05; 0.916 −123.7 1.127 −0.02; 4.0e-5 −0.994 0.001 0.003];
DH = [0.0 0.0 0.0 6.4; 1.3 0.0 0.1 0.1; 1.3 0.0 0.1 0.1; 1.3 0.0 0.1 0.1; 1.3 0.0 0.1 0.1];

% Gains and delay times for the low pressure column:
KL = [0.792 −0.029 0.007 2.161 0.012; 0.790 −0.051 0.003 3.291 0.038];
% Note: The coefficients in the fourth column have been multiplied by −1
% since QRL = −QCH
DL = [0.1 0.1 0.1 0.0 1.4; 8.5 0.0 0.0 0.0 0.0];

for ku = 1:3              % Looping over all manipulated variables (m = 3)
    for k = 1:n           % Looping over all frequencies.

%       Computing the frequency response of each component part submatrix [see
%       Eqs. (21.56) and (21.57)].
        ph = KH.*exp(−DH*s(k))./(13*s(k)+1);
        pl = KL.*exp(−DL*s(k))./(17*s(k)+1);
%       Computing P(s) and Pd(s) at the current frequency [See Eqs. (21.60) and (21.61)]
        P = [ph(1,,1:2) 0; pl(:,1:4)*ph(2:5,1:2) pl(:,:5)];
        Pd = [ph(1,3:4); pl(:,1:4)*ph(2:5,3:4)];
%       Scaling:
        P(:,1) = P(:,1)*11;P(:,2) = P(:,2)*0.222;P(:,3) = P(:,3)*11;
        PD(:,1) = Pd(:,1)*18;Pd(:,2) = Pd(:,2)*0.2;
        u2 = inv(P)*Pd*dd';    % Computing DC
        for i_dir = 1:n        % Looping over d direction 0 → 180
            z(i_dir,k) = norm(u2(ku,i_dir));
        end
    end
%       End of frequency loop.
```

```
        v = [0.1 0.2 0.3 0.4 0.5 0.6 0.7 0.8 0.9 1.0];
        figure
        cs = contour(ome,phi,z);
        clabel(cs)
        title(['Disturbance Cost for Input',num2str(ku)]);
        xlabel('log(w)');
        ylabel('Direction [deg]');
end     % End of manipulated-variable loop
```

This script generates the DC contour maps in Figure 21.39 for each manipulated variable separately. Note that there is no bandwidth limitation to perfect disturbance rejection in any of the control variables. ∎

21.7 SUMMARY

In this chapter, the methods for shortcut C&R analysis, using the results of steady-state simulations, have been described. The methods require the use of software for the solution of material and energy balances in process flowsheets (e.g., ASPEN PLUS, HYSYS.Plant) and for controllability and resiliency analysis (i.e., MATLAB). The reader is now prepared to tackle small- to medium-scale problems, and in particular should

1. Be able to generate a linear model of a chemical process in one of its standard forms, using either the equations expressed in a MATLAB function, or the solution of the material and energy balances computed by a process simulator.
2. Be able to compute the frequency-dependent process transfer functions using MATLAB, given a linear model in one of its standard forms.
3. Be able to generate the C&R measures of relative-gain array (RGA) and disturbance cost (DC), given the process transfer functions, using MATLAB.
4. Be able to select the appropriate pairings for a decentralized control system for the process using the static and dynamic RGAs and appropriate resiliency measures.
5. Be able to perform C&R analysis to select between alternative process configurations, given the results of process simulations.

Several examples have been selected to show how the methods are used to screen alternative flowsheets in Stage 2 of the design process (Table 20.1). In the first example (Section 21.3), dynamic C&R analysis enables the most resilient heat-integrated distillation configuration to be selected. In Case Study 21.1, two designs for an exothermic reactor, involving either one or two CSTR(s) in series, show that while the latter is more economical (assuming steady-state operation), the former is more resilient to disturbances. In Case Study 21.2, a steady-state analysis of two heat-exchanger network configurations leads to the conclusion that while a design equipped with bypasses may be subject to significant constraints leading to poor resiliency, a design without them may lead to poor dynamic performance. Here, dynamic C&R analysis is crucial. Finally, Case Study 21.3 involves a recycle process and shows the benefits of C&R analysis in the detailed design stage (Stage 3 in Table 20.1).

As shown in the case studies, it is recommended that dynamic simulation be employed to verify the results obtained by C&R analysis. This simulation is routinely performed using HYSYS.Plant and ASPEN DYNAMICS, as demonstrated in this chapter. The reader is referred to the book *Plantwide Dynamic Simulators in Chemical Processing and Control* (Luyben, 2002) for many additional examples in which dynamic simulation assists plantwide controllability analysis.

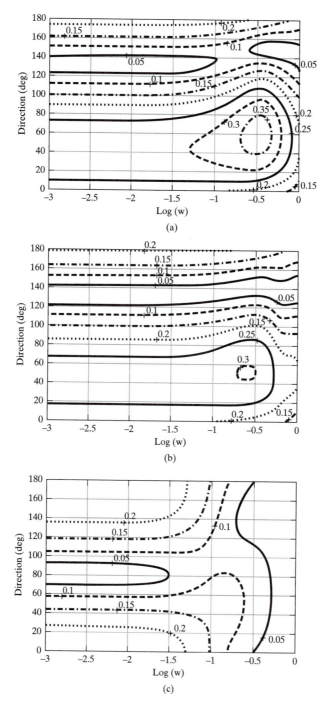

Figure 21.39 DC contour maps for the LSF configuration to dehydrate methanol: (a) L_H; (b) Q_{RH}; (c) L_L. The bounds on the disturbances are $\pm 20\%$ from their nominal values. The DC contour maps for each manipulated variable are computed separately. See Figure 21.17 for the DC contour maps for the SC, FS, and LSR configurations.

REFERENCES

Bequette, B.W., *Process Dynamics: Modeling, Analysis, and Simulation*, Prentice-Hall, Englewood Cliffs, New Jersey (1998).

Bristol, E.H., On a New Measure of Interactions for Multivariable Process Control, *IEEE Trans. Auto. Control*, **AC-11**, 133–134 (1966).

Chiang, T., and W.L. Luyben, Comparison of the Dynamic Performances of Three Heat-integrated Distillation Configurations, *Ind. Eng. Chem. Res.*, **27**, 99–104 (1988).

Denn, M.M., and R. Lavie, Dynamics of Plants with Recycle, *Chem. Eng.*, **24**, 55–59 (1982).

Lewin, D.R., A Simple Tool for Disturbance Resiliency Diagnosis and Feedforward Control Design, *Comput. Chem. Eng.*, **20**(1), 13–25 (1996).

Lewin, D.R., and D. Bogle, Controllability Analysis of an Industrial Polymerization Reactor, *Comput. Chem. Eng.*, **20**(S), S871-S876 (1996).

Lewin, D.R., W.D. Seider, J.D. Seader, E. Dassau, J. Golbert, D. Goldberg, M. Fucci, and R.B. Nathanson, *Using Process Simulators in the Chemical Engineering Curriculum: A Multimedia Guide for the Core Curriculum*, Version 2.0, Multimedia CD-ROM, John Wiley & Sons, New York (2003).

Luyben, W.L., *Process Modeling, Simulation and Control for Chemical Engineers*, 2nd ed., McGraw-Hill, New York (1990).

Luyben, W.L., *Plantwide Dynamic Simulators in Chemical Processing and Control*, Marcel Dekker, New York (2002).

Luyben, W.L., B.D. Tyreus, and M.L. Luyben, *Plantwide Process Control*, McGraw-Hill, New York (1999).

Mathisen, K.W., S. Skogestad, and E.A. Wolff, Bypass Selection for Control of Heat Exchanger Networks, *Comput. Chem. Eng.*, **16**(S), S263–S272 (1993).

McAvoy, T.J. *Interaction Analysis*, Instrument Society of America, Research Triangle Park, North Carolina (1983).

Morari, M. Design of Resilient Processing Plants III, A General Framework for the Assessment of Dynamic Resilience, *Chem. Eng. Sci.*, **38**, 1881–1891 (1983).

Morari, M., and E. Zafiriou, *Robust Process Control*, Prentice-Hall, Englewood Cliffs, New Jersey (1989).

Naot, I., and D.R. Lewin, Analysis of Process Dynamics in Recycle Systems Using Steady State Flowsheeting Tools, *Proc. 4th IFAC Symposium on Dynamics and Control of Chemical Reactors, Distillation Columns and Batch Processes (DYCORD'95)*, Helsingor, Danish Automation Society, Copenhagen (1995).

Ogunnaike, B A., and W.H. Ray, *Process Dynamics, Modeling and Control*, Oxford University Press, New York (1994).

Perkins, J.D. The Interaction between Process Design and Process Control, *Proc. IFAC Symposium on Dynamics and Control of Chemical Reactors and Distillation Columns (DYCORD'89)*, 195–203 (1989).

Prett, D.M., and M. Morari, *Shell Process Control Workshop*, pp. 355–360, Butterworth, Stoneham, Massachusetts (1986).

Rivera, D.E., S. Skogestad, and M. Morari, Internal Model Control. 4. PID Controller Design, *Ind. Eng. Chem. Res.*, **25**, 252–265 (1986).

Rotstein, G.E., and D.R. Lewin, Simple PI and PID Tuning for Open Loop Unstable Systems, *Ind. Eng. Chem. Res.*, **30**, 1,864–1,869 (1991).

Russo, L.P., and B.W. Bequette, Operability of Chemical Reactors: Multiplicity Behavior of a Jacketed Styrene Polymerization Reactor, *Chem. Eng. Sci.*, **53**(1), 27–45 (1998).

Sandelin, P.M., K.E. Haggblom, and K.V. Waller, Indirect Two-Point Control Through One-Point Control of Distillation, in J.E. Rijnsdorp, J.F. MacGregor, B.D. Tyreus, and T. Takamatsu, eds., *Dynamics and Control of Chemical Reactors, Distillation Columns and Batch Processes*, IFAC Symposia Series 1990, No. 7, Pergamon Press, Oxford, 143–148 (1990).

Seborg, D.E., T.F. Edgar, and D.A. Mellichamp, *Process Dynamics and Control*, Wiley, New York (1989).

Seferlis, P., and J. Grievink, "Plant Design Based on Economic and Static Controllability Criteria," *Proc. of the 5th Int. Conf. of Foundations of Computer-aided Process Design*, pp. 346–350, AIChE, New York (1999).

Shinskey, F.G., *Distillation Control*, pp. 83–89, 2nd ed., McGraw-Hill, New York (1984).

Shinskey, F.G., *Process Control Systems*, 3rd ed., McGraw-Hill, New York (1988).

Skogestad, S. *Studies on Robust Control of Distillation Columns*, Ph.D. thesis, California Institute of Technology (1987).

Skogestad, S., and M. Morari, The Effect of Disturbance Directions on Closed Loop Performance, *Ind. Eng. Chem. Res.*, **26**, 2,029–2,035 (1987).

Solovyev, B.E., and D.R. Lewin, "A Steady-state Process Resiliency Index for Non-linear Processes," *Proc. DYCOPS'6*, Jejudo Island, Korea (2001).

Stephanopoulos, G., *Chemical Process Control*, Prentice-Hall, Englewood Cliffs, New Jersey (1984).

Weitz, O., and D.R. Lewin, Dynamic Controllability and Resiliency Diagnosis Using Steady-State Process Flowsheet Data, *Comput. Chem. Eng.*, **20**(4), 325–335 (1996).

Weitz, O., Integration of Controllability Measures into Process Design, M.Sc. thesis, Technion, Haifa, Israel (1994).

Wolff, E.A., K.W. Mathisen, and S. Skogestad, Dynamics and Controllability of Heat Exchanger Networks, *Proc. COPE-91*, 117–128 (1991).

Ziegler, J.G., and N.B. Nichols, Optimum Settings for Automatic Controllers, *Trans. A.S.M.E.*, **64**, 759–768 (1942).

EXERCISES

21.1 The following RGA matrix has been obtained for a MIMO process:

$$\Lambda = \begin{bmatrix} 0.8 & & \\ -4.0 & 4.3 & \\ & & -4.6 \end{bmatrix}$$

Note that the missing elements are unavailable. If the process is to be controlled using a decentralized control system, what are the most promising pairings?

a. u_1–y_1, u_2–y_2, u_3–y_3.

b. u_3–y_1, u_2–y_2, u_1–y_3.

c. u_1–y_1, u_3–y_2, u_2–y_3.

d. There is not enough information to decide.

21.2 a. Consider a two-stream blender

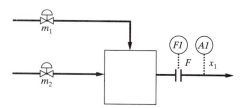

where m_1 and m_2 are the mass flow rates of species 1 and 2, F is the total flow rate, and x_1 is the mass fraction of species 1 in the effluent stream. Use the relative gain array to select the control loop pairings for the effluent composition:

1. $x_1 = 0.8$
2. $x_1 = 0.3$

b. When blending pure streams of species 1, 2, and 3

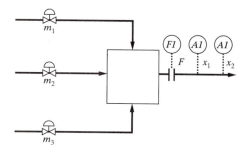

pair the control loops for operation at $F = 1$, $x_1 = 0.1$, and $s_1 = 0.2$.

21.3 Two liquid phases are separated using the continuous decanter shown in Figure 21.40. The output variables, which must be controlled, are: F_1, the volumetric feed rate, P_1, the operating pressure, and I, the dispersion interface level in the decanter. The positions of the three control valves, m_1, m_2, and m_3 are the manipulated variables. A linear model is available to describe the process:

$$\begin{bmatrix} F_1 \\ P_1 \\ I \end{bmatrix} = \begin{bmatrix} 2.7 & 8.4 & 8.4 \\ 0.38 & -0.56 & -0.56 \\ 0 & 12/s & -0.35/s \end{bmatrix} \begin{bmatrix} m_1 \\ m_2 \\ m_3 \end{bmatrix}$$

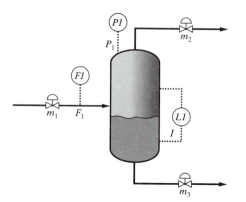

Figure 21.40 Continuous separation of two liquid phases.

The equipment manufacturer has suggested the pairings F_1–m_1, P_1–m_2, and I–m_3. Are these the most appropriate to use?

$$\text{Hint:} \begin{bmatrix} 2.7 & 8.4 & 8.4 \\ 0.38 & -0.56 & -0.56 \\ 0 & 12/s & -0.35/s \end{bmatrix}^{-1}$$

$$= \begin{bmatrix} 0.12 & 1.8 & 0 \\ 0.0023 & -0.016 & 0.081s \\ 0.079 & -0.56 & -0.081s \end{bmatrix}$$

21.4 In a recent publication, a decentralized control system was proposed for an experimental reactor involving heat integration between two sections of the reactor. The manipulated variables available for control are the heat duties to the two sections, Q_p and Q_c. The control system is intended to regulate the operating temperatures in the two sections, T_p and T_c. The authors developed a detailed nonlinear model of the process that, after linearization, gives

$$\begin{bmatrix} T_p \\ T_c \end{bmatrix} = \begin{bmatrix} \dfrac{295s + 37}{133s^2 + 48s + 1} & \dfrac{30}{133s^2 + 48s + 1} \\ \dfrac{33}{133s^2 + 48s + 1} & \dfrac{148s + 35}{133s^2 + 48s + 1} \end{bmatrix} \begin{bmatrix} Q_p \\ Q_c \end{bmatrix}$$

They tuned simple PI controllers and found that the closed-loop response of the overall system became faster and *less oscillatory* when they increased the controller gains. Use the dynamic RGA to explain this observation.

21.5 Reproduce the DC map in Figure 21.11 for the Shell process.

21.6 Three component parts were given by Weitz (1994) for the FS configuration of the heat-integrated distillation columns in Figure 20.2. His linearized models are, for the high-pressure column

$$\begin{bmatrix} x_{DH} \\ x_{BH} \\ Q_{CH} \end{bmatrix} = \frac{1}{11s + 1} \times$$

$$\begin{bmatrix} 0.018 & -1.471 & 0.003 & 0.170e^{-4.8s} \\ 0.047e^{-1.2s} & -7.219 & 0.041e^{-0.2s} & 1.449e^{-0.2s} \\ -0.001 & -0.861 & 0.0003 & -0.028 \end{bmatrix} \begin{bmatrix} L_H \\ Q_{RH} \\ F_H \\ x_{FH} \end{bmatrix}$$

And for the low-pressure column

$$\begin{bmatrix} x_{DL} \\ x_{BL} \end{bmatrix} = \frac{1}{16s + 1} \times$$

$$\begin{bmatrix} -1.112 & 0.0185 & 0.001 & 0.168e^{-6.9s} \\ -6.745 & 0.048e^{-1.4s} & 0.034e^{-0.3s} & 1.483e^{-0.3s} \end{bmatrix} \begin{bmatrix} Q_{RL} \\ L_L \\ F_L \\ x_{FL} \end{bmatrix}$$

and for the feed splitter (pure gain),

$$\begin{bmatrix} F_H \\ x_{FH} \\ F_L \\ x_{FL} \end{bmatrix} = \begin{bmatrix} 11.72 & 0.490 & 0 \\ 0 & 0 & 1 \\ -11.72 & 0.510 & 0 \\ 0 & 0 & 1 \end{bmatrix} \begin{bmatrix} F_H/F_L \\ F \\ x_F \end{bmatrix}$$

It can be assumed that all of the inputs are nominally at 50% of their full ranges. The nominal values of the inputs are taken as $L_H = L_L = 11$ kmol/min, $Q_{RH} = 0.205 \times 10^6$ kcal/min, $F_H/F_L = 0.49$. The maximum disturbance magnitudes are taken as $|\Delta F| = 18$ kmol/min and $|\Delta x_F| = 0.2$ ($\pm 20\%$ of full range). Using these models, and noting the interconnections between the component parts in Figure 21.13a, reproduce the DC contour maps in Figure 21.17 for the FS configuration.

21.7 Two component parts were given by Weitz (1994) for the LSR configuration of the heat-integrated distillation columns in Figure 20.2. His linearized models are, for the high-pressure column

$$\begin{bmatrix} x_{BH} \\ x_{DH} \\ Q_{CH} \end{bmatrix} = \frac{1}{14s + 1} \times$$

$$\begin{bmatrix} 1.136e^{-0.2s} & -0.047e^{-0.2s} & 0.013e^{-0.2s} & 0.022e^{-1.4s} & -3.425 \\ 0.154e^{-6.3s} & -0.027 & 0.001 & 0.023 & -1.551 \\ -0.045 & -0.013 & 0.0002 & -0.0008 & -0.872 \end{bmatrix}$$

$$\times \begin{bmatrix} x_{BL} \\ T_{BL} \\ B_L \\ L_H \\ Q_{RH} \end{bmatrix}$$

and for the low-pressure column

$$\begin{bmatrix} x_{DL} \\ x_{BL} \\ T_{BL} \\ B_L \end{bmatrix} = \frac{1}{18s + 1} \times$$

$$\begin{bmatrix} 0.021 & -1.012 & 0 & 0.131e^{-8.9s} \\ 0.010e^{-1.4s} & -1.772 & 0.005e^{-0.1s} & 1.297e^{-0.1s} \\ -0.272e^{-1.4s} & 50.05 & -0.144e^{-0.1s} & -36.96e^{-0.1s} \\ 0.913e^{-1.4s} & -112.5 & 0.998e^{-0.1s} & -1.085e^{0.1s} \end{bmatrix}$$

$$\times \begin{bmatrix} L_L \\ Q_{RL} \\ F \\ x_F \end{bmatrix}$$

It can be assumed that all of the inputs are nominally at 50% of their full ranges. The nominal values of the inputs are taken as $L_H = L_L = 11$ kmol/min, $Q_{RH} = 0.205 \times 10^6$ kcal/min, $F_H/F_L = 0.49$. The maximum disturbance magnitudes are taken as $|\Delta F| = 18$ kmol/min and $|\Delta x_F| = 0.2$ ($\pm 20\%$ of full range). Using these models, and noting the interconnections between the component parts in Figure 21.15c, reproduce the DC contour maps in Figure 21.17 for the LSR configuration.

21.8 A product P is produced by two sequential exothermic reactions, $A \rightarrow B \rightarrow P$, with an additional endothermic reaction of B leading to an unwanted product X. These reactions are carried out in a jacketed CSTR, whose material and energy balances are

$$\dot{C}_A = q(C_{A0} - C_A) - k_1(T)C_A$$
$$\dot{C}_B = -qC_B + k_1(T)C_A - k_2(T)C_B - k_3(T)C_B$$
$$\dot{C}_P = -qC_P + k_2(T)C_B$$
$$\dot{T} = q(T_0 - T) - \frac{1}{\rho c_P}[k_1(T)\Delta H_1 C_A + (k_2(T)\Delta H_2 + k_3(T)\Delta H_3)C_B]$$
$$\quad - \frac{UA}{\rho c_P V}(T - T_J)$$
$$\dot{T}_J = q_J(T_{J0} - T_J) + \frac{UVA}{\rho c_P}(T - T_J)$$

with reaction rate constants

$$k_i(T) = k_{i0}\exp\left(\frac{-E_i}{T}\right), i = 1,2,3$$

The controlled variables are the concentration of P in the reactor effluent, C_P, and the reactor temperature, T. The manipulated variables are the feed flow rate, q, and the jacket coolant flow rate, q_J. The process disturbances are the feed concentration of A, C_{A0}, and the feed temperature, T_0. Additional information is given in Table 21.13.

Using the model and Table 21.13, compute the steady-state RGA and DC. You may assume that the disturbances in the feed concentrations are limited to within ± 1M and those of the feed temperature to ± 5 K, and that the manipulated variables are nominally midway between their lower and upper bounds. Based on these computations, answer the following questions: a. What are the appropriate pairings to use for decentralized control? b. What is the worst possible combination of disturbances in T_0 and C_{A0}?

Table 21.13 Process Information for Exercise 21.8

Variable		Value	Variable		Value
C_P	(M)	1.00	k_{10}	(min^{-1})	1.169×10^{10}
T	(K)	353.15	k_{20}	(min^{-1})	1.445×10^{11}
q	(min^{-1})	0.15	k_{30}	(min^{-1})	1.689×10^{11}
q_J	(min^{-1})	0.10	E_1	(K)	9,000
C_{A0}	(M)	5.00	E_2	(K)	9,500
T_0	(K)	343.15	E_3	(K)	9,800
T_{J0}	(K)	288.15	ΔH_1	(kJ/mol)	-40
$\dfrac{UA}{V}$	$\left(\dfrac{\text{kJ·L}}{\text{min·K}}\right)$	0.225	ΔH_2	(kJ/mol)	-20
ρc_P	(kJ/L-K)	1.00	ΔH_3	(kJ/mol)	120

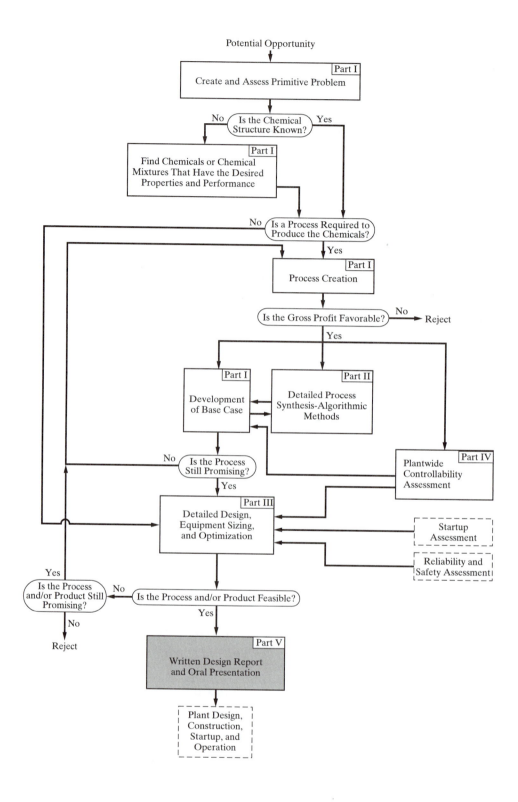

Part Five

DESIGN REPORT

In this part, which is comprised of Chapter 22, the contents of the written design report are presented and recommendations are made concerning its completion. Emphasis is placed on the need to document the design throughout the design process, including the data used, the flowsheets considered, the material and energy balances, the detailed calculations for the process units, the cost estimates, and the profitability analysis. Special instructions are provided for the documentation of product designs. To accomplish these tasks, the design team needs to keep a repository of its materials. From time to time, it needs to report to its supervisor and, at universities, to its faculty advisor, industrial consultants, and fellow students. When documented properly, much of the material presented, together with the background calculations and drawings, can be used in the final design report. Hence, an objective of Part Five is to help the design team set milestones for the completion of aspects of the design as well as its design report.

In addition, the oral design presentation is discussed and suggestions are made for the organization of the presentation, the media used, methods for rehearsing, and written handouts. Also discussed, are typical criteria for evaluation and the usage of videotapes.

To see how Part Five relates to the other four parts of this text and to the entire design process, see the figure at the left.

Chapter 22

Written Reports and Oral Presentations

22.0 OBJECTIVES

At the completion of its design work, whether it be for a process or a product, the design team is required to prepare a detailed report, documenting the details of the design and how it was produced, projecting its profitability, and making a recommendation as to whether or not management should make an investment in the process or the product. Management will also need a marketing report, which may, but probably will not, be part of the design report. While marketing may seem foreign to chemical engineering, far more often than not, marketing mistakes, rather than technical mistakes, are responsible for the failure of a new process or product.

A *process design report* identifies the key assumptions in the design and their potential impact on the performance of the process and its anticipated profitability. This is particularly important for designs completed by undergraduate students at universities, where facilities and time are rarely available for laboratory work or pilot-plant testing.

It is these uncertainties, especially when data are lacking, that engineers encounter throughout their careers. Even when laboratories and pilot plants are available, engineering judgments are needed to determine when the investment of money and time is justified to organize an experimental program. In this respect, engineers are asked regularly to estimate the profitability of processes about which they have too little information. For these reasons, design teams usually expend considerable effort in trying to eliminate as much of the uncertainty as possible (by locating data in the literature, conducting process simulations, etc.). Invariably, however, uncertainties remain, and it is important that the design report identifies the uncertainties and, when it recommends that a process be constructed, makes recommendations as to how the uncertainties should be resolved.

A *product design report* documents all of the steps leading up to the design of the product. This includes a discussion of the need for a new product, a summary of all the possible ways generated by the design team to satisfy the need, the rationale behind the selection of the best ideas and ultimately the one selected, and the design of the new product. In industry, the report might also include the development of the manufacturing process as well as the performance of a prototype of the design. In universities, undergraduate students may not have the time to produce a prototype of the product and determine its performance. On the other hand, a product design developed by a student design team in a previous year could be given to a new design team in the current year for the development and assessment of a prototype.

The design team should view its design report as an opportunity to showcase its most creative engineering efforts for management. Wherever possible, a process design should highlight the engineering work that it believes will lead to greater economies than are achievable using alternative or conventional technologies. A product design should emphasize the superiority of the new product over available products with similar functions.

For most student design teams, the design report is the first extensive report of their professional careers. It is the culmination of a major engineering effort and, when done well, is deserving of considerable attention by other students, faculty, industrial consultants, and perspective employers. In this respect, the professional reputation of the design team depends, in part, on how well the design problem has been analyzed, how ingeniously the process has been laid out or the product developed, and how thoroughly the engineering calculations and design work have been done. The efforts of a design team are judged almost entirely by the quality of the engineering report provided to their supervisor, which describes the work that has been accomplished. Of particular importance to management is the strength of and justification for the recommendations made in the report. Almost always, the report will be accompanied by an oral presentation by the leader of the design team and perhaps some team members, where questions can be asked by management.

There is perhaps a tendency to view the preparation of the design report and the oral presentation to management as activities reserved for the completion of a design project. Although, indeed, the level of activity in writing builds steadily toward the end, especially as the design becomes more promising, an objective of this chapter is to present the many reasons for documenting the results gradually, as the design project proceeds. In fact, a close look at Figure 1.2, and the discussions throughout the book, especially in Sections 1.1, 1.2, and 3.5, shows that up-to-date documentation is very important to the success of a design team, especially as the composition of the team changes.

After studying this chapter, the reader should

1. Understand the template that prescribes the sections in most design reports, and have a good appreciation of the materials to be included in each section.
2. Be prepared to coordinate the preparation of the design report with the other members of the design team, beginning early in the design process, and to recognize the important milestones, that is, those portions of the report that are best prepared before work begins on the next step in the design process.
3. Understand the role and format of a typical oral presentation, including the alternative media for the presentation, and related topics such as the need for rehearsals and the desirability of a written handout.

22.1   CONTENTS OF THE WRITTEN REPORT

This section begins with a template of items to be included in most design reports and is followed by a discussion of several techniques found to be helpful in their preparation, as well as recommendations for the page format to permit the design report to be bound for distribution.

Opening Sections of the Report

This listing is presented in outline form to identify, at a glance, the sections that are normally included in the sequence shown. The first four items are common to both process design and product design reports.

1. *Letter of Transmittal.* This letter, on professional letterhead, is normally directed to the supervisor who requested that the design work be done. It should be signed by all members of the design team.
2. *Title Page.* In addition to the title, in uppercase, the authors and their affiliations are listed, as well as the publication date. The title should be short, but very descriptive.

3. *Table of Contents.* All sections in the report should be listed, including the page numbers on which they begin. Hence, all pages in the report, *without exception*, must be numbered. This applies to text pages, for which the word processor will probably have provided the page numbers, as well as tables, figures, and appendixes, whose pages may have to be numbered manually. Note that unnumbered pages are not readily found by the reader, who may resent the time wasted thumbing through the report to find pages that are missing or not numbered.

4. *Abstract.* The abstract is a brief description, in one or two paragraphs, of the design report, its key conclusions, special features, and assumptions. These include projections of any applicable economic measures of goodness (e.g., the return on investment and the net present value) and recommendations to management.

Remaining Sections of a Process Design Report

5. *Introduction.* The introduction contains some or all of the following sections.

   a. A description of the product to be manufactured, including its chemical formula, and a discussion of the role of this product in the industry and its significance in national and international trade.

   b. A survey of the methods used in manufacturing this product, including the process being described in this report, giving the raw materials, the principal chemical reactions, byproducts, and intermediates.

   c. An explanation of the choice of the production method. This often involves a description of partially synthesized flowsheets and the reasons why these flowsheets were rejected in favor of the design selected.

   d. A discussion of the choice of the production level and plant location.

   e. A discussion of the reasons for entering the market at this time.

   f. An overview of the environmental issues encountered, including the toxicity of the chemicals, and the potential safety problems.

6. *Process Flow Diagram and Material Balances.* This is the detailed process flow diagram discussed in Section 3.5 and shown for a vinyl chloride process in Figure 3.19. All of the streams are numbered clearly and all of the process units are labeled. At some point on the arc for each stream, the temperature and pressure should appear, or the information should be tabulated (e.g., see Table 3.5). Note that, as mentioned in Section 3.5, many software packages are available to simplify the preparation of flow diagrams, most notably those associated with the process simulators.

   In addition, the drawing should contain a *material balance block*, similar to the one shown for the the vinyl chloride process in Table 3.5, that is, a table showing for each numbered stream:

   a. Total flow rate

   b. Flow rate of each chemical species

   c. Temperature

   d. Pressure

and other properties of importance (density, enthalpy, etc.). It is desirable that the flow diagram and the material balance block appear on a single sheet for continuous reference, preferably $8\frac{1}{2}$ by 11 in., so that it can be bound easily with the remainder of the report. Most commonly, this combination is prepared by computer, often using Microsoft VISIO, although a hand-drawn flow diagram is normally acceptable. If the flow diagram is hand drawn, it should be of a larger size, perhaps 20 in. by 30 in. Such flow diagrams should be drawn on paper in ink, or pencilled on vellum or other erasable, translucent media, so that a reproduction (such as an Ozalid or similar black-on-white print) may be prepared, and folded and included in the pocket on the inside back cover of the report. The symbols on the drawing should follow a standard list, such as those provided in Figure 3.20 and by Peters et al. (2003), Sandler and Luckiewicz (1993), and Ulrich (1984). All entries should be made with straightedges and templates.

7. *Process Description.* This section provides an explanation of the flow diagram. It best begins, however, with reference to a block flow diagram, similar to that in Figure 3.18, which shows just the process steps that involve chemical reactions and the separation of chemical mixtures. Then, a more detailed description is presented that refers to all steps in the process that are shown in the process flow diagram (e.g., Figure 3.19). The detailed description describes the function of each equipment item and discusses the reasons for each particular choice. Note that the details of each major equipment item are presented below in Section 9, on unit descriptions. To aid the reader, however, the discussion of each item in Section 7 should be accompanied by a reference to the page number in Section 9. As in the introduction, when this flow diagram has been selected from among alternatives, it is appropriate to present the alternative flow diagrams and process descriptions, and to describe the reasons for the final choice.

8. *Energy Balance and Utility Requirements.* In describing most chemical processes, it is desirable to have a section that discusses the energy requirements of the process, and the measures adopted to improve the plant economics by energy and mass conservation, usually through the application of the methods described in Chapter 9 on second-law analysis, Chapter 10 on heat and power integration, and Chapter 11 on mass integration. In this section, all of the heating, cooling, power, and other utility and mass-separating-agent demands should be identified (with numerical values provided), and the methods of satisfying these demands shown. A list should be provided of each demand (e.g., 500,000 Btu/hr to heat stream 5 from 80 to 200°F) and the vehicle for its satisfaction (e.g., 500,000 Btu/hr from stream 15 as it is cooled from 250 to 100°F). When power generated by a turbine is used to drive a compressor and pumps, these integrations should be listed as well. Methods used to minimize the need for solvents and other mass-separating agents, as well as to minimize wastes, should be described.

9. *Equipment List and Unit Descriptions.* In this section, every process unit in the flow diagram is described in terms of its specifications and the design methodologies (e.g., the methods for estimating the heat transfer coefficients, the graphical design of a distillation tower by means of the McCabe–Thiele method, and the recommendations of industrial consultants) and the data employed (e.g., to characterize the reaction kinetics and vapor–liquid equilibria). The important approximations should be discussed, as well as any difficulties encountered in performing the design calculations (e.g., in converging equilibrium-stage calculations with a simulator). In addition, the materials of construction should be indicated, together with the reasons for their selection.

Each process unit described in Section 9 should refer to the page number in the appendix on which the design calculations appear or are described. Note that the latter calculations are usually printed neatly, and when done by computer, the printed output is carefully annotated. In addition, the description for each process unit should refer to a corresponding specification sheet, discussed below, which is assembled with the other specification sheets in Section 10. Finally, the descriptions should refer to the installed and operating costs for the process unit in cost summaries, discussed below.

The identification of each process unit (e.g., Unit No. E-154, the condenser on an ethanol still) should be very clear, so that the concerned reader is able, without confusion, not only to relate each unit description to the corresponding specification sheet, its costs in the cost summaries, and its design calculations in the appendix, but also to locate that additional information readily and to check it when necessary.

The process units described in Section 9 should include (a) storage facilities for the feed, product, byproduct, and intermediate chemicals, (b) spare equipment items (pumps, adsorption towers, etc.) required to avoid shutdowns in the event of operating difficulties, and equipment for startup, which is often not needed during normal operation.

The descriptions are accompanied by an equipment list, which includes the unit number, unit type, brief function, material of construction, size, and operating conditions of temperature and pressure.

10. *Specification Sheets.* Specification sheets are required to guide purchasing agents in locating vendors of desired equipment and to enable vendors to prepare bids. These sheets provide the design specifications for each of the process units in the process flow diagram, as referred to in the unit descriptions. A typical example is shown in Figure 22.1.

DISTILLATION COLUMN

Identification: **Item** *Distillation Column*
Item No. *T-700*
No. required 1

Date: *9 April 1997*

By: *SFG*

Function: Separate Benzoic Acid and Benzaldehyde from VCH, Styrene, and other organics.

Operation: Continuous

Materials handled:

	Feed	Feed 2	Liquid Dist.	Bottoms	Vapor Dist.
Quantity (lb/hr):	161,527		153,022	6947	1558
Composition:					
Butadiene	4 PPB		2 PPB	trace	236 PPB
VCH	0.059		0.061	2 PPM	0.109
Styrene	0.861		0.899	0.087	0.630
Butene	10 PPB		5 PPB	trace	604 PPB
Cis-Butene	29 PPB		16 PPB	trace	2 PPM
Trans-Butene	9 PPB		5 PPB	trace	545 PPB
n-butane	3 PPB		1 PPB	trace	171 PPB
Isobutylene	7 PPB		3 PPB	trace	454 PPB
Isobutane	trace		trace	trace	9 PPB
Ethyl Benzene	0.039		0.041	96 PPM	0.041
Benzoic Acid	0.011		trace	0.244	trace
Benzaldehyde	0.028		31 PPM	0.647	10 PPM
H_2O	0.004		0.002	trace	0.205
N_2	139 PPM		2 PPM	trace	0.014
CO_2	6 PPB		1 PPB	trace	559 PPB
O_2	150 PPB		5 PPB	trace	15 PPM
Tar	902 PPM		trace	0.021	trace
Stabilizer					
Temperature (°F):	70.0		126.3	255.9	126.3

Design Data: Number of trays: 23
Pressure: 3.2 psig
Functional height: 70.5 ft
Material of construction: Carbon-steel
Recommended inside diameter: 21.0 ft
Tray efficiency: 0.70
Feed stage: 13
Feed 2 stage:
Side stream stage: 1

Molar reflux ratio: 10
Tray spacing: 3.0 ft
Skirt height: 14.5 ft

Utilities: Cooling water at 1.09 MM lb/hr and 370.52 M lb/hr 100 # stream
Controls:
Tolerances:
Comments and drawings: See Process Flow Sheet, 7 and Appendix F, 222-4.

Figure 22.1 Typical specification sheet for a process unit.

11. ***Equipment Cost Summary.*** In this section a table is prepared, containing the estimated purchase price of every equipment unit in the process flow diagram, identified according to the unit number and unit type on the process flow diagram and in the equipment list. The sources of the prices should be identified (graphical or tabulated cost data, a quotation from a specific manufacturer, etc.).

12. ***Fixed-Capital Investment Summary.*** In this section, the fixed-capital investment is related to the purchase cost of the equipment items. If desired, the equipment list and the list of equipment purchase costs can be combined. The methods for estimating the fixed capital investment, beginning with the purchase costs, should be clearly stated. If a factored cost estimate is used, the overall factor or individual equipment factors should be noted.

13. ***Other Important Considerations.*** In most design reports, the following considerations may deserve separate sections. Often, they are sufficiently important to warrant coverage apart from any discussion in the other parts of the report. These include those aspects of the design that address

    a. Environmental problems and methods used to eliminate them.
    b. Safety and health concerns, including a HAZOP (hazard and operability) study and a HAZAN (hazard analysis).
    c. Process controllability and instrumentation, including a piping and instrumentation diagram (P&ID).
    d. Startup, including additional equipment and costs.
    e. Plant layout.

To the extent that these matters influence the choice of particular or additional items of equipment, as well as operating strategies, at least some discussion should be included in Sections 5–12. This section is intended to allow for a more thorough discussion of these subjects than might be appropriate elsewhere, and to enable the design team to draw attention to their importance in developing the design.

14. ***Operating Cost and Economic Analysis.*** This section begins with a presentation of the annual costs of operating the proposed plant, that is, the cost sheet, as discussed in Section 17.2 and shown in Table 17.1. In addition to the total production cost on the cost sheet, it should provide an estimate of the cost per unit weight of the product (e.g., $ per lb, kg, ton, or tonne). Note that when cash flows are computed for different production rates from year to year, a separate cost sheet is required for each unique production rate. Note also that, in addition to appearing on the cost sheet, the utilities for each equipment unit and their costs should be summarized in a separate table.

    Next, the working capital is presented, with a discussion of how it was estimated. Then the total capital investment is presented.

    This section concludes with a presentation of the calculations to obtain several of the profitability measures. Normally, this includes one or more of the approximate measures, such as return on investment (ROI) and venture profit (VP), and one or more of the rigorous methods that involve cash flows, such as net present value (NPV) and investor's return on investment (IRR). The latter is also referred to as the discounted cash flow rate of return (DCFRR). In all cases, it is important to indicate clearly the depreciation schedule and, for the rigorous methods, to provide a table that shows the calculation of the annual cash flows, as shown in Example 17.29, as well as plots of cash flow of the type shown in the same example. Finally, the design team should present its judgment of the profitability of the proposed plant.

15. ***Conclusions and Recommendations.*** The principal conclusions of the design study should be presented, together with a clear statement of the recommendations, accompanied by justifications, for management. At this point, before the remaining sections of the report are discussed, it is important to emphasize that an engineering supervisor

may find it necessary to check the calculations of the engineers in the design team. For this purpose, Sections 9–12, and 14, as well as the associated sections of the appendix, are very important. References to the specific pages in each of these sections for every equipment item are equally important. Neither the supervisor responsible for the work of the design team, nor the faculty member who grades the design report, will regard with favor references to various sections of the report, including the appendix, that are absent or difficult to locate. The same is true of an industrial supervisor who causes such a report to be created.

16. *Acknowledgments.* Most design teams obtain considerable assistance and advice from industrial consultants, equipment vendors, librarians, fellow students, faculty, and the like. This section provides an opportunity to acknowledge their contributions with an expression of appreciation and thanks.

17. *Bibliography.* All works referred to in the design report, including the appendix, should be listed in this section. It is recommended that the references appear in the form shown in the Reference sections near the end of each chapter in this textbook.

18. *Appendix.* The following items are typically included in the appendix, whose pages should be numbered sequentially with the body of the report.

   a. The design procedures and detailed calculations for all of the equipment items in Section 9 must be included here. These are normally *not* typed, but must be sufficiently neat to be easily read and understood. Photocopies of legible calculation sheets, even bearing erasures, are adequate.

   b. Computer programs developed for the design should be listed with sufficient documentation to enable the principal sections to be identified. This can normally be accomplished through the use of comment statements at the beginning of each section, including definitions of the key variables.

   c. Relevant portions of the computer output (the variables on each stage of a distillation column, a graph showing the variables as a function of the stage number, etc.) should be included here. It is important that the output be sufficiently well annotated to permit the reader to read it intelligently. In some cases, hand-written annotations are helpful and adequate.

   d. Pertinent printed material (e.g., materials provided by equipment vendors that describe their products) should be included here.

At the risk of stating the obvious, it cannot be emphasized too strongly that the appendix is not a repository in which large quantities of computer printout, pertinent or not, are included to increase the weight and thickness of the report. Unless the information in the appendix can readily be located by appropriate references in Sections 5–14, a responsible supervisor may doubt the results that appear in the foregoing sections. This can only affect adversely the evaluation of the report and the quality of the proposed design.

Remaining Sections of a Product Design Report

No one outline can apply to all product design reports. However, the following list may provide a starting point.

5. *Introduction.* The introduction can be very short, just long enough to inform the reader about the nature of the product and the customer needs that it will satisfy. The remaining sections will present the necessary background and approach that was used to develop the product.

6. *Existing Similar Products.* This section is based on discussions among members of the design team, bolstered by interviews with major customers and consumers of the existing products that are similar to the new product being presented in the report. This section should answer the questions:

     a. What are the main functions of the existing products?

     b. What product improvements would be welcomed and what has hindered the improvements?

     c. How and where are the existing products marketed?

     d. Are new markets possible if the nature of the product could be improved?

     e. What patents protect the existing products or possible future products?

7. *Product Specifications.* This section lists the specifications for the new product and delineates the differences between the new product and the existing products.

8. *Generation of Concepts.* This section presents a list of all the ideas and concepts for the new product, as generated by the design team. It is preferable that the list be organized into several categories. Then the categories are compared in an attempt to select the best one or two categories for further consideration.

9. *Selection of the Best Concept(s).* This section follows from the previous section in an attempt to narrow the list of concepts to just a few, giving reasons why some concepts are dropped and others retained. Usually one, or at the most three, concepts will be selected for further evaluation. The selection is best based on supporting calculations, presented in an appendix, to prove the workability of the concept(s). This section should answer the question of possible patent protection for the new concept(s).

10. *Architectural Design of the Product(s).* This section presents the design of the new product(s), complete with dimensioned sketches and reference to design calculations presented in an appendix.

11. *Prototype.* This section presents the development of a working model of the product(s), testing procedures, and an assessment of the superiority of the product(s) over current competitors.

12. *Manufacture.* This section presents proposed methods for manufacturing the product(s), ensuring that product specifications are met.

13. *Human Factors.* This section discusses any potential adverse effects of the new product(s) on humans or the environment.

14. *Marketing and Cost Estimates.* This section discusses the potential market for the new product(s) and presents estimates of the selling price.

The remaining sections are essentially identical to Sections 15–18 of the process design report.

Preparation of the Written Report

Coordination of the Design Team

As mentioned in the introduction to this chapter, it is important for a design team to document its work throughout the design process. In this regard, each member is normally assigned responsibilities for a portion of the design work, as well as for its documentation. In industry, the assignments are usually coordinated by the head of the design team, who is normally appointed by the project supervisor. At a university, it is also recommended that a member of a student design team be appointed the team leader. The team leader schedules meetings to review progress of the team, to plan its next steps, make assignments, and set due dates. The faculty advisor is often very helpful in advising the team as it reviews its progress and plans its next steps.

Project Notebook

When carrying out a design, the design team normally maintains a project notebook, most likely a loose-leaf binder, in which important sources of information are placed. These include articles from the literature, data from the laboratory or the literature, design calculations, and computer programs and printed outputs. This repository of information is updated regularly and is particularly helpful during the meetings of the design team, especially when visitors, such as the team's faculty advisor and industrial consultants, are present.

Milestones

Since no two design projects follow exactly the same sequence of steps, it is not possible to suggest a timetable with specific milestones to be met by all design teams. Rather, in this subsection, it should suffice to identify the milestones, with emphasis on the steps to be accomplished and the portions of the design report that can be written. It is up to the team leader to prepare the timetable so that the final completion date can be met. The following steps pertain to a process design report. Similar steps, not given here, can be formulated for a product design report.

a. ***Assessment of the primitive design problem and literature search.*** These are the first steps in carrying out the design. As the specific problems are created and the preliminary database develops, the design notebook is augmented. This is an excellent time to write a draft of the Introduction (Section 5), which discusses many of these findings.

b. ***Complete the block flow diagram and the detailed process flow diagram showing the material balances.*** Most design teams spend considerable time in the process creation steps, identifying alternative process flow diagrams and creating the synthesis tree, as discussed in Section 3.4. While these steps, and the application of the algorithmic methods for process synthesis (which are usually carried out in parallel), are very important in leading to the most profitable processes, it is crucial not to spend too much time generating alternatives. Fairly early in the design process, the team should begin to focus its attention on the base-case design, as discussed in Section 3.5. This involves the preparation of a detailed process flow diagram (see Figure 3.19) and the completion of the material balances. As this is completed, the design team should prepare a draft of Sections 6 and 7 of the report. Should the base-case design be modified, the section is revised accordingly to show how the modifications improve upon the original design.

c. ***Complete the heat integration.*** In many cases, an attempt to achieve a high degree of heat and power integration is not undertaken until after mass integration is complete and the reactor(s) and separation equipment have been designed. After heat and power integration is complete and the heat exchangers, pumps, and compressors are installed in the base-case design, it is recommended that Section 8, on the energy balance, be completed.

d. ***Complete the detailed equipment design.*** After this step is completed, Sections 9 and 10, on the unit descriptions and the specification sheets, should be written. Note that it helps to complete hand calculations neatly so that they can be inserted into the appendix without any additional work. Furthermore, it is recommended that the important sections of the computer outputs be removed and annotated when necessary for insertion into the appendix.

e. ***Complete the fixed-capital investment and the profitability analysis.*** After these steps are completed, Sections 11, 12, and 14 should be written.

For the novice design team, it is hoped that the preceding pointers will help to simplify both the preparation of the design report and the design process. Although many merely follow common sense, they are included to help the design team set milestones to achieve throughout the design process.

Word Processing and Desktop Publishing

The advent of the word processor has had a major impact on the preparation of the design report. Because sections of text can be cut and pasted with ease, it is possible to write drafts of many sections, as discussed previously. As the base-case design is modified, new sections can be composed and added easily to the previously prepared sections, which can usually be included with minor modifications. For technical writing, Word, WordPerfect, PageMaker, and LaTeX are the most commonly used word processors. Except for highly mathematical manuscripts, the former two word processors are preferred.

Many design reports have on the order of 100 pages that include the 18 sections discussed earlier. Since there are many cross-references between the sections, it can be very helpful to add headers to the pages that identify the section numbers and titles. Furthermore, in addition to the table of contents, an index can be very helpful when the reader is searching for coverage of a specific topic.

Editing

No matter how careful an author is, it is difficult to compose concise text without redundant terms and the use of words that add little, if any, meaning. Most novice designers and writers, examine their manuscripts carefully for spelling errors, with the help of the spelling checkers in their word processors. They also seek to confirm that their statements are technically correct. However, many are inexperienced in the art of editing.

To obtain a more tightly structured document, it is recommended that the design team read its text carefully with the objectives of improving the grammatical constructions (eliminating split infinitives, avoiding the use of long strings of adjectives, etc.), avoiding the usage of redundant terms, and eliminating terms that add no meaning to the sentence. This step is important even for the most experienced writers, who can take advantage of recent versions of word processors that include grammar checkers. Checks are made and suggestions sometimes given for:

1. Incomplete sentences.
2. Use of passive voice when active voice would give more punch.
3. Improper use of who, whom, which, and that.
4. Capitalization.
5. Hyphenation.
6. Punctuation.
7. Subject and verb agreement.
8. Possessives and plurals.
9. Sentence structure.
10. Wordiness.

It must be noted, however, that the grammar checkers are not always correct and suggested corrections should, therefore, not always be accepted.

Page Format

At many companies and universities, the design reports are bound for storage in technical libraries and repositories. When this is the case, to save space on the bookshelves and simplify the usage of the reports, the following guidelines are recommended in the preparation of a manuscript for binding.

a. The pages of the report, including the appendix, should be numbered at the bottom center of each page.

b. The pages of the report should be printed back-on-back (two-sided), with the odd page numbers appearing on the right-hand page, as shown below.

c. All pages, including the appendix, should have left and right margins that are at least 1 in. wide, as shown below.

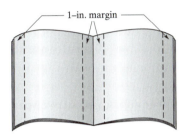

d. Sheets that appear sideways (broadside) should be mounted so that their tops face the left margin, as shown below. Remember that, for sideways sheets, the top and bottom (which become the left and right side of the page when rotated, must have 1-in. margins.

e. Black ink should be used for the printed calculations, to ensure that the pages will photocopy adequately.

f. When a large flow diagram is prepared by hand, as discussed in Section 6, it cannot be bound into the report, which is printed on $8\frac{1}{2}$ in. 11 in. pages. Such a flow diagram should be folded for insertion into a cover pocket, which is pasted onto the inside back cover of the report after the binding is completed.

g. The complete manuscript should be submitted in a file folder for binding.

Sample Design Reports

Samples of design reports are available in the libraries and repositories of technical reports maintained by companies and universities. In a few cases, they are available by Interlibrary Loan; for example, the design reports prepared by students at the University of Pennsylvania since 1993. Note that titles of the problem statements that led to these reports are reproduced in Appendix II of this textbook, with full problem statements included in the file, Design Problem Statements.pdf, on the CD-ROM that accompanies this textbook.

22.2 ORAL DESIGN PRESENTATION

It is probably most common for the oral design report in industry to be presented to the immediate supervisor of a design team, together with managers who are responsible for deciding upon the prudence of investing funds in the proposed design. Similar presentations, to a somewhat different audience (including industrial consultants, faculty, and fellow students) are prepared at universities, usually to provide the students with an experience similar to that which they are likely to encounter in industry. It should be noted, however, that only occasionally do young engineers have the opportunity to attend a meeting where their work and ideas are presented to the decision makers among their employers, and especially to make the presentation in person.

Typical Presentation

A typical oral design presentation by a team comprised of three students at a university is scheduled for 30 min with an additional 10 min for questions and discussion. This provides sufficient time to

1. Introduce the design problem.
2. Provide an overview of the proposed product or process (emphasizing alternative designs that were rejected).
3. Discuss sections of the proposed process (emphasizing the strengths of the design) or aspects of the proposed product.
4. Present the results of the economic analysis.
5. Discuss other considerations.
6. Summarize the design and make recommendations.

Normally, each student speaks for 10 min, although it is not uncommon to split the presentation into as many as six or seven segments, with each member of the design team covering those portions of the design with which he or she is most familiar.

Media for the Presentation

Overhead Projector

Until recently, the overhead projector was the most actively used vehicle for displaying the key concepts, graphs, figures, and tables that accompany a design presentation. In some cases, two overhead projectors permitted the presenters to describe concepts that benefit from the simultaneous display of two complementary figures. Although most engineers have extensive experience in the use of these projectors, it continues to be helpful to remind presenters of the importance of using sufficiently large fonts and maintaining appropriate borders to enable the entire transparency to be displayed clearly and at once.

Computer Projection Software

Rapidly gaining favor with speakers in recent years, with the availability of LCD pads and computer projection devices, are computerized projection facilities. To prepare and display the images, several software packages have been developed, including PowerPoint. This software is capable of displaying animated sequences, halftones, and videos. In most cases, the quality is significantly improved relative to the use of overhead projectors. Students using this technique of presentation should be warned to carefully check out the system just prior to the presentation. Computer-projector connections have not yet reached the point of being foolproof.

Preparation of Exhibits

To avoid duplicate work effort, it is recommended that design teams prepare all of the figures and tables for their written reports in such a way that they are displayed properly by an overhead projector or computer projection software. This requires that the figures and tables be prepared with sufficiently large fonts and the information be placed less compactly than when an oral presentation is not required.

Rehearsing the Presentation

One of the most difficult tasks a design team encounters is the organization of a 30-min presentation to summarize the most salient features of an extensive written report. It is especially challenging because the members of the team have usually been so involved in the details of the design calculations that they often find it difficult to summarize the really important results without overemphasizing the details. For this reason, and to help the team see the forest through the trees, it is important to rehearse the presentation in the presence of a colleague or teacher. In the best situation, this person will have attended many design presentations in the past, and will be well positioned to recommend that certain topics be expanded upon while others are deemphasized or eliminated entirely.

In one format for the rehearsal, the team makes a complete presentation without any interruptions, with the critic sitting toward the back of the room, to check that the exhibits can be seen and that the speakers can be heard easily. In addition, the critic takes notes and records the time that each speaker begins his or her presentation. Then, when the presentation is completed, the critic reviews the timing and offers some general comments. Often the design team makes a brief pass through its exhibits to enable the critic to offer more specific criticisms. The critic often has a major impact on the organization of the report and, more specifically, in helping the design team to achieve a well-balanced presentation.

Written Handout

In some situations, design teams find it easier to make their points through the preparation of a small written handout. Often, this includes the detailed process flow diagram, similar to that shown in Figure 3.19, or a design sketch of a product.

Evaluation of the Oral Presentation

When preparing an oral design presentation, it helps the speakers to have an appreciation of the criteria by which it will be judged. Similarly, when serving on a team to evaluate oral presentations, it is important for the evaluators to understand the criteria and to apply them fairly, especially when they play an important role in the preparation of a course grade and the selection of an award winner.

For process designs, one possible list of items to be evaluated is shown in Figure 22.2. The list can be modified for product designs. Included are the quality of the process description, the descriptions of the process units, and the discussion of the economic analysis. These are at the heart of the design presentation and deserve the most attention. The next item, novelty, is more difficult to judge, as some design problems provide more of an opportunity to be novel than others. This is recognized by most judges, who attempt to rate the creativity of the design work in the context of the design problem and the opportunities it provides to develop novel solutions. The next items address the organization of the presentation and its execution. Then, the quality of the exhibits and visual aids are evaluated. Finally, the overall presentation is rated, which includes a recommendation for a grade when the design report is the work of a student design team.

Name of presenter(s):
Title of Presentation:
Date of Presentation:
Name of Examiner/Appraiser:

Content	**Noteworthy**	**Acceptable**	**Needs Improving**
Process Description			
Unit Description			
Economics			
Novelty of Design			
Totals			

Presentation – Organization	**Noteworthy**	**Acceptable**	**Needs Improving**
Core Message			
Clear Objective			
Overall Structure			
Visible Logic			
Totals			

Presentation – Execution	**Noteworthy**	**Acceptable**	**Needs Improving**
Confident, enthusiastic, forceful, convincing			
Controlled pace/natural finish			
Voice quality (clear, calm, understandable)			
Frequent eye contact			
Totals			

Visual Aids	**Noteworthy**	**Acceptable**	**Needs Improving**
Interesting, relevant			
Easy to read			
Totals			

Figure 22.2 Oral design presentation evaluation form.

Videotapes and DVDs

Increasingly, oral design presentations are recorded on videotape or DVD to provide a record of the presentations, as well as to enable each design team to critique its own presentation with a view toward improving the next one. In most cases, a portable camcorder, mounted on a tripod, is adequate to capture the bulk of the presentation.

22.3   AWARD COMPETITION

At many universities, an award is presented to the design team that prepares a design judged to be the most outstanding. Normally, the criteria are a combination of those discussed for both the written and oral design reports. However, since the written reports become available using Interlibrary Loans, and the best reports are often submitted for regional competitions in which the judges select from among reports that originate from other universities, it is common to place more emphasis on the written reports.

Usually, a small awards committee, comprised of academic and industrial members, is appointed to make the judgment. It begins by reading the reports of those design teams whose oral presentations were judged to be among the best.

At many universities, the design award is presented to the design team at the commencement exercises, either for the Chemical Engineering Department, the Engineering School, or the entire university. It often involves a small stipend and a certificate or plaque.

Finally, it is important to mention the annual National Student Design Competition prepared by AIChE members from industry for the AIChE Student Chapters. The design contest is timed to be completed by the end of the spring semester, after which the awardees are selected to receive their awards at the Annual Meeting of the AIChE, usually in November, and to make oral presentations at the associated Student Chapter Meeting.

22.4 SUMMARY

In this chapter, readers have been presented with a template and associated milestones that must be completed, for guidance in the preparation of the written design report. No exercises are included because the template is intended to be used by design teams when writing their written reports.

Furthermore, readers have learned how to organize an oral design presentation. In this connection, they have become familiar with the alternative media for the presentation, with the reasons for rehearsing the presentation and the methods used to evaluate presentations.

REFERENCES

Peters, M.S., K.D. Timmerhaus and R. West, *Plant Design and Economics for Chemical Engineers*, 5th ed., McGraw-Hill, New York (2003).

Sandler, H.J., and E.T. Luckiewicz, *Practical Process Engineering*, XIMIX, Philadelphia, Pennsylvania (1993).

Ulrich, G.D., *A Guide to Chemical Engineering Process Design and Economics*, Wiley, New York (1984).

Appendix **I**

Residue Curves for Heterogeneous Systems

Beginning with Eq. (7.19), which also applies for heterogeneous systems, the liquid mole fractions, x_j, are replaced by the overall liquid mole fractions, x_j^o. These are accompanied by the equations that define the vapor–liquid and liquid–liquid equilibrium constants, K_j^{VL} and K_j^{LL}, respectively, and the component mass balances, to give

$$\frac{dx_j^o}{d\hat{t}} = x_j^o - y_j \qquad\qquad j = 1, \ldots, C \qquad\qquad \textbf{(A-I.1a)}$$

$$y_j = x_j^I K_j^{VL}\{T, P, \underline{x}^I, \underline{y}\} \quad j = 1, \ldots, C \qquad\qquad \textbf{(A-I.1b)}$$

$$x_j^{II} = x_j^I K_j^{LL}\{T, P, \underline{x}^I, \underline{x}^{II}\} \quad j = 1, \ldots, C \qquad\qquad \textbf{(A-I.1c)}$$

$$x_j^o = \alpha x_j^I + (1 - \alpha)x_j^{II} \quad j = 1, \ldots, C \qquad\qquad \textbf{(A-I.1d)}$$

$$\sum_{j=1}^{C} x_j^I - \sum_{j=1}^{C} x_j^{II} = 0 \qquad\qquad\qquad \textbf{(A-I.1e)}$$

$$\sum_{j=1}^{C} y_j = 1 \qquad\qquad\qquad\qquad \textbf{(A-I.1f)}$$

where x_j^I and x_j^{II} are the mole fractions of species j in the first and second liquid phases, respectively, and α is the mole fraction of the first liquid phase in the total liquid. To trace a residue curve from some starting composition, this system of differential-algebraic equations is solved by numerical integration. Equations (A-I.1b)–(A-I.1f) are solved to determine the compositions in vapor–liquid–liquid equilibrium as $\hat{t}$ is advanced in the integration.

Appendix II*

 Design Problem Statements

A-II.0   CONTENTS AND INTRODUCTION

*Appendix II appears in the file, Design Problem Statements.pdf, on the CD-ROM that accompanies this text. Only the titles of the design problem statements are listed here.

This appendix contains the problem statements for 50 design projects, each prepared for design teams of three students at the University of Pennsylvania by chemical engineers in the local chemical industry. At Penn, each team selects its design project during the first lecture course in the fall, and spends the spring semester completing the design. In the spring, each group meets regularly with its faculty advisor and industrial consultants, including the individual who provided the problem statement, to report on its progress and gain advice.

The problem statements in the PDF on the CD-ROM are in their original forms, as they were presented to the student design teams on the date indicated. Some provide relatively little information, whereas others are fairly detailed concerning the specific problems that need to be solved to complete the design. The reader should recognize that, in nearly every case, as the design team proceeded to assess the primitive problem statement and carry out a literature search, the specific problems it formulated were somewhat different than stated herein. Still, these problem statements should be useful to students and faculty in several respects. For students, they should help to show the broad spectrum of design problems that chemical engineers have been tackling in recent years. For the faculty, they should provide a basis for similar design projects to be created for their courses.

In formulating design problem statements, the industrial consultants strive to create process opportunities that lead to designs that are timely, challenging, and offer a reasonable likelihood that the final design will be attractive economically. Every effort is made to formulate problems that can be tackled by chemical engineering seniors without unduly gross assumptions and for which good sources of data exist for the reaction kinetics and thermophysical and transport properties. In this respect, this was accomplished in each of the problems included herein; furthermore, successful designs were completed by a student design team for most of these problems.

As seen in the contents, the projects have been assigned to one of the following areas, in some cases arbitrarily: Petrochemicals, Petroleum Products, Gas Manufacture, Foods, Pharmaceuticals, Polymers, and Environmental.

Credit is given to each formulator on his problem statement. In addition, the names of the contributors are listed below with many thanks, as their contributions in preparing these design problems have been crucial to the success of the design course.

Rakesh Agrawal	Air Products and Chemicals
E. Robert Becker	Environex, Wayne, PA
David D. Brengel	Air Products and Chemicals
Robert M. Busche	Bio-en-gene-er Associates, Wilmington, DE
Leonard A. Fabiano	CDI Corporation (formerly ARCO Chemical and Lyondell)
Brian F. Farrell	Air Products and Chemicals
Mike Herron	Air Products and Chemicals
F. Miles Julian	DuPont
Ralph N. Miller	DuPont
Robert Nedwick	Pennsylvania State University (formerly ARCO Chemical and Lyondell)
Frank Petrocelli	Air Products and Chemicals
Mark R. Pillarella	Air Products and Chemicals
Matthew J. Quale	Mobil Technology Company
William B. Retallick	Consultant, West Chester, PA
David G. R. Short	University of Delaware (formerly DuPont)
Peter Staffeld	Exxon/Mobil
Albert Stella	AlliedSignal (formerly General Electric)
Bjorn D. Tyreus	DuPont
Kamesh G. Venugopal	Air Products and Chemicals
Bruce Vrana	DuPont
Andrew Wang	Air Products and Chemicals
Steve Webb	Air Products and Chemicals
John Wismer	Atochem North America
Jianguo Xu	Air Products and Chemicals

Appendix **III**

Materials of Construction

The selection of materials of construction, based on strength, corrosivity, and cost of fabrication, is vital to process design and economic evaluation. The most common materials for process equipment, which include metals, glass, plastics, and ceramics, are listed in Table A-III.1, together with typical applications. Much more extensive tables are given by M. S. Peters and K. D. Timmerhaus in *Plant Design and Economics for Chemical Engineers,* fourth ed. (McGraw-Hill, New York, 1991); by G. D. Ulrich in *A Guide to Chemical Engineering Process Design and Economics* (John Wiley & Sons, New York, 1984); and in Section 28 of the seventh edition of *Perry's Chemical Engineers' Handbook* (McGraw-Hill, New York, 1997). Table A-III.1 should be used only for preliminary process design and economic evaluation. For final process design, corrosion and strength data as a function of temperature are needed for the expected chemical compositions within the process. Equipment vendors can also assist in the final selection of materials. In general, carbon steel is used whenever possible because of its low cost and ease of fabrication.

Table A-III.1 Materials of Construction for Process Equipment

Material	Maximum Temperature, °C (°F)	Typical Applications
Carbon steel (e.g. SA-285C) Cast iron (not strong) Ductile iron (stronger)	400 (750)	Cooling-tower water, boiler feed water, steam, air, hydrocarbons, glycols, mercury, molten salts, acetone
Low alloy (Cr-Mo) steel (e.g. SA-387B	500 (930)	Same as carbon steel, hydrogen
Stainless steels	700 (1,300)	Aqueous salt solutions, aqueous nitric acid, aqueous basic solutions, food intermediates, alcohols, ethers, freons, hydrogen, hydrogen sulfide, molten salts, molten metals
Aluminum	150 (300)	Aqueous calcium hydroxide, hydrogen, oxygen
Copper and copper alloys, aluminum bronze, brass, bronze	150 (300)	Aqueous sulfate and sulfite solutions, hydrogen, nitrogen, alcohols and other organic chemicals, cooling-tower water, boiler feed water
Nickel-based alloys, e.g., Hastelloy, Inconel, Monel, Incoloy, Carpenter 20	400 (750)	Aqueous nitric and organic acids, flue gases, chlorine, bromine, halogenated hydrocarbons, ammonia, sulfur dioxide, sulfur trioxide, organic solvents, brackish water and seawater
Titanium-based alloys	400 (750)	Aqueous solutions, carbon dioxide, organic solvents

Table A-III.1 Materials of Construction for Process Equipment (*Continued*)

Material	Maximum Temperature, °C (°F)	Typical Applications
Conventional plastics (polyethylene, polypropylene, ABS)	50–120 (120–250)	Aqueous solutions at near-ambient temperatures
Fluorocarbon plastics	250 (480)	Almost everything except halogens and halogenated chemicals
Rubber lining	250 (480)	Aqueous salt solutions and aqueous basic solutions at near-ambient temperatures
Glass lining	250 (480)	Aqueous sulfuric acid solutions, almost everything except fluorine and hydrogen fluoride
Ceramics	2,000 (3,630)	Almost all aqueous solutions, except hydrogen fluoride and sodium hydroxide, at near-ambient temperatures; most gases, except fluorine and hydrogen fluoride; most solvents; water
Graphite	2,000 (3,630)	Aqueous salt and base solutions, organic solvents Cl_2, HCl, H_2, H_2S, N_2, Hg, hydrocarbons, molten salts

Author Index

Subject Index